Optical Fiber Telecommunications VIB

Systems and Networks

Dedication

To the memory of Dr. Tingye Li
(July 7, 1931 – December 27, 2012)

A pioneer, luminary, friend, mentor, and champion of our field.
We will miss him dearly.

From the optical communications community

题词

光通信领域的同仁们：
请向我们的先驱、引路人、朋友、导师、倡导者和勇士，

厉鼎毅博士
(1931年7月7日 – 2012年12月27日)

致以深切的怀念！

Optical Fiber Telecommunications VIB

Systems and Networks

Sixth Edition

Ivan P. Kaminow

Tingye Li

Alan E. Willner

AMSTERDAM • BOSTON • HEIDELBERG • LONDON
NEW YORK • OXFORD • PARIS • SAN DIEGO
SAN FRANCISCO • SINGAPORE • SYDNEY • TOKYO

Academic Press is an Imprint of Elsevier

Academic Press is an imprint of Elsevier
The Boulevard, Langford Lane, Kidlington, Oxford OX5 1GB, UK
225 Wyman Street, Waltham, MA 02451, USA

Sixth edition 2013

Library of Congress Cataloging-in-Publication Data
A catalog record for this book is availabe from the Library of Congress

British Library Cataloguing in Publication Data
A catalogue record for this book is available from the British Library

ISBN: 978-0-12-396960-6

For information on all Academic Press publications
visit our web site at books.elsevier.com

Contents

CHAPTER 5 Digital Signal Processing (DSP) and Its Application in Optical Communication Systems 163

Polina Bayvel, Carsten Behrens, and David S. Millar

CHAPTER 9 Optical OFDM and Nyquist Multiplexing 381

Juerg Leuthold and Wolfgang Freude

CHAPTER 10 Spatial Multiplexing Using Multiple-Input Multiple-Output Signal Processing 433

Peter J. Winzer, Roland Ryf, and Sebastian Randel

CHAPTER 12 Multimode Communications Using Orbital Angular Momentum... 569

Jian Wang, Miles J. Padgett, Siddharth Ramachandran, Martin P.J. Lavery, Hao Huang, Yang Yue, Yan Yan, Nenad Bozinovic, Steven E. Golowich, and Alan E. Willner

CHAPTER 15 ROADM-Node Architectures for Reconfigurable Photonic Networks 683

Sheryl L. Woodward, Mark D. Feuer, and Paparao Palacharla

CHAPTER 20 Recent Advances in High-Frequency (>10 GHz) Microwave Photonic Links 853

Charles H. Cox, III and Edward I. Ackerman

CHAPTER 21 Advances in 1-100 GHz Microwave Photonics: All-Band Optical Wireless Access Networks Using Radio Over Fiber Technologies 873

Gee-Kung Chang, Yu-Ting Hsueh, and Shu-Hao Fan

CHAPTER 25 Modern Undersea Transmission Technology 1041

Jin-xing Cai, Katya Golovchenko, and Georg Mohs

Preface—Overview of OFT VI A & B

Optical Fiber Telecommunications VI (*OFT VI*) is the sixth installment of the *OFT* series. Now 34 years old, the series is a compilation by the research and development community of progress in the field of optical fiber communications. Each edition reflects the current state of the art at the time. As editors, we started with a clean slate and selected chapters and authors to elucidate topics that have evolved since *OFT V* or that have now emerged as promising areas of research and development.

SIX EDITIONS

Installments of the series have been published roughly every 5–8 years and chronicle the natural evolution of the field:

- In the late 1970s, the original *OFT* (Chenoweth and Miller, 1979) was concerned with enabling a simple optical link, in which reliable fibers, connectors, lasers, and detectors played the major roles.
- In the late 1980s, *OFT II* (Miller and Kaminow, 1988) was published after the first field trials and deployments of simple optical links. By this time, the advantages of multi-user optical networking had captured the imagination of the community and were highlighted in the book.
- *OFT III* (Kaminow and Koch, 1997) explored the explosion in transmission capacity in the early-to-mid-1990s, made possible by the erbium-doped fiber amplifier (EDFA), wavelength-division-multiplexing (WDM), and dispersion management.
- By 2002, *OFT IV* (Kaminow and Li, 2002) dealt with extending the distance and capacity envelope of transmission systems. Subtitle nonlinear and dispersive effects, requiring mitigation or compensation in the optical and electrical domains, were explored.
- *OFT V* (Kaminow, Li, and Willner, 2008) moved the series into the realm of network management and services, as well as employing optical communications for ever-shorter distances. Using the high-bandwidth capacity in a cost-effective manner for customer applications started to take center stage.
- The present edition, *OFT VI* (Kaminow, Li, and Willner, 2013), continues the trend of photonic integrated circuits, higher-capacity transmission systems, and flexible network architectures. Topics that have gained much interest in increasing performance include coherent technologies, higher-order modulation formats, and space-division-multiplexing. In addition, many of the topics from earlier volumes are brought up to date and new areas of research which show promise of impact are featured.

Although each edition has added new topics, it is also true that new challenges emerge as they relate to older topics. Typically, certain devices may have adequately solved transmission problems for the systems of that era. However, as systems become more complex, critical device technologies that might have been considered a "solved problem" would now have new requirements placed upon them and need a fresh technical treatment. For this reason, each edition has grown in sheer size, i.e. adding the new and, if necessary, re-examining the old.

An example of this circular feedback mechanism relates to the fiber itself. At first, systems simply required low-loss fiber. However, long-distance transmission enabled by EDFAs drove research on low-dispersion fiber. Further, advances in WDM and the problems of nonlinear effects necessitated development of non-zero-dispersion fiber. Cost and performance considerations today drive research in plastic fibers, highly bendable fibers, few-mode fibers, and multicore fibers. We believe that these cycles will continue.

Perspective of the past 5 years

OFT V was published in 2008. At that point, our field was still emerging from the unprecedented upheaval circa 2000, at which time worldwide telecom traffic ceased being dominated by the slow-growing voice traffic and was overtaken by the rapidly growing Internet traffic. The *irrational* investment exuberance and subsequent depression-like period of oversupply (i.e. the "bubble-and-bust") wreaked havoc on our field. We are happy to say that, by nearly all accounts, the field continues to gain strength again and appears to have entered a stage of *rational* growth. Demand for bandwidth continues to grow at a very healthy rate. Capacity needs are real, and are expected to continue in the future.

We note that optical fiber communications is firmly entrenched as part of the global information infrastructure. For example: (i) there would be no Internet as we know it if not for optics, (ii) modern data centers may have as many as 1,000,000 lasers to help interconnect boards and machines, and (iii) Smartphones would not be so smart without the optical fiber backbone.

A remaining question is how deeply will optical fiber penetrate and complement other forms of communications, e.g. wireless, access and on-premises networks, Interconnections, satellites, etc. The odds are that, indeed, optics will continue to play a significant role in assisting all types of future communications. This is in stark contrast to the voice-based future seen by *OFT*, published in 1979, which occurred before the first commercial intercontinental or transatlantic cable systems deployed in the 1980s. We now have Tbit/s systems for metro and long-haul networks. It is interesting and exciting to contemplate what topics, concerns, and innovations might be contained in the next edition of the series, *OFT VII*.

In this edition, *OFT VI*, we have tried to capture the rich and varied technical advances that have occurred in our field. Innovations continue to abound! We hope our readers learn and enjoy from all the chapters.

We wish sincerely to thank Tim Pitts, Charlie Kent, Susan Li, Jason Mitchell of Elsevier and Hao Huang of USC for their gracious and invaluable support throughout the publishing process. We are also deeply grateful to all the authors for their laudable efforts in submitting their scholarly works of distinction. Finally, we wish to thank the many people whose insightful suggestions were of great assistance.

Below are brief highlights of the different chapters in the two volumes.

OFT VI Volume A: Components and Subsystems
1A. Advances in Fiber Distributed-Feedback Lasers
Michalis N. Zervas

This chapter covers advances in fiber distributed-feedback (DFB) lasers and their potential use in modern coherent optical telecommunication systems. In particular, it describes novel DFB cavity designs and configurations and considers their impact on the laser performance. Special emphasis is given to the fiber parameters that define the power scalability and stability, the polarization performance, as well as the linewidth and phase-noise characteristics. The wavelength coverage and tunability mechanisms are also discussed. The chapter finally reviews the use of fiber DFB lasers in non-telecom applications, such as advanced optical fiber sensors, and concludes with an outlook of the fiber laser technologies and their future prospects.

2A. Semiconductor Photonic Integrated Circuit Transmitters and Receivers
Radhakrishnan Nagarajan, Christopher Doerr, and Fred Kish

This chapter covers the field of semiconductor photonic integrated circuits (PIC) used in access, metro, long-haul, and undersea telecommunication networks. Although there are many variants to implementing optical integration, the focus is on monolithic integration where multiple semiconductor devices, up to many hundreds in some cases, are integrated onto the same substrate. Monolithic integration poses the greatest technical challenge and the biggest opportunity for bandwidth and size scaling. The PICs discussed here are based on the two most popular semiconductor material systems: Groups III–V indium phosphide-based devices and Group IV silicon-based devices. The chapter also covers the historical evolution of the technology from the decades-old original proposal to the current-day Tbit/s class, coherent PICs.

3A. Advances in Photodetectors and Optical Receivers
Andreas Beling and Joe C. Campbell

This chapter reviews the significant advances in photodetectors that have occurred since *Optical Fiber Telecommunications V*. The quests for higher-speed p-i-n detectors and lower-noise avalanche photodiodes (APDs) with high gain-bandwidth product remain.

To a great extent, high-speed structures have coalesced to evanescently coupled wave-guide devices; bandwidths exceeding 140 GHz have been reported. A primary APD breakthrough has been the development of Ge on Si separate-absorption-and-multiplication devices that achieve long-wavelength response with the low-noise behavior of Si. For III–V compound APDs, ultra-low noise has been achieved by strategic use of complex multilayer multiplication regions that provide a more deterministic impact ionization. However, much of the excitement and innovation have focused on photodiodes that can be incorporated into InP-based integrated circuits and photodetectors for Si photonics.

4A. Fundamentals of Photonic Crystals for Telecom Applications— Photonic Crystal Lasers
Susumu Noda

Photonic crystals, in which the refractive index changes periodically, provide an exciting tool for the manipulation of photons and have made substantial progresses in recent years. This chapter first introduces research activities that are geared toward realizing the ultimate nanolaser using the photonic bandgap effect. Important aspects of this effort are in the achievement of spontaneous emission suppression and strong optical confinement using a photonic nanocavity. During the process of implementation of this goal, interesting phenomena, which can be classified as Quantum Anti-Zeno effect, have been observed. The rest of the chapter focuses on the current state of research in the field of broad-area coherent photonic crystal lasers using the band-edge effect, which occupies a position opposite to that of nanolasers discussed above. The main characteristics of these lasers will be discussed, including their high-power operation, the generation of tailored beam patterns, the surface-emitting laser operation in the blue-violet region, and even the beam-steering functionality.

5A. High-Speed Polymer Optical Modulators
Raluca Dinu, Eric Miller, Guomin Yu, Baoquan Chen, Annabelle Scarpaci, Hui Chen, and Corey Pilgrim

Recent advances in thin-film-polymer-on-silicon (TFPS) technology have provided the foundation to support commercial devices manufactured at production levels. A fundamental understanding of the material systems and fabrication techniques has been demonstrated, and will provide a stable platform for future developments to support next-generation applications. The chapter focuses on high-speed polymer-based optical modulators and on the molecular engineering of chromophores. The design of electron donor, bridge, electron acceptor, and isolating groups are discussed. Finally, the current commercial technologies are presented.

6A. Nanophotonics for Low-Power Switches

Lars Thylen, Petter Holmström, Lech Wosinski, Bozena Jaskorzynska, Makoto Naruse, Tadashi Kawazoe, Motoichi Ohtsu, Min Yan, Marco Fiorentino, and Urban Westergren

Switches and modulators are key devices in ubiquitous applications of photonics: telecom, measurement equipment, sensor, and the emerging field of optical interconnects in high-performance computing systems. The latter could accomplish a breakthrough in offering a mass market for these switches. This chapter deals with photonic switches and the quest for the partly interlinked properties of low-power dissipation in operation and nanostructured photonics. It first summarizes some of the most important existing and emerging materials for nanophotonics low-power switches, and describes their physical mechanisms, operation mode, and characteristics. The chapter then focuses on basic operation and power dissipation issues of electronically controlled switches, which in many important cases by using a simple model are operated by charging and discharging capacitors and thus changing absorption and/or refraction properties of the medium between the capacitor plates. All optical switches are also discussed and some present devices are presented.

7A. Fibers for Short-Distance Applications

John Abbott, Scott Bickham, Paulo Dainese, and Ming-Jun Li

This chapter first reviews the current use of multimode fibers (MMF) with short-wavelength VCSELs for short-distance applications. Standards are in place for 100 Gbit/s applications based on 10 Gbit/s optics and are being developed for ~25 Gbit/s optics. Then it briefly introduces the theory of light propagation in multimode fibers. The actual performance of an MMF link (the bit error rate and inter-symbol interference) depends on both the fiber and the laser. Effective model bandwidth, which includes both fiber and laser effects, will be discussed, and the method of characterizing fiber with the differential-mode-delay measurement and the laser with the encircled flux measurement will be summarized as well. Bend-insensitive multimode fiber is then presented, explaining how the new fiber achieves high bandwidth with low bend loss. New fibers for short-distance consumer applications and home networking are discussed. Finally, fibers designed for high-performance computing are reviewed, including multicore fibers for optical interconnects.

8A. Few-Mode Fiber Technology for Spatial Multiplexing

David W. Peckham, Yi Sun, Alan McCurdy, and Robert Lingle Jr.

This chapter gives an overview of design and optimization of few-mode optical fibers (FMF) for space-division-multiplexed transmission. The design criteria are outlined, along with performance limitations of the traditional step-profile and graded-index profiles. The trade-offs between number of usable optical modes (related to total channel capacity), differential group delay, differential mode attenuation, mode coupling, and the impact on multiple-input and multiple-output (MIMO) receiver complexity are outlined. Improved fiber designs are analyzed which maximize channel capacity with foreseeable next-generation receiver technology. FMF measurement technology is overviewed.

9A. Multi-Core Optical Fibers
Tetsuya Hayashi

Spatial division multiplexing attracts lots of attention for tackling the "capacity crunch," which is anticipated as a problem in the near future, and therefore various types of optical fibers and multiplexing methods have been intensively researched in recent years. This chapter introduces the multi-core fibers for spatial division multiplexed transmission. It describes various characteristics specific to the multi-core fibers, which have been elucidated theoretically and experimentally in recent years. Though there are many important factors, many pages are devoted especially to the description of inter-core crosstalk, which is crucial when signals are transmitted over each core independently. The chapter also describes other characteristics related to the improvement of core density.

10A. Plastic Optical Fibers and Gb/s Data Links
Yasuhiro Koike and Roberto Gaudino

As high-speed data processing and communication systems are required, plastic optical fibers (POFs) become promising candidates for optical interconnects as well as optical networking in local area networks. This chapter presents an overview of the evolution of POF, reviewing the technical achievements of both fiber design and system architectures that today allow using POF for Gb/s data links. In particular, the chapter presents the different POF materials such as polymethyl methacrylate (PMMA), perfluorinated polymers, types such as step-index POF and graded-index POF, as well as the POF production process, describing the resulting optical characteristics in terms of attenuation, dispersion, and bandwidth. The main applications of POF in industrial automation, home networking, and local area networks are also discussed.

11A. Integrated and Hybrid Photonics for High-Performance Interconnects
Nikos Bamiedakis, Kevin A. Williams, Richard V. Penty, and Ian H. White

Optical interconnection technologies are increasingly deployed in high-performance electronic systems to address challenges in connectivity, size, bandwidth, latency, and cost. Projected performance requirements lead to formidable cost and energy efficiency challenges. Hybrid and integrated photonic technologies are currently being developed to reduce assembly complexity and to reduce the number of individually packaged parts. This chapter provides an overview of the important challenges that photonics currently face, identifies the various optical technologies that are being considered for use at the different interconnection levels, and presents examples of demonstrated state-of-the-art optical interconnection systems. Finally, the prospects and potential of these technologies in the near future are discussed.

12A. CMOS Photonics for High-Performance Interconnects
Jason Orcutt, Rajeev Ram, and Vladimir Stojanović

For many applications, multicore chips are primarily constrained by the latency, bandwidth, and capacity of the external memory system. One of the most significant challenges is how to effectively connect on-chip processors to off-chip memories. This chapter introduces optical interconnects as a possible solution to the emerging performance wall in high-density supercomputer applications, arising from limited bandwidth and density of on-chip interconnects and chip-to-chip (processor-to-memory) electrical interfaces. The chapter focuses on the translation of system- and link-level performance metrics to photonic component requirements. The topics to be developed include network topology, photonic link components, circuit and system design for photonic links.

13A. Hybrid Silicon Lasers
Brian R. Koch, Sudharsanan Srinivasan, and John E. Bowers

The term "hybrid silicon laser" refers to a laser that has a silicon waveguide and a III–V material that is in close optical contact. In this structure, the optical confinement can be easily transferred from one material to the other and intermediate modes exist for which the light is contained in both materials simultaneously. In hybrid silicon lasers, the optical gain is provided by the electrically pumped III–V material and the optical cavity is ultimately formed by the silicon waveguide. This type of laser can be heterogeneously integrated with silicon components that have superior performance compared to III–V components. These lasers can be fabricated in high volumes as components of complex photonic integrated circuits, largely with CMOS-compatible processes. These traits are expected to allow for highly complex, non-traditional photonic integrated circuits with very high yields and relatively low manufacturing costs. This chapter discusses the theory of hybrid silicon lasers, wafer-bonding techniques, examples of experimental results, examples of system demonstrations based on hybrid silicon lasers, and prospects for future devices.

14A. VCSEL-Based Data Links
Julie Sheridan Eng and Chris Kocot

Vertical cavity surface emitting laser (VCSEL)-based data links are attractive due to their low-power dissipation and low-cost manufacturability. This chapter reviews the foundations for this technology, as well as the device and module design challenges of extending the data rate beyond the current level. The chapter begins with a review of data communications from the business perspective, and continues with a brief discussion of the current and future standards. This is followed by a survey of recent advances in VCSELs, including data links operating at 28 Gbit/s. Recent efforts on ultra-fast data links are reviewed and the advantages of the different approaches are discussed. The chapter also examines key design aspects of optical transceiver modules and focuses on novel applications in high-performance computing using both multi-mode and single-mode fiber optics. The importance of the device/component-level and system-level modeling is highlighted, and some modeling examples are shown with comparison to measured data. The chapter concludes with a comparison of the VCSEL-based data links with other competing technologies, including silicon photonics and short-cavity edge-emitting lasers.

15A. Implementation Aspects of Coherent Transmit and Receive Functions in Application-Specific Integrated Circuits
Andreas Leven and Laurent Schmalen

One of the most challenging components of an optical coherent communication system is the integrated circuits (ICs) that process the received signals or condition the transmit signals. This chapter discusses implementation aspects of these ICs and their main building blocks, as data converters, baseband signal processing, forward error correction, and interfacing. This chapter also highlights selected implementation details for some baseband signal processing blocks of a coherent receiver. The latest generation of coherent ICs also supports advanced forward error correction techniques based on soft decisions. The circuits for encoding and decoding low-density parity-check (LDPC) codes are introduced and evaluation of different forward error correction schemes based on a set of recorded measurement data is presented in this chapter.

16A. All-Optical Regeneration of Phase-Encoded Signals
Joseph Kakande, Radan Slavík, Francesca Parmigiani, Periklis Petropoulos, and David Richardson

This chapter reviews the general principles and approaches used to regenerate phase-encoded signals of differing levels of coding complexity. It first reviews different approaches and nonlinear processes that may be used to perform the regeneration of phase-encoded signals. The primary focus is on parametric effects, which as explained previously can operate directly on the optical phase. The chapter then proceeds to review progress on regenerating the simplest of phase modulation formats, namely DPSK/BPSK- and for which the greatest progress has been made to date. In the following, the progress in regenerating more complex modulation format signals—in particular (D)QPSK and other M-PSK signals—is discussed. The chapter also reviews the choice of nonlinear components available to construct phase regenerators. Finally, it reviews the prospects for regenerating even more complex signals including QAM and mixed phase-amplitude coding variants.

17A. Ultra-High-Speed Optical Time Division Multiplexing

Leif Katsuo Oxenløwe, Anders Clausen, Michael Galili, Hans Christian Hansen Mulvad, Hua Ji, Hao Hu, and Evarist Palushani

The attraction of optical time division multiplexing (OTDM) technology is the promise of achieving higher bit rates per channel than electronics could provide, thus alleviating the so-called electronic speed bottleneck. In this chapter, the state-of-the-art OTDM systems are presented, with a focus on experimental demonstrations. This chapter especially highlights demonstrations at 640–1280 Gbaud per polarization based on a variety of materials and functionalities. Many essential network functionalities are available today using a plethora of available materials, so now it is time to look at new network scenarios that take advantage of the serial nature of the data, e.g. try to come up with practical schemes for ultra-high bit rate optical data packets in supercomputers or within data centers.

18A. Technology and Applications of Liquid Crystal on Silicon (LCoS) in Telecommunications

Stephen Frisken, Ian Clarke, and Simon Poole

Liquid crystal is now the dominant technology for flat-screen displays and has been used in telecom systems since the late 1990s. More recently, the adoption of liquid crystals in Wavelength Selective Switches—with the control of light on a pixel-by-pixel basis—has been enabled by developments in Liquid Crystal on Silicon (LCoS) backplane technologies derived from projection displays. This chapter presents the principles of operation of liquid crystals, focusing in particular on how they operate within an LCoS chip. It then explains in detail the design and operation of an LCoS-based wavelength selective switch (WSS), with particular emphasis on the key optical parameters that determine performance in an optical communications network. In the final section, the chapter briefly describes the broad scope of new opportunities that arise from the intrinsic performance and flexibility of LCoS as a switching medium.

OFT VI Volume B: Systems and Networks

1B. Fiber Nonlinearity and Capacity: Single-Mode and Multimode Fibers

René-Jean Essiambre, Robert W. Tkach, and Roland Ryf

This chapter presents the trends in optical network traffic and commercial system capacity, discusses fundamentals of nonlinear capacity of single-mode fibers, and indicates that improvements in the properties of single-mode fibers only moderately increase the nonlinear fiber capacity. This leads to the conclusion that fiber capacity can be most effectively grown by increasing the number of spatial modes. This chapter also discusses nonlinear propagation in multimode fiber, a complex field still largely unexplored. It gives a basic nonlinear propagation equation derived from the Maxwell equation, along with simplified propagation equations in the weak- and strong-coupling approximations, referred to as generalized Manakov equations. Finally, the chapter presents experimental observations of two inter-modal nonlinear effects, inter-modal cross-phase modulation, and inter-modal four-wave mixing, over a few-km-long few-mode fiber. Important differences between intra-modal and inter-modal nonlinear effects are also discussed.

2B. Commercial 100-Gbit/s Coherent Transmission Systems
Tiejun J. Xia and Glenn A. Wellbrock

This chapter provides a global network service provider's view on technology development and product commercialization of 100-Gbit/s for optical transport networks. Optical channel capacity has been growing over the past four decades to address traffic demand growth and will continue this trend for the foreseeable future to meet ever-increasing bandwidth requirements. In this chapter, optical channels are reclassified into three basic design types. Commercial 100-Gbit/s channel development experienced all three types of channel designs before eventually settling on the single-carrier polarization-multiplexed quadrature-phase-shift keying (PM-QPSK) format using coherent detection, which appears to be the optimal design in the industry. A series of 100-Gbit/s channel related field trials was performed in service providers' networks to validate the technical merits and business advantages of this new capacity standard before its deployment. Introduction of the 100-Gbit/s channel brings new opportunities to boost fiber capacity, accommodates increases in client interface speed rates, lowers transmission latency, simplifies network management, and speeds up the realization of next-generation optical add/drop functions.

3B. Advances in Tb/s Superchannels
S. Chandrasekhar and Xiang Liu

Optical superchannel transmission, which refers to the use of several optical carriers combined to create a channel of desired capacity, has recently attracted much research and development in an effort to increase the capacity and cost-effectiveness of wavelength-division multiplexing (WDM) systems. Using superchannels avoids the electronic bottleneck via optical parallelism and provides high per-channel data rates and better spectral utilization, especially in transparent mesh optical networks. This chapter reviews recent advances in the generation, detection, and transmission of optical superchannels with channel data rates on the order of Tbit/s. Multiplexing schemes such as optical orthogonal-frequency-division-multiplexing (O-OFDM) and Nyquist-WDM are described, in conjunction with modulation schemes such as OFDM and Nyquist-filtered single-carrier modulation. Superchannel transmission performance is discussed. Finally, networking implications brought by the use of superchannels, such as flexible-grid WDM, are also discussed.

4B. Optical Satellite Communications
Hamid Hemmati and David Caplan

Current satellite-based communication systems are increasingly capacity-limited. Based on radio frequency or microwave technologies, current state-of-the-art satellite communications (Satcom) are often constrained by hardware and spectrum allocation limitations. Such limitations are expected to worsen due to the use of more sophisticated data-intensive sensors in future interplanetary, deep-space, and manned missions, an increased demand for information, and the demand for a bigger return on space-exploration investment. This chapter presents the recent advances in optical satellite communications technologies. Lasercom link budgets, the first step in designing a lasercom system, are discussed. The chapter then reviews the major challenges facing laser beam propagation through the atmosphere, including atmospheric attenuation, scattering, radiance, and turbulence. It also discusses mitigation approaches. The rest of the chapter focuses on optical transceiver technologies for satellite communications systems. Finally, space and ground terminals in optical satellite communications are discussed.

5B. Digital Signal Processing (DSP) and its Application in Optical Communication Systems
Polina Bayvel, Carsten Behrens, and David S. Millar

The key questions in current optical communications research are how to maximize both capacity and transmission distance in future optical transmission networks by using spectrally efficient modulation formats with coherent detection and how digital signal processing can aid in this quest. There is a clear trade-off between spectral efficiency and transmission distance, since the more spectrally efficient modulation formats are more susceptible to optical fiber nonlinearities. This chapter illustrates the application of nonlinear back-propagation to mitigate both linear and nonlinear transmission impairments in a range of modulation formats at varying symbol rates, wavelength spacing, and signal bandwidth. The basics of coherent receiver structure and digital signal processing (DSP) algorithms for chromatic dispersion compensation, equalization, and phase recovery of different modulation formats employing amplitude, phase, and polarization are reviewed and the effectiveness of the nonlinearity compensating DSP based on digital back-propagation is explored. This chapter includes a comprehensive literature review of the key experimental demonstrations of nonlinearity compensating DSP.

6B. Advanced Coding for Optical Communications
Ivan B. Djordjevic

This chapter represents an overview of advanced coding techniques for optical communication. Topics include the following: codes on graphs, coded modulation, rate-adaptive coded modulation, and turbo equalization. The main objectives of this chapter are as follows: (i) to describe different classes of codes on graphs of interest for optical communications, (ii) to describe how to combine multilevel modulation and channel coding, (iii) to describe how to perform equalization and soft-decoding jointly, and (iv) to demonstrate efficiency of joint demodulation, decoding, and equalization in dealing with various channel impairments simultaneously. The chapter describes both binary and nonbinary LDPC codes, their design, and decoding algorithms. A field-programmable gate array (FPGA) implementation of decoders for binary LDPC codes is discussed. In addition, this chapter demonstrates that an LDPC-coded turbo equalizer is an excellent candidate to simultaneously mitigate chromatic dispersion, polarization mode dispersion, fiber nonlinearities, and I/Q-imbalance. In the end, the information capacity study of optical channels with memory is provided for completeness of presentation.

7B. Extremely Higher-Order Modulation Formats
Masataka Nakazawa, Toshihiko Hirooka, Masato Yoshida, and Keisuke Kasai

This chapter reviews recent progress on coherent quadrature amplitude modulation (QAM) and orthogonal frequency-division multiplexing (OFDM) transmission with higher-order multiplicity, which is aiming at ultra-high spectral efficiency approaching the Shannon limit. Key technologies are the coherent detection with a frequency-stabilized fiber laser and an optical PLL circuit. Single-carrier 1024 QAM and 256 QAM-OFDM transmissions are successfully achieved, demonstrating a spectral efficiency exceeding 10 bit/s/Hz. Such an ultra-high spectrally efficient transmission system would also play a very important role in increasing the total capacity of WDM systems and improving the tolerance to chromatic dispersion and polarization mode dispersion as well as in reducing power consumption. The chapter also describes a novel high-speed, spectrally efficient transmission scheme that combines the OTDM and QAM techniques, in which a pulsed local oscillator (LO) signal obtained with an optical phase-lock loop (OPLL) enables precise demultiplexing and demodulation simultaneously. An optimum OTDM and QAM combination would provide the possibility for realizing long-haul Tbit/s/channel transmission with a simple configuration, large flexibility, and low-power consumption.

8B. Multicarrier Optical Transmission
Xi Chen, Abdullah Al Amin, An Li, and William Shieh

This chapter is an overview of multicarrier transmission and its application to optical communication. Starting with an introduction to historical perspectives in the development of optical multicarrier technologies, the chapter presents different variants of optical multicarrier transmission, including electronic and optical fast Fourier transform (FFT)-based realizations. In the next section, several problems of fiber nonlinearity in optical multicarrier transmission systems are highlighted and an analysis of fiber capacity under nonlinear impairments is presented. The applications of multicarrier techniques to long-haul systems, access networks, and free-space optical communication systems are also discussed. Finally, this chapter summarizes several possible directions for research into the implementation of multicarrier technologies in optical transmission.

9B. Optical OFDM and Nyquist Multiplexing
Juerg Leuthold and Wolfgang Freude

New pulse shaping techniques allow for optical multiplexing with the highest spectral efficiencies. This chapter introduces the general theory of orthogonal pulse shaping followed by a discussion that places more emphasis on the orthogonal frequency-division multiplexing (OFDM) and Nyquist frequency-division multiplexing schemes. Subsequently, the chapter shows that the rectangular-shaped pulses used for OFDM can mathematically be treated by the Fourier transform. This leads to the theory of the time-discrete Fourier transform (DFT) and to a discussion of practical implementations of the DFT and its inverse in the optical domain. The chapter concludes with exemplary implementations of OFDM transceivers that either rely on direct pulse shaping or use the DFT approaches.

10B. Spatial Multiplexing Using Multiple-Input Multiple-Output Signal Processing
Peter J. Winzer, Roland Ryf, and Sebastian Randel

In order to further scale network capacities and to avoid a looming "capacity crunch," *space* has been identified as the only known physical dimension yet unexploited for optical modulation and multiplexing. Space-division multiplexing (SDM) may use uncoupled or coupled cores of multi-core fiber, or individual modes of multimode waveguides. If crosstalk rises to levels where it cannot be treated as a transmission impairment any more, multiple-input multiple-output (MIMO) digital signal processing (DSP) techniques have to be used to manage crosstalk in highly integrated SDM systems. This chapter reviews the fundamentals and practical experimental aspects of MIMO-SDM. First, it discusses the importance of selectively addressing all modes of a coupled-mode SDM channel at transmitter and receiver in order to achieve reliable capacity gains. It shows that reasonable levels of mode-dependent loss (MDL) are acceptable without much loss of channel capacity. The chapter then introduces MIMO-DSP techniques as an extension of familiar algorithms used in polarization-division multiplexed (PDM) digital coherent receivers and discusses their functionality and scalability. Finally, the design of mode multiplexers that allows for the mapping of the individual transmission signals onto an orthogonal basis of waveguide mode is reviewed and its performance in experimental demonstrations is discussed.

11B. Mode Coupling and its Impact on Spatially Multiplexed Systems
Keang-Po Ho and Joseph M. Kahn

Mode coupling is the key to overcoming challenges in mode-division multiplexed transmission systems in multimode fiber. This chapter provides an in-depth description of mode coupling, including its physical origins, its effect on modal dispersion (MD) and mode-dependent loss (MDL) or gain, and the resulting impact on system performance and implementation complexity. Strong mode coupling reduces the group delay spread from MD, minimizing the complexity of digital signal processing used for compensating MD and separating multiplexed signals. Likewise, strong mode coupling reduces the variations of MDL that arise from transmission fibers and inline optical amplifiers, thus maximizing average channel capacity. When combined with MD, strong mode coupling creates frequency diversity, which reduces the probability of outage caused by MDL and enables outage capacity to approach average capacity. The statistics of strongly coupled MD and MDL depend only on the number of modes and the variances of MD or MDL, and they can be derived from the eigenvalue distributions of certain random matrices.

12B. Multimode Communications Using Orbital Angular Momentum
Jian Wang, Miles J. Padgett , Siddharth Ramachandran, Martin P.J. Lavery, Hao Huang, Yang Yue, Yan Yan, Nenad Bozinovic, Steven E. Golowich, and Alan E. Willner

Laser beams with a helical phase front, such as Laguerre-Gaussian beams, carry orbital angular momentum (OAM). Based on the fact that different OAM beams can be inherently orthogonal with each other, OAM multiplexing was introduced to provide an additional degree of freedom in optical communications, and further increase the capacity and spectral efficiency in combination with advanced multilevel modulation formats and conventional multiplexing technologies. This chapter provides a comprehensive review of multimode communications using OAM technologies. The fundamentals of OAM are introduced first, followed by the techniques for OAM generation, multiplexing/demultiplexing, and detection. The chapter then presents recent research into free-space communication links and fiber-based transmission links using OAM multiplexing with optical signal processing using OAM (data exchange, add/drop, multicasting, monitoring, and compensation). Future challenges for OAM communications are then discussed.

13B. Transmission Systems Using Multicore Fibers
Yoshinari Awaji, Kunimasa Saitoh, and Shoichiro Matsuo

As the simplest form of space-division multiplexing (SDM), multi-core fiber (MCF) transmission technologies have been widely studied. Many types of MCFs exist, but the most common is "Uncoupled MCF" in which each individual core is assumed to be an independent optical path. The key issue in these systems is how to suppress the inter-core crosstalk and the coupling/decoupling mechanism. Currently, many MCF varieties, coupling methods, splicing techniques, and transmission schemes have been proposed and demonstrated, and despite the fact that many of the component technologies are still in the development stage, MCF systems already present the capability for huge transmission capacities. In this chapter, these component technologies and the early experimental trials of MCF transmission are reviewed. First, an overview of medium- to long-haul MCF transmission and theories is provided. Second, coupling technologies between MCF-SMF and MCF-MCF are reviewed. Finally, several experimental demonstrations, including transmission exceeding 100 Tbit/s and over 1000 km, are described.

14B. Elastic Optical Networking
Ori Gerstel and Masahiko Jinno

Service provider (SP) networks are undergoing major changes. These changes imply that the optical layer will have to be low-cost, flexible, and reconfigurable. To properly address this challenge, flexible and adaptive networks equipped with flexible transceivers and network elements that can adapt to the actual traffic demands are needed. The combination of adaptive transceivers, a flexible grid, and intelligent client nodes enables a new "elastic" networking paradigm, allowing SPs to address the increasing needs of the network without frequently overhauling it. This chapter starts by looking at the challenges faced by the optical layer in the future. These challenges are fueled by the insatiable appetite for more bandwidth, coupled with a reduced ability to forecast and plan for such growth. Different enabling technologies, including flexible spectrum reconfigurable optical add/drop multiplexers (ROADM), bit rate variable transceivers, and the extended role of network control systems are reviewed. The concept of elastic optical network (EON) is envisioned and the benefits are highlighted by further comparing the EON to a fixed WDM system.

15B. ROADM-Node Architectures for Reconfigurable Photonic Networks
Sheryl L. Woodward, Mark D. Feuer, and Paparao Palacharla

The deployment of reconfigurable optical add/drop multiplexers (ROADMs) is gradually transforming a transport layer made of point-to-point optical links into a highly interconnected, reconfigurable photonic mesh. To date, the widespread use of ROADMs has been driven by the cost savings and operational simplicity they provide to quasi-static networks (i.e. networks in which new connections are frequently set up but rarely taken down). However, new applications exploiting the ROADMs' ability to dynamically reconfigure a photonic mesh network are now being investigated. This chapter reviews the attributes and limitations of today's ROADMs and other node hardware. It also surveys proposals for future improvements, including colorless, non-directional, and contentionless add/drop ports. The application of reconfigurable networks is also discussed with emphasis on the backbone network of a major communications service provider. Finally, the chapter assesses which of these new developments is most likely to bring added value in the short and long future.

16B. Convergence of IP and Optical Networking
Kristin Rauschenbach and Cesar Santivanez

Rapidly increasing network demand based on unpredictable services has driven research into methods to provide intelligent provisioning, efficient restoration and recovery from failures, and effective management schemes that reduce the amount of "hands-on" activity to plan and run the network. Integrating the service-oriented IP layer together with the efficient transport capabilities of the optical layer is a cornerstone of this research. Converged IP-optical networks are being demonstrated in large multi-carrier and multi-vendor venues. Research is continuing on making this convergence more efficient, flexible, and scalable. This chapter reviews the current key technologies that contribute to the convergence of IP and optical networks, and describes control and management plane technologies, techniques, and standards in some detail. Current research challenges and future research directions are also discussed.

17B. Energy-Efficient Telecommunications
Daniel C. Kilper and Rodney S. Tucker

For many years, advances in telecommunications have been driven by the need for increased capacity and reduced cost. Recently, however, concerns about the rising energy use of telecommunications networks have brought the issue of energy efficiency into the mix for both equipment vendors and network operators. This chapter provides an overview of energy consumption in telecommunications networks. This chapter identifies the key contributors to energy consumption and the trends in the growth of energy consumption. The chapter also compares the performance of state-of-the-art equipment with theoretical lower bounds on energy consumption and points to opportunities for improving the energy efficiency of core metro and access networks. The potential of significantly improving energy efficiency in telecommunications is envisioned.

18B. Advancements in Metro Regional and Core Transport Network Architectures for the Next-Generation Internet
Loukas Paraschis

The expanding role of Internet-based service delivery, and its underlying infrastructure of internetworked data centers, is motivating an evolution to an IP next-generation network architecture with a flatter hierarchy of more densely interconnecting networks. This next-generation Internet is required to cost-effectively scale to Zettabytes of bandwidth with improved operational efficiency, in an environment of increasing traffic variability, dynamism, forecast unpredictability, and uncertainty of future traffic types. This chapter explores the implications of this change in the metro regional and core transport network architectures, and the important advancements in optical, routing, and traffic engineering technologies that are enabling this evolution. The chapter accounts particularly for the increasingly important role of optical transport, and photonics technology innovations.

19B. Novel Architectures for Streaming/Routing in Optical Networks
Vincent W.S. Chan

Present-day networks are being challenged by dramatic increases in the data rate demands of emerging applications. New network architectures for streaming/routing large "elephant" transactions will be needed to reduce costs and improve power efficiency. This chapter examines a number of possible optical network transport mechanisms, including optical packet switching, burst switching, and flow switching and describes the necessary physical layer, routing, and transport layer architectures for these transport mechanisms. Performance comparisons are made based on capacity utilization, scalability, costs, and power consumption.

20B. Recent Advances in High-Frequency (>10 GHz) Microwave Photonic Links
Charles H.Cox, III and Edward I.Ackerman

The transmission of multi-band radio signals through optical fibers has attracted great attention recently due to its potential for cellular backhaul networks, mobile cloud computing, and wireless local area networks. As wireless services and technologies evolve into multi-gigabit radio access networks, their speed is increased, but the wireless coverage of a single access point is inevitably and dramatically reduced. As a result, the importance of >10 GHz radio-over-fiber techniques has been emphasized for the capability of expanding wireless coverage feasibility, and in the meantime reducing system complexity and operation expenditure, especially in the high-speed millimeter-wave regime. This chapter introduces the radio-over-fiber technique and its challenge to handle optical millimeter-wave generation, transmission, and converged multi-band systems. By exploring real-world system implementation and characterization, the unique features and versatile applications of radio-over-fiber technologies are investigated and reviewed to reach next-generation converged optical and wireless access networks.

21B. Advances in 1-100 GHz Microwave Photonics: All-Band Optical Wireless Access Networks Using Radio Over Fiber Technologies
Gee-Kung Chang, Yu-Ting Hsueh, and Shu-Hao Fan

With the growing bandwidth demand for the last mile and last meter in the access network, radio-over-fiber (RoF) technology at millimeter-wave (mmW) band has been viewed as one of the most promising solutions to providing ubiquitous multi-gigabit wireless services with simplified and cost-effective base stations (BSs) and low-loss, bandwidth-abundant fiber optic networks. This chapter first outlines the general methods and types of optical mmW generation, and summarizes their advantages and disadvantages. Owing to ultra-wide bandwidth and protocol transparent characteristics, a RoF system can be utilized to simultaneously deliver wired and multi-band wireless services for both fixed and mobile users. In the rest of this chapter, several multi-band 60-GHz RoF systems are reviewed, including mmW with baseband, microwave, mmW with commercial wireless services in low RF regions, and 60-GHz sub-bands.

22B. PONs: State of the Art and Standardized
Frank Effenberger

This chapter aims to describe the current state of the passive optical network (PON) technology, including both state-of-the-art systems that are currently under research in the laboratory and "standardized" systems that have been or soon will be described as an industry norm. A short introduction to the PON topic is given, to set the scene and provide the basic motivation for why PON is so important to fiber access. Then, each of the major technologies is reviewed, including time division multiplexing, video overlay, wavelength-division multiplexing, frequency-division multiplexing, and hybrid multiplexing. The focus of each review is at a system level to present a wide view of the whole range, and comparisons are made to different technologies.

23B. Wavelength-Division-Multiplexed Passive Optical Networks (WDM PONs)
Y.C. Chung and Y. Takushima

Wavelength-division multiplexed passive optical network (WDM PON) has long been considered as an ultimate solution for a future optical access network capable of providing practically unlimited bandwidth to each subscriber. On the other hand, it is still considered to be too expensive for mass deployment. To solve this problem and to meet the ever-increasing demand for bandwidth, there have been numerous efforts to improve the competitiveness of WDM PON. This chapter reviews the current status and future direction of these WDM PON technologies. It first reviews various colorless light sources, which are critical for the cost-effective implementation of the optical network units (ONUs), and several representative network architectures proposed for WDM PONs. The chapter then reviews the recent research activities for the realization of high-speed (>10 Gb/s) and long-reach WDM PONs. Various fault-monitoring and protection techniques are also reviewed, as they may be increasingly important in future high-capacity WDM PONs.

24B. FTTX Worldwide Deployment
Vincent O'Byrne, Chang Hee Lee, Yoon Kim, and Zisen Zhao

Since the early 2000s, Fiber-to-the-X, where X refers to different meanings for to different operators, has taken off around the world and is seen as the main method to meet the continued growth in the broadband needs of residential and business customers. This chapter covers two types of architectures, including the shared network among many users and the point-to-point network, and the standing of the various technologies for access space. The status of FTTX and some of the issues that operators are facing around the world are discussed. The chapter then reviews technologies that have been deployed to date and the new technologies that are under consideration to meet their customers' residential and business needs in the future.

25B. Modern Undersea Transmission Technology
Jin-xing Cai, Katya Golovchenko, and Georg Mohs

Much progress has been made over the last few years in undersea optical fiber telecommunication systems. Most importantly, coherent receivers have become practical, enabling polarization multiplexing and higher-order modulation formats with increased spectral efficiency. This chapter provides an overview of the progress in undersea transmission technology. After a brief general introduction to undersea systems and their unique challenges and design constraints, the principles of coherent transmission technologies are outlined. These include polarization multiplexing, linear equalizers, and multiple bits per symbol. The chapter then describes the use of strong optical filtering to help to improve spectral efficiency, and it reviews the techniques to mitigate the effects of inter-symbol interference. Higher-order modulation formats that can further increase spectral efficiency by increasing the number of bits per symbol are then introduced. The implications of the receiver sensitivity degradation and the mitigation techniques are discussed.

Ivan P. Kaminow
(Bell Labs, retired)
University of California, Berkeley, CA, USA

Tingye Li
(Bell Labs and AT&T Labs, deceased)
Boulder, CO, USA

Alan E. Willner
University of Southern California,
Los Angeles, CA, USA

Fiber Nonlinearity and Capacity: Single-Mode and Multimode Fibers

René-Jean Essiambre, Robert W. Tkach, and Roland Ryf

Bell Laboratories, Alcatel-Lucent, 791 Holmdel-Keyport Road, Holmdel, NJ 07733, USA

1.1 INTRODUCTION

The vast majority of all communications on the planet goes through a worldwide network of interconnected fused-silica optical fibers forming the backbone of optical networks. This fact results from important intrinsic advantages of optical fibers. A first critical property of fused-silica fiber is its wideband frequency region ($\sim$40 THz) of low transmission loss ($\sim$0.2–0.35 dB/km) centered around a carrier frequency of $\sim$200 THz. This broad low-loss frequency range allows a larger quantity of information to be transmitted [1] than in the narrower frequency bands available for other types of communication systems such as wireless (tens of MHz), digital subscriber lines (DSLs) ($\sim$a few tens of MHz) and satellite communications (a few hundreds of MHz). A second advantage of optical fibers is that they provide tight spatial confinement of a few tens of micrometers that enables the use of multiple independent channels with great spatial density.

A fused-silica optical fiber is a medium that differs in a fundamental manner from other transmission media—it exhibits the optical Kerr effect [2,3], a nonlinear phenomenon that introduces distortions that increase with signal power. As a result, there exists a maximum quantity of information that can be transmitted through optical fibers (see [4–7] and references therein for single-mode fibers, SMFs). This maximum nonlinear capacity is often referred to as the "nonlinear Shannon capacity limit" that we abbreviate in this chapter to simply "nonlinear capacity limit." The nonlinear capacity limit of a SMF depends on some of the fiber physical properties, such as loss and nonlinear coefficients and chromatic dispersion [7]. The nonlinear capacity limit of a 500-km-long system using the standard SMF (SSMF) is estimated to be between 70 Tb/s for a C-band system having 4 THz optical bandwidth and 175 Tb/s for 10 THz optical bandwidth for an extended C- and L-band system (see Section 1.5.2) [4].

Current backbone optical networks are exclusively based on SMFs and further capacity increase will eventually require installing new systems in parallel

Optical Fiber Telecommunications VIB. http://dx.doi.org/10.1016/B978-0-12-396960-6.00001-8

when we closely approach the nonlinear capacity limit of SMFs [4]. An alternative approach to parallel fibers is to consider fibers supporting multiple spatial modes to increase capacity [8]. Systems based on multimode fibers (MMFs) or multicore fibers (MCFs) have the potential to be of lower cost by making use of optical and electronic integrations while achieving greater spatial densities and reducing management complexity.

This chapter starts by providing some statistics on traffic demand in optical networks and the capacity scaling over time of commercial optical communication systems. These observations, in combination with the knowledge of a nonlinear capacity limit, suggest that a fiber capacity crunch may be looming [8]. This section is followed by a brief review of the basic results of information theory. We then describe the stochastic nonlinear Schrödinger equation (SNSE), the equation that governs nonlinear propagation in SMFs. This is followed by calculations of nonlinear capacity limit estimates for SSMF and for advanced fibers having improved transmission characteristics. An analytical formula of nonlinear capacity is also presented.

We then introduce a set of coupled partial differential equations (PDEs) describing nonlinear propagation of polarization-division multiplexed (PDM) signals in SMFs along with nonlinear capacity estimates for these systems. The next section focuses on MMFs and MCFs. We first present an elementary analysis of MMFs and MCFs capacities in the absence of fiber nonlinearity. The rest of the chapter focuses on nonlinear effects in MMFs and MCFs, with an emphasis on MMFs and few-mode fibers (FMFs). The impact of nonlinearity in fibers supporting multiple spatial modes is still an area with many unknowns, despite the fact that nonlinear effects in fibers were first observed using FMFs (see Ref. [9] for a historical review). The chapter concludes by reporting experimental observations of two important nonlinear effects between spatial modes: inter-modal cross-phase modulation (IM-XPM) and inter-modal four-wave mixing (IM-FWM).

1.2 NETWORK TRAFFIC AND OPTICAL SYSTEMS CAPACITY

The capacity of optical communication systems has seen an incredible growth since their inception four decades ago, increasing by more than a factor of 100,000 to reach today's commercial systems capacities of nearly 10 Tb/s. The technological implementation of such a data carrying capacity in a single fiber was unimaginable in the early days of optical communications, and would have seemed far beyond the needs of society if it had been envisioned. But since the mid-1990s, the advent of the Internet and its associated applications has driven a growth rate in data traffic that challenges the ability of optical technology to keep up. Figure 1.1 shows the capacity of commercial optical communication systems versus their year of introduction (including a projected point at 17.6 Tb/s in 2013). Also shown is a curve corresponding to North American core network traffic based on 2009 traffic levels and measured growth rates [10]. The traffic curve is dominated by voice before 2000 and by data traffic after 2004. Plotting these two quantities on the same graph yields

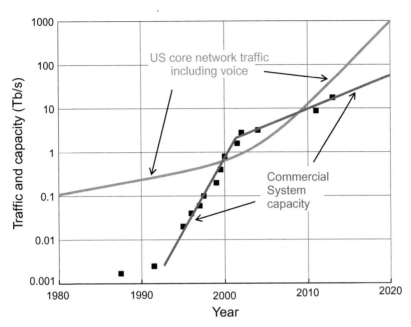

FIGURE 1.1 Commercial system capacities (squares and red curve) and total network traffic including voice (blue curve). The growth of network traffic currently exceeds the growth of systems capacity. (For interpretation of the references to color in this figure legend, the reader is referred to the web version of this book.)

a striking observation: Traffic growth of roughly 50% per year is greatly outstripping the growth in the capacity of optical communications systems of less than 20% per year. Thus, while the entire traffic in the North American core network could be carried on a single fiber in 2008, in 2011 more than two fibers were required. If the growth trends continue as shown, the required number of fibers (and systems) will double every 3 years. While this situation may seem rosy for system vendors, it is frightening to their customers.

Even these projected growth rates for system capacity are likely to be overly optimistic. If we examine the historical trends for system capacity we see a period of extremely rapid growth in the mid-1990s arising from the introduction of wavelength-division multiplexing (WDM). In those years, most of the capacity growth was achieved by simply expanding the number of WDM channels and concomitantly the occupied bandwidth of the erbium-doped fiber amplifiers (EDFAs). After the first systems with capacities of several hundred Gb/s were introduced, the bandwidth of the amplifiers was fully occupied. Further increases in capacity have come from increased efficiency in the use of the spectrum. This "spectral efficiency" (SE) is a familiar concept used in many other fields of communication, and particularly in wireless communication systems. It is expressed in bits per second per Hz. Figure 1.2

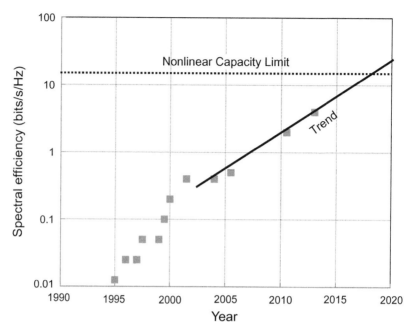

FIGURE 1.2 Spectral efficiencies for the commercial WDM systems of Figure 1.1. The dashed horizontal line at the top is the nonlinear capacity limit estimate of Section 1.5.2.

shows the SE of the same commercial systems shown in Figure 1.1, plotted versus year. Up through the year 2000 all systems shown used on-off-keying encoding one bit per time slot on the presence or absence of a pulse of light. Bit rates for WDM systems began at 2.5 Gb/s in the 1990s, increased to 10 Gb/s in 1996, and to 40 Gb/s in 2003. Those increases in bit rate (and bandwidth of the signals) at fixed WDM channel spacing resulted in increased SE. But the introduction of 40 Gb/s rates resulted in signal bandwidths that became comparable to optical channel spacings, then at 50 GHz. Further increases in SE were obtained through the use of PDM signals, i.e. sending two independent signals on the two polarizations of light, and through more complex modulation formats. Today's high-capacity systems use quaternary phase-shift keying (QPSK) employing four possible phases on the pulse of light in a given pulse time slot. Combined with PDM signaling yields 4 bits of data for each time slot. Channel spacings of 50 GHz and bit rates of 100 Gb/s yield a net SE of 2 bits/s/Hz. The systems that will be introduced in the next few years will increase SE further by using multiple pulse amplitudes as well as phases to create 16 possible combinations and will achieve a net SE of 4 bits/s/Hz. However the increased number of possible states requires an improved signal-to-noise ratio (SNR) to ensure accurate discrimination of the possibilities. Increasing SNR requires increasing signal power that in turns increases the impact of fiber nonlinearity.

As mentioned in the Introduction, there has been a significant effort to understand the impact of fiber nonlinearity on the capacity of fiber optical systems. This complex problem is the subject of this chapter. However, the effect can be broadly understood in simple terms. Shannon's theory of channel capacity states that the capacity of a communication channel scales with the bandwidth of the channel times logarithm of 1 plus the SNR as seen in Eq. (1.1). Thus there is no maximum capacity as long as signal power can be increased. But, in a typical optical network, fiber nonlinearity generates distortions that increase faster than the signal power. This eventually sets a limit on the signal power that may be used. This manifests itself as an eventual decrease in performance as signal power is increased and the appearance of a maximum capacity for any given bandwidth or, equivalently, a maximum SE. In Figure 1.2, the dashed horizontal line roughly indicates this maximum SE corresponding to typical system parameters. Note that the continuation of the trend of improving SE would cross this limit in roughly 5–7 years. We have already seen that the current trend of improvement in system SE lags behind the rate of traffic growth, but even that rate of improvement will soon be limited by fundamental concerns.

1.3 INFORMATION THEORY

We present in this section the basic notions of Shannon's information theory for the additive white Gaussian noise (AWGN) channel. We also establish relations between information theory and the language of optical communication.

1.3.1 Basic concepts

Shannon introduced the concept of channel "capacity" in his landmark paper in 1948 [11]. He defined the channel as "the medium used to transmit the signal from transmitter to receiver." In a pragmatic approach intended to deal with various models used to represent a physical system, it has become common to define the channel as "that part of a communication system that the designer is unable or un-willing to change" [12,13]. A representation of a modern channel is displayed in Figure 1.3 as a part of a communication from the origin and a destination.

A channel can be defined as supporting only real signals or supporting complex signals that have two quadratures. In this chapter, we adopt the convention of a channel that supports complex signals. It is interesting to point out that, in optical communication, the term "channel" is commonly used to refer to a signal of limited bandwidth used as part of a WDM system. To avoid any confusion, the term "WDM channels" is used for these optical signals.

The capacity of a channel in the sense of Shannon can be described as the asymptote of the rates of transmission of information that can be achieved with arbitrarily low error rate. Even though its formulation has great generality, in his seminal paper [11], Shannon gave a great importance to the AWGN channel [14–16]. Over the years, Shannon's theory has been adapted to include other propagation effects that

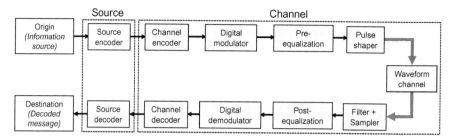

FIGURE 1.3 Representation by blocks of typical functions that allow efficient transmission of information. The channel and source functions are surrounded by dotted boxes. The thick arrows represent analog waveforms of the "physical world" channel.

are present in different communication channels such as the wireless [17,18], DSLs [19,20], and satellite [21,22] communication channels.

In this section, we review the capacity relations for a *single channel* with AWGN. Note that a SMF supports one guided transverse mode having two states of polarization, each supporting a complex signal. Therefore, a single-mode fiber supports two (complex) channels in the sense of Shannon. For a bandlimited channel, it is convenient to quote the SE defined as the capacity C per unit of bandwidth B. From Shannon's theory, the SE of a single AWGN channel is given by the widely known relation [14,16],

$$\text{SE} \equiv \frac{C}{B} = \log_2{(1 + \text{SNR})}, \tag{1.1}$$

where the SNR is defined as $\text{SNR} = P/N$ where P is the average signal power and N is the average noise power in a bandwidth equal to the baud rate R_s of a time-division multiplexed (TDM) signal. The minimum bandwidth that such a signal can have without suffering from inter-symbol interference (ISI) [23,24] when sampled at the optimum sampling point is equal to the symbol rate R_s [11,25–27]. Such signal can be generated by superposing delayed pulses of the form $\sin(t)/t$ associated with each symbol [23,4]. The symbol is encoded in the amplitude and the phase of each pulse. For this minimum-bandwidth signal the noise power is given by $N = N_0 R_s$, where N_0 is the noise power spectral density. The SNR of Eq. (1.1) can be written in a few different forms, [23,24,28,29]

$$\text{SNR} \equiv \frac{P}{N} = \frac{P}{N_0 R_s} = \frac{E_s}{N_0}, \tag{1.2}$$

where we used the relation $P = E_s R_s$, where E_s is the energy per symbol. At low SNR, Eq. (1.1) can be approximated by

$$\text{SE} \approx \frac{1}{\ln 2}\left(\text{SNR} - \frac{\text{SNR}^2}{2}\right). \tag{1.3}$$

An interesting observation from Eq. (1.3) is that both SE and SNR go simultaneously to zero. This implies that one cannot achieve high SE at low SNR on a single channel. Using Eq. (1.2) one can define an SNR *per bit*, SNR_b, as [16,23,24]

$$\frac{E_b}{N_0} \equiv SNR_b = \frac{SNR}{SE}, \tag{1.4}$$

where we use the relation $E_b = E_s/SE$ that relates the energy per bit to the energy per symbol. Using Eq. (1.4), one can rewrite Eq. (1.1) as

$$SNR_b = \frac{2^{SE} - 1}{SE}. \tag{1.5}$$

At high SE, $SNR_b \sim 2^{SE}/SE$. One can rewrite Eq. (1.5) as a series expansion around SE$=0$. It is given by

$$SNR_b = \sum_{q=0}^{Q} (\ln 2)^{q+1} SE^q / (q+1)!. \tag{1.6}$$

Expansion to second order and at low SE gives,

$$SNR_b \approx \ln 2 + \frac{(\ln 2)^2}{2} SE + \frac{(\ln 2)^3}{6} SE^2. \tag{1.7}$$

One notes that, in contrast to the relation between SE and SNR of Eq. (1.3), the SNR per bit, SNR_b, does not go to zero with the SE, but assumes a minimum value

$$SNR_b^{min} = \ln 2, \tag{1.8}$$

or $\sim -1.59\,dB$ [30] (see Figure 1.4). Therefore, there is a minimum SNR or energy per bit to transmit information over the AWGN channel [14,16] and this minimum occurs when operating at low SE.

It is interesting to define the following ratio:

$$\Delta SNR_b \equiv SNR_b / SNR_b^{min}, \tag{1.9}$$

which can be interpreted as an "excess" energy per bit at which a system operates above the minimum energy per bit required for the AWGN channel. A schematic representation of the quantity ΔSNR_b is shown in Figure 1.4 for a system operating at ~ 8.5 bits/s/Hz. Also shown are various approximations for the SE.

Finally, one can write the SE as a function of SNR_b to first order in SNR_b as,

$$SE \approx \frac{2}{(\ln 2)^2} \left(SNR_b - SNR_b^{min} \right). \tag{1.10}$$

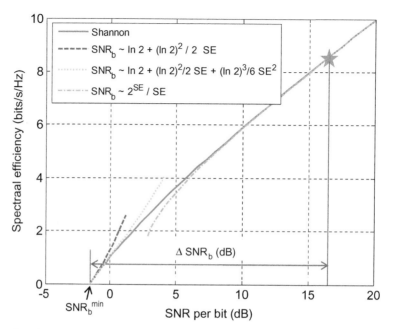

FIGURE 1.4 SE versus SNR per bit along with a few approximations of SNR$_b$. Also defined is the "excess" SNR per bit, ΔSNR$_b$, for a system operating at a certain SE.

1.3.2 Link to optical communication

In optical communication, the *optical* SNR (OSNR) is traditionally utilized to represent the ratio of powers between signal and noise. It is defined as [31]

$$OSNR = \frac{P}{2 N_{ASE} B_{ref}},$$

(1.11)

where P is the average signal power, N_{ASE} is the spectral density of amplified spontaneous emission (ASE) per polarization and B_{ref} is a *fixed* reference bandwidth of 0.1 nm (12.5 GHz at 1550 nm). The power P is the sum of the signal power in both states of polarization. Note that there are no assumptions of the signal carrying data in the definition of OSNR. This is because this quantity is commonly used to characterize continuous-wave (CW) lasers that are unmodulated. The OSNR also considers the noise in both polarization states independent of whether a polarization-division multiplexed (PDM) signal is used or not. The OSNR is defined to easily relate to spectral measurements using an optical spectrum analyzer (OSA) that measure all the power in a given bandwidth, typically 0.1 nm in the early models available.

From Eqs. (1.2) and (1.11), the relation between the OSNR and SNR is given by [4]

$$OSNR = \frac{p R_s}{2 B_{ref}} SNR,$$

(1.12)

where the parameter $p = 1$ when only a single polarization of the signal is used and $p = 2$ for a PDM signal. The noise power spectral densities N_{ASE} and N_0 have canceled out since they represent the same physical quantity.

1.4 SINGLE-MODE FIBERS: SINGLE POLARIZATION

We consider in this section the nonlinear transmission and nonlinear capacity of a singly polarized signal over SMFs. We first present nonlinear capacity estimates for SSMF and then of advanced SMFs with greatly improved physical properties, in abstraction of their physical realizability. Finally, we discuss an analytical formula of capacity that fits well the numerical capacity results that have been presented.

1.4.1 Stochastic nonlinear Schrödinger equation

Nonlinear propagation through optical fibers having loss continuously compensated by gain is described by a PDE, often referred to as the stochastic nonlinear Schrödinger equation (SNSE) [2,4],

$$\frac{\partial E}{\partial z} + \frac{\iota}{2}\beta_2 \frac{\partial^2 E}{\partial t^2} - \iota \gamma |E|^2 E = \iota \mathcal{N}, \tag{1.13}$$

where $E(z, t)$ is the optical field at a given location z and time t, γ the fiber nonlinear coefficient and $\mathcal{N}(z,t)$ is the field that represents the noise, or in this case the ASE, and $\beta_2 \equiv \beta_2(\omega_0) = d^2\beta/d\omega^2|_{\omega=\omega_0}$ is the group-velocity dispersion (GVD) parameter evaluated at ω_0, an arbitrary angular optical frequency generally conveniently chosen to be the signal center frequency. The relation between β_2 and the more commonly used chromatic dispersion D is

$$D = -\frac{2\pi c}{\lambda^2}\beta_2, \tag{1.14}$$

where c is the speed of light in vacuum.

The field $E(z, t)$ is assumed to be singly polarized and the compensation of the fiber loss in Eq. (1.13) is assumed to be by ideal distributed Raman amplification [32]. The fiber nonlinear coefficient γ is defined as [2]

$$\gamma = \frac{n_2 \omega_s}{c A_{\text{eff}}}, \tag{1.15}$$

where n_2 is the fiber nonlinear refractive index [3], $\omega_s = 2\pi v_s$ is the angular optical frequency at the signal wavelength with v_s being the optical frequency and A_{eff} the fiber effective area defined as [2, Eq. (2.3.29)]

$$A_{\text{eff}} = \frac{\left(\iint_{-\infty}^{\infty} |F(x,y)|^2 dx\, dy\right)^2}{\iint_{-\infty}^{\infty} |F(x,y)|^4 dx\, dy}, \tag{1.16}$$

where $F(x,y)$ is the mode field transverse distribution.

If we assume that ASE can be modeled by an AWGN source [33,34], Shannon's information theory can be applied to evaluate the impact of ASE noise on capacity. Besides being uncorrelated in time, the ASE field $\mathcal{N}(z,t)$ is also assumed to be uncorrelated at different locations in the fiber. It therefore possesses the following autocorrelation [35–37]

$$\mathcal{E}[\mathcal{N}(z,t)\mathcal{N}^*(z',t')] = \alpha L h \nu_s K_T \delta(z-z')\delta(t-t'), \qquad (1.17)$$

where $\mathcal{E}[\cdot]$ is the expectation value operator and δ the Dirac functional. The parameter K_T in Eq. (1.17) originates from Raman amplification and is given by $K_T = 1 + \eta(T, \nu_s, \nu_p)$ where $\eta(T, \nu_s, \nu_p)$ is the phonon occupancy factor [38, Eq. (7)], [39]. At room temperature, $K_T \sim 1.13$.

For ideal distributed Raman amplification, the noise spectral density per state of polarization can be written as [30,32],

$$N_{\text{ASE}} = \alpha L h \nu_s K_T, \qquad (1.18)$$

where L is the system length and h the Planck constant. The energy of the noise per symbol and per polarization is simply $N_{\text{ASE}} R_s$. The fiber loss coefficient α is often expressed in dB using $\alpha_{\text{dB}} = 10 \log_{10}(e)\alpha$, where e is the Euler's number.

1.4.2 Nonlinear capacity of standard single-mode fiber

We will now present some nonlinear capacity estimate results of the standard single-mode fiber (SSMF) based on Refs. [4,7]. We refer to these capacity results as "estimates" because a few assumptions have entered the capacity calculations and that they can either lower or increase capacity, resulting in no definitive capacity bound.

The parameters of the SSMF considered are given in Table 1.1. The calculated nonlinear capacity estimate curve for 500 km SSMF transmission using ideal distributed amplification and digital nonlinear back-propagation on the WDM channel of interest is given in Figure 1.5. The symbol rate $R_s = 100$ Gbaud and the channel

Table 1.1 Standard single-mode fiber parameters.

Parameter	Symbol	SSMF
Chromatic dispersion	D	17 ps/(nm-km)
Dispersion slope	S	0.07 ps/(nm²-km)
Loss coefficient	α_{dB}	0.2 dB/km
Nonlinear refractive index	n_2	2.5×10^{-20} m²/W
Effective area	A_{eff}	80 μm²
Nonlinear coefficient	γ	1.27 (W-km)$^{-1}$
Wavelength	λ_s	1550 nm
Optical frequency	ν_s	193.41 THz

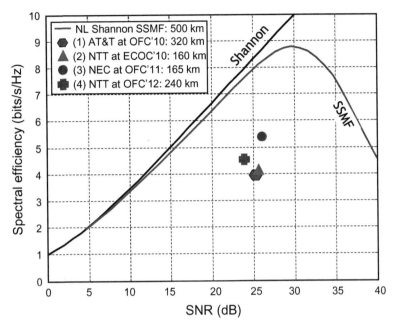

FIGURE 1.5 Nonlinear capacity for transmission over 500 km of SSMF (see Table 1.1) for single-polarization signals and current record capacity experiments. The experiments (1)–(4) are from Refs. [41–44], respectively.

spacing $\Delta f = 102$ GHz. Even though only this baud rate is presented, similar results were obtained for different baud rates for the same ratio $R_s/\Delta f$. The capacity curve shown is for a zero dispersion slope ($S=0$, see [40, Eq. (6.24)] for the definition of S) but calculations using different values of S produce virtually identical curves. Also shown are the latest record SE experiments for transmission over at least 100 km [41–44]. The experimental capacity records are within a factor 1.6–2 from the nonlinear capacity limit estimate of close to 9 bits/s/Hz (for single polarization). Note that the highest SE experiments have been achieved for distances shorter than 500 km and that a reduction in SE is expected if the distance were to be extended to 500 km.

Calculations of nonlinear fiber capacity estimates have been performed for various distances and the nonlinear capacity limits for each distance are displayed in Figure 1.6. The nonlinear capacity limit for 8000 km is about 5.5 bits/s/Hz, a value more than half the nonlinear capacity for 500 km, even though the distance is increased by a factor of 16. This seemingly surprising result can be understood by the fact that higher capacities are achieved using denser constellations (larger constellation size $\mathcal{M}$), and that the larger the constellation, the more sensitive a signal is to nonlinear distortions. These larger constellations does not allow raising the signal power significantly, even when the transmission distance is shortened dramatically.

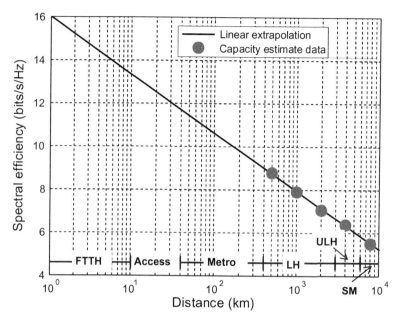

FIGURE 1.6 Nonlinear capacity limit as a function of distance for SSMF (Table 1.1) along with a linear extrapolation of data.

Figure 1.6 shows the calculated nonlinear capacity limit estimate or maximum achievable SE versus distance. These calculated SE values fit well a linear relation when the distance is plotted on a logarithmic scale, so a linear fit to the SE is also shown (see Section 1.4.4 for more details). The types of optical networks corresponding to the various distances are displayed at the bottom of the plot. One should note that the maximum SE of submarine (SM), ultra-long-haul (ULH), and long-haul (LH) systems does not vary much considering the difference of reach of these systems. The maximum SE continues to increase as the system reach decreases to metropolitan (metro) and access networks. Finally, fiber-to-the-home (FTTH) has the highest maximum SE, but, remarkably, only three times the maximum SE of the longest optical systems, the submarine lines. This is a rather small increase in maximum SE considering that there is a difference of four orders of magnitude in distance between FTTH and submarine systems. This underscores the difficulties associated to increasing SE for a nonlinear medium like the optical fiber.

1.4.3 Advanced single-mode fibers

Given an optical system of a certain length, the nonlinear fiber capacity limit depends mainly on the fiber properties. We discuss below the impact of three fiber parameters on the nonlinear capacity limit: the fiber loss coefficient α, the fiber nonlinear coefficient γ, and fiber dispersion D.

1.4.3.1 Fiber loss

Figure 1.7 shows the dependence of the maximum nonlinear fiber capacity on the fiber loss coefficient α_{dB}. One observes that the maximum nonlinear capacity does not increase dramatically with a large reduction of α_{dB}. For instance, reducing α_{dB} from 0.2 to 0.05 dB/km increases capacity only from ~8 to ~9 bits/s/Hz. Even though the gain in nonlinear capacity is small, fibers with lower loss can reduce the cost of systems by allowing a large spacing between pump stations for Raman amplification and larger amplifier spacing for periodic optical amplification.

The typical loss coefficient of SSMF of 0.2 dB/km is shown in Figure 1.7 as well as the record lowest fiber loss coefficient of 0.1484 dB/km achieved in a pure silica-core fiber (PSCF) [45]. A striking feature of Figure 1.7 is that even though lowering the fiber loss coefficient below 0.15 dB/km represents a tremendous challenge, the increase in fiber capacity is surprisingly limited.

1.4.3.2 Fiber nonlinear coefficient

The impact on the nonlinear capacity limit of the nonlinear coefficient γ is shown in Figure 1.8. A reduction of the fiber nonlinear coefficient has a similar effect on capacity as a reduction in the fiber loss coefficient. As in the case of the fiber loss coefficient, it is striking to see that a reduction in the nonlinear coefficient by a factor of 1000, such as what can be nearly achieved by using hollow-core fibers (HCFs) [46,47], would increase the nonlinear capacity limit by a mere ~30%.

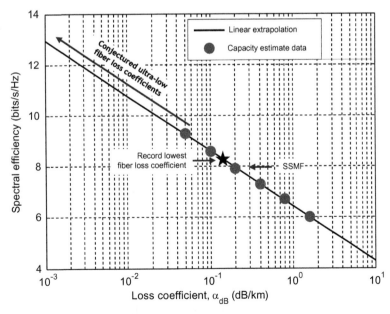

FIGURE 1.7 Nonlinear capacity limit as a function of the fiber loss coefficient. $L = 1000$ km, $\gamma = 1.27$ (W-km)$^{-1}$, and $D = 17$ ps/(nm-km).

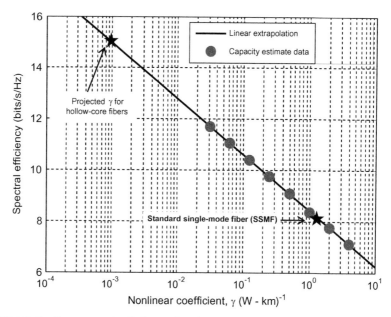

FIGURE 1.8 Nonlinear capacity limit as a function of the fiber nonlinear coefficient.
$L=500\,km$, $\alpha_{dB}=0.15\,dB/km$, and $D=17\,ps/(nm\text{-}km)$.

Clearly, capacity scales very slowly with the nonlinear coefficient γ. Because they operate at lower SE due to higher noise levels, the impact of a reduction of γ has a higher relative impact on capacity in systems based on discrete EDFAs than for systems based on ideal distributed Raman amplification. However, for both system types, very large changes in γ are necessary to significantly increase capacity. Perhaps more importantly, reducing the nonlinear coefficient does not reduce the noise level of a system and the gain in capacity would be achieved by increasing signal power in proportion to the reduction in the nonlinear coefficient. Clearly, the 1000 times increase in signal power required to achieve a 30% increase in the nonlinear capacity limit of a fiber with a 1000 times lower nonlinear coefficient is dramatically energy inefficient and unrealistic. Installing parallel systems or using fibers supporting multiple transverse modes such as MMFs and MCFs as described in Section 1.6 appears a more appealing option.

1.4.3.3 Fiber dispersion

The impact of the fiber chromatic dispersion D on the nonlinear capacity limit is shown in Figure 1.9. The range of chromatic dispersion simulated covers most commercial fibers. A scaling of the chromatic dispersion D produces smaller variations of the nonlinear capacity limit than for the loss and nonlinear coefficients. This shows that a variation of chromatic dispersion has a lower impact on capacity than the two other fiber parameters considered. Note that we are assuming that we are not too

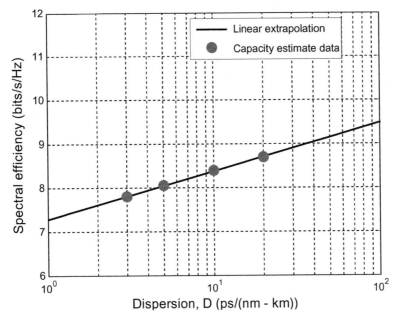

FIGURE 1.9 Nonlinear capacity limit as a function of the fiber chromatic dispersion. $L = 500\,\text{km}$, $\alpha_{\text{dB}} = 0.2\,\text{dB/km}$, $\gamma = 1.27\,(\text{W-km})^{-1}$.

close to the zero dispersion value where nonlinear effects such as four-wave mixing (FWM) can be significantly enhanced [48].

1.4.4 Analytic formula of fiber capacity

The dominant mechanism responsible for the nonlinear Shannon capacity limit for single polarization signals was established from full simulations to be cross-phase modulation (XPM) [4]. Based on this knowledge, a theory that accounts for XPM in the highly dispersive regime, often referred to as pseudo-linear transmission, has been developed in Ref. [49]. The analytic nonlinear capacity limit C and SE can be expressed as [49]

$$\text{SE} = \frac{C}{\Delta f} = \log_2 \left(1 + \left[\frac{n_{\text{sp}} \hbar \omega_0 \alpha L R_s}{P} + 4 \frac{\gamma^2 \mathcal{P}^2 L}{|R_s^2 \beta_2|} \right]^{-1} \right), \tag{1.19}$$

where

$$\mathcal{P} = \left[\sum_{n=-N_{\text{ch}}/2(n \neq 0)}^{N_{\text{ch}}/2} \frac{\kappa}{2\pi} \frac{R_s}{|\Delta f_n|} \right]^{1/2} P \tag{1.20}$$

and where $\kappa = 1$ for a Gaussian constellation and

$$\kappa = \frac{n^4/5 + n^3/2 + n^2/3 - 1/30}{(n^2/3 + n/2 + 1/6)^2} - 1, \tag{1.21}$$

for equally spaced n-ring constellations with equal probability of occupation on each ring [4]. The values of κ for $n = 2, 3, 4, 16$ are 0.36, 0.5, 0.5733, 0.7433, and 0.8 for an infinite number of rings.

Figure 1.10 shows the nonlinear capacity curves from the full nonlinear simulations for various distances [4, Fig. 35] and the corresponding analytic capacity curves from Eq. (1.19). The parameters used for the analytic capacity Eq. (1.19) not already specified in Table 1.1 or before are: spontaneous emission factor $n_{sp} = 1$, symbol rate $R_s = 100$ Gbaud, system length L as indicated in Figure 1.10, average signal power P, constellation shaping factor $\kappa = 0.74$ for a 16-ring constellation, number of channels $N_{ch} = 5$, channel spacing $\Delta f = 100$ GHz and frequency separation between the central and neighboring channels $\Delta f_n = n\Delta f$. These parameters give $\mathcal{P} \sim 0.59P$ and the analytical curves of Figure 1.10. The fit between the analytical capacity formula (1.19) and the numerical results is rather good as seen in the figure. The main source of deviation is thought to originate from not taking into account explicitly the

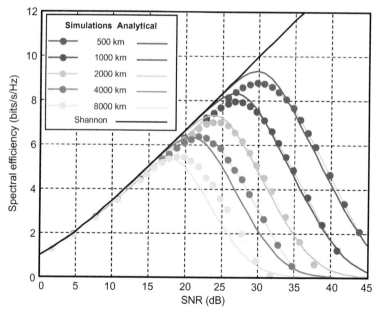

FIGURE 1.10 Maximum SE versus SNR for various distances. Filled circles represent the results of numerical simulations while lines without circles are from the analytic formula of Eq. (1.19).

reduction in capacity from the ring constellation shape in the linear part of C where a Gaussian constellation is assumed.

The maximum value of the nonlinear capacity of Eq. (1.19) or nonlinear capacity limit is given by [49],

$$\mathrm{SE_{max}} = \log_2 \left\{ 1 + \frac{1}{3L} \left[\frac{\left(n_{\mathrm{sp}} \hbar \omega_0 \alpha \gamma \right)^2 \mathcal{S}_B}{|\beta_2|} \right]^{-\frac{1}{3}} \right\}, \tag{1.22}$$

where $\mathcal{S}_B$ is a constant independent of the fiber. At high SNR and on a log scale for the fiber parameters, $\mathrm{SE_{max}}$ behaves as $\propto L^{-1}, \propto \alpha^{-2/3}, \propto \gamma^{-2/3}$, and $\propto |\beta_2|^{-1/3}$, a scaling consistent with the behavior observed in Figures 1.6–1.9. Finally, one should mention that other nonlinear capacity formulas based on FWM have been derived [50,51] that produce similar but non-identical results.

1.5 SINGLE-MODE FIBERS: POLARIZATION-DIVISION MULTIPLEXING

The term "single" in single-mode fibers (SMFs) refers to a single *spatial* mode. Because there are two dimensions of polarization for the signal, SMFs support two independent modes. In the absence of nonlinear effects, one can transmit independent signals that are impaired by independent sources of noise. In the *linear regime*, the SE achievable is therefore twice the value derived by Shannon for a single channel, or

$$\mathrm{SE}^{(2)} = 2 \log_2 \left(1 + \mathrm{SNR} \right), \tag{1.23}$$

which is the Shannon capacity for two independent complex channels. However, as for single-polarization signals, nonlinear effects are expected to limit capacity of PDM signals. The calculation of a nonlinear capacity estimate of PDM systems requires the knowledge of the set of nonlinear propagation equations that captures nonlinear interactions between the two states of polarization of a PDM signal. This is reported in the next section.

1.5.1 Nonlinear propagation: stochastic Manakov equations

Optical fibers exhibit small residual and essentially random birefringence due to the small ellipticity of the fiber core or to small difference between the refractive indices between different polarization states. The presence of such small random birefringence produces rapid random dephasing between the x and y-polarization components. This dephasing results in a redistribution of the nonlinear effects and a set of coupled NSEs, generally referred to as the Manakov equations [52,53], can be derived to model nonlinear propagation. This results in the following set of coupled PDEs,

$$\frac{\partial E_x}{\partial z} + \frac{\iota}{2}\beta_2\frac{\partial^2 E_x}{\partial t^2} - \iota\gamma\frac{8}{9}\left(|E_x|^2 + |E_y|^2\right)E_x = \iota\mathcal{N}_x, \tag{1.24}$$

$$\frac{\partial E_y}{\partial z} + \frac{\iota}{2}\beta_2\frac{\partial^2 E_y}{\partial t^2} - \iota\gamma\frac{8}{9}\left(|E_y|^2 + |E_x|^2\right)E_y = \iota\mathcal{N}_y, \tag{1.25}$$

where $E_x(z,t)$ and $E_y(z,t)$ are the x and y components of the optical field and $\mathcal{N}_x(z,t)$ and $\mathcal{N}_y(z,t)$ are the x and y components of the noise field. Each noise field component $\mathcal{N}_x$ and $\mathcal{N}_y$ obeys Eq. (1.17) individually. The Manakov equations are considered to accurately describe the nonlinear effects between the two polarization states in fibers that almost invariably exhibit some small random residual birefringence [53]. The Manakov equations (1.24) and (1.25) can be written in a more compact form as,

$$\frac{\partial \mathbf{E}}{\partial z} + \iota\frac{\beta_2}{2}\frac{\partial^2 \mathbf{E}}{\partial t^2} - \iota\gamma\frac{8}{9}|\mathbf{E}|^2\mathbf{E} = \iota\mathcal{N}, \tag{1.26}$$

where $\mathbf{E}(z,t)$ is a 1×2 column vector of the form $\mathbf{E}(z,t) = [E_x(z,t)\,E_y(z,t)]^{\mathrm{T}}$ where $E_x(z,t)$ and $E_y(z,t)$ are the x and y polarization components of the field. The noise field vector $\mathcal{N}(z,t)$ is defined similarly to $\mathbf{E}(z,t)$ as $\mathcal{N}(z,t) = [\mathcal{N}_x(z,t)\mathcal{N}_y(z,t)]^{\mathrm{T}}$.

1.5.2 Capacity of PDM systems

To fully exploit the two polarization states or modes of SMFs, one should use polarization-division multiplexing (PDM). Figure 1.11 shows a calculated nonlinear capacity estimate using PDM over 500 km of SSMF. The SSMF parameters are as reported in Table 1.1. Independent ring constellations have been used for each polarization component of the PDM signal. As a reference, the nonlinear capacity curve for single polarization of Figure 1.5 is displayed. Also shown is twice this nonlinear capacity that represents the capacity that would be obtained in the absence of nonlinear interactions between the two polarization components of the PDM signal. The Shannon limit for two independent channels of Eq. (1.23) is also shown. As can be seen in Figure 1.11, the nonlinear capacity limit of the PDM signal is slightly lower than twice the single-polarization nonlinear capacity, suggesting that nonlinear interactions between polarization components may cause a nonlinear capacity reduction in PDM systems. The most likely nonlinear interaction responsible for this capacity reduction is cross polarization modulation (XPolM) [2,53]. Those are the terms involving $|E_y|^2$ in Eq. (1.24) and $|E_x|^2$ in Eq. (1.25).

If one considers a nonlinear capacity limit estimate of $\sim$17.5 bits/s/Hz in Figure 1.11 (peak of PDM curve), the capacity limit of a 32-nm (4 THz) C-band system is 70 Tb/s. For an extended 80-nm (10 THz) (C+L)-band system it is 175 Tb/s. For the entire 300-nm low-loss window from $\lambda = 1300$ nm to 1600 nm (43 THz),

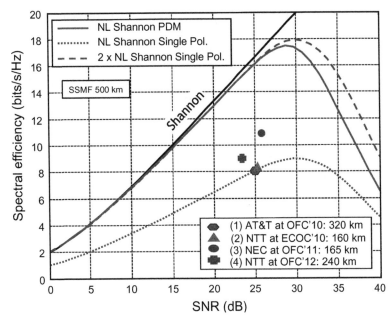

FIGURE 1.11 Nonlinear Shannon capacity for polarization-division multiplexing (PDM) signals for transmission over 500 km of SSMF. Single-polarization nonlinear capacity curve, and twice that curve, are shown as references. Also displayed are current record capacity experiments. The experiments (1)–(4) are from Refs. [41–44].

assuming a 0.2 dB/km of loss coefficient, the capacity would be 0.75 Pb/s. The capacity of C-band systems of 70 Tb/s should be achieved toward the end of the decade as shown in Figure 1.1 while the 175-Tb/s capacity a few years after 2020. Finally, a further small increase of the nonlinear capacity limit can be envisioned by considering four-dimensional constellations [54] optimized for nonlinear systems.

1.6 MULTICORE AND MULTIMODE FIBERS

The study of nonlinear capacity limit of SMFs has led to the realization that we are less than a decade away from closely approaching this limit at the current rate of capacity growth of commercial systems [4,55]. It is reasonable to expect that it will become increasingly difficult to achieve an increase in SE as we approach the nonlinear capacity limit. This is because the last bits/s/Hz will require, for instance, to nearly achieve ideal distributed Raman amplification, have modulation formats with large optimized constellation, a forward-error correction (FEC) closely approaching the linear Shannon limit, Nyquist pulse shaping with very steep roll-off, etc. As a result of the nonlinear capacity results, it naturally follows that considering fibers supporting multiple spatial modes appears to be the most efficient way to increase capacity per fiber strand.

In this section, we first describe different types of fibers that support multiple spatial modes. We follow by a calculation of the capacity scaling with the number of Shannon channels, considered to be linear and independent. A fiber supporting M spatial modes has $2M$ Shannon channels. Because the fields in different modes overlap in space, there is the possibility of nonlinear interactions between them. Even though the first publications on nonlinear interactions in MMFs started in the early seventies, this topic still remains largely unexplored due to the complexity of the interactions and the lack of simple PDEs describing the nonlinear evolution of the various fields in such fibers. In this section, we present a set of reduced PDEs that can be used, under certain approximations, to model nonlinear propagation in fibers supporting multiple spatial modes. Finally, we present two nonlinear effects observed in MMFs, inter-modal XPM and inter-modal FWM.

1.6.1 Types of multicore and multimode fibers

Figure 1.12 shows cross-sections of a SMF and of fibers supporting multiple spatial modes. Fiber 1 is the SMF represented for comparison. Fibers 2–4 are MCFs with a different number of cores placed on an hexagonal grid [56–65]. The hexagonal pattern of cores allows for the most compact placement of cores when all

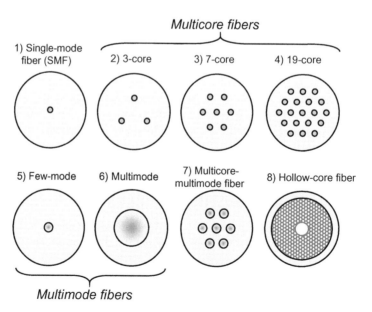

FIGURE 1.12 Cross-sections of a single-mode fiber and fibers supporting spatial multiplexing. Filled areas with darker shades of gray have a higher index of refraction (solid lines are shown as delimiters only).

nearest-neighbor cores are separated by the same distance (this is related to the problem of sphere packing in two dimensions [54]). Fibers that contain a single large core that can support multiple spatial modes are called MMFs [66–69]. If the number of spatial modes supported is between 2 and ~10, one generally refers to these fibers as FMFs. Fiber 5 depicts a FMF while fiber 6 represents a MMF. An hybrid between MCF and MMF is possible and an example is displayed as fiber 7. Finally, fiber 8 represents a hollow-core fiber (HCF), a type of photonic bandgap fiber (PBF).

A MCF can be designed to have independent cores or coupled cores. For independent cores, the goal is to create uncoupled parallel channels [58,59,70–72] by preventing linear coupling between cores. This approach results in independent channels in the fiber and is covered in other chapters in this book.

A second design of MCF does not impose any restrictions on the linear coupling between cores, allowing large coupling to occur. One refers to these fibers as coupled-core MCFs [73–75]. As a result of coupling, data launched in each core generally exit the fiber from multiple cores. Fortunately, one can recover the data by applying multiple-input multiple-output (MIMO) digital signal processing (DSP) at the receiver [17] after coherent detection. Such MIMO processing enables data recovery with low penalty as described in the chapter by Winzer et al. in this book.

By making the core of a SMF larger, the fiber eventually supports a few spatial modes and becomes a FMF [76,77] (see chapter by Peckham et al. in this book). By increasing the core further, the FMF supports a larger number of modes and becomes a MMF. The number of spatial modes guided in a MMF increases rapidly with the core diameter and the number of spatial modes that are guided can exceed a few hundred. For an *ideal* MMF, spatial modes overlap in space but remain orthogonal. In practice, coupling between spatial modes is introduced by macro- [67,78,79] and micro-bending [67,80–82,78,83]. In a similar manner to the MCFs with linear coupling, MIMO techniques can be applied in combination to coherent detection to recover the data in realistic MMFs.

A different type of fiber that can support multiple spatial modes are photonic bandgap fibers (PBFs). These fibers can be engineered to dramatically alter the propagation properties of a fiber. An example of PBF of interest is the HCF shown in Figure 1.12. A HCF can be designed to support a single or multiple spatial modes [84,85]. Because the core does not contain material, most of the guided fields propagate in air and experience an ultra-low nonlinear coefficient [84]. It has been conjectured that fiber loss coefficients possibly as low as ~0.05 dB/km at a wavelength of $2 \, \mu$m could be achieved [86,87]. The impact of reduced loss and reduced nonlinearity on capacity for single-mode operation has been discussed in Section 1.4.3.

1.6.2 Capacity scaling with the number of modes

In this section, we discuss how capacity scales with the number of modes in the absence of nonlinear effects. We consider M_S channels (in the sense of Shannon) and compare to a single channel in terms of SE [17,18,28,29,88–93]. An optical fiber supporting M linearly polarized (LP) spatial modes [94], each supporting two polarization states, has $M_S = 2 \times M$ Shannon channels. If one transmits over only one polarization mode of the fiber, then $M_S = M$. We focus here only on the gain in SE. Considerations of gain in energy per bit for systems using multiple channels can be found in Refs. [7,28,29,90,95–100].

For M_S channels, the bit rate $R_b^{(M_S)}$ is defined as the sum over all modes and can be simply written as

$$R_b^{(M_S)} = M_S \widetilde{R}_c \log_2(\mathcal{M}) R_s, \tag{1.27}$$

where $\widetilde{R}_c$ is the encoder rate and $\mathcal{M}$ is the constellation size. The encoder rate is the ratio of input to output bits at the encoder and takes a value between 0 and 1. We consider the simplest case of M_S independent parallel channels in order to estimate the potential benefits of spatial multiplexing in fibers. Nonlinear effects are likely to limit the capacity of fibers supporting multiple spatial modes and are treated in the next sections.

The signal and noise powers in the mth channel out of the M_S channels are denoted by P_m and N_m, respectively. The *total* signal power in all channels is $P^{(M_S)} = \sum_{m=1}^{M_S} P_m$. The SNR in each channel m is $\text{SNR}_m = P_m / N_m$ and the *average* SNR is given by

$$\text{SNR}^{(M_S)} \equiv \frac{\sum_{m=1}^{M_S} P_m}{\sum_{m=1}^{M_S} N_m}. \tag{1.28}$$

The capacity of a single AWGN channel is given by Eq. (1.1). In the case of independent channels, Eq. (1.1) applies to each channel and one can write

$$\text{SE}_m = \log_2(1 + \text{SNR}_m) \tag{1.29}$$

for each m of the M_S channels. The total SE, $\text{SE}(M_S)$, is simply given by,

$$\text{SE}^{(M_S)} = \sum_{m=1}^{M_S} \text{SE}_m = \sum_{m=1}^{M_S} \log_2(1 + \text{SNR}_m). \tag{1.30}$$

Interesting relations can be obtained when $\text{SNR}_m \ll 1$ for all the M_S channels, where $\text{SE}^{(M_S)}$ in Eq. (1.30) simplifies to

$$\text{SE}^{(M_S)} \approx \frac{1}{\ln 2} \sum_{m=1}^{M_S} \text{SNR}_m = \frac{1}{\ln 2} \sum_{m=1}^{M_S} \frac{P_m}{N_m}. \tag{1.31}$$

If one assumes that the noise present in all channels is identical to the noise produced in the reference single channel, i.e. $N_m = N$ for all values of m, a realistic first-order approximation for spatial multiplexing in fibers, Eq. (1.31) simplifies to

$$\text{SE}^{(M_S)} \approx \frac{1}{\ln 2} \sum_{m=1}^{M_S} \frac{P_m}{N} = \frac{P^{(M_S)}}{N \ln 2}. \tag{1.32}$$

Interestingly, Eq. (1.32) means that in the low SNR per channel regime, it is the total power injected in the fiber that determines the total capacity, irrespective of the power distribution among modes in as much as the low value of SNR per mode SNR_m is not violated.

We now consider the case for which SNR_m is not necessarily small. For the case of M_S identical channels the total SE of the M_S channels, $SE^{(M_S)}$, is simply M_S times the SE of each channel. If we further assume that the signal and noise powers in each channel are identical, i.e. $P_m = P_c$ and $N_m = N$, $SE^{(M_S)}$ can be written simply as,

$$SE^{(M_S)} = M_S \log_2 \left(1 + \frac{P_c}{N}\right) = M_S \log_2 \left(1 + \rho \, SNR\right), \tag{1.33}$$

where $SNR = P/N$ is the SNR in a reference single channel and $\rho = P_c/P$ is the ratio of the power in each channel P_c to P. The reason to express $SE^{(M_S)}$ in terms of the ratio P_c/P is that, in fibers, P_c and P have maximum values determined by fiber nonlinearity and fiber types. In the limit of an infinite number of channels M_S with a fixed $P_c = P/M_S$, $SE^{(\infty)} \to SNR/\ln 2$, which is identical to the single-channel result of Eq. (1.3) to first order in SNR.

We can express the gain in SE as $G_{SE} = SE^{(M_S)}/SE$. Considering now channels with identical noise per channel, i.e. $N_m = N$, and with identical signal powers P_c, one can write the gain in SE as,

$$G_{SE} = \frac{M_S \log_2(1 + \rho SNR)}{\log_2(1 + SNR)}. \tag{1.34}$$

Equation (1.34) indicates that three independent parameters affect G_{SE}: the number of channels M_S, the ratio of powers $\rho = P_c/P$ and the SNR of the reference channel P/N. When the power per channel P_c equals the single-channel power P, $G_{SE} = M_S$ in Eq. (1.34), as it should be for M_S independent channels.

The maximum value that P_c can assume will be determined by the nonlinear effects in the fiber used for spatial multiplexing. It is likely that the value of P_c will depend on the fiber design and it is not known at this point in time how to precisely determine P_c. Therefore, it becomes interesting to consider the impact of different ratios of the power per channel P_c to the single-channel power P on the gain in SE of fibers supporting spatial multiplexing. From Eq. (1.34), one can see that the gain in SE per mode, G_{SE}/M_S, is independent of M_S.

Figure 1.13 shows the gain in SE per mode as a function of the ratio P_c/P for various values of SNR. The reference power is P and a negative gain corresponds to a reduction in SE relative to the SE of the reference power. As seen in Figure 1.13, the highest values of SNR are the least affected by a reduction or increase in signal power per channel P_c. This can be understood by the fact that a reduction of a few dBs in power when operating at high SNR on the SE curve changes the SE by only a small amount relative to the original SE, thereby producing lower gain or loss in SE. Note that this behavior is independent of the number of channels M_S.

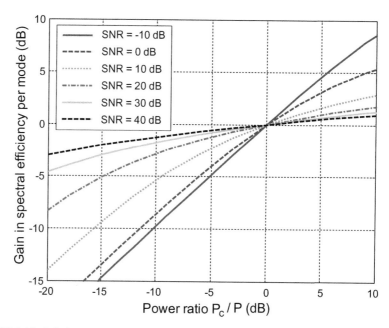

FIGURE 1.13 Gain in SE *per mode*, G_{SE}/M_S, **as a function of the ratio of the signal power per mode** P_c **to the single channel power** P. **These curves are independent of the number of modes** M_S.

1.6.3 Generalized Manakov equations for multimode fibers

A set of coupled PDEs describing nonlinear propagation in MMFs can be derived from the Maxwell equations under certain approximations [101–103] for propagation of ultrashort pulses (for instance, see Eq. (1.6) of Ref. [102]). In the context of optical communication, ultrafast phenomena (faster than $\sim$1 ps) have less impact on nonlinear propagation due to the long duration of the transmitted symbols ($>$1 ps) in the foreseeable future of optical communication systems. In this context, we neglect the non-instantaneous part of the Kerr nonlinearity [2] to simplify the nonlinear propagation PDEs derived in Refs. [101–103]. By further assuming an LP-modes representation [94] of the fields, we can write the following set of nonlinear propagation PDEs [104],

$$
\frac{\partial \mathbf{E}_p}{\partial z} - \iota(\boldsymbol{\beta}_{0p} - \beta_g)\mathbf{E}_p + \left(\boldsymbol{\beta}_{1p} - \frac{1}{v_g}\right)\frac{\partial \mathbf{E}_p}{\partial t} + \iota\frac{\boldsymbol{\beta}_{2p}}{2}\frac{\partial^2 \mathbf{E}_p}{\partial t^2}
$$

$$
= \iota \sum_{lmn=1}^{M} f_{lmnp}\frac{\gamma}{3}\left[\left(\mathbf{E}_n^{\mathrm{T}}\mathbf{E}_m\right)\mathbf{E}_l^* + 2\left(\mathbf{E}_l^{\mathrm{H}}\mathbf{E}_m\right)\mathbf{E}_n\right],
$$

(1.35)

where $\mathbf{E}_p(z,t)$ is defined similarly as explained below Eq. (1.26) but for each of the p spatial fiber modes. It is a 1×2 column vector $\mathbf{E}_p(z,t) = [E_{px}(z,\omega)E_{py}(z,\omega)]^T$ where $E_{px}(z,t)$ and $E_{py}(z,t)$ are the x and y polarization components of the field of the pth spatial mode. Other parameters are derived from the propagation constant matrix,

$$\boldsymbol{\beta}_p = \begin{pmatrix} \beta_{px} & 0 \\ 0 & \beta_{py} \end{pmatrix}. \tag{1.36}$$

with $\boldsymbol{\beta}_{np} = \partial^n \boldsymbol{\beta}_p / \partial \omega^n |_{\omega_0}$ where $n=0$, 1, 2. The constants $\boldsymbol{\beta}_{0p}, \boldsymbol{\beta}_{1p}$, and $\boldsymbol{\beta}_{2p}$ are 2×2 diagonal matrices representing the propagation constant, inverse group velocity, and GVD at the angular frequency ω_0 of each polarization component of the pth spatial mode, respectively. β_g and v_g are the references propagation constant and group velocity, respectively. We assume here that the two polarization components of a spatial mode may have different phase and group velocities but we assume that they have identical GVD. Stated differently, the phase and group velocities are polarization-dependent while the GVD is considered identical for the two polarization states. We omitted the noise terms in Eq. (1.35) for simplicity.

The nonlinear coefficient γ is as defined in Eq. (1.15). The nonlinear coefficient tensor f_{lmnp} in Eq. (1.35) provides a relative weight among the various nonlinear terms between the spatial modes. It is defined as,

$$f_{lmnp} = \frac{A_{\text{eff}}}{(I_l I_m I_n I_p)^{1/2}} \iint F_l^* F_m F_n F_p^* \, dx \, dy, \tag{1.37}$$

where the indices l, m, n, and p run through the number M of spatial modes. The parameters I_l, I_m, I_n, and I_p represent the constants of normalization for the modal fields F_l, F_m, F_n, and F_p. They are given by

$$\iint F_m(x,y) F_p^*(x,y) dx \, dy = I_m \delta_{mp}, \tag{1.38}$$

where δ_{mp} is the Kronecker delta. Note that the value of γ in Eq. (1.15) is generally calculated using the fundamental mode of the MMF. This requires using the effective area of the same mode in the definition of the f_{lmnp} in Eq. (1.37).

It is interesting to observe that, in Eq. (1.35), the nonlinear term in brackets is summed over the three indices l, m, n, each running from 1 to M, the number of spatial modes. The number of these nonlinear terms grows like M^3, for each of the M PDEs that are modeling the M spatial modes. For a relatively small number of modes like 6 modes, the number of terms in brackets is $6^3 = 216$. For $M=20$ it is 8000! Such a large number of terms makes it difficult to isolate specific nonlinear phenomena and understand nonlinear propagation in MMFs.

In the case of SMFs, the original nonlinear propagation equations could be simplified to give the Manakov equations (see Section 1.5.1) by considering random polarization evolution due to the random birefringence that virtually all fibers exhibit.

A similar approach has been considered for MMFs, and a set of simplified nonlinear equations of the Manakov type can be derived under certain assumptions [104–106]. A first set of generalized Manakov equations have been derived assuming strong linear mode coupling between all spatial modes [104,107]. This is referred to as the strong-coupling regime. Another set of generalized Manakov equations have also been derived for the somewhat opposite case where there is weak or negligible linear coupling between any pair of spatial modes [105]. This is referred to as the weak-coupling regime. We discuss both cases below.

The derivation of the generalized Manakov equations for the weak-coupling regime starts from Eq. (1.35). The fields in each spatial mode are then assumed to experience independent random birefringence along the fiber length. This random birefringence couples the two polarization components of each spatial mode but do not couple distinct spatial modes to each other. This is referred to as the weak-coupling regime. The resulting propagation equations are then averaged over all realizations of the random coupling matrices between the two polarizations of each LP mode. We then arrive at the generalized Manakov equations describing the evolution of the field with distance. The resulting generalized Manakov equations in the weak-coupling regime can be written as [104],

$$
\frac{\partial \mathbf{E}_p}{\partial z} + \langle \delta \beta_{0p} \rangle \mathbf{E}_p + \langle \delta \beta_{1p} \rangle \frac{\partial \mathbf{E}_p}{\partial t} + \iota \frac{\beta_{2p}}{2} \frac{\partial^2 \mathbf{E}_p}{\partial t^2}
$$
$$
= \iota \gamma \left(f_{pppp} \frac{8}{9} |\mathbf{E}_p|^2 + \sum_{m \neq p} f_{mmpp} \frac{4}{3} |\mathbf{E}_m|^2 \right) \mathbf{E}_p, \tag{1.39}
$$

where

$$
\langle \delta \boldsymbol{\beta}_{0p} \rangle = \frac{1}{2} (\beta_{px}(\omega_0) + \beta_{py}(\omega_0)) - \beta_g \tag{1.40}
$$

$$
\langle \delta \boldsymbol{\beta}_{1p} \rangle = \frac{1}{2} \left(\frac{\partial \beta_{px}}{\partial \omega} \bigg|_{\omega_0} + \frac{\partial \beta_{py}}{\partial \omega} \bigg|_{\omega_0} \right) - \frac{1}{v_g}. \tag{1.41}
$$

Equations (1.40) and (1.41) can be interpreted as the averaged "β_0" and "β_1" for each mode p. It is interesting to note that the difference in phase and group velocities is preserved in Eqs. (1.40) and (1.41). This is a consequence of weak coupling between distinct spatial modes.

One can also derive a set of generalized Manakov equations for the case of strong mode coupling. The procedure to follow is the one outlined above in Eq. (1.39) except that now all the M spatial modes are assumed to randomly couple. After averaging over all possible random coupling states, the generalized Manakov equations in the strong-coupling regime can be written as [104,107],

$$
\frac{\partial \mathbf{E}_p}{\partial z} + \iota \frac{\bar{\beta}_2}{2} \frac{\partial^2 \mathbf{E}_p}{\partial t^2} = \iota \gamma \kappa \sum_m |\mathbf{E}_m|^2 \mathbf{E}_p, \tag{1.42}
$$

where $\bar{\beta}_2 = \sum_{\rho=1}^{M} \beta_{2\rho}/(M)$ is the average dispersion across all M spatial modes. In this regime of strong coupling, all fields components E_{px} and E_{py} with $p = 1, \ldots, M$ obey a similar equations for the linear part of propagation, i.e. the left-hand side of Eq. (1.42). One notes that the group velocity of individual modes is not present in Eq. (1.42). This is because, in the strong coupling regime, all modes couple back and forth and therefore converge toward identical average phase and group velocities. This is in contrast to the weak-coupling regime of Eq. (1.39) which retains the individual linear propagation characteristics of each mode.

The nonlinear coefficient κ in Eq. (1.42) is given by,

$$\kappa = \sum_{k \leqslant l}^{M} \frac{32}{2^{\delta_{kl}}} \frac{f_{kkll}}{6M(2M+1)}. \tag{1.43}$$

How much linear coupling there is between the different spatial modes of MMFs depends on the fiber design, the fabrication process and environmental stress. In general, there can be different values of linear coupling between different pairs of spatial modes. Therefore, it becomes important to consider the case of arbitrary levels of linear coupling between modes in the derivation of a generalized Manakov equation. This represents the natural next step in the development of the nonlinear propagation equations for MMFs.

Because nonlinear modeling of MMFs and FMFs is still under development, it is important to experimentally investigate the nonlinear effects present in these fibers, and especially the inter-modal nonlinear effects that are new to optical fiber communication. The rest of this chapter is devoted to experimental demonstrations of two inter-modal nonlinear effects: inter-modal cross-phase modulation (IM-XPM) and inter-modal four-wave mixing (IM-FWM). These nonlinear effects are expected to play a critical role in determining the system performance of fibers supporting multiple spatial modes and setting their nonlinear capacity limits.

1.6.4 Description of a few-mode fiber

We describe in this section a 4.7-km-long graded-index FMF (GI-FMF) used to perform an experimental investigation of inter-modal nonlinear effects. The fiber supports three spatial modes, LP01, LP11a, and LP11b modes, with two orthogonal states of polarization for each mode. The fiber has a loss coefficient of 0.226 dB/km and no noticeable mode-dependent loss. By selectively launching each spatial mode in the fiber using a phase-plate-based mode coupler [76], we observed strong linear coupling between the LP11a and LP11b modes while the coupling between the LP01 and LP11 modes was measured to be −20 dB.

The GI-FMF was designed to have a small difference in the group velocity v_g between the LP01 and LP11 modes across the whole C-band, which is advantageous to reduce the complexity of the digital signal processing at the receiver during MIMO transmission [108] (see chapters on MIMO). The group velocities of the LP01 and

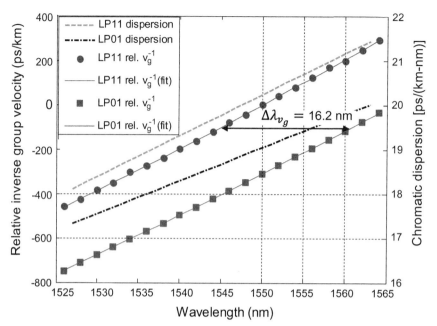

FIGURE 1.14 Measurement of relative average inverse group velocity of the LP01 and LP11 modes of a GI-FMF as a function of wavelength. The dispersion of the LP01 and LP11 modes is also shown for reference.

LP11 modes were measured using a time-of-flight technique [109] based on a 100 ps test pulse that was selectively launched into the corresponding mode.

Figure 1.14 shows the measured inverse group velocity (IGV) and chromatic dispersion of the LP01 and LP11 modes. The IGV is given by $1/v_g$, where the group velocity at the wavelength λ_s corresponds to $v_g = (d\beta/d\omega)^{-1}|_{\omega=\omega_s} = 1/\beta_1(\omega_s)$. The group velocity of the LP11 modes at 1550 nm is used as a reference. The chromatic dispersion D is as defined in Eq. (1.14). From these measurements, we can observe that the two LP modes have slightly different values of chromatic dispersion: 18.7 and 19.8 ps/(nm-km) at 1545 nm for the LP01 and LP11 modes, respectively. There is however a sizeable difference in IGV which value varies from 290 ps/km, at 1525 nm, to 330 ps/km, at 1565 nm. Such differential IGV, combined with chromatic dispersion, leads to groups of wavelengths belonging to different spatial modes to propagate at the same group velocity for a certain wavelength separation. For the GI-FMF considered, the wavelength separation is $\Delta\lambda_{v_g} = \lambda_{s,01} - \lambda_{s,11}$, where $\lambda_{s,01}$ and $\lambda_{s,11}$ are the signal wavelengths of the waves in the LP01 and LP11 modes, respectively. When $\Delta\lambda_{v_g}$ approaches a value between 15.8 and 16.2 nm depending on the location in the C-band, the two spatial modes are group-velocity matched. The effective areas and the nonlinear coefficients of the GI-FMF modes were calculated based on the fiber refractive index profile. The effective area of the fundamental mode LP01 is $A_{\text{eff}} = 59.6 \, \mu\text{m}^2$, which corresponds to $\gamma = 1.77 \, (\text{W-km})^{-1}$. The nonlinear coefficients f_{lmnp} calculated according to Eq. (1.37) are reported in Table 1.2 for the LP01, LP11a, and LP11b modes.

The nonlinear coefficient tensor f_{lmnp} has 81 elements for a fiber with 3 spatial modes. Table 1.2 reports all 21 non-zero elements which can take 4 different values. We used the following classification for the nonlinear coefficients: Self-phase modulation (SPM)-type for all coefficients of the form f_{mmmm}, IM-XPM-type for all coefficients of the form f_{mmnn} and f_{mnmn} where m and n are different modes, and IM-FWM-type for all other combinations of indices that give a non-zero value of f_{lmnp}.

The largest coefficients are observed for nonlinear interactions that are related to SPM-type of terms, whereas the smallest coefficients are related to SPM-type of terms, whereas the smallest coefficients are related to processes that only involve both LP11a and LP11b modes. On the other hand, there are many more of these processes than SPM. Because the LP11a and LP11b modes are degenerate, there is no

Table 1.2 Non-zero elements of the nonlinear coefficient f_{lmnp} for a few-mode fiber with 3 spatial modes, calculated for LP01, LP11a, and LP11b modes.

Nonlinear term type	LP mode indices	f_{lmnp}
SPM	LP01, LP01, LP01, LP01	1
	LP11a, LP11a, LP11a, LP11a	0.747
	LP11b, LP11b, LP11b, LP11b	
IM-XPM	LP01, LP01, LP11a, LP11a	0.496
	LP11a, LP11a, LP01, LP01	
	LP01, LP01, LP11b, LP11b	
	LP11b, LP11b, LP01, LP01	
	LP11a, LP01, LP11a, LP01	
	LP11b, LP01, LP11b, LP01	
	LP01, LP11a, LP01, LP11a	
	LP01, LP11b, LP01, LP11b	
	LP11a, LP11a, LP11b, LP11b	0.249
	LP11b, LP11b, LP11a, LP11a	
	LP11a, LP11b, LP11a, LP11b	
	LP11b, LP11a, LP11b, LP11a	
IM-FWM	LP01, LP11a, LP11a, LP01	0.496
	LP01, LP11b, LP11b, LP01	
	LP11a, LP01, LP01, LP11a	
	LP11b, LP01, LP01, LP11b	
	LP11a, LP11b, LP11b, LP11a	0.249
	LP11b, LP11a, LP11a, LP11b	

unique orthogonal basis for the vector space they span. The conventional definition of the LP11a and LP11b modes proposed by Gloge [94] results in field amplitude distributions $F(x, y)$ that are real, and have spatial intensity patterns that are clearly different.

A second orthogonal base for the LP11 mode has the form,

$$F_{\text{LP11}\pm} = R_{\text{LP11}}(r)e^{\pm i\phi}, \tag{1.44}$$

where R_{LP11}, as $F_{\text{LP11}\pm}$, is also a solution for the radial field distribution of the LP11 mode, and r and φ are the polar coordinates. This basis has the property of having the same intensity pattern for both degenerate modes. This can be seen in Figure 1.15, which shows the intensity and phase distributions of various LP modes. The conventional notation of the LP modes is then related to the LP11$\pm$ basis by the simple linear transformation,

$$F_{\text{LP11}+} = (F_{\text{LP11a}} - i F_{\text{LP11b}})/\sqrt{2}, \tag{1.45}$$

$$F_{\text{LP11}-} = i(F_{\text{LP11a}} + i F_{\text{LP11b}})/\sqrt{2}, \tag{1.46}$$

which can be shown to be unitary.

The nonlinear coefficient tensor f_{lmnp} depends on the basis chosen for the LP11 modes. Recalculating f_{lmnp} using LP01, LP11+, and LP11− as basis, the number of non-zero coefficients is reduced from 21 to 19, which can now take three different values. The results are reported in Table 1.3.

Compared to Table 1.2, we notice that the SPM-type and the IM-XPM-type coefficients for the LP11 mode are now equal.

Is there a basis of LP11 modes that is more representative of the experimental observations? It depends on numerous factors. In most experiments, LP11a and LP11b are launched, but because the LP11 modes are not the true modes of the fiber, they will not propagate as isolated modes but will rapidly mix with typical beating

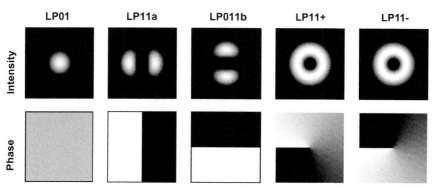

FIGURE 1.15 Different LP mode bases. Top row: Theoretical intensity distributions of a GI-FMF (highest intensity in white). Bottom row: Phase distribution.

Table 1.3 Non-zero nonlinear coefficients for a few-mode fiber with 3 spatial modes, calculated for the LP01, LP11+, and LP11− modes.

Nonlinear term type	LP mode indices	f_{lmnp}
SPM	LP01, LP01, LP01, LP01	1
	LP11+, LP11+, LP11+, LP11+	0.498
	LP11−, LP11−, LP11−, LP11−	
IM-XPM	LP01, LP01, LP11+, LP11+	0.496
	LP11+, LP11+, LP01, LP01	
	LP01, LP01, LP11−, LP11−	
	LP11−, LP11−, LP01, LP01	
	LP01, LP11+, LP01, LP11+	
	LP01, LP11−, LP01, LP11−	
	LP11+, LP01, LP11+, LP01	
	LP11−, LP01, LP11−, LP01	
	LP11+, LP11+, LP11−, LP11−	0.498
	LP11−, LP11−, LP11+, LP11+	
	LP11+, LP11−, LP11+, LP11−	
	LP11−, LP11+, LP11−, LP11+	
IM-FWM	LP01, LP11+, LP11−, LP01	0.496
	LP01, LP11−, LP11+, LP01	
	LP11+, LP01, LP01, LP11−	
	LP11−, LP01, LP01, LP11+	

length ranging from a few centimeters to meters. In addition, minute variations of the fiber index profile along the fiber length, macro- and micro-bending also cause a strong mixing between the LP11a and LP11b modes, which will not allow the LP11 modes to propagate as unperturbed modes. The exact conditions of propagation determine the effective nonlinear coefficient of the nonlinear interaction considered.

To determine whether inter-modal nonlinear effects can be observed in "real" fibers having non-idealized characteristics, we experimentally investigate inter-modal nonlinear interactions in the FMF described in this section. We consider two configurations of waves represented schematically in Figure 1.16. The first configuration shown in Figure 1.16a is designed to measure IM-XPM while the second is designed to observe IM-FWM.

1.6.5 Inter-modal cross-phase modulation

Inter-modal cross-phase modulation (IM-XPM) has been explored by using ultrashort pulses [110] or in MMFs with a large number of spatial modes [111]. Here we focus on a FMF with signal bandwidth of a few tens of GHz and clearly show the effect and compare it to intra-modal XPM.

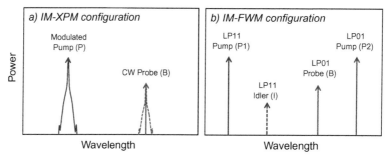

FIGURE 1.16 Schematic representation of wave configurations in the study of inter-modal nonlinearities: (a) one modulated pump and one CW probe for the study of IM-XPM and (b) two CW pumps and one CW probe to study IM-FWM. The full lines represent the waves at the fiber input while the dashed lines represent new waves observed at the fiber output.

For the measurement of IM-XPM, a non-return-to-zero (NRZ) modulated pump is injected in one spatial mode, the LP11a mode, and co-propagates with a CW probe launched in a different spatial mode, the LP01. In the wavelength region where IM-XPM between the pump and the probe occurs, a broadening of the probe is expected as depicted schematically by the dashed curve at the probe location in Figure 1.16a. But before presenting the experimental observations, let us consider a simple analytical model describing the impact of XPM between two waves, a high-power modulated pump and one low-power CW probe.

The nonlinear phase generated on a CW probe (labeled B) by a modulated pump (labeled P) can be expressed by the transformation of the probe field $E_B(z,t)$ as [112,113],

$$E_B(z,t) = E_B(0, t - z/v_{gB})e^{-\alpha z/2}\, e^{i\phi_B(z,t)}, \tag{1.47}$$

where α is the fiber loss coefficient and v_{gB} the group velocity of the probe. The phase modulation $\phi_B(z,t)$ imprinted on the CW probe by the modulated pump can be written as,

$$\phi_B(z,t) = 2\gamma_B \int_0^z \left| E_P\left(0, t - \frac{z}{v_{gP}} + d_{BP}\, z'\right) \right|^2 e^{-\alpha z'}\, dz', \tag{1.48}$$

where γ_B is the nonlinear coefficient experienced by the probe from the presence of the pump, $d_{BP} = v_{gB}^{-1} - v_{gP}^{-1}$ where v_{gP} is the group velocity of the pump. Assuming that the pump has a sinusoidal power modulation of the form $|E_P(0,t)|^2 = P_0 + P_m \cos(\omega t)$, the XPM-induced phase shift on the probe by the pump modulation *alone* is given by,

$$\phi_B(z,t) = 2\gamma_B P_m(1 - e^{-\alpha L})\sqrt{\eta_{\text{XPM}}}/\alpha, \tag{1.49}$$

where the XPM efficiency coefficient η_{XPM} is given by,

$$\eta_{\text{XPM}} = \frac{\alpha^2}{\omega^2 d_{BP}^2 + \alpha^2} \left[1 + \frac{4\sin^2(\omega d_{BP} L/2)e^{-\alpha L}}{(1 - e^{-\alpha L})^2} \right]. \tag{1.50}$$

Note that η_{XPM} assumes its maximum value of 1 for the same group velocity between probe and pump, i.e. when $d_{BP} = 0$. The value of η_{XPM} decreases when there is a mismatch in group velocity between pump and probe. The phase modulation of the probe $\phi_B(z,t)$ is proportional to the pump power and the nonlinear coefficient γ_B.

To experimentally demonstrate IM-XPM, we use the GI-FMF 4.7-km-long described in Section 1.6.4 that supports three spatial modes: LP01, LP11a, and LP11b. The results of an experimental comparison of inter-modal and intra-modal XPM are shown in Figure 1.17 where the relative spectral broadening of the probe as a function of the pump wavelength when pump and probe are in the same or in different modes is presented. The spectral broadening is defined as the ratio of the standard deviation of the power spectrum of the probe in the presence of the pump to the standard deviation in the absence of the pump. The first two configurations (a and b in Figure 1.17) are for pump and probe in the same mode, either LP01 or LP11. The probe wavelength is set to $\lambda_{\text{probe}} = 1545.045$ nm for these two same mode configurations and spectral broadening is observed when the pump is at the same wavelength as the probe. The third configuration (c) is for the probe in the LP01 mode and the pump in the LP11 mode. The probe wavelength has been moved to

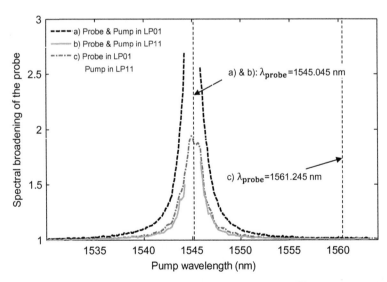

FIGURE 1.17 Spectral broadening of the probe measured for three different pump-probe configurations of the 4.7 km-GI-FMF of Section 1.6.4.

$\lambda_{probe} = 1561.245$ nm to make the broadening curves coincide in pump wavelength. For this third configuration, spectral broadening occurs with the probe wavelength λ_{probe} that is 16.2 nm longer than the pump wavelength λ_{pump}, corresponding to the wavelength separation $\Delta\lambda_{v_g} = 16.2$ nm for which the group velocities of the LP01 and LP11 modes match as shown in Figure 1.14. For all three configurations, the maximum spectral broadening occurs when the average group velocities of the pump and the probe closely match and the spectral broadening becomes negligible when the pump is more than 7 nm away from the wavelength of maximum spectral broadening. This behavior is consistent with Eq. (1.50).

From Figure 1.17, one observes that the maximum spectral broadening occurs when both pump and probe are launched into the LP01 mode (configuration a). The two other configurations show smaller and nearly identical spectral broadenings. Even though a more complete nonlinear propagation study is necessary to predict the spectral broadening, one can get insights by considering the nonlinear coefficients reported in Tables 1.2 and 1.3. If one considers that the LP11a and LP11b modes are the most relevant modal representation for nonlinear interactions, the corresponding nonlinear terms in Table 1.2 are the ones involving {LP01, LP01, LP01, LP01}, {LP11a, LP11a, LP11a, LP11a}, and {LP01, LP01, LP11a, LP11a}, for the waves configurations a, b, and c of Figure 1.17, respectively. If the LP11+ and LP11− modes are considered, the LP11a should be replaced by the LP11+ above and the nonlinear coefficient taken from Table 1.3. For LP11a, one has $f_a = 1$, the nonlinear coefficients are $f_b = 0.747$ and $f_c = 0.496$ for the three configurations while for LP11+, one has $f_a = 1$, $f_b = 0.498$, and $f_a = 0.496$. The measured spectral broadenings of configurations b and c being nearly identical suggest that the LP11+ and LP11− modes are more suitable for estimating the effect of IM-XPM. For a more precise description of the effects of IM-FWM, a more accurate model based on the set of Eq. (1.35) should be studied.

1.6.6 Inter-modal four-wave mixing

Nonlinear effects in MMFs, and in particular, IM-FWM, were first explored by Stolen and co-workers [114–116]. They focused on partially degenerate IM-FWM where only two waves were launched into the FMF to generate a third wave, a configuration of waves still being studied today [117,118]. Here, we focus on non-degenerate IM-FWM by using three distinct CW waves in two different spatial modes as shown schematically in Figure 1.16b. The fiber used is the FMF described in Section 1.6.4. After nonlinear propagation, an idler is generated when a sufficient phase matching condition between all the waves is achieved.

Any nonlinear interactions in a FMF must first satisfy energy conservation which determines the frequency of the generated fourth wave. Additionally, for FWM to build up, a second condition needs to be achieved called phase matching. One can find at least two nonlinear processes that satisfy energy conservation between three waves in a nonlinear medium. The optical angular frequencies of a fourth wave ω_4 generated from FWM between three waves for these two processes are,

$$(\text{PROC1}) \quad \omega_4 = \omega_1 - \omega_2 + \omega_3, \tag{1.51}$$

$$(\text{PROC2}) \quad \omega_4 = \omega_1 + \omega_2 - \omega_3, \tag{1.52}$$

where the superscripts are used to label the waves, and PROC1 and PROC2 labels the FWM processes seen in Figure 1.18 when identifying the waves (1)–(4) by P1, B, P2, and I, respectively, as shown in Figure 1.16b. The phase matching conditions between these waves that correspond to these two processes are

$$(\text{PROC1}) \quad \beta^{(4)}(\omega_4) = \beta^{(1)}(\omega_1) - \beta^{(2)}(\omega_2) + \beta^{(3)}(\omega_3), \tag{1.53}$$

$$(\text{PROC2}) \quad \beta^{(4)}(\omega_4) = \beta^{(1)}(\omega_1) + \beta^{(2)}(\omega_2) - \beta^{(3)}(\omega_3). \tag{1.54}$$

Ignoring any polarization dependence, one can write the constant of propagation of each wave $\beta^{(i)}(\omega)$ in a Taylor series centered around an angular optical frequency ω_0 as,

$$\beta^{(i)}(\omega) = \beta_0^{(i)} + \beta_1^{(i)}(\omega - \omega_0) + \frac{\beta_2^{(i)}}{2}(\omega - \omega_0)^2 + \cdots, \tag{1.55}$$

where terms of order higher than two are neglected, i.e. third-order and higher-order dispersion terms are not taken into account.

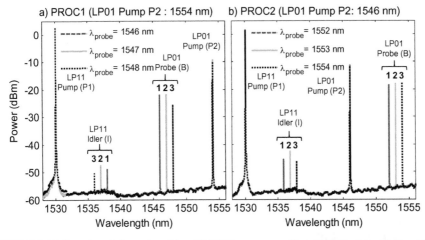

FIGURE 1.18 Experimental spectra of two IM-FWM processes. Three spectra are superposed on each graph. Only four waves as depicted in Figure 1.16b are present at a time–two pumps and one of three probe positions labeled 1–3 and the corresponding idler. (For interpretation of the references to color in this figure legend, the reader is referred to the web version of this book.)

We now assign each wave to the spatial mode they are launched into. We use the notation $\beta_{n,\text{mode}}$ where n is the expansion coefficient index of the constant of propagation and "mode" is either "01" or "11" for the corresponding LP mode and $\omega_{\text{wave,mode}}$. The subscript wave is either P1, B, or P2, I, as mentioned below Eq. (1.52). Assigning the waves P1 and I to the LP11 mode and the waves B and P2 to the LP11 mode in the two processes PROC1 and PROC2 of Eqs. (1.53) and (1.54), we arrive at the same phase matching condition for both processes given by

$$
\beta_{1,11} + \beta_{2,11} \left(\frac{\Delta\omega_{P1,11} + \Delta\omega_{I,11}}{2} \right) = \beta_{1,01} + \beta_{2,01} \left(\frac{\Delta\omega_{B,01} + \Delta\omega_{P2,01}}{2} \right).
$$

(1.56)

One can interpret Eq. (1.56) in the following way: the phase matching of this type of IM-FWM is achieved when the group velocities evaluated at the average frequencies of the two waves present in each mode are equal. Note that Eq. (1.56) is independent of the β_0 of each spatial mode so that fluctuations of β_0 [119] common to both modes cancel out.

Figure 1.18 shows the experimental observation of the two processes from Eqs. (1.51) and (1.52). The powers launched into the GI-FMF are 16 dBm for the LP11 pump, 18 dBm for the LP01 pump and 6 dBm for the LP01 probe. The waves are launched in the GI-FMF according to Figure 1.16b, with the pump P1 launched in the LP11a mode. Since there is strong linear coupling between the LP11a and LP11b modes, the power spectra shown are the sum of the power spectra measured at the LP11a and LP11b modes. This insures that the entire power in the LP11 mode is captured. One can identify the spectra of Figures 1.18a and b as corresponding to the processes PROC1 of Eq. (1.51) and PROC2 of Eq. (1.52), respectively.

Finally, it is interesting to point out that IM-FWM can be phase matched in the presence of dispersion in each spatial mode as can be shown from Eq. (1.56). This is in sharp contrast to SMFs that requires zero dispersion to fulfill phase matching conditions [112]. Consequently, in SMF with non-zero dispersion, FWM is never fully phase matched and is highly suppressed. In MMFs, IM-FWM can actually be fully phase matched and therefore potentially be more important in systems using spatial multiplexing in fibers than FWM in SMFs. Further details can be found in Refs. [120–123].

1.7 CONCLUSION

We presented the trends in optical network traffic and commercial system capacity. We discussed fundamentals of nonlinear capacity of single-mode fibers and described that improvements in the properties of single-mode fibers only moderately increase the nonlinear fiber capacity. It leads us to the conclusion that fiber capacity can be most effectively grown by increasing the number of spatial modes. We outlined the capacity scaling of multimode systems in the absence of fiber nonlinearity as a reference. We then discuss nonlinear propagation in multimode fiber, a complex field still largely unexplored. We wrote down a basic set of nonlinear propagation equations derived from Maxwell's equations, along with simplified propagation equations in the weak- and strong-coupling approximations, referred to as generalized Manakov equations. Finally, we presented experimental observations of two inter-modal nonlinear effects, inter-modal cross-phase modulation, and inter-modal four-wave mixing, over a few-km-long few-mode fiber. Important differences between intra-modal and inter-modal nonlinear effects have been discussed. Understanding nonlinear propagation in multimode and multicore fibers is required to eventually establish a nonlinear capacity limit for these fibers.

Acknowledgments

We would like to thank many colleagues and collaborators, in particular S. Mumtaz, A. Mecozzi, M.A. Mestre, G.P. Agrawal, G. Kramer, J. Foschini, A. Gnauck, P.J. Winzer, S. Randel, A. Chraplyvy, A. Tulino, M. Magarini and collaborators from OFS and Sumitomo electric company as well as many other colleagues inside and outside Bell Laboratories.

References

[1] K.C. Kao, G.A. Hockham, Dielectric-fiber surface waveguides for optical fiber, Proc. IEE. 113 (1966) 1151–1158 also reprinted in Proc. IEE 133 (1986) 191–198.

[2] G.P. Agrawal, Nonlinear Fiber Optics. fourth ed., Elsevier Science & Technology, San Diego, 2006.

[3] R.W. Boyd, Nonlinear optics. third ed., Academic Press, 2008.

[4] R.-J. Essiambre, G. Kramer, P.J. Winzer, G.J. Foschini, B. Goebel, Capacity limits of optical fiber networks, J. Lightwave Technol. 28 (2010) 662–701.

[5] A. Ellis, J. Zhao, D. Cotter, Approaching the non-linear Shannon limit, J. Lightwave Technol. 28 (4) (2010) 423–433.

[6] I. Djordjevic, W. Ryan, B. Vasic, Optical channel capacity, in Coding for optical channels, Springer Verlag, 2010, pp. 353–398 (Chapter 10).

[7] R.-J. Essiambre, R.W. Tkach, Capacity trends and limits of optical communication networks, Proc. IEEE 100 (5) (2012) 1035–1055.

[8] A.R. Chraplyvy, The coming capacity crunch, in: Proc. Eur. Conf. Opt. Commun. (ECOC) Plenary talk, 2009.

[9] R.H. Stolen, The early years of fiber nonlinear optics, J. Lightwave Technol. 26 (9) (2008) 1021–1031.

[10] R. Tkach, Scaling optical communications for the next decade and beyond, Bell Labs Tech. J. 14 (4) (2010) 3–9.

[11] C.E. Shannon, A mathematical theory of communication, Bell Syst. Tech. J. 27 (1948) 379–423 and 623–656.

[12] J.M. Wozencraft, B. Reiffen, Sequential Decoding, MIT Press, Wiley, 1961.

[13] J.L. Massey, Channel models for random-access systems, NATO Advances Studies Institutes Series E142, 1988.

[14] R.G. Gallager, Information Theory and Reliable Communication, John Wiley and Sons, New York, 1968.

[15] D.J.C. MacKay, Information Theory, Inference and Learning Algorithms, Cambridge, University Press, 2003.

[16] T.M Cover, J.A Thomas, Elements of Information Theory. second ed., John Wiley and Sons, 2006.

[17] G.J. Foschini, Layered space-time architecture for wireless communication in a fading environment when using multi-element antennas, Bell Labs Tech. J. 1 (1996) 41–59.

[18] I.E. Telatar, Capacity of multi-antenna Gaussian channels, Eur. Trans. Telecommun. 10 (6) (1999) 585–595.

[19] I. Kalet, S. Shamai, On the capacity of a twisted-wire pair: Gaussian model, IEEE Trans. Commun. 38 (3) (1990) 379–383.

[20] J.J. Werner, The HDSL environment, IEEE J. Sel. Area. Comm. 9 (6) (1991) 785–800.

[21] J.P. Gordon, Quantum effects in communications systems, Proc. IRE 50 (1962) 1898–1908.

[22] D. Boroson, A survey of technology-driven capacity limits for free-space laser communications, Proc. SPIE 6709 (2007) 670918.

[23] J.G. Proakis, M. Salehi, Digital Communications. fifth ed., Mc Graw Hill, 2007.

[24] S. Haykin, Communication Systems. fifth ed., Wiley, 2009.

[25] H. Nyquist, Certain factors affecting telegraph speed, Bell Syst. Tech. J. 3 (1924) 324–346.

[26] H. Nyquist, Certain topics of telegraph transmission theory, Trans. Am. Inst. Electr. Eng. 47 (1928) 617–644.

[27] R.V.L. Hartley, Certain factors affecting telegraph speed, Bell Syst. Tech. J. 7 (1928) 535–563.

[28] D. Tse, P. Viswanath, Fundamentals of Wireless Communication, Cambridge University Press, 2005.

[29] A. Goldsmith, Wireless Communications, Cambridge University Press, 2005.

[30] R.-J. Essiambre, G.J. Foschini, G. Kramer, P.J. Winzer, Capacity limits of information transmission in optically-routed fiber networks, Bell Labs Tech. J. 14 (4) (2010) 149–162.

[31] G.P. Agrawal, Fiber-Optic Communication Systems. third ed., Wiley-Interscience, 2010.

[32] L.F. Mollenauer, J.P. Gordon, Solitons in Optical Fibers: Fundamentals and Applications, Academic Press, 2006.

[33] J.P. Gordon, W.H. Louisell, L.R. Walker, Quantum fluctuations and noise in parametric processes II, Phys. Rev. 129 (1) (1963) 481–485.

[34] J.P. Gordon, L.R. Walker, W.H. Louisell, Quantum statistics of masers and attenuators, Phys. Rev. 130 (2) (1963) 806–812.

[35] A. Mecozzi, Limits to long-haul coherent transmission set by the Kerr nonlinearity and noise of the in-line amplifiers, J. Lightwave Technol. 12 (11) (1994) 1993–2000.

[36] C.W. Gardiner, Handbook of Stochastic Methods: for Physics, Chemistry and the Natural Science. third ed., Springer, 2004.

[37] C.W. Gardiner, P. Zoller, Quantum Noise: A Handbook of Markovian and Non-Markovian Quantum Stochastic Methods with Applications to Quantum Optics. third ed., Springer, 2004.

[38] N.W. Ashcroft, N.D. Mermin, Solid State Physics, Brooks Cole, 1976.

[39] J. Bromage, Raman amplification for fiber communications systems, J. Lightwave Technol. 22 (1) (2004) 79–93.

[40] R.-J. Essiambre, G. Raybon, B. Mikkelsen, Pseudo-linear transmission of high-speed TDM signals: 40 and 16 Gb/s, in: I. Kaminow, T. Li (Eds.), Optical Fiber Telecommunications IV, Academic Press, 2002, pp. 232–304 (Chapter 6).

[41] X. Zhou, J. Yu, M.-F. Huang, Y. Shao, T. Wang, L. Nelson, P. Magill, M. Birk, P.I. Borel, D.W. Peckham, R. Lingle Jr., 64-Tb/s (640×107-Gb/s) PDM-36 QAM transmission over 320 km using both pre- and post-transmission digital equalization, in: Proc. Opt. Fiber Commun. Conf. (OFC), 2010, Paper PDPB9.

[42] A. Sano, T. Kobayashi, A. Matsuura, S. Yamamoto, S. Yamanaka, E. Yoshida, Y. Miyamoto, M. Matsui, M. Mizoguchi, T. Mizuno, 100×120-Gb/s PDM 64-QAM transmission over 160 km using linewidth-tolerant pilotless digital coherent detection, in: Proc. Eur. Conf. Opt. Commun. (ECOC), 2010, Paper PD2.4.

[43] D. Qian, M. Huang, E. Ip, Y. Huang, Y. Shao, J. Hu, T. Wang, 101.7-Tb/s (370×294-Gb/s) PDM-128QAM-OFDM transmission over 3×55-km SSMF using pilot-based phase noise mitigation, in: Proc. Opt. Fiber Commun. Conf. (OFC), 2011, Paper PDPB5.

[44] A. Sano, T. Kobayashi, S. Yamanaka, A. Matsuura, H. Kawakami, Y. Miyamoto, K. Ishihara, H. Masuda, 102.3-Tb/s (224×548-Gb/s) C-and extended L-band all-Raman transmission over 240 km using PDM-64QAM single carrier FDM with digital pilot tone, in: Proc. Opt. Fiber Commun. Conf. (OFC), 2010, Paper PDP5C.3.

[45] K. Nagayama, M. Kakui, M. Matsui, I. Saitoh, Y. Chigusa, Ultra-low-loss (0.1484 dB/km) pure silica core fibre and extension of transmission distance, IEE Electron. Lett. 38 (20) (2002) 1168–1169.

[46] D.G. Ouzounov, F.R. Ahmad, D. Müller, N. Venkataraman, M.T. Gallagher, M.G. Thomas, J. Silcox, K.W. Koch, A.L. Gaeta, Generation of megawatt optical solitons in hollow-core photonic band-gap fibers, Science 301 (5640) (2003) 1702.

[47] A.R. Bhagwat, A.L. Gaeta, Nonlinear optics in hollow-core photonic bandgap fibers, Opt. Express 16 (7) (2008) 5035–5047.

[48] D. Marcuse, Single-channel operation in very long nonlinear fibers with optical amplifiers at zero dispersion, J. Lightwave Technol. 9 (3) (1991) 356–361.

[49] A. Mecozzi, R.-J. Essiambre, Nonlinear Shannon limit in pseudo-linear coherent systems, J. Lightwave Technol. 30 (12) (2012) 2011–2024.

[50] W. Shieh, X. Chen, Information spectral efficiency and launch power density limits due to fiber nonlinearity for coherent optical ofdm systems, Photon. J. 3 (2) (2011) 158–173.

[51] G. Bosco, P. Poggiolini, A. Carena, V. Curri, F. Forghieri, Analytical results on channel capacity in uncompensated optical links with coherent detection, Opt. Express 19 (26) (2011) B440–B451.

[52] S.V. Manakov, Contribution to the theory of two-dimensional stationary self-focusing of electromagnetic waves, Sov. Phys. JETP 38 (2) (1974) 248–253.

[53] D. Marcuse, C.R. Menyuk, P.K.A. Wai, Application of the Manakov-PMD equation to studies of signal propagation in optical fibers with randomly varying birefringence, J. Lightwave Technol. 15 (9) (1997) 1735–1746.

[54] J.H. Conway, N.J.A. Sloane, Sphere Packings, Lattices, and Groups. third ed., Springer, Verlag, 2010.

[55] R.-J. Essiambre, G.J. Foschini, P.J. Winzer, G. Kramer, Capacity limits of fiber-optic communication systems, in: Proc. Opt. Fiber Commun. Conf. (OFC), 2009, Paper OThL1.

[56] S. Inao, T. Sato, S. Sentsui, T. Kuroha, Y. Nishimura, Multicore optical fiber, in: Proc. Opt. Fiber Commun. Conf. (OFC), 1979, Paper WB1.

[57] M. Koshiba, K. Saitoh, Y. Kokubun, Heterogeneous multi-core fibers: proposal and design principle, IEICE Electron. Express 6 (2) (2009) 98–103.

[58] B. Zhu, T.F. Taunay, M.F. Yan, J.M. Fini, M. Fishteyn, E.M. Monberg, F.V. Dimarcello, Seven-core multicore fiber transmissions for passive optical network, Opt. Express 18 (11) (2010) 11117–11122.

[59] J. M. Fini, T. Taunay, B. Zhu, M. Yan, Low cross-talk design of multi-core fibers, in: Proc. Conf. Lasers and Electro-optics, 2010, Paper CTuAA3.

[60] T. Hayashi, T. Taru, O. Shimakawa, T. Sasaki, E. Sasaoka, Low-crosstalk and low-loss multicore fiber utilizing fiber bend, in: Proc. Opt. Fiber Commun. Conf. (OFC), 2011, Paper OWJ3.

[61] T. Hayashi, T. Taru, O. Shimakawa, T. Sasaki, E. Sasaoka, Ultra-low-crosstalk multi-core fiber feasible to ultra-long-haul transmission, in: Proc. Opt. Fiber Commun. Conf. (OFC), 2011, Paper PDPC2.

[62] S. Chandrasekhar, A.H. Gnauck, X. Liu, P.J. Winzer, Y. Pan, E.C. Burrows, T.F. Taunay, B. Zhu, M. Fishteyn, M.F. Yan, J.M. Fini, E.M. Monberg, F.V. Dimarcello, WDM/SDM transmission of 10×128-Gb/s PDM-QPSK over 2688-km 7-core fiber with a per-fiber net aggregate spectral-efficiency distance product of 40,320 km b/s/Hz, Opt. Express 20 (2) (2012) 706–711.

[63] T. Hayashi, T. Taru, O. Shimakawa, T. Sasaki, E. Sasaoka, Characterization of crosstalk in ultra-low-crosstalk multi-core fiber, J. Lightwave Technol. 30 (4) (2012) 583–589.

[64] J.M. Fini, B. Zhu, T.F. Taunay, M.F. Yan, K.S. Abedin, Crosstalk in multicore fibers with randomness: gradual drift vs. short-length variations, Opt. Express 20 (2) (2012) 949–959.

[65] J.M. Fini, B. Zhu, T.F. Taunay, M.F. Yan, K.S. Abedin, Statistical models of multicore fiber crosstalk including time delays, J. Lightwave Technol. 30 (12) (2012) 2003–2010.

[66] D. Gloge, Optical power flow in multimode fibers, Bell Syst. Tech. J. 51 (8) (1972) 1767–1783.

[67] D. Gloge, Bending loss in multimode fibers with graded and ungraded core index, Appl. opt. 11 (11) (1972) 2506–2513.

[68] D. Gloge, E.A.J. Marcatili, Multimode theory of graded-core fibers, Bell Syst. Tech. J. 52 (9) (1973) 1563–1578.

[69] D. Marcuse, Theory of dielectric optical waveguides, in: P.F. Liao, P.L. Kelley (Eds.), Academic Press, 1991.

[70] Y. Kokubun, M. Koshiba, Novel multi-core fibers for mode division multiplexing: proposal and design principle, IEICE Electron. Express 6 (8) (2009) 522–528.

[71] K. Imamura, K. Mukasa, T. Yagi, Investigation on multi-core fibers with large Aeff and low micro bending loss, in: Proc. Opt. Fiber Commun. Conf. (OFC), 2010, Paper OWK6.

[72] J. Sakaguchi, B. Puttnam, W. Klaus, Y. Awaji, N. Wada, A. Kanno, T. Kawanishi, K. Imamura, H. Inaba, K. Mukasa, R. Sugizaki, T. Kobayashi, M. Watanabe, 305-Tb/s space division multiplexed transmission using homogeneous 19-core fiber, J. Lightwave Technol. 31 (4) (2012) 554–562.

[73] R. Ryf, R.-J. Essiambre, S. Randel, A.H. Gnauck, P.J. Winzer, T. Hayashi, T. Taru, T. Sasaki, MIMO-based crosstalk suppression in spatially multiplexed 56-Gb/s PDM-QPSK signals in strongly-coupled 3-core fiber, Photon. Technol. Lett. 23 (7) (2011) 1469–1471.

[74] R. Ryf, M.A. Mestre, A.H. Gnauck, S. Randel, C. Schmidt, R.-J. Essiambre, P.J. Winzer, R. Delbue, P. Pupalaikis, A. Sureka, T. Hayashi, T. Taru, T. Sasaki, Space-division multiplexed transmission over 4200-km 3-core microstructured fiber, in: Proc. Opt. Fiber Commun. Conf. (OFC), 2012, Paper PDP5C.2.

[75] C. Xia, N. Bai, I. Ozdur, X. Zhou, G. Li, Supermodes for optical transmission, Opt. Express 19 (2011) 16653–16664.

[76] R. Ryf, S. Randel, A.H. Gnauck, C. Bolle, A. Sierra, S. Mumtaz, M. Esmaeelpour, E.C. Burrows, R.-J. Essiambre, P.J. Winzer, D.W. Peckham, A. McCurdy, R. Lingle Jr , Mode-division multiplexing over 96 km of few-mode fiber using coherent 6×6 MIMO processing, J. Lightwave Technol. 30 (4) (2012) 521–531.

[77] S. Randel, R. Ryf, A.H. Gnauck, M.A. Mestre, C. Schmidt, R.-J. Essiambre, P.J. Winzer, R. Delbue, P. Pupalaikis, A. Sureka, Y. Sun, X. Jiang, R. Lingle Jr., Mode-multiplexed 6×20-GBd QPSK transmission over 1200-km DGD-compensated few-mode fiber, in: Proc. Opt. Fiber Commun. Conf. (OFC), 2012, Paper PDP5C.5.

[78] R. Olshansky, Mode coupling effects in graded-index optical fibers, Appl. Opt. 14 (4) (1975) 935–945.

[79] D. Marcuse, Curvature loss formula for optical fibers, J. Opt. Soc. Am. 66 (3) (1976) 216–220.

[80] D. Marcuse, Coupled mode theory of round optical fibers, Bell Syst. Tech. J. 52 (6) (1973) 817–842.

[81] W.B. Gardner, Microbending loss in optical fibers, Bell Syst. Tech. J. 54 (2) (1975) 457–465.

[82] L. Jeunhomme, J.P. Pocholle, Mode coupling in a multimode optical fiber with microbends, Appl. Opt. 14 (10) (1975) 2400–2405.

[83] L. Su, K. Chiang, C. Lu, Microbend-induced mode coupling in a graded-index multimode fiber, Appl. Opt. 44 (34) (2005) 7394–7402.

[84] R.F. Cregan, B.J. Mangan, J.C. Knight, T.A. Birks, P.S.J. Russell, P.J. Roberts, D.C. Allan, Single-mode photonic band gap guidance of light in air, Science 285 (5433) (1999) 1537–1539.

[85] M. Nielsen, C. Jacobsen, N. Mortensen, J. Folkenberg, H. Simonsen, Low-loss photonic crystal fibers for transmission systems and their dispersion properties, Opt. Express 12 (7) (2004) 1372–1376.

[86] P.J. Roberts, F. Couny, H. Sabert, B.J. Mangan, D.P. Williams, L. Farr, M.W. Mason, A. Tomlinson, T.A. Birks, J.C. Knight, P.S.J. Russell, Ultimate low loss of hollow-core photonic crystal fibres, Opt. Express 13 (1) (2005) 236–244.

[87] M.N. Petrovich, F. Poletti, J.P. Wooler, A.M. Heidt, N.K. Baddela, Z. Li, D.R. Gray, R. Slavk, F. Parmigiani, N.V. Wheeler, J.R. Hayes, E. Numkam, L. Grüner-Nielsen, B. Pálsdóttir, R. Phelan, B. Kelly, M. Becker, N. MacSuibhne, J. Zhao, F.C. Garcia Gunning, A.D. Ellis, P. Petropoulos, S.U. Alam, D.J. Richardson, First demonstration of 2 μm data transmission in a low-loss hollow core photonic bandgap fiber, in: Proc. Eur. Conf. Opt. Commun. (ECOC), 2012, Paper Th.3.A.5.

[88] H.R. Stuart, Dispersive multiplexing in multimode optical fiber, Science 289 (5477) (2000) 281–283.

[89] H. Bölcskei, D. Gesbert, A.J. Paulraj, On the capacity of OFDM-based spatial multiplexing systems, IEEE Trans. Commun. 50 (2) (2002) 225–234.

[90] S. Verdú, Spectral efficiency in the wideband regime, IEEE Trans. Inform. Theory 48 (6) (2002) 1319–1343.

[91] R.C.J. Hsu, A. Tarighat, A. Shah, A.H. Sayed, B. Jalali, Capacity enhancement in coherent optical MIMO (COMIMO) multimode fiber links, IEEE Commun. Lett. 10 (3) (2006) 195–197.

[92] A. Tarighat, R. Hsu, A. Shah, A. Sayed, B. Jalali, Fundamentals and challenges of optical multiple-input multiple-output multimode fiber links, IEEE Commun. Mag. 45 (5) (2007) 57–63.

[93] I.B. Djordjevic, M. Arabaci, L. Xu, T. Wang, Spatial-domain-based multidimensional modulation for multi-Tb/s serial optical transmission, Opt. Express 19 (7) (2011) 6845–6857.

[94] D. Gloge, Weakly guiding fibers, Appl. Opt. 10 (10) (1971) 2252–2258.

[95] A. Lozano, A.M. Tulino, S. Verdú, Multiple-antenna capacity in the low-power regime, IEEE Trans. Inform. Theory 49 (10) (2003) 2527–2544.

[96] S.K. Jayaweera, An energy-efficient virtual MIMO architecture based on V-BLAST processing for distributed wireless sensor networks, in: Proc. IEEE Conf. Sensor Ad Hoc Commun. Network, 2003, pp. 299–308.

[97] S. Cui, A.J. Goldsmith, A. Bahai, Energy-efficiency of MIMO and cooperative MIMO techniques in sensor networks, IEEE J. Sel. Area. Comm. 22 (6) (2004) 1089–1098.

[98] S. Cui, A.J. Goldsmith, A. Bahai, Energy-constrained modulation optimization, IEEE Trans. Wireless Commun. 4 (5) (2005) 2349–2360.

[99] P.J. Winzer, Energy-efficient optical transport capacity scaling through spatial multiplexing, Photon. Technol. Lett. 23 (13) (2011) 851–853.

[100] I.B. Djordjevic, Energy-efficient spatial-domain-based hybrid multidimensional coded modulations enabling multi-Tb/s optical transport, Opt. Express 19 (17) (2011) 16708–16714.

[101] M. Kolesik, J.V. Moloney, Nonlinear optical pulse propagation simulation: From Maxwell's to unidirectional equations, Phys. Rev. E 70 (3) 2004, Paper 036604.

[102] F. Poletti, P. Horak, Description of ultrashort pulse propagation in multimode optical fibers, J. Opt. Soc. Am. B 25 (10) (2008) 1645–1654.

[103] F. Poletti, P. Horak, Dynamics of femtosecond supercontinuum generation in multimode fibers, Opt. Express 17 (8) (2009) 6134–6147.

[104] S. Mumtaz, R.-J. Essiambre, G.P. Agrawal, Nonlinear propagation in multimode and multicore fibers: generalization of the Manakov equations, J. Lightwave Technol. 31 (3) (2013) 398–406.

[105] S. Mumtaz, R.-J. Essiambre, G.P. Agrawal, Birefringence effects in space-division multiplexed fiber transmission systems: generalization of Manakov equation, IEEE Summer Topical, 2012, Paper MC3.5.

[106] S. Mumtaz, R.-J. Essiambre, G.P. Agrawal, Reduction of nonlinear impairments in coupled-core multicore optical fibers, IEEE Summer Topical, 2012, Paper MC3.3.

[107] A. Mecozzi, C. Antonelli, M. Shtaif, Nonlinear propagation in multi-mode fibers in the strong coupling regime, Opt. Express 20 (11) (2012) 11673–11678.

[108] S. Randel, R. Ryf, A. Sierra, P.J. Winzer, A.H. Gnauck, C.A. Bolle, R.-J. Essiambre, D.W. Peckham, A. McCurdy, R. Lingle Jr, 6×56-Gb/s mode-division multiplexed transmission over 33-km few-mode fiber enabled by 6×6 MIMO equalization, Opt. Express 19 (17) (2011) 16697–16707.

[109] L. Cohen, Comparison of single-mode fiber dispersion measurement techniques, J. Lightwave Technol. 3 (5) (1985) 958–966.

[110] G. Millot, S. Pitois, P.T. Dinda, M. Haelterman, Observation of modulational instability induced by velocity-matched cross-phase modulation in a normally dispersive bimodal fiber, Opt. Lett. 22 (22) (1997) 1686–1688.

[111] G. Rademacher, S. Warm, K. Petermann, Nonlinear interference in mode multiplexed multimode fibers, IEEE Summer Topical, 2012, Paper MC3.4.

[112] R.W. Tkach, A.R. Chraplyvy, F. Forghieri, A.H. Gnauck, R.M. Derosier, Four-photon mixing and high-speed WDM systems, J. Lightwave Technol. 13 (5) (1995) 841–849.

[113] T.K. Chiang, N. Kagi, M.E. Marhic, L.G. Kazovsky, Cross-phase modulation in fiber links with multiple optical amplifiers and dispersion compensators, J. Lightwave Technol. 14 (3) (1996) 249–260.

[114] R.H. Stolen, E.P. Ippen, Raman gain in glass optical waveguides, Appl. Phys. Lett. 22 (6) (1973) 276–278.

[115] R.H. Stolen, J.E. Bjorkholm, A. Ashkin, Phase-matched three-wave mixing in silica fiber optical waveguides, Appl. Phys. Lett. 24 (7) (1974) 308–310.

[116] R.H. Stolen, Phase-matched-stimulated four-photon mixing in silica-fiber waveguides, IEEE J. Quantum Electron. 11 (3) (1975) 100–103.

[117] J. Cheng, M. Pedersen, K. Charan, C. Xu, L. Grüner-Nielsen, D. Jacobsen, High-efficiency intermodal four-wave mixing in a higher-order-mode fiber, in: Proc. Conf. Lasers Electrooptics, 2012, Paper CTh3G.6.

[118] L. Rishøj, Y. Chen, P. Steinvurzel, K. Rottwitt, S. Ramachandran, High-energy fiber lasers at non-traditional colours, via intermodal nonlinearities, in: Proc. Conf. Lasers Electro-optics, 2012, paper CTu3M.6.

[119] J.M. Fini, B. Zhu, T.F. Taunay, M.F. Yan, Statistics of crosstalk in bent multicore fibers, Opt. Express 18 (14) (2010) 15122–15129.

[120] R.-J. Essiambre, R. Ryf, M.A. Mestre, A.H. Gnauck, R.W. Tkach, A.R. Chraplyvy, S. Randel, Y. Sun, X. Jiang, R. Lingle Jr., Inter-modal nonlinear interactions between well separated channels in spatially-multiplexed fiber transmission, in: Proc. Eur. Conf. Opt. Commun. (ECOC), 2012, Paper Tu.1.C.4.

[121] R.-J. Essiambre, M.A. Mestre, R. Ryf, A.H. Gnauck, R.W. Tkach, A.R. Chraplyvy, S. Randel, Y. Sun, X. Jiang, R. Lingle Jr., Demonstration of broadband inter-modal four-wave mixing in graded-index few-mode fibers, in: Proc. Opt. Fiber Commun. Conf. (OFC), 2013, Paper OM3B.2.

[122] R.-J. Essiambre, R. Ryf, M.A. Mestre, A.H. Gnauck, R.W. Tkach, A.R. Chraplyvy, Y. Sun, X. Jiang, R. Lingle Jr., Experimental observation of inter-modal cross-phase modulation in few-mode fibers, Photon. Technol. Lett. 25 (6) (2013) 535–538.

[123] R.-J. Essiambre, R. Ryf, M. A. Mestre, A. H. Gnauck, R. W. Tkach, A. R. Chraplyvy, Y. Sun, X. Jiang, R. Lingle Jr, Experimental investigation of inter-modal four-wave mixing in few-mode fibers, Photon. Technol. Lett. 25 (6) (2013) 539–541.

Commercial 100-Gbit/s Coherent Transmission Systems

Tiejun J. Xia and Glenn A. Wellbrock

Verizon, Richardson, TX, USA

2.1 INTRODUCTION

An optical transmission system is a part of the transport layer in a service provider's network. The transmission system carries information on optical channels, which have certain protocols, such as SONET or OTN containers to encapsulate the user data and provide network management functions. The optical channels are carried across the network via an optical fiber infrastructure. The channel capacity, which governs how much information a channel is able to carry across the network in a unit time, and the number of channels in a strand of fiber determine the total transmission capacity a fiber link provides. The channel capacity has been growing to keep up with traffic demands over the past several decades. Initial optical transport networks mainly served telephony services, which required very little bandwidth. However, introduction of the Internet and fiber-based video services, which need much higher bandwidth than conventional phone calls, has become the primary application thus making IP (Internet Protocol) the dominant traffic type a transport network carries. The optical transport network therefore now carries mainly the traffic from data equipment, such as IP Routers, MPLS (Multi-Protocol Label Switching) switches, or Ethernet switches. Normally the equipment hands the data traffic to the transport network via Ethernet ports and the transport network aggregates it before transporting it to the desired destination and delivering it back to the far end data equipment. Therefore, the capacity of transport equipment must support the need of the data equipment. Figure 2.1a shows the global Internet Protocol traffic growth from the beginning of the Internet [1]. The Internet traffic surpassed TDM (Time Domain Multiplexing) traffic, mainly phone calls, in 2002 [2]. By 2015 global IP traffic is predicted to be about eight orders of magnitude higher than it was in 1990. Given that IP traffic has such a high growth while the TDM traffic has only a few percentage points of annual growth rate, the majority of the traffic these days is almost all in the IP data format, at least from a customer or end user's perspective. In 2011 the global IP traffic reached 30 exabytes per month. It represents most of the traffic that the transport network must handle, but not all of it. IP or Ethernet transport is not defined in such a way to support long inter-office reaches.

The implementation history of optical transport channel capacity standards and Ethernet port speed standards is shown in Figure 2.1b. The Ethernet port speed has

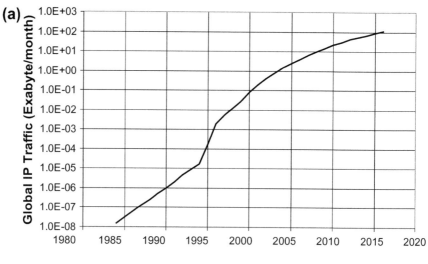

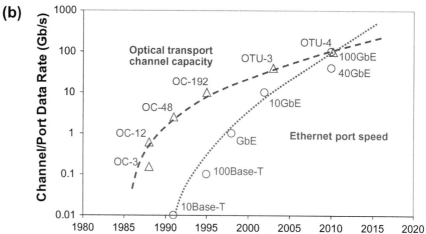

FIGURE 2.1 (a) Global IP traffic demands, data beyond 2011 are estimated [1]; (b) standards of transport channel capacities and Ethernet port speeds.

grown from 10 Mb/s to 100 Gb/s with a 10× incremental (except for the 40 GbE standard) while the optical channel capacity standards have grown from 155 Mb/s (OC-3) to 100 Gb/s (OTU4) with a 4× incremental (except the 100 Gb/s standard). For the first time, both optical channel capacity and data equipment port speed match each other at 100 Gb/s in 2010. It can be seen from Figure 2.1b that the transport channel capacity has always been larger than or at least equal to the port speed of data equipment. That is understandable since an optical channel, as a "container" for data packets, should be equal to or larger than the size of its "content." The trends of the capacity increases for transport equipment and the speed increases for data equipment as shown in the figure, however, indicate the data port speed may exceed the transport channel capacity in the future (beyond 100 Gb/s). Given that the "container" must be equal to or larger than the "content," the trend of transport channel capacity must follow the data port speed trend, instead of its historical trend. If this happens the 100 Gb/s standard becomes a turning point for transport channel capacity designs, unless inverse multiplexing is considered for data mapping between Ethernet ports and transport channels.

This chapter is going to review optical channel development and the road leading to commercially available coherent 100G systems. Various optical channel design approaches will be analyzed from a new angle by sorting them into three basic channel types. Network service providers' efforts to introduce 100G transmission systems to their networks will be presented as well along with the network architecture changes made possible with the introduction of the coherent receivers. The benefits of 100G transmission systems to network service providers will be discussed. Finally the next step in optical channel designs beyond 100G will be envisioned. To simplify this discussion, notations such as 10G, 100G, 1T, etc. may be used in this chapter to represent data rates around 10 Gb/s, 100 Gb/s, 1 Tb/s, etc.

2.2 OPTICAL CHANNEL DESIGNS

As the channel capacity of commercially available optical transport systems reaches the unprecedented level of 100 Gb/s [3], it is interesting to review the history of optical channel capacity evolution and speculate what is going to happen after this. Here, an "optical channel" is referred to as a unit of capacity, into which optical transport systems load traffic data and carry it from one node to another node via light paths. The "channel capacity" is referred to as the amount of traffic the channel is able to carry in a unit of time. For example, an OTU2 (Optical Transport Unit 2) optical channel in the Optical Transport Network (OTN) protocol is able to carry about 10-Gb/s of traffic across a transport network, in which "10-Gb/s" is called its "channel capacity."

In the early 1970s, when optical fiber communication was in its infancy, the optical channel capacity was very low by today's standards. Goell reported an experiment of a 6.3-Mb/s optical signal repeater [4] targeting digital communication with newly developed low-loss optical fiber [5] in 1973. In the experiment the optical signal was generated by direct modulation of a light-emitting diode at the 900 nm wavelength. Pin diodes and an avalanche detector were used as the receiver. Bit error rate (BER) curves were measured in the experiment. In the same year, White and Chin reported an investigation of a 100-Mb/s fiber communication system, in which the modulated light from a 1.5-mW light-emitting diode was coupled into a 60-μm multimode optical fiber, and the BER curve and the eye diagram were recorded [6]. Ozeki and Ito reported a 200-Mb/s experimental optical communication system, in which a 20% coupling efficiency for the light source to a multimode optical fiber was achieved, and an avalanche photodiode with a rise and fall time of 200 ps was used as the receiver [7]. Also in the same year, Miller, Li, and Marcatili reported a 300-Mb/s optical communication result [8]. Since then, after almost 40 years of development, optical channel capacities have experienced a steady and fast growth with experiments of more than 10 Tb/s have been reported [9], which is about six orders of magnitude increase in channel capacity compared with the channel capacity in the first optical fiber communication experiment mentioned above.

The optical channel capacity has increased several orders of magnitude and the optical channel design has been evolved into various flavors. There are many ways to design an optical channel. In general, construction of an optical channel may consist of two steps: modulation of an optical carrier or a group of optical carriers and multiplexing of the modulated optical carriers. At least six dimensions of an optical carrier can be modulated to carry information: amplitude, phase, time, space, frequency, and polarization. In practice, the most common characteristics for optical modulation are amplitude and phase similar to what is done in radio communication environments. Modulation of other carrier attributes is possible, for example, time modulation has been reported in which a bit can be placed in one of 16 time slots to represent different data patterns [10]; however, these modulation techniques are rare and have not been commercialized to date. These same characteristics can be considered for multiplexing as well, but there are limitations in the optical environment for certain modulated signals, for example, a polarization modulated signal would have a hard time to be polarization multiplexed. Optical time domain multiplexing (OTDM) was a heavily studied approach for optical channel designs and used for many "hero" experiments. With OTDM the channel capacity can be much higher than the data rate generated by an optical modulator, since multiple optical modulators are used to build an optical channel via the TDM process. Space domain multiplexing (SDM), which is also emerging as a new way to construct optical channels, has attracted a lot of attention in the industry for its potential to increase channel capacity and fiber capacity by using multi propagation modes in a waveguide and/or multi waveguides in a strand of fiber, but yet to be materialized. Given that multiplexing in optical frequency used to be the approach to combine optical channels in fiber, now multiplexing optical carriers in the frequency domain is becoming a new way to design optical channels, particularly for channels with capacities higher than 100 Gb/s, such as "superchannels" [11]. Polarization multiplexing can be employed for nearly all modulation techniques (two polarizations instead of just one) and is clearly the method being pursued for current generation channel capacities to double the bit rate. When combinations of various modulation and multiplexing approaches are considered in optical channel designs there could be a lot of differently flavored optical channels.

It is helpful to put these different optical channel options into just a few distinguishable channel types so that studying the evolution of optical channel designs can be made easier. Consequently, a new optical channel classification method has been proposed [12]. With this new method optical channels can be sorted into just three basic channel types, based on how the multiplexing process is accomplished. The three basic channel types are defined as the following.

Type 1 optical channel is defined as only one optical carrier and no multiplexing, therefore it is denoted as Single Carrier No Multiplexing (SCNM) format. For example, an NRZ 10-Gb/s commercially available channel is a Type 1 channel, since it contains only one optical carrier and no multiplexing process.

A Type 2 optical channel still involves only one optical carrier but has more than one level of multiplexing, hence it can be denoted as SCWM (Single Carrier With Multiplexing). A Polarization Multiplexed Binary Phase Shift Keying (PM-BPSK) channel is an example of a Type 2 channel, in which polarization multiplexing is used in the channel design. In this polarization multiplexed channel, the data is loaded into two independent polarization states of one optical carrier. However, when the channel propagates along the fiber, the instant polarization state of the channel is determined by the phases of the two streams of data, which were originally loaded into two polarizations at the transmitter, but are scrambled due to inherent characteristics of light propagating through glass fiber. An optical time domain multiplexing (OTDM) channel is a typical Type 2 channel, too. In an OTDM channel one pulsed optical carrier is split into several copies and each copy is loaded with traffic data via an optical modulator. The traffic data loaded optical signals are then combined together without overlapping each other in time domain thus the channel can have a data rate much higher than the modulation rate. However, there is only one optical carrier for the channel. Here, "optical carrier" means a light signal with a representative optical frequency. For example, the light signal carrying an OTU2 optical channel running at 1553.33 nm can be considered as the optical carrier of the channel with a representative optical frequency of 193.00 THz. Space domain multiplexing (SDM) can be employed in Type 2 channels as well, if an optical carrier is used for multiple optical signals, belonging to one channel, which are distributed in space.

Type 3 involves multiplexing in the optical frequency domain such that multiple optical carriers are presented in the channel; therefore this type of channel is denoted as Multiple Carriers With Multiplexing (MCWM). Of course frequency multiplexing is not the only method applicable for this type of channel, other characteristics of the optical carrier can also be used in the multiplexing processes. When a channel contains multiple optical carriers, it is referred to as a "superchannel" [11]. Various flavors of superchannels belong to the Type 3 family. When a channel is allowed to contain multiple optical carriers, the boundary between a single channel capacity and fiber capacity is blurred.

Figure 2.2 shows a schematic illustration of the three basic channel types defined in this chapter. A binary Return-to-Zero (RZ) On-Off Keying (OOK) modulation format is used as an example of optical signals. An Electrical Time Domain Multiplexing (ETDM) channel, an OTDM channel, and a superchannel are used as examples to show the differences between them. Signal wave formats in both frequency domain and time domain are also shown in the figure to help illustrate the differences.

Since Type 1 channels do not involve signal multiplexing in optical domain, all signal multiplexing must be accomplished in the electrical domain [13], so it is often called an ETDM channel. As shown in Figure 2.2a the data being transported is fed into an electrical processor via the data tributaries. Multiple lower-rate data streams

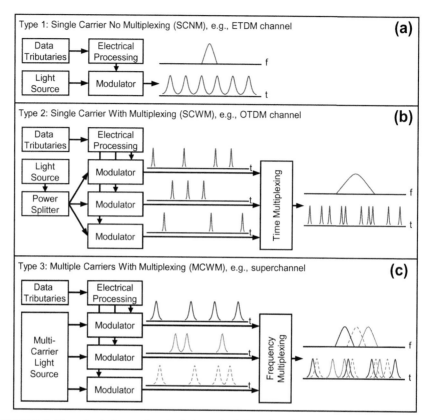

FIGURE 2.2 Three basic types of optical channels defined in this chapter: (a) single carrier no multiplexing (SCNM), (b) single carrier with multiplexing (SCWM), and (c) multiple carriers with multiplexing (MCWM) channels.

are electrically multiplexed to form a drive signal for the modulator. The light emitted from the source, usually a laser, is modulated by the drive signal. At the output of the transmitter the optical channel produces one modulated carrier in the optical domain and a serial bit stream in the time domain. For Type 1 channels the channel capacity is limited by the working bandwidth of the modulator and the number of modulation levels. Increasing the channel capacity can be accomplished by increasing the modulation frequency, which is equivalent to requiring a higher modulator bandwidth, and increasing the levels of modulation. Until just recently, the modulation format was mainly binary On-Off Keying (OOK). The OOK scheme works just fine for channel capacities up to 10 Gb/s. Beyond 10 Gb/s, multiple levels of modulation have been introduced to ease the requirement for higher modulator bandwidth. For example, in a 43 Gb/s channel the required modulator bandwidth is about 30 GHz if a binary modulation format is used. This is nearly the limit of a commercially available lithium niobate modulator

(LiNbO$_3$) modulator [14]. If a four-level modulation format were used, the bandwidth requirement would be reduced to half this rate. The historical channel capacity records of Type 1 channels [15–28] are shown in Figure 2.3. This type of channel showed impressive growth until the 1990s, after that the growth curve flattened out reflecting technical challenges in further increasing channel capacity using this technique. The highest channel capacity of a binary ETDM channel is 107 Gb/s [25]. Type 1 channels could increase channel capacity by using multiple levels of modulations [26]; however, polarization multiplexing (a Type 2 technique) is an easier way to double the channel capacity without the need to increase bandwidth and seems like a better choice to increase channel capacity beyond 40 Gb/s. Consequently, most high capacity channel designs in recent years have adopted polarization multiplexing and the effort to increase channel capacity for Type 1 channels has diminished.

Type 2 channel designs use only one optical carrier just like Type 1 channel designs, but the optical signal of the channel is generated via one or more optical multiplexing processes giving designers more freedom choosing approaches to increase the channel capacity which is then no longer limited by the bandwidth of a single modulator. Given that the multiplexing process in Type 2 channels is to directly combine multiple modulated data streams onto the same wavelength, the approaches to perform this combination must be able to avoid optical interference between the data streams, unless of course the relative phase relationship between the data streams is well controlled. There are three common approaches to perform optical multiplexing for signals with the same wavelength. They are multiplexing in polarization, in time, and in space.

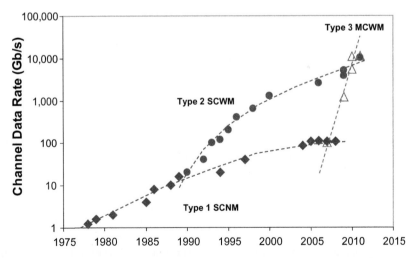

FIGURE 2.3 Channel capacity records for the three basic types of optical channels defined in this chapter, where SCNM—single carrier no multiplexing; SCWM—single carrier with multiplexing; MCWM—multi-carrier with multiplexing.

Again, until just recently only one polarization state was used to carry data traffic when in fact signals loaded into two orthogonal polarization states are independent from each other and therefore using both polarization states to carry data can double the channel capacity without increasing the spectral bandwidth. The process of combining two different signal streams into two different polarizations is called "polarization multiplexing (PM)." This is referred to as "dual polarization (DP)." It is relatively easy to generate a PM channel at the transmitter, but quite complex to extrapolate a PM channel at a receiver. At the transmitter a PM channel can be constructed by just putting two data streams with orthogonal polarizations together through a Polarization Beam Combiner (PBC). At the receiver the data streams must be separated back out again into individual data signals. This is a bit tricky, however, since the polarization states of the signals will inevitably be scrambled as they propagate in the fiber. There are basically only two approaches to separate the data back out, optically or with a significant amount of digital signal processing. With an optical approach the two data streams in a channel are separated by a Polarization Beam Splitter (PBS) before they hit the two independent receivers. For example, signals belong to two polarizations in polarization multiplexed 20-Gb/s channels which were separated with a PBS in front of the receivers with the help of a polarization controller [29]. This approach is quite simple from a component point of view; however, the real challenge is to guarantee that the signals in the incoming channel align to the axes of the PBS accurately making the polarization tracking controller (especially in a situation where polarization states change rapidly) very complex. The data streams in each of the two polarization states can also be separated in the digital domain. Actually, a polarization multiplexed channel behaves just like one optical signal to a photo detector, since the detector cannot distinguish between the signals belonging to the two different polarizations at all. However, the full information of the polarization multiplexed channel—its amplitude, phase, and polarization states as a function of time—can be obtained with a coherent detection receiver. With a coherent receiver, the original signals in both polarizations can be recovered by canceling the polarization scrambling and linear impairments experienced during propagation in the fiber. This recovery can be accomplished by digitizing the received signal and processing it in the digital domain [30]. This design for high capacity channels (polarization multiplexing) is now a common practice using coherent detection and digital signal processing as the main choice for signal recovery.

Optical time domain multiplexing (OTDM) is another way to construct a Type 2 channel. A short pulse clock stream can be duplicated and the multiple copies of the clock stream can be used as the light source of the channel. Data is loaded into each copy of the clock stream with modulators to form multiple data streams and the data streams are combined in time domain. Since the pulse width is short enough, the bits from different data streams do not overlap each other in the time domain. With this OTDM approach, the channel capacity can be much higher than a single modulator is able to generate. The pulse width of an OTDM signal, however, must be smaller than the final bit interval of the channel. Therefore to perform an optical time domain multiplexer, the modulation must be a Return-to-Zero (RZ) format. For example, short optical pulses generated with a pulse carver [31] or optical solitons generated by a fiber laser [32] were used for OTDM channel generation. In one report optical pulses with a pulse width as short as 380 fs were used to generate a 1.28-Tb/s OTDM channel, in which the bit interval was 780 fs [33]. The OTDM approach has been used in the past for many hero experiments, but is still in the research stage as there have been no commercial products based on this technology observed by the authors.

Space domain multiplexing (SDM) is an emerging approach. In optical fiber transmission, SDM requires multiplexing signals in multiple fiber waveguides and/or multiple propagation modes in a single waveguide. Here a waveguide can be a fiber core or one defined in a complex fiber structure. In most cases SDM is used for multiplexing multiple channels, not for multiplexing multiple data streams to form a channel. There are high exceptions though for this technique in the future. A 240-Gb/s channel was built with a six-fold SDM, in which six data streams with the same wavelength in six propagation modes in a single strand of fiber were multiplexed. Each data stream was a 40-Gb/s signal with a QPSK modulation format [34]. SDM with 19 fiber cores has also been reported [35]. The concept of SDM has been generally restricted to short reach optical applications so far, where a channel is generated by multiplexing data streams in multiple strands of fiber [36]. Much like other approaches, this could be applied to backbone networks as well as the technology matures.

The channel capacity records of Type 2 channels [33,37–46] are shown in Figure 2.3 as well. Due to the availability of short pulsed laser sources used for TDM in the 1990s, Type 2 channel capacities had a sharp increase in that time period. After that Type 2 channel capacity increases have been accomplished by adding multiple levels of modulation and polarization multiplexing, in addition to introduction of shorter laser pulses.

The Type 3 optical channel design breaks down the limitation of using only one optical carrier for an optical channel. Multiple optical carriers are allowed to be multiplexed to form a single optical channel. In this way the channel capacity is nearly no longer a limitation since all possible multiplexing approaches can be utilized to construct a "superchannel." This is the primary reason why the concept of optical frequency multiplexing has been introduced [47], so the channel capacities of this type have quickly caught up with the channel capacities of the other two types of channels [9,48–50], as shown in Figure 2.3. For example, a channel of 11.2 Tb/s, which is the highest single channel capacity reported so far, has been presented with a Type 3 channel design [50]. This channel used 112 optical carriers, generated from a single laser source via a multi-stage optical carrier generation process. Each of the carriers has a PM-QPSK modulation format with a symbol rate of 28 Gbaud. The industry term for Type 3 channels, which have multiple optical carriers, is a "superchannel" [11]. Compared to Type 1 and Type 2 channels, Type 3 channels have significantly more freedom in design. The number of carriers in a Type 3 channel can be small or large, depending on the needs. Superchannels with as little as two optical carriers [51] or as many as 112 carriers, as mentioned above, have been reported.

In commercial optical transport networks, optical channels with a capacity of 10 Gb/s or below mainly use the single polarization NRZ modulation format (Type 1 channels). The designs for 40G channels have many flavors. There are directly modulated ETDM 40G channels (Type 1), for example, NRZ 40G channel [23], RZ-DQPSK

40G channel [52], or 16QAM 40G channel [53] have been reported. There are Type 2 40G channels with optical multiplexing, for example, RZ-4xOTDM 40G channel [32], PM-DPSK 40G channel [54], and PM-QPSK 40G channel with coherent detection [30] have been reported. It appears that there is no need to build a 40-Gb/s channel with Type 3 design since one optical carrier with current generation modulator speeds is enough for various 40G channel flavors. For 100-Gb/s channels, however, all three types of channel designs have been studied in the industry. The detail of 100-Gb/s channel development is the focus of the following sections.

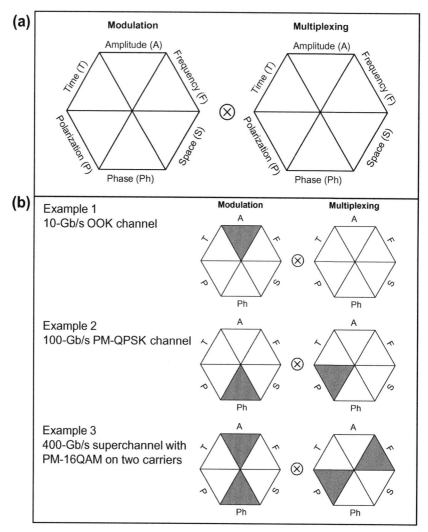

FIGURE 2.4 (a) Graphic representation of optical carrier characteristics for channel designs; (b) examples using the graphic representation to particular channel designs.

It is convenient to use a graph to represent a type of channel or a particular channel design, based on the above analysis on optical channels. The options for both modulation and multiplexing in channel designs can be represented by a hexagon. Each side of the hexagon represents a characteristic of the optical carriers such as amplitude (A), Phase (Ph), time (T), space (S), polarization (P), or frequency (F), as shown in Figure 2.4a. A tensor product symbol is used to represent a combination of various modulation formats and various multiplexing approaches in the channel designs. Three examples chosen to represent the three basic types of channels are shown in Figure 2.4b. A 10-Gb/s OOK signal is a typical Type 1 channel. It uses amplitude modulation and has no multiplexing, therefore only the amplitude triangle in the modulation hexagon is highlighted. A 100-Gb/s PM-QPSK channel is a typical Type 2 channel. It has only phase modulation and polarization multiplexing, so that the phase triangle in the modulation hexagon and the polarization triangle in the multiplexing hexagon are highlighted. A Type 3 channel is represented by a 400-Gb/s superchannel with PM-16QAM on two optical carriers. This channel has both amplitude modulation and phase modulation as well as polarization multiplexing and frequency multiplexing. As a result, the four corresponding triangles are highlighted to represent the channel design. With this graphic representation the modulation and multiplexing scheme of a channel can be seen clearly.

2.3 100G CHANNEL—FROM WISH TO REALITY

Even though optical channel capacity has passed the 10 Tb/s level in research reports [9], the highest channel capacity available for commercial deployment is 100 Gb/s today. The standard for 100 Gb/s transport channel capacity, OTU4 in ITU-T G.709, and that for 100 Gb/s Ethernet port speed, 100 GbE in IEEE 802.3ba, was finalized in 2010; the research work focusing on 100-Gb/s channel capacity was initiated many years ago. In the early 1990s, high-speed modulation and high-speed electrical signal processors were not available and at that time 100-Gb/s channels were mainly constructed by using OTDM approaches (Type 2 channels). For example, a 100-Gb/s channel was generated via $8\times$ OTDM and polarization multiplexing, with a 6.3-Gb/s modulated signal as the speed [39]. The bandwidth limitation of optoelectronic devices also prevented the detection and processing of the high capacity channel electrically in the early development stage. To handle OTDM 100G channels, in many cases, all-optical signal processing technology was used. For example, all-optical processing was a key enabling technology for data packet header recognition and packet switching of an experimental 100-Gb/s channel [55]. The OTDM approach for 100G channel designs was not successfully commercialized, partially due to the introduction of Dense Wavelength Division Multiplexing (DWDM) technology in the late 1990s and partially due to transmission difficulties of short optical pulses in fiber. At the time when DWDM technology was introduced, even though each DWDM channel might not be able to provide a channel capacity as high as 100 Gb/s and a typical DWDM channel capacity was 2.5 Gb/s or 10 Gb/s, the overall capacity of a DWDM system could easily surpass the capacity a single OTDM 100 Gb/s channel could provide and therefore the OTDM approach to build 100G systems did not attract enough attention from the industry.

After turning to the new millennium ETDM technology once looked promising for 100G channels as a result of significant progresses in optical and electrical components. These components had much wider working bandwidths than before. A 107 Gb/a binary NRZ channel was generated directly from a modulator with a 107 Gb/s electrical driving signal [25]. This type of binary 100G channel successfully traveled over 1000 km fiber in labs [56]. Such ETDM 100G channels have also been tested in live networks several times. For the first time an NRZ 107-Gb/s channel transmitted over 160-km field installed fiber using vestigial sideband modulation. The trial results showed a high tolerance toward narrowband optical filtering by using the modulation format [57]. Eight 107-Gb/s NRZ channels with a channel spacing of 100 GHz were demonstrated over 500-km of installed standard SMF in a service provider's network. In the trial a spectral efficiency of 1 b/s/Hz was achieved for a link composed with spans of high losses and realistic PMD values [58,59].

Since both the electrical and optical components are working at their extreme conditions for an ETDM NRZ 100G channel, the binary ETDM approach may not be suitable for commercial 100G products. In addition, using a binary modulation format is not spectrally efficient in principle. To reduce the symbol rate and ease the requirements for the component bandwidth, multi-level modulation was introduced to construct ETDM 100G channels. Differential QPSK (DQPSK) modulation format has been fully studied for 100G channel designs. The DQPSK has four modulation levels, therefore each symbol represents two data bits and the symbol rate is half of the data rate. In experiments a 107 Gb/s DQPSK channel was designed with a symbol rate of 53.5 Gbaud [60]. The signal traveled over 2000 km in a lab environment successfully. To verify the performance of the channel in a real network, a field trial was conducted with this channel in a global service provider's long haul network [61–63]. Figure 2.5 shows the configuration of field trial.

In this first 100G field trial with live traffic, a 504-km long haul network route between Tampa and Miami, Florida, equipped with a long haul transport system was chosen. The system is a Raman-pumped DWDM system with alternating 65-GHz/35-GHz DWDM passbands in the extended L-band. Nine working DWDM channels (10 Gb/s) and the 100G channel under test were on the same fiber. The 10G channels were bi-directional and the 100G channel only propagated in one direction. The frequencies of the 10G channels were between 186.70 THz and 188.70 THz and the 100G channel was added at the ROADM in Tampa as an alien wavelength at 188.60 THz (100 GHz away from the nearest 10G channel). The 100G channel was dropped at a ROADM in Miami. The 10G channels were monitored during the trial by the network operation center to ensure there was no impact to the customer traffic due to the introduction of the 100G channel. As shown in Figure 2.5, at the 100G transmitter, an OC192 signal, which contained live HDTV traffic, was tapped optically from the service provider's national video service network and fed to the client port of the 100G equipment. There, the OC192 signal passed through G.709

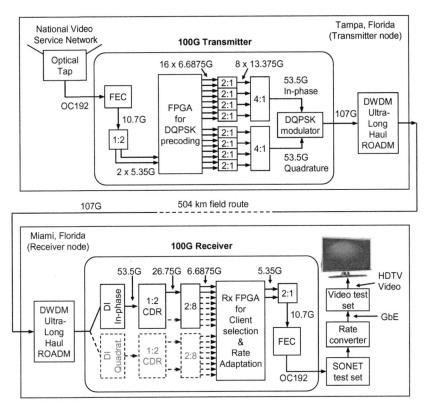

FIGURE 2.5 Configuration of DQPSK 100G channel field trial with live traffic [61–63].

compliant enhanced forward error correction (FEC) and framing to produce OTU2 frames. The OTU2 signal at 10.7 Gb/s entered an FPGA at 2×5.35 Gb/s. Within the FPGA, the 10G client was immediately duplicated to obtain a true 107-Gb/s data bus, on which the necessary DPQSK precoding and some simple 100G framing was performed, resulting in an output bus of 16×6.6875-Gb/s. These signals were 8:1 multiplexed in two groups to represent the 53.5-Gb/s in-phase (I) and quadrature (Q) components of the 107-Gb/s DQPSK signal. The 107-Gb/s RZ-DQPSK optical signal was generated by a nested LiNbO$_3$ modulator, fed into the ROADM, and transmitted over 504 km to the receiver. There, the signal was dropped using another ROADM and fed into the 100G receiver. The receiver consisted of a Bragg grating tunable dispersion compensator, followed by a delay interferometer with a 67-GHz free spectral range and low polarization dependence. Balanced detection was followed by a 1:2 electronic data and clock recovery demultiplexer. After further demultiplexing to a 6.6875-Gb/s data bus, the signal entered another FPGA for extraction of a single OTU2 signal (2×5.35 Gb/s). This signal entered an FEC decoder to arrive at the original OC192 signal containing the live HDTV video traffic. The OC192 signal was fed into standard transport equipment to map four of the STS-1s into a GbE channel, which was then fed into a video test set to extract different HDTV channels for display. Neither SONET errors nor video signal defects were observed on the 100G wavelength during the trial, and all of the other 10G channels with live traffic on the fiber remained error free. This field trial demonstrated the viability of carrying live traffic on a 100G wavelength over a deployed 10G/40G optical line system without changing any of the embedded networking infrastructure or control software.

In spite of successful research/development and field trials on Type 1 100G channels, the industry was still searching for a more optimal 100G channel design among all three types of channels. Without multiplexing options Type 1 channel designs for 100 Gb/s suffered either high symbol rates, which means low spectral efficiency and difficult propagation conditions in fiber, or high modulation levels, which means more stringent requirements for optical signal to noise ratio (OSNR). Therefore, while Type 1 channel designs were trumpeting their successes in 100G channel development the commercial 100G design still hesitated moving forward in this direction.

There was a short time period when Type 3 designs caught 100G channel design-ers' attention. Multi-carrier OFDM channel design was once considered to support 100G channels. For example, a nine-carrier OFDM channel with a channel capacity of 109 Gb/s traveled over 400 km of standard single mode fiber (SSMF) successfully [64]. A dual-carrier PM-QPSK channel design was also used for 100G transport systems. Since two optical carriers were used, the symbol rate of the signal from a modulator point of view was dropped to one eighth of the channel data rate, therefore mature 10G technology could be re-used for the 100G channel. A 111-Gb/s channel with two closely spaced optical carriers working together to carry 100-Gb/s data traf-fic was reported [65] and early deployed commercial 100G systems were based on dual-carrier PM-QPSK modulation format as well [51].

Toward the end of the first decade in the new century, the telecom industry quickly moved toward a consensus on the optimal 100G channel design. The modu-lation format selected by the industry is the single-carrier PM-QPSK with coherent detection [66], which is a Type 2 channel by the definition defined in this chapter. With this design the symbol rate is a quarter of the data rate, around 30 Gbaud, which fits well within the bandwidth of commercially available optical modulators. The coherent detection adds some complexity, but it is a real game changer given that the full information of an optical signal, including real-time amplitude, phase, and polarization of the electrical field, are able to be detected which would be impos-sible using direct detection. With full information of the electrical field, the receiver tolerance to noise can be greatly improved and fiber propagation impairments can be removed using signal processing. Coherent detection was heavily studied in the early days of optical communication to enhance receiver sensitivity [51]. However, at that time optical amplification proved to be a cheaper and easier way to improve resilience to noise than coherent detection, therefore coherent detection did not gain much industry momentum until the benefits of optical amplifiers were exhausted in fiber transmission designs. Newly developed high-speed analog-to-digital converters (ADC) with about a 60-GS/s sampling rate help in digitizing the coherently detected 100G channel for digital signal processing (DSP). The powerful DSP is able to correct almost all distortions of the signal that accumulated during propagation. Linear effects, such as chromatic dispersion (CD), polarization mode dispersion (PMD), frequency drifting, and polarization scrambling, can be monitored, detected, and corrected [67]. The DSP is even able to partially correct nonlinear effects, such as nonlinear phase noise [68–74]. Typical tolerance of a commercial 100G system to CD is around 50,000 ps/nm and that to PMD is around 100 ps instant differential group delay (DGD). Due to strong CD compensation capability a terrestrial coherent 100G system does not need optical CD compensators anymore, which were very common in traditional long haul transport systems. Without CD compensation modules in the transmission line, channels with different optical frequencies have huge walk-off between each other, so the cross-phase modulation (XPM) can be reduced significantly. Strong FEC algorithms in most DSP chips give the 100G optical channel a net coding gain as high as 10 dB [75] or more which means the OSNR requirement for error-free operation can be relaxed for the same amount of dB. The FEC feature is key to allow the coherent 100G channels to have a reach similar to that of the traditional 10G channels in backbone networks. Now the industry has converged on its effort toward this design for 100G channels. Table 2.1 shows the typical specifications of commercial 100G systems.

The industry took a long way to come to the final choice of commercial 100G channel design. The technology choice started with Type 2 designs (e.g. OTDM channels), then went to Type 1 designs (e.g. ETDM channels), then swung to Type 3 designs (e.g. two-carrier channels), and eventually came back to a Type 2 design (e.g. PM-QPSK channels with coherent detection). Figure 2.6 shows a typical design for the transmitter and the receiver of a commercial 100G PM-QPSK channel, in which data traffic is mapped to I and Q driving signals for two polarization states in the

Table 2.1 Typical specifications of commercial 100G systems.

Item	Value
Optical band used	C-band
Optical bandwidth	4.5 THz
Number of 100G channels	90
Total system capacity	9 Tb/s
Channel spacing	50 GHz
Modulation format	PM-QPSK
Protocol and FEC overhead ratio	25%
Reach distance	2500 km
Chromatic dispersion tolerance	50,000 ps/nm
PMD tolerance (instant DGD)	100 ps

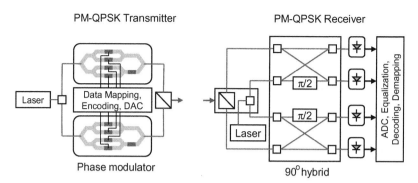

FIGURE 2.6 Schematic transmitter and receiver structure of a PM-QPSK 100G channel with coherent detection.

transmitter; FEC is added in the encoding process and DAC helps to modify the spectral shape of the modulated signals. At the receiver a laser is used as a local oscillator for the coherent detection. Two 90° hybrid devices help to detect the full electrical field of the incoming optical signal. After the detector array obtains the signals, ADC converts the detected signals into a digitized format. Then the DSP does a series of data processing to recover the original data traffic in two polarizations.

2.4 INTRODUCTION OF 100G CHANNELS TO SERVICE PROVIDER NETWORKS

As the traffic demand continues growing as shown in Figure 2.1a, telecom network service providers have planned to introduce the newly developed coherent 100G transport systems into their networks to meet the demand. History shows us that network service providers have made good use of every stage of a new channel capacity available from equipment developers. Figure 2.7 shows the timeline for increases in fiber link capacity in service provider's networks. In the early 1990s, a capacity of a few hundred Mb/s per link and only one channel per strand of fiber in a transport network was typical. As email became a new communication tool in the middle 1990s, the fiber capacity gradually increased to a few Gb/s, and this growth continued to address the demand that people needed to start accessing the Internet. Into the later 1990s fiber capacity continued to grow with the deployment of 10-Gb/s channels and WDM techniques to multiplex and amplify a small number of wavelengths (4–8) on a single fiber pair. In early 2000s Internet usage had become commonplace but networking kept pace with the introduction of DWDM (dense wavelength division multiplexing) techniques that could support forty, eighty, or even more wavelengths allowing fiber capacities to be near Tb/s. This extensive fiber capacity increase helped the transport network support continually increasing user demands. In the late 2000s, the introduction of 40G channels gave the capacity of the networks another boost. By 2010, video

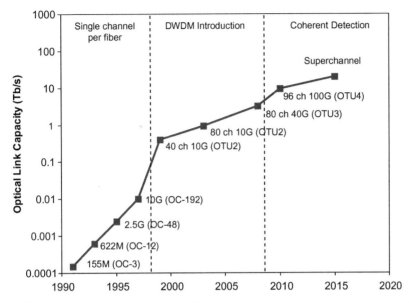

FIGURE 2.7 Typical channel capacities and fiber capacities of transport networks, when they began to be popular.

sharing on the Internet by applications such as YouTube and other video on demand (VoD) services again began to stress existing network capacity. The introduction of the newly developed coherent 100Gb/s channels has provided another 10× increase in fiber capacity to approximately 10Tb/s per fiber. This should address near term capacity requirements, but moving forward, cloud computing and other bandwidth hungry applications will continue to consume network resources, and new optical techniques to increase channel capacity and optical link capacity will be introduced progressively.

The coherent 100G PM-QPSK system selected by the industry is able to run at the same channel spacing (50GHz) as a 10G commercial system does in existing networks, therefore the 100G system can provide enough capacity for network service providers to support customer demands in the near term without a network overbuild. With the new 100G system, service providers expect that the cost per bit declines at the same rate as or a faster pace than the decline rate of serves prices service providers can charge their customers, so that service providers are able to remain competitive. Before telecom service providers introduce commercial coherent 100G systems into their networks, normally a series of technology trials must be conducted in their existing networks to check the performance of the new technology. The main purpose of the technology trials is to guarantee the 100G channel behaves well in existing fiber network infrastructures. Fiber routes in the field may have high transmission attenuation, high PMD values, multiple connections and splicing points, various fiber types, etc. While most lab experiments are conducted with fiber loop configurations, a linear configuration in field trials is more preferred to mimic optical links in real networks. Field trials give network service providers proper expectation for the performance of the systems, which will be installed in networks. Issues found in these trials can also be sent back to the system developers for further product improvement.

In one field trial a 112-Gb/s coherent channel transmitted over 1730 km deployed DWDM link in a service provider's network [76]. A carrier suppressed RZ and differential PM-QPSK modulation format was used for the channel in the trial. The trial results show that the coherent 100G channel is able to serve long haul routes. The plug and play performance of the equipment and robustness to chromatic dispersion and PMD impairments was demonstrated in the trial. Co-propagating the 100G channel with adjacent 10-Gb/s signals without touching the fiber infrastructure proved one viable migration path to next generation networks. It is a requirement for service providers to maintain the networks scalable and cost effective while increasing channel capacity and fiber capacity to have next generation multi-terabit networks. In another field trial a real-time, single carrier, coherent 100G PM-QPSK upgrade of an existing 10G/40G terrestrial system was demonstrated in a service provider's network [77]. The field experiment shows the performance of the 100G channel sufficient for error-free operation after FEC over installed 900 km and 1800 km fiber links. The experiment proves that flexible and seamless 100 Gb/s channel upgrades to existing 10G and 40G DWDM systems are possible and practical. Yet another coherent 100G channel field trial was performed on dispersion shifted fiber (DSF) links [78]. The trial involved eighty 127-Gb/s channels propagating on a deployed fiber link. L-band spectrum was used to avoid zero dispersion point in spectrum, different from using C-band for SSMF or NZDSF for more common cases. The 100G channels, with 50-GHz channel spacing, traveled over 458-km DSF successfully with L-band EDFA-only. Sufficient Q-margins were still left for the 80 channels after the 458-km transmission. This field trial demonstrated that a 10-Tb/s-class capacity DWDM system is feasible under the condition of small local dispersion by deploying coherent detection and high overhead (20%) coding gain FEC. This trial represented the highest fiber capacity in the field at the time the trial was conducted.

The purpose of introducing 100G channels into transport networks is to carry large IP data traffic across IP networks, therefore, an "end-to-end" transport trial, i.e. a complete data transport trial from data equipment to data equipment, via a coherent 100G channel transmission over a long distance, is particularly meaningful to service providers. One such field trial, which involved a global network service provider, a data equipment developer, a transport equipment developer, and a client interface developer, has been reported [3]. In this trial a 112-Gb/s single carrier real-time coherent PM-QPSK channel from a transponder carried native IP packet traffic over 1520-km field deployed fiber, with 100GE router cards and 100G CFP interfaces. This trial shows the feasibility of interoperability between multi-suppliers' equipment for 100G transport. This field trial, which fully emulated a practical near-term deployment scenario, confirmed that all key components needed for deployment of 100-Gb/s Ethernet technology are maturing at the time the trial was conducted (early 2010). The detailed configuration of the trail is shown in Figure 2.8. A 10GE test set generates

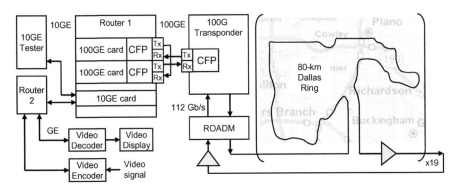

FIGURE 2.8 Configuration of end-to-end 100G transport field trial [3].

10GE traffic for Router 1 and the test set is used for analyzing packet throughput as well. Another router (Router 2) is used to accept a GbE signal containing a video signal via a video encoder and to send the video signal to a video display via a video decoder after the signal transverses the trial path. Router 2 connects to Router 1 with another 10GE link, which contains the video traffic. Router 1 routes both 10GE data streams to one of the 100GE cards and routes back the 10GE data streams from the other 100GE card to the corresponding 10GE ports. The 100G CFP interfaces are used to connect 100GE cards and the 100G transponder. The transmitter port of the CFP in the first 100GE card is connected to the receiver port of the CFP in the transponder and the receiver port of the second 100GE card is connected to the transmitter port of the CFP in the transponder. The receiver port of the CFP in the first 100GE card and the transmitter port of the CFP in the second 100GE card are connected with a fiber jumper to close the loop. The 100G transponder sends the 112-Gb/s optical signal to the fiber route equipped with a long haul DWDM system. Both directions of the inline amplifiers have been used for the trial to save on equipment needed.

With these successful 100G system field trials, telecom network service providers and other network operators have been convinced that the single optical carrier PM-QPSK with coherent detection is the most promising 100G channel solution, at least for the time being. Now commercial 100G systems are available to the customers of the equipment developers and the customers are going to enjoy the 10× fiber capacity jump in their networks.

2.5 IMPACT OF COMMERCIAL 100G SYSTEM TO TRANSPORT NETWORK

To understand the impact of introducing commercial 100G systems to transport networks, it is helpful to discuss the basic functions of a transport network first. A service provider's typical backbone transport network is illustrated in Figure 2.9.

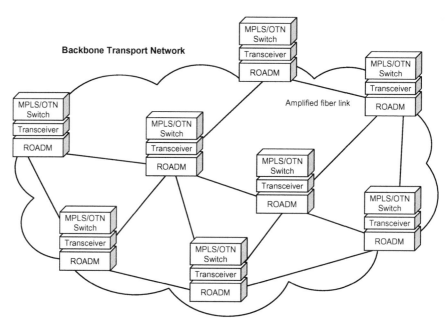

FIGURE 2.9 Typical backbone transport network architecture.

The network is composed of multiple nodes, which are connected by amplified fiber links. The main job of the network is to carry traffic of various services across the world. The functions of each node in the transport network can be represented by the functional blocks as shown in Figure 2.9. The electrical switching function block includes MPLS switches and OTN switches. MPLS switches are for packet traffic aggregation and distribution while OTN switches are for TDM traffic aggregation and distribution. The transceiver function block is to load traffic into optical channels using OTN protocols, and the ROADM is for wavelength level switching and optical channel add/drop. A network service provider may provide IP services, Ethernet services, and private line services. In the backbone network, IP and Ethernet switched services require data packet handling which is accomplished with an MPLS switch. The data traffic at the ingress port receives an MPLS label and is sent to the proper egress port where the transceiver encapsulates it into OTN channels and the ROADM connects it to the right fiber pair. At an intermediate node the data packets in an OTN channel are sent to the MPLS switch, switched to proper exit ports, and carried by OTN channels to continue their journey. Finally the data packets are delivered at the egress point. Private line services normally require only TDM switching when the traffic goes across the network. At the edge of the network the traffic of private line services is aggregated into OTN channels. At intermediate nodes the contents of the OTN channels are sent to the OTN switch to perform sub-channel grooming. Finally the traffic of private line services is delivered at the edge of the network. It is preferred to use more optical switching and less electrical switching in transport network for cost savings [79]. To reduce the burden to switch transit traffic at intermediate nodes in the electrical domain, optical express links are used whenever it is possible. For example, if the traffic demand of IP services between Node A and Node Z is above the capacity of an OTN channel for most of the time, an optical express channel can be set up between Node A and Node Z. The traffic carried by this channel does not need to be switched by either the MPLS switch or the OTN switch at intermediate nodes and the end-to-end light path of the channel can be set up by ROADMs only.

Adding coherent 100G systems into transport networks brings service providers some unprecedented benefits. First, the 100G channel has a channel capacity 10 times that of a 10G channel and the 100G channel occupies the same 50-GHz optical channel bandwidth like a 10G channel; in principle, a transport network immediately gains a $10\times$ capacity increase by introducing 100G systems. At the initial stage of 100G system deployment a lot of 10G channels can be multiplexed into 100G channels via muxponders or OTN switches, therefore many wavelength slots can be freed up in the original DWDM systems. That not only releases a lot of optical bandwidth for future capacity increases, but also reduces the number of optical channels in the network the service provider needs to manage. A reduction in the channel count also greatly simplifies network maintenance tasks.

Introducing 100G systems into transport networks also provides the capability to accommodate higher-speed data ports in customers' data equipment. For example, 100GbE ports now are available on many high bandwidth IP routers and MPLS switches. A service provider who does not have a 100G transport system will find it difficult to serve these customers.

The coherent 100G system provides lower latency transmission compared with traditional non-coherent transmission. Since the 100G channel is equipped with electrical dispersion compensation, the traditional optical Dispersion Compensation Module (DCM) is no longer needed for the 100G system. Optical dispersion compensation modules have been used widely in backbone networks for 10G and some 40G systems. These systems use direct detection and a DCM is a must to compensate the accumulated dispersion in fiber, since the receivers have very small dispersion tolerances. Removing DCM modules from DWDM transmission systems brings service providers another benefit—latency reduction. It is quite typical that an optical DCM adds about 15% latency to a transmission link, when the transmission fiber is SSMF and the compensation module is dispersion compensation fiber (DCF) based. With the optical DCM-free 100G system service providers are able to send the traffic to the destination faster than before. For example, the latency to send a 10G channel from New York to Chicago is about 7.14 ms (if a highway mileage is used to calculate the latency, 1267 km plus 15%). With the new 100G system, the latency is reduced to 6.21 ms. The one-way latency saving is 0.93 ms. Low latency services are attractive to latency-sensitive customers, such as financial institutes, and reduction in latency brings competitive advantages to network service providers.

Coherent detection also significantly simplifies the ROADM design. The local oscillator in the coherent 100G channel receiver is effectively a channel selector in the optical spectral domain. When a host of channels arrive at the receiver, the receiver can pick the channel it wants to receive and treats the other channels as noise by tuning the frequency of the local oscillator to the central frequency of the desired channel. This receiver capability helps to reduce the complexity in ROADM designs, especially for the next generation ROADM, which may have features such as color-less, directionless, contentionless (C/D/C), and even gridless. Figure 2.10a shows a schematic C/D/C ROADM design considering the advantage of coherent detection. A multi-cast switch is used as an add/drop module, see Figure 2.10b. Since the coherent receiver is channel selective, the add/drop module can send the dropped signal, which may contain multiple channels, to a receiver. The receiver uses the local oscillator to pick the channel sent to it and ignore all other channels. Coherent detection was able to pick one channel from as many as 70 channels and was demonstrated [80] in a field trial. Only with the C/D/C function can network service providers fully manage the network remotely, since only one add/drop location is needed to consolidate all added and dropped channels with the C/D/C function implemented. A channel from a local transceiver is able to be configured to have any color or wavelength and go to any direction without restriction, as long as the network is able to accept the channel.

To improve channel performance, particularly the reach distance, various FEC algorithms have been considered in channel designs. The effect of FEC is represented

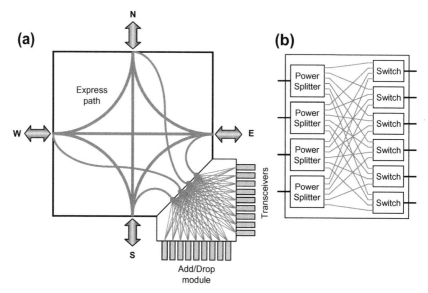

FIGURE 2.10 (a) ROADM with colorless, directionless, and contentionless (C/D/C) functions; (b) multi-cast switch as add/drop module.

by net coding gain (NCG), which is commonly defined as the difference of OSNR requirements for 10^{-15} BER at the receiver with and without the FEC. NCG of a channel depends on the channel overhead ratio and whether a hard decision (HD) or a soft decision (SD) is used in signal digitization and decoding at a receiver. Here hard decision means the received signal is digitized into a number of levels equal to that in the transmitter, and soft decision means the received signal is digitized to a number of levels more than that in the transmitter so more information is provided to the decoder. Higher overhead ratio leads to more coding gain. For example, the theoretical coding gain for a hard decision with a 7% overhead ratio is 10 dB while a 25% overhead ratio is 11.6 dB. Also the coding gain limit may increase for another 1 dB or so if the hard decision is replaced by a soft decision [81]. In 100G channel designs various FEC overhead ratios have been considered as well as both hard decisions and soft decisions. In one report 164 100-Gb/s PM-QPSK channels traveled over 2550 km distance, in which a hard decision FEC with an overhead ratio of 7% was used [82]. FEC overhead ratios of 15%, 19.25%, and 33.75% [83–85] were reported, respectively.

When a channel data rate is mentioned in the industry, however, there is some confusion in its exact meaning, since only one unit "bits per second (b/s)" has been used to demote various data rates. A channel may include the payload, the protocol overhead, and the FEC overhead. When channel rates are referred in technical reports the meaning of channel rates varies. For example, in one report the channel data rate was referred to as 128 Gb/s [84], in which a 19.25% FEC overhead has been counted, and a "net rate" of 107.3 Gb/s was mentioned when the FEC overhead was discounted. In another report a channel data rate of 100 Gb/s was mentioned [82]. In fact the total channel rate was 110.8 Gb/s, which included both protocol overhead (7%) and FEC overhead (4%). The confusion in the exact meaning of channel rates was not significant when the FEC overhead ratio was small. But recently larger FEC overheads, such as the one with a more than 30% FEC overhead [85], have been used to improve channel performance as techniques such as soft-decision FEC are introduced. Since the FEC overhead contains redundant information for error correction purposes only, an "information rate" of a channel may need to be defined by excluding the FEC overhead while the total channel rate may be called the "gross rate" of the channel. Both gross rate and information rate are important to network service providers since the gross rate determines the channel bandwidth that the network must support while the information rate describes actual channel capacity the service providers may use. Furthermore, a "payload rate" may be defined as well by excluding both protocol overhead and FEC overhead and indicates how much customer traffic can be loaded into the channel for the service providers. Table 2.2 shows the three proposed channel rates with their units. For example, the gross rate of standard OTU4 channel is 111.81 Gb/s. With a protocol overhead of 0.42% and a standard FEC overhead of 6.7%, the information rate and the payload rate of the channel are 104.79 Gb/s and 104.36 Gpb/s, respectively. The channel rate definitions can be extended to

Table 2.2 Proposed names of channel rates and their units with OTU4 channel as an example.

Proposed Name of Channel Rate	Inclusion	Proposed Unit	Example (OTU4)
Gross data rate	Payload, protocol overhead, FEC overhead	bits per second (b/s)	111.81 Gb/s
Information data rate	Payload, protocol overhead	information bits per second (ib/s)	104.78 Gib/s
Payload data rate	Payload only	payload bits per second (pb/s)	104.36 Gpb/s

expressions of fiber capacity and spectral efficiency as well. With these definitions the expressions of channel data rates can be much clearer and simpler with less confusion.

2.6 OUTLOOK BEYOND COMMERCIAL 100G SYSTEMS

The next most likely step toward improving commercial 100G systems is to change the modulation format of the channel to increase the channel capacity at about the same cost. The standard 100G channel modulation is PM-QPSK. Since the modulator in the 100G transceiver has the ability to modulate both amplitude and phase anyway and the modulator driving signal can be modified easily with the DSP and the digital-to-analog converter (DAC), the same design should be able to support modulation formats higher than QPSK, such as 8QAM, 16QAM, etc. PM-16QAM modulation format has attracted a lot of attention lately in the industry. With the same symbol rate a PM-16QAM channel doubles the channel rate to 200G. Even though higher level modulation requires higher OSNR, the 100G PM-QPSK channel is able to support several thousand kilometers of transmission which is not always needed. The 200G PM-16QAM channel should be sufficient for metro and regional transmission distances.

It is not clear at this moment if the mentioned 200G PM-16QAM channel will be a new channel capacity standard given that 400G and 1T are the two most discussed channel capacity standards beyond 100G. Either way, new channel designs will not be able to follow the traditional design principle for transport networks (one channel contains only one optical carrier), at least for a reasonable transmission distance. Due to the bandwidth limitation in components the symbol rate of optical channels will not increase much based on that for the commercial 100G system. If the symbol rate is limited, the logical way to increase the channel capacity with a single optical carrier is to increase modulation levels. Increasing modulation levels, however, significantly reduces the transmission distance. Therefore, the standard 100G PM-QPSK design may be the last high capacity channels using only one optical carrier, which is a Type 2 channel based on the definition of channel types in this chapter. Both the 400G channel and the 1T channel under discussion in the industry will require more than one optical carrier, For example, the 400G channel could be a two-carrier PM-16QAM superchannel (each carrier has a net data rate of 200G) and the 1T channel could be a four-carrier PM-16QAM superchannel (each carrier has a net data rate of 250G) [86]. The history of optical channel designs shows that Type 1 dominated channel designs of 10G, Type 1 and Type 2 have been used for 40G, and Type 2 and Type 3 are the choices for commercial 100G. Beyond 100G Type 3 probably will be the only choice.

In conventional ROADM-based transport networks, optical channels are switched following the light paths that belong to each channel. Considering frequency filtering effects during channel switching in ROADMs, usually a frequency gap is reserved between channels. A superchannel contains more than one optical carrier. Since these optical carriers belong to the same channel, they will be switched as a group in ROADMs. No gap is needed between these carriers. That means the carriers can be packed more tightly than when each of them represents an individual channel. The net spectral efficiency of the current commercial 100G system is 2 b/s/Hz, so that a higher spectral efficiency is expected if superchannels are constructed in the network, even with the same symbol rate and the same modulation format. A field trial shows that with the same modulation format of PM-QPSK as the 100G commercial system uses, multi-carrier superchannels with 400G or 1T channel capacities have a spectral efficiency about 50% higher than that of the commercial 100G system [87].

There are several different ways to construct a superchannel. Figure 2.11 shows four different flavors of superchannels under investigation. The generic waveform

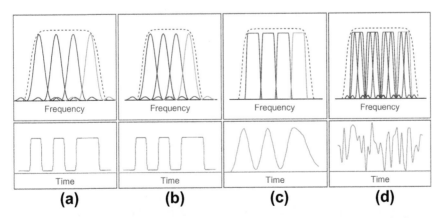

FIGURE 2.11 Various superchannel flavors in frequency domain, and in time domain, in which only one optical carrier is shown, (a) generic WDM superchannel, (b) all-optical OFDM superchannel, (c) Nyquist superchannel, and (d) Classic OFDM superchannel.

of one carrier in each superchannel flavor is included as well to show the carrier behavior in the time domain. Figure 2.11a shows a generic WDM superchannel, which is built using multiple single carrier channels with minimal shaping of the optical spectrum. Each optical carrier maintains a digital bit form in the time domain. For example, a 400G superchannel can be created by just reducing the spectral gaps between four 100G PM-QPSK channels. The performance of the superchannel directly depends on the spectral spacing between the carriers. Figure 2.11b shows an all-optical OFDM superchannel, in which the optical carriers are arranged such that adjacent carrier spacing is exactly the symbol rate of the carriers, therefore the carriers satisfy the condition of "orthogonal to each other." Figure 2.11c shows a Nyquist superchannel, which has the spectral shape of its carriers close to a rectangular spectrum and a bandwidth equal to the symbol rate. Its Fourier transform in the time domain is a sync function, which means that there will be no inter-symbol interference (ISI) in the same carrier and not much spectral overlapping between the carriers either. However, bit overlap does happen as shown in the time domain. Figure 2.11d shows a classic OFDM superchannel, which has many optical carriers generated by a conventional OFDM design. Each optical carrier is generated by a group of electrical subcarriers, which are orthogonal to each other. Since each optical carrier is composed of many lower baud rate electrical carriers the optical spectra are very much square shaped, which edge slope is determined by the bandwidth of the electrical subcarriers. Therefore there is little spectral overlap between the optical carriers. Because the phases of the electrical carriers are independent of each other, the optical carrier behaves like a random signal in time domain.

To further enhance spectral efficiency, the spectral gap between superchannels is kept to a minimum as well leaving just enough for superchannel switching at the ROADM layer. Thus, the total spectral width of a superchannel, including the spectral width occupied by the carriers and the superchannel gap, most likely is no longer an integral multiple of the current DWDM 50-GHz grid in common transport systems. Therefore the concept of flexible channel bandwidths was introduced to ROADM-based networks [88]. With the flexible channel bandwidth design, the spectral granularity could be smaller than the standard 50 GHz used in the current 100G systems, see Figure 2.12a. The bandwidths 25 GHz, 12.5 GHz, or even 6.25 GHz have been proposed for a new spectral granularity standard. Introduction of the concept of flexible channel spectral width also brings in another concept of non-uniform channel spacing assignment. Different from the traditional uniform channel spacing assignment, in which all channels in a fiber link have the same channel spacing, this new concept proposes that channels belonging to an optical link could have different channel bandwidths. For example, a 400G PM-QPSK superchannel may be built with four optical carriers. Each carrier is a PM-QPSK signal with a symbol rate of 28 Gbaud, so that the gross data rate of each carrier is 112 Gb/s. If each carrier occupies only 31 GHz in spectrum (calculated as 1.1 times its symbol rate, as a common practice in the industry) and the superchannel gap reserved is 12.5 GHz, the total spectral width of the 400G superchannel is 136.5 GHz. If the spectral granularity

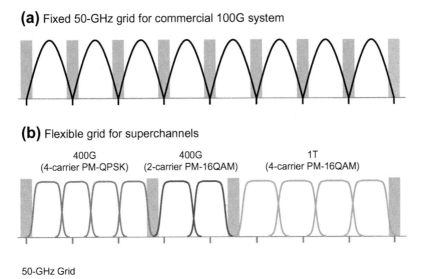

(a) Fixed 50-GHz grid for commercial 100G system

(b) Flexible grid for superchannels

| 400G | 400G | 1T |
| (4-carrier PM-QPSK) | (2-carrier PM-16QAM) | (4-carrier PM-16QAM) |

50-GHz Grid

FIGURE 2.12 (a) Commercial 100G system with fixed 50 GHz frequency grid; (b) examples of superchannels with flexible spectral grid.

(or unit) is set to 12.5 GHz for a DWDM network, the superchannel should occupy 11 units, i.e. 137.5 GHz, making the spectral efficiency about 45% higher than using four 100G individual channels and 10% higher than using conventional 50 GHz granularity to host the superchannel, see Figure 2.12b. If a 400G channel has a two-carrier PM-16QAM design with a symbol rate of 32 Gbaud, the bandwidth occupied by the superchannel is 82.9 GHz following the same convention in the previous calculation. In this case seven spectral units of 12.5 GHz will be used and the total bandwidth is 87.5 GHz, see Figure 2.12b again. The last example is a four-carrier PM-16QAM 1T channel with a 40 Gbaud symbol rate. Based on the calculation method mentioned above, the channel occupies 14 spectral units of 12.5 GHz, resulting in a net spectral efficiency of 5.7 b/s/Hz, which is shown in Figure 2.12b as well.

Introducing flexible spectral bandwidth does increase spectral efficiency; however, it also increases network management complexity. Smaller spectral granularity also requires better wavelength accuracy control in lasers, wavelength selective switches (WSS), and local oscillators. Non-uniform channel bandwidth assignment adds complexity in bandwidth reservation for light paths at network level as well. Therefore a balance should be maintained between spectral efficiency optimization and network operational simplicity. The industry is still searching for the best balance point.

So far this chapter has been focused on channel capacity or channel data rates. Fiber capacity, which is the total capacity a strand of fiber is able to provide, is just as important as the channel capacity to network service providers. Given that a commercial 100G system is able to provide about 10 Tb/s fiber capacity, results from research and development labs have reached milestones of 100 Tb/s per fiber [35,89–91]; but these results do include using newly developed multi-core fiber. The historical single core fiber capacity records [26,89,91–96] are shown in Figure 2.13. The single core fiber capacity of these hero experiments has increased about ten times in a decade. This capacity improvement came from progresses in polarization multiplexing, coherent detection, high level modulation formats, wide optical bandwidth, and high symbol rates. In recent years, space division multiplexing (SDM) has been introduced in optical communication research to boost capacity of fiber in backbone networks. A seven-core fiber, as an implementation of SDM, has been developed and high capacity experiments have been reported [90,97]. Multi-mode fiber has also been proposed for backbone network applications as another version of SDM [34,98,99]. The highest fiber capacity reported so far is a little bit more than 300 Tbd/s. It is generated by a 19-core fiber [35].

The first field trail greater than 10 Tb/s transmission on standard single mode fiber was reported in 2012 [100]. In the trial a 21.7-Tb/s transmission experiment was conducted successfully using 22 optical superchannels and a flexible band WDM. 1503 km of field installed fiber with EDFAs and Raman amplifiers supported the superchannels. All channels were multi-subcarrier superchannels generated using the all-optical OFDM technique and DP-8QAM or DP-QPSK modulation of each subcarrier. A novel dynamic modulation format selector was also used to combat uneven OSNR distribution within the transmitted spectrum. A coherent receiver was used to select the optical carrier of interest by tuning the local oscillator frequency. This trial proved that it is feasible to transport more than 20 Tb/s on an installed standard fiber infrastructure at long haul distances with C-band only.

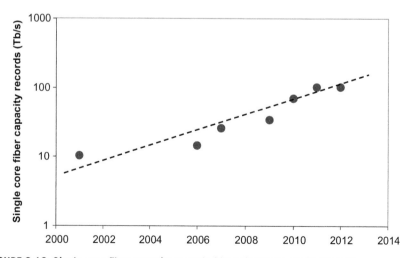

FIGURE 2.13 Single core fiber capacity records (the references are in the text).

You are probably wondering at this point just how much capacity a strand of fiber is able to provide. An analytical formula to calculate the maximum capacity primarily based on the Shannon limit and fiber nonlinear effects has been derived [101]. In addition to the main factors, fiber characteristics, such as attenuation, transmission distance, dispersion, signal symbol rate, signal power, and channel wavelength are parameters in the formula. Figure 2.14 shows the highest possible single core fiber capacity as a function of transmission distance derived from the formula. The conditions used in Figure 2.14 are the following: Standard single mode fiber, 0.2 dB/km fiber loss, 1550 nm central wavelength, 10 THz total optical bandwidth (C+L band), and 32 Gbaud symbol rate. The commercial 100G system and several high capacity experiment results [91,95,102,103] are shown in the figure as well. With this figure, results of different transmission distances may be compared in a relatively "fair" fashion. Several "equal performance" lines have been drawn in the figure. It is assumed that the data points on the same line are considered to have the same performance in terms of capacity. For example, even though the second experimental result data point (ii) shows a higher capacity than that of the third experimental result data point (iii), actually the result of the third data point (iii) is better than that of the second one (ii), since the third data point is closer to the capacity limit than the second one. The capacity of the commercial 100G system sits in the low portion of the figure since only C-band and PM-QPSK are considered in the system design.

To further enhance capacity of a commercial transmission system, three directions can be considered as shown in Figure 2.15. Increasing spectral efficiency can help, but noise reduction in the amplified fiber link would be critical to maintain the

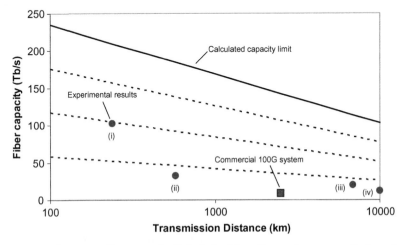

FIGURE 2.14 Single core fiber capacity limit derived from the maximum capacity formula [101] with the results of the commercial 100G system and several high capacity experiments (the conditions of the figure are in the text).

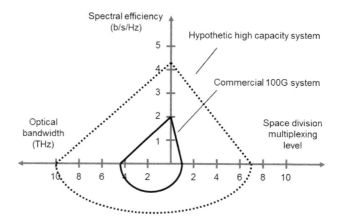

FIGURE 2.15 Three directions to further enhance capacity of commercial transmission systems. The solid line represents the commercial 100G system and the dotted line represents a hypothetic high capacity system.

necessary signal to noise ratio (SNR). New amplifiers, such as phase-sensitive amplifiers [104], may help in this area. Increasing optical bandwidth seems a natural choice to further enhance capacity since most commercial systems use only a single optical band currently. Using both C-band and L-band would double the capacity without upgrading the fiber infrastructure. This would be a huge benefit to network service providers as large scale fiber infrastructure upgrade is very costly. It is true that introducing SDM would help increase capacity, either by using multi-core fiber or multi-mode fiber. However, replacing fiber in the field with newly developed multi-core fiber or multi-mode fiber, as mentioned above, is very expensive to service providers. To reach the same effect of SDM, using a bundle of standard single core fibers may be a viable alternative choice. The commercial 100G system is shown in Figure 2.15 by a contour. It only occupies a small area in the figure. That means there is still a lot of room to grow capacity. A hypothetical high capacity system is shown in the figure as well. It is assumed to be the two-carrier PM-16QAM 400G superchannel system mentioned above, a 10 THz optical bandwidth, and the channels are loaded into a seven-core fiber. The net capacity of the system is about 320 Tb/s, which is about 35 times that of the current commercial 100G system.

2.7 SUMMARY

Optical channel capacity has been growing over the past four decades to address traffic demands and will need to continue this trend for the foreseeable future to meet ever-increasing bandwidth requirements. There are several different optical channel designs being considered to best serve optical transport networks. In this chapter the channel designs are re-classified into three basic channel types: Type 1—single optical carrier with no multiplexing; Type 2—single optical carrier with multiplexing; and Type 3—multiple optical carriers with multiplexing. With this channel type categorization it is easier to analyze and compare various channel designs. In research labs channel capacities have been achieved beyond 10 Tb/s per channel with Type 2 and Type 3 designs, while only a little over 200 Gb/s has been achieved for Type 1 channel designs. To date, the highest commercially available optical transport systems employ channel capacity of 100 Gb/s which did take quite a bit of time to become a deployable solution. Many different channel designs were considered including OTDM, binary ETDM, single polarization DQPSK, multi-carrier OFDM, multi-carrier PM-QPSK, etc. before eventually settling on the single-carrier PM-QPSK with coherent detection, which appears to be the optimal design in the industry. Before 100G systems were deployed, leading network service providers had to perform a series of field trials in their networks to validate the technical merits as well as business advantages of the new capacity standard. Deployment of 100G into a service provider's network brings new opportunities to boost fiber capacity, accommodate increases in client interface speed rates, lower transmission latency, simplify network management, and speed up the realization of C/D/C ROADM functions. Looking forward, channel designs for capacities beyond 100G may have to undergo a paradigm shift due to component symbol rate limitations and the optical signal propagation of current generation fiber. The commercial 100G channel design: with coherent PM-QPSK, may be the last design with a single optical carrier for channels in optical transport networks, at least in the long distance transmission space. Higher capacity channels, for example, 400G or 1T, most likely will use a multiple-optical-carrier superchannel design. The individual optical carriers of these superchannels will likely be packed much tighter than today so that the total capacity of a fiber link can be further enhanced.

References

[1] Cisco, Visual Networking Index, 2012.

[2] S.K. Korotky, Traffic trends: drivers and measures of cost-effective and energy-efficient technologies and architectures for backbone optical networks, in: OFC/NFOEC 2012, OM2G.1.

[3] T.J. Xia, G. Wellbrock, E.B. Basch, S. Kotrla, W. Lee, T. Tajima, K. Fukuchi, M. Cvijetic, J. Sugg, Y. Ma, B. Turner, C. Cole, C. Urricariet, End-to-end native IP data 100G single carrier real time DSP coherent detection transport over 1520-km field deployed fiber, in: OFC/NFOEC 2010, PDPD4.

[4] J. Goell, A repeater with high input impedance for optical-fiber transmission systems, IEEE J. Quant. Electron. 9 (6) (1973) 641–642.

[5] R.D. Maurer, Glass fibers for optical communications, Proc. IEEE 61 (4) (1973) 452–462.

[6] G. White, G.M. Chin, A 100-Mb/s fiber-optic communication channel, Proc. IEEE 61 (5) (1973) 683–684.

[7] T. Ozeki, T. Ito, A 200 Mbit/s PCM DH-GaAlAs laser communication experiment, IEEE J. Quant. Electron. 9 (6) (1973) 692.

[8] S.E. Miller, T. Li, E.A.J. Marcatili, Research toward optical-fiber transmission systems. Part II: Devices and systems considerations, Proc. IEEE 61 (12) (1973) 1726–1751.

[9] D. Hillerkuss, T. Schellinger, R. Schmogrow, M. Winter, T. Vallaitis, R. Bonk, A. Marculescu, J. Li, M. Dreschmann, J. Meyer, S. Ben Ezra, N. Narkiss, B. Nebendahl, F. Parmigiani, P. Petropoulos, B. Resan, K. Weingarten, T. Ellermeyer, J. Lutz, M. Möller, M. Huebner, J. Becker, C. Koos, W. Freude, J. Leuthold, Single source optical OFDM transmitter and optical FFT receiver demonstrated at line rates of 5.4 and 10.8 Tbit/s, in: OFC/NFOEC 2010, PDPC1.

[10] X. Liu, T.H. Wood, R. Tkach, S. Chandrasekhar, Demonstration of record sensitivity in an optically pre-amplified receiver by combining PDM-QPSK and 16-PPM with Pilot-assisted digital coherent detection, in: OFC/NFOEC 2011, PDPB1.

[11] R. Antosik, Super-channel architectures for in-service capacity expansion of CWDM/DWDM systems, in: ICTON 2003, We.C.8.

[12] T.J. Xia, Optical channel capacity—from Mb/s to Tb/s and beyond, Opt. Fiber Technol. 17 (5) (2011) 328–334.

[13] M. Chacinski, U. Westergren, B. Stoltz, R. Driad, R.E. Makon, V. Hurm, A.G. Steffan, 100 Gb/s ETDM transmitter module, IEEE J. Sel. Top. Quant. Electron. 16 (5) (2010) 1321–1327.

[14] J. Li, C. Schubert, R.H. Derksen, R.E. Makon, V. Hurm, A. Djupsjöbacka, M. Chacinski, U. Westergren, H.-G. Bach, G.G. Mekonnen, A.G. Steffan, R. Driad, H. Walcher, J. Rosenzweig, 112 Gb/s field trial of complete ETDM system based on monolithically integrated transmitter & receiver modules for use in 100GbE, in: ECOC2010, P4.03.

[15] J. Yamada, M. Saruwatari, K. Asatani, H. Tsuchiya, A. Kawana, K. Sugiyama, T. Kimura, High-speed optical pulse transmission at 1.29-μm wavelength using low-loss single-mode fibers, IEEE J. Quant. Electron. 14 (11) (1978) 791–800.

[16] J.-I. Yamada, S. Machida, T. Kimura, H. Takata, Dispersion-free single-mode fiber transmission experiments up to 1.6 Gbit/s, Electron. Lett. 15 (10) (1979) 278–279.

[17] J.I. Yamada, S. Machida, T. Kimura, 2 Gbit/s optical transmission experiments at 1.3 μm with 44 km single-mode fiber, Electron. Lett. 17 (13) (1981) 479–480.

[18] A.H. Gnauck, B.L. Kasper, R.A. Linke, R.W. Dawson, T.L. Koch, T.J. Bridges, E.G. Burkhardt, R.T. Yen, D.P. Wilt, J.C. Campbell, K.C. Nelson, L.G. Cohen, 4-Gb/s transmission over 103 km of optical fiber using a novel electronic multiplexer/demultiplexer, J. Lightwave Technol. 3 (5) (1985) 1032–1035.

[19] A.H. Gnauck, J.E. Bowers, J.C. Campbell, 8 Gbit/s transmission over 30 km of optical fibre, Electron. Lett. 22 (11) (1986) 600–602.

[20] S. Fujita, N. Henmi, I. Takano, M. Yamaguchi, T. Torikai, T. Suzaki, S. Takano, H. Ishihara, M. Shikada, A 10 Gb/s-80 km optical fiber transmission experiment using a directly modulated DFB-LD and a high speed InGaAs-APD, in: OFC1988, PD16.

[21] A.H. Gnauck, C.A. Burrus, S.-J. Wang, N.K. Dutta, 16 Gbit/s transmission over 53 km of fibre using directly modulated 1.3 μm DFB laser, Electron. Lett. 25 (20) (1989) 1356–1358.

[22] A.R. Chraplyvy, A.H. Gnauck, R.W. Tkach, R.M. Derosier, 160-Gb/s (8 × 20 Gb/s WDM) 300-km transmission with 50-km amplifier spacing and span-by-span dispersion reversal, in: OFC 1994, PD19.

[23] W. Bogner, E. Gottwald, A. Schopflin, C.J. Weiske, 40 Gbit/s unrepeatered optical transmission over 148 km by electrical time division multiplexing and demultiplexing, Electron. Lett. 33 (25) (1997) 2136–2137.

[24] K. Schuh, B. Junginger, E. Lach, A. Klekamp, E. Schlag, 85.4 Gbit/s ETDM receiver with full rate electronic clock recovery circuit, in: ECOC 2004, Th4.1.1.

[25] P.J. Winzer, G. Raybon, M. Duelk, 107-Gb/s optical ETDM transmitter for 100G Ethernet transport, in: ECOC 2005, Th4.1.1.

[26] A. Sano, H. Masuda, Y. Kisaka, S. Aisawa, E. Yoshida, Y. Miyamoto, M. Koga, K. Hagimoto, T. Yamada, T. Furuta, H. Fukuyama, 14-Tb/s (140 × 111-Gb/s PDM/WDM) CSRZ-DQPSK transmission over 160 km using 7-THz bandwidth extended L-band EDFAs, in: ECOC 2006, Th4.1.1.

[27] K. Schuh, B. Junginger, E. Lach, G. Veith, 1 Tbit/s (10 × 107 Gbit/s ETDM) NRZ transmission over 480 km SSMF, in: OFC/NFOEC 2007, PDP23.

[28] X. Zhou, J. Yu, M. Du, G. Zhang, Tb/s (20′107 Gb/s) RZ-DQPSK straight-line transmission over 1005 km of standard single mode fiber (SSMF) without Raman amplification, in: OFC/NFOEC 2008, OMQ3.

[29] A.R. Chraplyvy, A.H. Gnauck, R.W. Tkach, J.L. Zyskind, J.W. Sulhoff, A.J. Lucero, Y. Sun, R.M. Jopson, F. Forghieri, R.M. Derosier, C. Wolf, A.R. McConnick, 1-Tb/s transmission experiment, IEEE Photon. Technol. Lett. 8 (9) (1996) 1264–1266.

[30] S.J. Savory, G. Gavioli, R.I. Killey, P. Bayvel, Transmission of 42.8 Gbit/s polarization multiplexed NRZ-QPSK over 6400 km of standard fiber with no optical dispersion compensation, in: OFC/NFOEC 2007, OTuA1.

[31] Y. Miyamoto, K. Yonenaga, A. Hirano, N. Shimizu, M. Yoneyama, H. Takara, K. Noguchi, K. Tsuzuki, 1.04-Tbit/s DWDM transmission experiment based on alternate-polarization 80-Gbit/s OTDM signals, in: Proceedings of the European Conference on Optical Communication, vol. 3, 1998, pp. 53–57.

[32] I. Morita, M. Suzuki, N. Edagawa, S. Yamamw, S. Akiba, Single-channel 40 Gbit/s, 5000 km straight-line soliton transmission experiment using periodic dispersion compensation, in: ECOC 1996, TuD.3.1.

[33] M. Nakazawa, T. Yamamoto, K.R. Tamura, 1.28 Tbit/s-70 km OTDM transmission using third- and fourth-order simultaneous dispersion compensation with a phase modulator, Electron. Lett. 36 (24) (2000) 2027–2029.

[34] R. Ryf, A. Sierra, R.-J. Essiambre, A.H. Gnauck, S. Randel, M. Esmaeelpour, S. Mumtaz, P.J. Winzer, R. Delbue, P. Pupalaikis, A. Sureka, T. Hayashi, T. Taru, T. Sasaki, Coherent 1200-km 6 × 6 MIMO mode-multiplexed transmission over 3-core microstructured fiber, in: ECOC 2011, Th.13.C.1.

[35] J. Sakaguchi, B.J. Puttnam, W. Klaus, Y. Awaji, N. Wada, A. Kanno, T. Kawanishi, K. Imamura, H. Inaba, K. Mukasa, R. Sugizaki, T. Kobayashi, M. Watanabe, 19-Core fiber transmission of 19 × 100 × 172-Gb/s SDM-WDM-PDM-QPSK signals at 305 Tb/s, in: OFC/NFOEC 2012, PDP5C.1.

[36] ITU-T, Interfaces for the optical transport network (OTN), ITU-T Recommendation G.709/Y.1331, December 2009.

[37] M. Nakazawa, K. Suzuki, E. Yamada, Y. Kimura, 20 Gbit/s soliton transmission over 200 km using erbium-doped fibre repeaters, Electron. Lett. 26 (19) (1990) 1592–1593.

[38] K. Iwatsuki, K. Suzuki, S. Nishi, M. Saruwatari, 40 Gbit/s optical soliton transmission over 65 km, Electron. Lett. 28 (19) (1992) 1821–1822.

[39] S. Kawanishi, H. Takara, K. Uchiyama, T. Kitoh, M. Saruwatari, 100 Gbit/s, 50 km, and nonrepeated optical transmission employing all-optical multi/demultiplexing and PLL timing extraction, Electron. Lett. 29 (12) (1993) 1075–1077.

[40] K. Iwatsuki, K. Suzuki, S. Nishi, M. Saruwatari, 60 Gb/s × 2 ch time/polarization-multiplexed soliton transmission over 154 km utilizing an adiabatically compressed, gain-switched, DFB-LD pulse source, IEEE Photon. Technol. Lett. 6 (11) (1994) 1377–1379.

[41] S. Kawanishi, H. Takara, T. Morioka, O. Kamatani, M. Saruwatari, 200 Gbit/s, 100 km time-division-multiplexed optical transmission using supercontinuum pulses with prescaled PLL timing extraction and all-optical demultiplexing, Electron. Lett. 31 (10) (1995) 816–817.

[42] S. Kawanishi, H. Takara, T. Morioka, O. Kamatani, K. Takiguchi, T. Kitoh, M. Saruwatari, Single channel 400 Gbit/s time-division-multiplexed transmission of 0.98 ps pulses over 40 km employing dispersion slope compensation, Electron. Lett. 32 (10) (1996) 916–918.

[43] M. Nakazawa, E. Yoshida, T. Yamamoto, E. Yamada, A. Sahara, TDM single channel 640 Gbit/s transmission experiment over 60 km using 400 fs pulse train and walk-off free, dispersion flattened nonlinear optical loop mirror, Electron. Lett. 34 (9) (1998) 907–908.

[44] C. Schubert, R. Ludwig, S. Ferber, C. Schmidt-Langhorst, B. Huettl, H.G. Weber, Single channel transmission beyond 1 Tbit/s, in: LEOS 2006, ThX1.

[45] C. Schmidt-Langhorst, R. Ludwig, D.-D. Groß, L. Molle, M. Seimetz, R. Freund, C. Schubert, Generation and coherent time-division demultiplexing of up to 5.1 Tb/s single-channel 8-PSK and 16-QAM signals, in: OFC/NFOEC 2009, PDPC6.

[46] Thomas Richter, Evarist Palushani, Carsten Schmidt-Langhorst, Markus Nolle, Reinhold Ludwig, J.K. Fischer, Colja Schubert, Single wavelength channel 10.2 Tb/s TDM-data capacity using 16-QAM and coherent detection, in: OFC/NFOEC 2011, PDPA9.

[47] R. Dischler F. Buchali, Transmission of 1.2 Tb/s continuous waveband PDM-OFDM-FDM signal with spectral efficiency of 3.3 bit/s/Hz over 400 km of SSMF, in: OFC/NFOEC 2009, PDPC2.

[48] A. Sano, H. Masuda, E. Yoshida, T. Kobayashi, E. Yamada, Y. Miyamoto, F. Inuzuka, Y. Hibino, Y. Takatori, K. Hagimoto, T. Yamada, Y. Sakamaki, 30 × 100-Gb/s all-optical OFDM transmission over 1300 km SMF with 10 ROADM Nodes, in: ECOC 2007, PD 1.7.

[49] S. Chandrasekhar, X. Liu, B. Zhu, D.W. Peckham, Transmission of a 1.2-Tb/s 24-carrier no-guard-interval coherent OFDM superchannel over 7200-km of ultra-large-area fiber, in: ECOC 2009, PD2.6.

[50] J. Yu, Z. Dong, X. Xiao, Y. Xia, S. Shi, C. Ge, W. Zhou, N. Chi, Y. Shao, Generation, transmission and coherent detection of 11.2 Tb/s (112 × 100 Gb/s) single source optical OFDM superchannel, in: OFC/NFOEC 2011, PDPA6.

[51] K. Roberts, 100G—Key technology enablers of 100 Gbit/s in carrier, in: OFC/NFOEC 2011, NWA1.

[52] S. Chandrasekhar, X. Liu, D. Kilper, C.R. Doerr, A.H. Gnauck, E.C. Burrows, L.L. Buhl, 0.8-bit/s/Hz terabit transmission at 42.7-Gb/s using hybrid RZ-DQPSK and NRZ-DBPSK formats over 16 × 80 km SSMF spans and 4 bandwidth-managed ROADMs, in: OFC/NFOEC 2007, PDP28.

[53] R. Freund, H. Louchet, M. Gruner, L. Molle, M. Seimetz, A. Richter, 80 Gbit/s/λ polarization multiplexed star-16QAM WDM transmission over 720 km SSMF with electronic distortion equalization, in: OECC 2009, WP2.

[54] J.-X. Cai, O.V. Sinkin, C.R. Davidson, D.G. Foursa, A.J. Lucero, M. Nissov, A.N. Pilipetskii, W.W. Patterson, N.S. Bergano, 40 Gb/s transmission using polarization division multiplexing (PDM) RZ-DBPSK with automatic polarization tracking, in: OFC/NFOEC 2008, PDP4.

[55] T.J. Xia, Y. Liang, K.H. Ahn, J.W. Lou, O. Boyraz, Y.H. Kao, X.D. Cao, S. Chaikamnerd, J.K. Andersen, M.N. Islam, All-optical packet-drop demonstration using 100-Gb/s

words by integrating fiber-based components, IEEE Photon. Technol. Lett. 10 (1) (1998) 153–155.

[56] P.J. Winzer, G. Raybon, C.R. Doerr, 10×107 Gb/s electronically multiplexed NRZ transmission at 0.7 bits/s/Hz over 1000 km non-zero dispersion fiber, in: ECOC 2006, Tu1.5.1.

[57] S.L. Jansen, R.H. Derksen, C. Schubert, X. Zhou, M. Birk, C.-J. Weiske, M. Bohn, D. van den Borne, P.M. Krummrich, M. Möller, F. Horst, B.J. Offrein, H. de Waardt, G.D. Khoe, A. Kirstädter, 107-Gb/s full-ETDM transmission over field installed fiber using vestigial sideband modulation, in: OFC 2007, OWE3.

[58] S. Vorbeck, D. Breuer, K. Schuh, B. Junginger, E. Lach, W. Idler, A. Klekamp, G. Veith, M. Schneiders, P. Wagner, C. Xie, D. Werner, H. Haunstein, M. Paul, A. Ehrhardt, R.-P. Braun, 8×107 Gbit/s serial WDM field trial over 500 km SSMF, in: OECC 2009, WP3.

[59] K. Schuh, B. Junginger, E. Lach, W. Idler, G. Veith, S. Vorbeck, M. Schneiders, P. Wagner, D. Werner, H. Haunstein, D. Breuer, R.-P. Braun, M. Paul, A. Ehrhardt, 8×107 Gbit/s NRZ-VSB DWDM field transmission over 500 km SSMF, in: ITG Symposium on Photonic Networks, 2009.

[60] P.J. Winzer, G. Raybon, C.R. Doerr, L.L. Buhl, T. Kawanishi, T. Sakamoto, M. Izutsu, K. Higuma, 2000-km WDM transmission of 10×107-Gb/s RZ-DQPSK, in: ECOC 2006, Th4.1.3.

[61] T.J. Xia, G. Wellbrock, W. Lee, G. Lyons, P. Hofmann, T. Fisk, B. Basch, W. Kluge, J. Gatewood, P.J. Winzer, G. Raybon, T. Kissel, T. Carenza, A.H. Gnauck, A. Adamiecki, D.A. Fishman, N.M. Denkin, C.R. Doerr, M. Duelk, T. Kawanishi, K. Higuma, Y. Painchaud, C. Paquet, Transmission of 107-Gb/s DQPSK over Verizon 504-km Commercial LambdaXtreme Transport System, in: OFC/NFOEC'2008, NMC2.

[62] P.J. Winzer, G. Raybon, H. Song, A. Adamiecki, S. Corteselli, A.H. Gnauck, D.A. Fishman, C.R. Doerr, S. Chandrasekhar, L.L. Buhl, T.J. Xia, G. Wellbrock, W. Lee, B. Basch, T. Kawanishi, K. Higuma, Y. Painchaud, 100-Gb/s DQPSK transmission: from laboratory experiments to field trials, J. Lightwave Technol. 26 (20) (2008) 3388–3402.

[63] G. Wellbrock, T.J. Xia, W. Lee, G. Lyons, P. Hofmann, T. Fisk, B. Basch, W. Kluge, J. Gatewood, P.J. Winzer, G. Raybon, H. Song, A. Adamiecki, S. Corteselli, A.H. Gnauck, D.A. Fishman, T. Kawanishi, K. Higuma, Y. Painchaud, Field trial of 107-Gb/s channel carrying live video traffic over 504 km in-service DWDM route, in: 21th IEEE/LEOS Annual Meeting, WH1, 2008.

[64] F. Buchali, R. Dischler, A. Klekamp, M. Bernhard, D. Efinger, Realisation of a real-time 12.1 Gb/s optical OFDM transmitter and its application in a 109 Gb/s transmission system with coherent reception, in: ECOC 2009, PD2.1.

[65] H. Masuda, E. Yamazaki, A. Sano, T. Yoshimatsu, T. Kobayashi, E. Yoshida, Y. Miyamoto, S. Matsuoka, Y. Takatori, M. Mizoguchi, K. Okada, K. Hagimoto, T. Yamada, S. Kamei, 13.5-Tb/s (135×111-Gb/s/ch) no-guard-interval coherent OFDM transmission over 6,248 km using SNR maximized second-order DRA in the extended L-Band, in: OFC/NFOEC 2009, PDPB5.

[66] OIF, 100G Ultra Long Haul DWDM Framework Document, June 30, 2009.

[67] M. Tomizawa, DSP aspects for deployment of 100G-DWDM systems in carrier networks, in: OFC/NFOEC 2012, NTh1I.1.

[68] A. Pilipetski, Nonlinearity management and compensation in transmission systems, in: OFC/NFOEC 2010, OTuL1.

[69] E. Yamazaki, A. Sano, T. Kobayashi, E. Yoshida, Y. Miyamoto, Mitigation of nonlinearities in optical transmission systems, in: OFC/NFOEC 2011, OThF1.

[70] X. Zhou, E.F. Mateo, G. Li, Fiber nonlinearity management—from carrier perspective, OFC/NFOEC 2011, NThB4.

[71] A. Awadalla, M. O'Sullivan, D. Yevick, Efficient pre-compensation algorithm for self phase modulation (SPM), in: IEEE/PS Summer Topical Meeting 2010, TuC4.4.

[72] L.B. Du, A.J. Lowery, Practical XPM compensation method for coherent optical OFDM systems, IEEE Photon. Technol. Lett. 22 (5) (2010) 320–322.

[73] E. Ip, N. Bai, T. Wang, Complexity versus performance tradeoff for fiber nonlinearity compensation using frequency-shaped, multi-subband backpropagation, in: OFC/NFOEC 2011, OThF4.

[74] X. Li, X. Chen, G. Goldfarb, E. Mateo, I. Kim, F. Yaman, G. Li, Electronic post-compensation of WDM transmission impairments using coherent detection and digital signal processing, Opt. Express 16 (2) (2008) 880–888.

[75] B. Li, K. J. Larsen, D. Zibar, I.T. Monroy, Over 10 dB net coding gain based on 20% overhead hard decision forward error correction in 100G optical communication systems, in: ECOC 2011, Tu.6.A.3.

[76] R.-P. Braun, D. Fritzsche, A. Ehrhardt, L. Schürer, P. Wagner, M. Schneiders, S. Vorbeck, C. Xie, Z. Zhao, W. Wan, P. Liu, Q. Zhou, P. Hostalka, 112 GBit/s PDM-CSRZ-DQPSK field trial over 1730 km deployed DWDM-link, in: ACP 2010.

[77] M. Birk, P. Gerard, R. Curto, L. Nelson, X. Zhou, P. Magill, T.J. Schmidt, C. Malouin, B. Zhang, E. Ibragimov, S. Khatana, M. Glavanovic, R. Lofland, R. Marcoccia, G. Nicholl, M. Nowell, F. Forghieri, Field trial of a real-time, single wavelength, coherent 100 Gbit/s PM-QPSK channel upgrade of an installed 1800 km Link, in: OFC/NFOEC 2010, PDPD1.

[78] T. Kobayashi, S. Yamanaka, H. Kawakami, S. Yamamoto, A. Sano, H. Kubota, A. Matsuura, E. Yamazaki, M. Ishikawa, K. Ishihara, T. Sakano, E. Yoshida, Y. Miyamoto, M. Tomizawa, S. Matsuoka, 8-Tb/s(80 × 127Gb/s) DP-QPSK L-band DWDM transmission over 457-km installed DSF links with EDFA-only amplification, in: OECC 2010, PD2.

[79] S. Elby, Bandwidth flexibility and high availability, Service Provider Summit, in: OFC/NFOEC 2009.

[80] R. Pastorelli, S. Piciaccia, A.D. Torre, F. Forghieri, C. Fludger, J. Geyer, T. Duthel, J. Schiessl, S. Gehrke, P. Presslein, T. Kupfer, C. Schulien, DWDM transmission of 70 100Gb/s CP-DQPSK channels over 2000km of uncompensated SMF with real-time DSP and coherent channel selection, in: OFC/NFOEC 2012, NTh1I.3.

[81] OIF, 100G Forward Error Correction White Paper, May 2010.

[82] G. Charlet, J. Renaudier, H. Mardoyan, P. Tran, O.B. Pardo, F. Verluise, M. Achouche, A. Boutin, F. Blache, J.-Y. Dupuy, S. Bigo, Transmission of 16.4-bit/s capacity over 2550 km using PDM QPSK modulation format and coherent receiver, J. Lightwave Technol. 27 (3) (2009) 153–157.

[83] G. Zhang, L.E. Nelson, Y. Pan, M. Birk, C. Skolnick, C. Rasmussen, M. Givehchi, B. Mikkelsen, T. Scherer, T. Downs, W. Keil, 3760km, 100G SSMF transmission over commercial terrestrial DWDM ROADM systems using SD-FEC, in: OFC/NFOEC 2012, PDP5C.4.

[84] S. Chandrasekhar, A.H. Gnauck, X. Liu, P.J. Winzer, Y. Pan, E.C. Burrows, B. Zhu, T.F. Taunay, M. Fishteyn, M.F. Yan, J.M. Fini, E.M. Monberg, F.V. Dimarcello, WDM/SDM transmission of 10 × 128-Gb/s PDM-QPSK over 2688-km 7-core fiber with a per-fiber net aggregate spectral-efficiency distance product of 40,320 km b/s/Hz, in: ECOC 2011, Th.13.C.4.

[85] S. Zhang, M.-F. Huang, F. Yama, E. Mateo, D. Qian, Y. Zhang, L. Xu, Y. Shao, I.B. Djordjevic, T. Wang, Y. Inada, T. Inoue, T. Ogata, Y. Aoki, 340 × 117.6 Gb/s PDM-16QAM OFDM transmission over 10,181 km with soft-decision LDPC coding and nonlinear compensation, in: OFC/NFOEC 2012, PDP5C.4.

[86] T.J. Xia, S. Gringeri, M. Tomizawa, High capacity optical transport networks, IEEE Commun. Mag. 50 (11) (2012) 170–178.

[87] H. Murai, M. Kagawa, H. Tsuji, K. Fujii, 80-Gb/s error-free transmission over 5600 km using a cross absorption modulation based optical 3R regenerator, IEEE Photon. Technol. Lett. 17 (9) (2005) 1965–1967.

[88] S. Gringeri, B. Basch, V. Shukla, R. Egorov, T.J. Xia, Flexible architectures for optical transport nodes and networks, IEEE Commun. Mag. 48 (7) (2010) 40–50.

[89] D. Qian, M.-F. Huang, E. Ip, Y.-K. Huang, Y. Shao, J. Hu, T. Wang, 101.7-Tb/s (370 × 294-Gb/s) PDM-128QAM-OFDM transmission over 3 × 55-km SSMF using pilot-based phase noise mitigation, in: OFC/NFOEC 2011, PDPB5.

[90] J. Sakaguchi, Y. Awaji, N. Wada, A. Kanno, T. Kawanishi, T. Hayashi, T. Taru, T. Kobayashi, M. Watanabe, 109-Tb/s (7 × 97 × 172-Gb/s SDM/WDM/PDM) QPSK transmission through 16.8-km homogeneous multi-core fiber, in: OFC/NFOEC 2011, PDPB6.

[91] A. Sano, T. Kobayashi, S. Yamanaka, A. Matsuura, H. Kawakami, Y. Miyamoto, K. Ishihara, H. Masuda, 102.3-Tb/s (224 × 548-Gb/s) C- and extended L-band all-Raman transmission over 240 km using PDM-64QAM single carrier FDM with digital pilot tone, in: OFC/NFOEC 2012, PDP5C.3.

[92] S. Bigo, Y. Frignac, G. Charlet, W. Idler, S. Borne, H. Gross, R. Dischler, W. Poehlmann, P. Tran, C. Simonneau, D. Bayart, G. Veith, A. Jourdan, J.-P. Hamaide, 10.2 Tbit/s (256 × 42.7 Gbit/s PDM/WDM) transmission over 100 km TeraLight™ fiber with 1.28 bit/s/Hz spectral efficiency, in: OFC 2001, PD25.

[93] A.H. Gnauck, G. Charlet, P. Tran, P.J. Winzer, C.R. Doerr, J.C. Centanni, E.C. Burrows, T. Kawanishi, T. Sakamoto, K. Higuma, 25.6-Tb/s C+L-band transmission of polarization-multiplexed RZ-DQPSK signals, in: OFC/NFOEC 2007, PDP19.

[94] A.H. Gnauck, G. Charlet, P. Tran, P.J. Winzer, C.R. Doerr, J.C. Centanni, E.C. Burrows, T. Kawanishi, T. Sakamoto, K. Higuma, 25.6-Tb/s WDM transmission of polarization-multiplexed RZ-DQPSK signals, J. Lightwave Technol. 26 (1) (2008) 79–84.

[95] X. Zhou, J. Yu, M.-F. Huang, Y. Shao, T. Wang, P. Magill, M. Cvijetic, L. Nelson, M. Birk, G. Zhang, S. Ten, H.B. Matthew, S.K. Mishra, 32 Tb/s (320 × 114 Gb/s) PDM-RZ-8QAM Transmission over 580 km of SMF-28 ultra-low-loss fiber, in: OFC/NFOEC 2009, PDPB4.

[96] A. Sano, H. Masuda, T. Kobayashi, M. Fujiwara, K. Horikoshi, E. Yoshida, Y. Miyamoto, M. Matsui, M. Mizoguchi, H. Yamazaki, Y. Sakamaki, H. Ishii, 69.1-Tb/s (432 × 171-Gb/s) C- and extended L-band transmission over 240 km using PDM-16-QAM modulation and digital coherent detection, in: OFC/NFOEC 2010, PDPB7.

[97] B. Zhu, T. Taunay, M. Fishteyn, X. Liu, S. Chandrasekhar, M. Yan, J. Fini, E. Monberg, F. Dimarcello, K. Abedin, P.W. Wisk, D.W. Peckham, P. Dziedzic, Space-, wavelength-, polarization-division multiplexed transmission of 56-Tb/s over a 76.8-km seven-core fiber, in: OFC/NFOEC 2011, PDPB7.

[98] M. Salsi, C. Koebele, D. Sperti, P. Tran, P. Brindel, H. Mardoyan, S. Bigo, A. Boutin, F. Verluise, P. Sillard, M. Bigot-Astruc, L. Provost, F. Cerou, G. Charlet, Transmission at 2 × 100 Gb/s, over two modes of 40 km-long prototype few-mode fiber, using LCOS based mode multiplexer and demultiplexer, in: OFC/NFOEC 2011, PDPB9.

[99] C. Koebele, M. Salsi, L. Milord, R. Ryf, C. Bolle, P. Sillard, S. Bigo, G. Charlet, 40 km transmission of five mode division multiplexed data streams at 100 Gb/s with low MIMO-DSP complexity, in: ECOC 2011, Th.13.C.3.

[100] T.J. Xia, G.A. Wellbrock, Y.-K. Huang, M.-F. Huang, E. Ip, P.N. Ji, D. Qian, A. Tanaka, Y. Shao, T. Wang, Y. Aono, T. Tajima, 21.7 Tb/s field trial with 22 DP-8QAM/QPSK optical superchannels over 1,503-km of installed SSMF, in: OFC/NFOEC 2012, PDP5C.6.

[101] R.-J. Essiambre, A. Mecozzi, Capacity limits in single-mode fiber and scaling for spatial multiplexing, in: OFC/NFOEC 2012, OW3D.1.

[102] J.-X. Cai, Y. Cai, C. Davidson, A. Lucero, H. Zhang, D. Foursa, O. Sinkin, W. Patterson, A. Pilipetskii, G. Mohs, N. Bergano, 20 Tbit/s capacity transmission over 6,860 km, in: OFC/NFOEC 2011, PDPB4.

[103] D. Qian, M.-F. Huang, S. Zhang, P.N. Ji, Y. Shao, F. Yaman, E. Mateo, T. Wang, Y. Inada, T. Ogata, Y. Aoki, Transmission of 115 × 100G PDM-8QAM-OFDM channels with 4 bits/s/Hz spectral efficiency over 10,181 km, in: ECOC 2011, Th.13.K.3.

[104] J. Kakande, F. Parmigiani, R. Slavík, P. Petropoulos, D.J. Richardson, Phase sensitive amplifiers for regeneration of phase encoded optical signal formats, in: ICTON 2011, Tu.A1.6.

Advances in Tb/s Superchannels

3

S. Chandrasekhar and Xiang Liu

Bell Labs, Alcatel-Lucent, 791 Holmdel-Keyport Road, Holmdel, NJ 07733, USA

3.1 INTRODUCTION

To satisfy the ever-increasing capacity demand in optical fiber communications, the data rate carried by each wavelength channel in wavelength-division multiplexing (WDM) systems has been increasing exponentially [1–3]. The dramatic increase in serial interface rates and WDM capacities of optical networks is shown in Figure 3.1 [3]. 100-Gb/s per-channel data rates have been available in commercial systems since mid-2010 [3], and 200-Gb/s per channel data rates have been recently demonstrated. With the introduction of optical superchannels, which avoid the electronic bottleneck via optical parallelism, optical transmission with per-channel data rates beyond 100 Gb/s and up to several Terabits/s (Tb/s) has been experimentally demonstrated [4–7]. The term "superchannel" was first coined by Chandrasekhar and X. Liu et al. [4] for multiple single-carrier-modulated signals arranged under the orthogonal frequency-division multiplexing (OFDM) conditions [8,9], although the use of optical OFDM (O-OFDM) to group multiple modulated bands together was previously demonstrated in the context of OFDM signals [5,6] and in the context of 2-carrier transmission [10]. The superchannel concept was later generalized to any collection of optical signals that are (1) modulated and multiplexed together with high spectral efficiency (SE) at a same originating site, (2) transmitted and routed together over a same optical link, and (3) received at a same destination site. To achieve high-SE multiplexing, "Nyquist-WDM," "quasi-Nyquist-WDM," and offset quadrature amplitude modulation (o-QAM) [11–18] have been introduced as promising alternatives of O-OFDM, as we will describe later.

There are five key benefits of using superchannels in WDM systems as follows:

1. Higher per-channel data rate, to meet the demand of ever-increasing serial interface rates, by avoiding the electronic bottleneck imposed by optoelectronic converters (O/E), electro-optical converters (E/O), digital-to-analog converters (DAC), and analog-to-digital converters (ADC).
2. Higher spectral efficiency in WDM transmission, especially in transparent optical networks based on reconfigurable optical add/drop multiplexers

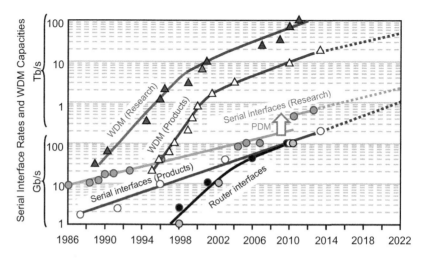

FIGURE 3.1 Serial interface rates and WDM capacities of optical networks versus year. (Courtesy of P.J. Winzer [3].)

(ROADMs), by reducing the percentage of wasted optical spectrum between channels.

3. Increased efficiency in digital signal processing (DSP), by processing the subchannels within each superchannel separately.

4. Better leverage of photonic integrated circuits (PICs) and application-specific integrated circuits (ASICs).

5. Supporting software-defined optical transmission, enabled by DSP at both transmitter and receiver, to improve system throughput and flexibility.

Tb/s superchannels have also led to a rethinking of spectral bandwidth allocation beyond the current fixed grid architectures, leading to the so-called flexible grid architecture that allows for more efficient utilization of the optical spectrum. In this chapter, we review recent progress on Tb/s superchannel transmission and discuss and compare various superchannel implementations based on O-OFDM and Nyquist WDM. This chapter is organized as follows. Section 3.2 describes the superchannel principle and various terminologies used for superchannels. Section 3.3 presents common modulation schemes used to form a superchannel. Section 3.3 presents two main multiplexing schemes, O-OFDM and quasi-Nyquist WDM. Section 3.5 discusses the detection schemes for receiving a superchannel. Section 3.6 addresses the transmission performance of superchannels based on different modulation formats. Section 3.7 briefly discusses the network implications of using Tb/s superchannels with various optical bandwidths. Finally, Section 3.8 concludes this chapter, with an outlook of future perspectives on superchannel-enabled optical transmission systems and networks.

3.2 SUPERCHANNEL PRINCIPLE

Over the last several decades, the field of WDM transmission has evolved from sparsely populated channels, as in coarse WDM (CWDM), to very high density WDM, as in the case of superchannel transmission. It is therefore instructive to classify WDM systems based on the channel bandwidth allocation (or channel spacing) Δf relative to the modulation symbol rate of the channel B. In Table 3.1, different classes of WDM systems are thus defined. One can clearly see that the recent progress in high spectral efficiency systems using advanced modulation formats with coherent detection has opened new regimes, identified as "quasi-Nyquist" WDM (for $1 \leq \Delta f/B \leq 1.2$), "Nyquist-WDM" (for $\Delta f/B = 1$), and "super-Nyquist" WDM (for $\Delta f/B < 1$), respectively. Here, the term "Nyquist" is borrowed from the well-known Nyquist-Shannon sampling theorem, named after Nyquist [19] and Shannon [20], which essentially states that if a function $x(t)$ contains no frequencies higher than $B/2$ Hz, then it is completely determined by giving its ordinates at a series of points spaced $1/B$ seconds apart [20]. This is a fundamental result in the field of information theory, particularly in telecommunications and signal processing. More specifically, Nyquist-WDM here means that the optical spectral bandwidth of an optical signal modulated at a symbol rate of B can be limited to B without losing its fidelity, and similar such signals can be packed at a frequency spacing equal to B in the ideal case, or $\Delta f/B = 1$. Packing two signals closer than the Nyquist frequency, termed as "super-Nyquist" WDM or "faster-than-Nyquist" WDM [21], will cause crosstalk, which can be compensated, albeit with added DSP complexity and reduced optical signal-to-noise (OSNR) performance [14]. The definition makes no assumptions on how a channel is modulated, or any physical impairment associated with placing channels close together, such as crosstalk from overlapping spectral content. As an example, optical prefiltering has been employed to mitigate crosstalk in the demonstration of "quasi-Nyquist" WDM [12] and "super-Nyquist" WDM [14]. Alternatively, electronic pre-filtering has also been employed for such demonstrations [16,17].

Table 3.1 Definitions of various classes of WDM.

Definition	Condition ($\Delta f/B$)	Example
Coarse WDM	>50	10 Gb/s on 20 nm
WDM	>5	10 Gb/s on 100 GHz
DWDM	$1.2 < \Delta f/B \leq 5.0$	28-Gbaud PDM-QPSK on 50 GHz
"Quasi-Nyquist" WDM	$1.0 < \Delta f/B \leq 1.2$	28-Gbaud PDM-16QAM on 33 GHz
"Nyquist" WDM	$\Delta f/B = 1$	28-Gbaud PDM-QPSK on 28 GHz
"Super-Nyquist" WDM	$\Delta f/B < 1$	28-Gbaud PDM-QPSK on 25 GHz

Δf is the allocated channel bandwidth (or channel spacing) and B is the channel symbol rate.

A special case of "Nyquist" WDM is the one that additionally satisfies the O-OFDM conditions, as described below, allowing for crosstalk-free reception of symbol-rate spaced channels without using optical or electrical pre-filtering [4,8,22–25]. The O-OFDM conditions that must be met for multiplexing multiple modulated carriers to form a superchannel can be enumerated [8] as follows.

1. The carrier spacing must equal the symbol rate with sufficient accuracy (inversely proportional to the duration of each processing block at the receiver). This implies that the carriers on which the modulation is imprinted need to be frequency locked. Experimental evidence for such a need is illustrated later in Section 3.4.2.1 under seamless multiplexing.
2. The modulated symbols on the carriers need to be time aligned at the point of de-multiplexing. (This follows from Figure 2 in Ref. [8].)
3. Typically, the frequency-domain response of the modulated symbols is a sinc function. This implies that sufficient bandwidth is needed at the transmitter and the receiver to modulate each subcarrier. At the receiver, there must also be sufficient oversampling speed to capture most of the sinc function for each of the modulated subcarriers.

The transmitter and receiver bandwidth requirements for meeting the O-OFDM conditions can be relaxed by using OFDM to modulate each signal [26], which effectively reduces the bandwidth of each orthogonal optical subcarrier. In this case, one has N_e electronic subcarriers for each of the N_o optical subcarriers. The bandwidth requirements can also be relaxed by introducing a guard interval (GI) between adjacent modulated symbols, albeit at the expense of reduced WDM SE [27]. While O-OFDM requires that the orthogonality conditions be met by the modulated signals that construct a superchannel, quasi-Nyquist-WDM relaxes the multiplexing requirement by allocating a guard band (GB) between adjacent modulated signals, albeit at the expense of reduced SE.

Generally, the synthesis of Terabit/s superchannels is a two-step process involving both modulation and multiplexing. In the first step (for modulation), one needs to pick a modulation format with the appropriate optical and electronic hardware to realize what we term a single-band transmitter (SB-TX), generating a signal, whose data rate is usually limited by the electronic bottleneck, e.g. resulting from limited bandwidth of components such as optical modulator, modulator driver, photo-detector,

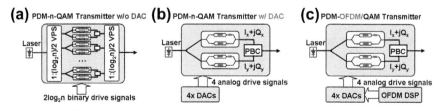

FIGURE 3.2 Three common types of single-band transmitters (SB-TX). VPS: variable power splitter; DAC: digital-to-analog converter; DSP: digital signal processor.

trans-impedance amplifier (TIA), DAC, and ADC. Here SB means that there is no optical frequency domain parallelization. In the second step (for multiplexing), the outputs of the SB-TXs are multiplexed in parallel to generate a superchannel having a desired total data rate. The multiplexing can be based on O-OFDM or Nyquist-type WDM. The detection of a superchannel can be categorized as multi-step detection [26] and single-step detection [28,29]. In multi-step detection, a superchannel is divided into multiple subchannels that are detected separately, e.g. via multiple digital coherent detection front-ends. The details on the modulation, multiplexing, and detection schemes will be presented in the following sections.

3.3 MODULATION

There are three common modulation schemes used to construct a SB-TX for high-level constellations, as shown in Figure 3.2 [30]. The first scheme (a) uses an array of PDM I/Q modulators (PDM-IQMs) that are driven by binary drive signals. To generate a PDM-n-QAM signal, $\log_2(n)/2$ PDM-IQMs are needed, together with two $1{:}\log_2(n)/2$ variable power splitters (VPSs) [31]. The second scheme (b) uses a single PD-IQM that is driven by four analog drive signals, corresponding to the I and Q components of two polarization states of the signal. Four DACs are needed. Digital signal processing (DSP) may be used for pre-equalization and pulse shaping at the transmitter. The third scheme (c) is based on OFDM with QAM subcarrier modulation, which additionally requires OFDM DSP.

It is of value to compare the above high-SE transmission schemes. Table 3.2 compares the three SB-TX schemes. SB-(a) has benefits of (1) not requiring a DAC, (2) not requiring DSP, and (3) generating signals with low peak-to-average-power ratio (PAPR) and with low optical loss. However, it requires more than one PD-IQM for

Table 3.2 Comparison among modulation schemes.

Single-Band Transmitter Type	Number of MZMs Needed	DAC Needed? (Sampling Rate)	DSP Needed?	Spectral Width[a]	PAPR
(a) PDM-n-QAM w/o DAC	$2\log_2 n$	No	No	~2B	Low
(b) PDM-n-QAM w/ DAC	4	Yes (2B preferred)	Optional	~2B w/o DF~B w/ DF	Low
(c) OFDM	4	Yes (1.2~1.5B)	Yes	~B	High[b]

[a] The spectral width is measured as null-to-null bandwidth.
[b] The high PAPR of OFDM can be reduced by more DSP, e.g. via DFT-spread-OFDM. TX: transmitter; B: signal baud; DF: digital filter.

$n > 4$, so photonic integration of multiple PDM-IQMs would make this scheme more attractive. SB-(b) and SB-(c) have the advantage of needing only one PD-IQM, but they require high-speed DACs. Hybrid options are possible as well to trade DAC complexity with parallel-optics complexity. Compared to SB-(c), SB-(b) offers lower PAPR but prefers a slightly higher DAC sampling speed. For single-carrier modulation, DSP can also be applied to tightly confine the signal spectrum through Nyquist filtering with small roll-off factors. With tight spectral filtering, the PAPR of a single-carrier modulated signal increases as compared to the unfiltered case. The high PAPR in SB-(c) can be reduced by the discrete Fourier transform (DFT)-spread technique [32–35].

There are two emerging modulation formats worth mentioning. The signal tolerance to inter-symbol interference (ISI) can be improved by offsetting the in-phase and quadrature components of a quadrature-amplitude modulation (QAM)-based subcarrier by half modulation period in time, leading to the offset-QAM format (o-QAM) [18,36,37]. When the OFDM subcarriers are modulated on a one-dimensional constellation, e.g. via BPSK, the subcarrier spacing can be half of the modulation symbol rate, resulting in the fast-OFDM scheme [38], which uses inverse discrete cosine transforms rather than inverse discrete Fourier transform for subcarrier multiplexing. Fast-OFDM recently found interesting applications in intensity-modulation and direct-detection systems [39].

3.4 MULTIPLEXING

3.4.1 Overview of multiplexing schemes

There are three common schemes for constructing a multi-band transmitter (MB-TX) that consists of multiple SB-TXs, as shown in Figure 3.3 [30]. The first scheme (a) is based on O-OFDM, which requires a set of frequency-locked carriers [4–7]. The second scheme (b) is to use optical filters (OFs) to perform the filtering needed to support quasi-Nyquist-WDM [11,12]. The third scheme (c) is to use digital filters (DFs), instead of OFs, to perform the filtering needed to support quasi-Nyquist-WDM [16,17]. In the third scheme, DSP and DAC are both needed. The implementation of the DF can be a root-raised-cosine (RRC) filter. In a more general sense, the inverse fast Fourier transform (IFFT) used in OFDM modulation can also be regarded as a DF that naturally produces a well-confined square-like signal spectrum with a sharp roll-off.

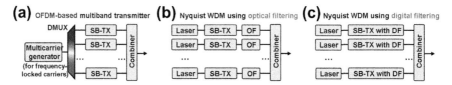

FIGURE 3.3 Three common types of multi-band transmitters (MB-TX). DMUX: wavelength de-multiplexer; OF: optical filter; DF: digital filter. (After Ref. [30]. © 2011 OSA.)

Table 3.3 Comparison among multiplexing schemes.

Multiplexing	O-OFDM		Quasi-Nyquist-WDM	
Carrier modulation	SC	OFDM	SC w/OF	SC w/DF
Frequency locking	Needed	Needed	Not needed	Not needed
TX bandwidth	>~2B	~B	~B	~B
ADC sampling rate	>~4B	~1.5B	1.5~2B	1.5~2B
Extra components	No	DAC, TX-DSP	Optical filters	DAC, TX-DSP

SC: single-carrier modulation; TX-DSP: transmitter digital signal processor.

Table 3.3 compares several common multiplexing schemes. Single-carrier modulation in conjunction with the orthogonality conditions described earlier, as used in Ref. [4], does not require DAC and transmitter DSP, but requires the sampling speed of the receive-side ADC to be much larger than the modulation speed of each modulated carrier. On the other hand, OFDM modulation on each optical carrier, combined with seamless band multiplexing, as used in Refs. [5,6,26], has the benefits of (1) not requiring the tight OFs and (2) lower requirements on transmitter bandwidth and ADC speed. Quasi-Nyquist multiplexing has the advantage of not requiring frequency-locked carriers, so independent lasers can be used. Confinement of the signal spectrum using DF has the advantage of not requiring bulky optical filters, although a moderate amount of additional DSP is needed to implement the DF at the transmitter. Also, DF usually produces sharper spectrum roll-offs than OF, thereby allowing the modulated carriers to be packed closer. Overall, the use of digital filtering to tightly confine the signal spectrum to enable high-SE multiplexing in the superchannel formation is becoming a preferred solution.

3.4.2 Seamless multiplexing

3.4.2.1 O-OFDM with single-carrier-modulated signals

One common type of seamless multiplexing is based on packing single-carrier-modulated signals under the O-OFDM condition [4,8,9]. This type of multiplexing was also referred to as no-guard-interval (NGI) coherent optical OFDM [10], alluding to the aspect that there is no time-domain guard interval (GI) added after the multiplexing, and all-optical OFDM [40,41], alluding to the aspect that no digital signal processing (e.g. digital FFT/IFFF) is used for the multiplexing. A GI can be added at the modulation stage to mitigate the negative impact of limited modulator bandwidth [42], but at the expense of reduced power efficiency and spectral utilization. The need to have frequency locked multi-carriers spaced closely at the symbol rate, as previously defined for a superchannel, is experimentally confirmed and shown in Figure 3.4,

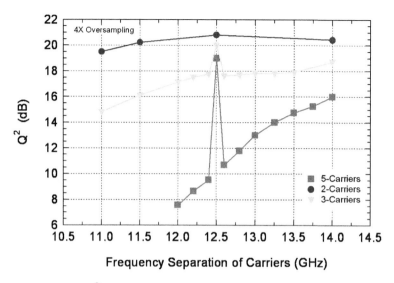

FIGURE 3.4 Measured Q^2 (at full OSNR) as a function of the frequency separation of the carriers for a 2-, 3-, and 5-carrier NGI-CO-OFDM system with 12.5-Gbaud PDM-QPSK modulation on each carrier [9].

where the measured Q^2, at full optical signal-to-noise ratio (OSNR), is plotted as a function of the frequency separation of the carriers for three superchannels, respectively, with 2, 3, and 5 carriers, each modulated with 12.5-Gbaud PDM-QPSK. As the number of carriers increases, the carrier separation frequency has to be locked to the symbol rate very precisely.

A superchannel transmitter design based on O-OFDM of multiple single-carrier-modulated signals is shown in Figure 3.5. A group of frequency-locked carriers is

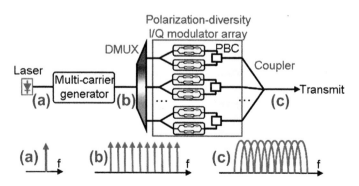

FIGURE 3.5 Schematic of a multi-carrier superchannel transmitter based on O-OFDM of multiple single-carrier-modulated signals. Optical spectra at locations (a)–(c) are illustrated. DMUX: wavelength de-multiplexer; PBC: polarization beam combiner [9].

first separated by a wavelength demultiplexer (DMUX), before being individually modulated with PDM-OFDM by an *I/Q* modulator array (*I/Q* modulators followed by polarization-beam combiners). To achieve orthogonality among the modulated carriers, all the carriers, in addition to being spaced at the modulation symbol rate, need to be synchronously modulated or symbol aligned. With orthogonality among the carriers, no frequency guard band is needed between adjacent carriers. The modulated carriers are then combined by an optical coupler. The optical spectra right after the laser, the multi-carrier generator, and the coupler are also illustrated in the insets (a), (b), and (c), respectively, of Figure 3.5. An extension of the scheme shown in Figure 3.5 is to include DACs in the drive path to the data modulators to generate superchannels with high-level modulation formats. Suitable DSP is needed to map the binary data stream intended for transmission to the constellation of the selected modulation format. Photonic integration of all or most of the optical elements in the multi-carrier transmitter is expected to be essential to enable future cost-effective generation of such Tb/s superchannels.

Figure 3.6 shows the measured optical spectrum of a 1.2-Tb/s superchannel based on O-OFDM of 24 PDM-QPSK signals [4]. The spectrum of the 24 frequency-locked optical carriers before modulation is also shown. The penalty associated with the seamless multiplexing was shown to be negligible. This 1.2-Tb/s superchannel was transmitted over a record transmission distance of 7200 km with a mean bit error ratio (BER) of $<3 \times 10^{-3}$, a typical threshold of hard-decision (HD) forward-error correction (FEC) codes.

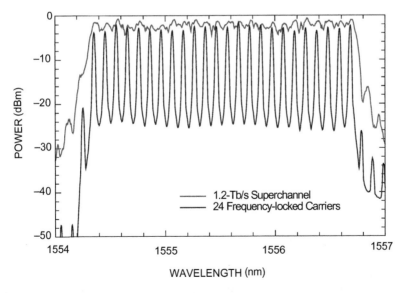

FIGURE 3.6 Measured optical spectrum of a 1.2-Tb/s superchannel based on O-OFDM of 24 PDM-QPSK signals [4].

3.4.2.2 O-OFDM with OFDM-modulated signals

Another common type of seamless multiplexing is based on packing OFDM-modulated signals under the O-OFDM condition [5–7,26]. In essence, each OFDM-modulated signal can be seen as a set of subcarriers that satisfy the O-OFDM condition with a subcarrier spacing Δf_{SC}, and multiple such OFDM signals can be arranged to maintain the O-OFDM condition, by synchronizing the OFDM symbols in these signals and putting the GB between any two adjacent OFDM signals to be $m \cdot \Delta f_{SC}$, where m is a positive integer. When m is set to 1, the GB between two adjacent OFDM signals becomes exactly the subcarrier spacing of the OFDM signals, leading to seamless multiplexing of the OFDM signals, or zero-wasted spectrum in the multiplexing process.

In conventional OFDM [43–47], a time-domain GI, e.g. in the form of a cyclic prefix (CP), is inserted in the time domain between adjacent OFDM symbols to accommodate for fiber chromatic dispersion (CD) induced inter-symbol interference (ISI). The larger the chromatic dispersion, the longer is the GI needed, leading to an increased overhead and a reduced spectral efficiency. Recently, reduced-GI OFDM (RGI-OFDM) was proposed to reduce the needed GI in the presence of large CD, by digitally compensating the CD effect prior to the receiver-side OFDM signal processing scheme [26,48]. The RGI-OFDM can be efficiently implemented by a reconfigurable static frequency-domain equalizer as the CD effect is determined by the fiber link and can be viewed as stationary in practice as long as the fiber link is unchanged. Also, the CD effect causes a well-defined quadratic phase response on the signal, which can be readily equalized. A much reduced GI or CP between adjacent OFDM symbols is used to accommodate ISI with short memory, such as that induced by transmitter bandwidth limitations or polarization-mode dispersion (PMD). This approach enables the reduction of the GI from >20% for conventional OFDM to only ~2% for RGI-OFDM in a typical long-haul 100-Gb/s OFDM system [26].

Figure 3.7 shows an exemplary transmitter setup for forming a superchannel based on O-OFDM of multiple RGI-OFDM signals in a lab context [48]. An external cavity laser (ECL) at 1548.3 nm with a linewidth of ~100 kHz was used as the laser source. Twenty frequency-locked optical carriers with 6.46-GHz spacing were generated by cascading a 5-comb generator, based on a Mach-Zehnder modulator (MZM) driven by a 25.94-GHz sine-wave with ~$3V_\pi$ amplitude, and a 4-comb generator, based on a nested MZM whose two branches were respectively driven by 3.24 and 9.73-GHz sine-waves with ~$1V_\pi$ amplitudes. A wavelength-selective switch (WSS) was configured to have a 3-dB bandwidth of 120 GHz to reject the unwanted harmonics generated by the 5-comb generator. The 20 frequency-locked carriers were then modulated by a PDM-IQM to generate a PDM-32QAM-OFDM superchannel. The x- and y-polarization components of the PDM signal were independently modulated to better emulate a real transmitter. Four independent drive patterns were stored in two synchronized arbitrary waveform generators (AWGs), each having two 10-GS/s DACs. Pseudo-random bit sequences (PRBS) of length $2^{15} - 1$ were used as the payload data. The IFFT size used for OFDM was 128, and the GI was two samples, resulting in a small GI-overhead of 1.56%. Each polarization component of an OFDM symbol

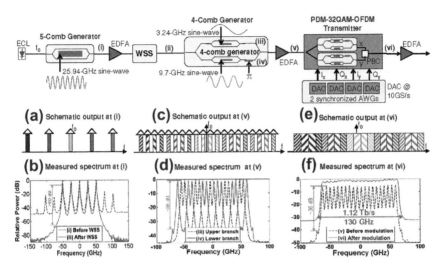

FIGURE 3.7 Schematic of the experimental setup of a 1.12-Tb/s superchannel transmitter based on seamless multiplexing of multiple OFDM signals under the O-OFDM condition. Insets (a–f) show the schematic outputs (in the frequency domain) and the measured optical spectra at different stages in the transmitter. EDFA: Erbium-doped fiber amplifier; PBC: polarization-beam combiner. (After Ref. [48]. © 2011 OSA.)

contained 78 32-QAM data subcarriers (SCs), four pilot SCs, one unfilled DC SC, and 45 unfilled edge SCs. The spectral bandwidth of each modulated subchannel was 6.48 GHz ($=83/128 \times 10$ GHz), and the 20 frequency-locked 6.48-GHz-spaced input carriers enabled *seamless* superchannel formation with a total bandwidth of 130 GHz, as shown in inset (f). Three correlated dual-polarization training symbols (TSs) [49] were used for every 697 payload OFDM symbol, resulting in a small TS-overhead of 0.43%. Excluding 7% overhead for forward-error correction, the net payload data rate of the superchannel was 1.12 Tb/s ($=10$ GHz $\times 10$ b/s/Hz $\times 78/130 \times 697/700 \times 20/1.07$), corresponding to a net intra-channel SE (ISE, defined later in Section 3.6) of 8.61 b/s/Hz ($=1.12$ Tb/s/129.7 GHz). This superchannel could likely be put on a 150-GHz grid with <-40 dB crosstalk to neighbors, as indicated in inset (f), to achieve a WDM SE of 7.43 b/s/Hz.

3.4.3 Multiplexing with guard band

3.4.3.1 Multiplexing OFDM-modulated signals with guard band

Seamless multiplexing requires frequency-locked optical carriers, which may be difficult to generate with high tone-to-noise ratio and with low unwanted harmonics. In addition, these frequency-locked carriers need to be separated before modulation by independent modulators. Spectral guard-bands (GBs) can be introduced between OFDM signals to allow for frequency-unlocked independent lasers to be the carriers of a superchannel, at the expense of slightly reduced SE. Figure 3.8 shows the optical spectrum of a superchannel based on multiplexing of eight OFDM-modulated signals with guard band. For each OFDM signal, the FFT size, the sampling rate, and the spectral bandwidth of the filled subcarriers are 128, 30 GS/s, and 19.5 GHz, respectively. The GB used is 1 GHz, representing a spectral overhead of ~5%, yielding a $\Delta f/B$ of 1.05, a "quasi-Nyquist" WDM superchannel. The subcarrier modulation is PDM-16QAM. The arrangements of the OFDM subcarriers and symbols are similar to that described in the previous section. The net payload data rate of the superchannel is 1.07 Tb/s ($=30$ GHz $\times$ 8 b/s/Hz $\times$ 78/130 $\times$ 697/700 $\times$ 8/1.07), and the overall spectral bandwidth of the superchannel is ~165 GHz. It is feasible to transmit such 1-Tb/s superchannels on a 175-GHz grid (or a 200-GHz grid to allow ROADM filtering) to achieve a WDM SE of ~5.7 b/s/Hz (or 5 b/s/Hz).

It is important to estimate the crosstalk penalty as a function of the GB bandwidth for various FFT sizes and modulation formats. Figure 3.9 shows the experimentally measured BER performance of one of the eight OFDM signals of the 1.07-Tb/s superchannel (in the back-to-back configuration) as a function of the GB width. For comparison, the BER performance of O-OFDM-based seamless multiplexing by using frequency-locked carriers is also shown (as the dashed line). When the guard band is larger than ~1 GHz, or ~5% of the OFDM signal bandwidth, the performance of the superchannel with guard-banded multiplexing becomes better than that with O-OFDM-based seamless multiplexing, which we attribute to the worse back-to-back OSNR of the O-OFDM case due to the high loss of the frequency-comb generator used, resulting in a reduced signal OSNR.

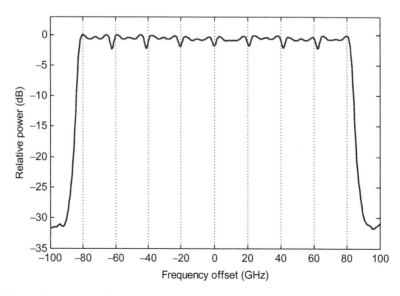

FIGURE 3.8 Measured optical spectrum of a 1.07-Tb/s superchannel that multiplexes eight OFDM signals with a GB of 1 GHz.

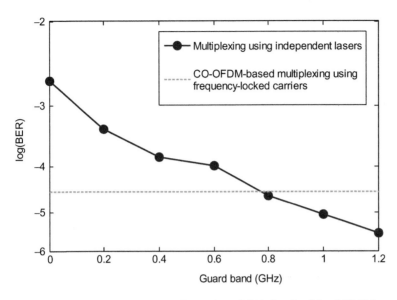

FIGURE 3. 9 Measured BER performance of one of the OFDM signals of the 1.07-Tb/s superchannel as a function of the guard band width.

The crosstalk penalty in guard-banded multiplexing of OFDM signals can be analytically estimated. For each OFDM subcarrier, we have a square pulse in the time domain and hence a sinc waveform in the frequency domain. The edge subcarriers of an OFDM signal are most contaminated by the non-orthogonal subcarriers of other OFDM signals. The amount of the worst-case crosstalk imposed on these edge subcarriers can be expressed as

$$X(\text{dB}) = 10 \log \left(\sum_{n=0}^{\infty} \left[\pi \left(\frac{\text{GB}}{\Delta f_{\text{SC}}} + n \right) \right]^{-2} \right)$$

$$= 10 \log \left(\sum_{n=0}^{\infty} \left[\pi \left(\frac{\text{GB} \cdot N_{\text{FFT}}}{B_{\text{OFDM}}} + n \right) \right]^{-2} \right), \qquad (3.1)$$

where Δf_{SC}, N_{FFT}, and B_{OFDM} are the OFDM subcarrier spacing, FFT/IFFT size used in OFDM, and the spectral bandwidth of the OFDM signal, respectively. Figure 3.10 shows the crosstalk as a function of the ratio between the GB width and the OFDM signal bandwidth ($\text{GB}/B_{\text{OFDM}}$). The crosstalk-induced penalty depends on modulation format and the targeted BER. A comprehensive study on this subject has recently been reported by Winzer et al. [50]. To limit the OSNR penalty at BER $= 10^{-3}$ to be within 3 dB, the allowed crosstalk amounts are approximately 10 dB, 17.5 dB, and 24 dB for QPSK, 16-QAM, and 64-QAM, respectively. Assuming that we can allow the worst-case edge subcarriers to have 3-dB OSNR penalty at BER $= 10^{-3}$, the GB-to-bandwidth ratio $\text{GB}/B_{\text{OFDM}}$ needs to be larger than $\sim$1.2%,

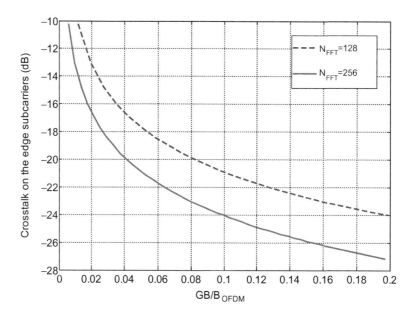

FIGURE 3.10 Calculated worst-case crosstalk on the edge subcarriers of an OFDM signal as a function of the ratio between the GB width and the OFDM signal bandwidth.

~5%, and ~20% for QPSK, 16-QAM, and 64-QAM subcarrier modulation, respectively. This indicates that for OFDM signals with $N_{FFT} > 128$ and with n-QAM subcarrier modulation where $n \leq 16$, spectrally efficient multiplexing of these OFDM signals with only ~5% spectral overhead is feasible. For higher-level QAM formats such as 64-QAM, larger GB or additional spectral shaping of the OFDM signal to suppress its sidebands is needed.

The above analysis can be also applied to estimate the needed GB between superchannels that are based on O-OFDM of single-carrier-modulated signals, by replacing N_{FFT} with the number of single-carrier-modulated signals and replacing B_{OFDM} with the superchannel bandwidth. Obviously, when the number of single-carrier-modulated signals is reduced, the GB-to-bandwidth ratio needs to be increased, as recently shown through numerical simulations by Bosco [51].

3.4.3.2 Multiplexing Nyquist-filtered single-carrier signals with guard band

As discussed previously, quasi-Nyquist-filtered single-carrier signals can have well-confined spectra with a small roll-off factor, β, commonly defined as the ratio between the excess signal bandwidth and the Nyquist bandwidth of the signal ($B/2$). Without considering laser frequency drift, the needed GB to avoid crosstalk between adjacent Nyquist-filtered signals is thus $\beta \cdot B/2$. Smaller roll-off is desired to multiplex signals at higher SE, but the smaller the roll-off, the larger the DSP complexity as longer filters are needed to perform the Nyquist filtering. A numerical study on the robustness of a Nyquist-WDM-based superchannel to optical filtering and to crosstalk induced by adjacent superchannels has been reported [51]. In a recent Nyquist-WDM experiment, a roll-off factor as small as 0.01 has been demonstrated, leading to a record single-mode fiber capacity of 102.3 Tb/s [52].

DFT-spread OFDM [32–35] is also commonly known as single-carrier frequency-division multiple access (SC-FDMA) [53], in which discrete Fourier transform (DFT) and inverse discrete Fourier transform (IDFT) are used to perform the spectral shaping. In SC-FDMA, the modulation is done in the time domain while the equalization is performed in the frequency domain. Multiple SC-FDMA signals can be multiplexed to form a superchannel, and the needed GB to confine the crosstalk penalty to a certain level can be estimated by using the analysis presented in the previous section.

3.5 DETECTION

In conventional WDM systems, wavelength channels are first de-multiplexed before being received. For superchannels, however, the modulated carriers inside each superchannel are typically too closely spaced to be separated by WDM filters without incurring a filtering penalty. As sharp filtering functions can be readily generated in the digital domain, digital coherent detection enables banded-detection of a superchannel [4,26,54–56], which consists of the following steps:

1. Splitting the superchannel into M copies;
2. Mixing these M copies in M polarization-diversity optical hybrids with M different optical local oscillators (OLOs);
3. Performing digital coherent detection of each of the M copies, with an RF bandwidth that is slightly larger than half of the occupied optical spectral bandwidth of the modulated carrier(s) intended for detection;
4. Digitally filtering each modulated carrier and recovering the data carried by that carrier.

Note that the tight confinement of the spectral content of each modulated carrier in a superchannel, e.g. through transmitter DF, is very beneficial as it reduces the oversampling ratio requirement at the receiver, leading to relaxed ADC sampling speed requirement and more efficient digital signal processing for channel recovery.

It is possible to simultaneously detect more than one carrier per digital sampling at the receiver. At 112 Gb/s, a 2-carrier signal was shown to be detected with low-sampling-rate ADC to reduce both hardware complexity and receiver DSP load [54]. It was also shown that an oversampling factor, defined as the ratio between the sampling rate and the symbol rate of the carrier modulation, as small as 1.4 is sufficient [55]. In Refs. [4] and [56], the simultaneous detection of two and three subcarriers, in the presence of all 24 subcarriers, was demonstrated, respectively.

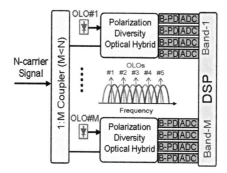

FIGURE 3.11 Schematic of the banded detection of a multi-carrier signal. Inset: illustration of the detection of a 10-carrier signal with five detection bands, each containing two modulated carriers. B-PD: balanced photo-detector [9].

Figure 3.11 shows the schematic of the banded detection of a superchannel containing N single-carrier-modulated signals. There are M (M<N) detection bands, each of which contains more than one modulated signal to be recovered. The inset illustrates the detection of a 10-carrier superchannel with five detection bands, each containing two modulated carriers. Note that the needed optical hardware (including the optical hybrids and OLOs) is reduced in proportion to the number of carriers per detection band, by the use of faster ADCs and DSPs. In the case of the reception of a 448-Gb/s superchannel based on O-OFDM of ten RGI-OFDM signals [26], two banded detections were used to recover all the ten signals, corresponding to an oversampling ratio of 1.67. It was also shown that multi-carrier superchannels can be detected with oversampling ratios below two with negligible penalties. Figure 3.12 shows the schematic of the DSP block diagram for recovering the (N/M) carriers when detecting the first band of an N-carrier signal using M equally sized detection bands [4]. First, electronic dispersion compensation (EDC) is performed,

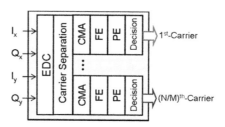

FIGURE 3.12 Schematic of the DSP block diagram for recovering the (N/M) carriers when detecting the 1st band of an N-carrier signal using M equally sized detection bands. EDC: electronic dispersion compensation; CMA: constant-modulus algorithm-based channel compensation; FE: frequency estimation and compensation; PE: phase estimation and compensation [9].

e.g. by using the well-known frequency-domain equalization approach. Then, carrier separation is conducted by shifting each carrier to the baseband and passing each shifted carrier through two cascaded T/2 and T/4 delay-and-add filters, where T is the modulation symbol period. Constant modulus algorithm (CMA)-based blind equalization with a 15-tap T/4-spaced adaptive FIR filter can be used for polarization demultiplexing and further signal equalization. Other DSP processes include frequency estimation (FE), phase estimation (PE), and data recovery. Finally, the bit error ratio (BER) of each of the detected carriers is measured.

For a superchannel with OFDM-modulated signals, the DSP at the receiver is strongly linked to the DSP at the transmitter, particularly with the frame structure, training symbols, and pilot subcarriers. Concepts such as correlated dual-polarization (CDP) training symbols [49,57] and intra-symbol frequency-domain averaging (ISFA) [58] have enabled reliable and efficient reception of OFDM signals after long-distance transmission. OFDM-based superchannels are well suited for banded detection for the following reasons. First, when OFDM modulation is used, a square-like optical spectrum with sharp roll-off is naturally obtained, reducing the needed guard band compared to quasi-Nyquist WDM if independent lasers are used as the multiplexing scheme. Second, when CO-OFDM multiplexing is used, the orthogonality among the modulated carriers can be exploited to achieve crosstalk-free demultiplexing of these carriers. Finally, the banded detection makes the frequency range to be covered in the subsequent receiver signal processing much smaller than the bandwidth of the superchannel. This dramatically reduces the complexity of electronic CD compensation, as the size of the CD equalizer scales quadratically with the processing frequency range. In effect, the banded processing of a superchannel increases the overall receiver DSP efficiency. A novel filter-bank-based efficient processing of RGI-OFDM signals and Nyquist-WDM signals was recently proposed by Tolmachev and Nazarathy [59,60].

3.6 SUPERCHANNEL TRANSMISSION

A useful parameter that specifies the spectral efficiency of the superchannel is the intrachannel SE (ISE). For a superchannel whose carriers are multiplexed by PDM n-point quadrature-amplitude modulation (n-QAM), the maximum ISE that can be achieved without a coherent crosstalk penalty is

$$ISE_{max} = 2 \log_2(n), \tag{3.2}$$

where the factor of 2 on the right-hand side accounts for PDM. The actual ISE for an OFDM-based superchannel can be expressed as

$$ISE = ISE_{max}/(1 + O_{FEC})/(1 + O_{GI}), \tag{3.3}$$

where O_{FEC} is the overhead used for forward error correction (FEC) and O_{GI} is the OFDM signal processing related overhead, used for guard intervals, training symbols, and/or pilot subcarriers.

3.6.1 Transmission based on single-carrier modulation and O-OFDM multiplexing

Table 3.4 shows the transmission performance reported for various superchannels in experimental demonstrations. An early demonstration [22], called coherent wavelength division multiplexing (CoWDM), was based on non-return-to-zero (NRZ) signaling, with 42.66-GHz carrier spacing and 42.66-Gb/s on-off keyed (OOK) data modulation on the carriers, with appropriate phase control applied to minimize crosstalk. Subsequently, two-carrier optical OFDM was demonstrated [40] where differential quadrature phase shift keying (DQPSK) was used to generate a 100-Gb/s superchannel, with the two carriers spaced 25-GHz apart, and each modulated at 25 Gbaud. A variant of this approach with four carriers, each modulated using the duobinary format, was also reported [61]. All three of the above demonstrations used direct detection (DD) to recover information from each carrier. All subsequent demonstrations have used coherent detection. O-OFDM using two carriers, each modulated with single-carrier PDM-QPSK, was demonstrated for 100-Gb/s long-haul transmission [54]. The underlying O-OFDM principles elucidated in [8] were used to demonstrate 1.2-Tb/s 24-carrier superchannel generation, detection, and transmission [4]. This was also the first time ultra-large area fiber (ULAF) was used for Terabit/s superchannel transmission. The ULAF had an average fiber loss, dispersion, and dispersion slope at 1550 nm of 0.185 dB/km, 19.9 ps/nm/km, 0.06 ps/nm^2/km, respectively. The effective area was 120 μm^2, which allowed for high signal launch powers without suffering very much from fiber nonlinearities. As can be seen in Table 3.4, several multi-level modulation formats (QPSK, 8-QAM, and 16-QAM) have been used to synthesize superchannels, and a wide range of ISEs have been demonstrated.

Table 3.4 Experimental demonstrations of superchannels based on single-carrier modulation and O-OFDM.

Ref.	Modulation Format	Super-channel Data Rate (Gb/s)	Composition	Intra-channel SE (b/s/Hz)	Reach (km)	ISEDP (km × b/s/Hz)
22	NRZ-OOK (DD)	288	7 × 41.3-Gb/s	0.93	1200	1116
40	DQPSK (DD)	100	2 × 25-Gb/s	1.87	1300	2431
61	Duobinary (DD)	100	4 × 25-Gb/s	0.93	100	93
54	PDM-QPSK	112	2 × 56-Gb/s	3.74	10,093	37,748
4	PDM-QPSK	1200	24 × 50-Gb/s	3.74	7200	26,928
23	PDM-QPSK	11,200	112 × 100-Gb/s	3.57	640	2285
24	PDM-QPSK	1150	23 × 50-Gb/s	3.33	10,000	33,300
25	PDM-8QAM	1200	16 × 100-Gb/s	4.71	1600	7536
41	PDM-16QAM	1500	15 × 100-Gb/s	7.00	1200	8400

In Table 3.4, we also compare another figure of merit, namely, the intrachannel spectral efficiency and distance product (ISEDP). Large ISEDP values reflect the superior transmission characteristics of high SE superchannels. As the complexity of a modulation format increases, the ISEDPs have correspondingly reduced due to higher OSNR requirements. Record ISEDP values have been achieved by using PDM-QPSK modulation with ISEs ranging from 3.33 to 3.74 b/s/Hz.

3.6.2 Transmission based on OFDM modulation and O-OFDM multiplexing

In Table 3.5, we list experimental demonstrations of superchannels using RGI-OFDM formats as well as quasi-Nyquist filtered single-carrier formats, using digital

Table 3.5 Experimental demonstrations of superchannels based on OFDM for the O-OFDM optical subcarriers, as well as quasi-Nyquist-WDM and guard-banded OFDM.

Ref.	Format (Subcarrier Modulation)	Super-channel Raw Data Rate	Composition	Intra-channel SE (b/s/Hz)	Reach (km)	ISEDP (km b/s/Hz)
26	RGI-OFDM (PDM-16QAM)	448 Gb/s	10 × 45-Gb/s (2-band detection)	7.00	2000	14,000
63	RGI-OFDM (PDM-16QAM)	485 Gb/s	10 × 48.5-Gb/s (single detection)	6.20	4800	29,760
28	RGI-OFDM (PDM-32QAM)	606 Gb/s	10 × 60.6-Gb/s (single detection)	7.76	1600	12,416
29	RGI-OFDM (PDM-64QAM)	728 Gb/s	10 × 72.8-Gb/s (single detection)	8.00	800	6400
16	Quasi-Nyquist-WDM PDM-2/64QAM	504 Gb/s	5 × 100-Gb/s	8.40	1200	10,080
17	Quasi-Nyquist-WDM PDM-64QAM	538 Gb/s	8 × 67-Gb/s	8.96	1200	10,752
62	Quasi-Nyquist-WDM PDM-64QAM	12.84 Tb/s	100 × 128.4-Gb/s	7.90	320	2528
35	DFT-Spread-OFDM (PDM-16QAM)	1630 Gb/s	16 × 2 × 34-Gb/s	5.50	2500	13,750
64	Guard-banded OFDM (PDM-16QAM)	1864 Gb/s	8 × 233-Gb/s	5.75	5600	32,000

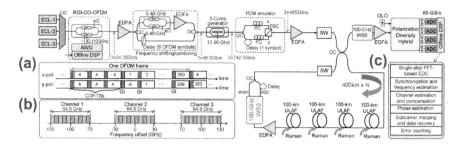

FIGURE 3.13 Schematic of the experimental setup [63]. Insets: (a) OFDM frame structure; (b) optical spectra of three 485-Gb/s WDM channels before and after 4800-km transmission; (c) block diagram of the receiver offline DSP.

signal processing techniques at the transmitter as described earlier. It is evident that achieving ISEs beyond 5 b/s/Hz is generally easier using either RGI-OFDM or the quasi-Nyquist-filtering based modulation as compared to unfiltered single-carrier modulation. In addition, the demonstrated ISEDP values in Table 3.5 are generally larger for approximately the same ISE as compared to the values demonstrated in Table 3.5. The reason for both these observations is related to the relative ease of generating high-quality crosstalk-free subcarriers that are multi-level modulated when DAC and transmitter-side signal processing are used. For quasi-Nyquist WDM, the intentionally allocated guard bands between channels alleviate crosstalk impairments, albeit at the expense of slightly reduced SE.

We illustrate the performance of RGI-OFDM superchannels with the experiment reported in Ref. [63] and depicted in Figure 3.13. The same setup was successfully used to demonstrate three different modulation formats, show optical parallelization concepts, demonstrate banded detection of the entire superchannel with one digital sampling, and evaluate the concatenation performance of reconfigurable optical add/ drop multiplexers (ROADMs). A brief description follows.

At the transmitter, three 100-GHz-spaced 485-Gb/s superchannels based on RGI-OFDM modulation and CO-OFDM multiplexing were generated. The WDM channels were launched into a transmission loop consisting of four Raman-amplified 100-km ULAF spans. To assess the signal performance in optically routed networks with ROADMs, we used a 100-GHz WSS in the loop to separate and recombine the even and odd channels. For each WSS passband, the 0.1-dB and 35-dB bandwidths were ~72 and ~125 GHz, respectively. The optical spectra of the three WDM channels, measured by an optical spectrum analyzer with 0.1-nm resolution before and after 4800-km transmission, are shown as inset (b) in Figure 3.12. Evidently, no optical filtering-induced spectral clipping is observed, even after 12 WSS passes. At the receiver, each WDM channel was sequentially filtered out by a second 100-GHz WSS for performance evaluation. An optical local oscillator (OLO) was tuned to the center frequency of the channel under test.

With the use of four 80-GS/s ADCs with 32.5-GHz RF bandwidth, each 485-Gb/s signal with an optical bandwidth of 64.8 GHz can be completely sampled through a single coherent detection, without having to resort to banded detection. The digitized waveforms were processed offline. The DSP blocks are shown in inset (c) of Figure 3.13.

With the use of transmitter DSP, the modulation format used for subcarrier modulation in OFDM can be easily changed, leading to a so-called *software-defined* transmission link. Figure 3.14 shows the recovered constellations of the OFDM subcarriers when modulated by 16-QAM [63], 32-QAM [28], and 64-QAM [29], achieving superchannel data rates of 485, 606, and 728 Gb/s, respectively. The distances transmitted over ULAF fiber with all-Raman amplification were 4800, 1600, and 800 km, respectively.

It is interesting to note that the three formats had a spectral occupancy of between 60 and 65 GHz, demonstrating the adaptive nature of multilevel formats to support high data rates without sacrificing optical spectrum.

With the prevalence of soft-decision forward error correction (SD-FEC) in next generation transport systems, it is important not just to look at pre-FEC error rates (as is permissible for hard-decision FEC if errors are uncorrelated) but also to investigate the probability density function (pdf) of the signal after transmission to see if the noise distribution is, in fact, Gaussian, as the net coding gain (NCG) for the FEC is typically derived using the additive white Gaussian noise (AWGN) assumption [65]. Figure 3.15a shows the pdf of the I and Q components of both polarizations of the center 485-Gb/s PDM-16QAM RGI-CO-OFDM superchannel in the back-to-back configuration, which closely follows the Gaussian distribution. Figure 3.15b shows the pdf after 4800-km transmission at the optimal signal power, which also closely follows the Gaussian distribution. In addition, careful analysis of the pdfs for all 16 constellation points showed that all pdfs are identical and individually obey circularly symmetric complex Gaussian statistics. This indicates that soft-decision

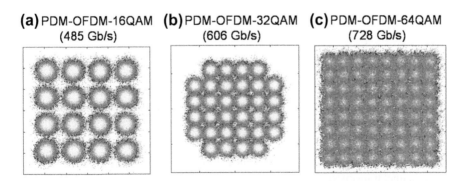

(a) PDM-OFDM-16QAM　**(b)** PDM-OFDM-32QAM　**(c)** PDM-OFDM-64QAM
(485 Gb/s)　　　　　　　(606 Gb/s)　　　　　　　(728 Gb/s)

FIGURE 3.14 Constellations of the OFDM subcarriers (recovered in the back-to-back configuration) when modulated by 16-QAM [63], 32-QAM [28], and 64-QAM [29], respectively, achieving superchannel data rates of 485, 606, and 728 Gb/s.

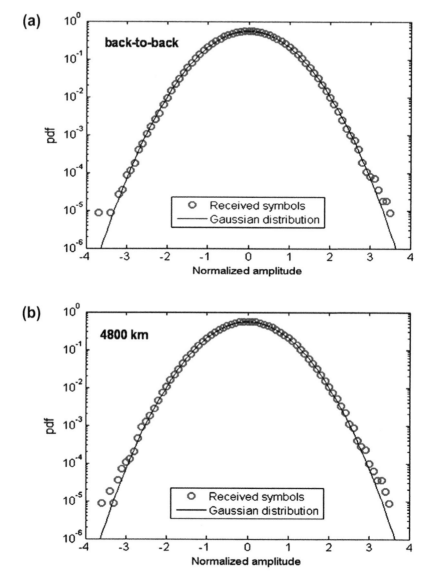

FIGURE 3.15 (a) Probability density function (pdf) of the signal distortion in the back-to-back configuration; (b) pdf after 4800-km transmission at 1 dBm per superchannel [63].

FEC can be effective, even in the nonlinear transmission regime, for OFDM-based superchannel transmission.

More recently, a guard-banded OFDM superchannel with 1.5-Tb/s net data rate, consisting of eight all-electronically-generated 30-Gbaud PDM-OFDM-16QAM signals spaced at 32.8 GHz, was transmitted over 56 100-km ULAF spans (5600 GHz km), achieving a record spectral-efficiency-distance product of over 32,000 km b/s/Hz for terrestrial Tb/s-class superchannel transmission with >5 b/s/Hz net spectral efficiency [64]. Soft-decision forward error correction was implemented using offline processing as an inner code, resulting in a corrected bit-error ratio below the threshold of an outer hard-decision (HD) FEC decoder. This demonstration was enabled by high-speed 50-GS/s DACs and ADCs allowing for a record 233-Gb/s all-electronically-generated and detected OFDM signal, low-nonlinearity ULAF, Raman amplification, low-overhead OFDM, optimized one-step-per-span nonlinear compensation, and SD-FEC.

3.6.3 Transmission based on Nyquist-WDM

The transmission performance of Tb/s Nyquist-WDM superchannels has been studied experimentally [12,13] and numerically [15,66]. Bosco et al. comprehensively studied the transmission performance of Tb/s Nyquist-WDM with PDM-BPSK, PDM-QSPK, PDM-8QAM, and PDM-16QAM formats through numerical simulations [66]. Figure 3.16 shows the maximum reach of Nyquist-WDM superchannels at BER $\leq 4 \times 10^{-4}$ as a function of capacity in the C-band and as a function of spectral efficiency for PM-BPSK, PM-QPSK, PM-8QAM, and PM-16QAM. (Although this study was done using a BER threshold of 4×10^{-3}, longer reaches are achievable when the FEC threshold is higher, as in current generation SD-FEC.) For a

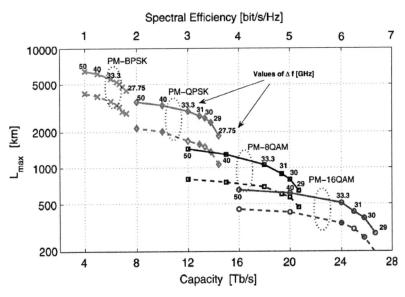

FIGURE 3.16 Maximum reach with BER $\leq 4 \times 10^{-3}$ versus capacity in the C-band (bottom axis) and spectral efficiency (top axis) for Nyquist-WDM superchannel with 27.75-Gbaud PM-BPSK (crosses), PM-QPSK (diamonds), PM-8QAM (squares), and PM-16QAM (circles). Lines are obtained connecting simulation results. Solid lines refer to SSMF and dashed lines refer to NZDSF. The corresponding values of Δf in GHz are shown at the SSMF points. (After Ref. [66]. © 2011 IEEE.) Note that longer transmission distances are feasible when the BER threshold is increased, e.g. by using stronger forward error correction codes.

given modulation format, the reach decreases as the frequency spacing between the Nyquist-filtered signals approaches the modulation symbol rate (27.75 GHz). For the same modulation format and the same frequency spacing, the transmission distance in standard single-mode fiber (SSMF) is longer than that in non-zero dispersion-shifted fiber (NZDF) due to the larger dispersion and smaller nonlinear coefficient of SSMF. More remarkably, as the constellation size doubles, e.g. from BPSK to QPSK or from 8-QAM to 16-QAM, the transmission distance is almost halved. This observation was also reported by Essiambre et al. in 2010 [67]. There are several other observations from the study. At the optimum transmitted power, the value of the maximum reachable distance is approximately equal to 2/3 of the maximum reachable distance in the linear regime, for all formats and fibers, and the OSNR penalty due to fiber nonlinearity at the optimal signal launch power is ~ 1.76 dB. The optimal signal power spectral density is essentially independent of the modulation format. This is also consistent with the simulation results presented in Ref. [67]. These observations can be explained by analyses to be presented in the following section.

3.6.4 Optimization of the spectral-efficiency-distance-product

For future optical transport systems, it is desirable to increase the SE for higher network capacity. On the other hand, higher SE leads to less tolerance to linear and nonlinear impairments and thus shorter transmission distances, as shown experimentally and numerically in the previous sections. What really matters in many cases is the cost per end-to-end transported information bit (CPB), which is preferred to be as low as possible. A useful metric that closely relates to the CPB is the spectral-efficiency-distance-product (SEDP), because the SEDP is linked to the number of transponders used in a system: usually the higher the SEDP, the less need for optical-to-electronic-to-optical (O/E/O) conversion, and thus the lower the CPB (assuming that O/E/O conversion cost is the major portion of the system cost, which is usually true in WDM systems). In this section, we compare the commonly used PDM-n-QAM modulation formats based on their respective achievable SEDPs. We present simple closed-form expressions for the achievable SEDP. Given the popular use of digital signal processing at both the transmitter and receiver of a Tb/s superchannel link, the optimal format that provides the highest SEDP for a given optical link could be readily implemented.

Assuming that the signal distortions are Gaussian distributed, which is a reasonably accurate assumption for dispersion-uncompensated high-speed transmission links [68–71] we define an effective signal-to-noise ratio (SNR) as the ratio between the signal power and the sum of the linear noise resulting from ASE and the nonlinear noise, given by

$$\text{SNR} = \frac{P}{\sigma_L^2 + \sigma_{NL}^2} = \frac{1}{aP^{-1} + bP^2}. \tag{3.4}$$

where P, δ_L^2, and δ_{NL}^2 are respectively the signal power, linear noise power, and nonlinear noise power, and a and b are parameters that depend on link conditions. The SNR is maximized to

$$\text{SNR}_{\text{max}} = a^{-2/3}b^{-1/3}/(2^{1/3} + 2^{-2/3}) \tag{3.5}$$

at the optimum power $P_{opt} = \left(\frac{a}{2b}\right)^{1/3}$. Evidently, the SNR penalty due to fiber non-linearity (as compared to the case without fiber nonlinearity) at the optimal signal launch power is $-10\log(2/3) \approx 1.76\,\text{dB}$. This is in good agreement with that obtained through numerical simulations for dispersive coherent transmission [66].

Assuming that the transmission link consists of uniform fiber spans with the same physical characteristics and seamless O-OFDM multiplexing of signals with the same modulation format, the maximum effective SNR can be expressed by [69]

$$\text{SNR}_{max} = \frac{(8\pi\alpha|\beta_2|)^{1/3}}{3[3n_0^2\gamma^2 N_s h_e \ln(B/B_0)]^{1/3}}, \tag{3.6}$$

where α, β_2, and γ are the fiber coefficients for loss, dispersion, and nonlinearity, respectively, N_s is the number of fiber spans of the optical link, n_0 is the power spectral density of the optical amplified spontaneous (ASE) noise, which is proportional to $N_s \cdot 10^{\alpha L_s}$ (where L_s is the length of each fiber span), B is the bandwidth of the WDM signals, B_0 is the larger of the CD-induced walk-off bandwidth and the phase estimation bandwidth and h_e is the nonlinear multi-span noise enhancement factor [70]. The maximum SNR is obtained at the optimum signal power spectral density

$$I_{max} = \left(\frac{8n_0\pi\alpha|\beta_2|}{3[3\gamma^2 N_s h_e \ln(B/B_0)]}\right)^{1/3}. \tag{3.7}$$

For dispersion-uncompensated optical transmission (DUMT) or dispersion compensation module (DCM) free transmission, which is a popular approach enabled by the capability of electronic dispersion compensation in digital coherent receivers, $h_e \approx 1$, we have

$$\text{SNR}_{max,DUMT} \propto \frac{[\alpha|\beta_2|/\ln(B/B_0)]^{1/3}}{N_s 10^{2\alpha L_s/3}\gamma^{2/3}}, \tag{3.8}$$

and

$$I_{max,DUMT} \propto 10^{\alpha L_s/3}[\alpha|\beta_2|/\ln(B/B_0)]^{1/3}\gamma^{-2/3}. \tag{3.9}$$

The maximum signal qualify factor (Q^2) of a signal with n-QAM modulation format has the following dependence on link parameters

$$Q^2_{max,DUMT} \propto \frac{\text{SNR}_{max,DUMT}}{n} \propto \frac{[\alpha|\beta_2|/\ln(B/B_0)]^{1/3}}{n N_s 10^{2\alpha L_s/3}\gamma^{2/3}}. \tag{3.10}$$

The maximum achievable SE obtained by using PDM-n-QAM for the link is achieved by setting the maximum signal quality factor equal to the quality factor corresponding to the underlying FEC threshold, Q^2_{FEC},

$$\text{SE}_{max,DUMT} = 2\log_2\left(\frac{s\,(\alpha|\beta_2|)^{1/3}}{N_s 10^{2\alpha L_s/3}\gamma^{2/3}Q^2_{FEC}}\right), \tag{3.11}$$

where s is a scaling factor. The above analytical results suggest the following key transmission characteristics in DUMT.

1. To increase spectral efficiency by 2 bit/s/Hz, the dispersion coefficient ($|\beta_2|$) needs to be increased by a factor of 8, the nonlinear coefficient (γ) needs to be decreased by a factor of $2^{1.5}$ (or ~ 2.83), or the number of fiber spans (N_s) needs to be reduced by a factor of 2 [69]. These analytical findings are in reasonable agreement with those obtained through numerical simulation [72].

2. The optimum signal power spectral density is dependent on fiber parameters, but is independent of the transmission distance and the modulation format [66].

3. The optimum signal quality factor Q^2 scales inversely proportional to the transmission distance. The physical reason behind this is that in DUMT, the nonlinear noises (or distortions) generated by the fiber spans are de-correlated [71], so the accumulation of the nonlinear noise during transmission scales the same way as the linear noise resulting from the ASE. Reducing the FEC threshold (Q_{FEC}^2) by 3 dB leads to doubled transmission distance, showing the importance of using powerful FEC codes with higher coding gains, e.g. soft-decision (SD) FEC codes, to enable longer transmission distance.

The above findings from the analytical study are in reasonably good agreement with the results obtained by experiments and numerical simulations for DUMT of superchannels based on O-OFDM and Nyquist-WDM [66]. It is useful to find the SE or modulation format that provides the minimum CPB of a transmission system of a given size. When the system cost is dominated by the optical transponders or the O/E/O conversion, the minimum CPB is achieved when the SEDP of the transmission system is maximized. The effective SEDP of a maximum system length L_{Link} using PDM-n-QAM modulation can be expressed as [73]

$$\text{SEDP} = 2\log_2(n) \cdot \min(L_{\text{Link}}, L_{\text{Achievable}}), \tag{3.12}$$

where $L_{\text{Achievable}}$ is the transmission distance that is achievable by the signal based on fiber parameters of the transmission system and can be written as

$$L_{\text{Achievable}} = N_s L_s = \frac{s(\alpha|\beta_2|)^{1/3} L_s}{n 10^{2\alpha L_s/3} \gamma^{2/3} Q_{\text{FEC}}^2}. \tag{3.13}$$

Figure 3.17 shows the calculated SEDP as a function of SE in systems using superchannels based on PDM-n-QAM modulation and 100-km ULAF spans [63]. The calculation is based on Eqs. (3.11) and (3.12), and the scaling factor s in Eq. (3.11) was obtained using the results reported in Ref. [63]. The FEC BER threshold and the overall spectral overhead (for FEC and multiplexing) are assumed to be 2×10^{-2} and 29%, respectively [63]. The key implications from the results are (i) there is an optimum SE, SE_{opt}, at which the system SEDP is maximized (for lowest CPB); (ii) the SE_{opt} depends on the system size L_{link}; and (iii) for long-haul systems where $L_{\text{link}} > 1000\,\text{km}$, the SE_{opt} is smaller than 10. A useful takeaway message is that for core optical networks with long-haul reach, there is not much value (in terms of CPB) to push the SE beyond 10 by using superchannels modulated with a constellation size larger than PDM-64QAM. Moreover, when the implementation penalty of higher-level QAM and practical system margin are taken into consideration, the

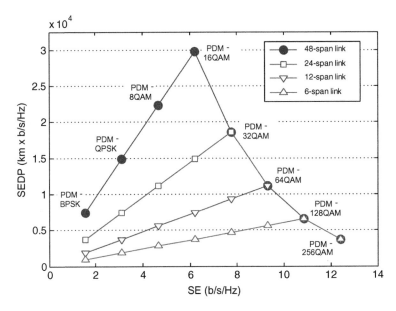

FIGURE 3.17 Calculated SEDP as a function of SE in systems using superchannels and 100-km ULAF spans. The FEC BER threshold and the overall spectral overhead (for FEC and multiplexing) are assumed to be 2×10^{-2} and 29%, respectively.

optimum system SE is expected to be pushed downwards, e.g. to be between 4 and 8 b/s/Hz.

The above analysis is based on several assumptions, such as homogeneous fiber spans, negligible signal-to-noise nonlinear interaction, and ideal transceivers having zero implementation penalty. More general and comprehensive studies on the ultimate capacities of fiber transmission systems can be found in Refs. [67,72,74–76].

3.7 NETWORKING IMPLICATIONS

In modern dense WDM (DWDM) systems, transparent optical routing based on ROADMs is an essential technology that enables network flexibility and efficiency. In current DWDM systems, a fixed channel grid, as defined by the International Telecommunications Union (ITU), is used. The channel spacing is typically 50 GHz, as shown in Figure 3.18a. For future systems with 1-Tb/s and 400-Gb/s superchannels, the optical bandwidth needed for each channel is expected to be more than 50 GHz. This calls for a new DWDM channel allocation scheme where the channel bandwidth is flexible, or adjustable in order to support these high-data-rate superchannels, as illustrated in Figure 3.18b. This new type of DWDM can be called flexible DWDM or elastic DWDM [77–79]. In the recent 448-Gb/s

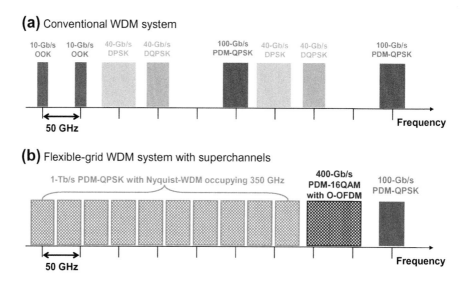

FIGURE 3.18 Illustrations of a conventional WDM channel allocation (a) and a flexible-grid WDM channel allocation suitable for supporting future 1-Tb/s and 400-Gb/s superchannels (b). The data rates labeled are for net information bit rates.

RGI-OFDM demonstration [26], it has been shown that the superchannel is capable of passing five 80-GHz-grid ROADMs with negligible filtering penalty. For maximum system SE, the center of these superchannels may not coincide with the ITU 50-GHz grid because of their nonstandard optical bandwidth requirements. On the other hand, it may cause too much architectural changes to completely abandon the well-established ITU grid. So, a plausible compromise would be to use a finer ITU grid, e.g. the 25-GHz grid or the 12.5-GHz grid, but allow the channel bandwidth to be flexible, e.g. ranging from 50 to 350 GHz, to efficiently support 100-Gb/s, 400-Gb/s, and 1-Tb/s channels [9,80].

Currently, a flexible WDM grid architecture with a bandwidth granularity of 12.5 GHz and a center frequency granularity of 6.25 GHz has been standardized by the International Telecommunication Union (ITU) as the ITU G.694.1 standard [81] and illustrated in Figure 3.19. According to the new ITU standard, the allowed frequency slots have a nominal central frequency (in THz) defined by: $193.1 + n \times 0.00625$, where n is a positive or negative integer including 0, and 0.00625 is the nominal central frequency granularity in THz, and a slot width defined by: $12.5 \times m$, where m is a positive integer and 12.5 is the slot width granularity in GHz. Any combination of frequency slots is allowed as long as no two frequency slots overlap [81]. The reason that the nominal central frequency granularity needs to be 6.25 GHz is to be able to place a slot that has a width of an even multiple of 12.5 GHz next to one with a width of an odd multiple of 12.5 GHz without a spectral gap or guard band. The flexible WDM architecture is also supported by the recent availability of flexible bandwidth WSSs [82].

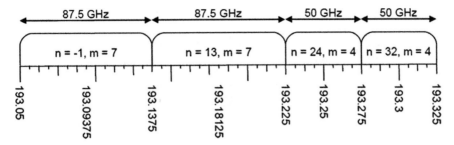

FIGURE 3.19 A flexible grid allocation example in accordance with the ITU G.694.1 standard. (After Ref. [81].)

The benefits of the flexible WDM grid architecture on the networking layer have also been studied recently [83–87]. It was shown that the introduction of the elasticity and adaptation in the optical domain yields significant spectral savings and leads to increased network capacity and enhanced survivability in the 400-Gb/s and 1-Tb/s era [83]. The blocking performance of spectrum efficient superchannels in dynamic flexible grid networks was studied, and the results demonstrate that increased spectral efficiency and flexible superchannel assignment do translate into network efficiency gains [84]. The energy efficiency of flexible-grid WDM networks has also been studied [85–87]. It is expected that the flexible-grid WDM architecture would become a popular choice for future Tb/s-per-channel optical networks.

Another key benefit of superchannels is that it efficiently leverages the advances in large-scale photonic integrated circuits (PICs) to reduce optical circuit complexity, as compared to using discrete optical components, and offers maximum flexibility for an engineering design [88–90]. A compact coherent-receiver front end consisting of an integrated 4×40 arrayed waveguide grating (AWG) array following a polarization-diversity optical hybrid was recently demonstrated and used for complete demodulation of a 1.12-Tb/s multi-carrier superchannel having ten 112-Gb/s PDM-QPSK signals [91]. Figure 3.20 shows the schematic of the compact superchannel coherent receiver. The cyclic feature of the AWG array allows the receiver to receive modulated carriers that are not adjacently spaced. The scheme can be extended to multi-carrier signals whose carriers are closely spaced, e.g. under the O-OFDM condition, for high-spectral-efficiency Tb/s transmission applications [91]. Large-scale photonic circuit integration [92], both at the transmitter and at the receiver, are needed to enable efficient superchannel synthesis and reception. Furthermore, superchannel architectures can advantageously leverage the advances in large-scale application-specific integrated circuits (ASIC) capable of high-speed parallel signal processing [93], especially when PICs are used to increase the intimacy between photonics and electronics [90].

More recently, there has been vigorous research in the field of space division multiplexing (SDM) [94] to explore capacity enhancement using spatial modes/cores in

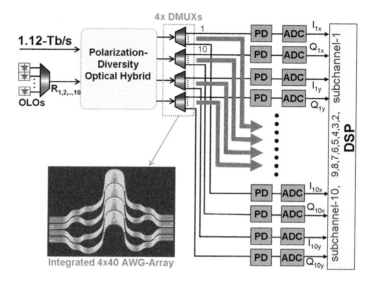

FIGURE 3.20 Schematic of a compact coherent receiver based on large-scale PICs for receiving a 1.12-Tb/s superchannel with 10 112-Gb/s PDM-QPSK signals [91]. Inset: layout of an integrated 4 × 40 AWG array. DMUX: wavelength demultiplexer; PD: photo-detector; OLO: optical local oscillator; DSP: digital signal processor.

fiber as another dimension. In that context, concepts of "spatial superchannels" have been proposed and demonstrated [95], opening opportunities for further expanding the application of superchannels.

3.8 CONCLUSION

We have reviewed the principle and recent progress of superchannel transmission with per-channel data rates of Tb/s and beyond. The generation, detection, nonlinear transmission, and networking of Tb/s-per-channel-class superchannels have been discussed. Superchannel transmission offers several key benefits such as higher serial interface rates, higher spectral efficiency and WDM capacity, increased efficiency in digital signal processing, and better leverage of large-scale PICs and ASICs. Enabled by digital signal processing at both transmitters and receivers, high spectral efficiency modulation based on Nyquist-filtered single-carrier and OFDM, together with multiplexing based on Nyquist-WDM and O-OFDM, can be realized. Software-defined optical transmission to optimize the system throughput depending on the link conditions also becomes feasible. The presence of Tb/s superchannels with different spectral bandwidths in a WDM system has led to the introduction of flexible grid WDM architectures to achieve more efficient utilization of the optical spectrum. It is expected that with the further advances in PICs, ASICs, efficient digital signal processing, flexible-bandwidth ROADM and WSS, and network-layer provisioning of flexible bandwidth WDM channels, Tb/s-per-channel-class super-channel transmission will play an important role in future high-capacity optical fiber networks, offering both increased performance and lowered cost per bit to meet the ever-increasing capacity demands of the Internet era.

Acknowledgments

The authors are especially grateful to their colleagues working in this field in Bell Laboratories, Alcatel-Lucent, for fruitful collaborations and valuable discussions. Among them are L.L. Buhl, E.C. Burrows, R. Essiambre, A.H. Gnauck, and P.J. Winzer. The authors also wish to thank A.R. Chraplyvy, R.W. Tkach, and P.J. Winzer for their support.

Glossary

ADC	analog-to-digital conversion (converter)
ASIC	application-specific integrated circuit
BER	bit error rate
BPSK	binary phase-shift keying
CD	chromatic dispersion
CMA	constant modulus algorithm
DAC	digital-to-analog converter
DSP	digital signal processor (processing)
EDC	electronic dispersion compensation
FEC	forward error correction
IFWM	intrachannel four-wave mixing
ISI	intersymbol interference
MZM	Mach-Zehnder modulator
OFDM	orthogonal frequency-division multiplexing
OLO	optical local oscillator
O-OFDM	optical OFDM
OSNR	optical signal-to-noise ratio
PIC	photonic integrated circuit
PMD	polarization-mode dispersion
QAM	quadrature amplitude modulation
QPSK	quadrature phase-shift keying
ROADM	reconfigurable optical add/drop multiplexer
SNR	signal-to-noise ratio
SPM	self phase modulation
WDM	wavelength-division multiplexing
WSS	wavelength selective switch

References

[1] A.R. Chraplyvy, The coming capacity crunch, in: ECOC Plenary Talk, 2009.

[2] R.W. Tkach, Scaling optical communications for the next decade and beyond, Bell Labs Tech. J. 14 (2010) 3–10.

[3] P.J. Winzer, Beyond 100G Ethernet, IEEE Commun. Mag. 48 (7) (2010) 26–30.

[4] S. Chandrasekhar, X. Liu, B. Zhu, D.W. Peckham, Transmission of a 1.2-Tb/s 24-carrier no-guard-interval coherent OFDM superchannel over 7200-km of ultra-large-area fiber, in: ECOC'09, Post-deadline paper PD2.6.

[5] Y. Ma, Q. Yang, Y. Tang, S. Chen, W. Shieh, 1-Tb/s per channel coherent optical OFDM transmission with subwavelength bandwidth access, in: OFC'09, Post-deadline paper PDPC1.

[6] R. Dischler, F. Buchali, Transmission of 1.2 Tb/s continuous waveband PDM-OFDM-FDM signal with spectral efficiency of 3.3 bit/s/Hz over 400 km of SSMF, in: Proceedings of OFC 2009, March 22–26, 2009, Paper PDPC2.

[7] B. Zhu, X. Liu, S. Chandrasekhar, D.W. Peckham, R. Lingle, Jr., Ultra-long-haul transmission of 1.2-Tb/s multicarrier no-guard interval CO-OFDM superchannel using ultra-large-area fiber, IEEE Photon. Technol. Lett. 22 (2010) 826–828.

[8] S. Chandrasekhar, X. Liu, Experimental investigation on the performance of closely spaced multi-carrier PDM-QPSK with digital coherent detection, Opt. Express 17 (2009) 12350–12361.

[9] S. Chandrasekhar, X. Liu, Terabit superchannels for high spectral efficiency transmission, in: Proceedings of ECOC'10, 2010, Paper Tu.3.C.5.

[10] A. Sano, E. Yamada, H. Masuda, E. Yamazaki, T. Kobayashi, E. Yoshida, Y. Miyamoto, R. Kudo, K. Ishihara, Y. Takatori, No-guard-interval coherent optical OFDM for 100-Gb/s long-haul WDM transmission, J. Lightwave Technol. 27 (2009) 3705–3713.

[11] G. Gavioli et al., Investigation of the impact of ultra-narrow carrier spacing on the transmission of a 10-carrier 1 Tb/s superchannel, in: Proceedings of OFC 2010, March 21–25, 2010, Paper OThD3.

[12] G. Gavioli, E. Torrengo, G. Bosco, A. Carena, S.J. Savory, F. Forghieri, P. Poggiolini, Ultra-narrow-spacing 10-channel 1.12 Tb/sD-WDM long-haul transmission over uncompensated SMF and NZDSF, IEEE Photon. Technol. Lett. 22 (19) (2010) 1419–1421.

[13] E. Torrengo et al., Transoceanic PM-QPSK Terabit superchannel transmission experiments at baud-rate subcarrier spacing, in: Proceedings of ECOC 2010, September 19–23, 2010, Paper We.7.C.2.

[14] J.-X. Cai et al., Transmission of 96 × 100G pre-filtered PDM-RZQPSK channels with 300% spectral efficiency over 10 608 km and 400% spectral efficiency over 4 368 km, in: Proceedings of OFC 2010, March 21–25, 2010, Post-deadline Paper PDPB10.

[15] G. Bosco, A. Carena, V. Curri, P. Poggiolini, F. Forghieri, Performance limits of Nyquist-WDM and CO-OFDM in high-speed PM-QPSK systems, IEEE Photon. Technol. Lett. 22 (2010) 1129–1131.

[16] X. Zhou et al., 1200 km tansmission of 50 GHz spaced, 5 × 504-Gb/s PDM-32-64 hybrid QAM using electrical and optical spectral shaping, in: OFC 2012, Paper OM2A.2.

[17] T. Kobayashi et al., 45.2 Tb/s C-band WDM transmission over 240 km using 538 Gb/s PDM-64QAM single carrier FDM signal with digital pilot tone, in: ECOC 2011, PD Th.13.C.6.

[18] S. Randel, A. Sierra, X. Liu, S. Chandrasekhar, P. Winzer, Study of multicarrier offset-QAM for spectrally efficient coherent optical communications, in: ECOC 2011, Paper Th.11.A.

[19] H. Nyquist, Certain topics in telegraph transmission theory, Trans. Am. Inst. Electr. Eng. 47 (1928) 617–644.

[20] C.E. Shannon, Communication in the presence of noise, Proc. Inst. Radio Engineers 37 (1) (1949) 10–21 Reprint as classic paper in Proc. IEEE 86(2) (1998).

[21] G. Colavolpe, T. Foggi, A. Modenini, A. Piemontese, Faster-than-Nyquist and beyond: how to improve spectral efficiency by accepting interference, Opt. Express 19 (2011) 26600–26609.

[22] A.D. Ellis, F.C.G. Gunning, B. Cuenot, T.C. Healy, E. Pincemin, Towards 1 TbE using coherent WDM, in: Proceedings of OECC/ACOFT 2008, Sydney, Australia, 2008, Paper WeA-1.

[23] J. Yu et al., Generation, transmission and coherent detection of 11.2 Tb/s (112 × 100 Gb/s) single source optical OFDM superchannel, in: OFC'2011, PDPA6.

[24] T.J. Xia et al, 10,000-km enhanced long-haul transmission of 1.15-Tb/s superchannel using SSMF only, in: OECC 2011, PD1.

[25] Y.-K. Huang et al., Terabit/s optical superchannel with flexible modulation format for dynamic distance/route transmission, in: OFC'2012, Paper OM3H.4.

[26] X. Liu, S. Chandrasekhar, B. Zhu, P. Winzer, A. Gnauck, D. Peckham, 448-Gb/s reduced-guard-interval CO-OFDM transmission over 2000 km of ultra-large-area fiber and five 80-GHz-grid ROADMs, J. Lightwave Technol. 29 (4) (2011) 483–490.

[27] D. Hillerkuss et al., 26 Tbit s^{-1} line-rate super-channel transmission utilizing all-optical fast Fourier transform processing, Nat. Photon. 5 (2011) 364–371.

[28] X. Liu et al., Single coherent detection of a 606-Gb/s CO-OFDM signal with 32-QAM subcarrier modulation using 4 × 80-Gsamples/s ADCs, in: ECOC2010 PD2.6.

[29] X. Liu et al., 728-Gb/s CO-OFDM transmission over 800-km ULAF using 64-QAM subcarrier modulation and single-step coherent detection with 4 × 80-Gsamples/s ADCs, in: ACP 2010 PD1.

[30] X. Liu, S. Chandrasekhar, High spectral-efficiency transmission techniques for systems beyond 100 Gb/s, in: 2011 OSA Summer Topical Meeting on Signal Processing in Photonics Communications (SPPCom), 2011, Paper SPMA1.

[31] X. Liu, X. Wei, US patent 7,558,487, filed on September 25, 2005.

[32] Y. Tang, W. Shieh, B. Krongold, DFT-spread OFDM for fiber nonlinearity mitigation, IEEE Photon. Technol. Lett. 22 (2010) 1250.

[33] X. Chen, A. Li, G. Gao, W. Shieh, Experimental demonstration of improved fiber nonlinearity tolerance for unique-word DFT-spread OFDM systems, Opt. Express 19 (2011) 26198–26207.

[34] Q. Yang, Z. He, Z. Yang, S. Yu, X. Yi, W. Shieh, Coherent optical DFT-Spread OFDM transmission using orthogonal band multiplexing, Opt. Express 20 (2012) 2379–2385.

[35] A. Li, X. Chen, G. Gao, W. Shieh, B.S. Krongold, Transmission of 1.63-Tb/s PDM-16QAM unique-word DFT-spread OFDM signal over 1,010-km SSMF, in: Optical Fiber Communication Conference, OSA Technical Digest, Optical Society of America, 2012, Paper OW4C.1.

[36] S. Randel, S. Corteselli, S. Chandrasekhar, A. Sierra, X. Liu, P. Winzer, T. Ellermeyer, J. Lutz, R. Schmid, Generation of 224-Gb/s multicarrier offset-QAM using a real-time transmitter, in: Optical Fiber Communication Conference 2012, Paper OM2H.2.

[37] J. Zhao, A.D. Ellis, Offset-QAM based coherent WDM for spectral efficiency enhancement, Opt. Express 19 (2011) 14617–14631.

[38] J. Zhao, A.D. Ellis, A novel optical fast OFDM with reduced channel spacing equal to half of the symbol rate per carrier, in: Optical Fiber Communication Conference, 2010, Paper OMR1.

[39] E. Giacoumidis, S.K. Ibrahim, J. Zhao, T.M. Tang, A.D. Ellis, I. Tomkos, Experimental and theoretical investigations of intensity-modulation and direct-detection optical fast-OFDM over MMF-links, Photon. Technol. Lett. 24 (2012) 52–54.

[40] A. Sano et al., 30 × 100-Gb/s all-optical OFDM transmission over 1300 km SMF with 10 ROADM nodes, in: Proceedings of ECOC'07, 2007, PD1.7.

[41] Y.-K. Huang et al., Transmission of spectral efficient super-channels using all-optical OFDM and digital coherent receiver technologies, IEEE J. Lightwave Technol. 29 (2011) 3838–3844.

[42] D. Hillerkuss et al., 26 Tbit-1 line-rate super-channel transmission utilizing all-optical fast Fourier transform processing, Nature Photon. 5 (2011) 364–371.

[43] J.S. Chow, J.C. Tu, J.M. Cioffi, A discrete multitone transceiver system for HDSL applications, IEEE J. Sel. Areas Commun. 9 (6) (1991) 895–908.

[44] A.J. Lowery, L. Du, J. Armstrong, Orthogonal frequency division multiplexing for adaptive dispersion compensation in long haul WDM systems, in: Optical Fiber Communication Conference, Anaheim, CA, 2006, Paper PDP39.

[45] W. Shieh, C. Athaudage, Coherent optical orthogonal frequency division multiplexing, Electron. Lett. 42 (2006) 587–589.

[46] W. Shieh, H. Bao, Y. Tang, Coherent optical OFDM: theory and design, Opt. Express 16 (2008) 841–859.

[47] S.L. Jansen, I. Morita, T.C. Schenk, H. Tanaka, Long-haul transmission of 16 × 52.5 Gbits/s polarization-division-multiplexed OFDM enabled by MIMO processing (Invited), J. Opt. Netw. 7 (2008) 173–182.

[48] X. Liu, S. Chandrasekhar, X. Chen, P.J. Winzer, Y. Pan, T.F. Taunay, B. Zhu, M. Fishteyn, M.F. Yan, J.M. Fini, E.M. Monberg, F.V. Dimarcello, 1.12-Tb/s 32-QAM-OFDM superchannel with 8.6-b/s/Hz intrachannel spectral efficiency and space-division multiplexed transmission with 60-b/s/Hz aggregate spectral efficiency, Opt. Express 19 (2011) B958–B964.

[49] X. Liu, F. Buchali, R.W. Tkach, Improving the nonlinear tolerance of polarization-division-multiplexed CO-OFDM in long-haul fiber transmission, J. Lightwave Technol. 27 (2009) 3632–3640.

[50] P. Winzer, A. Gnauck, A. Konczykowska, F. Jorge, J. Dupuy, Penalties from in-band crosstalk for advanced optical modulation formats, in: 37th European Conference and Exposition on Optical Communications, OSA Technical Digest (CD), Optical Society of America, 2011, Paper Tu.5.B.7.

[51] G. Bosco, Spectrally efficient transmission: a comparison between Nyquist-WDM and CO-OFDM approaches, in: 2012 OSA Summer Topical Meeting on Signal Processing in Photonics Communications (SPPCom), Paper SpW3B.1 2012.

[52] A. Sano, T. Kobayashi, S. Yamanaka, A. Matsuura, H. Kawakami, Y. Miyamoto, K. Ishihara, H. Masuda, 102.3-Tb/s (224 × 548-Gb/s) C- and Extended L-band All-Raman transmission over 240 km using PDM-64QAM single carrier FDM with digital pilot tone, in: Optical Fiber Communication Conference, OSA Technical Digest, Optical Society of America, 2012, Paper PDP5C.3.

[53] H.G. Myung, J. Lim, D.J. Goodman, Single carrier FDMA for uplink wireless transmission, IEEE Veh. Technol. Mag. 1 (3) (2006) 30–38.

[54] E. Yamada et al., Novel no-guard-interval PDM CO-OFDM transmission in 4.1 Tb/s (50 × 88.8-Gb/s) DWDM link over 800 km SMF including 50-GHz spaced ROADM nodes, in: Proceedings of OFC'08, 2008, PDP8.

[55] R. Kudo, 111 Gb/s no-guard-interval OFDM using low sampling rate analogue-to-digital converter, in: Proceedings of ECOC'09, 2009, Paper P4.09.

[56] X. Liu et al., Efficient digital coherent detection of a 1.2-Tb/s 24-carrier no-guard-interval CO-OFDM signal by simultaneously detecting multiple carriers per sampling, in: Proceedings of OFC'10, 2010, OWO2.

[57] C.J. Youn, X. Liu, S. Chandrasekhar, Y.-H. Kwon, J.-H. Kim, J.-S. Choe, D.-J. Kim, K.-S. Choi, E.S. Nam, Channel estimation and synchronization for polarization-division multiplexed CO-OFDM using subcarrier/polarization interleaved training symbols, Opt. Express 19 (2011) 16174–16181.

[58] X. Liu, F. Buchali, Intra-symbol frequency-domain averaging based channel estimation for coherent optical OFDM, Opt. Express 16 (2008) 21944–21957.

[59] A. Tolmachev, M. Nazarathy, Filter-bank based efficient transmission of Reduced-Guard-Interval OFDM, Opt. Express 19 (2011) B370–B384.

[60] A. Tolmachev, M. Nazarathy, Real-time-realizable filtered-multi-tone (FMT) modulation for layered-FFT Nyquist WDM spectral shaping, in: ECOC 2011, Paper We.10.P1.65.

[61] K. Yonenaga et al., 100 Gbit/s all-optical OFDM transmission using 4 × 25 Gbit/s optical duobinary signals with phase-controlled optical sub-carriers, in: Proceedings of OFC'08, 2001, JThA48.

[62] J. Yu et al., 30-Tb/s (3 × 12.84-Tb/s) signal transmission over 320 km using PDM 64-QAM modulation, in: OFC 2012, Paper OM2A.4.

[63] X. Liu et al., 3 × 485-Gb/s WDM transmission over 4800 km of ULAF and 12 × 100-GHz WSSs using CO-OFDM and single coherent detection with 80-GS/s ADCs, in: OFC'11, 2011, JThA37.

[64] X. Liu et al., 1.5-Tb/s Guard-banded superchannel transmission over 56 × 100-km (5600-km) ULAF using 30-Gbaud pilot-free OFDM-16QAM signals with 5.75-b/s/Hz net spectral efficiency, in: ECOC'12, Post-dealine Paper Th.3.C.5.

[65] J. Cho, C. Xie, P.J. Winzer, Analysis of soft-decision FEC on non-AWGN channels, Opt. Express 20 (7) (2012) 7915–7928.

[66] G. Bosco, V. Curri, A. Carena, P. Poggiolini, F. Forghieri, On the performance of Nyquist-WDM terabit superchannels based on PM-BPSK, PM-QPSK, PM-8QAM or PM-16QAM subcarriers, J. Lightwave Technol. 29 (1) (2011) 53–61.

[67] R.J. Essiambre, G. Kramer, P.J. Winzer, G.J. Foschini, B. Goebel, Capacity limits of optical fiber networks, J. Lightwave Technol. 28 (4) (2010) 662–701.

[68] A. Splett, C. Kurzke, K. Petermann, Ultimate transmission capacity of amplified optical fiber communication systems taking into account fiber nonlinearities, in: Proceedings of ECOC 1993, 1993, pp. 41–44.

[69] W. Shieh, X. Chen, Information spectral efficiency and launch power density limits due to fiber nonlinearity for coherent optical OFDM system, IEEE Photon. J. 3 (2) (2011) 158–173.

[70] X. Chen, W. Shieh, Closed-form expressions for nonlinear transmission performance of densely spaced coherent optical OFDM systems, Opt. Express 18 (2010) 19039–19054.

[71] F. Vacondio, O. Rival, C. Simonneau, E. Grellier, A. Bononi, L. Lorcy, J.-C. Antona, S. Bigo, On nonlinear distortions of highly dispersive optical coherent systems, Opt. Express 20 (2012) 1022–1032.

[72] R.J. Essiambre, R.W. Tkach, Capacity trends and limits of optical communication networks, Proc. IEEE 100 (2012) 1035–1055.

[73] X. Liu, S. Chandrasekhar, Advanced modulation formats for core networks, in: Proceedings of 2011 OptoElectronics and Communications Conference (OECC), 2011, pp. 399–400.

[74] P.P. Mitra, J.B. Stark, Nonlinear limits to the information capacity of optical fiber communications, Nature 411 (6841) (2001) 1027–1030.

[75] R.-J. Essiambre, G.J. Foschini, G. Kramer, P.J. Winzer, Capacity limits of information transport in fiber-optic networks, Phys. Rev. Lett. 101 (1) (2008) Paper 163901.

[76] D. Rafique, A.D. Ellis, Impact of signal-ASE four-wave mixing on the effectiveness of digital back-propagation in 112 Gb/s PM-QPSK systems, Opt. Express 19 (4) (2011) 3449–3454.

[77] S.V. Kartalopoulos, Elastic bandwidth, IEEE Circuits Dev. Mag. 18 (2002) 8–13.

[78] N. Kataoka, N. Wada, K. Sone, Y. Aoki, H. Miyata, H. Onaka, K. Kitayama, Field trial of data-granularity-flexible reconfigurable OADM with wavelength-packet-selective switch, J. Lightwave Technol. 24 (1) (2006) 88–94.

[79] M. Jinno, H. Takara, B. Kozicki, Y. Tsukishima, Y. Sone, S. Matsuoka, Spectrum-efficient and scalable elastic optical path network: architecture, benefits, and enabling technologies, IEEE Commun. Mag. 47 (2009) 66–75.

[80] S. Chandrasekhar, X. Liu, OFDM based superchannel transmission technology, J. Lightwave Technol., in press.

[81] ITU-T G.694.1, Spectral grids for WDM applications: DWDM frequency grid.

[82] S. Frisken, S. Poole, G. Baxter, Reconfiguration in transparent agile optical networks, Proc. IEEE 100 (5) (2012) 1056–1064.

[83] M. Jinno, H. Takara, Y. Sone, Elastic optical path networking: Enhancing network capacity and disaster survivability toward 1 Tbps era, in: Proceedings of 2011 OptoElectronics and Communications Conference (OECC), 2011, pp. 401–404.

[84] S. Thiagarajan, M. Frankel, D. Boertjes, Spectrum efficient super-channels in dynamic flexible grid networks—a blocking analysis, in: OFC/NFOEC 2011, 2011, Paper OTuI6.

[85] W. Shieh, OFDM for flexible high-speed optical networks, J. Lightwave Technol. 29 (10) (2011) 1560–1577.

[86] A. Klekamp, U. Gebhard, F. Ilchmann, Energy and cost efficiency of adaptive and mixed-line-rate IP over DWDM networks, J. Lightwave Technol. 30 (2) (2012) 215–221.

[87] A. Nag, T. Wang, B. Mukherjee, On spectrum-efficient green optical backbone networks, IEEE Global Telecommunications Conference (GLOBECOM 2011), 2011, pp. 1–5.

[88] S. Chandrasekhar, X. Liu, Enabling components for future high-speed coherent communication Systems, in: Optical Fiber Communication Conference 2011, 2011, Tutorial talk OMU5.

[89] G. Bennett, Superchannels to the rescue!, Lightwave 29 (2) (2012).

[90] T. Koch, III–V and silicon photonic integrated circuit technologies, in: OFC/NFOEC 2012, 2012, Tutorial talk OTh4D.

[91] X. Liu, D.M. Gill, S. Chandrasekhar, L.L. Buhl, M. Earnshaw, M.A. Cappuzzo, L.T. Gomez, Y. Chen, F.P. Klemens, E.C. Burrows, Y.-K. Chen, R.W. Tkach, Multi-carrier coherent receiver based on a shared optical hybrid and a cyclic AWG array for terabit/s optical transmission, IEEE Photon. J. 2 (3) (2010) 330–337.

[92] J. Rahn et al., 250 Gb/s real-time PIC-based super-channel transmission over a gridless 6000 km terrestrial link, in: OFC/NFOEC 2012, 2012, Paper PDP5D.5.

[93] I. Dedic, High-speed CMOS DSP and data converters, in: Optical Fiber Communication Conference, 2011, Paper OTuN1.

[94] P.J. Winzer, Optical networking beyond WDM, IEEE Photon. J. 4 (2) (2012) 647–651.

[95] M.D. Feuer et al., Demonstration of joint DSP receivers for spatial superchannels, Photonics Society Summer Topical, 2012, Paper MC4.2.

Optical Satellite Communications

4

Hamid Hemmati[a] and David Caplan[b]

[a]Jet Propulsion Laboratory (JPL), California Institute of Technology, CA, USA
[b]MIT Lincoln Laboratory, MA, USA

4.1 INTRODUCTION

Satellite-based communication systems of today are increasingly capacity limited. Based on radio frequency or microwave (generically RF) technologies, current state-of-the-art satellite communications (satcom) are often constrained by hardware and spectrum allocation limitations. Consequently, mobile payload sensors and instruments (on satellite or aircraft) are often implemented with restricted capacity to better match that of the host platform. Such limitations are expected to worsen as future interplanetary, deep-space, and manned missions use more sophisticated data-intensive sensors and as the demand for information—and a bigger return on the space-exploration investment—continues to increase.

The potential improvements of optical satellite communications stem from three primary factors: (1) *Reduced diffraction*—short optical carrier wavelength (relative to RF) reduces diffraction (free-space beam spreading), greatly improving efficiency of far-field power delivery using lower size, weight, and power (SWAP) antennas; (2) *Available bandwidth*—virtually unlimited and unregulated optical spectrum enables higher data rates, simplified implementations, and improved sensitivities through advanced modulation and coding; and (3) *Commercially available technologies*—the availability of a mature high-reliability high-performance telecommunication technology base.

4.1.1 Reduced diffraction

The short optical wavelengths $\sim$1000–1600 nm (200–300 THz frequency) used in free-space laser communications (also known as lasercom) are a factor of $\sim$10,000 smaller than state-of-the-art RF systems with $\sim$1 cm wavelengths (32 GHz), resulting in a corresponding reduction in far-field beam divergence $\theta = \lambda/D_{TX}$ [rad] [1,2], where λ is the carrier wavelength and D_{TX} is the transmitter (TX) aperture diameter. For a receiver (RX) aperture diameter of D_{RX}, the power delivered to the far field, (P_{RX}) is given by the antenna loss equation [2]

$$P_{RX} \approx \frac{P_{TX} A_{RX} A_{TX}}{\lambda^2 L^2} = P_{TX} \left(\frac{\pi D_{RX} D_{TX}}{4\lambda L} \right)^2, \tag{4.1}$$

Optical Fiber Telecommunications VIB. http://dx.doi.org/10.1016/B978-0-12-396960-6.00004-3

where the λ-dependent loss is significantly ($\sim 80\,\text{dB}$) larger for the longer RF wavelength. In practice, a system designer typically trades some of this resulting gain for smaller aperture sizes and/or lower-power TXs—which reduces SWAP and cost, and uses the remainder to increase the data rate (by orders of magnitude beyond today's conventional RF systems), or simply increase link margin.

4.1.2 Available bandwidth

For most satellite applications, the many THz of unregulated optical spectrum available over the free-space channel combined with wavelength division multiplexing (WDM) technologies (which enable efficient access to it) practically make the optical channel bandwidth (BW) virtually unlimited. This is due to the TX power requirements needed to span such long (typically power starved) links, which generally limit the data rates (and thus channel BW) to small fractions of the available spectrum. Consequently, lasercom systems have a significant advantage over BW-limited RF systems in terms of data rates they can support and BW-intensive modulation and coding techniques that may be employed to improve system performance.

4.1.3 Commercially available technologies

Driven by the rapid growth of the Internet and corresponding investment in telecom technologies, optical communications has provided unprecedented capacity in modern networks, with single-fiber capacities in excess of many Tb/s [3]. The new and enabling technologies developed during the optical communications revolution of the past decade can also be applied to lasercom. Satellite-based lasercom networks have the potential to surround the planet with flexible and agile wide-band connectivity [4–7] that could extend to the moon [8], Mars [9,10], and beyond—capable of bridging billion kilometer links at hundreds of Mb/s data rates [11–15]. Such capabilities require power-efficient TXs, photon-efficient RXs, and precise laser pointing capabilities, key topics covered in this chapter.

The tremendous telecom investment in optical technologies can be leveraged to accelerate the integration of cutting-edge commercial-off-the-shelf (COTS) telecom technologies—developed for and widely used throughout the telecom industry—into reliable lasercom designs. Furthermore, this cost-effective approach can provide access to components with field-tested heritage as well as Telcordia (formerly known as Belcore) qualification. This is of considerable value since these standards (e.g. [16–19]) often test to mechanical and thermal levels that meet or exceed many of the environmental requirements for space-based platforms.

4.1.4 Lasercom challenges

The primary challenges of lasercom include: precision (micro-radian to sub-micro-radian) laser beam pointing over vast distances spanning from Earth orbit (e.g. 40,000 km to GEO) to interplanetary (e.g. 400,000,000 km to Mars); inefficiency of current lasercom TXs; quantum-noise-limited detection, especially in the presence of additive background during periods of near-Sun pointing; and atmospheric degradation due to attenuation, turbulence, and weather. Future implementations of optical communications systems can be realized with 10–100 times the data rate of RF systems in the same size, mass, and power envelope, or can offer equivalent data rates with a significantly lower-SWAP package [14].

Figure 4.1 shows a generic space-to-ground optical link block diagram consisting of the transmit channel, the optical channel, and the receive channel. The ultimate capacity of the communications link is achieved when each of the three channels is optimized. Channel degradation drivers include laser TX extinction ratio (ER) and waveform distortion; code efficiency; detector jitter; detector blocking; detector efficiency; thermal noise (background noise, dark noise); diffraction-limited and/or seeing-limited point-spread functions (limits the ability to mitigate blocking and dark noise); system efficiency; and system BW. The state of technology for some of these parameters is briefly described below and expanded upon in subsequent sections.

4.1.4.1 Transmit channel

With many THz of optical spectrum available and the ability to access it using WDM technologies, channel BW practically imposes no limitations on lasercom capacity. While powerful forward error correction (FEC) codes within 1 dB of the Shannon limit have been developed (e.g. [20]) and demonstrated for both photon counting [21–25] and coherent RXs [26], realizing them with low-SWAP implementations or at high data rates is still a practical limitation. With limited BW expansion, coherent homodyne PSK has the best theoretical sensitivity (photon efficiency) [27], with 2 bits/photon sensitivity realizable with rate-½ coding (2× BW expansion). In the limit of large BW expansion, orthogonal modulation formats such as pulse-position modulation (PPM) in conjunction with ideal photon-counting detection and strong coding have the best theoretical sensitivity of >10-bits/photon efficiency [27–29]. However, practical limitations discussed later have limited such systems to ~1 bit/photon-class sensitivities. Other practical TX limitations include modulation ER and nonlinear spectral broadening which tend to impact low-duty cycle and high-peak power optical TXs [13].

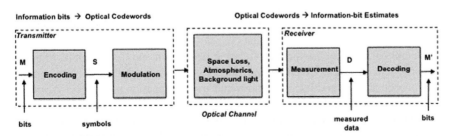

FIGURE 4.1 Block diagram example for a space-to-ground optical communications link.

4.1.4.2 Optical channel

Unlike RF communications, lasercom systems need clear line of sight, and can be completely blocked by obstructions such as attenuating clouds. Atmospheric effects such as fading and scintillation [30] can also lead to significant degradation due to wavefront distortion, which in turn can significantly reduce transmitted power delivered to the target RX and the power coupled into the RX, especially for single-mode RXs. Mitigation techniques include spatial [31,32] and temporal diversity [33,34], and adaptive optics (AO) systems, to remove the impact of channel fluctuations and improve coupling from TX to RX [35]. Background light inevitably accompanies the transmitted signal into the RX and can set severe limitations on achievable link capacity, especially for multi-mode photon-counting RXs. Few spatial modes coupled with matched optical filtering can minimize the impact of background on the communications channel. Dispersive (temporal) pulse spreading, a problem with fiber optics communications, is generally not a limitation in free-space optical communications since there is generally minimal dispersion in the free-space channel.

4.1.4.3 Receive channel

Limitations on the RX include suboptimal detection (photodetector detection efficiency, dark noise, timing jitter, and blocking loss for photon-counters); suboptimal decoding; noise and poorly matched filtering in optical or electronic amplifiers; and capacity efficiency limits. The area of the detector can be decreased (to reduce dark noise) only to the diffraction limit of the telescope for a beam that has traveled through the atmosphere, and the time resolution (minimum slot-width) can be decreased to the RX BW limit. Detector timing jitter is the random delay from the time a photon is incident on a photodetector to the time a photoelectron is detected. Jitter limits the data rate, ability to mitigate dark noise, and ability to decrease the slot-width without incurring losses.

Key characteristics of major space-borne lasercom technology demonstrations are summarized in Table 4.1. Included are downlink and crosslink examples, some in the multi-Gb/s data-rate regime, successfully linking *low-Earth orbit* (LEO) or *geosynchronous Earth orbit* (GEO) to Earth, LEO-to-GEO terminals, and GEO-to-airplanes [4–8]. These demonstrations have matured the lasercom technology to the point that operational lasercom systems from Earth orbit are planned [8]. There are also plans for experimental telecom links from beyond Earth orbit, including lunar (in 2013), GEO-relay ($\sim$2017), and interplanetary (perhaps by 2020) [36].

Notable differences between lasercom at near-Earth and, for example, Mars are primarily related to laser-beam pointing implications, and include a 10,000$\times$ dimmer laser beacon at Mars and 100$\times$ greater round-trip propagation time. This long propagation time precludes handshaking between flight and ground terminals that could be used to infer residual uncompensated host platform vibrations affecting beam pointing. Lasercom to Mars may also have as much as 10$\times$ larger point-ahead angle, and the possibility of simultaneous Sun angles interfering with proper detection of

Table 4.1 Characteristics of past demonstrations and planned lunar laser-com links (best-known values) LCE = Laser Communications Experiment, SILEX = Semiconductor *laser* Intersatellite Link Experiment, LUCE = Laser Utilizing Communications Experiment, LCT = Laser Communications Terminal, LLCD = Lunar Laser Communications Demonstration, NA = Not Available. See also [37,38].

Project Name	LCE	SILEX	LUCE	LCT	LLCD
Host spacecraft	ETS-VI	SPOT-4 ARTIMIS	OICETS (to Grnd)	TerraSARX/ NFIRE	LADEE
Demonstration year	1995	2001	2006	2008	2013
Orbits(s)	Near-GEO	LEO/GEO	LEO	LEO–LEO	Lunar
Downlink/Crosslink data rate (Mb/s)	1	2 (GEO)	50	5,625	622
Link range (1000 km)	~36	36	0.6	5.1	360–400
Flight aperture (cm)	7.5	25/12.5	26	12.5	10
Ground/Crosslink aperture (cm)	1.5	N/A	20–150	12.5	80
Downlink wavelength (nm)	830	819 (GEO)	843–853	1064	1550
Downlink modulation	Intensity	PPM2, GEO	NRZ	BPSK	PPM16
Uplink wavelength (nm)	514	847 (LEO)	797–808 815–825	1064	1568
Downlink power (mW)	14	60–37	100	700	500
Uplink data rate (Mb/s)	1	50 (LEO)	~2	5,625	20
Detection technique	Amplitude	Amplitude	Amplitude	Homodyne	Amplitude
Flt terminal mass (kg)	22	100	140	32	~30
Power consumption (W)	81	150	220	120	50–140
Reference	[4]	[5]	[6]	[7,39]	[8,40–42]

the beam beacon or overwhelming photon-starved detection. Interplanetary laser-com necessitates significantly higher vibration isolation than is typically required for near-Earth links, use of orthogonal modulation techniques such as PPM that can improve the link margin by as much as 6 dB (resulting in the requirement for high peak-to-average TX output power), and use of photon-counting detectors that can enhance link margin by at least another 6 dB.

4.2 LASERCOM LINK BUDGETS

The link budget is one of the first steps in designing a lasercom system and performs many critical functions such as:

- Predicting performance before the link is established;
- Determining if there is sufficient optical power to span the link, typically for a defined worst-case condition;
- Assisting decision making, trade-offs, and assessments, for example, the telescope aperture diameter or transmit laser power requirements needed to support a given bit-error rate (BER);
- Ultimately becoming a central, guiding agreement between all involved.

Figure 4.2 shows a generic block diagram of a laser communications system. To predict the performance of the system, characteristics of each component and assembly must be taken into account in the link budget. Clearly, less than optimized functioning in each block diagram element results in degraded link performance.

The link budget can be derived from a given set of system parameters and their tolerances, and is a straightforward addition and subtraction of gains and losses (when values are converted to decibels). Link budget accounts for an optical system and include gains from antennas (telescopes) and the laser transmitter, losses

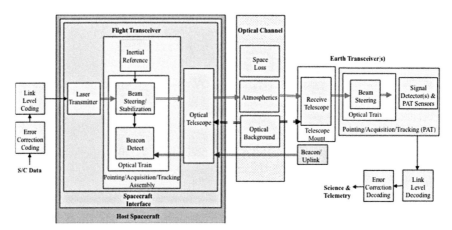

FIGURE 4.2 A generic end-to-end block diagram of a laser communications system showing different elements whose characteristics influence the overall link performance.

in propagation through the transmitting optical path, losses through space or atmosphere, and losses through the receiving optical train to the photodetector.

The primary interest in link budget analysis is the determination of the required power at the opposite terminal's detector. This value is derived from signal-to-noise ratio (SNR) or probability of bit error (PBE) analysis considered in Section 4.1. Any received power over and above the required power constitutes the link margin. Typically, separate link budgets are required for each of the communications, tracking, and acquisition subsystems, uplink, and downlink.

Link budget drivers may be categorized as:

1. Those fixed by the user's specifications (e.g. data rate, distance, and BER);
2. Those fixed by physics (beam divergence and space loss due to $1/r^2$ effect);
3. Variables of hardware design (e.g. laser power or aperture size); and
4. Variables of hardware quality (e.g. quality of optics).

In terms of implementation, major link budget drivers include:

1. The transmitter's beam divergence and the receiver's aperture diameter;
2. Receiver sensitivity: the minimum power the detector needs to receive for a given BER; and
3. Path-length losses, atmospheric attenuation, and atmospheric scintillation losses.

A positive link margin is required to successfully close a communications link. The link margin is simply the ratio of received power to the required power (by the receiver).

The *required signal* power is a complex function of (1) the target BER, (2) implementation losses, and (3) different noise terms.

The *received signal* can be estimated from the following simplified formula:

$$P_{RX} = (P_{TX}G_{TX}L_{TX}L_S) \cdot (L_{RX}L_{ABS}L_{FADE}L_{AO}) \cdot (L_P L_{TRK}) \cdot (G_R L_R L_{IMPL}) \quad (4.2)$$

The first term: $P_{TX}G_{TX}L_{TX}L_S$ represents the transmitter's factors, and corresponds to EIRP (*equivalent isotropic radiated power*) commonly used in RF systems;

The second term: $L_{RX}L_{ABS}L_{FADE}L_{AO}$ describes the medium that the beam goes through: for propagation entirely through vacuum, most terms drop off except for range losses;

The third term: $L_P L_{TRK}$ represents control terms driven by the host platform's stability; and

The last term: $G_R L_R L_{IMPL}$ represents the receiver's terms [43], where P_{RX} is the received power, P_{TX} is the transmit laser power, G_T is the transmitter gain, L_{TX} is the transmitter losses (e.g. optics imperfections); L_P represents the pointing losses induced by the host platform's vibrational frequencies; L_R is the range loss; L_S represents the Strehl loss due to wavefront aberrations stemming from optics imperfections; L_{ABS} is losses due to atmospheric attenuation; L_{FADE} represents losses due to atmospheric scintillation; L_{AO} is the aero-optics effect at the boundary layer of an airplane's window; G_R is receiver gain; L_R represents receiver losses stemming from optics imperfections; and L_{TRK} stems from losses due to tracking errors (e.g. receive platform jitter).

A simplified expansion of the key parameters of this equation by neglecting optical assembly and channel medium losses results in the antenna loss equation, Eq. (4.1).

Loss-term examples include *absorption* at refractors and surface coatings; *refraction* at coated and uncoated surfaces; *scattering* due to surface contamination; *signal splitting*, e.g. for initial acquisition; *wavefront error* due to manufacturing quality; *polarization* mismatches; and *vignetting* due to optical/mechanical segments getting in the way of each other.

Space loss (or beam spreading loss) is caused by diffraction at finite-aperture diameters, and is a dominating loss factor in long-range communications. For range length L, the dimensionless loss term is given by:

$$L_{SL} \approx (\lambda_{TX}/4\pi L)^2. \tag{4.3}$$

The *transmitter module's* efficiency is influenced by a large number of factors, including laser beam quality, transmit aperture area, efficiency of optics assembly (optics and optical/wavefront quality losses), and the type of modulation or coding scheme applied to the transmit beam. Very efficient modulation and coding schemes are now available, such that efficiency can approach within a fraction of 1 dB of (Shannon's) capacity. The *transmit telescope (antenna) gain* is a major factor in link analysis, and for an aperture with diameter D is given by:

$$G_T = 2/\alpha^2 [\exp(\alpha^2) - \exp(\alpha^2 \gamma^2)]^2 \cdot (\pi D/\lambda)^2, \tag{4.4}$$

where $\alpha \approx 1.12 - 1.3\gamma^2 + 2.12\gamma^4$ is the optimal truncation ratio of the transmit laser's Gaussian beam as it leaves the telescope, and $\gamma = d/D$, with d being the diameter of the secondary mirror (obscuration). Assuming that σ is the RMS (root-mean-square) path difference for an aberrated optical element relative to an ideal (perfect) optical design, the Strehl ratio (S) is a measure of the wavefront aberration losses and is given by:

$$S = \exp[-(2\pi\sigma/\lambda)^2]. \tag{4.5}$$

The received signal and background power, and the receiver sensitivity, are influenced by a large set of parameters, including whether it is a direct detection or a coherent detection scheme; receiver aperture area; receiver optics efficiency; receiver pointing losses; receive Strehl ratio; polarization mismatch losses (if pertinent); atmospheric attenuation and scintillation-induced losses; and detector characteristics (e.g. different noise parameters). The receiver's *telescope gain* is expressed as:

$$G_R = (1 - \gamma^2) \cdot (\pi D_R/\lambda)^2. \tag{4.6}$$

Table 4.2 shows the rolled-up version of a downlink budget analysis, where all calculated gain factors are summed and subtracted from the total of the loss parameters [44].

Table 4.2 Rolled-up version of a Mars downlink communications link budget.

Link Parameter	Nominal			Worst		
	dB	Other	Units	dB	Other	Units
Signaling						
PPM Order		16			16	
Laser transmitter						
Average laser power	6.02	4	W	6.02	4	
Transceiver						
Far-field antenna gain	112.98	7.0	μrad	112.98	7.0	μrad
Transmitter efficiency	−5.47			−5.88		
Range and atmosphere						
Space loss	−354.14	0.42	AU	−354.14	0.42	AU
Atmospheric transmission loss	−0.26			−0.91		
Ground receiver						
Receiver gain	147.52	11.8	m	147.52	11.8	m
Receiver efficiency	−5.27			−6.07		
Detection/ Implementation						
Detection/Implementation losses	−5.71			−5.7103		
Mean signal flux		1.22E+08	phe/s		1.42E+08	ph/s
Background, noise, and atmospherics						
Mean background flux	70.8279	1.21E+07	phe/s	76.18	4.15E+07	ph/s
Link performance						
Poisson channel capacity		2.96E+08	b/s		3.45E+08	b/s
Throughput		267.00	Mb/s		267.00	Mb/s
Link margin	4			2		

4.3 LASER BEAM PROPAGATION THROUGH THE ATMOSPHERE

Laser communications beams traversing through the (free space) optical channel will experience a number of atmospheric effects, which individually or jointly can reduce the communications link margin to the point of total signal loss. These effects include:

1. Atmospheric attenuation, which is primarily atomic and molecular absorption of the beam along the path, for example, due to precipitation present in the beam path (see Figure 4.3);
2. Atmospheric scattering, for example, due to clouds, fog, rain, or snow, resulting in contrast reduction at the receiver's photodetector due to loss of a portion of the beam;
3. Atmospheric radiance, resulting in contrast reduction, increased noise, or saturation at the photodetector due to introduction of background light; and
4. Atmospheric turbulence (refractive index fluctuation effects), which has multiple properties including global beam deflection (tilt and jitter), and phase degradation (wavefront aberrations and scintillation).

4.3.1 Atmospheric attenuation

Horizontal propagation atmospheric attenuation over a path length L can be described by transmittance as:

$$T = e^{-\mu(\lambda) \cdot L},\tag{4.7}$$

where $\mu(\lambda)$ is the wavelength-dependent attenuation coefficient with contributions from atomic absorption and scattering, and molecular absorption and scattering, illustrated in Figure 4.3. Note that there are low losses at both 1064 nm and 1550 nm (telecom band) where there are a variety of good laser sources and detectors available.

4.3.2 Atmospheric radiance

Sky radiance is present during both daytime and nighttime laser communications as a major source of noise, that can adversely affect the received signal-to-noise ratio, (most notably) of multimode tracking systems and photon-counting receivers. The power (P_R) at a receiver whose field of view is smaller than the angular extent of an extended source (e.g. Sun, moon, or planets) is given by $P_R = I \, O_{TL} \, A\Omega \, \Delta\lambda$, where I is the source irradiance (W/cm^2 Sr μm), O_{TL} is the receiver's optical transmission loss, A represents the receiver's area (cm^2), Ω is the receiver's field of view (solid angle, Sr), and $\Delta\lambda$ is the bandpass of the filter in the receiver path (μm).

The magnitude of sky radiance is driven by the receiver's line-of-sight Sun angle, the receiver's altitude, aerosol concentration in the atmosphere, and cloud density (both sources of light scattering), and can vary by a factor of 10–100 throughout the day.

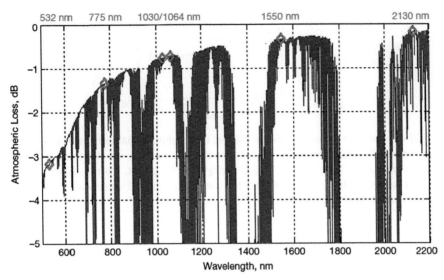

FIGURE 4.3 Representative atmospheric loss in dB vs. wavelength for ground-to-space propagation at 70° zenith angle, from a location 2 km above sea level.

4.3.3 Atmospheric turbulence

Solar heating of Earth's surface and the resulting heat transfer to the adjacent air causes atmospheric inhomogeneity and random temperature variations (gradients) along the beam path. This phenomenon creates atmospheric zones of differing densities that act as lenses, causing the laser beam to deviate from its intended path. As such, the atmosphere behaves akin to a collection of moving prisms of different sizes and refractive indices in the beam path, causing phase and irradiance fluctuations.

Turbulence results in spatial and temporal intensity fluctuations, angle-of-arrival fluctuations, focusing and defocusing, large-scale beam steering, image dancing, beam wander, and speckle. Beam steering is the result of angular deviation of the received beam from the line-of-sight path. This can result in a focused beam that falls outside of the receive telescope, and therefore misses the receiver. Beam spreading (or defocusing) is the result of small-angle scattering, resulting in increased beam divergence and reduced spatial power density at the receiver. Beam scintillation is the small-scale (destructive) interference at the beam's cross-section that can result in fast, millisecond-class fluctuations of the spatial power density at the photodetector. Image dancing is the result of the received beam's angular variations, where the focal spot size vacillates at the RX's photodetector.

The net effect is deep signal fades (e.g. with attenuation >10 dB) at the receiver, lasting from 1 to ~1000 µs, as shown in Figure 4.4. Turbulence depends on temperature (solar loading affected by time of day), receiver site attitude and pressure, wind speed, and relative beam size. Turbulence effects can vary by as much as an order of

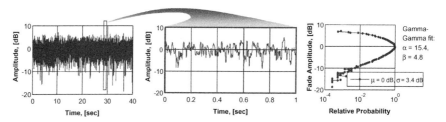

FIGURE 4.4 (Left) 40 second turbulence-induced fade profile, normalized to a 0 dB mean transmission. (Center) Zoom-in one-second view of the fade profile, showing msec-class features of the fade characteristics. (Right) Histogram of the profile, showing large >10 dB but infrequent fades that are well matched to a gamma-gamma distributed profile with $\alpha = 15.4$ and $\beta = 4.8$ [34].

magnitude during a 24-h period. The worst turbulence typically occurs at midday when the temperature is highest, whereas the least-turbulent atmosphere occurs usually just before dawn and dusk.

Turbulent atmosphere can be described by the Kolmogorov model [45,46]. The atmospheric refractive index structure function for distance (r) of two structural points is given by:

$$D_n(r) = \langle n(r_1) - n(r_2)]^2 \rangle = \langle [n(r_1) - n(r_1 + r)]^2 \rangle = C_n^2 r^{2/3}, \qquad (4.8)$$

where the term C_n^2 is the index of refraction structure parameter quantifying the turbulence strength. C_n^2 of $10^{-13}\,\text{m}^{-2/3}$ is considered strong turbulence, while C_n^2 of $10^{-15}\,\text{m}^{-2/3}$ is considered weak turbulence. The value of C_n^2 is dependent on a number of factors, for example, wavelength, geographic location, altitude, weather, and season. The longer the wavelength or the higher the altitude, the lower the C_n^2 magnitude.

The atmospheric coherence length parameter r_0 (also known as Fried's parameter) is a function of C_n^2, for a slant path of range L, and wave number $k\,(= 2\pi\lambda)$, and is given by [15]

$$r_0 = (0.423 K^2 C_n^2 L)^{-3/5}, \qquad (4.9)$$

while for a horizontal path, r_0 is given by:

$$r_0 = (0.158 K^2 C_n^2)^{-3/5}. \qquad (4.10)$$

As an example, the (r_0 dependent) atmospheric seeing effect deteriorates the diffraction limit of a ground telescope of diameter D and focal length f, receiving a laser beam of wavelength λ, from $2.44f\lambda/D$ to $(2.44f\lambda/D)(D/r_0) = 2.44f\lambda/r_0$ [47]. This results in a blurred spot size at the focal point of the telescope, as opposed to a well-defined (diffraction-limited) spot. A consequence of this effect is that without mitigation, the focused spot cannot be efficiently coupled to a single-mode fiber. Straightforward single-mode coupling is desirable since it can greatly simplify the backend receiver configuration, and enable use of single-mode RX technologies such as coherent and EDFA-based preamplified receivers.

The magnitude of beam wander (σ^2, also known as tilt variance, or jitter) for a beam traversing the atmosphere can be calculated from [15]:

$$\sigma^2 = 0.17(D/r_0)^{5/3}(\lambda/D)^2. \qquad (4.11)$$

4.3.4 Turbulence mitigation approaches

As indicated earlier, turbulence causes a number of undesirable effects on both laser beam uplink to the spacecraft and downlink from the spacecraft. For each scenario, a number of turbulence-mitigation effects have been developed as outlined below.

4.3.4.1 Aperture averaging effects on downlink beam

The number of turbulence cells captured by the receiver aperture is approximately given by the area ratio $N = (D/r_0)^2$. As the term "aperture averaging" implies, for a ground aperture receiving multiple independent turbulence cells, the overall signal summation represents an average intensity. For direct detection, the larger the diameter of a ground receiver aperture, the greater the averaging/integration effect of the non-uniform irradiance of the downlink beam [48]. With this effect, as the ground telescope aperture diameter increases, the photodetector at its focal spot experiences a decreased level of signal fluctuations. For coherent detection through the atmosphere, since the wavefront is coherent over a distance r_0, the effective ground aperture diameter is limited by r_0 and increasing the aperture diameter beyond this does not increase the power coupled into the single mode. As described below, the adaptive optics technique can largely mitigate the latter effect.

Figure 4.5a shows the aperture averaging effect on mitigating turbulence as a function of aperture diameter.

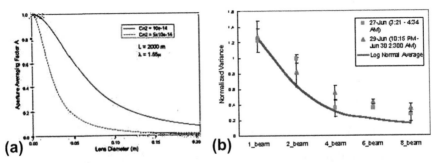

FIGURE 4.5 (a) Effect of aperture averaging as a function of aperture diameter and (b) normalized variance of the flight detector as a function of number of beams.

4.3.4.2 Multibeaming for uplink

Transmitting multiple laser beams that are mutually incoherent relative to each other and separated by a distance greater than r_0 (single to transmitting a single beam) reduces the effects of atmospheric scintillation and beam wander [31,49].

Figure 4.5b shows normalized signal variance as received by the flight detector as a function of the number of beams. Up to four beams, signal variance decreases approximately as $1/N$, where N is the number of beams, and approximately as $1/N^{1/2}$ when N exceeds four. (See also discussion of the LLCD uplink transmitter in Section 4.6.2.)

4.3.4.3 Adaptive optics technique for downlink

The adaptive optics (AO) technique requires, at the minimum, incorporating a wavefront sensor and a deformable mirror to the aft optics section of the ground receiver telescope. A portion of the downlink signal is diverted to the wavefront sensor (WFS) for analysis and feedback to a deformable mirror (DM) to dynamically compensate for the wavefront distorted by the atmosphere [35]. AO also mitigates optical signal fades and precisely couples both ends of the communications link, thereby optimizing the signal-to-noise ratio. A major benefit of AO technology is that it allows coupling of a received laser beam (which has traversed through the atmosphere) into a single-mode fiber. Common limitations of AO include availability of a sufficiently strong signal (turbulence and resulting fades are not too strong); WFS, DM, and processor with high enough bandwidth; and adequate spatial correction bandwidth.

4.3.4.4 Coding of downlink and downlink transmitter

Recently developed block codes are showing considerable promise for mitigating the turbulence effects on both downlink and uplink beams. Similar to the spatial-diversity-based multibeaming approach described above, temporal diversity using a combination of forward error correction coding and interleaving has also been shown to be an effecitve method of mitigating the effects of fading and scintillation. Introducing an interleaver that spreads atmospheric fades over many codewords of an error correction code may mitigate the losses. In this approach each codeword sees a large number of uncorrelated fades (essentially averaging the performance). A portion of the fades is thereby mitigated via interleaver gain. In the limit of an infinitely large interleaver, the fading mitigation gain can be as large as 13 dB. However, in practice, typical gains with a finite interleaver are on the order of a few dBs [33,34,50–53].

4.4 OPTICAL TRANSCEIVERS FOR SPACE APPLICATIONS

It is well known that free-space optical (FSO) communications can provide cost-effective high-speed connectivity suitable for long-haul intersatellite and interplanetary links due to significantly reduced diffraction losses relative to incumbent RF technologies [9,12,13,15,38,54,55]. These benefits are enhanced by improvements in receiver sensitivity, which can extend link distances, or simply reduce size, weight, and power (SWAP). These are often key design drivers in space-based FSO transmitter (TX) subsystems, but are often quite relevant for fiber-optic communications as well. Over the past ~20 years, state-of-the-art RX sensitivities have gone from hundreds of photons/bit (PPB) to the single PPB regime, a notable >20 dB improvement.

The many THz of excess channel bandwidth in the optical regime can be used to further improve receiver sensitivities in FSO links, while still supporting Gb/s class data rates over ~10^6 km distances such as the Earth-moon link (e.g. [8,56]). Power-efficient implementation is often a primary design driver, particularly for the TX power amplifier [57,58], but also for modulation and waveform generation hardware. This is especially true for small nanosatellite (<10 kg) and microsatellite (100 kg) based designs, or for large-channel WDM designs, where power requirements generally grow with the number of wavelengths. Other important metrics include reliability, especially in the space environment, complexity, and performance over the free-space channel, which may include background noise and atmospheric channel effects discussed in Section 4.3. In Section 4.4.1 we will present a brief overview of modulation formats and receiver sensitivities with and without forward error correction (FEC) coding. In Section 4.4.2, we will discuss current transmitter technologies, specifically addressing features of master-oscillator power amplifier transmitters and average-power-limited waveform generation. Section 4.4.3 addresses performance and implementation considerations for photon-counting, coherent, and preamplified receivers. Systems based on these technologies show the most promise for next generation solutions for satellite-based optical communications [59].

4.4.1 Overview of FSO modulation formats and sensitivities

While spectral-efficiency has long been a key design parameter in the telecom industry, FSO links often have excess channel bandwidth available that may be used to improve performance when photon efficiency is the design driver.

Figure 4.6 illustrates the theoretical tradeoff between photon-efficiency and spectral-efficiency, with each capacity curve representing a lower bound on achievable error-free communications for a given RX type, modulation format, and type of decoding (hard or soft decision).

The upper-left-hand portion of Figure 4.6 features good spectral-efficiency (low BW expansion) and poor photon-efficiency, whereas the lower-right-hand region features good photon-efficiency at the cost of large bandwidth-expansion. In the lower right region, photon-counting RXs have the best theoretical sensitivity, with orthogonal modulation achieving −10 dB PPB sensitivity (10 bits/photon) with 1000× bandwidth expansion (in principle). See Pierce [28], Yamamoto and Haus [29], Shapiro et al. [61], and Boroson [60] for further discussion of the ultimate sensitivity limits of photonic communications. In the bandwidth-limited regime, coherent RXs have the best theoretical sensitivity, with coherent homodyne having

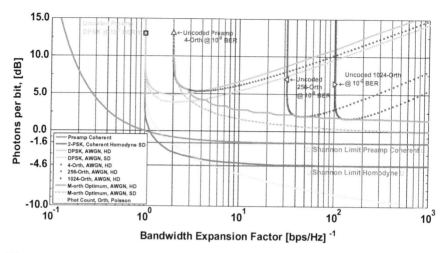

FIGURE 4.6 Theoretical trade between sensitivity (photon-efficiency) as measured in photons per bit in dB, and bandwidth expansion factor (inverse spectral-efficiency). Capacity curves are shown for coherent homodyne and preamplified coherent (equivalent to heterodyne) RXs, preamplified differential phase shift keying (DPSK) and *M*-ary orthogonal modulation (e.g. *M*-PPM or *M*-FSK), for both hard decision (HD) and soft decision (SD) decoding. Adapted from Caplan et al. [56]. See also [27,60].

a 3 dB *theoretical* advantage over heterodyne or preamplified coherent RXs in the limit of large bandwidth expansion. However, in practice, little of this sensitivity advantage is realized due to >2–3 dB additional implementation losses that are difficult to overcome in homodyne RXs (see, e.g. [26] for a detailed breakdown of experimental homodyne RX penalties). DPSK has often been a modulation format of interest for both free-space and fiber applications, since it is reasonably energy- and spectrally-efficient, and easier to implement than coherent RXs, due to relaxed linewidth and phase-locking requirements. For a quantum-limited optically-preamplified DPSK RX with optimal coding (~rate-½, 100% overhead), Shannon-limited performance (with hard decision) approaches 3 photons/bit (PPB) with ~0.5 b/s/Hz efficiency.

The increases in photon efficiency shown in Figure 4.6 via bandwidth expansion may be realized using both modulation and forward error correction coding (FEC). In practice, high-rate high-sensitivity DPSK demonstrations using 33% RS(255, 191), 24.6% turbo product code and 7% enhanced RS(255, 239) FEC have achieved between 7 and 9 PPB RX sensitivities at 2.5, 10, and 40 Gb/s data rates [34,62–64]. While low-overhead algebraic codes like 7% Reed-Solomon are commercially available at rates up to 40 Gb/s, the stronger, more advanced FEC relies on iterative, soft-decision decoding using low-density parity check (LDPC) or turbo codes to facilitate further improvements in receiver sensitivity [62]—and these are not yet readily available with 100% redundancy at Gbit/sec rates [65] (see, e.g. rate-½ SCPPM [8,20] and DVBS-2 [66]).

Examples of modulation formats that improve sensitivity while expanding channel bandwidth include *M*-ary orthogonal modulation formats such as pulse position modulation [21,22,25,33,68,69], frequency shift keying (FSK) [69–71], and hybrids described in more detail in [13].

Such *M*-ary orthogonal formats, with *M* orthogonal symbols convey

$$k = \log_2(M) \tag{4.12}$$

bits per symbol, and thus, the information content per photon and corresponding RX sensitivity improves as *k* grows. This comes with a cost of increased channel BW given by

$$\mathrm{BW_{Factor}} = \frac{M}{\log_2(M)} = \frac{M}{k}. \tag{4.13}$$

M-FSK modulation shown in Figure 4.7 is of particular interest relative to more commonly used *M*-PPM (Figure 4.8) because it has the same theoretical sensitivity advantage over binary formats but has lower peak power and may be implemented with slower, lower BW electronics—and is compatible with high-speed operation over both free-space and fiber-optic channels. For example, while *M*-PPM and *M*-FSK symbols illustrated below have the same average power, information per symbol, channel BW, and sensitivity, PPM requires *M/k* greater electrical BW relative to the data rate, whereas *M*-FSK may be implemented with reduced 1/*k* electrical BW requirements [70]. Moreover, the high peak-to-average power ratio of *M*-PPM can lead to nonlinear impairments [72] in TX power amplifiers [13,73] (used in FSO applications) and in long-distance fiber applications.

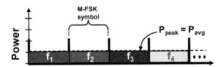

FIGURE 4.7 *M*-ary frequency shift keying (*M*-FSK) symbols consist of *M*-orthogonal frequencies (or wavelengths), shown here for *M*=4.

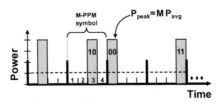

FIGURE 4.8 *M*-ary pulse position modulation (*M*-PPM) symbols consist of *M*-orthogonal time-slots (in this case *M*=4), with information conveyed by the photon arrival position.

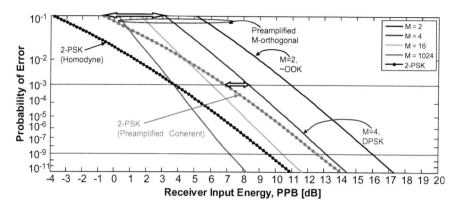

FIGURE 4.9 BER curves for homodyne and preamplified-coherent 2-PSK, preamplified DPSK and *M*-ary orthogonal modulation for *M*=2, 4, 16, and 1024. Adapted from Caplan [13]. See also [11].

Figure 4.9 shows the theoretical quantum-limited bit-error-rate (BER) performance for 2-PSK and preamplified *M*-orthogonal performance. For the *M*-ary orthogonal case, as symbol alphabet size *M* is increased from 2 to 1024, so does the information per symbol ($\log_2 M$) and sensitivity, though with diminishing returns as *M* grows. At *M*=1024 (10-bits/symbol) there is ~8.8 dB sensitivity improvement over the binary case (1-bit/symbol) at 10^{-9} BER.

Large *M*-orthogonal modulation also comes with additional implementation complexities. For example, as *M* grows large, there is increased need for good modulation extinction ratio (ER) in order to efficiently direct TX power to the intended symbol [67,74]. While fiber optic links may work well with ~8 dB ER (see, e.g. [76]) for binary on-off-keying (OOK) modulation, *M*-orthogonal modulation requires

$$|ER| > M + 15 \text{ dB}, \qquad (4.14)$$

which limits TX power loss to ~0.2 dB [13,53]. For example, 32-orthogonal requires ER>30 dB (see, e.g. [40,73]) and 1024-orthogonal requires >45 dB to avoid significant TX power penalty.

Note that the relative performance of preamplified coherent 2-PSK versus preamplified orthogonal formats varies significantly depending on the error rate of interest and corresponding FEC coding employed. For example, at high error rates, e.g. 10^{-1} BER (which is near the error-free cutoff point for rate-½ SCPPM and DVB-S2 FEC), the theoretical benefits of coherent 2-PSK are more pronounced, having ~1 dB better sensitivity than preamplified 1024-orthogonal. However, at 10^{-3} BER (near the cutoff of enhanced RS(255, 239) FEC), the 1024-orthogonal has a 3 dB advantage, albeit, requiring 100× more channel bandwidth. For an FSO application where sensitivity is the primary design driver, channel bandwidth is abundant, and 7% FEC is readily available, large alphabet orthogonal modulation may be an attractive option.

Similarly, in comparing preamplified DPSK to preamplified coherent 2-PSK, from a sensitivity versus complexity trade space, 2-PSK provides little benefit (only $\sim$1 dB) over DPSK at 10^{-3} BER (with conventional 7% coding) at a cost of increased line-width restrictions and optical phase-locking complexities. However, the trade looks more attractive for coherent PSK at 10^{-1} BER, where the sensitivity improvement grows to more than $\sim$3 dB.

As noted earlier, with the availability of strong rate-½ FEC that can support high data rates of interest, the dramatic improvements in coherent RX sensitivity (at high error rates) lead to the best bandwidth-limited sensitivities—with potential to provide significant benefit to both fiber-optic and free-space applications.

4.4.2 Transmitter technologies

Average power limited master oscillator power amplifier (MOPA) transmitters are the most widely used in high-rate high-sensitivity optical communications (see Figure 4.10). The MOPA transmitter can generate high-power high-fidelity low-jitter high-extinction pulse-shaped waveforms, which are especially desirable for achieving optimized performance in preamplified RXs [76,77].

MOPA transmitters are typically implemented using a CW distributed feedback (DFB) laser, followed by data and pulse-carving modulators, and a high-efficiency power amplifier (e.g. [57,58]). For EDFA and YDFA based-MOPAs, state-of-the-art wall-plug power amplifier efficiency is $\sim$13% [57] and 21% [58], respectively, and represent an upper bound limit on TX power efficiency when power amplifier requirements dominate. However, other elements in the TX (e.g. laser and modulator) often take an appreciable amount of electrical power, further reducing the overall TX efficiency. This effect is especially noticeable for lower power TXs (e.g. <$\sim$500 mW output). CW DFB lasers, for example, may take 2–5 W electrical power for current and temperature control (depending on the ambient temperature range), and external data modulators (and carvers) may require 3–6 W for driver-amplifiers and bias control. These power requirements are substantial relative to the 5 W needed for a $\sim$10% efficient 0.5 W EDFA (that is capable of supporting 622 Mbps data transmission over the 400,000 km Earth-moon link [8,40]).

Use of filtered direct-drive modulation of DFB laser sources [78–80] can potentially eliminate the external modulator and have demonstrated impressive performance in long-haul fiber-based demonstrations with low ($\sim$8 dB) modulation extinction waveforms [76]. However, as noted in Section 4.1, large $|ER| > M + 15$ dB

FIGURE 4.10 Master oscillator power amplifier (MOPA) transmitter, consisting of a modulated laser source followed by a power amplifier.

is needed to support *M*-orthogonal formats to avoid TX power losses and maintain near-theoretical RX performance [11,13]. In this regard, direct-drive techniques can also be used as a low-power means of generating the high-extinction waveforms (also needed for quantum-limited RX performance [59]), a capability that can offer significant SWAP savings in multi-wavelength WDM TXs that can support scalable high-sensitivity orthogonal formats such as *M*-FSK, *M*-PPM and hybrids [13]. For example, a filtered-direct-drive TX in [70] demonstrated high-fidelity 8-FSK waveform generation (with ER > 30 dB) using only ~20 mW drive power per wavelength, a power reduction of two orders of magnitude relative to conventional externally-modulated TXs.

4.4.2.1 *Average power limited (APL) multi-rate transceivers (TRXs)*

For power-starved free-space applications, multi-rate capability provides valuable architectural flexibility by extending the operational range of receiver power levels. This enables bandwidth on demand when conditions are favorable, fall-back modes, and operation over a variety of link conditions (e.g. distance and channel state) and TRX designs (e.g. range of TX power × TX aperture area × RX aperture area) [34,81,82]. For example, NASA's anticipated demand for deep-space communications at rates from 1s to 1000s of Mbps [10,83] could potentially be satisfied with a single TRX platform.

The average-power-limited properties mentioned earlier can support multi-rate capabilities for rates >~MHz-class using, for example, EDFA- or YDFA-based TXs with power-off time $\tau_{off} < {\sim}1\,\mu sec \ll \tau_s$, the upper state lifetime [85,86]. The performance and efficiency of such APL TXs is largely independent of modulation format, pulse-shape, and duty cycle. This enables aggressive pulse shaping and variable-duty cycle TX waveforms that can support high-sensitivity multi-rate communications with simple RX implementations [13].

As illustrated in Figure 4.11 for variable-duty cycle 2-PPM waveforms at the input to an APL optical amplifier, the primary difference between the multi-rate waveforms is the duty cycle (% on time); the same pulse-shape is used at all rates, but the off-time between pulses is increased as the rate is reduced. In this manner, a multi-rate RX may be implemented with (optical and electrical) filtering matched to a single pulse shape, and operate with near-theoretical performance over a wide range of rates [81,82]. Such characteristics are particularly well suited for *M*-ary PPM [67], since RX optical power (W) and RX sensitivity (photon/bit) both improve with increasing *M*.

The output of an APL TX with variable-duty cycle 2-PPM input waveforms is shown in Figure 4.12. As the duty cycle and data rate are reduced, the peak power increases so that the average power is maintained (a condition maintained if |ER| ≫ |DC|).

Pulse-carving may be used to further improve RX performance, generating, for example, Gaussian-like waveforms that are more tolerant to precise filter matching [76,77]. Moreover, burst-mode waveforms may be used to implement multi-rate DPSK waveforms that can be demodulated using a single, easy-to-implement

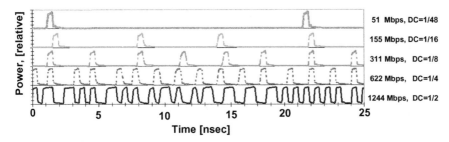

FIGURE 4.11 Measured peak-power-limited (PPL) variable-duty-cycle 2-PPM waveforms at the input to a high-gain average-power-limited EDFA. DC = duty cycle = % on time [81].

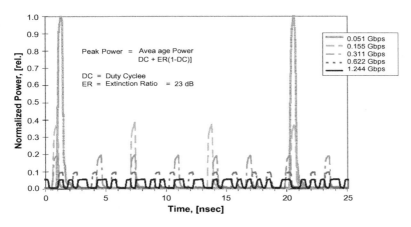

FIGURE 4.12 Measured variable-duty-cycle 2-PPM waveforms at the output of an average-power-limited high-gain EDFA [81]. Reprinted with permission (@ 1999 IEEE).

interferometer. For example, multi-rate burst-mode DPSK communications was demonstrated with near-theoretical performance at rates from 3.38 Gb/s to 3.2 Mb/s (a ~1000× range in channel rate) using a 3.38 GHz delay-line interferometer. Performance at all rates was ~1 dB from theory, also demonstrating that this variable-duty cycle burst-mode approach can accommodate low-rate (D)PSK without incurring linewidth penalties. In addition, extensions of this approach are applicable to higher-order differentially-encoded modulation formats such as QDPSK, as well as coherent M-PSK and QAM formats if the fade-free burst window is long enough to determine or phase-lock to the incoming optical signal [34].

Unlike terrestrial fiber-based communications which are often limited by channel nonlinearities, nonlinear limitations in FSO applications are restricted to the TX power amplifier or the short meter-class output fiber leading to the telescope. Such effects typically occur at ~400 W peak power levels, depending on the amplifier design, output fiber core size, and length. Other restrictions on variable-duty-cycle multi-rate communications include sufficiently large ER > ~30 dB—to avoid TX power loss, and short off-times τ_{off} < ~1 μsec—to avoid power loss due to amplifier gain dynamics [13,53,73].

4.4.3 Receiver technologies and performance

Photon-counting receiver architectures have been proposed [9,86] and realized at Mb/s [22] and ~Gb/s [25,87] data rates with the best demonstrated coded RX sensitivities near 1 photon/bit (PPB), with potential for improvement to multiple-bits/photon sensitivities and greatly simplified processing due to the digital nature of the counting process. However, these photon-counting RXs are presently limited to power-starved links with little or no background noise and <~Gb/s rates due to dark-count and reset-time constraints, which preclude their use with the sun in the field of view. Also, the suitability of key detector technologies, e.g. Geiger-mode avalanche photodiode (GM-APD) arrays [88] and superconducting single photon counting detectors SSPDs [89] and SSPD arrays [88,91] for use in the space environment, has not yet been established. Si-based GM-APDs, for example, have shown sensitivity to radiation [92], and SSPDs require cooling to cryogenic temperatures, requiring a significant overhead in SWAP. For ground-based receivers, however, where reliability and receiver SWAP are not a driving limitation, these technologies offer significant potential. Unlike coherent and preamplified RXs, these photon-counting detectors can receive multiple spatial modes (though this also increases background noise) and can efficiently collect signals distorted by the atmospheric channel without the need for wavefront correction. Furthermore, the net detection area can be efficiently scaled in distributed telescope arrays without the need to build large and costly telescopes [59,93].

As noted in Section 4.4.1, coherent homodyne RXs using PSK are spectrally efficient and provide among the best theoretical RX sensitivity in the bandwidth-limited regime when combined with strong ~rate-½ coding [27,60,94]. However, in practice the high-sensitivity potential of homodyne PSK has not yet been realized, in part due to challenging component and laser linewidth requirements, difficulties associated with phase-locking the local oscillator, and other excess losses (e.g. detection efficiency, coupling losses, thermal noise limitations). Until recently, the best reported *uncoded* PSK performance in the Gbit/s regime is ~35 PPB at ~6 Gb/s and ~80 PPB at ~8 Gb/s [55], providing little performance benefit over optically preamplified DPSK, e.g. [34,62,64,76], which is an easier to implement WDM-scalable approach. Nevertheless, coherent homodyne systems at 1064 nm wavelengths have been fielded in LEO orbit, successfully demonstrating LEO-to-LEO communications at a 5.6 Gbit/s data rate [7]. From an RX sensitivity perspective, however, the advantages of coherent PSK become much more apparent when combined with strong FEC [26]. Furthermore, preamplified coherent RXs have overcome most of the excess losses noted above, enabling them to approach theoretical performance as demonstrated by several recent high-sensitivity demonstrations [68,71,95].

Preamplified RXs can directly leverage the field-tested heritage of telecom-type 1.55 μm technologies that are compatible with operation in the space environment and have demonstrated the best sensitivities at high data rates (>~Gb/s) of 25–30 PPB for uncoded DPSK and 7–10 PPB with coding [62,64,76]. Moreover, it has been shown that nearly theoretical RX performance (within ~1 dB of quantum-limited performance for single-polarization RXs) may be achieved using a combination of pulse-shaping at the TX and matched optical filtering in the RX [34,76,77] for direct detection-based RXs. With the use of M-ary orthogonal modulation formats, direct-detection (non-coherent) RX sensitivities can approach the 1–2 PPB regime albeit with substantial bandwidth expansion [56,67]. In this case, the use of hybrid modulation formats including frequency, position, and polarization modulation along with WDM rate scaling can be used to access the many THz available in EDFAs, and overcome electronic bandwidth limitations [13,70]. More recently, as noted above, similar near-theoretical performance has also been demonstrated with preamplified coherent RXs using DSP-based (offline-processed) demodulation and coding [68,71,95].

Aside from sensitivity considerations, DSP-based coherent RXs clearly have notable auxiliary benefits in terrestrial fiber applications such as improving symbol constellation density, and electronically correcting for dispersion and PMD effects, in addition to extending the reach between repeaters [68,96]. However, since these auxiliary benefits are of little value for satellite-based FSO communications, it is unclear if the sensitivity improvements alone (especially when used with low-over-head FEC at 10^{-3} BERs) outweigh the additional costs (primarily additional SWAP and complexity) needed to implement DSP-based coherent RXs relative to traditional optical phase lock loop (OPLL)-based implementations. Recently, a lower-complexity alternative coherent RX design demonstrated an optically injection locked OPLL, with tracking rates suitable to accommodate LEO-ground Doppler wavelength shifts [97]. However, it remains to be seen if this promising approach can work in the low-SNR high-BER regime that is more attractive for FSO applications. These considerations and others, such as the ability to operate over the fading channel described in Section 4.3, are likely to have an important impact on future developments in this trade space.

A summary of reported high-sensitivity demonstrations versus data rate is shown in Figure 4.13, highlighting performance for a variety of RX types, modulation formats, and coding options discussed in this section, expanding upon previous summaries reported in 2007 [13] and 2009 [53]. Since then, there been have been several notable developments. On the photon-counting RX front, some of the most sensitive high-speed detectors (SSPDs [90,98,99]) have matured sufficiently to be incorporated in the LLCD downlink RX, supporting satellite-based moon-to-Earth lasercom at 622 Mb/s data rates [8,100]. Also, previously reported offline-processed ~single-PPB demodulation and decoding results at 781 Mb/s [25] have now progressed to real-time decoded implementations, though only at 155 Mb/s [101,102]. This illustrates both the predictive benefits of offline-processed proof-of-concept demonstrations and the practical challenges of making them operate in real-time.

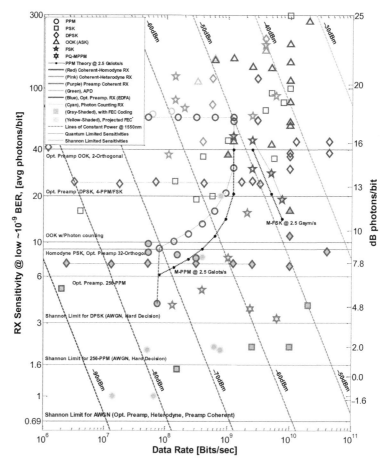

FIGURE 4.13 Summary of reported high-sensitivity optical communication demonstrations at low ~10^{-9} BER (or error-free operation) versus data rate. Modulation type is indicated by marker type, for instance circles represent PPM. RX type is indicated by color: coherent is red (dark red = homodyne and pink = heterodyne), purple is preamplified coherent, dark blue indicates optically preamplified, (usually single polarization) and light blue indicates photon-counting RXs. Coded demonstration results are shaded gray. *Yellow-shaded demonstrations have reported sensitivity at an FEC threshold (e.g. 10^{-3} BER).* In an attempt to maintain a more level comparison with results reported at ~10^{-9} BER, these have been projected to ~10^{-9} by factoring in the assumed FEC overhead (e.g. 7% at 10^{-3} BER, 19.25% at 1.5×10^{-2}). Also shown are the quantum-limited sensitivities for various modulation formats (uncoded), *M*-PPM at 2.5 Gslots/s, and FSK at 2.5 Gsym/s, as well as Shannon-limited sensitivities and lines of constant power (dashed diagonal). Adapted from Alexander [11] and Caplan [13]. See also [26,34,51,53,68,70,71,95]. (For interpretation of the references to color in this figure legend, the reader is referred to the web version of this book.)

Similarly, preamplified *M*-orthogonal demonstrations have evolved from offline demodulation and decoding [67] to real-time multi-rate implementations [51]. There have also been notable improvements to coherent RXs—realizing more of the theoretical potential—with near-capacity-approaching ∼single-PPB sensitivities (with offline decoding) [26,95a,b] to ∼few-PPB sensitivities (with offline demodulation and assumed decoding) [68,71,95].

Its important to note that the summary in Figure 4.13 includes both real-time hardware-based results as well as demonstrations achieved using offline processing as part of the demodulation and/or decoding. The real-time demonstrations provide a useful indication of what can be done with potential for use in near-term applications. It should be recognized that the increasing prevalence of offline processing (which can obscure real implementation complexities), makes fair comparisons more difficult; while offline processing is useful for proof-of-concept demonstrations, there may be a sizeable gap to practically useful implementations.

All told, the continuing improvements in RX sensitivities that leverage commercially available telecom-type WDM-scalable hardware make the prospects of cost-effective satellite-based communications of the future quite promising.

4.5 SPACE TERMINAL

4.5.1 Space environment

During launch, the flight terminal experiences vibrations and mechanical shock caused by propulsion and fairing opening [103]. The vibro-acoustic launch environment results from the propagation of sound waves generated during launch by high-velocity engine exhaust gases, resonant motion of internal engine components, and the vehicle's motion through the atmosphere. This environment can place severe stress on the flight lasercom terminal and affect its reliability. Table 4.3 provides an example of the frequency range of vibrations experienced by flight payloads [104,105], and Figure 4.14, shows typical acoustic noise frequency distribution for different launch vehicles [106].

Once in space, the flight terminal experiences a variety of thermal conditions, including extreme temperatures and thermal gradients; vacuum that eliminates convection cooling as an option; ionizing radiation from galactic cosmic rays, high-energy particles from the solar wind, and magnetically trapped charge particles; an

Table 4.3 Typical space-payload vibration levels.

Vibration Type	Frequency (Hz)	Response
Random vibration	10–2000	$0.01–1\,g^2/Hz$
Acoustic vibration	20–10,000	100–140 dB
Pyrotechnic shock	30–10,000	5–10,000 g

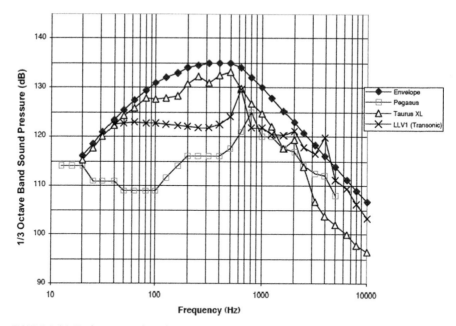

FIGURE 4.14 Typical acoustic noise requirement for different launch vehicles and an envelope of acoustic flight data.

electromagnetic field, including static electricity, discharges, emission/susceptibility, coronal arcing, and plasma bias [107,108]. The temperature of an Earth-orbiting satellite may vary significantly during an orbit. For example, when the satellite is facing the Sun, the temperature of the outside surfaces can exceed 140 °C, while in shadow, the temperature of the outside surfaces can reach −150 °C. Depending on the orbit, such temperature cycling may repeat several times per day. An important context for these temperature extremes is that typically most electronics are designed for maximum operating temperatures of −30 °C to +70 °C.

Approximate values for examples of space environment thermal loading conditions are shown in Table 4.4 [15].

Table 4.4 Thermal loading examples in a space environment.

Parameter	Value	Notes
Solar energy flux	1.3–1.45 kW/m^2	Above the atmosphere
Reflected solar energy	0.45 kW/m^2	Albedo (maximum)
Outgoing infrared	0.1–0.27 kW/m^2	
Cold sky temperature	2.7 K	
Temperature extremes	−140 °C to +150 °C	Uncontrolled
	−10 °C to +60 °C	Controlled

Earth's Van Allen belts contain electrons and protons in Earth's magnetic field in the range of 1000–6000 km, and are seen as low as 100 km. Galactic cosmic rays contain high-energy (hundreds of MeV) ions. Solar flares contain highly intense protons, electrons, and some heavy ion bursts.

Damage to laser communications components and assemblies by particles and rays includes that caused by trapped protons and electrons, solar protons from flares, and cosmic rays degrading materials and electronic components, causing single-event effects in semiconductor components; solar radiation (X-ray to infrared) causing materials degradation; plasma from magnetic substorms causing spacecraft charging; and atomic oxygen eroding exposed surfaces.

Single-event effects (SEE) include *single-event upset* (SEU), which is a soft failure and not harmful to hardware (e.g. bit flips); *single-event latchup* (SEL), which is functional and a hard failure, typically requiring power reset to the device; *single-event functionality interrupt* (SEFI), which is a recoverable failure; and *single-event dielectric rupture* (SEDR), which is a hard failure. *Displacement damage* due to protons and neutrons can cause bulk damage to semiconductor diode lasers, photodetectors, and analog devices. In the latter case, energetic particles displace an atom from its crystal lattice, permanently altering the device's electrical properties. *Total ionizing dose* (TID) stemming from electrons, protons, or gamma-ray bombardment causes gradual and cumulative parametric and sudden degradation, leading to malfunction.

Table 4.5 summarizes the radiation effects of particles and rays on laser communications components (opto-electronics and fiber-optics).

Table 4.5 Effects of ionizing radiation, rays and particles on laser communications components [15] (p. 336). White areas indicate little to no occurrence.

Type	Radiation Effect			
	Trapped Electrons	Trapped Protons	Solar Protons	Cosmic Rays
Glass (and fiber) darkening	■	■		
Diode laser power degradation	■	■		
Detector noise increase	■	■		
Total dosage damage	■	■	■	
Scintillation	■	■		
Single-event upsets		■	■	■
Displacement damage		■	■	■
Bremsstrahlung	■			

4.5.2 Pointing, acquisition, and tracking

Due to the small divergence (narrow beamwidth) of the transmitted laser beam, spatial acquisition and pointing is one of the most critical aspects of a lasercom system's operation. Inaccurate laser beam pointing results in large signal fades at the receiving site, severely degrading system performance. Lasercom links from airborne and spaceborne platforms typically require beam pointing on the order of 1/10th of the beamwidth [109]. This means pointing to the receiver's location with accuracy of a few microradians (μrad) to sub-μrad levels.

To accomplish this challenge, and to minimize signal loss due to pointing errors, a dedicated pointing-control assembly is typically required. The pointing, acquisition, and tracking assembly must have the capability to acquire the incoming beacon signal in the presence of large platform attitude uncertainty, and then accurately deliver (point) the transmit laser beam to its intended target.

The requirement for sub-μrad pointing is further compounded by the host spacecraft platform's angular microvibrations, which are generally the dominant source of beam mispointing. Disturbances that can affect the optical path include guidance and control activities, such as momentum dumping, reaction wheels, thrusters, and slewing to acquire a target, and retrorocket firings; and vibrations caused by motors, thermal gradients, sudden temperature change, and thermal snap. The extent of these microvibrations' frequencies spans from less than 0.1 Hz to hundreds of Hz, and their amplitude is typically much larger than the transmit beamwidth. A combination of passive, active, a hybrid of passive and active disturbance-rejection mechanisms, or inertially referenced stabilized platforms is often applied between the flight terminal and the host platform to largely prevent the platform disturbances from reaching the flight terminal (see, e.g. [42,110,111]).

High-frequency disturbances (vibrational modes) resulting from platform jitter that affect the lasercom telescope may lead to boresight errors (angular variations in the alignment of telescope TX and RX axes). Jitter parameters are normally root-sum-squared to achieve an estimate of pointing accuracy. Low-bandwidth disturbances are called bias and are added linearly. Contributions to overall pointing errors include the optical tracking sensor's noise equivalent angle, inertial sensor noise (if applicable), residual uncompensated platform motion, line-of-sight motion between the two terminals, errors in alignment between the two terminals, and errors in the computation of point-ahead angle.

Imprecise beam pointing results in major degradation of link margin and system performance due to large signal fades at the receive terminal. The power link budget typically allocates a pointing loss to account for statistical pointing-induced fades (PIF) and mispoint angles. The 3-sigma total pointing accuracy is defined as (3 × jitter+ bias). The angular width of a 1000-nm laser beam transmitted through a 30-cm-diameter telescope is approximately 3.3 μrad. In this example, a 2-dB pointing loss allocation with a 1-% PIF requires a total (3-sigma) pointing accuracy of 2 μrad. The allocated pointing loss is then divided into bias and jitter mispoint errors. A portion of these losses may be mitigated via coding and interleaving (see Section 4.3.4.4), whereby an interleaver spreads a PIF over many code-words to effectively deliver the signal to the ground station [112]. The flight lasercom terminal must be capable of tracking the receiving station such that the tracking jitter error is less than approximately 10% of the transmit beamwidth (roughly 0.33 μrad). Such a pointing requirement is orders of magnitude more precise than those for radio frequency telecommunications from space. The primary contributors to a flight lasercom transceiver's total pointing error are errors in position determination, inaccuracies in boresight calibration, and residual tracking errors not compensated for by the pointing subsystem control loop (due to inadequate compensation of the host platform vibrations by the pointing control loop).

Typically, the higher the bandwidth of the control loop, the higher the degree of platform jitter compensation and the lower the residual pointing error. An inertial navigation assembly, joined intimately with the flight transceiver, for tracking the ground-based receiver site from space mitigates the need for the flight subsystem to require image centroiding update rates as high as several kHz. Compensation of lower-frequency disturbances to maintain accurate beam pointing requires data from a reliable beacon signal and a dedicated pointing control subsystem. Examples of a suitable beacon include a laser originating from Earth, Sun-illuminated Earth images in the visible or infrared region of the spectrum, and precision star tracking.

Various PAT strategies may be required to successfully address the beam-pointing requirements of a given link. For precision beam pointing, typically a strong beacon signal (e.g. a laser beam from the vicinity of the receive terminal) is required to aid with removing the host platform's residual vibration, and maintaining receiver position within a few μrad. A coarse-pointing mechanism (e.g. a gimbal) may be required to perform large-angle motions. A fine-pointing mechanism (e.g. a fast-steering mirror) performs high-frequency and high-acceleration portions of pointing motions, and its main function is to compensate for and reject base motion and force and torque disturbances.

4.5.3 Flight optomechanics assembly

The optomechanics assembly includes the transmit and receive aperture(s) (telescope), aft optics behind the telescope to route the incoming and outgoing beams, acquisition and tracking sensors and associated actuators for beam pointing, a mechanical structure that integrates the optics, a coarse-pointing mechanism that compensates for the platform's attitude, and inertial sensors or isolation mechanisms.

Generic goals for the optomechanics assembly often include: maximizing aperture diameter while at the same time minimizing the mass and swept volume; having an optical system capable of supporting multiple wavelengths; transmit and receive channels with adequate field of view while maintaining good wavefront quality over the entire field of view; excellent background light rejection, e.g. via narrowband filtering with wide-enough field of view; and a structure with high fundamental frequency and high temporal and thermal stability.

Numerous parameters influence the design of the lasercom flight terminal (see, e.g. [4,8,10,113]). Examples are aperture diameter, background light rejection and baffling, beam divergence, bidirectional reflectance distribution function, effective focal length, field of view of the detectors, laser beam throughput, mass and volume, polarized light accommodation, redundancy, radiation hardness of the optics, secondary mirror obscuration, stray light control, Strehl ratio of the optics, transmit/receive isolation, and unidirectional or bidirectional link. The acquisition and tracking field of view and receiver detector field of view, as well as the transmit laser beam's divergence, are typically the major design drivers. There is a choice of common or separate transmit and receive aperture. A common aperture lends itself to precise alignment between the paths (as required for almost all long-range lasercom systems). The common aperture is often challenged with achieving good $>\sim 100\,dB$ transmit/receive isolation in order to prevent, for example, a $\sim 30\,dBm$ TX from blinding an RX with $-60\,dBm$ sensitivity. This can be accomplished via a good optical design (e.g. anti-reflective coatings, and clean, low-backscatter optics, beams stops, etc.) in combination with spectral- and polarization-based isolation schemes.

Stray (undesired) light is often problematic when receiving a faint beacon or communications signal. Possible sources are sunlight (background light) scattered by atmospheric constituents and falling within the optical filter bandpass, directly viewed or scattered background light entering the telescope, ghost reflections, and diffraction. Atmospheric background light and stray light are largely mitigated by the optical design. Stray-light control techniques include effective optical system baffling design; inclusion of filed stop(s) to prevent sunlight from entering the aft optics region; inclusion of a Lyot stop to block diffracted light from the secondary mirror and its support structure; and adherence to a specific cleanliness level for all surfaces, which entails specification of BRDF (bidirectional reflectance distribution function) for telescope optics and its verification after assembly and before launch.

Requirements on the structure used for the optomechanics assembly include ease of fabrication to achieve the required wavefront quality; thermal stability (both soaked and gradient temperatures); structural integrity, including fundamental resonance frequency stiffness, dynamic response, and fracture threshold; and temporal stability.

4.6 GROUND TERMINAL

4.6.1 Ground terminal—telescope and optomechanics assembly

The function of the ground terminal's optomechanics assembly is to collect downlink signal light from the flight terminal, filter out the background light, detect the photons, synchronize to the signal, and monitor atmospheric channel conditions. Representative elements and functions of the ground terminal are illustrated in Figure 4.15.

Major design drivers include the requirements to provide large antenna gain (collecting area), and daytime operation at low Sun angles and at low signal-to-background ratio conditions.

The downlink receive telescope's effective optical quality is degraded by several atmospheric turbulence-induced effects (discussed earlier in Section 4.3) due to turbulence just above the telescope entrance aperture. These effects include

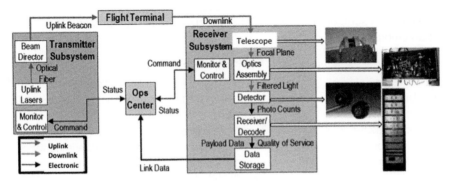

FIGURE 4.15 A block diagram of the ground terminal and its constituent assemblies.

scintillation, in the form of random fluctuations of laser power arriving at the telescope's focal plane; wavefront tip/tilt, in the form of angle-of-arrival fluctuations; and blurred focal image due to wavefront distortions. Clearly, higher-altitude ground station locations reduce the turbulence effects. But, depending on the link requirements, that is not always possible. The aperture-averaging effect, which becomes more pronounced as the aperture diameter increases, helps reduce the angle-of-arrival fluctuations as well as the focal-spot blurring. Blurred spot size is driven by the D/r_0 ratio, where D is the telescope diameter and r_0 is the atmospheric coherence length (or the Fired parameter).

The largest ground telescope aperture diameters are often desired, restricted only by cost and allocated volume. Both the primary mirror's surface quality and the telescope gimbal's blind-pointing accuracy and repeatability are major cost drivers. For near-Earth lasercom applications, often the specified telescope has diffraction-limited quality with about $\lambda/10$ optics, resulting in a point-spread function of 1 μrad at the telescope focal plane (without atmosphere). For direct-detected interplanetary lasercom applications, a near-diffraction-limited telescope with about 1–λ optics and corresponding 10-μrad point-spread function at the telescope focal plane will suffice (the focal image is blurred into a focal spot governed by the atmosphere).

Considering the importance of atmospheric effects on the downlink and uplink signals, use of a suite of atmospheric analysis instruments with the optical ground station will not only help to mitigate certain atmospheric effects, but can answer certain link disruptions, if they were to occur. Table 4.6 lists some of the atmospheric characterization instruments, the instruments that could be used, and the atmospheric parameters that they can measure.

The required telescope aperture may be formed by a single primary mirror (monolithic or segmented), or by an array of smaller-diameter telescopes followed by electrical signal multiplexing (see, e.g., [93]). Each option has advantages and disadvantages. The required number of opto-electronic receivers increases in proportion to the number of telescopes and may become the driving factor in the overall cost of an array of fully equipped optical receiving telescopes.

Table 4.6 A suite of atmospheric characterization instruments and their characteristics.

Instrument	Measure Values
Cloud imager	Cloud coverage and cloud optical depth
Star DIMM[a]	Nighttime r_0 (fried parameter)
Sun scintillometer	Daytime r_0
Ground scintillometer	Turbulence profile
Polaris photometer	Atmospheric losses, by monitoring Polaris irradiance
Sun photometer	Atmospheric transmission losses and sky background light
Weather station	Temperature, wind speed, pressure, humidity, particle profile

[a] *DIMM = Differential Image Motion Monitor.*

Proper stray-light rejection is often a major consideration for large-diameter optics. Stray-light performance due to telescope mirror surface contamination can be quantified via its bidirectional reflectance distribution function (BRDF) [114]. The BRDF is the ratio of directional scattered surface radiance ($W\,m^{-2}\,sr^{-1}$) to the value of incident surface irradiance ($W\,m^{-2}$). A typical lasercom telescope BRDF specification value is $0.003\ sr^{-1}$, not an overly challenging requirement.

Protection of the ground telescope from contamination and solar heating is another challenge. Narrow-bandpass optical filters on a membrane substrate with a single-piece diameter of nearly 2 m have recently been developed to meet this challenge [115]. The membrane substrate was selected to minimize overall weight and to minimize wavefront distortions that might be introduced by thicker substrates.

4.6.2 Ground terminal—uplink transmitter

Ground-based lasercom terminals often have the luxury of not being SWAP-constrained, and therefore, can support large optical powers as needed to satisfy the link requirements. This also enables operation in a relatively benign environment. An example of this is the LLCD uplink TX, which resides in a temperature-controlled trailer shown in Figure 4.16. This system uses a MOPA TX design and variable-duty cycle 4-PPM modulation described in Section 4.4, to deliver 10 and 20 Mbps data rates over the 400,000 km Earth-moon link [73].

In order to overcome the large propagation losses and mitigate atmospheric impairments, the uplink TX delivers up to an aggregate of 40 W average power to the uplink telescope using four parallel 10 W slices as shown in Figure 4.17. This parallel approach enables power scaling by simply increasing the number of TX slices and corresponding apertures. It also enables spatial-diversity, which when

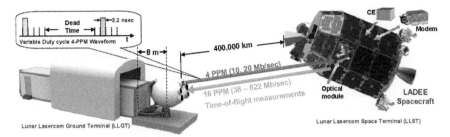

FIGURE 4.16 LLCD overview highlighting the bidirectional high-rate laser communications over the 400,000 km link between the Earth-based LLGT and the LLST onboard the **LADEE** spacecraft orbiting the moon [8]. The uplink uses a variable-duty cycle 4-PPM waveform [81,82,116] with rate-½ SCPPM forward error correction [20] at ~10 and ~20 Mbps. The UL TX resides in the control-room trailer operating nominally at room temperature and communicates with the LLCD space terminal Uplink RX [40] on board the **LADEE** spacecraft [119]. The 4 × 10 W UL-TX waveforms are delivered to four 15 cm aperture uplink telescopes [41] through <8-m single mode fiber. CE = control electronics [74].

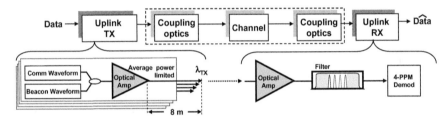

FIGURE 4.17 The LLCD Uplink TX is comprised of 4 × 10 W slices that together generate up to 40 W of WDM-based spatial diversity signals that mitigate the impact of atmospheric fading on both acquisition and communications. Each TX slice generates both Comm and Beacon (spatial tracking) signals through a shared EDFA-based high-power optical amplifier (HPA), with the Comm-to-Beacon power ratio adjusted to support the TX mode of operation, e.g. spatial acquisition or communications and fine-track [74].

combined with strong rate-½ SCPPM forward error correction coding [20] and interleaving, provides an effective solution for far-field atmospheric-turbulence-induced fading as discussed in Section 4.3. Precision wavelength control is used to separate the four TX-channels by $\Delta\lambda_{CH} = 1\,GHz \pm 0.25\,GHz(\pm 2\,pm)$, which is wide enough to prevent inter-channel coherent interference at the space-terminal RX, but packed tightly enough to fit within the narrow uplink RX optical filter [40,73,116].

Each transmitter slice generates a 1 kHz modulated beacon signal (Beacon) at 1568 nm used for spatial acquisition and tracking [42], and a multi-rate 4-PPM communication signal (Comm) at 1558 nm using variable 1:16 and 1:32 duty cycles at a fixed 311 MHz slot rate. By varying the dead-time (and hence the duty cycle), the TX supports 38.9 Mb/s and 19.4 Mb/s channel rates, respectively [40,116]. Both Beacon and Comm are amplified via a shared EDFA-based high-power optical amplifier (HPA). With the rate-½ FEC, the uplink data rates are ~20 Mb/s and ~10 Mb/s. The power ratio of Comm and Beacon is adjustable—depending on the mode of operation—permitting, for example, an increase of power to the Beacon prior to spatial acquisition, and subsequent shifting of power to Comm once signal-based fine-tracking has commenced [73].

4.6.3 Ground terminal—acquisition, pointing, and tracking assembly

The function of the pointing, acquisition, and tracking (PAT) assembly in the ground terminal is to acquire the link within a prespecified time period from the start of telescope blind pointing, once the incident downlink irradiance level is satisfied.

The narrow beamwidths of uplink and downlink beams necessitate a cooperative target and a dedicated PAT assembly within the ground terminal. The cooperative target supplies a laser beacon to help with co-alignment of the two terminals' optical axes. Typically, two sensors are used within the ground terminal: one for acquisition, the other for tracking. The acquisition sensor supplies the fast-steering mirror (FSM) with data to direct the downlink signal to within the field of view of the tracking sensor (often a quadrant detector). The FSM also helps to compensate for atmospheric-induced tip-tilt motion of the downlink beam. Proper PAT implementation requires adequately high bandwidth dynamics with real-time software response.

To acquire the flight terminal, either the ground terminal raster/spiral scans its uplink beacon towards the spacecraft's known location, or the flight terminal scans its beam in the direction of the ground terminal's location and across the uncertainty cone (driven by the attitude-control knowledge). Alternatively, the ground (or flight terminal) transmits a wide beam and gradually narrows its beamwidth as handshaking between the two terminals is accomplished successfully. Therefore, accurate knowledge of spacecraft ephemeris and GPS location of the ground station has to be supplied to the opposite terminal in order for each terminal to properly accomplish its pointing objective. For interplanetary communications, the round-trip times are too long for handshaking to occur. Therefore, both terminals are in effect open loop pointing, but rely on data-fusion from the laser beacon sensor, Earth image sensor, and inertial navigation sensor(s).

Due to cross-velocity between the spacecraft and Earth, point-ahead has to be implemented on both the downlink and the uplink beams. The magnitude of point-ahead is as much as 50 μrad for Earth orbiters and as much as 500 μrad for interplanetary communications. The point-ahead angle is typically implemented via a second fine-pointing mirror within the PAT assembly, with adequate pointing accuracy.

4.7 LIST OF ACRONYMS

APD	Avalanche photo diode
APL	Average power limited
BER	Bit error rate
BRDF	Bidirectional reflectance distribution function
BW	Bandwidth
CW	Continuous wave
DC	Duty cycle
DFB	Distributed feedback
DPSK	Differential phase shift keying
EDFA	Erbium doped fiber amplifier
ER	Extinction ratio
FEC	Forward error correction
FSO	Free space optical
FM	Frequency modulation
FSK	Frequency shift keying
FSM	Fast steering mirror
GEO	Geosynchronous Earth orbit (~40,000 km)
HD	Hard decision (FEC decoding)
HPA	High-power optical amplifier
GM-APD	Gieger-mode avalanche photo diode
LEO	Low Earth orbit (~200–2000 km)
LLCD	Lunar laser communications demonstration
MOPA	Master oscillator power amplifier
PAT	Pointing, acquisition, and tracking
PIF	Pointing-induced fades
PPB	Photon/bit
PPM	Pulse position modulation
PSK	Phase shift keying
RMS	Root mean square
RX	Receiver
SD	Soft decision (FEC decoding)
SSPD	Superconducting single photon counting detectors
SWAP	Size, weight, and power
TX	Transmitter
TRX	Transceiver
YDFA	Ytterbium doped fiber amplifier
WDM	Wavelength division multiplexing

Acknowledgments

Portions of the work described here were carried out at the Jet Propulsion Laboratory, California Institute of Technology, under contract with the National Aeronautics and Space Administration.

Portions of this work were sponsored by the National Aeronautic and Space Administration under Air Force Contract #FA8721-05-C-0002. Opinions, interpretations, conclusions, and recommendations are those of the authors and are not necessarily endorsed by the United States Government.

References

[1] R.M. Gagliardi, S. Karp, Optical Communication, second ed., John Wiley & Sons, Inc., New York, 1995.

[2] E. Hecht, Optics, second ed., Addison Wesley, 1987.

[3] R. Essiambre, G. Kramer, P.J. Winzer, G.J. Foschini, B. Goebel, Capacity limits of optical fiber networks, J. Lightwave Technol. 28 (2010) 662–701.

[4] Y. Arimoto, M. Toyoshima, M. Toyoda, T. Takahashi, M. Shikatani, K. Araki, Preliminary result on laser communication experiment using Engineering test satellite-VI (ETS-VI), in: Proceedings of the SPIE, 1995.

[5] T.T. Nielsen, G. Oppenhaeuser, In-orbit test result of an operational optical intersatellite link between ARTEMIS and SPOT4 SILEX, Proc. SPIE 4635 (April 2002).

[6] T. Jono, Y. Takayama, K. Shiratama, I. Mase, B. Demelenne, Z. Sodnik, A. Bird, M. Toyoshima, H. Kunimori, D. Giggenbach, N. Perlot, M. Knapek, K. Arai, Overview of the inter-orbit and the orbit-to-ground lasercom demonstration by OICETS, Proc. SPIE 6457 (2007).

[7] B. Smutny, H. Kaempfner, G. Muehlnikel, U. Sterr, B. Wandernoth, F. Heine, U. Hildebrand, D. Dallmann, M. Reinhardt, A. Freier, R. Lange, K. Boehmer, T. Feldhaus, J. Mueller, A. Weichert, P. Greulich, S. Seel, 5.6 Gbps optical intersatellite communication link, Proc. SPIE 7199 (2009).

[8] D.M. Boroson, B.S. Robinson, D.A. Burianek, D.V. Murphy, A. Biswas, Overview and status of the lunar laser communication demonstration, Proc. SPIE 8246 (2012).

[9] B.L. Edwards et al., Overview of the Mars laser communications demonstration project, in: American Institute of Aeronautics and Astronautics, Space 2003 Conference and Exposition, 2003.

[10] S.A. Townes et al., The Mars laser communication demonstration, in: IEEE Aerospace Conference, 2004.

[11] S.B. Alexander, Optical Communication Receiver Design, SPIE Opt. Eng. Press, Bellingham, Washington, USA, 1997.

[12] H. Hemmati, Deep Space Optical Communications, JPL Deep Space Communications and Navigation Series, Wiley, 2006.

[13] D.O. Caplan, Laser communication transmitter and receiver design, J. Opt. Fiber Commun. Res. 4 (2007) 225–362.

[14] H. Hemmati, D.M. Boroson, A. Biswas, Prospects for improvement of interplanetary laser communication data rates by 30 dB, Proc. IEEE 95 (October 2007) 2082–2092.

[15] H. Hemmati, Near-Earth Laser Communications, CRC Press, Florida, 2009 (Chapter 8).

[16] Generic reliability assurance requirements for optoelectronic devices used in telecommunications equipment, Telcordia GR-468-CORE, No. 2, December 2002.

[17] Generic requirements for optical fiber amplifiers, GR-1312-CORE, No. 2, December 1996.

[18] Generic reliability assurance requirements for fiber optic branching components, GR-1221-CORE, No. 1, December 1994.

[19] Test methods and procedures for microelectronics, Military Standard MIL-STD-883C, August 1983.

[20] B.E. Moision, J. Hamkins, Coded modulation for the deep-space optical channel: serially concatenated pulse-position modulation, Interplanetary Network Progress Report 42-161, 2005.

[21] B.S. Robinson, D.O. Caplan, M.L. Stevens, R.J. Barron, E.A. Dauler, S.A. Hamilton, 1.5-photons/bit photon-counting optical communications using Geiger-Mode Avalanche photodiodes, in: IEEE LEOS Summer Topical Meetings, 2005.

[22] P.I. Hopman, P.W. Boettcher, L.M. Candell, J.B. Glettler, R. Shoup, G. Zogbi, An end-to-end demonstration of a receiver array based free-space photon counting communications link, SPIE 6304 (September 2006).

[23] R. Shoup, Hardware implementation of a high-throughput 64-PPM serial concatenated turbo decoder, SPIE (Optical Information Systems IV) 6311 (2006).

[24] M.K. Cheng, M.A. Nakashima, B.E. Moision, J. Hamkins, Optimizations of a hardware decoder for deep-space optical communications, IEEE Trans. Circuits Syst. 55 (March 2008) 644–658.

[25] B.S. Robinson, A.J. Kerman, E.A. Dauler, R.J. Barron, D.O. Caplan, M.L. Stevens, J.J. Carney, S.A. Hamilton, J.K.W. Yang, K.K. Berggren, 781-Mbit/s photon-counting optical communications using superconducting NbN-nanowire detectors, Opt. Lett. 31 (2006) 444–446.

[26] M.L. Stevens, D.O. Caplan, B.S. Robinson, D.M. Boroson, A.L. Kachelmyer, Optical homodyne PSK demonstration of 1.5 photons per bit at 156 Mbps with rate-½ turbo coding, Opt. Express (June 2008).

[27] D.M. Boroson, A survey of technology-driven capacity limits for free-space laser communications, in: SPIE Proceedings, 2007.

[28] J.R. Pierce, Optical channels: practical limits with photon counting, IEEE Trans. Commun. COM-26 (1978) 1819–1821.

[29] Y. Yamamoto, H.A. Haus, Preparation, measurement and information capacity of optical quantum states, Rev. Mod. Phys. 58 (October 1986) 1001–1020.

[30] L.C. Andrews, A.J. Phillips, Laser Beam Propagation through Random Media, second ed., SPIE Press, Bellingham, WA, 2005.

[31] I.I. Kim, H. Hakakha, P. Adhikari, E. Korevaar, A.K. Majumdar, Scintillation reduction using multiple transmitters, in: Proceedings of the SPIE, 1997.

[32] R. Parenti, R.J. Sasiela, L.C. Andrews, R.L. Phillips, Modeling the PDF for the irradiance of an uplink beam in the presence of beam wander, in: SPIE, 2006.

[33] J.A. Mendenhall et. al., Design of an optical photon counting array receiver system for deep space communications, in: Proceedings of the IEEE, 2007.

[34] D.O. Caplan, H. Rao, J.P. Wang, D.M. Boroson, J.J. Carney, A.S. Fletcher, S.A. Hamilton, R. Kochhar, R.J. Magliocco, R. Murphy, M. Norvig, B.S. Robinson, R.T. Schulein, N.W. Spellmeyer, Ultra-wide-range multi-rate DPSK laser communications, in: CLEO, Paper CPDA8, 2010.

[35] R.K. Tyson, Bit-error rate for free-space adaptive optics laser communications, J. Opt. Soc. Am. A 19 (2002) 753–758.

[36] A. Biswas, H. Hemmati, S. Piazzolla, B. Moision, K. Birnbaum, K. Quirk, Deep-space optical terminal (DOT)—system engineering, JPL's IPN Progress, Report, vols. 42–183, November 2010.

[37] Z. Sodnik, B. Furch, H. Lutz, Optical intersatellite communication, IEEE J. Quant. Electron. 16 (September/October 2010) 1051–1057.

[38] F. Fidler, M. Knapek, J. Horwath, W.R. Leeb, Optical communications for high altitude platforms, IEEE J. Sel. Top. Quant. Electron. 16 (September/October 2010) 1058–1070.

[39] B. Smutny, R. Lange, H. Kampfner, D. Dallmann, G. Muhlnikel, M. Reinhardt, K. Saucke, U. Sterr, B. Wandernoth, In-orbit verification of optical inter-satellite communication links based on homodyne BPSK, in: Proc. SPIE (Free-Space Laser Communication Technologies XX), 2008.

[40] S. Constantine, L.E. Elgin, M.L. Stevens, J.A. Greco, K. Aquino, D.D. Alves, B.S. Robinson, Design of a high-speed space modem for the lunar laser communications demonstration, Proc. SPIE 7923 (January 2010).

[41] D. Fitzgerald, Design of a transportable ground telescope array for the LLCD, Proc. SPIE 7923 (2011).

[42] J.W. Burnside, S.D. Conrad, C.E. DeVoe, A.D. Pillsbury, Design of an inertially-stabilized telescope for the LLCD, in: Proceedings of the SPIE, 2011.

[43] H. Hemmati, Deep Space Optical Communiations, John Wiley & Sons, Interscience, New Jersey, 2006 (Chapter 2).

[44] A. Biswas, H. Hemmati, S. Piazzolla, B. Moision, K. Birnbaum, K. Quirk, Deep-space optical terminals (DOT) systems engineering, The Interplanetary Network Progress Report, vols. 42–183, November 15, 2010, pp. 1–38.

[45] R.J. Noll, Zernike polynomials and atmospheric turbulence, J. Opt. Soc. Am. 66 (March 1976) 207–2011.

[46] D.L. Fried, Statistics of a geometric representation of wavefront distortion, J. Opt. Soc. Am. 55 (1965) 1427–1435.

[47] L.C. Andrews, R. Phillips, C.Y. Hopen, Laser Beam Scintillation with Applications, SPIE Press, Bellingham, WA, 2001.

[48] J.H. Chumside, Aperture averaging of optical scintillations in the turbulent atmosphere, Appl. Opt. 15 (1991) 198201994.

[49] K.E. Wilson, J.R. Lesh, Overview of the ground-to-orbit lasercom demonstration (GOLD), in: Proceedings of the SPIE (Free-Space Laser Communication Technologies IX), 1997, pp. 1–9.

[50] J.D. Moores, F.G. Walther, J.A. Greco, S. Michael, W.E. Wilcox, A.M. Volpicelli, R.J. Magliocco, S.R. Henion, Architecture overview and data summary of a 5.4 km free-space laser communications experiment, in: Proceedings of the SPIE, 2009.

[51] N.W. Spellmeyer, S.L. Bernstein, D.M. Boroson, D.O. Caplan, A.S. Fletcher, S.A. Hamilton, R.J. Murphy, M. Norvig, H.G. Rao, B.S. Robinson, S.J. Savage, R.T. Schulein, M.L. Stevens, J.P. Wang, Demonstration of multi-rate thresholded preamplified 16-ary pulse-position-modulation, in: OFC, 2010.

[52] X. Zhu, J.M. Kahn, Communication techniques and coding for atmospheric turbulence channels, in: A. Majumdar, J.C. Ricklin (Eds.), Free-Space Laser Communications: Principles and Advances, Springer, 2007.

[53] D.O. Caplan, M.L. Stevens, B.S. Robinson, Free-space laser communications: global communications and beyond, European Conference on Optical Communications (ECOC), Elsevier, Austria, 2009.

[54] V.W.S. Chan, Optical satellite networks, J. Lightwave Technol. 21 (November 2003) 2811–2827.

[55] R. Lange, B. Smutny, Highly-coherent optical terminal design status and outlook, in: LEOS, 2005.

[56] D.O. Caplan, B.S. Robinson, M.L. Stevens, D.M. Boroson, S.A. Hamilton, High-rate photon-efficient laser communications with near single photon/bit receiver sensitivities, in: OFC, 2006.

[57] P. Wysocki, T. Wood, A. Grant, D. Holcomb, K. Chang, M. Santo, L. Braun, G. Johnson, High reliability 49 dB gain, 13W PM fiber amplifier at 1550 nm with 30 dB PER and record efficiency, in: OFC, Paper PDP17, 2006.

[58] N.W. Spellmeyer, D.O. Caplan, B.S. Robinson, D. Sandberg, M.L. Stevens, M.M. Willis, D.V. Gapontsev, N.S. Platonov, A. Yusim, A high-efficiency ytterbium-doped fiber amplifier designed for interplanetary laser communications, in: OFC, 2007.

[59] D.O. Caplan, M.L. Stevens, B.S. Robinson, S. Constantine, D.M. Boroson, Ultra-long distance free space laser communications, in: CLEO, 2007.

[60] D.M. Boroson, Optical Communications: A Compendium of Signal Formats, Receiver Architectures, Analysis Mathematics, and Performance Comparisons, *in MIT Lincoln Laboratory Report,* 2005.

[61] J.H. Shapiro, S. Guha, B.I. Erkmen, Ultimate channel capacity of free-space optical communications, J. Opt. Network 4 (August 2005).

[62] T. Mizuochi, Y. Myiata, T. Kobayashi, K. Ouchi, K. Kuno, K. Kubo, Forward error correction based on block turbo code with 3-bit soft decision for 10-Gb/s optical communication systems, IEEE J. Sel. Top. Quant. Electron. 10 (2004) 376–386.

[63] D.O. Caplan, J.C. Gottschalk, R.J. Murphy, N.W. Spellmeyer, M.L. Stevens, A.M.D. Beling, Performance of high-rate high-sensitivity optical communications with forward error correction coding, in: CLEO: Paper CPDD9, 2004.

[64] N.W. Spellmeyer, J.C. Gottschalk, D.O. Caplan, M.L. Stevens, High-sensitivity 40-Gb/s RZ-DPSK with forward error correction, IEEE Photon. Technol. Lett. 16 (June 2004) 1579–1581.

[65] K. Onohara, T. Sugihara, Y. Konishi, Y. Miyata, T. Inoue, S. Kametani, K. Sugihara, K. Kubo, H. Yoshida, T. Mizuochi, Soft-decision-based forward error correction for 100 Gb/s transport systems, IEEE J. Quant. Electron. 16 (September/October 2010) 1258–1267.

[66] ETSI, Digital video broadcasting (DVB); second generation framing structure, channel coding and modulation systems for broadcasting, interactive services, news gathering and other broadband satellite applications (DVB-S2), in: ETSI EN 302 307 V1.2.1, 2009.

[67] D.O. Caplan, B.S. Robinson, R.J. Murphy, M.L. Stevens, Demonstration of 2.5-Gslot/s optically-preamplified M-PPM with 4 photons/bit receiver sensitivity, in: Optical Fiber Conference (OFC), Paper PDP23, 2005.

[68] X. Liu, T.H. Wood, R.W. Tkach, S. Chandrasekbar, Demonstration of record sensitivities in optically preamplified receivers by combining PDM-QPSK and M-Ary pulse-position modulation, J. Lightwave Technol. 30 (February 2012) 406–413.

[69] S.B. Alexander, R. Barry, D.M. Castagnozzi, V.W.S. Chan, D.M. Hodsdon, L.L. Jeromin, J.E. Kaufmann, D.M. Materna, R.J. Parr, M.L. Stevens, D.W. White, 4-ary FSK coherent optical communication system, Electron. Lett. 26 (1990) 1346–1348.

[70] D.O. Caplan, J.J. Carney, S. Constantine, Parallel direct modulation laser transmitters for high-speed high-sensitivity laser communications, in: Conference on Lasers and Electro-Optics (CLEO), Paper PDPB12, 2011.

[71] K. Kikuchi, M. Osaki, Highly-sensitive coherent optical detection of M-ary frequency-shift keying signal, Opt. Express 19 (December 2011).

[72] G.P. Agrawal, Nonlinear Fiber Optics, second ed., Academic Press, Inc., New York, 1995.

[73] D.O. Caplan, J.J. Carney, R.E. Lafon, M.L. Stevens, Design of a 40 watt 1.55 μm uplink transmitter for Lunar Laser Communications, in: Proceedings of the SPIE, Free-Space Laser Communication Technologies XXIV, 2012.

[74] D.O. Caplan, A technique for measuring and optimizing modulator extinction ratio, in: CLEO, 2000.

[75] S. Chandrasekhar, A.H. Gnauck, G. Raybon, L.L. Buhl, D. Mahgerefteh, X. Zheng, Y. Matsui, K. McCallion, Z. Fan, P. Tayebati, Chirp-managed laser and MLSE-RX enables transmission over 1200 km at 1550 nm in a DWDM environment in NZDSF at 10 Gb/s without any optical dispersion compensation, Photon. Technol. Lett. 18 (July 2006) 1560–1562.

[76] D.O. Caplan, M.L. Stevens, J.J. Carney, R.J. Murphy, Demonstration of optical DPSK communication with 25 photons/bit sensitivity, in: CLEO, 2006.

[77] D.O. Caplan, W.A. Atia, A quantum-limited optically-matched communication link, in: OFC, 2001.

[78] R.S. Vodhanel, A.F. Elrefaie, M.Z. Iqbal, R.E. Wagner, J.L. Gimlett, S. Tsuji, Performance of directly modulated DFB lasers in 10-Gb/s ASK, FSK, and DPSK lightwave systems, J. Lightwave Technol. 8 (1990) 1379–1386.

[79] D. Mahgerefteh, P.S. Cho, J. Goldhar, H.I. Mandelberg, Penalty-free propagation over 600 km of nondispersion shifted fiber at 2.5 Gb/s using a directly laser modulated transmitter, in: CLEO, 1999.

[80] C.H. Lee, S.S. Lee, H.K. Kim, J.H. Han, Transmission of directly modulated 2.5-Gb/s signals over 250-km of nondispersion-shifted fiber by using a spectral filtering method, Photon. Technol. Lett. 8 (December 1996) 1725–1727.

[81] D.O. Caplan, M.L. Stevens, D.M. Boroson, J.E. Kaufmann, A multi-rate optical communications architecture with high sensitivity, in: LEOS, 1999.

[82] M.L. Stevens, D.M. Boroson, D.O. Caplan, A novel variable-rate pulse-position modulation system with near quantum limited performance, in: LEOS, 1999.

[83] C.D. Edwards, C.T. Stelzried, L.J. Deutsch, L. Swanson, NASA deep space telecommunications road map, in: Telecommunications and Mission Operations Progress Report 42-136, Jet Propulsion Laboratory, 1999.

[84] E. Desurvire, Erbium-Doped Fiber Amplifiers, John Wiley & Sons, New York, 1994.

[85] D.O. Caplan, P.W. Juodawlkis, J.J. Plant, M.L. Stevens, Performance of high-sensitivity OOK, PPM, and DPSK communications using high-power slab-coupled optical waveguide amplifier (SCOWA) based transmitters, in: OFC, 2006.

[86] D.M. Boroson, C.C. Chen, B.L. Edwards, The Mars laser communication demonstration project: truly ultralong-haul optical transport, in: OFC, 2005.

[87] B.S. Robinson, A.J. Kerman, E.A. Dauler, R.J. Barron, D.O. Caplan, M.L. Stevens, J.J. Carney, S.A. Hamilton, J.K. W. Yang, K.K. Berggren, High-data-rate photon-counting optical communications using a NbN nanowire superconducting detector, in: CLEO, 2006.

[88] K.A. McIntosh et al., Arrays of III-V semiconductor Geiger-mode avalanche photodiodes, in: LEOS, 2003.

[89] K. Rosfjord, J. Yang, E. Dauler, A. Kerman, V. Anant, B. Voronov, G. Gol'tsman, K. Berggren, Nanowire single-photon detector with an integrated optical cavity and anti-reflection coating, Opt. Express 14 (January 2006) 527–534.

[90] E.A. Dauler, B.S. Robinson, A.J. Kerman et al., Multi-element superconducting nanowire single-photon detector, IEEE Trans. Appl. Supercond. 17 (2007) 279–284.

[91] E.A. Dauler, A.J. Kerman, B.S. Robinson, J.K.W. Yang, B. Voronov, G.N. Goltsman, S.A. Hamilton, K.K. Berggren, Photon-number-resolution with sub-30-ps timing using multi-element superconducting nanowire single photon detectors, J. Mod. Opt. 56 (January/February 2009) 364–373.

[92] X. Sun et al., Space-qualified silicon avalanche-photodiode single-photon-counting modules, J. Mod. Opt. 51 (July 2004) 1333–1350.

[93] D.M. Boroson, R.S. Bondurant, D.V. Murphy, LDORA: a novel laser communication receiver array architecture, Proc. SPIE 5338 (2004) 16–28.

[94] B.I. Erkmen, B.E. Moision, K.M. Birnbaum, The classical capacity of single-mode free-space optical communication: a review, IPN Progress, Report, vols. 42–179, November 2009.

[95] (a) X. Liu, S. Chandrasekbar, T.H. Wood, R.W. Tkach, E.C. Burrows, P.J. Winzer, Demonstration of 2.7-PPB receiver sensitivity using PDM-QPSK with 4-PPM and unrepeatered transmission over a single 370-km unamplified ultra-large-area fiber span, in: OFC, 2011. (b) D.J. Geisler, T.M. Yarnall, W.E. Keicher, M.L. Stevens, A.S. Fletcher, R.R. Parenti, D.O. Caplan, S.A. Hamilton, Demonstration of 2.1 Photon-per-bit Sensitivity for BPSK at 9.94-Gb/s with Rate-1/2 FEC, in: OFC, 2013.

[96] G. Li, Recent advances in coherent optical communication, Adv. Opt. Photon. 1 (2009) 279–307.

[97] Y. Shoji, M.J. Fice, Y. Takayama, A.J. Seeds, A pilot-carrier coherent LEO-to-ground downlink system using an optical injection phase lock loop (OIPLL) technique, J. Lightwave Technol. 30 (August 2012) 2696–2706.

[98] K.M. Rosfjord, J.K.W. Yang, E.A. Dauler, V. Anant, K.K. Berggren, A.J. Kerman, Increased detection efficiencies of nanowire single-photon detectors by integration of an optical cavity and anti-reflection coating, in: CLEO, 2006.

[99] E. Dauler, R.J. Molnar, A. Kerman, High detection efficiency superconducting nanowire single-photon detectors, in: European Conference on Applied Superconductivity, Dresden, Germany, 2009.

[100] M.E. Grein, A.J. Kerman, E.A. Dauler, O. Shatrovoy, R.J. Molnar, D. Rosenberg, J. Yoon, C.E. DeVoe, D.V. Murphy, B.S. Robinson, D.M. Boroson, Design of a ground-based optical receiver for the lunar laser communications demonstration, in: International Conference on Space Optical Systems and Applications (ICSOS), 2011.

[101] M.M. Willis, B.S. Robinson, M.L. Stevens, B.R. Romkey, J.A. Matthews, J.A. Greco, M.E. Grein, E.A. Dauler, A.J. Kerman, D. Rosenberg, D.V. Murphy, D.M. Boroson, Downlink synchronization for the lunar laser communications demonstration, in: International Conference on Space Optical Systems and Applications (ICSOS), Santa Monica, CA, USA, 2011.

[102] M.M. Willis, A.J. Kerman, M.E. Grein, J. Kansky, B.R. Romkey, E.A. Dauler, D. Rosenberg, B.S. Robinson, D.V. Murphy, D.M. Boroson, Performance of a multimode photon-counting optical receiver for the NASA lunar laser communications demonstration, in: International Conference on Space Optical Systems and Applications (ICSOS) Corsica, France, 2012.

[103] S.L. Huston, Space Environments and Effects, NASA Publication #NASA/CR-2002-211785, June 2012.

[104] K.Y. Chang, Deep space 1 spacecraft vibration qualification testing, J. Sound Vib. (March 2001).

[105] P. Fortescue, G. Swinerd, J. Stark, Spacecraft Systems Engineering. fourth ed., John Wiley and Sons, 2011.

[106] Acoustic Noise Requirement, Practice No. PD-ED-1259, 1996.

[107] J.W. Howard Jr., D.M. Hardage, Spacecraft Environments Interactions: Space Radiation and its Effects on Electronic Systems, NASA Publication, #NASA/TP-1999-209373 1999.

[108] A.C. Tribble, The Space Environment: Implications for Spacecraft Design, Princeton University Press, 2003.

[109] V.W.S. Chan, Optical space communications, IEEE Sel. Top. Quant. Electron. 6 (November/December 2000) 959–975.

[110] E.A. Swanson, V.W.S. Chan, Heterodyne spatial tracking system for optical space communication, IEEE Trans. Commun. 34 (February 1986) 118–126.

[111] W.H. Farr, M.W. Regher, M.W. Wright, A. Sahasrabudhe, J.W. Gin, D.H. Nguyen, Overview and trades for DOT flight laser transceiver, Jet Propulsion Laboratory, 2011.

[112] H. Hemmati, Interplanetary laser communications, Optics and Photonic News, November 2007.

[113] R. Lange, B. Smutny, BPSK laser communication terminals to be verified in space, in: Milcom, 2004, pp. 441–444.

[114] V.L. Williams, R.T. Lockie, Optical contamination assessment by bidirectional reflectance-distribution function (BRDF) measurement, Opt. Eng. 18 (March/April 1979) 152–156.

[115] W.T. Roberts, Optical membrane technology for deep space optical communications filters, in: IEEE Aerospace Conference, 2005.

[116] M.L. Stevens, D.M. Boroson, A simple delay-line 4-PPM demodulator with near-optimum performance, Opt. Express 20 (5) (February 2012)

[117] G.T. Delory et al., The lunar atmosphere and dust environment explorer (LADEE), in: 40th Lunar and Planetary Science Conference, 2009.

Digital Signal Processing (DSP) and Its Application in Optical Communication Systems

5

Polina Bayvel[a], Carsten Behrens[a, b], and David S. Millar[a, c]

[a]*Optical Networks Group, Department of Electronic and Electrical Engineering, University College London (UCL), London, UK*
[b]*Telekom Innovation Laboratories, Berlin, Germany*
[c]*Mitsubishi Electric Research Laboratories, Cambridge, MA, USA*

5.1 INTRODUCTION

5.1.1 Maximizing capacity in optical transport networks

We now take for granted the ubiquity of the Internet, the ability to communicate instantly, delay-free and with seemingly limitless bandwidth. In fact, the unprecedented growth of optical fiber infrastructure in recent decades has underpinned this, making possible broadband communications, e-commerce, video-on-demand and streaming media, tele-presence, and high performance distributed computing. It has dramatically changed the whole landscape of public, business, and government activities, stimulating relentless traffic growth. Indeed over 99% of all data is carried over optical fibers. But is fiber capacity unlimited? Internet traffic continues to grow at a nearly exponential pace (see Figure 5.1), exposing current backhaul and core network capacities to unprecedented demand pressures. The search for a solution to the predicted "capacity crunch" [1] has led to new research in system design, including the use of digital signal processing, stimulated by the development of coherent detection and the use of higher-order modulation formats. Initially this was seeded by the promise of potential cost reductions with increase in spectral efficiency which would allow the use of 10 Gbit/s-based hardware for 40 Gbit/s transmission, then by the needs to solve the perceived capacity crunch, but ultimately by the realization of the power of DSP.

Although the coherent optical receiver was subject of much research in the late 1980s and early 1990s (see the excellent chapter by Kikuchi in the OFTVB, Chapter 3 [4]), no significant practical progress had been observed for about 20 years. The digital coherent receiver allows the straightforward detection of in-phase and quadrature components of both polarisations and, therefore, quadruples the spectrum available for modulation, and thus the capacity. It detects the entire optical field in the digital domain, that is both amplitude and phase, which allows the signal to be processed and manipulated by powerful DSP algorithms. The last 5 years has seen much exciting

Optical Fiber Telecommunications VIB. http://dx.doi.org/10.1016/B978-0-12-396960-6.00005-5

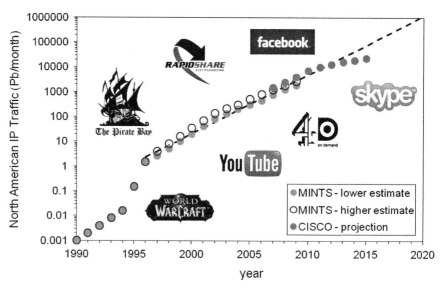

FIGURE 5.1 Exponential Internet traffic growth over the last two decades and important drivers [2,3]. IP traffic doubles nearly every 16 months.

research in this area, so prophetically predicted by Winzer and Essiambre, in the previous edition of OFT (Chapter 2 in OFTVB [5]). Virtually unlimited amounts of chromatic dispersion and PMD can now be compensated for digitally allowing for significant cost reductions [6], because dispersion compensating modules, optical filter compensators, and up to half the number of the EDFAs per link could be saved. Another advantage is the resultant reduction of inter-channel nonlinear effects due to the averaging out of nonlinear distortion, arising from the high values of accumulated dispersion in the absence of optical dispersion compensation. In addition, coherent detection offers better scaling characteristics at increased line rates compared to alternative solutions such as differential direct detection, and increased flexibility, because of its ability to select a wide range of wavelength channels by tuning the local oscillator. However, DSP algorithms require a large number of logic gates when implemented on an application specific integrated circuit (ASIC) in complementary metal oxide semiconductor technology (CMOS) as demonstrated by Sun et al. [7] (20 million gates in 90 nm CMOS). Under these circumstances power dissipation becomes an important figure of merit (see for example, [7]: 21 W for the ASIC and 140 W for the entire transceiver card).

Probably the most important advantage of coherent detection is the ability to detect higher-order modulation formats such as PDM-QPSK [8], PDM-8PSK [9], PDM-8QAM [10], PDM-16QAM [11], and PDM-36QAM [12]. These modulation formats use all the possible degrees of freedom offered by an optical wave to encode information, employing IQ modulators to access in-phase and quadrature components in both polarizations [13], utilizing the available bandwidth much more efficiently than binary modulation formats and, therefore, allowing to boost capacity without the requirement for installing new fiber. Figure 5.2 shows the overall capacity of some

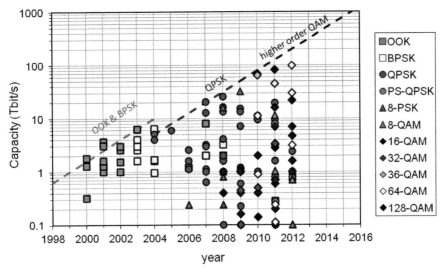

FIGURE 5.2 The maximum capacity of experimental optical communication systems has doubled every 18 months for more than a decade.

groundbreaking optical transmission system experiments of the last decade. Interestingly, maximum available capacity appears to double every 18 months, almost tracking the IP traffic growth that doubles every 16 months (Figure 5.1). In 2004 the first high-capacity QPSK demonstrations were reported (initially employing balanced detection [14]), achieving similar capacities to previous lab experiments using only binary modulation formats [15]. QPSK dominated high capacity systems for 5 years, when 8QAM entered the scene for the first time matching the available capacity of QPSK systems. Since then, further increases of capacity have only been possible by resorting to even more spectrally efficient formats like 36QAM and 128QAM. Note that a capacity of 305 Tb/s has recently been demonstrated by using PDM-QPSK and a 19-core fiber [16]. However, in this chapter, we have focused on only transmission over single-mode fibers, and have not considered the transmission under the spatial-division-multiplexed (SDM) regime.

The use of denser constellation diagrams makes higher-order modulation formats more susceptible to circularly symmetric Gaussian noise, generated by EDFAs along the transmission link. Even though the launch power per wavelength channel can be increased to improve the signal-to-noise ratio at the receiver, transmission is limited by nonlinear distortions due to the Kerr effect, which have a more severe impact on higher-order modulation formats. This leads to a dramatically reduced transmission reach, sacrificed at the expense of increased capacity. Figure 5.3 illustrates this dilemma by showing the capacity distance product of the key transmission experiments since the year 2000. While capacity doubles every 1.5 years as highlighted in Figure 5.2, equivalent capacity distance product doubles only every ~5 years, indicating that the increased capacity can only be delivered over much shorter distances.

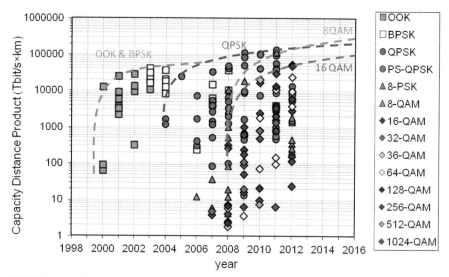

FIGURE 5.3 Achieved capacity × distance product for experimental optical communication systems during the last decade. Capacity × distance product doubles only every 5 years due to nonlinear limitations. (For interpretation of the references to color in this figure legend, the reader is referred to the web version of this book.)

So, the key question of current optical communications research is: how to maximize the transmission reach, while increasing the capacity of current optical transport networks by using spectrally efficient modulation formats and how can digital signal processing aid in this quest? While ASE noise cannot be compensated for due to its random nature, nonlinear distortions arising from the Kerr effect are deterministic and, therefore, predictable. Receiver-based nonlinear compensation can be performed in the digital domain by deploying powerful DSP algorithms, such as digital backpropagation (DBP). Several DBP regimes were considered, ranging from a fixed complexity per distance solution, which has been experimentally investigated to ideal multi-channel backpropagation with high complexity studied by means of extensive computer simulations. It is worthwhile to note that transmitter side nonlinear compensation has also been demonstrated [17]. The shortcoming of transmitter-based compensation is the need for feedback from the receiver in the presence of randomly varying processes such as PMD or noise, but there is some debate in relative merits of DSP, in split compensation between transmitter and receiver.

The rest of the chapter is organized as follows.

Section 5.2 describes the coherent receiver structure and DSP algorithms for chromatic dispersion compensation, equalization, and phase recovery of PDM-BPSK, PS-QPSK, PDM-QPSK, PDM-8PSK, PDM-8QAM, and PDM-16QAM. Additionally, it includes a section about receiver-based digital backpropagation with a literature review on the topic.

Section 5.3 shows how the DBP-based DSP can be applied in the nonlinear regime for a range of modulation formats at varying symbol-rates and channel spacings, while keeping the spectral efficiency constant for every modulation format. The system is assumed to be unconstrained in terms of algorithm complexity with the consideration of two scenarios for DBP—backpropagation of only the central channel and over a fixed bandwidth approach covering 100 GHz of the optical spectrum. Summary and outlook for the future are given in Section 5.4.

5.2 DIGITAL SIGNAL PROCESSING AND ITS FUNCTIONAL BLOCKS

5.2.1 Optical coherent receiver and digital signal processing functionality

The aim of DSP for optical communications is to process the transmitted digital data (in the electrical domain) so as to correctly detect it, ideally compensating or at least mitigating all impairments, both linear and nonlinear, deterministic and random. Although transmitter-based DSP is also possible, for example—electronic predistortion was one of the first recent examples of digital signal processing [18,19], it is most effective when the channel properties are known *a priori*. Conversely, in the case of the source of distortion randomly varying in the course of transmission, any transmitter-based DSP would require a feedback signal from the receiver, impractical because of the long delays. The case of split receiver and transmitter-based compensation is one which has not been investigated in great detail, but could offer some promise. However, the starting point for most DSP in optical communications follows coherent detection. The detector is a phase and polarization diverse coherent receiver. The transmitted signal interferes inside an optical hybrid with a LO-signal provided by another CW-laser (see Figure 5.4) converting both quadratures of X- and Y-polarization into the electrical domain. Both the LO- and transmitter laser have phase noise, which can be modeled as a random walk process of the variance $\sigma^2 = 2\pi \cdot \Delta\vartheta \cdot dt$ with $\Delta\vartheta$ representing the laser linewidth and dt is the time between two observations. Typical laser linewidths range from $\sim$100 kHz, corresponding to

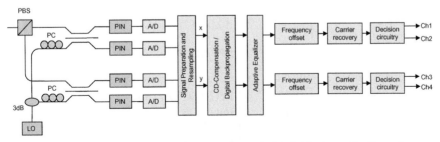

FIGURE 5.4 Coherent receiver using a fiber coupler as 90° hybrid, chromatic dispersion compensation, equalization, digital phase recovery, and differential decoding.

commercially available external cavity lasers (ECL), up to a few MHz for a conventional DFB laser. The LO-laser is usually free running within ~ 1 GHz of the optical frequency of the transmit laser, which is referred to as intradyne detection. In this case, the remaining frequency offset is compensated digitally, e.g., by applying a higher-order nonlinearity to the signal and estimating the offset from the spectrum.

A four port phase- and polarization-diversity hybrid may be used with single-ended detection, resulting in the four photocurrents given in (5.1):

$$
\begin{bmatrix} i_{XI}(t) \\ i_{XQ}(t) \\ i_{YI}(t) \\ i_{YQ}(t) \end{bmatrix} \sim \begin{bmatrix} \mathrm{Re}(E_X(t) \cdot E_{LO}^*) + \frac{1}{2}|E_X(t)|^2 + \frac{1}{4}|E_{LO}|^2 \\ \mathrm{Im}(E_X(t) \cdot E_{LO}^*) + \frac{1}{2}|E_X(t)|^2 + \frac{1}{4}|E_{LO}|^2 \\ \mathrm{Re}(E_Y(t) \cdot E_{LO}^*) + \frac{1}{2}|E_Y(t)|^2 + \frac{1}{4}|E_{LO}|^2 \\ \mathrm{Im}(E_Y(t) \cdot E_{LO}^*) + \frac{1}{2}|E_Y(t)|^2 + \frac{1}{4}|E_{LO}|^2 \end{bmatrix}. \tag{5.1}
$$

Each of the photocurrents consists of three terms: a coherent detection term (leftmost), a signal envelope term (center), and an LO envelope term (rightmost). Although the LO envelope term is constant (and may therefore be removed with AC coupling), the signal envelope term is time varying and must be minimized relative to the coherent detection term by making the LO much more powerful than the signal.

To overcome the constraints imposed on signal-LO power ratios, balanced photo-detection is often employed for coherent optical receivers. In this scenario, an 8 port optical hybrid is used, with a 180° phase shift between each quadrature pair. The pairs of outputs are then differentially amplified to eliminate the direct-detection components in the signal. The eight output ports of the hybrid are given by

$$
\begin{bmatrix} i_{XI+}(t) \\ i_{XI-}(t) \\ i_{XQ+}(t) \\ i_{XQ-}(t) \\ i_{YI+}(t) \\ i_{YI-}(t) \\ i_{YQ+}(t) \\ i_{YQ-}(t) \end{bmatrix} \sim \begin{bmatrix} \frac{1}{2}\mathrm{Re}(E_X(t) \cdot E_{LO}^*) + \frac{1}{4}|E_X(t)|^2 + \frac{1}{8}|E_{LO}|^2 \\ -\frac{1}{2}\mathrm{Re}(E_X(t) \cdot E_{LO}^*) + \frac{1}{4}|E_X(t)|^2 + \frac{1}{8}|E_{LO}|^2 \\ \frac{1}{2}\mathrm{Im}(E_X(t) \cdot E_{LO}^*) + \frac{1}{4}|E_X(t)|^2 + \frac{1}{8}|E_{LO}|^2 \\ -\frac{1}{2}\mathrm{Im}(E_X(t) \cdot E_{LO}^*) + \frac{1}{4}|E_X(t)|^2 + \frac{1}{8}|E_{LO}|^2 \\ \frac{1}{2}\mathrm{Re}(E_Y(t) \cdot E_{LO}^*) + \frac{1}{4}|E_Y(t)|^2 + \frac{1}{8}|E_{LO}|^2 \\ -\frac{1}{2}\mathrm{Re}(E_Y(t) \cdot E_{LO}^*) + \frac{1}{4}|E_Y(t)|^2 + \frac{1}{8}|E_{LO}|^2 \\ \frac{1}{2}\mathrm{Im}(E_Y(t) \cdot E_{LO}^*) + \frac{1}{4}|E_Y(t)|^2 + \frac{1}{8}|E_{LO}|^2 \\ -\frac{1}{2}\mathrm{Im}(E_Y(t) \cdot E_{LO}^*) + \frac{1}{4}|E_Y(t)|^2 + \frac{1}{8}|E_{LO}|^2 \end{bmatrix}. \tag{5.2}
$$

After differential amplification, this becomes the four-dimensional signal given by

$$
\begin{bmatrix} i_{XI}(t) \\ i_{XQ}(t) \\ i_{YI}(t) \\ i_{YQ}(t) \end{bmatrix} \sim \begin{bmatrix} \mathrm{Re}(E_X(t) \cdot E_{LO}^*) \\ \mathrm{Im}(E_X(t) \cdot E_{LO}^*) \\ \mathrm{Re}(E_Y(t) \cdot E_{LO}^*) \\ \mathrm{Im}(E_Y(t) \cdot E_{LO}^*) \end{bmatrix}. \tag{5.3}
$$

The received signal defined by Eq. (5.3) represents the ideal coherently received optical field, that is: no direct-detection terms, infinite common-mode rejection between differential pairs and perfectly matched optical path lengths in the hybrid resulting in an exact 90° difference between quadratures.

Following the detection of both the amplitude and phase of the optical signal, needed to completely describe it, the signal is digitized by analog-to-digital converters (ADCs), de-skewed and resampled at twice the symbol-rate to prepare it for subsequent digital signal processing. Figure 5.4 shows the following DSP blocks performing the key signal processing functions of chromatic dispersion compensation, CMA equalizer, digital carrier recovery, and differential decoding of the signal.

5.2.1.1 Chromatic dispersion compensation

The compensation of chromatic dispersion is generally (although not exclusively) carried out by using a finite-impulse-response (FIR) filter [20], which is the approach on which we focus here.

The propagation of an optical wave $A(z, t)$ inside a fiber, assuming only chromatic dispersion is present, can be described as follows:

$$\frac{\partial A(z,t)}{\partial Z} = j \frac{D\lambda^2}{4\pi c} \frac{\partial^2 A(z,t)}{\partial t^2}. \tag{5.4}$$

Equation (5.4) can be solved in the frequency domain by $H(z, w)$, which is then converted into the time domain to obtain the impulse response $h(z, t)$:

$$H(z, w) = \exp\left(-j \frac{D\lambda^2}{4\pi c} w^2\right) \Leftrightarrow h(z, t) = \sqrt{\frac{c}{jD\lambda^2 z}} \exp\left(j \frac{\pi c}{D\lambda^2 z} t^2\right). \tag{5.5}$$

To compensate for chromatic dispersion the sign of D must be reversed, which results in the impulse response of the compensating filter:

$$\tilde{h}(z, t) = \sqrt{\frac{jc}{D\lambda^2 z}} \exp\left(-j \frac{\pi c}{D\lambda^2 z} t^2\right). \tag{5.6}$$

$\tilde{h}(z, t)$ is infinite in time, non-causal and passes all frequencies for a finite sampling frequency, introducing aliasing. To apply this impulse response in an FIR filter the continuous time impulse response is approximated by a sampled impulse response, which can be implemented using a tapped delay-line (Figure 5.5).

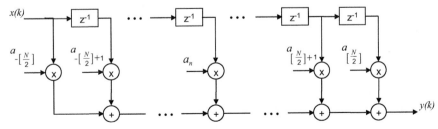

FIGURE 5.5 Structure of a FIR filter for compensating for chromatic dispersion in the time domain.

Additionally, the impulse response is truncated to an odd number of taps, with the following tap-weights:

$$
a_n = \sqrt{\frac{jcT^2}{D\lambda^2 z}} \exp\left(-j\frac{\pi cT^2}{D\lambda^2 z}n^2\right) \quad \text{for} -\left\lfloor\frac{N}{2}\right\rfloor \leqslant n \leqslant \left\lfloor\frac{N}{2}\right\rfloor \quad \text{with } N = 2\left\lfloor\frac{|D|\lambda^2 z}{2cT^2}\right\rfloor + 1.
$$

$$(5.7)$$

If it is assumed that a transmission system similar to the one investigated in Section 5.3 (1000 km SSMF with $D = 16$ ps/km/nm operating at 28 GBd) a FIR filter with 401 taps would be required to compensate for all of the accumulated chromatic dispersion in the link.

Indeed Equation 5.6 determines the minimum number of taps necessary to compensate for the maximum amount of delay over the signal bandwidth. To reduce ripple associated with FIR filters a large number of taps may be required. To improve performance with small amounts of dispersion, a windowed design using the frequency domain description of chromatic dispersion may be used. For an arbitrary number of taps N, we define our window as being able to represent the discrete set of frequencies f_n given by Eq. (5.8), where f_s is the sampling frequency. We may then describe the frequency response at these discrete frequencies given by the solution to the linear part of the NLSE as (5.9).

$$
f_n = n\Delta f: \text{ where } \left\lfloor-\frac{N}{2}\right\rfloor \leqslant n \leqslant \left\lfloor\frac{N}{2}\right\rfloor \quad \text{and} \quad \Delta f = f_s/N, \tag{5.8}
$$

$$
H(f_n) = \exp\left(\frac{j\pi D(\lambda f_n)^2}{c}\right). \tag{5.9}
$$

These filter coefficients may be used either in the frequency domain using an algorithm such as overlap-add or overlap-save to perform filtering, or the iFFT of the discrete frequency response may be calculated and filtering performed in the time domain using conventional linear convolution.

5.2.1.2 DSP for nonlinearity compensation-digital backpropagation (DBP)

Access to both the amplitude and field of the optical signal should allow for both linear and nonlinear impairments to be compensated. Digital backpropagation is a nonlinear compensation scheme which uses the knowledge about the physical channel and inverts it. In principle, all linear and nonlinear impairments could be compensated but in practice, the presence of randomly generated processes such as ASE noise and PMD precludes complete compensation. Despite this, DBP is the most general of DSP algorithms, and this chapter focuses on its implementation and application to a range of optical communication transmission systems, and the analysis of its efficacy.

An electrical field propagating in an optical fiber is usually described by the coupled nonlinear Schrödinger equation (CNLSE) [21], which does not account for the effect of randomly varying birefringence along the fiber. However, the Manakov approximation to the CNLSE does accurately model transmission over the length scales important for optical communications [22]; it assumes that fiber nonlinearity acts equally on both polarizations, since the birefringence scatters the state of polarization on a much smaller length scale than the nonlinear length. Throughout this chapter, the digital backpropagation algorithm is based on the application of the Manakov equation:

$$
\begin{aligned}
\frac{\partial}{\partial z} E_X &= -\frac{\alpha}{2} E_X + \frac{j\beta_2}{2} \frac{\partial^2}{\partial t^2} E_X - j\gamma \frac{8}{9}(|E_X|^2 + |E_Y|^2) E_X, \\
\frac{\partial}{\partial z} E_Y &= -\frac{\alpha}{2} E_Y + \frac{j\beta_2}{2} \frac{\partial^2}{\partial t^2} E_Y - j\gamma \frac{8}{9}(|E_Y|^2 + |E_X|^2) E_Y
\end{aligned}
\tag{5.10}
$$

with α denoting the fiber attenuation coefficient, β_2 the chromatic dispersion coefficient, and γ the nonlinear coefficient as well as X and Y for the two orthogonal polarizations. Since there is no analytical solution to this equation, the split-step method has to be applied to find an approximate solution for the inverse channel. Consequently, Eq. (5.10) can be split up into a linear and nonlinear part

$$
\frac{\partial \mathbf{E}}{\partial z} = (\widehat{D} + \widehat{N})\mathbf{E}
\tag{5.11}
$$

with $\mathbf{E} = [E_X\ E_Y]^T$ describing the optical field in both polarizations, $\widehat{D}$ the effect of chromatic dispersion, and $\widehat{N}$ describing the Kerr effect as well as the attenuation of the fiber:

$$
\widehat{D} = \frac{j\beta_2}{2} \frac{\partial^2}{\partial t^2},
\tag{5.12}
$$

$$
\widehat{N} = -j\gamma \frac{8}{9} \mathbf{E}^H \mathbf{E} - \frac{\alpha}{2}.
\tag{5.13}
$$

For a small step size h the solution to Eq. (5.8) can be approximated by:

$$
\mathbf{E}(z+h, T) \approx \exp\left(\widehat{D}\frac{h}{2}\right) \exp(\widehat{N} h_{\text{eff}}) \exp\left(\widehat{D}\frac{h}{2}\right) \mathbf{E}(z, T),
\tag{5.14}
$$

where $h_{\text{eff}} = (1-\exp(-\alpha h))/\alpha$ denotes the effective length of the step size. In Eq. (5.14) the symmetrical split-step method has been applied, leading to higher accuracy as the dispersive step is split into two equal steps [5]. It should be noted that the symmetrical split-step method increases the implemented hardware complexity but nonetheless allows to explore the maximum achievable performance of digital backpropagation.

The dispersive step is performed in the frequency domain by inverting the frequency response of a dispersive fiber:

$$
\widehat{D}(\omega) = -j\frac{\beta_2}{2}\omega^2,
\tag{5.15}
$$

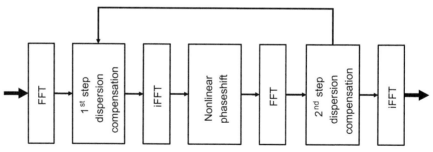

FIGURE 5.6 Block diagram of digital backpropagation.

while the nonlinear step is performed in the time domain applying an inverse nonlinear phase shift proportional to the total power in both polarizations:

$$\widehat{N}(t) = -j\gamma\varphi\frac{8}{9}(|E_Y|^2 + |E_X|^2)P_{\text{in}}10^{([\frac{s}{n}-1]\frac{\alpha L}{10})} \tag{5.16}$$

Here ϕ is a variable that converges with increasing number of steps toward 1 (in the noise free case) but has to be optimized for a realistic number of steps. P_{in} denotes the input power per span, while $10^{([\frac{s}{n}-1]\frac{\alpha L}{10})}$ accounts for the varying power along the span, with n the number of steps per span, s the index of the step within a span and L the span length in km and the fibres attenuation coefficient in dB/km. Although, digital backpropagation has been considered for various oversampling rates, to account for spectral broadening during the nonlinear step [23], in most cases 2 samples per symbol allow significant performance improvements to be acheived [24], whilst relaxing hardware requirements. The minimum sufficient sampling rate is a subject of current research. Recent work by Du and Lowery [25] demonstrated that similar performance can be achieved with two samples per symbol backpropagation by suppressing aliasing effects with a low-pass filter applied to the estimated nonlinear phase shift.

The symmetric split-step method for digital backpropagation can be implemented as shown in the signal flow model of Figure 5.6 and replaces the block compensating for chromatic dispersion in conventional digital coherent receivers.

5.2.1.3 Application of digital backpropagation-based DSP in optical transmission experiments

Assuming the predictions of the "capacity crunch" mentioned in the Introduction [1] will indeed take place, an increase in the optical transport capacity is essential. This could be achieved by using spectrally more efficient modulation formats. These, however, have increased OSNR requirements as well as a higher susceptibility to nonlinear distortions, reducing the reach when the capacity is increased. A solution to increase transmission reach is to compensate for deterministic nonlinear distortions by digitally backpropagating either part of or the *entire* received spectrum.

Table 5.1 Experimental demonstrations of digital backpropagation to compensate for intra-channel nonlinearities in a WDM environment.

Year	Modulation Format	Line Rate (Gbit/s)	Number of Channels	Channel Spacing (GHz)	Performance Improvement
2009 [26]	PDM-QPSK	112	19	100	2 dBQ
2009 [27]	PDM-QPSK	112	40	100	1.7 dBQ
2009 [28]	PDM-QPSK	112	72	50	0.25 dBQ
2010 [29]	PDM CO-OFDM (16-QAM)	224	7	50	0.5 dBQ
2010 [30]	PDM-QPSK	112	10	50	3% reach
2010 [30]	PDM-QPSK	112	10	100	24% reach
2011 [31]	PDM-16-QAM	224	3	50	10% reach
2011 [32]	PDM-16-QAM	112	3	50	24% reach
2011 [33]	PDM CO-OFDM (16-QAM)	448	3	80	25% reach
2011 [34]	PDM CO-OFDM (QPSK)	111	10	not specified	25% reach (1.3 dBQ)
2011 [35]	PDM-QPSK	112	40	100	1.3 dBQ
2011 [36]	PDM-QPSK	224	40	100	1.2 dBQ
2011 [37]	PDM-QPSK, NRZ aggressors	112	80	50	0.9 dBQ
2011 [38]	PDM-QPSK	112	40	100	1.6 dBQ
2012 [39]	PDM-8-QAM	112	7	50	17.8% reach
2012 [39]	PDM-8-PSK	112	7	50	20% reach
2012 [39]	PDM-QPSK	112	7	50	19.6% reach

Table 5.1 shows most of the experimental results using digital backpropagation to compensate for intra-channel nonlinearities. All these experiments have in common that a single channel (or band of subcarriers, in the case of OFDM) is coherently detected after being propagated in a WDM environment. The received portion of the spectrum is then digitally backpropagated to increase the margin or transmission reach as detailed in the last column. While DBP can lead to modest performance improvements of +2 dBQ [26] or +24% transmission reach [30] when a coarse frequency grid of 100 GHz is used, it is limited when the frequency spacing is reduced. In the case of a 50 GHz spacing, the improvement of transmission performance has been limited to less than 1 dBQ [28,29,37] or small increases in transmission reach of +3% [30]. Recent research, however, has shown that transmission reach can be increased by ~20% even on a 50 GHz grid [39].

It has been demonstrated that at least 7–9 WDM channels should be transmitted to capture all XPM distortions that restrict the efficiency of the DBP algorithm and reliably assess the performance under WDM conditions [31,47]. Consequently, the improvement of up to ~25% increase in reach obtained for three-channel PDM-16QAM [32] and PDM CO-OFDM [33] may be significantly reduced with additional WDM channels. Polarization mode dispersion has been found to limit the efficiency of DBP depending on the ratio between symbol slot and differential group delay (DGD) by up to 2 dBQ for single carrier and 6 dBQ for OFDM transmission [47]. In practice however, DGD is expected to be as small as 0.1 ps/$\sqrt{\text{km}}$ leading only to a negligible walk-off of ~$0.12 \times$ symbol slot in case of 28 GBd transmission over 2000 km. Even under high PMD conditions DBP has been demonstrated to be effective [37].

In the absence of uncompensated inter-channel nonlinearities the performance improvement gained from DBP is much higher as detailed in Table 5.2, which shows recent experimental results employing DBP to the entire transmitted optical field. Under these circumstances the DBP algorithm has been found

Table 5.2 Experimental demonstrations of full-field digital backpropagation in a single-channel environment.

Year	Modulation Format	Line Rate (Gbit/s)	Number of Channels	Channel Spacing (GHz)	Performance Improvement
2008 [24][a]	BPSK	6	3	7	121% reach
2009 [40]	PDM CO-OFDM (QPSK)	111	1		13% reach
2009 [41][a]	PDM-BPSK	12	3	7	16 dBQ
2009 [42]	PDM-QPSK	42.7	1		33% reach
2009 [42]	PDM-QPSK	85.4	1		50% reach
2010 [43]	PDM CO-OFDM (8-QAM)	61.7	1		2.2 dBQ
2010 [30]	PDM-QPSK	112	1		46% reach
2011 [44]	PDM-QPSK	42.7	1		1.6 dBQ
2011 [44]	PDM-16-QAM	85.4	1		1 dBQ
2011 [32]	PDM-16-QAM	112	1		67% reach
2011 [31]	PDM-16-QAM	224	1		12% reach
2011 [34]	PDM CO-OFDM (QPSK)	111	1		13.3% reach (0.5 dBQ)
2011 [45]	PDM-QPSK	43	1		1.9 dBQ
2012 [46]	PS-QPSK	112	1		20.7% reach
2012 [39]	PDM-8-QAM	112	1		69.7% reach
2012 [39]	PDM-8-PSK	112	1		59.3% reach
2012 [39]	PDM-QPSK	112	1		31.6% reach

[a]*All three channels are backpropagated.*

to be largely limited by non-deterministic nonlinear signal-ASE interactions [48]. Nevertheless, Goldfarb [47] and Yaman [41] demonstrated an impressive increase of +121% transmission reach and a performance improvement of 16dBQ, by using low dispersion fibers and under the condition that a very low symbol-rate is used so that all three transmitted WDM channels can be fitted into the electrical bandwidth of a single coherent receiver to be digitally backpropagated. Furthermore, it has been found that higher-order modulation formats show an increased benefit for modulation formats when comparing increase in transmission reach of +67% for PDM-16QAM [32] as well as +69.7% and +59.3% for PDM-8QAM and PDM-8PSK [39] on one side to +31.6% PDM-QPSK [39] and +20.7% for PS-QPSK [46] on the other side. Similar conclusions have been drawn as a result of simulation studies comparing PDM-QPSK and PDM-16QAM transmission [49] and multilevel QAM formats [50].

A key parameter in the digital signal processing is the number of computational DBP steps which has been shown to be related to the spectral width of the received signal (symbol-rate) and the chromatic dispersion parameter of the fiber [49]. Up to a symbol-rate of 28 GBd, 1 step per transmitted span is widely established as providing a good trade-off between algorithm complexity and performance improvement [26,27,40,44,51]. However, at higher symbol-rates such as 56 GBd more steps per span are necessary to provide optimum performance as demonstrated in [36]. In an attempt to reduce complexity of the DBP algorithm, the correlation of the nonlinear distortion incident on neighboring symbols has been exploited by filtering the calculated nonlinear phase shift [25]. This method has been implemented in the frequency domain [25] as well as in the time domain [35], reducing the number of required steps by 75% without sacrificing performance.

5.2.1.4 Alternative nonlinearity compensation schemes

A number of possible alternative solutions for nonlinear equalization have been investigated in the last decade, mainly focusing on reduced complexity.

- An example of a low complexity solution is a intensity-dependent *nonlinear phase shift* at the receiver, which can be regarded as digital backpropagation with a single step per link. This compensation method has been investigated in the analog [52] and digital domains [53], along with the introduction of non-rectangular decision boundaries based on a priori knowledge of the transmission link [54]. All of these approaches exploit the fact that optically compensated transmission links produce non-circular symmetric nonlinear distortions ("bean"-shaped-constellation diagrams), which can be easily equalized. However, in uncompensated transmission, distortion statistics tend to be circularly symmetric [55], rendering this approach less effective.
- *Maximum-likelihood sequence estimation* (MLSE) has been found to be the optimum nonlinear decoder in the presence of deterministic distortions such as chromatic dispersion and intra-channel nonlinearities [56]. It is based on

finding the most likely transmitted sequence by computing the cross-correlations between a set of expected sequences and the received one. The MLSE is usually implemented with the Viterbi algorithm [57] in which case the computational complexity scales with M^N where M is the symbol alphabet and N is the memory length related to the pulse spreading. Since modern coherent systems omit optical dispersion compensation and increased amounts of accumulated dispersion have been shown to reduce inter-channel nonlinear distortions, a high pulse spreading is the consequence and MLSE would incur an unacceptable complexity burden. Nevertheless, reduced complexity MLSE has been demonstrated for uncompensated 10.7 Gbit/s IMDD transmission and high memory length [58] and in the case of reduced memory length due to optically compensated transmission [59]. Note that even though MLSE has been implemented for coherent detection of 112 Gbit/s PDM-QPSK in [60], it is used to reduce ISI due to aggressive filtering and is not capable of compensating for large amounts of nonlinearity, since the memory introduced by chromatic dispersion is compensated separately.

- Another approach to the compensation for deterministic nonlinearities is the *maximum a posteriori probability* (MAP) detector, which exploits the pattern dependency of nonlinear distortions. A training sequence is sent across the channel to initialize a look-up table at the receiver with statistical distributions of a certain memory length. Similarly to MLSE, the pattern with the highest correlation is chosen as the MAP decision, but minimizing the symbol error rate rather than the probability of a sequence error as in the case of MLSE [61]. Notable performance improvements have been demonstrated recently in lab experiments [62,63] and in simulation [64], albeit at the expense of significant DSP complexity.

- Volterra series transfer functions have been investigated to increase the speed of fiber transmission simulations [65]. These types of nonlinear transfer functions based on a generalization of the Taylor series have recently attracted much interest, because they allow to design nonlinear filters which are capable of compensating for nonlinear distortions [66–68]. However, in general DBP algorithms outperform nonlinear Volterra equalizers designed with the focus on low complexity [68].

- As already mentioned, the transmitter-side equivalent to digital backpropagation is *electronic predistortion*, which requires digital-to-analog converters (DACs) and signal processing capabilities at the transmitter [17,19,69]. In the case of coherent detection an ASIC would still be required at the receiver for the adaptive equalizer, carrier recovery, and FEC decoding.

5.2.1.5 Equalization

Multiple-input-multiple-output (MIMO) systems are widely used in optical communications [20] to combat multipath propagation effects such as polarization-mode-dispersion (PMD). This is sometimes referred to as a butterfly structure.

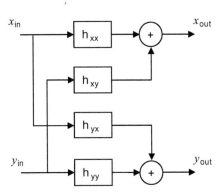

FIGURE 5.7 Butterfly structure to compensate for polarization-mode-dispersion.

It can be implemented to consist of four FIR filters ($\mathbf{h}_{xx}$, $\mathbf{h}_{xy}$, $\mathbf{h}_{yx}$, and $\mathbf{h}_{yy}$) (Figure 5.7). These filters have adaptive taps, due to the time-dependent nature of PMD. The transfer function of the sample $x_{out}(k)$, which depends on the input samples x_{in} and $y_{in}(k)$, is:

$$x_{out}(k) = \mathbf{h}_{xx}^T \mathbf{x}_{in} + \mathbf{h}_{xy}^T \mathbf{y}_{in} = \sum_{m=0}^{M-1} h_{xx}(m) x_{in}(k-m) + h_{xy}(m) y_{in}(k-m).$$

$$(5.17)$$

Generally, an adaptive equalizer tries to estimate the Jones matrix of the channel and apply the inverse of it to the signal. Additionally, it emulates a matched filter and, therefore, compensates for linear distortions incurred by filtering, leading to an optimization of the receiver sensitivity. An adaptive equalizer is also capable of compensating for chromatic dispersion. However, splitting up PMD- and chromatic dispersion compensation is desirable, since it leads to a much smaller footprint of the DSP and allows for the adaptation timescales for the partitioned polarization-dependent and independent blocks to be different, as discussed in depth in the seminal work by Savory [20]. If we assume a transmission system similar to the one described in the next section (1000 km SSMF with a PMD coefficient of $0.1 \, \text{ps}/\sqrt{\text{km}}$), the mean DGD adds up to 3.2 ps which is in the case of 28 GBd less than $0.1 \times$ symbol spacing resulting in three taps to be sufficient to track the mean DGD as opposed to more than 400 taps for chromatic dispersion compensation. Since PMD exhibits a Maxwellian distribution around the mean DGD value, some 15 equalizer taps can be used, sufficient to track most of the random polarization rotations experienced in the system.

To adapt the taps to the changing channel conditions, different update algorithms are necessary, depending on the properties of the modulation formats, as described below.

- In the case of PDM-QPSK and PDM-8PSK, the most widely adopted algorithm is the constant modulus algorithm (CMA [70]), which exploits the fact that the symbols lie on intensity rings, i.e. have a constant modulus. The LMS updating algorithms are given in Eq. (5.18), with the step size parameter μ, the polarization-dependent error signals (e_x, e_y), and the complex conjugates of the input sequences $(\overline{x}_{in}, \overline{y}_{in})$

$$
\begin{aligned}
\mathbf{h}_{xx} &\rightarrow \mathbf{h}_{xx} + \mu e_x x_{out}(k)\overline{x}_{in}, \\
\mathbf{h}_{xy} &\rightarrow \mathbf{h}_{xy} + \mu e_x x_{out}(k)\overline{x}_{in}, \\
\mathbf{h}_{yx} &\rightarrow \mathbf{h}_{yx} + \mu e_y y_{out}(k)\overline{x}_{in}, \\
\mathbf{h}_{yy} &\rightarrow \mathbf{h}_{yy} + \mu e_y y_{out}(k)\overline{y}_{in}.
\end{aligned}
\tag{5.18}
$$

The input samples have to be normalized to unit power to determine the error signals e_x and e_y (Figure 5.8). The equalizer tries to minimize the amplitudes of e_x and e_y in a mean squares sense to converge on the inverse channel matrix.

- In the case of 8QAM and 16QAM one must decide the radius to which the current symbol belongs, before calculating an error signal. The set of error signals of this so-called radially directed equalizer (RDE) is shown in Figure 5.8a and b [71].
- For PDM-BPSK a decision directed equalizer can help to improve convergence with respect to the standard CMA. After pre-convergence with the CMA and frequency offset removal, the receiver moves into a decision directed mode, similarly to the equalizer implemented for PDM-QPSK in [72]. In this case the outputs $x_{out}(k)$ and $y_{out}(k)$ of the butterfly structure (Figure 5.7) are combined with an additional phase correction term, resulting in new output values: $X(k) = \exp(-j\phi_x(k))x_{out}(k)$ and $Y(k) = \exp(-j\phi_y(k))y_{out}(k)$. The updating algorithms in (5.18) become:

$$
\begin{aligned}
\mathbf{h}_{xx} &\rightarrow \mathbf{h}_{xx} + \mu e_x \exp(j\varphi_x)\overline{x}_{in}, \\
\mathbf{h}_{xy} &\rightarrow \mathbf{h}_{xy} + \mu e_x \exp(j\varphi_x)\overline{y}_{in}, \\
\mathbf{h}_{yx} &\rightarrow \mathbf{h}_{yx} + \mu e_y \exp(j\varphi_y)\overline{x}_{in}, \\
\mathbf{h}_{yy} &\rightarrow \mathbf{h}_{yy} + \mu e_y \exp(j\varphi_y)\overline{y}_{in}.
\end{aligned}
\tag{5.19}
$$

With the corresponding error functions for each polarization:

$$
\begin{aligned}
e_x &= \text{sgn}(\text{Re}\{X\}) - X, \\
e_y &= \text{sgn}(\text{Re}\{Y\}) - Y,
\end{aligned}
\tag{5.20}
$$

where $\text{sgn}(\text{Re}\{x\})$ denotes the sign of the real part of the relevant symbol. The estimated phase values, averaged over the N following symbols, can be written as:

$$
\varphi_x = \frac{1}{N} \arg \sum_{i=1}^{N} \text{sgn}(\text{Re}\{X_i\}) \cdot X_i,
$$

$$
\varphi_y = \frac{1}{N} \arg \sum_{i=1}^{N} \text{sgn}(\text{Re}\{Y_i\}) \cdot Y_i.
\tag{5.21}
$$

- In the case of PS-QPSK a polarization-switched CMA is used, to ensure convergence [73,74]. A decision based on the energy of the symbol enables to identify the polarization with the QPSK constellation. This polarization is then equalized with a standard CMA equalizer, forcing the symbol to the radius $R=1$, while the other polarization's energy is being minimized with $R=0$. The resulting error functions are shown in Eq. (5.22):

$$e_x = R_x - |x_{\text{out}}|^2,$$
$$e_y = R_y - |y_{\text{out}}|^2. \tag{5.22}$$

Additionally, for every modulation format the bit-error rate is monitored to prevent the equalizer from converging on the same polarization, in which case the equalizer is re-initialized with a different tap weight until it has converged correctly.

5.2.1.6 Frequency offset compensation

Frequency offset may be estimated in the frequency domain by using an Mth power nonlinearity to remove modulation, and then finding the maximum power in the Fourier transform of the resultant signal. This peak will be located at the offset frequency, as shown in Eq. (5.23) where T_s is the symbol period and FFT denotes the discrete Fourier transform

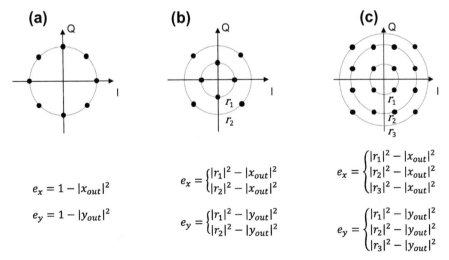

(a)

$$e_x = 1 - |x_{out}|^2$$
$$e_y = 1 - |y_{out}|^2$$

(b)

$$e_x = \begin{cases} |r_1|^2 - |x_{out}|^2 \\ |r_2|^2 - |x_{out}|^2 \end{cases}$$
$$e_y = \begin{cases} |r_1|^2 - |y_{out}|^2 \\ |r_2|^2 - |y_{out}|^2 \end{cases}$$

(c)

$$e_x = \begin{cases} |r_1|^2 - |x_{out}|^2 \\ |r_2|^2 - |x_{out}|^2 \\ |r_3|^2 - |x_{out}|^2 \end{cases}$$
$$e_y = \begin{cases} |r_1|^2 - |y_{out}|^2 \\ |r_2|^2 - |y_{out}|^2 \\ |r_3|^2 - |y_{out}|^2 \end{cases}$$

FIGURE 5.8 Constellation diagrams of 8PSK, 8QAM, and 16QAM, with equivalent error signals.

$$C(f) = \text{FFT}(x^4),$$
$$C(f_0) = \max(C(f)),$$
$$\Delta\varphi = \frac{2\pi f_0}{T_s}, \tag{5.23}$$
$$y(k) = x(k)\exp(-jk\Delta\varphi).$$

While computationally intensive, this method is accurate without requiring training or convergence, and may be used for short signals such as those captured in experiments where the sequence length may be on the order of 10^5 symbols [75].

5.2.1.7 Carrier phase recovery

The next DSP block is the digital phase estimation, required to recover the signal's carrier phase. A widely used carrier phase recovery scheme for PSK signals (e.g. for BPSK, QPSK, and 8PSK) is the feed-forward Mth power phase estimation [76] (or Viterbi and Viterbi algorithm [77]), the latter was used in the analysis of different transmission systems, described in the next part of the chapter. It operates as follows. The received complex samples are first raised to the Mth power to eliminate the phase modulation (Figure 5.9). Subsequently, the kth symbol $x(k)$ is added to its N predecessors and successors to average the estimated phase in order to combat the influence of noise. The argument, divided by M, leads to a phase estimate $\phi'(k)$ for $x(k)$:

$$\hat{\varphi}(k) = \frac{1}{M} \cdot \arg\left(\frac{1}{2N+1}\sum_{l=-N}^{N} x(k+l)^M\right). \tag{5.24}$$

Generally, this averaging can be seen as a low-pass filtering with the number of taps determining the bandwidth of the filter. Therefore, the optimum block-length $2N+1$ depends on the amount of noise that has been picked up by the signal and the symbol-rate, since a lower symbol-rate leads to a larger phase walk-off between neighboring symbols, increasing the laser phase noise variance $\sigma^2 = 2\pi \cdot \Delta v \cdot T_S$. The estimated phase must be unwrapped to combat the remaining phase ambiguity and avoid "cycle-slipping." The unwrapping and phase correction algorithm is given by:

$$a(k) = a(k-1) + \left\lfloor 1/2 - \frac{M}{2\pi}(\hat{\varphi}(k) - \hat{\varphi}(k-1))\right\rfloor,$$
$$\varphi'(k) = \hat{\varphi}(k) + a\frac{2\pi}{M}; \tag{5.25}$$
$$y(k) = x(k)e^{-j\varphi'(k)}$$

"Cycle-slipping" occurs when the true phase crosses a boundary while the phase estimate is not unwrapped, or conversely when the true phase does not cross a

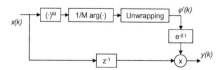

FIGURE 5.9 Carrier recovery using the Mth power scheme.

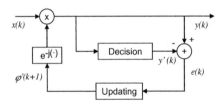

FIGURE 5.10 Digital phase-locked loop carrier recovery.

boundary while the phase estimate is unwrapped. This results in a rotation of the signal constellation and subsequent misdetection of the signal phase, resulting in burst
errors. Differential coding can mitigate the worst burst errors which could cause
signal failure; however, unwrapping is still highly desirable to improve performance.

To recover the phase of a polarization switched QPSK the Mth power phase estimator must be modified. The two parallel signal streams of a PS-QPSK signal, one
in each polarization, can be collapsed to a single QPSK stream by making decisions
on the energy in each symbol slot. The resulting QPSK signal can then be processed
with the standard Mth power scheme and the phase can be recovered [73].

Since QAM modulation is not restricted to the optical phase alone, the Mth
power algorithm as introduced above is not suitable for this kind of modulation.
Therefore, in this work, a decision directed phase-locked loop [71] has been
used to track the phase in the case of 8QAM and 16QAM modulation. As shown in
Figure 5.10, the PLL calculates the error between the $y(k)$ and the corresponding
hard decision $y'(k)$:

$$e(k) = y(k) - y'(k). \tag{5.26}$$

The decision boundaries are set at this stage, under the assumption that the phase-
noise is the dominant distortion [78]. The error information is used to update the
phase estimate for the following symbol:

$$\varphi'(k+1) = \varphi'(k) - \mu \text{Im}\{y(k)e^*(k)\}, \tag{5.27}$$

where * denotes the complex conjugate and μ is the step size parameter, which
was set to 0.1 [71]. Finally, the estimated phase is then applied to the next symbol
$y(k+1)=x(k+1)\exp(-j\phi(k+1)')$.

5.3 APPLICATION OF DBP-BASED DSP TO OPTICAL FIBER TRANSMISSION IN THE NONLINEAR REGIME

5.3.1 Nonlinearity compensation in optical communications

Having described the principles and the underlying blocks of the DSP, the next part
of the chapter is used to show how the backpropagating DSP is applied in the analysis of the optical transmission systems and what benefits it brings, especially in terms
of extending the transmission reach of long-haul and ultra-long-haul transmission

of higher-order modulation formats in single channel and WDM configurations, especially in the presence of nonlinearities. As demand for capacity continues to increase and limits to the available optical bandwidth are conceivable, higher-order modulation formats become a viable solution to use the available bandwidth more efficiently. However, with increasing constellation density, noise limitations become more stringent and signal power cannot be ramped up indefinitely due to nonlinear distortions arising from the Kerr effect.

The focus here is on the nonlinear compensation, based on the DSP of digital backpropagation. We consider both cases of backpropagating a single channel and also when enough receiver (both electrical and sampling) bandwidth is available to backpropagate more than one wavelength channel, in order to explore the fundamental limits of backpropagation. Although this might have previously been thought of as unconceivable, very recent work [79] has highlighted the feasibility of approaching this practically. Additionally, we focus on the nonlinear limitations of higher-order modulation formats for varying symbol-rates and channel spacing, and specifically on the trade-off between inter-channel nonlinearities, such as cross-phase modulation (XPM) and intra-channel nonlinearities, such as self-phase-modulation (SPM). This trade-off is key to the comparison described in this chapter.

5.3.2 Single-channel optical transmission performance

To investigate the relative contributions of intra- and inter-channel nonlinear effects to the nonlinear penalties, and how the DSP can be used to mitigate for these, polarization multiplexed single-channel transmission with varying symbol-rate 56 GBd, 28 GBd, 14 GBd, 7 GBd, and 3.5 GBd is considered here initially.

In every case, it was assumed that the payload of up to 400 Gbit/s (see net-bitrates per channel in Table 5.3) is transmitted with a 4% overhead for the Ethernet frame and 7% overhead for forward error correction (FEC) which adds up to a 12% overhead to the payload (e.g. 104 Gbit/s + 7% = 111.28 Gbit/s). Each WDM channel was modulated with 2^{15} symbols using a different random symbol sequence drawn from a uniform probability distribution rendering every symbol equally probable, as shown in Figure 5.11. Single-channel propagation has been modeled with eight temporal samples per symbol, to allow for sufficient bandwidth to accommodate the spectrum and excess bandwidth to cover nonlinearity induced spectral broadening.

QPSK is generated by an IQ modulator which enables the modulation of in-phase and quadrature components of the optical field with different binary signals, 8PSK was obtained by inserting another phase modulator which varies the phase between 0 and $\pi/4$, as determined by a third driving signal. PS-QPSK is generated with an IQ modulator to obtain a QPSK constellation and a polarization switching stage consisting of two parallel Mach-Zehnder modulators. The polarization switching stage is used to extinguish one polarization while the other one is in transmit state. To generate 8QAM the driving signal for one arm of an IQ modulator is attenuated and the in-built phase shifter is set to a constant phase-shift of $\pi/4$. A subsequent phase-modulator varies the optical phase between 0 and $\pi/4$ to obtain the desired 8-symbol constellation.

Table 5.3 Spectral efficiency and net-bitrate per channel for a given symbol-rate and modulation format is shown below.

		PS-QPSK	PDM-QPSK	PDM-8PSK	PDM-8QAM	PDM-16QAM
Spectral efficiency (bit/s/Hz)		1.5	2	3	3	4
Bit per symbol		3	4	6	6	8
Symbol-Rate (GBd)	Optical Filter BW (GHz)	Net-bitrate (Gbit/s)				
56	100	150	200	300	300	400
28	50	75	100	150	150	200
14	25	37.5	50	75	75	100
7	12.5	18.75	25	37.5	37.5	50
3.5	6.25	9.375	12.5	18.75	18.75	25

Finally, 16QAM was generated by driving an IQ modulator with 4-level driving signals, leading to 16 different symbols in the complex plane. In every case, the limited transmitter bandwidth was emulated by applying a fifth-order electrical Bessel filter with a 3 dB bandwidth of $0.8 \times$ symbol-rate and the laser linewidth of the transmitter was set to 100 kHz, corresponding to a conventional, commercially available external cavity laser (ECL). Note that the electrical bandwidth of the transmitter is usually dominated by the modulator bandwidth and the highest 3 dB electrical bandwidth needed would be $0.8 \times 56\,\text{GHz} = 44.8\,\text{GHz}$. Commercially available lithium niobate phase modulators [80] and IQ modulators [81] have a 3 dB electrical bandwidth around 30 GHz; however, due to the shallow roll-off of the frequency transfer function, they are suitable for modulation bandwidths beyond their 3 dB bandwidth. To reduce the number of optical components, all modulation formats have NRZ pulse shape, because RZ pulse shapes require another pulse carver and additional driving electronics, even though pulse carving has been shown to improve nonlinear transmission performance [82]. After modulation, the signals were polarization-multiplexed with another random symbol-pattern and finally passed through an optical interleaver consisting of second-order Gaussian filters with a 3 dB bandwidth, as shown in Table 5.3.

At the receiver, the signal is detected with a digital coherent receiver. The linewidths of transmitter- and LO-laser are both 100 kHz, while the frequency offset between the two was assumed to be 0 GHz. The limited receiver bandwidth was modeled with fifth-order Bessel filters, employing a 3 dB bandwidth of $0.8 \times$ symbol-rate and resampled to 2 samples/symbol. The specifications of a maximum receiver bandwidth of 44.8 GHz and a required ADC speed of 112 Gsample/s are reasonable assumptions for future digital coherent receivers, considering today's state-of-the-art oscilloscopes offer an electrical bandwidth of up to 45 GHz and up to 120 Gsample/s

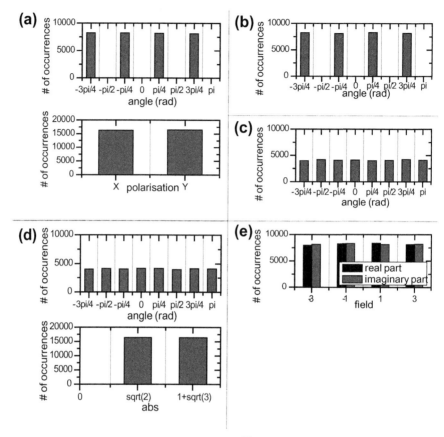

FIGURE 5.11 Uniform probability distributions of 2^{15} symbols for (a) PS-QPSK (phase and polarization), (b) QPSK, (c) 8PSK (phase), (d) 8QAM (phase and amplitude), and (e) 16QAM (real and imaginary part).

[83]. After digitization, digital signal processing algorithms are applied, as described in Section 5.2.1, the symbols are differentially decoded and Monte Carlo error counting is performed to determine the BER.

It is advisable to compare the simulations against theoretical values for the receiver sensitivity in the presence of additive white Gaussian noise. Figure 5.12 shows the BER vs. the signal-to-noise ratio (SNR) per bit, comparing Monte Carlo simulations (symbols) and analytical approximations to the BER performance for optimum and differential coding. The SNR per bit is related to the optical-signal-noise ratio (OSNR) as follows:

$$\text{SNR}_{\text{bit}} = \frac{2B_{\text{Ref}}}{F_B}\text{OSNR} \qquad (5.28)$$

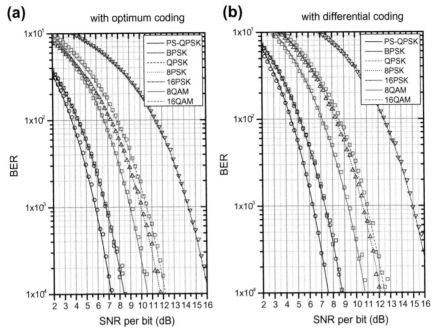

FIGURE 5.12 Receiver sensitivities for PS-QPSK, BPSK, QPSK, 8PSK, 16PSK, 8QAM, and 16QAM with (a) optimum coding and (b) differential coding. Lines show analytical equations (8QAM: numerical approximation) and symbols results of Monte Carlo simulations.

with B_{Ref} referring to the noise reference bandwidth (0.1 nm or 12.5 GHz in this case) and the overall bitrate F_B.

Excellent agreement between analytical approximation and simulations employing ideal matched filters with a root-raised cosine frequency response at the transmitter and receiver can be seen, especially in the high SNR region. At low SNR values, the assumption that errors occur only between neighboring symbols is no longer true. Most analytical approximations are based on the Marcum Q-function [76]:

$$Q(x) = \frac{1}{2}\text{erfc}\left(\frac{x}{\sqrt{2}}\right). \tag{5.29}$$

The following equations have been used to obtain the bit error probabilities of BPSK [76]:

$$\text{BER}_{\text{BPSK}} = Q\left(\sqrt{2 \cdot \text{SNR}_{\text{bit}}}\right). \tag{5.30}$$

QPSK [76]:

$$\text{BER}_{\text{QPSK}} = Q\left(\sqrt{2 \cdot \text{SNR}_{\text{bit}}}\right)\left(1 - \frac{1}{2}Q(\sqrt{2 \cdot \text{SNR}_{\text{bit}}})\right). \tag{5.31}$$

8PSK and 16PSK [76]:

$$\text{BER}_{\text{MPSK}} = \frac{2}{\log_2(M)} Q\left(\sqrt{2 \cdot \log_2(M) \cdot \sin^2(\frac{\pi}{M}) \cdot \text{SNR}_{\text{bit}}}\right) \quad (5.32)$$

as well as 16QAM [76]:

$$\text{BER}_{\text{16QAM}} = \left(1 - \frac{1}{\sqrt{16}}\right) Q\left(\sqrt{\frac{12}{15} \cdot \text{SNR}_{\text{bit}}}\right)\left(1 - \left(1 - \frac{1}{\sqrt{16}}\right) Q\left(\sqrt{\frac{12}{15} \cdot \text{SNR}_{\text{bit}}}\right)\right).$$

$$(5.33)$$

Note that an analytical approximation of the BER for PS-QPSK requires the numerical solution of an integral [84]:

$$\text{BER}_{\text{PS-QPSK}} = \frac{1}{2\sqrt{\pi}} \int_{-\infty}^{\infty} \text{erfc}(x)(3 - 3\text{erfc}(x)) + \text{erfc}^2(x)) \exp(-(x - 3 \cdot \text{SNR}_{\text{bit}})^2) dx.$$

$$(5.34)$$

Furthermore, 8QAM requires an entirely numerical approach, which is due to the more complex decision boundaries, implemented within the DSP, as shown in [78].

In the case of differentially coded field, as in Figure 5.12b, the analytical approximations in Eqs. (5.30)–(5.34) are not valid anymore, because in the event of a symbol error the following symbol will be erroneous as well. Indeed, the resulting BER is higher than in case of optimum coding, e.g. by a factor of 2 for phase shift keying as shown in Table 5.4. However, differential coding is still likely to be implemented in commercial systems, because it prevents error bursts due to cycle-slipping.

Table 5.5 summarizes the SNR per bit at the FEC rate of $\text{BER} = 3 \times 10^{-3}$ as inferred from Figure 5.12a for optimum coding and Figure 5.12b for differential coding.

Assuming that both polarizations of the optical field are used, the spectral efficiency is given and it is easy to see that the receiver sensitivity reduces monotonically with increasing spectral efficiency. Once the spectral efficiency of the modulation format increases, the energy must be increased to maintain the symbol points separate, given the same amount of additive white Gaussian noise. A notable exception to this is BPSK with a spectral efficiency of 2 bit/s/Hz, which does not use the four available dimensions (in-phase and quadrature components in the two polarizations) by using only one dimension per polarization.

The remainder of this section describes the application of the backpropagation-based DSP to a range of transmission systems using a range of application formats including PS-QPSK, PDM-QPSK, PDM-8PSK, PDM-8QAM, and PDM-16QAM.

Table 5.4 Correction factors for BER in the case of a differentially coded field.

M-PSK	8QAM	16QAM	PS-QPSK
2	1.4545	1.625	2.222

Table 5.5 Required SNR per bit at an FEC rate of BER $= 3 \times 10^{-3}$ for various modulation formats.

	PS-QPSK	BPSK	QPSK	8QAM	8PSK	16QAM	16PSK
Spectral efficiency (bit/s/Hz)	3	2	4	6	6	8	8
Optimum coding (dB)	4.9	5.8	5.8	8.0	8.9	9.4	13.2
Differential coding (dB)	5.5	6.4	6.4	8.4	9.6	9.9	13.9

Table 5.6 Fiber and link parameters.

α (dB/km)	0.2
D (ps/km/nm)	16
S (ps/km/nm^2)	0.06
γ (1/W/km)	1.2
PMD coefficient (ps/$\sqrt{k/m}$)	0.1
Span length (km)	80
Number of spans	13
EDFA noise figure (dB)	4.5

To illustrate the comparison, a sample transmission line is used and it is based on 13×80 km SSMF spans without any inline dispersion compensation. EDFAs, with a noise figure of 4.5 dB, are used to compensate for the fiber attenuation, giving a fixed output power of 17 dBm. Note that the noise was added at each amplifier along the link to model the interaction between ASE noise and nonlinearity, an important non-linear limitation especially in long and ultra-long haul transmission systems beyond 3000 km [85]. Signal propagation in the fiber can be modeled with the symmetrical split-step Fourier method including the effect of chromatic dispersion, dispersion slope, polarization mode dispersion, power dependence of the refractive index (Kerr effect), and nonlinear polarization scattering. In the described results, a step size of 100 m was used to ensure a valid representation of PMD via the waveplate model and accurate modeling of the peak nonlinear phase shift per step. Table 5.6 summarizes the link parameters used in the simulations described in this chapter.

Figure 5.13 shows the variation of BER vs. input launch power for PS-QPSK, PDM- PDM-QPSK, PDM-8PSK, PDM-8QAM, and PDM-16QAM, plotted for different baud rates. It can be seen that the linear parts of the curves at low input powers are shifted to the left by 3 dB as the symbol-rate is halved. This seems intuitively correct if one considers that in half the spectral width only half of the in-band

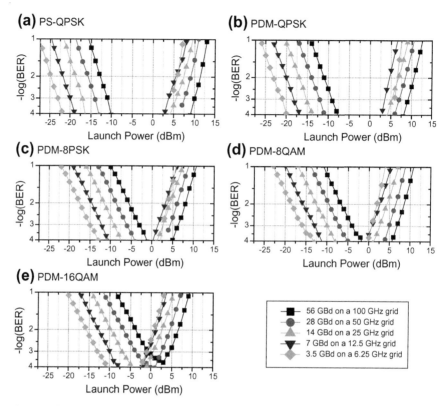

FIGURE 5.13 BER vs. input launch power in single-channel transmission of coherently detected PS-QPSK, PDM-QPSK, PDM-8PSK, PDM-8QAM, and PDM-16QAM, as a function of baud rate.

ASE-power is detected, so that only half the signal power is required to achieve the same SNR at the receiver. Figure 5.14 highlights this effect by showing SNR per bit for the transmission link under investigation with a separate launch power axis in blue for each spectral width.

However, the effect of varying in-band ASE-power can be neglected when plotting the received SNR per bit as a function of the power spectral density (PSD), defined by the launch power per channel normalized to the channel spacing (see Figure 5.14). In this case, different symbol-rates can be compared at the same SNR and curves of the same spectral efficiency and modulation format overlap in the linear transmission regime and have the same BER as shown in Figure 5.15. Therefore, the power spectral density was selected as the basis for comparing transmission performance at different symbol-rates.

Note that the PSD should not be given in dBm/Hz, because in this case the power would scale logarithmically while the frequency spacing scales linearly, again leading to a biased estimation of the linear and nonlinear performance. Instead the units of mW/GHz, as in this work, should be used.

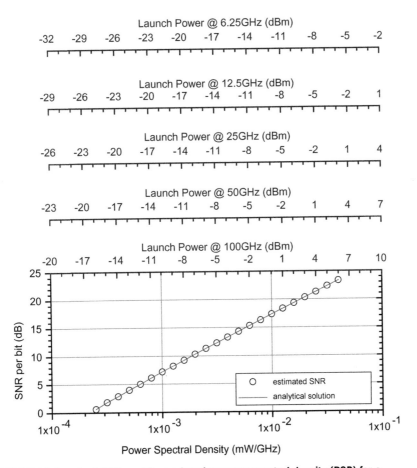

FIGURE 5.14 Received SNR per bit as a function power spectral density (PSD) for a 13 × 80 km standard single mode fiber link with EDFA amplification. Blue axes show the power per channel depending on the spectral width. (For interpretation of the references to color in this figure legend, the reader is referred to the web version of this book.)

Figure 5.16 shows the maximum power spectral density (PSD) at BER $= 3 \times 10^{-3}$ (corresponding to $-\log(\text{BER}) = 2.52$), selected as a figure of merit to compare the transmission performance of all modulation formats in the nonlinear regime. The maximum PSD was taken as a slice through the nonlinear part of the waterfall curves at BER $= 3 \times 10^{-3}$ in Figure 5.15. All the modulation formats show a clear improvement in nonlinear performance when the symbol-rate is reduced, which agrees very well with recent experimental results comparing PDM-QPSK and PS-QPSK at different symbol-rates [86,87] as well as simulation results obtained by Piyawanno et al. [88] comparing PDM-QPSK, PDM-16QAM, and PDM-64QAM. The improved transmission performance with lower symbol-rate can be attributed to a significantly reduced pulse overlap during transmission.

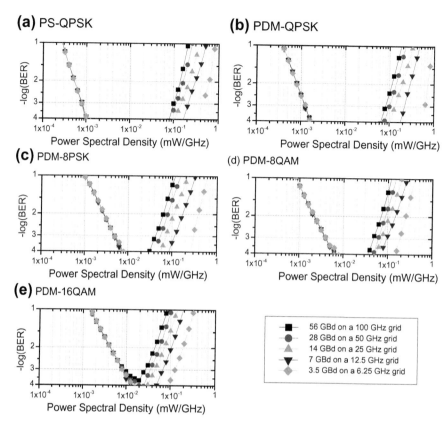

FIGURE 5.15 BER vs. power spectral density for single-channel transmission of coherently detected PS-QPSK, PDM-QPSK, PDM-8PSK, PDM-8QAM, and PDM-16QAM, for different values of baud rate.

Intra-channel four-wave mixing (IFWM) is the limiting nonlinear effect for single-channel phase shift keyed formats and its severity increases with the number of pulses that are involved in the nonlinear mixing process [89]. While a pulse propagating at 3.5 GBd only affects three neighboring pulses, this number increases to 745 pulses at 56 GBd.

Clearly, in the case of the single channel, the intra-channel self-phase modulation (ISPM) and intra-channel cross phase modulation (IXPM) depend on the signals pulse shape and, therefore, result in a constant nonlinear phase shift for every pulse for phase shift keyed formats. This phase shift will not degrade performance, since the digital phase recovery block within the DSP is able to mitigate for it. However, in the case of modulation formats with multiple intensity rings, like 8QAM and 16QAM, an increased de-rotation of the constellation rings due to different nonlinear phase shifts depending on the intensity can be observed [82]. This effect becomes more significant with reduced symbol-rate and, therefore, reduced pulse spread,

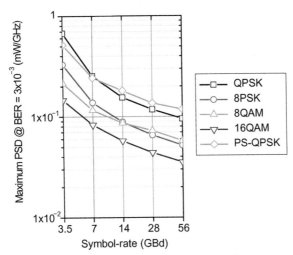

FIGURE 5.16 Maximum power spectral density @ BER $=3 \times 10^{-3}$ vs. symbol-rate for single-channel transmission of PDM-QPSK, PDM-8PSK, PDM-8QAM, PDM-16QAM, and PS-QPSK.

which can clearly be seen when comparing 8PSK to 8QAM. For modulation formats with multiple intensity rings, the de-rotation cannot be mitigated by the digital phase recovery, since the phase shift varies between adjacent symbols of different intensity. However, implementing a simple intensity-dependent phase shift at the receiver can improve performance in this case [53].

It can be seen that PS-QPSK shows better performance than all other formats, which can be explained with the absence of cross phase modulation (XPM) between the orthogonal polarizations, characteristic of every polarization multiplexed modulation format.

The severity of nonlinear distortions increases with increasing spectral efficiency, similarly to the case of linear distortions due to additive white Gaussian noise (see Figure 5.12). Recently, a number of authors, including, for example, Carena et al. [55] and Poggiolini et al. [90], have argued that in uncompensated transmission the nonlinear distortion can be reasonably well approximated with a Gaussian noise process whose variance is proportional to the square of the launch power.

5.3.3 Single-channel digital backpropagation

In this section, we describe the potential performance improvement that can be gained from applying optimum digital backpropagation-based DSP to a single-channel polarization multiplexed signal. The split-step Fourier method is implemented with the Manakov equation. The nonlinear step can be described by Eq. (5.16):

$$\widehat{N}(t) = -j\gamma\varphi\frac{8}{9}(|E_Y|^2 + |E_X|^2)P_{\text{in}}10^{(-[\frac{s}{n}-1]\frac{\alpha L}{10})},$$

where P_{in} denotes the input power per span, while $10^{(-\lceil \frac{s}{n} \rceil - 1]\frac{\alpha L}{10})}$ accounts for the varying power profile along the span, with n being the number of steps per span, s the index of the step within a span, L the span length in km, and α the fibers' attenuation coefficient in dB/km.

ϕ is a variable that converges with increasing number of computational steps toward 1, but has to be optimized for a realistic number of steps. ϕ and the number of steps per transmitted span have been optimized to minimize the error vector magnitude (EVM), which is defined as root mean square value of the Euclidian distance between detected symbol x_j and the closest member of the symbol alphabet s_j:

$$\text{EVM} = 100\% \cdot \sqrt{\frac{\frac{1}{N}\sum_{j=1}^{N}|s_j - x_j|^2}{\frac{1}{N}\sum_{j=1}^{N}|s_j|^2}}. \tag{5.34}$$

Here, the EVM is used as a percentage value. In this optimization, the BER is not an appropriate performance metric. This is because given a fixed number of symbols to count errors on, the accuracy at low values of BERs reduces due to the small number of counted errors.

The optimization ϕ and the number of steps per span were performed in the nonlinear transmission regime at 11 dBm, 9 dBm, 7 dBm, 5 dBm, and 7 dBm for 56 GBd, 28 GBd, 14 GBd, 7 GBd, and 3.5 GBd, respectively (see Figure 5.17). The accuracy of the digital backpropagation was monotonically increased by increasing the number of steps per span from 1 to 25, while varying ϕ between 0 and 1.

In general, the performance, likely to be achieved with digital backpropagation, improves as the number of steps is increased, as the approximation of instantaneously acting nonlinearity becomes more accurate as the dispersion per step is reduced. Of interest here is what is ultimately achievable with digital backpropagation with the choice of the optimum number of steps per span, that is the point at which a significant reduction of the EVM (<0.1% EVM) would no longer be achievable.

The optimum values are the same for every modulation format:

- 25 steps per span for 56 GBd,
- 7 steps per span for 28 GBd,
- 3 steps per span for 14 GBd,
- 1 step per span for 7 GBd,
- 1 step per span 3.5 GBd,

with $\phi = 0.9$–0.95.

Note that there is clearly a trade-off between the achievable performance improvement and hardware complexity. Recently Du and Lowery [25] showed that by applying a filtered nonlinear phase shift, the performance of the backpropagation algorithm can be improved, and the complexity was reduced by a factor of 4. A recent review of backpropagation-based DSP Schmauss et al. highlighted that further reductions are possible through further DSP optimization [91]. Another conclusion is that the optimum number of steps per span scales with DR_S^2 where D is the accumulated dispersion within a span and R_S is the symbol-rate. The faster

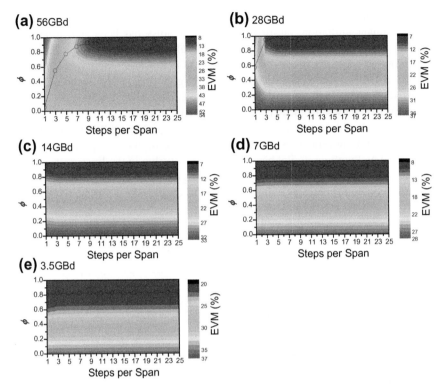

FIGURE 5.17 Backpropagation optimization for different symbol-rates of PDM-QPSK. The parameters φ and steps per span are optimized with EVM as a performance metric. Open symbols show the descent of the optimum parameters toward lowest EVM.

the low and high frequency components walk off from each other, the faster the waveform evolves during fiber propagation. Hence, an accurate representation of the waveform requires more samples, or steps per span, for signals with a higher bandwidth, or symbol-rate.

As mentioned earlier, in an ideal case the parameter ϕ converges toward 1 with increasing number of steps per span, so that the nonlinear phase shift in Eq. (5.16) becomes the original nonlinear term of the Manakov equation. However, it has to be noted that the digital backpropagation algorithm is limited by a lot of factors, related to transmission distortions and the non-ideal implementation of the coherent receiver:

- Gordon Mollenauer noise [48].
- Out-of-band nonlinear noise (XPM, XPolM) [31,47].
- PMD [47], PDL.
- Frequency Response of the Receiver (Optical Filter, Electrical Front end...).
- ADC sample rate [23].
- Quantization noise.
- LO-laser phase noise [92].

While in the full field DBP Gordon Mollenauer noise [48] has been identified as the dominant limitation; out-of-band nonlinear noise has been shown to dominate [31] the achievable efficiency of the DBP algorithm when only a portion of the transmitted spectrum is backpropagated. Note that if it is only possible to apply DBP-based DSP to the received channel, the neighboring WDM-channels may be added and dropped along the link, causing unpredictable nonlinear distortions [93].

Figure 5.18 shows the BER as a function of input power density for transmission of PS-QPSK, PDM-QPSK, PDM-8PSK, PDM-8QAM, and PDM-16QAM with and without optimum numbers of steps and full-field digital backpropagation. The curves, with and without digital backpropagation, overlap in the linear transmission regime, since the nonlinear phase shift is negligible and digital backpropagation compensates only for chromatic dispersion in this case.

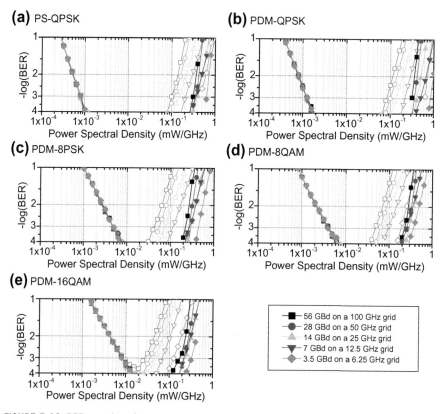

FIGURE 5.18 **BER as a function power spectral density for single-channel transmission of coherently detected PS-QPSK, PDM-QPSK, PDM-8PSK, PDM-8QAM, and PDM-16QAM. Open symbols denote transmission without nonlinear compensation, while filled symbols show transmission with optimum digital backpropagation.**

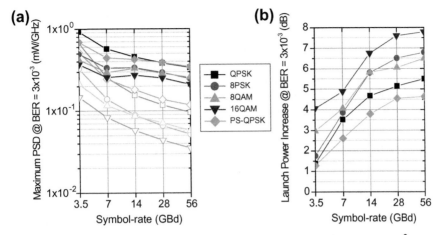

FIGURE 5.19 Figure (a) depicts the maximum power spectral density @ BER$=3\times10^{-3}$ vs. symbol-rate for single-channel transmission with (filled symbols) and without optimum digital backpropagation (open symbols). Figure (b) shows the resulting increase in maximum launch power @ BER$=3\times10^{-3}$.

Similarly to the previous section the nonlinear transmission performance can be compared more easily among different modulation formats by looking at the maximum power spectral density at BER$=3\times10^{-3}$. Figure 5.19a shows the maximum PSD for transmission with (filled symbols) and without optimum nonlinear compensation (open symbols). It can be seen that lower symbol-rates show an improved transmission performance than higher symbol-rates, irrespective of whether nonlinear compensation is applied or not. PDM-QPSK shows the best nonlinear performance throughout all the symbol-rates, while PS-QPSK is the only modulation format to match performance at 56 GBd and 28 GBd. PDM-8QAM and PDM-8PSK have a similar maximum PSD for 56 GBd, 28 GBd, and 14 GBd, while PDM-8PSK outperforms PDM-8QAM at lower symbol-rates. PDM-16QAM exhibits the worst nonlinear performance, among all the modulation formats.

However, more spectrally efficient modulation formats benefit the most (up to 7.75 dB for PDM-16QAM) from nonlinear compensation as shown in Figure 5.19b, which translates the maximum PSD into the equivalent increase in maximum launch power at BER$=3\times10^{-3}$. This effect can be explained as follows: since the denser constellations suffer more from nonlinear distortions due to the higher proximity of the constellation points, the improvement in BER tends to be more significant when this distortion is compensated for. Another observation is that higher symbol-rates benefit more from digital backpropagation. Nonlinear mixing processes between signal frequencies and ASE-noise, which is usually denoted as Gordon-Mollenauer noise, is the effect which ultimately limits digital backpropagation in a single-channel regime [48]. For a narrower spectrum, signal components show a lower phase variation across the spectrum, which facilitates mixing processes between neighboring signal components, which are already affected by Gordon-Mollenauer noise.

Multi-ring modulation formats such as PDM-8QAM and PDM-16QAM benefit more from digital backpropagation at lower symbol-rates, because the de-rotation between adjacent rings, which dominated the BER in the nonlinear regime, is mitigated by the digital backpropagation.

5.3.4 **WDM transmission**

In this section, the attention focuses on WDM-transmission of coherently detected PS-QPSK, PDM-QPSK, PDM-8PSK, PDM-8QAM, and PDM-16QAM. In WDM transmission systems with polarization multiplexing, SPM is not the only severe nonlinear distortion, since cross phase modulation and cross polarization modulation also limit transmission. These nonlinear impairments are known to grow in severity with reduced frequency spacing due to reduced walkoff between neighboring channels [94]. However, as described in the previous section, it was found that in the case of SPM-limited transmission the nonlinear degradation reduces with a narrower spectrum for every modulation format. Consequently, by changing the frequency spacing and symbol-rate intra-channel nonlinearities and inter-channel nonlinearities can be traded off against each other, leading to optimum transmission performance at a specific symbol-rate.

Similarly for the single-channel section, we illustrate the effectiveness of the DBP spectral efficiency per modulation format and we compared different symbol-rates in terms of the power spectral density (PSD). The use of the PSD facilitates comparison between different symbol-rates, because signals with the same signal-to-noise ratio have the same PSD, but not the same launch power per channel. The channel spacing is varied between 100 GHz and 12.5 GHz according to the symbol-rate, while the number of channels had to be increased from 9 in the case of 56 GBd up to 144 in the case of 3.5 GBd to ensure a full occupation of the optical bandwidth that has been investigated (Figure 5.20). Note that all WDM channels as well as X- and Y-polarization contain completely decorrelated symbol patterns based on pseudo-random symbol sequences with different seeds. This is particularly important when investigating ultra-dense frequency grids below 25 GHz, since WDM-channels walk off much slower from each other leading to correlated distortions. Note that WDM propagation was modeled with 16, 32, 64, 128, and 256 temporal samples per symbol for 56 GBd, 28 GBd, 14 GBd, 7 GBd, and 3.5 GBd, respectively. This poses a good trade-off between simulation time (in the order of weeks for 3.5 GBd) and accuracy, because it allows for sufficient bandwidth to accommodate the spectrum and excess bandwidth to cover nonlinearity induced spectral broadening.

To compare nonlinear performance of WDM-systems, it is useful to plot the BER as a function of the power spectral density (PSD), as in the previous section. Figure 5.21 shows the BER as a function of the PSD for all the compared modulation.

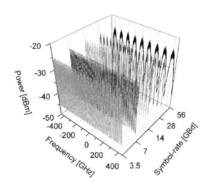

FIGURE 5.20 Optical power spectra for QPSK at 56 GBd (black), 28 GBd (red), 14 GBd (green), 7 GBd (blue), and 3.5 GBd (orange). (For interpretation of the references to color in this figure legend, the reader is referred to the web version of this book.)

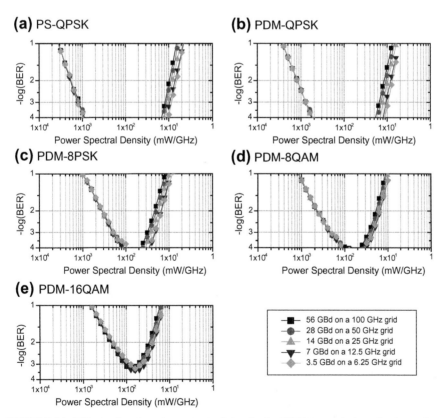

FIGURE 5.21 BER as a function power spectral density for WDM-transmission of coherently detected PS-QPSK, PDM-QPSK, PDM-8PSK, PDM-8QAM, and PDM-16QAM.

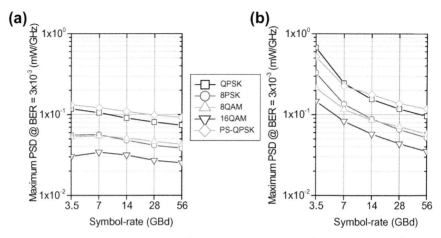

FIGURE 5.22 Figure (a) depicts the maximum power spectral density @ BER=3×10^{-3} vs. symbol-rate for WDM transmission, while figure (b) shows single-channel transmission.

In the linear transmission regime curves of different symbol-rates overlap because signals with the same PSD have the same SNR and consequently the same BER, similarly to the single-channel case (Figure 5.15).

To compare transmission performance in the nonlinear regime the maximum PSD at BER=3×10^{-3} is shown in Figure 5.22. In the case of WDM transmission (Figure 5.22a), all modulation formats follow the same trend of improved performance for lower symbol-rates down to 7 GBd. However, at lower symbol-rates the performance starts to be dominated by inter-channel XPM and FWM, as the gap between nonlinear performance of single channel and WDM-transmission widens (compare Figure 5.22a and b). Additionally, an increased impact on higher density constellations can be observed, which confirms the findings in [88].

It can be seen that PS-QPSK exhibits the best nonlinear performance with a maximum PSD of 0.13 mW/GHz at 3.5 GBd. PDM-QPSK shows nearly the same maximum PSD (0.12 mW/GHz at 3.5 GBd), while PDM-8PSK and PDM-8QAM tolerate only half the maximum PSD with 0.056 and 0.055 mW/GHz at 3.5 GBd and 7 GBd, respectively. The maximum PSD for PDM-16QAM is 0.034 mW/GHz at 7 GBd. The improvement that can be gained by reducing the symbol-rate with respect to 56 GBd ranges from 2 dB for QPSK and 1.6 dB for 8PSK to 1.5 dB for PS-QPSK, 1.3 dB for 16QAM, and 1 dB for 8QAM. The benefit of reducing the symbol-rate in reducing intra-channel nonlinear distortion as depicted for single-channel transmission in Figure 5.22b is offset by the increased inter-channel XPM and FWM resulting from the reduced channel spacing.

The next question to answer is how effective is the DBP and associated nonlinear compensation as the length of link is increased. The answer to this question is shown in Figure 5.23, where the transmission distance for a WDM-system as a function

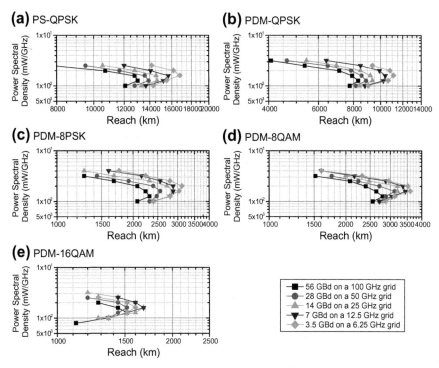

FIGURE 5.23 Achievable transmission reach @ BER = 3 × 10⁻³ as a function of power spectral density for WDM transmission of (a) PS-QPSK, (b) PDM-QPSK, (c) PDM-8PSK, (d) PDM-8QAM, and (e) PDM-16QAM.

of the input power spectral density, modulation format, and symbol-rate is plotted. A higher nonlinear tolerance is clearly observed at lower symbol-rates for all the modulation formats (similarly to a fixed link length in Figure 5.22a). This increased nonlinear performance translates into an equivalent increase in transmission distances, listed in Table 5.7.

Due to the highest linear and nonlinear tolerance, the longest transmission distance with up to 16,800 km for 3.5 GBd is achieved with PS-QPSK, while PDM-QPSK is second with up to 11,120 km at 3.5 GBd. PS-QPSK and PDM-QPSK have recently been experimentally investigated at 20 GBd on a 50 GHz grid [86] and at 15 GBd on a 25 GHz grid [87], confirming the higher transmission reach for PS-QPSK in this region. Furthermore, 28 GBd PDM-QPSK can be compared to the experimental results for an equivalent system in Ref [95]. The experimental maximum transmission distance was more than 1400 km lower (7382 km compared to 8800 km), due to additional implementation penalty and an increased amount of ASE noise. The direct comparison between PDM-8PSK and PDM-8QAM at 3.5 GBd shows a 480 km higher transmission reach of 3600 km for PDM-8QAM compared to 3120 km for

Table 5.7 Maximum transmission distance in kilometers, assuming FEC can correct for BER $= 3 \times 10^{-3}$ (largest transmission distances per modulation format are highlighted).

Symbol-Rate(GBd)	PS-QPSK	PDM-QPSK	PDM-8PSK	PDM-8QAM	PDM-16QAM
56	13,040	8,320	2,320	2,800	1,520
28	13,920	8,880	2,560	3,120	1,520
14	14,640	9,360	2,880	3,200	1,680
7	15,680	10,400	2,880	3,440	1,680
3.5	16,800	11,120	3,120	3,600	1,600

PDM-8PSK. The longest transmission distance for PDM-16QAM is observed at 7 and 14 GBd: 1680 km, in line with recent simulation results of Piyawanno et al. [88]. However, it is worth noting that all maximum transmission distances for 16QAM lie within three spans, ranging from 1520 to 1680 km. In the case of 28 GBd PDM-16QAM, the simulated maximum reach was more than 300 km higher compared to an equivalent system investigated in [31] (1520 km compared to 1200 km), again due to additional implementation penalty and higher levels of ASE-noise.

Generally, the symbol-rates with highest transmission distances correspond very well to the nonlinear performance on the fixed link displayed in Figure 5.22a. A notable exception is the relative performance of PDM-8PSK and PDM-8QAM, where PDM-8QAM shows worse nonlinear performance on the fixed link, but a higher maximum transmission reach at the lowest symbol-rate of 3.5 GBd. This inconsistency can be explained by taking into account that 8QAM has a 1.2 dB better linear performance at BER $= 3 \times 10^{-3}$ as shown in Figure 5.12b. This benefit in the linear region, combined with an improved nonlinear performance at higher symbol-rates, can perhaps explain the longer transmission distance, for every symbol-rate.

Note that the optimum symbol-rate for WDM transmission of a particular modulation format is not only determined by the trade-off between inter- and intra-channel nonlinearities, but is also influenced by practical considerations such as cost effectiveness (governed by the trade-off between the number of transceivers required and the operating speed of the electronics) and the most convenient WDM channel spacing for network routing.

The optimum launch power spectral density lies in the same, relatively narrow range between 0.013 and 0.02 mW/GHz for all modulation formats (see Table 5.8). The maximum transmission reach is correlated with the optimum power spectral density: the higher the optimum PSD the larger the tolerance to nonlinearities will be, when comparing symbol-rates of the same modulation format. However, when comparing transmission performance with different modulation formats this correlation is masked by different tolerances to ASE-noise, which affect the optimum PSD.

Table 5.8 Optimum power spectral density in mW/GHz, assuming FEC can correct for BER$=3\times10^{-3}$ (highest power spectral densities per modulation format are highlighted).

Symbol-Rate(GBd)	PS-QPSK	PDM-QPSK	PDM-8PSK	PDM-8QAM	PDM-16QAM
56	0.013	0.013	0.013	0.013	0.013
28	0.013	0.013	0.013	0.016	0.016
14	0.013	0.013	0.016	0.016	0.016
7	0.016	0.016	0.016	0.016	0.016
3.5	0.016	0.016	0.020	0.016	0.016

5.3.5 Digital backpropagation of the central channel

In this section we investigate the potential gain of the digital backpropagation algorithm based on the Manakov equation described in Section 5.2.1. As for the investigation of single-channel transmission in Section 5.3.3, we focus on optimum backpropagation of one channel at different symbol-rates, but now in a WDM environment.

The optimum number of steps per span and the value of optimization parameter ϕ are found to be similar as in the single-channel case (see Figure 5.17), which is unsurprising since both of these parameters depend only on the backpropagated spectrum and not on the adjacent channels. The optimum values are identical irrespective of the modulation format: 25 steps per span for 56 GBd, 7 steps per span for 28 GBd, 3 steps per span for 14 GBd, and 1 step per span for 7 GBd and 3.5 GBd with $\phi=0.9$–0.95.

Figure 5.24 shows the calculated BER for all modulation formats and symbol-rates, as a function of the PSD. It can be seen that curves in the linear region overlap as before, because nonlinear compensation does not improve performance in the regions which have a negligible nonlinear phase shift.

Figure 5.25 shows the maximum PSD @ BER$=3\times10^{-3}$ and resulting launch power increases in the nonlinear transmission regime. PS-QPSK performs best for every symbol-rate that has been investigated and shows the nearly same performance of 0.13–0.14 mW/GHz. PDM-QPSK shows same flat performance over all symbol-rates with a maximum PSD of 0.11–0.12 mW/GHz. Modulation formats with 3 bit/symbol like 8PSK and 8QAM show similar performance in the region of 0.06 mW/GHz for all symbol-rates. In the case of 16QAM the best nonlinear performance of 0.044 mW/GHz can be seen at the highest symbol-rate 56 GBd.

Figure 5.25b translates the improvement that can be gained by digital backpropagation from maximum PSD into launch power in dB. All modulation formats show a clear trend of improved benefit from digital backpropagation for higher symbol-rates similar to the single-channel case in Figure 5.22b. However, there is significantly less improvement to be gained, which can be attributed to the nonlinear phase shift from the adjacent channels (cross phase modulation (XPM)). Figure 5.26a illustrates

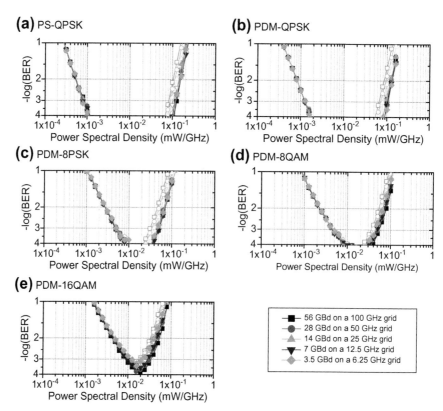

FIGURE 5.24 BER as a function of power spectral density for WDM-transmission of coherently detected PS-QPSK, PDM-QPSK, PDM-8PSK, PDM-8QAM, and PDM-16QAM. Open symbols denote transmission without nonlinear compensation (for comparison), while filled symbols show transmission with digital backpropagation of the central channel.

how the nonlinear distortion depends on the symbol-rate R_S. The nonlinear distortion due to SPM increases with the symbol-rate, while the XPM distortion reduces with increased frequency spacing/symbol-rate. Although SPM is fully compensated, since the full spectrum of the channel of interest is backpropagated (see Figure 5.26b–f), XPM is not compensated for and acts as additional distortion that limits the efficiency of digital backpropagation, so that the benefit is negligible in the case of a low symbol-rate of, e.g., 3.5 GBd on a 6.25 GHz grid. At 7 GBd the launch power can be increased by 0.4 dB and at 14 GBd by 0.6–0.8 dB. 16QAM shows a higher benefit than all other modulation formats at 28 GBd (1.6 dB compared to 1.1–1.3 dB) and at 56 GBd (2.4 dB compared to 1.8–1.9 dB), respectively.

It is interesting that increased improvement for spectrally more efficient modulation formats has been already observed in the single-channel case (see Section 5.3.3). This effect can be attributed to the higher density of constellation points—denser

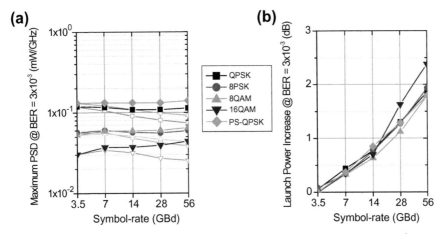

FIGURE 5.25 Figure (a) depicts the maximum power spectral density @ BER=3×10⁻³ vs. symbol-rate for WDM-transmission with (filled symbols) and without optimum digital backpropagation of the central channel (open symbols). Figure (b) shows the resulting increase in maximum launch power at @ BER=3×10⁻³.

constellations (i.e. where the constellation points are spaced closer together) suffer more from nonlinear distortions, therefore the improvement in BER tends to be more pronounced when this distortion is compensated for. However, this effect is heavily masked by the degrading impact of XPM which limits the efficiency of the digital backpropagation algorithm and can only be seen at higher symbol-rates for PDM-16QAM [93].

To highlight the impact of the improved nonlinear performance, the maximum achievable transmission distances for PS-QPSK, PDM-QPSK, PDM-8PSK, PDM-8QAM, and PDM-16QAM have been plotted in Figure 5.27, showing a comparison between transmission with and without digital backpropagation.

Although the performance improvement on the 13×80 km link at lower symbol-rates (see Figure 5.25b) is relatively small, the maximum achievable transmission reach increased significantly (up to 1840 km in the case of PS-QPSK). This can be explained by the fact that on all amplified transmission links degradations tend to accumulate in a logarithmic fashion (e.g. SNR increases with launch power in dB—see Figure 5.14). As a consequence, small differences in nonlinear performance after a few spans can lead to significant differences in maximum achievable reach.

It is worth noting that the symbol-rate with the best nonlinear performance on the fixed link is not necessarily the symbol-rate with the highest maximum reach, especially for PS-QPSK and PDM-QPSK. The disparity between the two sets of results can be explained by the high value of power spectral density at which the results are obtained. For the BER vs. PSD simulations on the 13×80 km link, changing SNR conditions are characteristic. First, the SNR increases in the linear regime and then reduces in the nonlinear regime, whereas maximum reach curves are taken at a fixed BER

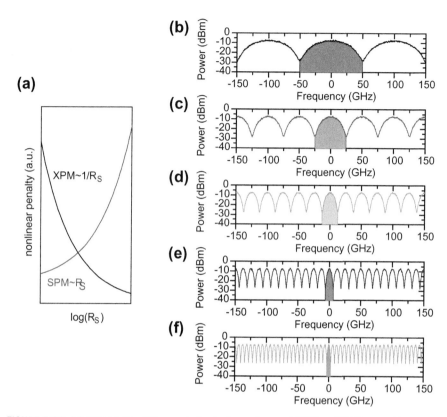

FIGURE 5.26 (a) Schematic of the symbol-rate dependency of SPM and XPM as well as backpropagated spectral content for (b) 56 GBd, (c) 28 GBd, (d) 14 GBd, (e) 7 GBd, and (f) 3.5 GBd.

and, therefore, at a fixed SNR. This can negatively affect equalization convergence and phase recovery algorithms, especially when an increased amount of nonlinear noise is present and the length of averaging windows has been designed for high SNR regions.

PS-QPSK achieves the highest transmission distance with 18,400 km at 3.5 GBd and PDM-QPSK performs second best with 12,240 km at 3.5 GBd as documented in Table 5.9. Similar to the results for the fixed link (see Figure 5.25b), an increased benefit of digital backpropagation for higher symbol-rates can be observed consistently for all modulation formats. In the case of PS-QPSK, digital backpropagation increases reach by 1600 km for 3.5 GBd, 1840 km for 7 GBd, 2480 km for 14 GBd, 2800 km for 28 GBd, and 3120 km for 56 GBd.

However, the achievable transmission distances are different for the different modulation formats. While PS-QPSK and PDM-QPSK have a maximum in the transmission reach at the lowest symbol-rate of 3.5 GBd (18,400 km and 12,240 km), a similar maximum transmission distance for all symbol-rates can be observed at

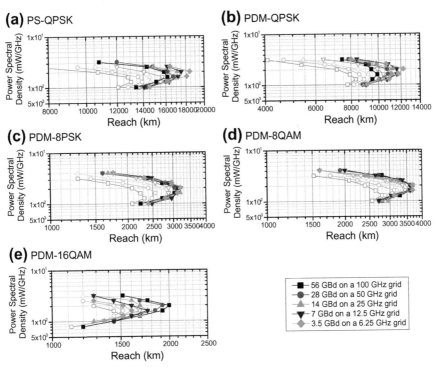

FIGURE 5.27 Achievable transmission reach @ BER = 3×10^{-3} with (filled symbols) and without digital backpropagation (open symbols). WDM- transmission of (a) PS-QPSK, (b) PDM-QPSK, (c) PDM-8PSK, (d) PDM-8QAM, and (e) PDM-16QAM is investigated at varying symbol-rates.

Table 5.9 Maximum transmission distance with digital backpropagation in kilometers, assuming FEC can correct for BER = 3×10^{-3} (largest transmission distances per modulation format are highlighted).

Symbol-Rate(GBd)	PS-QPSK	PDM-QPSK	PDM-8PSK	PDM-8QAM	PDM-16QAM
56	16,160	9,920	3,120	3,680	2,000
28	16,720	10,800	3,120	3,680	1,920
14	17,120	11,200	3,200	3,600	1,760
7	17,520	11,600	3,200	3,600	1,760
3.5	18,400	12,240	3,200	3,680	1,600

PDM-8PSK and PDM-8QAM ($\sim$3200 km and $\sim$3700 km). This trend is reversed with PDM-16QAM, for which the maximum transmission distance of 2000 km can be achieved at 56 GBd. This behavior can be explained by an increased nonlinear penalty induced by neighboring WDM channels modulated with higher order modulation formats such as 16QAM compared to, e.g., PDM-QPSK [93]. The increased nonlinear crosstalk is more detrimental to the efficiency of the digital backpropagation algorithm in the case of a narrow grid such as 6.25 GHz than in the case of a 100 GHz grid. This suggests that if the interfering channels had been modulated consistently with, e.g., PDM-QPSK (or another modulation format with a low spectral efficiency), digital backpropagation would provide a greater benefit for more spectrally efficient modulation formats at all symbol-rates. However, it is worth noting that this benefit comes at the cost of reduced capacity of the neighboring WDM channels. Rafique et al. [93] reported that the increased degradation induced by more spectrally efficient modulation formats is a result of the higher peak-to-average power ratio at low values of accumulated dispersion and could be mitigated by appropriate dispersion pre-compensation.

Improvements of the DBP algorithm can be compared for 28 GBd PDM-QPSK between simulation and the experiment described in [95]. In the case of the experiment it was possible to extend the maximum achievable transmission distance by +20% (corresponding to 1400 km from 7382 km to 8826 km) with a 1 step per span algorithm, while in the case of the simulation transmission reach was extended by +23% (corresponding to 2000 km from 8800 km to 10800 km) with a 7 steps per span algorithm. Comparing the simulation of 28 GBd PDM-16QAM to a similar experiment [32], it appears that in experimental transmission reach has been extended by only +13% (160 km from 1200 km to 1360 km) with a 1 step per span algorithm, as opposed to +26% (400 km from 1520 km to 1920 km) with a 7 steps per span algorithm. However, it should be noted that a simulation study [31] based on the experiment in [32] revealed that a +30% improvement would be more characteristic for this system. At this symbol-rate, a single step per span seems to be sufficient to enjoy the majority of the improvement available through DBP.

5.3.6 Multi-channel digital backpropagation

Clearly DBP for a single channel within a WDM comb has significant shortcomings. Given the improvements in the ADC speeds and progress in digital coherent receivers for larger bandwidths covering adjacent WDM-channels, it will become possible to compensate not only for intra-channel nonlinearities, but for inter-channel nonlinearities such as XPM as well. Let us now assume that for every symbol-rate investigated a coherent receiver with a 3 dB electrical bandwidth of $0.8 \times 56 = 44.8$ GHz (corresponding to the 56 GBd receiver) is available. In this case the full available bandwidth would be digitized at 112G samples and digitally backpropagated. For symbol-rates below 56 GBd, more than one channel is backpropagated and, therefore, cross phase modulation is compensated for, additionally to self-phase modulation. We illustrate this section by applying digital backpropagation, again based on the Manakov equation, as described in Section 5.2.1. The optimum number of steps per span is 25, which is identical to the value used for 56 GBd single channel and WDM transmission, since the backpropagated spectral width is 100 GHz in all cases. The overall power of the backpropagated waveform at the beginning of backpropagation is the sum of the power of all channels that have been digitized. However, due to the roll-off of the optical filter at the receiver, spectral content of the channels in the roll-off region is cut out. The optimization parameter ϕ must reflect this and is correspondingly smaller. After backpropagation the central channel was selected by resampling the signal to two samples per symbol and an equalizer, a phase recovery circuit, and a differential decoder is applied, before error counting.

Figure 5.28 shows the resultant BER for all modulation formats and symbol-rates as a function of the PSD. Again, curves in the linear region overlap, because nonlinear compensation does not improve performance in regions with negligible nonlinear phase shift.

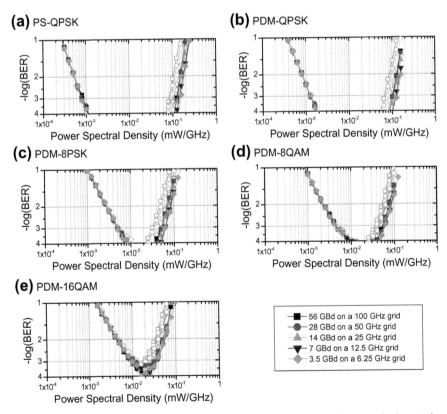

FIGURE 5.28 BER vs. power spectral density for WDM-transmission of coherently detected PS-QPSK, PDM-QPSK, PDM-8PSK, PDM-8QAM, and PDM-16QAM. Open symbols denote transmission without nonlinear compensation (for comparison), while filled symbols show transmission with digital backpropagation covering 100 GHz.

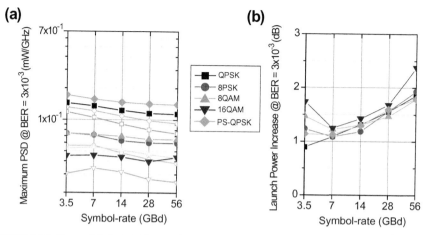

FIGURE 5.29 Figure (a) depicts the maximum power spectral density @ BER=3×10⁻³ vs. symbol-rate for WDM transmission with (filled symbols) and without optimum digital backpropagation covering 100 GHz (open symbols). Figure (b) shows the resulting increase in maximum launch power.

Figure 5.29 shows the maximum PSD @ BER$=3\times10^{-3}$ and resulting launch power increases in the nonlinear transmission regime. PS-QPSK performs best for every symbol-rate, similarly to the previous section, where only the central channel was backpropagated. However, when backpropagating a 100 GHz portion of the spectrum, 3.5 GBd shows the best performance with a maximum PSD of 0.17 mW/GHz compared to 0.13 mW/GHz, when only the central channel (6.25 GHz) is backpropagated. PDM-QPSK shows the same performance improvement toward lower symbol-rates with a maximum PSD of 0.14 mW/GHz at 3.5 GBd. PDM-8QAM and PDM-8PSK show similar performance in the region of 0.06–0.07 mW/GHz, with slightly better performance at lower symbol-rates. In the case of 16QAM the maximum PSD is nearly flat across the symbol-rates in the region of 0.044–0.045 mW/GHz.

Figure 5.29a shows the resultant increase in the launch power that can be gained by digital backpropagation from maximum PSD. As expected, all modulation formats show better performance at lower symbol-rates compared to digital backpropagation of the central channel (see Figure 5.25), since a larger proportion of the spectrum is backpropagated (see Figure 5.30b–f). However, the benefit from digital backpropagation remains different at different symbol-rates, even though the same spectrum of 100 GHz is backpropagated in every case.

As already discussed in the previous section, Figure 5.30a shows how the impact of SPM increases with symbol-rate, while the distortion due to XPM is reduced with increased frequency spacing or symbol-rate. However, because a fixed bandwidth of 100 GHz is backpropagated, not only SPM is compensated but at lower symbol-rates increasing amounts of XPM are compensated as well. The DBP-algorithm is now

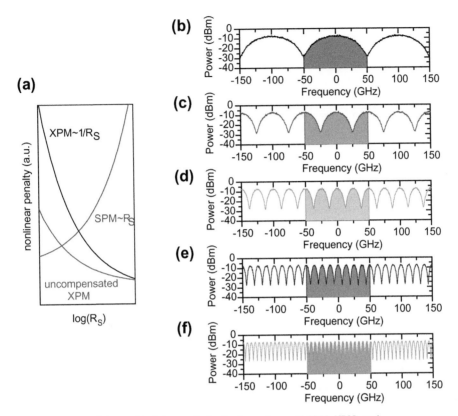

FIGURE 5.30 (a) Schematic of the symbol-rate dependency of SPM, XPM, and uncompensated XPM as well as backpropagated spectral content (100 GHz) for (b) 56 GBd, (c) 28 GBd, (d) 14 GBd, (e) 7 GBd, and (f) 3.5 GBd.

limited by XPM induced by the channels outside the backpropagated bandwidth. The influence of this uncompensated XPM increases with reduced symbol-rate and frequency spacing (Figure 5.30a), resulting, as before, in higher benefit for DBP at higher symbol-rates as displayed in Figure 5.29a.

Similarly to previous sections, the maximum achievable transmission reach has been calculated when 100 GHz worth of bandwidth is backpropagated.

Figure 5.31 shows the reach curves for (a) PS-QPSK, (b) PDM-QPSK, (c) PDM-8PSK, (d) PDM-8QAM, and (e) PDM-16QAM (filled symbols) comparing it to transmission without DBP (open symbols).

Table 5.10 summarizes the maximum achievable transmission distances for all modulation formats with 100 GHz DBP, highlighting the trend of increased reach for lower symbol-rates similarly to what has been observed without DBP (see Table 5.7). Again, PS-QPSK outperforms all other modulation formats by achieving a maximum transmission distance of 21,600 km at 3.5 GBd (+28.6% with respect to no

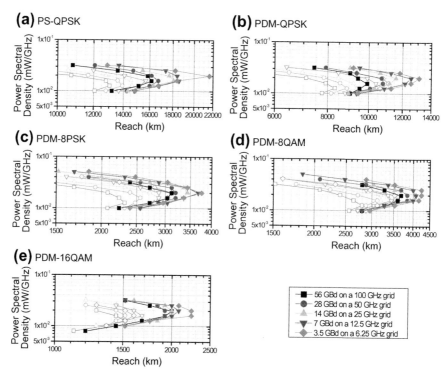

FIGURE 5.31 Achievable transmission reach @ BER = 3×10^{-3} with (filled symbols) and without 100 GHz digital backpropagation (open symbols). WDM-transmission of (a) PS-QPSK, (b) PDM-QPSK, (c) PDM-8PSK, (d) PDM-8QAM, and (e) PDM-16QAM is investigated at varying symbol-rates.

Table 5.10 Maximum transmission distance with 100 GHz digital backpropagation in km, assuming FEC can correct for BER = 3×10^{-3} (largest transmission distances per modulation format are highlighted).

Symbol-Rate(GBd)	PS-QPSK	PDM-QPSK	PDM-8PSK	PDM-8QAM	PDM-16QAM
56	16,160	9,920	3,120	3,680	2,000
28	16,800	11,040	3,200	3,840	1,920
14	18,080	11,600	3,440	4,080	2,000
7	18,560	12,560	3,680	4,080	2,080
3.5	21,600	13,120	3,760	4,240	2,240

DBP and +17.4% compared to central channel DBP), which is more than 7000 km more compared to the second best modulation format PDM-QPSK (+17.9% with respect to no DBP and +7.2% compared to central channel DBP) at the same symbol-rate. PDM-8QAM outperforms PDM-8PSK by more than 400 km irrespective of the symbol-rate. At 3.5 GBd PDM-8PSK gains +20.5% reach compared to no DBP and +17.5% compared to central channel DBP, while PDM-8QAM gains +17.8% and +15.2%, respectively. PDM-16QAM shows a +40% improvement compared to the case without DBP or central channel DBP.

5.4 SUMMARY AND FUTURE QUESTIONS

This chapter has focused on the application of digital backpropagation-based DSP in WDM optical communication systems with different modulation formats. There is a clear trade-off between intra- and inter-channel nonlinearities. With increasing symbol-rate inter-channel nonlinear distortions show a reduced impact, which is due to the wider frequency spacing. At the same time, however, inter-channel nonlinearities show a more severe impact with increasing symbol-rate, since the pulse overlap within the system increases with spectral width. The optimum symbol-rate can be found to balance these effects and shifts from 3.5 to 7 GBd when increasing the density of the constellation, e.g., from QPSK to 16QAM, indicating that higher-order modulation formats lead to more severe inter-channel distortions.

The simplest case is to apply the backpropagation only to the central channel. Higher symbol-rate signals such as 56 GBd and higher-order modulation formats such as PDM-16QAM have been found to benefit more from digital backpropagation. However, low symbol-rates such as 3.5 GBd and spectrally less efficient modulation formats such as PS-QPSK still show a higher maximum transmission reach in single-channel transmission. Even though deterministic nonlinearities are compensated for nonlinear interactions between signal and ASE noise limit the efficiency of the DBP algorithm. These tend to be less harmful for signals with a narrower spectrum (lower symbol-rate) since less ASE power is picked up along the transmission link. Adding more WDM channels to the signal leads to an increased amount of uncompensated inter-channel nonlinearities, which is particularly harmful in a narrower grid (lower symbol-rate). Under these conditions the optimum symbol-rate is the result of a trade-off between nonlinear signal-ASE interactions and uncompensated inter-channel nonlinearities. While less spectrally efficient modulation formats such as QPSK and PS-QPSK have their optima at 3.5 GBd, 8QAM and 8PSK show a similar transmission reach for all symbol-rates and 16QAM clearly favors 56 GBd, which is due to an increased influence of inter-channel nonlinearities induced by spectrally more efficient modulation formats.

In the case of the receiver with an electrical bandwidth of 100 GHz (relatively conservative compared to recently published work in [76] reporting a 228 GHz receiver!), capable of backpropagating adjacent channels, the trade-off between nonlinear signal-ASE interactions and uncompensated inter-channel nonlinearities is notably relaxed, especially at lower symbol-rates. This results in increased transmission distances for lower symbol-rates, irrespective of the modulation format.

However, for selecting the optimum symbol-rate in a real system one has to take into account practical considerations, such as cost effectiveness (governed by the trade-off between the number of transceivers required and the operating speed of the electronics) and the most convenient WDM channel spacing for network routing.

The achievable increases in transmission distances with nonlinearity-compensating DBP DSP are in the region of 20–30%, equivalent to around 1 dB. It has recently been argued that these increases are insignificant given the computational complexity. However, it can also be argued that a dB scale for assessing these benefits is an incorrect measure for distances, and a 30% increase—in transmission distance, going some way toward increasing a trans-Atlantic to trans-Pacific transmission distance is a worthwhile achievement. So—what next for DSP for optical communication?

DSP design for the first generation of digital coherent systems was driven primarily by the need to achieve adequate performance with reliable demodulation. Designs were heavily constrained by the processing resources available (due to the high bitrates required) which led to the use of highly compact algorithms which were originally developed for single-carrier wireless applications in the early 1980s. As DSP for coherent optical fiber communication has developed as a field, we have seen the development of new algorithms that allow us to exploit the well-known physical properties of the channel. Much research is currently under way into transmission systems which utilize different channels, such as few-mode fiber and multicore fiber, these new channels present new challenges and opportunities to the DSP designer, and will provide many future avenues for research. The increasing integration of transmitter DSP with FEC encoding, and of receiver DSP with FEC decoding, promise further opportunities to develop more statistically optimal receivers with the highly constrained resources available when operating in the terabit/s regime. As bit and baud rates continue to increase, the issue of maximizing capacity per fiber and per unit of spectrum becomes ever more important. Finally (and most importantly to the DSP designer), digital signal processing is now as fundamentally important to the field of optical fiber communications as is communication theory and nonlinear optics. The capabilities and constraints of DSP algorithms affect, and are affected by, every aspect of the design process within the system, from laser cavity dimensions and optical hybrid manufacturing tolerances to micro-structured fiber design and fiber system and link designs.

Clearly, there is much more to DSP than just backpropagation and that all backpropagation-based DSP is currently probably too complex. There is much research to do to find simplified DSP algorithms for a range of optical systems designs, such as the DSP is appropriate to both the link design and the nature of nonlinearity-caused impairment. And looking to the future, if the channel carrying capacity is to continue to increase, in line with the ever-growing bandwidth demands, fundamentally and radically new solutions too will be needed to help maintain the reduction of cost (and energy) per transmitted bit—on which all previous bandwidth growth has been predicated. This will need practical ways of overcoming the limits imposed by fiber nonlinearities. And so the search for new modulation formats combined with advanced DSP and coding, specifically tailored for the nonlinear channel, remains the goals in the quest to increase the fiber channel capacity, toward the Shannon limit and beyond.

Acknowledgments

The authors would like to acknowledge the invaluable contributions, discussions, and help of colleagues in the Optical Networks Group—Dr. Robert Killey, Dr. Seb Savory, Domanic Lavery, and Sean Kilmurray, in the course of this work and the financial support from the UK EPSRC, under the UNLOC Programme Grant and other projects, the Royal Society, and Huawei Technologies.

References

[1] A. Chraplyvy, The coming capacity crunch, in: Proceedings of the 35th European Conference on Optical Communication ECOC 2009, Vienna, 2009, p. Mo1.0.2.

[2] Minnesota Internet Traffic Studies (MINTS), University of Minnesota, August 2009. <http://www.dtc.umn.edu/mints/home.php>.

[3] Cisco Visual Networking Index, June 2011. <http://www.cisco.com/en/US/solutions/collateral/ns341/ns525/ns537/ns705/ns827/white_paper_c11-481360_ns827_Networking_Solutions_White_Paper.html>.

[4] K. Kikuchi, Coherent optical communications systems, in: I.P. Kaminow, T. Li, A.E. Willner (Eds.), Optical Fiber Telecommunications V B, Academic Press, 2008.

[5] P.J. Winzer, R.J. Essiambre, Advanced optical modulation formats, in: I.P. Kaminow, T. Li, A.E. Willner (Eds.), Optical Fiber Telecommunications V B, Academic Press, 2008.

[6] S.J. Savory, G. Gavioli, V. Mikhailov, R.I. Killey, P. Bayvel, Ultra Long-Haul QPSK transmission using a digital coherent receiver, in: Proceedings of the Digest of the IEEE LEOS Summer Topical Meetings, 2007, pp. 13–14.

[7] H. Sun, K.-T. Wu, K. Roberts, Real-time measurements of a 40 Gb/s coherent system, Opt. Express 16 (2008) 873–879.

[8] M. Salsi, H. Mardoyan, P. Tran, C. Koebele, E. Dutisseuil, G. Charlet, S. Bigo, 155 × 100 Gbit/s coherent PDM-QPSK transmission over 7200 km, in: Proceedings of the 35th European Conference on Optical Communication ECOC 2009, 2009, p. PD2.5.

[9] J. Yu, X. Zhou, M.-F. Huang, Y. Shao, D. Qian, T. Wang, M. Cvijetic, P. Magill, L. Nelson, M. Birk, S. Ten, H.B. Matthew, S.K. Mishra, 17 Tb/s (161 × 114 Gb/s) PolMux-RZ-8PSK transmission over 662 km of ultra-low loss fiber using C-band EDFA amplification and digital coherent detection, in: Proceedings of the 34th European Conference on Optical Communication ECOC 2008, 2008.

[10] X. Zhou, J. Yu, M.-F. Huang, T.S.Y., P. Magill, M. Cvijetic, L. Nelson, M. Birk, G. Zhang, S. Ten, H.B. Mattew, S.K. Mishra, 32 Tb/s (320 × 114 Gb/s) PDM-RZ-8QAM transmission over 580 km of SMF-28 ultra-low-loss fiber, in: Proceedings of the Conference on Optical Fiber Communication and the National Fiber Optic Engineers Conference OFC/NFOEC 2009, 2009.

[11] A.H. Gnauck, P.J. Winzer, S. Chandrasekhar, X. Liu, B. Zhu, D.W. Peckham, 10×224-Gb/s WDM transmission of 28-Gbaud PDM 16-QAM on a 50-GHz grid over 1200 km of fiber, in: Proceedings of the Conference on Optical Fiber Communication OFC '10, 2010, p. PDPB.

[12] X. Zhou, J. Yu, M. Huang, Y. Shao, T. Wang, L. Nelson, P. Magill, M. Birk, P.I. Borel, D.W. Peckham, R. Lingle, 64-Tb/s (640×107-Gb/s) PDM-36QAM transmission over 320 km using both pre- and post-transmission digital equalization, in: Proceedings of the Conference on Optical Fiber Communication and the National Fiber Optic Engineers Conference OFC/NFOEC 2010, 2010, p. PDPB9.

[13] S. Shimotsu, S. Oikawa, T. Saitou, N. Mitsugi, K. Kubodera, T. Kawanishi, M. Izutsu, Single side-band modulation performance of a LiNbO$_3$ integrated modulator consisting of four-phase modulator waveguides, Photon. Technol. Lett. 13 (2001) 364–366.

[14] N. Yoshikane, I. Morita, 160% spectrally-efficient 5.12 Tb/s (64×85.4 Gb/s RZ DQPSK) transmission without polarisation demultiplexing, in: Proceedings of the 30th European Conference on Optical Communication ECOC 2004, 2004.

[15] G. Charlet, J.-C. Antona, S. Lanne, S. Bigo, From 2,100 km to 2,700 km distance using phase-shaped binary transmission at 6.3 Tbit/s capacity, in: Proceedings of the Conference on Optical Fiber Communication and the National Fiber Optic Engineers Conference OFC/NFOEC 2003, 2003, pp. 329–330.

[16] J. Sakaguchi, B.J. Puttnam, W. Klaus, Y. Awaji, N. Wada, A. Kanno, T. Kawanishi, K. Imamura, H. Inaba, K. Mukasa, R. Sugizaki, T. Kobayashi, M. Watanabe, 19-core fiber transmission of $19 \times 100 \times 172$-Gb/s SDM-WDM-PDM-QPSK signals at 305 Tb/s, in: OFC 2012, p. PDP5C.1.

[17] K. Roberts, C. Li, L. Strawczynski, M. O'Sullivan, I. Hardcastle, Electronic precompensation of optical nonlinearity, Photon. Technol. Lett. 18 (2006) 403–405.

[18] R.I. Killey, P.M. Watts, V. Mikhailov, M. Glick, P. Bayvel, Electronic dispersion compensation by signal predistortion using digital processing and a dual-drive Mach-Zehnder modulator, J. Lightwave Technol. 17 (2005) 714–716.

[19] R. Waegemans, S. Herbst, L. Hohlbein, P. Watts, P. Bayvel, C. Fuerst, R.I. Killey, 10.7 Gb/s electronic predistortion transmitter using commercial FPGAs and D/A converters implementing real-time DSP for chromatic dispersion and SPM compensation, Opt. Express 17 (May) (2009) 8630–8640.

[20] S.J. Savory, Digital filters for coherent optical receivers, Opt. Express 16 (2008) 804–817.

[21] G.P. Agrawal, Nonlinear Fiber Optics, Academic Press, 1995.

[22] C.R. Menyuk, Application of multiple-length-scale methods to the study of optical fiber transmission, J. Eng. Math. 36 (1999) 113–136.

[23] E. Ip, J.M. Kahn, Compensation of dispersion and nonlinear effects using digital backpropagation, J. Lightwave Technol. 26 (2008) 3416–3425.

[24] G. Goldfarb, M.G. Taylor, G. Li, Experimental demonstration of fiber impairment compensation using the split-step finite-impulse-response filtering method, Photon. Technol. Lett. 20 (2008) 1887–1889.

[25] L.B. Du, A.J. Lowery, Improved single channel backpropagation for intra-channel fiber nonlinearity compensation in long-haul optical communication systems, Opt. Express 18 (August) (2010) 17075–17088.

[26] S. Oda, T. Tanimura, T. Hoshida, C. Ohshima, H. Nakashima, Z. Tao, J.C. Rasmussen, 112 Gb/s DP-QPSK transmission using a novel nonlinear compensator in digital coherent

receiver, in: Proceedings of the Conference on Optical Fiber Communication OFC 2009, 2009, pp. 1-3.

[27] T. Tanimura, T. Hoshida, S. Oda, T. Tanaka, C. Oshima, Z. Tao, J.C. Rasmussen, Systematic analysis on multi-segment dual-polarisation nonlinear compensation in 112Gb/s DP-QPSK coherent receiver, in: Proceedings of the 35th European Conference on Optical Communication ECOC 2009, Vienna, 2009, p. 9.4.5.

[28] G. Charlet, M. Salsi, P. Tran, M. Bertolini, H. Mardoyan, J. Renaudier, O. Bertran-Pardo, S. Bigo, 72×100 Gb/s transmission over transoceanic distance, using large effective area fiber, hybrid Raman-Erbium amplification and coherent detection, in: Proceedings of the National Fiber Optic Engineers Conference, 2009, p. PDPB6.

[29] X. Liu, S. Chandrasekhar, B. Zhu, P.J. Winzer, D.W. Peckham, 7×224-Gb/s WDM transmission of reduced-guard-interval CO-OFDM with 16-QAM subcarrier modulation on a 50-GHz grid over 2000km of ULAF and five ROADM passes, in: Proceedings of the 36th European Conference on Optical Communication ECOC 2010, Torino, 2010, p. Tu.3.C.2.

[30] S.J. Savory, G. Gavioli, E. Torrengo, P. Poggiolini, Impact of interchannel nonlinearities on a split-step intrachannel nonlinear equalizer, Photon. Technol. Lett. 22 (May) (2010) 673–675.

[31] C. Behrens, S. Makovejs, R.I. Killey, S.J. Savory, M. Chen, M. Bayvel, Pulse-shaping versus digital backpropagation in 224Gbit/s PDM-16QAM transmission, Opt. Express 19 (2011) 12879–12884.

[32] S. Makovejs, High-speed optical fibre transmission using advanced modulation formats, PhD Thesis, Electrical and Electronic Engineering, University College London, London, 2011.

[33] X. Liu, S. Chandrasekhar, B. Zhu, P.J. Winzer, A.H. Gnauck, D.W. Peckham, 448-Gb/s reduced-guard-interval CO-OFDM transmission over 2000km of ultra-large-area fiber and five 80-GHz-grid ROADMs, J. Lightwave Technol. 29 (February) (2011) 483–489.

[34] E. Yamazaki, A. Sano, T. Kobayashi, E. Yoshida, Y. Miyamoto, Mitigation of nonlinearities in optical transmission systems, in: Proceedings of the Conference on Optical Fiber Communication and the National Fiber Optic Engineers Conference OFC/NFOEC 2011, Los Angeles, 2011, p. OThF1.

[35] L. Li, Z. Tao, L. Dou, W. Yan, S. Oda, T. Tanimura, T. Hoshida, J.C. Rasmussen, Implementation efficient nonlinear equalizer based on correlated digital backpropagation, in: Proceedings of the Conference on Optical Fiber Communication and the National Fiber Optic Engineers Conference OFC/NFOEC 2011, Los Angeles, 2011, p. OWW3.

[36] M. Salsi, O. Bertran-Pardo, J. Renaudier, W. Idler, H. Mardoyan, P. Tran, G. Charlet, S. Bigo, WDM 200Gb/s single carrier PDM-QPSK transmission over 12,000km, in: Proceedings of the 37th European Conference on Optical Communication ECOC 2011, Geneva, 2011, p. Th.13.C.5.

[37] T. Tanimura, S. Oda, T. Hoshida, L. Li, Z. Tao, J.C. Rasmussen, Experimental characterisation of nonlinearity mitigation by digital back propagation and nonlinear polarization crosstalk canceller under high PMD condition, in: Proceedings of the Conference on Optical Fiber Communication and the National Fiber Optic Engineers Conference OFC/NFOEC 2011, Los Angeles, 2011, p. JWA20.

[38] W. Yan, Z. Tao, L. Dou, L. Li, S. Oda, T. Tanimura, T. Hoshida, J.C. Rasmussen, Low complexity digital perturbation back-propagation, in: Proceedings of the European Conference on Optical Communications ECOC 2011, Geneva, 2011, p. Tu.3.A.2.

[39] C. Behrens, D. Lavery, R.I. Killey, S.J. Savory, P. Bayvel, Long-haul WDM transmission of PDM-8PSK and PDM-8QAM with nonlinear DSP, in: Proceedings of the Conference on Optical Fiber Communication and the National Fiber Optic Engineers Conference OFC/NFOEC 2012, Los Angeles, 2012, p. OMA3A.4.

[40] E. Yamazaki, H. Masuda, A. Sano, T. Yoshimatsu, T. Kobayashi, E. Yoshida, Y. Miyamoto, R. Kudo, K. Ishihara, M. Matsui, Y. Takatori, Multi-staged nonlinear compensation in coherent receiver for 16340-km transmission of 111-Gb/s no-guard-interval Co-OFDM, in: Proceedings of the 35th European Conference on Optical Communication ECOC 2009, 2009.

[41] F. Yaman, G. Li, Nonlinear impairment compensation for polarization-division multiplexed WDM transmission using digital backward propagation, Photon. J. 1 (August) (2009) 144–152.

[42] D.S. Millar, S. Makovejs, V. Mikhailov, R.I. Killey, P. Bayvel, S.J. Savory, Experimental comparison of nonlinear compensation in long-haul PDM-QPSK transmission at 42.7 and 85.4 Gb/s, in: Proceedings of the 35th European Conference on Optical Communication ECOC 2009, 2009.

[43] L. Du, B. Schmidt, A. Lowery, Efficient digital backpropagation for PDM-CO-OFDM optical transmission systems, in: Proceedings of the Conference on Optical Fiber Communication and the National Fiber Optic Engineers Conference OFC/NFOEC 2010, 2010, p. OTuE2.

[44] D.S. Millar, S. Makovejs, C. Behrens, S. Hellerbrand, R.I. Killey, P. Bayvel, S.J. Savory, Mitigation of fiber nonlinearity using a digital coherent receiver, J. Sel. Top. Quant. Electron. 16 (2010) 1217–1226.

[45] T. Yoshida, T. Sugihara, H. Goto, T. Tokura, K. Ishida, T. Mizuochi, A study on statistical equalization of intra-channel fiber nonlinearity for digital coherent optical systems, in: Proceedings of the European Conference on Optical Communications ECOC 2011, Geneva, 2011, p. Tu.3.A.1.

[46] D. Lavery, C. Behrens, S. Makovejs, D.S. Millar, R.I. Killey, S.J. Savory, P. Bayvel, Long-haul transmission of PS-QPSK at 100 Gb/s using digital backpropagation, Photon. Technol. Lett. 24 (February) (2012) 176–178.

[47] E. Ip, Nonlinear compensation using backpropagation for polarization-multiplexed transmission, J. Lightwave Technol. 28 (March) (2010) 939–951.

[48] D. Rafique, A.D. Ellis, The impact of signal-ASE four-wave mixing in coherent transmission systems, in: Proceedings of the Conference on Optical Fiber Communication and the National Fiber Optic Engineers Conference OFC/NFOEC 2011, 2011, pp. OthO2.

[49] C. Behrens, R.I. Killey, S.J. Savory, M. Chen, P. Bayvel, Benefits of digital backpropagation in coherent QPSK and 16QAM fibre links, in: Communications and Photonics Conference and Exhibition (ACP), Shanghai, 2010, pp. 359–360.

[50] D. Rafique, J. Zhao, A.D. Ellis, Digital back-propagation for spectrally efficient WDM 112 Gbit/s PM m-ary QAM transmission, Opt. Express 19 (March) (2011) 5219–5224.

[51] S. Makovejs, E. Torrengo, D. Millar, R. Killey, S. Savory, P. Bayvel, Comparison of pulse shapes in a 224 Gbit/s (28 Gbaud) PDM-QAM16 long-haul transmission experiment, in: Proceedings of the Conference on Optical Fiber Communication and the National Fiber Optic Engineers Conference OFC/NFOEC 2011, Los Angeles, 2011, pp. OMR5.

[52] C. Xu, X. Liu, Postnonlinearity compensation with data driven phase modulators in phase-shift keying transmission, Opt. Lett. 27 (September) (2002) 1619–1621.

[53] K. Kikuchi, Electronic post-compensation for nonlinear phase fluctuations in a 1000-km 20-Gbit/s optical quadrature phase-shift keying transmission system using the digital coherent receiver, Opt. Express 16 (2008) 889–896.

[54] K.-P. Ho, J.M. Kahn, Electronic compensation technique to mitigate nonlinear phase noise, Photon. Technol. Lett. 22 (2004) 779–783.

[55] A. Carena, G. Bosco, V. Curri, P. Poggiolini, M. Tapia Taiba, F. Forghieri, Statistical characterization of PM-QPSK signals after propagation in uncompensated fiber links, in: Proceedings of the European Conference on Optical Communications ECOC 2010, Torino, 2010, p. P4.07.

[56] G.D. Forney, Maximum-likelihood sequence estimation of digital sequences in the presence of intersymbol interference, Trans. Inform. Theory IT-18 (May) (1972) 363–378.

[57] G.D. Forney, The Viterbi algorithm, Proc. IEEE 61 (March) (1973) 268–278.

[58] S.J. Savory, Y. Benlachtar, R.I. Killey, P. Bayvel, G. Bosco, P. Poggiolini, J. Prat, M. Omella, IMDD transmission over 1,040 km of standard single-mode fiber at 10 Gbit/s using a one-sample-per bit reduced-complexity MLSE receiver, in: Proceedings of the Conference on Optical Fiber Communication and the National Fiber Optic Engineers Conference OFC/NFOEC 2007, 2007, p. OThK2.

[59] S. Chandrasekhar, A.H. Gnauck, Performance of MLSE receiver in a dispersion-managed multispan experiment at 10.7 Gb/s under nonlinear transmission, Photon. Technol. Lett. 18 (December) (2006) 2448–2450.

[60] J.X. Cai, Y. Cai, C.R. Davidson, A. Lucero, H. Zhang, D.G. Foursa, O.V. Sinkin, W.W. Patterson, A. Philipetskii, G. Mohs, N.S. Bergano, 20 Tbit/s capacity transmission over 6,860 km, in: Proceedings of the Conference on Optical Fiber Communication and the National Fiber Optic Engineers Conference OFC/NFOEC 2011, Los Angeles, 2011, pp. PDPB4.

[61] J.G. Proakis, Adaptive equalization for TDMA digital mobile radio, IEEE Trans. Vehicular Technol. 40 (May) (1991) 333–341.

[62] J.-X. Cai, Y. Cai, C.R. Davidson, D.G. Foursa, A. Lucero, O. Sinkin, W. Patterson, A. Pilipetskii, G. Mohs, N.S. Bergano, Transmission of 96×100-Gb/s bandwidth-constrained PDM-RZ-QPSK channels with 300% spectral efficiency over 10610 km and 400% spectral efficiency over 4370 km, J. Lightwave Technol. 29 (February 2011) 491–497.

[63] Y. Cai, D.G. Foursa, C.R. Davidson, J.X. Cai, O. Sinkin, M. Nissov, A. Philipetskii, Experimental demonstration of coherent MAP detection for nonlinearity mitigation in long-haul transmissions, in: Proceedings of the Conference on Optical Fiber Communication and the National Fiber Optic Engineers Conference OFC/NFOEC 2010, 2010, p. OTuE1.

[64] J. Zhao, A.D. Ellis, Performance improvement using a novel MAP detector in coherent WDM systems, in: Proceedings of the 34th European Conference on Optical Communication ECOC 2008, 2008, p. Tu.1.D.2.

[65] K.V. Peddanarappagari, M. Brandt-Pearce, Volterra series approach for optimizing fiber-optic communications system designs, J. Lightwave Technol. 16 (1998) 2046–2055.

[66] Y. Gao, F. Zhang, L. Dou, Z. Chen, A. Xu, Intra-channel nonlinearities mitigation in pseudo-linear coherent QPSK transmission systems via nonlinear electrical equaliser, Opt. Commun. 282 (March) (2009) 2421–2425.

[67] F.P. Guiomar, J.D. Reis, A.L. Teixeira, A.N. Pinto, Mitigation of intra-channel nonlinearities using a frequency-domain Volterra series equaliser, Opt. Express 20 (January) (2012) 1360–1368.

[68] Z. Pan, B. Châtelain, M. Chagnon, D.V. Plant, Volterra filtering for nonlinearity impairment mitigation in DP-16QAM and DP-QPSK fiber optic communication systems, in: Proceedings of the Conference on Optical Fiber Communication and the National Fiber Optic Engineers Conference OFC/NFOEC 2011, Los Angeles, 2011, p. JThA40.

[69] C. Weber, J.K. Fischer, C.A. Bunge, K. Petermann, Electronic precompensation of intrachannel nonlinearities at 40 Gb/s, Photon. Technol. Lett. 18 (August) (2006) 1759–1761.

[70] D. Godard, Self-recovering equalization and carrier tracking in two-dimensional data communication systems, Trans. Commun. 28 (1980) 1867–1875.

[71] I. Fatadin, D. Ives, S.J. Savory, Blind equalization and carrier phase recovery in a 16-QAM optical coherent system, J. Lightwave Technol. 27 (2009) 3042–3049.

[72] S.J. Savory, G. Gavioli, R.I. Killey, P. Bayvel, Electronic compensation of chromatic dispersion using a digital coherent receiver, Opt. Express 15 (2007) 2120–2126.

[73] D.S. Millar, S.J. Savory, Blind adaptive equalization of polarization switched QPSK modulation, Opt. Express 19 (2011) 8533–8538.

[74] D.S. Millar, Digital signal processing for coherent optical fibre communications, PhD Thesis, Electrical and Electronic Engineering, University College London, London, 2011.

[75] D. Rife, R. Boorstyn, Single tone parameter estimation from discrete-time observations, Trans. Inform. Theory 20 (1974) 591–598.

[76] J.D. Proakis M. Salehi, Digital Communications, McGraw-Hill, 2008.

[77] A.J. Viterbi, A.M. Viterbi, Nonlinear estimation of PSK-modulated carrier phase with application to burst digital transmission, Trans. Inform. Theory 29 (1983) 543–551.

[78] E. Ip, J.M. Kahn, Carrier synchronization for 3- and 4-bit-per-symbol optical transmission, J. Lightwave Technol. 23 (2005) 4110–4124.

[79] N.K. Fontaine, G. Raybon, B. Guan, A. Adamiecki, P.J. Winzer, R. Ryf, A. Konczykowska, F. Jorge, J.-Y. Dupuy, L.L. Buhl, S. Chandrasekhar, 228-GHz coherent receiver using digital optical bandwidth interleaving and reception of 214-GBd (856-Gb/s) PDM-QPSK, in: Proceedings of the 38th European Conference on Optical Communications ECOC 2012, Amsterdam, The Netherlands, 2012.

[80] Covega Co., March 2011. <http://www.covega.com/products/pdfs/LN%20027-066%20 Rev%20F.pdf>.

[81] Fujitsu Optical Components Ltd., November 2011. <http://jp.fujitsu.com/group/foc/ downloads/services/100gln/ln100gdpqpsk-e-111102.pdf>.

[82] C. Behrens, R.I. Killey, S.J. Savory, M. Chen, P. Bayvel, Nonlinear distortion in transmission of higher-order modulation formats, Photon. Technol. Lett. 22 (2010)

[83] Le Croy Co. December 2011. <http://cdn.lecroy.com/files/pdf/lecroy_labmaster_9_zi_ datasheet.pdf>.

[84] E. Agrell, M. Karlsson, Power-efficient modulation formats in coherent transmission systems, J. Lightwave Technol. 27 (2009) 5115–5126.

[85] P. Serena, N. Rossi, A. Bononi, PDM-iRZ-QPSK vs. PS-QPSK at 100 Gbit/s over dispersion-managed links, Opt. Express 20 (March) (2012) 7895–7900.

[86] M. Sjödin, B.J. Puttnam, P. Johannisson, S. Shinada, N. Wada, P.A. Andrekson, M. Karlsson, Transmission of PM-QPSK and PS-QPSK with different fiber span lengths, Opt. Express 20 (March) (2012) 7544–7554.

[87] M. Sjödin, B.J. Puttnam, P. Johannisson, S. Shinada, N. Wada, P.A. Andrekson, M. Karlsson, Comparison of PS-QPSK and PM-QPSK at different data rates in a

25 GHz-spaced WDM system, in: Proceedings of the Conference on Optical Fiber Communication and the National Fiber Optic Engineers Conference OFC/NFOEC 2012, Los Angeles, 2012, p. OTu2A.3.

[88] K. Piyawanno, M. Kuschnerov, B. Spinnler, B. Lankl, Optimal symbol-rate for optical transmission systems with coherent receivers, in: Proceedings of the SPPCom 2010, Karlsruhe, Germany, 2010, p. SPWA2.

[89] K.-P. Ho, Phase Modulated Optical Communication Systems, Springer, 2005.

[90] P. Poggiolini, Analytical modeling of non-linear propagation in coherent systems, in: Proceedings of the 38th European Conference on Optical Communication ECOC 2012, Amsterdam, The Netherlands, 2012, p. Th.2.G.1.

[91] B. Schmauss, C.-Y. Lin, R. Asif, Progress in digital backpropagation, in: Proceedings of the 38th European Conference on Optical Communications ECOC 2012, Amsterdam, The Netherlands, 2012, p. Th.1.D.5.

[92] I. Fatadin, S.J. Savory, Impact of phase to amplitude noise conversion in coherent optical systems with digital dispersion compensation, Opt. Express 18 (2010) 16273–16278.

[93] D. Rafique, A.D. Ellis, Nonlinear penalties in long-haul optical networks employing dynamic transponders, Opt. Express 19 (May) (2011) 9044–9049.

[94] T.K. Chiang, N. Kagi, T.K. Fong, M.E. Marhic, L.G. Kazovsky, Cross-phase modulation in fiber links with multiple optical amplifiers and dispersion compensators, J. Lightwave Technol. 14 (1996) 249–259.

[95] C. Behrens, Mitigation of nonlinear impairments for advanced optical modulation formats, PhD Thesis, Electrical and Electronic Engineering, University College London, London, 2012.

Advanced Coding for Optical Communications

6

Ivan B. Djordjevic

University of Arizona, Department of Electrical and Computer Engineering, Tucson,
AZ 85721, USA

6.1 INTRODUCTION

As a response to the ever-increasing demands of telecommunication needs, the 100 G Ethernet has recently been standardized, and research effort has moved to 400 Gb/s and 1 Tb/s optical transport [1]. At those data rates, the performance of fiber-optic communication systems is degraded significantly due to intra- and inter-channel fiber nonlinearities, polarization-mode dispersion (PMD), and chromatic dispersion [2–4]. In order to deal with various channel impairments simultaneously novel advanced techniques in modulation, detection, coding and signal processing should be developed; and some important aspects on advanced coding techniques for optical communications are described in this chapter.

The state of the art in optical communication systems standardized by the International Telecommunication Union-Telecommunication Standardization Sector (ITU-T) employs concatenated Bose-Ray-Chaudhuri-Hocquenghem (BCH)/Reed-Solomon (RS) codes [8,9,46,49–53,55,56,61,62]. The RS(255, 239) in particular has been used in a broad range of long-haul communication systems, and it is commonly considered as the first-generation of FEC [10,11]. The elementary FEC schemes (BCH, RS, or convolutional codes) may be combined to design more powerful FEC schemes, e.g. RS(255, 239) + RS(255, 233). Several classes of concatenation codes are listed in ITU-T G975.1. Different concatenation schemes, such as the concatenation of two RS codes or the concatenation of RS and convolutional codes, are commonly considered as second generation (2G) of FEC [10,11] for optical communication.

Codes on graphs, such as turbo codes [5] and low-density parity-check (LDPC) codes [2,4,18–28], have revolutionized communications, and are becoming standard in many applications. LDPC codes, invented by Gallager in 1960s, are linear block codes for which the parity-check matrix has low density of 1s [18]. LDPC codes have generated great interests in the coding community recently [2–4,19–28,40–46]. It has been shown that an iterative LDPC decoder based on the *sum-product algorithm* (SPA) has been shown to achieve a performance as close as 0.0045 dB to the Shannon

Optical Fiber Telecommunications VIB. http://dx.doi.org/10.1016/B978-0-12-396960-6.00006-7

limit [45]. The codes on graphs, in particular TPCs and LDPC codes, can be called the third generation (3D) [10,11] of FEC for optical communications.

The purpose of this chapter is: (i) to describe different classes of codes on graphs of interest for optical communications, (ii) to describe how to combine multilevel modulation and channel coding, (iii) to describe how to perform equalization and soft decoding jointly, and (iv) to demonstrate efficiency of joint de-modulation, decoding and equalization in dealing with various channel impairments simultaneously.

We first describe briefly, in Section 6.2, the channel coding preliminaries: the basics of FEC, linear block codes, and definition of coding gain. The codes on graphs suitable for use in optical communications, namely, turbo codes, turbo-product codes (TPCs), and LDPC codes are described in Section 6.3. Due to their channel capacity achieving bit-error rate (BER) performance, while having a reasonable low complexity of decoding algorithm, in this chapter we are mostly concerned with LDPC codes. Turbo codes are described in Section 6.3.1, and turbo product-codes are described in Section 6.3.2. The basic concepts of LDPC codes and large girth quasi-cyclic (QC) LDPC code design are provided in Sections 6.3.3 and 6.3.4, respectively. Section 6.3.5 is devoted to decoding of binary LDPC codes and their BER performance evaluation. The main problem in decoder implementation for large girth binary LDPC codes is the excessive codeword length and fully parallel implementation on a single FPGA is quite a challenging problem. To solve this problem, we will describe nonbinary LDPC codes over $GF(2^m)$ of large girth, in Section 6.3.6. The next section is devoted to FPGA implementation of decoders for large-girth QC-LDPC codes. Then we describe, in Section 6.4, how jointly to perform multilevel de-modulation and soft-decoding through: (i) multilevel coding (Section 6.4.1), (ii) coded orthogonal frequency division multiplexing (OFDM) (Section 6.4.2) and (iii) nonbinary LDPC-coded modulation (Section 6.4.3). Section 6.4.4 is devoted to multi-dimensional coded-modulation, in particular to LDPC-coded four-dimensional coded-modulation. In Section 6.5, we describe rate-adaptive LDPC-coded modulation. Next, in Section 6.6, we discuss how to combine the *maximum a posteriori probability* (MAP) detector (equalizer) with an LDPC decoder, in so-called turbo equalization fashion. When used in combination with large girth LDPC codes as channel codes, this scheme represents a universal equalizer scheme for *simultaneous* mitigation of fiber nonlinearities, chromatic dispersion and PMD effects, and various receiver imperfections; and it is imperfectly compensated applicable to both direct and coherent detections. To further improve the overall BER performance, we perform the iteration of extrinsic LLRs between LDPC decoder and multilevel MAP equalizer. We use the extrinsic information transfer (EXIT) chart approach due to S. ten Brink [47] to match the LDPC decoders and multilevel MAP equalizer. Because the complexity of turbo equalizer grows exponentially as state memory and signal constellation sizes increase, in Section 6.6.5 we describe how to use this method in combination with digital back-propagation.

For completeness of presentation we also provide in Section 6.6 the independent identically distributed (IID) channel capacity study of fiber-optics channels with *memory*. This chapter has been written having a typical optical communication expert in mind. Coding experts should skip introductory material and go directly to Section 6.3.4.

6.2 LINEAR BLOCK CODES

A typical digital optical communication system, with direct detection, employing channel coding is shown in Figure 6.1. The discrete source generates the information in the form of sequence of symbols. The channel encoder accepts the message symbols and adds redundant symbols according to a corresponding prescribed rule. The channel coding is the act of transforming of a length-k sequence into a length-n codeword. The set of rules specifying this transformation is called the *channel code* [59–66], which can be represented as the following mapping: $C : M \rightarrow X$, where C is the channel code, M is the set of information sequences of length k, and X is the set of codewords of length n. The decoder exploits these redundant symbols to determine which message symbol was actually transmitted. Encoder and decoder consider whole digital transmission system as a discrete channel. Different classes of channel codes can be categorized into three broad categories: (i) *error detection* in which we are concerned only with detecting the errors occurring during transmission (examples include automatic request for transmission-ARQ), (ii) *forward error correction* (FEC), where we are interested in correcting the errors occurring during transmission, and (iii) hybrid channel codes that combine the previous two approaches. In this chapter we are concerned only with FEC.

The codes commonly considered in fiber-optics communications belong to the class of *block codes*. In an (n, k) *block code* the channel encoder accepts information in successive k-symbol blocks, adds $n - k$ redundant symbols that are algebraically

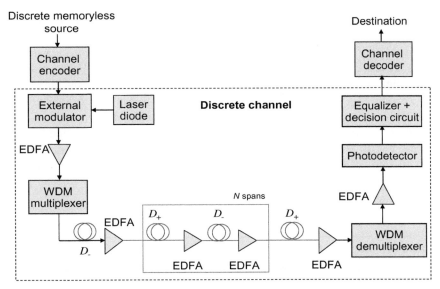

FIGURE 6.1 Block diagram of a point-to-point digital optical communication system with direct detection.

related to the k message symbols; thereby producing an overall encoded block of n symbols ($n > k$), known as a *codeword*. If the block code is *systematic*, the information symbols stay unchanged during the encoding operation, and the encoding operation may be considered as adding the $n - k$ generalized parity checks to k information symbols. Since the information symbols are statistically independent (a consequence of source coding or scrambling), the next codeword is independent of the content of the current codeword. The *code rate* of an (n, k) block code is defined as $R = k/n$, and *overhead* by $OH = (1/R - 1) \cdot 100\%$.

The *linear block code* (n, k), using the language of vector spaces, can be defined as a subspace of a vector space over finite field GF(q), with q being the prime power. Every space is described by its *basis*—a set of linearly independent vectors. The number of vectors in the basis determines the dimension of the space. Therefore, for an (n, k) linear block code the dimension of the space is n, and the dimension of the code subspace is k.

EXAMPLE 6.1

(n, 1) repetition code. The repetition code has two codewords $x_0 = (00 \ \ldots \ 0)$ and $x_1 = (11 \ \ldots \ 1)$.

Any linear combination of these two codewords is another codeword as shown below

$$x_0 + x_0 = x_0 \quad x_0 + x_1 = x_1 + x_0 = x_1 \quad x_1 + x_1 = x_0.$$

The set of codewords from a linear block code forms a group under the addition operation, because all-zero codeword serves as the identity element, and the codeword itself serves as the inverse element. This is the reason why the linear block codes are also called the group codes. The linear block code (n, k) can be observed as a k-dimensional subspace of the vector space of all n-tuples over the binary filed GF(2) $= \{0, 1\}$, with addition and multiplication rules given in Table 6.1. All n-tuples over GF(2) form the vector space. The sum of two n-tuples $a = (a_1 \ a_2 \ \ldots \ a_n)$ and $b = (b_1 \ b_2 \ \ldots \ b_n)$ is clearly an n-tuple and commutative rule is valid because $c = a + b = (a_1 + b_1 \ a_2 + b_2 \ \ldots \ a_n + b_n) = (b_1 + a_1 \ b_2 + a_2 \ \ldots \ b_n + a_n) = b + a$. The all-zero vector $\mathbf{0} = (0 \ 0 \ \ldots \ 0)$ is the identity element, while n-tuple a itself is the inverse element $a + a = 0$. Therefore, the n-tuples form the Abelian group

Table 6.1 Addition (+) and multiplication (·) rules in GF(2).

+	0	1	·	0	1
0	0	1	0	0	0
1	1	0	1	0	1

with respect to the addition operation. The scalar multiplication is defined by: $\alpha a = (\alpha a_1 \ \alpha a_2 \ \ldots \ \alpha a_n)$, $\alpha \in \mathrm{GF}(2)$. The distributive laws

$$\alpha(a + b) = \alpha a + \alpha b,$$

$$(\alpha + \beta)a = \alpha a + \beta a, \quad \forall \alpha, \beta \in \mathrm{GF}(2)$$

are also valid. The associate law $(\alpha \cdot \beta)a = \alpha \cdot (\beta a)$ is clearly satisfied. Therefore, the set of all n-tuples is a vector space over $\mathrm{GF}(2)$. It can be shown, in a fashion similar to that above, that all codewords of an $(n,\ k)$ linear block code form the vector space of dimensionality k. There exist k basis vectors (codewords) such that every codeword is a linear combination of basis ones.

EXAMPLE 6.2

The codewords of $(n, 1)$ repetition code are given by $C = \{(0\ 0\ \ldots\ 0),\ (1\ 1\ \ldots\ 1)\}$. Two codewords in C can be represented as linear combination of all-ones basis vector: $(11\ \ldots\ 1) = 1 \cdot (11\ \ldots\ 1)$, $(00\ \ldots\ 0) = 1 \cdot (11\ \ldots\ 1) + 1 \cdot (11 \ldots 1)$.

6.2.1 Generator matrix

Any codeword x from the $(n,\ k)$ linear block code can be represented as a linear combination of k basis vectors g_i $(i = 0,\ 1,\ \ldots,\ k-1)$ as given below:

$$x = m_0 g_0 + m_1 g_1 + \cdots + m_{k-1} g_{k-1} = m \begin{bmatrix} g_0 \\ g_1 \\ \ldots \\ g_{k-1} \end{bmatrix} = mG;$$

$$G = \begin{bmatrix} g_0 \\ g_1 \\ \ldots \\ g_{k-1} \end{bmatrix}, \quad m = (m_0 \ \ m_1 \ \cdots \ m_{k-1}),$$

(6.1)

where m is the message vector, and G is the generator matrix (of dimensions $k \times n$), in which every row represents a vector from the coding subspace. Therefore, in order to encode, the message vector m $(m_0,\ m_1,\ \ldots,\ m_{k-1})$ has to be multiplied with a generator matrix G to get $x = mG$, where x $(x_0,\ x_1,\ \ldots,\ x_{n-1})$ is a codeword.

EXAMPLE 6.3

Generator matrices for repetition $(n, 1)$ code G_{rep} and $(n, n-1)$ single-parity-check code G_{par} are given, respectively, as

$$G_{\mathrm{rep}} = [11 \ldots 1] \quad G_{\mathrm{par}} = \begin{bmatrix} 100 \ldots 01 \\ 010 \ldots 01 \\ \ldots \\ 000 \ldots 11 \end{bmatrix}.$$

By elementary operations on rows in the generator matrix, the code may be transformed into systematic form.

$$G_s = [I_k | P],$$ (6.2)

where I_k is unity matrix of dimensions $k \times k$, and P is the matrix of dimensions $k \times (n-k)$ with columns denoting the positions of parity checks

$$P = \begin{bmatrix} P_{00} & P_{01} & \cdots & P_{0,n-k-1} \\ P_{10} & P_{11} & \cdots & P_{1,n-k-1} \\ \cdots & & \cdots & \cdots \\ P_{k-1,0} & P_{k-1,1} & \cdots & P_{k-1,n-k-1} \end{bmatrix}.$$ (6.3)

The codeword of a systematic code is obtained by

$$x = [m|b] = m[I_k | P] = mG, \quad G = [I_k | P],$$ (6.4)

and the structure of systematic codeword is shown in Figure 6.2.

Therefore, during encoding the message vector stays unchanged and the elements of vector of parity checks b are obtained by

$$b_i = p_{0i} m_0 + p_{1i} m_1 + \cdots + p_{k-1,i} m_{k-1},$$ (6.5)

where

$$p_{ij} = \begin{cases} 1, & \text{if } b_i \text{ depends on } m_j, \\ 0, & \text{otherwise.} \end{cases}$$

During transmission an optical channel introduces the errors so that the received vector r can be written as $r = x + e$, where e is the error vector (pattern) with elements components determined by

$$e_i = \begin{cases} 1 & \text{if an error occured in the } i\text{th location,} \\ 0 & \text{otherwise.} \end{cases}$$

To determine whether the received vector r is a codeword vector, we are introducing the concept of a *parity-check matrix*.

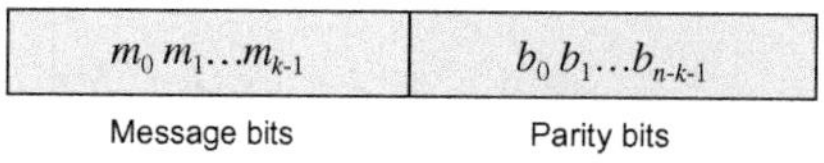

$m_0 \, m_1 ... m_{k-1}$	$b_0 \, b_1 ... b_{n-k-1}$
Message bits	Parity bits

FIGURE 6.2 Structure of a systematic codeword.

6.2.2 Parity-check matrix

Another useful matrix associated with the linear block codes is the parity-check matrix. Let us expand the matrix equation $x = mG$ in scalar form as follows:

$$x_0 = m_0,$$
$$x_1 = m_1,$$
$$\cdots$$
$$x_{k-1} = m_{k-1},$$
$$x_k = m_0 p_{00} + m_1 p_{10} + \cdots + m_{k-1} p_{k-1,0},$$
$$x_{k+1} = m_0 p_{01} + m_1 p_{11} + \cdots + m_{k-1} p_{k-1,1}, \qquad (6.6)$$
$$\cdots$$
$$x_{n-1} = m_0 p_{0,n-k-1} + m_1 p_{1,n-k-1} + \cdots + m_{k-1} p_{k-1,n-k-1}.$$

By using the first k equalities, the last $n-k$ equations can be rewritten as follows:

$$x_0 p_{00} + x_1 p_{10} + \cdots + x_{k-1} p_{k-1,0} + x_k = 0,$$
$$x_0 p_{01} + x_1 p_{11} + \cdots + x_{k-1} p_{k-1,0} + x_{k+1} = 0,$$
$$\cdots$$
$$x_0 p_{0,n-k+1} + x_1 p_{1,n-k-1} + \cdots + x_{k-1} p_{k-1,n-k+1} + x_{n-1} = 0. \qquad (6.7)$$

The matrix representation of (6.7) is given below:

$$\begin{bmatrix} x_0 & x_1 & \cdots & x_{n-1} \end{bmatrix} \begin{bmatrix} p_{00} & p_{10} & \cdots & p_{k-1,0} & 1 & 0 & \cdots & 0 \\ p_{01} & p_{11} & \cdots & p_{k-1,1} & 0 & 1 & \cdots & 0 \\ \cdots & & \cdots & & & \cdots & & \cdots \\ p_{0,n-k-1} & p_{1,n-k-1} & \cdots & p_{k-1,n-k-1} & 0 & 0 & \cdots & 1 \end{bmatrix} = x H^T = 0,$$

$$(6.8)$$

$$x = \begin{bmatrix} x_0 & x_1 & \cdots & x_{n-1} \end{bmatrix}, \quad H = \begin{bmatrix} P^T & I_{n-k} \end{bmatrix}_{(n-k) \times n},$$

where P is already introduced by (6.3). The H-matrix in (6.8) is known as the parity-check matrix. We can easily verify that:

$$GH^T = \begin{bmatrix} I_k & P \end{bmatrix} \begin{bmatrix} P \\ I_{n-k} \end{bmatrix} = P + P = 0, \qquad (6.9)$$

meaning that the parity-check matrix of an (n, k) linear block code H is a matrix of rank $n-k$ and dimensions $(n-k) \times n$ whose null-space is k-dimensional vector with basis being the generator matrix G.

EXAMPLE 6.4

Parity-check matrices for $(n, 1)$ repetition code H_{rep} and $(n, n-1)$ single-parity check code H_{par} are given, respectively, as:

$$H_{rep} = \begin{bmatrix} 100...01 \\ 010...01 \\ ... \\ 000...11 \end{bmatrix} \quad H_{par} = [11...1].$$

EXAMPLE 6.5

For Hamming $(7, 4)$ code the generator G and parity check H matrices are given, respectively, as

$$G = \begin{bmatrix} 1000|110 \\ 0100|011 \\ 0010|111 \\ 0001|101 \end{bmatrix} \quad H = \begin{bmatrix} 1011|100 \\ 1110|010 \\ 0111|001 \end{bmatrix}.$$

Every linear block code with generator matrix G and parity-check matrix H has a dual code with generator matrix H and parity-check matrix G. For example, $(n, 1)$ repetition and $(n, n-1)$ single-parity check codes are dual.

6.2.3 Coding gain

A very important characteristic of an (n, k) linear block code is so-called coding gain, which can be defined as the savings attainable in the energy per information bit to noise spectral density ratio (E_b/N_0) required to achieve a given bit error probability when coding is used compared to that with no coding. Let E_c denote the transmitted bit energy, and E_b denote the information bit energy. Since the total information word energy kE_b must be the same as the total codeword energy nE_c, we obtain the following relationship between E_c and E_b:

$$E_c = (k/n)E_b = RE_b. \tag{6.10}$$

The probability of error for BPSK on an AWGN channel, when coherent hard decision (bit-by-bit) demodulator is used, can be obtained as follows:

$$p = \frac{1}{2}\text{erfc}\left(\sqrt{\frac{E_c}{N_0}}\right) = \frac{1}{2}\text{erfc}\left(\sqrt{\frac{RE_b}{N_0}}\right), \tag{6.11}$$

where $\text{erfc}(x)$ function is defined by

$$\text{erfc}(x) = \frac{2}{\sqrt{\pi}}\int_x^{+\infty} e^{-Z^2}\, dZ.$$

By using the Chernoff bound, we obtain the following expression for hard decision decoding coding gain

$$\frac{(E_b/N_0)_{\text{uncoded}}}{(E_b/N_0)_{\text{coded}}} \approx R(t+1), \tag{6.12}$$

where t is the error correction capability of the code. The corresponding soft decision coding gain can be estimated by [46,49,50]

$$\frac{(E_b/N_0)_{\text{uncoded}}}{(E_b/N_0)_{\text{coded}}} \approx Rd_{\text{min}}, \tag{6.13}$$

and it is about 3 dB better than hard decision decoding (because the minimum distance $d_{\text{min}} \geq 2t+1$). In optical communications it is very common to use the Q-factor[1] as the figure of merit instead of SNR, which is related to the BER on an AWGN as follows:

$$\text{BER} = \frac{1}{2}\text{erfc}\left(\frac{Q}{\sqrt{2}}\right). \tag{6.14}$$

Let BER_{in} denote the BER at the input of FEC decoder, let BER_{out} denote the BER at the output of FEC decoder, and let BER_{ref} denote target BER (such as either 10^{-12} or 10^{-15}). The corresponding coding gain GC and net coding gain NCG are respectively defined as [11]

$$\text{CG} = 20\log_{10}\left[\text{erfc}^{-1}(2\text{BER}_{\text{ref}})\right] - 20\log_{10}\left[\text{erfc}^{-1}(2\text{BER}_{\text{in}})\right] \text{ [dB]}, \tag{6.15}$$

$$\text{NCG} = 20\log_{10}\left[\text{erfc}^{-1}(2\text{BER}_{\text{ref}})\right] - 20\log_{10}\left[\text{erfc}^{-1}(2\text{BER}_{\text{in}})\right] + 10\log_{10} R \text{ [dB]}. \tag{6.16}$$

6.3 CODES ON GRAPHS

The codes on graphs [70] of interest in optical communications include turbo codes, turbo-product codes, and LDPC codes. The turbo codes [5,10–14] can be considered as the generalization of the concatenation of codes in which, during iterative decoding, the decoders interchange the soft messages for a certain number of times. Turbo codes can approach channel capacity closely in the region of interest for wireless communications. However, they exhibit strong error floors in the region of interest for fiber-optics communications (see [4,5]); therefore, alternative iterative soft decoding approaches are to be sought. As recently shown in [2–4,7,10–12,17,19,21,22,25–28], turbo-product codes and LDPC codes can provide excellent coding gains and, when properly designed, do not exhibit error floor in the region of interest for fiber-optics communications.

6.3.1 Turbo codes

A turbo encoder comprises the concatenation of two (or more) convolution encoders, while corresponding decoders consist of two (or more) convolutional soft decoders

---

[1]The Q-factor is defined as $Q=(\mu_1-\mu_0)/(\sigma_1+\sigma_0)$, where μ_j and σ_j ($j=0,\ 1$) represent the mean and the standard deviation corresponding to the bits j.

in which extrinsic probabilistic information is iterated back and forth among soft decoders [4,5]. *Parallel turbo encoder*, shown in Figure 6.3a, consists of two rate ½ convolutional encoders arranged in parallel concatenation scheme. In a *serial turbo encoder*, shown in Figure 6.3b, the serially concatenated convolutional encoders are separated by K/R_0-interleaver, where R_0 is the code rate of outer encoder. The interleaver takes incoming block of bits and arranges them in pseudo-random fashion prior to being encoded by second encoder, so that the same information bits are not encoded twice by the same recursive systematic convolutional (RSC) code, in case identical RSC codes are used. The generator matrix of an RSC code can be written as follows: $G_{RSC}(x) = \begin{bmatrix} 1 & g_2(x)/g_1(x) \end{bmatrix}$, where $g_2(x)/g_1(x)$ denotes the transfer function of the parity branch of encoder (see Figure 6.3a). For example, the RSC encoder described by $G_{RSC}(x) = \begin{bmatrix} 1(1+x^4)/(1+x+x^2+x^3+x^4) \end{bmatrix}$ is shown in Figure 6.3c.

Iterative (turbo) decoder, shown in Figure 6.4, interleaves two soft-input/soft-output (SISO) decoders, exchanging the extrinsic information iteratively and cooperatively. The role of iterative decoder is to iteratively estimate the *a posteriori*

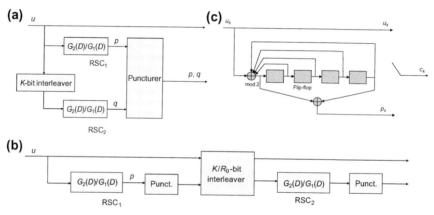

FIGURE 6.3 Turbo codes encoder configurations for: (a) parallel and (b) serial turbo codes. (c) RSC encoder with $G_{RSC}(x) = [1(1+x^4)/(1+x+x^2+x^3+x^4)]$.

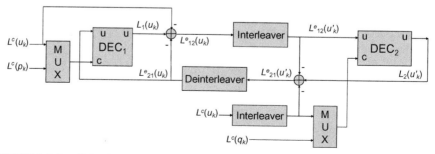

FIGURE 6.4 Parallel turbo decoder configuration. MUX: multiplexer, DEC: decoder.

probabilities (APPs) $\Pr(u_k|y)$, where u_k is kth data bit ($k=1, 2, \ldots, K$), and y is a received codeword plus noise, $y=c+n$. For iterative decoders, as illustrated in Figure 6.4, each component decoder receives extrinsic or soft information for each u_k from its companion decoder, which provides "the prior" information. The key idea behind extrinsic information is that decoder DEC$_2$ provides the soft information to DEC$_1$ for each u_k using only information not available to DEC$_1$. The knowledge of APPs allows the optimum decisions for information bits u_k to be made by the maximum *a posteriori* probability (MAP) rule:

$$\frac{p(u_k = +1|y)}{p(u_k = -1|y)} \overset{+1}{\underset{-1}{\gtrless}} 1 \Leftrightarrow \hat{u}_k = \text{sign}\,[L(u_k)], \quad L(u_k) = \log\left[\frac{p(u_k = +1|y)}{p(u_k = -1|y)}\right], \quad (6.17)$$

where $L(u_k)$ is the log-APP also known as log-likelihood ratio (LLR). By applying the Bayes' rule, we can write:

$$L(b_k) = \log\left[\frac{P(b_k = +1|y)}{P(b_k = -1|y)}\frac{p(y)}{p(y)}\right] = \log\left[\frac{P(y|b_k = +1|y)}{P(y|b_k = -1)}\frac{P(b_k = +1)}{P(b_k = -1)}\right]$$

$$= \log\left[\frac{P(y|b_k = +1)}{P(y|b_k = -1)}\right] + \log\left[\frac{P(b_k = +1)}{P(b_k = -1)}\right], \quad (6.18)$$

where $b_k \in \{u_k, p_k\}$, and the second term represents *a priori* information. The first term represents the *extrinsic* information, which is in iterative decoders obtained from corresponding companion decoder. Notice that for conventional decoders we typically have $P(b_k=+1)=P(b_k=-1)$. The key idea behind extrinsic information is to provide to the companion decoder only soft information about b_k not already available to it. Therefore, although initially the prior information is zero, it becomes nonzero after the first iteration. As shown in Figure 6.4, the extrinsic information to be sent from DEC$_1$ to DEC$_2$, denoted as $L_{12}^e(u_k)$, can be calculated by subtracting the channel reliability $L^c(u_k)$ and extrinsic information already received from DEC$_2$, $L_{21}^e(u_k)$, from DEC$_1$ output LLR $L_1(u_k)$. On the other hand, the extrinsic information to be sent from DEC$_2$ to DEC$_1$, $L_{21}^e(u_k)$, is obtained by subtracting the interleaved channel reliability $L^c(u_k)$ and interleaved extrinsic information already received from DEC$_1$ $L_{12}^e(u_k')$ from DEC$_2$s output LLR $L_2(u_k)$. Since the DEC$_2$ operates on interleaved sequence this extrinsic information needs to be deinterleaved before being sent to DEC$_1$. This exchange of extrinsic information is performed until the successful decoding or until the pre-determined number of iterations is reached. Decoders DEC$_i$ ($i=1, 2$) operate on trellis description of encoder, by employing the BCJR algorithm [6]. The BCJR algorithm can also be used for MAP detection and as such it will be described in the section on turbo equalization.

Since the resulting code rate of parallel turbo encoder is low, $R=1/3$, the puncturer deletes the selected bits in order to reduce the coding overhead, as shown in Figure 6.3a, and resulting code rate is $R=K/(K+P)$, where P is the number of parity bits remaining after puncturing. Notice that puncturing, on the other hand, reduces the minimum distance of original code, and leads to performance degradation and an early error floor as shown in [4,5]. Because of low code rate and early error floor, the turbo codes are not widely used in fiber-optics communications.

6.3.2 Turbo-product codes (TPCs)

A turbo-product code (TPC) (illustrated in Figure 6.5), also known as block turbo code (BTC) [15], is an $(n_1 n_2, \ k_1 k_2, \ d_1 d_2)$ code in which codewords form an $n_1 \times n_2$ array such that each row is a codeword from an $(n_1, \ k_1, \ d_1)$ code C_1, and each column is a codeword from an $(n_2, \ k_2, \ d_2)$ code C_2. With n_i, k_i, and d_i $(i = 1, \ 2)$ we denoted the codeword length, dimension, and minimum distance, respectively, of the ith component code. The soft bit reliabilities are iterated between decoders for C_1 and C_2. In fiber-optics communications, TPCs based on BCH component codes were intensively studied, e.g. [10,11,16,17]. The product codes were proposed by Elias [48], but the term "turbo" is used when two soft-input soft-output (SISO) decoders exchange the extrinsic information [15]. It is possible to show that the minimum distance of a product code is the product of minimum distances of component codes, $d = d_1 d_2$. The constituent codes are typically extended BCH codes, because with extended BCH codes we can increase the minimum distance for $d_1 + d_2 + 1$ compared to nominal BCH codes. The expressions for bit error probability P_b and codeword error probability P_{cw} under maximum likelihood (ML) decoding, for AWGN channel, can be estimated as follows [4]:

$$P_b \sim \frac{w_{\min}}{2 k_1 k_2} \text{erfc} \left(\sqrt{\frac{R d E_b}{N_0}} \right) \quad P_{cw} \sim \frac{A_{\min}}{2} \text{erfc} \left(\sqrt{\frac{R d E_b}{N_0}} \right), \quad (6.19)$$

where $w_{\min}$ is the minimum weight of all $A_{\min}$ TPC codewords at the minimum distance d. Notice that $w_{\min}$ and $A_{\min}$ are quite large compared to turbo codes, resulting in excellent BER performance of TPCs. Unfortunately, due to high complexity ML decoding is typically not used in TPC decoding, but simple Chase II-like decoding algorithms are used instead [10,15]. One such algorithm, which is independent on channel model, is described below. Let u_j be jth bit in a codeword $\boldsymbol{u} = [u_1 \ldots u_n]$, and r_j be corresponding received sample. The initial bit LLRs can be calculated by

$$L(u_j) = \log \frac{P(u_j = 0 | r_j)}{P(u_j = 1 | r_j)},$$

$$P(u_j | r_j) = \frac{P(r_j | u_j) P(u_j)}{P(r_j | u_j = 0) P(u_j = 0) + P(r_j | u_j = 1) P(u_j = 1)}, \quad (6.20)$$

where $P(r_j | \cdot)$ can be evaluated by estimation of histograms.

The constituent *SISO decoding* algorithms are based on modified Chase II decoding algorithm [4]:

1. Determine p least reliable positions starting from (6.20). Generate 2^p test patterns to be added to the hard-decision word obtained after (6.20).

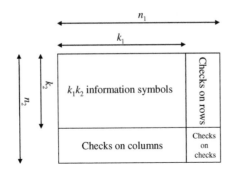

FIGURE 6.5 The structure of a codeword of a turbo-product code.

2. Determine the ith ($i = 1, \ldots, 2^p$) perturbed sequence by adding (modulo-2) the test pattern to the hard-decision word (on least reliable positions).
3. Perform the algebraic or hard decoding to create the set of candidate codewords. Simple syndrome decoding is suitable for high-speed transmission.
4. Calculate the candidate codeword c_i LLRs by

$$\Lambda[c_i = (c_i(1)c_i(2)\ldots c_i(n))] = \sum_{j=1}^{n} \log\left[\frac{e^{(1-c_i(j))^L(c_i(j))}}{1 + e^{L(c_i(j))}}\right]. \tag{6.21}$$

5. Calculate the extrinsic bit reliabilities for next decoding stage using

$$L_{ext}\left(u_j\right) = \lambda\left(u_j\right) - L\left(u_j\right), \quad \lambda\left(u_j\right) = \log\left[\frac{\sum_{c_i}^{(j)=0} \Lambda(c_i)}{\sum_{c_i}^{(j)=1} \Lambda(c_i)}\right]. \tag{6.22}$$

In calculating (6.21) and (6.22), the following "max-star" operator is applied recursively

$$\max{}^*(x, y) \triangleq \log(e^x + e^y) + \max(x, y) + \log\left[1 + e^{-|x-y|}\right]. \tag{6.23}$$

Given this description of SISO constituent decoding algorithm, the TPC decoder operates as follows. Let $L_{rc,j}^e$ denote the extrinsic information to be passed from row to column decoder, and let $L_{cr,j}^e$ denote the extrinsic information to be passed in opposite direction. Then, assuming that the column-decoder operates first, the **TPC decoder** performs the following steps:

1. *Initialization*: $L_{cr,j}^e = L_{rc,j}^e$ for all js.
2. *Column decoder*: Run the SISO decoding algorithm described above with the following inputs $L(u_j) + L_{rc,j}^e$ to obtain $\{L_{column}(u_j)\}$ and $\{L_{cr,j}^e\}$. The extrinsic information is calculated by Eq.(6.23). Pass the extrinsic information $\{L_{cr,j}^e\}$ to row decoder.

3. *Row decoder*: Run the SISO decoding algorithm with the following inputs $L(u_j) + L^e_{cr,j}$ to obtain $\{L_{row}(u_j)\}$ and $\{L^e_{rc,j}\}$. Pass the extrinsic information $\{L^e_{cr,j}\}$ to companion column decoder.
4. *Bit decisions*: Repeat the steps 2–3 until a valid codeword is generated or a predetermined number of iterations has been reached. Make the decisions on bits by $sgn[L_{row}(u_k)]$.

Unlike [10,11,15], because not any approximation is used in calculation of the extrinsic reliabilities, there is no need to introduce the scaling factors and the correction factors.

6.3.3 Low-density parity-check (LDPC) codes

If the parity-check matrix has a low density of 1s and the number of 1s per row and per column are both constant, the code is said to be a *regular LDPC* code. To facilitate the implementation at high speed, we prefer the use of regular rather than irregular LDPC codes. The graphical representation of LDPC codes, known as bipartite (Tanner) graph representation, is helpful in efficient description of LDPC decoding algorithms. A *bipartite (Tanner) graph* is a graph whose nodes may be separated into two classes (*variable* and *check* nodes), where *undirected edges* may only connect two nodes not residing in the same class. The Tanner graph of a code is drawn according to the following rule: check (function) node c is connected to variable (bit) node v whenever element h_{cv} in a parity-check matrix H is a 1. In an $m \times n$ parity-check matrix, there are $m = n - k$ check nodes and n variable nodes.

EXAMPLE 6.6

As an illustrative example, consider the H-matrix of the following code

$$H \begin{bmatrix} 1 & 0 & 1 & 0 & 1 & 0 \\ 1 & 0 & 0 & 1 & 0 & 1 \\ 0 & 1 & 1 & 0 & 0 & 1 \\ 0 & 1 & 0 & 1 & 1 & 0 \end{bmatrix}$$

For any valid codeword $x = [x_0 x_1 \ldots x_{n-1}]$, the checks used to decode the codeword are written as,

- Equation (c_0): $x_0 + x_2 + x_4 = 0$ (mod 2).
- Equation ($c1$): $x_0 + x_3 + x_5 = 0$ (mod 2).
- Equation (c_2): $x_1 + x_2 + x_5 = 0$ (mod 2).
- Equation (c_3): $x_1 + x_3 + x_4 = 0$ (mod 2).

The bipartite graph (Tanner graph) representation of this code is given in Figure 6.6(a).

The circles represent the bit (variable) nodes while squares represent the check (function) nodes. For example, the variable nodes x_0, x_2, and x_4 are involved in Eq. (c_0), and therefore connected to the check node c_0. A closed path in a bipartite

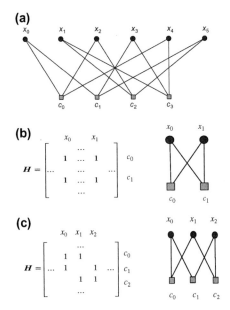

FIGURE 6.6 (a) Bipartite graph of (6,2) code described by H matrix above. Cycles in a Tanner graph: (b) cycle of length 4, and (c) cycle of length 6.

graph comprising l edges that closes back on itself is called a *cycle* of length l. The shortest cycle in the bipartite graph is called the *girth*. The girth influences the minimum distance of LDPC codes, correlates the extrinsic LLRs, and therefore affects the decoding performance. The use of large girth LDPC codes is preferable because the large girth increases the minimum distance and de-correlates the extrinsic info in the decoding process. To improve the iterative decoding performance, we have to avoid cycles of length 4, and preferably 6 as well. To check for the existence of short cycles, one has to search over H-matrix for the patterns shown in Figure 6.6b and c.

6.3.4 Quasi-cyclic (QC) binary LDPC code design

In this section, we describe a method for designing large girth QC LDPC codes; while an efficient and simple variant of sum-product algorithm (SPA) suitable for high-speed implementation, namely the min-sum-with-correction term algorithm, is described in Section 6.3.5.

6.3.4.1 Design of large girth quasi-cyclic LDPC codes

Based on Tanner's bound for the minimum distance of an LDPC code [71]

$$d \geqslant \begin{cases} 1 + \frac{w_c}{w_c-2}\left((w_c-1)^{\lfloor(g-2)/4\rfloor}-1\right), & g/2 = 2m+1, \\ 1 + \frac{w_c}{w_c-2}\left((w_c-1)^{\lfloor(g-2)/4\rfloor}-1\right) + (w_c-1)^{\lfloor(g-2)/4\rfloor}, & g/2 = 2m \end{cases}$$

$$(6.24)$$

(where g and w_c denote the girth of the code graph and the column weight, respectively, and where d stands for the minimum distance of the code), it follows that large girth leads to an exponential increase in the minimum distance, provided that the column weight is at least 3 ($\lfloor \ \rfloor$ denotes the largest integer less than or equal to the enclosed quantity). For example, the minimum distance of girth-10 codes with column weight $r=3$ is at least 10. The parity-check matrix of regular QC LDPC codes [73,75] can be represented by

$$\begin{bmatrix} I & I & I & \cdots & I \\ I & P^{s[1]} & P^{s[2]} & \cdots & P^{s[c-1]} \\ I & P^{2s[1]} & P^{2s[2]} & \cdots & P^{2s[c-1]} \\ \cdots & \cdots & \cdots & \cdots & \cdots \\ I & P^{(r-1)s[1]} & P^{(r-1)s[2]} & \cdots & P^{(r-1)s[c-1]} \end{bmatrix}, \quad (6.25)$$

where I is $B \times B$ (B is a prime number) identity matrix, P is $B \times B$ permutation matrix given by $P=(p_{ij})_{B \times B}$, $p_{i,i+1}=p_{B,1}=1$ (zero otherwise), and where r and c represent the number of block-rows and block-columns in (6.25), respectively. The set of integers S are to be carefully chosen from the set $\{0, 1, \ldots, B-1\}$ so that the cycles of short length, in the corresponding Tanner (bipartite) graph representation of (6.25), are avoided. According to Theorem 2.1 in [75], we have to avoid the cycles of length $2k$ ($k=3$ or 4) defined by the following equation

$$S[i_1]j_1 + S[i_2]j_2 + \cdots + S[i_k]j_k = S[i_1]j_2 + S[i_2]j_3 + \cdots + S[i_k]j_1 \bmod B, \quad (6.26)$$

where the closed path is defined by (i_1, j_1), (i_1, j_2), (i_2, j_2), (i_2, j_3), ..., (i_k, j_k), (i_k, j_1) with the pair of indices denoting row-column indices of permutation-blocks in (6.25) such that $l_m \neq l_{m+1}$, $l_k \neq l_1$ ($m=1, 2, \ldots, k$; $l \in \{i, j\}$). Therefore, we have to identify the sequence of integers $S[i] \in \{0, 1, \ldots, B-1\}$ ($i=0, 1, \ldots, r-1$; $r<B$) not satisfying Eq. (6.26), which can be done either by computer search or in a combinatorial fashion. For example, to design the QC LDPC codes in [43], we introduced the concept of the cyclic-invariant difference set (CIDS). The CIDS-based codes come naturally as girth-6 codes, and to increase the girth we had to selectively remove certain elements from a CIDS. The design of LDPC codes of rate above 0.8, column weight 3, and girth-10 using the CIDS approach is a very challenging and is still an open problem. Instead, in our recent paper [73], we solved this problem by developing an efficient computer search algorithm. We add an integer at a time from the set $\{0, 1, \ldots, B-1\}$ (not used before) to the initial set S and check if the equation (6.26) is satisfied. If the equation (6.26) is satisfied, we remove that integer from the set S and continue our search with another integer from set $\{0, 1, \ldots, B-1\}$ until we exploit all the elements from $\{0, 1, \ldots, B-1\}$. The code rate of these QC codes, R, is lower-bounded by

$$R \geqslant \frac{|S| B - r B}{|S| B} = 1 - r/|S|, \quad (6.27)$$

and the codeword length is $|S|B$, where $|S|$ denotes the cardinality of set S. For a given code rate R_0, the number of elements from S to be used is $\lfloor r/(1-R_0) \rfloor$. With this algorithm, LDPC codes of arbitrary rate can be designed.

EXAMPLE 6.7

By setting $B=2311$, the set of integers to be used in Eq.(6.25) is obtained as $S=\{1, 2, 7, 14, 30, 51, 78, 104, 129, 212, 223, 318, 427, 600, 808\}$. The corresponding LDPC code has rate $R_0 = 1 - 3/15 = 0.8$, column weight 3, girth-10, and length $|S|B = 15 \cdot 2311 = 34665$. In the example above, the initial set of integers was $S=\{1, 2, 7\}$, and the set of row to be used in Eq.(6.25) is $\{1, 3, 6\}$. The use of a different initial set will result in a different set from that obtained above.

EXAMPLE 6.8

By setting $B=269$, the set S is obtained as $S=\{0, 2, 3, 5, 9, 11, 12, 14, 27, 29, 30, 32, 36, 38, 39, 41, 81, 83, 84, 86, 90, 92, 93, 95, 108, 110, 111, 113, 117, 119, 120, 122\}$ If 30 integers are used, the corresponding LDPC code has rate $R_0 = 1 - 3/30 = 0.9$, column weight 3, girth-8, and length $30 \cdot 269 = 8070$.

6.3.5 Decoding of binary LDPC codes and BER performance evaluation

In this sub-section, we describe the min-sum with correction term decoding algorithm [74,76]. It is a simplified version of the original algorithm proposed by Gallager [18]. Gallager proposed a near-optimal iterative decoding algorithm for LDPC codes that computes the distributions of the variables in order to calculate the *a posteriori probability* (APP) of a bit v_i of a codeword $v = [v_0 v_1 \ldots v_{n-1}]$ to be equal to 1, given a received vector $y = [y_0 y_1 \ldots y_{n-1}]$. This iterative decoding scheme engages passing the extrinsic info back and forth among the c-nodes and the v-nodes over the edges to update the distribution estimation. Each iteration in this scheme is composed of two half-iterations. In Figure 6.7, we illustrate both the first and the second halves of an iteration of the algorithm. As an example, in Figure 6.7a, we show the message sent from v-node v_i to the c-node c_j. v_i-node collects the information from channel (y_i sample), in addition to extrinsic info from other c-nodes connected to v_i-node, processes them, and sends the extrinsic info (not already available info) to c_j. This extrinsic info contains the information about the probability $\Pr(c_i = b | y_0)$, where $b \in \{0, 1\}$. This is performed in all c-nodes connected to v_i-node. On the other hand, Figure 6.7b shows the extrinsic info sent from c-node c_i to the v-node v_j, which contains the information about $\Pr(c_i$ equation is satisfied $| y)$. This is done repeatedly to all the c-nodes connected to v_i-node.

After this intuitive description, we describe the min-sum-with-correction-term algorithm in more detail [74] because of its simplicity and suitability for high-speed implementation. Generally, we can either compute *a posteriori probability* (APP)

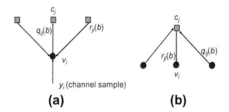

FIGURE 6.7 Illustration of the half-iterations of the sum-product algorithm: (a) first half-iteration: extrinsic info sent from *v*-nodes to *c*-nodes and (b) second half-iteration: extrinsic info sent from *c*-nodes to *v*-nodes.

$\Pr(v_i|\mathbf{y})$ or the APP ratio $l(v_i) = \Pr(v_i=0|\mathbf{y})/\Pr(v_i=1|\mathbf{y})$, which is also referred to as the likelihood ratio. In log-domain version of the sum-product algorithm, we replace these likelihood ratios with log-likelihood ratios (LLRs) due to the fact that the probability domain includes many multiplications which lead to numerical instabilities, whereas the computation using LLRs computation involves addition only. Moreover, the log-domain representation is more suitable for finite precision representation. Thus, we compute the LLRs by $L(v_i) = \log[\Pr(v_i=0|\mathbf{y})/\Pr(v_i=1|\mathbf{y})]$. For the final decision, if $L(v_i)>0$, we decide in favor of 0 and if $L(v_i)<0$, we decide in favor of 1. To further explain the algorithm, we introduce the following notations due to MacKay [72]:

$V_j = \{v\text{-nodes connected to } c\text{-node } c_j\}$,
$V_j\backslash i = \{v\text{-nodes connected to } c\text{-node } c_j\}\backslash\{v\text{-node } v_i\}$,
$C_i = \{c\text{-nodes connected to } v\text{-node } v_i\}$,
$C_i\backslash j = \{c\text{-nodes connected to } v\text{-node } v_i\}\backslash\{c\text{-node } c_j\}$,
$M_v(\sim i) = \{\text{messages from all } v\text{-nodes except node } v_i\}$,
$M_c(\sim j) = \{\text{messages from all } c\text{-nodes except node } c_j\}$,
$P_i = \Pr(v_i=1|y_i)$,
$S_i = \text{event that the check equations involving } c_i \text{ are satisfied}$,
$q_{ij}(b) = \Pr(v_i=b|S_i, \ y_i, \ M_c(\sim j))$,
$r_{ji}(b) = \Pr(\text{check equation } c_j \text{ is satisfied}|v_i=b, M_v(\sim i))$.

In the log-domain version of the sum-product algorithm, all the calculations are performed in the log-domain as follows:

$$L(v_i) = \log\left[\frac{\Pr(v_i = 0|y_i)}{\Pr(v_i = 1|y_i)}\right], \ L(r_{ji}) = \log\left[\frac{r_{ji}(0)}{r_{ji}(1)}\right], \ L(q_{ji}) = \log\left[\frac{q_{ji}(0)}{q_{ji}(1)}\right].$$

$$(6.28)$$

The algorithm starts with the initialization step where we set $L(v_i)$ as follows:

$$L(v_i) = (-1)^{y_i} \log\left(\frac{1-\varepsilon}{\varepsilon}\right), \quad \text{for BSC,}$$

$$L(v_i) = 2y_i/\sigma^2, \quad \text{for binary input AWGN,} \qquad (6.29)$$

$$L(v_i) = \log\left(\frac{\sigma_1}{\sigma_0}\right) - \frac{(y_i - \mu_0)^2}{2\sigma_0^2} + \frac{(y_i - \mu_1)^2}{2\sigma_1^2}, \quad \text{for BA-AWGN,}$$

$$L(v_i) = \log\left(\frac{\Pr(v_i = 0|y_i)}{\Pr(v_i = 1|y_i)}\right), \quad \text{for arbitrary channel,}$$

where ε is the probability of error in the binary symmetric channel (BSC), σ^2 is the variance of the Gaussian distribution of the AWGN, and μ_j and σ_j^2 ($j=0,\ 1$) represent the mean and the variance of Gaussian process corresponding to the bits $j=0,\ 1$ of a binary asymmetric (BA)-AWGN channel. After initialization of $L(q_{ij})$, we calculate $L(r_{ji})$ as follows:

$$L(r_{ji}) = L\left(\sum_{i' \in V_j \setminus i} b'_i\right) = L(\cdots \oplus b_k \oplus b_l \oplus b_m \oplus b_n \cdots) \quad (6.30)$$

$$= \cdots L_k \boxplus L_l \boxplus L_m \boxplus L_n \boxplus \cdots$$

where $\oplus$ denotes the modulo-2 addition, and use the square symbol in eq. (6.30) here denotes a pairwise computation defined by

$$L_1 \boxplus L_2 = \prod_{k=1}^{2} \text{sign}(L_k) \cdot \phi\left(\sum_{k=1}^{2} \phi(|L_k|)\right), \quad \phi(x) = -\log\tanh(x/2). \quad (6.31)$$

Upon calculation of $L(r_{ji})$, we update

$$L(q_{ij}) = L(v_i) + \sum_{j' \in C_i \setminus j} L(r_{j'i}), \quad L(Q_i) = L(v_i) + \sum_{j' \in C} L(r_{j'i}) \quad (6.32)$$

Finally, the decision step is as follows:

$$\hat{v}_i = \begin{cases} 1, & L(q_i) < 0, \\ 0, & \text{otherwise.} \end{cases} \quad (6.33)$$

If the syndrome equation $\hat{v} H^T = 0$ is satisfied or the maximum number of iterations is reached, we stop, otherwise, we recalculate $L(r_{ji})$ and update $L(q_{ij})$ and $L(Q_i)$ and check again. It is important to set the number of iterations high enough to ensure that most of the codewords are decoded correctly and low enough not to affect the processing time. It is important to mention that decoder for good LDPC codes requires fewer iterations to guarantee successful decoding. The **Gallager log-domain SPA** can be formulated as follows:

0. *Initialization:* For $j=0,\ 1,\ \ldots,\ n-1$; initialize the messages to be sent from v-node i to c-node j by channel LLRs, namely $L\left(q_{ij}\right) = L\left(c_i\right)$.

1. *c-node update rule:* For $j=0,1, \ldots, n-k-1$;

    compute $L\left(r_{ji}\right) = \left(\prod_{i' \in R_j \setminus i} \alpha_{i'j}\right) \phi \left[\sum_{i' \in R_j \setminus i} \phi\left(\beta_{i'j}\right)\right]$,

    where $\alpha_{ij} = \text{sign}\left[L\left(q_{ij}\right)\right]$, $\beta_{ij} = \left|L\left(q_{ij}\right)\right|$, and
    $\phi(x) = -\log \tan h\,(x/2) = \log\left[(e^x + 1) / (e^x - 1)\right]$.

2. *v-node update rule:* For $i=0, 1, \ldots, n-1$; set
    $L\left(q_{ij}\right) = L\left(c_i\right) + \sum_{j' \in C_i \setminus j} L\left(r_{j'i}\right)$ for all c-nodes for which $h_{ji}=1$.

3. *Bit decisions:* Update $L(Q_i)$ $(i=0, \ldots, n-1)$ by
    $L\left(Q_i\right) = L\left(c_i\right) + \sum_{j \in C_i} L\left(r_{ji}\right)$ and set $\hat{c}_i = 1$ when $L(Q_i)<0$
    (otherwise, $\hat{c}_i = 0$). If $\hat{c}\boldsymbol{H}^T = \boldsymbol{0}$ or pre-determined number of iterations
    has been reached then stop, otherwise go to step 1.

Because the c-node update rule involves log and tanh functions, it is computation-ally intensive, and there exist many approximations. The very-popular one is the *min-sum-plus-correction-term approximation* [20]. Namely, it can be shown that "box-plus" operator $\boxplus$ can also be calculated by

$$L_1 \boxplus L_2 = \prod_{k=1}^{2} \text{sign}(L_k) \cdot \min(|L_1|, |L_2|) + c(x, y), \qquad (6.34)$$

where $c(x, y)$ denotes the correction factor defined by

$$c(x, y) = \log[1 + \exp(-|x + y|)] - \log[1 + \exp(-|x - y|)], \qquad (6.35)$$

commonly implemented as a look-up table (LUT). (For an alternative reduced-complexity decoding algorithm an interested reader is referred to [77].)

6.3.5.1 BER performance of binary LDPC codes

The results of simulations for an AWGN channel model are given in Figure 6.8, where we compare the large girth LDPC codes (Figure 6.8a) against RS codes, concatenated RS codes, TPCs, and other classes of LDPC codes. Therefore, dif-ferent generations of FEC proposed for optical communications have been com-pared. In optical communications, it is a common practice to use the Q-factor as a figure of merit of binary modulation schemes instead of signal-to-noise ratio. In all simulation results in this section, we maintained the double precision. For the LDPC(16935, 13550) code, we also provided 3- and 4-bit fixed-point simulation results (see Figure 6.8a). Our results indicate that the 4-bit representation performs comparable to the double-precision representation whereas the 3-bit representa-tion performs 0.27 dB worse than the double-precision representation at the BER of 2×10^{-8}. The girth-10 LDPC(24015, 19212) code of rate 0.8 outperforms the concatenation RS(255, 239)+RS(255, 223) (of rate 0.82) by 3.35 dB and RS(255, 239) by 4.75 dB both at BER of 10^{-7}. The same LDPC code outperforms projective geometry (PG) (2, 2^6) based LDPC(4161, 3431) (of rate 0.825) of girth-6 by 1.49 dB at BER of 10^{-7}, and outperforms CIDS-based LDPC(4320, 3242) of rate 0.75 and girth-8 LDPC codes by 0.25 dB. At BER of 10^{-10}, it outperforms lat-tice-based LDPC(8547, 6922) of rate 0.81 and girth-8 LDPC code by 0.44 dB, and BCH(128, 113) × BCH(256, 239) TPC of rate 0.82 by 0.95 dB. The net coding gain

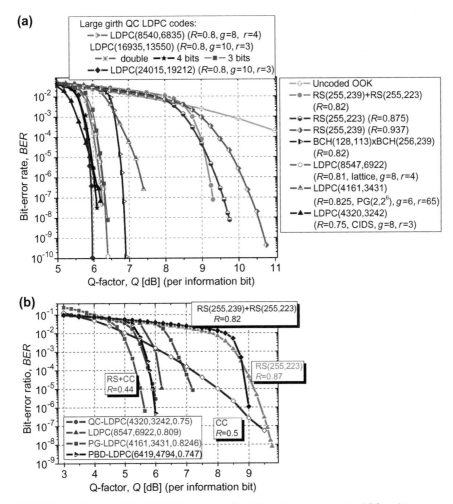

FIGURE 6.8 (a) Large girth QC LDPC codes against RS codes, concatenated RS codes, TPCs, and girth-6 LDPC codes on an AWGN channel model, and (b) LDPC codes versus convolutional, concatenated RS, and concatenation of convolutional and RS codes on an AWGN channel. Number of iterations in sum-product-with-correction-term algorithm was set to 25. (Modified from Ref. [2]; ©IEEE 2009; reprinted with permission.)

at BER of 10^{-12} is 10.95 dB. In Figure 6.8b, different LDPC codes are compared against RS(255, 223) code, concatenated RS code of rate 0.82 and convolutional code (CC) (of constraint length 5). It can be seen that LDPC codes, both regular and irregular, offer much better performance than hard-decision codes. It should be noticed that pairwised balanced design (PBD) [57,58] based irregular LDPC code of rate 0.75 is only 0.4 dB away from the concatenation of convolutional-RS codes (denoted in Figure 6.8b as RS + CC) of significantly lower code rate $R = 0.44$ at BER of 10^{-6}. As expected, irregular LDPC codes outperform regular LDPC codes.

6.3.6 Nonbinary LDPC codes

The parity-check matrix H of a nonbinary QC-LDPC code can be organized as an array of sub-matrices of equal size as in (6.36), where $H_{i,j}$, $0 \leq i < \gamma$, $0 \leq j < \rho$, is a $B \times B$ sub-matrix in which each row is a cyclic shift of the row preceding it. This modular structure can be exploited to facilitate hardware implementation of the decoders of QC-LDPC codes [21,22]. Furthermore, the quasi-cyclic nature of their generator matrices enables encoding of QC-LDPC codes to be performed in linear time using simple shift-register-based architectures [22,25]

$$
H = \begin{bmatrix}
H_{0,0} & H_{0,1} & \cdots & H_{0,\rho-1} \\
H_{1,0} & H_{1,1} & \cdots & H_{1,\rho-1} \\
\vdots & \vdots & \ddots & \vdots \\
H_{\gamma-1,0} & H_{\gamma-1,1} & \cdots & H_{\gamma-1,\rho-1}
\end{bmatrix}.
\tag{6.36}
$$

If we select the entries of H from the binary field GF(2), as it was done in previous sections, then the resulting QC-LDPC code is a binary LDPC code. On the other hand, if the selection is made from the Galois field of q elements denoted by GF(q), then we obtain a q-ary QC-LDPC code. In order to decode binary LDPC codes, as described above, an iterative message-passing algorithm is referred to as SPA. For nonbinary LDPC codes, a variant of the SPA known as the q-ary SPA (QSPA) is used [23]. When the field order is a power of 2, i.e. $q = 2^m$, where m is an integer and $m \geq 2$, a fast Fourier transform (FFT) based implementation of QSPA, referred to as FFT-QSPA, significantly reduces the computational complexity of QSPA. FFT-QSPA is further analyzed and improved in [24]. A mixed-domain FFT-QSPA implementation, in short MD-FFT-QSPA, aims to reduce the hardware implementation complexity by transforming multiplications in the probability domain into additions in the log domain whenever possible. It also avoids instability issues commonly faced in probability-domain implementations.

Following the code design we had discussed above, we generated (3,15)-regular, girth-8 LDPC codes over the fields GF(2^p), where $0 \leq p \leq 7$. All the codes had a code rate (R) of at least 0.8 and hence an overhead $OH = (1/R - 1)$ of 25% or less. We compared the BER performances of these codes against each other and against some other well-known codes, namely the ITU-standard RS(255, 239), RS(255, 223), and RS(255, 239) + RS(255, 223) codes; and BCH(128, 113) $\times$ BCH(256, 239) TPC. We used the binary AWGN (BI-AWGN) channel model in our simulations and set the maximum number of iterations to 50. In Figure 6.9a, we present the BER performances of the set of non-binary LDPC codes discussed above. Using the figure, we can conclude that when we fix the girth of a non-binary regular, rate-0.8 LDPC code at eight, increasing the field order above eight exacerbates the BER performance. In addition to having better BER performance than codes over higher order fields, codes over GF(4) have smaller decoding complexities when decoded using MD-FFT-QSPA algorithm since the complexity of this algorithm is proportional to the field order. Thus, we focus our attention on non-binary, regular, rate-0.8, girth-8 LDPC codes over GF(4) in the rest of the section.

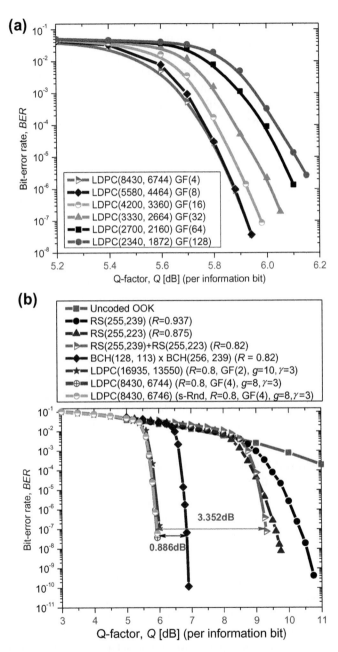

FIGURE 6.9 (a) Comparison of non-binary, (3,15)-regular, girth-8 LDPC codes over BI-AWGN channel. (b) Comparison of 4-ary (3,15)-regular, girth-8 LDPC codes; a binary, girth-10 LDPC code, three RS codes, and a TPC code. (Modified from Ref. [2]; ©IEEE 2009; reprinted with permission.)

In Figure 6.9b, we compare the BER performance of the LDPC(8430, 6744) code over GF(4) discussed in Figure 6.9a, against that of the RS(255, 239) code, RS(255, 223) code, RS(255, 239) + RS(255, 223) concatenation code, and BCH(128, 113) × BCH(256, 239) TPC. We observe that the LDPC code over GF(4) outperforms all of these codes with a significant margin. In particular, it provides an additional coding gain of 3.363dB and 4.401 dB at BER of 10^{-7} when compared to the concatenation code RS(255, 239) + RS(255, 223) and the RS(255, 239) code, respectively. Its coding gain improvement over BCH(128, 113) × BCH(256, 239) TPC is 0.886dB at BER of 4×10^{-8}. Finally, we computed the NCG of the 4-ary, regular, rate-0.8, girth-8 LDPC code over GF(4) to be 10.784dB at BER of 10^{-12}. We also presented in Figure 6.9b a competitive, binary, (3, 15)-regular, LDPC(16935, 13550) code proposed in [73]. We can see that the 4-ary, (3, 15)-regular, girth-8 LDPC(8430, 6744) code beats the bit-length-matched binary LDPC code with a margin of 0.089 dB at BER of 10^{-7}. More importantly, the complexity of the MD-FFT-QSPA used for decoding the non-binary LDPC code is lower than the min-sum-with-correction-term algorithm [20] used for decoding the corresponding binary LDPC code. When the MD-FFT-QSPA is used for decoding a (γ, ρ)-regular q-ary LDPC($N/\log q$, $K/\log q$) code, which is bit-length-matched to a (γ, ρ)-regular binary LDPC(N,K) code, the complexity is given by $(M/\log q)2\rho q(\log q + 1 - 1/(2\rho))$ additions, where $M = N - K$ is the number of check nodes in the binary code. On the other hand, to decode the bit-length-matched binary counterpart using min-sum-with-correction-term algorithm [20], one needs $15M(\rho - 2)$ additions. Thus, a (3, 15)-regular 4-ary LDPC code requires 91.28% of the computational resources required for decoding a (3, 15)-regular binary LDPC code of the same rate and bit length. For additional details on various classes of nonbinary LDPC codes and corresponding decoding algorithms, an interested reader is referred to [80–84].

6.3.7 FPGA implementation of decoders for large-girth QC-LDPC codes

In our study of LDPC decoders' implementation [2,26], we used the min-sum LDPC decoding algorithm, which represents a simplified version of the min-sum-with-correction-term algorithm introduced above, in which the correction term in (6.34) is omitted. Among various alternatives [2,26], we adopted a partially parallel architecture because it is a natural choice for quasi-cyclic codes. In this architecture, a processing element (PE) is assigned to a group of bit/check nodes instead of a single node. A PE mapped to a group of bit nodes is called a bit-processing element (BPE), and a PE mapped to a group of check nodes is called a check-processing element (CPE). BPEs (CPEs) process the nodes assigned to them in a serial fashion. However, all BPEs (CPEs) carry out their tasks simultaneously. In Figure 6.10a, we depict a convenient method for assigning BPEs and CPEs to the nodes in a QC-LDPC code that we introduced in [73]. This method is easy to implement because in addition to regular structure of parity-check matrices, it simplifies the memory addressing. The messages between BPEs and CPEs are exchanged via memory banks. In Table 6.2, we summarize the memory allocation,

(a)

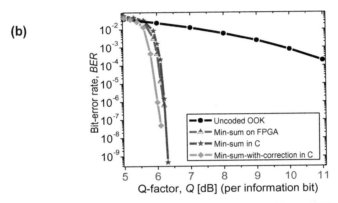

BPEs

		0	1	2	...	c-1
	0	I	I	I	...	I
	1	I	$p^{S[1]}$	$p^{S[2]}$	...	$p^{S[c-1]}$
CPEs	2	I	$p^{2S[1]}$	$p^{2S[2]}$	...	$p^{2S[c-1]}$
	...	...	...	...	...	...
	r-1	I	$p^{(r-1)S[1]}$	$p^{(r-1)S[2]}$	...	$p^{(r-1)S[c-1]}$

(b)

FIGURE 6.10 (a) Assignment of bit nodes and check nodes to BPEs and CPEs. *I* denotes the identity matrix of size $p \times p$ (p is a prime), *P* is the permutation matrix given by $P = (p_{ij})_{p \times p}$, $p_{i,i+1} = p_{p,1} = 1$ (zero otherwise), and *r* and *c* represent the number of block-rows and block-columns. The set of integers *S* are carefully chosen from the set $\{0,1,...,p-1\}$ so that the cycles of short length are avoided. (b) BER performance comparison of FPGA and software implementations. (Modified from Ref. [2]; ©IEEE 2009; reprinted with permission.)

where we used the following notation: MEM B and MEM C denote the memories used to store bit node and check node edge values; MEM E stores the codeword estimate; and MEM I stores the initial LLRs. In Figure 6.10b, we present BER performance comparison of FPGA and software implementations for a QC LDPC(16935, 13550) code. A close agreement between BER curves is observed. For additional details on this FPGA implementation an interested reader is referred to [78,79].

Table 6.2 Memory allocation of the implementation.

MEM Name	MEM B	MEM C	MEM E	MEM I
Data word (bits)	8	11	1	8
Address word (bits)	16	16	15	15
Block size (words)	50,805	50,805	16,935	16,935

6.4 CODED MODULATION

In this section, we describe how to combine modulation with channel coding, and describe several coded-modulation schemes: (i) multilevel coding [85,86], (ii) coded-OFDM [87], (iii) nobinary LDPC-coded modulation [22], and (iv) multidimensional coded modulation [29–34]. Using this approach, modulation, coding, and multiplexing are performed in a unified fashion so that, effectively, the transmission, signal processing, detection, and decoding are done at much lower symbol rates. At these lower rates, dealing with the nonlinear effects and PMD is more manageable, while the aggregate data rate per wavelength is maintained above 100 Gb/s.

6.4.1 Multilevel coding and block-interleaved coded modulation

M-ary PSK, M-ary QAM and M-ary DPSK achieve the transmission of $\log_2 M (=m)$ bits per symbol, providing bandwidth-efficient communication. In coherent detection for M-ary PSK, the data phasor $\varphi_l \in \{0, 2\pi/M, ..., 2\pi(M-1)/M\}$ is sent at each lth transmission interval. In direct detection, the modulation is differential, the data phasor $\varphi_l = \varphi_{l-1} + \Delta\varphi_l$ is sent instead, where $\Delta\varphi_l \in \{0, 2\pi/M, ..., 2\pi(M-1)/M\}$ is determined by the sequence of m input bits using an appropriate mapping rule. Let us now introduce the transmitter architecture employing LDPC codes as channel codes. If component LDPC codes are of different code rates but of the same length, the corresponding scheme is commonly referred to as multilevel coding (MLC). If all component codes are of the same code rate, corresponding scheme is referred to as the block-interleaved coded-modulation (BICM). The use of MLC allows us to adapt the code rates to the constellation mapper and channel. For example, for Gray mapping, 8-PSK and AWGN, it was found in [88] that optimum code rates of individual encoders are approximately 0.75, 0.5, and 0.75, meaning that 2 bits are carried per symbol. In MLC, the bit streams originating from m different information sources are encoded using different (n, k_i) LDPC codes of code rate $r_i = k_i/n$. k_i denotes the number of information bits of the ith $(i = 1,2, ..., m)$ component LDPC code, and n denotes the codeword length, which is the same for all LDPC codes. The mapper accepts m bits, $c = (c_1, c_2, ..., c_m)$, at time instance i from the $(m \times n)$ interleaver column-wise and determines the corresponding M-ary $(M = 2^m)$ constellation point $s_i = (I_i, Q_i) = |s_i|\exp(j\varphi_i)$ (see Figure 6.11a).

The receiver input electrical field at time instance i for an optical M-ary differential phase-shift keying (DPSK) receiver configuration from Figure 6.11b is denoted by $E_i = |E_i|\exp(j\varphi_i)$. The outputs of I- and Q-branches (upper- and lower-branches in Figure 6.11b) are proportional to $\text{Re}\{E_i E_{i-1}^*\}$ and $\text{Im}\{E_i E_{i-1}^*\}$, respectively. The corresponding coherent detector receiver architecture is shown in Figure 6.11c, where

$$S_i = |S|e^{j\varphi_{S,i}} \quad (\varphi_{S,i} = \omega_S t + \varphi_i + \varphi_{S,PN}) \tag{6.37}$$

is coherent receiver input electrical field at time instance i and

$$L = |L|e^{j\varphi_L} \quad (\varphi_L = \omega_L t + \varphi_{L,PN}) \tag{6.38}$$

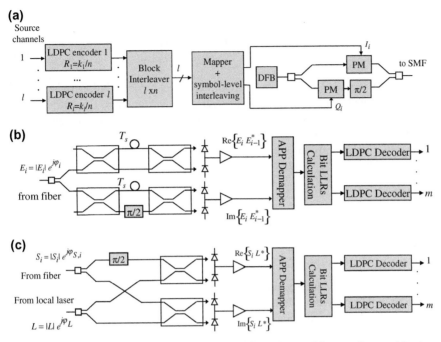

FIGURE 6.11 Block-interleaved LDPC-coded modulation scheme: (a) transmitter architecture, (b) direct detection architecture, and (c) coherent detection receiver architecture. $T_s = 1/R_s$, R_s is the symbol rate.

is the local laser electrical field. For homodyne balanced coherent detection, the frequency of the local laser (ω_L) is the same as that of the incoming optical signal (ω_L), so the balanced outputs of I- and Q-channel branches (upper- and lower-branches of Figure 6.11c) can be written as

$$
\begin{aligned}
v_I(t) &= R\,|S_k|\,|L|\cos\left(\varphi_i + \varphi_{S,PN} - \varphi_{L,PN}\right) \quad (i-1)\,T_s \le t < iT_s \\
v_Q(t) &= R\,|S_k|\,|L|\sin\left(\varphi_i + \varphi_{S,PN} - \varphi_{L,PN}\right) \quad (i-1)\,T_s \le t < iT_s
\end{aligned}
\tag{6.39}
$$

where R is photodiode responsivity while $\phi_{S,PN}$ and $\phi_{L,PN}$ represent the laser phase noise of transmitting and receiving (local) laser, respectively. The outputs at I- and Q-branches (in either coherent or direct detection case) are sampled at the symbol rate (we assume perfect synchronization), and the LLRs of symbols s_i ($i = 1, \ldots, M$) are calculated in an APP demapper block as follows:

$$
\lambda(s_i) = \log \frac{P(s_i|r)}{P(s_r|r)},
\tag{6.40}
$$

where $P(s|r)$ is determined by using Bayes' rule

$$P\left(s_i|r\right) = \frac{P\left(r|s_i\right) P\left(s_i\right)}{\sum_{s'} P\left(r|s'\right) P\left(s'\right)}. \tag{6.41}$$

Notice that $si = (I_i, Q_i)$ is the transmitted signal constellation point at time instance i, while $ri = (r_{I,i}, r_{Q,i})$, $r_{I,i} = v_I(t = iT_s)$, and $r_{Q,i} = v_Q(t = iT_s)$ are the samples of I- and Q-detection branches from Figure 6.11b and c. In the presence of fiber nonlinearities, $P(r|s)$ from (6.41) is to be estimated evaluating the histograms, by employing a sufficiently long training sequence. With $P(s_i)$ we denoted the *a priori* probability of symbol s_i, while s_r is a referent symbol. By substituting Eq. (6.41) into (6.40) we obtain:

$$\begin{aligned}\lambda\left(s_i\right) &= \log\left[\frac{P\left(s_i|r_i\right) P\left(s_i\right)}{P\left(r_i|s_r\right) P\left(s_r\right)}\right] = \log\left[\frac{P\left(r_i|s_i\right)}{P\left(r_i|s_r\right)}\right] + \log\left[\frac{P\left(s_i\right)}{P\left(s_r\right)}\right] \\ &= \log\left[\frac{P\left(r_i|s_i\right)}{P\left(r_i|s_r\right)}\right] + \lambda_a\left(s_i\right), \quad \lambda_a\left(s_i\right) = \log\left[\frac{P\left(s_i\right)}{P\left(s_r\right)}\right],\end{aligned} \tag{6.42}$$

where with $\lambda_a(s_i)$ we denoted the prior reliability of symbol s_i. Let us denote by c_j the jth bit in the observed symbol s_i binary representation $c = (c_1, \ldots, c_m)$. The prior symbol LLRs for the next iteration are determined by:

$$\lambda_a\left(s_i\right) = \log\frac{P\left(s_i\right)}{P\left(s_r\right)} = \log\frac{\prod_{j=1}^{l} P\left(c_j\right)}{\prod_{j=1}^{l} P\left(c_j = 0\right)} = \sum_{j=1}^{l}\log\frac{P\left(c_j\right)}{P\left(c_j = 0\right)}, \tag{6.43}$$

where we assumed that referent symbol is $s_r = (0\ldots0)$. Because

$$\log\frac{P(c_j)}{P(c_j = 0)} = \begin{cases} 0, & c_j = 0 \\ -L(c_j), & c_j = 1 \end{cases} = -c_j L(c_j); \quad L(c_j) = \log\frac{P(c_j = 0)}{P(c_j = 1)} \tag{6.44}$$

the prior symbol LLRs become

$$\lambda_a(s) = -\sum_{j=1}^{l} c_j L(c_j). \tag{6.45}$$

Finally, the prior symbol estimate can be obtained from

$$\lambda_a(\hat{s}) = -\sum_{j=1}^{l} c_j L_{D,e}(c_j), \tag{6.46}$$

where

$$L_{D,e}(\hat{c}_j) = L\left(c_j^{(t)}\right) - L\left(c_j^{(t-1)}\right). \tag{6.47}$$

In Eq. (6.47), we use $L(c_j^{(t)})$ to denote the LDPC decoder output in current iteration (iteration t). The bit LLRs $L(c_j)$ are determined from symbol LLRs by

$$L(\hat{c}_j) = \log \frac{\sum_{c:c_j=0} \exp[\lambda(s)] \exp\left(\sum_{c:c_k=0,k\neq j} L_a(c_k)\right)}{\sum_{c:c_j=1} \exp[\lambda(s)] \exp\left(\sum_{c:c_k=0,k\neq j} L_a(c_k)\right)}. \quad (6.48)$$

Therefore, the jth bit reliability is calculated as the logarithm of the ratio of a probability that $c_j=0$ and probability that $c_j=1$. In the nominator, the summation is done over all symbols s (with corresponding binary representation c) having 0 at the position j, while in the denominator over all symbols s having 1 at the position j. With $L_a(c_k)$ we denoted the prior (extrinsic) information determined from the APP demapper. The inner summation in (6.48) is performed over all bits of symbol s, selected in the outer summation, for which $c_k=0$, $k\neq j$. The bit LLRs are forwarded to LDPC decoders, which provide extrinsic bit LLRs for demapper according to (6.47), and are used as inputs to (6.46) as the prior information. The iteration between the APP demapper and LDPC decoder is performed until the maximum number of iterations is reached, or the valid codewords are obtained.

The results simulations, for 30 iterations in the sum-product algorithm and 10 iterations between the APP demapper and the LDPC decoder, and by employing the BICM and Gray mapping, are shown in Figure 6.12. Although the actual noise in the repeated systems is dominated by the ASE noise, in this calculation we observed the thermal noise dominated scenario, to be consistent with digital communication literature [46–56]. The NCG for 8-PSK at the BER of 10^{-9} is about 9.5 dB and a much larger NCG is expected at BERs below 10^{-12}. Block-interleaved LDPC-coded 8-PSK with coherent detection outperforms LDPC-coded 8-DPSK with direct detection by 2.23 dB at the BER of 10^{-9}. 8-DQAM outperforms 8-DPSK by 1.15 dB at the same BER. LDPC-coded 16-QAM slightly outperforms LDPC-coded 8-PSK, and significantly outperforms LDPC-coded 16-PSK. As expected, LDPC-coded BPSK and LDPC-coded QPSK (with Gray mapping) perform very closely, and they both outperform LDPC-coded OOK by almost 3 dB.

6.4.2 Polarization-multiplexed coded-OFDM

In this sub-section we describe how to combine coded modulation with OFDM, which is illustrated in Figure 6.13. We also describe how 1 Tb/s serial transport and beyond can be achieved based on coded multiband-OFDM. The transmitter configuration up to the mapper is identical to that already described in Figure 6.11. The two-dimensional (2D) signal constellation points (see Figure 6.13b) are split into two streams for OFDM transmitters corresponding to the x- and y-polarizations. The QAM constellation points are considered to be the values of the fast Fourier transform (FFT) of a multi-carrier OFDM signal. The OFDM symbol is generated as follows: N_{QAM} input QAM symbols are zero-padded to obtain N_{FFT} input samples for inverse FFT (IFFT), N_G non-zero samples are inserted to create the guard interval, and the OFDM symbol is multiplied by the Blackman-Harris window function.

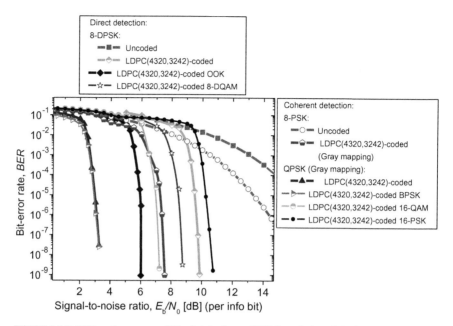

FIGURE 6.12 BER performance of block-interleaved LDPC-coded modulation schemes with both coherent detection and direct detection over the AWGN channel. E_b represents the average bit energy, and N_0 is the power spectral density. (Modified from Ref. [2]; ©IEEE 2009; reprinted with permission.)

For efficient chromatic dispersion and PMD compensation, the length of cyclically extended guard interval should be longer than the total spread due to chromatic dispersion and DGD. The cyclic extension is accomplished by repeating the last $N_G/2$ samples of the effective OFDM symbol part (N_{FFT} samples) as a prefix, and repeating the first $N_G/2$ samples as a suffix. After D/A conversion (DAC), the RF OFDM signal is converted into the optical domain using the dual-drive Mach-Zehnder modulator (MZM). Two MZMs are needed, one for each polarization. The outputs of MZMs are combined using the polarization beam combiner (PBC). One DFB laser is used as CW source, with x- and y-polarization separated by polarization beam splitter (PBS).

On receiver side, the polarization-detector soft estimates of symbols carried by the kth subcarrier in the ith OFDM symbol, $s_{i,k,x(y)}$, are forwarded to the APP demapper, which determines the symbol LLRs $\lambda_{x(y)}(q)$ ($q=0,\ 1,\ \ldots,\ 2^b-1$) of x-polarization (y-polarization) by

$$\lambda_{x(y)}(q) = -(\text{Re}[\tilde{s}_{i,k,x(y)}] - \text{Re}[\text{QAM}(\text{map}(q))])^2/(2\sigma^2) - (\text{Im}[\tilde{s}_{i,k,x(y)}] - \text{Im}[\text{QAM}(\text{map}(q))])^2/(2\sigma^2), \tag{6.49}$$

where Re[ ] and Im[ ] denote the real and imaginary part of a complex number, QAM denotes the QAM-constellation diagram, σ^2 denotes the variance of an equivalent Gaussian noise process originating from ASE noise, and map(q) denotes a

(a)

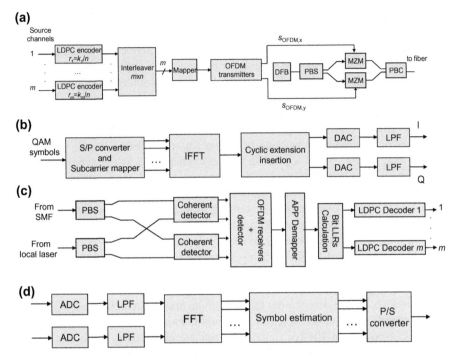

FIGURE 6.13 Polarization-division multiplexed LDPC-coded OFDM employing both polarizations: (a) transmitter architecture, (b) RF OFDM transmitter configuration, (c) receiver architecture, and (d) RF OFDM receiver configuration. DFB: distributed feedback laser, PBS(C): polarization beam splitter (combiner), MZM: dual-drive Mach-Zehnder modulator serving as *I/Q* modulator.

corresponding mapping rule. (*b* denotes the number of bits per constellation point.) Let us denote by $v_{j,x(y)}$ the *j*th bit in an observed symbol *q* binary representation $v = (v_1, v_2, \ldots, v_b)$ for *x*- (*y*-) polarization. The bit LLRs needed for LDPC decoding are calculated from symbol LLRs in fashion similar to Eq. (6.48). The extrinsic LLRs are iterated backward and forward until convergence or pre-determined number of iterations has been reached. The polarization-detector soft estimates can be obtained by employing: (i) polarization-time coding [89] similar to space-time coding proposed for use in MIMO wireless communication systems [90], (ii) using BLAST algorithm [91], (iii) by polarization interference cancelation scheme [91], or (iv) by carefully performed channel matrix inversion [92]. Notice that equation (6.49) is valid only in the quasi-linear regime, in the nonlinear regime we need to determine the likelihoods based on histogram method in a fashion similar to that given by Eq. (6.40).

Since the 100 GbE has been recently standardized, the next natural step would be the introduction of 1 Tb/s Ethernet (1 TbE). Moreover, it is believed that the 1 TbE standard should be standardized soon [35]. Coherent optical OFDM is one promising pathway toward achieving beyond 1 Tb/s optical transport [3,95]. Initial studies [3] indicate that the system *Q*-factor when multiband OFDM with orthogonal sub-bands is used is low (about 13.2 dB after 1000 km of SMF [3]). Such a low Q-factor represents a very tight margin in terms of 7% overhead for RS(255, 239) code, and the use of stronger LDPC codes is advocated in [3,95]. In this section, we describe one potential LDPC-coded multiband OFDM scheme suitable for beyond 1 Tb/s optical transport.

In Figure 6.14, we describe the conceptual diagram for multiplexing/demultiplexing enabling 1 Tb/s Ethernet based on polarization-division multiplexed (PDM) multiband OFDM. The frame corresponding to 1 Tb/s (see Figure 6.14a) is organized in five band groups, each carrying 200 Gb/s traffic, which is equivalent to one 100 GbE frame per polarization. Each group band contains five OFDM bands, with guard spacing $\Delta f_G = m \Delta f_{sc}$ (m is a positive integer), where Δf_{sc} is the subcarrier spacing. Because the central frequencies of neighboring OFDM bands are orthogonal to each other, we may simplify the separation of OFDM bands by anti-aliasing filters, as explained in [3]. Every particular OFDM band carries 40 Gb/s traffic, employing polarization multiplexing and RF multiplexing. The 40 Gb/s traffic can originate from 40 GbE, in which case there is no need for polarization multiplexing. If 40 Gb/s traffic originates from 10 GbE, we first perform RF multiplexing to 20 Gb/s and then polarization multiplexing to 40 Gb/s. The proposed multiplexing/demultiplexing scheme is compatible with 10 GbE, 40 GbE, 100 GbE, and 400 GbE.

In Figure 6.14b we describe our three-layer integrated circuit hierarchy. The lowest (baseband) level, at 40 Gb/s, can be straightforwardly implemented in CMOS ASIC technology, and 40 Gb/s signal can originate either from 40 GbE or 10 GbE as we explained above. The second layer, i.e. RF layer that is used to perform RF multiplexing, can quite readily be implemented in the mixed-circuit CMOS ASIC technology, thanks to recent progress in CMOS technology [36,37]. The third layer is the photonics layer, which can be implemented in photonic integrated circuit (PIC)

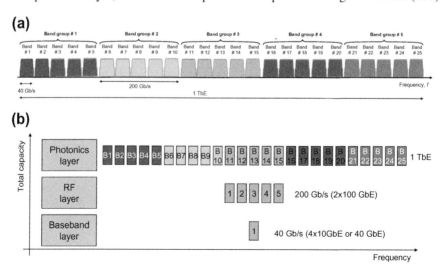

FIGURE 6.14 Conceptual diagram for multiplexing/demultiplexing for 1 Tb/s Ethernet: (a) organization of the frame and (b) three-layer integrated circuit hierarchy.

technology [38,39], and simply requires the integration of five-frequency locked lasers that are combined optically into 1 Tb/s optical transport signal.

In Figure 6.15, we show the single-band BER performance of both the uncoded and LDPC-coded polarization multiplexed OFDM, against the polarization diversity OFDM scheme, for different constellation sizes. For DGD of 1200 ps, the polarization-multiplexed scheme [92] performs comparable to the polarization-diversity OFDM scheme in terms of BER (the corresponding curves overlap each other), while offering two times higher spectral efficiency. The net coding gain increases as the constellation size grows. For $M = 4$ QAM based polarization multiplexed coded-OFDM the net NCG is 8.36 dB at BER of 10^{-7}, while for $M = 32$ QAM based LPDC-coded OFDM (of aggregate data rate 100 Gb/s) the NCG is 9.53 dB at the same BER.

6.4.3 Nonbinary LDPC-coded modulation

The use of coded modulation with non-binary LDPC codes provides several advantages over the previously proposed binary counterparts based on block-interleaved LDPC-coded modulation (BI-LDPC-CM), described in Section 6.4.1, which can be summarized as follows [22,25,27,28]: (i) m binary LDPC encoders/decoders needed for 2^m-QAM modulation (where $m > 1$) are collapsed into a single 2^m-ary encoder/decoder (reducing the overall computational complexity of the system), (ii) the use of block (de-) interleavers; binary-to-non-binary and vice versa conversion interfaces are eliminated (reducing the latency in the system), (iii) APP de-mapper and LDPC decoder are integrated into a single block, and thus the need for iterating extrinsic information between APP de-mapper and LDPC decoder is eliminated

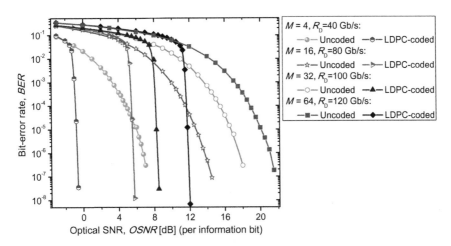

FIGURE 6.15 Single-band BER performance of polarization multiplexed coded-OFDM, for DGD of 1200 ps. R_D denotes the aggregate data rate. (Modified from Ref. [2]; ©IEEE 2009; reprinted with permission.)

(reducing the latency in the system). In addition to lowering both the complexity and the latency in an optical communication system, we demonstrated in [22,25] that the nonbinary coded modulation scheme provides significantly higher coding gains than its binary counterparts based on BI-LDPC-CM. The transmitter (Tx) and receiver (Rx) configurations of the non-binary LDPC-coded modulation (NB-LDPC-CM) scheme are shown in Figure 6.16. Two independent data streams are encoded using different non-binary LDPC codes of code rates $R_i = K_i/N$ ($i \in \{x, y\}$) where K_x (K_y) denotes the number of information symbols used in the non-binary LDPC code corresponding to x- (y-) polarization, and N denotes the codeword length, which is the same for both LDPC codes. The non-binary LDPC codes operate over finite fields $GF(q_i = 2^{m_i})$, $i \in \{x, y\}$. The Mapper x (y) accepts a non-binary symbol from the LDPC encoder x (y) at time instance l and determines the corresponding q_x-ary (q_y-ary) constellation point $s_{l,x} = (I_{l,x}, Q_{l,x}) = |s_{l,x}|\exp(j\varphi_{l,x})$ [$s_{l,y} = (I_{l,y}, Q_{l,y}) = |s_{l,y}|\exp(j\varphi_{l,y})$] with coordinates being used as the inputs of a dual-drive MZM$_x$ (MZM$_y$) or I/Q Modulator$_x$ (I/Q Modulator$_y$), as shown in Figure 6.16a. At the receiver side, the outputs at I- and Q-branches (in either x- or y-polarization receiver branch) are sampled at the symbol rate, while the symbol LLRs are calculated as follows:

$$\lambda(s) = \log[P(s|r)/P(s_0|r)], \tag{6.50}$$

(a)

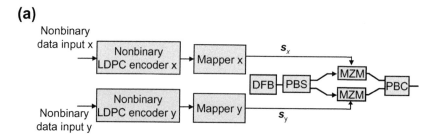

(b)

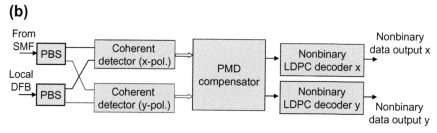

FIGURE 6.16 The NB-LDPC-CM scheme: (a) transmitter (Tx) and (b) receiver (Rx) configurations. PBS/C: polarization beam splitter/combiner. The $s_X = (s_{X,I}, s_{X,Q})$ and $s_Y = (s_{Y,I}, s_{Y,Q})$ denote the symbols transmitted in x- and y-polarizations, respectively; where we used subscripts I and Q to denote components corresponding to in-phase (I) and quadrature (Q) channels. MZM: dual-drive Mach-Zehnder modulator serving as I/Q modulator.

where $s = (I_l, Q_l)$ and $r = (r_I, r_Q)$ denote the transmitted signal constellation point and received symbol at time instance l (in either x- or y-polarization), respectively, and s_0 represents the referent symbol.

In order to evaluate the performance of NB-QC-LDPC-coded modulation techniques, we performed Monte Carlo simulations by observing the thermal noise dominated scenario, to be compatible with Section 6.4.1. In Figure 6.17, we present the BER performance results of quasi-cyclic (γ, ρ)-regular binary and nonbinary LDPC-coded schemes, modulated using 2^m-QAM where $m = 2$, 3, and 4, respectively. The corresponding parity-check matrix H of dimension $(N-K) \times N$, $K < N$, given by Eq. (6.36), has γ 1s in each column and ρ 1s in each row. We report BER performance when component LDPC codes of rates 0.8 (see Figure 6.17a) and 0.85 (see Figure 6.17a) are used. The nonbinary LDPC codes and binary LDPC codes are of comparable lengths. The number of decoding iterations for BI-LDPOC-CM is set to 25, and the number between APP demapper and LDPC decoders to 3. The number of decoding iterations for NB-LDPC-CM is set to 50. The LDPC code parameters and aggregate information rates, defined as $R_s \times R \times m \times 2$ (where R_s denotes the symbol rate, R is the code rate, m is the number of bits per symbol, and factor 2 corresponds to two orthogonal polarization states), are provided in Table 6.3. The symbol rate is set to $R_s = 50$ GS/s, and polarization-multiplexing is used. We further calculated the NCGs of both binary and nonbinary LDPC-coded modulation schemes at BER of 10^{-6} for different number of bits per symbol m, and the results are summarized in Table 6.4. It is interesting to notice that NCG improvement of nonbinary over binary coded-modulation schemes increases as the signal constellation size increases, and for 16-QAM the NCG improvement is 1.2 dB, when LDPC codes of rate 0.8 are used.

When an FFT-based decoding algorithm is employed, the computational complexity of an NB-LDPC-CM scheme using a (γ, ρ)-regular 2^m-ary LDPC(N, K) code with 2^m-QAM modulation is given by the numerator of the complexity ratio (CR) term in Eq. (6.51). On the other hand, when the min-sum-plus-correction-term algorithm [20] is used, the complexity of a corresponding BI-LDPC-CM scheme using $m(\gamma, \rho)$-regular binary LDPC(N,K) codes with 2^m-QAM modulation is given by the denominator in Eq. (6.51), where $M = N - K$ and $q = 2^m$. For example, if (3, 30)-regular, rate-0.9 LDPC codes are used in both schemes, we get CR = 42.62% and CR = 75.87% when QPSK and 8-QAM are employed, respectively

$$\text{CR} = \frac{2\rho q M(m + 1 - 1/(2\rho))}{\rho m M(\rho - 2)} = 2q[m + 1 - 1/(2\rho)]/[m(\rho - 2)]. \qquad (6.51)$$

In simulations above, we observed the thermal noise dominated scenario. In practice, to deal with various channel impairments simultaneously, such as residual chromatic dispersion, PMD, polarization dependent loss (PDL) and fiber nonlinearities, we will need to use turbo equalization, which is the subject of Section 6.6. The complexity of turbo equalizer is typically dominated by MAP equalizer, implemented based on BCJR algorithm [6], as described later. In turbo equalization, the APP demapper is substituted by MAP equalizer, and BI-LDPC-CM requires the iteration of extrinsic bit reliabilities between MAP equalizer and LPDC decoder and,

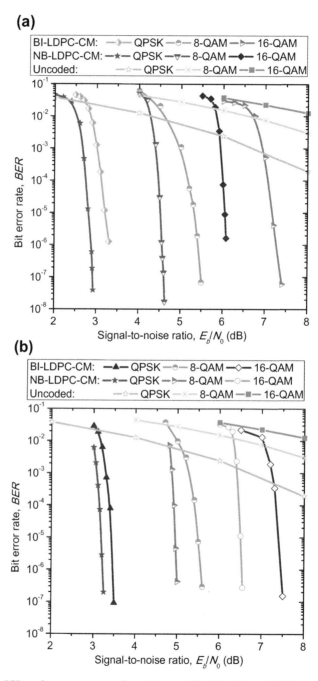

FIGURE 6.17 BER performance comparison between NB-LDPC-CM and BI-LDPC-CM schemes for component LDPC codes of rate: (a) 0.8 and (b) 0.85.

Table 6.3 LDPC code parameters and achievable aggregate information bit rates for polarization-multiplexed BI-LDPC-CM and NB-LDPC-CM schemes.[a]

Code Rate	(γ, ρ)	(N, K)	Aggregate Information Bit Rate (Gb/s)		
			QPSK	8-QAM	16-QAM
0.8	(3, 15)	(8550, 6840)	160	170	180
	(4, 21)	(8547, 6922)			
0.85	(3, 20)	(16,200, 13,770)	240	255	270
	(4, 27)	(16,200, 13,810)			

[a]NB-LDPC-CM (BI-LDPC-CM) scheme is the first (second) row for a given code rate.

Table 6.4 Coding gains (in dBs) determined at the BER of 10^{-6} for polarization-multiplexed BI-LDPC-CM and NB-LDPC-CM schemes.[a]

Code Rate	Modulation Format		
	QPSK	8-QAM	16-QAM
0.8	7.62	8.06	8.31
	7.19	7.21	7.11
0.85	7.27	7.64	7.87
	7.02	7.06	6.94

[a]NB-LDPC-CM (BI-LDPC-CM) scheme is the first (second) row for a given code rate.

therefore, multiple usage of BCJR algorithm. On the other hand, in NB-LDPC-CM, the BCJR algorithm is used only once meaning that complexity of NB-LDPC-CM is even lower than that described by equation (6.51). For more details on complexity when turbo equalization is used an interested reader is referred to [28].

6.4.4 Multidimensional coded modulation

In order to satisfy high-bandwidth demands of future optical networks and solve interoperability problems while keeping system cost and power consumption reasonably low, we proposed the use of multidimensional coded-modulation schemes in a series of articles [29–34]. The key idea behind this proposal is to exploit various degrees of freedom already available for the conveyance of information on a photon. Available degrees of freedom include frequency, time, phase, amplitude, and polarization. (Please refer to [93] for additional details on polarization shift keying theory.) Here we describe one approach based on subcarrier-multiplexed four-dimensional (4D) LDPC-coded modulation [33,34].

The 4D subcarrier-multiplexed system is composed of N 4D subsystems, as shown in Figure 6.18a. The $N \times m$ input bit streams from different information sources are divided into N groups of m streams per subcarrier. The m streams of each

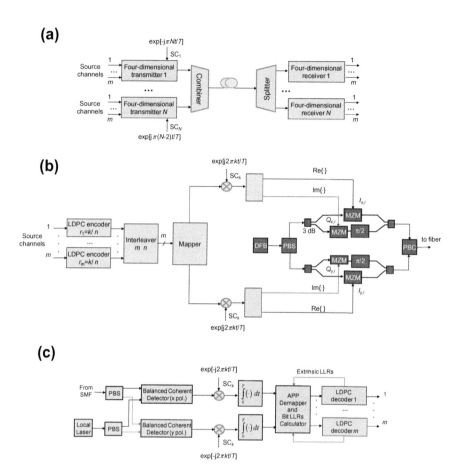

FIGURE 6.18 Four-dimensional coded optical subcarrier-multiplexed system: (a) system configuration, (b) 4D-transmitter configuration in Cartesian (*I/Q*) coordinates, and (c) 4D-receiver configuration.

subcarrier are used as input to a 4D transmitter. To each subsystem a unique subcarrier $\exp(j2\pi kt/T)$ is assigned, wherein subcarriers are orthogonal to each other (they contain the integer number of cycles within a symbol duration *T*). The outputs of the *N* 4D transmitters are combined together using a power combiner and sent over optical transmission system of interest.

The 4D transmitter configuration is shown in Figure 6.18b. The *m* bit independent streams of *k*th subcarrier are used as inputs of *m* LDPC encoders of code rate k_i/n (k_i represents the information word length of the *i*th encoder and *n* is the codeword length that is common for all encoders). The encoded data streams from these branches are forwarded to an $m \times n$ block interleaver where they are written row-wise. At a time instant *i*, the mapper reads *m* bits column-wise from interleaver to determine the corresponding 2^m-ary signal constellation point. The mapper is based on a simple look-up table with 2^m memory locations. It follows a given mapping rule to select the output voltages needed to control the modulator. The output of the mapper is in the 4D signal constellation, which is multiplied by complex exponential term $\exp(j2\pi kt/T)$. The corresponding signal constellation point of ith symbol interval can be represented in a vector form as follows:

$$
s_i^{(k)} = \begin{pmatrix} \Re(E_{x,i}) \\ \Im(E_{x,i}) \\ \Re(E_{y,i}) \\ \Im(E_{y,i}) \end{pmatrix} e^{j2\pi kt/T} = \begin{pmatrix} I_{x,i} \\ Q_{x,i} \\ I_{y,i} \\ Q_{y,i} \end{pmatrix} e^{j2\pi kt/T} = \begin{pmatrix} |s_{x,i}|\cos\theta_{x,i} \\ |s_{x,i}|\sin\theta_{x,i} \\ |s_{y,i}|\cos\theta_{y,i} \\ |s_{y,i}|\sin\theta_{y,i} \end{pmatrix} e^{j2\pi kt/T}.
$$

$$(6.52)$$

The first two coordinates are used as I and Q inputs of I/Q modulator corresponding to x-polarization, while the second two coordinates as the inputs of I/Q modulator corresponding to y-polarization. The x- and y-polarization streams are combined together by polarization beam combiner (PBC) as shown in Figure 6.18b. The superscript (k) is used to denote the kth subcarrier.

At the receiver side (see Figure 6.18a), the signal is split into N subcarrier branches and forwarded to the corresponding 4D receivers. The kth subcarrier 4D receiver is shown in Figure 6.18c. The optical signal is split into two orthogonal polarizations by using the polarization beam splitter (PBS) and is used as input into two balanced coherent detectors. The balanced coherent detectors provide the estimated in-phase and quadrature information for both polarizations to be used in the turbo-like decoding as explained below. The outputs of the detectors are demodulated by the subcarrier specified for the corresponding 4D receiver (see Figure 6.18c). The output samples are then forwarded to the APP demapper and a bit LLRs' calculator in order to provide the bit LLRs required for iterative LDPC decoding. The APP demapper calculates the symbol LLRs using the following equation:

$$
\lambda\left(s_i^{(k)}\right) = \log\left[\frac{P\left(s_0|r_i^{(k)}\right)}{P\left(s_i^{(k)}|r_i^{(k)}\right)}\right],
$$

$$(6.53)$$

where $P\left(s_i^{(k)}\mid r_i^{(k)}\right)$ is determined by Bayes' rule as:

$$
P\left(s_i^{(k)}|r_i^{(k)}\right) = \frac{P\left(r_i^{(k)}|s_i^{(k)}\right)P\left(s_i^{(k)}\right)}{\sum_{s'}P\left(r_i^{(k)}|s_i'^{(k)}\right)P\left(s_i'^{(k)}\right)}.
$$

$$(6.54)$$

The bit LLRs calculator on the other hand calculates the bit LLRs from the symbol LLRs, as follows:

$$
L\left(\hat{v}_j^{(k)}\right) = \log\left[\frac{\sum_{s_i:v_j=0}\exp\left(\lambda\left(s_i^{(k)}\right)\right)}{\sum_{s_i:v_j=1}\exp\left(\lambda\left(s_i^{(k)}\right)\right)}\right].
$$

$$(6.55)$$

In the above equations $s_i^{(k)}$ denotes the transmitted signal constellation point on kth subcarrier, $r_i^{(k)}$ denotes the received constellation point (on kth subcarrier), and s_0 denotes the referent constellation point. The $P(r_i^{(k)}|s_i^{(k)})$ denotes the conditional probability, and $P(s^{(k)})$ is the *a priori* symbol probability, while $\hat{v}_j^{(k)}$ ($j \in \{0, 1, \ldots, n-1\}$) is the jth bit of the codeword v. The bit LLRs are forwarded to LDPC decoders, which provide extrinsic bit LLRs for demapper and are used as inputs to (6.54) as prior information.

The turbo-like decoding process is used to reduce the number of iterations required by the LDPC decoder to reach convergence, and it is performed as follows. After the bit LLRs are calculated, the extrinsic LLRs of the demapper are forwarded to the LDPC decoder as the *a priori* probabilities to be used in the LDPC decoding process. The resulting extrinsic information of the LDPC decoder are sent back to the APP demapper to be used as the *a priori* reliabilities again. In the turbo-like decoding algorithm, the outer back and forth iterations are repeated until convergence is achieved unless a predefined number of iterations is reached. Once the iterations stop, the LDPC decoders will yield the decoded data to the m outputs.

To illustrate the high-potential of this scheme, in Figure 6.19 we provide the BER performance results when two orthogonal subcarriers are used and vertices of different 4D polytopes as signal constellation points. This scheme is compared with conventional PDM-QAM. For 4D polytopes with 96 vertices, the 64 vertices are used to carry 6 bits per symbol by using the first subcarrier, while the remaining 32 points are used to carry 5 bits/symbol on the second subcarrier. Therefore, 11 bits/symbol are transmitted, which represents a highly bandwidth-efficient scheme. The constellation points for 96-4D signal constellation based on cantellated tesseract are chosen as permutations of $\left(\pm 1, \pm 1, \pm \left(1 + \sqrt{2} \right), \pm \left(1 + \sqrt{2} \right) \right)$. On the other hand,

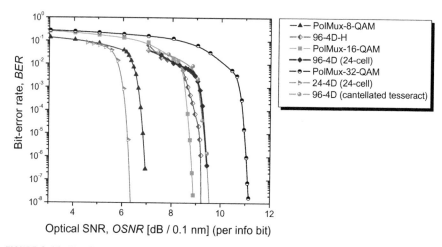

FIGURE 6.19 The 2-subcarrier-multiplexed 4D-LDPC-coded modulation scheme BER performance at baseband information rate of 40 Gb/s.

the constellation points for 96-4D signal constellation based on rectified 24-cell are given as permutations of $(0, \pm 1, \pm 1, \pm 2)$. The constellation points of 96-4D-H signal constellation are given in [29] as combination of 64 and 32 subcarrier-multiplexed-4D. Finally, the constellation points for 24-4D signal constellation based on 24-cell are given as follows: 8 points as permutations of $(\pm 1, 0, 0, 0)$ on the first subcarrier and 16 points as different combinations of $(\pm 1/2, \pm 1/2, \pm 1/2, \pm 1/2)$ on the second subcarrier. This constellation carries 7 bits per symbol. The 24-4D scheme outperforms the PDM-8-QAM by ~ 0.7 dB while having higher aggregate data rate of 280 Gb/s. The aggregate data rate of PDM-8-QAM is 240 Gb/s. The various 96-4D schemes have the aggregate data rate of 440 Gb/s and are compatible with 400 Gb/s Ethernet. The corresponding PDM-32-QAM has the aggregate data rate of 400 Gb/s. The worst 96-4D scheme outperforms the corresponding PDM-32-QAM by ~ 1.5 dB.

6.5 ADAPTIVE NONBINARY LDPC-CODED MODULATION

Current limitations of optical transport networks result from the heterogeneity of the infrastructure and consequential bottlenecks at domain-, layer, physical media-, and technology-boundaries and interfaces, respectively. For example, in optically routed networks, neighboring WDM channels transport the traffic that is random, and different lightwave paths experience different penalties due to deployment of reconfigurable optical add-drop muxes (ROADMs) and wavelength crossconnects (WXCs). In order to provide seamless integrated transport platforms, which can support heterogeneous networking, in this section, we describe the use of rate-adaptive non-binary LDPC-coded polarization-division multiplexed QAM with coherent detection, which we introduced in [25] (see also [4,27,28]). The use of different channel codes for different destinations would be costly to implement due to increased hardware complexity, unless unified encoding and decoding architectures can be used for all destinations. The structured quasi-cyclic LDPC codes provide us with this unique feature. By using the multilevel modulation and polarization-multiplexing, all related coding, signal processing, and transmission are performed at lower symbol rates (such as 33 GS/s), where dealing with fiber nonlinearities and PMD is more manageable, while the aggregate rate is kept beyond 100 Gb/s. The code rate of non-binary LDPC (NB-LDPC) code, for a given constellation size, is chosen in accordance with the channel conditions. When the channel conditions are favorable (corresponding to large SNR), higher code rate LDPC code is employed. Compared to the adaptive modulation approach, the proposed adaptive coding approach is very friendly from an implementation point of view because the symbol rate is kept constant so that all synchronization issues related to the variable symbol rate modulation adaptation are avoided. The use of non-binary coded multilevel modulation schemes is described because they offer lower decoding complexity and latency, and at the same time provide larger coding gains compared to their binary counterparts, as we have shown in Section 6.4.3.

We consider nonbinary LDPC code rates of rates $R=0.833$, 0.875, and 0.9. We test the performance of rate-adaptive LDPC-coded modulation scheme by using QPSK, 8-star-QAM, and 16-star-QAM modulation formats, which map $m=2$, 3, and 4 bits per symbol, respectively; and by observing the thermal noise dominate scenario to be consistent with Section 6.4.3. Therefore, the achievable information bit rates can be determined as $2mR_sR$ bit/s, and hence the lowest information bit rate is obtained when the lowest code rate ($R=0.833$) is used. Consequently, using $R_s=60\,\text{GS/s}$, the lowest information bit rates achievable using our scheme is given by 200 Gb/s, 300 Gb/s, and 400 Gb/s for QPSK, 8-star-QAM, and 16-star-QAM modulations, respectively.

We employ a closely related family of structured QC-LDPC codes over GF(2^m) in our rate-adaptive LDPC-coded scheme. All our codes are $(3,\rho)$-regular and of fixed length N symbols, and corresponding parity-check matrices are obtained by Eq. (6.36), by varying the size of sub-matrices B, as shown in Table 6.5. As it can be concluded from the table, the three different code rates considered in our scheme are 0.833, 0.875, and 0.9 with corresponding overheads (*OHs*) of 20%, 14.29%, and 11.11%, respectively, where overhead is defined by $OH=(1/R-1)\times100\%=(N/K-1)\times100\%$. To keep the complexity of decoder reasonable low, the column weight is fixed to $\gamma=3$ for all component codes. Hence, the code rate adaptation is obtained by changing either the sub-matrix size B or the row weight ρ.

To illustrate the efficiency of this scenario, in Figure 6.20 we present the BER performance curves for these codes when conventional QAM modulations are used under thermal noise dominated scenario. It is clear from the figure that for a given BER, a higher SNR is required for a higher rate code compared to a lower rate code to achieve the same target BER. The region of SNR should be properly quantized, and based on channel SNR corresponding LDPC code should be used. For simulation results related to the long-haul transmission an interested reader is referred to [27,28].

6.6 LDPC-CODED TURBO EQUALIZATION

In this section we describe an LDPC-coded turbo equalization (TE) scheme [2,4,6,111], as a universal scheme that can be used for simultaneous mitigation of: (i) fiber nonlinearities, (ii) imperfectly compensated PMD, (iii) PDL, (iii) imperfectly compensated chromatic dispersion, and (iv) *I/Q*-imbalance effects in multi-level coded-modulation schemes. (For review of various optical channel impairments an interested reader is referred to [68,69].)

6.6.1 MAP detection

Before we describe the LDPC-coded turbo equalization scheme, we provide the basic concepts of optimum detection of binary signaling in minimum probability of error sense [96,97]. Let x denote the transmitted sequence and y the received sequence.

Table 6.5 Parameters for a rate-adaptive nonbinary LDPC-coded scheme.

Code Rate	N	K	M	γ	ρ	B
0.833	13,680	11,400	2280	3	18	760
0.875	13,680	11,970	1710	3	24	570
0.9	13,680	12,312	1368	3	30	456

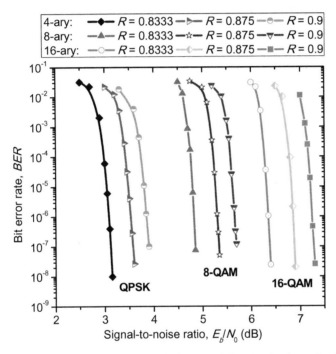

FIGURE 6.20 Performance of LDPC codes listed in Table 6.5 under the thermal noise dominated transmission scenario.

The *optimum receiver* assigns $\hat{x}_k$ to the value $x \in \{0, 1\}$ that maximizes the *a posteriori* probability (APP) $P(x_k = x|\mathbf{y})$ given the received sequence $\mathbf{y}$ as follows:

$$\hat{x}_k = \arg \max_{x \in \{0,1\}} P(x_k = x|\mathbf{y}). \tag{6.56}$$

The corresponding detection algorithm is commonly referred to as a maximum *a posteriori* probability (MAP) detection algorithm. In practice, it is common to use the logarithmic version of equation (6.56) as follows:

$$\hat{x}_k = \begin{cases} 0, L(x_k|\mathbf{y}) \leqslant 0 \\ 1, \text{otherwise} \end{cases} \quad L(x_k|\mathbf{y}) = \log \left[\frac{P(x_k = 0|\mathbf{y})}{P(x_k = 1|\mathbf{y})} \right], \tag{6.57}$$

where $L(x_k|y)$ is the conditional log-likelihood ratio (LLR). To calculate the $P(x_k=x|y)$ needed in either equation above we invoke the Bayes' rule:

$$P(x_k = x|y) = \sum_{\forall x:x_k=x} P(x|y) = \sum_{\forall x:x_k=x} \frac{P(y|x)P(x)}{P(y)}, \quad (6.58)$$

where $P(y|x)$ is conditional probability density function (PDF), and $P(x)$ is the *a priori* probability of input sequence x, when the symbols are independent factors as $P(x) = \prod_{i=1}^{n} P(x_i)$, where n is the codeword length. By substituting Eq. (6.58) into Eq. (6.57), the conditional LLR can be written as:

$$L(x_k|y) = \log\left[\frac{\sum_{\forall x:x_k=0} P(y|x) \prod_{i=1}^{n} P(x_i)}{\sum_{\forall x:x_k=1} P(y|x) \prod_{i=1}^{n} P(x_i)}\right] = L_{\text{ext}}(x_k|y) + L(x_k), \quad (6.59)$$

where the extrinsic information about x_k contained in y $L_{\text{ext}}(x_k|y)$ and the *a priori* LLR $L(x_k)$ are defined respectively as

$$L_{\text{ext}}(x_k|y) = \log\left[\frac{\sum_{\forall x:x_k=0} P(y|x) \prod_{i=1,i\neq k}^{n} P(x_i)}{\sum_{\forall x:x_k=1} P(y|x) \prod_{i=1,i\neq k}^{n} P(x_i)}\right], L(x_k) = \log\left[\frac{P(x_k=0)}{P(x_k=1)}\right].$$

$$(6.60)$$

From Eq. (6.60) it is clear that computation of conditional LLRs can be computationally extensive. One possible computation is based on BCJR algorithm [6], with log-domain version when applied to the multilevel modulation schemes being described in the following section.

6.6.2 Multilevel turbo equalization

The multilevel LDPC-coded turbo equalizer is composed of two ingredients: (i) the MAP detector based on multilevel BCJR detection algorithm [2,4,7,114,115], and (ii) the LDPC decoder. The transmitter configuration, for MLC, is already described in previous sections (e.g. Figures 6.11, 6.13, 6.16, and 6.18). The receiver configuration of LDPC-coded turbo equalizer is shown in Figure 6.21. The outputs of upper- and lower-coherent-detection-balanced branches, proportional to $\text{Re}\{S_iL^*\}$ and $\text{Im}\{S_iL^*\}$, respectively, are used as inputs of multilevel BCJR equalizer, where the local laser electrical field is denoted by $L=|L|\exp(j\phi_L)$ (ϕ_L denotes the laser phase noise process of the local laser) and incoming optical signal at time instance i is denoted by S_i.

The multilevel BCJR equalizer operates on a discrete dynamical trellis description of the optical channel. Notice that this equalizer is universal and applicable to any two-dimensional signal constellation such as M-ary PSK, M-ary QAM, or M-ary polarization-shift keying (PolSK), and both coherent and direct detections. This scheme can easily be generalized to any multidimensional scheme. This dynamical trellis is uniquely defined by the following triplet: the previous state, the next state, and the channel output. The state in the trellis is defined as $s_j=(x_{j-m}, x_{j-m+1}, \ldots, x_j, x_{j+1}, \ldots, x_{j+m})=x[j-m, j+m]$, where x_k denotes the index of the

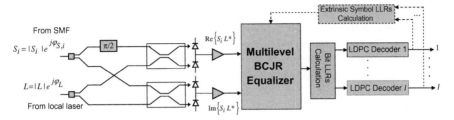

FIGURE 6.21 LDPC-coded turbo equalization scheme architecture.

symbol from the following set of possible indices $X = \{0,\ 1,\ \ldots,\ M-1\}$, with M being the number of points in corresponding M-ary signal constellation. Every symbol carries $l = \log_2 M$ bits, using the appropriate mapping rule (natural, Gray, anti-Gray, etc.) The memory of the state is equal to $2m+1$, with $2m$ being the number of *symbols* that influence the observed symbol from both sides. An example trellis of memory $2m+1=3$ for 4-ary modulation formats (such as QPSK) is shown in Figure 6.22. The trellis has $M^{2m+1}=64$ states ($s_0,\ s_1,\ \ldots,\ s_{63}$), each of which corresponds to the different 3-symbol patterns (symbol-configurations). The state index

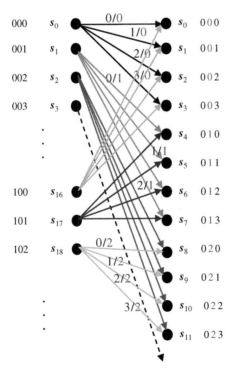

FIGURE 6.22 A portion of trellis for 4-level MAP detector with memory $2m+1=3$.

is determined by considering $(2m+1)$ symbols as digits in numerical system with the base M. For example, in Figure 6.22, the quaternary numerical system (with the base 4) is used. The left column in dynamic trellis represents the current states and the right column denotes the terminal states. The branches are labeled by two symbols, the input symbol is the last symbol in initial state, the output symbol is the central symbol of terminal state. Therefore, the current symbol is affected by both previous and incoming symbols. For the complete description of the dynamical trellis, the transition probability density functions (PDFs) $p(y_j|x_j) = p(y_j|s)$, $s \in S$ are needed; where S is the set of states in the trellis, and y_j is the vector of samples (corresponding to the transmitted symbol index x_j). The conditional PDFs can be determined from *collected histograms* or by using *instanton-Edgeworth expansion* method [113]. The number of edges originating in any of the left-column states is M, and the number of merging edges in arbitrary terminal state is also M.

The *forward metric* is defined as $\alpha_j(s) = \log\{p(s_j=s, y[1, j])\}$ $(j = 1, 2, ..., n)$; the *backward metric* is defined as $\beta_j(s) = \log\{p(y[j+1, n]|s_j=s)\}$; and the *branch metric* is defined as $\gamma_j(s', s) = \log[p(s_j=s, y_j, s_{j-1}=s')]$. The corresponding metrics can be calculated iteratively as follows:

$$\alpha_j(s) = \max_{s'}{}^* \left[\alpha_{j-1}(s') + \gamma_j(s', s)\right], \tag{6.61}$$

$$\beta_{j-1}(s') = \max_{s}{}^* \left[\beta_j(s) + \gamma_j(s', s)\right], \tag{6.62}$$

$$\gamma_j(s', s) = \log\left[p\left(y_j|x[j-m, j+m]\right)P\left(x_j\right)\right]. \tag{6.63}$$

The $\max^*$-operator used in Eqs. (6.61), (6.62) is defined by $\max^*(x, y) = \log(e^x + e^y)$, and it is efficiently calculated by [20] $\max^*(x, y) = \max(x, y) + c_f(x, y)$, where $c_f(x, y)$ is the correction factor, defined as $c_f(x, y) = \log[1 + \exp(-|x-y|)]$, which is commonly approximated or implemented using a look-up table. $p(y_j|x[j-m, j+m])$ is obtained, as already indicated above, by either collecting the histograms or by instanton-Edgeworth expansion method, and $P(x_j)$ represents *a priori* probability of transmitted symbol x_j. In the first outer iteration $P(x_j)$ is set to either $1/M$ (because equally probable transmission is observed) for an existing transition from trellis given in Figure 6.22, or to zero for a nonexisting transition. The outer iteration is defined as the calculation of symbol LLRs in multilevel BCJR equalizer block, the calculation of corresponding bit LLRs needed for LDPC decoding, the LDPC decoding, and the calculation of extrinsic symbol LLRs needed for the next iteration. The iterations within LDPC decoder, based on min-sum-with-correction-term algorithm [20,76], are called here inner iterations.

The initial forward and backward metrics values are set to

$$\alpha_0(s) = \begin{cases} 0, s = s_0 \\ -\infty, s \neq s_0 \end{cases} \quad \text{and} \quad \beta_n(s) = \begin{cases} 0, s = s_0 \\ -\infty, s \neq s_0 \end{cases}, \tag{6.64}$$

where s_0 is an initial state. Let $s'=x[j-m-1, j+m-1]$ represent the previous state, $s=x[j-m, j+m]$ the present state, $x=(x_1, x_2, ..., x_n)$—the transmitted word of symbols, and $y=(y_1, y_2, ..., y_n)$—the received sequence of samples. The LLR, denoting the reliability, of symbol $x_j=\delta$ $(j=1, 2, ..., n)$ can be calculated by

$$\Lambda(x_j = \delta) = \max_{(s',s):x_j=\delta}{}^* \left[\alpha_{j-1}(s') + \gamma_j(s',s) + \beta_j(s)\right]$$
$$- \max_{(s',s):x_j=\delta_0}{}^* \left[\alpha_{j-1}(s') + \gamma_j(s',s) + \beta_j(s)\right], \quad (6.65)$$

where δ represents the observed symbol ($\delta \in \{0, 1, ..., M-1\}/\{\delta_0\}$), and δ_0 is the referent symbol. The forward and backward metrics are calculated using Eqs. (6.61) and (6.62). The forward and backward recursion steps of 4-level BCJR MAP detector are illustrated in Figure 6.23a and b, respectively. In Figure 6.23a, s denotes

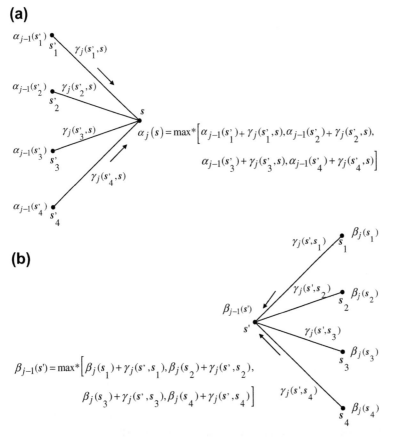

(a)

$$\alpha_j(s)=\max{}^*\big[\alpha_{j-1}(s'_1)+\gamma_j(s'_1,s),\alpha_{j-1}(s'_2)+\gamma_j(s'_2,s),$$
$$\alpha_{j-1}(s'_3)+\gamma_j(s'_3,s),\alpha_{j-1}(s'_4)+\gamma_j(s'_4,s)\big]$$

(b)

$$\beta_{j-1}(s')=\max{}^*\big[\beta_j(s_1)+\gamma_j(s',s_1),\beta_j(s_2)+\gamma_j(s',s_2),$$
$$\beta_j(s_3)+\gamma_j(s',s_3),\beta_j(s_4)+\gamma_j(s',s_4)\big]$$

FIGURE 6.23 Forward/backward recursion steps for $M=4$-level BCJR equalizer: (a) the forward recursion step and (b) the backward recursion step.

an arbitrary terminal state, which has $M=4$ edges originating from corresponding initial states, denoted as s'_1, s'_2, s'_3, and s'_4. Notice that the first term in branch metric is calculated only once, before the detection/decoding takes place, and stored. The second term, $\log(P(x_j))$, is recalculated in every outer iteration. The forward metric of state s in jth step ($j=1, 2, \ldots, n$) is updated by preserving the maximum term (in $\max^*$-sense) $\alpha_j - 1(s'_k) + \gamma_j(s, s'_k)$ ($k=1, 2, 3, 4$). The procedure is repeated for every state in column of terminal states of jth step. The similar procedure is used to calculate the backward metric of state s', $\beta_{j-1}(s')$, (in $(j-1)$th step), as shown in Figure 6.23b, but now proceeding in backward direction ($j=n, n-1, \ldots, 1$).

We further calculate bit LLRs from symbol LLRs in a similar fashion to that described in Section 6.4. To improve the overall performance of LDPC-coded turbo equalizer we perform the iteration of *extrinsic* LLRs between LDPC decoder and multilevel BCJR equalizer.

6.6.3 Performance of LDPC-coded turbo equalizer

As an illustration of the potential of the proposed scheme, the BER performance of an LDPC-coded turbo equalizer is given in Figure 6.24 for the dispersion map shown in Figure 6.25 (launch power of 0 dBm and single channel transmission). The nonlinear interaction of signal and ASE noise has been taken into account. EDFAs with a noise figure of 5 dB are deployed after every fiber section. The bandwidth of the optical filter is set to $3R_l$ and that of the electrical filter is set to $0.7R_l$, where $R_l = R_s/R$ with R_s being the symbol rate and R being the code rate (0.8). In Figure 6.24a, we present simulation results for QPSK transmission at the symbol rate of 50 Giga symbols/s. The symbol rate is appropriately chosen so that the effective aggregate information rate is 100 Gb/s. Through polarization-multiplexing the aggregate data rate of 200 Gb/s can be achieved. The figure depicts the uncoded BER and the BER after iterative decoding with respect to the number of spans, which was varied from 4 to 84. The propagation was modeled by solving the nonlinear Schrödinger equation using the split-step Fourier method. It can be seen from Figure 6.24a that when a 4-level BCJR equalizer of state memory $2m+1=1$ and an LDPC (16935, 13550) code of girth-10 and column weight 3 are used, we can achieve QPSK transmission at the symbol rate of 50 Giga symbols/s over 55 spans (6600 km) with a BER below 10^{-9}. On the other hand, for the turbo equalization scheme based on a 4-level BCJR equalizer of state memory $2m+1=3$ (see Figure 6.24a) and the same LDPC code, we are able to achieve even 8160 km at the symbol rate of 50 Giga symbols/s with a BER below 10^{-9}. Notice that in both cases the BCJR equalizer trellis detection depth was equal to the codeword length. The BER performance comparison of LDPC-coded turbo equalizer against large-girth LDPC codes and turbo-product codes for RZ-OOK system operating at 40 Gb/s (in effective information rate) is given in Figure 6.24b, for different trellis memories. LDPC-coded turbo equalizer with state memory $2m+1=7$ provides almost 12 dB improvement over the BCJR equalizer with state memory of $m=0$ at BER of 10^{-8}.

In order to apply the proposed multilevel turbo equalizations scheme to real 100 Gb/s systems, the practical circuit implementation study would be mandatory.

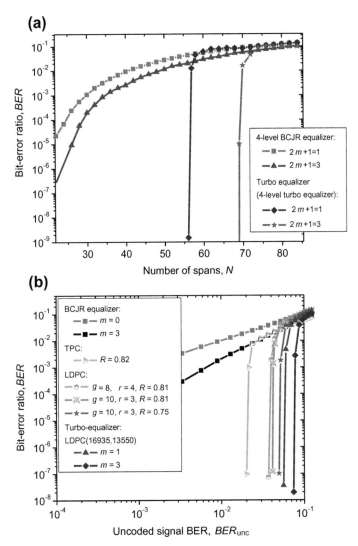

FIGURE 6.24 BER performance of LDPC-coded turbo equalizer in the presence of fiber nonlinearities for: (a) QPSK modulation format with aggregate data rate of 100 Gb/s and (b) RZ-OOK modulation format at 40 Gb/s. For both simulations, dispersion map shown in Figure 6.25 is used. (Modified from Ref. [2]; ©IEEE 2009; reprinted with permission.)

It is evident from Figure 6.22 that complexity of dynamic trellis grows exponentially, because the number of states is determined by M^{2m+1}, so that the increase in signal constellation leads to increase of the base, while the increase in channel memory assumption $(2m+1)$ leads to the increase of exponent. It is clear from Figure 6.24, that even small state memory assumption $(2m+1=3)$ leads to significant performance

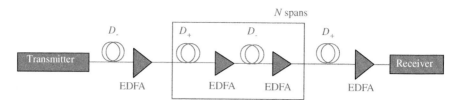

FIGURE 6.25 Dispersion map under study is composed of **N** spans of length $L = 120$ km, consisting of $2L/3$ km of D_+ fiber followed by $L/3$ km of D_- fiber, with pre-compensation of -1600 ps/nm and corresponding post-compensation. The fiber parameters are given in Table 6.6.

improvement with respect to the state memory $m = 0$. For larger constellations and/or larger memories the reduced-complexity *sliding-window* BCJR algorithm [112] is to be used instead. Namely, instead of detection of sequence of symbols corresponding to the length of codeword n, in sliding-window detector we can observe shorter sequences. Further, we do not need to memorize all branch metrics but several largest ones. In forward/backward metrics' update, we need to update only the metrics of those states connected to the edges with dominant branch metrics, and so on. Moreover, when $\max^*(x, y) = \max(x, y) + \log[1 + \exp(-|x - y|)]$ operation, required in forward and backward recursion steps, is approximated by $\max(x, y)$ operation, the forward and backward BCJR steps become the forward and backward Viterbi algorithms, respectively.

The LDPC-coded multilevel turbo equalizer can also be used for residual *PMD compensation*. Figure 6.26 shows the experimental setup for PMD compensation study in polarization multiplexed schemes with coherent detection [114]. In this example, we jointly perform detection and decoding of symbols transmitted in two orthogonal polarizations. The two orthogonal polarizations of a continuous wave laser source are separated by a polarization beam splitter and are modulated by two phase modulators (Covega) driven at 10 Gb/s (Anritsu MP1763C). (The symbol rate was determined by available equipment.)

A pre-coded test pattern was loaded into the pattern generator via personal computer with GPIB interface. A polarization beam combiner was used to combine the two modulated signals, followed by a PMD emulator (JDSU PE3) which introduced controlled amount of differential group delay (DGD) to the signal. Then the signal distorted by PMD was mixed with controlled amount of amplified spontaneous emission (ASE) noise with 3 dB coupler. Modulated signal level was maintained at 0 dB while the ASE power level was changed to obtain different optical signal-to-noise ratios (OSNRs). Next, the optical signal was pre-amplified, filtered (JSDU 2 nm band-pass filter), and coherently detected. The coherent detection is performed by mixing the received signal with signal from local laser with 3 dB coupler. The resulting signal is detected with a detector (Agilent 11982A) and an oscilloscope (Agilent DCA 86105A), triggered by the data pattern that was used to acquire the samples. To maintain constant power of -6 dBm at the detector, a variable attenuator was used.

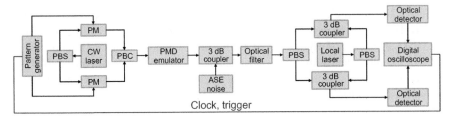

FIGURE 6.26 Experimental setup for polarization multiplexed BPSK study. CW Laser: continuous wave laser, PM: phase modulator.

Data was transferred via GPIB back to the PC. The PC also served as a multilevel turbo equalizer with offline processing. To avoid any imbalance of two independent symbols transmitted in two polarizations, we detect both the symbols simultaneously. Because the symbols transmitted in both polarizations are considered as one super-symbol, the BER performance of turbo equalizer is independent on power splitting ratio between principal states of polarization.

The experimental results for BER performance of the multilevel turbo equalizer are summarized in Figure 6.27. For the experiment, a quasi-cyclic LDPC(16935, 13550) code of girth 10 and column weight 3 was used as channel code. The number of extrinsic iterations between LDPC decoder and BCJR equalizer was set to 3, and the number of the intrinsic LDPC decoder iterations was set to 25. The state memory of

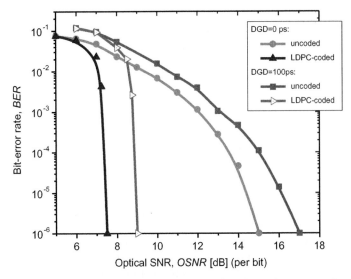

FIGURE 6.27 BER performance of multilevel turbo equalizer for PMD compensation. (Modified from Ref. [112]; ©IEEE 2009; reprinted with permission.)

$2m+1=3$ was sufficient for the compensation of the first order PMD with DGD of 100 ps. The OSNR penalty for 100 ps of DGD is 1.5 dB at BER of 10^{-6}. Coding gain for DGD of 0 ps is 7.5 dB at BER of 10^{-6}, and the coding gain for DGD of 100 ps is 8 dB. For more details on this experiment an interested reader is referred to [114]. For residual chromatic dispersion compensation by turbo equalization an interested reader is referred to [7].

6.6.4 Multilevel turbo equalizer robust to *I/Q*-imbalance and polarization offset

In this section, we describe a turbo equalization scheme robust to *I/Q*-imbalance and polarization offset [115]. In this scheme, we perform maximum *a posteriori* probability (MAP) detection of two independent symbols transmitted over two orthogonal polarizations. The MAP detection is based on four-dimensional (4D) *sliding-window* BCJR algorithm. The unique property of this scheme is that it considers the independent symbols transmitted in both polarizations as a super-symbol $s=(s_x, s_y)$, where s_x (s_y) is a QAM symbol transmitted in *x*-polarization (*y*-polarization). For experimental validation, we study the transmission of two independent DQPSK signals at symbol rate of 11 Giga symbol/s (11 GS/s) over *x*- and *y*-polarization channels (resulting in aggregate data rate of 44 Gb/s). We show that proposed scheme is much more robust to *I-Q* imbalance and polarization offset than conventional 2D schemes. The super-symbol coded-modulation approach is general and applicable to arbitrary signal constellation.

The general transmitter and receiver architectures for the proposed scheme are shown in Figure 6.28a and b, respectively. For the two orthogonal polarizations *x* and *y*, m_x and m_y independent input streams are encoded with different QC-LDPC codes of rates $R_i=k_i/n$ ($i \in \{x, y\}$). The output of the encoders in every polarization is followed by a $m_x \times n$ ($m_y \times n$ respectively) bit-interleaver. At every symbol interval m_x (m_y) bits, taken from the interleaver column-wise, are mapped into a 2^{m_x}-ary (2^{m_y}-ary) QAM signal constellation point by a QAM mapper. The constellation points are represented in Cartesian *I-Q* coordinates as $s_{i,x}=(I_x, Q_x)$ [$s_{i,y}=(I_y, Q_y)$]. The outputs of every mapper are used to drive two single-drive Mach-Zehnder modulators (MZMs). The signals from the two polarizations are then multiplexed before transmission. The transmitted super-symbol $s \in S$ consists of the symbols in the two polarizations $s=(s_x, s_y)$. At the receiver side Figure 6.28b, the signal is split in the corresponding polarizations and coherently detected. Let R denote the received sequence. The received symbol $r \in R$ consists of four components corresponding to the *I* and *Q* for every polarization $r = (r_x^{(I)}, r_x^{(Q)}, r_y^{(I)}, r_y^{(Q)})$ and every component is sampled at the symbol rate. The four components of the received vector are then passed on the DSP block followed by MAP equalizer and set of LDPC decoders. The 4D MAP decoder calculates symbol LLRs, which are used to determine bit LLRs needed in LDPC decoding. The initial symbol LLRs are obtained by using $\lambda(s)=\log[P(s|r)/P(s_0|r)]$ with s denoting the transmitted super-symbol constellation point and s_0 representing the reference symbol. The turbo-equalization principle is used to compensate

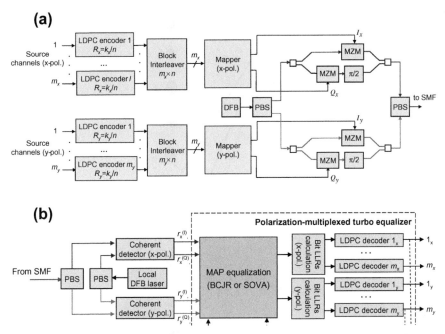

FIGURE 6.28 Polarization-division multiplexed LDPC-coded QAM: (a) transmitter architecture and (b) receiver architecture. PBS/C: polarization beam splitter/combiner. (Modified from Ref. [115]; ©IEEE 2010; reprinted with permission.)

for multiple channel impairments, such as fiber nonlinearities, PMD, and imbalance between the I and Q channels.

The 4D-MAP equalizer operates on an optical channel modeled as a nonlinear intersymbol interference (ISI) channel with memory $2m+1$. This model allows for channel description by a discrete dynamical trellis. An example of such trellis for QPSK transmission is similar to that shown in Figure 6.22, and is shown in Figure 6.29. It illustrates the dynamic trellis for two consecutive discrete moments in time for a channel with memory $2m+1=3$. The memory assumption signifies that every super-symbol s_i during transmission is influenced by the preceding m (s_{i-m}, s_{i-m+1}, ..., s_{i-1}) super-symbols and the next m (s_{i+1}, ..., s_{i+m}) super-symbols. A state s_i is defined as $s_i = (s_{i-m}, ..., s_i, s_{i+1}, ..., s_{i+m})$. Let's denote the QPSK symbols transmitted over x-polarization (y-polarization) with 0_x, 1_x, 2_x, and 3_x (0_y, 1_y, 2_y, and 3_y). Based on this notation the super-symbols are defined as: $s_0 = (0_x, 0_y)$, $s_1 = (0_x, 1_y)$, $s_2 = (0_x, 2_y)$, $s_3 = (0_x, 3_y)$, $s_4 = (1_x, 0_y)$, ..., $s_{15} = (3_x, 3_y)$. The ordered triple {previous state, channel output, next state} defines the trellis uniquely at any moment of time. The right column is the current instance of time, while the left column denotes the previous instance of time. The super-symbol to be detected is the middle super-symbol of the terminal state. All possible transitions from one moment in time to next one are indicated by the arrows. The numbers above the arrows indicate the transmitted super-symbol and the

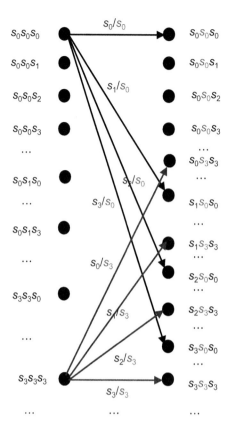

FIGURE 6.29 Dynamic trellis description of fiber-optic channel. (Modified from Ref. [115]; ©IEEE 2010; reprinted with permission.)

central super-symbol in terminal state, respectively. Complete characterization of the optical channel is achieved by determining the conditional probability density functions (PDFs) $p(r_j|s)$. These conditional PDFs are experimentally determined by propagating the sufficiently long training sequence and creating the histograms of the states.

The experimental setup used for the performance evaluation of proposed scheme is shown in Figure 6.30a. Continuous wave tunable laser output is modulated by *I/Q* modulator driven by two independent LDPC encoded sequences at 11 Gb/s, generated by Agilent N4901B and HP70340 pulse-pattern generators, preloaded by personal computer. The pulse carver is further used to perform NRZ-to-RZ conversion. The modulated signal is then split with PBS, whose outputs are decorrelated and then re-combined with PBC. The two polarization components can be considered, therefore, independent of each other. The purpose of this manipulation is to emulate the polarization-multiplexing transmitter. Controlled amount of ASE is then introduced to the signal. Modulated signal level was maintained constant while the ASE power

(a)

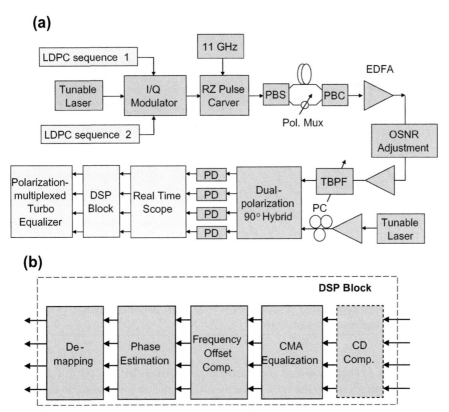

(b)

FIGURE 6.30 **Experimental setup for PDM-DQPSK. TBPF: tunable bandpass filter, PC: polarization controller, PD: photodetector. (Modified from Ref. [115]; ©IEEE 2010; reprinted with permission.)**

level was changed to obtain different optical signal-to-noise ratios (OSNRs). Signal is then passed to a coherent DQPSK receiver and four photodetectors are used before sampling with a digital sampling oscilloscope. The collected samples are then processed offline by a personal computer, which performs DSP and polarization-division multiplexed turbo equalization (PDM-TE). Figure 6.30b shows the DSP block. It performs initial chromatic dispersion compensation and I-Q imbalance compensation. The remaining channel distortions are compensated for by 4D PDM-TE.

Experimental results are summarized in Figure 6.31. The horizontal axis in Figure 6.31 represents the OSNR per bit. The code used for the experiment is a QC-LDPC code of large girth. Its parameters are (16935, 13550), the code rate is 0.8, girth is 10, and column weight is 3. The coding gain of the 4D PDM-TE based on this LDPC code with respect to BCJR equalizer is 7.4 dB at BER of 10^{-6}. Much higher gains are expected at lower BERs. This scheme outperforms the

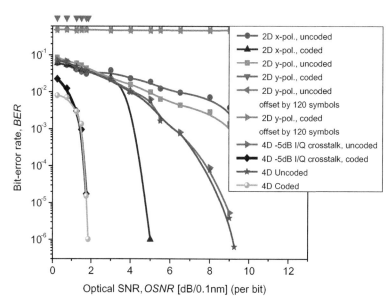

FIGURE 6.31 BER performance of super-symbol 4D scheme. (Modified from Ref. [115];
©IEEE 2010; reprinted with permission.)

corresponding 2D scheme for x-polarization by ~ 3.2 dB at BER of 10^{-6}. The top set
of curves corresponds to the misalignment of the signal in y-polarization compared
to the signal in x-polarization by 120 symbols processed by the conventional 2D
turbo equalizer, which considers symbols transmitted over both polarizations independently. The misalignment causes the conventional scheme to exhibit an error floor
phenomenon that cannot be handled by the best-known FEC codes. On the other
hand, the 4D PDM-TE shows robustness against polarization misalignment since
the BCJR sliding-window depth exceeds the misalignment delay. We further demonstrate the efficiency of 4D-TE scheme in I-Q imbalance compensation by showing the performance of PDM-TE in the presence of -5 dB crosstalk between I- and
Q-channels. BER performance degradation is small compared to the case when there
is no I-Q imbalance.

6.6.5 Multilevel turbo equalization with digital backpropagation

The LDPC-coded turbo equalizer described in previous sections is an excellent
equalizer to deal simultaneously with both linear and nonlinear fiber impairments.
However, the complexity of this equalizer grows exponentially as either channel
memory or constellation size increases. To solve this problem, we recently proposed
the use of coarse digital backpropagation (with reasonable small number of coefficients) to reduce the required channel memory and to compensate for remaining

channel impairments by turbo equalization scheme [112,128], which is shown in Figure 6.32.

The $m_x + m_y$ (index x (y) corresponds to x- (y-) polarization) independent data streams are encoded using different LDPC codes of code rates $R_i = K_i/N$ ($i \in \{x, \ y\}$) where K_x (K_y) denotes the number of information bits used in the binary LDPC code corresponding to x- (y-) polarization, and N denotes the codeword length, which is the same for both LDPC codes. The m_x (m_y) input bit streams from m_x (m_y) different information sources pass through identical LDPC encoders that use large-girth quasi cyclic LDPC codes with code rate R_x (R_y). The outputs of the encoders are then bit-interleaved by an $m_x \times N$ ($m_y \times N$) bit-interleaver where the sequences are written row-wise and read column-wise. The output of the interleaver is sent in one bit-stream, m_x (m_y) bits at a time instant i, to a mapper. The mapper maps each m_x (m_y) bits into a 2^{m_x}-ary (2^{m_y}-ary) IPM signal constellation point based on a look-up table, as explained above. The iterative polar modulation (IPM) [129] mapper x (y) constellation point s $_{i,x} = (I_{i,x}, \ Q_{i,x}) = |s_{i,x}| \exp(j\varphi_{i,x})$ $[s_{i,y} = (I_{i,y}, \ Q_{i,y}) = |s_{i,y}| \exp(j\varphi_{i,y})]$ coordinates are used as the inputs of an $I/Q\ MOD_x$ ($I/Q\ MOD_y$), as shown in Figure 6.32. At the receiver side, the outputs at I- and Q-branches in two polarizations are sampled at the symbol rate, while the symbol LLRs are calculated as follows $\lambda(s) = \log[P(s|\mathbf{r})/P(s_0|\mathbf{r})]$, where $s = (I_i, \ Q_i)$ and $\mathbf{r} = (r_I, \ r_Q)$ denote the transmitted signal constellation point and received symbol at time instance i (in either x- or y-polarization), respectively, and s_0 represents the reference symbol. To reduce the channel memory, while keeping complexity reasonably low, we use the coarse digital backpropagation with small number of coefficients. To compensate for remaining channel distortion we employ

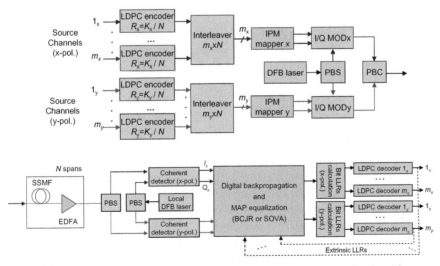

FIGURE 6.32 LDPC-coded PDM-IPM scheme. PBS/C: polarization beam splitter/combiner, MAP: maximum *a posteriori* probability, LLRs: log-likelihood ratios, IPM: iterative polar modulation (see Refs. [127–129]).

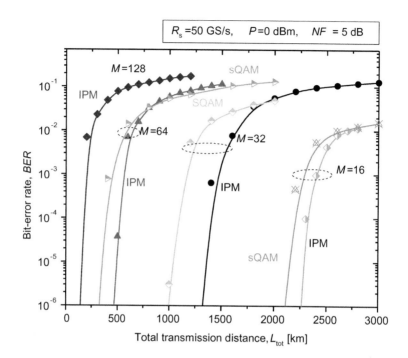

FIGURE 6.33 BER versus total transmission distance (L_{tot}). (Modified from Ref. [129]; ©IEEE 2010; reprinted with permission.)

the LDPC-coded turbo equalization. The bit reliabilities for LDPC decoders are calculated from symbol reliabilities, as we described previously. To improve BER performance we use the EXIT chart analysis [4,47] and iterate extrinsic reliabilities between MAP equalizer and LDPC decoders in turbo equalization fashion, until convergence or predetermined number of iterations has been reached.

In Figure 6.33, we report the BER results obtained for polarization-multiplexed (8547, 6922, 0.81)-coded modulation schemes with digital backpropagation and turbo equalization (for three outer MAP equalizer-LDPC decoder iterations and 25 LDPC decoder inner iterations), for symbol rate $R_s = 50$ GS/s and launch power $P = 0$ dBm. The dispersion map was composed of standard SMF (SSMF) only with EDFAs of noise figure $NF = 5$ dB being deployed every 100 km (for dispersion map please refer to Figure 6.36 in next section). We see that coded-IPM outperforms star-QAM for different signal constellation sizes, and allows longer transmission distances. The total transmission distance for different signal constellation sizes is found to be: 2250 km for $M = 16$ (aggregate rate $R_D = 400$ Gb/s), 1320 km for $M = 32$ ($R_D = 500$ Gb/s), 460 km for $M = 64$ ($R_D = 600$ Gb/s) and 140 km for $M = 128$ ($R_D = 700$ Gb/s).

6.7 INFORMATION CAPACITY OF FIBER-OPTICS COMMUNICATION SYSTEMS

Previous coded modulation and turbo equalization schemes have shown great robustness to various linear and nonlinear channel impairments so that the question of ultimate information capacity naturally arises. There have been numerous attempts to determine the channel capacity of a nonlinear fiber-optics communication channel [98–110]. The main approach, until recently, was to consider ASE noise as a predominant effect and to observe the fiber nonlinearities as the perturbation of linear case or as the multiplicative (signal dependent) noise. In this section, we describe how to determine the true fiber-optics channel capacity. Because in most practical applications the channel input distribution is uniform, we also describe how to determine the uniform information capacity, which represents the lower bound on channel capacity. This method consists of two steps: (1) approximating probability density functions (PDFs) for energy of pulses, which is done by one of the following approaches: (a) evaluation of histograms, (b) instanton approach [108], or (c) edgeworth expansion [113], and (2) estimating information capacities by applying a method originally proposed by Arnold and Pfitser [116–118].

6.7.1 Channel capacity of channels with memory

Let the input and output alphabets of the optical channel be finite and be denoted by $\{A\}$ and $\{B\}$ respectively; and the channel input and output be denoted by X and Y. For memoryless channels the noise behavior is generally captured by a conditional probability matrix $P\{b_j|a_k\}$ for all $b_j \in B$ and $a_j \in A$. For the channels with finite memory, such as the optical channel, the transition probability is dependent on the transmitted sequences up to the certain prior finite instance of time. For example, for channel described by Markov process the transition matrix has the following form $P\{Y_k=b|..., X_{-1}, X_0, X_1, ..., X_k\} = P\{Y_k=b|X_k\}$. We are interested in a more general description, which is due to McMillan [120] and Khinchin [121] (see also [122]). Let us consider a member of input ensemble x and its corresponding channel output y: $\{X\} = \{..., x_{-2}, x_{-1}, x_0, x_1, ...\}$, $\{Y\} = \{..., y_{-2}, y_{-1}, y_0, y_1, ...\}$. Let X denote all possible input sequences and Y denote all possible output sequences. By fixing a particular symbol at specific location we obtain so-called *cylinder* [121]. For example, cylinder $x^{4,1}$ is obtained by fixing the symbol a_1 at position x_4: $x^{4,1} = ..., x_{-1}, x_0, x_1, x_2, x_3, a_1, x_5, ...$. The output cylinder $y^{1,2}$ is obtained by fixing the output symbol b_2 at position 1: $y^{1,2} = ..., y_{-1}, y_0, b_2, y_2, y_3, ...$. To characterize the channel we have to determine the following transition probability $P(y^{1,2}|x^{4,1})$, that is the probability that cylinder $y^{1,2}$ was received given that cylinder $x^{4,1}$ was transmitted. Therefore, for all possible input cylinders $S_A \subset X$ we have to determine the probability that cylinder $S_B \subset Y$ was received given that S_A was transmitted. The channel is completely specified by: (i) input alphabet A, (ii) output alphabet B, and (iii) transition probabilities $P\{S_B|S_A\} = v_x$ for all $S_A \in X$ and $S_B \in Y$. Thus the channel is specified by the triplet: $[A, v_x, B]$. If the transition probabilities are invariant with respect to time shift T, that is, $v_{Tx}(TS) = v_x(S)$, then the channel is said to be *stationary*. If the distribution of Y_k depends only on the statistical properties of the sequence $\ldots, x_{k-1}, x_k$, we say that the channel is without *anticipation*. If furthermore the distribution of Y_k depends only on $x_{k-m}, \ldots, x_k$ we say that channel has the finite *memory* of m units.

The source and channel may be described as a new source $[C, \omega]$ with C being the product of input A and output B alphabets, namely $C = A \times B$, and ω is a corresponding probability measure. The joint probability of symbol $(x, y) \in C$, where $x \in A$ and $y \in B$, is obtained as the product of marginal and conditional probabilities: $P(x''' y) = P\{x\}P\{y|x\}$.

Let us further assume that both source and channel are stationary. The following description due to Khinchin [121,122] is useful in describing the concatenation of a stationary source and a stationary channel.

1. If the source $[A, \mu]$ (μ is the probability measure of the source alphabet) and the channel $[A, v_x, B]$ are stationary, the product source $[C, \omega]$ will also be stationary.
2. Each stationary source has an entropy, and therefore $[A, \mu]$, $[B, \eta]$ (η is the probability measure of the output alphabet), and $[C, \omega]$ each have the finite entropies.
3. These entropies can be determined for all n-term sequences $x_0, x_1, \ldots, x_{n-1}$ emitted by the source and transmitted over the channel as follows [121,122]:

$$H_n(X) \leftarrow \{x_0, x_1, \ldots, x_{n-1}\} \quad H_n(Y) \leftarrow \{y_0, y_1, \ldots, y_{n-1}\},$$
$$H_n(X, Y) \leftarrow \{(x_0, y_0), (x_1, y_1), \ldots, (x_{n-1}, y_{n-1})\}, \tag{6.66}$$

$$H_n(Y|X) \leftarrow \{(Y|x_0), (Y|x_1), \ldots, (Y|x_{n-1})\},$$
$$H_n(X|Y) \leftarrow \{(X|y_0), (X|y_1), \ldots, (X|y_{n-1})\}.$$

It can be shown that the following is valid:

$$H_n(X, Y) = H_n(X) + H_n(Y|X) \quad H_n(X, Y) = H_n(Y) + H_n(X|Y). \tag{6.67}$$

Equation (6.67) can be rewritten in terms of entropies per symbol:

$$\frac{1}{n}H_n(X, Y) = \frac{1}{n}H_n(X) + \frac{1}{n}H_n(Y|X),$$
$$\frac{1}{n}H_n(X, Y) = \frac{1}{n}H_n(Y) + \frac{1}{n}H_n(X|Y). \tag{6.68}$$

For sufficiently long sequences the following channel entropies exist:

$$\lim_{n \to \infty} \frac{1}{n}H_n(X, Y) = H(X, Y) \quad \lim_{n \to \infty} \frac{1}{n}H_n(X) = H(X),$$

$$\lim_{n \to \infty} \frac{1}{n}H_n(Y) = H(Y) \quad \lim_{n \to \infty} \frac{1}{n}H_n(X|Y) = H(X|Y),$$

$$\lim_{n \to \infty} \frac{1}{n}H_n(Y|X) = H(Y|X). \tag{6.69}$$

The mutual information exists and it is defined as

$$I(X,Y) = H(X) + H(Y) - H(X,Y). \qquad (6.70)$$

The *stationary information capacity* of the channel is obtained by maximization of mutual information over all possible information sources:

$$C(X,Y) = \max I(X,Y). \qquad (6.71)$$

Equipped with this knowledge, in the next section we will discuss how to determine the information capacity of fiber-optics channel with memory.

6.7.2 Calculation of information capacity of multilevel modulation schemes by forward recursion of BCJR algorithm

Here we address the problem of calculating of channel capacity of multilevel modulation schemes for an independent identically distributed (IID) information source, in literature also known as the achievable information rate (see [7,106,107] and references therein). The IID channel capacity represents a lower bound on channel capacity. To calculate the IID channel capacity, we model the whole transmission system as the dynamical ISI channel, in which m previous and next m symbols influence the observed symbol, which is shown in Figure 6.22. The optical communication system is characterized by the conditional probability density function (PDF) of the output complex vector of samples $y = (y_1, \ldots, y_n, \ldots)$, where $y_i = \mathrm{Re}\{y_i\}$, $\mathrm{Im}\{y_i\}) \in Y$, given the source sequence $x = (x_1, \ldots, x_n, \ldots)$, $x_i \in X = \{0, 1, \ldots, M-1\}$. The set X represents the set of indices of constellation points in corresponding M-ary two-dimensional signal constellation diagram (such as M-ary phase-shift keying (PSK), M-ary quadrature-amplitude modulation (QAM) or M-ary polarization-shift keying (PolSK)), while Y represents the set of all possible channel outputs. The $\mathrm{Re}\{y_i\}$ corresponds to the in-phase channel sample, and the $\mathrm{Im}\{y_i\}$ represents the quadrature channel sample.

The information rate can be calculated by:

$$I(Y; X) = H(Y) - H(Y|X), \qquad (6.72)$$

where $H(U) = E(\log_2 P(U))$ denotes the entropy of a random variable U and $E(\cdot)$ denotes the mathematical expectation operator. By using the Shannon-McMillan-Brieman theorem that states [123]:

$$E\left(\log_2 P(Y)\right) = \lim_{n\to\infty} (1/n)\log_2 P(y[1,n]), \qquad (6.73)$$

the information rate can be determined by calculating $\log_2(P(y[1, n]))$, by propagating the sufficiently long source sequence. By substituting Eq. (6.73) into Eq. (6.72) we obtain the following expression suitable for practical calculation of IID information capacity

$$I(Y; X) = \lim_{n\to\infty} \frac{1}{n}\left[\sum_{i=1}^{n}\log_2 P\left(y_i|y[1,i-1],x[1,n]\right) - \sum_{i=1}^{n}\log_2 P\left(y_i|y[1,i-1]\right)\right].$$

$$(6.74)$$

The first term in (6.74) can be straightforwardly calculated from conditional PDFs $P(y[j-m, j+m]|s)$. To calculate $\log_2 P(y_i|y[1, i-1])$ we use the forward recursion of the multilevel BCJR algorithm [7,112], described in previous section, wherein the *forward metric* $\alpha_j(s) = \log\{p(s_j = s, y[1, j])\}$ ($j = 1, 2, ..., n$), and the *branch metric* $\gamma_j(s', s) = \log[p(s_j = s, y_j, s_{j-1} = s'')]$ are defined as follows:

$$\alpha_j(s) = \max_{s'}{}^* \left[\alpha_{j-1}(s') + \gamma_j(s', s) - \log_2 M\right],$$
$$\gamma_j(s', s) = \log\left[p\left(y_j | x[j-m, j+m]\right)\right], \tag{6.75}$$

where the $\max^*$-operator is defined by $\max^*(x, y) = \log(e^x + e^y) = \max(x, y) + \log[1 + \exp(-|x - y|)]$. The ith term $\log_2 P(y_i|y[1, i-1])$ can be calculated iteratively by

$$\log_2 P(y_i|y[1, i-1]) = \max_{s}{}^* \alpha_i(s), \tag{6.76}$$

where $\max^*$-operator was applied for all $s \in S$ (S denotes the set of states in the trellis shown in Figure 6.22). Information capacity is defined as

$$C = \max I(Y; X), \tag{6.77}$$

where the maximization is performed over all possible input distributions. Because the optical channel has the memory, it is natural to assume that optimum input distribution will be with memory as well. By considering the stationary input distributions of the form $p(x_i|x_{i-1}, x_{i-2}, ...) = p(x_i|x_{i-1}, x_{i-2}, ..., x_{i-k})$, we can determine the transition probabilities of corresponding Markov model that maximizes the information rate in Eq. (6.76) by nonlinear numerical optimization [124,125].

This method is applicable to both memoryless channels and for channels with memory. In Figure 6.34, we provide the information capacities for different signal constellation sizes and two types of QAM constellations: square-QAM and star-QAM [126] (see also [105]), by observing a linear channel model. We also provide the information capacity for an optimum signal constellation, based on so-called iterative polarization quantization (IPQ), also known as IPPM, introduced in [127–129]. We can see that information capacity can be closely approached even with an IID information source providing that constellation size is sufficiently large. It is interesting to notice that star-QAM outperforms the corresponding square QAM for low and medium signal-to-noise ratios (SNR), while for high SNRs square QAM outperforms star QAM. The IPQ significantly outperforms both square-QAM and star-QAM.

After this generic description of IIID information capacity calculation for fiber-optics channel in next section we study the information capacity of fiber-optics communication systems with coherent detection. For direct detection results an interested reader is referred to [7] (see also [4]).

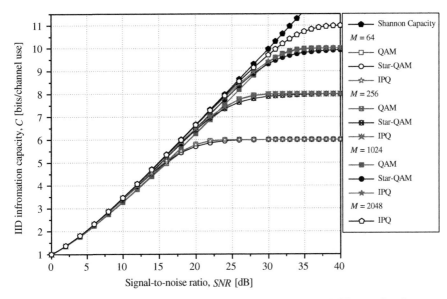

FIGURE 6.34 IID information capacities for linear channel model and different signal constellation sizes. (64-star-QAM contains 8 rings with 8 points each, 256-star-QAM contains 16 rings with 16 points, and 1024-star-QAM contains 16 rings with 64 points.) SNR is defined as E_s/N_0, where E_s is the symbol energy and N_0 is the power spectral density. (Modified from Ref. [129]; ©IEEE 2010; reprinted with permission.)

6.7.3 Information capacity of systems with coherent detection

In Figure 6.35, we show the IID information capacity against the number of spans (obtained by Monte Carlo simulations) for dispersion map shown in Figure 6.25 (and the fiber parameters are the same as in Table 6.6) and QPSK modulation format of aggregate data rate 100 Gb/s for two different memory assumptions. The transmitter and receiver configurations are shown in Figure 6.11.

We see that by using the LDPC code (of rate $R=0.8$) of sufficient length and large girth, we are able to achieve the total transmission distance of 8760 km for state memory $m=0$, and even 9600 km for state memory $m=1$. The transmission distance can further be increased by observing larger memory channel assumptions, which requires higher computational complexity for corresponding turbo equalizer. On the other hand, we can use backpropagation approach [105,119] to keep the channel memory reasonably low, and then apply the method described in this section. Notice that digital backpropagation method cannot account for the nonlinear ASE noise-Kerr nonlinearities interaction, and someone should use the method described in previous sub-section in information capacity calculation to account for this effect. In the same figure we show the IID information capacity, when digital backpropagation method is used, for dispersion map composed of standard SMF only with EDFAs of

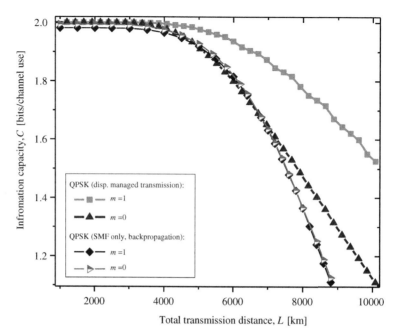

FIGURE 6.35 IID information capacity per single polarization for QPSK of aggregate data rate of 100 Gb/s against the transmission distance for dispersion map shown in Figure 6.25. (Modified from Ref. [112]; ©IEEE 2009; reprinted with permission.)

Table 6.6 Fiber parameters.

	D_+ **fiber**	D_- **fiber**
Dispersion [ps/(nm km)]	20	−40
Dispersion slope [ps/(nm² km)]	0.06	−0.12
Effective cross-sectional area (μm²)	110	50
Nonlinear refractive index (m²/W)	2.6×10^{-20}	2.6×10^{-20}
Attenuation coefficient (dB/km)	0.19	0.25

noise figure of 6 dB being deployed every 100 km, as shown in Figure 6.36. We see that digital backpropagation method helps to reduce the channel memory, since the improvement for $m=1$ over $m=0$ case is small.

In Figure 6.37 we show the IID information capacities for three different modulation formats: (i) MPSK, (ii) star-QAM (sQAM), and (iii) IPM; obtained by employing the dispersion map from Figure 6.36a. The symbol rate was 50 GS/s, and the launch power was set to 0 dBm. We see that IPQ outperforms star-QAM and significantly

(a)

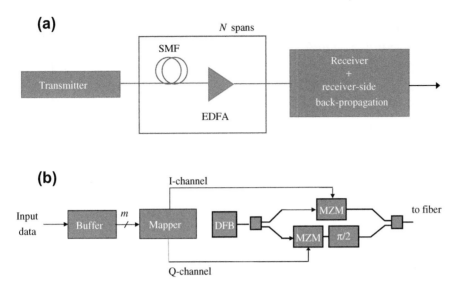

(b)

FIGURE 6.36 (a) Dispersion map composed of SMF sections only with receiver-side digital backpropagation and (b) transmitter configuration. The receiver configuration is shown in Figure 6.11.

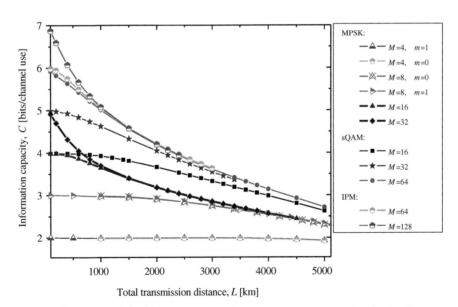

FIGURE 6.37 IID Information capacities per single-polarization for star-QAM (SQAM), MPSK and IPM for different constellation sizes and dispersion map from Figure 6.25. EDFAs *NF*= 6 dB.

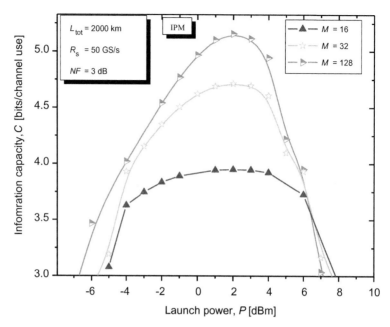

FIGURE 6.38 Information capacity per single-polarization against launch power P for total transmission distance of $L_{tot} = 2000$ km. EDFAs $NF = 3$ dB.

outperforms MPSK. For transmission distance of 5000 km the IID information capacity is 2.72 bits/symbol (the aggregate rate is 136 Gb/s per wavelength), for 2000 km it is 4.2 bits/symbol (210 Gb/s), and for 1000 km the IID information capacity is 5.06 bits/symbol (253 Gb/s per wavelength). For the completeness of presentation in Figure 6.38 we show the information capacity as a function of launch power for fixed total transmission distance $L_{tot} = 2000$ km and $NF = 3$ dB, for dispersion map shown in Figure 6.36a. We see that for optimum launch power of $P_{opt} = 2$ dBm for $M = 128$ we can extend the total transmission distance to 2000 km and achieve the channel capacity of $C_{opt} = 5.16$ bits/s/Hz, which is very close to the result reported by Essiambre et al. [105]. Notice that authors [105] use star-QAM of size 2048 and optimum dispersion map based on Raman amplifiers, while our dispersion map is based on SMF only with periodically deployed EDFAs.

6.8 CONCLUDING REMARKS

This chapter has provided an overview of advanced coding techniques for optical communications. The following topics have been covered: codes on graphs, coded modulation, rate-adaptive coded modulation, and turbo equalization. The main objectives of this chapter have been: (i) to describe different classes of codes on graphs of interest for optical communications, (ii) to describe how to combine multilevel modulation and channel coding, (iii) to describe how to perform equalization and soft-decoding jointly, and (iv) to demonstrate efficiency of joint de-modulation, decoding, and equalization in dealing with various channel impairments simultaneously. The following codes on graphs of interest for next-generation FEC for high-speed optical transport have been described: turbo codes, turbo-product codes, and LDPC codes. Both binary and nonbinary LDPC codes, their design and decoding algorithms, have been described. In addition, the FPGA implementation of decoders for binary LDPC codes have been discussed. It has been demonstrated that the LDPC-coded turbo equalizer can be used to simultaneously mitigate chromatic dispersion, PMD, fiber nonlinearities, and I/Q-imbalance. Finally, for completeness of presentation, the information capacity of optical channels with memory has been studied.

Other advanced coding schemes suitable for optical communications, not being described due to space limitations, include generalized LDPC codes [40,130], continuously interleaved Bose-Chaudhuri-Hocquenghem (CI-BCH) cyclic codes [131], concatenated RS-LDPC codes [131], convolutional LDPC codes [132,133], and their related family of staircase codes [134]. In order to increase the spectral efficiency, coded-modulation has been described above as an efficient way to combine coding with higher modulation schemes. Traditionally, the redundant bits of FEC are accommodated for by increasing the symbol rate of transmission. However, in situations where the bandwidth expansion is not a viable option, the nonbinary LDPC-coded modulation schemes without bandwidth expansion should be used as proposed in [135,136].

The exponential Internet traffic growth projections place enormous transmission rate demand on the underlying information infrastructure at every level, from the core to access networks. As a response to these high bandwidth demands, the IEEE has ratified its 40/100 Gb/s Ethernet Standard IEEE 802.3ba in June 2010 [141,142]. Deployment of 100 Gb/s Ethernet (GbE) has already started and it is expected to accelerate within a couple of years. To meet the ever-growing bandwidth demands, 1 Tb/s Ethernet (TbE) and beyond (4 TbE and 10 TbE) should be standardized in the near future. There are several technologies, such as DWDM, OTDM, multiband-OFDM, and polarization-multiplexed-QAM, that can potentially be used to deliver optical multi-TbE. However, some practical challenges still exist with respect to using these techniques as enabling technologies for optical TbE. For instance, PDM QAM requires huge QAM constellations to reach Tb/s-range with commercially available symbol rates, while high-speed DSP capability is still limited. The reality is that any bit rates of Terabit range require some extent of parallel processing and optical channel design. The envisioned growth in Internet traffic will place enormous demand not only on transmission rate at every level and the overall transmission capacity, but also on the energy that is needed for bandwidth creation and distribution. Recent studies indicate that the power consumed by the information and communication technologies, currently about 2–4% of the total carbon emissions, will be doubled by the end of this decade if current trend continues [137]. Therefore, the Internet is becoming constrained not only by speed and capacity, but also by the energy consumption

[138–140]. In order to solve the bandwidth efficiency and energy-efficiency problems simultaneously, several energy-efficient hybrid-coded-modulation schemes enabling multi-Tb/s serial optical transport have been introduced in [139,140,143–147].

Acknowledgments

This chapter was supported in part by the National Science Foundation (NSF) under Grant CCF-0952711; through NSF CIAN ERC Center for Integrated Access Network under grant EEC-0812072; and in part by NEC Labs.

References

[1] M. Cvijetic, Modulation and coding techniques, and optical networking technologies enabling multi terabit bandwidth delivery, in: Proceedings of the CLEO 2012, 6–11 May 2012, San Jose, CA, USA, Paper CTh3C.1.

[2] I.B. Djordjevic, M. Arabaci, L. Minkov, Next generation FEC for high-capacity communication in optical transport networks, IEEE/OSA J. Lightwave Technol. 27 (2009) 3518–3530 (Invited Paper).

[3] W. Shieh, I. Djordjevic, OFDM for Optical Communications, Elsevier, 2009.

[4] I. Djordjevic, W. Ryan, B. Vasic, Coding for Optical Channels, Springer, 2010.

[5] W.E. Ryan, Concatenated convolutional codes and iterative decoding, in: J.G. Proakis (Ed.), Wiley Encyclopedia of Telecommunications, John Wiley & Sons, 2003.

[6] L.R. Bahl, J. Cocke, F. Jelinek, J. Raviv, Optimal decoding of linear codes for minimizing symbol error rate, IEEE Trans. Inform. Theory IT-20 (2) (1974) 284–287.

[7] I.B. Djordjevic, L.L. Minkov, H.G. Batshon, Mitigation of linear and nonlinear impairments in high-speed optical networks by using LDPC-coded turbo equalization, IEEE J. Sel. Areas Comm. Optical Comm. and Network 26 (2008) 73–83.

[8] ITU, Telecommunication standardization sector: forward error correction for submarine systems, Rec. G.975, Geneva, 1996.

[9] ITU, Telecommunication standardization sector: forward error correction for high bit rate DWDM submarine systems, Rec. G. 975.1, 02/2004.

[10] T. Mizuochi et al., Forward error correction based on block turbo code with 3-bit soft decision for 10 Gb/s optical communication systems, IEEE J. Sel. Top. Quant. Electron. 10 (2004) 376–386.

[11] T. Mizuochi et al., Next generation FEC for optical transmission systems, in: Proceedings of the Optical Fiber Communication Conference (OFC 2003), vol. 2, 2003, pp. 527–528.

[12] O.A. Sab, FEC techniques in submarine transmission systems, in: Proceedings of the Optical Fiber Communication Conference (OFC 2001), vol. 2, 2001, pp. TuF1-1–TuF1-3.

[13] C. Berrou, A. Glavieux, P. Thitimajshima, Near Shannon limit error-correcting coding and decoding: turbo codes, in: Proceedings of the IEEE International Conference on Communication (ICC 1993), 1993, pp. 1064–1070.

[14] C. Berrou, A. Glavieux, Near optimum error correcting coding and decoding: turbo codes, IEEE Trans. Commun. 44 (1996) 1261–1271.

[15] R.M. Pyndiah, Near optimum decoding of product codes, IEEE Trans. Commun. 46 (1998) 1003–1010.

[16] O.A. Sab, V. Lemarie, Block turbo code performances for long-haul DWDM optical transmission systems, in: Proceedings of the OFC 2001, vol. 3, 2001, pp. 280–282.

[17] T. Mizuochi, Recent progress in forward error correction and its interplay with transmission impairments, IEEE J. Sel. Top. Quant. Electron. 12 (2006) 544–554.

[18] R.G. Gallager, Low Density Parity Check Codes, MIT Press, Cambridge, 1963.

[19] I.B. Djordjevic, S. Sankaranarayanan, S.K. Chilappagari, B. Vasic, Low-density parity-check codes for 40 Gb/s optical transmission systems, IEEE/LEOS J. Sel. Top. Quant. Electron. 12 (4) (2006) 555–562.

[20] J. Chen, A. Dholakia, E. Eleftheriou, M. Fossorier, X.-Y. Hu, Reduced-complexity decoding of LDPC codes, IEEE Trans. Commun. 53 (2005) 1288–1299.

[21] I.B. Djordjevic, B. Vasic, Nonbinary LDPC codes for optical communication systems, IEEE Photon. Technol. Lett. 17 (10) (2005) 2224–2226.

[22] M. Arabaci, I.B. Djordjevic, R. Saunders, R.M. Marcoccia, Non-binary quasi-cyclic LDPC based coded modulation for beyond 100-Gb/s transmission, IEEE Photon. Technol. Lett. 22 (6) (2010) 434–436.

[23] M.C. Davey, Error-correction using low-density parity-check codes, PhD Dissertation, University of Cambridge, Cambridge, UK, 1999.

[24] D. Declercq, M. Fossorier, Decoding algorithms for nonbinary LDPC codes over GF(q), IEEE Trans. Commun. 55 (4) (2007) 633–643.

[25] M. Arabaci, I.B. Djordjevic, R. Saunders, R.M. Marcoccia, Polarization-multiplexed rate-adaptive non-binary-LDPC-coded multilevel modulation with coherent detection for optical transport networks, Opt. Express 18 (3) (2010) 1820–1832.

[26] M. Arabaci, I.B. Djordjevic, An alternative FPGA implementation of decoders for quasi-cyclic LDPC codes, in: Proceedings of the TELFOR 2008, November 2008, pp. 351–354.

[27] M. Arabaci, I.B. Djordjevic, R. Saunders, R. Marcoccia, Non-binary LDPC-coded modulation for high-speed optical metro networks with back propagation, in: Proceedings of the SPIE Photonics West 2010, OPTO: Optical Communications: Systems and Subsystems, Optical Metro Networks and Short-Haul Systems II, 23–28 January 2010, San Francisco, California, USA, Paper No. 7621-17.

[28] M. Arabaci, Nonbinary-LDPC-coded modulation schemes for high-speed optical communication networks, PhD Dissertation, University of Arizona, November 2010.

[29] H.G. Batshon, I.B. Djordjevic, T. Schmidt, Ultra high speed optical transmission using subcarrier-multiplexed four-dimensional LDPC-coded modulation, Opt. Express 18 (19) (2010) 20546–20551.

[30] H.G. Batshon, I.B. Djordjevic, L. Xu, T. Wang, Multidimensional LDPC-coded modulation for beyond 400 Gb/s per wavelength transmission, IEEE Photon. Technol. Lett. 21 (16) (2009) 1139–1141.

[31] H.G. Batshon, I.B. Djordjevic, Beyond 240 Gb/s per wavelength optical transmission using coded hybrid subcarrier/amplitude/phase/polarization modulation, IEEE Photon. Technol. Lett. 22 (5) (2010) 299–301.

[32] H.G. Batshon, I.B. Djordjevic, L. Xu, T. Wang, Modified hybrid subcarrier/amplitude/phase/polarization LDPC-coded modulation for 400 Gb/s optical transmission and beyond, Opt. Express 18 (13) (2010) 14108–14113.

[33] I. Djordjevic, H.G. Batshon, L. Xu, T. Wang, Four-dimensional optical multiband-OFDM for beyond 1.4 Tb/s serial optical transmission, Opt. Express 19 (2) (2011) 876–882.

[34] I.B. Djordjevic, Four-dimensional coded optical OFDM for ultra-high-speed metro networks, in: SPIE Photonics West 2011, Optical Metro Networks and Short-Haul Systems III, 22–27 January 2011, The Moscone Center, San Francisco, California, USA, Paper No. 7959–2 (Invited Paper).

[35] J. McDonough, Moving Standards to 100 GbE and Beyond, IEEE Applications & Practice 45 (11) (2007) 6–9.

[36] B. Razavi, A 60-GHz CMOS receiver front-end, IEEE J. Solid-State Circ. 41 (2006) 17–22.

[37] C. Doan, S. Emami, A.M. Niknejad, R.W. Brodersen, Millimeter-wave CMOS design, IEEE J. Solid-State Circ. 40 (2005) 144–155.

[38] R. Nagarajan et al., Large-scale photonic integrated circuits, IEEE J. Sel. Top. Quant. Electron. 11 (2005) 50–64.

[39] D.F. Welch et al., Large-scale InP photonic integrated circuits: enabling efficient scaling of optical transport networks, IEEE J. Sel. Top. Quant. Electron. 13 (1) (2007) 22–31.

[40] I.B. Djordjevic, O. Milenkovic, B. Vasic, Generalized low-density parity-check codes for optical communication systems, IEEE/OSA J. Lightwave Technol. 23 (2005) 1939–1946.

[41] B. Vasic, I.B. Djordjevic, R. Kostuk, Low-density parity check codes and iterative decoding for long haul optical communication systems, IEEE/OSA J. Lightwave Technol. 21 (2003) 438–446.

[42] I.B. Djordjevic et al., Projective plane iteratively decodable block codes for WDM high-speed long-haul transmission systems, IEEE/OSA J. Lightwave Technol. 22 (2004) 695–702.

[43] O. Milenkovic, I.B. Djordjevic, B. Vasic, Block-circulant low-density parity-check codes for optical communication systems, IEEE/LEOS J. Sel. Top. Quant. Electron. 10 (2004) 294–299.

[44] B. Vasic, I.B. Djordjevic, Low-density parity check codes for long haul optical communications systems, IEEE Photon. Technol. Lett. 14 (2002) 1208–1210.

[45] S. Chung et al., On the design of low-density parity-check codes within 0.0045 dB of the Shannon Limit, IEEE Commun. Lett. 5 (2001) 58–60.

[46] S. Lin, D.J. Costello, Error Control Coding: Fundamentals and Applications, Prentice-Hall, Inc., USA, 1983.

[47] S. ten Brink, Convergence behavior of iteratively decoded parallel concatenated codes, IEEE Trans. Commun. 40 (2001) 1727–1737.

[48] P. Elias, Error-free coding, IRE Trans. Inform. Theory IT-4 (1954) 29–37.

[49] J.B. Anderson, S. Mohan, Source and Channel Coding: An Algorithmic Approach, Kluwer Academic Publishers, Boston, MA, 1991.

[50] F.J. MacWilliams, N.J.A. Sloane, The Theory of Error-Correcting Codes, Amsterdam, North Holland, The Netherlands, 1977.

[51] S.B. Wicker, Error Control Systems for Digital Communication and Storage, Prentice-Hall, Inc., Englewood Cliffs, NJ, 1995.

[52] G.D. Forney Jr, Concatenated Codes, MIT Press, Cambridge, MA, 1966.

[53] D.B. Drajic, An Introduction to Information Theory and Coding, second ed., Akademska Misao, Belgrade, Serbia, 2004 (in Serbian).

[54] S. Haykin, Communication Systems, John Wiley & Sons, Inc., 2004.

[55] J.G. Proakis, Digital Communications, McGraw-Hill, Boston, MA, 2001.

[56] R.H. Morelos-Zaragoza, The Art of Error Correcting Coding, John Wiley & Sons, Boston, MA, 2002.

[57] I. Anderson, Combinatorial Designs and Tournaments, Oxford University Press, 1997.

[58] D. Raghavarao, Constructions and Combinatorial Problems in Design of Experiments, Dover Publications, Inc., New York, 1988 (reprint).

[59] T.M. Cover, J.A. Thomas, Elements of Information Theory, John Wiley & Sons, Inc., New York, 1991.

[60] F.M. Ingels, Information and Coding Theory, Intext Educational Publishers, Scranton, 1971.

[61] I.S. Reed, G. Solomon, Polynomial codes over certain finite fields, SIAM J. Appl. Math. 8 (1960) 300–304.

[62] S.B. Wicker, V.K. Bhargva, Reed-Solomon Codes and their Applications, IEEE Press, New York, 1994.

[63] J.K. Wolf, Efficient maximum likelihood decoding of linear block codes using a trellis, IEEE Trans. Inform. Theory IT-24 (1) (1978) 76–80.

[64] B. Vucetic, J. Yuan, Turbo Codes-Principles and Applications, Kluwer Academic Publishers, Boston, 2000.

[65] G. Bosco, P. Poggiolini, Long-distance effectiveness of MLSE IMDD receivers, IEEE Photon. Technol. Lett. 18 (9) (2006) 1037–1039.

[66] D. Divsalar, F. Pollara, Turbo codes for deep-space communications, TDA progress, Report 42–120, February 15, 1995, pp. 29–39.

[67] M.E. van Valkenburg, Network Analysis, third ed., Prentice-Hall, Englewood Cliffs, 1974.

[68] R.-J. Essiambre, G. Raybon, B. Mikkelsen, Pseudo-linear transmission of high-speed TDM signals at 40 and 160 Gb/s, in: I.P. Kaminow, T. Li (Eds.), Optical Fiber Telecommunications IVB, Academic, San Diego, CA, 2002., pp. 233–304.

[69] G.P. Agrawal, Nonlinear Fiber Optics, Academic, San Diego, CA, 2001.

[70] F.R. Kschischang, B.J. Frey, H.-A. Loeliger, Factor graphs and the sum-product algorithm, IEEE Trans. Inform. Theory 47 (2001) 498–519.

[71] R.M. Tanner, A recursive approach to low complexity codes, IEEE Trans. Inform. Theory IT-27 (1981) 533–547.

[72] D.J.C. MacKay, Good error correcting codes based on very sparse matrices, IEEE Trans. Inform. Theory 45 (1999) 399–431.

[73] I.B. Djordjevic, L. Xu, T. Wang, M. Cvijetic, Large girth low-density parity-check codes for long-haul high-speed optical communications, in: Proceedings of the OFC/NFOEC, IEEE/OSA, San Diego, CA, 2008, Paper No. JWA53.

[74] W.E. Ryan, An introduction to LDPC codes, in: B. Vasic (Ed.), CRC Handbook for Coding and Signal Processing for Recording Systems, CRC Press, 2004.

[75] M.P.C. Fossorier, Quasi-cyclic low-density parity-check codes from circulant permutation matrices, IEEE Trans. Inform. Theory 50 (2004) 1788–1793.

[76] H. Xiao-Yu, E. Eleftheriou, D.-M. Arnold, A. Dholakia, Efficient implementations of the sum-product algorithm for decoding of LDPC codes, in: Proceedings of the IEEE Globecom, vol. 2, November 2001, pp. 1036–1036E.

[77] Y. Miyata, R. Sakai, W. Matsumoto, H. Yoshida, T. Mizuochi, Reduced-complexity decoding algorithm for LDPC codes for practical circuit implementation in optical communications, in: Proceedings of the Optical Fiber Communication Conference 2007 (OFC 2007), Paper No. OWE5.

[78] M. Arabaci, I.B. Djordjevic, An alternative FPGA implementation of decoders for quasi-cyclic LDPC codes, in: Proceedings of the TELFOR 2008, November 2008, pp. 351–354.

[79] Mitrion Users Guide, Mitrionics Inc., v1.5.0-001, 2008.

[80] M. Arabaci, I.B. Djordjevic, R. Saunders, R. Marcoccia, A class of non-binary regular girth-8 LDPC codes for optical communication channels, in: Proceedings of the OFC/NFOEC 2009, San Diego, CA, 22–26 March 2009, Paper No. JThA.

[81] M.C. Davey, Error-correction using low-density parity-check codes, PhD Thesis, University of Cambridge, 1999.

[82] C. Spagnol, W. Marnane, E. Popovici, FPGA Implementations of LDPC over GF(2^m) decoders, in: IEEE Workshop on Signal Processing Systems, Shanghai, China, 2007, pp. 273–278.

[83] A. Voicila, F. Verdier, D. Declercq, M. Fossorier, P. Urard, Architecture of a low-complexity non-binary LDPC decoder for high order fields, in: Proceedings of the ISIT, 2007, pp. 1201–1206.

[84] L. Lan, L. Zeng, Y.Y. Tai, L. Chen, S. Lin, K. Abdel-Ghaffar, Construction of quasi-cyclic LDPC codes for AWGN and binary erasure channels: a finite field approach, IEEE Trans. Inform. Theory 53 (2007) 2429–2458.

[85] I.B. Djordjevic, B. Vasic, Multilevel coding in M-ary DPSK/differential QAM high-speed optical transmission with direct detection, IEEE/OSA J. Lightwave Technol. 24 (2006) 420–428.

[86] I.B. Djordjevic, M. Cvijetic, L. Xu, T. Wang, Using LDPC-coded modulation and coherent detection for ultra high-speed optical transmission, IEEE/OSA J. Lightwave Technol. 25 (2007) 3619–3625.

[87] I.B. Djordjevic, B. Vasic, LDPC-coded OFDM in fiber-optics communication systems [Invited], OSA J. Opt. Network 7 (2008) 217–226.

[88] J. Hou, P.H. Siegel, L.B. Milstein, H.D. Pfitser, Capacity-approaching bandwidth-efficient coded modulation schemes based on low-density parity-check codes, IEEE Trans. Inform. Theory 49 (9) (2003) 2141–2155.

[89] I.B. Djordjevic, L. Xu, T. Wang, PMD compensation in coded-modulation schemes with coherent detection using Alamouti-type polarization-time coding, Opt. Express 16 (18) (2008) 14163–14172.

[90] E. Biglieri, R. Calderbank, A. Constantinides, A. Goldsmith, A. Paulraj, H.V. Poor, MIMO Wireless Communications, Cambridge University Press, Cambridge, 2007.

[91] I.B. Djordjevic, L. Xu, T. Wang, PMD compensation in multilevel coded-modulation schemes with coherent detection using BLAST algorithm and iterative polarization cancellation, Opt. Express 16 (19) (2008) 14845–14852.

[92] I.B. Djordjevic, L. Xu, T. Wang, Beyond 100 Gb/s optical transmission based on polarization multiplexed coded-OFDM with coherent detection, IEEE J. Sel. Areas Commun. Optical Commun. Network. 27 (3) (2009).

[93] S. Benedetto, P. Poggiolini, Theory of polarization shift keying modulation, IEEE Trans. Commun. 40 (1992) 708–721.

[94] VPITransmisionMaker. <http://www.vpiphotonics.com>.

[95] Y. Ma, Q. Yang, Y. Tang, S. Chen, W. Shieh, 1-Tb/s single-channel coherent optical OFDM transmission over 600-km SSMF fiber with subwavelength bandwidth access, Opt. Express 17 (11) (2009) 9421–9427.

[96] C. Douillard, M. Jézéquel, C. Berrou, A. Picart, P. Didier, A. Glavieux, Iterative correction of intersymbol interference: turbo equalization, Eur. Trans. Telecommun. 6 (1995) 507–511.

[97] M. Tüchler, R. Koetter, A.C. Singer, Turbo equalization: principles and new results, IEEE Trans. Commun. 50 (5) (2002) 754–767.

[98] E.E. Narimanov, P. Mitra, The channel capacity of a fiber optics communication system: perturbation theory, IEEE/OSA J. Lightwave Technol. 20 (3) (2002) 530–537.

[99] E. Narimanov, P. Patel, Channel capacity of fiber optics communications systems: WDM vs. TDM, in: Proceedings of the Conference on Lasers and Electro-Optics (CLEO'03), 2003, pp. 1666–1668.

[100] P.P. Mitra, J.B. Stark, Nonlinear limits to the information capacity of optical fiber communications, Nature 411 (2001) 1027–1030.

[101] K.S. Turitsyn, S.A. Derevyanko, I.V. Yurkevich, S.K. Turitsyn, Information capacity of optical fiber channels with zero average dispersion, Phys. Rev. Lett. 91 (20) (2003) 203–901.

[102] J. Tang, The multispan effects of Kerr nonlinearity and amplifier noises on Shannon channel capacity for a dispersion-free nonlinear optical fiber, IEEE/OSA J. Lightwave Technol. 19 (2001) 1110–1115.

[103] A. Mecozzi, M. Shtaif, On the capacity of intensity modulated systems using optical amplifiers, IEEE Photon. Technol. Lett. 13 (2001) 1029–1031.

[104] J.M. Kahn, K.-P. Ho, Spectral efficiency limits and modulation/detection techniques for DWDM systems, IEEE Sel. Top. Quant. Electron. 10 (2004) 259–272.

[105] R.-J. Essiambre, G.J. Foschini, G. Kramer, P.J. Winzer, Capacity limits of information transport in fiber-optic networks, Phys. Rev. Lett. 101 (2008) 16390-1–163901-4.

[106] I.B. Djordjevic, B. Vasic, M. Ivkovic, I. Gabitov, Achievable information rates for high-speed long-haul optical transmission, IEEE/OSA J. Lightwave Technol. 23 (11) (2005) 3755–3763.

[107] I.B. Djordjevic, L. Xu, T. Wang, On the channel capacity of multilevel modulation schemes with coherent detection, in: Proceedings of the Asia Communications and Photonics Conference and Exhibition (ACP) 2009, Shangai, China, 2–6 November 2009, Paper ThC4.

[108] M. Ivkovic, I.B. Djordjevic, B. Vasic, Calculation of achievable information rates of long-haul optical transmission systems using instanton approach, IEEE/OSA J. Lightwave Technol. 25 (5) (2007) 1163–1168.

[109] I. Djordjevic, N. Alic, G. Papen, S. Radic, Determination of achievable information rates (AIRs) of IM/DD systems and AIR loss due to chromatic dispersion and quantization, IEEE Photon. Technol. Lett. 19 (2007) 12–14.

[110] L.L. Minkov, I.B. Djordjevic, H.G. Batshon, L. Xu, T. Wang, M. Cvijetic, F. Kueppers, Demonstration of PMD compensation by LDPC-coded turbo equalization and channel capacity loss characterization due to PMD and quantization, IEEE Photon. Technol. Lett. 19 (2007) 1852–1854.

[111] L.L. Minkov, I.B. Djordjevic, L. Xu, T. Wang, F. Kueppers, Evaluation of large girth LDPC codes for PMD compensation by turbo equalization, Opt. Express 16 (2008) 13450–13455.

[112] I.B. Djordjevic, L.L. Minkov, L. Xu, T. Wang, Suppression of fiber nonlinearities and PMD in coded-modulation schemes with coherent detection by using turbo equalization, IEEE/OSA J. Opt. Commun. Network. 1 (2009) 555–564.

[113] M. Ivkovic, I. Djordjevic, P. Rajkovic, B. Vasic, Pulse energy probability density functions for long-haul optical fiber transmission systems by using instantons and Edgeworth expansion, IEEE Photon. Technol. Lett. 19 (20) (2007) 1604–1606.

[114] L.L. Minkov, I.B. Djordjevic, L. Xu, T. Wang, PMD Compensation in Polarization Multiplexed Multilevel Modulations by Turbo Equalization, IEEE Photon. Technol. Lett. 21 (23) (2009) 1773–1775.

[115] I.B. Djordjevic, L. Xu, L.L. Minkov, T. Wang, S. Zhang, Polarization-multiplexed LDPC-coded QAM robust to I-Q imbalance and polarization offset, in: Proceedings of the Communications and Photonics Conference and Exhibition (ACP 2010), 8–12 December, Shanghai, China, 2010, Paper No. SH 3.

[116] D. Arnold, A. Kavcic, H.-A. Loeliger, P.O. Vontobel, W. Zeng, Simulation-based computation of information rates: upper and lower bounds, in: Proceedings of the IEEE International Symposium on Information Theory (ISIT 2003), 2003, p. 119.

[117] D. Arnold, H.-A. Loeliger, On the information rate of binary-input channels with memory, in: Proceedings of the 2001 International Conference on Communications, Helsinki, Finland, 11–14 June 2001, pp. 2692–2695.

[118] H.D. Pfitser, J.B. Soriaga, P.H. Siegel, On the achievable information rates of finite state ISI channels, in: Proceedings of the Globecom 2001, San Antonio, TX, 25–29 November 2001, pp. 2992–2996.

[119] E. Ip, J.M. Kahn, Nonlinear impairment compensation using backpropagation, in: Optical Fibre, New Developments, In-Tech, Vienna, Austria, 2009.

[120] B. McMillan, The basic theorems of information theory, Ann. Math. Statistics 24 (1952) 196–219.

[121] A.I. Khinchin, Mathematical Foundations of Information Theory, Dover Publications, New York, 1957.

[122] F.M. Reza, An Introduction to Information Theory, McGraw-Hill, New York, 1961.

[123] T.M. Cover, J.A. Thomas, Elements of Information Theory, Wiley, New York, 1991.

[124] D.P. Bertsekas, Nonlinear Programming, second ed., Athena Scientific, 1999.

[125] E.K.P. Chong, S.H. Zak, An Introduction to Optimization, third ed., John Wiley & Sons, New York, 2008.

[126] W.T. Webb, R. Steele, Variable rate QAM for mobile radio, IEEE Trans. Commun. 43 (1995) 2223–2230.

[127] Z.H. Peric, I.B. Djordjevic, S.M. Bogosavljevic, M.C. Stefanovic, Design of signal constellations for Gaussian channel by iterative polar quantization, in: Proceedings of the 9th Mediterranean Electrotechnical Conference, 18–20 May, vol. 2, Tel-Aviv, Israel, 1998, pp. 866–869.

[128] I.B. Djordjevic, H.G. Batshon, L. Xu, T. Wang, Coded polarization-multiplexed iterative polar modulation (PM-IPM) for beyond 400 Gb/s serial optical transmission, in: Proceedings of the OFC/NFOEC 2010, 21–25 March, San Diego, CA, 2010, Paper No. OMK2.

[129] H.G. Batshon, I.B. Djordjevic, L. Xu, T. Wang, Iterative polar quantization based modulation to achieve channel capacity in ultra-high-speed optical communication systems, IEEE Photon. J. 2 (4) (2010) 593–599.

[130] I.B. Djordjevic, L. Xu, T. Wang, M. Cvijetic, GLDPC codes with Reed-Muller component codes suitable for optical communications, IEEE Commun. Lett. 12 (2008) 684–686.

[131] F. Chang, K. Onohara, T. Mizuochi, Forward error correction for 100G transport networks, IEEE Commun. Mag. 48 (3) (2010) S48–S55.

[132] R.M. Tanner, D. Sridhara, A. Sridharan, T.E. Fuja, D.J. Costello Jr, LDPC block and convolutional codes based on circulant matrices, IEEE Trans. Inform. Theory 50 (12) (2004) 2966–2984.

[133] D. Chang, F. Yu, Z. Xiao, N. Stojanovic, F.N. Hauske, Y. Cai, C. Xie, L. Li, X. Xu, Q. Xiong, LDPC convolutional codes using layered decoding algorithm for high speed

coherent optical transmission, in: Proceedings of the 2012 Optical Fiber Communication Conference and Exposition/the National Fiber Optic Engineers Conference (OFC/NFOEC), 4–8 March, 2012, Paper No. OW1H.4.

[134] B.P. Smith, A. Farhood, A. Hunt, F.R. Kschischang, J. Lodge, Staircase codes: FEC for 100 Gb/s OTN, J. Lightwave Technol. 30 (1) (2012) 110–117.

[135] M. Arabaci, I.B. Djordjevic, L. Xu, T. Wang, Nonbinary LDPC-coded modulation for high-speed optical fiber communication without bandwidth expansion, IEEE Photon. J. 4 (3) (2012) 728–734.

[136] M. Arabaci, I.B. Djordjevic, L. Xu, T. Wang, Nonbinary LDPC-coded modulation for rate-adaptive optical fiber communication without bandwidth expansion, IEEE Photon. Technol. Lett. 24 (16) (2012) 1402–1404.

[137] W. Vereecken, W. Van Heddeghem, M. Deruyck, B. Puype, B. Lannoo, W. Joseph, D. Colle, L. Martens, P. Demeester, Power consumption in telecommunication networks: overview and reduction strategies, IEEE Commun. Mag. 49 (6) (2011) 62–69.

[138] I.B. Djordjevic, Energy-efficient spatial-domain-based hybrid multidimensional coded-modulations enabling multi-Tb/s optical transport, Opt. Express 19 (17) (2011) 16708–16714.

[139] I.B. Djordjevic, Spatial-domain-based hybrid multidimensional coded-modulation schemes enabling multi-Tb/s optical transport, IEEE/OSA J. Lightwave Technol. 30 (14) (2012) 2315–2328.

[140] I.B. Djordjevic, L. Xu, T. Wang, Statistical physics inspired energy-efficient coded-modulation for optical communications, Opt. Lett. 37 (8) (2002) 1340–1342.

[141] P. Winzer, Beyond 100G Ethernet, IEEE Comm. Mag. 48 (2010) 26–30.

[142] T. Xia, G. Wellbrock, Y. Huang, E. Ip, M. Huang, Y. Shao, T. Wang, Y. Aono, T. Tajima, S. Murakami, M. Cvijetic, Field experiment with mixed line-rate transmission (112-Gb/s, 450-Gb/s, and 1.15-Tb/s) over 3,560 km of installed fiber using filterless coherent receiver and EDFAs only, in: Proc. OFC/NFOEC 2011, Paper PDPA3, LA, CA, March 6-10, 2011.

[143] S. Murshid, A. Khayrattee, Multiplexing of optical channels as a function of orbital angular momentum of photons, in: Proc. Frontiers in Optics (FiO) 2008/Laser Science XXIV (LS) Conf., OSA Technical Digest (CD) (Optical Society of America, 2008), paper FThE6, October 2008 Rochester, New York, USA.

[144] X. Chen, A. Li, J. Ye, A. Al Amin, W. Shieh, Reception of dual-LP11-mode CO-OFDM signals through few-mode compatible optical add/drop multiplexer, in: Proc. OFC/NFOEC, Postdeadline Papers (OSA, 2012), Paper PDP5B.4.

[145] R. Ryf, R. Essiambre, A. Gnauck, S. Randel, M.A. Mestre, C. Schmidt, P. Winzer, R. Delbue, Pe. Pupalaikis, A. Sureka, T. Hayashi, T. Taru, T. Sasaki, Space-division multiplexed transmission over 4200 km 3-core microstructured fiber, in: Proc. OFC/NFOEC, Postdeadline Papers (Optical Society of America, 2012), Paper PDP5C.5.

[146] J. Sakaguchi, B.J. Puttnam, W. Klaus, Y. Awaji, N. Wada, A. Kanno, T. Kawanishi, K. Imamura, H. Inaba, K. Mukasa, R. Sugizaki, T. Kobayashi, M. Watanabe, 19-core fiber transmission of $19 \times 100 \times 172$-Gb/s SDM-WDM-PDM-QPSK signals at 305 Tb/s, in: Proc. OFC/NFOEC, Postdeadline Papers (OSA, 2012), Paper PDP5C.1.

[147] P.M. Krummrich, Optical amplifiers for multi mode / multi core transmission, in: Proc. OFC/NFOEC (OSA, 2012), Paper OW1D.1.

[148] P.S. Randel, R. Ryf, A. Gnauck, M.A. Mestre, C. Schmidt, R. Essiambre, P. Winzer, R. Delbue, P. Pupalaikis, A. Sureka, Y. Sun, X. Jiang, R. Lingle, Mode-multiplexed 6×20-GBd QPSK transmission over 1200-km DGD-compensated few-mode fiber, in: Proc. OFC/NFOEC, Postdeadline Papers (OSA, 2012), Paper PDP5C.5.

Extremely Higher-Order Modulation Formats

Masataka Nakazawa, Toshihiko Hirooka, Masato Yoshida, and Keisuke Kasai

Research Institute of Electrical Communication, Tohoku University, 2-1-1 Katahira, Aoba-ku, Sendai 980-8577, Japan

7.1 INTRODUCTION

The capacity of the optical communication infrastructure in backbone networks has increased a 1000-fold over the last 20 years. This has been made possible by the development of the EDFA and WDM in the 1990s. Despite such rapid progress, the Internet traffic is still growing at an annual rate of 40%, driven by innovative new services such as real-time video streaming, 3-D and ultra-realistic transmission, and cloud computing. This means that in 20 years we will need Peta bit/s or even Exa bit/s optical communication. However, it is widely recognized that the maximum transmission capacity of a single fiber is rapidly approaching its limit, owing to optical power limitations imposed by the fiber fuse phenomenon and the finite transmission bandwidth determined by optical amplifiers.

In order to expand the total WDM capacity within a finite optical amplification bandwidth, it has been an important research subject to explore how to increase the spectral efficiency toward the Shannon limit. Figure 7.1 shows the progress made on fiber-optic capacity and spectral efficiency over the last 20 years. This is plotted based on record-breaking experimental demonstrations reported at OFC and ECOC [1]. This figure indicates that, in spite of relatively slow progress on the single-channel bit rate, a larger WDM capacity, supported by the increase in spectral efficiency, had been achieved by around 2005. This was basically made possible by employing a closer channel spacing. However, there has subsequently been a notable increase in spectral efficiency to 10 bit/s/Hz. This was a consequence of adopting coherent multi-level transmission.

In particular, *M*-ary quadrature amplitude modulation (QAM), i.e. the multi-level amplitude modulation of in-phase (I) and quadrature-phase (Q) carriers, is advantageous as regards expanding the spectral efficiency to >10 bit/s/Hz and even approaching the Shannon limit, as the increase in the multiplicity ($M = 2^N$) leads to an N-fold increase in the spectral efficiency of binary modulation. Since the first demonstration of coherent optical QAM [2], the multiplicity level has been increased from 64 [3] to 128 [4], 256 [5], 512 [6], and 1024 [7]. Higher-order multi-level modulation also enables us to realize a high-speed system with low-speed devices, and therefore helps

Optical Fiber Telecommunications VIB. http://dx.doi.org/10.1016/B978-0-12-396960-6.00007-9

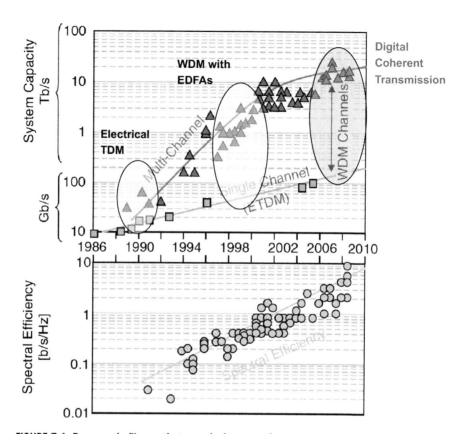

FIGURE 7.1 Progress in fiber-optic transmission capacity and spectral efficiency [1].

to enhance tolerance to chromatic dispersion (CD) and polarization-mode dispersion (PMD) as well as to reduce power consumption. Orthogonal frequency division multiplexing (OFDM) is another approach for highly spectral-efficient transmission. In OFDM transmission, the multi-carrier transmission of low-speed orthogonal sub-carriers enables us to improve both spectral efficiency and dispersion tolerance by adopting high-level subcarrier modulation formats and employing coherent detection.

Recently, higher-order QAM and OFDM have been widely adopted for achieving the spectral efficiency exceeding 10 bit/s/Hz. QAM with a higher multiplicity could play an important role to achieve further increases in the transmission capacity by taking full advantage of the finite bandwidth over the C- and L-bands, as well as to realize higher data rate transmissions under a fixed symbol rate. Indeed, QAM has been applied to WDM transmission with a record-breaking capacity beyond 100 Tbit/s per single core [8,9]. As an application of QAM to a transport system beyond 100 Gbit/s, 16 QAM has been employed for Pol-Mux 224 Gbit/s/ch [10–12] and 448 Gbit/s/ch [13] transmissions at a symbol rate of 28 and 56 Gsymbol/s, respectively. Recently, Pol-Mux 120–538 Gbit/s/ch transmission with 64 QAM at 10–44.8 Gsymbol/s has also been reported [14–16].

This chapter describes fundamental technologies and recent progress on coherent QAM transmission, with a special focus on challenges to extremely higher-order QAM such as 256–1024 levels. Section 7.2 reviews an analytical description of the bit error rate (BER) and spectral efficiency of QAM signals, and discusses the possibility of approaching the Shannon limit. Section 7.3 describes key components for higher-order QAM transmission, such as a coherent light source, an optical phase-locked loop (OPLL), an IQ modulator, and a digital demodulator. Based on the fundamental configuration for QAM transmission, we describe recent demonstrations of single-carrier 1024 QAM, 256 QAM-OFDM, and OTDM-RZ/32 QAM transmissions in Section 7.4.

7.2 SPECTRAL EFFICIENCY OF QAM SIGNAL AND SHANNON LIMIT

QAM is a modulation format that combines two carriers whose amplitudes are modulated independently with the same optical frequency and whose phases are 90° apart. These carriers are called in-phase carriers (I) and quadrature-phase carriers (Q). The QAM can assign 2^N states by using I and Q, which is called 2^N QAM. Figure 7.2 shows constellation maps for 16 (2^4) QAM and conventional OOK. As shown in Figure 7.2, a 2^N QAM signal processes N bits in a single channel, so it can realize N times the spectral efficiency of OOK.

The BER of M-ary QAM signals can be analytically calculated, and the result for $M = 2^N$ is given by [13]

$$P_b = \frac{\sqrt{M} - 1}{\sqrt{M} \log_2 \sqrt{M}} \text{erfc} \sqrt{\left[\frac{3 \log_2 M}{2(M - 1)}\right] E_b/N_0}. \tag{7.1}$$

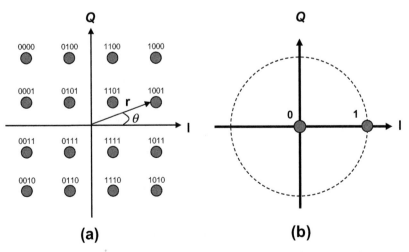

(a)   **(b)**

FIGURE 7.2 Constellation maps for 16 QAM (a) 16 QAM (4 × 4 = 2^4) and (b) OOK.

Here, E_b and N_0 are the energy and noise power per bit, thus E_b/N_0 corresponds to the signal-to-noise ratio (SNR) per bit. The specific forms of Eq. (7.1) for $M = 16, 64, 256$, and 1024 are written as follows:

$$P_b = \frac{3}{8}\text{erfc}\sqrt{\frac{2}{5}E_b/N_0} \quad (16 \text{ QAM}),\tag{7.2}$$

$$P_b = \frac{7}{24}\text{erfc}\sqrt{\frac{1}{7}E_b/N_0} \quad (64 \text{ QAM}),\tag{7.3}$$

$$P_b = \frac{15}{64}\text{erfc}\sqrt{\frac{4}{85}E_b/N_0} \quad (256 \text{ QAM}),\tag{7.4}$$

$$P_b = \frac{31}{160}\text{erfc}\sqrt{\frac{5}{341}E_b/N_0} \quad (1024 \text{ QAM}).\tag{7.5}$$

Equations (7.2–7.5) are plotted in Figure 7.3 as a function of E_b/N_0.

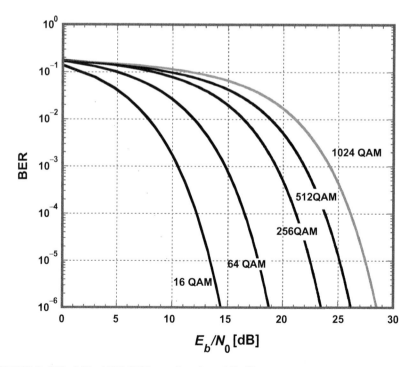

FIGURE 7.3 BER of 16–1024 QAM as a function of E_b/N_0.

The spectral efficiency of an *M*-ary QAM signal is shown in Figure 7.4 as a function of E_b/N_0. Here, the ultimate spectral efficiency is given by the Shannon limit:

$$\frac{C}{W} = \log_2\left(1 + \frac{E_b}{N_0}\frac{C}{W}\right),$$ (7.6)

which is known as the Shannon-Hartley theorem [18]. This figure indicates that, as the multiplicity *M* increases, the spectral efficiency of *M*-QAM approaches closer to the Shannon limit than other advanced modulation formats such as *M*-PSK or *M*-FSK.

The increase in *M*, however, requires a larger E_b/N_0 value under the same BER as shown in Figure 7.3. To realize a better BER performance with a lower E_b/N_0, the forward error correction (FEC) technique has been developed. Figure 7.5 shows the BER after applying FEC versus input Q value without FEC, Q_{in} [19]. Q_{in} is the SNR given by

$$Q_{in} = \frac{I_1 - I_0}{\sigma_1 + \sigma_0},$$ (7.7)

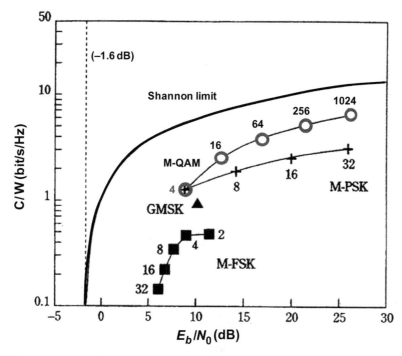

FIGURE 7.4 Spectral efficiency of *M*-ary QAM signal and the Shannon limit. E_b/N_0 at BER $= 10^{-4}$ is shown assuming synchronous detection.

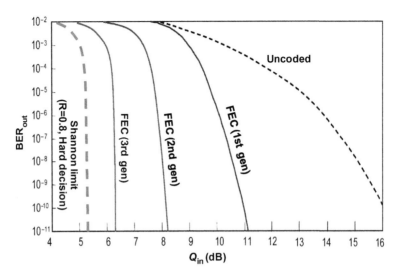

FIGURE 7.5 Relationship between BER after FEC and *Q* value without FEC [19].

where I_1 and I_0 are the mean values and σ_1 and σ_0 are the standard deviations of the bits corresponding to 1 and 0, respectively. Here, the Shannon limit describes the lowest Q_{in} value needed to achieve an infinitely low BER by employing FEC under a certain code rate R:

$$R = 1 - \mathrm{BER}_{in} \log_2 \mathrm{BER}_{in} + (1 - \mathrm{BER}_{in}) \log_2(1 - \mathrm{BER}_{in}), \qquad (7.8)$$

$$\mathrm{BER}_{in} = \frac{1}{2}\mathrm{erfc}\left(\frac{Q_{in}}{\sqrt{2}}\right), \qquad (7.9)$$

which is known as Shannon's second theorem or the noisy-channel coding theorem [18]. This provides the ultimate limit for the minimum Q value needed to achieve an infinitely low BER. Recently, third-generation FEC, namely the turbo block code with a soft decision, has been developed that enables us to realize BER performance very close to the Shannon limit. This indicates the distinct possibility of realizing ultrahigh spectral efficiency by combining QAM and FEC.

7.3 FUNDAMENTAL CONFIGURATION AND KEY COMPONENTS OF QAM COHERENT OPTICAL TRANSMISSION

The fundamental configuration of a QAM coherent optical transmission system is shown in Figure 7.6. As a coherent light source, a narrow linewidth CW fiber laser or laser diode such as an external-cavity laser is employed in order to meet the

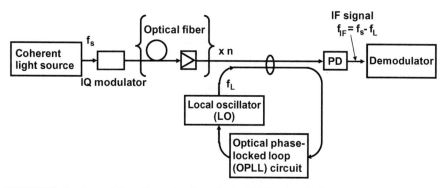

FIGURE 7.6 Fundamental configuration for coherent QAM transmission.

narrow linewidth requirement for high-order QAM. The optical QAM signal can be easily generated with an IQ modulator, consisting of two nested MZ modulators and a 90° phase shifter, driven by QAM baseband signals. A transmitted QAM signal is mixed with a local oscillator (LO) output and detected with a coherent receiver to convert the QAM data into an IF signal. For high-order QAM, an OPLL technique using a high-speed free-running laser as an LO plays a very important role as regards producing a stable IF frequency automatically. The IF signal is then A/D converted and accumulated in a digital signal processor (DSP). All digital signals are demodulated into I and Q data and finally into a binary sequence in the DSP.

The following sections present a detailed description of these key components for QAM coherent transmission, including the coherent light source, OPLL, optical IQ modulator, and digital demodulator.

7.3.1 Coherent light source

A stable optical frequency in the 1.5 μm region is indispensable as a light source for QAM coherent optical transmission. C_2H_2 molecules have been utilized as a frequency standard to stabilize the frequency of semiconductor and fiber lasers at 1.55 μm [20]. Fiber lasers are particularly attractive as a coherent light source because of their narrow linewidth. Here we describe a frequency-stabilized, polarization-maintained erbium fiber ring laser.

Figure 7.7 shows a configuration of the $^{13}C_2H_2$ frequency-stabilized fiber ring laser [21]. The frequency deviation of a fiber laser from the $^{13}C_2H_2$ absorption line is detected with a phase-sensitive detection circuit and fed back to PZT to control the laser frequency. Figure 7.8 shows the absorption characteristics of $^{13}C_2H_2$ molecules. Of these absorption lines, the P(10) linear absorption line was used as a frequency reference, whose center wavelength is 1538.8 nm and a spectral linewidth is 500 MHz.

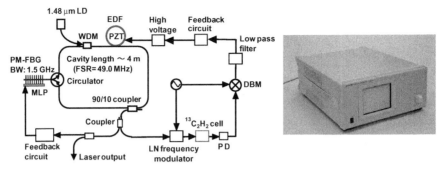

FIGURE 7.7 Configuration of a $^{13}C_2H_2$ frequency-stabilized fiber laser and its overview.

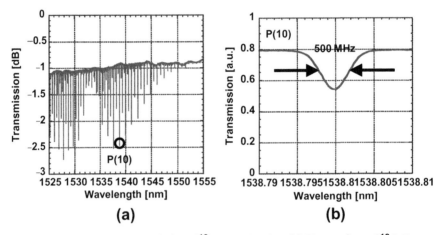

FIGURE 7.8 Absorption characteristics of $^{13}C_2H_2$ molecules. (a) Observations of $^{13}C_2H_2$ absorption lines for long span. (b) P(10) linear absorption line.

The laser linewidth was measured with a delayed self-heterodyne detection method [22]. Figure 7.9a shows the electrical spectrum of the heterodyne signal. The linewidth is only 4 kHz, which is almost the same as that of a free-running fiber laser. There is no additional linewidth broadening because of the adoption of the external frequency stabilization scheme. The frequency stability was evaluated from the square root of the Allan variance, which was estimated from the fluctuation of the beat note signals [23]. The result is shown in Figure 7.9b. The frequency stability is 2.5×10^{-11} for an integration time, τ of 1 s, and 6.3×10^{-12} for a τ of 100 s, which indicates excellent short- and long-term stabilities.

Other than fiber lasers, various types of narrow linewidth semiconductor lasers have been developed recently. Several examples are shown in Figure 7.9. The one shown in Figure 7.10a involves the adoption of waveguide ring resonators [24].

This makes it possible to realize a linewidth as narrow as 100 kHz as well as wide wavelength tunability. Figure 7.10b is an external cavity DBR laser with a linewidth of 4 kHz, which is commercially available [25]. Figure 7.10c is DFB-LD with external fiber ring cavity [26]. Here both edges of a DFB-LD are AR-coated and a fiber ring cavity is attached to form a long external cavity. It features a narrow linewidth associated with a fiber ring cavity as well as a low RIN value for the DFB-LD. Figure 7.10d is a DFB-LD with an external FBG mirror [27]. The frequency deviation

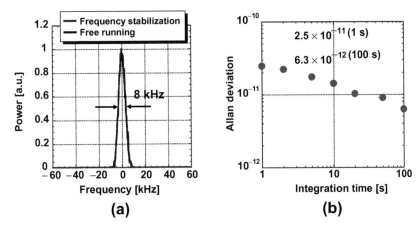

FIGURE 7.9 Electrical spectrum of delayed self-heterodyne signal (a) and the square root of the Allan variance of the frequency fluctuation of a $^{13}C_2H_2$ frequency-stabilized fiber laser.

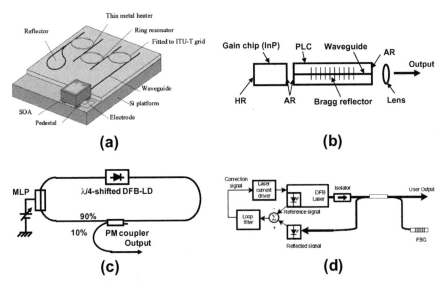

FIGURE 7.10 Narrow linewidth semiconductor lasers.

from the peak of the fiber Bragg grating (FBG) is monitored by comparing the laser output power and the power reflected at the FBG.

The external frequency modulation scheme adopted for the fiber laser has also been applied to the frequency stabilization of an external-cavity laser diode shown in Figure 7.10b [28]. In this case an FM-eliminated output beam with a linewidth of only 4 kHz and a RIN as low as -135 dB/Hz was obtained, and the frequency stability reached as high as 2.3×10^{-11} for $\tau = 1$ s. Such a low intensity noise is difficult to realize with a fiber laser owing to the small power fluctuation associated with the relaxation oscillation of erbium ions. However, the laser has a broader linewidth in the tail compared to fiber lasers.

7.3.2 Optical IQ modulator

The configuration of an optical IQ modulator is shown in Figure 7.11a [29]. It is composed of three Mach-Zehnder interferometers (MZIs), in which the sub-MZIs, MZ_A and MZ_B, are installed in each arm of the main MZI (MZ_C). I and Q optical data are generated individually with MZ_A and MZ_B using the I and Q components of the QAM baseband signal. They are combined with MZ_C with a dc bias DC_C so that a 90° phase shift is introduced between the two signals.

An optical IQ modulator has been realized not only with an LN modulator as shown in Figure 7.11b but also a monolithic InP modulator as shown in Figure 7.11c [30], which is beneficial for compact integration. A more complicated IQ modulator can be

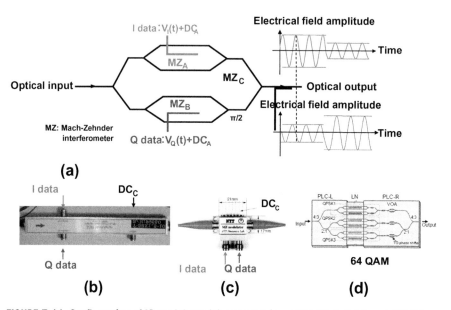

FIGURE 7.11 Configuration of IQ modulator (a) and overview of LN (b), InP (c) and PLC-LN (d) modulators.

fabricated using PLC-LN hybrid integration as shown in Figure 7.11d [31]. In this case, three IQ modulators are integrated, and by employing QPSK modulation at each modulator with a power ratio of 4:2:1, 64 QAM signal can be generated from a binary data.

In the LN-based IQ modulator, surface acoustic waves are generated by the piezo-electric effect in the LN crystal, which degrades the low-frequency response of the modulator [31]. To suppress the acoustic wave, we tapered the edge of the modulator and reduced its thickness. Figure 7.12a and b shows the E/O characteristics of IQ modulators with the conventional and new structures, respectively. The low-frequency response was successfully improved with the new structure. This improvement plays a very important role in increasing the multiplicity level in QAM transmission.

7.3.3 Coherent optical receiver and optical PLL

The precise optical carrier phase recovery is one of the fundamental functionalities in coherent optical transmission. Figure 7.13 shows three kinds of coherent optical receivers: heterodyne, homodyne, and intradyne receivers. Recent advances in digital signal processing (DSP) technologies have enabled an intradyne detection with a digital coherent receiver, in which the phase fluctuation between a signal and an LO is eliminated with a DSP [33]. This technique does not require an OPLL

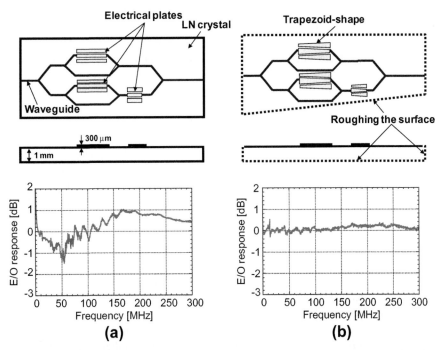

FIGURE 7.12 Improvement of IQ modulator. Schematic diagram of the modulator and its E/O characteristics (a) before and (b) after improvement.

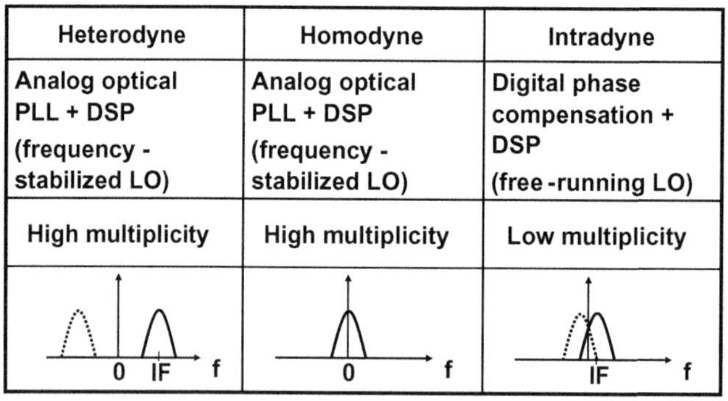

Heterodyne	Homodyne	Intradyne
Analog optical PLL + DSP (frequency - stabilized LO)	Analog optical PLL + DSP (frequency - stabilized LO)	Digital phase compensation + DSP (free -running LO)
High multiplicity	High multiplicity	Low multiplicity

FIGURE 7.13 Comparison of heterydone, homodyne, and intradyne receivers.

for phase synchronization and therefore enables us to realize coherent transmission using free-running lasers as a light source and an LO. Digital coherent receivers relax the requirements as regards the linewidth and frequency stability of a coherent light source, and the IF fluctuation can be reduced in the DSP without employing an OPLL. However, the IF stability thus obtained may not be sufficient for higher QAM multiplicity.

On the other hand, to realize an ultrahigh multiplicity, heterodyne or homodyne detection with an analog optical PLL circuit is more advantageous, which can automatically produce a stable IF frequency. In heterodyne or homodyne detection, the optical frequency difference between a transmitter and an LO must be kept constant in order to obtain a stable IF signal. Here, the use of a high-speed OPLL is a key technique for automatic frequency control. The linewidth of the IF signal is evaluated as

$$\sigma_\phi^2 = \frac{\delta f_T + \delta f_L}{2 f_c},$$

(7.10)

where δf_T and δf_L are the linewidths of the transmitter and LO, and f_c is the bandwidth of the feedback circuit [34]. This indicates that the reduction of the phase noise (linewidth) of the two lasers and the large bandwidth of the feedback circuit are very important factors as regards realizing a precise OPLL. Of the many available lasers, the fiber laser is suitable for an OPLL because of its low phase noise (narrow linewidth).

Figure 7.14 shows an OPLL configuration [35]. Here, a free-running fiber laser is used as an LO whose configuration is almost the same as that of the transmitter, except that an LN modulator was adopted in the laser cavity for the high-speed tracking of the IF signal. The LO linewidth is also approximately 4 kHz. The signal from the transmitter is heterodyne-detected with the LO signal. The phase of the beat signal (IF signal: $f_{IF} = |f_S - f_L|$) is compared with the phase of the reference signal from the synthesizer (f_{syn}) by the DBM and the difference between them is fed back to the LO through the feedback circuits. The phase noise of the OPLL is mainly dominated by the loop bandwidth. With this configuration, we could obtain an FM bandwidth for the LO of up to 1 GHz. The OPLL circuit contains two feedback circuits with different loop filter bandwidths. One is a broadband filter (~1 MHz) for fast frequency tuning with the LN modulator, the other is a narrow band filter (~10 KHz) for slow frequency tuning with the PZT.

Figure 7.15a and b shows the IF spectrum and the single sideband (SSB) phase noise spectrum. The linewidth of the spectrum was less than 10 Hz, which was below the measurement resolution as shown in Figure 7.15a. The phase noise variance

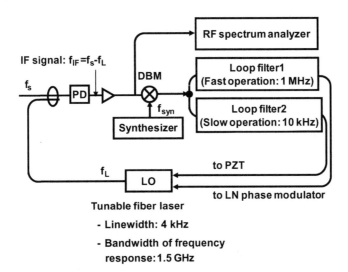

FIGURE 7.14 Configuration of OPLL for coherent transmission.

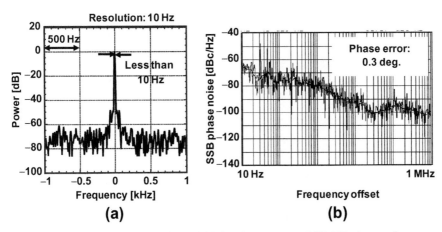

FIGURE 7.15 IF signal characteristics. (a) IF signal spectrum and (b) SSB phase noise spectrum.

(RMS) of the IF signal estimated by integrating the SSB noise spectrum was as low as 0.3°, which indicates that a stable OPLL operation is successfully achieved under a low phase noise condition. Such a low phase noise of the IF signal in spite of the relatively large PLL bandwidth is attributed to the narrow linewidth of the fiber laser.

The tolerance of the phase noise for 64, 256, and 1024 QAM signals can be estimated from the constellation maps. As shown in Figure 7.16, the angle between the two closest symbols is $2\delta\phi = 4.7$, 2.0, and 0.95° for 64, 256, and 1024 QAM, respectively, which correspond to the tolerable phase noise. Therefore, the RMS phase noise of 0.3° is sufficiently small for demodulating even a 1024 QAM signal.

The QAM signal and LO are coupled into a coherent receiver for demodulation. Figure 7.17 shows the configuration of a coherent receiver used for homodyne detection. The LO output is divided into two signals with a shift 90° phase and coupled with the QAM signal in a 90° optical hybrid circuit, so that the cosine (I) and sine (Q) components are obtained after balanced detection. A compact 90° optical hybrid

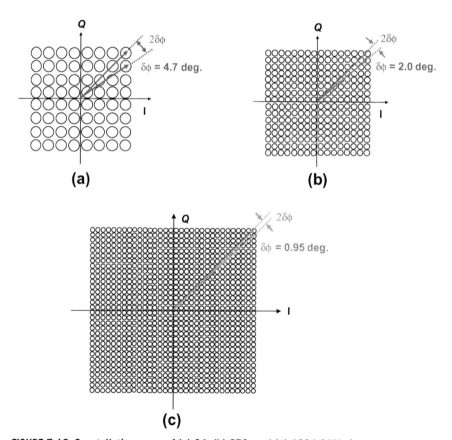

FIGURE 7.16 Constellation maps of (a) 64, (b) 256, and (c) 1024 QAM signals and the comparison of the tolerable phase noise.

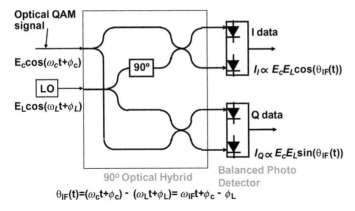

FIGURE 7.17 Schematic diagram of coherent receiver.

circuit and balanced photo-diodes are commercially available, and the integration of these components is under development [36].

7.3.4 Digital demodulator and equalizer

Figure 7.18 shows a schematic diagram of a digital demodulator. Here a polarization diversity configuration with a dual-polarization optical hybrid circuit is adopted to receive a polarization-multiplexed signal. The QAM data are first digitized with four A/D converters that are operated synchronously and fed into DSP where I and Q data are demodulated. In the DSP, there is an adaptive equalization to compensate for waveform distortions due to chromatic dispersion and fiber nonlinearity during transmission, or hardware imperfections in individual components such as a non-ideal frequency response or skew in optical or electrical devices. Polarization demultiplexing is also carried out in the equalizer. Finally the demodulated data are converted into binary data in the decoder.

Recent advances in DSP technologies have enabled not only the coherent detection of high-speed multi-level optical signals with high precision, but also compensation for linear and nonlinear transmission impairments in the electrical domain. A number of adaptive equalization techniques have been demonstrated to compensate

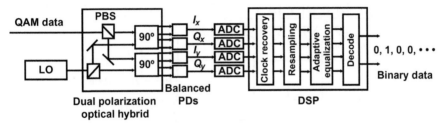

FIGURE 7.18 Schematic diagram of digital demodulator.

for linear impairments, which are categorized mainly into time- and frequency-domain approaches. A well known time-domain equalizer is a finite impulse response (FIR) filter [37]. A configuration of an adaptive FIR filter is shown in Figure 7.19. The output from the FIR filter is given by

$$y(n) = \sum_{k=-N}^{N} h(k)x(n+k), \qquad (7.11)$$

where $x(n)$ is the complex amplitude of the input signal to the filter, $h(k)$ is the tap coefficient, and $2N+1$ is the number of taps. Here, the filter bandwidth is given by $B = 1/T$, where T is the symbol period, and the frequency resolution of the FIR filter is $\Delta f = B/(2N+1)$. The tap coefficients can be adaptively determined by minimizing the error between $y(n)$ and the reference $d(n)$ such as with least mean square (LMS) algorithm [38].

In FIR filters, the frequency resolution is determined mainly by the finite number of filter taps, which cannot be increased largely in order to avoid complex calculations. Insufficient resolution in the equalizers becomes disadvantageous for higher-order QAM, as these signals typically contain non-negligible frequency components in lower frequency regime, such as below 10 MHz. In order to improve the frequency resolution without the expense of calculation complexity, frequency-domain equalization (FDE) schemes have been developed [39,40] and have recently received attention in digital coherent transmission [41]. FDE features an equalization capability with less computational complexity than FIR filters by virtue of the FFT operation. The principle of FDE is shown in Figure 7.20a. The distortion function of the transmission system in the frequency domain can be described as

$$F_{\text{distortion}}(f) = \frac{F_{\text{out}}(f)}{F_{\text{in}}(f)}, \qquad (7.12)$$

where $(F_{\text{in}} f)$ and $F_{\text{out}}(f)$ are the frequency spectra of the transmitted and received QAM data. Then we define the compensation function as $F_{\text{comp}}(f) = 1/F_{\text{distortion}}(f)$. Namely, the transfer function of the equalizer can be determined from the transmitted

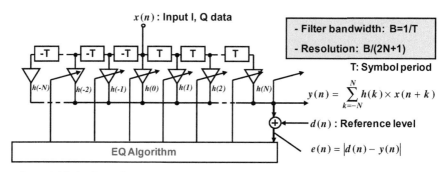

FIGURE 7.19 Configuration of an adaptive FIR filter.

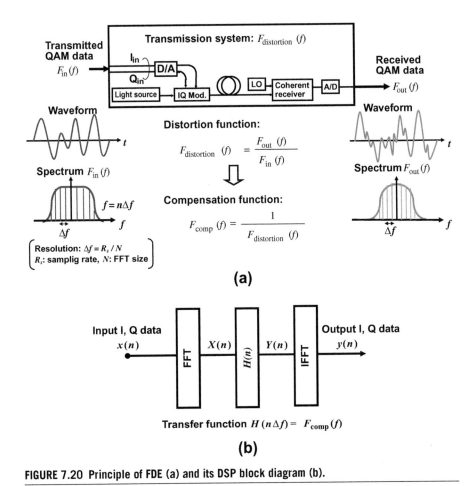

FIGURE 7.20 Principle of FDE (a) and its DSP block diagram (b).

and received QAM data that are converted to the frequency domain with FFT. Once the compensation function is determined, we can implement it in the DSP as shown in Figure 7.20b, where $H(n)$ is a digitized transfer function of $F_{comp}(f)$. The frequency resolution of FDE is given by the FFT size, N_{FFT}, and the sampling rate $R_s : \Delta f = \beta R_s / N_{FFT}$, where β is the number of samples per symbol.

Here we compared the capability of FIR and FDE for 256 QAM, 4 Gsymbol/s signal. Figure 7.21a and b, respectively, shows the back-to-back error vector magnitude (EVM) of the 256 QAM signal for various frequency resolution values when a digital FIR filter and FDE were adopted as equalizers, respectively. In Figure 7.21a, the resolution of the FIR filter Δf was varied by changing the number of FIR taps, N_{FIR}, which are related as $\Delta f = 4$ Gsymbol/s $/N_{FIR}$. The maximum N_{FIR} was 99, which corresponds to the resolution $\Delta f = 40$ MHz, in which the minimum EVM was 1.38%. A larger N_{FIR} results in a significant increase in computational complexity

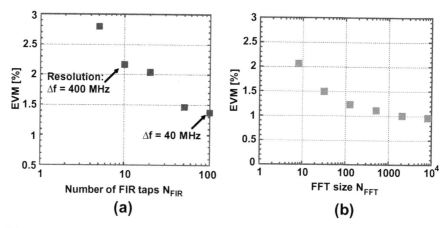

FIGURE 7.21 Dependence of the EVM of a 4 Gsymbol/s, 256 QAM signal on the frequency resolution in an equalizer with a digital FIR filter (a) and FDE (b).

and poor convergence of the tap coefficient calculation. On the other hand, the resolution of FDE is given by $\Delta f = 4$ Gsymbol/s$\times$(2sample/symbol)/N_{FFT}. With an FFT size of $N_{FFT} = 8192$, δf can be greatly reduced to 1 MHz without too much computational complexity. In this case, EVM was improved to 0.94% as shown in Figure 7.21b, which is a consequence of the higher resolution.

In addition to the linear impairments, fiber nonlinearity also becomes a major limiting factor especially for higher-order QAM transmission because of the power-dependent phase rotation caused by self-phase modulation (SPM) and cross-phase modulation (XPM) [42–44]. SPM can be canceled by simply multiplying a phase shift $\exp[-i\gamma M L_{eff}|A(t)|^2]$, where γ is the nonlinear coefficient, M is the number of spans, and L_{eff} is the effective span length defined as $L_{eff} = (1/\alpha)[1 - \exp(-\alpha L)]$, where α is the fiber loss and L is the span length. However, as the symbol rate increases, interplay between nonlinearity and dispersion becomes non-negligible, and hence the individual averaged compensation of dispersion and SPM is not sufficient to eliminate distortions. In order to compensate for linear and nonlinear impairments simultaneously, the backpropagation (BP) method [45,46] has been recently proposed, which computes an inverse fiber propagation using the nonlinear Schrödinger equation by using a split-step Fourier method. BP has been used to extend the transmission distance in 112 Gbit/s 16 QAM [47] and 120 Gbit/s 64 QAM [48] transmissions. In a polarization-multiplexed transmission, we employ a split-step Fourier analysis of the Manakov equation, which describes pulse propagation in the presence of dispersion, SPM, and XPM between the two orthogonal polarizations under a randomly varying birefringence [49]:

$$\begin{cases} -i\frac{\partial A_x}{\partial z} = -\frac{\beta_2}{2}\frac{\partial^2 A_x}{\partial t^2} + \frac{8}{9}\gamma |A_x|^2 A_x + \frac{8}{9}\gamma |A_y|^2 A_x + i\frac{\alpha}{2}A_x, \\ -i\frac{\partial A_y}{\partial z} = -\frac{\beta_2}{2}\frac{\partial^2 A_y}{\partial t^2} + \frac{8}{9}\gamma |A_y|^2 A_y + \frac{8}{9}\gamma |A_x|^2 A_y + i\frac{\alpha}{2}A_y, \end{cases} \tag{7.13}$$

where A_x and A_y represent the amplitude of the x and y polarization components, and α, β_2, and γ are the loss, dispersion, and nonlinear coefficients, respectively. In BP, we solve Eq. (7.13) with the reversed sign of α, β_2, and γ. BP has the potential to provide significantly improved performance as a result of enhanced OSNR by allowing a higher transmission power.

Recently, intensive efforts have been made to realize real-time digital coherent receivers for practical implementation of coherent QAM transmission. Real-time field programmable gate array FPGA intradyne coherent receivers for 16 QAM (0.625 [50] and 5 Gsymbol/s [51]) and a heterodyne receiver for 64 QAM (1 Gsymbol/s) [52] have been successfully demonstrated.

7.4 HIGHER-ORDER QAM TRANSMISSION EXPERIMENTS

In this section, we present recent demonstrations of a 1024 QAM single-carrier transmission (Section 7.4.1), which is currently the highest QAM multiplicity, and 256 QAM-OFDM transmission (Section 7.4.2) with a spectral efficiency as high as 14 bit/s/Hz. A novel OTDM RZ/QAM technique for realizing highly spectral-efficient, ultrahigh-speed coherent transmission beyond the limit of DSP is also described in Section 7.4.3, where the result of a 400 Gbit/s, 32 RZ/QAM transmission is presented in detail.

7.4.1 1024 QAM (60 Gbit/s) single-carrier transmission

As a first example of extremely higher-order QAM transmission, here we describe a recent demonstration of 1024 QAM single-carrier coherent optical transmission, in which a 60 Gbit/s polarization-multiplexed signal (3 Gsymbol/s) was transmitted over 160 km within an optical bandwidth of 4.05 GHz [53].

The experimental setup is shown in Figure 7.22. At the transmitter, coherent CW light emitted from a C_2H_2 frequency-stabilized fiber was modulated by an IQ modulator driven with a 3 Gsymbol/s, 1024 QAM baseband signal generated by an arbitrary waveform generator (AWG). The AWG was operated at 12 Gsample/s with a 10-bit resolution. The bandwidth of the 1024 QAM baseband signal was reduced with the adoption of a digital Nyquist raised-cosine filter [54].

It is well known in the microwave communication field that a Nyquist filter is very useful for reducing the bandwidth of a data signal without introducing inter-symbol interference [55,56]. Figure 7.23 shows the transfer function and impulse response of the raised-cosine Nyquist filter. The transfer function is given by

$$H(f) = \begin{cases} \frac{1}{2}\left\{1 - \sin\frac{\pi(f-0.5)}{\alpha}\right\}, & 0.5 - \frac{\alpha}{2} \leq |f| < 0.5 + \frac{\alpha}{2}, \\ 1, & |f| < 0.5 - \frac{\alpha}{2}, \\ 0, & |f| \geq 0.5 + \frac{\alpha}{2}, \end{cases} \tag{7.14}$$

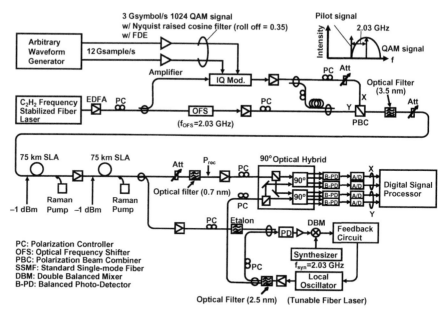

FIGURE 7.22 Experimental setup for 1024 QAM (60 Gbit/s) coherent transmission over 150 km.

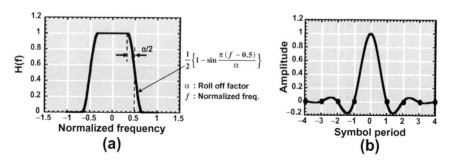

FIGURE 7.23 Transfer function (a) and impulse response (b) of the raised-cosine Nyquist filter.

where α is called a roll-off factor. As shown in Figure 7.23b, the impulse response becomes zero at the location of adjacent symbols. This indicates that the bandwidth can be reduced with the Nyquist filter while avoiding the intersymbol interference. When the roll-off factor is too small, the eye pattern is closed and the clock cannot be extracted at the receiver. Here we employed a raised-cosine Nyquist filter with $\alpha = 0.35$ so that the bandwidth of the QAM signal is reduced to 4.05 GHz.

In addition to the Nyquist filtering, digital pre-equalization with FDE was employed by converting the data into the frequency domain with a fast Fourier transform (FFT). The transfer function of the equalizer was determined in order to compensate for the non-ideal frequency response of individual components, mainly at the IQ modulator, as we described in Section 7.3.2. To compensate for the frequency response, a frequency resolution of less than 1 MHz is needed in the pre-equalization. Here the FFT size for FDE was set at 16,384. This enabled us to improve the frequency resolution to 0.73 MHz, compared with the 30 MHz resolution obtained with a conventional 99-tap FIR filter.

The optical QAM was polarization multiplexed with a polarization beam combiner to generate 60 Gbit/s data. In parallel with these processes, part of the laser output was divided in front of the IQ modulator, and its frequency was downshifted by 2.03 GHz against the carrier frequency. This signal was combined with the QAM data and used as a pilot tone in the receiver for OPLL as described in Section 7.3.3. The transmission link was composed of two 75 km spans of super large area (SLA) fiber with an effective area of 106 μm^2, whose loss (0.20 dB/km) was compensated for by using EDFAs and Raman amplifiers. The power launched into each span was set at −1 dBm, which was optimally chosen to maximize the OSNR and minimize the nonlinear impairments. The Raman amplifiers were backward pumped and provided 9.5 dB of the total gain of 16.0 dB at each span.

At the receiver, which was followed by a 0.7 nm optical filter and an EDFA preamplifier, the transmitted QAM data were combined with an LO and received by a polarization-diverse 90° optical hybrid and four balanced photodiodes. The detected signals were A/D converted at 40 Gsample/s and processed with an offline DSP. Figure 7.24 shows the RF spectrum of the demodulated signal at the DSP.

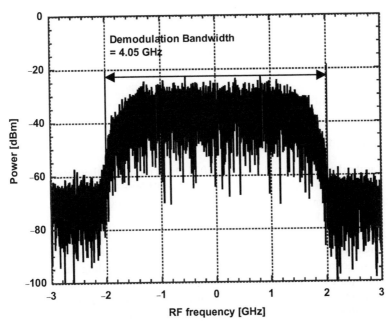

FIGURE 7.24 RF spectrum of demodulated 1024 QAM signal after 150 km transmission.

The demodulation bandwidth was set at 4.05 GHz due to the adoption of a Nyquist filter. In the DSP, we first compensated for dispersion, SPM, and XPM simultaneously with a BP as we mentioned in Section 7.3.4. Here the step size in terms of distance was set at 9.375 km. Finally, the compensated QAM signal was demodulated into binary data, and the bit error rate (BER) was evaluated.

We first measured the OSNR and phase noise of the transmitted data, and compared them with the requirement for 1024 QAM. To compare the experimental results with theoretical BER presented in Section 7.2, we need to convert the received optical power P_{rec} to E_b/N_0. The received optical power can be converted to OSNR and E_b/N_0 in the following way. The signal and noise power after a preamplifier with gain G and noise figure NF are given by $P_{sig} = G P_{rec}$ and $P_{ASE} = n_{sp}h\nu(G - 1)\Delta$, respectively, where $n_{sp} = NF/2$ is the spontaneous emission factor, h is Planck's constant, ν is the optical frequency, and $\Delta\nu$ is the detection bandwidth of the optical signal. Here, the power is defined in terms of single polarization. Therefore, the OSNR is obtained as

$$\text{OSNR} = \frac{P_{sig}}{P_{ASE}} = \frac{P_{rec}G}{n_{sp}h\nu(G - 1)\Delta\nu} \cong \frac{P_{rec}}{n_{sp}h\nu\Delta\nu} = \frac{P_{rec}}{(NF/2)h\nu\Delta\nu}, \quad (7.15)$$

where we assumed $G \gg 1$. In dB unit, Eq. (7.15) is written as

$$\begin{aligned}\text{OSNR[dB]} &= 10\log_{10} P_{rec} - \text{NF[dB]} + 3 - 10\log_{10}(h\nu\Delta\nu) \quad (7.16)\\ &= P_{rec}\,[\text{dBm}] - 30 - \text{NF[dB]} + 3 + 88 \\ &= 61 + P_{rec}\,[\text{dBm}] - \text{NF[dB]},\end{aligned}$$

where we used $h = 6.626 \times 10^{-34}$ J s, $\nu = 194$ THz, and $\Delta\nu = 12.5$ GHz [57]. From Eqs. (7.15) and (7.16), E_b/N_0 is given by

$$E_b/N_0 = \frac{P_{sig}/R}{N_0} = \frac{\Delta\nu}{R}\frac{P_{sig}}{N_0\Delta\nu} = \frac{\Delta\nu}{R}\frac{P_{sig}}{P_{ASE}} = \frac{\Delta\nu}{R}\text{OSNR}, \quad (7.17)$$

$$E_b/N_0\,[\text{dB}] = 10\log_{10}(\Delta\nu/R) + \text{OSNR}\,[\text{dB}]. \quad (7.18)$$

Here, R is the data rate per polarization. 1024 QAM requires a theoretical E_b/N_0 value as high as 24 dB to achieve an FEC threshold of BER $= 2 \times 10^{-3}$ as can be seen in Figure 7.3. This corresponds to OSNR $= 27.8$ dB, where we used $R = 30$ Gbit/s and $\Delta\nu = 12.5$ GHz in Eq. (7.17).

Figure 7.25a and b, respectively, shows the optical spectra of the 1024 QAM signal before and after a 150 km transmission. It can be seen that the optical bandwidth was accommodated within a bandwidth of 4.05 GHz including the pilot tone (not visible because the tone level was 20 dB below the data). The OSNR, which we measured with a 0.1 nm resolution before transmission, was 40 dB, and it had degraded slightly to 36.5 dB after a 150 km transmission. This indicates that the present OSNR is sufficient for the FEC threshold after a 150 km transmission.

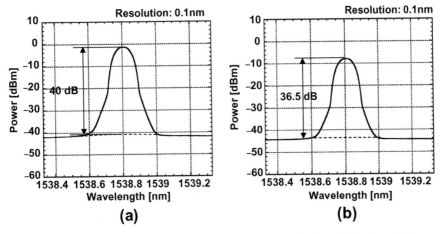

FIGURE 7.25 Optical spectra of 1024 QAM signal. (a) Back-to-back, and (b) after 150 km transmission.

Next we evaluated the phase noise tolerance of the transmitted 1024 QAM signal. Figure 7.26 shows the SSB noise power spectrum of a heterodyne beat note between the LO and the pilot tone after a 150 km transmission. By integrating this spectrum, the phase noise was estimated to be 0.46°, whereas it was 0.25° before transmission. This slight increase was mainly caused by OSNR degradation. On the other hand, the phase tolerance for 1024 QAM, determined by the phase difference between the two nearest symbols, is ±0.95° as shown in Figure 7.16c. This implies that the OPLL successfully achieved a residual phase noise in the IF signal that was within the tolerance for 1024 QAM even after a 150 km transmission.

Here we present the BER performance. Figure 7.27 shows the measured BER after a 150 km transmission as a parameter of the fiber launched power without and with a BP method. The optimum launched power was increased from −6 to −1 dBm by using the BP method. Figure 7.28a and b shows the constellation diagrams at an optimum launched power without and with BP. The error vector magnitude (EVM) of the constellation was improved from 1.30% to 1.12% by employing nonlinear compensation. Of the various linear and nonlinear transmission impairments, XPM between the two polarizations was a major cause of performance degradation. It should be noted that the individual equalization of dispersion, SPM, and XPM is insufficient for 1024 QAM, and simultaneous equalization with BP plays an important role.

The BER characteristics as a function of the received power are shown in Figure 7.29. The power penalty at a BER of 2×10^{-3} was approximately 7 dB for both polarizations, but both sets of polarization data achieved a BER lower than the FEC limit of 2×10^{-3}. The present result is scalable to a net spectral efficiency as high as 13.8 bit/s/Hz in a multi-channel transmission even when taking account of the 7% FEC overhead. The power penalty was mainly caused by the residual phase noise that remained after nonlinear compensation.

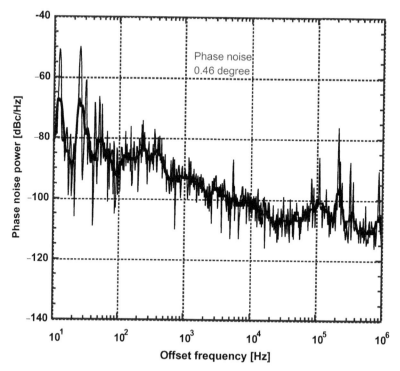

FIGURE 7.26 Single side-band (SSB) noise power spectrum of a heterodyne beat note between LO and pilot tone after 150 km transmission.

7.4.2 256 QAM-OFDM coherent transmission

Another attractive candidate for transmission with high spectral efficiency is orthogonal frequency division multiplexing (OFDM), where the multi-carrier transmission of low-speed orthogonal subcarriers enables us to improve the spectral efficiency [58]. The baseband OFDM signal can be represented as

$$S(t) = \sum_{n=0}^{N-1} \left\{ a_n(t) \cos\left(2\pi n \frac{t}{T}\right) - b_n(t) \sin\left(2\pi n \frac{t}{T}\right) \right\}, \qquad (7.19)$$

where $a_n(t)$ and $b_n(t)$ are the symbol data, T is the symbol period, and N is the number of subcarriers. The subcarrier spacing is set so that it is equal to the symbol rate $1/T$, which allows orthogonality between the subcarriers:

$$\frac{2}{T} \int_{t_0}^{t_0+T} \cos\left(2\pi n \frac{t}{T}\right) \cos\left(2\pi k \frac{t}{T}\right) dt = \begin{cases} 1, n = k, \\ 0, n \neq k. \end{cases} \qquad (7.20)$$

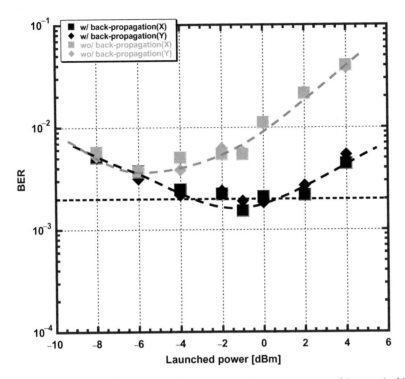

FIGURE 7.27 BER after 150 km transmission versus fiber launched power without and with digital nonlinear compensation using backpropagation method.

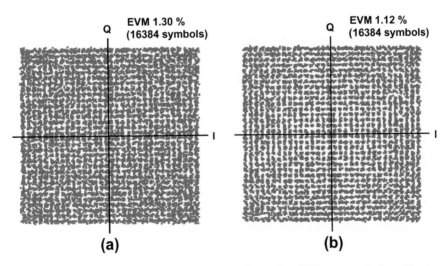

FIGURE 7.28 Constellation diagram of 1024 QAM signal after 150 km transmission without and with digital nonlinear compensation using backpropagation method.

A typical example of an OFDM signal spectrum and waveform is shown in Figure 7.30. Although the spectrum of each subcarrier overlaps, the oscillating tail of each subcarrier spectrum intersects at the zero level. The data can be easily demodulated in spite of the spectral overlap by extracting a subcarrier using the orthogonality. For example, the data $a_k(t)$ can be demodulated in the following way:

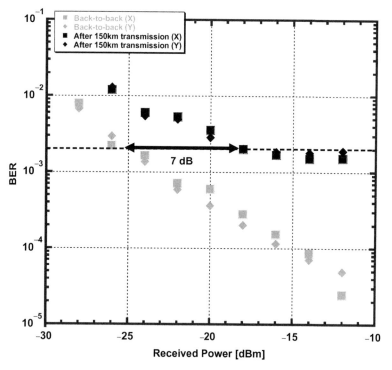

FIGURE 7.29 BER characteristics of 1024 QAM, 60 Gbit/s transmission.

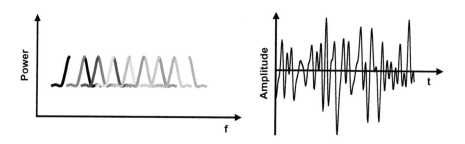

FIGURE 7.30 Typical example of an OFDM signal spectrum and waveform.

$$\frac{2}{T} \int_{t_0}^{t_0+T} S(t) \cos\left(2\pi k \frac{t}{T}\right) dt = \frac{2}{T} \sum_{n=0}^{N-1} \left\{ a_n(t_0) \int_{t_0}^{t_0+T} \cos\left(2\pi n \frac{t}{T}\right) \cos\left(2\pi n \frac{t}{T}\right) dt \right.$$

$$\left. - b_n(t_0) \int_{t_0}^{t_0+T} \sin\left(2\pi n \frac{t}{T}\right) \cos\left(2\pi n \frac{t}{T}\right) dt \right\} \quad (7.21)$$

$$= a_k(t_0).$$

Subcarrier multiplexing with a spectral overlap allows us to improve the spectral efficiency and thus the dispersion tolerance. By virtue of these features, a 1 Tbit/s per channel OFDM transmission has been demonstrated [59]. With this scheme, 1.2 Tbit/s transmission over 7200 km [60] and the back-to-back demonstration at 26 Tbit/s [61] have been reported. It should be noted, however, that OFDM is more sensitive to nonlinear effects than a single-carrier transmission owing to the large peak-to-average power ratio (PAPR). This occurs as a result of phase coherence among subcarrier signals.

Figure 7.31 shows a block diagram of baseband OFDM signal generation and demodulation. At the transmitter, as shown in Figure 7.32a, the binary data are first encoded with a multi-level format such as QPSK or QAM. A training symbol used for amplitude and phase equalization of subcarriers is then added and the data are converted into a parallel sequence and divided into N subcarriers using inverse fast Fourier transformation (IFFT). After adding a guard interval and employing D/A conversion, a baseband OFDM signal is obtained. At the receiver, the IF signal is synchronously detected, and after removing the guard interval and extracting the training symbol, the OFDM signal is demodulated via the fast Fourier transformation (FFT) of the subcarriers. The demodulated signal is finally converted to binary data.

To achieve higher spectral efficiency in OFDM, high-level subcarrier modulation formats and coherent detection play a very important role. A spectral efficiency of 8.0 bit/s/Hz has been realized with 32 QAM subcarrier modulation, where a 400 Gbit/s PDM-OFDM signal was transmitted with a channel spacing of 50 GHz [62]. By increasing the QAM multiplicity level further, a 101.7 Tbit/s WDM capacity has been demonstrated using a Pol-Mux 128 QAM-OFDM with a spectral efficiency of 11 bit/s/Hz [8]. More recently, 1024 QAM subcarrier modulation at 1 Msymbol/s has also been reported [63]. As an alternative format to QAM, 256-level interative polar modulation has been applied to subcarrier modulation, which is found to reduce the required OSNR by 1.2 dB compared to 256 QAM [64].

Here we describe 400 Gbit/s frequency-division-multiplexed, Pol-Mux 256 QAM-OFDM transmission with a spectral efficiency of 14 bit/s/Hz using an optical PLL technique [65]. Figure 7.32 shows the experimental setup. The C_2H_2 frequency-stabilized fiber laser output was divided into two paths, and one was coupled to a multi-carrier generator that consisted of two Mach-Zehnder modulators (MZM) and an optical frequency shifter (OFS1). Each MZM was driven with a 5.18 GHz sinusoidal signal, and five sidebands were generated. The OFS1 fed a frequency downshift of 2.59 GHz against the original frequency f_0. As a result, 10 optical sidebands separated at 2.59 GHz were generated from the multi-carrier generator. Then, each of the five sidebands from the MZM was modulated using an IQ modulator with an OFDM signal generated by an AWG at a sampling rate of 12 GSa/s. After data modulation, the OFDM signals were

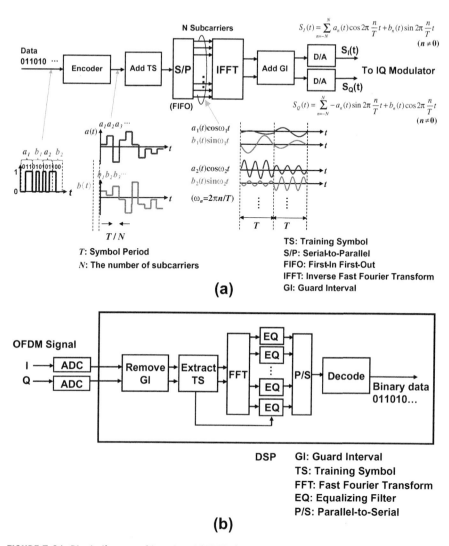

FIGURE 7.31 Block diagram of baseband OFDM signal generation (a) and demodulation (b).

combined with an optical coupler and then polarization-multiplexed with a polarization beam combiner (PBC). The other path from the laser output led to OFS2, which fed a frequency downshift of 1.3 GHz against f_0 for use as the pilot tone signal required for the optical phase tracking of the LO under optical PLL operation. The polarization of the pilot signal was aligned with one of the two polarization axes of the OFDM signal.

The AWGs generated the OFDM baseband signals using an FFT-based technique. An OFDM signal was encoded from binary data into a 256 QAM format, and training symbols (TS) and a cyclic prefix were added to the sequence. The data was then divided into 1714 subcarriers and pre-equalization was adopted for each subcarrier

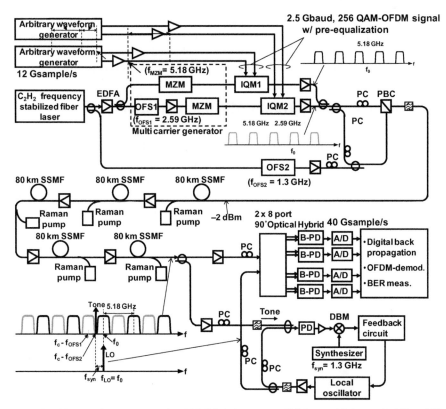

FIGURE 7.32 Experimental setup for 400 Gbit/s frequency-division-multiplexed, Pol-Mux 256 QAM-OFDM transmission over 400 km.

to compensate for the transfer function of the transmission system. Here, the OFDM symbol rate per subcarrier was 1.465 Msymbol/s, the FFT size was 8192, the cyclic prefix length was 128 samples per OFDM symbol as a 1/64 overhead, and four out of every 204 OFDM symbols were assigned as TSs. The net data rate per 2.59 GHz channel was $2 \times 8 \times 1714 \times (200/204) \times (64/65) \times 1.465 \text{ MHz} = 38.8 \text{ Gbit/s}$. After taking a 7% FEC overhead into account, the achieved spectral efficiency was 14 bit/s/Hz.

The transmission link consisted of 5×80 km SSMFs, in which Raman amplifiers and EDFAs were employed. At the receiver, the OFDM signal was homodyne-detected with a dual-polarization 90° optical hybrid using an LO signal from a frequency-tunable tracking fiber laser, whose phase was locked to the pilot signal. After detection with four balanced PDs (B-PD), the data were A/D-converted and accumulated in a high-speed digital oscilloscope at a sampling rate of 40 GSa/s and with a resolution of 8 bits. Then, the digital data were post-processed with a DSP in an offline condition. At the DSP, the linear and nonlinear fiber impairments were compensated with a BP method. Then, each OFDM band was extracted electronically by a digital filter. The OFDM signal was demodulated with an FFT and converted to a binary data sequence.

Figure 7.33a shows the BER performance, and (b) and (c) show the constellations before and after the 400 km transmission. Here we employed BP with eight steps per span. These results indicate that all 10 channels achieved a BER lower than the FEC limit of 2×10^{-3}, and the power penalty after 400 km transmission is 6 dB.

7.4.3 Ultrahigh-speed OTDM-RZ/QAM transmission

Coherent QAM transmission with ultrahigh multiplicity makes it possible to realize a high-speed system with low-speed devices and therefore helps to enhance the tolerance to CD and PMD as well as to reduce power consumption. However, these coherent transmission systems are limited to modest symbol rates owing to the speed and bandwidth limitations of such electrical components as analog-digital (A/D) converters and DSPs. In contrast, OTDM enables us to realize ultrahigh-speed transmission with a symbol rate beyond the limit available with electrical signal processing. Therefore, OTDM is very useful for increasing the channel rate even in a coherent transmission.

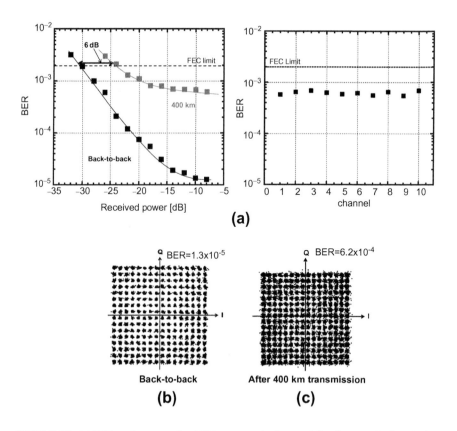

FIGURE 7.33 (a) BER performance for 400 km transmission, and (b, c) constellations before and after 400 km transmission.

If we combine QAM and OTDM as an ultimate technology for handling coherent pulses, i.e. by encoding data on both the amplitude and phase of ultrashort RZ optical pulses and multiplexing them with OTDM, we could obtain an enormous advantage in terms of both a high bit rate and high spectral efficiency. This scheme has been used to achieve demodulation performance in 640 Gbit/s-1073 km QPSK [65] and 10.2 Tbit/s-29 km 16 QAM transmission [67].

Figure 7.34 shows the basic configuration for OTDM RZ/QAM transmission. Here it is important to note that, in order to employ OTDM, the QAM signal must be generated in an RZ format, i.e. as an optical pulse. Since frequency-stabilized coherent pulse sources are not yet available, here we use a coherent CW laser followed by RZ pulse carving composed of an optical comb generator and appropriate optical filtering. After QAM modulation at 10 Gsymbol/s, the signal is OTDM multiplexed N times and launched into a transmission line. On the receiver side, we prepare a pulsed local oscillator for coherent detection, in which a CW-LO is carved into a pulse. Coherent detection with a pulse LO also functions as an OTDM demultiplexer.

Based on this configuration, a single-channel 400 Gbit/s, Pol-Mux, 10 Gsymbol/s, 4-OTDM-32 RZ/QAM transmission over 225 km was recently demonstrated by using an OPLL [68]. Figure 7.35 shows the experimental setup. The output signal of the continuous-wave (CW) C_2H_2 frequency-stabilized fiber laser was introduced into an optical comb generator consisting of an LN intensity modulator [69,70].

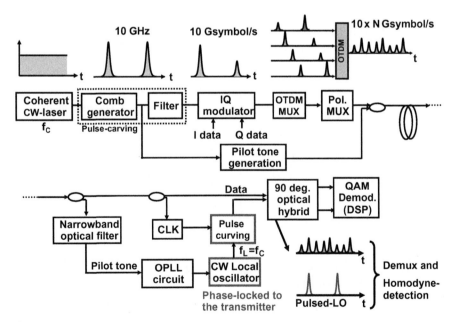

FIGURE 7.34 Basic configuration for ultrahigh-speed OTDM RZ/QAM coherent optical transmission.

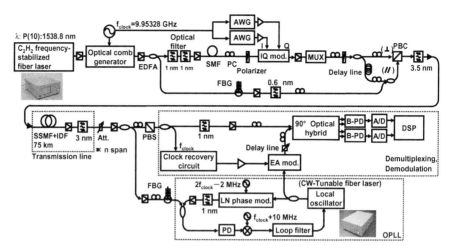

FIGURE 7.35 Experimental setup for 400 Gbit/s, Pol-Mux, OTDM, 32 RZ/QAM coherent transmission.

This modulator was driven at a frequency of $f_{clock} = 9.95328$ GHz and generated an optical comb with a -10-dB spectral width of 290 GHz (29 comb lines). The optical comb signal was divided into two arms. On one arm, we prepared a 9.95328 GHz optical pulse train with a 6-ps pulse with a spectral width of 0.59 nm that had passed through 1.0 nm optical filters a single-mode fiber (SMF). Figure 7.36a and b shows its optical spectrum and autocorrelation trace. Then, two AWGs (20 Gsample/s) were used to drive an IQ modulator to generate 10 Gsymbol/s, RZ-32 QAM signals. Then,

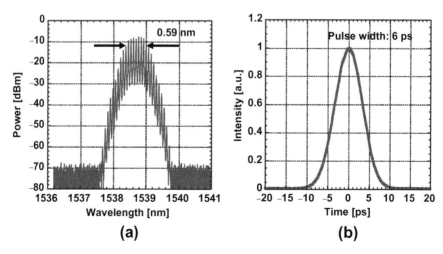

FIGURE 7.36 Optical spectrum (a) and autocorrelation waveform (b) of a 10 GHz RZ pulse generated with the optical comb generator.

the data signal was time-division-multiplexed four times and polarization-multiplexed with a polarization beam combiner, and thus a 400 Gbit/s data signal was generated.

On the other arm, the 13th harmonic signal of the comb was extracted by a fiber Bragg grating with a bandwidth of 1.4 GHz. This signal was used as a pilot tone signal for OPLL operation. Since the data spectrum is -30 dB lower from the peak when 130 GHz apart from the center, the use of the 13th harmonic signal as a pilot tone allows us to avoid the overlap between the data and pilot tone. They were combined and transmitted over a 225 km dispersion-managed fiber link. Each span consists of a 50-km standard SMF (SSMF) and a 25-km inverse dispersion fiber (IDF, -40 ps/nm/km) and has an average loss of 18 dB.

The receiver circuit was composed of two main parts. One was a homodyne-demodulation circuit. The other was an OPLL circuit whose configuration was the same as that in Sections 7.4.1 and 7.4.2, except for the inclusion of an LN phase modulator. Figure 7.37 shows the frequency relationship between the optical comb spectrum at the transmitter and the phase-modulated LO signal at the OPLL circuit. Here, f_{Trans} and f_{LO} denote the transmitter and LO frequency, respectively. In the OPLL circuit, a CW-LO signal was phase-modulated at $2f_{clock} - 1.67$ MHz. Here the driving frequency was set to $2 f_{clock}$, rather than f_{clock}, in order to obtain a sufficiently broad optical comb with high S/N per mode from an LN phase modulator. The offset of -1.67 MHz was chosen to give a 10 MHz offset to the reference frequency (namely, $f_{clock} + 10$ MHz $= 9.9633$ GHz). Without the offset, extra beat components from other modes (such as between the pilot and the seventh harmonic of the modulated LO) also enter into the PLL. With the $+10$ MHz offset, it is possible to avoid these extra beat components to fall inside the PLL bandwidth of 1 MHz.

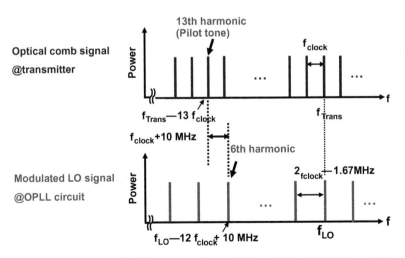

FIGURE 7.37 Frequency relationship between the optical comb signal at the transmitter and the modulated LO signal at OPLL.

The phase of the beat signal between the sixth harmonic of the modulated LO signal, whose frequency was $f_{LO}-6 \times (2f_{clock}-1.67 \text{ MHz}) = f_{LO}-12f_{clock} + 10 \text{ MHz}$, and the transmitted pilot signal ($f_{Trans}-13f_{clock}$) was compared with the reference phase from the synthesizer by the DBM. Since the beat frequency of these two modes is given by $(f_{LO}-f_{Trans}) + f_{clock} + 10 \text{ MHz}$, we define the reference frequency to be $f_{clock} + 10 \text{ MHz}$ for phase locking. The phase difference was fed back to the LO through the loop filter for phase locking. Thus, the LO frequency was phase-locked to the transmitter frequency, and homodyne detection could be easily realized between an RZ signal and a synchronized LO signal. The phase noise of the detected signal was $1.7°$ (10^{-1} MHz), which was sufficiently small to achieve 32 QAM.

In the homodyne detection circuit, an optical pulse train with a repetition rate of f_{clock}, which is used for a pulse LO, was generated from a CW-LO with an electro-absorption (EA) modulator. The EA modulator was driven by a clock signal obtained with a clock recovery circuit for the transmitted QAM data. The RZ-QAM data and the synchronized local pulse signal were incident on a 90° optical hybrid. The data was simultaneously demultiplexed and downconverted to a base-band signal. Here, the phase of the local pulses was adjusted by an optical delay line to make it possible to choose one of the tributaries. Finally, the detected signal was A/D-converted and accumulated in a high-speed digital scope (40 Gsample/s, 12 GHz bandwidth) and sent to a DSP circuit operating in an offline condition.

Figure 7.38 shows the BER characteristics of the 10 Gsymbol/s tributaries demultiplexed from 400 Gbit/s data signals for back-to-back, 150 and 225 km transmissions. After the 225 km transmission, all tributaries had BERs below the FEC limit of 2×10^{-3}, indicating an error-free transmission at a net data rate of 374 Gbit/s

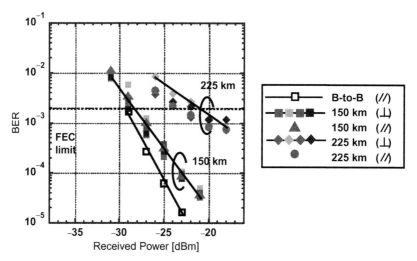

FIGURE 7.38 BER characteristics of 10 Gsymbol/s, tributaries demultiplexed from 400 Gbit/s, Pol-Mux, OTDM, 32 RZ/QAM signal.

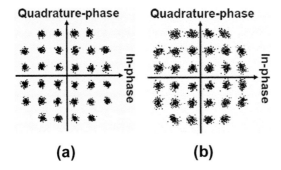

FIGURE 7.39 Constellation maps of 10 Gsymbol/s 32 RZ/QAM signal. (a) back-to-back and (b) after 150 km transmission.

including the 7% FEC overhead. Figures 7.39a and b show constellation maps for one of the tributaries of the orthogonal polarization state before and after a 150 km transmission. The constellation points broadened due to optical signal-to-noise ratio degradation after transmission.

Recently, the bit rate has been increased from 400 to 800 Gbit/s by increasing the OTDM multiplicity from 4 to 8 [71]. As the pulse width is shortened associated with higher OTDM multiplicity, only a small fraction of power in the broad bandwidth of an RZ data signal contributes to the demodulation at the coherent receiver, due to the limited receiver bandwidth. This causes a serious SNR degradation in one symbol. In order to overcome this, an RZ-CW conversion scheme [72] is newly introduced, in which the OTDM demultiplexed RZ data is converted to a signal modulated on a CW carrier. This enables the conversion of the broad spectrum to a narrow band, and thus allows a demodulation with high SNR.

7.5 CONCLUSION

We have described recent progress on coherent QAM and OFDM transmission with higher-order multiplicity, aimed at ultrahigh spectral efficiency approaching the Shannon limit. Key technologies are the coherent detection with a frequency-stabilized fiber laser and an optical PLL circuit. Single-carrier 1024 QAM and 256 QAM-OFDM transmissions were successfully achieved, demonstrating a spectral efficiency exceeding 10 bit/s/Hz. Such an ultrahigh spectrally efficient transmission system would also play very important roles as regards increasing the total capacity of WDM systems and improving tolerance to CD and PMD as well as in reducing power consumption. We also described a novel high-speed, spectrally efficient transmission scheme combining OTDM and the QAM technique, in which a pulsed LO signal obtained with OPLL enabled precise demultiplexing and demodulation simultaneously. An optimum OTDM and QAM combination would provide the possibility of realizing long-haul Tbit/s/ch transmission with a simple configuration, large flexibility, and low power consumption.

References

[1] P.J. Winzer, Modulation and multiplexing in optical communication systems, IEEE LEOS Newslett. 23 (1) (2009) 4–10.

[2] M. Nakazawa, M. Yoshida, K. Kasai, J. Hongou, 20 Msymbol/s, 64 and 128 QAM coherent optical transmission over 52 km using heterodyne detection with frequency-stabilised laser, Electron. Lett. 42 (12) (2006) 710–712.

[3] J. Hongo, K. Kasai, M. Yoshida, M. Nakazawa, 1-Gsymbol/s 64-QAM coherent optical transmission over 150 km, IEEE Photon. Technol. Lett. 19 (9) (2007) 638–640.

[4] M. Nakazawa, Challenges to FDM-QAM coherent transmission with ultrahigh spectral efficiency, in: European Conference on Optical Communication (ECOC 2008), Tu.1.E.1.

[5] M. Nakazawa, S. Okamoto, T. Omiya, K. Kasai, M. Yoshida, 256-QAM (64 Gb/s) coherent optical transmission over 160 km with an optical bandwidth of 5.4 GHz, IEEE Photon. Technol. Lett. 22 (3) (2010) 185–187.

[6] S. Okamoto, K. Toyoda, T. Omiya, K. Kasai, M. Yoshida, M. Nakazawa, 512 QAM (54 Gbit/s) coherent optical transmission over 150 km with an optical bandwidth of 4.1 GHz, in: European Conference on Optical Communication (ECOC 2011), PD 2.3.

[7] Y. Koizumi, K. Toyoda, M. Yoshida, M. Nakazawa, 1024 QAM (60 Gbit/s) single-carrier coherent optical transmission over 150 km, Opt. Express 20 (11) (2012) 12508–12514.

[8] D. Qian, M. Huang, E. Ip, Y. Huang, Y. Shao, J. Hu, T. Wang, 101.7-Tb/s (370 × 294-Gb/s) PDM-128QAM-OFDM Transmission over 3 × 55-km SSMF using pilot-based phase noise mitigation, in: Optical Fiber Communication Conference (OFC 2011), PDPB5.

[9] A. Sano, T. Kobayashi, S. Yamanaka, A. Matsuura, H. Kawakami, Y. Miyamoto, K. Ishihara, H. Masuda, 102.3-Tb/s (224 × 548-Gb/s) C- and extended L-band all-Raman transmission over 240 km using PDM-64QAM single-carrier FDM with digital pilot tone, in: Optical Fiber Communication Conference (OFC 2012), PDP5C.3.

[10] M.S. Alfiad, M. Kuschnerov, S.L. Jansen, T. Wuth, D. van den Borne, H. de Waardt, 11×224-Gb/s POLMUX-RZ-16QAM transmission over 670 km of SSMF with 50-GHz channel spacing, IEEE Photon. Technol. Lett. 22 (15) (2010) 1150–1152.

[11] A.H. Gnauck, P.J. Winzer, S. Chandrasekhar, X. Liu, B. Zhu, D.W. Peckham, 10 × 224-Gb/s WDM transmission of 28-Gbaud PDM 16-QAM on a 50-GHz grid over 1200 km of fiber, in: Optical Fiber Communication Conference (OFC 2010), PDPB8.

[12] P.J. Winzer, A.H. Gnauck, C.R. Doerr, M. Magarini, L.L. Buhl, Spectrally efficient long-haul optical networking using 112-Gb/s polarization-multiplexed 16-QAM, J. Lightwave Technol. 28 (4) (2010) 547–556.

[13] P.J. Winzer, A.H. Gnauck, S. Chandrasekhar, S. Draving, J. Evangelista, B. Zhu, Generation and 1,200-km transmission of 448-Gb/s ETDM 56-Gbaud PDM 16-QAM using a single I/Q modulator, in: European Conference on Optical Communication (ECOC 2010), PD2.2.

[14] S. Okamoto, T. Omiya, K. Kasai, M. Yoshida, M. Nakazawa, 140 Gbit/s coherent optical transmission over 150 km with a 10 Gsymbol/s polarization-multiplexed 128 QAM signal, in: Optical Fiber Communication Conference (OFC 2010), OThD5.

[15] A.H. Gnauck, P. Winzer, A. Konczykowska, F. Jorge, J. Dupuy, M. Riet, G. Charlet, B. Zhu, D.W. Peckham, Generation and transmission of 21.4 Gbaud PDM 64 QAM using a high power DAC driving a single I/Q modulator, OFC 2011, PDPB2.

[16] T. Kobayashi, A. Sano, A. Matsuura, Y. Miyamoto K. Ishihara, Nonlinear tolerant long-haul WDM transmission over 1200 km using 538 Gb/s/ch PDM-64QAM SC-FDM signals with pilot tone, in: OFC 2012, OM2A.5.

[17] J.G. Proakis, Digital Communications, fourth ed., McGraw Hill, 2000.

[18] C.E. Shannon, A mathematical theory of communication, Bell Syst. Tech. J. 27 (1948) 379–423, 623–656.

[19] T. Mizuochi, Recent progress in forward error correction for optical communication systems, IEICE Trans. Comm. E88-B (2005) 1934–1946.

[20] A. Onae, K. Okumura, K. Sugiyama, F.L. Hong, H. Matsumoto, K. Nakagawa, R. Felder, O. Acef, Optical frequency standard at 1.5 m based on Doppler-free acetylene absorption, in: Proceedings 6th Symposium on Frequency Standards and Metrology, 2002, p. 445.

[21] K. Kasai, A. Suzuki, M. Yoshida, M. Nakazawa, Performance improvement of an acetylene (C_2H_2) frequency-stabilized fiber laser, IEICE Electron. Express 3 (2006) 487–492.

[22] T. Okoshi, K. Kikuchi, A. Nakayama, Novel method for high resolution measurement of laser output spectrum, Electron. Lett. 16 (1980) 630–631.

[23] D.W. Allan, Statistics of atomic frequency standards, Proc. IEEE 54 (1966) 221–230.

[24] T. Takeuchi, M. Takahashi, K. Suzuki, S. Watanabe, H. Yamazaki, Wavelength tunable laser with silica-waveguide ring resonators, IEICE Trans. Electron. E92C (2) (2009) 198–204.

[25] L. Stolpner, S. Lee, S. Li, A. Mehnert, P. Mols, S. Siala, J. Bush, Low noise planar external cavity laser for interferometric fiber optic sensors, Proc. SPIE 7004 (2008) 700457.

[26] H.T. Quynhanh, A. Suzuki, M. Yoshida, T. Hirooka, M. Nakazawa, A λ/4-shifted distributed-feedback laser diode with a fiber ring cavity configuration having an OSNR of 85 dB and a linewidth of 7 kHz, IEEE Photon. Technol. Lett. vol. 20 (18) (2008) 1578–1580.

[27] J. Cliche, Y. Painchaud, C. Latrasse, M. Picard, I. Alexandre, M. Têtu, Ultra-narrow Bragg grating for active semiconductor laser linewidth reduction through electrical Feedback, in: Bragg Gratings, Photosensitivity, and Poling in Glass Waveguides (BGPP2007), BTuE2.

[28] K. Kasai, M. Nakazawa, FM-eliminated C_2H_2 frequency-stabilized laser diode with a RIN of −135 dB/Hz and a linewidth of 4 kHz, Opt. Lett. 34 (14) (2009) 2225–2227.

[29] S. Shimotsu, S. Oikawa, T. Saitou, N. Mitsugi, K. Kubodera, T. Kawanishi, M. Izutsu, Single side-band modulation performance of a $LiNbO_3$ integrated modulator consisting of four-phase modulator waveguides, IEEE Photon. Technol. Lett. 13 (2001) 364–366.

[30] N. Kikuchi, Y. Shibata, K. Tsuzuki, H. Sanjoh, T. Sato, E. Yamada, T. Ishibashi, H. Yasaka, 80-Gb/s low-driving-voltage InP DQPSK modulator with an n-p-i-n structure, IEEE Photon. Technol. Lett. 21 (2009) 787–789.

[31] H. Yamazaki, T. Yamada, T. Goh, Y. Sakamaki, A. Kaneko, 64QAM modulator with a hybrid configuration of silica PLCs and $LiNbO_3$ phase modulators, IEEE Photon. Technol. Lett. 22 (2010) 344–346.

[32] R.L. Jungerman, C.A. Flory, Low-frequency acoustic anomalies in lithium niobate Mach-Zehnder interferometers, Appl. Phys. Lett. 53 (1988) 1477–1479.

[33] D.-S. Ly-Gagnon, S. Tsukamoto, K. Katoh, K. Kikuchi, Coherent detection of optical quadrature phase-shift keying signals with carrier phase estimation, J. Lightwave Technol. 24 (2006) 12–21.

[34] K. Kikuchi, T. Okoshi, M. Nagamatsu, N. Henmi, Degradation of bit-error rate in coherent optical communications due to spectral spread of the transmitter and the local oscillator, J. Lightwave Technol. LT-2 (1984) 1024–1033.

[35] K. Kasai, J. Hongo, M. Yoshida, M. Nakazawa, Optical phase-locked loop for coherent transmission over 500 km using heterodyne detection with fiber lasers, IEICE Electron. Express 4 (2007) 77–81.

[36] C.R. Doerr, L. Zhang, P.J. Winzer, N. Weimann, V. Houtsma, T.-C. Hu, N.J. Sauer, L.L. Buhl, D.T. Neilson, S. Chandrasekhar, Y.K. Chen, Monolithic InP dual-polarization and dual-quadrature coherent receiver, IEEE Photon. Technol. Lett. 23 (2011) 694–696.

[37] A. Antoniou, Digital Signal Processing: Signals, Systems, and Filters, McGraw-Hill Professional (2005).

[38] S. Haykin, B. Widrow, Least-Mean-Square Adaptive Filters, John Wiley & Sons, 2003.

[40] M.V. Clark, Adaptive frequency-domain equalisation and diversity combining for broadband wireless communications, IEEE J. Sel. Areas Commun. 16 (8) (1998) 1385–1395.

[41] K. Ishihara, T. Kobayashi, R. Kudo, Y. Takatori, A. Sano, E. Yamada, H. Masuda, Y. Miyamoto, Frequency-domain equalisation for optical transmission systems, Electron. Lett. 44 (2008) 870–871.

[42] J.P. Gordon, L.F. Mollenauer, Phase noise in photonic communication systems using linear amplifiers, Opt. Lett. 15 (1990) 1351–1353.

[43] A.D. Ellis, J. Zhao, D. Cotter, Approaching the non-linear Shannon limit, J. Lightwave Technol. 28 (4) (2010) 423–433.

[44] R.-J. Essiambre, G. Kramer, P.J. Winzer, G.J. Foschini, B. Goebel, Capacity limits of optical fiber networks, J. Lightwave Technol. 28 (4) (2010) 662–701.

[45] C. Paré, A. Villeneuve, P.-A. Bélanger, N.J. Doran, Compensating for dispersion and the nonlinear Kerr effect without phase conjugation, Opt. Lett. 21 (7) (1996) 459–461.

[46] X. Li, X. Chen, G. Goldfarb, E. Mateo, I. Kim, F. Yaman, G. Li, Electronic post-compensation of WDM transmission impairments using coherent detection and digital signal processing, Opt. Express 16 (2008) 880–888.

[47] S. Makovejs, D.S. Millar, V. Mikhailov, G. Gavioli, R.I. Killey, S.J. Savory, P. Bayvel, Experimental investigation of PDM-QAM16 transmission at 112 Gbit/s over 2400 km, in: Optical Fiber Communication Conference (OFC 2010), OMJ6.

[48] T. Kobayashi, A. Sano, A. Matsuura, E. Yamazaki, E. Yoshida, Y. Miyamoto, T. Nakagawa, Y. Sakamaki, T. Mizuno, 120-Gb/s PDM 64-QAM transmission over 1280 km using multi-staged nonlinear compensation in digital coherent receiver, in: Optical Fiber Communication Conference (OFC 2011), OThF6.

[49] P.K.A. Wai, C.R. Menyuk, H.H. Chen, Stability of solitons in randomly varying birefringent fibers, Opt. Lett. 16 (16) (1991) 1231–1233.

[50] A. Al-Bermani, C. Wördehoff, S. Hoffmann, K. Puntsri, T. Pfau, U. Rückert, R. Noé, Realtime 16-QAM transmission with coherent digital receiver, in: Opto Electronics and Communications Conference (OECC 2010), 7B4-2.

[51] T. Pfau, N. Kaneda, S. Corteselli, A. Leven, Y.-K. Chen, Real-time FPGA-based intradyne coherent receiver for 40 Gbit/s polarization-multiplexed 16-QAM, in: Optical Fiber Communication Conference (OFC 2011), OTuN4.

[52] M. Yoshida, T. Omiya, K. Kasai, M. Nakazawa, Real-time FPGA-based coherent optical receiver for 1 Gsymbol/s, 64 QAM transmission, in: Optical Fiber Communication Conference (OFC 2011), OTuN3.

[53] Y. Koizumi, K. Toyoda, M. Yoshida, M. Nakazawa, 1024 QAM (60 Gbit/s) single-carrier coherent optical transmission over 150 km, Opt. Express 20 (11) (2012) 12508–12514.

[54] K. Kasai, J. Hongo, H. Goto, M. Yoshida, M. Nakazawa, The use of a Nyquist filter for reducing an optical signal bandwidth in a coherent QAM optical transmission, IEICE Electron. Express 5 (1) (2008) 6–10.

[55] H. Nyquist, Certain topics in telegraph transmission theory, AIEE Trans. 47 (1928) 617–644.

[56] S.D. Personick, Receiver design for digital fiber optic communication systems, I, Bell Syst. Tech. J. 52 (1973) 843–874.

[57] N.S. Bergano, C.R. Davidson, Circulating loop transmission experiments for the study of long-haul transmission systems using erbium-doped fiber amplifiers, J. Lightwave Technol. 13 (1995) 879–888.

[58] S. Hara, R. Prasad, Multicarrier Techniques for 4G Mobile Communications, Artech House, Boston, 2003.

[59] Y. Ma, Q. Yang, Y. Tang, S. Chen, W. Shieh, 1-Tb/s per channel coherent optical OFDM transmission with subwavelength bandwidth access, in: Optical Fiber Communication Conference (OFC 2009), PDPC1.

[60] S. Chandrasekhar, X. Liu, B. Zhu, D. Peckham, Transmission of a 1.2-Tb/s 24-carrier no-guard-interval coherent OFDM superchannel over 7200-km of ultra-large-area fiber, in: European Conference on Optical Communication (ECOC 2009), PD2.6.

[61] D. Hillerkuss, R. Schmogrow, T. Schellinger, M. Jordan, M. Winter, G. Huber, T. Vallaitis, R. Bonk, P. Kleinow, F. Frey, M. Roeger, S. Koenig, A. Ludwig, A. Marculescu, J. Li, M. Hoh, M. Dreschmann, J. Meyer, S. Ben Ezra, N. Narkiss, B. Nebendahl, F. Parmigiani, P. Petropoulos, B. Resan, A. Oehler, K. Weingarten, T. Ellermeyer, J. Lutz, M. Moeller, M. Huebner, J. Becker, C. Koos, W. Freude, J. Leuthold, 26 Tbits-1 line-rate super-channel transmission utilizing all-optical fast Fourier transform processing, Nat. Photon. 5 (2011) 364–371.

[62] H. Takahashi, K. Takeshima, I. Morita, H. Tanaka, 400-Gbit/s optical OFDM transmission over 80 km in. 50-GHz frequency grid, in: European Conference on Optical Communication (ECOC 2010), Tu.3.C.1.

[63] M.-F. Huang, D. Qian, E. Ip, 50.53-Gb/s PDM-1024QAM-OFDM transmission using pilot-based phase noise mitigation, in: Opto Electronics and Communications Conference (OECC 2011), PDP1.

[64] X. Liu, S. Chandrasekhar, T.H. Lotz, P.J. Winzer, H. Haunstein, S. Randel, S. Corteselli, B. Zhu, D. Peckham, Generation and FEC-decoding of a 231.5-Gb/s PDM-OFDM signal with 256-iterative-polar-modulation achieving 11.15-b/s/Hz intrachannel spectral efficiency and 800-km reach, in: Optical Fiber Communication Conference (OFC 2012), PDP5B.3.

[65] T. Omiya, K. Toyoda, M. Yoshida, M. Nakazawa, 400 Gbit/s Frequency-division-multiplexed and polarization-multiplexed 256 QAM-OFDM transmission over 400 km with a spectral efficiency of 14 bit/s/Hz, in: Optical Fiber Communication Conference (OFC2012), OM2A.7, March 2012.

[66] C. Zhang, Y. Mori, M. Usui, K. Igarashi, K. Katoh, K. Kikuchi, Straight-line 1,073-km transmission of 640-Gbit/s dual-polarization QPSK signals on a single carrier, in:

Proceedings European Conference on Optical Communication (ECOC), Vienna, Austria, Postdeadline Paper PD2.8, 2009.

[67] T. Richter, E. Palushani, C. Schmidt-Langhorst, M. Nölle, R. Ludwig, J.K. Fischer, C. Schubert, Single wavelength channel 10.2 Tb/s TDM-data capacity using 16-QAM and coherent detection, in: Optical Fiber Comomunication Conference (OFC 2011), PDPA9.

[68] K. Kasai, T. Omiya, P. Guan, M. Yoshida, T. Hirooka, M. Nakazawa, Single-channel 400-Gb/s OTDM-32 RZ/QAM coherent transmission over 225 km using an optical phase-locked loop technique, IEEE Photon. Technol. Lett. 22 (8) (2010) 562–564.

[69] T. Sakamoto, T. Kawanishi, M. Izutsu, Asymptotic formalism for ultraflat optical frequency comb generation using a Mach-Zehnder modulator, Opt. Lett. 32 (2007) 1515–1517.

[70] M. Sugiyama, M. Doi, F. Futami, S. Watanabe, H. Onaka, A low drive voltage LiNbO$_3$ phase and intensity integrated modulator for optical frequency comb generation and short pulse generation, in: European Conference on Optical Communication (ECOC 2004), Tu3.4.3.

[71] K. Kasai, D.O. Otuya, M. Yoshida, T. Hirooka, M. Nakazawa, Single-carrier 800-Gb/s 32 RZ/QAM coherent transmission over 225 km employing a novel RZ-CW conversion technique, IEEE Photon. Technol. Lett. 24 (5) (2012) 416–418.

[72] M. Nakazawa, K. Kasai, M. Yoshida, T. Hirooka, Novel RZ-CW conversion scheme for ultra multi-level, high-speed coherent OTDM transmission, Opt. Express 19 26 (2011) B574–B580.

Multicarrier Optical Transmission

8

Xi Chen, Abdullah Al Amin, An Li, and William Shieh

Department of Electrical and Electronic Engineering, The University of Melbourne,
VIC 3010, Australia

In this chapter, we present an overview of multicarrier transmission and its application to optical communication, which has been in the focus of research for the last few years. Among all the multicarrier communication techniques, orthogonal frequency-division multiplexing (OFDM) is most well known and has been already adopted in many radio-frequency (RF) communication standards. With the advent of coherent detection technologies, optical multicarrier techniques, mainly optical OFDM has become an attractive candidate for high-speed optical transmission, especially at the emerging rates of 100 Gb/s to 1 Tb/s. In the following sections, we highlight some of the historical perspectives in the development of optical multicarrier technologies, followed by basic mathematical formulations for OFDM, the most popular multicarrier technique. We next present different variants of optical multicarrier transmission, including electronic and optical FFT-based realizations. We also highlight the problem of fiber nonlinearity in optical multicarrier transmission systems and present an analysis of fiber capacity under nonlinear impairments. Furthermore, we discuss applications of multicarrier techniques to long-haul systems, access networks, and free-space optical communication systems. Finally, we summarize with some possible research directions in implementing multicarrier technologies in optical transmission.

8.1 HISTORICAL PERSPECTIVE OF OPTICAL MULTICARRIER TRANSMISSION

The concept of multicarrier transmission is attractive as an effective way of increasing data transmission rate by using many parallel carriers each carrying relatively slow data rate. It is an old concept which began in the form of subcarrier multiplexing (SCM), but with subcarrier spacing at more than multiple of symbol rate. With the invention of orthogonal frequency division multiplexing (OFDM) in the 1960s [1], and subsequent efficient realization by discrete Fourier transforms (DFT) [2], it emerged as an effective modulation format to combat inter-symbol-interference (ISI) from multi-paths or other dispersive effects. Since early 1990s, OFDM and its variant, discrete multi-tone (DMT), have been widely deployed in a number of wireless and cable transmission standards.

Even though the wavelength-division multiplexing (WDM) schemes in optical fiber transmission can also be considered as a form of multicarrier transmission [3], the application of multicarrier modulation within each optical channel to mitigate dispersion and gain high spectral efficiency (SE) is a relatively new trend in the optical communications community. Before the advent of coherent detection, each WDM channel used a single carrier (SC) modulation with simple generation and detection methods, but going to higher data rates such as 40 Gb/s or 100 Gb/s became problematic due to inter-symbol-interference (ISI) from chromatic and polarization mode dispersion (CD/PMD), which required precise dispersion management. A multicarrier technique called subcarrier multiplexing has been proposed to achieve high date rates but with low spectral efficiency and receiver sensitivity [4]. With the arrival of full-field optical signal capture by coherent detection and subsequent digital signal processing (DSP), many choices of advanced modulation formats have been explored. While SC-based optical transmission was quickly developed and has now become a commercial product for 40/100 Gb/s line cards [5], the strengths of multicarrier techniques to overcome linear dispersive effects like CD/PMD and the ability to achieve high SE are also recognized in the optical communications community. The increased interest in multicarrier transmission is evidenced by explosive growth of the number of publications on this topic from 2007 onwards.

8.1.1 Variations of optical multicarrier transmission methods

While some reports focused on the simple implementation of optical OFDM by directly modulating the OFDM waveform on optical signal intensity as early as 1996 [6] and more recently [7–9], others explored the flexibility of complex (intensity and phase combined) OFDM modulation combined with coherent detection [10,11]. These two methods are commonly known as direct detection optical (DDO-) OFDM and coherent optical (CO-) OFDM, respectively [12]. Together, they form an important class of optical multicarrier transmission where the waveform is electronically generated and demodulated by fast Fourier transform (FFT) in the digital domain. Because the signal is generated and detected by electronic digital-to-analog or analog-to-digital converters (DAC/ADC), respectively, the OFDM signal bandwidth is limited by the sampling rate of DAC/ADC.

Another class of optical multicarrier transmission relies on all-optically generated or demodulated orthogonal subcarriers, for which the signal bandwidth becomes the product of DAC/ADC sampling rates times the number of optical carriers employed. In this way, a large data rate (beyond 10s of Tb/s per channel) has been shown to be achievable. Some variations of this method include all-optical OFDM [13,14], no-guard interval OFDM [15,16], and coherent WDM [17].

In terms of their applications, optical multicarrier transmission has been demonstrated to be feasible for both short-reach and long-reach applications. Typically the short-reach, access network applications require a very cost-effective implementation rather than high SE [18,19]. On the other hand, the high-capacity long-haul applications can pay a premium for SE and reach [20,21] (typically over 1000 km). There has also been some study on free-space optical communication by multicarrier techniques [22–24].

8.1.2 Research trends in optical multicarrier transmission

In order to implement very high SE and capacity, a number of multiplexing methods have been proposed. The most straightforward method is to multiplex the multicarrier signal on two orthogonal polarizations of the single-mode fiber (SMF), which enables the doubling of data rate via a channel equalization method known as multiple input, multiple output (MIMO) [25]. MIMO is very well suited for multicarrier methods such as OFDM, where the linear interference from neighboring transmitters on the same frequency is reversed by a simple matrix multiplication [26]. A further extension of MIMO is the use of space-division multiplexing (SDM), whereby the signal travels from N transmitters to N receivers via an optical fiber link that can support N modes (spatial and polarization) [27–32]. Recently, this type of novel fiber has attracted much attention, and optical OFDM transmission is shown to be a very effective modulation format for such SDM systems.

Even though SC coherent optical systems are becoming commercially available in the data rates up to 100 Gb/s in the last few years, no such products have so far been developed for optical OFDM, and as such a question is asked about the future of this method in comparison to SC [33]. With the coherent detection and DSP methods, linear impairments have been shown to be mitigated along with the phase noise [12], which can affect OFDM because of its longer symbol lengths [34]. But another detrimental effect is the fiber nonlinearity which is exacerbated in multicarrier systems due to high peak-to-average ratio (PAPR). Similar to SC systems, such nonlinear effects could be compensated in multicarrier cases, and it has even been shown that for very high data rates, multicarrier systems could outperform SC ones [35].

As the demand is growing for further higher speeds (400 Gb/s or 1 Tb/s) amid lack of such high-bandwidth electronics, it has been accepted in the community that adoption of some form of multicarrier techniques is indispensable for these data rates. This multicarrier format for parallel transmission within a channel is called "superchannel" [36], which may also lead to grid-less or flexible-grid optical networks in the future [37,38]. In this way, multicarrier optical transmission methods continue to attract research attention for their versatility and flexibility to realize software-reconfigurable optical links and optical networks in the coming era.

8.2 OFDM BASICS

One of the central features that set OFDM apart from SC modulation is its uniqueness of signal processing. SC technique has been employed in optical communication systems for the last three decades. As a result, OFDM signal processing may seem unfamiliar to an optical engineer at first glance. However, OFDM technology provides an exceptionally scalable pathway for migration to higher data rates. Once the algorithms and hardware designs are developed for the current generation product, it is very likely that these skill sets can be incorporated for the next generation product. In this respect, OFDM is a future-proof technology, and subsequently various aspects of OFDM signal processing deserve careful perusal.

For conventional optical SC systems, as the transmission speed increases, the requirement for optimal timing sampling precision becomes critical. Excessive timing jitter would place the sampling point away from the optimal, incurring severe penalty. On the other hand, for optical OFDM systems, a precise time sampling is not necessary. As long as an appropriate "window" of sampling points is selected containing an uncontaminated OFDM symbol, it is sufficient to remove inter-symbol-interference (ISI). However, this tolerance to sampling point imprecision is traded off against the stringent requirement of frequency offset and phase noise in OFDM systems.

In this section, we will lay out various aspects of OFDM signal processing associated with (i) OFDM basics and mathematical aspects of OFDM, (ii) DFT implementation, and (iii) cyclic prefix for OFDM. Following the description of signal processing, we will discuss the spectral efficiency of optical OFDM.

8.2.1 Mathematical formulation of an OFDM signal

In a generic OFDM system [1], any signal $s(t)$ can be represented as

$$s(t) = \sum_{i=-\infty}^{+\infty} \sum_{k=1}^{N_{sc}} c_{ki} s_k(t - iT_s), \tag{8.1}$$

$$s_k(t) = \Pi(t) e^{j2\pi f_k t}, \tag{8.2}$$

$$\Pi(t) = \begin{cases} 1, & (0 < t \leqslant T_s), \\ 0, & (t \leqslant 0, t > T_s), \end{cases} \tag{8.3}$$

where c_{ki} is the ith information symbol at the kth subcarrier, s_k is the waveform for the kth subcarrier, N_{sc} is the number of subcarriers, f_k is the frequency of the subcarrier, T_s is the symbol period, and $\Pi(t)$ is the pulse-shaping function. The optimum detector for each subcarrier could use a filter that matches the subcarrier waveform, or a correlator matched to the subcarrier. Therefore, the detected information symbol c'_{ik} at the output of the correlator is given by

$$c'_{ki} = \frac{1}{T_s} \int_0^{T_s} r(t - iT_s) s_k^* \, dt = \frac{1}{T_s} \int_0^{T_s} r(t - iT_s) e^{-j2\pi f_k t} \, dt, \tag{8.4}$$

where $r(t)$ is the received time-domain signal. Typical multicarrier modulation uses non-overlapped band-limited signals and can be implemented with a bank of large number of oscillators and filters at both transmit and receive end [39]. The major disadvantage of this implementation is that it requires excessive bandwidth.

This is because in order to design the filters and oscillators cost-effectively, the channel spacing has to be multiple of the symbol rate, greatly reducing the spectral efficiency. OFDM was investigated as a novel approach employing spectrally overlapped yet orthogonal signal set [1]. This orthogonality originates from straightforward correlation between any two subcarriers, given by

$$\delta_{kl} = \frac{1}{T_s}\int_0^{T_s} s_k s_l^* \, dt = \frac{1}{T_s}\int_0^{T_s} \exp(j2\pi(f_k - f_l)t)dt$$

$$= \exp(j\pi(f_k - f_l)T_s)\frac{\sin(\pi(f_k - f_l)T_s)}{\pi(f_k - f_l)T_s}, \qquad (8.5)$$

It can be seen that if the following condition

$$f_k - f_l = m\frac{1}{T_s} \qquad (8.6)$$

is satisfied, then the two subcarriers are orthogonal to each other. This signifies that these orthogonal subcarrier sets, with their frequencies spaced at multiple of inverse of the symbol rate, can be recovered with the matched filters in (8.4) without inter-carrier-interference (ICI), in spite of strong signal spectral overlapping.

8.2.2 Discrete Fourier transform implementation of OFDM

A fundamental challenge with OFDM is that a large number of subcarriers are needed so that the transmission channel affects each subcarrier as a flat channel. This leads to an extremely complex architecture involving many oscillators and filters at both transmit and receive end. Weinstein and Ebert first revealed that OFDM modulation/demodulation can be implemented by using inverse discrete Fourier transform (IDFT)/discrete Fourier transform (DFT) [2]. This is evident by studying OFDM modulation (8.2) and OFDM demodulation (8.4). Let us temporarily omit the index "i," re-denote N_{sc} as N in (8.2) to focus our attention on one OFDM symbol, and assume that we sample $s(t)$ at every interval of T_s/N, and the mth sample of $s(t)$ from the expression (8.2) becomes

$$s_m = \sum_{k=1}^N c_k \cdot e^{j2\pi f_k \cdot \frac{(m-1)T_s}{N}}. \qquad (8.7)$$

Using orthogonality condition of (8.6), and the convention that

$$f_k = \frac{k-1}{T_s}, \qquad (8.8)$$

and substituting (8.8) into (8.7), we have

$$s_m = \sum_{k=1}^N c_k \cdot e^{j2\pi f_k \cdot \frac{(m-1)T_s}{N}} = \sum_{k=1}^N c_k \cdot e^{j2\pi \frac{(K-1)(m-1)}{N}} = F^{-1}\{c_k\}, \qquad (8.9)$$

where $\boldsymbol{F}$ stands for Fourier transform and $m \in [1, N]$. In a similar fashion, at the receive end, we arrive at

$$C'_k = \boldsymbol{F}\{r_m\}, \qquad (8.10)$$

where r_m is the received signal sampled at every interval of T_s/N. From (8.9) and (8.10) it follows that the discrete value of the transmitted OFDM signal $s(t)$ is merely a simple N-point IDFT of the information symbol c_k, and received information symbol c'_k is a simple N-point DFT of the receive sampled signal. It is worth noting that there are two critical devices we have assumed for the DFT/IDFT implementation which are: (i) DAC, needed to convert the discrete value of s_m to the continuous analog value of $s(t)$ and (ii) ADC, needed to convert the continuous received signal $r(t)$ to discrete sample r_m. There are two fundamental advantages of DFT/IDFT implementation of OFDM. First, because of existence of efficient IFFT/FFT algorithm, the number of complex multiplications for IFFT in (8.9) and FFT in (8.10) is reduced from N^2 to $\frac{N}{2}\log_2(N)$, increasing almost linearly with the number of subcarriers, N [40]. Second, a large number of orthogonal subcarriers can be generated and demodulated without resorting to much more complex RF oscillator and filter banks. This leads to a relatively simple architecture for OFDM implementation when large number of subcarriers are required. The corresponding architecture using DFT/IDFT and DAC/ADC is shown in Figure 8.1. At the transmit end, the input serial data bits are first converted into many parallel data pipes, each mapped onto corresponding information symbols for the subcarriers within one OFDM symbol, and the

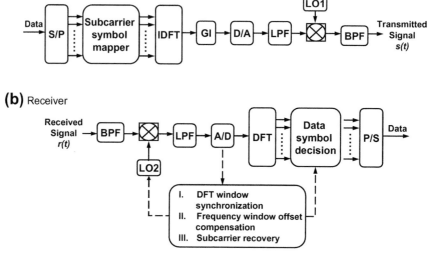

(a) Transmitter

(b) Receiver

S/P: Serial-to-parallel   GI: Guard Time Insertion D/A: Digital-to-Analog   (I) DFT: (Inverse) Discrete Fourier Transform LPF: Low Pass Filter BPF: Band Pass Filter

FIGURE 8.1 Conceptual diagram for (a) OFDM transmitter and (b) OFDM receiver.

digital time domain signal is obtained by using IDFT, which is subsequently inserted with guard interval and converted into real-time waveform through DAC. The guard interval is inserted to prevent ISI due to channel dispersion. The baseband signal can be up-converted to an appropriate RF passband with an IQ mixer/modulator. At the receive end, the OFDM signal is first down-converted to baseband with an IQ demodulator, and sampled with an ADC, and demodulated by performing DFT and baseband signal processing to recover the data.

It is worth noting that from (8.7), the OFDM signal s_m is a periodical function of f_k with a period of N/T_s. Therefore, any discrete subcarrier set with its frequency components spanning one period of N/T_s is equivalent. Namely, in Eqs. (8.7) and (8.8), the subcarrier frequency f_k and its index k can be generalized as

$$f_k = \frac{k-1}{T_s}, \quad k\epsilon \, [k_{min} + 1, k_{min} + N], \qquad (8.11)$$

where k_{min} is an arbitrary integer. However, only two subcarrier index conventions are widely used, which are $k \in [1, N]$ and $k \in [-N/2 + 1, N/2]$.

8.2.3 Cyclic prefix for OFDM

One of the enabling techniques for OFDM is the insertion of cyclic prefix [26,41]. Let us first consider two consecutive OFDM symbols that undergo a dispersive channel with a delay spread of t_d. For simplicity, assume each OFDM symbol includes only two subcarriers with the fast delay and slow delay spread at t_d, represented by "fast subcarrier" and "slow subcarrier," respectively. Figure 8.2a shows that inside each OFDM symbol, the two subcarriers, "fast subcarrier" and "slow subcarrier," are aligned upon the transmission. Figure 8.2b shows the same OFDM signals upon the reception where the "slow subcarrier" is delayed by t_d against the "fast subcarrier." We select a DFT window containing a complete OFDM symbol for the "fast subcarrier." It is apparent that due to the channel dispersion, the "slow subcarrier" has crossed the symbol boundary leading to the interference between neighboring OFDM symbols, which is known as inter-symbol-interference (ISI). Furthermore, because the OFDM waveform in the DFT window for "slow subcarrier" is incomplete, the critical orthogonality condition for the subcarriers (8.5) is lost, resulting in an inter-carrier-interference (ICI) penalty.

Cyclic prefix was proposed to resolve the channel dispersion induced ISI and ICI [26]. Figure 8.2c shows insertion of a cyclic prefix by cyclic extension of the OFDM waveform into the guard interval, Δ_G. As shown in Figure 8.2c, the waveform in the guard interval is essentially an identical copy of that in the DFT window, with time-shifted by "t_s" forward. Figure 8.2d shows the OFDM signal with the guard interval upon reception. Let us assume that the signal has traversed the same dispersive channel, and the same DFT window is selected containing a complete OFDM symbol for the "fast subcarrier" waveform. It can be seen from Figure 8.2d, a complete OFDM symbol for "slow subcarrier" is also maintained in the DFT window, because a proportion of the cyclic prefix has moved into the DFT window to replace the

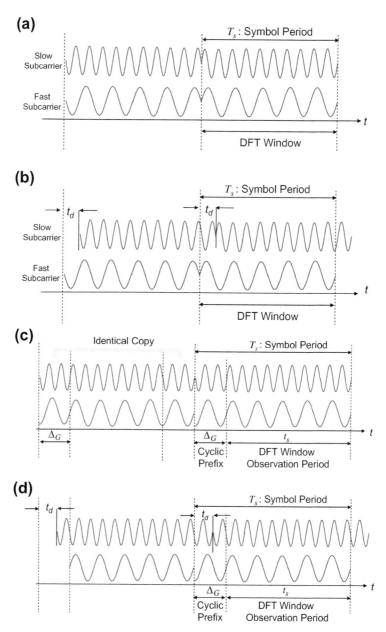

FIGURE 8.2 OFDM signals (a) without cyclic prefix at the transmitter, (b) without cyclic prefix at the receiver, (c) with cyclic prefix at the transmitter, and (d) with cyclic prefix at the receiver.

identical part that has shifted out. As such, the OFDM symbol for "slow subcarrier" is a near-identical copy of the transmitted waveform with an additional phase shift. This phase shift is dealt with through channel estimation and will be subsequently removed before symbol decision. Now we arrive at the important condition for ISI-free OFDM transmission, given by

$$t_d < \Delta_G. \tag{8.12}$$

It can be seen that to recover the OFDM information symbol properly, there are two critical procedures that need to be carried out: (i) selection of an appropriate DFT window, called DFT window synchronization and (ii) estimation of the phase shift for each subcarrier, called channel estimation or subcarrier recovery. Both signal processing procedures are actively pursued research topics, and their references can be found in these two books [26,41].

An elegant way to describe the cyclic prefix is to maintain the same expression as (8.2) for the transmitted signal $s(t)$, but to extend the pulse shape function (8.3) to the guard interval, given by

$$\Pi(t) = \begin{cases} 1, & (-\Delta_G < t \leqslant t_s) \\ 0, & (t \leqslant -\Delta_G, t > t_s). \end{cases} \tag{8.13}$$

The corresponding time-domain OFDM symbol is illustrated in Figure 8.3 that shows one complete OFDM symbol comprised of observation period and cyclic prefix. The waveform within the observation period will be used to recover the frequency-domain information symbols.

8.2.4 Spectral efficiency for optical OFDM

In direct-detection optical OFDM (DDO-OFDM) systems, the optical spectrum is usually not a linear replica of the RF spectrum. Therefore, the optical spectral efficiency is dependent on the specific implementation method. We will turn our attention to the optical spectral efficiency for coherent optical OFDM (CO-OFDM)

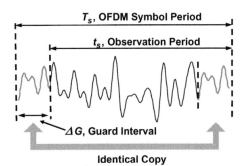

FIGURE 8.3 The time-domain OFDM signal for one complete OFDM symbol.

systems. In CO-OFDM systems, N_{sc} subcarriers are transmitted in every OFDM symbol period of T_s. Thus the total symbol rate R for CO-OFDM systems is given by

$$R = N_{sc}/T_s. \tag{8.14}$$

Figure 8.4a shows the spectrum of wavelength-division-multiplexed (WDM) channels each with CO-OFDM modulation, and Figure 8.4b shows the zoomed-in optical spectrum for each wavelength channel. We use the bandwidth of the first null to denote the boundary of each wavelength channel. The OFDM bandwidth, B_{OFDM}, is thus given by

$$B_{OFDM} = \frac{2}{T_s} + \frac{N_{sc} - 1}{T_s}, \tag{8.15}$$

where t_s is the observation period (Figure 8.3). Assuming a large number of subcarriers are used, the bandwidth efficiency of OFDM η is found to be

$$\eta = 2\frac{R}{B_{OFDM}} = 2\alpha, \quad \alpha = \frac{t_s}{T_s}. \tag{8.16}$$

The factor of 2 accounts for two polarizations in the fiber. Using a typical value of 8/9 for α, we obtain the optical spectral efficiency factor η of 1.8 Baud/Hz. The optical spectral efficiency becomes 3.6 b/s/Hz if QPSK modulation is used for each subcarrier. The spectral efficiency can be further improved by using higher-order QAM modulation [42,43]. To practically implement CO-OFDM systems, the optical spectral efficiency will be reduced because of the need for a sufficient guardband between WDM channels taking account of laser frequency drifts (about 2 GHz). This

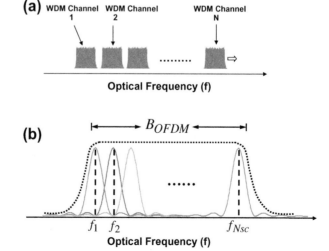

FIGURE 8.4 Optical spectra for (a) *N* wavelength-division-multiplexed CO-OFDM channels and (b) zoomed-in OFDM signal for one wavelength.

guardband can be avoided by ensuring orthogonality across the WDM channels by using frequency locked sources [44].

8.3 OPTICAL MULTICARRIER SYSTEMS BASED ON ELECTRONIC FFT

One of the major strengths of OFDM modulation format is its rich variation and ease of adaptation to a wide range of applications. This rich variation stems from the intrinsic advantages of OFDM modulation including dispersion robustness, ease of dynamic channel estimation and mitigation, high spectral efficiency, and capability of dynamic bit and power loading. Recent progress in optical OFDM is no exception. Despite the fact that OFDM has been extensively studied in the RF domain, it is rather surprising that the first report on optical OFDM in the open literature only appeared in 1996 by Pan et al. [6] where they presented in-depth performance analysis of hybrid AM/OFDM subcarrier-multiplexed (SCM) fiber-optic systems. The lack of interest in optical OFDM in the past is largely due to the fact that the digital signal processing power of CMOS integrated circuit (IC) had not reached the point where sophisticated OFDM signal processing could be performed economically.

In this section, we discuss optical multicarrier systems that rely on electronic FFT to process each subcarrier as opposed to using optical FFT. Based on the detection methods, we further classify such electronic FFT based multicarrier systems into two main categories, CO-OFDM and DDO-OFDM. In the remainder of the section, we will describe the fundamentals of these two optical multicarrier systems.

8.3.1 Coherent optical OFDM

CO-OFDM represents the ultimate performance in receiver sensitivity, spectral efficiency, and robustness against polarization dispersion, but requires high complexity in transceiver design. In the open literature, CO-OFDM was first proposed by Shieh and Athaudage [10], and the concept of the coherent optical MIMO-OFDM was formalized by Shieh et al. in [45]. The early CO-OFDM experiments were carried out by Shieh et al. for a 1000 km SSMF transmission at 8 Gb/s [46], and by Jansen et al. for 4160 km SSMF transmission at 20 Gb/s [47]. The principle and transmitter/receiver design for CO-OFDM are given below.

8.3.1.1 Principle of coherent optical OFDM

The synergies between coherent optical communications and OFDM are twofold. OFDM enables channel and phase estimation for coherent detection in a computationally efficient way. Coherent detection provides linearity in RF-to-optical (RTO) up-conversion and optical-to-RF (OTR) down-conversion, much needed for OFDM. Consequently, CO-OFDM is a natural choice for optical transmission in the linear regime. A generic CO-OFDM system is depicted in Figure 8.5. In general, a CO-OFDM system can be divided into five functional blocks including (i) RF OFDM

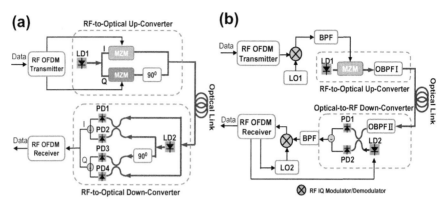

FIGURE 8.5 A CO-OFDM system in (a) direct up-down-conversion architecture and (b) intermediate frequency (IF) architecture.

transmitter, (ii) RTO up-converter, (iii) the optical channel, (iv) the OTR down-converter, and (v) the RF OFDM receiver. The detailed architecture for RF OFDM transmitter/receiver has already been shown in Figure 8.1, which generates/recovers OFDM signals either in baseband or an RF band. Let us assume for now a linear channel where optical fiber nonlinearity is not considered. It is apparent that the challenges for CO-OFDM implementation are to obtain a linear RTO up-converter and linear OTR down-converter. It has been proposed and analyzed that by biasing the Mach-Zehnder modulators (MZMs) at null point, a linear conversion between the RF signal and optical field signal can be achieved [10,48]. It has also been shown that by using coherent detection, a linear transformation from optical field signal to RF (or baseband electrical) signal can be achieved [10,48–50]. Now by putting together such a composite system cross RF and optical domain [10,46,47], a linear channel can be constructed where OFDM can perform its best role of mitigating channel dispersion impairment in both RF domain and optical domain. In this section, we use the term "RF domain" and "electrical domain" interchangeably.

8.3.1.2 Coherent detection for linear down-conversion and noise suppression

As shown in Figure 8.6, coherent detection uses a six-port 90° optical hybrid and a pair of balanced photo-detectors. The main purposes of coherent detection are: (i) to linearly recover the I and Q components of the incoming signal and (ii) to suppress or cancel the common mode noise. Using a six-port 90° hybrid for signal detection and analysis has been practiced in RF domain for decades [51, 52], and its application to single-carrier coherent optical systems can be also found in [49, 50]. In what follows, in order to illustrate its working principle, we will perform an analysis of down-conversion via coherent detection assuming ideal conditions for each component shown in Figure 8.6.

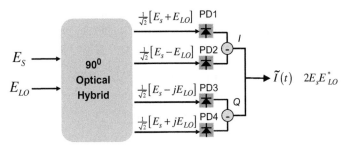

E_s: Incoming Signal E_{LO}: Local Oscillator Signal

PD: Photo-detector $\tilde{I}(t)$: Complex photocurrent

FIGURE 8.6 Coherent detection using an optical hybrid and balanced photo-detection.

The purpose of the four output ports of the 90° optical hybrid is to generate a 90° phase shift between I and Q components, and 180° phase shift between balanced detectors. Ignoring imbalance and loss of the optical hybrid, the output signals E_{1-4} can be expressed as

$$E_1 = \frac{1}{\sqrt{2}}[E_s + E_{LO}], \quad E_2 = \frac{1}{\sqrt{2}}[E_s - E_{LO}],$$

$$E_3 = \frac{1}{\sqrt{2}}[E_s - jE_{LO}], \quad E_4 = \frac{1}{\sqrt{2}}[E_s + jE_{LO}], \tag{8.17}$$

where E_s and E_{LO} are, respectively, the incoming signal and local oscillator (LO) signal. We further decompose the incoming signal into two components: (i) the received signal when there is no amplified spontaneous noise (ASE), $E_r(t)$ and (ii) the ASE noise, $n_o(t)$, namely

$$E_s = E_r + n_o. \tag{8.18}$$

We first study how the I component of the photo-detected current is generated, and the Q component can be derived accordingly. The I component is obtained by using a pair of photo-detectors, PD1 and PD2, in Figure 8.6, whose photocurrent I_{1-2} can be described as

$$I_1 = |E_1|^2 = \frac{1}{2}\{|E_s|^2 + |E_{LO}|^2 + 2\mathrm{Re}\{E_s E_{LO}^*\}\}, \tag{8.19}$$

$$I_2 = |E_2|^2 = \frac{1}{2}\{|E_s|^2 + |E_{LO}|^2 - 2\mathrm{Re}\{E_s E_{LO}^*\}\}, \tag{8.20}$$

$$|E_s|^2 = |E_r|^2 + |n_o|^2 + 2\mathrm{Re}\{E_r n_o^*\}, \tag{8.21}$$

$$|E_{LO}|^2 = I_{LO}(1 + I_{RIN}(t)), \tag{8.22}$$

where I_{LO} and $I_{RIN}(t)$ are the average power and relative intensity noise (RIN) of the LO laser, and "Re" or "Im" denotes the real or imaginary part of a complex signal. For simplicity, the photo-detection responsivity is set to unity. The three terms at the right-hand side of (8.21) represent signal-to-signal beat noise, signal-to-ASE beat noise, and ASE-to-ASE beat noise. Because of the balanced detection, using Eqs. (8.19) and (8.20), the I component of the photocurrent becomes

$$I_I(t) = I_1 - I_2 = 2\mathrm{Re}\{E_s E_{LO}^*\}. \tag{8.23}$$

Now the noise suppression mechanism becomes quite clear because the three noise terms in (8.21) and the RIN noise in (8.22) from a single detector are completely canceled via balanced detection. Nevertheless, it has been shown that coherent detection can be performed by using a single photo-detector, but at the cost of reduced dynamic range [53].

In a similar fashion, the Q component from the other pair of balanced detectors can be derived as

$$I_Q(t) = I_3 - I_4 = 2\mathrm{Im}\{E_s E_{LO}^*\}. \tag{8.24}$$

Using the results of (8.23) and (8.24), the complex detected signal $I(t)$ consisting of both I and Q components becomes

$$I(t) = I_I(t) + jI_Q(t) = 2E_s E_{LO}^*. \tag{8.25}$$

From (8.25), the linear down-conversion process via coherent detection becomes quite clear; the complex photocurrent $I(t)$ is in essence a linear replica of the incoming complex signal that is frequency down-converted by a local oscillator frequency. Thus with linear coherent detection at receiver and linear generation at transmitter, complex OFDM signals can be readily transmitted over the optical fiber channel.

8.3.2 Direct-detection optical OFDM

A direct-detection optical OFDM (DDO-OFDM) aims for simpler transmitter/receiver than CO-OFDM for lower costs. It has many variants which reflect the different requirements in terms of data rates and costs from a broad range of applications. For instance, the first report of the DDO-OFDM [6] takes advantage of the fact that the OFDM signal is more immune to the impulse clipping noise seen in CATV networks. Another example is single-side-band (SSB)-OFDM which has been recently proposed by Lowery et al. and Djordjevic et al. for long-haul transmission [7,9]. Tang et al. have proposed an adaptively modulated optical OFDM (AMOOFDM) that uses bit and power loading showing promising results for both multimode fiber and short-reach SMF fiber links [54–56]. The common feature for DDO-OFDM is use of a simple square-law photodiode at the receiver. DDO-OFDM can be divided into two categories according to how optical OFDM signal is being generated: (i) linearly mapped DDO-OFDM (LM-DDO-OFDM) where the optical OFDM spectrum is a replica of baseband OFDM, and (ii) nonlinearly mapped DDO-OFDM

(NLM-DDO-OFDM) where the optical OFDM spectrum does not display a replica of baseband OFDM. In what follows, we discuss the principles and design choices for these two classes of direct-detection OFDM systems.

8.3.2.1 Linearly mapped DDO-OFDM

As shown in Figure 8.7, the optical spectrum of an LM-DDO-OFDM signal at the output of the O-OFDM transmitter is a linear copy of the RF OFDM spectrum plus an optical carrier that is usually 50% of the overall power. The position of the main optical carrier can be one OFDM spectrum bandwidth away [7,57] or right at the end of the OFDM spectrum [58,59]. Formally, such type of DDO-OFDM can be described as

$$s(t) = e^{j2\pi f_0 t} + \alpha e^{j2\pi(f_0 + \Delta f)t} \cdot s_B(t), \tag{8.26}$$

where $s(t)$ is the optical OFDM signal, f_0 is the main optical carrier frequency, Δf is guardband between the main optical carrier and the OFDM band (Figure 8.7), and α is the scaling coefficient that describes the OFDM band strength related to the main carrier. $s_B(t)$ is the baseband OFDM signal given by

$$s_B = \sum_{k=-\frac{1}{2}N_{sc}+1}^{\frac{1}{2}N_{sc}} c_k e^{j2\pi f_k t}, \tag{8.27}$$

where c_k and f_k are, respectively, the OFDM information symbol and the frequency for the kth subcarrier. For explanatory simplicity, only one OFDM symbol is shown in (8.27). After the signal passes through fiber link with chromatic dispersion, the OFDM signal can be approximated as

$$r(t) = e^{j(2\pi f_0 t + \Phi_D(-\Delta f) + \phi(t))} + \alpha e^{j(2\pi(f_0 + \Delta f)t + \phi(t))} \cdot \sum_{k=-\frac{1}{2}N_{sc}+1}^{\frac{1}{2}N_{sc}} c_{ik} e^{(j2\pi f_k t + \Phi_D(f_k))},$$

$$\tag{8.28}$$

$$\Phi_D(f_k) = \pi \cdot c \cdot D_t \cdot f_k^2 / f_O^2, \tag{8.29}$$

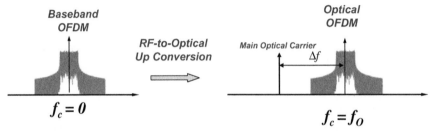

FIGURE 8.7 Illustration of linearly mapped DDO-OFDM (LM-DDO-OFDM) where the optical OFDM spectrum is a replica of the baseband OFDM spectrum.

where $\Phi_D(f_k)$ is the phase delay due to chromatic dispersion for the kth subcarrier. D_t is the accumulated chromatic dispersion in units of ps/pm, f_o is the center frequency of optical OFDM spectrum, and c is the speed of light in a vacuum. At the receiver, the photodetector can be modeled as a square-law detector and the resultant photo-current signal is

$$
I(t) \propto |r(t)|^2 = 1 + 2\alpha \text{Re} \left\{ e^{j2\pi\Delta ft} \sum_{k=-\frac{1}{2}N_{sc}+1}^{\frac{1}{2}N_{sc}} c_{ik} e^{(j2\pi f_k t + \Phi_D(f_k) - \Phi_D(-\Delta f))} \right\}
$$

$$
+ |\alpha^2| \sum_{k_1=-\frac{1}{2}N_{sc}+1}^{\frac{1}{2}N_{sc}} \sum_{k_2=-\frac{1}{2}N_{sc}+1}^{\frac{1}{2}N_{sc}} c_{k_2}^* c_{k_1} e^{(j2\pi(f_{k_1}-f_{k_2})t + \Phi_D(f_{k_1}) - \Phi_D(f_{k_2}))}.
$$

$$(8.30)$$

The first term is a DC component that can be easily filtered out. The second term is the fundamental term consisting of linear OFDM subcarriers that are to be retrieved. The third term is the second-order nonlinearity term that needs to be removed.

There are several approaches to minimize the penalty due to the second-order nonlinearity term:

a. *Offset SSB-OFDM:* Sufficient guardband is allocated such that the second-term and third-term RF spectra are non-overlapping. As such, the third term in Eq. (8.30) can be easily removed using a RF or DSP filter, as proposed by Lowery et al. in [7].

b. *Baseband optical SSB OFDM:* α coefficient is reduced as much as possible such that the distortion as a result of the third-term is reduced to an acceptable level. This approach has been adopted by Djordjevic et al. [9] and Hewitt et al. [58].

c. *Subcarrier interleaving:* From Eq. (8.30), it follows that if only odd subcarriers are filled, i.e. c_k is nonzero only for the odd subcarriers, the second-order intermodulation will be at even subcarriers, which are orthogonal to the original signal at the odd subcarrier frequencies. Subsequently, the third-term does not produce any interference. This approach has been proposed by Peng et al. [60].

d. *Iterative distortion reduction:* The basic idea is to go through a number of iterations of estimation of the linear term, and compute the second-order term using the estimated linear term, and removing the second-order term from the right hand side of Eq. (8.30). This approach has been proposed by Peng et al. [59].

There are advantages and disadvantages among all these four approaches. For instance, Approach B has the advantage of better spectral efficiency, but at the cost of sacrificing receiver sensitivity. Approach D has both good spectral efficiency and receiver sensitivity, but has a burden of computational complexity.

Figure 8.8 shows one offset SSB-OFDM proposed by Lowery et al. in [61]. They show that such DDO-OFDM can mitigate an enormous amount of chromatic dispersion up to 5000 km standard SMF (SSMF) fiber. The proof-of-concept experiment was demonstrated by Schmidt et al. from the same group for 400 km DDO-OFDM transmission at 20 Gb/s [57]. The simulated system is 10 Gb/s with 4-QAM modulation with a

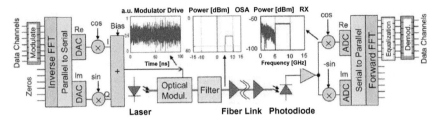

FIGURE 8.8 Direct-detection optical OFDM (DDO-OFDM) long-haul optical communication systems. After Ref. [61].

bandwidth of around 5 GHz [61]. In the electrical OFDM transmitter, the OFDM signal is up-converted to an RF carrier at 7.5 GHz generating an OFDM band spanning from 5–10 GHz. The RF OFDM signal is fed into an optical modulator. The output optical spectrum has the two side OFDM bands that are symmetric across the main optical subcarrier. An optical filter is then used to filter out one OFDM side band. This single-side band (SSB) is critical to ensure there is one-to-one mapping between the RF OFDM signal and the optical OFDM signal. The power of main optical carrier is optimized to maximize the sensitivity. At the receiver, only one photo-detector is used. The RF spectrum of the photocurrent is depicted as an inset in Figure 8.8. It can be seen that the second-order intermodulation, the third-term in Eq. (8.30), is from DC to 5 GHz, whereas the OFDM spectrum, the second term in Eq. (8.30), spans from 5 GHz to 10 GHz. As such, the RF spectrum of the intermodulation does not overlap with the OFDM signal, signifying that the intermodulation does not cause detrimental effects after proper electrical filtering.

8.3.2.2 Nonlinearly mapped DDO-OFDM (NLM-DDO-OFDM)

The second class of DDO-OFDM is nonlinearly mapped OFDM, which means that there is no linear mapping between the electric field (baseband OFDM) and the optical field. Instead, NLP-DD-OFDM aims to obtain a linear mapping between baseband OFDM and optical intensity. For simplicity, we assume generation of NLM-DDO-OFDM using direct modulation of a DFB laser, the waveform after the direct modulation can be expressed as [62]

$$E(t) = e^{j2\pi f_o t} A(t)^{1+jC}, \tag{8.31}$$

$$A(t) \equiv \sqrt{P(t)} = A_0 \sqrt{1 + \alpha \text{Re}(e^{j(2\pi f_{\text{IF}} t)} \cdot s_B(t))}, \tag{8.32}$$

$$s_B(t) = \sum_{k=-\frac{1}{2}N_{\text{sc}}+1}^{\frac{1}{2}N_{\text{sc}}} c_k e^{j2\pi f_k t}, \tag{8.33}$$

$$m \equiv \alpha \sqrt{\sum_{k=-\frac{1}{2}N_{\text{sc}}+1}^{\frac{1}{2}N_{\text{sc}}} |c_k|^2}, \tag{8.34}$$

where $E(t)$ is the optical OFDM signal, $A(t)$ and $P(t)$ are the instantaneous amplitude and power of the optical OFDM signal, c_k is the transmitted information
symbol for the kth subcarrier, C is the chirp constant for the direct modulated DFB
laser [62], f_{IF} is the IF frequency for the electrical OFDM signal for modulation,
m is the optical modulation index, α is a scaling constant to set an appropriate
modulation index m to minimize the clipping noise, and $s_B(t)$ is the baseband
OFDM signal. Assuming that the chromatic dispersion is negligible, the detected
current is

$$I(t) = |E(t)|^2 = |A|^2 = A_0(1 + \alpha\mathrm{Re}(e^{j(2\pi f_{IF}t)} \cdot s_B(t))). \qquad (8.35)$$

Equation (8.35) shows that the photocurrent contains a perfect replica of the OFDM
signal $s_B(t)$ with a DC current. We also assume that modulation index m is small
enough that clipping effect is not significant. Equation (8.35) shows that by using
NLM-DDO-OFDM with no chromatic dispersion, the OFDM signal can be perfectly
recovered. The fundamental difference between the NLM- and LM-DDO-OFDM
can be gained by studying their respective optical spectra. Figure 8.9 shows the optical spectra of NLM-DDO-OFDM using (a) direct modulation of a DFB laser with
the chirp coefficient C of 1 in (8.31) and modulation index m of 0.3 in (8.34) and (b)
offset SSB-OFDM. It can be seen that, in sharp contrast to SSB-OFDM, NLM-DDO-
OFDM has a multiple of OFDM bands with significant spectral distortion. Therefore, there is no linear mapping from the baseband OFDM to the optical OFDM.
The consequence of this nonlinear mapping is fundamental, because when any type
of the dispersion, such as chromatic dispersion, polarization dispersion, or modal
dispersion, occurs in the link, the detected photocurrent can no longer recover the
linear baseband OFDM signal. Namely, any dispersion will cause the nonlinearity
for NLM-DD-OFDM systems. In particular, unlike SSB-OFDM, the channel model
for direct-modulated OFDM is no longer linear under any form of optical dispersion.
Subsequently, NLM-DD-OFDM is only fit for short-haul application such as multimode fiber for local-area networks (LAN), or short-reach single-mode fiber (SMF)
transmission. This class of optical OFDM has attracted attention recently due to its

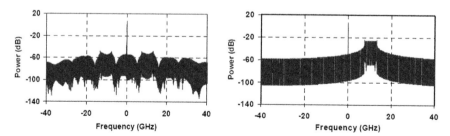

FIGURE 8.9 Comparison of optical spectra between (a) **NLM-DDO-OFDM** through direct-modulation
of DFB laser and (b) externally modulated offset SSB DDO-OFDM. The chirp constant C of 1 and
the modulation index m of 0.3 are assumed for direct-modulation in (a). Both OFDM spectrum
bandwidths are 5 GHz comprising 256 subcarriers.

low cost. Some notable works of NLM-DD-OFDM are experimental demonstrations and analysis of optical OFDM over multimode fibers [55,56,63] and compatible SSB-OFDM (CompSSB) proposed by Schuster et al. to achieve higher spectral efficiency than offset SSB-OFDM [64].

8.4 OPTICAL MULTICARRIER SYSTEMS BASED ON OPTICAL MULTIPLEXING

Multicarrier technique has been recognized as a powerful means of combating channel dispersion as illustrated in Sections 8.2 and 8.3, which is focused on electronic DSP implementation. However, there is an increasing gap between the electronic DSP bandwidth and the required baud rate in order to accommodate the exponential growth of the Internet traffic. It is widely believed that Tb/s-class Ethernet will emerge within the next decade, which cannot be directly realized using electronic DSP alone. Another layer of multiplexing seems to be inevitable to achieve Tb/s Ethernet transport. In the last few years, a wide variety of techniques have been proposed and demonstrated in the area of optical multiplexing for ultrahigh-speed and high spectral-efficiency transmission. They include: (i) all-optical OFDM where FFT is realized using optical circuits instead of electronic ones, (ii) optical superchannel where individual optical carriers are optically multiplexed as a Tb/s-class transport entity, and (iii) optical frequency-division multiplexing where multiple unlocked wavelengths are packed into one ITU wavelength slot. In this section, we capture the recent progress on optical multicarrier systems based on optical multiplexing utilizing these three techniques.

8.4.1 All-optical OFDM

All-optical OFDM attracts attention for its advantages of fast processing and low power consumption [65–68]. As the name of all-optical OFDM suggests, the IFFT/FFT digital signal processing to combine or split OFDM subcarriers is done optically. Figure 8.10 shows a typical architecture for all-optical OFDM transmission. For the transmitter, individual laser sources at equidistant frequencies serve as subcarriers. On each subcarrier an IQ modulator encodes the information for transmission [65]. The modulated subcarriers are then combined with an optical coupler to form the optical OFDM signal. Conversely, at the receiver side, an optical FFT circuit can be used to separate the subcarriers, which performs both serial-to-parallel conversion and FFT in the optical domain using a cascade of delayed interferometers (DIs) with subsequent time gates [65] or using an arrayed-waveguide grating router as reported in [66]. After the optical FFT, the separated subcarriers are optically amplified and coherently detected to perform demodulation and symbol decision.

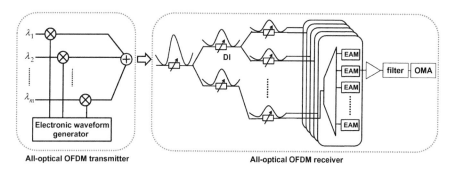

FIGURE 8.10 Schematic configuration for all-optical OFDM transmitter and receiver [65]. DI: Delayed interferometer, EAM: Electro-absorption modulator, OMA: Optical modulation analyzer.

By using the optical FFT/IFFT, the required electronic DSP only needs to process relatively narrow bandwidth for each low data-rate subcarrier as opposed to performing FFT across the entire signal spectrum, greatly reducing the bandwidth requirement and complexity of electronics. It has been also claimed that the power consumption of such all-optical OFDM systems is effectively reduced compared with conventional optical OFDM transceivers, since the power-hungry FFT algorithms are implemented in the optical domain.

8.4.2 Optical superchannel

The term "optical superchannel" is first given in [36] although similar techniques have already been proposed and demonstrated for Tb/s-class transport [17,69–72]. Optical superchannel commonly refers to a high data rate signal that originates from a single laser source and consists of multiple frequency-locked carriers which are synchronously modulated. By maintaining a suitable orthogonal condition among the modulated carriers, coherent crosstalk can be eliminated. Figure 8.11 illustrates an optical superchannel transceiver, and it has a similar configuration as the all-optical OFDM generator introduced in Section 8.3.1. The frequency locking and orthogonal condition are kept among all the optical tones. On the receiver side, the multiple signal bands can be detected either jointly or separately. After transmission, the superchannel signal is divided by a 1:M coupler (M can either equal to or be smaller than the number of bands). Each of the M parts is coherently detected and corresponding DSP is applied to the down-converted signal for signal recovery. Broadly speaking, coherent WDM (Co-WDM) can be also considered as one type of superchannel where the phase of multiple carriers is locked [17].

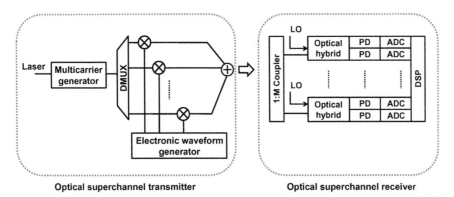

**FIGURE 8.11 Schematic diagram of an optical superchannel transmitter and receiver [36].
LO: Local oscillator, PD: Photodiode, ADC: Analog-to-digital converter, DSP: Digital signal
processing.**

8.4.3 Optical frequency division multiplexing

One could also use multiple frequencies within one ITU grid, provided they do not cause
significant interference. It was adopted in a commercial line card product [73], and we
call this method optical frequency-division multiplexing. In current systems, the optical
bandwidth of a path is determined in part by standards in adherence to the International
Telecommunication Union (ITU) Channel Grid [74], and in part by the technologies
of filters and wavelength-selective switches used to optically steer channels through a
network. Using N_{sc} different optical frequencies permits N_{sc} times the bit rate to be
transmitted within the filtered optical path. The maximum value of N_{sc} is limited by
the ratio of the optical filter bandwidth of the path to the information bandwidth of the
modulated optical carrier. The value of $N_{sc} = 2$ has been chosen for the commercially
available 100 Gb/s product [73]. The optical spectrum of this 100 Gb/s solution is shown
as the center channel (b) in Figure 8.12 accompanied by the spectra of single-carrier
10 Gb/s (a) and dual-polarization 40 Gb/s (c) channels. Each spectrum is centered on
a 50 GHz ITU channel. The concept of multiple frequency carriers can be extended to
arbitrary values of N_{sc}, provided that the system designer has freedom to vary the optical
filter width and channel positions. The extreme case is a continuum of low symbol-rate
carriers, densely filling the spectrum, such as OFDM.

8.5 NONLINEARITY IN OPTICAL MULTICARRIER
TRANSMISSION

Optical communication has recently witnessed the trend of the signal bandwidth
expansion, high spectral efficiency (SE), and ultra-long haul transmission [71,75].
As a result, fiber nonlinear noise becomes one of the major concerns for optical

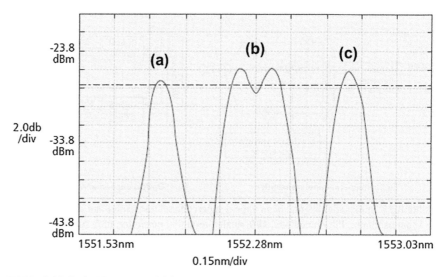

FIGURE 8.12 Optical frequency-division multiplexing comparison: (a) 10 Gb/s single-polarization single-carrier, (b) 100 Gb/s coherent dual-polarization dual-carrier, and (c) 40 Gb/s coherent dual-polarization single-carrier. All three channels are centered on the 50 GHz ITU grid [73].

transmission. Especially, there is a common belief that nonlinear impairments are more prominent in multicarrier systems such as optical OFDM due to its high PAPR than in single-carrier systems. In this section, we begin with the review of the recent progress on high SE transmission using optical multicarrier systems. We then discuss the optimal symbol rate for fiber nonlinearity in multicarrier systems. We also show analytical expressions for fiber nonlinearity noise and information spectral limit in multicarrier systems. Finally, we describe a few approaches for nonlinearity mitigation for multicarrier transmission.

8.5.1 High spectral-efficiency long-haul transmission

In the past few years, there have been impressive advances in experimental demonstrations of high SE multicarrier transmission employing narrow frequency guard interval and higher-order modulation format [20,71,76–78]. Table 8.1 summarizes the most recent records for high SE multicarrier transmission. It can be observed that while there is a steady advance in SE, there exists a trade-off between SE and reach. Generally speaking, high SE transmission systems can be achieved by either shrinking frequency guardband between wavelength channels, or utilizing higher-order modulation formats. However, both of these two approaches lead to increased sensitivity to fiber nonlinearity. Consequently, the nonlinearity impact and its mitigation strategy become a critical problem for SE multicarrier transmission.

Table 8.1 Experimental demonstrations of high spectral efficiency transmission.

Year	Modulation Format	SE (b/s/Hz)	Data Rate	Reach (km)	References
2009	QPSK CO-OFDM	3.3	1.2 Tb/s	600	[71]
2010	QPSK NGI-CO-OFDM	3.7	1.2 Tb/s	7200	[20]
2010	16-QAM CO-OFDM	7.0	520.8 Gb/s	240	[76]
2012	256-IPM CO-OFDM	11.2	231.5 Gb/s	800	[77]
2012	256-QAM CO-OFDM	14.0	400 Gb/s	400	[78]

8.5.2 Optimal symbol rate in multicarrier systems

It is known that the PAPR is one of the key characteristics which affect the performance of optical transmission due to fiber nonlinearity. It has been shown that multicarrier signals suffer from excessive nonlinear noise during fiber transmission due to their high PAPR. Though by using special algorithms such as hard-clipping the PAPR can be lowered at the transmitter. During transmission the PAPR can become very high again due to fiber dispersion. It is sensible to use not only PAPR reduction algorithms at the transmitter, but also strategies to maintain the low PAPR during transmission. For instance, for ultrahigh-speed systems such as 100-Gb/s and beyond, the fiber dispersion plays a critical role, inducing fast walk-off between subcarriers [79]. The PAPR of such a signal has a transient value during transmission due to fiber link dispersion, which renders the PAPR reduction at the transmitter ineffective. Nevertheless, if the PAPR mitigation approach is performed on a subband basis, due to the fact each subband has a much narrower bandwidth, the signal within each subband can be relatively undistorted over comparatively longer distances. This results in less inter-band and intra-band nonlinearity. In a nutshell, PAPR reduction on a subband basis will be more effective than on an entire spectrum basis.

Based on the above understanding, it is natural to predict there is a best trade-off of subband bandwidth within which the PAPR mitigation should be performed: On one hand, if the subband bandwidth is too broad, the PAPR reduction will not be effective due to the fiber dispersion. On the other hand, if the subbands are too narrow, the neighboring bands will interact just as narrowly spaced OFDM subcarriers, generating large inter-band crosstalk due to narrow subband spacing and incurring a large penalty. In the following, we describe the optimal subband allocation for optical OFDM signal transmission.

There are two mechanisms that may contribute to the optimal subband bandwidth. It relates to the four-wave mixing (FWM) efficiency which was derived in [80] and [81]. Due to the third-order fiber nonlinearity, the interaction of subcarriers at the frequencies of f_i, f_j, and f_k produces a mixing product at the frequency of $f_g = f_i + f_j - f_k$. The magnitude of the FWM product for N_s spans of the fiber link is given by [80]

$$P'_g = \frac{D_x^2}{9}\gamma^2 P_i P_j P_k e^{-\alpha L} \eta$$

$$\eta = \eta_1 \eta_2,$$

$$\eta_1 = \left| \frac{1 - e^{-\alpha L} e^{-j\Delta\beta_{ijk}L}}{j\Delta\beta_{ijk} + \alpha} \right|^2 \approx \frac{1}{(\Delta\beta_{ijk})^2 + \alpha^2} \qquad (8.36)$$

$$\eta_2 = \frac{\sin^2\left\{N_S \Delta\tilde{\beta}_{ijk}/2\right\}}{\sin^2 \Delta\tilde{\beta}_{ijk}/2}, \quad \Delta\tilde{\beta}_{ijk} = \Delta\tilde{\beta}_{ijk}L + \Delta\beta_{ijk,1}L_1$$

where D_x is the degeneration factor which equals 6 for non-degenerate FWM and 3 for degenerate FWM. $P_{i,j,k}$ is the input power at the frequency of $f_{i,j,k}$, α and L are, respectively, the loss coefficient and length of the fiber per span, respectively, γ is the third-order nonlinearity coefficient of the fiber, and L_{eff} is the effective fiber length given by $L_{\text{eff}} = (1 - e^{-\alpha L})/\alpha$. η is the FWM coefficient which has a strong dependence on the relative frequency spacing between the FWM components η_1 and η_2 are the phase mismatch in the transmission fiber, the subscript 1 stands for the parameters associated with the dispersion compensation fiber (DCF). η_1 is the intra-span FWM coefficient and η_2 originates from inter-span nonlinear interference. The derivation of η_1 and η_2 are shown in [89,90]. The FWM efficiency η will be discussed later in this section in more detail.

Figure 8.13a shows η_1 with varying fiber chromatic dispersions (CD) in a 10×100 km fiber link. It can be seen that the 3-dB bandwidth of η_1 is about 11, 8, 4.8 GHz for CDs of 3, 6, and 17 ps/nm/km, respectively. Figure 8.13b shows the

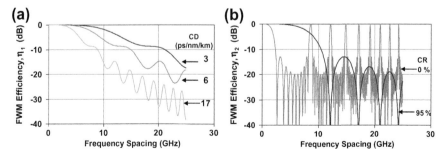

FIGURE 8.13 Four-wave mixing efficiency coefficients for a 10×100 km transmission link, with fiber loss coefficient of 0.2 dB/km. (a) due to the transmission fiber per span for different CD and (b) due to phase array effect. CD of 17 ps/nm/km. CR: chromatic dispersion compensation ratio.

FWM coefficient η_2 as a function of CD compensation ratio (CR) for a transmission fiber with CD of 17 ps/nm/km. For uncompensated systems (CR = 0%), the FWM 3-dB bandwidth is 1.8 GHz whereas for typical CD compensated systems with CR = 95%, the 3-dB bandwidth increases to 8 GHz. The idea behind the optimization of subband bandwidth is to maintain the FWM efficiency close to its maximum value within each subband while minimizing the intra-band FWM efficiency. Therefore, we could use the 3-dB bandwidth of FWM efficiency as the "ballpark" estimate of optimal subband bandwidth, and in that sense, Figures 8.13a and b give the approximate estimate of the optimal subband bandwidth. The 3-dB bandwidth increases with CD compensation, and therefore we anticipate that the optimal subband bandwidth of CD-uncompensated systems is narrower than CD-compensated systems.

We employ the DFT-Spread OFDM (DFT-S-OFDM) modulation to discuss the optimal symbol rate of optical multi-carrier transmission. The principle of DFT-S-OFDM is addressed in detail in [35,82]. A polarization division multiplexed 107-Gb/s coherent optical multiband DFT-Spread OFDM system is used in the simulation. The simulated transmission parameters are: fiber length of 100 km per span, $D_{SSMF} = 16$ ps/nm/km, $\alpha_{SSMF} = 0.2$ dB/km, $\gamma_{SSMF} = 1.3$ w^{-1} km^{-1}, noise figure of optical amplifiers of 6 dB, eight WDM channels with 50-GHz channel spacing, 64 number of subcarriers in each subband when the number of subbands is over 8, and QPSK modulation on each subcarrier. There is no dispersion compensation in this transmission simulation. A cyclic prefix ratio of 1/16 is used for all the cases. We simulate the link performance as a function of the optimal number of subbands, or equivalently, the optimal subband bandwidth for 107-Gb/s multiband CO-OFDM signal. Figure 8.14 shows the Q performance at fiber input powers of 4 dBm and 6 dBm for single-wavelength 107-Gb/s multi-band DFT-S-OFDM transmission. It can be seen that the optimal number of bands is close to 8, corresponding to subband bandwidth of 3.6 GHz in a 107-Gb/s multi-band DFT-S-OFDM transmission case using QPSK.

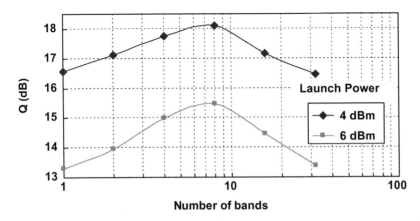

FIGURE 8.14 Q factor as a function of number of bands at 4- and 6-dBm launch powers with 107-Gb/s single-channel transmission over 10 × 100 km uncompensated SMF link.

8.5.3 The information spectral limit in multicarrier systems

By introducing the orthogonality among the multiple carriers and reducing the frequency guardbands between wavelength channels, the spectral efficiency can be maximized. In such systems, wavelength channels can be either continuously spaced without frequency guardband [20,83,84], or densely spaced with extremely small frequency guardband [72,76]. We now discuss the limits of information spectral efficiency in such multicarrier systems and summarize the outcome of a few theoretical works [85–88] where nonlinear launch power and information capacity are derived in analytical form. From these concise closed-form solutions, we can grasp the dependence of nonlinear transmission performance on some major system parameters such as chromatic dispersion and dispersion compensation ratio for dual-polarization CO-OFDM systems [89,90].

8.5.3.1 Derivation of analytical expressions for FWM noise in dual-polarization multicarrier transmission systems

1. FWM noise power density

As shown in [89,90], the nonlinear multiplicative noise spectral density I_{NL} is given by

$$
\begin{aligned}
I_{NL} &= \gamma^2 I^3 \frac{4}{\pi \alpha |\beta_2|} \left(\frac{2(N_s - 1 + e^{-\alpha \zeta L} - N_s e^{-\alpha \zeta L})e^{-\alpha \zeta L}}{(e^{-\alpha \zeta L} - 1)^2} + N_s \right), \\
&= \frac{3\gamma^2 N_s \ln(B/B_0) \cdot h_e}{8\pi \alpha |\beta_2|} I^3,
\end{aligned}
\tag{8.37}
$$

$$
h_e \equiv \frac{2(N_s - 1 + e^{-\alpha \zeta L N_s} - N_s e^{-\alpha \zeta L})e^{-\alpha \zeta L}}{N_s (e^{-\alpha \zeta L} - 1)^2} + 1,
\tag{8.38}
$$

$$
B_0 = \frac{\alpha}{\pi |\beta_2| B},
\tag{8.39}
$$

where γ is the fiber nonlinear coefficient, I is the launch power density, β_2 represents coefficient of chromatic dispersion, and ζ is the dispersion compensation ratio. (L) and N_s stand for the length of each fiber span and the number of transmission spans, respectively. α is the fiber loss coefficient. B represents the total width of signal bandwidth. We call h_e the noise enhancement factor accounting for the FWM noise interference among different fiber spans. We will discuss this interesting nonlinear enhancement factor h_e in more detail in the next section. This nonlinear noise power density I_{NL} of Eq. (8.37) can be expressed in a more concise form with the definition of nonlinear characteristic power density I_0 as follows:

$$
I_{NL} = \left(\frac{I}{I_0}\right)^2 I, \quad I_0 \equiv \frac{1}{\gamma} \sqrt{\frac{8\pi \alpha |\beta_2|}{3N_s h_e \ln(B/B_0)}}.
\tag{8.40}
$$

2. Signal-to-noise ratio and spectral efficiency limit in the presence of nonlinearity

The signal power in the presence of the nonlinear interference can be expressed as [90]

$$I = I \exp(-(I/I_0)^2) \cong I. \tag{8.41}$$

The noise can be considered as the summation of the white optical amplified spontaneous noise (ASE), n_0 and the FWM noise, and is given by [90]

$$n = 2n_0 + I(1 - \exp(-(I/I_0)^2)), \tag{8.42}$$

$$n_0 = N_s(G - 1)n_{sp}h\upsilon \cong 0.5N_s e^{\alpha L}h\upsilon \cdot NF, \tag{8.43}$$

where n_{sp} is the spontaneous emission noise factor equal to half of the noise figure of the optical amplifier NF in the ideal case, h is the Planck constant, and υ is the light frequency. The factor of 2 in Eq. (8.42) accounts for the unpolarized ASE noise. The signal power and noise power density in Eqs. (8.41) and (8.42) include the contribution from both polarizations. The signal-to-noise ratio (SNR) is thus given by

$$SNR = \frac{I \exp(-(I/I_0)^2)}{2n_0 + I(1 - \exp(-(I/I_0)^2))}. \tag{8.44}$$

For the SNR values larger than 10, Eq. (8.44) can be approximated as

$$SNR \cong \frac{I}{2n_0 + I(I/I_0)^2}. \tag{8.45}$$

The simplification is generally valid for the case of interests where the signal power density is much smaller than I_0.

It is verified in [87,88] that the FWM noise is of Gaussian distribution. Under the assumption of Gaussian noise distribution, the information spectral efficiency (defined as the maximum information capacity C normalized to bandwidth B) for dual polarization is readily given by [91]

$$S = 2\log_2(1 + SNR) = 2\log_2\left(1 + \frac{I\exp(-(I/I_0)^2)}{2n_0 + I\left(1 - \exp(-(I/I_0)^2)\right)}\right)$$

$$\cong 2\log_2\left(1 + \frac{I}{2n_0 + I(I/I_0)^2}\right). \tag{8.46}$$

From Eq. (8.46), the maximum spectral efficiency S_{opt} in the presence of fiber nonlinearity can be easily shown as

$$S_{opt} = 2\log_2\left(1 + \frac{1}{3}(I_0/n_0)^{2/3}\right). \tag{8.47}$$

3. Optimal launch power density, maximum Q, and nonlinear threshold of launch power density

In Eq. (8.46), the maximum possible spectral efficiency is obtained. However, in practice, the performance is always lower because of the practical implementation of modulation and coding. Next we derive a few important parameters that are important from the system design perspective. The first one is the maximum achievable Q

factor. Under the Gaussian noise assumption and QPSK modulation, the Q factor is equal to the SNR given by [89–91]

$$Q = \text{SNR} = \frac{I \exp(-(I/I_0)^2)}{2n_0 + I\left(1 - \exp(-(I/I_0)^2)\right)} \cong \frac{I}{2n_0 + I(I/I_0)^2}. \tag{8.48}$$

The optimum launch power density is another important parameter and is defined as the launch power density achieving where the maximum Q takes place. By simply differentiating Q of Eq. (8.48) over I, and setting it to zero, we obtain the optimum launch power density I_{opt} and the maximum Q as

$$I_{\text{opt}} = (n_0 I_0^2)^{1/3} = \left(\frac{8n_0\pi\alpha|\beta_2|}{3\gamma^2 N_s h_e \ln(B/B_0)}\right)^{1/3}, \tag{8.49}$$

$$Q_{\text{max}} = \frac{1}{3}\left(\frac{I_0}{n_0}\right)^{2/3} = \frac{(8\pi\alpha|\beta_2|)^{1/3}}{3(3n_0^2\gamma^2 N_s h_e \ln(B/B_0))^{1/3}}. \tag{8.50}$$

One of the inconveniences of using the expression in Eq. (8.49) is that it is dependent on the amplifier noise figure. The other commonly used term is nonlinear threshold launch power density that is defined as the maximum launch power density at which the BER due to the nonlinear noise can no longer be corrected by a certain type of forward-error-correction (FEC) code. For standard Reed-Solomon code RS (255, 239), the threshold Q is 9.8 (dB), or linear q_0 of 3.09. In Eq. (8.48), setting n_0 to zero and Q to q_0^2, we arrive at the nonlinear threshold power density

$$I_{\text{th}} = \frac{I_0}{q_0} = \frac{1}{q_0\gamma}\sqrt{\frac{8\pi\alpha|\beta_2|}{3N_s h_e \ln(B/B_0)}}, \tag{8.51}$$

where q_0 is the correctable linear Q for a specific FEC.

8.5.3.2 Application of the closed-form expressions
1. System Q factor and optimum launch power

Because the concise closed-form expressions are available, we can readily apply them to identify the system performance as a function of system parameters including fiber dispersion, number of spans, dispersion compensation ratio, and overall bandwidth. In this subsection, we will give examples of estimating the achievable system Q factor, optimum launch power, information spectral efficiency, and multi-span noise enhancement factor.

The significance of having closed-form formulas of Eqs. (8.49) and (8.50) for system Q factor and optimum launch power density is that they provide useful scaling rules for system designing. From Eqs. (8.49) and (8.50), it follows that for every $2\times(3\,\text{dB})$ increase in fiber dispersion, there is 1 dB increase in the optimal launch power density and the achievable Q; for every 3 dB increase in fiber nonlinear coefficient γ, there is 2 dB decrease in the optimal launch power and achievable Q.

As an illustrative example, we use the analytical expressions to generate the optimum launch power and achievable Q for a number of typical dispersion maps. We assume the following parameters: 16 wavelength channels, each covering 31-GHz bandwidth, giving total bandwidth B of 496 GHz; OFDM subcarrier frequency spacing of 85 MHz; QPSK modulation for each subcarrier; no frequency guardband between wavelength channels; 10-span of 100 km fiber link; fiber loss α of 0.2 dB/km; nonlinear coefficient $\gamma = 1.22$ w^{-1} km^{-1}; noise figure of the amplifier of 6 dB. Three transmission systems are investigated: (i) SSMF-type system with CD of 16 ps/nm/km without any dispersion compensation, abbreviated as "system I," (ii) CD of 16 ps/nm/km, but with dispersion 95% compensated per span, abbreviated as system II, and (iii) non-zero dispersion-shifted type fiber with CD of 4 ps/nm/km, abbreviated as "system III." As shown in Figure 8.15a, system I has the best performance due to large local dispersion and no per-span dispersion compensation. The advantage of system I over system II increases with the increase of the number of spans, for instance from 0 dB to 2.4 dB when the reach increases from single-span to 10 spans. The advantage of system I over system III is maintained at 1.7 dB independent of the number of spans. Figure 8.15b shows the optimal launch power versus number of spans. The optimum launch powers for non-compensated systems, systems I and III, are constant. This is because both the linear and nonlinear noises increase linearly with the number of the spans that leads to the optimum power independent of the number of spans. However, for the dispersion compensated system II, the optimum launch power density decreases with the number of spans due to the multi-span noise enhancement effect. Another interesting observation from Eqs. (8.49) and (8.50) is that both the optimal Q factor and launch power have very weak dependence on the overall system bandwidth: proportional to 1/3 power of logarithm of the overall bandwidth. It can be easily shown that for both systems I and III, the Q is decreased by only about 0.7 dB with the 10-fold increase of the overall system bandwidth from 400 GHz to 4000 GHz whereas system II incurs a larger decrease of the Q factor of 0.84 dB with the same bandwidth increase.

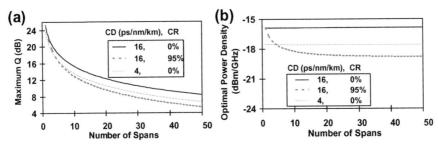

FIGURE 8.15 (a) The maximum Q factor and (b) the optimal launch power density versus number of spans with various dispersion maps. CD: chromatic dispersion. CR: (CD) compensation ratio.

2. Information spectral efficiency

The information spectral efficiency is important as it represents the ultimate bound of what we can achieve by employing all possible modulations (of course not limited to QPSK) and codes. For large SNR, we simplify Eq. (8.46) into

$$S = 2\log_2\left(1 + \frac{1}{3}(I_0/n_0)^{2/3}\right) \cong 2\log_2\left(\frac{1}{3}(8\pi\alpha|\beta_2|)^{1/3}(3\gamma^2 n_0^2 N_s h_e \ln(B/B_0))^{-1/3}\right).$$

(8.52)

Equation (8.52) clearly shows the challenges of improving spectral efficiency by redesigning the fiber system parameters: to increase spectral efficiency by 2 bit/s/Hz, the dispersion needs to be increased by a factor of 8, or the nonlinear coefficient γ be decreased by a factor of 2.8, or number of spans be reduced by a factor of 2, all of which are difficult to achieve. In a nutshell, it is of diminishing return to improve the spectral efficiency by modifying the optical fiber system parameters. The only effective method to substantially improve the spectral efficiency is to add more dimensions such as resorting to polarization multiplexing that leads to almost a factor of 2 improvement as discussed in the paper, or fiber mode multiplexing by at least a factor of 2 or more dependent on the capability of achievable digital signal processing (DSP). Figure 8.16 shows the achievable spectral efficiency for the three systems studied in Section 8.1. The only modification is that we assume 40 nm or 5 THz for the total bandwidth. The spectral efficiency for systems I, II, and III is, respectively, 9.90, 8.38, and 8.63 b/s/Hz. This shows a total capacity of 49.5 Tb/s can be achieved for 10×100 SSMF uncompensated EDFA-only dual-polarization systems within C-band.

8.5.4 Nonlinearity mitigation for multicarrier systems

Various techniques of nonlinearity mitigation have been proposed to improve the transmission performance. The commonly studied approaches are as follows:

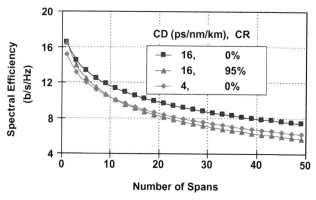

FIGURE 8.16 Information spectral efficiency as a function of the number of spans for various dispersion maps. The total bandwidth *B* is assumed to be 40 nm. The other OFDM and link parameters are the same as those for Figure 8.15.

(i) pre- and/or post-compensation where the nonlinear phase noise is compensated at the transmitter or/and receiver [45,92], (ii) joint cross-polarization nonlinearity cancelation, which is similar to approach (i), but broadens the compensation to nonlinear polarization rotation [93], (iii) nonlinear digital back-propagation (DBP) where the nonlinearity is unwrapped by back-propagating the received signal toward the transmitter digitally [94,95], (iv) Volterra nonlinear compensation where the nonlinearity is approximated as the Volterra series and the nonlinearity is compensated iteratively [96], and (v) DFT-spread OFDM at optimal symbol rate [35] as discussed in Section 8.5.2. In this subsection, we will focus on DBP which has drawn much attention for its flexible and comprehensive compensation of both intra- and inter-channel nonlinear effects.

With the exact knowledge of channel parameters, the deterministic nonlinear interactions among signals can be completely removed using DBP with fine enough back-propagation steps, but this requires huge digital processing power [45,95,97]. Signals propagating in generic optical fiber transmission systems can be described by the nonlinear Schrödinger equation (NLSE) given by

$$\frac{\partial E}{\partial z} = (\hat{N} + \hat{D})E, \quad \hat{N} = j\gamma E^2, \quad \hat{D} = -j\frac{\beta_2}{2}\frac{\partial^2}{\partial t^2} - \frac{\alpha}{2}, \quad (8.53)$$

where $\hat{N}$ and $\hat{D}$ are linear and nonlinear operators, and α, β_2, γ represent fiber loss, chromatic dispersion, and nonlinear coefficient, respectively. Equation (8.53) can be numerically solved using the symmetric split-step Fourier method (SSFM) [98] as follows

$$E(z+h,t) = \exp\left(\frac{h}{2}\hat{D}\right)\exp\left(\int_z^{z+h}\hat{N}(z')dz'\right) \times \exp\left(\frac{h}{2}\hat{D}\right)E(z,t), \quad (8.54)$$

where h is the step size. In the absence of noise, the transmitted signal can be calculated from the inverse NLSE:

$$\frac{\partial E}{\partial z} = (\hat{N}^{-1} + \hat{D}^{-1})E, \quad (8.55)$$

which is equivalent to passing the received signal through a "virtual" fiber with parameters of the opposite sign to the real fiber. By applying Eq. (8.55) to the received signal with appropriate step size, both linear and nonlinear effects incurred from transmission can be removed.

Although solving the inverse NLSE is computationally expensive, DBP offers good nonlinearity mitigation when transmission link parameters are known. Figure 8.17 shows the Q factors against launch power of 3200 km CO-OFDM transmission with DBP (solid round curve) using different number of steps [97]. We can see that more than 2-dB improvement in terms of nonlinear power tolerance can be achieved if the number of steps is larger than 8.

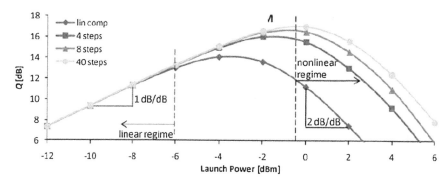

FIGURE 8.17 *Q* against launch power for a 56-Gb/s 3200 km transmission system using linear equalization or nonlinear (filtered BP) equalization [97].

8.6 APPLICATIONS OF OPTICAL MULTICARRIER TRANSMISSIONS

The various optical multicarrier transmission technologies introduced above have been explored for a wide range of potential applications. In this section, we give specific examples in three distinct areas of application, each of which has unique requirements in terms of data rate, spectral efficiency, reach, and complexity. These are: (i) long-reach and high-capacity systems, (ii) optical access systems, and (iii) indoor and free-space communication systems.

8.6.1 Long-reach and high-capacity systems

Long-reach applications have a specific demand to achieve high data capacity and spectral efficiency. As such, much research has been conducted to apply optical multicarrier transmission to long-haul and high-capacity systems in recent years [21,99–103]. Table 8.2 lists some of the recent achievements in long-reach systems (over 1000 km) with high transmission capacity (over 400 Gb/s). It is clear multicarrier technologies are capable of overcoming limitations due to fiber dispersion and nonlinearity, enabling high spectral efficiency and long-reach transmission in the post-100 Gb/s era.

In order to achieve such high-capacity transmission over longer distances, we need to adopt a few techniques to efficiently use bandwidth and reduce nonlinear impairments. For instance, in [99], a reduced-guard-interval (RGI) CO-OFDM technique is used to improve the spectral efficiency, and ultra large area fiber (ULAF) is used to reduce nonlinear impairments. In optical OFDM transceivers, GI (or the cyclic prefix mentioned in Section 8.2) is used to compensate fiber dispersion, the length of which increases with transmission distance and signal bandwidth. A lengthy GI then causes a large redundancy ratio (overhead) and reduces the channel capacity. The RGI-CO-OFDM method employs digital dispersion compensation and

Table 8.2 Recent achievements on long-reach (>1000 km) and high-capacity transmission (>400 Gb/s) using multicarrier optical transmission.

Modulation Format	Data Rate (Gb/s)	Composition	Reach (km)	Intrachannel SE (b/s/Hz)	References
RGI-CO-OFDM PDM-16QAM	448	10 × 45-Gb/s	2000	7.0	[99]
RGI-CO-OFDM PDM-16QAM	485	10 × 48.5-Gb/s	4800	6.2	[100]
RGI-CO-OFDM PDM-32QAM	606	10 × 60.6-Gb/s	1600	7.76	[101]
DFT-Spread OFDM PDM-16QAM	1630	16 × 64-Gb/s	2500	5.5	[21]
CO-OFDM PDM-16QAM	4704	40 × 117.6-Gb/s	10,180	4.7	[102]
CO-OFDM PDM-8QAM	11,500	115 × 100-Gb/s	10,180	4.0	[103]

therefore relatively short GI is needed. Another technique proposed for long-haul transmission is DFT-Spread OFDM [35], which improves performance of the long-reach transmission by reducing the PAPR of the signal. It also gives the capability to partition the whole bandwidth into smaller subbands with optimal symbol rate, which can minimize the nonlinearity penalty [21].

8.6.2 Optical access networks

Optical multicarrier techniques, such as optical OFDM, can be also a potential candidate for next-generation high-speed optical access networks [18,19,104]. The passive optical network which uses orthogonal frequency division multi-access is called OFDMA-PON. Figure 8.18 depicts an example of OFDMA-PON architecture. As shown in Figure 8.18 [18], at the optical line terminal (OLT), a bandwidth-sharing schedule is formed according to the demand from optical network units (ONU) side, and is distributed to all ONUs over pre-assigned subcarriers and/or timeslots. Different OFDM subcarriers are thus assigned to different ONUs. Since traffic is aggregated and de-aggregated electronically on an optical carrier, the architecture is compatible with the legacy fiber distribution network, which enables reuse of existing PON infrastructure and thus saving on deployment cost. At the ONUs, each ONU recovers its pre-assigned OFDM subcarriers and/or time slots in DSP. An orthogonal OFDM-based schedule for upstream transmission is likewise generated by the OLT and distributed to the ONUs. At the OLT, a complete OFDMA frame is assembled from the incoming sub-frames originating at different ONUs. It can be seen that such OFDMA-PON has the advantages of bandwidth flexibility and high

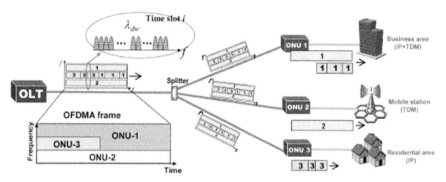

FIGURE 8.18 Example of OFDMA-PON architecture [18]. OLT: Optical line terminal, ONU: Optical network units, TDM: Time-division multiplexing.

spectral efficiency in addition to multi-user access capabilities, which make it an attractive approach for next-generation high-speed PON.

More recently, demonstration of PON utilizing OFDMA for downstream data transmission and achieving more than 100 Mb/s per-cell has been presented with low latency and relatively low cost [104]. This progress confirms that optical multicarrier could be a promising candidature for future optical access networks.

8.6.3 Indoor and free-space multicarrier optical systems

The current indoor and free-space systems being studied use intensity modulation and direct detection (IM/DD) for the simplicity of implementation. However, when optical waves propagate through air, they suffer from atmospheric turbulence causing fluctuations of both amplitude and phase. Such cases resemble wireless fading channels, and as a result optical multicarrier technologies are well suited for indoor and free-space systems. Because each subcarrier carries low rate data, it provides immunity to burst-errors due to intensity fluctuations [22–24].

The basic free-space OFDM transmitter and receiver configurations are shown in Figure 8.19a and c, respectively [22]. The corresponding free space link is shown in Figure 8.19b. An information bearing stream is first encoded (for instance, by an LDPC code) and then parsed into groups in the demultiplexer (DEMUX). The parsed stream is then mapped into a complex-valued signal constellation. The complex-valued signal points from all subchannels are considered as the values of the DFT of a multicarrier OFDM signal. The modulator and demodulator can be implemented by using the IFFT and FFT algorithms, respectively. After a D/A conversion and RF up-conversion, the OFDM signal is driven to the Mach-Zehnder modulator (MZM) and then transmitted over the free space link. At the receiver side, after photo-detection, RF down-conversion, and carrier synchronization, the received signal is demodulated by computing the FFT. The soft outputs of FFT demodulator are used to estimate the bit reliabilities that are fed into the LDPC decoder.

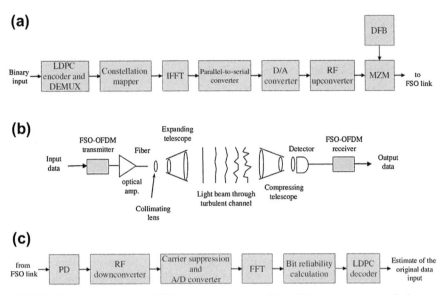

FIGURE 8.19 Basic free-space OFDM system: (a) transmitter, (b) free-space transmission link, and (c) receiver [22].

8.7 FUTURE RESEARCH DIRECTIONS FOR MULTICARRIER TRANSMISSION

Optical multicarrier transmission has become a fast progressing and vibrant research field. It is an exciting event that very advanced communication concepts in coding, modulation, reception, and channel equalization are being applied in the optical fiber communication, which traditionally used much simpler transmission and detection schemes compared to the wireless counterpart owing to the high-speed electronics involved. With the advent of extremely fast silicon DSP chips, it has now become possible to employ many of the advanced modulation schemes, enabling unprecedented data rates approaching Tb/s and beyond. Additionally, the marriage between high-speed electronics and photonics presents some tremendous challenges and opportunities in research. As concluding remarks for this chapter, here we lay out some examples of the research and development ideas that in our view will have significant ramifications in the field of optical multicarrier transmission:

1. As the channel rate goes beyond 1 Tb/s, the achievable capacity per fiber may become a bottleneck due to the fiber nonlinearity constraint [93]. To overcome the bottleneck, space division multiplexing (SDM)-based transmission over multi-core fiber (MCF) [32,105,109] or multi-mode fiber (MMF) [111–113], especially the few-mode fiber (FMF)-based transmission for optical multicarrier systems has been proposed recently [27–31]. FMF transmission, in conjunction with MIMO-OFDM, could be a promising technology to achieve

capacity higher than 100 Tb/s per fiber taking advantage of the mode-division multiplexing in the optical fiber. But SDM itself has many challenges, such as development of FMF fiber, few-mode fiber amplifiers, switches, add/drop multiplexers, etc. These engineering challenges are to be solved by innovative approaches in device technologies.

2. The traditional optical networks are rigidly designed to support only a fixed link data rate throughout the operational life. OFDM technique, as one realization of software-defined optical transmission (SDOT), provides many nabling functionalities for the future dynamically reconfigurable networks [37,38], the adaptation of channel rate according to the channel condition being one example. Rate-adaptive LDPC codes and adaptive bit loading can be employed to optimize the link capacity for difference reaches.

3. OFDMA has become an attractive multi-user access technique in which the subsets of subcarriers are assigned to individual users. OFDMA enables the flexible time and frequency domain resource partitioning. Additionally, OFDMA can seamlessly bridge the wireless and optical access networks via radio over fiber (RoF) systems. OFDMA is also shown to be a promising approach in offering resource management for passive-optical-networks (PON).

4. The last decade saw the dramatic resurgence of the research interests in optoelectronic integrated circuit (OEIC). Considering the extensive optical devices and digital signal processing involved in optical OFDM, we envisage the possibility of integration of many digital, RF, and optical components into one silicon IC which can perform all the main functionalities of an optical OFDM transceiver. Without doubt, the success of the OEIC will greatly influence the evolution of the optical multicarrier transmission systems.

References

[1] R.W. Chang, Synthesis of band-limited orthogonal signals for multichannel data transmission, Bell Sys. Tech. J. 45 (1966) 1775–1796.

[2] S.B. Weinstein, P.M. Ebert, Data transmission by frequency division multiplexing using the discrete Fourier transform, IEEE Trans. Commun. 19 (1971) 628–634.

[3] T. Ohara, H. Takara, T. Yamamoto et al., Over-1000-channel ultradense WDM transmission with supercontinuum multicarrier source, J. Lightwave Technol. 24 (6) (2006) 2311–2317.

[4] R.Q. Hui, B.Y. Zhu, R.X. Huang, C.T. Allen, K.R. Demarest, D. Richards, Subcarrier multiplexing for high-speed optical transmission, J. Lightwave Technol. 20 (2002) 417–427.

[5] H. Sun, K.T. Wu, K. Roberts, Real-time measurements of a 40 Gb/s coherent system, Opt. Express 16 (2) (2008) 873–879.

[6] Q. Pan, R.J. Green, Bit-error-rate performance of lightwave hybrid AM/OFDM systems with comparison with AM/QAM systems in the presence of clipping impulse noise, IEEE Photon. Technol. Lett. 8 (1996) 278–280.

[7] A.J. Lowery, J. Armstrong, Orthogonal-frequency-division multiplexing for dispersion compensation of long-haul optical systems, Opt. Express 14 (6) (2006) 2079–2084.

[8] J. Armstrong, A. Lowery, Power efficient optical OFDM, Electron. Lett. 42 (6) (2006) 370–372.

[9] I.B. Djordjevic, B. Vasic, Orthogonal frequency division multiplexing for high-speed optical transmission, Opt. Express 14 (2006) 3767–3775.

[10] W. Shieh, C. Athaudage, Coherent optical orthogonal frequency division multiplexing, Electron. Lett. 42 (2006) 587–589.

[11] S.L. Jansen, I. Morita, T.C. Schenk, N. Takeda, H. Tanaka, Coherent optical 25.8-Gb/s OFDM transmission over 4160-km SSMF, J. Lightwave Technol. 26 (1) (2008) 6–15.

[12] W. Shieh, I. Djordjevic, OFDM for Optical Communications. first ed., Academic Press, 2009.

[13] D. Hillerkuss, M. Winter, M. Teschke, A. Marculescu, J. Li, G. Sigurdsson, K. Worms, S.B. Ezra, N. Narkiss, W. Freude, J. Leuthold, Simple all-optical FFT scheme enabling Tbit/s real-time signal processing, Opt. Express 18 (9) (2010) 9324–9340.

[14] I. Kang, X. Liu, S. Chandrasekhar et al., Energy-efficient 0.26-Tb/s coherent-optical OFDM transmission using photonic-integrated all-optical discrete Fourier transform, Opt. Express 20 (2) (2012) 896–904.

[15] H. Masuda, E. Yamazaki, A. Sano, T. Yoshimatsu, T. Kobayashi, E. Yoshida, Y. Miyamoto, S. Matsuoka, Y. Takatori, M. Mizoguchi, and others, 13.5-Tb/s (135 × 111-Gb/s/ch) no-guard-interval coherent OFDM transmission over 6248 km using SNR maximized second-order DRA in the extended L-band, in: Optical Fiber Communication Conference (OFC), 2009, pp. 1–3.

[16] B.Y. Zhu, X. Liu, S. Chandrasekhar et al., Ultra-long-haul transmission of 1.2-Tb/s multicarrier no-guard-interval CO-OFDM superchannel using ultra-large-area fiber, IEEE Photon. Technol. Lett. 22 (11) (2012) 826–828.

[17] A.D. Ellis, F.C.G. Gunning, Spectral density enhancement using coherent WDM, IEEE Photon. Technol. Lett. 17 (2005) 504–506.

[18] N. Cvijetic, OFDM for next-generation optical access networks, J. Lightwave Technol. 30 (2012) 384–398.

[19] T.N. Duong, N. Genay, M. Ouzzif, J. Le Masson, B. Charbonnier, P. Chanclou, J.C. Simon, Adaptive loading algorithm implemented in AMOOFDM for NG-PON system integrating cost-effective and low-bandwidth optical devices, IEEE Photon. Technol. Lett. 21 (12) (2009) 790–792.

[20] S. Chandrasekhar, X. Liu, B. Zhu, D.W. Peckham, Transmission of a 1.2-Tb/s 24-carrier no-guard-interval coherent OFDM superchannel over 7200-km of ultra-large-area fiber, in: European Conference on Optical Communication (ECOC), 2009, p. PD2.6.

[21] A. Li, X. Chen, G. Gao, W. Shieh, B.S. Krongold, Transmission of 1.63-Tb/s PDM-16QAM unique-word DFT-Spread OFDM signal over 1,010-km SSMF, in: Optical Fiber Communication Conference (OFC), 2012, OW4C.1.

[22] I.B. Djordjevic, B. Vasic, M.A. Neifeld, LDPC coded OFDM over the atmospheric turbulence channel, Opt. Express 15 (2007) 6336–6350.

[23] N. Cvijetic, D. Qian, T. Wang, 10 Gb/s free-space optical transmission using OFDM, in: Optical Fiber Communication Conference (OFC), 2008, Paper OTHD2.

[24] O. González, R. Pérez-Jiménez, S. Rodŕguez, J. Rabadán, A. Ayala, OFDM over indoor wireless optical channel, IEE Proc. Optoelectron 152 (4) (2005) 199–204.

[25] W. Shieh, H. Bao, Y. Tang, Coherent optical OFDM: Theory and design, Opt. Express 16 (2) (2008) 841–859.

[26] S. Hara, R. Prasad, Multicarrier Techniques for 4G Mobile Communications, Artech House, Boston, 2003.

[27] A. Li, A.A. Amin, X. Chen, W. Shieh, Transmission of 107-GB/s mode and polarization multiplexed CO-OFDM signal over a two-mode fiber, Opt. Express 19 (9) (2011) 8808–8814.

[28] R. Ryf, S. Randel, A.H. Gnauck, C. Bolle, R. Essiambre, P. Winzer, D.W. Peckham, A. McCurdy, R. Lingle, Space-division multiplexing over 10 km of three-mode fiber using coherent 6×6 MIMO processing, in: Optical Fiber Communication Conference (OFC), 2011, Paper PDPB10.

[29] M. Salsi, C. Koebele, D. Sperti, P. Tran, P. Brindel, H. Mardoyan, S. Bigo, A. Boutin, F. Verluise, P. Sillard, M. Bigot-Astruc, L. Provost, F. Cerou, G. Charlet, Transmission at 2×100 GB/s, over two modes of 40 km-long prototype few-mode fiber, using LCOS based mode multiplexer and demultiplexer, in: Optical Fiber Communication Conference (OFC), 2011, Paper PDPB9.

[30] S. Randel, R. Ryf, A. Sierra, P.J. Winzer, A.H. Gnauck, C.A. Bolle, R.J. Essiambre, D.W. Peckham, A. McCurdy, R. Lingle, and others, 6×56-Gb/s mode-division multiplexed transmission over 33-km few-mode fiber enabled by 6×6 MIMO equalization, Opt. Express 19 (17) (2011) 16.

[31] A. Al Amin, A. Li, S. Chen, X. Chen, G. Gao, W. Shieh, Dual-LP11 mode 4×4 MIMO-OFDM transmission over a two-mode fiber, Opt. Express 19 (17) (2011) 16,672–16,679.

[32] Y. Sakaguchi, N. Awaji, A. Wada, T. Kanno, T. Kawanishi, T. Hayashi, T. Taru, M. Kobayashi, Watanabe, 109-Tb/s ($7 \times 97 \times 172 - $ Gb/s SDM/WDM/PDM) QPSK transmission through 16.8-km homogeneous multi-core fiber, in: Optical Fiber Communication Conference (OFC), 2011, Paper PDPB6.

[33] S.L. Jansen, I. Morita, K. Forozesh, S. Randel, D. van den Borne, H. Tanaka, Optical OFDM, a hype or is it for real?, in: European Conference on Optical Communication (ECOC), 2008, pp. 1–4.

[34] S. Randel, S. Adhikari, S.L. Jansen, Analysis of RF-pilot-based phase noise compensation for coherent optical OFDM systems, IEEE Photon. Technol. Lett. 22 (17) (2010) 1288–1290.

[35] Y. Tang, W. Shieh, B.S. Krongold, DFT-Spread OFDM for fiber nonlinearity mitigation, IEEE Photon. Technol. Lett. 22 (16) (2010) 1250–1252.

[36] S. Chandrasekhar, X. Liu, Terabit superchannels for high spectral efficiency transmission, in: European Conference and Exhibition on Optical Communication (ECOC), 2010, pp. 1–6.

[37] M. Jinno, H. Takara, B. Kozicki, Y. Tsukishima, Y. Sone, S. Matsuoka, Spectrum-efficient and scalable elastic optical path network: Architecture, benefits, and enabling technologies, IEEE Commun. Mag. (2009) 66–73.

[38] K. Christodoulopoulos, I. Tomkos, E.A. Varvarigos, Elastic bandwidth allocation in flexible OFDM-based optical networks, J. Lightwave Technol. 29 (9) (2011) 1354–1366.

[39] M.S. Zimmerman, A.L. Kirsch, AN/GSC-10 (KATHRYN) variable rate data modem for HF radio, AIEE Transaction 79 (1960) 248–255.

[40] P. Duhamel, H. Hollmann, Split-radix FFT algorithm, IET Electron. Lett. 20 (1984) 14–16.

[41] L. Hanzo, M. Munster, B.J. Choi, T. Keller, OFDM and MC-CDMA for Broadband Multi-User Communications, WLANs and Broadcasting, Wiley, New York, 2003.

[42] X. Yi, W. Shieh, Y. Ma, Phase noise on coherent optical OFDM systems with 16-QAM and 64-QAM beyond 10 Gb/s, in: European Conference on Optical Communication (ECOC), 2007, Paper 5.2.3.

[43] H. Takahashi, A.A. Amin, S.L. Jansen, I. Morita, H. Tanaka, 8 × 66. 8-Gbit/s coherent PDM-OFDM transmission over 640 km of SSMF at 5.6-bit/s/Hz spectral efficiency, in: European Conference on Optical Communication (ECOC), 2008, Paper Th.3.E.4.

[44] W. Shieh, Q. Yang, Y. Ma, 107 Gb/s coherent optical OFDM transmission over 1000-km SSMF fiber using orthogonal band multiplexing, Opt. Express 16 (9) (2008) 6378–6386.

[45] W. Shieh, X. Yi, Y. Ma, Y. Tang, Theoretical and experimental study on PMD-supported transmission using polarization diversity in coherent optical OFDM systems, Opt. Express 15 (2007) 9936–9947.

[46] W. Shieh, X. Yi, Y. Tang, Transmission experiment of multi-gigabit coherent optical OFDM systems over 1000 km SSMF fiber, Electron. Lett. 43 (2007) 183–185.

[47] S.L. Jansen, I. Morita, N. Takeda, H. Tanaka, 20-Gb/s OFDM transmission over 4160-km SSMF enabled by RF-Pilot tone phase noise compensation, in: Optical Fiber Communication Conference (OFC), 2007, Paper PDP15.

[48] Y. Tang, W. Shieh, X. Yi, R. Evans, Optimum design for RF-to-optical up-converter in coherent optical OFDM systems, IEEE Photon. Technol. Lett. 19 (2007) 483–485.

[49] D.S. Ly-Gagnon, S. Tsukarnoto, K. Katoh, K. Kikuchi, Coherent detection of optical quadrature phase-shift keying signals with carrier phase estimation, J. Lightwave Technol. 24 (2006) 12–21.

[50] S.J. Savory, G. Gavioli, R.I. Killey, P. Bayvel, Electronic compensation of chromatic dispersion using a digital coherent receiver, Opt. Express 15 (2007) 2120–2126.

[51] S.B. Cohn, N.P. Weinhouse, An automatic microwave phase measurement system, Microwave J. 7 (1964) 49–56.

[52] C.A. Hoer, K.C. Roe, Using an arbitrary six-port junction to measure complex voltage ratios, IEEE Trans. MTT MTT-23 (1975) 978–984.

[53] Y. Tang, W. Chen, W. Shieh, Study of nonlinearity and dynamic range of coherent optical OFDM receivers, in: Optical Fiber Communication Conference (OFC), 2008, Paper JWA65.

[54] M. Tang, K.A. Shore, 30-Gb/s signal transmission over 40-km directly modulated DFB-laser-based single-mode-fiber links without optical amplification and dispersion compensation, J. Lightwave Technol. 24 (6) (2006) 2318–2327.

[55] M. Tang, K.A. Shore, Maximizing the transmission performance of adaptively modulated optical OFDM signals in multimode-fiber links by optimizing analog-to-digital converters, J. Lightwave Technol. 25 (2007) 787–798.

[56] X.Q. Jin, J.M. Tang, P.S. Spencer, K.A. Shore, Optimization of adaptively modulated optical OFDM modems for multimode fiber-based local area networks, J. Opt. Netw. 7 (2008) 198–214.

[57] B.J.C. Schmidt, A.J. Lowery, J. Armstrong, Experimental demonstrations of 20 Gbit/s direct-detection optical OFDM and 12 Gbit/s with a colorless transmitter, in: Optical Fiber Communication Conference (OFC), 2007, Paper PDP18.

[58] D.F. Hewitt, Orthogonal frequency division multiplexing using baseband optical single sideband for simpler adaptive dispersion compensation, in: Optical Fiber Communication Conference (OFC), 2007, Paper OME7.

[59] W.R. Peng, X. Wu, V.R. Arbab, B. Shamee, J.Y. Yang, L.C. Christen, K.M. Feng A.E. Willner, S. Chi, Experimental demonstration of 340 km SSMF transmission using

a virtual single sideband OFDM signal that employs carrier suppressed and iterative detection techniques, in: Optical Fiber Communication Conference (OFC), 2008, Paper OMU1.

[60] W.R. Peng, X. Wu, V.R. Arbab, B. Shamee, L.C. Christen, J.Y. Yang, K.M. Feng, A.E. Willner, S. Chi, Experimental demonstration of a coherently modulated and directly detected optical OFDM system using an RF-Tone insertion, in: Optical Fiber Communication Conference (OFC), 2008, Paper OMU2.

[61] A.J. Lowery, L.B. Du, J. Armstrong, Performance of optical OFDM in ultralong-haul WDM lightwave systems, J. Lightwave Technol. 25 (2007) 131–138.

[62] G.P. Agrawal, Fiber-Optic Communication Systems, third ed., John Wiley & Sons Inc, New York.

[63] E. Jolley, H. Kee, P. Pickard, J. Tang, K. Cordina, Generation and propagation of a 1550 nm 10 Gbit/s optical orthogonal frequency division multiplexed signal over 1000 m of multimode fibre using a directly modulated DFB, in: Optical Fiber Communication Conference (OFC), 2005, Paper OFP3.

[64] M. Schuster, S. Randel, C.A. Bunge, S.C.J. Lee, F. Breyer, B. Spinnler, K. Petermann, Spectrally efficient compatible single-sideband modulation for OFDM transmission with direct detection, IEEE Photon. Technol. Lett. 20 (2008) 670–672.

[65] D. Hillerkuss, R. Schmogrow, T. Schellinger, M. Jordan, M. Winter, G. Huber, T. Vallaitis, R. Bonk, P. Kleinow, F. Frey, M. Roeger, S. Koenig, A. Ludwig, A. Marculescu, J. Li, M. Hoh, M. Dreschmann, J. Meyer, S. Ben Ezra, N. Narkiss, B. Nebendahl, F. Parmigiani, P. Petropoulos, B. Resan, A. Oehler, K. Weingarten, T. Ellermeyer, J. Lutz, M. Moeller, M. Huebner, J. Becker, C. Koos, W. Freude, J. Leuthold, 26 Tbit s-1 line-rate super-channel transmission utilizing all-optical fast Fourier transform processing, Nat. Photonics 5 (2011) 364–371.

[66] A.J. Lowery, L. Du, All-optical OFDM transmitter design using AWGRs and low-bandwidth modulators, Opt. Express 19 (17) (2011) 15,696–15,704.

[67] K. Lee, C.T.D. Thai, J.K.K. Rhee, All optical discrete Fourier transform processor for 100 Gbps OFDM transmission, Opt. Express 16 (6) (2008) 4023–4028.

[68] H.W. Chen, M.H. Chen, S.Z. Xie, All-optical sampling orthogonal frequency-division multiplexing scheme for high-speed transmission system, J. Lightwave Technol. 27 (21) (2009) 4848–4854.

[69] A. Sano, E. Yamada, H. Masuda, E. Yamazaki, T. Kobayashi, E. Yoshida, Y. Miyamoto, R. Kudo, K. Ishihara, Y. Takatori, No-guard-interval coherent optical OFDM for 100-Gb/s long-haul WDM transmission, J. Lightwave Technol. 27 (16) (2009) 3705–3713.

[70] W. Shieh, High spectral efficiency coherent optical OFDM for 1 Tb/s Ethernet transport, in: Optical Fiber Communication Conference (OFC), 2009, Paper OWW1.

[71] Y. Ma, Q. Yang, Y. Tang, S. Chen, W. Shieh, 1-Tb/s single-channel coherent optical OFDM transmission over 600-km SSMF fiber with subwavelength bandwidth access, Opt. Express 17 (2009) 9421–9427.

[72] R. Dischler, F. Buchali, Transmission of 1.2 Tb/s continuous waveband PDM-OFDM-FDM signal with spectral efficiency of 3.3 bit/s/Hz over 400 km of SSMF, in: Optical Fiber Communication Conference (OFC), 2009, Paper PDP C2.

[73] K. Roberts, D. Beckett, D. Boertjes, J.H. Berthold, C. Laperle, 100G and beyond with digital coherent signal processing, IEEE Commun. Mag. (2010) 62–69.

[74] ITU-T Rec. G.694.1, Spectral grids for WDM applications: DWDM frequency grid, June 2002.

[75] S. Chandrasekhar, X. Liu, Experimental investigation on the performance of closely spaced multi-carrier PDM-QPSK with digital coherent detection, Opt. Express 17 (24) (2009) 21350–21361.

[76] H. Takahashi, A. Al Amin, S.L. Jansen, Itsuro Morita, Hideaki Tanaka, Highly spectrally efficient DWDM transmission at 7.0 b/s/Hz using 8 × 65. 1-Gb/s coherent PDM-OFDM, J. Lightwave Technol. 28 (2010) 406–414.

[77] X. Liu, S. Chandrasekhar, T. Lotz, P.J. Winzer, H. Haunstein, S. Randel, S. Corteselli, B. Zhu, D.W. Peckham, Generation and FEC-decoding of a 231.5-Gb/s PDM-OFDM signal with 256-iterative-polar-modulation achieving 11.15-b/s/Hz intrachannel spectral efficiency and 800-km reach, in: Optical Fiber Communication Conference (OFC), 2012, PDP5B.3.

[78] T. Omiya, K. Toyoda, M. Yoshida, M. Nakazawa, 400 Gbit/s frequency-division-multiplexed and polarization-multiplexed 256 QAM-OFDM transmission over 400 km with a spectral efficiency of 14 bit/s/Hz, in: Optical Fiber Communication Conference (OFC), 2012, p. OMA2.7.

[79] M. Nazarathy, J. Khurgin, R. Weidenfeld, Y. Meiman, P. Cho, R. Noe, I. Shpantzer, V. Karagodsky, Phased-array cancellation of nonlinear FWM in coherent OFDM dispersive multi-span links, Opt. Express 16 (20) (2008) 15,777–15,810.

[80] K. Inoue, Phase-mismatching characteristic of four-wave mixing in fiber lines with multistage optical amplifiers, Opt. Lett. 17 (1992) 801–803.

[81] R.W. Tkach, A.R. Chraplyvy, F. Forghieri, A.H. Gnauck, R.M. Derosier, Four-photon mixing and high-speed WDM systems, J. Lightwave Technol. 13 (1995) 841–849.

[82] X. Chen, A. Li, G. Gao, W. Shieh, Experimental demonstration of improved fiber nonlinearity tolerance for unique-word DFT-spread OFDM systems, Opt. Express 19 (2011) 26198–26207.

[83] E. Yamada, A. Sano, H. Masuda, E. Yamazaki, T. Kobayashi, E. Yoshida, K. Yonenaga, Y. Miyamoto, K. Ishihara, Y. Takatori, T. Yamada, H. Yamazaki, 1 Tb/s (111 Gb/s/ch × 10ch) no-guard-interval CO-OFDM transmission over 2100 km DSF, in: Opto-Electronics Communications Conference/Australian Conference on Optical Fiber Technology (OECC), 2008, Paper PDP6.

[84] G. Goldfarb, G.F. Li, M.G. Taylor, Orthogonal wavelength-division multiplexing using coherent detection, IEEE Photon. Technol. Lett. 19 (2007) 2015–2017.

[85] A.J. Lowery, S. Wang, M. Premaratne, Calculation of power limit due to fiber nonlinearity in optical OFDM systems, Opt. Express 15 (2007) 13,282–13,287.

[86] M. Mayrock, H. Haunstein, Monitoring of linear and nonlinear signal distortion in coherent optical OFDM transmission, J. Lightwave Technol. 27 (2009) 3560–3566.

[87] P.P. Mitra, J.B. Stark, Nonlinear limits to the information capacity of optical fiber communications, Nature 411 (2001) 1027–1030.

[88] J. Tang, The channel capacity of a multispan DWDM system employing dispersive nonlinear optical fibers and an ideal coherent optical receiver, J. Lightwave Technol. 20 (2002) 1095–1101.

[89] X. Chen, W. Shieh, Closed-form expressions for nonlinear transmission performance of densely spaced coherent optical OFDM systems, Opt. Express 18 (2010) 19,039–19,054.

[90] W. Shieh, X. Chen, Information spectral efficiency and launch power density limits due to fiber nonlinearity for coherent optical OFDM systems, IEEE Photon. J. 3 (2011) 158–173.

[91] C.E. Shannon, A mathematical theory of communication, Bell Syst. Tech. J. 27 (1948)

[92] A.J. Lowery, Fiber nonlinearity pre- and post-compensation for long-haul optical links using OFDM, Opt. Express 15 (2007) 12965–12970.

[93] X. Liu, F. Buchali, R.W. Tkach, Improving the nonlinear tolerance of polarization-division-multiplexed CO-OFDM in long-haul fiber transmission, J. Lightwave Technol. 27 (2009) 3632–3640.

[94] E. Ip, J.M. Kahn, Compensation of dispersion and nonlinear impairments using digital backpropagation, J. Lightwave Technol. 26 (20) (2008) 3416–3425.

[95] E. Mateo, L. Zhu, G. Li, Impact of XPM and FWM on the digital implementation of impairment compensation for WDM transmission using backward propagation, Opt. Express 16 (2008) 16,124–16,137.

[96] R. Weidenfeld, M. Nazarathy, R. Noe, I. Shpantzer, Volterra nonlinear compensation of 100G coherent OFDM with baud-rate ADC, tolerable complexity and low intra-channel FWM/XPM error propagation, in: Optical Fiber Communication Conference (OFC), 2010, Paper OTuE3.

[97] B. Du, A.J. Lowery, Improved single channel back propagation for intra-channel fiber nonlinearity compensation in long-haul optical communication systems, Opt. Express 18 (16) (2010) 17075–17088.

[98] G.P. Agrawal, Nonlinear Fiber Optics, Academic Press, San Diego, California, 1989.

[99] X. Liu, S. Chandrasekhar, B. Zhu, P.J. Winzer, A.H. Gnauck, D.W. Peckham, Transmission of a 448-Gb/s reduced-guard-interval CO-OFDM signal with a 60-GHz optical bandwidth over 2000 km of ULAF and five 80-GHz-grid ROADMs, in: Optical Fiber Communication Conference (OFC), 2010, Paper PDPC2.

[100] X. Liu, S. Chandrasekhar, P. Winzer, B. Zhu, D.W. Peckham, S. Draving, J. Evangelista, N. Hoffman, C.J. Youn, Y. Kwon, E.S. Nam, 3 × 485-Gb/s WDM transmission over 4800 km of ULAF and 12 × 100-GHz WSSs using CO-OFDM and single coherent detection with 80-GS/s ADCs, in: Optical Fiber Communication Conference (OFC), 2011, Paper JThA37.

[101] X. Liu. S. Chandrasekhar, P.J. Winzer, S. Draving, J. Evangelista, N. Hoffman, B. Zhu, D.W. Peckham, Single coherent detection of a 606-Gb/s CO-OFDM signal with 32-QAM subcarrier modulation using 4 × 80-Gsamples/s ADCs, in: Opto-Electronics Communications Conference/Australian Conference on Optical Fiber Technology (OECC), 2010, Paper PD2.6.

[102] S. Zhang, M. Huang, F. Yaman, E. Mateo, D. Qian, Y. Zhang, L. Xu, Y. Shao, I. Djordjevic, T. Wang, Y. Inada, T. Inoue, T. Ogata, Y. Aoki, 40 × 117.6 Gb/s PDM-16QAM OFDM Transmission over 10,181 km with Soft-Decision LDPC Coding and Nonlinearity Compensation, in: Optical Fiber Communication Conference (OFC), 2012, Paper PDP5C.4.

[103] D. Qian, M. Huang, S. Zhang, P.N. Ji, Y. Shao, F. Yaman, E. Mateo, T. Wang, Y. Inada, T. Ogata, Y. Aoki, Transmission of 115 × 100G PDM-8QAM-OFDM channels with 4bits/s/Hz spectral efficiency over 10,181 km, in European Conference and Exhibition on Optical Communication (ECOC), 2011, Paper Th.13.K.3.

[104] N. Cvijetic, A. Tanaka, Y. Huang, M. Cvijetic, E. Ip, Y. Shao, T. Wang, 4+G mobile backhaul over OFDMA/TDMA-PON to 200 cell sites per fiber with 10 Gb/s upstream burst-mode operation enabling <1 ms transmission latency, in: Optical Fiber Communication Conference, OSA Technical Digest (Optical Society of America, 2012), Paper PDP5B.7.

[105] T. Hayashi, T. Taru, O. Shimakawa, T. Sasaki, E. Sasaoka, Ultra-low-crosstalk multi-core fiber feasible to ultra-long-haul transmission, in: Optical Fiber Communication Conference (OFC), 2011, Paper PDPC2.

[106] J. Sakaguchi, B.J. Puttnam, W. Klaus, Y. Awaji, N. Wada, A. Kanno, T. Kawanishi, K. Imamura, H. Inaba, K. Mukasa, R. Sugizaki, T. Kobayashi, M. Watanabe, 19-core fiber transmission of $19 \times 100 \times 172$-Gb/s SDM-WDM-PDM-QPSK signals at 305 Tb/s, in: Optical Fiber Communication Conference (OFC), 2012, Paper PDP5C.2.

[107] R. Ryf, R. Essiambre, A. Gnauch, S. Randel, M.A. Mestre, C. Schmidl, P. Winzer, R. Delbue, P. Pupalaikis, A. Sureka, T. Hayashi, T. Taru, T. Sasaki, Space-division multiplexed transmission over 4200 km 3-core microstructure fiber, in: Optical Fiber Communication Conference (OFC), 2012, Paper PDP5C.3.

[108] B. Zhu, T. Taunay, M. Fishteyn, X. Liu, S. Chandrasekhar, M. Yan, J. Fini, E. Monberg, F. Dimarcello, Space-, wavelength-, polarization-division multiplexed transmission of 56-Tb/s over a 76.8-km seven-core fiber, in: Optical Fiber Communication Conference (OFC), 2011, Paper PDPB7.

[109] B. Zhu, T. Taunay, M. Fishteyn, X. Liu, S. Chandrasekhar, M. Yan, J. Fini, E. Monberg, F. Dimarcello, 112-Tb/s space-division multiplexed DWDM transmission with 14-b/s/Hz aggregate spectral efficiency over a 76.8-km seven-core fiber, Opt. Express 19 (2011) 16665–16671.

[110] S. Berdagué, P. Facq, Mode division multiplexing in optical fibers, Appl. Opt. 21 (1982) 1950–1955.

[111] H.R. Stuart, Dispersive multiplexing in multimode optical fiber, Science 289 (2000) 281–283.

[112] B.C. Thomsen, MIMO enabled 40 Gb/s transmission using mode division multiplexing in multimode fiber, in: Optical Fiber Communication Conference (OFC), 2010, Paper OThM6.

[113] B. Franz, D. Suikat, R. Dischler, F. Buchali, H. Buelow, High speed OFDM data transmission over 5 km GI-multimode fiber using spatial multiplexing with 2×4 MIMO processing, in: European Conference and Exhibition On Optical Communication (ECOC), 2010, Tu3.C.4.

Optical OFDM and Nyquist Multiplexing

Juerg Leuthold [a,b] and Wolfgang Freude[c,d]

[a]*Swiss Federal Institute of Technology (ETH), Zurich, Switzerland*
[b]*Institute of Photonic Systems (IPS), Taiwan, China*
[c]*Karlsruhe Institute of Technology (KIT), Germany*
[d]*Institute of Photonics and Quantum Electronics (IPQ), Germany*

9.1 INTRODUCTION

Intricate pulse-shaping techniques have attracted the attention of the community. With these pulse-shaping techniques orthogonal frequency division multiplexing (OFDM) [1–11], Nyquist multiplexing [12–16], and other multiplexing schemes emerged [17]. Through these techniques transmission of information with highest spectral efficiencies has become possible. A recent transmission experiment shows that signal encoding with 15 bit/s/Hz and beyond is within reach [25]. Prerequisites for such multiplexing schemes are advanced optical transmitters that encode the information as complex-valued, specially shaped pulse envelopes of optical carriers. The shapes are selected such that signals are orthogonal in time and frequency. This allows for demultiplexing of appropriately generated signals even if they overlap in time and/or in the frequency domain. In addition, it has been suggested that such pulse-shaping can increase the nonlinear impairment tolerance [14,19–21].

A variety of pulse shapes have been investigated: rectangularly shaped [1], sinc-shaped [12], and raised-cosine [22] shaped pulses are just a few. The important question now is how optical transmitters can generate such pulses at the necessary speed.

In the past, pulse-shaping has been used excessively in medium and long-haul communication links. Pulse shapes were sculptured by so-called Mach-Zehnder interferometer (MZI) pulse carvers. With pulse carvers, one could trim non-return-to-zero (NRZ) signals into return-to-zero (RZ) or carrier-suppressed return-to-zero (CSRZ) signals. This was useful for reducing inter-symbol interference (ISI) in optical channels and helped in improving the transmission performance of the signal [23]. Yet, care had to be taken to ensure that signal spectra would not overlap significantly. This was necessary to ensure that demultiplexing with simple wavelength division multiplexing (WDM) filters would be possible without inter-channel interference (ICI). And indeed, in combination with advanced modulation formats that occupy a smaller spectral range for constant data rates, WDM systems with impressively dense optical channel spacing of down to 25 GHz have been built [24].

Currently, more versatile methods and pulse shapes are investigated, where digital signal processors, electronic filters or all-optical approaches generate the desired shape. Shaping signals with digital signal processing (DSP) is of increasing interest due to the availability of massively parallel high-speed electronics. We now can not only encode any complex value onto an optical carrier envelope forming a symbol during its associated time slot, but we even have sufficient processing bandwidth to shape it realtime within the symbol duration to a sufficient precision. Digital signal processing in electronics is extremely popular due to its extra-ordinary flexibility. Future software defined optical transmitters can adapt modulation formats, pulse shapes, and symbol rates at the push of a button [25]. Also, electronic implementations can easily combine hundreds of lower-speed tributaries into a single high-speed data stream. On the other hand, pulse forming and multiplexing in the optical domain is of interest for shaping and encoding optical carrier envelopes at speeds that are beyond present-day limits of electronics. All-optical analogue techniques have additional advantages over DSP, as quantization and the associated necessary clipping can be avoided. Further, all-optical processing has potential to be more energy-efficient than DSP. Gene rally, we observe that the "all-DSP" pulse-shaping approaches are most popular when lower-speed tributaries are combined into a super-channel at a moderate aggregate data rate. In turn, the "all-optical" approaches are attractive to combine high-bitrate channels spectrally efficient into multiple Tbit/s super-channels. In practice many implementations are neither "all-DSP" nor "all-optical" but use in part optical and in part DSP techniques. Thus one may imagine that many combinations of the schemes discussed below being implemented.

We exemplarily depict in Figure 9.1 four distinct multiplex implementations, where pulse-shaping plays an important role. All four approaches have been built and tested by our group. The two implementations on top show two "all-optical" or "optical" transmitter and receiver concepts. In these schemes the information is first encoded onto optical carriers before being multiplexed onto a fibre channel. The two schemes at the bottom rely exclusively on digital signal processing in the electronic domain. Here, the various tributaries are multiplexed in the electronic domain and the complete multi plex with all tributaries is modulated on an optical carrier. We now will briefly review the four implementations before discussing them thoroughly in the subsequent sections. It should be noted though that many variations of the four schemes exist. While the basic operations indicated in Figure 9.1 need to be implemented in all setups, the sequence of these operations might be interchanged. Operations performed in the electronic domain might also be implemented optically and vice versa. So, for instance, the "pulse-shaping" operation in Figure 9.1a is indicated to be performed in the optical domain. However, it could as well have been implemented in the electronic domain and be applied before the signal is upconverted into the optical domain by the E/O mixer.

A typical optical implementation of a transmitter (Tx) and a receiver (Rx) that exploits pulse-shaping is depicted in Figure 9.1a. The figure shows how the pulses of e.g. N complex symbols are first converted from the electronic to the optical domain (E/O-box) by an IQ-mixer ($\otimes$, also known as "optical IQ-modulator" in the optics

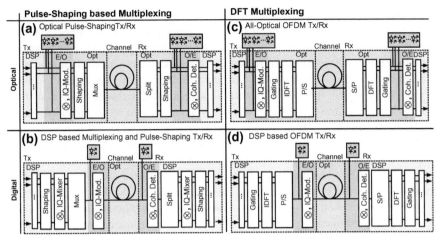

FIGURE 9.1 Transmitter-receiver concepts for multiplexing at highest spectral efficiencies. Subfigures (a) and (b) show transceiver concepts where symbols are encoded with a set of orthogonal pulse shapes such that they can be easily demultiplexed. Subfigures (c) and (d) show OFDM transceiver concepts where the orthogonality of the pulse shapes is exploited to encode and decode information using the inverse discrete Fourier transform (IDFT) and discrete Fourier transform (DFT) operation, respectively. The schemes (a) and (c) at the top and the schemes (b) and (d) at the bottom differ in the implementation of the multiplexing. In schemes (a) and (c), the symbols are first converted into the optical domain by IQ-modulators and then optically multiplexed. In schemes (b) and (d), the signals are modulated on different radio frequency carriers and multiplexed electronically. Subsequently, an optical IQ-modulator encodes the multiplex on a single optical carrier.

community). The form of the pulses is then carved to a particular shape (Shaping). Finally, the N optical signals are multiplexed (Mux) to the channel, which is transported in a single optical fibre. At the receiver the optical signal is first split into N identical copies (Split). Optical filters can then retrieve the signals with the respective orthogonal shape (Shaping). The respective sub-channels are downconverted by tuning the optical local oscillator frequency of a coherent receiver (O/E conversion, here symbolized by another IQ-mixer symbol, $\otimes$) to the frequency of the respective optical carrier. A recent example of such a transmission system has been given in, e.g. Ref. [26]. In this paper, data with a line rate of 32 Tbit/s were encoded on a single laser and transmitted over a fibre length of 227 km. For the said paper dual-polarization 16 QAM data were first Nyquist sinc-shaped and then encoded onto a total of 325 optical carriers (local oscillators) that were all derived from a single laser. Ref. [26] is also an example where pulse-shaping is performed in the digital domain prior to the O/E conversion. In the text below we will give an example where optical pulse-shaping is performed after O/E conversion. Ref. [27] shows an OFDM transmitter that is implemented in the spirit of Figure 9.1a.

A DSP-based approach of a transceiver that exploits pulse-shaping for efficient multiplexing is depicted in Figure 9.1b. In this implementation a number of N complex symbols is shaped in the electronic domain (Shaping). The shaped tributaries are mixed on N individual radio frequency (RF) subcarriers ($\otimes$), each of which has a different frequency. The modulated RF carriers are multiplexed (Mux) to form a broadband RF signal. It is only then that this electronic signal with all of its N tributaries is converted into the optical domain (E/O conversion) by an optical IQ-mixer ($\otimes$). The DSP-based receiver operates similarly. The optical signal is first detected in a coherent detector and mapped to the baseband with all its tributaries ($\otimes$). The electronic signal is split into N signal streams (Split). The individual subcarriers are subsequently selected by tuning an electronic local oscillator to the frequency of the respective subcarrier by means of an analogue electronic or DSP-based IQ-mixer ($\otimes$). A DSP processor then separates the carrier from its neighbours in a digital filter, exploiting the orthogonality of the pulse shapes (Shaping). A recent example illustrates the concept [28]. In this reference, several tributaries with different symbol rates are Nyquist sinc-pulse shaped and multiplexed to a larger bitrate in the electronic domain with a Nyquist frequency division multiplexing (FDM) scheme. In this example there are no guard bands between the tributary channels. Details on the practical implementation of 100 Gbit/s real-time Nyqist pulse-shaping transmitters were given in Ref. [14]. The potential of such a pulse forming technique has recently been demonstrated, when a Nyquist sinc-pulse transmitter generated a signal with a spectral efficiency of 18 bit/s/Hz and a net spectral efficiency of 15 bit/s/Hz [18].

The schemes depicted in Figure 9.1c and d visualize an orthogonal frequency division multiplexing (OFDM) transmitter and a corresponding receiver. In OFDM, sinusoidals with complex amplitudes and equally spaced frequencies inside a rectangular (rect-shaped) time window are multiplexed to form the OFDM signal. It can be shown that both multiplexing and demultiplexing operations can be performed efficiently by an inverse discrete Fourier transform (IDFT) and a discrete Fourier transform (DFT). The rectangular pulse shape in the time domain is automatically given by the finite length of the IDFT/DFT window.

The scheme in Figure 9.1c implements the IDFT and DFT optically for multiplexing and demultiplexing optical tributaries. The information is first encoded on N optical subcarriers ($\otimes$). Each of the optical signals can carry a data rate larger than 100 Gbit/s with a suitable choice of the modulation format. The optical signals then need to be tailored to the proper length (Gating) before being processed by an optical IDFT-processor. Subsequently, the N parallel data streams are serialized (P/S). The receiver retrieves the data with the inverse processes, including serial-to-parallel (S/P) conversion, DFT optical processing, optical gating, and coherent detection. The potential of this technology has been illustrated in a recent implementation published in Ref. [29]. In this paper, 325 optical carriers were encoded with data and superimposed to form an optical OFDM signal with an aggregate data rate of 26 Tbit/s. The tributaries were detected using an all-optical DFT in the receiver.

An exemplary electronic implementation of an OFDM transmitter and a receiver is shown in Figure 9.1d. In this scheme, an IDFT is computed for a number of N

data tributaries, which are S/P-converted and then encoded on an optical signal (O/E mixer ⊗). In the receiver, the inverse process includes O/E down-conversion, P/S-conversion, and DFT processing with gating to retrieve the transmitted data. An example of such a 100 Gbit/s real-time OFDM transmitter implementation can be found in Ref. [30].

The chapter is organized as follows. In Section 9.2 we review the theory of orthogonal functions in order to shape pulses for efficient orthogonal multiplexing at the highest spectral efficiency. The emphasis will be on rectangular-shaped pulses for OFDM, sinc-shaped pulses for Nyquist-WDM, and raised-cosine pulses for OFDM with relaxed modulator requirements. However, the theory introduced is more general. Implementation of various pulse-shaping techniques in the digital, the optical, and the electronic domain will be discussed. In Section 9.3 we review the pulse techniques that take advantage of DFT for efficient multiplexing. This will lead us to OFDM and various forms of implementation in the electronic and optical domain. In Section 9.4 we will discuss transmitter and receiver implementations, and an example is discussed in more detail. The chapter concludes with Section 9.5. The appendix (Section 9.6) gives a list of symbols and mathematical relations.

9.2 ORTHOGONAL SHAPING OF TEMPORAL OR SPECTRAL FUNCTIONS FOR EFFICIENT MULTIPLEXING

In this section the theory of orthogonal temporal or spectral shaping is discussed using a generic transmission system as displayed in Figure 9.2. It consists of a transmitter (Tx), a transmission channel, and a receiver (Rx). For this description it is irrelevant whether the information is mapped on radio frequency (RF) or optical carriers using an electrical or an optical IQ-mixer. We therefore do not differentiate between an electrical mixer, an electro-optical mixer (IQ-modulator) or an opto-electrical mixer (also known as a coherent detector) such as in Figure 9.1. However, for being definite, and because an optical fibre transmission channel is assumed, we imagine optical subcarriers.

Before looking at the detailed description of the subsystems involved, we explain the idea of "orthogonal temporal or spectral shaping." By this we mean that temporal or spectral shapes can be separated in time or frequency, respectively, even if the shapes under consideration overlap strongly, provided that certain orthogonality conditions are met. These conditions are defined in the following Subsection 9.2.1. Afterwards, a formalism to describe transmitter, channel, and receiver will be developed.

9.2.1 Definitions of orthogonality

The term "orthogonal" (*Greek* ὀρθός, straight, and γωνία, angle) refers to the fact of two quantities being perpendicular to each other and therefore independent, as with an orthogonal coordinate system where the axes represent independent directions in

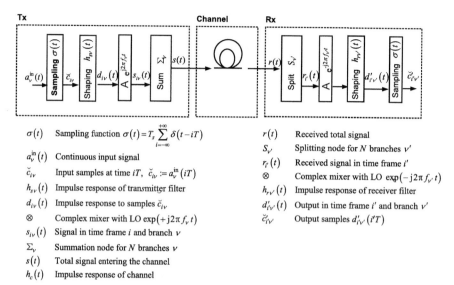

$\sigma(t)$ Sampling function $\sigma(t) = T_s \sum_{i=-\infty}^{+\infty} \delta(t - iT)$

$a_v^{in}(t)$ Continuous input signal

$\breve{c}_{iv}$ Input samples at time iT, $\breve{c}_{iv} := a_v^{in}(iT)$

$h_{sv}(t)$ Impulse response of transmitter filter

$d_{iv}(t)$ Impulse response to samples $\breve{c}_{iv}$

$\otimes$ Complex mixer with LO $\exp(+j2\pi f_v t)$

$s_{iv}(t)$ Signal in time frame i and branch v

Σ_v Summation node for N branches v

$s(t)$ Total signal entering the channel

$h_c(t)$ Impulse response of channel

$r(t)$ Received total signal

$S_{v'}$ Splitting node for N branches v'

$r_{i'}(t)$ Received signal in time frame i'

$\otimes$ Complex mixer with LO $\exp(-j2\pi f_{v'} t)$

$h_{rv'}(t)$ Impulse response of receiver filter

$d'_{i'v'}(t)$ Output in time frame i' and branch v'

$\breve{c}'_{i'v'}$ Output samples $d'_{i'v'}(i'T)$

FIGURE 9.2 Generic transmission system with transmitter (Tx), channel, and receiver (Rx). (Left) Tx with *N* branches *v*. Each branch consists of a sampler creating a time series of complex data coefficient $\breve{c}_{iv}$ from a continuous complex baseband input signal $a_v^{in}(t)$ at sampling time *iT*, an impulse-forming filter with impulse response $h_{sv}(t)$, and a complex (in-phase/quadrature, IQ) mixer with local oscillator (LO) described by an analytic signal $\exp(j2\pi f_v t)$, which up-converts the baseband filter output $d_{iv}(t)$ to the sub-channel frequency f_v. The summing node Σ_v adds the up-converted signals $s_{iv}(t)$ in time frame *i* for all branches *v*. The resulting total signal *s(t)* consists of concatenated time frames and enters the linear channel. (Right) Rx with *N* branches *v'*. The total received signal *r(t)* that leaves the channel enters a splitting node S_v and is distributed with equal amplitudes to the receiver branches *v'*. The received signal $r_{i'v'}(t)$ in time frame *i'* and branch *v'* is frequency down-converted with an IQ mixer and an LO $\exp(-j2\pi f_{v'} t)$. After a receiver filter with impulse response $h_{rv'}(t)$ and output signal $d'_{i'v'}(t)$, a sampler extracts at time $t = i'T + t_{gv'}$ the complex data coefficient $\breve{c}'_{i'v'}$.

space. A similar meaning is implied if we speak of orthogonal modulation techniques, which encode information by modifying the electric field of a carrier independently in amplitude, phase, frequency, or polarization. In general, linearly independent functions $\Psi_m(t)$ are called orthonormal if the inner product $\langle \Psi_{m'} | \Psi_m \rangle$ on the interval $[a, b]$ normalized by a weight function $\rho(t)$ is (Kronecker symbol Eq. (9.59))

$$\langle \Psi_{m'} | \Psi_m \rangle_\rho = \int_a^b \Psi_{m'}^*(t) \Psi_m(t) \, \rho(t) dt = \delta_{mm'}, \quad m, m' \in \mathbb{Z}. \tag{9.1}$$

If the inner product equals a constant different than one, the functions are simply named orthogonal. The following orthonormality relations are important,

$$\frac{1}{T} \int_{-\infty}^{+\infty} \text{rect}\left(\frac{t}{T} - m\right) \text{rect}\left(\frac{t}{T} - m'\right) dt = \delta_{mm'},$$

$$\frac{1}{T} \int_{-\infty}^{+\infty} \text{sinc}\left(\frac{t}{T} - m\right) \text{sinc}\left(\frac{t}{T} - m'\right) dt = \delta_{mm'},$$

$$\frac{1}{T} \int_{T_0-T/2}^{T_0+T/2} \exp\left(+j2\pi m\frac{t}{T}\right) \exp\left(-j2\pi m'\frac{t}{T}\right) dt = \delta_{mm'},$$

$$\frac{1}{F} \int_{F_0-F}^{F_0+F} \cos\left(\frac{\pi}{2}\frac{f}{F}\right) \exp\left(+j2\pi m\frac{f}{F}\right) \cos\left(\frac{\pi}{2}\frac{f}{F}\right) \exp\left(-j2\pi m'\frac{f}{F}\right) df = \delta_{mm'}.$$

$$(9.2)$$

The functions $\text{rect}(t/T)$ and $\text{sinc}(t/T)$ are defined in Eqs. (9.62) and (9.65), respectively. The first line in Eq. (9.2) states that rectangular non-overlapping impulses do not interfere, which is very obvious. The second line tells us that strongly overlapping sinc-pulses are also free of inter-symbol interference (ISI), as long as each impulse maximum falls on zeros of neighbouring sinc-pulses. The third line in Eq. (9.2) formulates the orthonormality of harmonic functions having frequency differences that are integer multiples of the basic frequency $1/T$. This relation applies for orthogonal frequency division multiplexing (OFDM), where inside a symbol period T many harmonics with different complex weights can be discriminated, if the weights are constant during T. Similar to sinc-pulses, sinc-like pulses [22] $\text{sincl}(t/T)$ as, for example, defined in Eq. (9.66) share the property that impulse maxima and zeros of neighbouring sincl-pulses can be made to coincide, but such pulses are not necessarily orthogonal in the sense of Eq. (9.2) and as such cannot be detected ISI-free when using a matched filter [31,32]. However, when employing pulses shaped with a root raised-cosine filter [33] having a transfer function according to Eq. (9.64), a matched receiver filter can be found. The last line of Eq. (9.2) establishes in the frequency domain that pulses with a root raised-cosine spectrum can be detected ISI-free when using a matched filter. We revisit related problems in Section 9.2.5.4.

There are many more orthogonality relations that would open the route to other promising multiplexing techniques. Zhao and Ellis [34], for instance, defined an orthogonality relation for fast orthogonal frequency division multiplexing (FODM) with half the standard OFDM carrier spacing $1/T$. We restrict ourselves to the most basic cases.

If a function $\psi_\nu[n]$ is defined in an interval $[0, T]$ only on a discrete set of N points $t_n = nT/N$, a discrete form of the scalar product Eq. (9.1) defines orthonormality,

$$\langle \psi_{\nu'}|\psi_\nu\rangle_{1/N} = \frac{1}{N} \sum_{n=0}^{N-1} \psi_{\nu'}^*[n]\,\psi_\nu[n] = \delta_{\nu\nu'}, \quad n, \nu, \nu' \in \mathbb{N}_0. \qquad (9.3)$$

As a second subscript we use the Greek letter ν (nu).

As an example, the discrete set of T-periodic harmonic functions obeys the ortho-normality relation

$$\frac{1}{N} \sum_{n=0}^{N-1} \exp\left(+j2\pi \frac{vn}{N}\right) \exp\left(-j2\pi \frac{v'n}{N}\right) = \delta_{vv'}. \tag{9.4}$$

Equation (9.4) is obviously true for $v = v'$. If $v \neq v'$ holds, the sum formula for a geometric progression must be employed, $\sum_{n=0}^{N-1} x^n = (1 - x^N)/(1 - x) = 0$ with $x = \exp\left[j2\pi (v - v')/N\right]$.

9.2.2 Transmitter

The schematic of a generic transmitter is to be seen in Figure 9.2. Only one out of N branches v is depicted. A continuous complex data signal $a_v^{in}(t)$ enters a sampling circuit in the Tx-branch v. The data signal $a_v^{in}(t)$ is sampled at times $t = iT$ ($i \in \mathbb{Z}$) with the sampling function $\sigma(t) = T_s \sum_{i=-\infty}^{+\infty} \delta(t - iT)$. The samples $a_v^{in}(t)\sigma(t) = a_v^{in}(t)T_s \sum_{i=-\infty}^{+\infty} \delta(t - iT)$ are the complex data coefficients $\check{c}_{iv} := a_v^{in}(iT)$. For simplicity, we left out the steps of quantization and coding. However, the coefficients $\check{c}_{iv}$ can be thought of as representing complex constellation points belonging to the constellation diagram of an advanced modulation format. Subsequently, the samples are fed into a filter having an impulse response $h_{sv}(t)$. The resulting (possibly non-causal) baseband impulse responses are centered at $t = iT$, repeat with the clock rate $f_T = 1/T$, and determine the shape of the baseband signal $d_{iv}(t)$ for each time frame i,

$$d_{iv}(t) = \int_{-\infty}^{+\infty} a_v^{in}(t')T_s\delta(t' - iT)h_{sv}(t - t')dt' = \check{c}_{iv}T_sh_{sv}(t - iT),$$
$$\check{c}_{iv} := a_v^{in}(iT). \tag{9.5}$$

The complex coefficients $\check{c}_{iv}$ represent the baseband data signal in branch v at time $t = iT$. A complex mixer (in-phase/quadrature mixer, IQ mixer) converts the baseband signal to the frequency f_v of a complex subcarrier represented by the analytic signal $\exp(j2\pi f_v t)$,

$$f_v = vF_s + f_c, \quad v = 0, 1, \ldots, N - 1, \quad F_s = 1/T_s, \quad \text{channel frequency } f_c. \tag{9.6}$$

As a result, a signal $s_{iv}(t)$ at subcarrier frequency f_v is generated,

$$s_{iv}(t) = \check{c}_{iv}T_sh_{sv}(t - iT)\exp(j2\pi f_v t). \tag{9.7}$$

It enters the node Σ_v where the subcarrier signals from all branches are summed (or multiplexed) to form the signal $s_i(t)$. Finally, the infinitely many consecutive time frames i constitute the signal $s(t)$ that is sent across the channel,

$$s(t) = \sum_{i=-\infty}^{+\infty} s_i(t), \quad s_i(t) = \sum_{v=0}^{N-1} s_{iv}(t). \tag{9.8}$$

9.2.3 Channel

The optical singlemode fibre channel is assumed to be linear and lossless. It has a length L and is modeled by its transfer function $\check{h}_c(f)$, where the propagation constant $\beta(\omega)$ of the fundamental mode is represented by a Taylor series at the channel frequency $f_c = \omega_c/(2\pi)$ up to the mth order ($m \in \mathbb{N}_0$),

$$\check{h}_c(f) = \exp[-\mathrm{j}\,\beta(\omega)L], \beta(\omega) \approx \beta_c^{(0)} + (\omega - \omega_c)\beta_c^{(1)} + \frac{(\omega - \omega_c)^2}{2!}\beta_c^{(2)} + \dots,$$

$$\beta_c^{(m)} = \frac{\mathrm{d}^m \beta}{\mathrm{d}\omega^m}\bigg|_{\omega=\omega_c}. \tag{9.9}$$

If the channel was distortionless, i.e. if $\beta_c^{(m \geqslant 2)} = 0$, then the carrier envelope would be transmitted unchanged, but with a group delay $t_g = \beta_c^{(1)}L$, while the carrier itself would be retarded in phase by $\varphi = -\beta_c^{(0)}L$. The effect of the channel then is to delay the signal impulse response $h_{sv}(t)$ of the transmitter by the group delay t_{gv}, and to retard the carrier phase by φ_{gv}, both of which are known for every v (or have to be estimated by the receiver).

The channel will further introduce envelope distortions due to dispersion if for any $m \geqslant 2$ the condition $\beta_c^{(m)} \neq 0$ is fulfilled. To simplify the mathematical formulation, we incorporate these linear distortions due to transmission into each of the transmitter's sub-channel baseband impulse responses so that the sent signal impulse response $h_{sv}(t)$ will be modified to be $h'_{sv}(t)$. The received signal $r(t)$ then reads

$$r(t) = \sum_{i=-\infty}^{+\infty} r_i(t), \quad r_i(t) = T_s \sum_{v=0}^{N-1} \check{c}_{iv}h'_{sv}\left(t - (iT + t_{gv})\right) \exp[\mathrm{j}\,(2\pi f_v t - \varphi_v)]. \tag{9.10}$$

9.2.4 Receiver

In the receiver, the data coefficients $\check{c}_{iv}$ are to be extracted. A schematic of a generic receiver that performs this operation is shown at the right-hand side of Figure 9.2. Again, only one out of N branches v' is depicted. The received signal $r(t)$ is distributed by a splitting node $S_{v'}$ with equal magnitudes and phases to N receiver branches v'. We first down-convert the received signal $r_{i'}(t)$ in branch v' for each time frame i' with a complex mixer and an LO $\exp(-\mathrm{j}2\pi f_{v'}t)$. Then we convolve the down-converted signal with the receiver's impulse response $h_{rv'}(t)$

$$d'_{i'v'}(t) = \sum_{i=-\infty}^{+\infty} \int_{-\infty}^{+\infty} r_i(t') \exp(-\mathrm{j}2\pi f_{v'}t')h_{rv'}(t - t')\mathrm{d}t'. \tag{9.11}$$

The received branch signal $d'_{i'v'}(t)$ in time slot i' comprises the transmitted signal at subcarrier frequency v', but it could also be perturbed by interferences from sub-channels with subcarrier frequencies $v \neq v'$ other than the present receiving branch v' (inter-channel interference, ICI), and by neigbouring data inside the same branch v' but belonging to different time slots $i \neq i'$ (inter-symbol interference, ISI). We rewrite Eq. (9.11) to express this fact by an interference term $I_{i,v'v}(t)$,

$$d'_{i'v'}(t) = \sum_{i=-\infty}^{+\infty} \sum_{v=0}^{N-1} \exp(-\mathrm{j}\varphi_v) I_{i,v'v}(t). \tag{9.12}$$

The interference term $I_{i,v'v}(t)$ is represented either by integrating weighed impulse responses $h'_{sv}(t), h_{rv'}(t)$ in time, or by integrating in the frequency domain the weighed transfer functions $\check{h}'_{sv}(f), \check{h}_{rv'}(f)$, which are the Fourier transforms Eq. (9.67) of the respective impulse responses,

$$I_{i,v'v}(t) = \check{c}_{iv} T_s \int_{-\infty}^{+\infty} h'_{sv}\left(t' - \left(iT + t_{gv}\right)\right) h_{rv'}(t - t') \exp[-\mathrm{j}2\pi(f_{v'} - f_v)t']\mathrm{d}t'$$

$$= \check{c}_{iv} T_s \exp[-\mathrm{j}2\pi(f_{v'} - f_v)t]$$

$$\times \int_{-\infty}^{+\infty} \check{h}'_{sv}(f)\check{h}_{rv'}(f - (f_{v'} - f_v)) \exp\left[\mathrm{j}2\pi f\left(t - \left(iT + t_{gv}\right)\right)\right] \mathrm{d}f. \tag{9.13}$$

At sampling time $t = i'T + t_{gv'}$ we detect $d'_{i'v'}(i'T + t_{gv'})$ and name it received complex data coefficient $\check{c}'_{i'v'}$ in receiver branch v',

$$\check{c}'_{i'v'} := d'_{i'v'}(i'T + t_{gv'}) = \sum_{i=-\infty}^{+\infty} \sum_{v=0}^{N-1} \exp(-\mathrm{j}\varphi_v) I_{i,v'v}(i'T + t_{gv'}). \tag{9.14}$$

9.2.5 Avoiding inter-channel and inter-symbol interference

The goal of the data transmission system Figure 9.2 is to assure the identity of the received data coefficient $\check{c}'_{i'v'}$ in sub-channel v' with its sent counterpart $\check{c}_{iv}$ for the case that the receiver's sub-channel frequency $f_{v'}$ and the optimum sampling time $t = i'T + t_{gv'}$ were chosen correctly. In this subsection we show how the receiver filters then depend on the respective pulse-shaping technique in the sender.

As already remarked, the received $\check{c}'_{i'v'}$ could be perturbed by interferences from sub-channels $v \neq v'$ other than the present receiving branch v' targets to (ICI), and by neighbouring data inside the same sub-channel v' but belonging to different time slots $i \neq i'$ (ISI). If the data are to be received correctly, the interference term in Eq. (9.14) must reduce to

$$I_{i,v'v}(i'T + t_{gv'}) = \check{c}_{iv}\delta_{ii'}\delta_{vv'}. \tag{9.15}$$

The receiver will then detect

$$\check{c}'_{i'v'} = \check{c}_{i'v'} \exp(-\mathrm{j}\varphi_{v'}). \tag{9.16}$$

The phase term $\exp(-\mathrm{j}\varphi_{v'})$ as well as the correct sampling time parameters i' and $t_{gv'}$ have to be found by channel estimation.

In conventional systems ICI and ISI are avoided by choosing non-overlapping impulses with reasonably large temporal guard intervals between channels (as is customary for time division multiplexing, TDM), or by choosing non-overlapping rect-shaped channel spectra combined with reasonably large time slots (as is customary for wavelength division multiplexing, WDM).

In multi-carrier systems, Eq. (9.13) shows that more is possible. ICI and ISI can also be avoided by choosing non-overlapping rectangular impulse responses with overlapping sinc-shaped channel spectra (orthogonal frequency division multiplexing, OFDM), or by deciding for non-overlapping rect-shaped channel spectra with overlapping sinc-shaped impulse responses (Nyquist pulses, also termed orthogonal time division multiplexing, OTDM). In either cases a matched receiver filter [31,32] $\check{h}_{rv}(f) = \check{h}_{sv}^{'*}(f)$ happens to be the best solution with respect to an optimum signal-to-noise power ratio.

If the nominally rectangular or sinc-shaped transmitter impulse responses are either intentionally pre-distorted by the transmitter, or unintentionally changed by the transmission channel, the goal formulated in Eq. (9.15) still holds, but for avoiding ICI and ISI then the *product* of received impulse response $h_{sv}'(t)$ and receiver impulse response $h_{rv'}(t)$ (or the *product* of the corresponding transfer functions $\check{h}_{sv}'(f)$ and $\check{h}_{rv'}(f)$) along with the proper sub-channel frequency and sampling time must be chosen properly. Yet, even when properly combined sender and receiver impulse responses are found and meet the requirement of orthogonality, they might not be the optimum as far as the signal-to-noise power ratio is concerned. It then depends on the special situation whether the influence of noise or the perturbations by interference have the stronger weight, and whether an optimum compromise between both effects can be found.

The following subsections first describe the two most basic cases to meet the goal of ICI-free and ISI-free reception. After that, the problem of temporal and spectral deviations from these basic cases is treated briefly.

9.2.5.1 Orthogonal frequency division multiplexing (OFDM)

For OFDM we choose rectangular transmitter impulse responses, which are assumed to arrive at the receiver undistorted, $h_{sv}'(t) = h_{sv}(t)$, but delayed by t_{gv}, have a maximum symbol period T and are sent at a clock rate $f_T = 1/T \leqslant F_s$, see Figure 9.3. The impulses from different time slots i do not overlap. Centered at $t = 0$ we postulate a region with a width named the symbol duration $T_s < T$, where the impulse response in each sub-channel branch v is constant, while its behavior outside the symbol duration, i.e. in the region $T_s/2 < |t| < T/2$, remains unspecified for the time being,

$$h_{sv}(t) = \begin{cases} 1/T_s & \text{for } |t| < T_s/2, \\ \text{unspecified} & \text{for } T_s/2 < |t| < T/2, \\ 0 & \text{for } T/2 < |t|. \end{cases} \qquad (9.17)$$

OFDM

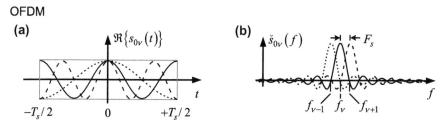

FIGURE 9.3 OFDM pulses in (a) time domain $s_{iv}(t)$ and (b) frequency domain, $\check{s}_{iv}(f)$. For a simplified display, the channel frequency in Eq. 9.6 is set to $f_c = 0$, so that $f_v = vF_s$ holds. The (green) box marks the finite symbol time. (a) The (green) graph shows the real parts of three on-off keyed sinusoidal subcarriers s_{0v}, the sum of which represents one specific time-domain OFDM symbol of duration T_s centred at time $t = iT$ with $i = 0$. (b) The (green) graph shows the corresponding spectrum $\check{s}_{0v}(f)$ of the subcarriers centred at frequencies f_{v-1}, f_v, f_{v+1} with a spectral separation of $|f_{v+1} - f_v| = F_s = 1/T_s$. (Figure derived from Ref. [14]) (For interpretation of the references to colour in this figure legend, the reader is referred to the web version of this book.)

Also, the receiver's filter impulse response for interference-free reception is non-zero only in a region $|t| < T_s/2$, see Eq. (9.62),

$$h_{rv}(t) = \frac{1}{T_s}\mathrm{rect}\left(\frac{t}{T_s}\right) = \begin{cases} 1/T_s & \text{for } |t| < T_s/2, \\ 0 & \text{for } |t| > T_s/2. \end{cases} \quad (9.18)$$

In this case and if the receiver sampling time $t = i'T + t_{gv'} = iT + t_{gv}$ with $i' = i$ and $t_{gv'} = t_{gv}$ has been estimated correctly, the interference term $I_{i,v'v}(i'T + t_{gv'})$ reduces with the help of the orthonormality condition Eq. (9.2) to

$$I_{i,v'v}(i'T + t_{gv'}) = \check{c}_{iv}T_s \int_{-\infty}^{+\infty} h'_{sv}\left(t' - (iT + t_{gv})\right)h_{rv'}\left((i'T + t_{gv'}) - t'\right)$$
$$\times \exp[-\mathrm{j}2\pi(f_{v'} - f_v)t']\mathrm{d}t'$$
$$= \check{c}_{iv}T_s \int_{-\infty}^{+\infty} \frac{1}{T_s}\delta_{ii'}\frac{1}{T_s}\mathrm{rect}\left(\frac{(i'T + t_{gv'}) - t'}{T_s}\right)$$
$$\times \exp[-\mathrm{j}2\pi(f_{v'} - f_v)t']\mathrm{d}t'$$
$$= \check{c}_{iv}\frac{1}{T_s}\int_{i'T+t_{gv'}-T_s/2}^{i'T+t_{gv'}+T_s/2} \delta_{ii'}\exp[-\mathrm{j}2\pi(f_{v'} - f_v)t']\mathrm{d}t' = \check{c}_{iv}\delta_{ii'}\delta_{vv'}. \quad (9.19)$$

According to Eq. (9.14), the output of receiver branch v' restores at sampling time $t = i'T + t_{gv'}$ the true complex data $\check{c}_{i'v'} = d_{i'v'}(i'T + t_{gv'})$ which the transmitter had sent to the receiver branch $v' = v$ at time $t = i'T$,

$$\check{c}'_{i'v'} = d'_{i'v'}(i'T + t_{gv'}) = \sum_{i=-\infty}^{+\infty}\sum_{v=0}^{N-1} \exp(-\mathrm{j}\varphi_v)\check{c}_{iv}\delta_{ii'}\delta_{vv'} = \check{c}_{i'v'}\exp(-\mathrm{j}\varphi_{v'}). \quad (9.20)$$

As remarked before, the phase term $\exp(-j\varphi_{v'})$ and the correct sub-channel group delay $t_{gv'} = t_{gv}$ have to be estimated in a separate step.

In Figure 9.3a an OFDM symbol with duration T_s is depicted [14,30] where the three subcarriers f_{v-1}, f_v, f_{v+1} in time frame $i = 0$ have the same data weights $\check{c}_{0v} = 1$. The subcarriers' spectra are frequency-shifted sinc-functions (non-causal in the frequency domain). They overlap strongly and have a characteristic width of F_s between maximum and first zero. The spectral maxima are separated by $|f_{v+1} - f_v| = F_s = 1/T_s$. This spacing is fixed by the symbol duration T_s (not to be mixed up with the symbol repetition period of $T \geqslant T_s$). At any spectral maximum f_v all neigbouring spectra centered at $f_{v'} \neq v$ are zero.

We now choose the transmitter impulse response to be constant not only for the symbol duration T_s, but for the whole symbol repetition period $T = T_s + T_g$ by adding a so-called guard interval T_g, see Figure 9.4. This guard interval realizes a cyclic prefix (CP) before and a cyclic suffix (CS) after the symbol. Both, CP and CS are of length $T_g/2$,

$$h_{sv}(t) = \frac{1}{T_s}\mathrm{rect}\left(\frac{t}{T}\right) = \begin{cases} 1/T_s & \text{for } |t| < T/2 \\ 0 & \text{for } |t| > T/2 \end{cases}, \quad T = T_s + T_g. \quad (9.21)$$

The so far unspecified region in Eq. (9.17) is now defined. The duration of the receiver impulse response, however, must not change and is kept at T_s. As a consequence, the receiver impulse response window is now allowed to shift inside the wider transmitter impulse response window (width T) by $\pm T_g/2$, so that the optimum receiver sampling point $t = iT + t_{gv} = i'T + t_{gv'}$ can deviate by $\pm T_g/2$,

$$I_{i,v'v}\left(i'T + \left(t_{gv'} \pm T_g/2\right)\right) = \check{c}_{iv}T_s \int_{-\infty}^{+\infty} \frac{1}{T_s}\mathrm{rect}\left(\frac{t' - (iT + t_{gv})}{T}\right)$$
$$\times \frac{1}{T_s}\mathrm{rect}\left(\frac{(i'T + (t_{gv'} \pm T_g/2)) - t'}{T_s}\right)$$
$$\times \exp[-j2\pi(f_{v'} - f_v)t']\mathrm{d}t'$$
$$= \check{c}_{iv}\frac{1}{T_s}\int_{i'T+(t_{gv}\pm T_g/2)-T_s/2}^{i'T+(t_{gv'}\pm T_g/2)+T_s/2} \delta_{ii'}$$
$$\times \exp[-j2\pi(f_{v'} - f_v)t']\mathrm{d}t'$$
$$= \check{c}_{iv}\delta_{ii'}\delta_{vv'}. \quad (9.22)$$

The sent data can be recovered without any ICI or ISI, but now with a tolerance in the receiver's sampling time, which means that for the same sampling time different group delays can be accepted within $\pm T_g/2$, such improving for instance the tolerance with respect to chromatic dispersion of the group delay t_{gv} at subcarrier frequencies f_v,

$$\check{c}'_{i'v'} = d'_{i'v'}\left(i'T + \left(t_{gv'} \pm T_g/2\right)\right) = \check{c}_{i'v'}\exp(-j\varphi_{v'}). \quad (9.23)$$

If the total group delay spread Δt_g across all sub-channels is not larger than the guard interval, $\Delta t_g \leqslant T_g$, then no individual group delays have to be estimated and sampling can be done at the same time for all sub-channels. However, due to the

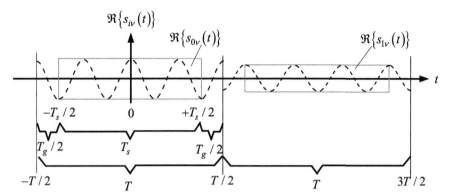

FIGURE 9.4 **OFDM transmitter impulse response with a guard interval T_g. The real part of two consecutive OFDM symbols consisting of only one subcarrier frequency f_v and having different coefficients $\check{c}_{iv}$ is shown. The symbol repetition period T is larger than the symbol duration T_s.**

frequency-dependent phase shift $\varphi_{v'}$, the rotation of each received constellation point $\check{c}_{i'v'} \exp(-j\varphi_{v'})$ has to be compensated.

9.2.5.2 *Orthogonal time division multiplexing (OTDM, sinc-shaped Nyquist pulses)*

For transmitting Nyquist pulses [12,14] we choose (non-causal) sinc-shaped transmitter impulse responses leading to rect-shaped spectra, see Eq. (9.24), and a characteristic temporal width of T_s between maximum and first zero, Figure 9.24. The pulses are again assumed to arrive at the receiver undistorted, $h'_{sv}(t) = h_{sv}(t)$, but delayed by t_{gv}. They have a width that is chosen to equal the clock period $T = T_s$,

$$h_{sv}(t) = \frac{1}{T_s}\mathrm{sinc}\left(\frac{t}{T_s}\right), \quad \check{h}_{sv}(f) = \mathrm{rect}\left(\frac{f}{F_s}\right), \quad T_s = \frac{1}{F_s} = T. \quad (9.24)$$

The maxima of the pulses in different time frames i are separated by $|t_{i+1} - t_i| = T_s = 1/F_s, t_i = iT_s$, so that the clock rate $f_T = 1/T_s = F_s$ equals the Nyquist bandwidth F_s. The sent impulses $s_{iv}(t)$ overlap strongly. At any pulse maximum all neighbouring pulses are zero. The rect-shaped spectra $\check{s}_{iv}(f)$ have the Nyquist bandwidth F_s. Due to the temporal time shift iT_s of subsequent sinc-pulses, the associated spectra are modulated with a spectral harmonic function $\exp(j2\pi f i T_s)$, see Figure 9.5b and Eq. (9.76). If in a Nyquist-WDM (N-WDM) system sinc-pulses are sent at different subcarrier frequencies, their spectra must not overlap. In particular, the associated sinc-pulses need not have the same width for different subcarrier frequencies, and therefore the subcarrier frequencies need not be equidistantly spaced. However, to simplify the notation, we assume equal T_s for all subcarrier frequencies f_v, so that adjacent centres of individual spectra are separated by $|f_{v+1} - f_v| = F_s = 1/T_s$.

Sinc-Shaped Nyquist

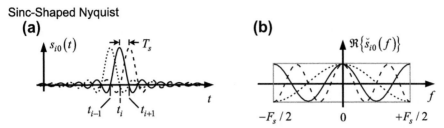

FIGURE 9.5 Sinc-shaped Nyquist pulses in time and in frequency domain. The (green) boxes mark the finite symbol windows in frequency domain. For a simplified display, the channel frequency in Eq. 9.6 is set to $f_c = 0$, so that $f_\nu = \nu F_s$ holds. (a) The graph shows three sinc-shaped Nyquist pulses $s_{i0}(t)$ ("temporal subcarriers") centred at times t_{i-1}, t_i, t_{i+1} with a temporal separation of $|t_{i+1} - t_i| = T_s = 1/F_s$. (b) The graph shows the corresponding real parts of the spectra $\check{s}_{i0}$ with three spectral sinusoidals, the sum of which represents one specific frequency-domain Nyquist symbol centred at $f_0 = 0$. (Figure modified from [14]) (For interpretation of the references to colour in this figure legend, the reader is referred to the web version of this book.)

In the receiver we need a filter transfer function $\check{h}_{r\nu'}(f)$ that provides the minimum ISI and ICI. We minimize the interference term $I_{i,\nu'\nu}(t)$ with its integral formulated in the frequency domain, Eq. (9.13). For the receiver filter we select the same rectangular shape as the one defined in Eq. (9.24) for the transmitter filter transfer function $\check{h}_{s\nu}(f)$, which according to our assumption remains unchanged even after transmission,

$$h_{r\nu'}(t) = \frac{1}{T_s}\operatorname{sinc}\left(\frac{t}{T_s}\right), \quad \check{h}_{r\nu'}(f) = \operatorname{rect}\left(\frac{f}{F_s}\right). \tag{9.25}$$

Because $\check{h}'_{s\nu}(f) = \check{h}_{s\nu}(f)$ and $\check{h}_{r\nu'}(f - (f_{\nu'} - f_\nu))$ in Eq. (9.13) overlap only if $f_{\nu'} = f_\nu$, the condition $\nu' = \nu$ must be met for a non-zero $I_{i,\nu'\nu}(t)$. With the ortho-normality condition Eq. (9.2) we find the result

$$I_{i,\nu'\nu}(i'T_s + t_{g\nu'}) = \check{c}_{i\nu}T_s \exp[-j2\pi(f_{\nu'} - f_\nu)t] \int_{-\infty}^{+\infty} \operatorname{rect}\left(\frac{f}{F_s}\right)\operatorname{rect}\left(\frac{f - (f_{\nu'} - f_\nu)}{F_s}\right)$$

$$\times \exp[j2\pi f\left((i'T_s + t_{g\nu'}) - (iT_s + t_{g\nu})\right)]\mathrm{d}f$$

$$= \check{c}_{i\nu}T_s\delta_{\nu\nu'} \int_{-F_s/2}^{+F_s/2} \exp[j2\pi f\left(i' - i\right)T_s]\mathrm{d}f$$

$$= \check{c}_{i\nu}\delta_{ii'}\delta_{\nu\nu'}. \tag{9.26}$$

Spectral orthogonality is maintained if spectra do not overlap according to the assumptions, so that ICI is eliminated. Temporal orthogonality and therefore disappearing ISI is guaranteed if sampling times are chosen to be $t = iT_s + t_{g\nu} = i'T_s + t_{g\nu'}$, i.e. if sampling takes place at the respective central maxima of the received sinc-pulses.

9.2.5.3 Sinc-like spectra and sinc-like pulses

Ideal sinc-shaped OFDM spectra and ideal sinc-shaped Nyquist pulses cannot be realized in practice, but reception without significant ICI and ISI can be achieved nonetheless. To this end we revisit the interference term $I_{i,v'v}(t)$ from Eq. (9.13). The receiver sampling time is understood to be $t = i'T + t_{gv}$, i.e. we assume that the proper group delay in receiver branch v' has been already found by channel estimation, so that $t_{gv'} = t_{gv}$ holds and only the correct temporal frame i' has to be fixed. We saw that for OFDM it was the rect-shaped non-overlapping transmitter impulse response (including the channel) and the rect-shaped receiver filter impulse response which guaranteed ICI-free and ISI-free reception. For OTDM this goal was achieved with rect-shaped non-overlapping transmitter spectra (including the channel) and rect-shaped receiver filter transfer functions. However, the condition for ISI- and ICI-free detection only requires that the respective temporal or spectral *products*

$$h'_{sv}\left(t' - \left(iT + t_{gv}\right)\right) h_{rv'}(t - t') \quad \text{for OFDM,}$$

$$\check{h}'_{sv}(f)\check{h}_{rv'}\left(f - (f_{v'} - f_v)\right) \quad \text{for Nyquist pulses} \tag{9.27}$$

must retain a rect-shape. Therefore, any deviation of $h'_{sv}(t)$, respectively, $\check{h}'_{sv}(f)$ from a rect-shape can be compensated by a proper choice of $h'_{sv}(t)$, respectively, $\check{h}_{sv}(f)$ so that

$$h'_{sv}\left(t' - \left(iT + t_{gv}\right)\right) h_{rv'}(t - t') = \frac{1}{T_s^2}\text{rect}\left(\frac{t' - (iT + t_{gv})}{T_s}\right)\delta_{ii'} \quad \text{for OFDM,}$$

$$\check{h}'_{sv}(f)\check{h}_{rv'}\left(f - (f_{v'} - f_v)\right) = \text{rect}\left(\frac{f}{F_s}\right)\delta_{vv'} \quad \text{for Nyquist sinc-pulses.} \tag{9.28}$$

With Eq. (9.28) and the orthonormality relation Eq. (9.2), the interference term Eq. (9.13) reduces again to the relation $I_{i,v'v}(i'T + t_{gv'}) = \check{c}_{iv}\delta_{ii'}\delta_{vv'}$ as in Eqs. (9.15), (9.19), and (9.26), so that reception is free of ICI and ISI. For OFDM signals, the successful compensation of a distorted transmitted signal was demonstrated [35] recently.

Neither form of Eq. (9.28) establishes the usual matched-filter [31,32] condition $\check{h}_{rv'}(f) = \check{h}'^*_{sv}(f)$ for optimal reception in case that additive wideband Gaussian noise (AWGN) superimposes the signal. In contrast, we see that, for instance, a drop in the transmitted amplitude spectrum (including the channel) could be compensated by a corresponding bulge in the receiver filter's amplitude spectrum. This is in contrast to the matched-filter requirement where the amplitude spectra meet the condition $|\check{h}'_{sv}(f)| = |\check{h}'^*_{rv'}(f)|$. In any case, the phase spectra of transmitter (including the channel) and receiver filter must be complex conjugate to each other if Eq. (9.27) is to be met for realizing ICI-free and ISI-free reception.

It should be emphasized that only for a limited number of transmitter and receiver impulse responses interference-free reception coincides with optimal matched-filter reception. Well-known examples are rect-shaped (OFDM) and sinc-shaped pulses (Nyquist pulses). Pulses with a root raised-cosine spectrum are as such not ISI-free, but can be detected ISI-free when using a matched filter. In the following section these topics are discussed in more detail.

9.2.5.4 Orthogonality revisited: ICI-free and ISI-free reception

It is often claimed that orthogonality in the sense of virtually error-free detection is maintained if spectral maxima (OFDM, Figure 9.3b) or temporal impulse maxima (OTDM, Figure 9.5b) coincide with zeros of neigbouring spcctra (ICI-free channels) or zeros of impulses (ISI-free symbols), respectively. And indeed, for rectangular OFDM symbols which lead to sinc-shaped spectra, and for sinc-shaped OTDM pulses which lead to rect-shaped spectra, spectral maxima and temporal impulse maxima coincide with zeros of neighbouring spectra and impulses, respectively. But in this spirit, also non-rectangular finite-length pulses with ICI-free sinc-like spectra and non-rectangular finite-width spectral shapes with ISI-free sinc-like pulses are admissible, the only requirement being that pulses with ICI-free spectra must not overlap in time, and that spectra of ISI-free pulses must not overlap in frequency.

As a consequence, error-free detection allegedly would be possible without complying to orthogonality when integrating Eq. (9.13). This means:

- Signals with sinc-like overlapping spectra, where spectral maxima coincide with zeros of neighbouring spectra, could thus be detected ICI-free by recording the complex value of the spectral maximum for each sub-channel. The associated receiver filters would need to have a bandwidth approaching zero. However, frequent spectral sampling inside a small bandwidth is not conceivable: The filter bandwidth (corresponding to the filter's ring time) must be at least equal to the symbol rate (corresponding to the symbol period). For ICI-free detection filters are needed which guarantee Eq. (9.15), leading again to received spectra being orthogonal in the sense of Eq. (9.2). In addition, non-overlapping pulses guarantee ISI-free detection.
- In a similar manner, signals with overlapping sinc-like pulses, where temporal maxima coincide with zeros of neighbouring pulses, could thus be detected ISI-free by recording the complex value of each impulse maximum in a virtually zero time interval. And indeed, frequent sampling inside short time windows is possible and technically feasible. This is part of the success story of equalizer filters, and it is part of the success story of time division multiplexing (TDM) (or "OTDM" where in this case the "O" stands for "optical" rather than "orthogonal"). In publications, the condition for ISI-free reception has become known as the "Nyquist ISI criterion" [36]. In addition, the associated spectrum must not overlap for ICI-free detection.

In Figure 9.6 an example of an ISI-free temporal signal can be seen. The figure shows the sinc-like sincl-function Eq. (9.66) together with a true sinc-function [22]

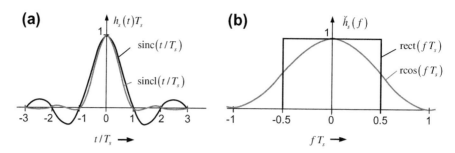

FIGURE 9.6 Impulse responses $h(t)$ and transfer functions $\check{h}(f)$ of finite length that still provide disappearing inter-symbol interference (ISI). (a) Sinc-function Eq. (9.65) and sinc-like function sincl Eq. (9.66). (b) Spectral rect-function Eq. (9.62) and spectral rcos-function (raised cosine) Eq. (9.63).

Eq. (9.65) and the respective Fourier transforms, the raised-cosine rcos-spectrum [22] Eqs. (9.63) and (9.74), together with a rect-shaped spectrum Eqs. (9.62) and (9.72).

The temporal pulses belonging to the root raised-cosine spectrum [32] Eq. (9.64) are as such not ISI-free, because the zeros of the pulses are not equidistantly spaced by T_s. However, if the transmitter pulses are shaped with a root raised-cosine filter $|\check{h}'_{sv}(f)| = |\check{h}^*_{rv'}(f)|$ (including the channel), thereby generating a pulse train where the ith pulse has the spectrum $\mathrm{rrcos}(fT_s)\exp(-j2\pi f i T_s)$, and the receiver uses a matched filter $\check{h}_{rv'}(f) = \check{h}'^*_{sv}(f)$ for detection, the orthogonality condition established in the last line of Eq. (9.2) leads to ISI-free reception.

9.2.6 Pulse-shaping in the digital, electrical, and optical domain—a comparison

An essential part of the generic transmitter setup of Figure 9.2 is the impulse-shaping filter $h_{sv}(t)$ which must be employed for each subcarrier v. As previously explained, there are many shaping options as long as the receiving filter complies with, e.g., Eq. (9.28) to establish orthogonality and to fulfill Eq. (9.15) for interference-free reception. Pulse-shaping techniques can decrease the sub-channel spacing and improve the spectral efficiency, for instance in the case of Nyquist-WDM. Here, sub-channels of possibly different spectral width can be positioned adjacent to each other without a guard band. With pulse-shaping, also the transmission performance in the presence of nonlinearities can be improved.

When looking at the position of the pulse-shaping filter $h_{sv}(t)$ in Figure 9.2 and at its effect on the received signal Eq. (9.13), it is obvious that filtering could be done also in the digital domain by manipulating the data $\check{c}_{iv}$ themselves, or in the electrical domain as depicted in Figure 9.7, by shaping $d_{iv}(t)$, or even in the optical domain by filtering $s_{iv}(t)$. Each of the methods has its own merits.

We demonstrated and compared the various pulse-shaping techniques recently [37]. The setup is depicted in Figure 9.7. We start with single-polarization (SP) non-return-to-zero (NRZ) quadrature phase shift keying (QPSK) pulses on a single carrier with a wavelength $\lambda_1 = 1.55\,\mu\text{m}$ (frequency $f_\nu = f_1 = c/\lambda_1$, $\nu = 1$, vacuum speed of light c). The subcarrier subscript ν will be omitted in the following. To generate the NRZ-QPSK data, we use a software-defined optical transmitter [25] (QPSK Tx). The DAC operate at sampling rates up to 30 GSa/s with a physical DAC resolution of 6 bit and an analogue electrical bandwidth $f_{\text{DAC}} > 18$ GHz. The respective pulse-shaping for the three schemes is implemented as follows:

- The digital filters (marked green in Figure 9.7) are realized in the FPGA. The additional electrical filters (red) are then used to remove the digitally generated image spectra (aliasing) when sinc-pulses are generated in the digital domain.
- When the sinc-shape is approximated in the electrical domain, the electrical filters alone shape the electrical drive-signals that are fed to the IQ modulator, a nested LiNbO_3 Mach-Zehnder IQ modulator (MZM) with a modulation bandwidth of $f_{\text{MZM}} > 25$ GHz, which in turn encodes QPSK data on an external cavity laser (ECL, wavelength λ_1, linewidth 100 kHz).
- When shaping the impulses in the optical domain, the DAC outputs are directly fed to the IQ modulator. A Finisar WaveShaper after the modulator serves as shaping filter (marked blue in Figure 9.7).

A variable optical attenuator (VOA) adjusts the optical power launched into the first erbium-doped fibre amplifier (EDFA) and thus varies the optical signal-to-noise power ratio (OSNR$_{\text{ref}}$ in a reference bandwidth of 0.1 nm). A schematic optical power spectrum centered at the ECL wavelength λ_1 is shown as an inset. The spectrum actually drops toward the band edges. In the case of digital filtering, this is due to the frequency response of the DAC and the anti-alias filter, an influence which

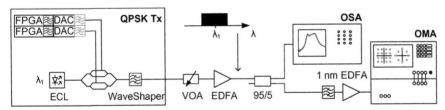

FIGURE 9.7 Setup for testing different pulse-shaping techniques. A pair of field programmable gate arrays (FPGA) drives two high-speed digital-to-analogue converters (DAC). A first pulse-shaper is realised in the digital domain (green). It is realised within the FPGAs. Electrical anti-alias filters (red) remove the alias spectra for the digital pulse-shaper. The pulse-shaper can as well be realised in the electrical domain by solely applying the electrical filters (red) on the QPSK signals. In the case of optical pulse-shaping a Finisar WaveShaper (blue) is employed. An external cavity laser (ECL) fed into an optical IQ modulator converts the electrical signals to the optical domain. We use a variable optical attenuator (VOA) together with an erbium-doped fibre amplifier (EDFA) to adjust the optical signal-to-noise ratio (OSNR). A 95/5 splitter directs the signals to an optical spectrum analyser (OSA) and to a coherent receiver (OMA, Agilent optical modulation analyser). (For interpretation of the references to colour in this figure legend, the reader is referred to the web version of this book.)

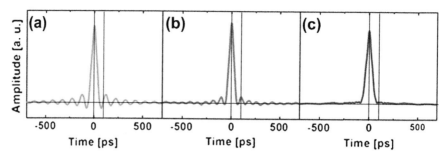

FIGURE 9.8 Different impulse responses measured from received spectra with (a) digital, (b) electrical, and (c) optical pulse-shapers. As expected, the digital pulse-shaper approximates a sinc-function Figure 9.3b closely. The electrical pulse-shaper still produces sinc-typical side lobes, whereas the optical pulse-shaper matches a sinc-function worst. (Figure modified from Ref. [37])

could have been compensated for by digital pre-conditioning. For the electrical and optical filters this spectral drop cannot be avoided in practice, and the spectral cut-off cannot be as sharp as for the digital filter. However, this non-ideal spectral shape can be compensated by the receiver filter $h_r(t)$ as has already been explained in Section 9.2.5.

In order to give a better idea of how accurately, for instance, a sinc-shaped impulse can be realised when employing digital, electrical, and optical pulse-shapers, we measured the optical spectrum $\check{r}(f) = \check{s}(f)$ (no channel, $\check{h}_c(f) = 1$) at the receiver input, Figure 9.7, the electrical spectrum $\check{x}'(f)$ after the IQ demodulator (no electrical receiver filter, $\check{h}_r(f) = 1$), and the spectrum of the digital data (no digital receiver filter).

From these spectra we derived the impulse responses as depicted in Figure 9.8a–c. The digital pulse-shaper in Figure 9.8a approximates a sinc-shaped impulse response best. The electrical pulse-shaper still produces sinc-typical side lobes, but they are smaller and decay more rapidly, Figure 9.8b. The optical pulse-shaper yields the worst sinc-approximation, Figure 9.8c, because optical filters produce only slopes with limited steepness.

9.3 OPTICAL FOURIER TRANSFORM BASED MULTIPLEXING

OFDM signal generation and reception bears a close relationship to the discrete Fourier transform (DFT) and its inverse (IDFT). More precisely, the mathematical expression of an OFDM signal to be transmitted is identical to an IDFT if the OFDM signal is sampled at proper discrete times t_n. We will see that the complex data coefficients $\check{c}_{i\nu}$ of an OFDM signal are the spectral Fourier coefficients for an IDFT. Conversely, upon reception these complex data coefficients $\check{c}_{i\nu}$ can be retrieved by performing the DFT on the OFDM signal.

To see this, we first define a transmitted OFDM symbol according to Eqs. (9.7), (9.8) and then perform the discretization in time. Our OFDM signal is composed of N subcarriers $f_\nu = \nu F_s + f_c$ spaced F_s apart. Each of the subcarriers is weighed with a complex data coefficient $\check{c}_{i\nu}$. For the time being and for simplicity's sake, we disregard any guard interval letting $T = T_s$, and we assume equal and rect-shaped impulse responses in each transmitter sub-channel, $h_{s\nu}(t) = h_s(t)$. To avoid unnecessary complications with unimportant phase factors, we redefine for the moment the transmitter impulse response window to *start* at time $t = 0$, and *not to be centred* at $t = 0$ as in Eq. (9.21),

$$h_s(t) = \frac{1}{T_s}\mathrm{rect}\left(\frac{t}{T_s} - \frac{1}{2}\right) = \begin{cases} 1/T_s & \text{for } 0 < t < T_s, \\ 0 & \text{else.} \end{cases} \tag{9.29}$$

Any discrete times $t_n = iT_s + nT_s/N\,(n = 0, 1, \ldots, N - 1)$ are then counted from the beginning of the symbol window, so that negative subscripts n are avoided. According to Eqs. (9.7), (9.8), and (9.29), the ith OFDM symbol $s_i(t)$ then is

$$s_i(t) = \underbrace{\mathrm{rect}\left(\frac{t - iT_s}{T_s} - \frac{1}{2}\right)\sum_{\nu=0}^{N-1}\check{c}_{i\nu}\exp(j2\pi\nu F_s t)}\exp(j2\pi f_c t) \tag{9.30}$$

Tx baseband OFDM symbol with duration T_s

In Figure 9.9 an exemplary signal $s_i(t)$ is displayed in the bottom subfigure. It consists of a fast oscillating carrier with channel frequency f_c (red) and the baseband OFDM symbol with duration T_s that is represented by the carrier envelope (blue) as defined by a sub-annotation in Eq. (9.30). As remarked earlier and in contrast to Figure 9.9, no guard interval was taken into account in Eq. (9.30).

In practice one does not work with continuous waveforms but generates and detects signals by time-discrete sampling. We now sample the OFDM symbol N times within a symbol duration T_s, i.e. we sample at times t_n. Also, the subcarrier spacing now

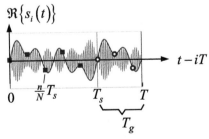

FIGURE 9.9 Exemplary OFDM signal. A sinusoidal carrier time function with frequency f_c is plotted (red). The baseband OFDM symbol (blue envelope) has a duration T_s. The OFDM symbol is repeated (cyclically extended) for the duration of the guard interval T_g. The symbol rate is $1/T$. Frequently, the text refers to this figure assuming no guard interval, $T = T_s$. (For interpretation of the references to colour in this figure legend, the reader is referred to the web version of this book.)

must be chosen such that $F_s = 1/T_s$ holds and consequently $\exp(j2\pi\nu F_s i T_s) = 1$ results. Eq. (9.30) then becomes

$$s_i(t_n) = \underbrace{\text{rect}\left(\frac{t_n - iT_s}{T_s} - \frac{1}{2}\right) \overbrace{\sum_{\nu=0}^{N-1} \check{c}_{i\nu} \exp\left(j2\pi\frac{\nu n}{N}\right)}^{=c_{in}} \exp(j2\pi f_c t)}_{\text{Tx baseband OFDM symbol with duration } T_s}. \quad (9.31)$$

It can now easily be recognised that the term with the super-annotation c_{in} has the form of an IDFT for the coefficients $\check{c}_{i\nu}$ such as in Eq. (9.80). The signal $s_i(t)$ sampled at times t_n within the ith time frame is thus a complex carrier with frequency f_c, modulated with the constant Fourier coefficient c_{in} within the symbol duration T_s. To visualise Eq. (9.31) we mark the samples inside the ith time frame of Figure 9.9 with blue rectangles (■). For completeness' sake the samples in a potential guard interval are marked with blue open circles (○).

Subsequently, the signal is sent across the channel. To keep things simple we replace the fibre channel by a direct connection, so that $r(t) = s(t)$.

At the receiver, a sampling detector will receive the signal $r_i(t) = s_i(t)$ within the ith time frame and detect the coefficients c_{in} that are encoded on the carrier frequency. In a first step, an IQ mixer down-converts the transmitted signal to the baseband. The task is then to retrieve the weights $\check{c}_{i\nu}$ of Eq. (9.30) or (9.31). This can be achieved by a DFT.

Subsequently, we will discuss and compare the various implementations. For a better understanding of optical Fourier processing, we briefly look in Section 9.3.1 at the electronic transmitter/receiver concept. The next two Sections 9.3.2 and 9.3.3 describe the functioning of a receiver and a transmitter with optical Fourier transform processors that operate at the speed of light. By making a transition from electrical to all-optical FFT processing, one may benefit from higher-speed and lower-power consumption. Section 9.3.4 then illustrates how such processors are constructed.

9.3.1 Electronic Fourier transform processing

We now describe the electronic processing in an electronic receiver upon reception of a signal $r_i(t) = s_i(t)$ as described by Eq. (9.30) and in an electronic transmitter.

We begin with the discussion of the receiver, which should ultimately retrieve the coefficients $\check{c}_{i\nu}$ from the received signal $r_i(t)$. The first step is to down-convert $r_i(t)$ with IQ mixers all using the same local oscillator frequency f_c (contrary to the procedure described in Figure 9.2 where a different local oscillator frequency for each subcarrier was employed) by performing the operation $d'_i(t) = r_i(t)\exp(-j2\pi f_c t)$. The resulting complex baseband signal $d'_i(t)$ is then measured by electronic sampling at times $t = t_n = iT_s + nT_s/N$. This gives

$$d'_i(iT_s + nT_s/N) =: c_{in}, \quad c_{in} = \sum_{\nu=0}^{N-1} \check{c}_{i\nu} \exp\left(j2\pi\frac{n\nu}{N}\right), \quad n = 0, 1, \ldots, N-1. \quad (9.32)$$

As a result one obtains the N baseband samples $d_i'(iT_s + nT_s/N)$ from within each OFDM symbol. Each sample is represented by a complex coefficient c_{in} that results from an inverse discrete Fourier transform (IDFT) (see Eq. (9.80)) of the complex data coefficients $\check{c}_{iv}$.

In a second step, the data $\check{c}_{iv}$ are to be retrieved. For this one needs to infer the amount to which each subcarrier v contributes to the baseband representation of the ith OFDM symbol. This analysis is done numerically with a discrete Fourier transform (DFT) (see Eq. (9.79)) in order to recover the original data coefficients,

$$\check{c}_{iv} = \frac{1}{N} \sum_{n=0}^{N-1} c_{in} \exp\left(-j2\pi \frac{nv}{N}\right).$$

$\qquad(9.33)$

On the electronic transmitter side a similar procedure can be designed: An inverse discrete Fourier transform (IDFT) Eq. (9.32) produces inside the ith time frame N samples c_{in} from N input data $\check{c}_{iv}$. After parallel-to-serial conversion (P/S), a digital-to-analogue converter (DAC) generates the continuous signal $s_i(t)$ for the ith time frame. Consecutive time frames i form the signal $s(t)$, which is transmitted through the channel and eventually received.

A basic disadvantage of transmitters and receivers employing the DFT (mostly in the form of the numerically efficient fast Fourier transform [38], FFT) comes from the fact that parallel data have to be serialized at the transmitter, and that the receiver has to take samples during each time frame for parallel processing. In addition, the processing must be completed inside the duration T_s of one time frame, and this N-point FFT must be repeated at the symbol rate $F_s = 1/T_s$. For large symbol rates the speed of the available electronic devices represents a bottleneck, which cannot be easily overcome.

9.3.2 The optical Fourier transform receiver

We now discuss how an optical Fourier transform receiver can retrieve the coefficients $\check{c}_{iv}$ from the received signal $r_i(t) = s_i(t)$, Eq. (9.30), disregarding the channel. Contrary to the electronic receiver, in an optical DFT receiver we apply an *optical discrete Fourier transform* (ODFT) first and only then down-convert the OFDM data, Figure 9.10a.

The first step for an ODFT of order N is to derive the complex data streams $\hat{r}_{iv'}(t)$ from N signals $r_{in}(t)$ taken from different temporal positions within the ith symbol of the received signal $r_i(t)$. This serial-to-parallel conversion together with the final sampling is equivalent to the sampling in Eq. (9.32). The operation then to be performed on the delayed data streams is

$$\hat{r}_{iv'}(t) = \frac{1}{N} \sum_{n=0}^{N-1} r_{in}(t) \exp\left(-j2\pi \frac{nv'}{N}\right).$$

$\qquad(9.34)$

An ODFT to calculate Eq. (9.34) is depicted in Figure 9.10a. It consists of a delay stage, a processor stage, and a sampling stage. The delay stage is needed to

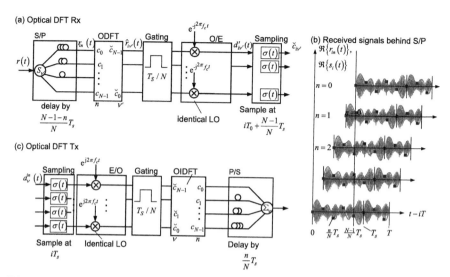

FIGURE 9.10 Schematic of receiver and transmitter employing the discrete optical (inverse) Fourier transform. In contrast to the setup in Figure 9.2, all local oscillators (LO) for the Tx as well as for the Rx can be derived from a single Tx-LO and a single Rx-LO, respectively. (a) Receiver with delays $(N - 1 - n)T_s/N$ after splitting node S_ν. (b) Symbol in time frame i with optical carrier frequency f_c (red). If the rapidly oscillating carrier is removed by mixing with $\exp(-j2\pi f_c)$, the baseband signal consisting of $N = 5$ subcarriers remains (blue). It is determined by $N = 5$ complex coefficients c_n, the values of which are measured at sampling times $t_{in} = iT + nT_s/N$ relative to the beginning of the symbol time frame $iT \leqslant t < iT + T_s$. In the region $iT + T_j \leqslant t < (i + 1)T$ the beginning of the symbol is repeated (cyclic "suffix" CS). Likewise, the end of the symbol can be repeated in front of the symbol (cyclic "prefix" CP), or a CP can be combined with a CS. Frequently, the text refers to this figure assuming neither CP nor CS, $T = T_s$. (c) Transmitter with delays nT_s/N before summing node Σ_ν. (For interpretation of the references to colour in this figure legend, the reader is referred to the web version of this book.)

provide the N input streams $r_{in}(t)$ each of which is delayed by temporal increments $(N - 1 - n)T_s/N$. With this choice of delays the earliest sample $r_{i0} = r_i(iT_s)$ with $n = 0$, see topmost subfigure in Figure 9.10b, is delayed by $(N - 1)T_s/N$, and the latest sample $r_{i,N-1} = r_i(iT_s + (N - 1)T_s/N)$ is not delayed at all, bottom subfigure in Figure 9.10b. After this serial-to-parallel conversion with delays, the ODFT inputs are computed from the received signal $r_i(t) = s_i(t)$, Eq. (9.30),

$$r_{in}(t) = \text{rect}\left(\frac{t - iT_s}{T_s} - \frac{1}{2} - \frac{N - 1 - n}{N}\right) \exp\left[j2\pi f_c\left(t - \frac{N - 1 - n}{N}T_s\right)\right]$$

$$\times \underbrace{\sum_{v=0}^{N-1} \check{c}_{iv} \exp\left[j2\pi v F_s\left(t - \frac{N - 1 - n}{N}T_s\right)\right]}_{=c_{in} \text{ for } t=t_\sigma=iT_s+(N-1)T_s/N}. \tag{9.35}$$

For $r(t) = s(t)$ (no channel), one specific OFDM symbol and its delayed copies are displayed in Figure 9.10b. For illustration's sake a cyclic suffix (CS) is included for the interval $T > T_s$, but we assume $T = T_s$ (no CS) in the following. The rapidly oscillating carrier f_c (red) is modulated with the baseband OFDM symbol (blue). At time $t = t_\sigma = iT_s + (N-1)T_s/N$ all optical samples c_{in}, see sub-annotation in Eq. (9.35) and observing $F_sT_s = 1$, have made it into the ODFT processor unit of Figure 9.10a.

In the ODFT processor stage, all inputs $r_{in}(t)$ are summed up and appropriate phases as per Eq. (9.34) are applied to the terms in order to produce the optical ODFT outputs $\hat{r}_{iv'}$. Proper outputs are valid only in an interval shorter than T_s/N. The gating thus removes invalid signal parts where samples from neigbouring sampling intervals $n' \neq n$ would interfere. However, the gating may be performed also later on and can be executed in the sampling stage.

In a next step the down-conversion is performed. We use again the *same optical LO* for all branches. As with the electronic DFT processors, this is in contrast to the receiver concept presented in Figure 9.2. After each mixer we obtain $d'_{iv'}(t) = \hat{r}_{iv'}(t) \exp(-j2\pi f_c t)$. Substituting Eq. (9.35) in Eq. (9.34) and observing the condition $F_sT_s = 1$, we find the baseband signal $d'_{iv'}(t)$,

$$
\begin{aligned}
d'_{iv'}(t) = &\frac{1}{N}\sum_{n=0}^{N-1} \text{rect}\left(\frac{t - iT_s}{T_s} - \frac{1}{2} - \frac{N-1-n}{N}\right) \exp\left[-j2\pi f_c T_s \frac{N-1-n}{N}\right] \\
&\times \sum_{v=0}^{N-1} \check{c}_{iv} \exp\left[j2\pi v F_s \left(t - \frac{N-1}{N}T_s\right)\right] \exp\left(j\, 2\pi \frac{n(v-v')}{N}\right).
\end{aligned}
$$
(9.36)

If we now take samples at times $t = t_\sigma = iT_s + (N-1)T_s/N$, the rect-function is always 1 as well as the phase factor $\exp[j2\pi v F_s(t_\sigma - (N-1)T_s/N)] = \exp[j2\pi v F_s T_s i] = 1$. Equation (9.36) then simplifies to

$$
d'_{iv'}(t_\sigma) = \exp\left(-j2\pi f_c T_s \frac{N-1}{N}\right) \frac{1}{N}\sum_{n=0}^{N-1}\sum_{v=0}^{N-1} \check{c}_{iv} \exp\left(j2\pi n \frac{v - (v' - f_c T_s)}{N}\right),
$$

$$
t_\sigma = iT_s + \frac{N-1}{N}T_s.
$$
(9.37)

If $f_c T_s$ is not an integer, i.e. if the carrier frequency f_c does not happen to be an integer multiple $M \times N$ of the subcarrier spectral separation F_s by a normalized frequency offset $\Delta f_c T_s$, then the individual phases $-2\pi n v'/N$ of the ODFT processor or the frequency f_c must be properly tuned such that $v' - f_c T_s = v''$ becomes an integer and ICI is avoided. This is not really difficult because $f_c T_s$ is of order 2×10^4 ($f_c = 200$ THz, $T_s = 100$ ps), and a frequency offset of $\Delta f_c = \pm 2.5$ GHz already leads to a tolerable normalized offset [11] of $\Delta f_c T_s = \pm 0.25$. In the following we assume that this "tweaking" effectively removes the term $j2\pi n f_c T_s/N$ in Eq. (9.37).

Finally, applying the orthonormality relation Eq. (9.4), the sum over n amounts to $N\delta_{\nu\nu'}$, which in turn reduces the sum over ν to a single term. Therefore, the sampled data $d'_{i\nu'}(t)$ at $t = t_\sigma$ directly lead to the original data coefficients $\check{c}_{i\nu'}$,

$$d'_{i\nu'}(t_\sigma) = \check{c}_{i\nu'} \exp\left(-j2\pi f_c T_s \frac{N-1}{N}\right), \quad t_\sigma = iT_s + \frac{N-1}{N}T_s. \tag{9.38}$$

For avoiding cross-talk, the sampling window must be smaller than T_s/N. The phase factor $\exp\left(-j2\pi f_c T_s (N-1)/N\right)$ that is common to all data points $\check{c}_{i\nu'}$ would rotate the constellation diagram, but this can be compensated by an appropriate estimation in the receiver.

9.3.3 The optical Fourier transform transmitter

To realize an OFDM transmitter with signals such as in Figure 9.3, an *optical inverse discrete Fourier transform* (OIDFT) processor can replace the transmitter of Figure 9.2. For implementing an OIDFT, Eqs. (9.34)–(9.38) have to be applied in the reverse sequence. An implementation of an OIDFT is depicted in Figure 9.10c. In each branch ν, sample the input signals modulate the N identical optical carriers $\exp(j2\pi f_c t)$ with impulses carrying the complex data coefficients $\hat{r}_{i\nu}$. The impulse width must be smaller than T_s/N and thus generates a broad optical spectrum. The optical inverse discrete Fourier processor analyses the modulated optical carrier $\exp(j2\pi f_c t)$ in the different branches ν by weighing the data $\check{c}_{i\nu} \exp(j2\pi f_c t)$ with phase factors $\exp(j2\pi n\nu/N)$ and by summing all contributions according to Eq. (9.31). This results in N signals r_{in} representing data samples c_{in} on an optical carrier $\exp(j2\pi f_c t)$ inside the OFDM symbol time frame $iT_s \leqslant t < (i+1)T_s$. These samples are parallel-to-serial converted (P/S) with optical delay lines and form the transmitted signal as depicted in Figure 9.10b bottom graph.

Instead of using a narrow sampling window of width smaller than T_s/N, the continuous laser source could be replaced by a mode-locked laser emitting pulses with a width smaller than T_s/N and at a rate $F_s = 1/T_s$. The sampling window could then be increased to T_s, and even a CP could be incorporated through widening the window size by a guard interval T_g to $T = T_s + T_g$. These techniques were recently compared by Wang et al. [39] and Lowery [40].

9.3.4 Optical Fourier transform processors

So far we treated the optical Fourier transform processor as a black box postulating only the introduced phase changes and the summing action of Eq. (9.36). Now we want to go into more physical details.

The optical Fourier transform is basically a gift of nature. It actually can be shown that within a paraxial approximation the fields in the front and in the back focal plane of a lens are related by the spatial Fourier transform. This will be demonstrated in Subsection 9.3.4.1. More precisely: If a monochromatic boundary field in plane $z = \text{const}$ has a radial extension r_M much smaller than the observation distance

d in point $P(x, y, z = \sqrt{d^2 - x^2 - y^2})$, i.e. if $d \gg r_M^2/\lambda$ holds, the field on a sphere with radius d centred at the origin $O(x = y = z = 0)$ is the two-dimensional spatial Fourier transform [41] of the boundary field. The Fourier variables are the spatial frequencies $\xi = \sin \gamma_x/\lambda = x/(\lambda d)$, and $\eta = \sin \gamma_y/\lambda = y/(\lambda d)$ of Eqs. (9.53), (9.54), (9.69) and (9.70), where γ_x and γ_y measure the angles included by line segment $\overline{OP}$ with the x- and y-axes, respectively. The actual calculations involved are lengthy, especially in vectorial form [42], so that usually a paraxial approximation is used.

It is interesting to note that wave propagation realizes only the forward Fourier transform, and that applying the forward Fourier transform two times in sequence leads to a spatial inversion. This is obvious when regarding the effect of a confocal system of two lenses. Further, the Fourier kernel of the spatial Fourier transform has the opposite sign as the kernel of the temporal Fourier transform. This fact is connected to the choice of the notation $\exp[j(\omega t - k_0 z)]$ in Eqs. (9.57) and (9.58) for a homogeneous monochromatic plane wave propagating with a vacuum propagation constant $k_0 = \omega/c$ "to the right" (vacuum speed of light c), because temporal and spatial Fourier transforms are just special cases of expanding a spatio-temporal field into plane waves. As remarked before, two forward Fourier transforms in sequence result in a spatial (temporal) inversion.

9.3.4.1 Free-space propagation—the Fourier transform lens

We consider in Figure 9.11 a two-dimensional opaque screen with transparent aperture slots at $x = x_n$ in plane $z = 0$ extending indefinitely in y-direction (perpendicularly to the drawing plane). The plane $z = 0$ represents the front focal plane of a thin cylindrical lens with focal length l positioned at $z = 0$. The monochromatic scalar field [41,42] $\psi(x) = c_n \Delta x \delta(x - x_n)$ emitted from the line objects (the aperture slots) have a wavelength λ and the spatial dependency of a cylindrical wave. The intensity of the wave thus decays linearly with distance or, in other words, the amplitude decays with the square root of $\lambda d = \lambda \sqrt{z^2 + (x - x_n)^2}$, and the phase factor is $\exp(j\pi/4)\exp(-j2\pi d/\lambda)$. The quantity $\Delta x > 0$ normalises the line objects and will be explained later on in more detail. In a plane $z = l$ just in front of the lens, we find the field [41]

$$\psi_{z=l}(x) = c_n \frac{\Delta x}{\sqrt{\lambda d}} \exp\left(j\frac{\pi}{4}\right) \exp\left(-j\frac{2\pi}{\lambda}\sqrt{l^2 + (x - x_n)^2}\right)$$

$$\approx c_n \frac{\Delta x}{\sqrt{\lambda l}} \exp\left(j\frac{\pi}{4}\right) \exp\left(-j2\pi\frac{l}{\lambda}\right) \exp\left(-j\pi\frac{x^2}{\lambda l}\right) \qquad (9.39)$$

$$\times \exp\left(-j\pi\frac{x_n^2}{\lambda l}\right) \exp\left(+j2\pi\frac{x_n x}{\lambda l}\right).$$

The approximation in Eq. (9.39) implies a paraxiality condition, where the maximum significant extension x_M of the field at $z = 0$ is much smaller than the focal length l, i.e. $x_M \ll l$. Under this condition, the squared transverse distances $x^2, x_n^2 \ll l^2$ are

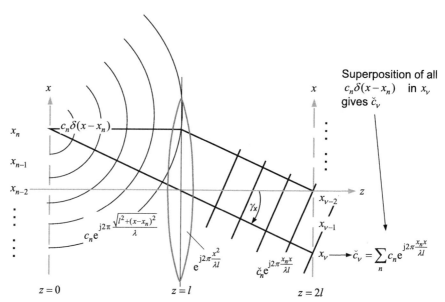

FIGURE 9.11 Fourier transform lens with focal length l in paraxial approximation. Line sources with complex weights c_n located in the front focal plane $z = l$ of a thin cylindrical lens emit cylindrical waves, which are transformed by the lens to plane waves propagating at an angle γ_x measured from the lens' z-axis. Only in the back focal plane $z = 2l$ the x-dependent phase factors of the plane waves are $\exp(j2\pi x_n x/(\lambda l))$ and therefore correspond to a Fourier transform kernel. The superposition of all input cylindrical waves realises the Fourier transform in terms of the paraxially approximated spatial frequencies $\xi = \sin \gamma_x/\lambda \approx x/(\lambda l)$. Input and output field samples are defined by $x_n = n\Delta x$ and $x_\nu = -\nu\Delta x$, respectively.

small, the amplitude factor simplifies, the square root can be expanded up to first order, and the ratio $x/l = \tan \gamma_x \approx \sin \gamma_x$ approximates the angle γ_x as indicated in Figure 9.11.

Next, the thin lens at $z = l$ contributes a phase factor [43]

$$\Psi_{\text{lens}}(x) = \exp\left(+j\pi \frac{x^2}{\lambda l}\right). \tag{9.40}$$

The wave emerging from the lens $\psi_{z=l}(x)\Psi_{\text{lens}}(x)$ is simply a tilted plane wave with a propagation constant $k_0^2 = k_x^2 + k_z^2$ that equals the vacuum propagation constant $k_0 = \omega/c = 2\pi/\lambda$. Its x-component amounts to $k_x \approx 2\pi x_n/(\lambda l)$, its z-component $k_z = k_0\sqrt{1 - k_x^2/k_0^2}$ is paraxially approximated by $k_z \approx k_0\left(1 - \frac{1}{2}x_n^2/l^2\right)$.

$$\psi_{z=l}(x) \approx c_n \frac{\Delta x}{\sqrt{\lambda d}} \exp\left(j\frac{\pi}{4}\right) \exp(-jk_z l) \exp\left(-jk_0\frac{x_n^2}{l^2}l\right) \exp(+jk_x x). \tag{9.41}$$

In a plane $z = l + l'$ and at point x after the lens we then observe the plane wave

$$\psi'_{z=l+l'}(x) \approx c_n \frac{\Delta x}{\sqrt{\lambda d}} \exp\left(j\frac{\pi}{4}\right) \exp(-jk_z(l + l')) \exp\left(-jk_0 \frac{x_n^2}{l^2} l\right) \exp(+jk_x x). \tag{9.42}$$

Only in the back focal plane $z = 2l$ of the lens, i.e. when choosing $l' = l$, the x-independent but x_n-dependent phase factor becomes 1. To see this we replace (and then k_x) k_z by their paraxial approximations as outlined above and find

$$\Psi'(x) := \psi'_{z=2l}(x) \approx c_n \frac{\Delta x}{\sqrt{\lambda l}} \exp\left(j\frac{\pi}{4}\right) \exp\left(-j2\pi \frac{2l}{\lambda}\right) \exp\left(+j2\pi \frac{x_n x}{\lambda l}\right). \tag{9.43}$$

Equation (9.43) is the spatial impulse response of the lens system with respect to its front and back focal planes. The phase factor $\exp(+j2\pi x_n x/(\lambda l))$ is determined by the shift of the spatial input impulse and is 1 if $x_n = 0$. As expected, the output field for this case has the same complex amplitude $c_n \Delta x \exp(j\pi/4) \exp(-j2\pi 2l/\lambda)/\sqrt{\lambda l}$ for all spatial frequencies $\xi = k_x/(2\pi) = \sin \gamma_x/\lambda \approx x/(\lambda l)$ in a range $|\xi| \leqslant 1/\lambda$. For an input field with N line objects (this sampled input field is an application of Huygens' principle, $\Psi(x) = \sum_{n=0}^{N-1} c_n \Delta x \delta(x - x_n)$), the continuous output field reads

$$\psi'(x) = \frac{\Delta x}{\sqrt{\lambda l}} \exp\left(j\frac{\pi}{4}\right) \exp\left(-j2\pi \frac{2l}{\lambda}\right) \sum_{n=0}^{N-1} c_n \exp\left(+j2\pi \frac{x_n x}{\lambda l}\right). \tag{9.44}$$

Next, we make the transition to discrete values $x = x_\nu$ of the output coordinate and define the discrete positions x_n in the input plane and the discrete positions x_ν in the output plane using the spatial increment Δx,

$$x_n = n\Delta x, \quad x = x_\nu = -\nu\Delta x, \quad n, \nu \in N_0, \quad \Delta x = +\sqrt{\lambda l/N}. \tag{9.45}$$

The output field $\psi'(x_\nu) \propto \check{c}_\nu$ at positions $x = x_\nu$ then represents an optical discrete Fourier transform (ODFT)

$$\Psi'(x_\nu) = \frac{\sqrt{\lambda l}}{\Delta x} \exp\left(j\frac{\pi}{4}\right) \exp\left(-j2\pi \frac{2l}{\lambda}\right) \check{c}_\nu, \quad \check{c}_\nu = \frac{1}{N} \sum_{n=0}^{N-1} c_n \exp\left(-j2\pi \frac{n\nu}{N}\right). \tag{9.46}$$

For a monochromatic signal, Eq. (9.46) realizes the Fourier sum of coefficients c_n appropriately weighed with phase factors as postulated in Eq. (9.33). It is remarkable that spacing Δx and common phase factor both depend on the signal's optical wavelength λ and on the λ-dependent focal length l. Therefore some tuning of the lens' phase factor Eq. (9.40) will be desired, if the subcarriers spread over a large spectral range. This phase tweaking has to be seen in the context of the comments after Eq. (9.37).

A spatial inversion of the output samples by choosing $x = x_\nu = +\nu\Delta x$ would result in an optical inverse discrete Fourier transform (OIDFT),

$$\Psi'(x_\nu) = \frac{\sqrt{\lambda l}}{\Delta x} \exp\left(j\frac{\pi}{4}\right) \exp\left(-j2\pi \frac{2l}{\lambda}\right) \check{c}_\nu,$$

$$\check{c}_\nu = \frac{1}{N} \sum_{n=0}^{N-1} c_n \exp\left(+j2\pi \frac{n\nu}{N}\right) \quad x_\nu = +\nu\Delta x. \tag{9.47}$$

If the output field Eq. (9.46) or (9.47) is input to a second, identical Fourier transform lens, the output field in the back focal plane of the second lens is the spatially reversed copy of the field in the front focal plane of the first lens, apart from an additional phase factor $j \exp(-j2\pi 4l/\lambda)$.

If we provide the spatial input data as serial-to-parallel converted samples of a temporal signal, the spatial Fourier transform acts as a temporal DFT processor when its output data are parallel-to-serial re-converted in the correct sequence.

9.3.4.2 Guided-wave propagation—FFT with butterfly topology

The fast Fourier transform [38] (FFT) performs the same input-output operation as the DFT, but avoids redundancies. The FFT breaks a DFT down into many smaller DFTs. When the algorithm is visualized in a graph, the crossings of its edges resemble a "butterfly" topology. With an FFT algorithm the DFT complexity of order $C_{\mathrm{DFT}} = N^2$ can typically be reduced to a complexity of order $C_{\mathrm{FFT}} = N \log_2 N$. A one-to-one implementation of an FFT following the scheme with a delay stage and a processor stage such as shown in Figure 9.10a was first suggested by Marhic [44] in 1987. Siegman [45] in 2001 and 2002 also suggested the FFT. It was for the first time applied to optical OFDM by Sanjoh et al. in Ref. [46] in 2002. Implementations of a 2×2 FFT were reported by Huang et al. [47], and a 4×4 FFT was shown by Takiguchi [48] in 2009 at the same meeting. Later on an impressive fully integrated 8×8 FFT was demonstrated by Takiguchi et al. [49] in 2010. While a 2×2 FFT is simple and only comprises a two-arm delay interferometer, the 8×8 FFT already has a considerable complexity. In Figure 9.12 an 8-point FTT is plotted with all delays, couplers, and phase-shifters. It can be seen that FFT implementations based on butterfly diagrams scale quite unfortunately with the order of the FFT. An FFT of the order N already requires precise tuning of $N \log_2 N$ phase-shifters and the integration of many couplers.

9.3.4.3 Guided-wave propagation—FFT with cascaded delay interferometers

In 1998 Li patented so-called Fourier filters with serial-to-parallel converters [51]. They were discussed in detail later on by Madsen and Zhao [52]. With these filters narrow spectral slicing could be performed. These filters were finally put in a wider context in connection with the wavelet transform by Cincotti [53,54] in 2002 and 2004 who also integrated the serial-to-parallel converter into the FFT network.

In 2010 detailed analyses were performed by Hillerkuss et al. [50], where the important aspects of temporally sampling the signals and exploiting a guard interval were elaborated. The theory also discusses errors and options for simplifying Nth order FFT schemes by approximating an FFT of an arbitrary order by one, two, and three stages of delay interferometers followed by conventional optical filters. As a result, a high line rate 10.8 Tbit/s optical OFDM transmission experiment using an all-optical FFT receiver was demonstrated [55]. In 2011, the record in OFDM transmission on a single laser was set to 26 Tbit/s, again by Hillerkuss et al. [29]. In both of these experiments the all-optical FFT based on the approximation with delay interferometers and filters was used to demultiplex the 325 subcarriers.

Finally, in 2011 Cincotti [56] published a generalization of the ODFT and an extension towards optical hybrids and the fractional Fourier transform, which had been defined by McBride and Kerr [57] in 1987, based on work of Namias [58] in 1980. It is interesting to note that also the so-called fractional optical Fourier transform can be employed [59] in the context of OFDM with chirped subcarriers.

In the following, we discuss the OFFT with cascaded delay interferometers following Hillerkuss et al. [50]. We start our discussion with the "butterfly" FFT approach and subsequently show that it is identical to cascading delay interferometers.

Our task is the Fourier analysis of N serial samples of a time signal, so we first need a S/P conversion as in Figure 9.10a, Figure 9.12 on Figure 9.13a. As said before, the FFT efficiently implements the DFT Eqs. (9.33) and (9.34), now for a number of $N = 2^p$ time samples (p is assumed to be an integer). If the coefficients c_{in} represent a time-series of N equidistant signal samples for a fixed i over an interval T_s, then the quantities $\check{c}_{iv}$ are the wanted complex spectral data coefficients in time slot i. The

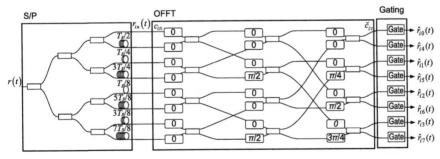

FIGURE 9.12 Direct implementation of an eight-point optical FFT. The butterfly-type FFT processor comprises a serial-to-parallel converting delay stage, a processor stage, and a gating stage. The unit processes signals with a symbol duration (T_s). Rectangular sharp-edged boxes stand for optical directional couplers with an (unphysical) amplitude split ratio of 1. Fibre loops symbolise time delays by fractions of the symbol duration (T_s). Rounded boxes with numbers mark the respective phase shifts φ according to $\varphi = -2\pi n\nu/N$ in Eqs. (9.33) and (9.46). The figure is derived from [50]. (The phase $\pi/4$ was erroneously placed in the seventh box from top and has been corrected to be in the fourth box from top instead. Thanks go to A.J. Lowery, Monash University, Australia.)

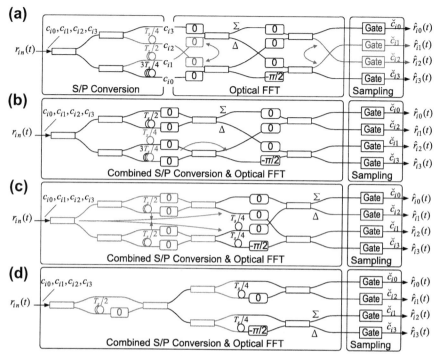

FIGURE 9.13 Exemplary four-point optical FFT for symbols of duration T_s and simplification steps for integrating the S/P conversion into the FFT circuit. (a) Traditional FFT implementation with separate S/P conversion stage. (b) S/P conversion and FFT structure consisting of two delay interferometers (DI) with the same differential delay; the additional $T_s/4$ delay is moved out of the second DI. (c) Two identical DI can be replaced by a single DI followed by signal splitters. (d) Resulting low-complexity scheme with combined S/P conversion and FFT. In this picture, rectangular sharp-edged boxes stand for optical directional couplers with an (unphysical) amplitude split ratio of 1. We follow the convention that the field in the upper output arm (Σ) results from summing the complex amplitudes of the two coupler inputs, while the field in lower output arm (Δ) stands for the difference (upper input field) − (lower input field). "Fibre loops" symbolise time delays by fractions of the symbol duration T_s. Rounded boxes with numbers mark the respective phase shifts φ according to $\varphi = -2\pi n\nu/N$ in Eqs. (9.33) and (9.34). (Modified from Ref. [50] © 2010 OSA.)

FFT typically decimates a DFT of size N into two interleaved DFT of size $N/2$ with recursive stages [38] so that the N-point DFT is given by

$$\check{c}_{i\nu} = \begin{cases} E_{i\nu} + \exp\left[-j\dfrac{2\pi}{N}\nu\right]O_{i\nu} & \text{if } \nu < \dfrac{N}{2} \\[4mm] E_{i,\nu-N/2} - \exp\left[-j\dfrac{2\pi}{N}\left(\nu-\dfrac{N}{2}\right)\right]O_{i,\nu-N/2} & \text{if } \nu \geqslant \dfrac{N}{2} \end{cases} \tag{9.48}$$

The quantities E_{iv} and O_{iv} are the *even* and *odd* DFT of size $N/2$ for even and odd inputs $c_{i,2m}$ and $c_{i,2m+1}$ ($m = 0, 1, 2, \ldots, N/2 - 1$), respectively. In Figure 9.13a, to the right of the S/P conversion, the direct implementation of the FFT for $N = 4$ is shown using optical directional couplers (empty sharp-edged boxes ■) and phase shifters (rounded boxes ⬭) with the phase shift $\varphi = -2\pi n v/N$ (Eqs. (9.33) and (9.34)) written into it. One may identify four butterfly structures and eight phases. For simplicity, all couplers are assumed to have an (unphysical) output amplitude split ratio of 1. We further follow the convention that the field in the couplers' upper output arm (marked by Σ) results from summing the complex amplitudes of the two coupler inputs according to the relation (upper input field) + (lower input field), while the field in the lower output arm (marked by Δ) stands for the difference relation (upper input field) − (lower input field). For the S/P conversion stage, the coupler's input fields at the (not depicted) lower arms are assumed to be zero, so both outputs reproduce input amplitude and phase. In Figure 9.13a, the gating is done in the optical domain for keeping the graph simple, but the gates could have been positioned also after the frequency down-conversion stage in the form of electronic sampling gates.

The standard S/P conversion-OFFT scheme in Figure 9.13a has a complexity of $C_{\text{SPC-OFFT-std}} = N - 1 + (N/2) \log_2 N$ couplers. The different phases have to be adjusted accurately with respect to all other phase shifters. This renders the scheme impractical if N is to be scaled up. However, the standard S/P conversion-OFFT scheme can be simplified by merging S/P conversion and OFFT stages.

In a first step we re-order the delays in the S/P conversion stage as indicated in Figure 9.13a and re-label the outputs accordingly. This way the OFFT input stage consists of two parallel delay interferometers (DI) with the same free spectral range (FSR) but different absolute delays, Figure 9.13b. By moving the common delay $T_s/4$ in both arms of the lower DI to its outputs, one obtains two identical DIs that are fed with the same input signal, Figure 9.13c. This redundancy can be eliminated by replacing the two DIs with one DI, and by splitting its output, Figure 9.13d. These simplification rules (a mathematical proof is available [50]) can be iterated for OFFT of any size N. The new optical OFFT processor consists of $N - 1$ cascaded DI with a complexity of only $C_{\text{SPC-OFFT-DI}} = 2(N - 1)$ couplers, which is always smaller (or equal for $N = 2$) than the standard complexity $C_{\text{SPC-OFFT-std}}$. The greatest advantage is that only the phase differences of $N - 1$ DI need stabilization, and that no inter-DI phase adjustment is required. This opens the route to integrate concatenated DIs on an optical chip. An implementation of such an 8-point OFFT is depicted in Figure 9.14.

If the arrangement of Figure 9.14 is operated in the reverse direction, and if the signs of all phases are set to their negative values, we have an OIFFT with combined parallel-to-serial converter. It is a direct implementation of the OIFFT-P/S converter transmitter as depicted in Figure 9.10c.

9.3.4.4 FFT approximated by conventional filters

In this subsection we discuss approximations to further simplify the implementations of the optical FFT. While FFT implementations going back to the butterfly topology are difficult to implement, the cascaded delay-interferometer approach is much

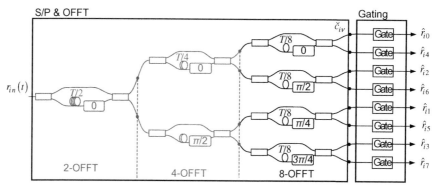

FIGURE 9.14 Implementation of an eight-point optical FFT using cascaded delay interferometers. Each DI cascade increases the order of the FFT by a factor 2. To derive a single Fourier coefficient $\hat{r}_{i\nu}$, one only needs three stages of DI, and a sampling gate to define the samples inside an OFDM symbol. Figure modified from [50].

simpler and quite doable. Yet, for a many-point FFT with $N > 4$ also this solution renders difficult.

An approximation to an FFT is a filter that as closely as possible approximates the ideal frequency response of the FFT. To find an approximation to the FFT it is useful to study the FFT DI cascade from Figure 9.14 with its subsequent gating in the optical or electrical domain. It will be shown below that in most cases a good approximation to an FFT of any order may be found by a simple single-stage DI together with a bandpass filter and a subsequent sampling gate [50]. Of course, the DI filter might be replaced by other filters—such as ring filters [60]. Ring filters were also suggested to extract single subcarriers from an OFDM signal [60]. They might provide a good approximation as well. To show this we follow the line of arguments given in Ref. [50].

The frequency response of the ideal eight point FFT of Figure 9.14 can easily be visualised. We plot the time and frequency response of the DI cascade in Figure 9.15a. It can be seen that the first DI cascade (red) is further refined by the second DI cascade (blue), which in turn is further refined by the third DI cascade (black) to finally provide the filter response of the total cascade (green). Simulations with a pseudorandom signal show that such a filter can derive the Fourier coefficient almost ideally. The filter provides open eyes for the Fourier coefficients and the plot in the graph at the bottom shows an almost ideal signal quality Q^2—above 30 dB. The frequency response of Figure 9.15a gives a hint as of how the filter response can be approximated. Since the first DI stage has the largest impact, we retain the first DI stage with the smallest delay and replaced the two subsequent DI cascades by a fourth-order Gaussian filter. It can be seen that when the filter passband is properly chosen, one can still achieve a satisfactory signal quality. If the DI cascade is replaced by the 4th-order super-Gaussian filter only, the signal quality degrades. In some instances this might still provide a sufficient signal quality. However, to retain at least one DI in the cascade seems to be a good

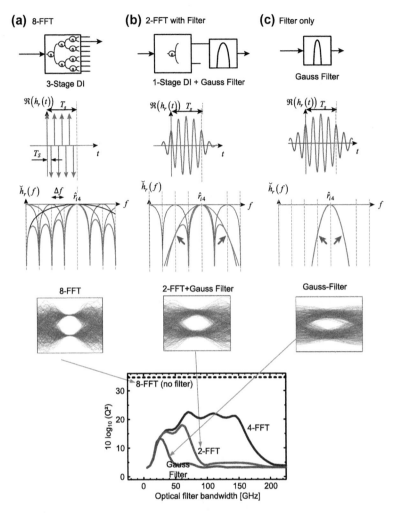

FIGURE 9.15 Three implementations of an eight-point FFT. (a) Ideal FFT based on three stages of cascaded delay interferometers (DI). The time and frequency response of the DI cascade are plotted below. The colours in the frequency response show how each DI cascade further refines the frequency response. The first DI cascade (red), the second cascade (blue), and the third cascade (black) provide the final response (green). The signal quality of the eye derived by this ideal FFT diagram is open and theoretical signal qualities are close to ideal. (b) When this ideal frequency response is approximated with a single-stage DI followed by a Gauss filter, then ISI and ICI crosstalk degrade the signal quality. Yet, the eye at the bottom still shows sufficient quality. (c) When the FFT is approximated by a Gaussian filter, the eye degrades and the signal quality decreases. We tested all kinds of optical filter passbands for the simulations of a 20 GBd signal on 25 GHz subcarrier spacings. To derive a reasonably good performance one needs a Gaussian filter with a 3 dB passband close to 25 GHz, see Ref. [50]. (For interpretation of the references to colour in this figure legend, the reader is referred to the web version of this book.)

compromise between signal quality and ease of implementation. It is noteworthy to mention that the maximum gate duration behind the FFT filter is increasing when reducing the number of DI stages. The duration depends on the filters that are cascaded, though.

All simulations in Figure 9.15 have been performed for 20 GBd NRZ on-off keying signals at a channel spacing of $F_s = 25$ GHz.

Approximations to the ideal FFT have been used many times so far. Experimentally, we approximated the 325 FFT by a dual-stage DI cascade and a Gaussian filter for extracting the Fourier coefficients from a 10 Tbit/s signal, and we used a single DI stage scheme followed by a narrow filter and an electro-absorption modulator as a sampling gate to derive the 325 Fourier coefficients from a 26 Tbit/s signal. Other groups used DI stages as well for separating subcarriers from within an OFDM signal [61].

The combination of a DI cascade and simple optical filters to perform OFDM demultiplexing is also known in the electrical domain. For instance in Ref. [9], cascaded delay-and-add elements were used in order to extract single sub-channels out of an optically filtered subband from within the OFDM spectrum.

9.3.4.5 Mixed guided and free-space propagation—the arrayed waveguide grating router

A guided-wave serial-to-parallel converter and a DFT Fourier lens can be combined in an arrayed waveguide gra ting router [62] (AWGR). In this function the device has been discussed by Lowery [63]. In Figure 9.16 the AWGR input slab coupler illuminates—when correctly designed—all grating waveguides with equal amplitude

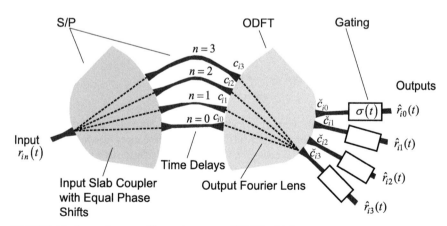

FIGURE 9.16 Arrayed waveguide grating router (AWGR) with input slab coupler that distributes the input equally to the inputs of the grating waveguides, which act as serial-to-parallel converters similar to the receiver schematic in Figure 9.10a. The output slab coupler functions as the Fourier lens, the inputs to which (plane $z = 0$ in Figure 9.11) are subscripted with n, while the outputs (plane $z = 2l$ in Figure 9.11) are identified by numbers corresponding to v in Figure 9.11. (Modified from Ref. [63] © 2010 OSA.)

and phase. The increasing lengths of grating waveguides introduce time delays incremented by T_s/N for providing a S/P conversion. The S/P conversion outputs n connects to the output slab coupler, which is designed for phase shifts as in Eq. (9.46) (Madsen and Zhao [52], Eq. (28) in Section 4.4), i.e. the output slab coupler acts as a Fourier lens. Samplers at the output ports complete the S/P conversion-ODFT setup, which has the identical functionality as the S/P conversion-OFFT scheme of Figure 9.12.

A disadvantage of the AWGR (which it shares with the Fourier transform lens discussed in Section 9.3.4.1) is the required precision of the $N-1$ phase shifts to be realized by the Fourier lens.

Transmitters according to Figure 9.10c employing the OIDFT combined with a parallel-to-serial converter in the form of an arrayed waveguide grating router (AWGR) were discussed by Lee et al. [64] and Lowery et al. [40]. Lowery even found ways of inserting a cyclic prefix [65].

Alternative implementations of the AWGR concept but using MMIs have been discussed in [66,67].

9.4 ENCODING AND DECODING OF OFDM SIGNALS

In this section we discuss implementations of optical OFDM transmitters and receivers. Finally, we exemplarily give account of a recent all-optical OFDM transmission experiment.

9.4.1 OFDM transmitter

OFDM transmitters may be implemented in the electronic domain or in the optical domain, a technique that allows for orthogonal multiplexing of almost any modulation. In this section we confine ourselves to the multiplexing of QAM signals. Yet, it needs to be understood that the same transmitters are also able to encode simple on-off keying, differential phase-shift keying, and other modulation formats.

Electronic OFDM transmitters are preferred when many lower bitrate subcarrier signals are to be multiplexed to a total data rate of a few 100 Gbit/s. A complete real-time OFDM transceiver operating at up to 11.25 Gbit/s has already been shown [68] and more recently the first 100 Gbit/s OFDM real-time transmitter has been demonstrated [30]. It is expected that electronic approaches will reach Tbit/s capacity in the future. Electronic implementations typically generate OFDM signals by carrying out a serial-to-parallel conversion of the input data stream and then performing the IDFT electronically as outlined in Section 9.3.1. After performing the IDFT, the signal is usually clipped before being passed on to digital-to-analogue (DAC) converters. The real and the complex part of the signal (the so-called inphase and qudrature part) are then separately fed to an optical IQ-modulator in order to encode the electrical signal on an optical carrier. Electronic implementations may flexibly adapt the order of the FFT and the symbol rate to changing channel conditions and customer requirements.

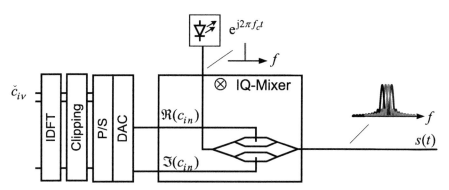

FIGURE 9.17 Electrical OFDM transmitter. The electronic part consists of an IDFT processor, a clipping circuit, a parallel-to-serial converter (P/S), and a digital-to-analogue converter (DAC). An optical IQ-modulator (IQ-mixer) then encodes the analogue signal on a narrow linewidth laser (local oscillator, LO). The signal spectra before encoding and after encoding the information are shown as insets.

With such transmitters, OFDM signals with very few up and to 1024 subcarriers have been implemented in optical communication systems [69]. Electronic OFDM transmitters are, however, limited by the speed of the digital signal processing circuit and by the electrical power consumption. A typical electronic OFDM setup is depicted in Figure 9.17.

Optical or "all-optical" OFDM transmitters are attractive for multiplexing higher-bit-rate tributaries into Tbit/s-capacity super-channels. Our theoretical discussion in the sections on OFDM and FFT led to two distinct implementations of OFDM transmitters. The first concept is based on a direct implementation of the rectangular-pulse-shaping technique discussed in Section 9.2. Alternatively, an optical OIDFT implementation may be applied as per Section 9.3. The two schemes are shown in Figure 9.18.

The first scheme is following the direct implementation of OFDM as outlined in Section 9.2. In this transmitter scheme an OFDM signal is generated by rectangular pulse-shaping of the individual OFDM subcarriers, and only then performing the summation of all subcarriers by an $N{:}1$ combiner. In Figure 9.18a we use a WDM-MUX instead of an $N{:}1$ combiner. To avoid the immense loss of a $N{:}1$ combiner, one would typically combine even and odd subcarrier channels in a WDM-MUX and only then combine them with a 2:1 coupler. The individual frequencies of the frequency comb of a single mode-locked laser (MLL) can be used as local oscillators as they have a precise frequency spacing. Alternatively, many equally spaced lasers may be used. Since OFDM only requires frequency locking, but no synchronization of the initial optical phase in a symbol, of the individual subcarriers, one may use a bank of well-stabilized lasers. To the best of our knowledge, such a transmitter was first used by Sano [27] to generate a 3 Tbit/s OFDM signal.

(a) OFDM Tx by Pulse-Shaping of Subcarriers

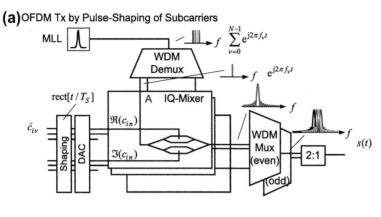

(b) OFDM Tx by Optical IFFT Filtering

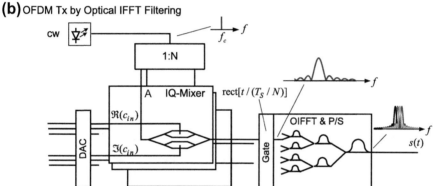

FIGURE 9.18 OFDM transmitter schemes. (a) Implementation of OFDM transmitter where each subcarrier is pulse-shaped and encoded with information before being multiplexed into a single fibre channel. (b) Implementation by coding the information onto _N_ local oscillators and then performing the optical IFFT.

The second scheme is a direct implementation of one of the all-optical DFT schemes outlined in Section 9.3. In Figure 9.18b, we have been following the all-optical IDFFT approach based on delay-interferometers. In this scheme, a local oscillator CW source is split into N local oscillators of the same frequency. Each of these copies is encoded with complex data by means of IQ-modulators. The signals are fed into an OIFFT processor. Here, the implementation based on delay interferometers is used. The N subcarriers are first shaped by a gate into signals of duration T_S/N before being IFFT processed in the delay-interferometer cascade. In practical implementations, a MLL would most likely replace the CW-source to perform the gating at the source rather than within the OIFFT processor.

While for lower-speed OFDM transmitters the electronic implementation of Figure 9.16 is quite common [30,68,70,69], the transmitter of Figure 9.18a is usually used for generating super-channels at highest data rates [27,29,70,9,71].

9.4.2 OFDM receivers

Lower bitrate OFDM signals may be received with a single coherent receiver, and all the subcarrier information may then be derived by digital signal processing (DSP) following our discussion in Section 9.3.1.

For higher bitrate OFDM signals, DSP may be used to a larger or lesser extent as shown in Figure 9.19a. This scheme illustrates how a Tbit/s super-channel can be split in up to N signals such that subcarriers or groups of subcarriers can be coherently received and subsequently processed with DSP. Such electronic receivers require sufficient oversampling. Electronic processing may then be applied by exploiting the orthogonality of rect-shaped functions such as outlined in Section 9.2 by evaluating Eq. (9.14), or by performing the DFT such as discussed in Section 9.3, Eq. (9.26), and outlined by Eq. (9.33). In electronic receivers, dispersion is typically undone by electronic digital signal processing prior to evaluating Eq. (9.14), (9.26) or the DFT of Eq. (9.35). Examples of implementations may, for example, be found in Refs. [9,71].

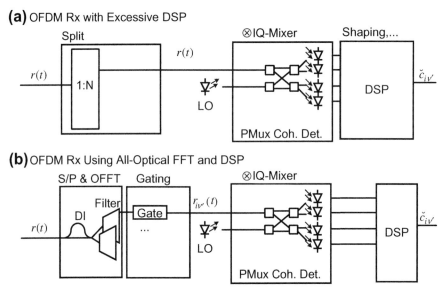

FIGURE 9.19 OFDM receiver concepts for the detection of higher-bit rate signals.
(a) In a DSP-based OFDM receiver, the incoming signal is split in up to *N* signals. Individual carriers or groups of carriers are detected with a local oscillator in the centre of the detected frequency band. Demultiplexing is typically performed in the electronic domain by applying the FFT. (b) All-optical OFDM receiver, where the FFT is performed in the optical domain. The subcarrier data can then directly be detected with a receiver that is adequate to detect the respective signal encoded onto each subcarrier. For QAM-signals this will be a coherent detector.

Alternatively, OFDM demultiplexing could be performed in the optical domain. Any of the all-optical schemes discussed in the previous sections could be applied. As an example, Figure 9.19b shows an all-optical OFDM receiver that performs the simplified OFFT with a single delay interferometer stage. In this scheme dispersion is typically first compensated in the optical domain by using dispersion compensating fibre. Subsequently, the optical FFT is performed. As a result, the complex data coefficients—still encoded on their respective optical carrier—are produced. These signals then can be detected by coherent receivers. Advantages of all-optical schemes are as follows. First, the FFT is performed passively in the optical domain. Such a scheme saves energy as it relies to a lesser extent on DSP. Second, all-optical detection schemes do not necessarily require coherent detection. In fact, a coherent receiver is only needed for advanced modulation formats where information is phase encoded. If on-off keying signals or differential-phase shift keying signals have been encoded, they can directly be detected with their respective direct-detection or self-coherent receiver as required by the respective modulation format. In addition, the factor 2 in oversampling is avoided, and the scheme is inherently format transparent. Examples of implementations may, for example, be found in Refs. [29,47,50].

9.4.3 OFDM transmission—an example of an all-optical implementation

As an example we show the all-optical encoding and decoding of a 26 Tbit/s superchannel onto a single mode-locked laser (MLL) source [29]. It is further shown that the quality of the signal is good enough for transmission over 50 km of SMF fibre in combination with a standard dispersion compensation module.

The experimental setup of the all-optical OFDM transmitter, receiver, and transmission link is shown in Figure 9.20. In the OFDM transmitter, the subcarriers are generated as a frequency comb with 12.5 GHz spacing which is derived from a mode-locked laser. Odd and even subcarriers are modulated separately with 16 QAM at 10 GBd and combined to form the all-optical OFDM signal. Additionally, the OFDM signal is polarisation multiplexed. The signal is then transmitted over 50 km of SMF-28 and the dispersion is compensated using a standard dispersion compensation module. The OFDM signal is then decoded in an all-optical OFDM receiver using the optical FFT [50]. The modulated subcarriers are then demodulated in a coherent receiver.

Here the symbol rate is not equal to the subcarrier spacing. We used a cyclic prefix to increase the tolerance to dispersion and other transmission effects [11]. In this case, each OFDM symbol is extended periodically. In our implementation we chose to implement a cyclic prefix using a reduced symbol rate (10 GBd), effectively introducing a cyclic prefix of 20 ps length.

To generate the local oscillators, we use a frequency comb derived from a mode-locked laser (MLL). The complete setup of the comb generator consists of the mode-locked laser (MLL) itself, an optical amplifier, a highly nonlinear photonic crystal fibre (HNLF), and a wavelength selective switch (WSS), and it employs spectral

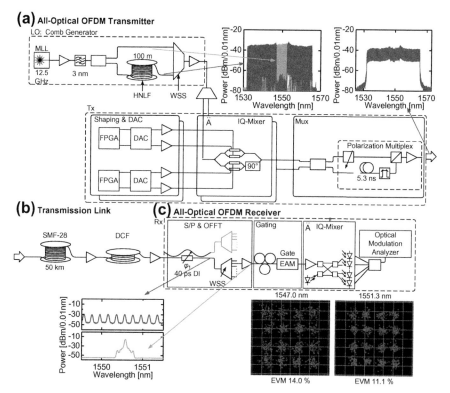

FIGURE 9.20 All-optical OFDM transmitter, link, and receiver setup. (a) The all-optical OFDM transmitter consists of a comb generator, a subcarrier modulation setup, and a polarization multiplexing stage. In the comb generator a mode-locked laser (MLL—12.5 GHz) is spectrally broadened in a highly nonlinear fibre (HLNF) and combined with the original MLL output in a wavelength selective switch (WSS). The WSS also provides equalizing for a uniform frequency comb. The 325 spectral lines are separated into even and odd subcarriers in an optical disinterleaver (DIL) and subsequently encoded with 16 QAM at 10 GBd. Even and odd channels are combined and polarization multiplexed to generate the OFDM channel. The signal is transmitted through a 50 km span of SMF-28 with subsequent standard dispersion compensation. The signal is decoded using an all-optical OFDM receiver consisting of a delay interferometer (DI), a WSS, and an electro absorption modulation (EAM). Only the first DI was implemented, and later DI have been replaced by a narrowband filter, sacrificing some performance. Ref. [29].

slicing to generate a broad frequency comb. The MLL (Time Bandwidth Products—ERGO XG) at a repetition rate of 12.5 GHz with a pulse width of 0.9 ps is amplified and filtered using a 3 nm band pass filter. The amplified pulse train is split, and one path is spectrally broadened through the Kerr nonlinearities of the HNLF (dispersion at 1480–1620 larger than 0 ps/nm/km, dispersion at 1480–1620 smaller than 1.5 ps/nm/km, mode field diameter at 1550 nm, nonlinear coefficient at 1550 nm is 11/W/km).

Due to the nature of the spectral broadening, the broadened spectrum will exhibit instably positioned minima in the centre of the spectrum. To compensate for this, the centre of the spectrum is replaced by the original MLL spectrum, and the resulting spectrum is equalised in the wavelength selective switch (Finisar Waveshaper) to generate a flat frequency comb.

The generated frequency comb is amplified and separated into odd and even subcarriers in an optical disinterleaver with a suppression of >35 dB. Odd and even subcarriers are modulated separately using two software defined 16 QAM transmitters [25]. In the 16 QAM transmitters, field programmable gale arrays (FPGA, Xiliux Virlex 5) generate the digital multilevel waveforms for a 16 QAM signal with a pseudo-random bit sequence (PRBS, $2^{15} - 1$) in real-time. The digital signal is fed to high-speed digital-to-analogue converters (Micram VEGA DAC25), and the generated multilevel drive signal is encoded on the optical carriers. The modulated subcarriers are combined in an optical 2×2 coupler to form the desired OFDM signal. The two outputs of the coupler are then delayed with respect to each other by 5.3 ns for decorrelation and subsequently multiplexed on two orthogonal polarisations.

The signal with an optical power of 9.5 dBm is launched into a 50 km transmission link of SMF-28 with a total span loss of 10.17 dB. After transmission, the signal is amplified, and the dispersion of the link is compensated using standard dispersion compensation modules with a total insertion loss of 4.8 dBm. The residual dispersion of the link is 17 ps/nm.

The receiver comprises an all-optical FFT circuit and an optical modulation analyser (Agilent N4391A) which performs real-time coherent detection and an EVM analysis. Here we implement the simplified optical FFT using a single delay interferometer to perform a two-point optical FFT, a wavelength selective switch (Finisar Waveshaper) to select the desired frequency band, and an optical gate which is synchronised to the transmitter clock at 10 GHz (electro-absorption modulator—EAM CIP 40G-PS-EAM-1550) to select the point in time in which the FFT is synchronised to the OFDM symbols. The recovered subcarriers are then received and demodulated in an optical modulation analyser (OMA—Agilent N4391A) with an electrical bandwidth of 16 GHz (Agilent Infiniium DSOX93204A). In the OMA, the signal is equalised with a 21 tap equalizer. For each polarisation, the error vector magnitude is calculated and averaged over 10 received sequences of 1024 symbols. This measurement is carried out for all subcarriers of the OFDM signal.

To evaluate the link performance, we plotted the error vector magnitudes (EVMs) for both polarisations in all 325 subcarriers as measured with the Agilent modulation analyser. The EVM is calculated from 2^{10} consecutive received symbols and averaged over 10 received sequences. The results are shown in Figure 9.21a for all measured 325 subcarriers. Typical constellation diagrams are depicted in Figure 9.20b. The symbols have a clear and distinct shape. To quantify the quality of the received signals, we have performed bit-error rate (BER) measurements as derived in Ref. [72] and renormalised them to be compatible with the EVM definition used in the optical modulation analyser. It can be seen that the EVM for all subcarriers after

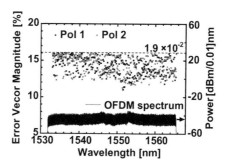

FIGURE 9.21 Error vector magnitude (EVM) for all 325 measured subcarriers. The EVM is shown for both polarizations on all subcarriers and is clearly below the horizontal line indicating the third generation FEC limit with a bit error ratio BER = 1.9×10^{-2}. The spectrum for all subcarriers is shown below the EVM plot. See Ref. [29].

transmission are between 11% and 16% with an average EVM of 14.2%. Estimations by Ref. [72] indicate that all subcarriers are below the 1.9×10^{-2} third-generation FEC limit [73] with 25% overhead. The demonstrated line rate of 26 Tbit/s therefore corresponds to a data rate of 20.8 Tbit/s. Degradation of the EVM due to transmission is negligible.

9.5 CONCLUSION

We have discussed multiplexing and demultiplexing schemes exploiting orthogonality of pulse shapes. A thorough theory has been introduced and used to discuss both OFDM and Nyquist pulse-shaping. The theory has then been further elaborated to work out a formalism to describe OFDM multiplexing and demultiplexing using by means of the discrete Fourier transform and the inverse Fourier transform. In a next step, electronic and all-optical implementations have been introduced and reviewed. The chapter concludes with a section on implementations of OFDM transmitters and receivers. Finally, the potential of all-optical schemes has been exemplified by discussing a recent 26 Tbit/s super-channel OFDM transmission experiment.

9.6 MATHEMATICAL DEFINITIONS AND RELATIONS

The following mathematical definitions and relations are used throughout the text and are referred to where required for understanding. In addition, Table 9.1 specifies the notation for temporal and spatial Fourier transforms and their inverse transforms for the chosen positive time dependency $\exp(j\omega t)$ (see Figure 9.20).

Table 9.1 Notation and formulae

Time	t	(9.49)
Frequency	f	(9.50)
Angular frequency	$\omega = 2\pi f$	(9.51)
Cartesian spatial coordinates	x, y, z	(9.52)
Spatial frequencies	ξ, η, ζ	(9.53)
Angular spatial frequencies	$k_x = 2\pi\xi, \quad k_y = 2\pi\eta, \quad k_z = 2\pi\zeta$	(9.54)
Imaginary unit	$j = \sqrt{-1}$	(9.55)
Complex conjugate of $u = p + jq$	$u^* = p - jq \quad$ (real p, q)	(9.56)
Plane wave propagation in homogeneous medium with refractive index n, positive time dependency	$\exp[j(\omega t - (k_x x + k_y y + k_z z))],$ $k_x^2 + k_y^2 + k_z^2 = n^2 \dfrac{\omega^2}{c^2}$	(9.57)
Plane wave propagation along $+z$, propagation constant β, positive time dependency	$\exp[j(\omega t - \beta z)]$	(9.58)
Kronecker symbol $\delta_{mm'}, m, m' \in \mathbb{Z}$	$\delta_{mm'} = \begin{cases} 1 & \text{for } m = m' \\ 0 & \text{else} \end{cases}$	(9.59)
Dirac function $\delta(t)$	$\Psi(0) = \displaystyle\int_{-\infty}^{+\infty} \delta(t)\Psi(t)dt, \quad \delta(t) = 0 \quad \text{for } t \neq 0$	(9.60)
Heaviside function $H(t)$	$\displaystyle\int_{0}^{+\infty} \Psi(t)dt = \int_{-\infty}^{+\infty} H(t)\Psi(t)dt, \quad H(t) = \begin{cases} 1 \text{ for } t > 0 \\ 0 \text{ for } t < 0 \end{cases}$	(9.61)
rect-function $\mathrm{rect}\left(\dfrac{t}{T}\right)$	$\displaystyle\int_{-T/2}^{+T/2} \Psi(t)dz = \int_{-\infty}^{+\infty} \mathrm{rect}\left(\dfrac{t}{T}\right)\Psi(t)dt,$ $\mathrm{rect}\left(\dfrac{t}{T}\right) = \begin{cases} 1 \text{ for } |t| < T/2 \\ 0 \text{ for } |t| > T/2 \end{cases}$	(9.62)
Raised-cosine (RC) function $\mathrm{r\,cos}\left(\dfrac{t}{T}\right)$	$\mathrm{r\,cos}\left(\dfrac{t}{T}\right) = \begin{cases} \cos^2\left(\dfrac{\pi}{2}t/T\right) & \text{for } |t| < T \\ 0 & \text{else} \end{cases}$	(9.63)
Root RC (RRC) function $\mathrm{rr\,cos}\left(\dfrac{t}{T}\right)$	$\mathrm{rr\,cos}\left(\dfrac{t}{T}\right) = \begin{cases} \cos\left(\dfrac{\pi}{2}t/T\right) & \text{for } |t| < T \\ 0 & \text{else} \end{cases}$	(9.64)

(continued)

Table 9.1 (Continued)

Sinc-function $\mathrm{sinc}\left(\dfrac{t}{T}\right)$	$\mathrm{sinc}\left(\dfrac{t}{T}\right) = \begin{cases} 1 & \text{for } t = 0 \\ \dfrac{\sin(\pi t/T)}{\pi t/T} & \text{else} \end{cases}$	(9.65)
Sinc-like function $\mathrm{sincl}\left(\dfrac{t}{T}\right)$	$\mathrm{sincl}\left(\dfrac{t}{T}\right) = \begin{cases} 1 & \text{for } t = 0 \\ \dfrac{\sin(2\pi t/T)}{2\pi t/T - (2\pi t/T)^3/\pi^3} & \text{else} \end{cases}$	(9.66)
Continuous temporal Fourier transform (FT), positive time dependency	$\check{\Psi}(f) = \displaystyle\int_{-\infty}^{+\infty} \Psi(t)\exp(-j2\pi ft)\,dt$	(9.67)
Continuous temporal inverse Fourier transform (IFT), positive time dependency	$\Psi(t) = \displaystyle\int_{-\infty}^{+\infty} \check{\Psi}(f)\exp(+j2\pi ft)\,df$	(9.68)
Continuous spatial Fourier transform (SFT)	$\widetilde{\Psi}(\xi, \eta) = \displaystyle\int_{-\infty}^{+\infty}\int_{-\infty}^{+\infty} \Psi(x, y)\exp[+j2\pi(\xi x + \eta y)]\,dx\,dy$	(9.69)
Continuous spatial inverse Fourier transform (SIFT)	$\Psi(x, y) = \displaystyle\int_{-\infty}^{+\infty}\int_{-\infty}^{+\infty} \widetilde{\Psi}(\xi, \eta)\exp[-j2\pi(\xi x + \eta y)]\,d\xi\,d\eta$	(9.70)
FT of rect-function	$\displaystyle\int_{-\infty}^{+\infty} \mathrm{rect}\left(\dfrac{t}{T}\right)\exp(-j2\pi ft)\,dt = T\mathrm{sinc}\,(fT)$	(9.71)
FT of sinc-function	$\displaystyle\int_{-\infty}^{+\infty} \mathrm{sinc}\left(\dfrac{t}{T}\right)\exp(-j2\pi ft)\,dt = T\mathrm{rect}\,(fT)$	(9.72)
FT of rcos-function	$\displaystyle\int_{-\infty}^{+\infty} r\cos\left(\dfrac{t}{T}\right)\exp(-j2\pi ft)\,dt = T\mathrm{sincl}(fT)$	(9.73)
FT of sincl-function	$\displaystyle\int_{-\infty}^{+\infty} \mathrm{sincl}\left(\dfrac{t}{T}\right)\exp(-j2\pi ft)\,dt = T\mathrm{rcos}\,(fT)$	(9.74)
Shift in frequency ("mixing" with analytic signal acting as complex carrier)	$\check{\Psi}(f - f_0) = \displaystyle\int_{-\infty}^{+\infty} \Psi(t)\exp(+j2\pi f_0 t)\exp(-j2\pi ft)\,dt$	(9.75)
Shift in time	$\Psi(t - t_0) = \displaystyle\int_{-\infty}^{+\infty} \check{\Psi}(f)\exp(-j2\pi ft_0)\exp(+j2\pi ft)\,df$	(9.76)
Fourier coefficients $\check{c}_\nu$ for T-periodic function $\Psi(t)$	$\check{c}_\nu = \dfrac{1}{T}\displaystyle\int_{-T/2}^{+T/2} \Psi(t)\exp\left(-j2\pi\nu\dfrac{t}{T}\right)\,dt$	(9.77)
Fourier series for T-periodic function $\Psi(t)$	$\Psi(t) = \displaystyle\sum_{\nu=-\infty}^{+\infty} \check{c}_\nu\exp\left(+j2\pi\nu\dfrac{t}{T}\right)$	(9.78)

Table 9.1 (Continued)

Discrete Fourier transform (DFT)	$\check{c}_v = \dfrac{1}{N} \displaystyle\sum_{n=0}^{N-1} c_n \exp\left(-j2\pi\,\dfrac{nv}{N}\right)$	(9.79)
Inverse discrete Fourier transform (IDFT)	$c_n = \displaystyle\sum_{v=0}^{N-1} \check{c}_v \exp\left(+j2\pi\,\dfrac{nv}{N}\right)$	(9.80)
Sampling function	$\sigma(t) = T \displaystyle\sum_{i=-\infty}^{+\infty} \delta\,(t - iT) = \displaystyle\sum_{i=-\infty}^{+\infty} \exp(j2\pi\,it/T)$	(9.81)
Fourier transform of sampling function	$\check{\sigma}(f) = T \displaystyle\sum_{i=-\infty}^{+\infty} \delta\,(f - iF), \quad F = 1/T$	(9.82)
Inner product	$(\Psi_1 \cdot \Psi_2) \equiv \langle\Psi_1|\Psi_2\rangle := \displaystyle\int_{-\infty}^{+\infty} \Psi_1^*\,(t')\,\Psi_2\,(t')\,dt'$	(9.83)
Convolution	$(\Psi_1 {*} \Psi_2)\,(t) := \displaystyle\int_{-\infty}^{+\infty} \Psi_1\,(t')\,\Psi_2\,(t - t')\,dt'$ $= \displaystyle\int_{-\infty}^{+\infty} \check{\Psi}_{1,2}(f)\,\exp(j2\pi\,ft)\,df$	(9.84)
Cross-correlation. Identical with convolution, if $\Psi_2(t) = \Psi_2^*\,(-t)$, $\check{\Psi}_2(f) = \check{\Psi}_2^*(f)$	$(\Psi_1 \otimes \Psi_2)\,(t) := \displaystyle\int_{-\infty}^{+\infty} \Psi_1\,(t')\,\Psi_2^*\,(t' - t)\,dt'$ $= \displaystyle\int_{-\infty}^{+\infty} \check{\Psi}_1(f)\check{\Psi}_2^*(f)\,\exp(j2\pi\,ft)\,df$	(9.85)

References

[1] R.W. Chang, Synthesis of band-limited orthogonal signals for multichannel data transmission, Bell Syst. Tech. J. 45 (1966) 1775–1796.

[2] W. Shieh, C. Athaudage, Coherent optical orthogonal frequency division multiplexing, Electron. Lett. 42 (2006) 587–589.

[3] A.J. Lowery, L. Du, J. Armstrong, Orthogonal frequency division multiplexing for adaptive dispersion compensation in long haul WDM systems, in: Optical Fiber Communication Conference, Anaheim, USA, Paper PDP39, 2006.

[4] I.B. Djordjevic, B. Vasic, Orthogonal frequency division multiplexing for high-speed optical transmission, Opt. Express 14 (2006) 3767–3775.

[5] N. Cvijetic, L. Xu, T. Wang, Adaptive PMD compensation using OFDM in Long-Haul 10 Gb/s DWDM systems, in: Optical Fiber Communication Conference, Anaheim, CA, Paper OTuA5, 2007.

[6] W. Shieh, Q. Yang, Y. Ma, 107 Gb/s coherent optical OFDM transmission over 1000-km SSMF fiber using orthogonal band multiplexing, Opt. Express 16 (2008) 6378–6386.

[7] S.L. Jansen, I. Morita, T.C. Schenk, H. Tanaka, Long-haul transmission of 1652.5 Gbits/s polarization-division-multiplexed OFDM enabled by MIMO processing, J. Opt. Network. 7 (2) (2008) 173–182.

[8] E. Yamada, A. Sano, H. Masuda, E. Yamazaki, T. Kobayashi, E. Yoshida, K. Yonenaga, Y. Miyamoto, K. Ishihara, Y. Takatori, T. Yamada, H. Yamazaki, 1 Tb/s (111 Gb/s/ch 10ch) no-guard-interval CO-OFDM transmission over 2100 km DSF, in: Opto-Electronics Communications Conference/Australian Conference on Optical Fiber Technology, Australia, Sydney, Paper PDP6, 2008.

[9] S. Chandrasekhar, X. Liu, B. Zhu, D.W. Peckham, Transmission of a 1.2-Tb/s 24-carrier no-guard-interval coherent OFDM superchannel over 7200-km of ultra-large-area fiber, Proceedings of European Conference on Optical Communication, Paper PD2.6, 2009.

[10] Q. Dayou, H. Ming-Fang, I. Ezra, H. Yue-Kai, S. Yin, H. Junqiang, W. Ting, 101.7-Tb/s (370294-Gb/s) PDM-128QAM-OFDM transmission over 355-km SSMF using pilot-based phase noise mitigation, Optical Fiber Communication Conference, March 2011, Optical Society of America, 2011, Paper PDPB5.

[11] W. Shieh, I. Djordjevic, OFDM for Optical Communications, Academic Press, 2009 ISBN: 978-0-12-374879-9.

[12] H. Nyquist, Certain topics in telegraph transmission theory, Trans. Am. Inst. Elec. Eng. (A.I.E.E.) 47 (2) (1928) 617–644<http://dx.doi.org/10.1109/T-AIEE.1928.5055024>.

[13] G. Bosco, V. Curri, A. Carena, P. Poggiolini, F. Forghieri, On the performance of Nyquist-WDM terabit superchannels based on PM-BPSK, PM-QPSK, PM-8QAM or PM-16QAM subcarriers, J. Lightwave Technol. 29 (2011) 53–61.

[14] R. Schmogrow, M. Winter, M. Meyer, D. Hillerkuss, S. Wolf, B. Baeuerle, A. Ludwig, B. Nebendahl, S. Ben-Ezra, J. Meyer, M. Dreschmann, M. Huebner, J. Becker, C. Koos, W. Freude, J. Leuthold, Real-time Nyquist pulse generation beyond 100 Gbit/s and its relation to OFDM, Opt. Express 20 (January) (2012) 317–337.

[15] R. Cigliutti, A. Nespola, D. Zeolla, G. Bosco, A. Carena, V. Curri, F. Forghieri, Y. Yamamoto, T. Sasaki, P. Poggiolini, Ultra-long-haul transmission of 16×112 Gb/s spectrally-engineered DAC-generated nyquist-WDM PM-16QAM channels with $1.05\times$ (symbol-rate) frequency spacing, in: Optical Fiber Communication Conference (OFC), 2012, Paper OTh3A.3.

[16] M. Yan, Z. Tao, W. Yan, L. Li, T. Hoshida, J.C. Rasmussen, Experimental comparison of no-guard-interval-OFDM and Nyquist-WDM superchannels, in: Optical Fiber Communication Conference (OFC), 2012, Paper OTh1B.2.

[17] A.D. Ellis, F.C.G. Gunning, Spectral density enhancement using coherent WDM, IEEE Photon. Technol. Lett. 17 (2) (2005) 504–506.

[18] R. Schmogrow, D. Hillerkuss, S. Wolf, B. Bäuerle, M. Winter, P. Kleinow, B. Nebendahl, T. Dippon, P.C. Schindler, C. Koos, W. Freude, J. Leuthold, 512QAM Nyquist sinc-pulse transmission at 54 Gbit/s in an optical bandwidth of 3 GHz, Opt. Express 20 (March) (2012) 6439–6447.

[19] X. Zhou, L.E. Nelson, P. Magill, B. Zhu, D.W. Peckham, 8×450-Gb/s, 50-GHz-spaced, PDM-32QAM transmission over 400 km and one 50 GHz-grid ROADM, in: Optical Fiber Communication Conference (OFC), 2011, Paper PDPB3.

[20] B. Châtelain, C. Laperle, K. Roberts, X. Xu, M. Chagnon, A. Borowiec, F. Gagnon, J. Cartledge, D.V. Plant, Optimized pulse shaping for intra-channel nonlinearities mitigation in a 10 Gbaud dual-polarization 16-QAM system, in: Optical Fiber Communication Conference (OFC), 2011, Paper OWO5.

[21] C. Behrens, S. Makovejs, R.I. Killey, S.J. Savory, M. Chen, P. Bayvel, Pulse-shaping versus digital backpropagation in 224 Gbit/s PDM-16QAM transmission, Opt. Express 19 (2011) 12879–12884.

[22] S.D. Personick, Receiver design for digital fiber optic communication systems, I, Bell Syst. Tech. J. 52 (1973) 843–874, Eq. (25) iv.

[23] P.J. Winzer, R.-J. Essiambre, Advanced optical modulation formats, Proc. IEEE 94 (5) (2006) 952–985.

[24] A. Sano, H. Masuda, T. Kobayashi, M. Fujiwara, K. Horikoshi, E. Yoshida, Y. Miyamoto, M. Matsui, M. Mizoguchi, H. Yamazaki, Y. Sakamaki, H. Ishii, 69.1-Tb/s (432 × 171-Gb/s) C- and extended L-band transmission over 240 km using PDM-16-QAM modulation and digital coherent detection, Optical Fiber Communication Conference, March 2010, Optical Society of America, 2010, Paper PDPB7.

[25] R. Schmogrow, D. Hillerkuss, M. Dreschmann, M. Huebner, M. Winter, J. Meyer, B. Nebendahl, C. Koos, J. Becker, W. Freude, J. Leuthold, Real-time software-defined multiformat transmitter generating 64QAM at 28 GBd, IEEE Photon. Technol. Lett. 22 (21) (2010) 1601–1603.

[26] D. Hillerkuss, R. Schmogrow, M. Meyer, S. Wolf, M. Jordan, P. Kleinow, N. Lindenmann, P.C. Schindler, A. Melikyan, X. Yang, S. Ben-Ezra, B. Nebendahl, M. Dreschmann, J. Meyer, F. Parmigiani, P. Petropoulos, B. Resan, A. Oehler, K. Weingarten, L. Altenhain, T. Ellermeyer, M. Moeller, M. Huebner, J. Becker, C. Koos, W. Freude, J. Leuthold, Single-laser 32.5 Tbit/s Nyquist WDM transmission, J. Opt. Commun. Network. 4 (2012) 715–723.

[27] A. Sano, H. Masuda, E. Yoshida, T. Kobayashi, E. Yamada, Y. Miyamoto, F. Inuzuka, Y. Hibino, Y. Takatori, K. Hagimoto, T. Yamada, Y. Sakamaki, 30 × 100-Gb/s all-optical OFDM transmission over 1300 km SMF with 10 ROADM nodes, in: European Conference on Optical Communication (ECOC), 2007, Paper PDP1.7.

[28] R. Schmogrow, S. Wolf, B. Baeuerle, D. Hillerkuss, B. Nebendahl, C. Koos, W. Freude, J. Leuthold, Nyquist frequency division multiplexing for optical communications, in: Conference on Lasers and Electro-Optics (CLEO'12), San Jose, CA, USA, Paper CTh1H.2, May 2012.

[29] D. Hillerkuss, R. Schmogrow, T. Schellinger, M. Jordan, M. Winter, G. Huber, T. Vallaitis, R. Bonk, P. Kleinow, F. Frey, M. Roeger, S. Koenig, A. Ludwig, A. Marculescu, J. Li, M. Hoh, M. Dreschmann, J. Meyer, S. Ben Ezra, N. Narkiss, B. Nebendahl, F. Parmigiani, P. Petropoulos, B. Resan, A. Oehler, K. Weingarten, T. Ellermeyer, J. Lutz, M. Moeller, M. Huebner, J. Becker, C. Koos, W. Freude, J. Leuthold, 26 Tbit s^{-1} line-rate super-channel transmission utilizing all-optical fast Fourier transform processing, Nature Photon. 5 (2011) 364–371.

[30] R. Schmogrow, M. Winter, D. Hillerkuss, B. Nebendahl, S. Ben-Ezra, J. Meyer, M. Dreschmann, M. Huebner, J. Becker, C. Koos, J. Becker, W. Freude, J. Leuthold, Real-time OFDM transmitter beyond 100 Gbit/s, Opt. Express 19 (June) (2011) 12740–12749.

[31] S. Stein, J.J. Jones, Modern Communication Principles—with Application to Digital Signaling, McGraw Hill, New York, 1967 (Chapter 13).

[32] J.G. Proakis, M. Salehi, Communication Systems Engineering, second ed., Prentice Hall, Upper Saddle River, NJ, 2002 (Section 7.5.2).

[33] J.G. Proakis, Digital Communication, fourth ed., McGraw Hill, New York, 2000 (Chapter 9).

[34] J. Zhao, A.D. Ellis, Novel optical fast OFDM with reduced channel spacing equal to half of the symbol rate per carrier, in: OFC, San Diego, Paper OMR1, March 2010.

[35] R. Schmogrow, B. Baeuerle, D. Hillerkuss. B. Nebendahl, C. Koos, W. Freude, J. Leuthold, Raised-cosine OFDM for enhanced out-of-band suppression at low subcarrier counts, in: Conference on Signal Processing in Photonics Communications

(SPPCom'12), Cheyenne Mountain Resort, Colorado Springs, Colorado, USA, Paper SpTu2A.2, June 2012.

[36] K.-D. Kammeyer, Nachrichten-Übertragung. fourth ed., Vieweg+Teubner Verlag, 2008.

[37] R. Schmogrow, S. Ben-Ezra, P.C. Schindler, B. Nebendahl, C. Koos, W. Freude, J. Leuthold, Pulse-shaping in the digital, electrical, and optical domain—A comparison, submitted to J. Lightwave Technol. 2013.

[38] J.W. Cooley, J.W. Tukey, An algorithm for the machine calculation of complex Fourier series, Math. Comput. 19 (1965) 297–301.

[39] Z. Wang, K.S. Kravtsov, Y.-K. Huang, P.R. Prucnal, Optical FFT/IFFT circuit realization using arrayed waveguide gratings and the applications in all-optical OFDM systems, Opt. Express 19 (2011) 4501–4512.

[40] A.J. Lowery, L. Du, All-optical OFDM transmitter design using AWGRs and low-bandwidth modulators, Opt. Express 19 (2011) 15696–15704.

[41] G. Grau, W. Freude, Optische Nachrichtentechnik. Eine Einführung (Optical Communications), Springer-Verlag, Berlin, 1991, Eq. (2.21) and Appendix C1 (since 1997 out of print; corrected electronic reprint available).

[42] W. Freude, G.K. Grau, Rayleigh-Sommerfeld and Helmholtz-Kirchhoff integrals: Application to the scalar and vectorial theory of wave propagation and diffraction, IEEE J. Lightwave Technol. 13 (1995) 24–32 Eq. (10) and Appendices.

[43] J.W. Goodman, Introduction to Fourier optics, second ed., McGraw-Hill, New York, 1996, Eq. (5)–(10).

[44] M.E. Marhic, Discrete Fourier transforms by single mode star networks, Opt. Lett. 12 (1987) 63–65.

[45] A.E. Siegman, Fiber Fourier optics, Opt. Lett. 26 (2001) 1215–1217 and 27 (2002) 381.

[46] H. Sanjoh, E. Yamada, Y. Yoshikuni, Optical orthogonal frequency division multiplexing using frequency/time domain filtering for high spectral efficiency up to 1 bit/s/Hz, OFC, Anaheim, Paper ThD1, March 2002.

[47] Y.-K. Huang, D. Qian, R.E. Saperstein, P.N. Ji, N. Cvijetic, L. Xu, T. Wang, Dual-polarization 2×2 IFFT/FFT optical signal processing for 100-Gb/s QPSK-PDM all-optical OFDM, in: OFC, San Diego, Paper OTuM4, March 2009.

[48] K. Takiguchi, M. Oguma, T. Shibata, H. Takahashi, Optical OFDM demultiplexer using silica PLC based optical FFT circuit, in: Proceedings Optical Fiber Communication Conference (OFC'2009), San Diego, Paper OWO3, March 2009.

[49] K. Takiguchi, M. Oguma, H. Takahashi, A. Mori, Integrated-optic eight-channel OFDM demultiplexer and its demonstration with 160 Gbit/s signal reception, Electron. Lett. 46 (8) (2010) 575–576.

[50] D. Hillerkuss, M. Winter, M. Teschke, A. Marculescu, J. Li, G. Sigurdsson, K. Worms, S. Ben Ezra, N. Narkiss, W. Freude, J. Leuthold, Simple all-optical FFT scheme enabling Tbit/s real-time signal processing, Opt. Express 18 (2010) 9329–9340.

[51] Y.P. Li, Dense waveguide division multiplexers implemented using a first stage Fourier filter, United States Patent, Patent No. 5,852,505, Date of Patent: December 22, 1998.

[52] C.K. Madsen, J.H. Zhao, Optical Filter Design and Analysis—A Signal Processing Approach, Wiley-Interscience, New York, 1999 (Section 4.2 "Cascade filters").

[53] G. Cincotti, Fibre wavelet filters, IEEE J. Quant. Electron. QE-38 (2002) 1420–1427.

[54] G. Cincotti, Full optical encoders/decoders for photonic IP routers, J. Lightwave Technol. 22 (2004) 337–342.

[55] D. Hillerkuss, T. Schellinger, R. Schmogrow, M. Winter, T. Vallaitis, R. Bonk, A. Marculescu, J. Li, M. Dreschmann, J. Meyer, S. Ben Ezra, N. Narkiss, B. Nebendahl, F.

Parmigiani, P. Petropoulos, B. Resan, K. Weingarten, T. Ellermeyer, J. Lutz, M. Möller, M. Hübner, J. Becker, C. Koos, W. Freude, J. Leuthold, Single source optical OFDM transmitter and optical FFT receiver demonstrated at line rates of 5.4 and 10.8 Tbit/s, in: OFC 2010, Postdeadline Paper PDPC1, 2010.

[56] G. Cincotti, Generalized fiber Fourier optics, Opt. Lett. 36 (2011) 2321–2323.

[57] A.C. McBride, F.H. Kerr, On Namias's fractional Fourier transforms, IMA J. Appl. Math. 39 (1987) 159–175.

[58] V. Namias, The fractional order Fourier transform and its application to quantum mechanics, J. Inst. Maths. Applics. 25 (1980) 241–265.

[59] G. Cincotti, Coherent optical OFDM systems based on the fractional Fourier transform, in: 14th International Conference on Transparent Optical Networks (ICTON'12), University of Warwick, Coventry, UK, Paper We.A1.1, July 2–5, 2012.

[60] M. Nazarathy, D.M. Marom, W. Shieh, Optical comb and filter bank (De)Mux enabling 1 Tb/s orthogonal sub-band multiplexed CO-OFDM free of ADC/DAC limits, in: Proceedings European Conference on Optical Communication (ECOC'2009), Vienna, Paper P3.12, September 2009.

[61] T. Kobayashi, A. Sano, E. Yamada, E. Yoshida, Y. Miyamoto, Over 100 Gb/s electro-optically multiplexed OFDM for high-capacity optical transport network, J. Lightwave Technol. 27 (16) (2009) 3714.

[62] M.K. Smit, C. Van Dam, PHASAR-based WDM-devices: principles, design and applications, IEEE J. Sel. Top. Quant. Electron. 2 (2) (1996) 236–250.

[63] A.J. Lowery, Design of arrayed-waveguide grating routers for use as optical OFDM demultiplexers, Opt. Express 18 (13) (2010) 14129–14143.

[64] K. Lee, C.T.D. Thai, J.-K.K. Rhee, All optical discrete Fourier transform processor for 100 Gbps OFDM transmission, Opt. Express 16 (6) (2008) 4023–4028.

[65] A.J. Lowery, Inserting a cyclic prefix using arrayed-waveguide grating routers in all-optical OFDM transmitters, Opt. Express 20 (9) (2012) 9742–9754.

[66] I. Kang et al., All-optical OFDM transmission of 7 × 5-Gb/s data over 84-km standard single-mode fiber without dispersion compensation and time gating using a photonic-integrated optical DFT device, Opt. Express 19 (2011) 9111–9117.

[67] K. Takiguchi, T. Kitoh, M. Oguma, Y. Hashizume, H. Takahashi, Integrated-optic OFDM demultiplexer using multi-mode interference coupler-based optical DFT circuit, in: Optical Fiber Communication Conference, OSA Technical Digest, Paper OM3J.6, Optical Society of America, 2012.

[68] R.P. Giddings, X.Q. Jin, E. Hugues-Salas, E. Giacoumidis, J.L. Wei, J.M. Tang, Experimental demonstration of a record high 11.25 Gb/s real-time optical OFDM transceiver supporting 25 km SMF end-to-end transmission in simple IMDD systems, Opt. Express 18 (2010) 5541–5555.

[69] B. Inan, S. Adhikari, O. Karakaya, P. Kainzmaier, M. Mocker, H. von Kirchbauer, N. Hanik, S. Jansen, Real-time 93.8-Gb/s polarization-multiplexed OFDM transmitter with 1024-point IFFT, Opt. Express 19 (26) (2011) B64–B68.

[70] Y. Ma, Q. Yang, Y. Tang, S. Chen, W. Shieh, 1-Tb/s single-channel coherent optical OFDM transmission over 600-km SSMF fiber with subwavelength bandwidth access, Opt. Express 17 (2009) 9421–9427.

[71] H. Masuda, E. Yamazaki, A. Sano, T. Yoshimatsu, T. Kobayashi, E. Yoshida, Y. Miyamoto, S. Matsuoka, Y. Takatori, M. Mizoguchi, K. Okada, K. Hagimoto, T. Yamada, S. Kamei, 13.5-Tb/s (135 × 111-Gb/s/ch) no-guard-interval coherent OFDM transmission over 6,248 km using SNR maximized second-order DRA in the extended L-band, in: National

Fiber Optic Engineers Conference, OSA Technical Digest (CD), Paper PDPB5, Optical Society of America, 2009.

[72] R. Schmogrow, B. Nebendahl, M. Winter, A. Josten, D. Hillerkuss, S. Koenig, J. Meyer, M. Dreschmann, M. Huebner, C. Koos, J. Becker, W. Freude, J. Leuthold, Error vector magnitude as a performance measure for advanced modulation formats, IEEE Photon. Technol. Lett. 24 (2012) 1–63.

[73] T. Mizuochi, Recent progress in forward error correction and its interplay with transmission impairments, IEEE J. Sel. Top. Quant. Electron. 12 (2006) 544–554.

Spatial Multiplexing Using Multiple-Input Multiple-Output Signal Processing

10

Peter J. Winzer, Roland Ryf, and Sebastian Randel

Bell Labs, Alcatel-Lucent, Holmdel, NJ 07733, USA

10.1 OPTICAL NETWORK CAPACITY SCALING THROUGH SPATIAL MULTIPLEXING

10.1.1 The capacity crunch

The amount of traffic carried on backbone networks has been growing exponentially over the past two decades, at about 30 to 60% per year (i.e. between 1.1 and 2 dB per year),[1] depending on the nature and penetration of services offered by various network operators in different geographic regions [1,2]. The increasing number of applications relying on machine-to-machine traffic and cloud computing could accelerate this growth to levels typical within data centers and high-performance computers [3,4]: According to Amdahl's rule of thumb [5,6], the interface bandwidth of a balanced computer architecture is proportional to its processing power. Since cloud services are increasingly letting the network take the role of an interface between distributed data processing nodes, the required network bandwidth for such applications may scale with data processing capabilities, at close to 90% (or 2.8 dB) per year [7]. Noncacheable real-time multimedia applications such as high-definition telepresence and immersive communications [8,9] will further drive the need for more network bandwidth.

For over two decades, the demand for communication bandwidth has been economically met by wavelength-division multiplexed (WDM) optical transmission systems, researched, developed, and abundantly deployed since the early 1990s [10]. At first, WDM capacities increased at around 80% per year, predominantly through improvements in optoelectronic device technologies. By the early 2000s, lasers had reached GHz frequency stabilities, optical filters had bandwidths allowing for 50-GHz WDM channel spacings, and 40-Gb/s optical signals filled up these frequency slots. At this remarkable point in time where "optical and electronic bandwidths met,"

---

[1]Following [1], we conveniently express traffic growth in decibels, i.e. a 30% growth corresponds to a growth of $10 \log_{10}(1.3) = 1.14 \text{ dB}$.

optical communications had to shift from physics toward communications engineering to increase spectral efficiencies, i.e. to pack more information into the limited (~5-THz) bandwidth of single-band optical amplifiers. Consequently, the last decade has seen a vast adoption of concepts from radio-frequency communications, such as advanced modulation formats, coherent detection, sophisticated digital signal processing (DSP), and powerful error correction coding. The associated evolution of experimentally achieved interface rates (per wavelength and polarization), modulation symbol rates, spectral efficiencies, and per-fiber capacities is visualized in Figure 10.1 and reviewed, e.g., in [11] and the references cited therein.

Today's commercial WDM systems transmit close to 10 Tb/s of traffic at 100 Gb/s per wavelength [12]. In research, interface rates of 320 Gb/s have been demonstrated using quadrature amplitude modulation (QAM) at a symbol rate of 80 GBaud [13], polarization multiplexed to yield a 640-Gb/s single-carrier channel. Interface rates of 1 Tb/s and beyond have been achieved through optical parallelism using multi-carrier [14–16] and orthogonal frequency division multiplexed (OFDM) optical superchannels [17]. Regarding aggregate per-fiber capacities, and as shown in Figure 10.1a, 100-Tb/s transmission over single-mode fiber has been reported [18,19], but capacity increases in WDM research have slowed down to about 20% (0.8 dB) per year since ~2002, with a similar trend seen in commercial systems with a time delay of about 5 years [1,20]. While spectral efficiencies have been able to keep up with a ~1-dB

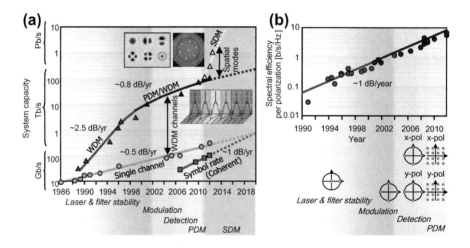

FIGURE 10.1 (a) Evolution of experimentally achieved single-channel bit rates (single-carrier, single-polarization, electronically multiplexed; green circles), symbol rates in digital coherent detection (purple squares), and aggregate per-fiber capacities (triangles) using wavelength-division multiplexing (WDM; red), polarization-division multiplexing (PDM; blue), and space-division multiplexing (SDM; yellow). (b) Evolution of experimentally achieved per-polarization spectral efficiencies. (Figure reproduced from [11].) (For interpretation of the references to color in this figure legend, the reader is referred to the web version of this book.)

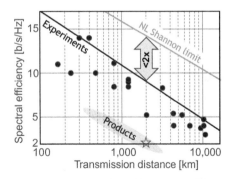

FIGURE 10.2 Trade-off between dual-polarization spectral efficiency and transmission reach, showing the nonlinear Shannon limit of [21] together with experimentally achieved results (circles). The ellipse indicates a range into which commercial systems might fall, and the asterisk represents an approximate commercially deployed optical transmission platform [12]. (Figure reproduced from [11].)

annual growth rate, cf. Figure 10.1b, capacities of conventional single-mode fiber systems over meaningful transmission distances are not expected to grow much further.

The evident capacity saturation is explained in recent studies on the nonlinear Shannon capacity of optical networks [21], setting an upper bound on the maximum achievable spectral efficiency for a given transmission distance and fiber type.[2] The resulting trade-off between spectral efficiency and system reach (including noise, fiber nonlinearities, as well as current technological shortfalls) is summarized in Figure 10.2. Record experimental results (circles) are shown together with the nonlinear Shannon limit of [21], scaled by a factor of two to represent polarization-division multiplexed (PDM) systems. Both curves trace straight lines on a logarithmic scale for the transmission distance L, since the signal-to-noise ratio (SNR) delivered by the line system to the receiver is inversely proportional to L, and the spectral efficiency is given by [23]

$$SE = \log_2(1 + SNR) \approx \log_2(SNR) \propto \log_2(1/L), \tag{10.1}$$

at reasonably high SNR which maximizes the performance of optical transport systems [21]. Importantly, we note that experimental records have approached the nonlinear Shannon limit to within a factor of less than two. The asterisk in Figure 10.2 represents Alcatel-Lucent's 1830 optical transmission platform, operating at a spectral efficiency of $\sim$2 b/s/Hz over $\sim$2000 km of fiber [12] and commercially deployed since mid-2010; the ellipse indicates a range into which commercial systems based on existing technologies and an approximate installed fiber with appropriate SNR margins might fall. At an annual traffic growth rate between 30 and 60%, and over the same, geography-enforced distances, optical transport systems surpassing the Shannon limit would have to be commercially available between 2015 and 2018.

---

[2]Advances in low-loss or low-nonlinearity fiber will not be able to change this picture significantly [22].

This observation leads to the notion of an imminent "*optical networks capacity crunch*" [24].

10.1.2 Spatial multiplexing

Realizing that WDM spectral efficiencies over the required transmission distances are no longer scalable, alternative solutions have to be developed. One option would be to trade spectral efficiency for reach by concatenating several high-spectral-efficiency systems. However, due to the unfavorable scaling of Eq. (10.1), which relates the SE to the *logarithm* of system reach, a large number of regenerators are needed to achieve a significant gain in SE [23]. For example, taking today's experimental records as a baseline for system performance, a target spectral efficiency of 20 b/s/Hz over 2000 km would require around 1500 1.3-km regeneration spans using, e.g., PDM 4096-QAM. Alternatively, a *parallel* approach may be taken. This requires the use of so far unexploited physical dimensions using multiple optical amplification bands or, likely more scalably, multiple parallel optical paths, referred to as *spatial multiplexing* or *space-division multiplexing* (*SDM*): With just three parallel paths at 7 bits/s/Hz each (e.g., using PDM 32-QAM), the desired aggregate SE is achieved with a total of three transponders per wavelength. The almost-three orders of magnitude difference in transponder count between the two solutions clearly point to SDM as the preferred solution for network capacity growth. A similarly large advantage of parallel transmission solutions over regenerated systems is obtained if not only transponders but also line system components are accounted for [23].

That going parallel in either amplification bands or space are the only capacity scaling options becomes clear from Figure 10.3, showing all known physical

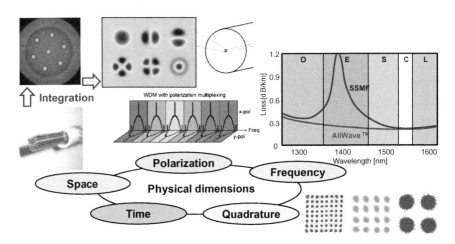

FIGURE 10.3 Spatial multiplexing exploits the only known physical dimension that has not yet been used in optical transport systems. Implementations include fiber bundles, multi-core, and multi-mode fiber, with increasing levels of integration.

dimensions that can be used to construct orthogonal signal spaces for modulation and multiplexing in optical communications [11,25]:

- The *time* dimension is used for modulation by transmitting one modulation symbol after the other, one per symbol duration T_S; the time dimension is used for multiplexing by allocating different time slots to different channels (time-division multiplexing, TDM).
- The *quadrature* dimension, i.e. the real and imaginary parts of the optical field, also referred to as the sine and cosine or the in-phase and quadrature components of this bandpass signal, are used to construct quadrature amplitude modulation (QAM) constellations to convey more than one bit per symbol.
- The *polarization* dimension is used for polarization-division multiplexing (PDM), allowing for a twofold increase in system capacity.
- The *frequency* dimension is used for multiplexing wavelength channels in WDM systems or to form optical superchannels or OFDM signals within the context of a single WDM channel. Since commercial WDM systems usually operate in a single amplification band (such as the C-band or the L-band), building multi-band systems would in principle be an option to scale system capacities by up to a factor of ~ 10. However, inherent problems associated with efficient and low-noise multi-band amplification as well as the fiber fuse effect, which arises on installed fiber at optical power levels significantly exceeding $1\,\mathrm{W}$ [26,27], may limit this option to very modest capacity gains and hence is unlikely to provide a long-term sustainable path forward.
- The *space* dimension is commonly used to scale short-reach interface rates [3] in a low-power, low-cost, and highly integrated manner. (For example, the 100GBASE-SR10 standard for 100G Ethernet specifies a 100-m interface using $10 \times 10.3\,\mathrm{Gb/s}$ over a multi-mode 20-fiber ribbon.) Exploiting the space dimension for optical transport networks through SDM holds the promise of getting network capacity growth back onto a 2–2.5-dB/year track, as impressively indicated by the three most recent data points at 112-Tb/s [28,29], 305-Tb/s [30], and 1-Pb/s [31] shown in Figure 10.1a.

Deploying SDM in its most trivial form by using M parallel optical line systems is a scalable but not yet an economically sustainable path forward, since it still does not reduce the cost or energy per bit compared to today's systems: M parallel systems carry M times the capacity at M times the cost, energy consumption, and footprint requirements compared to a single system. Commercially successful SDM technologies will be expected to scale capacity with a similar cost, energy, and footprint reduction as WDM (at $\sim 20\%$ per year [32]), leveraging *integration*, and sharing of system components among channels. Integration may take place on a system and network level, both from a capital (CAPEX) and operational (OPEX) expenditure point of view. Regarding the latter, Ref. [33] showed that it could be cheaper in certain scenarios to deploy a multi-core system just because of the associated reduced amplifier OPEX, even if sufficient single-mode fiber strands would be available to install M conventional line systems in parallel. Further, integration can take place on a transponder [34] and DSP level, on an optical amplifier level [35,36], and on a fiber level [37,38]. Importantly, apart from being more cost and energy effective than separate individual systems, integrated SDM solutions will have to allow for smooth network upgrades, reusing as much as possible the deployed WDM infrastructure. Initial global efforts in SDM research are reviewed in [39,40].

10.1.3 Crosstalk management in SDM systems

Since integration generally comes at the expense of *crosstalk* among parallel paths, proper crosstalk management is an important aspect of all SDM systems. Whether a certain level of crosstalk can be treated as a system impairment or needs to be actively compensated for depends on the underlying modulation format [41]: Figure 10.4 shows the crosstalk induced optical SNR (OSNR) penalty at a bit error ratio (BER) of 10^{-3} for single-polarization quadrature phase shift keying (QPSK), 16-QAM, and 64-QAM. The solid black curves represent a simple theoretical model, and circles denote measurement results. As the back-to-back implementation penalty in this 21.4-GBaud experiment increases from 0.9 dB (QPSK) to 1.8 dB (16-QAM) and 4.0 dB (64-QAM), the crosstalk tolerance also shrinks compared to theory. For a 0.5-dB crosstalk penalty, QPSK shows a tolerance of about −20 dB, 16-QAM tolerates about −26 dB, and 64-QAM tolerates only about −33 dB of crosstalk. In the context of SDM systems, these crosstalk numbers represent tolerable end-to-end crosstalk requirements, including crosstalk within transponders, optical amplifiers, transmission fibers, splices and connectors, as well as all other optical networking elements such as spatial and spectral optical crossconnects. For example, the 10-km span of 19-core fiber used for 305-Tb/s QPSK transmission in [30] showed a residual crosstalk level as high as −16 dB at the long-wavelength end of the exploited spectrum, revealing the difficulties of building highly scalable low-crosstalk optical components and transmission systems.

Recently, low-crosstalk multi-core fiber for SDM has been reported [29,42,43], and impressive SDM system experiments have been performed, including record

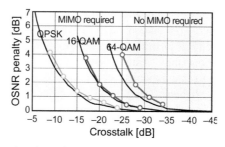

FIGURE 10.4 Tolerance of various higher-order modulation formats to in-band crosstalk (solid curves: theory; circles: experiments at 21.4 GBaud [41]). As soon as end-to-end crosstalk exceeds an acceptable system margin, MIMO techniques need to be employed.

per-fiber capacities of 109 Tb/s [28], 112 Tb/s [29], 305 Tb/s [30], and 1 Pb/s [31] over a few tens of km, as shown in Figure 10.1. Recent SDM experiments going beyond the nonlinear Shannon limit of single-mode fiber in terms of aggregate per-fiber spectral efficiency and transmission reach (cf. Figure 10.2) include 19-core transmission at 30.5 b/s/Hz over 10.1 km [30], 12-core transmission at 91.4 b/s/Hz over 52 km [31], as well as 7-core transmission at 15 b/s/Hz over 2688 km [44], at 42.2 b/s/Hz over 845 km [45], and 60 b/s/Hz over 76.8 km [46].

If crosstalk rises to levels where it induces penalties beyond acceptable system margins, *multiple-input-multiple-output (MIMO)* techniques, originally developed for wireless systems [47], have to be used to accommodate crosstalk by joint coherent detection and digital signal processing. As an example, the two highlighted areas in Figure 10.4 represent the case of 16-QAM with a 0.5-dB margin allocation for crosstalk. As soon as end-to-end crosstalk levels reach -26 dB, MIMO detection and processing is required, independent of the exact crosstalk level in this high-crosstalk regime.

Using MIMO techniques, a reliable (low-outage) capacity gain in a system with M coupled paths (or "modes") can be obtained, provided that the following key conditions are met [48,49]:

- The transmitter is able to uniquely map signals onto a complete orthonormal set of M (spatial and polarization) modes whose propagation is supported by the transmission fiber. This set of modes does not necessarily have to be the set of true fiber modes but can be any suitable linear combination of modes, as discussed in Section 10.4.
- The transmission fiber performs mostly unitary mode coupling, i.e. rotations of the M signals in mode space, without introducing excessive mode-dependent loss (MDL) or mode-dependent gain (MDG). As we shall see below, unitary transformations can be undone at the receiver without loss of information.
- The receiver is able to coherently detect a suitable complete orthonormal set of M modes. These M signals are then used to reconstruct the transmitted information using MIMO processing, as discussed in Section 10.3.

If the transmission properties of each mode in terms of noise and fiber nonlinearities are comparable to those of a single-mode reference system, an M-fold capacity gain can be achieved. The above requirements distinguish the type of MIMO-SDM described in this chapter from earlier work on mode-division multiplexing or mode-group multiplexing [50–56].

Recently, several impressive experimental demonstrations of coupled-mode MIMO-SDM transport have been reported, including transmission of six coupled spatial and polarization modes over up to 4,200-km microstructured fiber [57] as well as 1200-km few-mode fiber [58], MIMO-SDM using up to 12 modes [59], and discrete [60–62] as well as distributed Raman [63] amplification of few-mode signals.

In the remainder of this chapter, we describe key aspects of MIMO-SDM theory and implementation and highlight some of the challenges associated with this technology.

10.2 COHERENT MIMO-SDM WITH SELECTIVE MODE EXCITATION

10.2.1 Signal orthogonality

Communication signals can be multiplexed onto and can be uniquely demultiplexed from a common transmission channel if they are *orthogonal* in at least one of the physical dimensions discussed in the context of Figure 10.3.

10.2.1.1 Orthogonality in time and frequency

Orthogonality in *time and frequency* between two (scalar) waveforms $x_1(t)$ and $x_2(t)$ requires their inner product in signal space to vanish [25,64]:

$$\langle x_1(t)|x_2(t)\rangle = \int_{-\infty}^{\infty} x_1(t)x_2^*(t)\mathrm{d}t = \int_{-\infty}^{\infty} X_1(f)X_2^*(f)\mathrm{d}f = 0. \quad (10.2)$$

Here, t and f denote time and frequency, $X(f) = \mathcal{F}\{x(t)\}$ is the Fourier transform of $x(t)$, and * denotes the complex conjugate. Important examples for orthogonal waveforms in time and frequency are shown in Figure 10.5. (We will link these considerations to orthogonality in space in Section 10.2.1.2.) In particular, signals that are nonoverlapping either in time (b) or in frequency (c) are orthogonal, irrespective of their shape. Further, as shown in (a), T_S-time-shifted copies of certain temporally overlapping pulses $x(t)$ are orthogonal if $\mathcal{F}^{-1}\{|X(f)|^2\}$ has nulls at integer multiples of the symbol rate T_S, which follows from Eq. (10.2) with

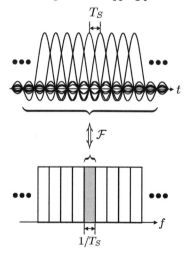

(a) Orthogonal overlapping pulses

(c) Orthogonal non-overlapping spectra

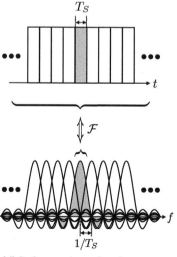

(b) Orthogonal non-overlapping pulses

(d) Orthogonal overlapping spectra

FIGURE 10.5 Examples for signal orthogonality in time and frequency.

$X_2(f) = X_1(f) \exp(j2\pi f k T_S)$. This condition is equivalent to Nyquist's criterion for no inter-symbol interference (ISI) [64,65]. An important class of pulse shapes satisfying this criterion are square-root raised cosine pulses. By the same token, using $x_2(t) = x_1(t) \exp(j2\pi t k R_S)$ in Eq. (10.2) reveals that R_S-frequency-shifted copies of $X(f)$ are orthogonal provided that $\mathcal{F}\{|x(t)|^2\}$ has nulls at integer multiples of $R_S = 1/T_S$, as illustrated in Figure 10.5(d). This condition is at the heart of pulse shaping in orthogonal frequency division multiplexing (OFDM) [66].

Once a set of orthogonal signals $x_i(t)$ supported by the communication channel is established as an orthonormal basis, multiple transmitters can transmit their information a_i, each by *uniquely* addressing one of the channel's basis functions, i.e. by sending $a_i x_i(t)$. Provided that the communication channel is able to maintain signal orthogonality, a receiver can demultiplex any such signal by forming the inner product between the aggregate received waveform $x(t) = \sum_i a_i x_i(t)$ and the respective basis function,

$$a_k = \langle x(t)|x_k(t)\rangle = \int_{-\infty}^{\infty} x(t)x_k^*(t)\mathrm{d}t = \int_{-\infty}^{\infty} X(f)X_k^*(f)\mathrm{d}f. \qquad (10.3)$$

For example, orthogonality in a WDM system is established through spectrally nonoverlapping frequency slots. Each transmitter sends its signal confined to within its frequency slot. Each receiver optically filters the aggregate WDM spectrum to extract the intended frequency slot, which is equivalent to calculating the inner product according to Eq. (10.3). If the transmitted spectrum spills into adjacent frequency slots, and unless frequency orthogonality is established through OFDM techniques, orthogonality is degraded, which is referred to as WDM crosstalk in this context. A degradation of orthogonality in the time domain is usually referred to as ISI.

10.2.1.2 Orthogonality in space and polarization

Orthogonality in *space and polarization* between a first spatial waveguide mode[3] with transverse electric field distribution $\vec{E}_{1,t}(\vec{r})$ and a second spatial waveguide mode with transverse magnetic field distribution $\vec{H}_{2,t}(\vec{r})$ is generally established as [67,68]

$$\iint\limits_{-\infty}^{\infty} [\vec{E}_{1,t}(\vec{r}) \times \vec{H}_{2,t}^*(\vec{r})]\vec{e}_z dA = 0, \qquad (10.4)$$

where A denotes the transverse ($\vec{r} = (x, y)^T$) plane relative to the propagation direction (unit vector $\vec{e}_z$). Any spatial field distribution can be expanded into a superposition of modes with expansion coefficients given by the integral in Eq. (10.4) [67]. It can further be shown that

$$[\vec{E}_{1,t}(\vec{r}) \times \vec{H}_{2,t}^*(\vec{r})]\vec{e}_z = \frac{\beta_2}{\omega\mu} \vec{E}_{1,t}(\vec{r}) \cdot \vec{E}_{2,t}^*(\vec{r}) + \frac{j}{\omega\mu} \vec{E}_{1,t}(\vec{r}) \cdot \begin{pmatrix} \partial_x E_{2,z}^*(\vec{r}) \\ \partial_y E_{2,z}^*(\vec{r}) \end{pmatrix}, \qquad (10.5)$$

---

[3]A waveguide mode is an electromagnetic field distribution that, apart from a scalar multiplication factor, does not change its shape upon transmission along the waveguide [67,68].

or alternatively that

$$[\vec{E}_{1,t}(\vec{r}) \times \vec{H}_{2,t}^*(\vec{r})]\vec{e}_z = \frac{\beta_2}{\omega\epsilon}\vec{H}_{1,t}(\vec{r}) \cdot \vec{H}_{2,t}^*(\vec{r}) - \frac{j}{\omega\epsilon}\vec{H}_{2,t}^*(\vec{r}) \cdot \begin{pmatrix} \partial_x H_{1,z}(\vec{r}) \\ \partial_y H_{1,z}(\vec{r}) \end{pmatrix}. \quad (10.6)$$

Here, β_2 is the propagation constant of mode 2, ω is the optical carrier frequency, and μ and ϵ are permeability and dielectric constant, respectively. The partial derivative with respect to x and y is denoted by ∂x and ∂y, respectively. The second term on the right-hand side of Eq. (10.5) vanishes for transversal electric (TE) modes, while the second term on the right-hand side of Eq. (10.6) vanishes for transversal magnetic (TM) modes. For modes with small z-components of electric or magnetic fields, such as LP pseudo-modes in the weakly guiding approximation of circular dielectric waveguides (i.e. optical fibers), the second terms may be neglected. In all these cases, the orthogonality relation (10.4) can either exactly or approximately be turned into the more familiar *spatial overlap integral*

$$\langle \vec{x}_1(t) | \vec{x}_2(t) \rangle = \int_{-\infty}^{\infty} dt \iint_{-\infty}^{\infty} dA\, \vec{x}_1(t,\vec{r}) \cdot \vec{x}_2(t,\vec{r}) = \int_{-\infty}^{\infty} df \iint_{-\infty}^{\infty} dA\, \vec{X}_1(f,\vec{r}) \cdot \vec{X}_2(f,\vec{r}),$$

$$(10.7)$$

where $\vec{x}(t)$ may denote either electric or magnetic field, leading to the unified notion of an "optical field." Equation (10.7) generalizes the familiar inner product of Eq. (10.2) to the space and polarization dimension. Two optical fields with orthogonal polarizations are hence orthogonal, as are two fields whose spatial profiles don't overlap, such as the fields of separate optical fibers or the fields that are confined to uncoupled cores of multi-core fibers. In analogy to time and frequency dimensions, and as discussed in the context of Figure 10.5, spatially *overlapping* optical fields distributions can also be orthogonal, provided that their overlap integral vanishes, which is the case, e.g., between waveguide modes with even and odd symmetries, such as the LP$_{01}$ and the LP$_{11}$ "modes" discussed in Section 10.4.

As with multiplexing in time and frequency, multiplexing in space and polarization requires the transmitters to selectively address a mode out of an orthogonal set of basis functions. Any spill-over manifests itself as performance degrading modal crosstalk. Similarly, the receiver has to calculate the inner product (i.e. the overlap integral) between the aggregate received field and the modal basis function it wants to receive. Owing to reciprocity considerations, a device that is able to selectively excite a certain waveguide mode at the transmitter will also extract that same mode at the receiver when operated in reverse. Orthogonal mode multiplexers (MMUXs) will be discussed in Section 10.4.

10.2.2 MIMO system capacities and outage

Neglecting inter- and intra-modal fiber nonlinearities, a coupled-mode SDM system in its simplest form can be represented by the linear matrix channel [49]

$$\vec{r} = \sqrt{E_0}\sqrt{L}\mathbf{H}\vec{s} + \vec{n}, \quad (10.8)$$

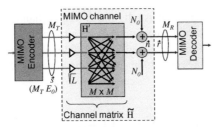

FIGURE 10.6 Basic MIMO system model for $M_T = 3$, $M_R = 2$, and $M = 4$.

as visualized in Figure 10.6; we use bold print to indicate stochastic processes and random variables. The SDM fiber supports a set of M orthogonal (spatial and polarization) modes. We use the term "mode" in the physical sense (cf. footnote 3 on page 437) to refer to a complete orthogonal set of waveguide modes, in contrast to wireless MIMO literature, which typically uses the term to refer to the eigenmodes of the channel matrix. Upon propagation, the fiber modes may be subject to coupling, differential gain or loss, and differential delay. We assume that the transmitter is able to selectively excite $M_T \leqslant M$ waveguide modes, and that the receiver is able to coherently extract $M_R \leqslant M$ modes. The average signal energy transmitted per symbol period and *per channel mode* is E_0, hence the total transmit energy across all modes is $M_T E_0$ per symbol and $\langle ||\vec{s}||^2 \rangle = M_T$. This definition differs from the constant total power constraint often used in wireless MIMO, where spatial waterfilling may then be used to optimally distribute power among the modes [69]. In an optical SDM system, the total transmit power is neither constrained by battery power nor by regulations pertaining to the use of the radio-frequency spectrum; rather, fiber nonlinearity is likely to set an upper bound on the optical power *per mode* [70] such that we are faced with a per-mode power constraint as opposed to an overall power constraint across all modes. Circularly symmetric complex Gaussian noise $\vec{n}$ with power spectral density N_0 per mode is added at the receiver. (We will discuss the case of distributed noise loading in Section 10.2.2.4.)

The SDM waveguide is described by an $M \times M$ matrix $\widetilde{\mathsf{H}}$, which we normalize as $\widetilde{\mathsf{H}} = \sqrt{L}\mathsf{H}'$ by factoring out the mode-average net propagation loss (or gain) L,

$$L = \frac{1}{M}\mathrm{tr}\{\widetilde{\mathsf{H}}\widetilde{\mathsf{H}}^\dagger\} = \frac{1}{M}\sum_{i=1}^{M}\widetilde{\lambda}_i; \tag{10.9}$$

$\mathrm{tr}\{\cdot\}$ denotes the trace of a matrix, $\widetilde{\lambda}_i$ are the M eigenvalues of $\widetilde{\mathsf{H}}\widetilde{\mathsf{H}}^\dagger$, and $\widetilde{\mathsf{H}}^\dagger$ is the conjugate transpose of $\widetilde{\mathsf{H}}$. Apart from noise, the MIMO channel is then given by the $M_R \times M_T$ matrix H spanning the subspace of H' addressed by the transponders.

We next assume that each individual instantiation H of the ensemble of channel matrices **H** is *known to the receiver* (e.g. through appropriate training sequences) but is *unknown to the transmitter* (since the receiver-to-transmitter feedback delays in optical transport networks are usually much longer than the kHz channel dynamics due to acoustic or thermal variations). Without access to channel state information, the transmitter sends uncorrelated signals of equal power on all transmit modes and at best achieves the open-loop Bell Labs Layered Space Time (BLAST) capacity[4] [47]

$$C = \sum_{i=1}^{d} \log_2 \left(1 + \lambda_i \frac{E_0 L}{N_0} \right), \quad (10.10)$$

where λ_i are the $d \leqslant \min\{M_T, M_R\} \leqslant M$ nonzero eigenvalues of $HH^\dagger$ and d is referred to as the rank of the MIMO channel. The term

$$\text{SNR} = \frac{E_0 L}{N_0} \quad (10.11)$$

represents the mode-average SNR at the receiver when all M channel modes are excited and detected ($M_T = M_R = M$) or, equivalently, the SNR measured as the ratio of the total received signal power to the total received noise power for equal power launched into all modes. We proceed to discuss capacities of several important optical MIMO-SDM channels [49].

10.2.2.1 Uncoupled channel

For the special case of M transmitters, M receivers, and M uncoupled channels, H is the M-dimensional identity matrix I_M, whose M eigenvalues are unity. Hence, Eq. (10.10) yields the capacity

$$C = M C_S = M \log_2(1 + \text{SNR}), \quad (10.12)$$

which is M times the corresponding single-mode capacity C_S, as expected.

10.2.2.2 Unitary mode coupling channel

Assuming again M transmitters and M receivers, but letting the M-dimensional channel randomly couple the signals in a unitary manner such that $H = U$ (corresponding to a random rotation in mode space), we have $HH^\dagger = UU^\dagger = UU^{-1} = I_M$, leading to the same capacity as in the uncoupled case,

$$C = M C_S = M \log_2(1 + \text{SNR}). \quad (10.13)$$

This shows that unitary mode coupling has no effect on the achievable MIMO transmission capacity. A familiar example are PDM systems operating on single-mode

---

[4]As is common in MIMO literature, we use the term "capacity" in this context to refer to the "capacity per channel use," which equals the "spectral efficiency" for signals using minimum-bandwidth Nyquist pulses. The overall fiber transmission capacity is obtained by multiplying this quantity by the employed optical bandwidth.

fiber. Such systems represent 2×2 MIMO channels with random polarization rotations. In the linear regime, and without impairments such as polarization-dependent loss (PDL), their capacity is exactly twice the capacity of a single-polarization system. (In the nonlinear propagation regime, their capacity is almost twice that of a single-polarization system, as discussed in Chapter 1 of this book.) Obviously, to recover the data at the receiver, appropriate DSP has to be used to rotate the received mode space back to its original transmit orientation. DSP techniques for M-mode MIMO-SDM will be discussed in Section 10.3.

10.2.2.3 Under-addressed channel

Another channel of interest [49,71] is the $M \times M$ random unitary channel of Figure 10.6, where we only use M_T transmitters and M_R receivers such that $d = \min\{M_T, M_R\} < M$. In contrast to wireless MIMO systems, the assumption of constant per-mode transmit power lets the SDM capacities of this channel be symmetric in M_T and M_R, i.e. an $M_T \times M_R$ system shows the same performance as an $M_R \times M_T$ system. If the transmitter is able to address all M modes but the receiver can only detect M_R modes, the capacity is given by $M_R C_S$. Conversely, if the transmitter is able to only address M_T modes but the receiver can detect all M modes, the capacity is given by $M_T C_S$ [49].

In all other cases, capacity loss has to be accepted. To analyze these reduced MIMO capacities due to under-addressing the SDM channel, we first acknowledge that the resulting MIMO capacity will be an inherently random quantity. For example, and with reference to Figure 10.6 for $M = 4$, $M_T = 3$, and $M_R = 2$, some random channel instantiation may couple 2 out of the 3 transmit signals to the 2 undetected receive modes, leaving only a single transmit mode for information transmission. At best, the receiver is able to extract 2 modes from the channel. In general, the minimum potential capacity of an under-addressed channel is given by $(M_R + M_T - M)C_S$ [71], and the maximum potential capacity of an under-addressed channel is given by dC_S [49], d being the rank of the channel, cf. Eq. (10.10). The statistical distribution of the resulting (random) MIMO capacity C is shown in Figure 10.7a, based on 100,000 random realizations of the 4×4 (unitary) channel matrix; the generation of uniform ensembles of M-dimensional unitary matrices is described in [72,49]. Most of the time, capacities approaching $dC_S = 2C_S$ are attained, but sometimes it may happen that the capacity almost drops to $(M_T + M_R - M)C_S = C_S$. This worst-case situation may occur very rarely, though, and designing the entire system for the worst case may result in unattractively low SDM capacities. Depending on the nature and dynamics of the channel, one may take one of the following approaches:

- *Slowly varying frequency-flat channel:* Assuming that the channel characteristics are constant across the signal bandwidth and change slowly compared to the burst error correction capability of the underlying code, as visualized in the top row of Figure 10.7b, the transmitter codes for (and transmits at) a certain capacity C_T that may occasionally lead to uncorrectable

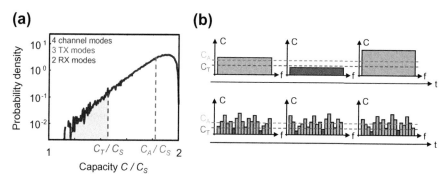

FIGURE 10.7 (a) Histogram of MIMO capacities C, **normalized to the single-mode capacity** C_S. **The shaded area represents the probability that** C **is smaller than** C_T, **leading to system outage if the system is designed to code for capacity** C_T. **The dashed line at** C_A **denotes the average capacity that would be obtained on a frequency-selective or rapidly time varying channel. (b) Visualization of slowly-varying temporal capacity evolution over a frequency-flat channel (top) and a strongly frequency-selective channel (bottom); green boxes denote channel capacities above the actually transmitted capacity, while red boxes denote outages. (For interpretation of the references to color in this figure legend, the reader is referred to the web version of this book.)**

blocks of bits. C_T is chosen such that the *probability* that a random channel instantiation only supports a capacity $C < C_T$ is acceptably small. These rare cases will then inherently lead to *system outage*. Hence, outage considerations become an integral part of this kind of SDM system design, similar to what is commonly done in the context of polarization-mode dispersion (PMD) [73–75]. The shaded area in Figure 10.7a represents the probability P_{out} that the channel capacity will fall below C_T to result in system outage,

$$P_{\text{out}} = \int_0^{C_T} p_C(C)dC, \tag{10.14}$$

where $p_C(C)$ is the probability density function (PDF) of the random channel capacity C. Note that in contrast to some wireless MIMO systems, fiber-optic transport networks typically require exceptionally low outage probabilities, typically below 10^{-5}, corresponding to 99.999% ("five nines") in system availability.

- *Rapidly varying frequency-flat channel:* As in the above case, channel variations are assumed to be constant across the signal bandwidth, but the channel varies rapidly compared to the burst error correction capabilities of the underlying code. This method is similar to what has been proposed to combat PMD outage in single-mode fiber systems [76,77] and results in a system that supports the *average* channel capacity $C_A = \langle C \rangle$, which is usually larger than the achievable capacity at low outage, cf. Figure 10.7. The method can be implemented at the expense of latency by code design, or at the expense of optical complexity by either introducing dynamic mode scrambling elements into the transmission line, or by switching between different sets of transmit and receive mode sets [49].

- *Frequency-selective channel:* In this case, the channel characteristics vary rapidly across the signal's bandwidth, as visualized in the bottom row of Figure 10.7b. This situation is encountered, e.g., for channels with significant differential group delay (DGD) between modes, as the number of uncorrelated frequency bins within the signal bandwidth is approximately given by the ratio of the signal bandwidth to the channel's coherence bandwidth (which is inversely proportional to the channel's delay spread) [78,79]. If each signal frequency component (e.g. each independent subcarrier of an OFDM signal) experiences a different channel instantiation, the signal ultimately sees the average channel capacity $C_A = \langle C \rangle$. In the idealized limiting case of a highly frequency selective channel, this average capacity is then *guaranteed*, i.e. if the transmitter codes for C_A, the receiver will always be able to extract C_A without experiencing outage [49,80].

In practice, the above three limiting cases may not be representative of any given real system and a hybrid approach may be necessary.

Figure 10.8 shows the 10^{-4} outage capacities C_T for static flat-fading (a) and the average capacities C_A for frequency-selective fading (b) in multiples of the single-mode capacity C_S; we assume channels supporting $M = 4, 8, 16, 32, 64,$ and 128 modes, and symmetric transponder pairs ($M_T = M_R = M_{TR}$) at 20-dB SNR. The dashed line $C_T = M_{TR} C_S$ represents crosstalk-free performance as an upper bound

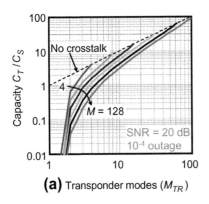
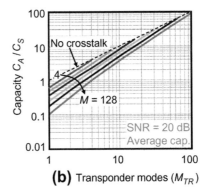

FIGURE 10.8 Achievable SDM capacity C_T relative to the single-mode capacity C_S at 10^{-4} outage (a) and average capacity C_A (b), both at 20 dB SNR, as a function of the number of modes M_{TR} processed by a fixed mode set transponder. The number of modes M supported by the waveguide parameterizes the curves.

to any MIMO system, only to be reached in the unitary MIMO case if the transponders are able to address *all* channel modes, as shown above. It can be seen that for a given transponder complexity (i.e. for fixed M_{TR}) it is generally preferable in terms of aggregate SDM capacity to use a waveguide that supports $M = M_{TR}$ modes. Deploying "future-proof" waveguides whose mode count exceeds current transponder capabilities results in a significant start-up capacity penalty. For example, operating a waveguide supporting $M = 32$ modes with transponders that only address $M_{TR} = 8$ modes results in a capacity of around $4C_S$. This capacity can also be achieved using only *half* the amount of transponders (and consequently much less MIMO signal processing) on a waveguide supporting just 4 modes. Hence, one should always strive to match the number of coupled waveguide modes to the number of modes that the transponders can address and are able to MIMO-process.

10.2.2.4 Channel with distributed noise loading

With reference to Figure 10.9a, we now consider an $M \times M$ SDM system operating on a matched M-mode waveguide composed of K concatenated segments with segment matrices H_i. Since we only consider the $M \times M$ case here, we have $H_i' = H_i$. Noise power N_i' is added at the end of each segment. As before, we factor out the net mode-average gain or loss L_i according to Eq. (10.9). In the linear regime covered

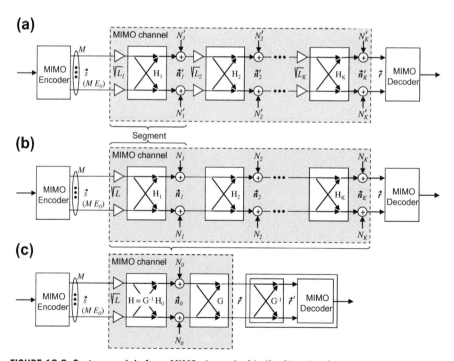

FIGURE 10.9 System models for a MIMO channel with distributed noise.

here, it is immaterial whether the mode-independent gain or loss within a segment physically occurs before or after the mode coupling described by the segment matrix. If a segment is identified with a net transparent amplification span, we have $L_i = 1$, and the physical segment loss is reflected in the amount of noise N_i' associated with optical amplification to reach per-span transparency.

We next define an equivalent system according to Figure 10.9b by replacing the individual segment losses by a lumped loss of $L = \prod_{i=1}^{K} L_i$ at the input to the channel and adjusting the noise powers at each segment such as to match the actual noise found in the original system, i.e. $N_i = N_i' \prod_{k=i+1}^{K} L_k$. Note that for an ensemble of random, nonunitary segment matrices, the overall net mode-average loss (or gain) L_Σ for each instantiation of the concatenated system will generally differ from our lumped loss variable L, with some channel instantiations showing more and some showing less net mode-average loss than L, as discussed in Ref. [49]. However, for most cases of practical interest, the *ensemble average* of the mode-average loss of the concatenated system is closely approximated by L.

If all noise sources are statistically independent, we find for the noise correlation matrix at the receiver:

$$\mathcal{R}_{\vec{n}} = \langle \vec{n}\vec{n}^\dagger \rangle = \left(N_1 H_K \cdots H_3 H_2 H_2^\dagger H_3^\dagger \cdots H_K^\dagger + N_2 H_K \cdots H_3 H_3^\dagger \cdots H_K^\dagger + \cdots + N_{K-1} H_K H_K^\dagger + N_K I_M \right).$$

$$(10.15)$$

If only modal crosstalk occurs within each fiber segment (i.e. if the segments H_i are all *unitary*), the correlation equals that of spatio-temporally white noise,

$$\mathcal{R}_{\vec{n}} = \sum_{i=1}^{K} N_i I_M = N_0 I_M. \qquad (10.16)$$

Hence, a concatenation of unitary segments with distributed noise has the same capacity as a unitary channel that is noise-loaded at the receiver by the same amount of noise, $N_0 = \sum_{i=1}^{K} N_i$.

If the segment matrices are not unitary, the channel can be written as

$$\vec{r} = \sqrt{E_0}\sqrt{L}H_0\vec{s} + G\vec{n}_0, \qquad (10.17)$$

where $H_0 = H_K H_{K-1} \ldots H_2 H_1$ is the overall channel matrix for the signal, and the white noise $\vec{n}_0$ of power density N_0 is colored by a matrix G such that $\mathcal{R}_{\vec{n}} = GG^\dagger$, as shown in Figure 10.9c. Placing the deterministic matrix G^{-1} as a whitening filter inside the receiver leaves the MIMO capacity unchanged but produces the equivalent white-noise MIMO channel

$$\vec{r}' = \sqrt{E_0}\sqrt{L}G^{-1}H_0\vec{s} + \vec{n}_0, \qquad (10.18)$$

as shown in Figure 10.9c; its open-loop BLAST capacity is given by Eq. (10.10) with $H = G^{-1}H_0$. The matrix G can be found from the measured noise correlation matrix $\mathcal{R}_{\vec{n}}$ by decomposing $\mathcal{R}_{\vec{n}}$ into the form $U\Theta U^\dagger$, where U is a unitary matrix and Θ is a diagonal matrix containing the eigenvalues θ_i of $\mathcal{R}_{\tilde{n}}$. The matrix G is then given as $U\Theta^{\frac{1}{2}}U^\dagger$, where $\Theta^{\frac{1}{2}}$ is diagonal with elements $\sqrt{\theta_i}$, and G^{-1} is given as $U\Theta^{-\frac{1}{2}}U^\dagger$, where $\Theta^{-\frac{1}{2}}$ is diagonal with elements $1/\sqrt{\theta_i}$.

10.2.2.5 Channel with mode-dependent loss (MDL)

Based on the above results, we next investigate MIMO capacity statistics for a channel with K segments and M modes, all of which can be addressed by transmitter and receiver. Each segment is composed of a random mode coupling element followed by a random MDL element. Without loss of generality, we assume the mode-average optical loss L_i for each individual segment to be unity. Hence, each segment matrix is composed of a random unitary matrix $\mathbf{U}_i$, followed by a random diagonal matrix $\mathbf{V}_i$ whose M real-valued, positive elements $\sqrt{v_{kk}}$ satisfy $\sum_{k=1}^{M} v_{kk} = M$ to ensure that $L_i = 1$. For each MDL segment matrix $\mathbf{V}_i$, the numbers v_{kk} are drawn from a uniform distribution (on a linear scale) such that the ratio of maximum to minimum matrix element equals the pre-specified value for the per segment MDL,

$$\text{MDL}_S = \frac{\max\{v_{kk}\}}{\min\{v_{kk}\}}. \tag{10.19}$$

We assume that all K segments have equal per-segment MDLs (i.e. the intervals from which the above random diagonal matrix elements are chosen are identical for all segments). Both the mode-average loss L_Σ and the MDL of the concatenated system MDL_Σ are random variables. The former is given by the trace of $\mathbf{HH}^\dagger$ and the latter by the ratio of largest to smallest eigenvalue of $\mathbf{HH}^\dagger$. For the two-mode case, i.e. in the context of polarization-dependent loss (PDL) in single-mode fiber, analytical relations of various sorts have been derived to describe the relationship between mode-average loss and mode-dependent loss as well as their statistical properties (see, e.g., [81–90]). An extension of the statistics of two-mode parameters to M modes is also given, e.g. in [91–94]. Numerically obtained statistics of the concatenated system's mode-average loss and mode-dependent loss as well as statistics of the concatenated system's MIMO capacity are given in [49].

Evaluating the average MIMO capacity for the frequency-selective channel, normalized to M times the single-mode capacity as a function of the per segment MDL for $K = 2, 16, 64$, and 128 segments results in the curves shown in Figure 10.10a for $M = 32$ modes, both with noise loading at the receiver (red, circles) and with distributed noise loading (blue, squares). The solid red lines and the dashed blue lines give corresponding analytical results [49] using the numerically exact values for mean and standard deviation of C. As expected, the SDM channel capacity drops with the per segment MDL as well as with the number of segments. For 128 segments, a per segment MDL of 1 dB is tolerable to still achieve over 90% of the ideal channel capacity. Further, we note that distributed noise loading gives slightly better capacities than noise loading at the receiver, which is due to the reduced noise enhancement by receive-side (spatial) equalization. This is in analogy to the better performance of single-mode receivers when noise loading is performed prior to narrow bandpass filtering [20]: If an

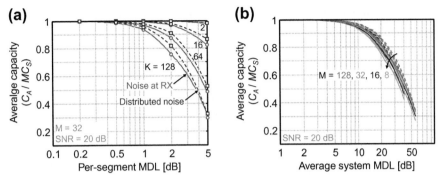

FIGURE 10.10 Average system capacity as a function of per segment MDL and number of segments K. (a) Systems with $M = 32$ modes and 20 dB SNR. Red circles and blue squares are numerically exact results for receive-side and distributed noise loading, respectively. Solid red and dashed blue lines represent an approximation given in [49]. (b) MDL performance of systems with $M = \{8, 16, 32, 128\}$ modes and $K = \{2, 16, 64, 128\}$ segments for 20 dB SNR and distributed noise loading as a function of the *aggregate* system MDL. (For interpretation of the references to color in this figure legend, the reader is referred to the web version of this book.)

optical filter is placed after all noise has been added, the filter acts on signal *and* noise, which lets the receive-side equalizer restore approximately white-noise conditions during signal equalization. On the other hand, an optical filter placed before any noise is added only affects the signal but not the noise, which results in noise enhancement upon signal equalization within the receiver.

To further test the scaling of SDM capacity with K, we plot the average capacity curves for the frequency-selective channel and distributed noise loading as a function of the *aggregate* average system MDL, $\langle \mathbf{MDL}_\Sigma \rangle_{[dB]}$ in Figure 10.10b. This lets all curves with equal M essentially collapse, as expected from the scaling relations discussed in [49]. Aggregate average system MDLs of 10 dB are seen to be acceptable for less than a 10% hit in SDM capacity, impressively illustrating the robustness of optical MIMO to MDL effects.

The curves shown in Figure 10.10, which represent the *average capacity* of a frequency-selective channel, can be compared to the curves in Figures 9 and 10 of [49], which apply to the 10^{-4} *outage capacity* of a frequency-flat channel. No significant performance difference in the tolerance to MDL is found between the two cases.

10.3 MIMO DSP

The above analyses of MIMO capacities represent limiting cases of transmission performance that may be approached assuming ideal digital signal processing (DSP) and coding. In this section, we extend the more familiar receiver DSP structure of a PDM system (which represents 2×2 MIMO), to an $M \times M$ MIMO system of M coupled spatial and polarization modes.

10.3.1 General receiver DSP functional blocks

Figure 10.11 visualizes the general 2×2 MIMO-DSP structure of a PDM coherent receiver (see, e.g., [95–97]) with its most basic functional blocks: In-phase (I) and quadrature (Q) components of the received x' and y' polarizations (in the local oscillator's x'-y'-coordinate system) are first converted into the digital domain typically through over-sampled analog-to-digital converters (ADCs) and combined into two complex sample streams, $I_{x'} + jQ_{x'}$ and $I_{y'} + jQ_{y'}$ that fully represent the received optical signal field. After applying front-end corrections that compensate for various imperfections of the optoelectronic receiver front-end, such as phase errors of the optical 90° hybrid or sampling skews between the ADCs, chromatic dispersion (CD) of the transmission link is compensated by digitally applying the inverse CD filter function (an allpass filter with quadratic phase [95,98]). Since the transmitter's x-y coordinate system is randomly rotated by the fiber's 2×2 Jones matrix relative to the receiver's x'-y'-coordinate system, the DSP next needs to apply the inverse Jones matrix to the vector $[x' y']$ to recover the two signal components multiplexed onto x and y polarizations at the transmitter. This inverse Jones matrix is represented by the shaded butterfly structure in Figure 10.11. In order not only to compensate frequency-independent matrices with scalar elements $H_{xx}, H_{xy}, H_{yx}, H_{yy}$ but also to correct for frequency-dependent quasi-unitary polarization rotations due to polarization-mode dispersion (PMD), the matrix elements $H_{xx}, H_{xy}, H_{yx}, H_{yy}$ are implemented as individual filters, typically with a few tens of filter taps. The complex filter coefficients are typically found using adaptive algorithms [95–97], such as the blind constant modulus algorithms (CMA), or various forms of blind decision-directed algorithms,

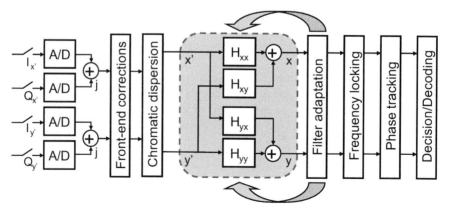

FIGURE 10.11 Functional blocks of a conventional polarization-diversity intradyne coherent receiver implementing 2 × 2 MIMO-DSP.

such as radius directed multi-modulus algorithm or the least-mean square (LMS) algorithm. Rapid coefficient acquisition can also be obtained with the help of data-aided algorithms that make use of specific training symbols inserted into the transmit data stream [99,100]. At the output of the butterfly filter, the transmitter's original [xy] coordinate system is re-established. Subsequent frequency and phase tracking lock the phase of the symbol constellation before hard-decision or soft-decision decoding can take place.

Extending this system to an $M \times M$ MIMO receiver, we note that the only functional block that changes in this DSP structure is the butterfly equalizer: Instead of undoing the action of a 2×2 channel, the equalizer now has to deal with an $M \times M$ channel based on an appropriate *channel estimation* or *filter adaptation* algorithm. In the former case, the channel matrix is estimated and inverted to obtain the equalizer coefficients based on the zero-forcing (ZF) criterion or the minimum-mean-square error (MMSE) criterion. In the latter case, gradient-based coefficient adaptation algorithms are used to directly adapt the filter coefficients to approach the MMSE solution [65].

10.3.2 Channel estimation

A detailed understanding of the communication channel is essential for the design of efficient equalization and coding schemes. In this context, channel estimation performed in the digital domain can serve as a useful tool that allows to visualize the channel state and to study channel statistics. Channel estimation can also be used to obtain a set of equalizer coefficients from a received training sequence. These coefficients can then be applied to equalize subsequent payload data. In the following, we describe the least-squares *(LS)* scheme and a stochastic gradient scheme based on the LMS algorithm as two ways to perform channel estimation for the fiber-optic MIMO channel.

10.3.2.1 Least squares channel estimation

Assume an $M \times M$ MIMO channel with a linear time invariant (LTI) channel matrix H that we want to estimate from M (possibly oversampled) complex-valued received waveforms, each represented by N consecutive received samples. We can write this received data as an $M \times N$ receive matrix

$$
R = \begin{pmatrix}
r_1[1] & r_1[2] & \cdots & r_1[N] \\
r_2[1] & r_2[2] & \cdots & r_2[N] \\
\vdots & \vdots & \ddots & \vdots \\
r_M[1] & r_M[2] & \cdots & r_M[N]
\end{pmatrix},
\tag{10.20}
$$

where $r_m[k]$ is the sampled waveform at the mth receiver, and $m \in \{1, 2, \ldots, M\}$. We next define the channel matrix as

$$H = \begin{pmatrix} h_{1,1}[1\ldots L] & h_{1,2}[1\ldots L] & \cdots & h_{1,M}[1\ldots L] \\ h_{2,1}[1\ldots L] & h_{2,2}[1\ldots L] & \cdots & h_{2,M}[1\ldots L] \\ \vdots & \vdots & \ddots & \vdots \\ h_{M,1}[1\ldots L] & h_{M,2}[1\ldots L] & \cdots & h_{M,M}[1\ldots L] \end{pmatrix}, \quad (10.21)$$

where each matrix element $h_{n,m}[1\ldots L] = (h_{n,m}[1], h_{n,m}[2], \ldots, h_{n,m}[L])$ is the L-tap discrete-time impulse response from the nth transmitter to the mth receiver. The size of the channel matrix H is $M \times ML$. Finally, from the M transmit sequences $s_m[k]$ with $k \in \{1, 2, \ldots, N\}$ of properly oversampled training symbols, and for a certain pattern delay k_0, we can define the $ML \times N$ matrix

$$S = \begin{pmatrix} s_1[k_0] & s_1[k_0 - 1] & \cdots & s_1[k_0 - N + 1] \\ s_1[k_0 - 1] & s_1[k_0 - 2] & \cdots & s_1[k_0 - N] \\ \vdots & \vdots & \ddots & \vdots \\ s_1[k_0 - L + 1] & s_1[k_0 - L] & \cdots & s_1[k_0 - N - L + 2] \\ s_2[k_0] & s_2[k_0 - 1] & \cdots & s_2[k_0 - N + 1] \\ s_1[k_0 - 2] & s_2[k_0 - 2] & \cdots & s_2[k_0 - N] \\ \vdots & \vdots & \ddots & \vdots \\ s_2[k_0 - L + 1] & s_2[k_0 - L] & \cdots & s_2[k_0 - N - L + 2] \\ \vdots & \vdots & \vdots & \vdots \\ \vdots & \vdots & \vdots & \vdots \\ s_M[k_0] & s_M[k_0 - 1] & \cdots & s_M[k_0 - N + 1] \\ s_M[k_0 - 1] & s_M[k_0 - 2] & \cdots & s_M[k_0 - N] \\ \vdots & \vdots & \ddots & \vdots \\ s_M[k_0 - L + 1] & s_M[k_0 - L] & \cdots & s_M[k_0 - N - L + 2] \end{pmatrix} \quad (10.22)$$

with its Toeplitz submatrices. This allows us to express the action of the MIMO system in compact matrix notation as

$$R = HS, \quad (10.23)$$

where the M^2 convolutions are now captured by a single matrix multiplication. If the matrices R and S are both known and are properly aligned in time through the parameter k_0, we can apply the LS method, to obtain the channel estimate

$$\hat{H}_{LS} = \arg\min_{H} \left(\|R - HS\|_2^2 \right) = RS^\dagger (SS^\dagger)^{-1}. \quad (10.24)$$

The matrix $S^\dagger(SS^\dagger)^{-1}$ is also referred to as pseudoinverse of S. For a fixed training pattern, the pseudoinverse can be determined ahead of time, stored in memory, and later used for estimating different MIMO channels. Regarding the selection of training patterns, one needs to take care that the training patterns of the M transmitters are uncorrelated. In practice, the LS method provides a very useful tool to analyze the properties of experimental fiber-optic MIMO systems, even though the assumption of an LTI channel is typically not strictly fulfilled in the presence of phase noise introduced by the transmit laser and the local oscillator.

10.3.2.2 Least mean squares channel estimation

Alternatively, the channel matrix can be estimated using the LMS algorithm. In this case the system can be expressed as $\vec{r}[k] = \hat{H}_{\mathrm{LMS}}[k]\vec{s}[k + k_0]$, where $\vec{r}[k] = (r_1[k], r_2[k], \ldots, r_M[k])^T$ corresponds to the M received waveforms at sampling instant k, $\hat{H}_{\mathrm{LMS}}[k]$ is the estimated channel matrix at time instant k, and $\vec{s}[k + k_0]$ is the time-aligned and properly upsampled vector of transmit data. The LMS algorithm provides an approximation of the MMSE solution by adapting the channel estimate $\hat{H}_{\mathrm{LMS}}[k]$ in the direction of the minimum instantaneous squared error

$$\vec{\varepsilon}[k] = \left| \vec{r}[k] - \hat{H}_{\mathrm{LMS}}[k]\vec{s}[k + k_0] \right|^2, \tag{10.25}$$

using the update rule

$$\hat{H}_{\mathrm{LMS}}[k + 1] = \hat{H}_{\mathrm{LMS}}[k] + 2\mu\vec{\varepsilon}[k]\vec{s}^{\dagger}[k + k_0], \tag{10.26}$$

where μ is the adaptation gain. A benefit of LMS-based channel estimation is that it can track variations of the channel matrix across the training sequence and that it can even be combined with a phase recovery scheme in order to separate the effect of laser phase noise from the channel matrix. However, LMS-based channel estimation typically results in a reduced accuracy compared to the LS scheme.

10.3.3 Adaptive MIMO equalization

Crosstalk from mode coupling in conjunction with modal delay spread (MDS), defined as the temporal width of the overall system's $M \times M$ channel matrix H, can set a hard limit on the transmission distance of SDM systems if no means are undertaken to compensate for it. We will show in this section that MIMO equalization can compensate for these impairments to a large extent if the communication channel is designed in a proper way. As we already pointed out above, adaptive 2×2 MIMO equalization is already widely used in single-mode optical communications in order to separate the two transmitted polarization-multiplexed data streams. If the SDM transmission link exhibits a sufficiently low level of MDL, the same concept can be extended to few-mode fiber transmission, resulting in a matrix of $M \times M$ equalizers.

Extending our discussions on MDL in Section 10.2.2.5, we first introduce the channel matrix in terms of the z transform as $H(z) = HD(z)$ where the $ML \times M$ delay matrix has the form

$$D(z) = \begin{pmatrix} z^0 & 0 & \cdots & 0 \\ z^{-1} & 0 & \cdots & 0 \\ \vdots & \vdots & \ddots & \vdots \\ z^{-L+1} & 0 & \cdots & 0 \\ 0 & z^0 & \cdots & 0 \\ 0 & z^{-1} & \cdots & 0 \\ \vdots & \vdots & \ddots & \vdots \\ 0 & z^{-L+1} & \cdots & 0 \\ 0 & 0 & \cdots & z^0 \\ 0 & 0 & \cdots & z^{-1} \\ \vdots & \vdots & \ddots & \vdots \\ 0 & 0 & \cdots & z^{-L+1} \end{pmatrix},$$

and obtain the frequency dependent channel matrix as $H(\omega) = H(z = e^{j\omega/f_s})$ over the frequency interval $-\pi f_s \leqslant \omega < \pi f_s$ (sampling rate f_s). Now, we can derive the MDL of the estimated channel matrix by carrying out a singular value decomposition according to

$$H(\omega) = U(\omega)\Lambda(\omega)V(\omega)^\dagger, \tag{10.27}$$

where $U(\omega)$ and $V(\omega)$ are frequency dependent unitary matrices of sizes $M \times M$, respectively. The $M \times M$ diagonal matrix $\Lambda(\omega)$ has the singular values $\lambda_m(\omega)$ on the main diagonal from which we obtain the MDL as

$$MDL = 10 \log_{10}\left(\frac{\max\{|\lambda_m(\omega)|^2\}}{\min\{|\lambda_m(\omega)|^2\}}\right). \tag{10.28}$$

Let us now consider a linear MIMO equalizer that is operating on a twofold oversampled input signal (oversampling ratios different from two are also feasible but require a proper modification of the DSP).

This oversampled implementation of the linear equalizer has the advantage that the optimal symbol-spaced sampled MMSE filter can be approximated using a fully digital adaptation scheme [65]. We transmit a sequence of symbols $\vec{a}[k] = (a_1[k], a_2[k], \ldots, a_M[k])^T$ with symbol index $k \in \{1, 2, \ldots, \infty\}$ through the MIMO channel, and receive the sequence $\vec{r}[k'] = (r_1[k'], r_2[k'], \ldots, r_M[k'])^T$ with sample index $k' \in \{1, 2, \ldots, \infty\}$. We further consider a linear equalizer with L_e taps at two samples per symbol, whose output $\vec{u}[k] = (u_1[k], u_2[k], \ldots, u_M[k])^T$ with $k \in \{1, 2, \ldots, \infty\}$ is calculated once for every transmit symbol. We can then obtain the output of the linear equalizer from

$$\vec{u}[k] = W\vec{\rho}[2k], \tag{10.29}$$

where W is an $M \times ML_e$ matrix of complex-valued equalizer coefficients and the received samples are expressed by the ML_e-element vector $\vec{\rho}[k'] = (r_1[k'], r_1[k'+1], \ldots, r_1[k'+L_e-1], r_2[k'], r_2[k'+1], \ldots, r_2[k'+L_e-1], \ldots, r_M[k'], r_M[k'+1], \ldots, r_M[k'+L_e-1])^T$. In the case where the channel has no MDL, the channel matrix H is unitary, i.e. $\lambda_m(\omega)$ for $m \in \{1, 2, \ldots, M\}$, and if the noise is white, all mode coupling and inter-symbol interference can be fully compensated using the coefficient matrix $W = H^{-1} = H^\dagger$. In the presence of weak or moderate MDL, i.e. less than approximately 10 dB (compare Figure 10.10b), a linear MIMO equalizer is still able to recover the transmitted signal with a small penalty. However, if the channel has strong MDL, linear equalization will lead to considerable noise enhancement and advanced schemes like decision-feedback equalization (DFE) or maximum-likelihood sequence estimation (MLSE) are required in order to recover the transmit signal [65].

In analogy to the general MIMO-DSP structure of Figure 10.11, the MIMO equalizer is typically followed by a carrier phase estimation scheme, such as the Viterbi and Viterbi scheme for QPSK constellations [101], that provides an estimate of the carrier phase offset $\vec{\phi}[k]$. In MIMO systems an interesting option is to apply a joint phase recovery over all modes in order to improve accuracy [102,103].

As the channel is expected to be slowly time-varying compared to the symbol rate, gradient-based adaptive algorithms are attractive in order to adapt the equalizer coefficients. A popular adaptation scheme is the CMA that updates the coefficients based on the error criterion

$$\vec{e}[k] = |\vec{u}[k]|^2 - |\vec{a}[k]|^2. \tag{10.30}$$

For QPSK modulation, we have $|\vec{a}[k]| = 1$. The equalizer coefficients are updated according to

$$W[k+1] = W[k] + 4\mu \left(\vec{e}[k] \cdot \vec{u}[k] \right) \vec{\rho}^\dagger[2k], \tag{10.31}$$

where $\cdot$ denotes element-wise multiplication. A benefit of the CMA algorithm is that it operates fully blindly and that no information about the carrier phase is required. This makes the algorithm attractive for real-time implementation. However, in an $M \times M$ MIMO system, the CMA can mis-converge in a way that different sub-equalizers converge to the same solution, resulting in the same information signal at multiple equalizer outputs while entirely dropping other information signals. Advanced algorithms like the independent component analysis (ICA) can be applied to solve this issue [104,105a] and to achieve fully blind convergence.

An alternative to the CMA is the LMS algorithm, where the adaptation error is defined as

$$\vec{e}_{\text{LMS-DD}}[k] = \left(\vec{a}[k] - \vec{u}[k] \cdot \exp(-j\vec{\phi}[k]) \right) \cdot \exp(j\vec{\phi}[k]), \tag{10.32}$$

and the coefficients are updated according to

$$W_{\text{LMS}}[k+1] = W_{\text{LMS}}[k] + 2\mu \vec{e}[k] \vec{\rho}^\dagger[2k]. \tag{10.33}$$

Compared to the CMA, the LMS algorithm requires additional information about the carrier phase offset $\vec{\phi}[k]$ and about the transmit symbol $\vec{a}[k]$. The former issue can be solved by feeding back the output of the carrier phase estimator to the equalizer adaptation algorithm. This approach, however, can cause difficulties in a high-speed real-time DSP implementation, as the required feedback delay becomes rather long. The problem that information about the transmitted symbol $M/2$ is required at the receiver can be solved by applying hard-decision after the carrier recovery to obtain an estimate of the transmitted symbol. In this so-called decision-directed mode, the LMS algorithm can track variations of the channel blindly. However, the decision-directed LMS algorithm usually fails to converge during the initial coefficient acquisition phase, as the transmit symbol and the carrier phase offset cannot be estimated reliably in the presence of strong ISI. An attractive way to achieve initial coefficient convergence in a reliable way is the use of training symbols that are inserted within the transmit sequence and that are known to the receiver [99,100]. Using this data-aided LMS scheme allows to achieve convergence of the equalizer coefficients after a certain number of training symbols. If we assume a system working at 28 GBaud and a 50,000-symbol training sequence for coefficient acquisition that is inserted once every millisecond, the required overhead is still less than 0.2%.

10.3.4 MIMO equalizer complexity

In the previous subsection we have shown that a linear MIMO equalizer is able to compensate for mode coupling and MDS in case of weak or moderate MDL. With respect to the practical implementation of MIMO-SDM systems, the overall DSP-complexity is an important constraint. As a reference for the following complexity analysis, we consider an SDM system with $M/2$ uncoupled single-mode waveguides terminated by receivers of the type shown in Figure 10.11. A 2×2 MIMO equalizer is applied for each waveguide in order to uncouple the polarizations and to compensate for polarization-mode dispersion such that the total number of sub-equalizers equals $4 \cdot M/2 = 2M$. If we compare this to a MIMO-SDM system with $M/2$ coupled waveguides supporting 2 polarizations each, we require a single $M \times M$ MIMO equalizer of the form shown in Figure 10.12, with a total of M^2 sub-equalizers. A coupled-mode MIMO-SDM system with $M/2$ spatial paths thus requires $M/2$ as many sub-equalizers as the corresponding uncoupled SDM system.

In addition to the larger number of required equalizers, a MIMO-SDM system will typically need a larger number of taps per sub-equalizer. To roughly estimate the resulting complexity increase for MIMO-SDM systems, we note that in single-mode fiber systems, the required equalizer memory for compensating PMD is on the order of 100 ps, while the required equalizer memory for long-haul MIMO-SDM transmission can grow up to, e.g., 10 ns [57,58], a factor of 100 larger. If the linear equalizer is implemented in the time-domain using a tapped-delay-line structure, the associated complexity is approximately one complex multiplication per sample per equalizer tap. In our example, this would mean that a MIMO-SDM equalizer is about $100M/2 = 50M$ more complex than the equalizer for the equivalent uncoupled SDM receiver. However, a frequency-domain implementation of the sub-equalizers scales much more favorable than a tapped-delay-line structure [105b,106,107a,107b]; a common way to implement the sub-equalizer in an efficient

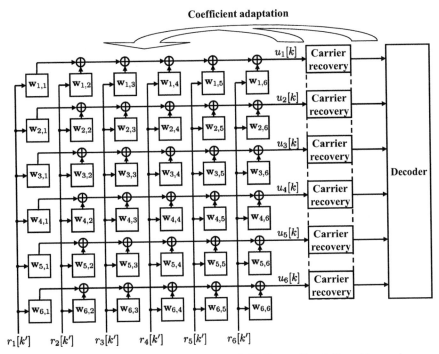

FIGURE 10.12 Schematic representation of a 6 × 6 MIMO equalizer followed by carrier recovery and joint decoder. The dashed line represents an optional joint carrier recovery scheme.

way is the overlap-and-save (or overlap-and-discard) method [108]. In this case, the complexity of a single sub-equalizer is of order

$$\mathcal{O}\left(\frac{2\alpha_{\text{FFT}} N_{\text{FFT}} \log_2(N_{\text{FFT}}) + N_{\text{FFT}}}{N_{\text{FFT}} - (L_e - 1)}\right), \tag{10.34}$$

where N_{FFT} is the FFT-size and α_{FFT} depends on the way the FFT is implemented (e.g. $\alpha_{\text{FFT}} = 1/2$ for a radix-2 implementation) [109]. Figure 10.13 shows the relative complexity of different equalizer structures. It becomes clear that the complexity of the overlap-save algorithm grows only slowly as a function of the equalizer length. Albeit this is only a rough estimate (e.g. since the complexity of the adaptation scheme is not taken into account), we note that the complexity of a coupled-mode MIMO-SDM equalizer scales linearly with $M/2$ but only slowly as a function of the equalizer memory as compared to the equalizer for an uncoupled SDM system.

If the complexity of a full $M \times M$ MIMO-SDM system is too large to be handled efficiently, be it due to DSP complexity, due to the required number of on-chip ADCs, or due to purely optical constraints such as the possibility of building optical fibers with reasonably low MDS or optical amplifiers with reasonably small

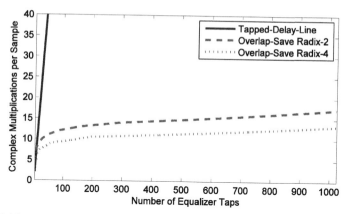

FIGURE 10.13 Number of complex multiplications per sample for different equalizer implementation schemes versus number of equalizer taps L_e.

modal gain variations, one can partition an SDM system of multiplicity $M/2$ into a set of M_u nominally uncoupled optical paths, each supporting $M_c/2$ coupled spatial modes. The resulting DSP then consists of M_u independent $M_c \times M_c$ MIMO-DSP engines that could be implemented on separate ASICs if needed, as visualized in Figure 10.14. An example of such a system could be a fiber whose LP "modes" show negligible coupling [59] or a multi-core fiber with M_u nominally uncoupled cores, each of which supports M_c propagation modes [31,58,110], allowing for a capacity

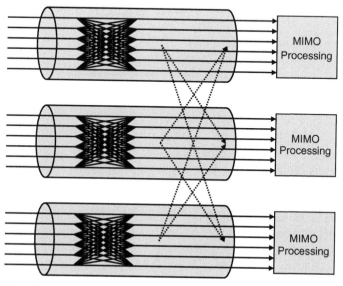

FIGURE 10.14 Schematic representation of three weakly coupled waveguides, each with $M_c = 6$ strongly coupled modes. MIMO processing is applied per waveguide.

increase by a factor $M_u M_c/2$ compared to a single-mode fiber. In a system of this kind, the uncompensated residual crosstalk between the M_u individual MIMO systems becomes a major design criterion, in analogy to the crosstalk considerations for weakly coupled single-mode systems (cf. Figure 10.4).

In [58] the crosstalk tolerance for a system of M_u, nominally uncoupled waveguides each supporting 3 spatial modes and performing 6×6 MIMO equalization, was studied, showing that for QPSK transmission a crosstalk of about -20 dB from the nominally uncoupled waveguides can be tolerated, similar to the single-mode case of Figure 10.4.

10.4 MODE MULTIPLEXING COMPONENTS

As described in Section 10.2.2, the ability to selectively excite and extract a complete orthogonal set of fiber modes is a critical requirement for MIMO-SDM performance. This is in complete analogy to a PDM system, where the transmitter needs to be able to couple its two signals distinctly into two orthogonal polarizations and the receiver also needs to be able to coherently extract a pair of orthogonal polarization states. If the transmitter does not couple its two signals into two essentially orthogonal polarization states, some irreversible crosstalk will occur that manifests itself in a performance degradation. Therefore, optical components that selectively couple M single-mode signals into M spatial fiber modes (or into unitary combinations thereof) without much nonunitary crosstalk are essential for MIMO-SDM systems. Devices that achieve this functionality are referred to as mode multiplexers (MMUXs).

10.4.1 Mode multiplexer characteristics

A mode multiplexer is fully characterized by its $M \times M$ transfer matrix $\widetilde{H}_{MMUX}$ (see Fig. 10.6 for channel matrix definitions), where each matrix element can be calculated by the overlap integral defined in Eq. (10.7), evaluated for each of the M optical field distributions generated by the MMUX with respect to each of the M optical field distributions of the fiber modes. (An exemplary calculation is given below.)

The impact of the MMUX on MIMO capacity can be determined using the methods presented in Section 10.2.2. Starting with the $M \times M$ MMUX transfer matrix $\widetilde{H}_{MMUX}$, we first factor out its mode-average loss L according to Eq. (10.9). We call this the MMUX's mode-average insertion loss. For an MMUX used at the transmitter, its mode-average insertion loss can be compensated for by increasing the transmit power without implications on system performance. Used in-line at amplifiers or other networking elements, its mode-average insertion loss directly impacts the corresponding span loss, and therefore contributes to the system's noise performance.

After factoring out the MMUX's mode-average insertion loss, we arrive at the normalized MMUX matrix H_{MMUX}. If the MIMO-SDM system consists solely of an MMUX and a mode demultiplexing receiver, the matrix H_{MMUX} by itself represents the MIMO channel. Knowing that a unitary MIMO channel does not degrade system performance, we first acknowledge that a transmit-side MMUX may map its M single-mode inputs onto any unitary transformation of fiber modes without loss of MIMO-SDM system performance. This fact also holds true for any arbitrary (linear) MIMO channel H_{Ch} following a unitary MMUX, since the overall MIMO channel capacity is then determined by the eigenvalues of $H_{Ch}H_{MMUX}H_{MMUX}^{\dagger}H_{Ch}^{\dagger} = H_{Ch}H_{Ch}^{\dagger}$, which are the same as those without a unitary MMUX. Furthermore, we note that the transmit-side MMUX, by definition preceding any noise loading, has no effect on the noise correlation matrix of Eq. (10.15).

If the MMUX is nonunitary, i.e. if it exhibits nonunitary crosstalk, at least two of the M eigenvalues λ_i of $H_{MMUX}H_{MMUX}^{\dagger}$ will differ from unity, resulting in the MDL channel model given by Eq. (10.19). The impact of MDL on MIMO system performance is discussed in Section 10.2.2.5 (Figure 10.10).

Note that a mode multiplexer is used here as a general term; in particular the same MMUX device can be used both on the transmit side and at the receive side, where it operates in reverse as a mode demultiplexer. When used as a mode demultiplexer at the receiver, the MMUX characteristics are less stringent, as this component will usually appear after all noise has been added to the signal. Hence, as long as the mode demultiplexer's matrix is invertible, and as long as its MDL is small enough not to lead to other sources of post-demultiplexing receiver noise that suddenly become important (such as thermal noise or quantization noise from signal digitization), its characteristics have no impact on system performance. Mode demultiplexers placed in-line (e.g. at amplifier sites or in connection with other networking elements) will add MDL and loss to the system, just like an equivalent MMUX at those locations.

10.4.2 Mode multiplexer design

A large variety of fibers has been proposed for SDM including few-mode fibers (FMFs), multi-core fibers (MCFs) with single-mode cores, multi-core fibers with few-mode cores, and photonic crystal fibers (PCFs). Each fiber asks for its optimized mode multiplexer. Furthermore, for a given fiber type, multiple strategies to implement mode multiplexers exist, depending on the crosstalk management strategy used in the transmission system: If crosstalk in an SDM system is not actively compensated but is rather treated as an impairment (cf. Figure 10.4), a direct one-to-one correspondence between the input single-mode channels and the fiber's spatial modes needs to be established in order for the receiver to uniquely detect a mode. This is the case, for example, in MCFs using nominally uncoupled cores, where each signal is uniquely coupled to one of the fiber cores. On the other hand, if MIMO processing is being used, any unitary mode combination may be excited at the transmitter, as discussed above. This offers more degrees of freedom in MMUX design.

Because the modes in a fiber are orthogonal according to Eq. (10.7), it should theoretically be possible to build a loss-less MMUX, in analogy to polarization multiplexing, where a polarizing beam splitter is used to losslessly multiplex two orthogonal polarizations onto a single fiber, or in analogy to WDM, where a diffraction grating may be used to losslessly combine WDM channels into a single fiber. In practice, this is not always possible, however. Generally, we can divide MMUX designs into two main categories:

- *Broadcast and select mode shapers:* Here, the MMUX shapes M individual spatial mode patterns of the waveguide. If these patterns overlap in space, they need to be passively combined by means of beam combiners before projection onto the waveguide's end facet. Since the light leaving the unused port of a beam combiner is inherently lost, this class of MMUXs tends to be lossy and scale poorly with M. Operated as mode demultiplexers in reverse, this class of MMUXs first splits the light coming from a SDM fiber into M equal copies and later selectively filters each path by a suitable mode converter. Examples for this architecture are phase-plate based designs [111] discussed in Section 10.4.3.1, Liquid Crystal on Silicon (LCoS)-based designs [112,113], single-stage thin hologram-based designs [114], as well as long-period fiber gratings (LPG) [115–117].

- *Mode transformers:* Here, the MMUX transforms its M single-mode inputs into M optical field patterns that are individually projected onto the waveguide's end facet without prior passive combining. Conversely, when operated as a demultiplexer, the light coming from a SDM fiber is mode-selectively split into M single-mode fibers. Depending on whether the mode transformation occurs abruptly or in a distributed fashion, we distinguish between:

    - *Single-step mode transformers:* Here, mode conversion happens abruptly in the waveguide coupling plane. An example is the spot coupler [118] discussed in Section 10.4.3.2. Since the abrupt mode conversion is generally not perfectly matched to the waveguide modes, it may induce loss through the excitation of radiation modes, or it may induce mode dependent loss through nonorthogonal mode mapping.

    - *Distributed mode transformers:* Here, mode conversion happens over a certain length within a volume rather than within a single coupling plane. MMUXs of this kind have the potential of being lossless for arbitrary values of M. Examples for this architecture are photonic lanterns [119] discussed in Section 10.4.3.2, volume holograms [120], and mode sorters [121].

In the following sections we review a selection of promising mode multiplexing components that are relevant for MIMO-SDM transmission systems.

10.4.3 Mode couplers for few-mode fibers

Few-mode fibers (FMFs) are single-core fibers where the refractive index profile of the core is designed to support a specific number of fiber modes (see Chapter 8 in Volume A). Refractive index profiles currently under study are step-index profiles, depressed-cladding index profiles, graded-index profiles, and ring-core profiles.

Mode multiplexers are often making use of symmetry properties of the modes. It is therefore instructive to consider the first 6 modes of a step-index fiber in detail.

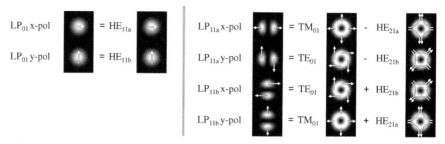

FIGURE 10.15 Grayscale intensity representation of the modes and the LP pseudo-modes of a step-index fiber supporting six spatial and polarization modes. The arrows indicate the direction of the electric field. Phase jumps are shown by opposite directions of the electric field vector.

These modes, which are represented graphically in Figure 10.15, comprise the fundamental HE_{11} mode (which is twofold degenerate in polarization, resulting in the HE_{11a} and HE_{11b} modes), the TE_{01} mode, the TM_{01} mode, and the HE_{21} mode (twofold degenerate into HE_{21a} and HE_{21b}). The HE_{11} modes are linearly polarized (LP) with a single radial maximum and no angular dependence, which motivates denoting them also as LP_{01}. In the limit of weakly guiding index profiles [122], the three higher-order modes TE_{01}, TM_{01}, and HE_{21} become almost degenerate as well. As shown in Figure 10.15, certain linear combinations of these modes are linearly polarized with one radial intensity maximum and with one set of positive maximum and negative minimum optical field amplitudes (resulting in two angular intensity maxima). This motivates grouping these modes into what is called the LP_{11} pseudo-mode. (In contrast to a true waveguide mode, whose electric field distribution is invariant with propagation distance, cf. footnote 3 on page 437, all but the LP_{0n} "modes" change their spatial distribution, owing to the small differences in propagation constant $\Delta\beta$ of their true waveguide mode constituents. The resulting beat length $L_B = 2\pi/\Delta\beta$ is in the range of a few cm up to meters for typical silica fibers, which causes the three constituent modes to strongly mix during transmission [123]).

10.4.3.1 Phase-plate based mode multiplexer

Experimentally, it is simpler to generate beams that resemble the LP "modes," because lasers usually generate linearly polarized light which can be readily phase and amplitude shaped to match an LP profile. In contrast, shaping the true waveguide modes making up LP_{11}, for example, requires polarization changes across the transversal coordinate (cf. Figure 10.15). Furthermore, LP "modes" have very characteristic phase profiles, consisting of one or more regions with constant phase and a π-phase jump between these regions. This suggests a simple way to generate LP "modes" by using binary phase plates, whereby the phase plates, introduced in the optical path between the feeding single-mode fiber and the FMF, match the phase profile of the target LP "mode" [124]. Such an MMUX design is shown in Figure 10.16 for a

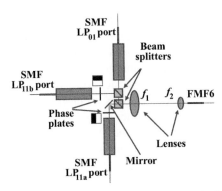

FIGURE 10.16 Phase-plate based mode multiplexer for a few-mode fiber that supports six spatial and polarization modes.

FMF that supports six spatial and polarization modes, equivalent to the first two LP "modes," LP_{01} and LP_{11}.

The strategy of phase-only mode shaping works well for fibers with six spatial and polarization modes (LP_{01} and LP_{11}). For fibers supporting 12 spatial and polarization modes ($LP_{01}, LP_{11}, LP_{21}$, and LP_{02}), the method introduces a crosstalk of 12 dB between the LP_{01} and the LP_{02} mode. This crosstalk dominates even when more modes are added. It can be reduced if in addition to binary phase plates the intensity profile is shaped as well, for example by using appropriate amplitude plates.

Since each phase plate shapes only a single spatial mode of the fiber, multiple phase plates are required to construct an MMUX. After phase shaping, the generated mode profiles are combined using beam combiners, which introduce significant loss since their second output ports remain unused. The loss scales proportional with the number of spatial modes. The same scaling law applies in general to all single-stage thin hologram-based couplers [125]. For six spatial modes, the combiner loss would amount to about 10 dB, which limits the practical use of phase-plate based MMUXs to a fairly small number of spatial modes.

Regarding the performance characteristics of the MMUX shown in Figure 10.16, we first note that it consists of 3 independent paths containing different phase plates. No phase plate is required for the LP_{01} mode of the FMF, since it matches the LP_{01} mode of the feeding single-mode fiber (SMF) up to a magnification factor M_{SMF}, defined as the magnification between the SMF and FMF end facets. For the LP_{11} "mode," the phase plate consists of two half planes with a constant phase difference of π. Rotating the phase plate by 90° produces the second LP_{11} "mode" (cf. Figure 10.15). The spatial Fourier transform of the LP "modes" has the interesting property of being self-similar [126], i.e. the field's phase structure is maintained by focusing/collimating the beam while the intensity profile retains similar characteristics (a spot stays a spot; a ring stays a ring but with changed geometric proportions). Hence, the mode-selecting phase plates can be placed either in the Fourier plane of the related FMF end facet or in the plane of the fiber end facet.

In order to calculate the elements of the MMUX transfer matrix $\widetilde{\mathbf{H}}_{MMUX}$ and extract its mode-average insertion loss and MDL, we first evaluate the overlap integrals η_{01} and η_{11} between the LP_{01} mode of a feeding standard SMF and the LP_{01} as well as the LP_{11} "modes" of a step index FMF with normalized frequency $V = 3.92$ and core diameter $d = 17\,\mu m$ according to Eq. (10.7). These are shown as a function of the magnification factor M_{SMF} in Figure 10.17 for an MMUX arrangement where the phase plates are placed in the imaging plane. The quantities $\tilde{\eta}_{01}$ and $\tilde{\eta}_{11}$ denote the case where the phase plates are located in the Fourier plane of the FMF. The coupling efficiency for coupling into the LP_{01} mode is essentially the same for both coupling geometries and shows a maximum coupling very close to 0 dB for a magnification of 1.7. The maximum coupling efficiencies when coupling into the LP_{11} mode are $-1\,dB$ and $-1.1\,dB$ for the phase plates located in the Fourier plane and image plane, respectively, albeit at different magnification factors. The arrangement with phase plates located in the Fourier plane has a slightly lower loss, but the image plane arrangement offers the advantage of being more tolerant toward errors in the magnification factor. By choosing the optimum beam combining ratio, which corrects for the difference in mode conversion loss due to the additional coupling loss for the LP_{11} "mode" by phase-only mode shaping, a theoretical minimal loss of 5.5 dB can be obtained. Although smaller losses are desirable, the loss is acceptable considering the simplicity of the optical arrangement.

In an experimental implementation [111], three identical collimators with a nominal beam diameter of 500 μm were imaged onto the end facet of a FMF by means of two lenses, the first with a focal length of 75 mm and the second, an aspheric lens, with a focal length of 3.9 mm placed in front of the FMF. This resulted in a magnification factor of 2.6. The phase plates were made of 0.5-mm thick Borosilicate glass, and a photolithographic process was used to create the phase pattern, which was

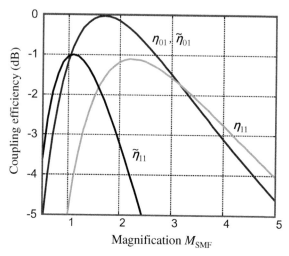

FIGURE 10.17 Theoretical coupling efficiency for phase-plate based coupler.

subsequently time etched into the glass, in order to achieve a 1.7-μm thickness difference. The three beams were combined using two splitters with a non-ideal splitting ratio of 37/63. A coupling loss of 8.3 dB, 10.6 dB, and 9.0 dB was measured for the LP_{01} and the two LP_{11} modes, respectively. Typically, large mode selectivity of >28 dB can be observed for an MMUX pair connected by a short (few meters long) FMF, which can be exploited not just for SDM transmission, but also as a characterization tool for SDM components.

10.4.3.2 Spot-based mode multiplexer

A second strategy to build mode multiplexers is based on the spatial sampling of the amplitude, phase, and polarization in different transverse locations on the end facet of a FMF. Ideally, the sampling spots excite orthogonal combinations of propagating fiber modes, described by a unitary mode coupling matrix. The optimum positions of the sampling spots (the "spot configuration") depend on the symmetry of the fiber modes. As discussed in Section 10.4.1, configurations that provide small MDL are of particular interest.

The intensity distributions of the true waveguide modes in a FMF all show radial symmetry (cf. Figure 10.15), if the appropriate linear combination is chosen for the degenerate modes, and will therefore appear as one or more "rings," which strongly suggests that the spots of the mode couplers should also be arranged on one or more circles. Also, the numbers of spots to be placed on each circle are preferably chosen according to the symmetry of the modes of the FMF [127].

Figure 10.18 shows practical arrangements for the first few LP modes of a FMF. The spot arrangements are constructed based on circles, were each circle corresponds to a particular radial mode number ν of the FMF, starting with the most

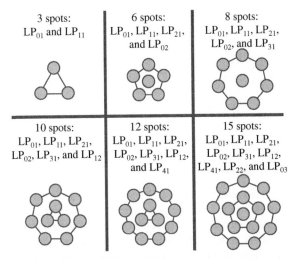

3 spots: LP_{01} and LP_{11}	6 spots: LP_{01}, LP_{11}, LP_{21}, and LP_{02}	8 spots: LP_{01}, LP_{11}, LP_{21}, LP_{02}, and LP_{31}
10 spots: LP_{01}, LP_{11}, LP_{21}, LP_{02}, LP_{31}, and LP_{12}	12 spots: LP_{01}, LP_{11}, LP_{21}, LP_{02}, LP_{31}, LP_{12}, and LP_{41}	15 spots: LP_{01}, LP_{11}, LP_{21}, LP_{02}, LP_{31}, LP_{12}, LP_{41}, LP_{22}, and LP_{03}

FIGURE 10.18 Example of different spot arrangements to couple into the modes of a FMF in the sequence they appear by increasing the normalized frequency of the FMF.

external circle corresponding to $v = 1$. The spots are then added and distributed on the corresponding circle according to their radial number; a single spot is added for the nondegenerate LP modes, and two spots are added for each twofold degenerate mode, respectively. In the case where there is only one mode with a certain radial mode number, the ring will collapse into a spot. Following this rule, we obtain patterns that fill the fiber cross-section homogeneously. The patterns are not unique, in particular it is possible, for example, to rotate the spots on one ring with respect to a second ring. Note that the obtained patterns differ from the more common hexagonal pattern which is the arrangement with largest packing density.

In the following we will analyze in detail the three-spot arrangement for a fiber supporting six spatial and polarization modes. The spot configuration consists of a single ring, and the 3 spots are arranged as an equilateral triangle centered with respect to the fiber core. Figure 10.19a shows graphically how the spots are related to the LP "modes" of the fiber.

Further, in the following calculations, the spots are not allowed to spatially overlap, but are clipped into regions corresponding to $120°$ angular sectors, demarcated by lines in Figure 10.19a. The truncation effect becomes relevant when the width of the spots becomes comparable to the distance between the spots, at which point the clipped light will dominate the MMUX insertion loss. Experimentally this is equivalent, e.g., to the case where the spots are created by combining collimated beams using a reflector with a shape of a pyramid with a triangular base. Alternatively the spots can be brought close together by using clipping mirrors [128] or planar waveguide circuits (PLCs) [129].

Making use of the form for the degenerate LP_{11} "modes" that has radial symmetry in intensity ($LP_{11c} = LP_{11a} + jLP_{11b}$, see also column 4 and 5 of Figure 10.19a), the

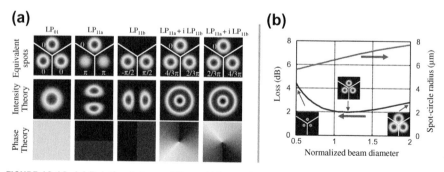

FIGURE 10.19 (a) Relation between LP_{01} and LP_{11} modes and 3-spot configurations. The first row shows the spot intensities in false-colors; the spot phases are indicated by labels next to the spots. The intensity and phase profiles of the targeted LP pseudo modes are shown in false-color in rows 2 and 3, respectively. (b) Insertion loss and radius of the circle where the spots are located, as a function of the beam diameter normalized to a standard SMF mode-field diameter. (For interpretation of the color in this figure, the reader is referred to the web version of this book.)

coupling matrix of the 3-spot coupler that describes the transformation from the 3 individual spots into the fiber modes LP_{01}, LP_{11c}, and LP_{11c}^* can be written as

$$H_{\text{MMUX}} = \begin{pmatrix} a_0 & a_0 & a_0 \\ a_1 & a_1 e^{i2\pi/3} & a_1 e^{-i2\pi/3} \\ a_1 & a_1 e^{-i2\pi/3} & a_1 e^{i2\pi/3} \end{pmatrix}, \qquad (10.35)$$

and the matrix elements can then be determined by evaluating only two overlap integrals to determine a_0 and a_1, which correspond to the overlap integral between a single spot and the LP_{01} and LP_{11c} mode, respectively. It can be shown that Eq. (10.35) is unitary up to a constant loss factor, and therefore MDL $= 0$, if the condition $a_0 = a_1$ is fulfilled. In practice $a_0 = a_1$ can always be achieved by varying the radius of the circle on which the spots are located. In fact if the radius is made small, more light is coupled into the LP_{01} mode and a_0 will become dominant, whereas a larger radius will couple more light into the LP_{11} mode, and a_1 will become larger than a_0. The theoretical insertion loss for a 3-spot coupler with MDL $= 0$ is reported in Figure 10.19b as a function of the spot width normalized to the mode size of a SMF. The calculation reported in Figure 10.20b predicts that insertion losses <2 dB are possible, and low-loss 3-spot couplers with MDL $= 0$ can be realized with a wide range of spot fill-factors, as long as the overall magnification between the fiber modes and the spot pattern is chosen correctly. In practice, insertion losses as low as 4 dB have been reported in [128] using discrete collimators and clipping mirrors, cf. arrangement in Figure 10.20b.

For larger numbers of spots, the MDL for the optimum spot arrangement is in general not zero [118]. It also becomes more difficult to implement the coupler using discrete collimators, and integrated technologies like photonic lanterns [119] or PLCs [129] are preferable. The two approaches are shown in Figure 10.21a and b, respectively. The photonic lantern offers the advantage of completely eliminating

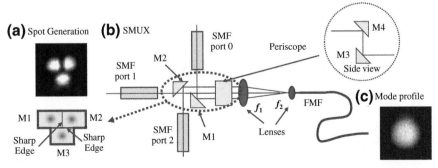

FIGURE 10.20 (a) Functional arrangement of the mirrors M_1, M_2, and M_3 used to generate 3 spots. (b) Setup for 3-spot coupler. (c) Intensity profile at the end facet of the transmission fiber when 3 spots are launched.

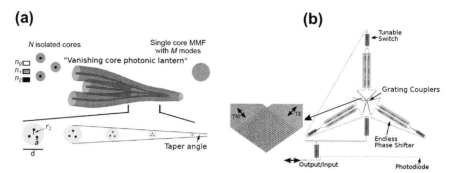

FIGURE 10.21 (a) Principle of the photonic lantern: Multiple single-mode cores are tapered until they form a multi-mode waveguide that couples into the FMF [119]. (b) Integrated-optics spot-based mode-coupler with all-optical MIMO processing [129].

the MDL by adapting the spot profile to the mode-profile of the FMF by using an adiabatically tapered waveguide structure.

10.4.4 Mode couplers for multi-core fibers

In uncoupled multi-core and coupled-core multi-core fibers, the intensity distributions of the individual core modes are typically well separated, and coupling to the individual cores can be performed using a coupler arrangement based on individual collimators [130], as presented in Figure 10.20, for example, by using integrated optical couplers [34] or tapered couplers [29]. For strongly coupled-core arrangements showing considerable differences between the intensity profile of entire multi-core waveguide structure (the super-modes) as compared to the superposition of the individual core modes, the fiber starts behaving more like a few-mode fiber and similar coupling strategies as presented in Section 10.4.3.2 can be applied, and the super-modes can be directly excited.

10.5 OPTICAL AMPLIFIERS FOR COUPLED-MODE TRANSMISSION

As discussed in Section 10.1.2, in order for SDM to be economic and energy efficient, integrated few-mode and multi-core optical amplifiers are a key requirement [35]. In particular, sharing pump power between SDM channels and reducing the number of optical components compared to M parallel single-mode amplifiers is key to reduce system cost and energy consumption. This is in analogy to optical amplification in WDM systems, where the optical pump power as well as the amplifier hardware is shared among all WDM channels.

10.5.1 Optical amplifiers for few-mode fibers

Few-mode fibers offer a significant advantage for optical amplification because of the strong intensity overlap between the modes, which allows to share the pump power across multiple fiber modes in a very effective way. Optical amplification in FMFs has been shown based on the Raman effect in passive FMFs as well as on rare earth doped FMFs. In general, Raman-based few-mode amplification is easier to understand, because it is solely described by pump and signal intensity distributions and does not require taking into account any doping profiles. Hence, algebraic solutions can be derived for practically relevant system configurations. Doped few-mode fiber amplifiers, on the other hand, require complex numerical simulations to arrive at quantitative results.

10.5.1.1 Raman amplification in few-mode fibers

Distributed Raman amplification does not require doped fibers, but utilizes the transmission fiber as an amplifying medium [131]. The Raman process requires in general higher pump powers than needed for doped amplifiers, and the optimal pump wavelength is around 1455 nm for maximum amplification at a signal wavelength of 1550 nm. In few-mode fiber, the pump power can be coupled into different fiber modes just like the transmitted signal, and the net gain experienced by each signal is determined by

$$\frac{dS_m}{dz} = -\alpha_S S_m + \gamma_R \left(\sum_n f_{n,m}(P_n^+ + P_n^-) \right) S_m, \quad \text{and} \qquad (10.36)$$

$$\frac{dP_n^\pm}{dz} = \mp \alpha_P P_n^\pm \mp \frac{\lambda_S}{\lambda_P} \gamma_R \left(\sum_n f_{n,m} S_m \right) P_n^\pm, \qquad (10.37)$$

where S_m is the signal power present in mode m at position z along the fiber, $P_n^\pm$ is the power at the pump wavelength in mode n, P_n^+ and P_n^- denote co-propagating and counter-propagating pumps, respectively, γ_R is the Raman gain coefficient, related to the cross-section of spontaneous Raman scattering, and α_S and α_P are the absorption coefficients at wavelengths λ_S and λ_P, respectively. The intensity overlap integrals are defined as

$$f_{n,m} = \frac{\displaystyle\iint_{-\infty}^{+\infty} I_n(x, y) I_m(x, y) dx\, dy}{\displaystyle\iint_{-\infty}^{+\infty} I_n(x, y) dx\, dy \iint_{-\infty}^{+\infty} I_m(x, y) dx\, dy}, \qquad (10.38)$$

where we assume that the wavelength dependence of the mode profile $I_n(x, y)$ is negligible. Example calculations for intensity overlap integrals are given in Table 10.1.

Equations (10.36) and (10.37) can be solved analytically in the undepleted pump approximation [131]. The solution is given by

$$\frac{S_m(z)}{S_m(0)} = \exp(-\alpha_S z + A_m^+(1 - e^{-\alpha_P z}) + A_m^- e^{-\alpha_P L}(e^{\alpha_P z} - 1)), \quad (10.39)$$

$$A_m^+ = \frac{\gamma_R}{\alpha_P} \sum_n f_{n,m} P_n^+(0), \text{ and} \quad (10.40)$$

$$A_m^- = \frac{\gamma_R}{\alpha_P} \sum_n f_{n,m} P_n^-(L_f), \quad (10.41)$$

where L_f is the length of the fiber, and the coefficients $A_m^\pm$ describe the exponential mode-dependent gain (MDG).

Low MDG can be obtained by optimizing the pump power distribution across the fiber modes. In order to correctly predict the amplification produced by the pump power in a particular mode, it is important to base the calculation on the true waveguide modes of the fiber, e.g. on HE_{11}, TE_{01}, TM_{01}, and HE_{21} instead of the LP pseudo-modes whenever there is LP degeneracy (cf. Figure 10.15). For example, coupling into the LP_{11a} mode with an unpolarized pump laser will equally excite *all* true waveguide modes forming both LP_{11a} and LP_{11b} "modes." Since Raman amplification takes place over fiber lengths considerably exceeding the beat length L_B between the true waveguide modes, the spatial pump pattern within the fiber will change multiple times to provide homogeneous gain for all waveguide modes forming the LP_{11} mode. This is especially true for backward pumping. The gain between LP_{01} and LP_{11} signals can then be equalized by adjusting the relative pump power between LP_{01} and LP_{11} "modes" at the pump wavelength. The exact power ratio can be calculated by evaluating the intensity overlap integrals defined in Eq. (10.38), which are summarized for a weakly guiding step-index FMF with $V = 5$ in Table 10.1.

The MDG can then be minimized by varying the pump power configurations $P_n^+(0)$ and $P_n^-(L)$ and finding the minimum value for the error function $[\max(A_m^\pm) - \min(A_m^\pm)]$, evaluated over all modes m of interest. In order to minimize the MDG along the whole fiber, the configurations for forward and backward

Table 10.1 Intensity overlap integrals $f_{n,m}$ (in $10^9/m^2$) for a FMF supporting 12 spatial- and polarization-modes.

	LP_{01}	LP_{11}	LP_{21}	LP_{02}
	HE_{11}	TE_{01}, TM_{01}, HE_{21}	HE_{31}, EH_{11}	HE_{12}
LP_{01} : HE_{11}	6.24	4.12	2.85	4.62
LP_{11} : TE_{01}, TM_{01}, HE_{21}	4.12	4.36	3.81	2.33
LP_{21} : HE_{31}, EH_{11}	2.85	3.81	3.88	2.12
LP_{02} : HE_{12}	4.62	2.33	2.12	6.15

amplification have to be optimized individually. However, the resulting optimum configuration will be the same for both directions.

For a FMF supporting six spatial and polarization modes (i.e. LP_{01} and LP_{11}), the optimal pump power configuration with perfectly equalized gain is obtained when 10% of the pump power is coupled into LP_{01} and 90% into one of the LP_{11} "modes;" this is close to the experimental condition explored in [63], where equalized gain in a depressed cladding FMF was observed by coupling 100% into one of the two degenerate LP_{11} "modes." For a FMF supporting 12 spatial and polarization modes (i.e. LP_{01}, LP_{11}, LP_{21}, and LP_{02}), the residual MDG can be reduced to 0.13 dB for each 10 dB of Raman gain when 69.6% of pump power is launched into LP_{21} and 30.4% into LP_{02}. Note that no power is launched into LP_{01} and LP_{11} in this case.

Because of the strong intensity overlap between the modes, the pump power is effectively shared between the signal modes, which results in a significant advantage in power efficiency [35]. In order to quantify this advantage, we compare the total pump power required to amplify $M/2$ SMFs with the same effective area as the LP_{01} mode of a FMF, where M is the number of spatial and polarization modes used in the FMF. The equivalent multiple SMF system requires 2.1 and 3.2 times the total Raman pump power of the FMF system for $M = 6$ and $M = 12$, respectively. This represents a considerable power advantage (which will, however, be reduced with pump depletion).

Based on this method, low MDG has been demonstrated using Raman amplification in six spatial- and polarization-mode fibers, with gains of up to 10 dB [63]. The experimental results for gain and equivalent noise figure are summarized in Figure 10.22.

10.5.1.2 Optical amplification in doped few-mode fiber

Optical amplification in Erbium-doped multimode fibers has been reported in 1991 by Nykolak et al. [132], but a strong mode-dependent gain (MDG) was observed. Minimizing the MDG for in-line amplification is essential to guarantee consistent performance. In contrast to Raman amplification, where the MDG only depends on the

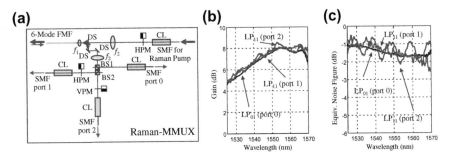

FIGURE 10.22 (a) Experimental arrangement of a phase-plate based mode multiplexer with backward Raman pump coupler. (b) On–off gain of the Raman amplification. (c) Equivalent noise figure at the FMF end [63].

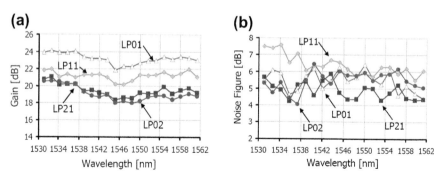

FIGURE 10.23 (a) Gain and (b) noise figure as a function of wavelength for a few-mode fiber supporting 12 spatial and polarization modes [62].

intensity overlap integrals, two additional drivers for the MDG are present in doped fibers, namely the transversal doping profile and the nonlinear nature of the optical active inversion in the amplifier. The latter depends on both the pump power and on the modal power content of the signal to be amplified. Detailed studies on MDG in Erbium-doped FMF have been presented in [133,134] (see Figure 10.23[62]).

In order to reduce the MDG in doped fibers, two main strategies have been adopted: The first approach optimizes the pump power distribution across the fiber modes [133], similar to the case of Raman amplification. The second approach is based on optimizing the doping profile [134–136]. The MDG in 6 and 12 spatial- and polarization-mode fibers has recently been brought under control and reduced to <4 dB variation across the C-band, by using ring shaped doping profiles, which reduces the gain of the otherwise dominant LP_{01} mode [62,135]. Figure 10.23 shows the experimental result presented in [62] for a 12 spatial- and polarization-mode fiber, where a combination of a ring- and a step-based doping profile is used to equalize the MDG.

10.5.2 Optical amplifier for multi-core fibers

In uncoupled-core and coupled-core multi-core fibers, optical amplifiers are straight forward to implement by using Raman amplification or multi-core rare earth-doped fibers [137,138]. Investigations have so far focused on integration, size reduction, pumping strategies like cladding pumping [137], and pump distribution schemes [138–140].

10.6 SYSTEMS EXPERIMENTS

Transmission experiments using mode division multiplexing in multi-mode fibers were first conducted in 1982 by Berdague and Facq [50], but due to the large number of modes supported by their fiber and without access to coherent detection and

MIMO techniques, experimental results were inherently limited in bit rate and transmission distance. MIMO was first experimentally introduced to optical fiber transmission in 2000 by Stuart [51] to exploit spatial diversity in multi-mode fiber in combination with direct detection. This was followed by numerous studies and experiments to increase the performance of local-area multi-mode fibers through mode group multiplexing [53–56]. Coherent MIMO techniques have also been proposed [141], however the lack of a *selective* access to *all* modes of the fiber limited the practicality of this method. The use of few-mode fiber [112,142,143], together with the selective excitation and detection of all fiber modes followed by full coherent MIMO signal processing, was first proposed and experimentally demonstrated by the authors [48,144]. This approach was shown to essentially reach a per-mode transmission performance comparable to that of single-mode fiber. Several impressive MIMO-SDM transmission systems of this kind have since been demonstrated [59,61,63,105a,111].

10.6.1 Single-span MIMO-SDM transmission over few-mode fiber

The first experimental demonstration of 6×6 MIMO transmission over few-mode fiber was presented in [144]. The initial single-span transmission distance of 10 km was subsequently increased to 33 and 96 km by improving the coupler alignment [100,111]. The experimental setup is shown in Figure 10.24.

The system was based on the phase-plate based MMUXs described in Section 10.4.3.1 and used 10 km of low-DGD depressed cladding few-mode fiber (DC-FMF). The fiber was designed to cut off the LP_{21} and to minimize the DGD between the LP_{01} and LP_{11} "modes," resulting in a DGD $<2.6 \pm 0.1$ ns for a 96-km long fiber over a wavelength range of 1530–1565 nm. In comparison, the DGD of a step-index (SI) profile FMF with similar parameters is more than two orders of magnitude

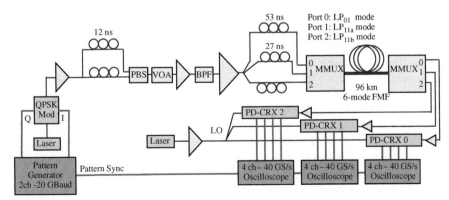

FIGURE 10.24 Experimental setup of [144]. QPSK-Mod: QPSK Modulator, PBS: Polarization beam splitter, VOA: Variable optical attenuator, BPF: Bandpass filter, LO: Local oscillator, PD-CRX: Polarization-diversity coherent receiver.

larger. The fiber loss was 0.205 dB/km at 1550 nm for all modes, and the effective areas were 155 and 159 μm^2 for the LP_{01} and LP_{11} "modes", respectively. The chromatic dispersion was 18 ps/(nm km) for both LP_{01} and LP_{11} "modes".

The transmission performance of this FMF was evaluated by launching three delay-decorrelated PDM-QPSK signals at 20 GBaud, into the three ports of the MMUX. This resulted in a spatial superchannel with a single-wavelength aggregate bit rate of 240 Gb/s. On the receiving end, the FMF was terminated by another MMUX acting as a mode demultiplexer, whose outputs were fed into three polarization-diversity coherent receivers (CRXs), followed by digitization at 40 GS/s using 12 ports of a synchronized high-speed digital oscilloscope system. The off-line DSP algorithms subsequently equalized the 6×6 MIMO channel matrix as described in Section 10.3 to recover the 6 independent data streams. Measurements performed on a 96-km span of the same FMF showed a penalty between measurement and theory of <2 dB at a BER of 10^{-3} when using 120-tap filters for the MIMO-DSP. This excellent performance shows that mode coupling in 96 km of this FMF can be almost completely compensated with low impact on system performance.

In order to better understand the coupling between the fiber modes, it is instructive to examine the 6×6 matrix of impulse responses of the MIMO channel using the channel estimation techniques presented in Section 10.3.

Figure 10.25 shows the squared magnitudes of the estimated 6×6 impulse responses as a two-dimensional array corresponding to each combination of input and output of the 6×6 MIMO channel. In this representation, the columns correspond to the ports of the transmit-side MMUX, and the rows correspond to the ports of the receive-side MMUX, respectively. In order to show the impulse responses only due to mode coupling, the mode-average chromatic dispersion of 96·18 ps/nm as well as the carrier frequency offset were electronically compensated prior to estimating the impulse response matrix. This results in sharp peaks that clearly identify the main mode coupling locations.

Figure 10.25 can be divided into four regions designated A, B, C, and D: Region A is the 2×2 array located in the top left corner that shows the coupling between the two polarizations of the fundamental mode (LP_{01x} and LP_{01y}), as also observed in a single-mode fiber. Region B is formed by the 4×4 array on the bottom right corner that shows the coupling between the degenerate LP "modes" LP_{11ax}, LP_{11ay}, LP_{11bx}, and LP_{11by}. The two remaining off-diagonal regions C and D describe the crosstalk between LP_{01} and LP_{11} "modes." We observe sharp and strong coupling peaks within regions A and B, and typically 100 to 1000 times weaker, 2.6-ns wide distributed coupling in regions C and D. The width of the distributed coupling in regions C and D corresponds to the DGD of 96 km of this FMF and represents distributed coupling at various locations along the fiber. If the light travels mostly in the LP_{01} "mode", it arrives earlier, whereas it arrives delayed by the DGD if it travels mostly in the slower LP_{11} "mode". Also, regions A and B show weaker distributed coupling next to the strong coupling peaks. This weaker distributed coupling is caused by light that couples back and forth between LP_{01} and LP_{11}. For LP_{01} "modes" (region A) the distributed coupling, whose width is also consistent with the DGD of 96 km of FMF, is located on the right of the main pulse, whereas it is located on the left of the main pulse for the LP_{11} "modes"

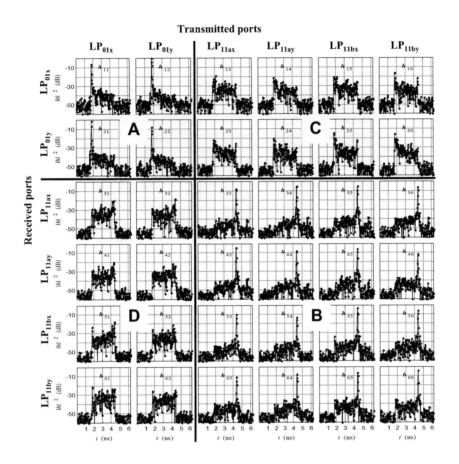

FIGURE 10.25 Squared magnitude of the PDM-SDM 6 × 6 impulse responses for 96 km of 6-mode FMF.

(region B). Finally, Figure 10.25 also confirms the excellent alignment of the MMUX: Any misalignment in the coupler would create a narrow crosstalk peak either at the beginning or at the end of the distributed coupling in regions C and D. The channel estimation gives a very clear picture of the crosstalk introduced by the MMUX and the propagation through the FMF and allows for a better understanding of the observed performance of the MIMO DSP.

Modal crosstalk can also be directly observed by imaging the end facet of the FMF to an InGaAs camera. The results are shown in Figure 10.26a and b for a fiber length of 96 and 33 km, respectively. The intensity profiles after 33-km FMF are in good agreement with the corresponding theoretical intensity profiles reported in Figure 10.26c, whereas after 96 km the images become "blurry" and coupling between the modes becomes evident. Crosstalk measurements at the MMUX for this kind of FMF of 2-m, 33-km, and 96-km length also confirmed a crosstalk of −18 and −11 dB between

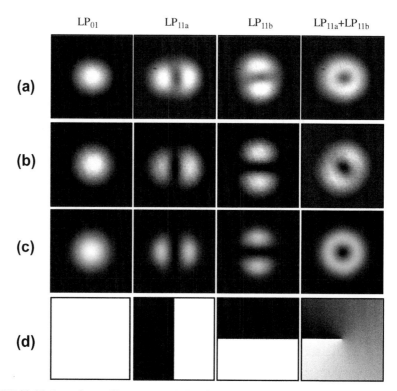

FIGURE 10.26 Intensity profiles measured after (a) 96 km and (b) 33 km of 6-mode FMF when launching either LP$_{01}$ or LP$_{11}$ "modes". (c) Theoretical mode intensity profiles and (d) theoretical phase profile of a 6-mode FMF. (e) Schematic setup of the MMUX.

LP$_{01}$ and LP$_{11}$, for a distance of 33 and 96 km, respectively. The accumulated crosstalk after 96 km is significant, and makes SDM transmission impossible without the use of MIMO-DSP.

Single-span transmission distances have since been considerably increased, up to 209 km in an experiment using backward-pumped distributed Raman amplification [63], by using spot-based couplers [128], and also using a DGD compensated hybrid fiber span [145].

10.6.2 Multi-span MIMO-SDM transmission over few-mode fiber

Within the boundary conditions typical of research environments, FMF transmission beyond a single amplification span is best demonstrated using a re-circulating loop, as shown in Figure 10.27, allowing to emulate tens of amplification spans with the equipment needs of a single span. However, in contrast to single-mode re-circulating loops, a FMF loop requires one set of amplifiers and switches for *each* supported spatial mode.

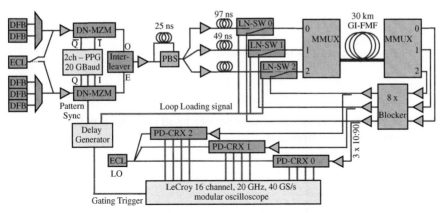

FIGURE 10.27 Experimental setup for MIMO loop transmission experiment. PBS: Polarization beam splitter. Triangles denote EDFAs.

Transmitter and receiver in Figure 10.27 are similar to the setup described in Figure 10.24. The triple re-circulating loop is loaded and closed using three Lithium Niobate switches (LN-SWs). Three Erbium-doped optical amplifiers (EDFAs) and wavelength blockers are used to amplify and gain-equalize the wavelength channels of the individual spatial modes after the MMUX. A critical part of the experimental setup is the exact adjustment of the relative loop lengths to within ~100 ps (corresponding to ~2 cm of fiber). In addition, the accumulated DGD in the fiber span, which is the main driver for impulse response broadening, has to be kept small to reduce the number of equalizer taps necessary for the MIMO-DSP. This is achieved by using a DGD compensated span, where multiple GI-FMFs with DGDs of opposite sign are spliced together to form a span with almost no residual DGD.

The fiber had a loss of 0.24 dB/km, and a dispersion of 18.5 ps/nm/km for both LP_{01} and LP_{11} "modes." The effective areas were 64 μm^2 for LP_{01} and 67 μm^2 for LP_{11} "modes," respectively. Multiple fiber spools with different lengths were measured with an intensity modulated pulse, as described in [111]. We used two spools with lengths of 25 and 5 km, respectively. One spool was additionally shortened to correct the overall DGD to <50 ps. The segments were then connected using a commercial splicer, where particular care was taken to minimize the crosstalk introduced by splice imperfections. The wavelength dependence of the overall DGD of the composite fiber was <100 ps over a wavelength range of 1525–1570 nm, thus providing good performance across the entire C-band.

Results for 6 × 6 MIMO transmission over the 30-km long span are shown in Figure 10.28 in terms of the Q-factors versus distance for all six spatial modes at different launch powers of −3 dBm, 0 dBm, and 2 dBm per spatial mode. At −3 dBm of launched power, a reach of 1200 km is achieved, assuming a forward-error-correction limit of 10^{-2} and an equalizer with 400 taps. For higher launch powers, the transmission distance is limited by the onset of fiber nonlinearity.

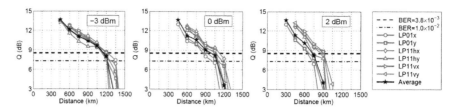

FIGURE 10.28 Q-factor versus transmission distance for all six modes at three different launch powers per spatial mode [58].

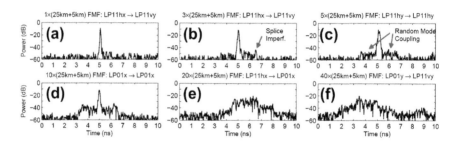

FIGURE 10.29 Exemplary impulse responses of a phase-plate based transmission experiment using a DGD-compensate few-mode fiber [58].

The impulse response of the MIMO channel, shown for various modes and distances in Figure 10.29, shows that DGD compensation is only partially working, because for long transmission distances the crosstalk in the fibers and particularly at the splice points between the individual fiber segments is no more negligible. It can be shown that the major driver for impulse response broadening is the maximal DGD excursion in the span [146], as long as the total accumulated crosstalk is significantly smaller than the power that stays in the mode in which it was originally launched. For even longer distances, the coupling is complete and the impulse response will take on a Gaussian shape whose width is expected to grow with the square root of the distance, as predicted in [92].

10.6.3 MIMO-SDM in coupled multi-core fiber

Uncoupled multi-core fibers offer a simple way to implement SDM over an optical fiber [28–46], but the minimum distance between the cores (and hence the spatial information density within the fiber) is limited by crosstalk considerations (cf. Section 10.1.3). If coupling between cores is allowed because MIMO-DSP is being provided, cores can be placed in closer vicinity, allowing for an increased spatial density of the SDM paths within the fiber cross-section. The first demonstration of 6×6 MIMO transmission over 24 km of coupled 3-core fiber (CCF) was reported in [147].

The distance between the cores was $38\,\mu m$, which is significantly smaller than what has been demonstrated for uncoupled multi-core fibers. More recently, the distance between the cores of a new 3-core CCF was further reduced to $29.4\,\mu m$ [148]. The fiber had a length of 60 km, a large effective area of $(129 \pm 2)\,\mu m^2$, and an attenuation of 0.181 dB/km for all cores. The coupling between cores was so strong that light launched into a core is almost equally distributed among the cores after 60 km of fiber. The presence of such strong coupling between cores suggests that the modes of the fiber can no longer be considered as simple superpositions of the individual core modes, but the modes of the whole structure, the "supermodes," have to be considered [149,150].

A cross-section of the 3-core CCF is shown in Figure 10.30a together with the calculated linearly polarized super-modes (upper row) and their far-fields (bottom row). The modes are named in analogy to the symmetry-equivalent LP modes in a FMF. Including polarization, a total of six independent channels, the same as for the 6-mode FMF, are available.

In order to study MIMO transmission over 3-core CCFs, we built a spot-based MMUX consisting of three individual collimators that are projected onto the end facet of the 3-core CCF. This way, low loss (<2 dB) and low crosstalk (<-40 dB) could be obtained. A similar re-circulating loop set up as shown in Figure 10.27 was used for transmission, and the results are reported in Figure 10.31. A maximum transmission distance of 4200 km was obtained for a per-core launch power of -3 dBm at a Q-factor >7.2 dB (equal to a BER of 10^{-2}) representing the limit tolerable for state-of-the-art 20% overhead hard-decision FEC. The required number of MIMO-DSP equalizer taps grows as a function of distance, starting with 60 taps after 60 km and growing up to 400 taps at 4200 km. This experiment clearly demonstrates that CCFs represent a promising long-haul transmission technology when MIMO-DSP is being used.

As in the case for a FMF supporting six spatial modes, the impulse responses matrix of the 3-core CCF also consists of a 6×6 impulse responses (cf. Figure 10.29). However, in contrast to the FMF, where the DGD between the LP_{01} and LP_{11} "mode" causes the impulse response of the MIMO channel to exhibit two prominent peaks separated by the DGD, the impulse response of the 3-core CCF after 60-km

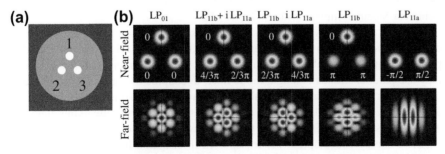

FIGURE 10.30 (a) Fiber cross-section of the 3-core CCF. (b) Linearly polarized super-modes (upper row) and corresponding far-fields (bottom row) of the 3-core CCF.

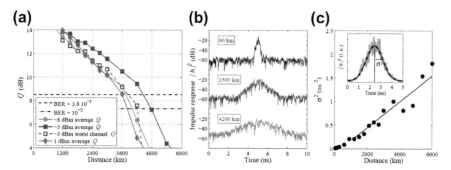

FIGURE 10.31   (a) Transmission results for 6 × 6 MIMO transmission in 3-core CCF. (b) Impulse response after 60-, 1500-, and 4200-km transmission distance. (c) Variance of the impulse response width as function of distance.

transmission only shows a single peak with a width of 600 ps. Also, all 36 impulse responses look similar and only one representative impulse response is hence shown in Figure 10.31b, where the squared magnitude of the impulse response h as obtained from LS channel estimation and after digital chromatic dispersion compensation is shown for a selection of transmission distances. Figure 10.31c shows the complete evolution of the variance of the width of $|h|^2$ obtained by a Gaussian fit. The linear fit suggests a complete randomization of the time delays between the cores during propagation, as well as a growth of the MDS as a function of the square-root of distance, as predicted for strong coupling in [92].

10.7 CONCLUSION

Wavelength-division multiplexing (WDM) has been the workhorse of data networks since the early 1990s, enabling ubiquitous and affordable data services with unabated exponential traffic growth. Today, commercial WDM systems can carry close to 10 Tb/s over a single fiber, and research experiments have reached the 100-Tb/s mark. Over the past few years, however, WDM capacities have approached the nonlinear Shannon limit to within a factor of 2.

In order to further scale network capacities and to avoid a looming "capacity crunch," *space* has been identified as the only known physical dimension yet unexploited for optical modulation and multiplexing. *Space-division multiplexing* (SDM) may use parallel strands of single-mode fiber, uncoupled or coupled cores of multi-core fiber, or individual modes of multi-mode waveguides. In this context, *integration* at various levels (including optical amplifiers, transponders, networking elements, and transmission fiber CAPEX and OPEX) is essential to continue the reduction in cost and energy consumption per transported information bit that has allowed the Internet to thrive. Integration, however, inherently comes at the expense of crosstalk. If crosstalk rises to levels where it cannot be treated as a transmission impairment any more, multiple-input multiple-output (MIMO) digital signal processing (DSP) techniques have to be used to manage crosstalk in highly integrated SDM systems.

At the beginning of an exciting new era in optical communications, we reviewed fundamentals as well as practical experimental aspects of MIMO-SDM: We discussed the importance of selectively addressing *all* modes of a coupled-mode SDM channel at transmitter and receiver in order to achieve reliable capacity gains and showed that reasonable levels of mode-dependent loss (MDL) are acceptable without much loss of channel capacity. We then introduced MIMO-DSP techniques as an extension of familiar algorithms used in polarization-division multiplexed (PDM) digital coherent receivers and discussed their functionality and scalability. Finally, we reviewed the design of mode multiplexers (MMUXs) that allow for the mapping of the individual transmission signals onto an orthogonal basis of waveguide modes, and discussed their performance in experimental demonstrations. Although MIMO-SDM has been experimentally proven to be a feasible technique for reliable high-capacity long-haul fiber transmission beyond the nonlinear Shannon limit of single-mode fiber, significant research, development, and standardization efforts have to be invested to enable a smooth upgrade path from today's single-mode systems to MIMO-SDM based optical communication networks.

Acknowledgments

We acknowledge valuable discussions with S. Bigo, S. Chandrasekhar, G. Charlet, D. Chizhik, A.R. Chraplyvy, C.R. Doerr, R.-J. Essiambre, N. Fontaine, G.J. Foschini, A.H. Gnauck, X. Liu, H. Kogelnik, S.K. Korotky, G. Raybon, M. Salsi, R.W. Tkach, and C. Xie from Bell Labs, Alcatel-Lucent, D. DiGiovanni, J. Fini, R. Lingle, D. Peckham, T.F. Taunay, and B. Zhu from OFS Labs, and M. Hirano, T. Sasaki, and Y. Yamamoto from Sumitomo Electric.

References

[1] R.W. Tkach, Scaling optical communications for the next decade and beyond, Bell Labs Tech. J. 14 (4) (2010) 3–9.

[2] S.K. Korotky, Traffic trends: drivers and measures of cost-effective and energy-efficient technologies and architectures for backbone optical networks, in: Proc. OFC, OM2G.1, 2012.

[3] M.A. Taubenblatt, Optical interconnects for high-performance computing, J. Lightwave Technol. 30 (4) (2012) 448–458.

[4] Y.A. Vlasov, Silicon CMOS-integrated nano-photonics for computer and data communications beyond 100G, IEEE Commun. Mag. 50 (2) (2012) S67–S72.

[5] J. Gray, P. Shenoy, Rules of thumb in data engineering, Microsoft Research Technical Report MS-TR-99-100, 2000.

[6] J.L. Hennessy, D.A. Patterson, Computer Architectures: A Quantitative Approach, Morgan Kaufmann, San Francisco, CA, 2003.

[7] <http://top500.org/lists/2010/06/performance>

[8] J.G. Apostolopoulos et al., The road to immersive communication, Proc. IEEE 100 (4) (2012) 974–990.

[9] M. Dou et al., Room-sized informal telepresence system, Proc. IEEE Virt. Real. (2012) 15–18.

[10] H. Kogelnik, On optical communication: Reflections and perspectives, Proc. ECOC (2004) Mo1.1.1.

[11] P.J. Winzer, High-spectral-efficiency optical modulation formats, J. Lightwave Technol. 30 (24) (2012) 3824–3835.

[12] Alcatel-Lucent 1830 Photonic Service Switch, Brochures and data sheets, <http://www.alcatel-lucent.com>

[13] G. Raybon et al., All-ETDM 80-Gbaud (640-Gb/s) PDM 16-QAM generation and coherent detection, Photon. Technol. Lett. 24 (15) (2012) 1328–1330.

[14] J. Renaudier et al., Spectrally efficient long-haul transmission of 22-Tb/s using 40-Gbaud PDM-16QAM with coherent detection, in: Proc. OFC, OW4C.2, 2012.

[15] G. Raybon et al., 1-Tb/s dual-carrier 80-GBaud PDM-16QAM WDM transmission at 5.2 b/s/Hz over 3200 km, in: Proc. IPC, PD1.2, 2012.

[16] X. Liu et al., 1.5-Tb/s guard-banded superchannel transmission over 56×100-km (5600-km) ULAF using 30-Gbaud pilot-free OFDM- 16QAM signals with 5.75-b/s/Hz net spectral efficiency, in: Proc. ECOC, Th.3.C.5, 2012.

[17] S. Chandrasekhar, X. Liu, OFDM based superchannel transmission technology, J. Lightwave Technol. 30 (24) (2012) 3816–3823.

[18] D. Qian et al., 101.7-Tb/s (370x294-Gb/s) PDM-128QAM-OFDM transmission over 3×55-km SSMF using pilot-based phase noise mitigation, in: Proc. OFC, PDPB5, 2011.

[19] A. Sano et al., 102.3-Tb/s (224×548-Gb/s) C- and extended L-band all-Raman transmission over 240km using PDM-64QAM single carrier FDM with digital pilot tone, in: Proc. OFC, PDP5C.3, 2012.

[20] P.J. Winzer, Beyond 100G Ethernet, IEEE Commun. Mag. 48 (2010) 26–30.

[21] R.-J. Essiambre et al., Capacity limits of optical fiber networks, J. Lightwave Technol. 28 (4) (2010) 662–701.

[22] R.-J. Essiambre, R.W. Tkach, Capacity trends and limits of optical communication networks, Proc. IEEE 100 (5) (2012) 1035–1055.

[23] P.J. Winzer, Energy-efficient optical transport capacity scaling through spatial multiplexing, Photon. Technol. Lett. 23 (13) (2011) 851–853.

[24] A.R. Chraplyvy, The coming capacity crunch, ECOC, plenary talk, 2009.

[25] P.J. Winzer, R.-J. Essiambre, Advanced optical modulation formats, in: I. Kaminow, T. Li, A. Willner (Eds.), Optical Fiber Telecommunication V B, Academic Press, New York, 2008, pp. 23–94 (Chapter 2)

[26] T. Morioka, Ultrafast optical technologies for large-capacity TDM/WDM photonic networks, in: IEEE Photonics Society Summer Topical Meetings, 2010, pp. 121–122.

[27] IEC Technical Report IEC 61292–4, Optical amplifiers—Part 4: Maximum permissible optical power for the damage-free and safe use of optical amplifiers, including Raman amplifiers, Edition 2.0, 2010.

[28] J. Sakaguchi et al., 109-Tb/s ($7 \times 97 \times 172$-Gb/s SDM/WDM/PDM) QPSK transmission through 16.8-km homogeneous multi-core fiber, in: Proc. OFC, PDPB6, 2011.

[29] B. Zhu et al., 112-Tb/s Space-division multiplexed DWDM transmission with 14-b/s/Hz aggregate spectral efficiency over a 76.8-km seven-core fiber, Opt. Express 19 (17) (2011) 16665–16671.

[30] J. Sakaguchi et al., 19-core fiber transmission of $19 \times 100 \times 172$-Gb/s SDM-WDM-PDM-QPSK signals at 305Tb/s, in: Proc. OFC, PDP5C.1, 2012.

[31] H. Takara et al., 1.01-Pb/s (12 SDM/222 WDM/456 Gb/s) crosstalk-managed transmission with 91.4-b/s/Hz aggregate spectral efficiency, in: Proc. ECOC, Th.3.C.1, 2012.

[32] R.S. Tucker, Green optical communications—part I: Energy limitations in transport, IEEE J. Sel. Top. Quantum Electron 17 (2) (2011) 245–260.

[33] S.K. Korotky, Price-points for components of multi-core fiber communication systems in backbone optical networks, J. Opt. Commun. Netw. 4 (5) (2012) 425–435.

[34] C.R. Doerr, T.F. Taunay, Silicon photonics core-, wavelength-, and polarization-diversity receiver, Photon. Technol. Lett. 23 (9) (2011) 597–599.

[35] P.M. Krummrich, Optical amplification and optical filter based signal processing for cost and energy efficient spatial multiplexing, Opt. Express 19 (17) (2011) 16636–16652.

[36] K.S. Abedin et al., Amplification and noise properties of an erbium-doped multicore fiber amplifier, Opt. Express 19 (17) (2011) 16715–16721.

[37] T. Morioka, New generation optical infrastructure technologies: EXAT initiative towards 2020 and beyond, in: Proc. OECC, FT4, 2009.

[38] Y. Kokubun, M. Koshiba, Novel multi-core fibers for mode division multiplexing: Proposal and design principle, IEICE Electron. Expr. 6 (8) (2009) 522–528.

[39] G. Li, X. Liu, Focus issue: Space multiplexed optical transmission, Opt. Express 19 (17) (2011) 16574.

[40] T. Morioka et al., Enhancing optical communications with brand new fibers, IEEE Commun. Mag. 50 (2012) s31.

[41] P.J. Winzer et al., Penalties from in-band crosstalk for advanced optical modulation formats, in: Proc. ECOC, Tu.5.B.7, 2011.

[42] K. Imamura et al., Trench assisted multi-core fiber with large Aeff over $100\,\mu m^2$ and low attenuation loss, in: Proc. ECOC, Mo.1.LeCervin.1, 2011.

[43] T. Hayashi et al., Design and fabrication of ultra-low crosstalk and low-loss multi-core fiber, Opt. Express 19 (17) (2011) 16576–16592.

[44] S. Chandrasekhar et al., WDM/SDM transmission of 10×128-Gb/s PDM-QPSK over 2688-km 7-core fiber with a per-fiber net aggregate spectral-efficiency distance product of 40,320 km b/s/Hz, in: Proc. ECOC, Th.13.C.4, 2011.

[45] A.H. Gnauck et al., WDM transmission of 603-Gb/s superchannels over 845 km of 7-core fiber with 42.2 b/s/Hz spectral efficiency, in: Proc. ECOC, Th.2.C.2, 2012.

[46] X. Liu et al., 1.12-Tb/s 32-QAM-OFDM superchannel with 8.6-b/s/Hz intrachannel spectral efficiency and space-division multiplexing with 60-b/s/Hz aggregate spectral efficiency, in: Proc. ECOC, Th.13.B.1, 2011.

[47] G.J. Foschini, Layered space-time architecture for wireless communication in a fading environment when using multi-element antennas, Bell Labs Tech. J. 1 (1996) 41–59.

[48] P.J. Winzer, G.J. Foschini, Outage calculations for spatially multiplexed fiber links, in: Proc. OFC, OThO5, 2011.

[49] P.J. Winzer, G.J. Foschini, MIMO capacities and outage probabilities in spatially multiplexed optical transport systems, Opt. Express 19 (17) (2011) 16680–16696.

[50] S. Berdague, P. Facq, Mode division multiplexing in optical fibers, Appl. Opt. 21 (11) (1982) 1950–1955.

[51] H.R. Stuart, Dispersive multiplexing in multimode optical fiber, Science 289 (5477) (2000) 281–283.

[52] S. Murshid et al., Spatial domain multiplexing: a new dimension in fiber optic multiplexing, Opt. Laser Technol. 40 (8) (2008) 1030–1036.

[53] A. Tarighat et al., Fundamentals and challenges of optical multiple-input-multiple-output multimode fiber links, IEEE Commun. Mag. 45 (5) (2007) 57–63.

[54] S. Schoellmann et al., Experimental realisation of 3×3 MIMO system with mode group diversity multiplexing limited by modal noise, in: Proc. OFC, JWA68, 2008.

[55] M. Nazarathy, A. Agmon, Coherent transmission direct detection MIMO over short-range optical interconnects and passive optical networks, J. Lightwave Technol. 26 (14) (2008) 2037–2045.

[56] B. Franz et al., High speed OFDM data transmission over 5 km GI-multimode fiber using spatial multiplexing with 2×4 MIMO processing, in: Proc. ECOC, Tu.3.C.4, 2010.

[57] R. Ryf et al., Space-division multiplexed transmission over 4200 km 3-core microstructured fiber, in: Proc. OFC, PDP5C.2, 2012.

[58] S. Randel et al., Mode-multiplexed 6×20-GBd QPSK transmission over 1200-km DGD-compensated few-mode fiber, in: Proc. OFC, PDP5C.5, 2012.

[59] R. Ryf et al., 12×12 MIMO transmission over 130-km few-mode fiber, Proc. Frontiers in Optics, paper FW6C.4, 2012.

[60] Y. Yung et al., First demonstration of multimode amplifier for spatial division multiplexed transmission systems, in: Proc. ECOC, Th.13.K.4, 2011.

[61] E. Ip et al., $88 \times 3 \times 112$-Gb/s WDM transmission over 50 km of three-mode fiber with inline few-mode fiber amplifier, in: Proc. ECOC, Th.13.C.2, 2011.

[62] M. Salsi et al., A six-mode Erbium-doped fiber amplifier, in: Proc. ECOC, Th.3.A.6, 2012.

[63] R. Ryf et al., Mode-equalized distributed Raman amplification in 137-km few-mode fiber, in: Proc. ECOC, Th.13.K.5, 2011.

[64] J.G. Proakis, M. Salehi, Digital Communications. fifth ed., Mc Graw Hill, New York, 2007.

[65] J.R. Barry, E.A. Lee, D.G. Messerschmitt, Digital Communication. third ed., Kluwer Academic Publishers, 2004.

[66] P. Tan, N.C. Beaulieu, Analysis of the effects of Nyquist pulse-shaping on the performance of OFDM systems with carrier frequency offset, Eur. Trans. Telecomm. 20 (2009) 9–22.

[67] H. Kogelnik, Theory of optical waveguides, in: N. Fraser (Ed.), Guided-Wave Optoelectronics, Springer Verlag, 1988.

[68] D. Marcuse, Light Transmission Optics, Van Nostrand Reinhold Company, 1972.

[69] T.M. Cover, J.A. Thomas, Elements of Information Theory. second ed., John Wiley and Sons, New York, 2006.

[70] C. Koebele et al., Nonlinear effects in long-haul transmission over bimodal optical fibre, in: Proc. ECOC, Mo.2.C.6, 2010.

[71] R. Dar et al., The underaddressed optical multiple-input, multiple-output channel: capacity and outage, Opt. Lett. 37 (15) (2012) 3150–3152.

[72] F. Mezzadri, How to generate random matrices from the classical compact groups, Notices of the AMS 54 (2007) 592–604.

[73] H. Kogelnik, L.E. Nelson, R.M. Jopson, Polarization mode dispersion, in: I.P. Kaminow, T. Li (Eds.), Optical Fiber Telecommunications IV B, Academic Press, San Diego, 2002, pp. 725–861 (Chapter 15)

[74] M. Brodsky, N.J. Frigo, M. Tur, Polarization mode dispersion, in: I.P. Kaminow, T. Li, A.E. Willner (Eds.), Optical Fiber Telecommunications V A, Academic Press, New York, 2008, pp. 605–670 (Chapter 17)

[75] C. Xie, Polarization-mode-dispersion impairments in 112-Gb/s PDM-QPSK coherent systems, in: Proc. ECOC, Th.10.E.6, 2010.

[76] B. Wedding, C.N. Haslach, Enhanced PMD mitigation by polarization scrambling and forward error correction, in: Proc. OFC, WAA1, 2001.

[77] X. Liu et al., Demonstration of broad-band PMD mitigation in the presence of PDL through distributed fast polarization scrambling and forward-error correction, Photon. Technol. Lett. 17 (5) (2005) 1109–1111.

[78] H. Bölcskei et al., On the capacity of OFDM-based spatial multiplexing systems, IEEE Trans. Commun. 50 (2) (2002) 225–234.

[79] A.J. Paulraj et al., An overview of MIMO communications—A key to Gigabit wireless, Proc. IEEE 92 (2) (2004) 198–218.

[80] K.-P. Ho, J.M. Kahn, Frequency diversity in mode-division multiplexing systems, J. Lightwave Technol. 29 (24) (2011) 3719–3726.

[81] M. Shtaif, Performance degradation in coherent polarization multiplexed systems as a result of polarization dependent loss, Opt. Express 16 (18) (2008) 13918–13932.

[82] A. Nafta et al., Capacity limitations in fiber-optic communications systems as a result of polarization dependent loss, Opt. Lett. 34 (23) (2009) 3613–3615.

[83] E. Meron et al., Use of space-time coding in coherent polarization-multiplexed systems suffering from polarization dependent loss, Opt. Lett. 35 (21) (2010) 3547–3549.

[84] A. Mecozzi, M. Shtaif, The statistics of polarization-dependent loss in optical communication systems, Photon. Technol. Lett. 14 (3) (2002) 313–315.

[85] Y. Fukada, Probability density function of polarization dependent loss (PDL) in optical transmission system composed of passive devices and connecting fibers, J. Lightwave Technol. 20 (6) (2002) 953–964.

[86] M. Yu et al., Statistics of polarization-dependent loss, insertion loss, and signal power in optical communication systems, Photon. Technol. Lett. 14 (12) (2002) 1695–1697.

[87] A. Mecozzi, M. Shtaif, Signal-to-noise-ratio degradation caused by polarization-dependent loss and the effect of dynamic gain equalization, J. Lightwave Technol. 22 (8) (2004) 1856–1871.

[88] L.E. Nelson et al., Statistics of polarization dependent loss in an installed long-haul WDM system, Opt. Express 19 (7) (2011) 6790–6796.

[89] A. Steinkamp et al., Polarization mode dispersion and polarization dependent loss in optical fiber systems, Proc. SPIE 5596 (2004) 243–254.

[90] A. El Amari et al., Statistical prediction and experimental verification of concatenations of fiber optic components with polarization dependent loss, J. Lightwave Technol. 16 (3) (1998) 332–339.

[91] C. Antonelli et al., Stokes-space analysis of modal dispersion in fibers with multiple mode transmission, Opt. Express 20 (11) (2012) 11718–11733.

[92] K.P. Ho, J.M. Kahn, Statistics of group delays in multimode fiber with strong mode coupling, J. Lightwave Technol. 29 (21) (2011) 3119–3128.

[93] K.-P. Ho, J.M. Kahn, Mode-dependent loss and gain: statistics and effect on mode-division multiplexing, Opt. Express 19 (17) (2011) 16612–16635.

[94] S. Warm, K. Petermann, Outage capacity for spliced mode multiplexed multi-mode fiber links, in: Proc. OFC, JW2A.39, 2012.

[95] S.J. Savory, Digital filters for coherent optical receivers, Opt. Express 16 (2) (2008) 804–817.

[96] M.G. Taylor, Algorithms for coherent detection, in: Proc. OFC, OThL4, 2010.

[97] S.J. Savory, Digital coherent optical receivers: Algorithms and subsystems, J. Sel. Top. Quant. Electron. 16 (5) (2010) 1164–1179.

[98] G. Agrawal, Nonlinear Fiber Optics, fourth edition, Elsevier, San Diego, CA, 2006.

[99] M. Kuschnerov et al., Data-aided versus blind single-carrier coherent receivers, IEEE Photon. J. 2 (3) (2010) 387–403.

[100] S. Randel et al., 6 × 56-Gb/s mode-division multiplexed transmission over 33-km few-mode fiber enabled by 6 × 6 MIMO equalization, Opt. Express 19 (17) (2011) 16697–16707.

[101] E. Ip, J.M. Kahn, Feedforward carrier recovery for coherent optical communications, J. Lightwave Technol. 25 (9) (2007) 2675–2692 and Addendum to Feedforward carrier recovery for coherent optical communications, J. Lightwave Technol. 25 (13) (2009) 2552–2553

[102] N. Bai et al., Mode-division multiplexed transmission with inline few-mode fiber amplifier, Opt. Express 20 (3) (2012) 2668–2680.

[103] M.D. Feuer et al., Demonstration of joint DSP receivers for spatial superchannels, in: IEEE Photonics Society Summer Topical Meetings, MC4.2, 2012.

[104] C.B. Papadias, Globally convergent blind source separation based on a multiuser kurtosis maximization criterion, IEEE Trans. Signal Proc. 48 (12) (2000) 3508–3519.

[105] (a) V.A.J.M. Sleiffer et al., 73.7 Tb/s (96 × 3 × 256-Gb/s) mode-division multiplexed DP-16QAM transmission with inline MM-EDFA, in: Proc. ECOC, Th.3.C.4, 2012. (b) J.J. Shynk, Frequency-domain and multirate adaptive filtering, IEEE Signal Proc. Mag. 9 (1) (1992) 14–37.

[106] J. Kahn, K.P. Ho, Mode coupling effects in mode-division-multiplexed systems, in: IEEE Photonics Society Summer Topical Meetings, TuC 3.4, 2012.

[107] (a) B. Inan et al., DSP requirements for MIMO spatial multiplexed receivers, in: IEEE Photonics Society Summer Topical Meetings, MC 4.4, 2012. (b) N. Bai et al., Adaptive frequency-domain equalization for the transmission of the fundamental mode in a fewmode fiber, Opt. Express 20 (21) (2012) 24010–24017.

[108] W.H. Press, S.A. Teukolsky, W.T. Vetterling, B.P. Flannery, Numerical Recipes in C, New York, Cambridge, 1992.

[109] J. Leibrich, W. Rosenkranz, Frequency domain equalization with minimum complexity in coherent optical transmission, in: Proc. OFC, OWV1, 2010.

[110] C. Xia, Hole-assisted few-mode multi-core fiber for high density space division multiplexing, Photon. Technol. Lett. 24 (21) (2012) 1914–1917.

[111] R. Ryf et al., Mode-division multiplexing over 96 km of few-mode fiber using coherent 6 × 6 MIMO processing, J. Lightwave Technol. 30 (4) (2012) 521–531.

[112] M. Salsi et al., Transmission at 2 × 100Gb/s, over two modes of 40km-long prototype few-mode fiber, using LCOS based mode multiplexer and demultiplexer, in: Proc. OFC, PDPB9, 2011.

[113] J. Carpenter, T.D. Wilkinson, All optical degenerate mode-group multiplexing using a mode selective switch, in: IEEE Photonics Society Summer Topical Meetings, 2012, pp. 236–237.

[114] R. Ryf et al., Optical coupling components for spatial multiplexing in multi-mode fibers, in: Proc ECOC, Th.12.B.1, 2011.

[115] C.D. Poole et al., Optical fiber-based dispersion compensation using higher order modes near cutoff, J. Lightwave Technol. 12 (10) (1994) 1746–1758.

[116] I. Giles, Fiber LPG mode converters and mode selection technique for multimode SDM, Photon. Technol. Lett. 24 (21) (2012) 1922–1925.

[117] L. Grüner-Nielsen, J.W. Nicholson, Stable mode converter for conversion between LP01 and LP11 using a thermally induced long period grating, in: IEEE Photonics Society Summer Topical Meetings, WC1.2, 2012.

[118] R. Ryf, Spot-based mode coupler for mode-multiplexed transmission in few mode fiber, in: IEEE Photonics Society Summer Topical Meetings, 2012, pp. 199–200.

[119] N.K. Fontaine et al., Evaluation of photonic lanterns for lossless mode-multiplexing, in: Proc. ECOC, Th.2.D.6, 2012.

[120] S.K. Case, M.K. Han, Multi-mode holographic waveguide coupler, Opt. Commun. 15 (2) (1975) 306–307.

[121] G.C.G. Berkhout et al., Efficient sorting of orbital angular momentum states of light, Phys. Rev. Lett. 105 (2010) 153601.

[122] D. Gloge, Weakly guiding fibers, Appl. Opt. 10 (10) (1971) 2252–2258.

[123] H. Kogelnik, P.J. Winzer, Modal birefringence in weakly guiding fibers, J. Lightwave Technol. 30 (14) (2012) 2240–2245.

[124] W.Q. Thornburg et al., Selective launching of higher-order modes into an optical fiber with an optical phase shifter, Opt. Lett. 19 (7) (1994) 454–456.

[125] J. Carpenter, T.D. Wilkinson, Holographic mode generation for mode division multiplexing, in: Proc. OFC, JW2A.42, 2012.

[126] V.A. Soifer, M.A. Golub, Laser Beam Mode Selection by Computer Generated Holograms, CRC Press, 1994.

[127] R.J. Black, L. Gagnon, Optical Waveguide Modes: Polarization, Coupling and Symmetry, McGraw-Hill Professional, 2009.

[128] R. Ryf et al., Low-loss mode coupler for mode-multiplexed transmission in few-mode fiber, in: Proc. OFC, PDP5B.5, 2012.

[129] N.K. Fontaine et al., Space-division multiplexing and all optical MIMO demultiplexing using a photonic integrated circuit, in: Proc. OFC, PDP5B.1, 2012.

[130] W. Klaus, Free-space coupling optics for multi-core fibers, in: IEEE Photonics Society Summer Topical Meetings, WC3.3, 2012.

[131] J. Bromage, Raman amplification for fiber communications systems, J. Lightwave Technol. 22 (1) (2004) 79–93.

[132] G. Nykolak et al., An Erbium-doped multimode optical fiber amplifier, Photon. Technol. Lett. 3 (12) (1991) 1079–1081.

[133] N. Bai et al., Multimode fiber amplifier with tunable modal gain using a reconfigurable multimode pump, Opt. Express 19 (17) (2011) 16601–16611.

[134] E. Ip, Gain equalization for few-mode fiber amplifiers with more than two propagating mode groups, in: IEEE Photonics Society Summer Topical Meetings, 2012, pp. 224–225.

[135] S. Alam, Modal gain equalization in a few moded erbium-doped fiber amplifier, in: IEEE Photonics Society Summer Topical Meetings, 2012, pp. 218–219.

[136] D. Askarov, J.M. Kahn, Design of multi-mode erbium-doped fiber amplifiers for low mode-dependent gain, in: IEEE Photonics Society Summer Topical Meetings, 2012, pp. 220–221.

[137] K.S. Abedin et al., Cladding-pumped erbium-doped multicore fiber amplifier, Opt. Express 20 (18) (2012) 20191–20200.

[138] H. Takahashi et al., First demonstration of MC-EDFA-repeatered SDM transmission of 40 × 128-Gbit/s PDM-QPSK signals per core over 6,160-km 7-core MCF, in: Proc. ECOC, Th.3.C.3, 2012.

[139] K. Tsujikawa, Optical fiber amplifier employing a bundle of reduced cladding erbium-doped fibers for multi-core fiber transmission, in: IEEE Photonics Society Summer Topical Meetings, 2012, pp. 228–229.

[140] Y. Yamauchi, A highly-efficient remotely-pumped multi-core EDFA transmission system with a novel hybrid wavelength/space-division multiplexing scheme, in: Proc. OECC, 2012, pp. 479–480.

[141] A.R. Shah et al., Coherent optical MIMO (COMIMO), J. Lightwave Technol. 23 (8) (2005) 2410–2419.

[142] A. Li et al., Reception of mode and polarization multiplexed 107Gb/s CO-OFDM signal over a two-mode fiber, in: Proc. OFC, PDPB8, 2011.

[143] N. Hanzawa et al., Demonstration of mode-division multiplexing transmission over 10 km two-mode fiber with mode coupler, in: Proc. OFC, OWA4, 2011.

[144] R. Ryf et al., Space division multiplexing over 10 km of three-mode fiber using coherent 6 × 6 MIMO processing, in: Proc. OFC, PDPB10, 2011.

[145] R. Ryf et al., Spot based mode couplers for mode multiplexed transmission in few mode fiber, Photon. Technol. Lett. 24 (21) (2012) 1973–1976.

[146] R. Ryf et al., Mode-multiplexed transmission over a 209-km dgd-compensated hybrid few-mode fiber span, Photon. Technol. Lett. 24 (21) (2012) 1965–1968.

[147] R. Ryf et al., MIMO-based crosstalk suppression in spatially multiplexed 3 × 56-Gb/s PDM-QPSK signals for strongly-coupled 3-core fiber, Photon. Technol. Lett. 23 (20) (2011) 1469–1471.

[148] R. Ryf et al., Coherent 1200-km 6 × 6 MIMO mode-multiplexed transmission over 3-core microstructured fiber, in: Proc. ECOC, Th.13.C.1, 2011.

[149] C. Xia et al., Supermodes for optical transmission, Opt. Express 19 (17) (2011) 16653–16664.

[150] R. Ryf et al., Impulse response analysis of coupled-core 3-core fibers, in: Proc. ECOC, Mo.1.F.4, 2012.

Mode Coupling and its Impact on Spatially Multiplexed Systems

11

Keang-Po Ho[a] and Joseph M. Kahn[b]

[a]*Silicon Image, Sunnyvale, CA 94085, USA*
[b]*E.L. Ginzton Laboratory, Department of Electrical Engineering, Stanford University, Stanford, CA 94305, USA*

11.1 INTRODUCTION

Recent developments in spatially multiplexed optical communication systems demand a deeper understanding of mode coupling effects in fibers. Spatial multiplexing is being considered for long-haul systems using coherent detection [1–6] or short-range systems using direct detection [7–9]. It increases transmission capacity by multiplexing several data signals in the cores of multicore fibers (MCFs) or in the modes of multimode fibers (MMFs), in which case, it is often called mode-division multiplexing (MDM). Index perturbations in fibers, whether intended or not, can induce coupling between signals in different modes, and can cause propagating fields to evolve randomly. Mode coupling may be classified as weak or strong, depending on whether the total system length is comparable to, or much longer than, a length scale over which propagating fields remain correlated. Mode coupling can affect MDM systems in several important ways.

First, mode coupling, whether occurring in transmission fibers [10,11] or in modal (de)multiplexers [12], leads to crosstalk between spatially multiplexed signals. In direct-detection systems, mode coupling must either be avoided by careful design of all these components, or mitigated by adaptive optical signal processing [13–15]. In systems using coherent detection, any linear crosstalk between modes can be compensated fully by multi-input multi-output (MIMO) digital signal processing (DSP) [1–4], but DSP complexity increases with an increasing number of modes. If mode coupling can be restricted to occur only within mode groups having nearly degenerate propagation constants [7,8], then DSP complexity may be reduced by processing each mode group separately [1].

Second, mode coupling substantially affects the end-to-end group delay (GD) spread of a system [10,16,17], which substantially affects the complexity of MIMO DSP. The GD spread determines the temporal memory required in MIMO time-domain equalization (TDE) of single-carrier modulation [1,18–20], while it determines the fast Fourier transform (FFT) block length in MIMO frequency-domain equalization (FDE) of single-carrier modulation [18] or in orthogonal

Optical Fiber Telecommunications VIB. http://dx.doi.org/10.1016/B978-0-12-396960-6.00011-0

frequency-division multiplexing (OFDM) [2,18]. When using FDE or OFDM with an optimized FFT block length, the DSP complexity per two-dimensional information symbol depends very weakly on the GD spread [18]. Nevertheless, the overall DSP circuit size and convergence time of an adaptive FDE scale roughly linearly with the GD spread or FFT block length [18], and unless strong mode coupling is used to significantly reduce the GD spread, long-haul spatially multiplexed systems may not be technically feasible.

Third, transmission fibers [21,22] and inline optical amplifiers [23–25] can introduce mode-dependent loss or gain, which we collectively refer to as MDL. MDL causes random variations of the powers of signals propagating in different modes. These power variations may affect the various frequency components of each signal differently, and may change over time. Like multipath fading in a MIMO wireless system [26,27], these power variations cause MIMO system capacity to become a random variable. As a result, the mean capacity may be reduced. Moreover, at a given point in time, the instantaneous capacity may drop below the transmission rate, causing system outage [28,29]. Strong mode coupling can reduce the power variations and the associated capacity fluctuations. In conjunction with modal dispersion (MD), mode coupling creates frequency diversity that can further reduce the capacity fluctuations and thus reduce the outage probability [30].

Mode coupling can be described by field coupling models [31, ch. 3], which account for complex-valued modal electric field amplitudes, or by power coupling models [31, ch. 5], a simplified description that accounts only for real-valued modal powers. Early MMF systems used incoherent light emitting diode sources, and power coupling models were used widely to describe important properties, including steady-state modal power distributions and fiber impulse responses [32–34]. Most recent MMF systems use lasers, but power coupling models are still used to describe important effects, including reduced GD spreads in plastic MMFs [16] and a filling-in of the impulse response [35]. By contrast, virtually all practical SMF systems have used laser sources. The study of random birefringence and mode coupling in SMF, which leads to polarization-mode dispersion (PMD), has always used field coupling models, which predict the existence of principal states of polarization (PSPs) [36–38]. PSPs are polarization states that undergo minimal dispersion, and which form the basis of optical compensation of PMD in direct-detection SMF systems [39,40]. In recent years, field coupling models have been applied to MMF, predicting minimally dispersive principal modes (PMs) [17,41], which are the basis for optical compensation of MD in direct-detection MMF systems [13,14]. With heightened recent interest in spatial multiplexing, field coupling models have been applied to study crosstalk in MCFs [42,43], the statistics of coupled GDs in MMF [10,19], and the statistics and system impact of coupled MDL in MMF [28–30].

In this chapter, we review mode coupling effects, how they are modeled, their effect on key fiber properties such as MD and MDL, and their impact on the performance and complexity of MDM systems. The remainder of this chapter is as follows. In Section 11.2, we review modes in optical fibers and the physical sources of coupling between modes. We then compare field- and power-coupling models, present a matrix propagation model used throughout the chapter, and discuss regimes of weak and strong mode coupling. In Section 11.3, we study MD in the weak- and strong-coupling regimes, especially the statistics of strongly coupled GDs and the GD spread. In Section 11.4, we study MDL, describing its statistics in the strong-coupling regime for both narrow- and wide-band systems. In Section 11.5, we briefly describe MDM in direct-detection systems. In Section 11.6, we discuss long-haul MDM systems using coherent detection, addressing the impact of MDL on average capacity and outage probability, and the effect of MD on DSP complexity.

11.2 MODES AND MODE COUPLING IN OPTICAL FIBERS

In this section, we discuss mode coupling, including its physical origins, models used to describe it, and regimes of weak and strong coupling. We begin the section by a brief review of modes in optical fibers.

11.2.1 Modes in optical fibers

Optical fibers are cylindrical waveguides comprising one (or possibly several) cores surrounded by a cladding having a slightly lower refractive index, as shown in Figure 11.1 [44–47]. A mode is a solution of a wave equation describing a field distribution that propagates in a fiber without changing, except for an overall scaling that describes amplitude and phase changes [44–49]. Throughout this chapter, we assume that a fiber supports D propagating modes, including spatial and polarization degrees of freedom. In fibers of circular cross section, in the limit of weak guidance (small index difference between core and cladding), the total number of propagating modes can

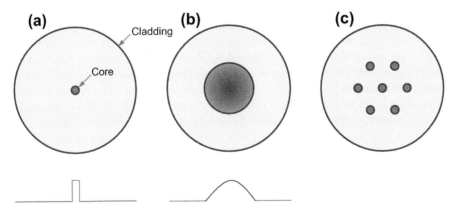

FIGURE 11.1 Different fiber types: cross sections (upper row) and typical refractive index profiles (lower row): (a) step-index single-mode fiber, (b) graded-index multimode fiber (MMF), and (c) multicore fiber.

assume a value $D = 2, 6, 10, 12, 16, 20, 24, 30, \ldots$ [49,50, Table 2.2]. In these weakly guiding fibers, there exist groups of modes having degenerate propagation constants or modal indices (including both phase and group indices) [8,49,50, Table 2.2]. When the refractive index difference between core and cladding is non-zero, the degeneracies within each mode group are generally broken slightly. Nevertheless, in typical glass fibers, the propagation coefficient differences within a mode group are usually small compared to the propagation coefficient differences between mode groups.

When the core radius and core-cladding index difference are sufficiently small, a fiber supports propagation of only one spatial mode in two polarizations ($D = 2$), and is called an SMF. An SMF having a step-index profile is shown in Figure 11.1a; other index profiles are also possible [48,50, Section 3.5]. Signals propagating in SMF are subject to chromatic dispersion (CD) arising both from material dispersion and waveguide dispersion [48,51,52], and are subject to PMD arising from random coupled birefringence [36–40], but are not subject to MD. In SMF, a change of the input launch conditions affects the launched power, but does not affect the spatial distribution at the fiber output, provided the fiber is sufficiently long (typically at least 1 m).

When the core radius and core-cladding index difference becomes sufficiently large, a fiber supports multiple spatial modes, and is called an MMF. Assuming the number of propagating modes is large, it is given approximately by $D \approx (2\pi/\lambda)^2 \int_0^R [n^2(r) - n_c^2]^{1/2} r \, dr$, where $n(r)$ describes the radial variation of the core index, n_c is the cladding index, R is the core radius, and λ is the wavelength [47]. Typically, the number of propagating modes is roughly proportional to the core-cladding index difference times $(R/\lambda)^2$.

MMF is subject to MD, whereby different spatial modes propagate with different modal group indices or GDs. MD causes signal distortion in direct-detection links. MD is severe in fibers with step-index profiles, while it is reduced substantially in fibers using nearly parabolic graded-index profiles, as shown in Figure 11.1b. Conventional glass MMFs for data centers or local-area networks have core diameters of 50 or 62.5 μm and support over one hundred modes at an 850-nm wavelength. Various optical multimode categories OM1 through OM4 [53] support increasing bandwidth × distance products, with bit rates of 10 Gb/s over distances up to 500 m. In an MMF, the output spatial distribution and the impulse response depend on the input launch conditions. In particular, excitation of higher-order modes can be enhanced by launching a beam with a transverse offset from the center axis [7,32,33,54] or with an angular offset [54]. Plastic optical fibers having core diameters of 0.5–1 mm [55,56] are used in short-range links. In these plastic fibers, index inhomogeneity causes strong mode coupling, which significantly reduces the GD spread and increases the supportable bit rate [16,57,58]. However, high losses in plastic optical fibers (33 dB/km is the minimum value given in [56]) limit link distances to under 100 m [56].

The plurality of modes propagating in MMF has long been viewed as a negative effect that limits the bit rate × distance product [33,53]. Attempts have been made to exploit multiple modes to increase capacity in short-range direct-detection links [7–9], but have not yet become practical, in part, because they did not address the problem of MD. More recently, MDM has yielded capacity increases in long-haul systems using coherent detection [1–4], by exploiting MIMO signal processing to compensate both MD and modal crosstalk. Analogously, multipath fading in wireless communications was long considered a negative effect [59, chs. 13–14,60, Section 5.5.1.2], but multipath is now exploited to increase system throughput in MIMO wireless systems [26,27].

As an alternative to MDM in MMF, spatial multiplexing can be performed in a MCF [5,6], as shown in Figure 11.1c. A MCF can be designed for very small crosstalk between cores [61,62], making it analogous to an array of SMFs embedded in the same cladding. Alternatively, a MCF can be designed with strong coupling between cores [42,43], making it equivalent to a special type of MMF in which super-modes propagate [63]. Our treatment of coupled modes in MMF is applicable to super-modes in strongly coupled MCF, where the total number of modes D is equal to twice the number of cores (assuming single-mode cores).

11.2.2 Mode coupling and its origins

In an ideal fiber, modes propagate without cross-coupling. In a real fiber, perturbations, whether intended or unintended, can induce coupling between spatial and/or polarization modes. Throughout this chapter, we consider only coupling between forward-propagating modes, since it has a dominant effect on the system properties of interest, including MD and MDL.

In multimode transmission fibers, unintended mode coupling can arise from several sources. These include manufacturing variations causing non-circularity of the core, roughness at the core-cladding boundary, variations in the core radius, or variations in the index profile in graded-index fibers. They also include stresses induced by the jacket [64,65], or by thermal mismatches between glasses of different compositions. Finally, mode coupling can arise from micro-bending, macro-bending, or twists [66–68].

Most unintended random perturbations of transmission fibers are having longitudinal power spectra that are lowpass, e.g. scaling as $\langle |F(\Delta\beta)|^2 \rangle \propto \Delta\beta^{-4}$ to $\Delta\beta^{-8}$ [11]. This notation is defined in Sections 11.2.3.1 and 11.2.3.2. As explained there, such lowpass perturbations can strongly couple modes having nearly equal propagation constants (small $\Delta\beta$), while they weakly couple modes having highly unequal propagation constants (large $\Delta\beta$). Therefore, in glass MMFs, nearly degenerate polarization modes or spatial modes within the same group are fully coupled after distances of order 300 m [32,33]. By contrast, even in fibers with low GD spread, spatial modes in different groups are only partially coupled after distances of order 100 km [1,69,70]. Hence, if full, strong mode coupling is desired, it may be necessary to introduce intentional perturbations to the fiber having longitudinal power spectra designed to couple different mode groups. These would be analogous to the spinning used to reduce the differential GD in SMF with PMD [48,71–74]. In Section 11.6, it is shown that strong mode coupling may be required to ensure the practicality of long-haul MDM systems using coherent detection.

In long-haul MDM systems, MDL in transmission fibers is likely to be weak [1], so MDL is expected to arise mainly from inline optical amplifiers [23–25]. In order to minimize the impact of MDL from inline amplifiers, it is sufficient to have a correlation length equal to the amplifier spacing [28]. If mode coupling in transmission fibers proves insufficient, a mode scrambler can be placed in each amplification node. Mode scramblers can be implemented using micro-bending [66–68] or by splicing together unlike fiber types [75,76].

In long-haul MDM systems, a modal multiplexer need not map each input data stream to a distinct uncoupled mode, but need only provide an approximately unitary mapping from the input data streams to those modes, which can imply significant mode coupling [12,77] (analogous considerations apply to a demultiplexer). Because such coupling occurs only at the beginning or end of a link, it cannot by itself reduce the correlation length below the total link length, i.e. it is not sufficient to achieve strong mode coupling.

11.2.3 Mode coupling models

Mode coupling can be described by field coupling models [31, chs. 3–4,78], which account for complex-valued modal electric fields, or by power coupling models [31, ch. 5,11,79,80], a simplified description that accounts only for real-valued modal powers. Both classes of models describe a coupling of energy between modes and how the coupling depends on the mode fields and on a perturbations inducing mode coupling. The transverse dependences of the mode fields and perturbation determine selection rules, while the longitudinal dependence of the perturbation governs phase matching.

Field coupling models, since they describe complex field amplitudes, can describe changes to the eigenmodes and corresponding eigenvalues of certain operators of interest that are caused by mode coupling. In the case of PMD in SMF, field coupling models describe how the PSPs and their GDs are affected by mode coupling [36–40]. PSPs form the basis for optical PMD compensation [39,40]. Likewise, in MMF, field coupling models describe PMs and how their GDs are affected by mode coupling [10,13,17,41], and explain an observed polarization dependence of the impulse response [13,81]. PMs form the basis for optical signal processing to avoid or compensate MD [13,15]. As described below, field coupling models can describe various other important properties of MMF and MDM systems.

Power coupling models also describe an exchange of energy among modes and how it depends on the mode fields and on perturbations inducing mode coupling [11,33,79]. Because they describe coupling as a diffusion process (non-negative real modal powers are coupled by non-negative real coefficients), these models do not describe the changes in eigenmodes and their associated eigenvalues caused by mode coupling. Likewise, power coupling models do not describe changes of eigenmodes or coupling coefficients with optical frequency [30,82].

Both field coupling and power coupling models appeared early in the study of MMF [31]. Early MMF systems used spatially and temporally incoherent light emitting diodes, and power coupling models were used to describe steady-state power distributions [11] and impulse responses [32,33]. Although more recent MMF systems use lasers, power coupling models have been used to explain reduced GD spreads in plastic fibers [16,57] and a filling-in of impulse responses in glass fibers [35].

Our brief review of field and power coupling models follows that in [31]. For simplicity, we expand propagating fields in a basis of ideal modes of an unperturbed fiber [31, Section 3.2]. This approach is suitable for perturbations of refractive index or geometry that represent small deviations from an ideal fiber. An alternate approach is to expand propagating fields in terms of local normal modes, which are eigenmodes that depend on the local refractive index and geometry [31, Section 3.3]. While more complicated, this approach is suitable for geometrical perturbations, such as bends or tapers, which represent large deviations from an ideal fiber, provided they vary slowly along the fiber's length. Also for simplicity, we assume weak guidance and assume the fields can be described as purely transverse. While including longitudinal components generally improves accuracy in detailed calculations, by ignoring them, we are able to obtain correct first-order insights in many problems. Important exceptions are known, e.g. longitudinal components are required to model coupling between orthogonal polarizations caused by elliptical core deformations [31, Section 4.5].

11.2.3.1 Field coupling models

We assume that light propagates along the z direction and let (x, y) define a transverse plane. The (square of the) unperturbed refractive index profile is $n_0^2(x, y)$, which is assumed to be independent of z. By solving a wave equation that depends on $n_0^2(x, y)$, we obtain a set of D orthonormal ideal propagating modal fields $\mathbf{E}_\mu(x, y), \mu = 1, \ldots, D$, which have propagation constants $\beta_\mu, \mu = 1, \ldots, D$. In this basis, any propagating field can be expanded as $\mathbf{E}(x, y, z) = \sum_{\mu=1}^{D} A_\mu(z)\mathbf{E}_\mu(x, y)$, where $A_\mu(z), \mu = 1, \ldots, D$ are complex-valued coefficients describing the amplitude and phase of each propagating mode. The quantities $\mathbf{E}(x, y, z), \mathbf{E}_\mu(x, y)$ and $A_\mu(z), \mu = 1, \ldots, D$ are all functions of angular frequency ω, but we suppress the frequency dependence for simplicity throughout this section.

If the fiber's refractive index profile (or geometry) is perturbed, since the $\mathbf{E}_\mu(x, y)$ are eigenmodes of the unperturbed fiber, they are coupled by the perturbation. If loss is neglected, propagation and coupling are described by the field coupling equations [31]:

$$\frac{dA_\mu}{dz} = -j\beta_\mu A_\mu + \sum_{v \neq \mu} C_{\mu v}(z)A_v \quad \mu = 1, \ldots, D. \tag{11.1}$$

On the right-hand side, the first term describes uncoupled propagation, and the second term describes coupling. As an alternative, the propagating field may be expanded as $\mathbf{E}(x, y) = \sum_\mu A_\mu(z)\mathbf{E}_\mu(x, y)\exp(-j\beta_\mu z)$, which leads to coupling equations similar to (11.1) without the first term of the right-hand side [78].

In order to illustrate some key principles governing spatial mode coupling, we assume the refractive index is modified by a small perturbation that can be factored to separate the transverse and longitudinal dependences. The (square of the) total index becomes:

$$n^2(x, y, z) = n_0^2(x, y) + \delta n^2(x, y) f(z). \tag{11.2}$$

Assuming a perturbation of the form (11.2), on the right-hand side of (11.1), the complex-valued field coupling coefficient is proportional to:

$$C_{\mu\nu}(z) \propto \int_{-\infty}^{\infty} \int_{-\infty}^{\infty} \delta n^2(x, y) \mathbf{E}_{\mu}^*(x, y) \cdot \mathbf{E}_{\nu}(x, y) dx dy \cdot f(z), \tag{11.3}$$

where the proportionality constant is given in [31, Section 3.2]. The first factor in (11.3) is an overlap integral describing how the transverse dependence of the index perturbation couples different modes depending on their field distributions, and it determines selection rules for mode coupling. For example, a perturbation independent of (x, y) cannot couple modes of opposite parities, while a perturbation depending linearly on x or y can couple only modes of opposite parities. It is important to realize that the form of the perturbation (11.2) is hardly general. Being isotropic, it cannot couple spatial modes in orthogonal polarizations. A perturbation of the dielectric tensor is required to describe birefringence caused by stress anisotropy, which is a major cause of polarization coupling in nominally circular fibers.

In order to explain the effect of the second factor $f(z)$ in (11.3), it is instructive to solve the coupled-mode Eq. (11.1). Instead of trying to solve them in general, we examine two special cases [31], which yield considerable insight.

As a first example, we assume that at $z=0$, only one mode ν is excited, and that as the modes propagate, the other modes $\mu \neq \nu$ are only weakly excited:

$$A_\nu(0) = 1, \quad A_\mu(0) = 0, \quad |A_\mu(z)| \ll |A_\nu(z)|, \quad \mu \neq \nu. \tag{11.4}$$

After propagating to $z = L$, the amplitude of mode $\mu \neq \nu$ is:

$$A_\mu(L) \approx e^{-j\beta_\mu L} \int_0^L C_{\mu\nu}(z) e^{-j(\beta_\nu - \beta_\mu)z} dz. \tag{11.5}$$

The right-hand side of (11.5) comprises two factors. The first describes the propagation phase. The second describes how the longitudinal dependence of the index perturbation couples different modes depending on their propagation constants, and defines conditions for phase matching. The integral in (11.5) is the Fourier transform of $C_{\mu\nu}(z)$ given by (11.3), which is proportional to the Fourier transform of $f(z)$, denoted as $F(\Delta\beta)$. The Fourier transform is computed over the interval $(0, L)$ and evaluated at $\Delta\beta = \beta_\nu - \beta_\mu$, the difference between the two propagation constants. As a consequence, nearly degenerate modes (small $\Delta\beta$) may be coupled efficiently by a lowpass $f(z)$, whereas strongly non-degenerate modes (large $\Delta\beta$) are coupled efficiently only by a broadband or resonant $f(z)$ that has significant Fourier components at the large value of $\Delta\beta$. Analogous principles are used in the design of spin profiles for manufacturing low-PMD SMF [73,74]. This example has assumed that the index perturbation is perfectly separable and that only one mode is excited strongly. Nevertheless, even when these conditions are not satisfied strictly, the dependence of pairwise mode coupling implied by the second factor in (11.5) often remains valid qualitatively. Note that the coupling coefficient in (11.5) also includes the integral over (x, y) defined in (11.3), and hence is subject to the selection rules described above.

As a second example, we consider a system with two modes and a perturbation independent of z which, according to (11.3), induces a constant coupling coefficient. The field coupling Eq. (11.1) becomes:

$$\frac{dA_1}{dz} = -j\beta_1 A_1 + C_{12}A_2, \quad \frac{dA_2}{dz} = -j\beta_2 A_2 - C_{12}^* A_1. \tag{11.6}$$

Assuming initial conditions $A_1(0) = 1$ and $A_2(0) = 0$, the solution becomes

$$A_1(z) = \exp\left(-j\frac{\beta_1 + \beta_2}{2}z\right)\left(\cos \gamma z - j\frac{\Delta\beta}{2\gamma}\sin \gamma z\right), \tag{11.7}$$

$$A_2(z) = \frac{C_{12}}{\gamma}\exp\left(-j\frac{\beta_1 + \beta_2}{2}z\right)\sin \gamma z, \tag{11.8}$$

where

$$\gamma^2 = \left(\frac{\Delta\beta}{2}\right)^2 + |C_{12}|^2 \quad \text{and} \quad \Delta\beta = \beta_1 - \beta_2.$$

The solutions (11.7) and (11.8) describe fields propagating with an average propagation constant $\frac{1}{2}(\beta_1 + \beta_2)$. A fraction of the total energy is coupled back and forth between the two modes, as described by the factors $\cos \gamma z$ and $\sin \gamma z$. The fraction of coupled energy depends on the ratio $|C_{12}/\Delta\beta|$, becoming small for $|C_{12}| \ll \Delta\beta$, and approaching unity for $|C_{12}| \gg \Delta\beta$. This observation helps explain how even weak perturbations can fully couple nearly degenerate modes, such as the two polarizations in SMF [83].

11.2.3.2 Power coupling models

For many purposes, such as computing the redistribution of average power among modes [11,79,80], it is sufficient to describe evolution of the modal powers $P_\mu(z) = \langle |A_\mu(z)|^2 \rangle, \mu = 1, \ldots, D$, which are non-negative and real. The brackets $\langle \rangle$ denote an ensemble average. Their evolution is described by the power coupling equations [31, ch. 5]:

$$\frac{dP_\mu}{dz} = -\alpha_\mu P_\mu + \sum_{\nu \neq \mu} h_{\mu\nu}(P_\nu - P_\mu), \quad \mu = 1, \ldots, D. \tag{11.9}$$

On the right-hand side, the first term describes loss by power attenuation coefficients α_μ, and the second term describes coupling by non-negative real coefficients $h_{\mu\nu}$. Let us assume a perturbed index profile of the form (11.2), except that the longitudinal dependence $f(z)$ is generally considered a stationary random process. The coupling coefficients are:

$$h_{\mu\nu} = \left\langle \left| \int_0^L C_{\mu\nu}(z) \, e^{-j(\beta_\nu - \beta_\mu)z} dz \right|^2 \right\rangle, \tag{11.10}$$

which is the power spectrum of the coupling coefficient $C_{\mu\nu}(z)$ (11.3), which is proportional to the power spectrum of $f(z)$, denoted by $\langle |F(\Delta\beta)|^2 \rangle$. The power spectrum is evaluated at $\Delta\beta = \beta_\mu - \beta_\nu$, the difference between the two propagation constants. This describes phase matching conditions, and can be interpreted as a stochastic version of (11.5), which involves the Fourier transform of $C_{\mu\nu}(z)$ or $f(z)$. Note that the coupling coefficients (11.10) also include the squared modulus of the integral over (x, y) defined in (11.3), and hence is subject to selection rules, as described above.

In the simple case of two modes, the power coupling equations become

$$\frac{dP_1}{dz} = -\alpha P_1 + h_{12}(P_2 - P_1), \frac{dP_2}{dz} = -\alpha P_2 + h_{12}(P_1 - P_2), \tag{11.11}$$

where we have assumed equal attenuation coefficients for the two modes. If the initial conditions are $P_1(0) = 1$ and $P_2(0) = 0$, the solution is

$$P_1(z) = \tfrac{1}{2} \exp(-\alpha z) \left[1 + \exp(-2h_{12}z) \right],$$
$$P_2(z) = \tfrac{1}{2} \exp(-\alpha z) \left[1 - \exp(-2h_{12}z) \right]. \tag{11.12}$$

Unlike the field coupling result for a fixed, uniform perturbation, for any non-zero coupling coefficient, given sufficient propagation distance, the optical power distribution in the two modes always becomes uniform. This apparent inconsistency can be resolved by observing that, in practice, even a nominally fixed, uniform perturbation is subject to random fluctuations (especially of optical phase) that vary over time and space. Hence, in modeling field coupling, a long fiber should be subdivided into short segments over which the perturbation is very nearly uniform, and modeled for an ensemble of different random phase shifts between the segments. When this procedure is carried out, the field coupling model reproduces the ensemble-average result obtained using the power coupling model.

In the remainder of this chapter, only field coupling models are used to analyze MDM systems, as these models can account for the evolution of complex-valued modal amplitudes, enabling study of how mode coupling affects the eigenmodes and eigenvalues of coupled systems, such as the GDs or gains/losses of coupled modes. Power coupling models can provide partial explanations of some relevant phenomena (e.g. in Section 11.3.1), but it is not clear how power coupling models can be extended to provide the quantitative predictions required for MDM systems.

11.2.3.3 Matrix propagation model

As mentioned above, a field at angular frequency ω propagating in a fiber can be represented as $\mathbf{E}(x, y, z, \omega) = \sum_{\mu=1}^D A_\mu(z, \omega)\mathbf{E}_\mu(x, y, \omega)$, where the fields $\mathbf{E}_\mu(x, y, \omega)$, $\mu = 1, \ldots, D$, are the orthonormal eigenmodes of an unperturbed fiber (this expression remains valid even if the fields are not strictly transverse). The eigenmodes $\mathbf{E}_\mu(x, y, \omega)$ are fixed, so it is convenient to represent the field by a vector $\mathbf{A}(z, \omega) = (A_1(z, \omega), \ldots, A_D(z, \omega))^T$, which we often write simply as $\mathbf{A}(\omega)$. Ignoring noise for now, linear propagation through an optical system can generally be described by a matrix equation:

$$\mathbf{A}^{(\text{out})}(\omega) = \mathbf{M}(\omega)\mathbf{A}^{(\text{in})}(\omega), \qquad (11.13)$$

where $\mathbf{A}^{(\text{in})}(\omega)$ and $\mathbf{A}^{(\text{out})}(\omega)$ describe the input and output fields and $\mathbf{M}(\omega)$ is a propagation operator, which is described by a $D \times D$ matrix.

In preparation for the treatment of MD and MDL in Sections 11.3 and 11.4, we discuss how to model a cascade of K passive fiber sections enumerated by $k = 1, \ldots, K$. Assume the kth section has length $L^{(k)}$ which, for now, has no specific relationship to the mode coupling correlation length. Uncoupled propagation of the modes in the kth section is described by the input-output relationship

$$A_\mu^{(\text{out})} = \exp\left[-\frac{\alpha_\mu^{(k)}}{2} L^{(k)} - j\beta_\mu^{(k)}(\omega) L^{(k)} \right] A_\mu^{(\text{in})}, \quad \mu = 1, \ldots, D, \quad (11.14)$$

where $\alpha_\mu^{(k)}$ are the power attenuation coefficients, which may be mode-dependent [21,22] but are assumed independent of frequency over the bandwidth of interest, and $\beta_\mu^{(k)}(\omega)$ are the propagation constants, which are generally mode-dependent [31, chs. 1,44,46,47,84].

To describe uncoupled power gain in the kth section, we define the mode-averaged attenuation constant $\bar{\alpha}^{(k)} = \frac{1}{D} \sum_\mu \alpha_\mu^{(k)}$. We quantify MDL by the uncoupled gains $g_\mu^{(k)} = -\left(\alpha_\mu^{(k)} - \bar{\alpha}^{(k)} \right) L^{(k)}, \mu = 1, \ldots, D$, where $\sum_\mu g_\mu^{(k)} = 0$. The vector

$$\mathbf{g}^{(k)} = \left(g_1^{(k)}, g_2^{(k)}, \ldots, g_D^{(k)} \right) \qquad (11.15)$$

describes the uncoupled MDL in the kth section.

To describe uncoupled MD and CD in the kth section, we perform a Taylor series expansion of the propagation constant $\beta_\mu^{(k)}(\omega)$ and keep the terms linear and quadratic in ω. MD is described by the uncoupled GDs $\beta_{1,\mu}^{(k)} L^{(k)}, \mu = 1, \ldots, D$, where $\beta_{1,\mu}^{(k)} = d\beta_\mu^{(k)}(\omega)/d\omega$. The mode-averaged uncoupled GD is $\bar{\beta}_1^{(k)} L^{(k)} = \frac{L^{(k)}}{D} \sum_\mu \beta_{1,\mu}^{(k)}$. We quantify MD by the uncoupled GDs $\tau_\mu^{(k)} = \left(\beta_{1,\mu}^{(k)} - \bar{\beta}_1^{(k)} \right) L^{(k)}, \mu = 1, \ldots, D$, where $\sum_\mu \tau_\mu^{(k)} = 0$. The vector

$$\boldsymbol{\tau}^{(k)} = \left(\tau_1^{(k)}, \tau_2^{(k)}, \ldots, \tau_D^{(k)} \right) \qquad (11.16)$$

describes the uncoupled modal GDs in the kth section.

CD includes contributions from both material dispersion and waveguide dispersion. As wavelength increases, any mode has more power in the cladding, where the refractive index is lower, so waveguide dispersion is always negative. Waveguide dispersion is enhanced for modes near cutoff [85–87]. When all modes are well above cutoff, which is desirable for low loss and good bend tolerance, mode-dependent CD is reduced [88].

In the kth section, CD is described by the parameters

$$\beta_{2,\mu}^{(k)} = d^2\beta_\mu^{(k)}(\omega)/d\omega^2 = \left(-\lambda^2/2\pi c\right) D_\mu^{(k)}, \quad \mu = 1,\ldots,D,$$

where $D_\mu^{(k)} = d\left(1/v_{g,\mu}^{(k)}\right)/d\lambda$ is the dispersion coefficient of mode μ. The mode-average CD parameter is $\bar{\beta}_2^{(k)} = \frac{1}{D}\sum_\mu \beta_{2,\mu}^{(k)}$, and mode-dependent CD is described by $\Delta\beta_{2,\mu}^{(k)} = \beta_{2,\mu}^{(k)} - \bar{\beta}_2^{(k)}$, where $\sum_\mu \Delta\beta_{2,\mu}^{(k)} = 0$.

Including MDL, MD, and mode-dependent CD, uncoupled propagation in the kth section is described by the diagonal matrix:

$$\Lambda^{(k)}(\omega) = \begin{pmatrix} e^{\frac{1}{2}g_1^{(k)} - j\omega\tau_1^{(k)} - \frac{j}{2}\omega^2\Delta\beta_{2,1}L^{(k)}} & & 0 \\ & \ddots & \\ 0 & & e^{\frac{1}{2}g_D^{(k)} - j\omega\tau_D^{(k)} - \frac{j}{2}\omega^2\Delta\beta_{2,D}L^{(k)}} \end{pmatrix}. \tag{11.17}$$

If it is necessary to include mode-averaged gain and dispersion, the diagonal matrix (11.17) can be multiplied by a constant factor:

$$\exp\left(-\frac{1}{2}\bar{\alpha}^{(k)}L^{(k)} - j\omega\bar{\beta}_1^{(k)}L^{(k)} - \frac{j}{2}\omega^2\bar{\beta}_2 L^{(k)}\right). \tag{11.18}$$

Including mode coupling, propagation in the kth section is modeled by a product of three $D \times D$ matrices:

$$\mathbf{M}^{(k)}(\omega) = \mathbf{V}^{(k)}\Lambda^{(k)}(\omega)\mathbf{U}^{(k)*}, \quad k = 1,\ldots,K, \tag{11.19}$$

where $*$ denotes Hermitian transpose. The $D \times D$ matrices $\mathbf{U}^{(k)}$ and $\mathbf{V}^{(k)}$ are frequency-independent unitary matrices representing mode coupling at the input and output of the kth section, respectively.

Finally, propagation through a cascade of K fiber sections is represented by a total propagation matrix $\mathbf{M}^{(t)}(\omega)$, which is the product of K matrices [10,17,19,28–30]:

$$\mathbf{M}^{(t)}(\omega) = \mathbf{M}^{(K)}(\omega)\mathbf{M}^{(K-1)}(\omega)\cdots\mathbf{M}^{(2)}(\omega)\mathbf{M}^{(1)}(\omega). \tag{11.20}$$

The matrix model (11.19) and (11.20) can describe signal propagation through any cascade of linear elements, as illustrated in Figure 11.2. For example, an optical amplifier can be described by a section in which $\bar{\alpha}^{(k)} < 0$, such that the factor $\exp\left(-\bar{\alpha}^{(k)}L^{(k)}\right) > 1$ describes the mode-averaged power gain. (De)multiplexers with or without crosstalk can be described easily.

Addition of noise by optical amplifiers cannot be described by the matrix model (11.17) and (11.19), and is discussed in Section 11.4.5.

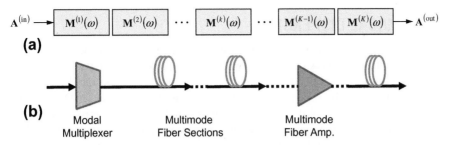

(a)

(b)

Modal Multiplexer

Multimode Fiber Sections

Multimode Fiber Amp.

FIGURE 11.2 (a) Matrix model for an MMF system described by the product of a cascade of matrices. (b) Each matrix in the cascade may represent a section of MMF, a modal multiplexer or demultiplexer, a multimode optical amplifier or other components.

11.2.3.4 Regimes of mode coupling

Different regimes of mode coupling are familiar from the study of randomly coupled birefringence and PMD [37,89], and apply as well to mode coupling in MMF [16,17,82]. These regimes are especially relevant in studying key effects such as MD (Section 11.3) or MDL (Section 11.4), where operators describing MD or MDL are defined, whose eigenvalues describe physical properties of interest, i.e. coupled modal GDs or coupled modal gains.

As stated above, modal fields are assumed to be strongly correlated over distances less than or equal to a correlation length, and weakly correlated over distances far larger than the correlation length. This correlation length is a generalization of the polarization correlation length defined in the study of PMD [83,90–92].

In the weak-coupling regime, the correlation length is comparable to, or slightly shorter than, the total system length. In this regime, signal propagation can be modeled using a small number of sections K, where each section should be slightly longer than the correlation length. If a larger number of sections K is employed, the unitary matrices $\mathbf{U}^{(k)}$ and $\mathbf{V}^{(k-1)}$ defined in (11.19) should be correlated to reflect correlation between sections. In the weak-coupling regime, the spread of eigenvalues describing quantities of interest, such as modal GDs or modal gains, scales linearly with the number of sections K or the total system length $L_t = \sum_k L^{(k)}$, and each coupled eigenmode is a linear combination of a small number of uncoupled modes.

In the strong-coupling regime, the correlation length is far shorter than the total system length. In this regime, signal propagation must be modeled using a large number of sections K, where each section should be slightly longer than the correlation length. The unitary matrices $\mathbf{U}^{(k)}$ and $\mathbf{V}^{(k-1)}$ should be statistically independent to ensure independence between sections. In the strong-coupling regime, the spread of eigenvalues describing quantities of interest, such as modal GDs or modal gains, scales with the square-root of the number of sections K or the square-root of the total system length L_t, and each coupled eigenmode is a linear combination of many uncoupled modes.

These regimes of mode coupling are further illustrated and compared in Section 11.3.1.

11.3 MODAL DISPERSION

In coherent MDM systems, MD poses no fundamental performance limitations. MIMO equalizer complexity increases with the GD spread [18–20], but if laser phase noise is sufficiently small and the channel does not change too rapidly, MD can be compensated with no penalty. To date, most long-haul coherent MDM systems [1–4,70] have operated in the weak-coupling regime. But as transmission distances and the number of multiplexed modes are increased, equalizer complexity is likely to become prohibitive unless strong mode coupling is exploited to reduce the GD spread.

In this section, field coupling models are used to study the effect of mode coupling on MD. PMs, which are coupled modes having well-defined GDs, are described. Of greatest relevance to long-haul MDM systems, the statistics of the GDs of the PMs in the strong-coupling regime are analyzed.

11.3.1 Coupled modal dispersion

The effect of mode coupling on MD was studied for the two-mode case in the context of PMD in SMF [36–40,89–91], where coupled polarization states having well-defined GDs were called PSPs. The theory was extended to the multimode case in [41], where coupled modes having well-defined GDs were called PMs. Following [41], we assume MDL is negligible (the combined impact of MD and MDL is studied in Sections 11.4.6 and 11.6.2). We further neglect mode-averaged loss or gain to simplify the notation here, although that assumption was not made in [38,41]. In this case, the total propagation operator defined in (11.20) becomes unitary: $\mathbf{M}^{(t)*}(\omega)\mathbf{M}^{(t)}(\omega) = \mathbf{I}$, where $\mathbf{I}$ is a $D \times D$ identity matrix. We restrict attention to optical signals occupying a narrow bandwidth near angular frequency ω. We define a set of D input PMs as a set of input field patterns defined by vectors $\mathbf{A}_\mu^{\mathrm{PM,(in)}}$, $\mu = 1, \ldots, D$. These input PMs are linear combinations of the uncoupled modes that have special properties. If an input PM is launched, the field at the fiber output is described by the corresponding output PM $\mathbf{A}_\mu^{\mathrm{PM,(out)}} = \mathbf{M}^{(t)}(\omega)\mathbf{A}_\mu^{\mathrm{PM,(in)}}$. Each of the input PMs is defined such that if we fix the input field pattern to be $\mathbf{A}_\mu^{\mathrm{PM,(in)}}$ and vary ω slightly, the output field pattern $\mathbf{A}_\mu^{\mathrm{PM,(out)}}$ remains unchanged to first order in ω. One can define a Hermitian GD operator:

$$\mathbf{G} = j\mathbf{M}_\omega^{(t)}\mathbf{M}^{(t)*}(\omega), \tag{11.21}$$

where $\mathbf{M}_\omega^{(t)} = d\mathbf{M}^{(t)}(\omega)/d\omega$ denotes differentiation with respect to ω. As shown in [41], the D input PMs are mutually orthogonal eigenmodes of the GD operator whose eigenvalues are the coupled GDs:

$$\mathbf{G}\mathbf{A}_\mu^{\mathrm{PM,(in)}} = \tau_\mu \mathbf{A}_\mu^{\mathrm{PM,(in)}}, \quad \mu = 1, \ldots, D. \tag{11.22}$$

In general, each input PM differs from the corresponding output PM, since the operators $\mathbf{M}^{(t)}$ and $\mathbf{G}$ do not commute.

For signals occupying a sufficiently narrow bandwidth near angular frequency ω, the overall input-output relationship of a fiber can be expressed as:

$$\mathbf{M}^{(t)}(\omega) = \mathbf{V}^{(t)}\mathbf{\Lambda}^{(t)}(\omega)\mathbf{U}^{(t)*}, \tag{11.23}$$

where $\mathbf{U}^{(t)}$ and $\mathbf{V}^{(t)}$ are unitary matrices that are independent of frequency (to first order), whose columns represent the input PMs and the output PMs, respectively, in the basis of ideal modes. The matrix

$$\mathbf{\Lambda}^{(t)}(\omega) = \begin{pmatrix} e^{-j\omega\tau_1} & & 0 \\ & \ddots & \\ 0 & & e^{-j\omega\tau_D} \end{pmatrix} \tag{11.24}$$

describes propagation of the PMs. The matrix (11.24) describes no crosstalk, since it is diagonal, and describes differential delay but no distortion, since its phase depends linearly on ω. Many factorizations of the form (11.23) are possible mathematically, but this choice is unique in yielding frequency-independent $\mathbf{U}^{(t)}$ and $\mathbf{V}^{(t)}$ and a diagonal $\mathbf{\Lambda}^{(t)}(\omega)$ with well-defined GDs.

In [17], the effect of spatial- and polarization-mode coupling on the modal GDs in graded-index MMF was studied. A fiber of fixed total length L_t was divided into a fixed number of sections K, and each section was given a random curvature to induce spatial-mode coupling and birefringence. Successive sections were rotated by a random angle to induce polarization-mode coupling. By changing the standard deviation (STD) of the curvature σ_K, different regimes of mode coupling can be obtained. Figure 11.3 shows the coupled GDs of the PMs in a 1-km-long, graded-index MMF supporting $D = 110$ modes in two polarizations. In the weak-coupling regime ($\sigma_K < 0.3$), modes form groups having GD degeneracies of 2, 4, 6, ..., and the GD spread scales linearly with K or L. In the medium-coupling regime ($0.3 < \sigma_K < 2$), these GD degeneracies are broken, and each coupled PM is a linear combination of a small number of uncoupled modes. In the strong-coupling regime ($2 < \sigma_K < 20$), the GD spread is reduced substantially, scaling in proportion to $\sqrt{K}$ or $\sqrt{L_t}$, and each coupled PM is a linear combination of many uncoupled modes. The reduction of the GD spread, while not commonly observed in glass MMFs of 1-km length, is analogous to that observed in plastic MMF. The regimes of mode coupling were explained previously using a power coupling model in [16]. The power coupling model, while intuitively appealing, may not be capable of providing the quantitative predictions required for MDM systems.

The field pattern of each PM varies with optical frequency ω, and is correlated over a coherence bandwidth that depends, in principle, on the uncoupled modes and GDs and the mode coupling [30,82]. In the absence of coupling and MDL, the PMs are essentially equivalent to the ideal modes [41], so the coherence bandwidth is extremely large, of the order of THz. In [17], the coherence bandwidths of the PMs

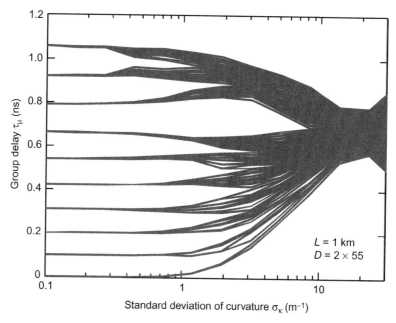

FIGURE 11.3 PM group delays (GDs) versus the standard deviation (STD) of curvature determining the mode coupling strength for a MMF of 1-km length, supporting $D = 110$ propagating modes. The GDs exhibit regimes of weak, moderate, and strong mode coupling. Adapted from [17].

in graded-index MMF were studied. For a 1-km fiber with weak coupling, coherence bandwidths were found to be about 300 GHz for a parabolic index profile, and smaller than 10 GHz for a non-parabolic profile. In the strong-coupling regime, following arguments similar to those given in Section 11.4.6 for coupled MDL, the coherence bandwidths of PMs should be of the order of $1/\sigma_{gd}$, where σ_{gd} is the overall STD of coupled GD, given by (11.45).

When a modulated signal occupies a bandwidth that is not small compared to the coherence bandwidth, higher-order MD effects occur [30,82]. The overall input-output relationship becomes more complicated than (11.23); in particular, the phase of the propagation operator is no longer a linear function of ω, as in (11.24). Higher-order MD effects include a filling-in between peaks in the pulse response of a fiber (observed in [35]), a nonlinear relationship between input and output intensity waveforms, polarization- and spatial mode-dependent CD, and enhanced depolarization of modulated signals, all of which have analogs in higher-order PMD [93,94]. Higher-order MD limits the signal bandwidth over which frequency-independent optical signal processing can avoid distortion from MD in direct-detection links [95] or avoid crosstalk in direct-detection MDM systems (see Section 11.5).

In long-haul coherent MDM systems, MD poses no fundamental performance limitations. The MIMO equalizer complexity increases with the GD spread due to MD [18–20], but in systems with sufficiently small laser phase noise and sufficiently slow channel variation over time, an arbitrarily large GD spread and arbitrary-order MD can be compensated with no penalty. Most long-haul coherent MDM systems demonstrated to date [1–4] have operated in the weak-coupling regime. Nevertheless, as shown below, strong coupling reduces GD spread and thus minimizes MIMO signal processing complexity, and it mitigates MDL. Since strong coupling is beneficial for long-haul coherent MDM systems, we analyze the statistics of strongly coupled GDs in the remainder of this section, and we analyze the statistics of strongly coupled MDL in Section 11.4.

11.3.2 Group delay statistics in strong-coupling regime

In the strong-coupling regime, MD in an MMF can be modeled by random matrices, and the distribution of the GDs can be characterized by studying the statistics of the eigenvalues of those random matrices [10,96].

11.3.2.1 Zero-trace Gaussian unitary ensemble

In the strong coupling regime, in the matrix model of Section 11.2.3.3, a MMF is modeled using many independent sections. Using the diagonal matrix (11.17) and neglecting MDL and mode-dependent CD, uncoupled propagation in the kth section can be modeled by the diagonal matrix:

$$\mathbf{\Lambda}^{(k)}(\omega) = \begin{pmatrix} e^{-j\omega\tau_1^{(k)}} & & 0 \\ & \ddots & \\ 0 & & e^{-j\omega\tau_D^{(k)}} \end{pmatrix}, \tag{11.25}$$

where the vector $\boldsymbol{\tau}^{(k)}$ (11.16) describes the uncoupled modal GDs.

In the absence of MDL, $\mathbf{\Lambda}^{(k)}(\omega)$, $\mathbf{U}^{(k)}$ and $\mathbf{V}^{(k)}$ are all unitary matrices, such that $\mathbf{M}^{(k)}(\omega)\mathbf{M}^{(k)*}(\omega) = \mathbf{I}$. With strong random mode coupling, following the discussion of Section 11.2.3.4, $\mathbf{U}^{(k)}$ and $\mathbf{V}^{(k)}$ can be assumed to be independent random unitary matrices, such that both input and output eigenvectors are independently oriented from one section to the next.

The model here is valid regardless of whether or not the modal delay vectors in each section $\boldsymbol{\tau}^{(k)} = \left(\tau_1^{(k)}, \tau_2^{(k)}, \ldots, \tau_D^{(k)}\right)$, $k = 1, \ldots, K$, (11.16) have the same statistical properties. The vector $\boldsymbol{\tau}^{(k)}$ may even be a deterministic vector, identical for each section. As in Section 11.2.3.3 and without loss of generality, we assume that $\sum_\mu \tau_\mu^{(k)} = 0$, ignoring the mode-averaged delay of each section that does not lead to MD.

In the kth section, the input-output relationship is

$$\mathbf{A}_k^{(\text{out})} = \mathbf{M}^{(k)}(\omega)\mathbf{A}_k^{(\text{in})}$$

and $\mathbf{A}_k^{(in)} = \mathbf{M}^{(k)*}(\omega)\mathbf{A}_k^{(out)}$. Using (11.22), the coupled GDs correspond to the eigenvalues of the GD operator $\mathbf{G}$ given by (11.21). With only a single section, we may verify that

$$\mathbf{G}^{(k)} = j\mathbf{M}_\omega^{(k)}\mathbf{M}^{(k)*} = \mathbf{V}^{(k)}\mathbf{T}^{(k)}\mathbf{V}^{(k)*}, \tag{11.26}$$

where $\mathbf{T}^{(k)} = \text{diag}\left[\tau_1^{(k)}, \tau_2^{(k)}, \dots, \tau_D^{(k)}\right]$ is a diagonal matrix of the uncoupled GDs in the kth section. With the constraint $\sum_\mu \tau_\mu^{(k)} = 0$, we have $\text{tr}\left(T^{(k)}\right) = \text{tr}\left(j\mathbf{M}_\omega^{(k)}\mathbf{M}^{(k)*}\right) = 0$. The GD operator matrices $\mathbf{G}^{(k)} = j\mathbf{M}_\omega^{(k)}\mathbf{M}^{(k)*}$, $k = 1, \dots, K$, are all Hermitian matrices with real eigenvalues given by $\mathbf{T}^{(k)}$.

Considering the overall propagation matrix $\mathbf{M}^{(t)}$, the overall PMs and their GDs correspond to the eigenvectors and eigenvalues of the overall GD operator $\mathbf{G} = j\mathbf{M}_\omega^{(t)}\mathbf{M}^{(t)*}(\omega)$. Because of the chain rule

$$\mathbf{M}_\omega^{(t)} = \mathbf{M}^{(K)}\cdots\mathbf{M}^{(2)}\mathbf{M}_\omega^{(1)} + \mathbf{M}^{(K)}\cdots\mathbf{M}_\omega^{(2)}\mathbf{M}^{(1)} + \cdots + \mathbf{M}_\omega^{(K)}\cdots\mathbf{M}^{(2)}\mathbf{M}^{(1)}, \tag{11.27}$$

we obtain

$$\mathbf{G} = j\mathbf{M}_\omega^{(K)}\mathbf{M}^{(K)*} + j\mathbf{M}^{(K)}\mathbf{M}_\omega^{(K-1)}\mathbf{M}^{(K-1)*}\mathbf{M}^{(K)*} + \\ \cdots + \mathbf{M}^{(K)}\cdots\mathbf{M}^{(2)}\mathbf{M}_\omega^{(1)}\mathbf{M}^{(1)*}\mathbf{M}^{(2)*}\cdots\mathbf{M}^{(K)*}. \tag{11.28}$$

The overall matrix $\mathbf{G}$ given by (11.28) is the summation of K random matrices. All those K random matrices are statistically identical to those of (11.26). Their eigenvectors are independent of each other. The first matrix $\mathbf{M}_\omega^{(K)}\mathbf{M}^{(K)*}$ has the same eigenvectors as $\mathbf{V}^{(K)}$. The second matrix $j\mathbf{M}^{(K)}\mathbf{M}_\omega^{(K-1)}\mathbf{M}^{(K-1)*}\mathbf{M}^{(K)*}$ has eigenvectors from $\mathbf{M}^{(K)}\mathbf{V}^{(K-1)}$. Both matrices $\mathbf{V}^{(K)}$ and $\mathbf{M}^{(K)}\mathbf{V}^{(K-1)}$ are unitary matrices that are obviously independent of each other. The matrix $\mathbf{M}^{(K)}\mathbf{V}^{(K-1)}$ is frequency-dependent. All K random matrices in (11.28) are independent, owing to the different directions of their independent eigenvectors. Excepting the first matrix $j\mathbf{M}_\omega^{(K)}\mathbf{M}^{(K)*}$, all other matrices are frequency-dependent, due to the frequency-dependent factor in $\mathbf{\Lambda}^{(k)}(\omega)$ (11.25).

The matrix elements of $\mathbf{G}$, $g_{\mu,\nu}$, $\mu, \nu = 1, \dots, D$, are the summation of K independent random variables. When the number of independent sections K is large, the matrix $\mathbf{G}$ is a random matrix whose entries are Gaussian random variables from the Central Limit Theorem. Because all K component matrices in (11.28) are Hermitian, $\mathbf{G}$ is also a Hermitian matrix. The diagonal elements $g_{\mu,\mu}$, $\mu = 1, \dots, D$, are all real Gaussian random variables. All off-diagonal elements $g_{\mu,\nu}$, $\mu \neq \nu$, $\mu, \nu = 1, \dots, D$, are complex Gaussian random variables with independent real and imaginary parts.

If the D vectors in $\mathbf{V}^{(k)}$ are assumed to be independent of each other, it can be shown that the variance of the matrix elements is

$$\sigma_g^2 = \frac{1}{D^2}\sum_{k=1}^K \left\|\mathbf{\tau}^{(k)}\right\|^2 = \frac{1}{D}\sum_{k=1}^K \sigma_{\tau^{(k)}}^2, \tag{11.29}$$

where $\sigma^2_{\tau(k)}$, $k = 1..., K$, are variances of the GDs in each section. If all K sections have statistically identical GD profiles, we have

$$\sigma^2_g = \frac{K}{D}\sigma^2_{\tau}, \tag{11.30}$$

where σ^2_{τ} are the GD variances in all sections.

The D vectors in $\mathbf{V}^{(k)}$ are not independent of each other, however, as the Dth vector is determined by other $D-1$ vectors due to the unitarity condition. The elements of the vector $\boldsymbol{\tau}^{(k)}$ also sum to zero due to the zero-trace condition. The variance of matrix elements σ^2_g is an averaged variance in which the diagonal elements have smaller variance than the non-diagonal elements.

In random matrix theory, the matrix $\mathbf{G}$ is described as zero-trace (or traceless) Gaussian unitary ensemble. Typically, a Gaussian unitary ensemble does not have any constraint aside from the variance of its matrix elements. However, in (11.28), all matrix components have zero-trace so that

$$\mathrm{tr}(\mathbf{G}) = 0. \tag{11.31}$$

In the strong mode coupling regime, the GDs in a MMF are statistically described by the eigenvalues of the zero-trace Gaussian unitary ensemble.

Assuming a Gaussian unitary ensemble of $\mathbf{A}$ without trace constraint, the diagonal elements are real Gaussian random variables with variance σ^2_a. The off-diagonal elements of $\mathbf{A}$ are complex Gaussian random variables with variance σ^2_a. Equivalently, the real and imaginary parts of the off-diagonal elements are independent of each other with variance $\sigma^2_a/2$. The corresponding zero-trace Gaussian unitary ensemble is equivalently the Gaussian random matrix $\mathbf{A} - \mathrm{tr}(\mathbf{A})\mathbf{I}/D$. The off-diagonal elements have variance σ^2_a but the diagonal elements have variance $\sigma^2_a(1 - 1/D)$. The variance of all matrix elements has an arithmetic average $\sigma^2_a(1 - 1/D^2)$.

For the zero-trace Gaussian unitary ensemble $\mathbf{G}$, numerical simulation shows that the diagonal elements have equal variance $\sigma^2_g(1 - 1/D)$ and the off-diagonal elements have equal variance $\sigma^2_g[1 + 1/D/(D - 1)]$, and the average variance is σ^2_g. If the zero-trace Gaussian unitary ensemble $\mathbf{G}$ is related to a Gaussian unitary ensemble $\mathbf{A}$ via $\mathbf{G} = \mathbf{A} - \mathrm{tr}(\mathbf{A})\mathbf{I}/D$, the Gaussian unitary ensemble $\mathbf{A}$ without trace constraint should have a variance $\sigma^2_g D^2/(D^2 - 1)$. This variance correction is required because the statistics of zero-trace Gaussian unitary ensemble are always derived based on the corresponding Gaussian unitary ensemble without trace constraint.

11.3.2.2 Group delay distribution

In the regime of strong mode coupling, at each single frequency, the PMs and their GDs are given by the eigenvectors and eigenvalues of the zero-trace Gaussian unitary ensemble (11.28). As above, the zero-trace Gaussian unitary ensemble $\mathbf{G}$ may be related to a Gaussian unitary ensemble without trace constraint $\mathbf{A}$ as $\mathbf{G} = \mathbf{A} - \mathrm{tr}(\mathbf{A})\mathbf{I}/D$. Without loss of generality, after normalization, the elements of $\mathbf{A}$ may be assumed to be zero-mean independent identically distributed Gaussian random variables with variance $\sigma_a^2 = 1/2$, similar to the classic normalization of [96, Section 3.3]. The diagonal elements of $\mathbf{A}$ are real with a variance of $\sigma_a^2 = 1/2$. The off-diagonal elements of $\mathbf{A}$ are complex Gaussian distributed with independent real and imaginary parts, each having a variance of 1/4.

The joint distribution for a Gaussian unitary ensemble without the zero-trace constraint is well known. The unordered joint probability density function of the eigenvalues of a $D \times D$ Gaussian unitary ensemble is [96, Section 3.3,97]:

$$p_{\mathrm{nc}}(x) = C_{D2}^{-1} \prod_{D \geqslant \mu \geqslant \nu \geqslant 1} (x_\mu - x_\nu)^2 \exp\left(-\sum_{\mu=1}^{D} x_\mu^2\right), \tag{11.32}$$

where

$$C_{D2} = \frac{\pi^{D/2}}{2^{D(D-1)/2}} \prod_{n=1}^{D} n!.$$

With suitable normalization, the random variables $x_\mu, \mu = 1, \ldots, D$, yield the GDs of the MMF with strong mode coupling. With the zero-trace constraint (11.31), the joint probability density function becomes

$$p_{\mathrm{zt}}(x) = \tilde{C}_{D2}^{-1} \delta\left(\sum_{\mu=1}^{D} x_\mu\right) \prod_{D \geqslant \mu \geqslant \nu \geqslant 1} (x_\mu - x_\nu)^2 \exp\left(-\sum_{\mu=1}^{D} x_\mu^2\right). \tag{11.33}$$

The constant $\tilde{C}_{D2}$ is not the same as C_{D2} and is found equal to $\tilde{C}_{D2} = C_{D2}/\sqrt{\pi D}$ later in this section.

The eigenvalue distribution for zero-trace Gaussian unitary ensemble $\mathbf{G}$ is given by

$$p_D(x_1) = \tilde{C}_{D2}^{-1} \int_{-\infty}^{+\infty} \cdots \int_{-\infty}^{+\infty} \delta\left(\sum_{\mu=1}^{D} x_\mu\right) \prod_{D \geqslant \mu \geqslant \nu \geqslant 1} (x_\mu - x_\nu)^2 \exp\left(-\sum_{\mu=1}^{D} x_\mu^2\right) dx_2 \cdots dx_D. \tag{11.34}$$

The $D \times D$ Vandermonde determinant gives [98, Section 4.6]

$$\det\left[x_\mu^{\nu-1}\right]_{\mu,\nu=1,2,\ldots,D} = \prod_{D \geqslant \mu \geqslant \nu \geqslant 1} (x_\mu - x_\nu),$$

where $\det[\cdot]$ denotes a determinant. Following the method of [96, Section 6.2], and directly from the properties of determinants, the Vandermonde determinant can be expressed in terms of Hermite polynomials as

$$\det\left[x_\mu^{\nu-1}\right]_{\mu,\nu=1,2,\ldots,D} = \det\left[\frac{1}{2^{\nu-1}} H_{\nu-1}(x_\mu)\right]_{\mu,\nu=1,2,\ldots,D}$$

$$= \det\left[\frac{1}{2^{\nu-1}} H_{\nu-1}(x_\mu + c_\nu)\right]_{\mu,\nu=1,2,\ldots,D}, \tag{11.35}$$

where $H_n(x)$ are Hermite polynomials and c_v are constants. The leading terms in the entries of the determinants (11.35) are all $x_\mu^{\nu-1}$. Using the Hermite polynomials, we obtain

$$p_{nc}(x) = \frac{1}{D!} \det[K_D(x_\mu, x_v)]_{\mu,v=1,2,\dots,D},$$ (11.36)

where

$$K_D(x, y) = \sum_{n=0}^{D-1} \frac{1}{2^n n! \sqrt{\pi}} H_n(x) H_n(y) \exp\left(-\frac{x^2 + y^2}{2}\right).$$ (11.37)

Without trace constraint, the eigenvalue distribution is given by $K_D(x,x)/D$ and $K_D(x,x)$ is called the correlation function [96, Section 6.2].

With zero-trace constraint, the joint density becomes

$$p_{zt}(x) = \frac{C_{D2}}{\tilde{C}_{D2}} \delta\left(\sum_{\mu=1}^{D} x_\mu\right) \det[k_D(x_\mu, x_v)]_{\mu,v=1,2,\dots,D}.$$ (11.38)

Similar to [99,100] but using Fourier instead of Laplace transform,

$$p_{zt}(x) = \frac{C_{D2}}{2\pi \tilde{C}_{D2} D!} \int_{-\infty}^{+\infty} \exp\left(i\omega \sum_{\mu=1}^{D} x_\mu\right) \det[K_D(x_\mu, x_v)]_{\mu,v=1,2,\dots,D} \, d\omega.$$ (11.39)

With some algebra, we obtain

$$p_{zt}(x) = \frac{C_{D2}}{2\pi \tilde{C}_{D2} D} \int_{-\infty}^{+\infty} \exp\left(-\frac{D}{4}\omega^2\right) \det\left[K_D\left(x_\mu + \frac{i\omega}{2}, x_v + \frac{i\omega}{2}\right)\right]_{\mu,v=1,2,\dots,D} \, d\omega.$$ (11.40)

In the above expression, the argument inside the Hermite polynomial changes from x_μ to $x_\mu + i\omega/2$ using the relation (11.35).

Similar to the method used to find the eigenvalue distribution, the probability density of the eigenvalues is given by

$$p_D(x) = \frac{C_{D2}}{2\pi \tilde{C}_{D2} D} \int_{-\infty}^{+\infty} K_D\left(x + \frac{i\omega}{2}, x + \frac{i\omega}{2}\right) \exp\left(-\frac{D}{4}\omega^2\right) d\omega$$ (11.41)

or

$$p_D(x) = \frac{C_{D2}}{2\pi \tilde{C}_{D2} D} \int_{-\infty}^{+\infty} \exp\left(-\frac{D}{4}\omega^2\right) \sum_{n=0}^{D-1} \frac{1}{\sqrt{\pi} 2^n n!} H_n^2\left(x + \frac{i\omega}{2}\right) d\omega.$$ (11.42)

We may first integrate (11.42) over x before integrating over ω to obtain $\tilde{C}_{D2} = C_{D2}/\sqrt{\pi D}$. The integration (11.42) becomes

$$p_D(x) = \frac{1}{2\sqrt{\pi D}} \int_{-\infty}^{+\infty} \exp\left(-\frac{D}{4}\omega^2 - \left(x + \frac{i\omega}{2}\right)^2\right) \sum_{n=0}^{D-1} \frac{1}{\sqrt{\pi} 2^n n!} H_n^2\left(x + \frac{i\omega}{2}\right) d\omega.$$

(11.43)

Using the substitution $s = i\omega$, the integration is very similar to the Mellin inversion formula for the Laplace transform [101] with an integration from $x - i\infty$ to $x + i\infty$. Based on the Laplace transform, we obtain

$$p_D(x) = \frac{\exp\left(-\frac{D}{D-1}x^2\right)}{\sqrt{\pi D(D-1)}} \sum_{n=0}^{D-1} \frac{1}{2^n n!} H_n^2\left(\frac{t}{2\sqrt{D-1}}\right)\Bigg|_{t^k \leftarrow (-1)^k H_k\left(\frac{Dx}{\sqrt{D-1}}\right)}.$$

(11.44)

In the probability density (11.44), the summation gives a $2(D-1)$-degree polynomial in t. The power t^k is algebraically substituted by the Hermite polynomial $(-1)^k H_k\left(Dx/\sqrt{D-1}\right)$.

The GD distribution (11.44) is valid for MMFs with various numbers of modes. The GD has zero mean and a variance $\frac{1}{2}(D - 1/D)$. Figure 11.4 compares $Dp_D(x)$ with the correlation function $K_D(x, x)$. Scaling the distribution (11.44) by a factor D enables comparison to $K_D(x, x)$. Figure 11.4 is plotted for $D = 2, 6, 12, 20, 30,$ and 64 modes. As the distribution is always symmetric with respect to the center, Figure 11.4 shows only the positive side of the curves.

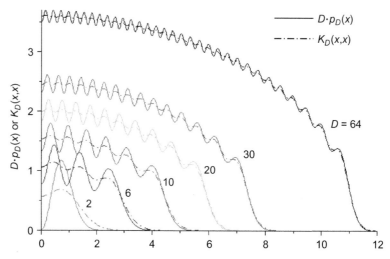

FIGURE 11.4 Probability density of the normalized GD (11.44) scaled by D, $Dp_D(x)$ shown as solid curves for $D = 2, 6, 12, 20, 30,$ and 64. The correlation functions $K_D(x, x)$ given by (11.37) are shown as dash-dotted curves for comparison.

Especially for large number of modes, the general curves of (11.44) are similar to the case without trace-constraint of $K_D(x, x)$. The number of peaks is the same as the number of modes D. Each peak corresponds to values of x where individual eigenvalues tend to be concentrated. There are more ripples with zero-trace constraint (11.31) as compared to the case without the constraint. Beyond the theory of this section, further theoretical study finds that the mean of each eigenvalue x_μ, assumed ordered, is not changed by the zero-trace constraint (11.31). However, the variance of the eigenvalues x_μ is always reduced by $\frac{1}{2}D^{-1}$. The variance reduction yields more localized eigenvalues.

As the probability density function (11.44) is derived from a Gaussian unitary ensemble in which all matrix elements have a variance of 1/2, if the elements of zero-trace Gaussian unitary ensemble **G** (11.28) have average variance σ_g^2, the Gaussian unitary ensemble **A** without zero-trace constraint has variance $\sigma_g^2 D^2/(D^2 - 1)$. Combining all normalization factors together, given a zero-trace Gaussian unitary ensemble **G** (11.28) with average variance σ_g^2, the GD variance is $\sigma_{gd}^2 = D\sigma_g^2$. Assuming all fiber sections have identical statistical properties and using (11.30) for σ_g^2, the overall GD variance is

$$\sigma_{gd}^2 = K\sigma_\tau^2. \tag{11.45}$$

Regardless of number of modes, the overall STD of GD among modes is always $\sigma_{gd} = \sqrt{K}\sigma_\tau$ and increases with the square-root of the number of sections. Equivalently, the overall GD increases with the square-root of system length. This result is consistent with the similar theory of PMD [36,37,89] and the results of [16,17] with strong mode coupling. Note that this is only relevant for fibers with strong mode coupling. As mentioned above, strong mode coupling is desirable to reduce the GD spread caused by MD.

The matrix **G** (11.28) is the summation of independent random matrices. The number of those random matrices is the same as the number of independent sections. Equivalently, the summation (11.28) may be considered the concatenation relationship for MD in MMF. For the case in which the number of independent sections in the fiber link is small, the matrix **G** may be calculated numerically to find the empirical GD distribution.

11.3.2.3 Few-mode fibers

Two-mode fiber is the simplest case, and may correspond to the two polarization modes in a SMF, i.e. the well-known PMD problem [36–38]. The purpose here is not to derive new properties of PMD, but to verify that the general random matrix model is applicable to PMD.

Using the density (11.44) for two-mode fiber, we obtain

$$p_2(x) = \sqrt{\frac{2}{\pi}}4x^2 \exp(-2x^2). \tag{11.46}$$

As in the PMD literature, we define $x_{1,2} = \pm\tau/2$ with τ as the differential GD with probability density

$$p_2(\tau) = \sqrt{\frac{2}{\pi}}\tau^2 \exp\left(-\frac{\tau^2}{2}\right), \quad \tau \geqslant 0, \tag{11.47}$$

which is the well-known Maxwellian distribution with normalized mean differential GD $\bar{\tau} = 2\sqrt{2/\pi} \approx 1.60$. Random matrix models specialized to the two-mode case were used to derive the Maxwellian distribution in [37,89,102].

For fibers with up to eight modes, direct integration of (11.34) has been used to find the GD distribution analytically, as given in [10,28].

Figure 11.5 compares the analytical distribution from (11.44) with simulation results for fibers with $D = 6$ modes. The simulation is conducted similar to that in [10]. The fiber has $K = 256$ independent sections. In each section, the six modes are chosen to have deterministic delays of $\pm\tau$, three of positive sign and three of negative sign, where $\tau = 1/\sqrt{K}$, to ensure that the overall STD of GD is unity, as from (11.45). This particular choice may not correspond to a physical fiber, but simplifies the simulation. The random unitary matrices $\mathbf{U}^{(k)}$, and $\mathbf{V}^{(k)}$, $k = 1, \ldots, K$, are first initialized by 6×6 random complex Gaussian matrices and then converted to unitary matrices using the Gram-Schmidt process [98, Section 5.2.8]. All sections have independent unitary matrices $\mathbf{U}^{(k)}$ and $\mathbf{V}^{(k)}$. A total of 300,000 eigenvalues are used in the curves shown in Figure 11.5.

In Figure 11.5, the simulation results show excellent agreement with the analytical eigenvalue probability density $p_6(x)$. Although the modes in each section have only two GDs, with strong mode coupling, a probability density function having six peaks is obtained. In the strong-coupling regime, similar results would be obtained using

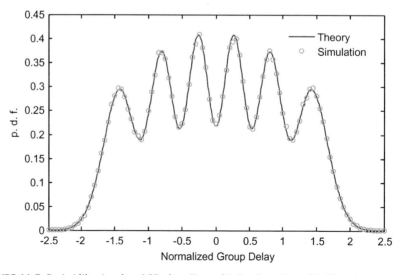

FIGURE 11.5 Probability density of GDs in a fiber with $D = 6$ modes, with GD variance normalized to unity. The curve represents the theoretical probability density (11.44) and the circles are from simulation.

any uncoupled GD vector in each section, provided that the six uncoupled GDs sum to zero and have a variance of $1/K$. To provide a more physically realistic example, four modes may have modal delays of $+\tau_1$ and the other two modes may have modal delays of $-2\tau_1$ with $\tau_1 = 1/\sqrt{2K}$. Similar results are obtained if the GDs in each section are, for example, $\pm\tau_2$, where τ_2 is random and follows a distribution with second moment $1/K$.

The simulation results in Figure 11.5 verify that zero-trace Gaussian unitary ensemble can be used to model the GD of few-mode fibers.

11.3.2.4 Many-mode fibers

With a large number of modes, a Gaussian unitary ensemble without the zero-trace constraint is described by a semicircle distribution with radius $\sqrt{2D}$ [96, Section 4.2]. With the normalization used in Sections 11.3.2.2 and 11.3.2.3, the variance of the eigenvalues is $D/2$. This semicircle law was first derived by Wigner for large random matrices [103,104]. The Wigner semicircle law is universally valid for many different types of large random matrices [105,106].

A Gaussian unitary ensemble, even with the zero-trace constraint, should follow the semicircle distribution [107,108]. As an alternative to considering **G** as a Gaussian unitary ensemble, a more straightforward derivation may use the Central Limit Theorem for free random variables.

In free probability theory, free random variables are equivalent to statistically independent large random matrices [109,110]. The Central Limit Theorem for the summation of free random variables gives the semicircle distribution [109,111]. The matrix **G** (11.28) is the summation of many independent random matrices.

The Central Limit Theorem for free random variables states the following: let $\mathbf{X}_k$, $k = 1, \dots, K$, be identically distributed independent zero-mean free random variables with unit variance, the summation

$$\mathbf{Y}_K = \frac{\mathbf{X}_1 + \mathbf{X}_2 + \cdots \mathbf{X}_K}{\sqrt{K}} \tag{11.48}$$

is described by semicircle distribution with radius of two and unit variance:

$$p_Y(r) = \begin{cases} \frac{1}{2\pi}\sqrt{4 - r^2} & |r| < 2, \\ 0 & \text{otherwise}, \end{cases} \tag{11.49}$$

as K approaches infinity.

In the Central Limit Theorem of free random variables, when free random variables are represented by large random matrices, the distribution of the free random variables is equivalent to the distribution of the eigenvalues of the random matrices. When the theorem is applied to **G** given by (11.28), if the variance of the zero-mean GD per section is σ_τ^2 for all K sections, the eigenvalues of **G** are described by a semicircle distribution with radius $2\sqrt{K}\sigma_\tau$ and variance $K\sigma_\tau^2$. Equivalently, the GD of the MMF has a semicircle distribution with variance $K\sigma_\tau^2$. The normalization used in this section based on the eigenvalues of $\mathbf{X}_k$ and $\mathbf{Y}_K$ in (11.48) is customary in free probability theory. However, the normalization used in Section 11.3.2.2 is based on the matrix elements of **G**, similar to that in Mehta [96].

Figure 11.6 compares the simulated marginal probability density of GD in fibers having $D = 16, 64$, and 512 modes to the semicircle distribution. In Figure 11.6a, the probability density for $D = 16$ modes from (11.44) is also shown for comparison. Each MMF is comprised of $K = 256$ sections. In each section, the uncoupled GDs are deterministic, with the first $D/2$ modes with a delay of τ and the other $D/2$ modes with a delay of $-\tau$. For normalization purposes, $\tau = 1/\sqrt{K}$ is chosen to facilitate comparison with a semicircle distribution with radius 2 and unit variance. The simulated curves are obtained from 1,600,000 and 640,000 and 102,400 eigenvalues for $D = 16$ and 64 and 512, respectively. The model here is valid as long as the component matrices in (11.28) may be modeled as free random variables [111].

In Figure 11.6, the simulated distributions match the semicircle distribution well for $D = 64$ and 512 modes. For a fiber having $D = 16$ modes, the distribution is close to a semicircle distribution, but has an obvious periodic structure with 16 peaks, consistent with the probability density (11.44). The ripples become less obvious as D increases from 16 to 64 to 512. Upon close examination of the curve for $D = 64$, the ripples seem periodic, consistent with the ripples in Figure 11.4.

The GD relationship $\sigma_{gd}^2 = K\sigma_\tau^2$ (11.45) remains valid when the number of modes D is very large. With a large number of modes, the relationship (11.45) can be derived directly from free probability theory.

The semicircle distribution, which describes the GDs in fibers with an infinite number of modes has strict upper and lower limits, so such fibers have a strictly bounded GD spread, given by $4\sigma_{gd}$. In fibers with large but finite number of modes D, it will be sufficient to accommodate a GD spread just slightly larger than $4\sigma_{gd}$. In the next section, the statistics of the GD spread of fibers with a finite number of modes is studied.

11.3.3 Statistics of group delay spread

As shown in Section 11.6.3.3, the complexity of MIMO equalization depends on the GD spread of the fiber [18,19]. This GD spread is equivalent to the difference between the maximum and minimum eigenvalues of a zero-trace Gaussian unitary ensemble. Comparing $\mathbf{G}$ and $\mathbf{A}$ related by $\mathbf{G} = \mathbf{A} - \mathrm{tr}(\mathbf{A})\mathbf{I}/D$, the difference between maximum and minimum eigenvalues is the same for both $\mathbf{G}$ and $\mathbf{A}$ with and without zero-trace constraint. The GD spread may be studied by the Gaussian unitary ensemble without trace constraint. As discussed earlier, the average variance of the matrix elements of $\mathbf{A}$ is a factor of $D^2/(D^2 - 1)$ larger than that of $\mathbf{G}$. The difference between the maximum and minimum eigenvalues for a Gaussian unitary ensemble $\mathbf{A}$ without trace constraint is studied here.

Many works have studied the maximum eigenvalue of Gaussian unitary ensemble, especially when the dimension of $\mathbf{G}$ approaches infinity. For many-mode fibers, the GD spread is approximately $4\sigma_{gd}$, as mentioned in Section 11.3.2.4.

For a finite-dimensional Gaussian unitary ensemble, the GD spread may be studied using the Fredholm determinant [96, Section 20.1], which was first derived to study integral equations. The solution of an integral equation gives numerical

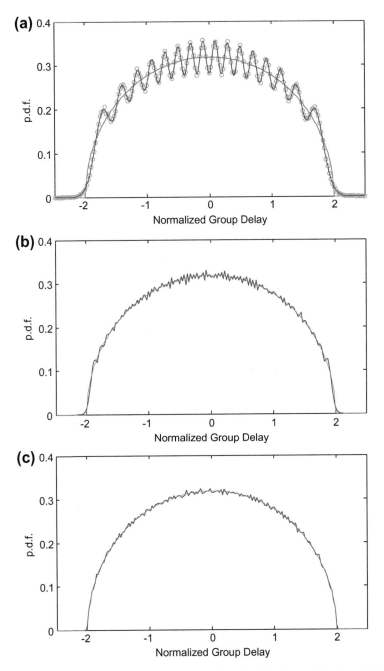

FIGURE 11.6 GD distribution compared to semicircle distribution. (a) $D = 16$: solid line with ripples from (11.44), circles from simulation. (b) $D = 64$: solid line with ripples from simulation. (c) $D = 512$: solid line with ripples from simulation. Adapted from [10].

values of the Fredholm determinant [112,113]. The joint probability density for the maximum and minimum eigenvalues is given by [114]

$$F^{(D)}(x, y) = \Pr(\lambda_{\max} \leqslant x, \lambda_{\min} \geqslant y) = E_2(0, J) = \det[1 - K_D(x, y)|_J],$$

$$(11.50)$$

where $J = (-\infty, y) \bigcup (x, +\infty)$ specifies a region, $E_2(0, J)$ denotes no eigenvalue in the region J, and $\det\left[1 - K_D(x, y)|_J\right]$ is the Fredholm determinant for the kernel $K_D(x, y)$ given by (11.37). The corresponding integral equation involves the kernel $K_D(x, y)$ integrated over the complement of J. The function $F^{(D)}(x, y)$ can be found by numerically solving the corresponding integral equation [113,114].

The GD spread distribution has a cumulative distribution of

$$\Pr(\lambda_{\max} - \lambda_{\min} \leqslant x) = -\int_{-\infty}^{+\infty} \frac{\partial F^{(D)}(t, y)}{\partial y}\bigg|_{t=y+x} dy. \qquad (11.51)$$

Figure 11.7 shows the complementary cumulative distribution $\Pr(\lambda_{\max} - \lambda_{\min} > x)$ for fibers with $D = 6$ and 12 modes. In the curves of Figure 11.7, the overall GD

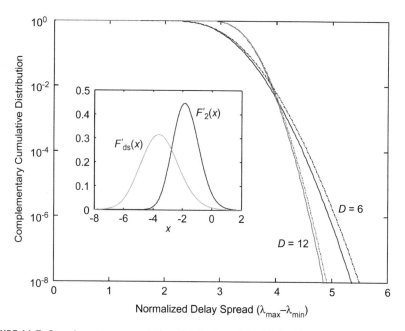

FIGURE 11.7 Complementary cumulative distribution of the GD for fibers with $D = 6$ and $D = 12$ modes, with GD variance normalized to unity. The solid curves are obtained by evaluating the Fredholm determinant (11.51) using a toolbox from [114]. The dash-dotted curves are derived using the approximation (11.54). The inset shows the Tracy-Widom distributions $F_2'(x)$ and the GD spread distribution $F_{ds}'(x)$.

variance σ_{gd}^2 is normalized to unity. Figure 11.7 is consistent with the empirical simulations in [18]. The complementary cumulative probability $\text{Pr}\,(\lambda_{max} - \lambda_{min} > x)$ corresponds to the event that the normalized GD spread of the fiber exceeds a value x, which typically describes the normalized temporal memory length of an equalizer for compensating MD. Here, $\text{Pr}\,(\lambda_{max} - \lambda_{min} > x)$ is referred to here as the "GD spread outage probability," although system outage does not necessarily occur when the fiber GD spread exceeds the equalizer length.

For many-mode fibers, the distribution of the GD spread can be approximated very accurately. For very large matrices, the maximum eigenvalue is described by the celebrated Tracy-Widom distribution [115,116]. For the Gaussian unitary ensemble described by the distribution (11.32), the largest eigenvalue is approximately $\sqrt{2D}$. The Tracy-Widom distribution for a Gaussian unitary ensemble is given by

$$\lim_{D \to \infty} \text{Pr}\left(\frac{\lambda_{max} - \sqrt{2D}}{2^{-1/2}D^{-1/6}} \le x\right) = F_2(x), \tag{11.52}$$

where λ_{max} denotes the maximum eigenvalue and $F_2(x)$ is the cumulative density function for the Tracy-Widom distribution, with numerical values given in [113,114].

For many-mode fibers, as shown in [117], the maximum and minimum eigenvalues are independent of each other. The GD spread for a large Gaussian unitary ensemble is given by

$$\lim_{D \to \infty} \text{Pr}\left(\frac{\lambda_{max} - \lambda_{min} - 2\sqrt{2D}}{2^{-1/2}D^{-1/6}} \le x\right) = F_{ds}(x) = \int_{-\infty}^{+\infty} F_2(x - t)\frac{d}{dt}F_2(t)dt \tag{11.53}$$

as the summation of two independent Tracy-Widom random variables. The inset of Figure 11.7 shows the Tracy-Widom distribution $F_2'(x)$ and the corresponding GD spread distribution $F_{ds}'(x)$.

Figure 11.7 also shows the approximate complementary cumulative distribution computed using (11.53):

$$\text{Pr}(\lambda_{max} - \lambda_{min} > x) \approx 1 - F_{ds}\left[D^{2/3}(x\sqrt{1 - D^{-2}} - 4)\right] \tag{11.54}$$

for fibers with $D = 6$ and 12 modes with GD variance normalized to unity. The approximation (11.54) follows directly from (11.53) for all values of D and takes into account that the GD variance for the statistics corresponding to (11.53) is $\frac{1}{2}(D - 1/D)$. The GD spread outage probabilities in Figure 11.7 govern the temporal memory length required for MD equalization. For typical outage probabilities in the range of 10^{-4}–10^{-6}, the equalizer length should be about 4 to 6 times σ_{gd}, the STD of GD.

The approximation based on the sum of two independent Tracy-Widom random variables always over-estimates the GD spread outage probability, but is sufficiently accurate for most engineering purposes for $D \ge 12$. The approximation (11.54) shows that as $D \to \infty$, the delay spread is upper-bounded by four times the STD of GD. In this limit, the GD follows a Wigner semicircle distribution from Section 11.3.2.4, which has a finite support equal to four times the STD.

The Tracy-Widom distribution has a mean of -1.77, and the normalized GD spread has a mean of approximately

$$\frac{4 - 3.54D^{-2/3}}{\sqrt{1 - D^{-2}}}. \tag{11.55}$$

The mean GD spread (11.55) approaches 4 as D increases. The Tracy-Widom distribution has an STD of 0.902 and the normalized GD spread has an STD of approximately

$$\frac{1.28D^{-2/3}}{\sqrt{1 - D^{-2}}}. \tag{11.56}$$

The GD spread STD (11.56) decreases to zero as $D^{-2/3}$ as D increases.

In Section 11.6.3.3 below, the GD spread outage probabilities in Figure 11.7 are used in quantifying the complexity of equalizers for compensating MD.

11.4 MODE-DEPENDENT LOSS AND GAIN

Multimode transmission fibers [21,22] or passive components may introduce loss that is mode-dependent, while multimode optical amplifiers [23–25] may introduce mode-dependent gain. Throughout this chapter, these effects are referred to collectively as MDL. MDL is likely to be relatively unimportant in short-range systems without amplifiers, but long-haul MDM systems with many cascaded amplifiers may accumulate significant amounts of MDL. Unlike MD, MDL is fundamentally a performance-limiting factor. In the extreme case, MDL is equivalent to a reduction in the number of propagating modes, leading to a proportional decrease in overall data rate or channel capacity.

Like the MD studied in Section 11.3, MDL can be studied using the matrix propagation model of Section 11.2.3.3. The overall propagation matrix $\mathbf{M}^{(t)}$ (11.20) describes coupled propagation of signals from input to output in an MMF, while $\mathbf{M}^{(t)*}$ describes propagation from output to input. We can define a round-trip propagation operator $\mathbf{M}^{(t)*}\mathbf{M}^{(t)}$ whose eigenmodes are the electric field profiles that, when propagated from the input to the output and then back to the input, remain unchanged except for a scaling describing gain or loss. The eigenvalues of $\mathbf{M}^{(t)*}\mathbf{M}^{(t)}$ describe the power gains or losses of these eigenmodes, and the square-root of the eigenvalues describe the scaling of the electric field amplitudes. MDL can obviously be defined equivalently in terms of the operator $\mathbf{M}^{(t)}\mathbf{M}^{(t)*}$. Since its eigenvalues describe modal power gains, we may refer to $\mathbf{M}^{(t)*}\mathbf{M}^{(t)}$ as a modal gain operator.

For polarization-dependent loss (PDL) in SMF ($D = 2$), the two eigenmodes are the modes having maximum and minimum gains. For MDL in MMF, the eigenmodes represent directions along which the gain is an extremum, which is either a maximum or minimum when approached from different orientations. For $D = 3$, MDL may be visualized in a three-dimensional space using a triaxial ellipsoid with three unequal semi-axes, where the distance from the origin to a point on the surface represents modal gain. The directions of maximal and minimal gain are obvious. The triaxial ellipsoid has an extremum, which can be either a maximum or minimum, depending on the direction from which it is approached. This extreme point corresponds to the third eigenmode. Without MDL, the ellipsoid degenerates to a sphere, and any three orthogonal directions may be chosen as eigenmodes.

Like the GD spread caused by MD, the variation of loss caused by MDL is minimized by strong mode coupling. In the remainder of this section, we study the statistics of MDL in the strong-coupling regime. In this regime, the propagation operator of an MMF can be represented by the product of many independent random matrices of the form (11.19) which converges to a limiting form whose statistics depend only on the number of modes and the accumulated MDL. In this regime, the accumulated MDL increases with the square-root of the number of independent matrices or the square-root of the total system length. However, the overall MDL exhibits a nonlinear dependence on the accumulated MDL.

11.4.1 Statistics of strongly coupled mode-dependent gains and losses

Based on the Central Limit Theorem for the product of random matrices [28], MDL follows a limiting distribution that depends only on the number of modes and a single additional parameter. In the regime of small MDL, the following two results are valid.

Proposition I: In the strong-coupling regime, when the overall MDL is small, the distribution of the overall MDL (measured in units of the logarithm of power gain or decibels) is identical to the eigenvalue distribution of zero-trace Gaussian unitary ensemble.

Proposition II: In the strong-coupling regime, when the overall MDL is small, the STD of the overall MDL σ_{mdl} depends solely on the square-root of the accumulated MDL variance ξ (often referred to simply as accumulated MDL) via:

$$\sigma_{mdl} = \xi\sqrt{1 + \frac{1}{12}\xi^2}. \tag{11.57}$$

If the MMF comprises K independent, statistically identical sections, each with MDL variance σ_g^2, the accumulated MDL is $\xi = \sqrt{K}\sigma_g$. In (11.57), σ_{mdl} and ξ are measured in units of the logarithm of power gain. Quantities measured in units of log power gain can be converted to decibels by multiplying by $\gamma = 10/\ln 10 \approx 4.34$, i.e. $\sigma_{mdl}(dB) = \gamma\sigma_{mdl}$ (log power gain). Unless noted otherwise, all expressions in this chapter assume that both ξ and σ_{mdl} are expressed in log power gain units.

Proposition I describes the shape of the MDL distribution, which is the same as (11.44) or Figure 11.4, whose variance depends on the variance of a Gaussian unitary ensemble used to describe MDL, just as such an ensemble was used to describe MD in Section 11.3. MDL has the same statistics as MD.

Propositions I and II have not been proven rigorously. Using numerical simulation, we have found that in the small-MDL region, the shape of the overall MDL distribution essentially depends only on the dimension of the zero-trace Gaussian unitary ensemble, which is the number of modes. For systems with overall MDL in the range of practical interest, $\sigma_{mdl} \leqslant 12$ dB or $\xi = \sqrt{K}\sigma_g \leqslant 10$ dB, the overall MDL (measured in decibels or log power gain) has a distribution very close to the eigenvalue distribution of zero-trace Gaussian unitary ensemble.

In Proposition II, the STD of the overall MDL depends solely on the accumulated MDL $\xi = \sqrt{K}\sigma_g$. For MD in MMF (see Section 11.3) or for PMD in SMF [36,37,89], the overall modal GD spread depends *linearly* on $\sqrt{K}\sigma_\tau$ (11.45). For MDL in MMF, the overall MDL depends *nonlinearly* on $\xi = \sqrt{K}\sigma_g$, as described by (11.57). The nonlinear relationship in Proposition II cannot be proven rigorously in the general case, but related results can be derived analytically in the limit of many modes. Proposition II has been tested by comparing (11.57) to numerical simulations. The approximation (11.57) is highly accurate for two-mode fibers with $\xi = \sqrt{K}\sigma_g$ up to 10 dB, and the region of its validity increases with an increasing number of modes.

In the region with $\xi = \sqrt{K}\sigma_g$ larger than 10 dB, which is beyond the range of practical interest, the overall MDL measured in log power gain units has the same statistical properties as the eigenvalues of the summation of two matrices [118]

$$\xi \mathbf{G} + \kappa_D \xi^2 \mathbf{F}, \tag{11.58}$$

where $\mathbf{G}$ is a zero-trace Gaussian unitary ensemble similar to that used to describe MD in Section 11.3, except it has unit eigenvalue variance, $\mathbf{F}$ is a deterministic uniform matrix, and $\kappa_D = \frac{1}{2}D/(1 + D)$ is a constant lying between 1/3 and 1/2, depending on number of modes D. The uniform matrix $\mathbf{F}$ has its eigenvalues deterministically and uniformly distributed between ± 1. For $D = 2$, an example is $\mathbf{F} = $ diag$[1,-1]$ or any other Hermitian matrix having the same eigenvalues. From the theory of MD in Section 11.3.2.1, the summation (11.58) represents the concatenation of two MMFs: the first with strong mode coupling represented by $\mathbf{G}$ and the second with deterministic and uniform MD represented by $\mathbf{F}$.

Comparing (11.57) with (11.58), the Gaussian unitary ensemble $\mathbf{G}$ gives the linear term ξ (or the 1 inside the square-root) to the overall MDL. With $\kappa_\infty = 1/2$, the uniform matrix $\mathbf{F}$ gives the nonlinear factor ($\xi^2/12$ inside the square-root). To a certain extent, Propositions I and II approximate the zero-trace uniform matrix $\mathbf{F}$ using a zero-trace Gaussian unitary ensemble. The approximation in Propositions I and II is sufficiently accurate for practical purposes.

Using $\kappa_D = \frac{1}{2}D/(1 + D)$ and (11.58), the accuracy of Proposition II may be improved slightly for few-mode fibers. The improved Proposition II is

$$\sigma_{mdl} = \xi \sqrt{1 + \frac{\xi^2}{12(1 - D^{-2})}}, \tag{11.59}$$

which gives more accurate values of overall MDL for $D = 2, 3$ but gives results similar to (11.57) for $D \geqslant 4$. The improvement in (11.59) is useful only in the large-MDL regime.

11.4.2 Model for mode-dependent loss and gain

In the strong-coupling regime, a MMF is divided into K independent sections, with each section modeled as a random matrix, similar to that in Section 11.3. The length of each section should be at least equal to the correlation length, such that each section can be considered independent of the others [10,28–30].

The fiber system is described by a product of matrices of the form (11.20) with each matrix given by the product of three matrices of the form (11.19). Mode-dependent CD is neglected. Initially, in modeling MDL at one frequency, the frequency dependence of MDL is neglected, so in the diagonal matrix (11.17), the modal delay vector (11.16) is set to zero, and only the modal gain vector $\mathbf{g}^{(k)} = \left(g_1^{(k)}, g_2^{(k)}, \dots, g_D^{(k)} \right)$ (11.15) is non-zero. Hence, uncoupled propagation in the kth section is described by

$$
\Lambda^{(k)} = \begin{pmatrix} e^{\frac{1}{2}g_1^{(k)}} & & 0 \\ & \ddots & \\ 0 & & e^{\frac{1}{2}g_D^{(k)}} \end{pmatrix}
\tag{11.60}
$$

In the overall propagation operator $\mathbf{M}^{(t)}$ given by (11.20), overall MDL is described by the singular values of $\mathbf{M}^{(t)}$ or, equivalently, by the eigenvalues of $\mathbf{M}^{(t)}\mathbf{M}^{(t)*}$ or $\mathbf{M}^{(t)*}\mathbf{M}^{(t)}$, which are both Hermitian matrices. Mathematically, the eigenvalues of $\mathbf{M}^{(t)}\mathbf{M}^{(t)*}$ are the squares of the singular values of $\mathbf{M}^{(t)}$. The singular values of $\mathbf{M}^{(t)}$ describe electric field gains, while the eigenvalues of $\mathbf{M}^{(t)}\mathbf{M}^{(t)*}$ describe power gains. As described earlier, $\mathbf{M}^{(t)*}\mathbf{M}^{(t)}$ or $\mathbf{M}^{(t)}\mathbf{M}^{(t)*}$ can be considered a round-trip propagation or modal gain operator.

As in MIMO wireless systems [26,27], at any single frequency, using singular value decomposition (SVD), the overall matrix $\mathbf{M}^{(t)}$ can be decomposed into D spatial channels:

$$
\mathbf{M}^{(t)} = \mathbf{V}^{(t)} \Lambda^{(t)} \mathbf{U}^{(t)*},
\tag{11.61}
$$

where $\mathbf{U}^{(t)}$ and $\mathbf{V}^{(t)}$ are input and output unitary beam-forming matrices, and we have defined

$$
\Lambda^{(t)} = \begin{pmatrix} e^{\frac{1}{2}g_1^{(t)}} & & 0 \\ & \ddots & \\ 0 & & e^{\frac{1}{2}g_D^{(t)}} \end{pmatrix}
\tag{11.62}
$$

Here, $\mathbf{g}^{(t)} = \left(g_1^{(t)}, g_2^{(t)}, \dots, g_D^{(t)} \right)$ is a vector of the logarithms of the eigenvalues of $\mathbf{M}^{(t)}\mathbf{M}^{(t)*}$, which quantifies the overall MDL of the MIMO system. Our goal here is to study the statistics of the overall MDL described by $\mathbf{g}^{(t)}$.

In order to describe MDL properly, we need to explain the difference between accumulated MDL and overall MDL. Accumulated MDL refers to the sum of the uncoupled MDL values in all K sections comprising a fiber. The variance of the accumulated MDL is:

$$
\xi^2 = \sigma_{g^{(1)}}^2 + \sigma_{g^{(2)}}^2 + \cdots + \sigma_{g^{(k)}}^2,
\tag{11.63}
$$

where $\sigma_{g(k)}^2, k = 1, \cdots, K$, are the variances of the uncoupled MDL vectors in the individual sections. The accumulated MDL is similar to the variance of the elements of the MD operator in (11.27). If the individual sections are statistically identical, we have $\xi = \sqrt{K}\sigma_g$, where σ_g^2 is the variance of the uncoupled MDL in each section. This accumulated MDL $\xi = \sqrt{K}\sigma_g$ increases with the square-root of number of sections or the square-root of the total system length.

Overall MDL refers to the end-to-end coupled MDL of a fiber comprising K independent sections. The overall MDL is computed from the gain vector $\mathbf{g}^{(t)}$ that appears in (11.62). The gains in $\mathbf{g}^{(t)}$ are assumed to be ordered as $g_1^{(t)} \geqslant g_2^{(t)} \geqslant \cdots \geqslant g_D^{(t)}$, and are assumed to sum to zero: $g_1^{(t)} + g_2^{(t)} + \cdots + g_D^{(t)} = 0$. When taken together, the D elements of $\mathbf{g}^{(t)}$ have zero mean; equivalently, an element $g_\mu^{(t)}$ chosen randomly from $\mathbf{g}^{(t)}$ has zero mean.

Two statistical parameters are especially useful for characterizing MDL: the STD of overall MDL σ_{mdl} and the mean of the maximum MDL difference $\left\langle g_1^{(t)} - g_D^{(t)} \right\rangle$. The maximum MDL difference is similar to the GD spread that is the difference between the maximum and minimum eigenvalues in Section 11.3.3; equivalently it is the condition number of the overall matrix $\mathbf{M}^{(t)}$ (11.20) expressed on a logarithmic scale.

The mean of the maximum MDL difference, $\left\langle g_1^{(t)} - g_D^{(t)} \right\rangle$, quantifies the gain difference between the strongest and weakest modes. In a two-mode fiber, $\left\langle g_1^{(t)} - g_2^{(t)} \right\rangle$ is commonly referred to as the mean PDL.

The STD of overall MDL is computed over all D elements of the gain vector $\mathbf{g}^{(t)}$; equivalently, it is the square-root of $\sigma_{\mathrm{mdl}}^2 = \left\langle \left(g_\mu^{(t)} \right)^2 \right\rangle$, where $g_\mu^{(t)}$ is an element chosen randomly from $\mathbf{g}^{(t)}$. The STD of overall MDL σ_{mdl} has the same units as the gains in $\mathbf{g}^{(t)}$.

The accumulated MDL variance ξ^2, given by (11.63), does not equal the variance of overall MDL because of the nonlinearity inherent in Proposition II, given by (11.57). Much of the complexity of PDL and MDL arises from the nonlinearity of (11.57).

11.4.3 Properties of the product of random matrices

Characterizing the statistics of the singular values of $\mathbf{M}^{(t)}$, or those of the eigenvalues of $\mathbf{M}^{(t)}\mathbf{M}^{(t)*}$, is the key to understanding the performance of MDM systems, as in MIMO wireless systems [26,27]. The analysis here is complicated by the fact that $\mathbf{M}^{(t)}$ is the product of random matrices by (11.20), rather than the sum of random matrices (11.28). Since the early works [119,120], there have been many studies on the statistics of the products of random matrices, but most addressed the Lyapunov exponent of the products [120–122]. We are interested here in the statistics of the eigenvalues of a product of matrices, not the Lyapunov exponent.

Because matrix multiplication is not commutative, i.e. $\mathbf{AB}$ is not generally equal to $\mathbf{BA}$, even for square matrices, the logarithm of the product of two matrices, log $\mathbf{AB}$, is not equal to the sum of $\log \mathbf{A}$ and $\log \mathbf{B}$. Unlike the product of positive random variables (that do commute), which has its central limit as the log-normal distribution, the product of positive-definite random matrices (those with positive eigenvalues) does not generally have its central limit as the exponent of a Gaussian unitary ensemble. The Central Limit Theorem for the summation of random matrices is not necessary helpful for the understanding of the products of random matrices.

For any matrix $\mathbf{X}$ and a very small number δ, we have $\log(\mathbf{I} + \delta\mathbf{X}) \approx \delta\mathbf{X}$, where $\mathbf{I} + \delta\mathbf{X}$ is intended to describe a matrix $\mathbf{M}^{(k)}$ when the gain vector $\mathbf{g}^{(k)}$ has small norm. If both matrices $\mathbf{A}$ and $\mathbf{B}$ are positive-definite and both $\log \mathbf{A}$ and $\log \mathbf{B}$ are small, $\log \mathbf{AB} \approx \log \mathbf{A} + \log \mathbf{B}$. As an approximation, the product of positive-definite random matrices with small logarithm has a central limit as the exponential of a Gaussian ensemble. When applied to the overall product matrix $\mathbf{M}^{(t)}$, if all gain vectors $\mathbf{g}^{(k)}$ are small, the matrix $\mathbf{M}^{(t)}$ is the exponential of the Gaussian ensemble. The approximation used here is similar to the results of Berger [123]. This small-gain approximation yields Proposition I, but not Proposition II. If we made the approximation $\log(\mathbf{I} + \delta\mathbf{X}) \approx \delta\mathbf{X}$, Proposition II would be a linear relationship, unlike the nonlinear relationship of both (11.57) and (11.59). Of course, the approximation (11.57) is linear when ξ is far less than unity. Numerical simulation shows that Proposition I remains valid for ξ up to 10 dB (about 2.3 in log power gain units), a regime in which, equivalently speaking, $\log \mathbf{A}$ and $\log \mathbf{B}$ are not very small. The nonlinearity in (11.57) may be understood as related to the second-order term in the approximation $\log(\mathbf{I} + \delta\mathbf{X}) \approx \delta\mathbf{X} - \frac{1}{2}\delta^2\mathbf{X}^2$. The factor of 1/2 in this approximation is difficult to relate to the factor of 1/12 in (11.57), however. Also, the more accurate model has the factor 1/12 related to the matrix dimension in (11.59).

For random matrices of the form (11.19), at a single frequency, we are interested in the statistics of the singular values of the product (11.20) in the decomposition (11.61). Approximate results were obtained for the 2×2 case in the study of PDL in SMF [124,125], and for very large matrices in the study of free random variables [109,110]. In both cases, for positive-definite matrices $\mathbf{A}$ and $\mathbf{B}$, the product $\mathbf{AB}$ may be interpreted as $\mathbf{A}^{1/2}\mathbf{BA}^{1/2*}$, similar to free probability theory. The repetition of $\mathbf{A}^{1/2}\mathbf{BA}^{1/2*}$ with $\mathbf{A} = \mathbf{M}^{(2)}\mathbf{M}^{(2)*}$ and $\mathbf{B} = \mathbf{M}^{(1)}\mathbf{M}^{(1)*}$ from $k = 2$ to K yields $\mathbf{M}^{(t)}\mathbf{M}^{(t)*}$.

The simplest possible case, a two-mode fiber modeled using 2×2 matrices, describes PDL in SMF, where approximate analytical results were derived for the small-PDL regime [124] and extended to the large-PDL regime [126]. Both [124,125] showed that for low PDL, the PDL (measured in decibels or log power gain units) has a Maxwellian distribution, the same as the distribution of the GD in SMF with PMD [89,102]. As shown in (11.47) of Section 11.3.2.3, the Maxwellian distribution is the eigenvalue distribution for a zero-trace 2×2 unitary Gaussian ensemble. The zero-trace 2×2 unitary Gaussian ensemble may be obtained by the summation of many zero-trace 2×2 Hermitian matrices. Numerical simulations confirm that Propositions I and II are correct for an accumulated MDL ξ smaller than 10 dB.

Using the notation here, the exact PDL distribution [126] is

$$p_2(x) = 3\sqrt{\frac{6}{\pi}} \frac{x \sinh x}{\xi^3} \exp\left(-\frac{3x^2}{2\xi^2} - \frac{\xi^2}{6}\right), \quad x \geqslant 0, \tag{11.64}$$

which can be approximated accurately by a Maxwellian distribution even in the high-PDL regime.

The exact distribution (11.64) is a non-central chi distribution with three degrees of freedom [127]. The Maxwellian distribution is the "central" chi distribution with three degrees of freedom [127]. The chi distribution is not as well known as the chi-square distribution [59, Section 2.3]. In $D = 2$, PDL has a non-central Maxwellian distribution but with very specific noncentrality parameter of $\xi/3$.

The exact PDL distribution (11.64) is the same as that of the exact MDL model (11.58) with $\kappa_2 = 1/3$, corresponding to concatenation of a random Maxwellian distributed PMD with root-mean-square differential GD (DGD) ξ and deterministic PMD with DGD $\xi^2/3$ that is analyzed in [102]. The large-PDL model of [126] is identical to the exact model (11.58) [118]. The exact model gives $\sigma_{mdl} = \xi\sqrt{1 + \xi^2/9}$, the same as (11.59) for $D = 2$, which is very close to the approximation (11.57), supporting Proposition II.

From the above discussion of the PDL, the comparison of the results of [124,125] with that in Section 11.3.2.3 supports Proposition I. By comparing (11.64) from [126] with the similar distribution derived from (11.58), the PDL of SMF is exactly the same as the exact model of (11.58), which reduces to Propositions I and II in the small-MDL regime.

Free random variables are equivalent to large matrices of the form (11.19) [109,110], and the statistics of free random variables are the statistical properties of the eigenvalues of the large matrices. The Central Limit Theorem for the summation of independent free random variables gives a semicircle distribution from Section 11.3.2.4, similar to the distribution of the eigenvalues of a large class of large random matrices [105,106]. In free probability theory, the semicircle distribution serves a function analogous to the normal distribution, which is the central limit for the summation of random variables in traditional probability theory [128, Section 5.10.4].

In traditional probability theory, the product of independent positive random variables has a central limit as the lognormal distribution. The lognormal distribution can model shadowing in wireless systems [60, Section 11.3.2.9] or distortion by stimulated Raman scattering in optical communication systems [129]. As shown in [130], the log-semicircle distribution is found to be the central limit of the product of positive free random variables if the free random variables have small variance; in the notation used here, this corresponds to ξ, defined in (11.63) having a small value.

From the results of [130], equivalently speaking, when the MDL is small and in the strong-coupling regime, the overall MDL has a log-semicircle distribution. The modal GDs follow a semicircle distribution in the limit of a large number of modes in Section 11.3.2.4. MDL (measured in decibels or log power gain) has the same distribution as the modal GDs in the limit of a large number of modes. Expression (11.58) is basically equivalent to the results in [130], which were derived from the free probability theory.

For MMF with larger number of modes, the overall MDL from (11.57) is the same as the more exact model from (11.58). We may conclude that Proposition II is valid in both small and large MDL regime for MMF with large number of modes.

11.4.4 Numerical simulations of mode-dependent loss and gain

Numerical simulation verifies the approximation in Propositions I and II and the more exact model (11.58). PDL in two-mode fiber has well-known exact results in [126], consistent with Propositions I and II in the small-PDL regime and described exactly by the model of (11.58) is all regimes. For many-mode fibers, analytical results from free-probability theory are consistent with the exact model of (11.58) and Proposition II in all regimes, and Proposition I in the low-MDL regime. Results for six-mode fibers are shown to match theory in Section 11.4.4.2.

11.4.4.1 Many-mode fibers

For fibers with a large number of modes, the modal GDs have the same statistical properties as the eigenvalues of a large zero-trace Gaussian unitary ensemble from Section 11.3.2.4. The eigenvalues of large random matrices have a semicircle distribution, as shown by Wigner [103,104] and verified numerically in Figure 11.6. The summation of free random variables also gives a semicircle distribution, as explained in Section 11.3.2.4.

Figure 11.8a shows the simulated eigenvalue distribution of a fiber with 64 modes. Similar to the simulation in Figure 11.6b, the fiber has $K = 256$ sections and all unitary matrices are generated by the method described in Section 11.3.2.3 or [28]. Each curve of Figure 11.8a is constructed using 64,000 eigenvalues. Each MMF section has uncoupled modal gains $g_{\mu}^{(k)} = \pm\alpha$, where $\alpha = \sigma_g$. The first 32 modes have gains of $+\alpha$, and the last 32 modes have gains of $-\alpha$, such that the gains sum to zero. The x-axis of each curve in Figure 11.8a is normalized by the simulated STD of the overall MDL σ_{mdl}.

In Figure 11.8a, the simulated eigenvalue distribution is very close to the semicircle distribution up to $\xi = 15\,dB$. The exact model (11.58) is also shown in Figure 11.8a as dashed lines. The distribution given by (11.58) is not available analytically for very large ξ but the moment generating function is given by [118, 130]

$$\exp\left(\frac{1}{2}\xi^2 s\right) {}_1F_1(1-s; 2; -\xi^2 s),\tag{11.65}$$

where ${}_1F_1(a,b;z)$ is the confluent hypergeometric function. The curves of Figure 11.8a are shown as the inverse Fourier transform of (11.65) with $s = i\omega$.

For ξ less than $15\,dB$, there is no significant difference between the semicircle distribution and the exact model. Even for ξ up to $20\,dB$, the simulation results are

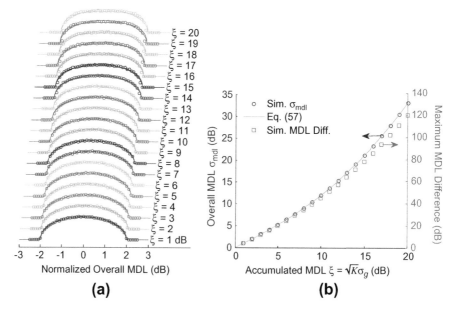

FIGURE 11.8 Mode-dependent loss (MDL) in fiber with $D = 64$ modes. (a) Comparing simulated distribution of the overall MDL (circles) with the semicircle distribution (solid curves) and exact model from (11.65) (dashed curves). The x-axis of each curve is normalized by the simulated STD of the overall MDL σ_{mdl}. (b) Comparing simulated STD of the overall MDL (circles) to approximation (11.57) (solid curves). The simulated mean maximum MDL difference is also shown (squares). Adapted from [28].

closer to the exact model but the difference between the exact and semicircle distribution is small.

Figure 11.8b compares the simulated STD of the overall MDL σ_{mdl} as a function of $\xi = \sqrt{K}\sigma_g$ to the approximation for σ_{mdl} given by (11.57) for a fiber with $D = 64$ modes. The approximation (11.57) agrees with simulated results within 0.01 dB for ξ up to 10 dB. For ξ from 10 to 20 dB, the overall MDL approximation (11.57) is always smaller than the simulated results and the discrepancy between the simulations and the approximation (11.57) increases to 0.15 dB. The discrepancy may arise from our simulating matrices of size $D = 64$ instead of infinitely large matrices. The discrepancy may also be caused by numerical uncertainty. Nevertheless, the approximation (11.57) can be considered highly accurate.

Figure 11.8b also shows the simulated maximum MDL difference, which is the mean of the maximum gain difference $\left\langle g_1^{(t)} - g_{64}^{(t)} \right\rangle$. In the range of ξ up to 15 dB, where the simulated distribution is close to the semicircle distribution (as seen in Figure 11.8a), the STD of overall MDL σ_{mdl} is as high as 33.4 dB and the maximum MDL difference is as high as 81 dB. Practical MDM systems should have MDL well below those values.

We conclude that for fibers with 64 modes, Proposition I is valid for ξ up to 15 dB, corresponding to σ_{mdl} up to 21.2 dB. The overall MDL approximation (11.57) from Proposition II is valid for values of ξ up to 20 dB. Over the range of validity of Proposition II, there is no observable difference between the exact model (11.58) and simulation results.

11.4.4.2 Six-mode fibers

The previous section confirms numerically that Propositions I and II are valid for fibers with $D = 64$ modes. The theoretical comparisons in Section 11.4.3 also found that Propositions I and II are valid for PDL in SMF, i.e. $D = 2$ modes. The eigenvalues of small zero-trace Gaussian unitary ensembles can be derived analytically by direct integration [10,28] or by algebraic substitution, as in (11.44).

Figure 11.9 compares results simulated for $D = 6$ modes with the approximate distribution (11.44), shown by solid lines. The approximate distribution (11.44) according to Proposition I is scaled by the overall MDL (11.57) according to Proposition II. The exact model (11.58), with distribution given by [118], is also shown by dashed lines. The MMF has $K = 256$ sections. Each curve is constructed using 300,000 eigenvalues.

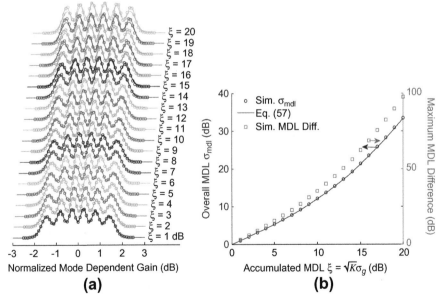

(a)   **(b)**

FIGURE 11.9 MDL in MMF with $D = 6$ modes. (a) Comparing simulated distribution of the overall MDL (circles) with the six-peak distribution from Figure 11.5 (solid curves). The *x*-axis of each curve is normalized by the simulated STD of the overall MDL σ_{mdl}. Also shown as dashed curves are the distributions derived from the exact model (11.58). (b) Comparing simulated STD of the overall MDL (circles) to approximation (11.57) (solid curves). The simulated mean maximum MDL difference is also shown (squares).

In Figure 11.9a, for a fiber with $D = 6$ modes, we observe that the simulated distribution matches the probability density (11.44) until $\xi = 13\,\text{dB}$. The exact model (11.58) is valid up to $\xi = 20\,\text{dB}$. As explained in Section 11.3, the number of peaks is the same as the number of modes, and each peak corresponds to values where the gain is concentrated.

In Figure 11.9b, for a fiber with $D = 6$ modes, we observe that the approximation for σ_{mdl} given by (11.57) is always smaller than the simulated results. The discrepancy is up to $0.05\,\text{dB}$ for ξ up to $10\,\text{dB}$, but increases to $0.20\,\text{dB}$ for ξ up to $20\,\text{dB}$. The exact model should provide a more accurate value of overall MDL, but the difference is already very small.

The numerical simulations of Figure 11.8 and Figure 11.9 show that Propositions I and II are valid in the small-MDL regime. With Proposition II and to a large extent even the exact model (11.58), the overall MDL depends solely on the accumulated MDL ξ. In strong-coupling regime, similar to the MD relationship (11.45), MDL depends on the square-root of the number of independent sections or the square-root of system length. With strong coupling, due to averaging effects, both MD and MDL depend on the square-root of the system length, i.e. strong coupling reduces both MD and MDL.

11.4.5 Spatial whiteness of received noise

In multimode systems using optical amplifiers, the amplified spontaneous emission noises generated in different uncoupled modes should be statistically independent. The MDL described in the previous sections affects both signal and noise, however, potentially making the noise spatially non-white. In system simulation using the matrix cascade of (11.20), it is straightforward to include multiple noise sources to capture this spatial correlation, as in [29]. It is helpful to observe, however, that if the number of noise sources is very large, the noises in different modes should have nearly the same power, and should be nearly uncorrelated with each other. This spatial whiteness is explained here.

Consider a system comprising K spans. For simplicity, suppose the Kth (last) fiber section contains a noise source, and that in the local uncoupled modes of the Kth section, it contributes independent noises with powers $\sigma_{K,\mu}^2, \mu = 1, \ldots, D$. At the fiber output, the electric fields from this noise source are described by a vector $\mathbf{V}^{(K)}\text{diag}\left[\sigma_{K,1}, \sigma_{K,2}, \ldots, \sigma_{K,D}\right]\mathbf{n}$, where $\mathbf{n}$ is a D-dimensional Gaussian noise vector having independent identically distributed (i.i.d.) elements with unit variance. At the fiber output, the noise correlation matrix is:

$$\mathbf{V}^{(K)}\text{diag}\left[\sigma_{K,1}^2, \sigma_{K,2}^2, \ldots, \sigma_{K,D}^2\right]\mathbf{V}^{(K)*}. \tag{11.66}$$

For a given realization of the random unitary matrix $\mathbf{V}^{(K)}$, the output noises may be neither independent nor identically distributed. Taking an expectation over all random matrices $\mathbf{V}^{(K)}$ yields:

$$\left\langle\mathbf{V}^{(K)}\text{diag}\left[\sigma_{K,1}^2, \sigma_{K,2}^2, \ldots, \sigma_{K,D}^2\right]\mathbf{V}^{(K)*}\right\rangle = \frac{\sigma_{K,1}^2 + \sigma_{K,2}^2 + \cdots + \sigma_{K,D}^2}{D}\mathbf{I}, \tag{11.67}$$

which is a constant times the identity matrix, and hence describes a noise vector with i.i.d. elements.

Considering a noise source in the kth fiber section, its output noise contribution is

$$\mathbf{M}^{(k)} \cdots \mathbf{M}^{(k+1)} \mathbf{V}^{(k)} \mathrm{diag} \left[\sigma_{k,1}, \sigma_{k,2}, \ldots, \sigma_{k,D}\right] \mathbf{n}.$$

The SVD of $\mathbf{M}^{(K)} \cdots \mathbf{M}^{(k+1)} \mathbf{V}^{(k)} \mathrm{diag} \left[\sigma_{k,1}, \sigma_{k,2}, \ldots, \sigma_{k,D}\right]$ may be assumed to yield $\tilde{\mathbf{V}}_n^{(k)} \tilde{\mathbf{\Lambda}}_n^{(k)} \tilde{\mathbf{U}}_n^{(k)}$, with output noise correlation matrix equal to $\tilde{\mathbf{V}}_n^{(k)} (\tilde{\mathbf{\Lambda}}_n^{(k)})^2 \tilde{\mathbf{V}}_n^{(k)*}$, which is of the same form as (11.66), with $\tilde{\mathbf{V}}_n^{(k)}$ independent of $\mathbf{V}^{(K)}$.

When the total number of noise sources is very large, by the law of large numbers [128, Section 7.4], the overall noise correlation matrix converges to a form similar to (11.66). The law of large numbers is applicable provided the number of noise sources is large and the $\tilde{\mathbf{V}}_n^{(k)}$ for each k indexing a noise source are independent, random unitary matrices. Based on the law of large numbers, at the fiber output, the noises in the different modes are i.i.d. Ideally, the receiver uses $\mathbf{V}^{(t)*}$ to diagonalize the channel, as in (11.61). After diagonalization, provided the number of noise sources is large and the $\mathbf{V}^{(t)*} \tilde{\mathbf{V}}_n^{(k)}$ for all values of k indexing noise sources are independent, random unitary matrices, the law of large numbers is applicable, and the noises in the different diagonal spatial channels are i.i.d. Because of the relationship between $\mathbf{V}^{(t)}$, $\tilde{\mathbf{V}}_n^{(k)}$, and the individual $\mathbf{U}^{(k)}$ and $\mathbf{V}^{(k)}$, the matrices $\mathbf{V}^{(t)*} \tilde{\mathbf{V}}_n^{(k)}$ are random unitary matrices, except in special cases.

Typically, the law of large numbers is concerned with averages. The overall noise correlation matrix is a summation over the correlation matrices corresponding to all the independent noise sources. The normalized cross-correlation of the noise is characterized by the ratios between its off-diagonal elements and its diagonal elements that are variances proportional to the number of noise sources. Thus, the normalized cross-correlation is implicitly an "average," to which the law of large numbers is applicable.

Figure 11.10 shows simulations quantifying the spatial non-whiteness of the output noise as a function of $1/\sqrt{K}$, where K is the number of noise sources. These simulations, for $D = 8$ modes, are similar to those in [28]. All K noise sources are spatially white, i.e. they contribute i.i.d. noises with equal variance in all D uncoupled modes. The accumulated MDL $\xi = \sqrt{K} \sigma_g$ is held constant at either 5 or 10 dB.

Given a realization of the $\mathbf{M}^{(k)}$, $k = 1, \ldots, K$, described by (11.19) with $\mathbf{\Lambda}^{(k)}$ given by (11.60) and a realization of the K noise sources, a realization of the D output noises is obtained; the absolute square of its discrete Fourier transform yields a realization of a spatial noise spectrum. Taking an ensemble average of the spatial noise spectrum over realizations of the noise sources yields a spatial spectral distribution, which quantifies the spatial whiteness of the output noise. If the output noises are uncorrelated (and thus i.i.d., since they are jointly Gaussian), the spatial spectral distribution is white, whereas if the noises are correlated, the spatial spectral distribution is non-white. The spatial spectral distribution is also the discrete Fourier transform of the spatial autocorrelation sequence of the D output noises [128, Section 9.3.4]. The procedure for calculating the spatial spectral distribution from the spatial autocorrelation sequence is similar to the serial correlation test for randomness described in [131, Section 3.3.2.K]. Figure 11.10

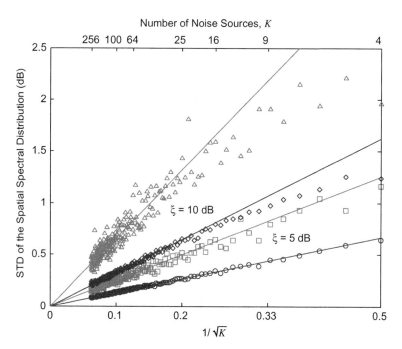

FIGURE 11.10 Spatial non-whiteness of the output noise, quantified by the STD of the spatial spectral distribution as a function of $1/\sqrt{K}$, where K is the number of noise sources, for $D = 8$ modes. The markers are simulation results, and the solid lines are fitted numerically. Upper and lower data sets for the corresponding accumulated MDL represent the maximum and mean of the STD over 100 random realizations at each value of K. The upper two data sets are for accumulated MDL $\xi = 10\,$dB and the lower two are for accumulated MDL $\xi = 5\,$dB. Adapted from [28]. (For interpretation of the color in this figure, the reader is referred to the web version of this book.)

quantifies non-whiteness by the STD of the spatial spectral distribution, presenting both the worst-case (maximum) and mean of the STD among 100 realizations of $\mathbf{M}^{(k)}$ for each value of K. The number of noise sources K ranges from 4 to 256.

In Figure 11.10, the spatial spectral distribution is observed to approach a white spectrum as K increases, with the STDs decreasing with the square-root of K, as given by the law of large numbers [128, Section 5.10.3]. The straight lines in Figure 11.10 indicate the best-fit slope of the STDs as function of $1/\sqrt{K}$. For different values of accumulated MDL $\xi = \sqrt{K}\sigma_g$, the slope is found to be approximately proportional to the value of overall MDL σ_{mdl} (11.57). While the simulated worst-case STD at any particular value of K is not statistically significant, slopes fitted to the worst-case STDs are clearly about twice as large as those fitted to the mean STDs. If a mean STD of 0.5 dB is desired, K needs to be larger than 8 and 42 for $\xi = 5$ and 10 dB, respectively. If a worst-case STD of 0.5 dB is desired, K needs to be larger than 25 and 160 for $\xi = 5$ and 10 dB, respectively. In any case, the spatial spectral distribution exhibits far smaller variations than the signal power variations caused by MDL.

Figure 11.10 confirms that spatial whiteness is a good assumption for systems with many noise sources, such as the long-haul systems considered in this chapter. This assumption may not be accurate for systems with a small number of noise sources, however.

11.4.6 Frequency-dependent mode-dependent loss and gain

In general, MDL depends on frequency if the diagonal matrix in each section, $\mathbf{\Lambda}^{(k)}(\omega)$ (11.17), includes both non-zero MDL vector $\mathbf{g}^{(k)}$ (11.15) and non-zero modal GDs vector $\boldsymbol{\tau}^{(k)}$ (11.16). The overall propagation matrix $\mathbf{M}^{(t)}(\omega)$, given by (11.20), is frequency-dependent. This frequency dependence of MDL gives frequency diversity. Frequency-dependent MDL has a coherence bandwidth that should be inversely proportional to the STD of GD, σ_{gd}. If MDM signals occupy a bandwidth far larger than the MDL coherence bandwidth, because of statistical averaging, signals at frequencies with large MDL are complemented by signals at frequencies with small MDL. This frequency diversity is described here, and its system implications are explained in Section 11.6.2.

Similar to MIMO wireless systems [26,27], at any single frequency, using SVD, the overall matrix $\mathbf{M}^{(t)}(\omega)$ can be decomposed into D spatial channels:

$$\mathbf{M}^{(t)}(\omega) = \mathbf{V}^{(t)}(\omega)\mathbf{\Lambda}^{(t)}(\omega)\mathbf{U}^{(t)}(\omega)^*, \tag{11.68}$$

where $\mathbf{U}^{(t)}(\omega)$ and $\mathbf{V}^{(t)}(\omega)$ are frequency-dependent input and output unitary beamforming matrices, respectively, and

$$\mathbf{\Lambda}^{(t)}(\omega) = \begin{pmatrix} e^{\frac{1}{2}g_1^{(t)}(\omega)} & & 0 \\ & \ddots & \\ 0 & & e^{\frac{1}{2}g_D^{(t)}(\omega)} \end{pmatrix}. \tag{11.69}$$

Here, $\mathbf{g}^{(t)}(\omega) = \left(g_1^{(t)}(\omega), g_2^{(t)}(\omega), \ldots, g_D^{(t)}(\omega) \right)$ is a frequency-dependent vector of the logarithms of the eigenvalues of $\mathbf{M}^{(t)}(\omega)\mathbf{M}^{(t)}(\omega)^*$, which quantifies the overall MDL of an MDM system. The decomposition (11.68) is the same as that of (11.61) at each individual frequency and the diagonal matrix (11.69) is the similar to (11.62).

The MDL given by the SVD (11.68) is frequency-dependent in general. In the special case that there is no MD, such that $\sigma_{gd} = \sqrt{K}\sigma_\tau$ is equal to zero, the MDL is independent of frequency as assumed in Section 11.4.3. Assuming non-zero σ_{gd}, the correlation of the MDL at two frequencies depends on the frequency separation. If the frequency separation is small, the phase factors for the uncoupled modes appearing in (11.17) are similar, leading to similar MDL values at the two frequencies. If the frequency separation is large, the values of $\mathbf{M}^{(t)}(\omega)$ at the two frequencies are independent, leading to independent MDL at the two frequencies.

Considering the simplest case of two modes, Figure 11.11 illustrates the frequency dependence of MDL in the regimes of small and large GD spread, quantified by σ_{gd},

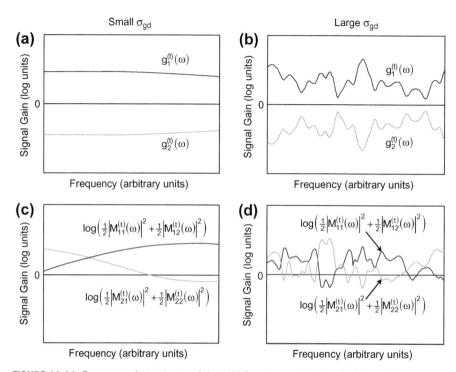

FIGURE 11.11 Frequency dependence of the MDL in a two-mode fiber for (a) small σ_{gd} and (b) large σ_{gd}, where σ_{gd} is the STD of GD. Output powers of signals launched into two orthogonal reference modes for (c) small σ_{gd} and (d) large σ_{gd}. Adapted from [30].

the STD of GD. Over the frequency range shown, the gains $g_1^{(t)}(\omega)$ and $g_2^{(t)}(\omega)$ (and thus the MDL) vary slowly for small σ_{gd} and rapidly for large σ_{gd}, as shown in Figure 11.11a and b, respectively. For signals launched into two orthogonal reference modes, the output powers (in logarithmic units) are

$$\log\left(\frac{1}{2}\left|M_{11}^{(t)}(\omega)\right|^2 + \frac{1}{2}\left|M_{12}^{(t)}(\omega)\right|^2\right) \text{ and } \log\left(\frac{1}{2}\left|M_{21}^{(t)}(\omega)\right|^2 + \frac{1}{2}\left|M_{22}^{(t)}(\omega)\right|^2\right),$$

respectively. Over frequency, these output powers vary slowly for small σ_{gd} and rapidly for large σ_{gd}, as shown in Figure 11.11c and d, respectively. For MDM signals spanning the frequency range shown, Figure 11.11c and d would correspond to regimes of low diversity order and moderate-to-high diversity order, respectively.

The correlation properties of MDL should depend on the normalized frequency separation $\vartheta = \Delta\omega\sigma_{gd}/2\pi$, where $\Delta\omega$ is the angular frequency separation. For small normalized frequency separation, $\vartheta \ll 1$, the MDLs at the two frequencies are identical, while for large normalized frequency separation, $\vartheta \gg 1$, the MDLs at the two

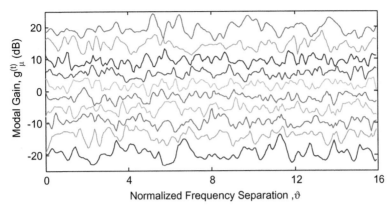

FIGURE 11.12 Modal gains, $g_\mu^{(t)}\mu = 1, ..., D$, as a function of normalized frequency separation for a fiber with $D = 10$ modes. Adapted from [30].

frequencies are independent. The coherence bandwidth of MDL should be of the same order as the reciprocal of the overall STD of GD, $1/\sigma_{gd}$; hence, the normalized coherence bandwidth should be of order unity.

Figure 11.12 shows simulations of the gain vector $\mathbf{g}^{(t)}(\omega)$ defined in (11.69) as a function of normalized frequency separation ϑ. The MMF has $D = 10$ modes and an accumulated MDL $\xi = 10\,\mathrm{dB}$. The MMF comprises $K = 256$ statistically identical sections, the same as in Section 11.3.2. The gain vector in each section $\mathbf{g}^{(k)}$ is the same as that in Section 11.4.4. The GD vector $\boldsymbol{\tau}^{(k)}$ in each section is generated as a Gaussian random vector whose entries sum to zero, using the method described in the Appendix of [28]. Each curve in Figure 11.12 corresponds to one of the elements of the vector $\mathbf{g}^{(t)}(\omega)$ as a function of normalized frequency separation ϑ. The x-axis of Figure 11.12 is the normalized frequency separation with respect to the first frequency.

Figure 11.12 illustrates how the correlation of the MDL depends on frequency separation, similar to Figure 11.11a and b. The gain of each mode is a smooth, continuous curve, so each modal gain is highly correlated for small frequency separations. Conversely, each modal gain is uncorrelated for large frequency separations. Figure 11.12 also shows that the highest and lowest modal gains are subject to larger variations than the intermediate modal gains because the outer peaks of the probability density function exhibit a larger spread than the inner peaks.

Figure 11.13 shows the correlation coefficients of the elements of the modal gain vector $\mathbf{g}^{(t)}(\omega)$ as a function of normalized frequency separation. The simulation parameters are the same as in Figure 11.12, but the correlation coefficients are calculated with 23,000 realizations of modal gain curves, each similar to Figure 11.12. In Figure 11.13, the correlation coefficient is calculated for each gain coefficient after conversion to a decibel scale. In Figure 11.13, the 10 curves are observed to cluster into five pairs, which are for the gain coefficients $g_\mu^{(t)}$ and $g_{D-\mu+1}^{(t)}, \mu = 1, ..., 5$.

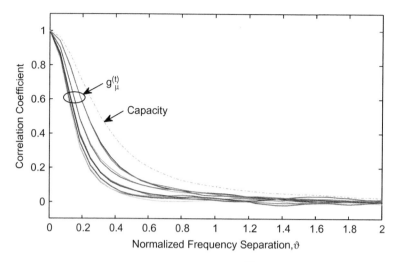

FIGURE 11.13 Correlation coefficients of modal gains, $g_\mu^{(t)} \mu = 1, ..., D$, as a function of normalized frequency separation for a fiber with $D = 10$ modes. Also shown is the correlation coefficient of the average channel capacity, which is used in Section 11.6.2, assuming a SNR of 20 dB and assuming CSI is not available at the transmitter. Adapted from [30].

The correlation coefficients are observed to decrease with an increase of μ. Referring to Figure 11.12, the highest and lowest two curves ($\mu = 1, g_1^{(t)}, g_{10}^{(t)}$) exhibit the largest (and similar) correlation over frequency, while the middle two curves ($\mu = 5, g_5^{(t)}, g_6^{(t)}$) exhibit the smallest (and similar) correlation over frequency.

In Figure 11.13, it is difficult to uniquely define a single coherence bandwidth for all the modal gains, because they decay at different rates, and do not decay fully to zero at large normalized frequency separation (this may arise, at least in part, from numerical errors). Considering the highest and lowest gains with the largest correlations, the normalized one-sided coherence bandwidths are 0.25 or 0.67 for correlation coefficients of 50% or 10%, respectively. At 10% correlation coefficient, the normalized one-sided coherence bandwidth ranges from 0.32 to 0.67 for the different gain coefficients. At a normalized frequency separation of unity, the correlation coefficients range from 0 to 4.7% for the different gain coefficients.

Figures 11.11–11.13 illustrate the frequency dependence of MDL. For wide-band MDM systems, frequencies having large MDL may be statistically averaged with frequencies having small MDL, at a frequency separation beyond the coherence bandwidth. The system benefits of this frequency diversity are quantified in Section 11.6.2.

Because the MDL is frequency-dependent, the potential exists for the received noise power spectrum to become non-white over frequency. Using arguments similar to those in Section 11.4.5 above, it can be shown that as the number of noise sources K becomes large, the received noise becomes white over frequency.

11.5 DIRECT-DETECTION MODE-DIVISION MULTIPLEXING

In short-range links, MDM using direct detection may be possible. Signals may be intensity-modulated or modulated by differential phase-shift keying. The entire end-to-end system must be designed with great care to avoid crosstalk between multiplexed signals and signal distortion by MD, or else these effects must be avoided by adaptive optical signal processing. If signal distortion and crosstalk are allowed to occur, they cannot be compensated fully by linear electrical signal processing, because of nonlinearity caused by higher-order MD [30,82], as explained in Section 11.3.1.

Direct-detection MDM has been proposed numerous times since the early 1980s. Some of the first proposals were based on multiplexing signals at different angles [132] or at different radial offsets [7], and attempted to exploit the relatively weak coupling between different mode groups over short distances, especially in step-index fibers with high GD differences between different mode groups. More recent works have pursued various approaches to multiplexing signals into weakly coupled mode groups [8,9,14,15,133,134]. As of this writing, however, we are not aware of any direct-detection MDM systems that have achieved significantly higher throughput than conventional MMF links.

In systems where MDL is negligible or small, MDM using optical signal processing to exploit PMs [135] can avoid crosstalk and signal distortion from MD in principle. Figure 11.14 shows a conceptual link design, assuming a single wavelength channel. To explain the operating principle, assuming the signal spans a narrow band near angular frequency ω, we can represent signal propagation by $\mathbf{M}^{(t)}(\omega) = \mathbf{V}^{(t)}\boldsymbol{\Lambda}^{(t)}(\omega)\mathbf{U}^{(t)*}$ (11.23). Recall that the columns of $\mathbf{U}^{(t)}$ and $\mathbf{V}^{(t)}$ describe the input and output PMs, respectively. In Figure 11.14, a modal multiplexer at the transmitter maps input signals into different input PMs, implementing $\mathbf{U}^{(t)}$, while

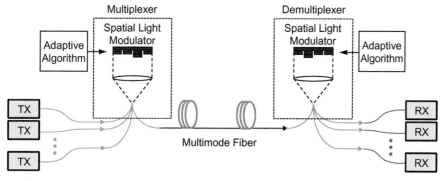

FIGURE 11.14 Schematic direct-detection mode-division multiplexed (MDM) system that multiplexes transmitted signals in input PMs, and demultiplexes received signals from output PMs. Adaptive optical components are used in both multiplexer and demultiplexer, with spatial light modulators shown as an example.

a modal demultiplexer at the receiver maps output signals from the corresponding output PMs, implementing $\mathbf{V}^{(t)*}$. Propagation of the multiplexed signals is described by $\mathbf{\Lambda}^{(t)}$, given by (11.24), which causes no crosstalk and no distortion, as explained in Section 11.3.1. By performing only memoryless optical signal processing, we have avoided signal distortion, obviating the need for signal processing that has memory. If the coherence bandwidth of the PMs (see Section 11.3.1) is sufficiently large, it may be possible to use the modal multiplexer and demultiplexer for multiple wavelength-division-multiplexed channels, similar to [136].

The modal multiplexer and demultiplexer in Figure 11.14 must be adjusted to track changes in the PMs caused by environmental perturbations. Spatial light modulators (SLMs) [13–15,133–137] or photonic integrated circuits [138–140] can be employed, but some algorithm is needed to determine the device settings. If the fiber propagation operator is known, the SVD $\mathbf{M}^{(t)}(\omega) = \mathbf{V}^{(t)}\mathbf{\Lambda}^{(t)}(\omega)\mathbf{U}^{(t)*}$ given by (11.23) may be computed, and then the input and output SLMs can be set to implement $\mathbf{U}^{(t)}$ and $\mathbf{V}^{(t)*}$, respectively. In practice, it may be difficult to measure the propagation operator and compute the SVD, so it may be preferable to adjust the multiplexer and demultiplexer using an adaptive algorithm that optimizes their settings based on measurement of eye openings, signal-to-noise-plus-interference ratios, or some other metrics. Standard adaptive algorithms [141] are not readily applicable. An adaptive algorithm for setting the SLMs is described in [135]. Adaptation of the modal multiplexer at the transmitter can be problematic in long-haul systems, where the round-trip signal propagation delay can be tens of milliseconds, which is longer than the time scale on which fiber properties can change.

An important outstanding question concerns how to implement the adaptive multiplexer and demultiplexer so as to minimize loss and crosstalk. For multiplexing (or demultiplexing) D modes, one can use D separate SLM regions to track the input (or output) PMs, and use beamsplitters for signal combining (or splitting). A multiplexer or demultiplexer would have a loss of at least $1/D$, for a combined loss of at least $1/D^2$. Alternatively, these devices can be implemented using the principle of a multiplexed correlator [15] or photonic integrated circuit [138–140], but it is not clear how the resulting loss and crosstalk scale with the number of multiplexed modes D.

Higher-order MD, described in Section 11.3.1 above, poses a fundamental limitation to direct-detection MDM systems. Although PMs are independent of frequency to first order, their frequency dependence can become important for wideband signals or long-haul propagation [30,82]. When $\mathbf{U}^{(t)}$ or $\mathbf{V}^{(t)}$ become frequency-dependent, SLMs or similar frequency-independent devices cannot realize them, and it may be too difficult or costly to implement frequency-dependent adaptive optics. Higher-order MD can cause nonlinear signal distortion and crosstalk that cannot be compensated effectively by linear electrical signal processing after a direct-detection receiver.

MDL poses another fundamental limitation to direct-detection MDM systems. Even in short links without optical amplifiers, MDL can be caused by offset connectors. MDL can make the signal-to-noise ratios different for various multiplexed signals. When signals are multiplexed into orthogonal input PMs, MDL can cause the received signals to become non-orthogonal, making it impossible to simultaneously avoid MD and crosstalk [142].

For $D = 2$, Figure 11.14 corresponds to polarization multiplexing in SMF by adaptively multiplexing into input PSPs and demultiplexing from output PSPs, which is fundamentally limited by higher-order PMD [143,144] and by PDL. To avoid the difficulties associated with an adaptive transmitter, it is far easier to multiplex signals into any pair of orthogonal polarization states, and use adaptive optical PMD compensation at the receiver prior to polarization demultiplexing (as a further simplification, PMD compensation may be replaced by simple polarization tracking, making the system subject to degradation by PMD [145–147]). By contrast, for $D > 2$, using a fixed orthogonal launch is probably not practical, because receiver-based adaptive optical compensation of MD becomes increasingly complicated as D increases.

11.6 COHERENT MODE-DIVISION MULTIPLEXING

As described in Section 11.5 above, in direct-detection MDM, unless the entire end-to-end system can be designed to achieve minimal crosstalk and distortion, it may be necessary to multiplex into input PMs at the transmitter and demultiplex from output PMs at the receiver using adaptive optics. Assuming frequency-independent multiplexers and demultiplexers, the usable signal bandwidth should not exceed the coherence bandwidth of the PMs, or higher-order MD will cause nonlinear distortion and crosstalk.

By contrast, in coherent MDM, the transmitter needs only multiplex signals into some orthogonal set of modes, and the receiver needs only to project the received signal onto some orthogonal set of modes [12,77] and perform dual-quadrature detection of each mode, providing sufficient statistics for signals to be separated using MIMO DSP [1–4,12,18,20]. In a coherent system, an arbitrarily large GD spread and arbitrary-order MD can be compensated with no penalty, provided laser phase noise is sufficiently small and the channel does not change too rapidly.

A coherent system may use either single- or multi-carrier modulation [1,2,18,148]. Multi-carrier modulation is typically implemented using OFDM, as we assume throughout this section. To compare these options briefly, single-carrier modulation is expected to incur less degradation from fiber nonlinearity, and typically requires a simpler transmitter but slightly more complex DSP at the receiver. OFDM is expected to incur more degradation from fiber nonlinearity and requires more complex DSP at the transmitter, but uses a simpler receiver architecture with lower computational complexity.

In this section, we address important issues governing the complexity and performance of coherent MDM systems, using the analyses of MD and MDL from Sections 11.3 and 11.4. In Section 11.6.1, we study how MDL can reduce average channel capacity. In Section 11.6.2, we show that strong mode coupling and MD lead to frequency diversity that can reduce outage probability. In Section 11.6.3, we discuss MIMO DSP methods for multi- or single-carrier MDM systems, and study how MD affects the implementation complexity of MIMO DSP.

11.6.1 Average channel capacity of narrowband systems

Long-haul MDM systems are assumed here to use inline optical amplification and coherent detection. The dominant noise at the receiver is assumed to arise from amplified spontaneous emission. The received noise power spectral density is assumed to be the same in each mode, as was assumed in wireless communications [26,27]. This assumption of spatially white noise was justified analytically and verified numerically in Section 11.4.5, but differs from the assumption made in [29]. The signal-to-noise ratio (SNR) ρ_t is defined as the received signal power (total over all D modes) divided by the received noise power (per mode). Throughout this subsection, the channel capacity at a single frequency is considered, assuming the signal has a very narrow bandwidth, as in [28,29].

Our definition of SNR follows the convention used in wireless MIMO systems [26] and is compatible with a conservative assumption that the total signal power in all D modes is constrained independent of D, while the noise power per mode is independent of D. Nevertheless, our results, if interpreted correctly, do not depend on this assumption in any way. For example, one might reason that the total signal power constraint scales in proportion to the number of modes D for a fixed time-averaged nonlinear phase shift, since D scales approximately in proportion to the core area from Section 11.2.1 and [47]. To accommodate such a constraint, one may scale the SNR as $\rho_t = D \cdot \rho_1$, where ρ_1 is the SNR per mode. Note, however, that both MD and CD may reduce nonlinear interactions between different modes [149–152].

In an MMF without MDL, such that all modes have equal gain, the power received in each mode is equal, and the channel capacity is equal to

$$C = D \log_2 \left(1 + \frac{\rho_t}{D}\right). \tag{11.70}$$

Throughout this chapter, channel capacity is expressed on a per-unit-bandwidth basis (i.e. in terms of spectral efficiency), and has units of b/s/Hz. When the SNR is much greater than the number of modes, the capacity (11.70) increases almost linearly with the number of modes. When the SNR is much less than the number of modes, the channel capacity is approximately proportional to SNR and independent of the number of modes. In the limit of an infinite number of modes, the capacity is given asymptotically by $C_\infty = \rho_t \log_2 e$, which is independent of the number of modes.

In MMF with MDL, channel state information (CSI) represents the information about a channel, described by a propagation operator $\mathbf{M}^{(t)}(\omega)$ (11.20), required by a transmitter to allocate transmit power optimally. CSI is defined more precisely in Section 11.6.1.2. In a coherent MDM system, the availability of CSI at the transmitter is an essential factor governing channel capacity. In the absence of CSI, an increase in MDL always leads to a decrease in capacity. In the extreme limit of MDL, the fiber supports propagation in only one mode. If CSI is available to the transmitter, only the surviving mode is used for transmission, and the channel capacity is $\log_2(1 + \rho_t)$, which is (11.70) with $D = 1$. If CSI is not available and the surviving mode is not known to the transmitter, the transmitter must allocate equal power to all modes, and the channel capacity is $\log_2(1 + \rho_t/D)$. As demonstrated below, channel capacity can be improved greatly by the availability of CSI. In some situations, when CSI is available, the channel capacity with MDL may exceed the capacity (11.70) without MDL.

Because MDL is a statistical phenomenon, as explained in Section 11.4, the channel capacity by itself is a random variable. In narrowband MDM systems, it is more relevant to study outage capacity, as in [28,29], than average capacity. As shown in Section 11.6.2, it is likely that practical MDM systems will need to operate in a wideband regime in which strong mode coupling and MD provide frequency diversity, such that the overall outage capacity approaches the average channel capacity, which is independent of bandwidth (when expressed on a per-unit-bandwidth basis). With this in mind, here we study only the average channel capacity of narrowband systems.

11.6.1.1 Average channel capacity without channel state information

When CSI is not available, the transmitter allocates equal power to each mode. Given the number of modes D and a realization of the gain vector $\mathbf{g}^{(t)}$ (11.15), the instantaneous channel capacity is

$$C = \sum_{i=1}^{D} \log_2\left[1 + \frac{\chi}{D}\exp\left(g_i^{(t)}\right)\right].\tag{11.71}$$

Taking the average of (11.71), the average channel capacity is:

$$C = D\int_{-\infty}^{+\infty} \log_2\left[1 + \frac{\chi}{D}\exp(\sigma_{\text{mdl}}x)\right] p_D(x)dx,\tag{11.72}$$

where $p_D(x)$ is the probability density defined in (11.44) with unit variance, σ_{mdl} is the STD of overall MDL, and the constant χ is determined by the constraint:

$$\chi = \frac{\rho_t}{\int_{-\infty}^{+\infty} \exp(\sigma_{\text{mdl}}x)\, p_D(x)dx}.\tag{11.73}$$

If the noise power per mode is normalized to unity, the constant χ is the total transmitted power and χ/D in (11.72) is the transmitted power per mode. Because of Proposition I of Section 11.4.1, the probability density $p_D(x)$ is the same as (11.44). However, scaling is required, as the probability density given by (11.44) has a variance of $\frac{1}{2}(D - D^{-1})$ instead of unity.

From Section 11.4, the MDL (measured in units of decibels or log power gain) has zero mean. Measured on a linear scale, the overall power loss/gain of a K-section MMF, summed over all D modes, is equal to $\cosh^K \sigma_g$. This overall gain gives a Lyapunov exponent, defined in [120–122], of $\log \cosh \sigma_g$. The factor χ, given by (11.73), normalizes the overall gain to unity, as measured on a linear scale. For a system with MDL, the channel gains are not constant, but are random variables. The SNR ρ_t is the mean (statistical average) SNR, and the factor χ given by (11.73) can be interpreted as the ratio of the mean SNR to the mean gain of the channel.

The product $\chi \exp(\sigma_{\text{mdl}} x)$ in (11.72) can be interpreted as the SNR of a channel realization with normalized gain x.

Analytical expressions are available for the constant χ, given by (11.73), but are too complicated for practical calculations. There is no known analytical expression for the capacity (11.72) except in the limit of very large D. Numerical integration of (11.72) may be employed to find the channel capacity. For fibers with a large number of modes, MDL is log-semicircle distributed, and the constant χ becomes

$$\chi = \frac{\rho_t \sigma_{\text{mdl}}}{2I_1(\sigma_{\text{mdl}})},$$

where $I_1(\cdot)$ is the modified Bessel function of the first kind. If D is far larger than $\chi \cdot \exp(2\sigma_{\text{mdl}})$, where $2\sigma_{\text{mdl}}$ is the upper limit of the semicircle distribution, the channel capacity approaches C_∞. In this limit, each mode is allocated a very small power, and the overall capacity is proportional to the total power.

Figure 11.15 shows the average capacity with MDL but without CSI, as a function of the accumulated MDL $\xi = \sqrt{K}\sigma_g$, for fibers with various numbers of modes D. As CSI is unavailable, equal power is allocated to each mode. The mean SNR is $\rho_t = 10$ dB. In Figure 11.15, the theoretical channel capacity is calculated by using (11.57) to

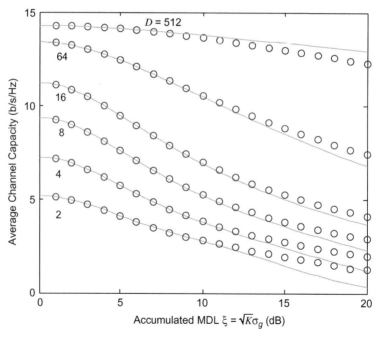

FIGURE 11.15 Average channel capacity for MDM in an MMF as a function of accumulated MDL ξ for fibers with various numbers of modes D, assuming CSI is not available. Equal power is allocated to each mode, and the SNR is $\rho_t = 10$ dB. Theoretical results are shown as curves and simulated results are shown as circles. Adapted from [28].

find the STD of overall MDL σ_{mdl}, and the probability distributions (11.45) (scaled to unit variance) are used in both (11.72) and (11.73) to compute the average channel capacity. The simulations use channel matrices generated by the same methods used to generate Figure 11.9.

In Figure 11.15, in the absence of CSI, average channel capacity always decreases with increasing ξ, particularly in fibers with smaller numbers of modes D. For values of ξ up to 10 dB, the average channel capacities from theory and simulation match very well. For a two-mode fiber, theoretical and simulated capacities match within 5% up to $\xi = 11$ dB. For a 512-mode fiber, theoretical and simulated capacities match within 5% up to $\xi = 19$ dB. The discrepancy between theory and simulation decreases with an increase in the number of modes.

The total launched power, equivalent to χ given by (11.73), is assumed not to change with the random realization of MDL. Satisfying this assumption requires that there be no amplifier saturation caused by MDL in each stage. For example, consider a two-mode case with unit total input power, unit noise level (in the absence of saturation), unit SNR, and gains of 0.5 and 1.5 (measured in linear units). In the absence of saturation, the output powers may vary from 0.5 to 1.5, depending on the alignment of the input signal to the eigenmodes of MDL (i.e. the modes of minimum and maximum gain). When saturation occurs, however, the maximum output powers may be limited to values less than the nominal value of 1.5, while noise levels decrease proportionally to maintain the mean SNR of unity. The model given here does not take account of these effects. If there are many optical amplifier stages or the system has a wide bandwidth, similar to the arguments of Section 11.4.5 based on the law of large numbers, a gain reduction in one stage (or at one frequency) should be compensated well by a gain enhancement in other stages (or at other frequencies). However, saturation effects for MDM systems are not well studied at this writing.

Due to statistical averaging, the outage capacity approaches the average channel capacity for a wideband system with strong mode coupling, as shown in Section 11.6.2.

11.6.1.2 Average channel capacity with channel state information

Ideally, in the operation of a MIMO system, the receiver estimates the overall channel matrix $\mathbf{M}^{(t)}$, computes the channel decomposition $\mathbf{M}^{(t)} = \mathbf{V}^{(t)}\mathbf{\Lambda}^{(t)}\mathbf{U}^{(t)*}$ (11.61), sends a description of the precoding matrix $\mathbf{U}^{(t)}$ and the gain vector $\mathbf{g}^{(t)}$ to the transmitter, and uses the received beam-forming vector $\mathbf{V}^{(t)}$ in decoding received signals [26]. The matrix $\mathbf{U}^{(t)}$ and the gain vector represent CSI that is to be fed back from the receiver to the transmitter. Transmit power and information bits are allocated to spatial channels based on the gain vector $\mathbf{g}^{(t)}$.

In a long-haul system, the feedback process described above may become impractical if the MMF changes on a time scale shorter than or comparable to the round-trip propagation delay, which can be tens of milliseconds. When feedback becomes impossible, space-time codes [153–155] or error-correction codes across the spatial channels can provide diversity. In these cases, all spatial channels are allocated equal power, yielding the capacity computed in Section 11.6.1.1. As described below in Section 11.6.3.3, receivers for systems with or without CSI have the same architecture, apart from the means to estimate CSI to feed back to the transmitter.

When CSI is available at the transmitter, transmit power may be allocated to spatial channels in an optimal way. If the noise power per mode is normalized to unity and the overall gain vector is $\mathbf{g}^{(t)}$ known, the optimal transmit powers are given by

$$\left[\eta - e^{-g_\mu^{(t)}}\right]^+, \ \mu = 1, \ldots, D,$$

where $[\ ]^+$ denotes limiting to non-negative values, i.e. $[x]^+ = \max(0, x)$. The constant η is chosen to satisfy the total power constraint:

$$\sum_{\mu=1}^{D}\left[\eta - e^{-g_\mu^{(t)}}\right]^+ = \chi, \tag{11.74}$$

with χ given by (11.73). The average channel capacity is

$$C = \left\langle \sum_{\mu=1}^{D} \log_2\left(1 + \left[\eta e^{g_\mu^{(t)}} - 1\right]^+\right)\right\rangle, \tag{11.75}$$

which is an expectation over random realizations of the MDL.

A brute-force method to compute the channel capacity (11.75) involves generating many random realizations of the overall matrix (11.20) and computing their singular values to evaluate the average capacity [29]. Each realization of (11.20) is the product of $3K$ matrices. Obviously, the total number of matrices can be reduced to $2K+1$ by combining $\mathbf{U}^{(k+1)}$ and $\mathbf{V}^{(k)}, k = 1, \ldots, K - 1$, into single random unitary matrices.

Assuming values of MDL sufficiently small to be of practical interest, Propositions I and II are valid (see Section 11.4.1). In this regime, a more efficient method of computing (11.75) is based on Proposition I, namely, that the joint probability density for the vector $g^{(t)}$ is given by the eigenvalue distribution of a zero-trace Gaussian unitary ensemble. Hence, a zero-trace random Gaussian Hermitian matrix may be generated using the method described in the Appendix of [28], and the eigenvalues of the random matrix can be used to compute the capacity (11.75). Also, instead of generating new matrices for different values of ξ, Proposition II can be exploited. A single matrix may be generated, and the eigenvalues can be scaled using the approximation for σ_{mdl} given by (11.57).

Figure 11.16 shows the average capacity for a system with MDL and with CSI, as a function of the accumulated MDL $\xi = \sqrt{K}\sigma_g$, for fibers with various numbers of modes D. The SNR is $\rho_t = 10$ dB, which represents a statistical average. The theoretical curves are obtained using 100,000 zero-trace random Gaussian-Hermitian matrices. Simulated results are obtained using brute-force generation of channel matrices by the same methods used for Figures 11.8 and 11.9.

In Figure 11.16, we see that for fibers with $D = 2$, 4, or 8 modes, even with CSI, the channel capacity decreases with increasing MDL. With $D = 16$ modes,

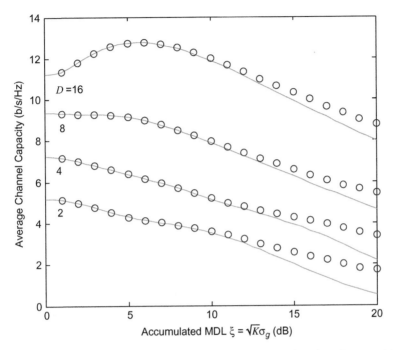

FIGURE 11.16 Average channel capacity for MDM in an MMF as a function of accumulated MDL ξ for fibers with various numbers of modes _D_. CSI is available, and transmit power is allocated optimally to each mode, and the SNR is $\rho_t = 10\,$dB. Theoretical results are shown as curves and simulated results are shown as circles. Adapted from [28].

at small MDL, the capacity increases with increasing MDL, although it eventually decreases with a further increase in MDL. More generally, MDL can increase capacity when the SNR measured in linear units is small relative to the number of modes D. This may seem counter-intuitive, but can be made plausible by the following example for $D=2$ modes. Assume that the total transmitted power is unity, the noise power per mode is unity, and the mean SNR is $\rho_t = 1$. Hence, the mean channel gain is $0\,$dB, corresponding to unity in linear units. Without MDL, using (11.70), the channel capacity is $2\log_2(1 + 0.5) = 1.17\,$b/s/Hz. Now suppose that MDL is present, and that in a particular channel realization, the modal gains are $-3.01\,$dB and $+1.76\,$dB, which correspond to 0.5 and 1.5 in linear units (the mean of these two values is unity, as in the absence of MDL). When CSI is not available, the transmitter allocates equal power of 0.5 to each mode. The capacity is $\log_2(1 + 0.5 \cdot 0.5) + \log_2(1 + 1.5 \cdot 0.5) = 1.13\,$b/s/Hz, slightly smaller than that without MDL. When CSI is available, the transmitter may allocate all the power to the stronger mode. The capacity becomes $\log_2(1 + 1.5 \cdot 1) = 1.32\,$b/s/Hz, slightly larger than that without MDL.

In Figure 11.16, for a two-mode fiber, theoretical and simulated average capacities agree within 5% up to $\xi = 11$ dB. For a 16-mode fiber, theoretical and simulated average capacities agree within 5% up to $\xi = 17$ dB. The discrepancy decreases with an increasing number of modes. While the more exact MDL model (11.58) gives slightly more accurate results, values of accumulated MDL $\xi > 10$ dB are likely too large for practical systems.

Although mathematically correct, the results of Figure 11.16 are difficult to reconcile with the saturation effects of optical amplifiers. The constraint (11.74) is nearly linear in χ, independent of η. Amplifier saturation effects may lead to inter-dependence between χ and η. Even for wideband systems and systems with many amplifier stages, it is not expected that amplifier saturation effects will fully average out between frequencies or stages.

In the remainder of this chapter, unless noted otherwise, we assume that CSI is not available.

11.6.2 Wideband systems and frequency diversity

In Section 11.4.6, Figures 11.11–11.13 demonstrate that MDL is frequency-dependent, and is strongly correlated only over a finite coherence bandwidth. If MDM signals occupy a bandwidth far larger than the coherence bandwidth, the outage channel capacity should approach the ensemble average channel capacity computed in Section 11.6.1 due to statistical averaging. For typical values of MDL and SNR, the coherence bandwidth of the capacity is found to be approximately equal to $1/\sigma_{gd}$, the reciprocal of the STD of GD. The difference between the average capacity and the outage capacity is found to decrease with the square-root of a diversity order that is given approximately by the ratio of the signal bandwidth to the coherence bandwidth of the channel capacity.

As explained earlier, the channel capacity for narrowband MDM systems is a random variable. Figure 11.17 shows the simulated distribution of the channel capacity of a narrowband system with $D = 10$ modes at an SNR $\rho_t = 20$dB, corresponding to an SNR per mode $\rho_1 = 10$dB, assuming CSI is not available at the transmitter. The system parameters are the same as those in Figure 11.13 of Section 11.4.6. The distribution in Figure 11.17 is constructed using about 5,900,000 channel capacity values. In Figure 11.17, the average channel capacity, near the peak of the distribution, is about 17.2 b/s/Hz, while the capacity for 10^{-3} outage probability is about 14.3 b/s/Hz. In a wideband MDM system, at any single frequency, the channel capacity has the same distribution as that in Figure 11.17.

Figure 11.13 of Section 11.4.6 shows the correlation coefficients of the modal gains versus normalized frequency separation, illustrating how the gains at nearby frequencies are highly correlated. Figure 11.13 also shows the correlation coefficient of the channel capacity. The capacity is computed as in Figure 11.17, assuming an SNR $\rho_t = 20$dB, no CSI available at the transmitter, and equal power allocated to all modes. The normalized one-sided coherence bandwidths of the channel capacity are 0.31 and 0.92 for correlation coefficients of 50% and 10%,

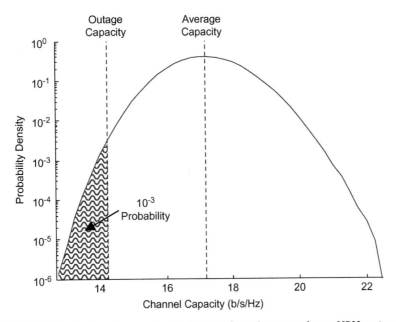

FIGURE 11.17 Distribution of channel capacity at a single frequency for an MDM system with $D = 10$ modes and an SNR of 20 dB, assuming CSI is not available. The average capacity and outage capacity at 10^{-3} outage probability are indicated. Adapted from [30].

respectively. A normalized frequency separation of unity gives a correlation coefficient of 8.8%.

If MDM signals occupy a bandwidth much greater than the coherence bandwidth of the capacity, statistical averaging over frequency should cause the outage channel capacity to approach the average channel capacity. Figure 11.18 shows the outage capacity as a function of SNR, for signals occupying different bandwidths. All parameters are as in Figure 11.13 and Figure 11.17 and the outage probability is 10^{-3}. A normalized signal bandwidth is defined as $b = B_{\text{sig}} \sigma_{\text{gd}}$, where B_{sig} is the signal bandwidth (measured in Hz). In Figure 11.18, signals occupy normalized bandwidths from $b = 0$ (a single frequency) to $b = 8$. The average capacity is also shown for comparison. Figure 11.18 shows that as the normalized bandwidth increases, the outage capacity approaches the average capacity.

Statistical averaging over frequency is a consequence of the law of large numbers. For example, consider two OFDM subchannels at frequencies whose separation far exceeds the coherence bandwidth of the capacity, e.g. at two frequencies well separated in Figure 11.12b. We assume that the subchannels have capacities C_1 and C_2; these are independent random variables following a common distribution (e.g. that in Figure 11.17). A channel comprising the two subchannels has an overall capacity $\frac{1}{2}(C_1 + C_2)$, as the capacity is computed on a per-unit-frequency basis. If

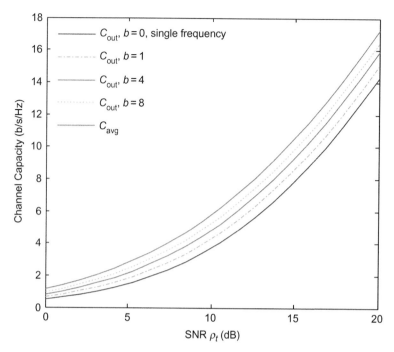

FIGURE 11.18 Outage channel capacity at 10^{-3} outage probability versus SNR for various normalized bandwidths, for an MDM system with $D = 10$ modes, assuming CSI is not available. The average channel capacity is shown for comparison. Adapted from [30].

each subchannel capacity has variance σ_C^2, the overall channel capacity has variance $\frac{1}{2}\sigma_C^2$, i.e. it is reduced by a factor of two. The channel comprising two independent subchannels has a diversity order of two. More generally, given a channel spanning a finite bandwidth B_{sig}, we define the diversity order in terms of a reduction of the variance of capacity: a diversity order equal to F_D corresponds to a reduction of the variance of capacity from σ_C^2 to σ_C^2/F_D. With this definition, the diversity order may be any real number not smaller than unity.

It would be useful to be able to estimate the diversity order F_D directly from the statistics of the frequency-dependent gain vector $\mathbf{g}^{(t)}(\omega)$, rather than having to compute the frequency-dependent capacity and characterize its statistics. The method of principal component analysis [156] may be used to calculate the diversity order directly from the frequency correlation coefficients of the channel capacity. Principal component analysis is very similar to the discrete Karhunen-Loève transform [157]. The conversion from correlation coefficients to diversity order is explained in detail in [30].

Numerical simulations of MDM systems similar to those in Figure 11.18 have been performed, with number of modes $D = 10$ and normalized bandwidth b ranging from 0 to 16. Regardless of the normalized bandwidth b, the distribution of the channel capacity is found to retain approximately the same shape as that of Figure 11.17, but the variance is reduced from σ_C^2 to approximately σ_C^2/F_D. The mean channel capacity does not change with diversity order. The outage capacity as a function diversity order is found to follow

$$C_{\text{out},F_D} \approx C_{\text{avg}} - \frac{1}{\sqrt{F_D}} \left(C_{\text{avg}} - C_{\text{out},1} \right), \qquad (11.76)$$

where $C_{\text{out},1}$ is the single-frequency outage capacity following the distribution in Figure 11.17. The relationship (11.76) is found to be independent of the outage probability, provided the outage capacities $C_{\text{out},1}$ and C_{out,F_D} refer to the same outage probability. The relationship (11.76) is found to be valid even if the shape of the distribution of capacity deviates from that shown in Figure 11.17. If the capacity distribution is assumed to be Gaussian, as in [29], the relationship between outage capacity and diversity order F_D at any particular outage probability can be computed analytically. However, when plotted on a semi-logarithmic scale, as in Figure 11.17, the capacity distribution is observed to deviate noticeably from a Gaussian distribution, i.e. it is slightly asymmetric. This non-Gaussianity is consistently observed at all SNRs and all diversity orders.

Figure 11.19 shows the outage capacity reduction ratio, defined as

$$\frac{C_{\text{avg}} - C_{\text{out},F_D}}{C_{\text{avg}} - C_{\text{out},1}}, \qquad (11.77)$$

as a function of $1/\sqrt{F_D}$, where F_D is the diversity order computed using principal component analysis from the correlation coefficients of the channel capacity shown in Figure 11.13. The simulation parameters used for Figure 11.19 are the same as those of Figure 11.13 and Figure 11.18, i.e. the MDM system uses $D = 10$ modes, and CSI is not available at the transmitter. Values of the diversity order F_D are only computed for SNR $\rho_t = 20\,\text{dB}$, as in Figure 11.13, but Figure 11.19 shows values of (11.77) computed at SNR $\rho_t = 10$ and $20\,\text{dB}$, illustrating that the diversity order F_D is valid over a wide range of SNR values. The correlation coefficients in Figure 11.13 are subject to numerical error, as they never go to zero even for large frequency separations. To limit numerical error, diversity orders in Figure 11.19 are computed only using values of the correlation coefficients from Figure 11.13 that are larger than 1%.

Based on (11.76), the outage capacity reduction ratio (11.77) should approximately equal $1/\sqrt{F_D}$, and the plots in Figure 11.19 should be straight lines with unit slope. In Figure 11.19, the best-fit slope is found to be 1.06. Figure 11.19 clearly shows that $C_{\text{avg}} - C_{\text{out},F_D}$ approaches zero as the diversity order F_D increases. The observed dependence of the difference between average and outage capacities on $1/\sqrt{F_D}$ is a direct consequence of the law of large numbers. The outage and average capacities converge slowly with an increase in diversity order F_D. The diversity order F_D must be four to decrease the capacity difference to half that without diversity, and must be 100 to decrease the difference to 10% of that without diversity.

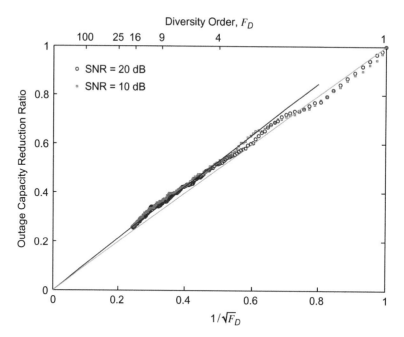

FIGURE 11.19 The outage capacity reduction ratio, given by (11.77), as a function of $1/\sqrt{F_D}$, where F_D is the diversity order. The lines have slopes of 1 (theoretical) and 1.06 (best-fit), respectively. The MDM system uses $D = 10$ modes, and CSI is not available. Adapted from [30]. (For interpretation of the color in this figure, the reader is referred to the web version of this book.)

The ratio of outage to average capacities, $C_{out,1}/C_{avg}$, decreases with an increase of MDL or a reduction of SNR. In the limit of a very high SNR and an MDL smaller than the SNR, the channel capacity without CSI (11.71) is approximately equal to $D \log_2 \chi/D + \log_2 e \sum_{i=1}^{D} g_i^{(t)} = D \log_2 \chi/D$, which is independent of the frequency-dependent gain vector $\mathbf{g}^{(t)}(\omega)$. The average SNR ρ_t needs to be large enough that even the weakest mode has sufficiently high SNR. At high SNR, the channel capacity is independent of $\mathbf{g}^{(t)}(\omega)$ even for a system with CSI.

The combined matrix model (11.20) shows that the GD statistics do not depend on the mode-averaged CD. In the case of spatial-mode-dependent CD, the higher-order frequency dependence of the GD statistics may be modified slightly. The general approach presented here should remain valid. The STD of GD becomes a frequency-dependent $\sigma_{gd}(\omega)$ to include the effect of spatial-mode-dependent CD. The normalized frequency separation may be modified to $\vartheta = (2\pi)^{-1} \int_0^{\Delta\omega} \sigma_{gd}(\omega_0 + s)ds$, where ω_0 is a reference frequency and $\Delta\omega$ is the frequency difference. The normalized bandwidth b is always the normalized frequency separation between the lowest and highest frequencies.

Numerical simulation of Figure 11.19 uses a maximum normalized bandwidth of $b = 16$. The diversity order F_D is approximately equal to the normalized bandwidth for systems with $b \geqslant 4$. As shown in Section 11.6.3.3 below, receiver DSP complexity scales in proportion to the normalized bandwidth b. Complexity constraints may permit systems to have normalized bandwidths b up to the order of 100 or larger. Assuming practical MDL values, the outage channel capacity of a wideband system with $b = 100$ should approach the average channel capacity of a narrowband system. With such high-order frequency diversity, average channel capacity is a good metric for characterizing system performance.

11.6.3 **Signal processing for mode-division-multiplexing**

In coherent MDM systems, DSP is used at the receiver, and possibly at the transmitter, to perform various functions [1–4,18,20]. In the transmitter, DSP may be used for precoding, and is used to perform modulation in OFDM systems. In the receiver, DSP is used for timing synchronization, frequency and phase synchronization, MIMO channel estimation and equalization, and other functions.

Figure 11.20 schematically shows key transmitter and receiver functions in an OFDM system, while Figure 11.21 shows corresponding functions in a single-carrier system, assuming an FDE-based receiver. These figures show only those functions that are relevant to the discussion in this section, and omit other important functions. For simplicity, only one channel in a wavelength-division-multiplexed system is shown. "Tx" includes digital-to-analog conversion and electrical-to-optical conversion and "Rx" includes combination of the signal and a local oscillator in an optical hybrid, optical-to-electrical conversion, and analog-to-digital conversion.

In Figures 11.20 and 11.21, the transmitter may optionally include a precoder. If CSI is available, precoding may be used with power and bit allocation (often

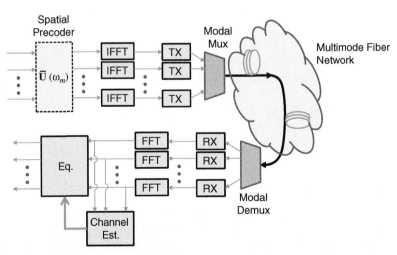

FIGURE 11.20 Major functions of OFDM transmitter and receiver in an MDM system.

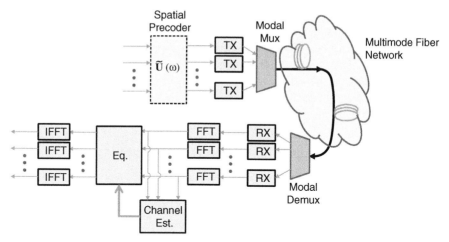

FIGURE 11.21 Major functions of single-carrier transmitter and receiver in an MDM system, assuming the receiver uses frequency-domain equalization (FDE).

known as "bit loading") to enhance performance, as discussed in Section 11.6.1.2. In this case, assuming the overall propagation matrix is decomposed using SVD to $\mathbf{M}^{(t)}(\omega) = \mathbf{V}^{(t)}(\omega)\mathbf{\Lambda}^{(t)}(\omega)\mathbf{U}^{(t)}(\omega)^*$ (11.68), the ideal precoder is $\mathbf{U}^{(t)}(\omega)$. Alternatively, precoding may be used to implement space-time coding [153–155] for spatial diversity. In general, the actual precoder employed is denoted by $\widetilde{\mathbf{U}}^{(t)}(\omega)$. If no precoder is used, we can make the assignment $\widetilde{\mathbf{U}}^{(t)}(\omega) = \mathbf{I}$. Even if CSI is available, the actual precoder $\widetilde{\mathbf{U}}^{(t)}(\omega)$ may not be equal to $\mathbf{U}^{(t)}(\omega)$ because of estimation error in the receiver due to additive noise, quantization noise or round-off error, or because of delay in the CSI feedback and estimation process. CSI may be estimated for a group of adjacent frequencies, resulting in estimation error at some frequencies within the group.

We assume that the overall propagation matrix $\mathbf{M}^{(t)}(\omega)$ (11.20) includes the modal multiplexer and demultiplexer. Including the precoder and mode-averaged CD, and ignoring mode-averaged gain, the overall channel response becomes:

$$\mathbf{H}(\omega) = \mathbf{M}^{(t)}(\omega)\widetilde{\mathbf{U}}^{(t)}(\omega)\exp\left(-j\frac{1}{2}\bar{\beta}_2(\omega)^2 L_t\right),\qquad(11.78)$$

where $L_t = \sum_k L^{(k)}$ is the total system length.

11.6.3.1 Multi-carrier signals

Figure 11.20 shows key functions in an OFDM transmitter and receiver. In the transmitter, for the mth subchannel at angular frequency ω_m, the precoder is $\widetilde{\mathbf{U}}^{(t)}(\omega_m)$. In the receiver, after downconverting all D spatial channels and performing FFTs of each one, each subchannel may be compensated using a memoryless one-tap equalizer [158–160]. Assuming a minimum mean-square error (MMSE) equalizer [59, Section 15.1–2], the equalizer response in the mth subchannel is:

$$\left[\mathbf{H}^*(\omega_m)\mathbf{H}(\omega_m) + \mathbf{S}_n(\omega_m)\right]^{-1}\mathbf{H}^*(\omega_m), \tag{11.79}$$

where $\mathbf{S}_n(\omega_m)$ is the power spectrum of the sampled noise. From Section 11.4.5, $\mathbf{S}_n(\omega_m)$ is likely to be approximately proportional to $\mathbf{I}$, but may exhibit variations among subchannels. In this section, $\mathbf{S}_n(\omega_m) \propto \mathbf{I}$ is assumed for each subchannel.

Frequency-dependent MDL causes the channel response $\mathbf{H}(\omega)$ to become non-unitary over space and frequency. At the output of the equalizer (11.79), the SNR may vary widely across different spatial and frequency subchannels. Strong error-correction coding across subchannels exploits frequency diversity and provides coding gain. Many modern strong codes are decoded using soft information, typically the log-likelihood ratio [161,162], which is the logarithm of the ratio of the probabilities that a bit is equal to 1 or 0. If $\mathbf{x}(\omega_m)$ is the D-dimensional transmitted vector for the mth subchannel, the received signal is $\mathbf{y}(\omega_m) = \mathbf{H}(\omega_m)\mathbf{x}(\omega_m) + \mathbf{n}(\omega_m)$, where $\mathbf{n}(\omega_m)$ is the additive noise in the subchannel. Assume that $\mathbf{b}_l, l = 1, \ldots, M$, are all possible transmitted signal vectors, where $M = 4^D$ for quadrature phase-shift keying (QPSK) signals in each mode, for example. The log-likelihood ratio is given by

$$\log \frac{\sum_{l,\text{bit}=1} \exp\left(-\frac{\|\mathbf{y}(\omega_m)-\mathbf{H}(\omega_m)\mathbf{b}_l\|^2}{\mathbf{S}_n(\omega_m)}\right)}{\sum_{l,\text{bit}=0} \exp\left(-\frac{\|\mathbf{y}(\omega_m)-\mathbf{H}(\omega_m)\mathbf{b}_l\|^2}{\mathbf{S}_n(\omega_m)}\right)}, \tag{11.80}$$

where the numerator sums over all possible transmit vectors in which a given bit is equal to 1 and the denominator sums over those in which it is equal to 0. The log-likelihood ratio in the form (11.80) is difficult to obtain. Using the log-max approximation, only the maximum terms in both numerator and denominator are used. Including only the maximum terms and ignoring some common factors, the log-likelihood ratio becomes

$$2\left[\max_{l,\text{bit}=1} \text{Re}\left\{\frac{\mathbf{b}_l^*\mathbf{H}^*(\omega_m)\mathbf{y}(\omega_m)}{\mathbf{S}_n(\omega_m)}\right\} - \max_{l,\text{bit}=0} \text{Re}\left\{\frac{\mathbf{b}_l^*\mathbf{H}^*(\omega_m)\mathbf{y}(\omega_m)}{\mathbf{S}_n(\omega_m)}\right\}\right]. \tag{11.81}$$

In a typical system, the approximation (11.81) is very close to the exact expression (11.80). To evaluate (11.81), we only need to calculate

$$2\mathbf{S}_n^{-1}(\omega_m)\mathbf{H}^*(\omega_m)\mathbf{y}(\omega_m) \tag{11.82}$$

in each OFDM subchannel and compare it with the possible $\mathbf{b}_l$ to obtain the log-like-lihood ratio. In an OFDM receiver, the equalizer just needs to multiply the received signal vector by the channel matrix (11.82), which is a very simple operation.

The possible transmitted $\mathbf{b}_l$ depends on CSI and the bit loading procedure. Because the same CSI should be available at both transmitter and receiver, the bit loading table may be calculated by the same procedure at both transmitter and receiver. However, the CSI may not be fed back to the transmitter correctly. More robust schemes require that the bit loading table or some indicator be transmitted as side information in the OFDM signals. In addition to the overhead of this side information, the round-trip delay of the feedback may be too long for a time-varying channel. For this reason, OFDM wireless systems typically do not use bit loading. In practice, if error correction coding is used across frequency and spatial subchannels, as in coded OFDM [163,164], the extra complexity of bit loading is usually not justified. The complexity of an OFDM receiver remains nearly the same regardless of whether CSI is fed back to the transmitter.

11.6.3.2 Single-carrier signals

Figure 11.21 shows key functions in a single-carrier transmitter and receiver, assuming an FDE-based receiver. As shown in [18,20], the DSP complexity for TDE is typically far higher.

A single-carrier transmitter, like its multi-carrier counterpart, may use a precoder $\widetilde{\mathbf{U}}^{(t)}(\omega)$. The single-carrier precoder $\widetilde{\mathbf{U}}^{(t)}(\omega)$ is frequency-dependent, and its implementation is not yet widely known. The precoder $\widetilde{\mathbf{U}}^{(t)}(\omega)$ should be designed to have minimal GD spread over the signal bandwidth, allowing it to be implemented in the time domain. If the GD spread of $\widetilde{\mathbf{U}}^{(t)}(\omega)$ is too large, a frequency-domain implementation is required, similar to the FDE in a receiver. Theoretically, any precoder described by

$$\widetilde{\mathbf{U}}^{(t)}(\omega)\mathrm{diag}\left[e^{-j\omega\tau_1^{(p)}}, e^{-j\omega\tau_2^{(p)}}, \ldots, e^{-j\omega\tau_D^{(p)}}\right] \tag{11.83}$$

for any delay vector $\boldsymbol{\tau}^{(p)} = \left(\tau_1^{(p)}, \tau_2^{(p)}, \ldots, \tau_D^{(p)}\right)$ will yield equivalent system performance. In designing the precoder, the vector $\boldsymbol{\tau}^{(p)}$ should be chosen to minimize the GD spread of (11.83).

In the receiver, all D spatial channels are optically downconverted and sampled, broken into blocks, and converted to the frequency domain using FFTs. Assuming MMSE equalization, the blocks are equalized by

$$\left[\mathbf{H}^*(\omega)\mathbf{H}(\omega) + \mathbf{S}_n(\omega)\right]^{-1}\mathbf{H}^*(\omega) \tag{11.84}$$

and converted back to time domain using IFFTs. In (11.84), $\mathbf{H}(\omega)$ is given by (11.78) and $\mathbf{S}_n(\omega)$ is the power spectrum of the sampled noise. Throughput may be maximized using overlap-add or overlap-save convolution [165,166], while receiver complexity is minimized by inserting a cyclic prefix [167] or unique word [168–170] at the transmitter. Transmission of the cyclic prefix or unique word reduces throughput and wastes energy, but the unique word can be used as pilots to aid in timing, frequency, or phase synchronization. Overlap-add or overlap-save convolution is assumed in the complexity analysis in Section 11.6.3.3.

When MDL causes the channel response $\mathbf{H}(\omega)$ to become non-unitary over space and frequency, each spatial output of the single-carrier equalizer (11.84) represents a linear combination of different frequency components, providing the averaging required for frequency diversity. In single-carrier systems, strong error-correction coding, typically decoded with soft information, such as log-likelihood ratios analogous to (11.80), provides coding gain, but is not essential to frequency diversity, unlike OFDM systems [171].

When CSI is not available at the transmitter, in the presence of significant frequency-dependent MDL, coded OFDM generally outperforms single-carrier modulation with the linear equalizer (11.84). Performance equivalent to OFDM can be achieved using single-carrier modulation with a decision-feedback equalizer (DFE) or a Tomlinson-Harashima precoder (THP) [172,173]. Unfortunately, both DFE and THP are extremely difficult to implement at very high speeds. Of course, if MDL is small and the overall channel (11.78) is nearly unitary, DFE or THP are not required for single-carrier modulation to have the same performance as OFDM.

Note that the above comparison between single-carrier and OFDM signals is in the context of linear propagation with MDL. In the presence of fiber nonlinearity, single-carrier signals are expected to tolerate a higher launched power and achieve a higher SNR [148], probably outweighing any performance advantage of OFDM in the linear regime.

11.6.3.3 Signal processing complexity

A primary goal of MDM is reducing energy consumption per bit [174]. Hence, the computational complexity and resulting energy consumption of DSP are of critical importance. In this section, we discuss the complexity of FDE in single- or multi-carrier systems. TDE in single-carrier systems is not considered here, as its complexity is typically far higher than FDE [18,20].

The received signal is sampled at a rate $r_{os} R_s$, where r_{os} is an oversampling ratio, usually between 1.5 and 2, and R_s represents the symbol rate for single-carrier modulation or the occupied bandwidth (an effective symbol rate) for OFDM. In an optimized architecture, only a fraction $1/r_{os}$ of the frequency-domain subchannels require equalization. In the OFDM system of Figure 11.20, only a fraction $1/r_{os}$ of the subchannels are modulated with nonzero amplitudes. In the single-carrier FDE receiver of Figure 11.21, it is only necessary to compute the IFFT over the middle subcarriers, which comprise a fraction $1/r_{os}$ of the total number of subchannels. In the FDE, the IFFT also downsamples the signal by a factor $1/r_{os}$ to the symbol rate R_s.

Receiver computational complexity depends strongly on the GD spread of the equivalent channel $\mathbf{H}(\omega)$, given by (11.75). The GD spread arises mainly from CD and MD. The duration of the impulse response of CD, measured in samples, is given by [18,175,176]

$$N_{cd} = \left\lceil 2\pi \left| \bar{\beta}_2 \right| L_t \left(r_{os} R_s \right)^2 \right\rceil, \tag{11.85}$$

where $\bar{\beta}_2$ is the mode-averaged CD coefficient described in Section 11.2.3.3, L_t is the total system length, and $\lceil x \rceil$ denotes the smallest integer larger than x. Assuming the strong-coupling regime, the duration of the impulse response of MD, measured in samples, is

$$N_{md} = \left\lceil \sigma_{gd} u_D(p) r_{os} R_s \right\rceil, \tag{11.86}$$

where σ_{gd} is the STD of coupled GD defined in (11.45) and $u_D(p)$ is a function of the number of modes D and a GD spread outage probability p [18], which is the probability that a random realization of the impulse response has duration longer than (11.86). The value of $u_D(p)$ is given by the complementary cumulative GD spread distribution, as shown in Figure 11.7. For suitably small values of p ($p = 10^{-6}$ is considered here), $u_D(p) = 4.94$ and 4.60 for $D = 6$ and 12, respectively, and decreases toward 4.00 as $D \to \infty$, since the support of the semicircle distribution approaches $4\sigma_{gd}$. In the duration (11.86), the product $\sigma_{gd} R_s$ approximately equals the normalized bandwidth b defined in Section 11.6.2 (equality is exact for OFDM signals).

Instead of choosing the FFT size to minimize computational complexity as in [18], we choose $N_{FFT} = 2^{\lceil 3 + \log_2(N_{cd} + N_{md}) \rceil}$, where the number 3 (corresponding to a factor of 8) is chosen as a compromise to reduce complexity without requiring an excessively large value of FFT size N_{FFT}. In an OFDM system, the IFFT in the transmitter has the same length as the FFT in the receiver. In a single-carrier receiver using FDE, the FFT size can be a factor r_{os} larger than the IFFT size. For simplicity, we consider each FFT (or IFFT) to require approximately $\frac{1}{2} N_{FFT} \lceil \log_2 N_{FFT} \rceil$ complex multiplications. In practice, the required number of complex multiplications depends on the FFT architecture, but the operations required per sample are always proportional to $\log_2 N_{FFT}$.

The OFDM receiver of Figure 11.20 may compensate MD and CD together using D complex multiplications, according to (11.82). Here, we assume that MD and CD are compensated separately using $D + 1$ complex multiplications. Assuming a sufficient cyclic prefix, and considering both transmitter and receiver, the required number of complex multiplications per two-dimensional symbol for OFDM is

$$CM_{OFDM} = 1 + D + r_{os} \log_2 N_{FFT}. \tag{11.87}$$

In the single-carrier receiver of Figure 11.21, the total number of complex multiplications per two-dimensional symbol is

$$CM_{FDE} = \frac{(1 + D)N_{FFT} + \frac{1}{2} r_{os} N_{FFT} \log_2 N_{FFT} + \frac{1}{2} N_{FFT} \lceil \log_2(N_{FFT}/r_{os}) \rceil}{N_{FFT} - N_{cd} - N_{md} + 1}, \tag{11.88}$$

where $D(N_{FFT} - N_{cd} - N_{md} + 1)/r_{os}$ is the number of symbols processed per FFT block, assuming overlap-add or overlap-save convolution. Single-carrier FDE requires fewer FFT operations than OFDM signals, but achieves lower computational efficiency due to overlap convolution.

Figure 11.22 shows the number of complex multiplications per two-dimensional symbol as calculated by (11.88). The total system length is $L_t = 2000$ km and the mode-averaged CD is $\bar{\beta}_2 = -22.5$ ps²/km (corresponding to 17.5 ps/nm/km at the wavelength of 1550 nm). Assuming fibers with numerical aperture NA $= 0.15$ and a graded-index depressed-cladding index profile, using values from [18], the STD of the uncoupled MD is

$$\left[\frac{1}{D} \sum_{\mu} \left(\beta_{1,\mu} - \bar{\beta}_1 \right)^2 \right]^{1/2} = 277 \text{ and } 383 \text{ ps/km}$$

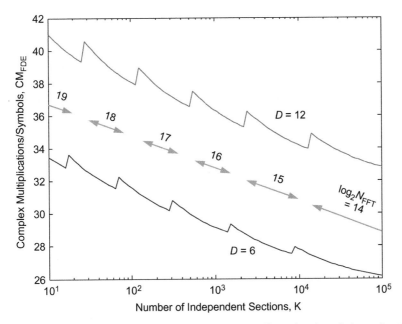

FIGURE 11.22 Number of complex multiplications per two-dimensional symbol as a function of the number of independent fiber sections for $D = 6$ and 12 modes for single-carrier modulation using FDE. Also indicated are the approximate ranges for different FFT block lengths.

for $D = 6$ and 12, respectively, corresponding to the value of σ_τ per kilometer in (11.45). The symbol rate is $R_s = 32$ GHz. An oversampling ratio $r_{os} = 1.5$ is assumed. In practice, although $r_{os} = 1.5$ cannot be achieved exactly using power-of-two FFT sizes, a value close to $r_{os} = 1.5$ such that N_{FFT}/r_{os} is an integer can be achieved using an appropriate FFT architecture. Figure 11.22 shows the number of complex multiplications per symbol as a function of the number of independent sections K. As K ranges from 10 to 10^5, the length of each section ranges from 200 km to 2 m. Over this wide range, the computational complexity is relatively insensitive to K. As K increases and the GD spread N_{md} decreases, the FFT size N_{FFT} decreases, but this reduces both the numerator and denominator of (11.88).

In Figure 11.22, the computational complexity exhibits small jumps when the FFT size N_{FFT} changes. The values of N_{FFT}, which are indicated in Figure 11.22, can be very large, which may have adverse implications for system practicality. In OFDM systems, large values of N_{FFT} can cause sensitivity to carrier frequency offset and phase noise-induced inter-carrier interference [171] and may require phase noise cancelation [177,178]. In single-carrier systems, equalization-enhanced phase noise increases with the GD spread from MD [179], similar to the effect of CD in SMF systems [180–182]. The convergence times of adaptive equalizers [18], and DSP memory requirements, increase at least linearly with N_{FFT}, and both can become prohibitive for large values of N_{FFT}.

For the particular type of MMF considered in Figure 11.22, the above issues are mitigated if the number of independent sections K is larger than 500–1000, corresponding to a mode-coupling correlation length shorter than 2–4 km. Within this range, the MD duration N_{md} (11.86) is about 6 to 9 times CD duration N_{cd} (11.85), N_{FFT} is less than 2^{16}, and compensating MD and the CD requires less than 30 or 37 complex multiplications per symbol for $D = 6$ and 12 modes, respectively. The corresponding normalized bandwidth in this range is always larger than $b = 500$. For comparison, compensating CD only requires $N_{FFT} = 2^{13}$, and about 19 complex multiplications per symbol.

As mentioned in Section 11.2.2 above, a short correlation length can be achieved by intentionally introducing perturbations in the MMF that enhance mode coupling, analogous to the spinning used to reduce DGD in SMF with PMD [48,71–74]. It would be helpful to employ MMFs having lower values of uncoupled GD spread than those considered here. As of this writing, however, these have been demonstrated only for $D = 6$ modes [1,183,184].

The overall complexity for MIMO DSP in MDM is expected to scale at least in proportion to $D \cdot N_{FFT}$. Given the high symbol rates employed, careful attention to DSP architecture optimization and application-specific integrated circuit implementation is required. In particular, FFT architectures must be chosen carefully, considering the available options. These include: radix-2 FFTs based on the Cooley-Tukey algorithm [185], which are often the most straightforward; split-radix FFTs [186,187], which minimize the total number of operations (multiplications and additions); Winograd FFTs [188], which minimize the number of multiplications; and prime-factor FFTs [189,190], which can implement FFTs of flexible size while avoiding twiddle factor multiplications. Also, buffer sizes and read/write bandwidths must be considered carefully, including those for performing FFTs [191,192] and those for storing channel matrix coefficients for LLR computations (11.82) or FDE (11.84). At each frequency in (11.82) or (11.84), a total of D^2 complex-valued channel coefficients must be stored.

11.7 CONCLUSION

Fields propagating in a multimode fiber with mode coupling are strongly correlated over distances shorter than a correlation length, and are weakly correlated over longer distances. In a transmission system, mode coupling may be classified as weak or strong, depending on whether the total system length is comparable to or much longer than the correlation length. In the weak-coupling regime, MD and MDL accumulate linearly with system length, while in the strong-coupling regime they accumulate with the square-root of system length. For a given system length, both MD and MDL are minimized by strong mode coupling.

Strong mode coupling may be essential for the practicality of long-haul MDM systems, which employ inline optical amplifiers and use coherent detection and

MIMO DSP to compensate MD and to separate multiplexed signals. A system's end-to-end GD spread governs the MIMO DSP complexity. Achieving sufficiently low GD spread may require intentionally perturbing the MMF to ensure strong mode coupling, analogous to the spinning used to reduce the GD spread in SMF with PMD. MDL arising from inline amplifiers can cause random signal fluctuations analogous to multipath fading in wireless systems. This MDL can reduce average channel capacity and may cause outage in narrowband systems. Strong mode coupling can minimize MDL and maximize average capacity. In combination with sufficient MD, strong coupling leads to frequency diversity, which reduces outage probability in wideband systems, allowing outage capacity to approach average capacity.

The operator describing strongly coupled MD is equivalent to a zero-trace Gaussian unitary ensemble, and the statistics of the GDs are the same as the statistics of the eigenvalues of the ensemble. The STD of coupled GD scales as the square-root of system length. The probability density of the GDs can be calculated analytically for few-mode fibers, and approaches a Wigner semicircle distribution for many-mode fibers. The statistics of the coupled GD spread, which governs DSP complexity, can be computed using the Fredholm determinant or approximated using the Tracy-Widom distribution. Strongly coupled MDL, when described on a logarithmic scale in the low-MDL regime of practical interest, follows the same statistics as strongly coupled GDs. Accumulated MDL scales with the square-root of system length. The overall MDL is determined from the accumulated MDL by a nonlinear relationship.

Acknowledgments

The authors are grateful for the contributions of S.Ö. Arık to Section 11.6.3.3. The research of J.M. Kahn was supported, in part, by National Science Foundation Grant No. ECCS-1 101 905 and Corning, Inc.

References

[1] R. Ryf, S. Randel, A.H. Gnauck, C. Bolle, A. Sierra, S. Mumtaz, M. Esmaeelpour, E.C. Burrows, R.-J. Essiambre, P.J. Winzer, D.W. Peckham, A.H. McCurdy, R. Lingle, Mode-division multiplexing over 96 km of few-mode fiber using coherent 6×6 MIMO processing, J. Lightwave Technol. 30 (4) (2012) 521–531.

[2] A. Al Amin, A. Li, S. Chen, X. Chen, G. Gao, W. Shieh, Dual-LP$_{11}$ mode 4×4 MIMO-OFDM transmission over a two-mode fiber, Opt. Express 19 (17) (2011) 16672–16678.

[3] S. Randel, R. Ryf, A. Sierra, P. Winzer, A.H. Gnauck, C. Bolle, R.-J. Essiambre, D.W. Peckham, A. McCurdy, R. Lingle, 6×56-Gb/s mode-division multiplexed transmission over 33-km few-mode fiber enabled by 6×6 MIMO equalization, Opt. Express 19 (17) (2011) 16697–16707.

[4] C. Koebele, M. Salsi, D. Sperti, P. Tran, P. Brindel, H. Margoyan, S. Bigo, A. Boutin, F. Verluise, P. Sillard, M. Bigot-Astruc, L. Provost, F. Cerou, G. Charlet, Two mode transmission at 2×100Gb/s, over 40km-long prototype few-mode fiber, using LCOS based mode multiplexer and demultiplexer, Opt. Express 19 (17) (2011) 16593–16600.

[5] B. Zhu, T.F. Taunay, M. Fishteyn, X. Liu, S. Chandrasekhar, M.F. Yan, J.M. Fini, E.M. Monberg, F.V. Dimarcello, 112-Tb/s Space-division multiplexed DWDM transmission with 14-b/s/Hz aggregate spectral efficiency over a 76.8-km seven-core fiber, Opt. Express 19 (17) (2011) 16665–16671.

[6] J. Sakaguchi, B.J. Puttnam, W. Klaus, Y. Awaji, N. Wada, A. Kanno, T. Kawanishi, K. Imamura, H. Inaba, K. Mukasa, R. Sugizaki, T. Kobayashi, M. Watanabe, 19-Core fiber transmission of $19 \times 100 \times 172$-Gb/s SDM-WDM-PDM-QPSK signals at 305Tb/s, in: OFC '12, Paper PDP5C.

[7] S. Berdagué, P. Facq, Mode division multiplexing in optical fibers, Appl. Opt. 21 (11) (1982) 1950–1955.

[8] S. Murshid, B. Grossman, P. Narakorn, Spatial domain multiplexing: A new dimension in fiber optic multiplexing, Opt. Laser Technol. 40 (2008) 1030–1036.

[9] C.P. Tsekrekos, A.M.J. Koonen, Mode-selective spatial filtering for increasing robustness in mode group diversity multiplexing link, Opt. Lett. 32 (9) (2007) 1041–1043.

[10] K.-P. Ho, J.M. Kahn, Statistics of group delays in multimode fiber with strong mode coupling, J. Lightwave Technol. 29 (21) (2011) 3119–3128.

[11] R. Olshansky, Mode-coupling effects in graded-index optical fibers, Appl. Opt. 14 (4) (1975) 935–945.

[12] R. Ryf, M.A. Mestre, A.H. Gnauck, S. Randel, C. Schmidt, R.-J. Essiambre, P.J. Winzer, R. Delbue, P. Pupalaikis, A. Sureka, Y. Sun, X. Jiang, D.W. Peckham, A. McCurdy, R. Lingle, Low-loss mode coupler for mode-multiplexed transmission in few-mode fiber, in: OFC '12, Paper PDP5B.

[13] X. Shen, J.M. Kahn, M.A. Horowitz, Compensation for multimode fiber dispersion by adaptive optics, Opt. Lett. 30 (20) (2005) 2985–2987.

[14] H.S. Chen, H.P.A. van den Boom, A.M.J. Koonen, 30-Gb/s 3×3 optical mode group-division-multiplexing system with optimized joint detection, IEEE Photon. Technol. Lett. 23 (18) (2011) 1283–1285.

[15] J. Carpenter, T. Wilkinson, All optical mode-multiplexing using holography and multimode fiber couplers, J. Lightwave Technol. 30 (12) (2012) 1978–1984.

[16] A.F. Garito, J. Wang, R. Gao, Effects of random perturbations in plastic optical fibers, Science 281 (5379) (1998) 962–967.

[17] M.B. Shemirani, W. Mao, R.A. Panicker, J.M. Kahn, Principal modes in graded-index multimode fiber in presence of spatial- and polarization-mode coupling, J. Lightwave Technol. 27 (10) (2009) 1248–1261.

[18] S.Ö. Arik, D. Askarov, J.M. Kahn, Effect of mode coupling on signal processing complexity in mode-division multiplexing, J. Lightwave Technol. 31 (3) (2013) 423–431.

[19] C. Antonelli, A. Mecozzi, M. Shtaif, P.J. Winzer, Stokes-space analysis of modal dispersion in fibers with multiple mode transmission, Opt. Express 20 (11) (2012) 11718–11733.

[20] B. Inan, B. Spinnler, F. Ferreira, D. van de Borne, A. Lobato, S. Adhikari, V.A.J.M. Slieffer, M. Kuschnerov, N. Hanik, S.L. Jansen, DSP requirements for mode-division multiplexed receivers, Opt. Express 20 (9) (2012) 10859–10869.

[21] R. Olshansky, D.A. Nolan, Mode-dependent attenuation of optical fibers; Excess loss, Appl. Opt. 15 (4) (1976) 1045–1047.

[22] A.R. Mickelson, M. Eriksrud, Mode-dependent attenuation in optical fibers, J. Opt. Soc. Am. 73 (10) (1983) 1282–1290.

[23] N. Bai, E. Ip, Y.-K. Huang, E. Mateo, F. Yaman, M.-J. Li, S. Bickham, S. Ten, J. Liñares, C. Montero, V. Moreno, X. Prieto, V. Tse, K.M. Chung, A.P.T. Lau, H.-Y. Tam, C. Lu, Y. Luo, G.-D. Peng, G. Li, T. Wang, Mode-division multiplexed transmission with inline few-mode fiber amplifier, Opt. Express 20 (3) (2012) 2668–2680.

[24] Y. Jung, S. Alam, Z. Li, A. Dhar, D. Giles, I.P. Giles, J.K. Sahu, F. Poletti, L. Grüner-Nielsen, D.J. Richardson, First demonstration and detailed characterization of a multimode amplifier for space division multiplexed transmission systems, Opt. Express 19 (26) (2011) B952–B957.

[25] D. Askarov, J.M. Kahn, Design of multi-mode Erbium-doped fiber amplifiers for low mode-dependent gain, in: IEEE Summer Topical on Spatial Multiplexing '12.

[26] D. Tse, P. Viswanath, Fundamentals of Wireless Communication, Cambridge University Press, 2005.

[27] A.M. Tulino, S. Verdú, Random Matrix Theory and Wireless Communications, Now, 2004.

[28] K.-P. Ho, J.M. Kahn, Mode-dependent loss and gain: Statistics and effect on mode-division multiplexing, Opt. Express 19 (17) (2011) 16612–16635.

[29] P.J. Winzer, G.J. Foschini, MIMO capacities and outage probabilities in spatially multiplexed optical transport systems, Opt. Express 19 (17) (2011) 16680–16696.

[30] K.-P. Ho, J.M. Kahn, Frequency diversity in mode-division multiplexing systems, J. Lightwave Technol. 29 (24) (2011) 3719–3726.

[31] D. Marcuse, Theory of Dielectric Optical Waveguide, second ed., Academic Press, 1991.

[32] L. Raddatz, I.H. White, D.G. Cunningham, M.C. Nowell, An experimental and theoretical study of the offset launch technique for the enhancement of the bandwidth of multimode fiber links, J. Lightwave Technol. 16 (3) (1998) 324–331.

[33] P. Pepeljugoski, S.E. Golowich, A.J. Ritger, P. Kolsar, A. Risteski, Modeling and simulation of next-generation multimode fiber links, J. Lightwave Technol. 21 (5) (2003) 1242–1254.

[34] K.M. Patel, S.E. Ralph, Enhanced multimode fiber link performance using a spatially resolved receiver, IEEE Photon. Technol. Lett. 14 (3) (2002) 393–395.

[35] K. Balemarthy, A. Polley, S.E. Ralph, Electronic equalization of multikilometer 10-Gb/s multimode fiber links: mode-coupling effects, J. Lightwave Technol. 24 (12) (2006) 4885–4894.

[36] J.P. Gordon, H. Kogelnik, PMD fundamentals: polarization-mode dispersion in optical fibers, Proc. Natl. Acad. Sci. 97 (9) (2000) 4541–4550.

[37] H. Kogelnik, R.M. Jopson, L.E. Nelson, Polarization-mode dispersion, in: I. Kaminow, T. Li (Eds.), Optical Fiber Telecommunications IVB, Academic Press, 2002.

[38] C.D. Poole, R.E. Wagner, Phenomenological approach to polarization dispersion in long single-mode fibers, Electron. Lett. 22 (19) (1986) 1029–1030.

[39] R. Noé, D. Sandel, M. Yoshida-Dierolf, S. Hinz, V. Mirvoda, A. Schöpflin, C. Glingener, E. Gottwald, C. Scheerer, G. Fischer, T. Weyrauch, W. Haase, Polarization mode dispersion compensation at 10, 20, and 40 Gb/s with various optical equalizers, J. Lightwave Technol. 17 (9) (1999) 1602–1616.

[40] H. Sunnerud, C. Xie, M. Karlsson, R. Samuelsson, P.A. Andrekson, A comparison between different PMD compensation techniques, J. Lightwave Technol. 20 (3) (2002) 368–378.

[41] S. Fan, J.M. Kahn, Principal modes in multimode waveguides, Opt. Lett. 20 (2) (2006) 135–137.

[42] Y. Kokubun, M. Koshiba, Noval multi-core fiber for mode division multiplexing: proposal and design principle, IEICE Electron. Express 6 (8) (2009) 522–528.

[43] A.W. Snyder, Coupled-mode theory for optical fibers, J. Opt. Soc. Am. 62 (11) (1972) 1267–1277.

[44] E. Snitzer, Cylindrical dielectric waveguide modes, J. Opt. Soc. Am. 51 (5) (1961) 491–498.

[45] K.C. Kao, G.A. Hockham, Dielectric-fibre surface waveguides for optical frequencies, Proc. IEE 113 (7) (1966) 1151–1158.

[46] D. Gloge, Weakly guiding fibers, Appl. Opt. 10 (10) (1971) 2252–2258.

[47] D. Gloge, E.A.J. Marcatilli, Multimode theory of graded-core fibers, Bell Syst. Tech. J. 52 (9) (1973) 1563–1578.

[48] M.-J. Li, D.A. Nolan, Optical transmission fiber design evolution, J. Lightwave Technol. 26 (9) (2008) 1079–1092.

[49] A.W. Snyder, J.D. Love, Optical Waveguide Theory, Chapman & Hall, 1983.

[50] G. Keiser, Optical Fiber Communications, third ed., McGraw-Hill, 2000.

[51] D. Gloge, Dispersion in weakly guiding fibers, Appl. Opt. 10 (11) (1971) 2442–2445.

[52] C. Lin, H. Kogelnik, L.G. Cohen, Optical-pulse equalization of low-dispersion transmission in single-mode fibers in the 1.3–1.7-m spectral region, Opt. Lett. 5 (11) (1980) 476–478.

[53] ISO/IEC 11801, Information technology—Generic cabling for customer premises, 2002.

[54] M. Webster, L. Raddatz, I.H. White, D.G. Cunningham, A statistical analysis of conditioned launch for gigabit Ethernet links using multimode fiber, J. Lightwave Technol. 17 (9) (1999) 1532 1541.

[55] Y. Koike, T. Ishigure, E. Nihei, High-bandwidth graded-index polymer optical fiber, J. Lightwave Technol. 13 (7) (1995) 1475–1489.

[56] Y. Koike, S. Takahashi, Plastic optical fibers: technologies and communication links, in: I.P. Kaminow, T. Li, A.E. Willner (Eds.), Optical Fiber Telecommunications VA, Elsevier Academic, 2008 (Chapter 16).

[57] A. Polley, S.E. Ralph, Mode coupling in plastic optical fiber enables 40-Gb/s performance, IEEE Photon. Technol. Lett. 19 (16) (2007) 1254–1256.

[58] R.F. Shi, C. Koeppen, G. Jiang, J. Wang, A.F. Garito, Origin of high bandwidth performance of graded-index plastic fibers, Appl. Phys. Lett. 72 (25) (1999) 3625–3627.

[59] J.G. Proakis, M. Salehi, Digital Communications, fifth ed., McGraw-Hill, 2008.

[60] T.S. Rapport, Wireless Communications: Principals & Practice, Prentice Hall, 1996.

[61] M. Koshiba, K. Saitoh, Y. Kobubun, Heterogeneous multi-core fibers: proposal and design principle, IEICE Electron. Express 6 (2) (2009) 98–103.

[62] T. Hayashi, T. Taru, O. Shimakawa, T. Sasaki, E. Sasaoka, Design and fabrication of ultra-low crosstalk and low-loss multi-core fiber, Opt. Express 19 (17) (2011) 16576–16592.

[63] C. Xia, N. Bai, I. Ozdur, X. Zhou, G. Li, Supermodes for optical transmission, Opt. Express 19 (17) (2011) 16653–16664.

[64] K. Nagano, S. Kawakami, S. Nishida, Change of the refractive index in an optical fiber due to external forces, Appl. Opt. 17 (13) (1978) 2080–2085.

[65] Y. Mitsunaga, Y. Katsuyama, Y. Ishida, Thermal characteristics of jacketed optical fibers with initial imperfection, J. Lightwave Technol. 2 (1) (1984) 18–24.

[66] M. Tokuda, S. Seikai, K. Yoshida, N. Uchida, Measurement of baseband frequency response of multimode fibre by using a new type of mode scrambler, Electron. Lett. 13 (5) (1977) 146–147.

[67] M. Ikeda, Y. Murakami, K. Kitayama, Mode scrambler for optical fiber, Appl. Opt. 16 (4) (1977) 1045–1049.

[68] L. Su, K.S. Chiang, C. Lu, Microbend-induced mode coupling in a graded-index multimode fiber, Appl. Opt. 44 (34) (2005) 7394–7402.

[69] K. Kitayama, S. Seikai, N. Uchida, Impulse response prediction based on experimental mode coupling coefficient in a 10-km long graded-index fiber, IEEE J. Quant. Electron. 16 (3) (1980) 356–362.

[70] S. Randel, R. Ryf, A.H. Gnauck, M.A. Mestre, C. Schmidt, R.-J. Essiambre, P.J. Winzer, R. Delbue, P. Pupalaikis, A. Sureka, Y. Sun, X. Jiang, R. Lingle, Mode-multiplexed 6 20-GBd QPSK transmission over 1200-km DGD-compensated few-mode fiber, in: OFC '12, Paper PDP5C.

[71] D.A. Nolan, X. Chen, M.-J. Li, Fiber with low polarization-mode dispersion, J. Lightwave Technol. 22 (4) (2004) 1066–1077.

[72] A.J. Barlow, J.J. Ramskov-Hansen, D.N. Payne, Birefringence and polarization mode-dispersion in spun single-mode fibers, Appl. Opt. 20 (17) (1981) 2962–2968.

[73] M.-J. Li, D.A. Nolan, Fiber spin-profile designs for producing fibers with low polarization mode dispersion, Opt. Lett. 23 (1998) 1659–1661.

[74] A. Galtarossa, L. Palmieri, A. Pizzinat, Optimized spinning design for low PMD fibers: An analytical approach, J. Lightwave Technol. 19 (10) (2001) 1502–1512.

[75] W.F. Love, Novel mode scrambler for use in optical-fiber bandwidth measurements, in: OFC '79, Paper ThG2.

[76] Y. Koyamada, T. Horiguchi, M. Tokuda, N. Uchida, Theoretical analysis of optical fiber mode exciters constructed with alternate concatenation of step-index and graded-index fibers, Electron. Commun. J., pt. I 68 (3) (1985) 66–74.

[77] M. Saffman, D.Z. Anderson, Mode multiplexing and holographic demultiplexing communication channels on a multimode fiber, Opt. Lett. 16 (5) (1991) 300–302.

[78] A. Yariv, Coupled-mode theory for guided-wave optics, IEEE J. Quant. Electron. 9 (9) (1973) 919–933.

[79] D. Gloge, Optical power flow in multimode fibers, Bell Syst. Tech. J. 51 (8) (1972) 1767–1780.

[80] D. Marcuse, Losses and impulse response in parabolic index fibers with random bends, Bell Syst. Tech. J. 52 (8) (1973) 1423–1437.

[81] S.S.-H. Yam, F.-T. An, M.E. Marhic, L.G. Kazovsky, Polarization sensitivity of 40 Gb/s transmission over short-reach 62.5 m multimode fiber, in: OFC '04, Paper FA-5.

[82] M.B. Shemirani, J.M. Kahn, Higher-order modal dispersion in graded-index multimode fiber, J. Lightwave Technol. 27 (23) (2009) 5461–5468.

[83] C.R. Menyuk, P.K.A. Wai, Polarization evolution and dispersion in fibers with spatially varying birefringence, J. Opt. Soc. Am. B 11 (7) (1994) 1288–1296.

[84] H. Kogelnik, P.J. Winzer, Modal birefringence in weakly guiding fibers, J. Lightwave Technol. 30 (14) (2012) 2240–2245.

[85] C.D. Poole, J.M. Wiesenfeld, A.R. McCormick, K.T. Nelson, Broadband dispersion compensation by using the higher-order spatial mode in a two-mode fiber, Opt. Lett. 17 (14) (1992) 985–987.

[86] A.H. Gnauck, L.D. Garrett, Y. Danziger, U. Levy, M. Tur, Dispersion and dispersion-slope compensation of NZDSF over the entire C band using higher-order-mode fibre, Electron. Lett. 36 (23) (2000) 1946–1947.

[87] A. Huttunen, P. Törmä, Optimization of dual-core and microstructure fiber geometries for dispersion compensation and large mode area, Opt. Express 13 (2) (2005) 627–635.

[88] G. Yabre, Comprehensive theory of dispersion in graded-index optical fibers, J. Lightwave Technol. 18 (3) (2000) 166–177.

[89] G.J. Foschini, C.D. Poole, Statistical theory of polarization dispersion in single mode fibers, J. Lightwave Technol. 9 (11) (1991) 1439–1456.

[90] C.D. Poole, Statistical treatment of polarization dispersion in single-mode fiber, Opt. Lett. 13 (8) (1988) 687–689.

[91] P.K.A. Wai, C.R. Menyuk, Polarization mode dispersion, decorrelation, and diffusion in optical fibers with randomly varying birefringence, J. Lightwave Technol. 14 (2) (1996) 148–157.

[92] A. Galtarossa, L. Palmieri, M. Schiano, T. Tambosso, Measurement of biregringence correlation length in long, single-mode fibers, Opt. Lett. 26 (13) (2001) 962–964.

[93] G.J. Foschini, L.E. Nelson, R.M. Jopson, H. Kogelnik, Probability densities of second-order polarization mode dispersion including polarization dependent chromatic fiber dispersion, IEEE Photon. Technol. Lett. 12 (3) (2000) 293–295.

[94] G.J. Foschini, L.E. Nelson, R.M. Jopson, H. Kogelnik, Statistics of second-order PMD depolarization, J. Lightwave Technol. 19 (12) (2001) 1882–1886.

[95] R.A. Panicker, A.P.T. Lau, J.P. Wilde, J.M. Kahn, Experimental comparison of adaptive optics algorithms in 10-Gb/s multimode fiber transmission, J. Lightwave Technol. 27 (24) (2009) 5783–5789.

[96] M.L. Mehta, Random Matrices, third ed., Elsevier Academic, 2006.

[97] J. Ginibre, Statistical ensembles of complex, quaternion, and real matrices, J. Math. Phys. 6 (3) (1965) 440–450.

[98] H.G. Golub, C.F. van Loan, Matrix Computations, third ed., Johns Hopkins, 1996.

[99] S.N. Majumdar, O. Bohigas, A. Lakshminarayan, Exact minimum eigenvalues distribution of entangled random pure state, J. Stat. Phys. 131 (1) (2008) 33–49.

[100] Y. Chen, D.-Z. Liu, D.-S. Zhou, Smallest eigenvalue distribution of the fixed-trace Laguerre beta-ensemble, J. Phys. A: Math. Theor. 43 (2010) 315303.

[101] J.L. Schiff, The Laplace Transform—Theory and Applications, Springer, 1999 (Chapter 4).

[102] M. Karlsson, Probability density functions of the differential group delay in optical fiber communication systems, J. Lightwave Technol. 19 (3) (2001) 324–331.

[103] E. Wigner, Characteristic vectors of bordered matrices with infinite dimensions, Ann. Math. 62 (3) (1955) 548–564.

[104] E. Wigner, On the distribution of the roots of certain symmetric matrices, Ann. Math. 67 (2) (1958) 325–328.

[105] T. Tao, V.H. Vu, From the Littlewood-Offord problem to the circular law: universality of the spectral distribution of random matrices, Bull. Am. Math. Soc. 46 (3) (2009) 337–396.

[106] L. Erdös, J. Ramíez, B. Schlein, T. Tao, V.H. Vu, H.T. Yau, Bulk universality for Wigner Hermitian matrices with subexponential decay, Math. Res. Lett. 17 (4) (2010) 667–674.

[107] F. Götze, M. Gordin, Limit correlation functions for fixed trace random matrix ensembles, Commun. Math. Phys. 281 (1) (2008) 203–229.

[108] D.-Z. Liu, D.-S. Zhou, Some universal properties for restricted trace Gaussian orthogonal, unitary and symplectic ensembles, J. Stat. Phys. 140 (2) (2010) 268–288.

[109] D. Voiculescu, K. Dykema, A. Nica, Free Random Variables, CRM Monograph Series vol. 1, American Mathematical Society, 1992.

[110] A. Nica, R. Speicher, Lectures on the Combinatorics of Free Probability, London Mathematical Society Lecture Note Series, vol. 335, Cambridge Univ. Press, 2006.

[111] D. Voiculescu, Limit laws for random matrices and free products, Invent. Math. 104 (1) (1991) 201–220.

[112] C.A. Tracy, H. Widom, Fredholm determinants, differential equations and matrix models, Commun. Math. Phys. 163 (1) (1994) 33–72.

[113] F. Bornemann, On the numerical evaluation of Fredholm determinants, Math. Comput. 79 (270) (2010) 871–915.

[114] F. Bornemann, On the numerical evaluation of distribution in random matrix theory: A review, Markov Proc. Relat. Fields 16 (4) (2010) 803–866.

[115] C.A. Tracy, H. Widom, Level-spacing distributions and the Airy kernel, Phys. Lett. B 305 (1–2) (1993) 115–118.

[116] C.A. Tracy, H. Widom, Level-spacing distributions and the Airy kernel, Commun. Math. Phys. 159 (1) (1994) 151–174.

[117] F. Bornemann, Asymptotic independence of the extreme eigenvalues of Gaussian unitary ensemble, J. Math. Phys. 51 (2010) 023514.

[118] K.-P. Ho, Exact model for mode-dependent gains and losses in multimode fiber, J. Lightwave Technol. 30 (23) (2012) 3603–3609.

[119] R. Bellman, Limit theorems for non-commutative operations I, Duke Math. J. 21 (3) (1954) 491–500.

[120] H. Furstenberg, H. Kesten, Products of random matrices, Ann. Math. Stat. 31 (2) (1960) 457–469.

[121] J.E. Cohen, C.M. Newman, The stability of large random matrices and their products, Ann. Probab. 12 (2) (1984) 283–310.

[122] A. Crisanti, G. Paladin, A. Vulpiani, Products of Random Matrices in Statistical Physics, Springer, 1993.

[123] M.A. Berger, Central limit theorem for products of random matrices, Trans. Am. Math. Soc. 285 (2) (1984) 777–803.

[124] A. Mecozzi, M. Shtaif, The statistics of polarization-dependent loss in optical communication systems, IEEE Photon. Technol. Lett. 14 (3) (2002) 313–315.

[125] P. Lu, L. Chen, X. Bao, Statistical distribution of polarization dependent loss in the presence of polarization mode dispersion in single mode fibers, IEEE Photon. Technol. Lett. 13 (5) (2001) 451–453.

[126] A. Galtarossa, L. Palmieri, The exact statistics of polarization-dependent loss in fiber-optic links, IEEE Photon. Technol. Lett. 15 (1) (2003) 57–59.

[127] P.W. Hooijmans, Coherent Optical System Design, John Wiley & Sons, 1994 (Section A.4).

[128] G.R. Grimmett, D.R. Stirzaker, Probability and Random Processes, second ed., Oxford, 1992.

[129] K.-P. Ho, Statistical properties of stimulated Raman crosstalk in WDM systems, J. Lightwave Technol. 18 (7) (2000) 915–921.

[130] K.-P. Ho, Central limits for the products of free random variables. <http://arxiv.org/abs/1101.5220>.

[131] D.E. Knuth, The Art of Computer Programming II: Seminumerical Algorithms, third ed., Addison Wesley, 1998.

[132] U. Levy, H. Kobrinsky, A.A. Friesem, Angular multiplexing for multichannel communication in a single fiber, IEEE J. Quant. Electron. QE-17 (11) (1981) 2215–2224.

[133] A. Amphawan, Holographic mode-selective launch for bandwidth enhancement in multimode fiber, Opt. Express 19 (10) (2011) 9056–9065.

[134] J. Carpenter, T.D. Wilkinson, Holographic offset launch for dynamic optimization and characterization of multimode fiber bandwidth, J. Lightwave Technol. 30 (10) (2012) 1437–1443.

[135] E. Alon, V. Stojanović, J.M. Kahn, S.P. Boyd and M.A. Horowitz, Equalization of modal dispersion in multimode fiber using spatial light modulators, in: GlobeCom '04.

[136] R.A. Panicker, J.P. Wilde, J.M. Kahn, D.F. Welch, I. Lyubomirsky, 10×10 Gb/s DWDM transmission through 2.2 km multimode fiber using adaptive optics, IEEE Photon. Technol. Lett. 19 (15) (2007) 1154–1156.

[137] B. Franz, H. Bülow, Experimental evaluation of principal mode groups as high-speed transmission channels in spatial multiplex systems, IEEE Photon. Technol. Lett. 24 (16) (2012) 1363–1365.

[138] C.R. Doerr, Proposed architecture for MIMO optical demultiplexing using photonic integration, IEEE Photon. Technol. Lett. 23 (21) (2011) 1573–1575.

[139] N.K. Fontaine, C.R. Doerr, M.A. Mestre, R. Ryf, P. Winzer, L. Buhl, Y. Sun, X. Jiang, R. Lingle, Space-division multiplexing and all-optical MIMO demultiplexing using a photonic integrated circuit, in: OFC '12, Paper PDP5B.

[140] H. Bülow, Optical-mode demultiplexing by optical MIMO filtering of spatial samples, IEEE Photon. Technol. Lett. 24 (12) (2012) 1045–1047.

[141] S.O. Haykin, Adaptive Filter Theory, fourth ed., Prentice Hall, 2001.

[142] A.J. Juarez, C.A. Bunge, S. Warm, K. Petermann, Perspectives of principal mode transmission in mode-division-multiplex operation, Opt. Express 20 (13) (2012) 13810–13823.

[143] M. Shtaif, A. Mecozzi, M. Tur, J.A. Nagel, A compensator for the effects of high-order polarization mode dispersion in optical fibers, IEEE Photon. Technol. Lett. 12 (4) (2000) 434–436.

[144] Q. Yu, L.-S. Yan, Y. Xie, M. Hauer, A.E. Willner, Higher order polarization mode dispersion compensation using a fixed time delay followed by a variable time delay, IEEE Photon. Technol. Lett. 13 (8) (2001) 863–865.

[145] S.G. Evangelides, L.F. Mollenauer, J.P. Gordon, N.S. Bergano, Polarization multiplexing with soliton, J. Lightwave Technol. 10 (1) (1992) 28–35.

[146] L. Nelson, H. Kogelnik, Coherent crosstalk impairments in polarization multiplexed transmission due to polarization mode dispersion, Opt. Express 7 (10) (2000) 350–361.

[147] Z. Wang, C. Xie, X. Ren, PMD and PDL impairments in polarization division multiplexing signals with direct detection, Opt. Express 17 (10) (2009) 7993–8004.

[148] E. Ip, J.M. Kahn, Fiber impairment compensation using coherent detection and digital signal processing, J. Lightwave Technol. 28 (4) (2010) 502–519.

[149] K. Koebele, M. Salsi, G. Charlet, S. Bigo, Nonlinear effects in mode-division-multiplexed transmission over few-mode optical fiber, IEEE Photon. Technol. Lett. 23 (18) (2011) 1316–1318.

[150] X. Chen, A. Li, G. Gao, A. Al Amin, W. Shieh, Characterization of fiber nonlinearity and analysis of its impact on link capacity limit of two-mode fibers, IEEE Photon. J. 4 (2) (2012) 455–460.

[151] A. Mecozzi, C. Antonelli, M. Shtaif, Nonlinear propagation in multi-mode fibers in the strong coupling regime, Opt. Express 20 (11) (2012) 11673–11678.

[152] S. Mumataz, R.-J. Essiambre, G.P. Agrawal, Nonlinear propagation in multimode and multicore fibers: Generalization of the Manakov equations, J. Lightwave Technol. 31 (3) (2013) 398–406.

[153] B. Vucetic, J. Yuan, Space-Time Coding, Wiley, 2003.

[154] S.M. Alamouti, A simple transmit diversity technique for wireless communications, IEEE J. Sel. Areas Commun. 16 (8) (1998) 1451–1458.

[155] V. Tarokh, H. Jafarkhani, A.R. Calderbank, Space-time block coding for wireless communications: Performance results, IEEE J. Sel. Areas Commun. 17 (3) (1999) 451–460.

[156] I.T. Jolliffe, Principal Component Analysis, second ed., Springer, New York, 2002.

[157] M. Vetterli, J. Kovačević, Wavelets and Subband Coding, Prentice Hall, 1995 (Section 7.1.1).

[158] Z. Tong, Q. Yang, Y. Ma, W. Shieh, 21.4 Gbit/s transmission over 200 km multimode fiber using coherent optical OFDM, Electron. Lett. 44 (23) (2008) 1373–1374.

[159] W. Shieh, H. Bao, Y. Tang, Coherent optical OFDM: theory and design, Opt. Express 16 (2) (2008) 841–859.

[160] A.J. Lowery, L.B. Du, J. Armstrong, Performance of optical OFDM in ultralong-haul WDM lightwave systems, J. Lightwave Technol. 25 (1) (2007) 131–138.

[161] T. Richardson, R. Urbanke, Modern Coding Theory, Cambridge Univ. Press, 2008.

[162] S. Lin, D.J. Costello, Error Control Coding, second ed., Prentice Hall, 2004.

[163] H. Sari, G. Karam, I. Jeanclaude, Transmission techniques for digital terrestrial TV broadcasting, IEEE Commun. Mag. (1995) 100–109 February.

[164] W.Y. Zou, Y. Wu, COFDM: An overview, IEEE Trans. Broadcast. 41 (1) (1995) 1–8.

[165] M.S. Faruk, K. Kikuchi, Adaptive frequency-domain equalization in digital coherent optical receivers, Opt. Express. 19 (13) (2011) 12789–12798.

[166] J.J. Shynk, Frequency-domain and multirate adaptive filtering, IEEE Sig. Process. Mag. 9 (1) (1992) 15–37.

[167] D. Falconer, S. Lek Ariyavisitakul, A. Benyamin-Seeyar, B. Eidson, Frequency domain equalization for single-carrier broadband wireless systems, IEEE Commun. Mag. 40 (4) (2002) 48–66.

[168] H. Witschnig, T. Mayer, A. Springer, A. Koppleer, L. Maurer, M. Huemer, R. Weigel, A different look on cyclic prefix for SC/FDE, in: IEEE PIMRC '02.

[169] J. Coon, M. Sandell, M. Beach, J. McGeehan, Channel and noise variance estimation and tracking algorithms for unique-word based single-carrier systems, IEEE Trans. Wireless Commun. 5 (6) (2006) 1488–1496.

[170] K. Kambara, H. Nishimoto, T. Nishimura, T. Ohgane, Y. Ogawa, Subblock processing in MMSE-FDE under fast fading environments, IEEE J. Sel. Areas Commun. 26 (3) (2008) 359–365.

[171] Z. Wang, X. Ma, G.B. Giannakis, OFDM or single-carrier block transmission? IEEE Trans. Commun. 52 (3) (2004) 380–394.

[172] N. Zervos, I. Kalet, Optimized decision feedback equalization versus optimized orthogonal frequency division multiplexing for high-speed data transmission over the local cable network, in: ICC '89.

[173] N. Benvenuto, S. Tomasin, On the comparison between OFDM and single carrier modulation with a FDE using frequency-domain feedforward filter, IEEE Trans. Commun. 50 (6) (2002) 947–955.

[174] P.J. Winzer, Energy-efficient optical transport capacity scaling through spatial multiplexing, IEEE Photon. Technol. Lett. 23 (13) (2011) 851–853.

[175] S.J. Savory, Digital coherent optical receivers: Algorithms and subsystems, IEEE J. Sel. Top. Quant. Electron. 16 (5) (2010) 1164–1179.

[176] S.J. Savory, Digital filters for coherent optical receivers, Opt. Express 16 (2) (2008) 804–817.

[177] J. Shentu, K. Pantu, J. Armstrong, Effects of phase noise on performance of OFDM systems using an ICI cancellation scheme, IEEE Trans. Broadcast. 49 (2) (2003) 221–224.

[178] S. Wu, Y. Bar-Ness, OFDM systems in the presence of phase noise: consequences and solutions, IEEE Trans. Commun. 52 (11) (2004) 1988–1996.

[179] W. Shieh, Interaction of laser phase noise with differential-mode-delay in few-mode fiber based MIMO systems, in: OFC '12.

[180] W. Shieh, K.-P. Ho, Equalization-enhanced phase noise for coherent-detection systems using electronic digital signal processing, Opt. Express 16 (20) (2008) 15718–15727.

[181] C. Xie, WDM coherent PDM-QPSK systems with and without inline optical dispersion compensation, Opt. Express 17 (6) (2009) 4815–4823.

[182] K.-P. Ho, A.P.T. Lau, W. Shieh, Equalization-enhanced phase noise induced timing jitter, Opt. Lett. 36 (4) (2011) 585–587.

[183] S. Randel, R. Ryf, A.H. Gnuack, M.A. Mestre, C. Schmidt, R.-J. Esiambre, R.J. Winzer, R. Delbue, P. Pupalaikis, A. Sureka, Y. Sun, X. Jiang, R. Lingle, Mode-multiplexed 6 × 20-GBd QPSK transmission over 1200-km DGD-compensated few-mode fiber, in: OFC '12, Paper PDP5C.5.

[184] L. Grüner-Nielsen, Y. Sun, J.W. Nicholson, D. Jakobsen, R. Lingle, B. Pálsdóttir, Few mode transmission fiber with low DGD, low mode coupling and low loss, in: OFC '12, Paper PDP5A.

[185] J.W. Cooley, J.W. Tukey, An algorithm for the machine calculation of complex Fourier series, Math. Comput. 19 (1965) 297–301.

[186] P. Duhamel, H. Hoolmann, Split-radix FFT algorithm, Electron. Lett. 20 (1) (1984) 14–16.

[187] H.V. Sorensen, M.T. Heideman, C.S. Burrus, On computing the split-radix FFT, IEEE Trans. Acoust. Speech Signal Process. 34 (1) (1986) 152–156.

[188] S. Winograd, On computing the discrete Fourier transform, Math. Comput. 32 (141) (1978) 175–199.

[189] I.J. Good, The interaction algorithm and practical Fourier analysis, J. Roy. Stat. Soc. B 20 (1958) 361–372 addendum, vol. 22, pp. 373–375, 1960.

[190] L.H. Thomas, Using a computer to solve problems in physics, Applications of Digital Computers, Ginn, 1963.

[192] Y.-W. Lin, H.-Y. Liu, C.-Y. Lee, A 1-GS/s FFT/IFFT processor for UWB applications, IEEE J. Solid-States Circ. 40 (8) (2005) 1726–1735.

Multimode Communications Using Orbital Angular Momentum

12

Jian Wang[a,b], Miles J. Padgett[c], Siddharth Ramachandran[d], Martin P.J. Lavery[c], Hao Huang[b], Yang Yue[b], Yan Yan[b], Nenad Bozinovic[d], Steven E. Golowich[e], and Alan E. Willner[b]

[a]*Wuhan National Laboratory for Optoelectronics, Huazhong University of Science and Technology, Wuhan 430074, Hubei, China*

[b]*Department of Electrical Engineering, University of Southern California, Los Angeles, California 90089, USA*

[c]*Department of Physics and Astronomy, University of Glasgow, UK*

[d]*Photonics Centre and ECE Department, Boston University, Boston, MA, USA*

[e]*Lincoln Laboratory, Massachusetts Institute of Technology, Cambridge, MA, USA*

12.1 PERSPECTIVE ON ORBITAL ANGULAR MOMENTUM (OAM) MULTIPLEXING IN COMMUNICATION SYSTEMS

The unabated exponential growth of broadband, mobile data, and cloud-based services has raised the challenges to find approaches to expending the capacity of networks. Miscellaneous multiplexing technologies together with advanced multi-level modulation formats have been widely employed to tackle the capacity crunch. A fundamental and straightforward way is to exploit different degrees of freedom or dimensions of a light beam. Generally speaking, the known physical dimensions of a light beam include frequency (wavelength), complex optical field (amplitude and phase), time, polarization, and space (transverse spatial distribution) [1]. For instance, the frequency or wavelength is used for wavelength-division multiplexing (WDM) with each frequency carrying an independent data channel, the complex optical field, i.e. real and imaginary parts or amplitude and phase components of an optical field, which is used to offer advanced multi-level quadrature amplitude modulation (QAM) formats in which multi-bit information is encoded in one symbol, the time is used for time-division multiplexing (TDM) by aggregating multiple low-speed data channels into high-speed data streams in the time domain, and the polarization is used for polarization-division multiplexing (PDM) to gain twice another increase of the capacity. Considering that the well-established WDM, TDM, and PDM have almost reached their scalability limits, it is highly desired to find additional multiplexing technologies, i.e. space-division multiplexing (SDM). Two candidates for SDM attracting lots of interest are known as multi-core fiber (MCF) and few-mode fiber (FMF) in fiber optical communications.

Optical Fiber Telecommunications VIB. http://dx.doi.org/10.1016/B978-0-12-396960-6.00012-2

SDM relies on the physical dimension of transverse spatial distribution which is termed the multiplexing of spatially separated optical fields in a MCF or few linearly polarized (LP) modes in an FMF. The concept of SDM can be further extended both in fiber and free-space optical communications. Actually, the transverse spatial distribution of a light beam can take many forms, which provides great flexibility in SDM. A special type of "twisted" light beam is noticed when looking into the transverse phase structure. A conventional propagating laser beam typically has an approximately flat phase front. However, in the 1990s, it was shown that propagating light beams can contain an interesting property known as orbital angular momentum (OAM) [2]. The OAM-carrying light beams have a helical phase front which is twisting along the direction of propagation, so-called "twisted" light beams. The twisted phase front into a corkscrew shape results in a doughnut-like intensity profile due to phase singularity at the center of light beams. Remarkably, OAM is a fundamental physical quantity which has previously been paid more attention to wide applications ranging from optical manipulation to quantum information processing [3,4]. A laudable goal would be to further introduce OAM as an additional degree of freedom into optical communications, taking into account the fact that different "twisted" light beams (multiple OAM modes) with varied twisting rates or OAM values are inherently orthogonal with each other. It is expected that OAM multiplexing can be employed to increase the capacity and spectral efficiency in combination with advanced multi-level modulation formats and conventional multiplexing technologies.

In this chapter, we tend to provide a comprehensive review of multimode communications using OAM. The fundamentals of OAM are introduced first followed by the techniques for OAM generation, multiplexing/demultiplexing, and detection. We then present recent research efforts to free-space communication links and fiber-based transmission links using OAM multiplexing together with optical signal processing using OAM (data exchange, add/drop, multicasting, monitoring and compensation. Future challenges of OAM communications are discussed at the end.

12.2 FUNDAMENTALS OF OAM

Since the mid-19th century the fact that light beams carry a momentum in addition to their energy has been recognized as a fundamental property of electromagnetic fields, described by Maxwell's equations [5]. This momentum means that when a light beam is absorbed by a surface, it exerts a radiation pressure, P, with a value $P = I/c$, where I is the intensity of the incident light beam and c is the velocity of light beam.

While studying Maxwell's equations, Poynting derived an expression for light carrying angular momentum. Poynting considered the rotation of the electromagnetic field, which occurs for circular polarized light beams, comparing this rotation to the mechanical rotation of a revolving shaft, which suggested that circularly polarized light does also carry angular momentum. In his formulation he showed that the angular momentum to energy ratio is $1/\omega$, where ω is the angular frequency of the light [6]. Noting that this effect is small, Poynting himself felt that a mechanical observation would be extremely unlikely. It took until the 1930s before an experiment was carried out by Beth demonstrating this angular momentum in the laboratory [7]. He showed that circularly polarized light transmitted through the half wave-plate caused the wave-plate to rotate. Key to this experiment was that the wave-plate was suspended from a quartz fiber to reduce the frictional forces. The rotation arose because the handiness of circularly polarized light was reversed by the wave-plate, and hence the angular momentum was also reversed. This change in the light's momentum resulted in a torque on the wave-plate, causing it to rotate.

These early works stem from a purely classical interpretation of a light beam, and do not require a quantized electromagnetic field. However, it is common for modern physicists to describe the interaction of light and matter in terms of photons rather than electromagnetic waves. One can quantify the momentum carried by a light beam as the momentum per photon, where the magnitude is expressed as $\hbar k$ and $\hbar$ for linear and angular momentum, respectively, where $\hbar$ is Plank's constant h divided by 2π and k is the wavenumber. When the momentum and energy ratios above are considered, these quantized expressions lead to the same results as those obtained under the classical interpretation, known to Maxwell and Poynting.

In the quantum interpretation of a light beam, one can consider circular polarization as arising from an individual photon spinning in either the left- or right-handed direction. Such spin is referred to as spin angular momentum (SAM), of value $\sigma\hbar$, where $\sigma=\pm 1$ depending on sense of circular polarization. However, even in the 1930s there was an indication that more complex momentum states could be carried by light. Within a photon picture, photons are produced during an atomic transition, and for the case of dipole transitions the change in angular momentum is $\hbar$. Hence, emitted light that is circularly polarized is consistent with momentum conservation. In 1932 Darwin considered the case of higher order transitions where the momentum transfer is greater than $\hbar$ [8]. He suggested that in the case where the light is emitted a short radius away from the center of mass of the atom, the linear momentum of the emitted photon would lead to an additional torque acting on the center of mass. This extra angular momentum is what we would now call orbital angular momentum.

Beyond higher order transitions, in 1992 Allen et al. while working in Leiden considered the OAM present in Laguerre-Gaussian (LG) laser beams [2]. Laguerre-Gaussian beams are characterized by having helical phase fronts which by necessity contain a phase singularity at their center. These phase singularities had been studied since the 1970s when there was the subject of much research in both acoustic and optical fields [9–12]. However, in none of this early work had the angular momentum associated with helically phased beams been identified. LG beams are described by two integer values ℓ and $\hbar$, and in the waist plane of the beam has the complex amplitude

$$\psi_{\ell,p}(r,\phi) = \frac{C_{\ell p}}{\omega_0} \cdot \left(\frac{r\sqrt{2}}{\omega_0}\right)^{|\ell|} \cdot L_p^{|\ell|}\left(\frac{2r^2}{\omega_0^2}\right) \cdot \exp\left(\frac{-r^2}{\omega_0^2}\right) \cdot \exp\left(i\ell\phi\right), \quad (12.1)$$

where ω_0 is the waist size, $L_p^{|\ell|}$ is the Laguerre polynomial for the integer ℓ and p, and $C_{\ell P}$ is the amplitude normalization term [13,14]. These two integers can take a wide range of values, as it is unbounded and p can be zero or take any positive integer value.

The key term in this complex amplitude is $\exp(i\ell\phi)$ determining the azimuthal phase profile of these beams. From Eq. (12.1), one can derive expressions for the energy and momentum within such a beam, described by the spatially dependent form of the Poynting vector, as shown in Figure 12.1. Within a ray optical picture, the Poynting vector determines the direction of a local ray emanating from a particular position on the wavefront. Helically phased beams have a skew of these local rays and as each ray has a linear momentum, when averaged over the whole beam, it leads to an azimuthal component of the beam momentum. This azimuthal component results in an angular momentum in the direction of propagation. If one considers the helical phase front, at a fixed radius, in azimuthal coordinates, it is simply a phase ramp of constant gradient. At a given radius from the beam axis, such a phase ramp has a base length of $2\pi r$ and a height of $\ell\hbar$, hence the skew angle of the local rays is

$$\beta = \frac{\ell\lambda}{2\pi r} = \frac{\ell}{kr}, \tag{12.2}$$

where k is the wavenumber associated with the beam. If one considers the linear momentum, $\mathbf{P}$, carried by a single skewed local ray, incident on surface at an angle, β, the OAM carried by the beam is

$$\mathbf{L} = \mathbf{r} \times \mathbf{P} = \mathbf{r} \times \hbar k \sin\beta = \mathbf{r} \times \frac{\ell\hbar}{r}. \tag{12.3}$$

Allen and co-works reasoned that such beams will carry an OAM to energy ratio of ℓ/ω, giving an OAM of $\ell\hbar$ per photon.

We note that unlike the SAM which has only two orthogonal values $\sigma = \pm 1$, OAM has an unbounded number of states described by ℓ. However, one should also

FIGURE 12.1 The local ray direction, shown as arrows, is perpendicular to the wavefront, shown as a green surface, of a light beam. (a) For a beam with $\ell = 0$, all the vectors are parallel with the beam axis. (b) (c) For higher order beams there is an azimuthal component of the Poynting vector. (b) $\ell = 1$. (c) $\ell = 2$. The angular deviation between the $\ell = 0$ case and higher orders is the skew angle of the rays. (For interpretation of the references to color in this figure legend, the reader is referred to the web version of this book.)

recognize that the low-loss transmission of high OAM beam requires optical transmission systems with high Fresnel number.

An experiment, analogous to Beth's approach to measuring spin angular momentum, was postulated by Allen et al., where two cylindrical lenses are used reverse to the handiness of the phase profile, hence changing the momentum carried by the beam. Such a change in momentum should result in a torque being applied to the lenses resulting in a rotational motion, as shown in Figure 12.2. However, attempts to show this effect at the macroscopic level have proved challenging.

Even though the torque induced by orbital angular momentum is too small to rotate large objects, in the microscopic regime the story is quite different. Soon after the initial work by Allen et al., He et al. considered the transfer of orbital angular momentum to trapped microscopic absorbing particles within optical tweezers (Figure 12.3) [15]. Optical tweezers trap particles due to the intensity gradient in the focused light beam. Due to this gradient force, dielectric particles are attracted to the center of the region of high field strength, which then allows the trapping and movement of micro-scale particles [16]. When trapping using a helically phased beam, with a beam waist similar to the size of the particle, the center of attraction corresponds to the location of the phase singularity at the center of the beam.

These early studies into the torque applied by beams carrying OAM used black micron-sized CuO particles, which were seen to spin when trapped using a plane polarized LG beam. In later studies, the torque due to OAM was combined with that due to the SAM of circular polarized light and by changing the relative sense of the two components the particle rotation could be sped up, slowed down, or even stopped [17,18].

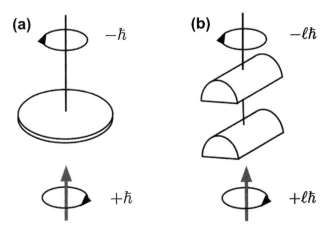

FIGURE 12.2 (a) Shows the rotational torque applied to a $\pi/2$ wave-plate as circularly polarized light is transmitted through the plate as the polarization state changes from left-handed into right-handed. **(b)** Suspended cylindrical lenses undergo a rotation when a helical beam passed through the lenses as the OAM value is converted from $\ell\hbar$ into $-\ell\hbar$.

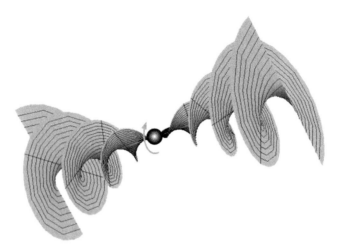

FIGURE 12.3 Focused helically-phased beam trapping a dielectric bead. A partially absorbing bead absorbs the OAM resulting in torque.

Aside from the obvious mechanical uses of beams carrying OAM, one of the next key studies was the conservation of OAM during the process of second harmonic generation (SHG). In SHG, a nonlinear crystal effectively combines photons of a particular energy to generate a new photon with twice that energy. In this process it was observed that OAM is conserved such that the second harmonic output had twice the OAM of the input light [19]. Such second-order processes, like SHG, are reversible. In the reverse of SHG the energy of a single pump photon is split into two new photons, a process called parametric down-conversion (Figure 12.4) [20]. In 2001 Mair et al. tested the conservation of OAM for down-conversion, which is commonly used for the preparation of entangled photons [21]. The work by Mair et al. showed not only that OAM was conserved, but that one down-converted photon could take one of many values, when the OAM of this photon is summed with its entangled partner, the OAM is still conserved. Their overall results were indicative of quantum entanglement, allowing investigations into this entanglement in higher dimensional Hilbert spaces.

Any state of complete polarization can be described as a superposition of two orthogonal states of circular polarization. A Poincare sphere is a common method to visualize these superpositions, where the north and south poles correspond to left and right circularly polarized light. Linear polarizations are equal superposition of left and right states, hence are represented around the equator of the Poincare sphere, where the orientation of the linear polarization is determined by the relative phase between the components in the superposition. An analogous sphere can be constructed for LG modes, for a two-state subset of the unbounded LG state-space, where the north and south poles correspond to $\pm\ell$, respectively [22]. Equal superpositions result in petal-type laser modes, where a change of the relative phase of the component modes leads to rotation of the mode. This analogy between spin angular moment and orbital

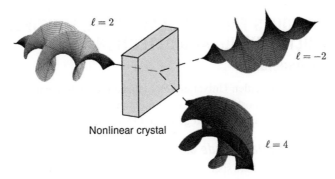

FIGURE 12.4 A photon carrying a particular OAM, when passed through a nonlinear crystal, will generate two photons each with half the energy of the input photon. Momentum is conserved in this process, hence the sum of the OAM of both photons will be the OAM of the input photon. These photons can take one of many possible values, above the specific case of $\ell=4$ and $\ell=-2$ is depicted.

angular momentum allows many tests of quantum mechanics, extensively studied for polarization, to be generalized for subsets of OAM state-spaces.

The discrete and unbounded nature of OAM, where ℓ can theoretically take any integer value, has made this optical property an area of interest for many researchers attempting to transmit ever-larger amounts of data across optical communications networks. Free-space links utilizing OAM are currently receiving the most interest, where several schemes have already been developed [23–25], and surely more are likely to be developed in the near future. In 2004, Gibson et al. demonstrated the detection of eight modes over a range of 15 meters, as at this range atmospheric turbulence only has a small effect on the mode quality of the transmitted beams [23]. For longer range links, compensation for atmospheric turbulence is an important consideration, as the OAM arises from the spatial phase structure of the beam and phase aberrations like those from turbulence degrade the mode quality [26–28]. More recently we presented a scheme using OAM multiplexing along with other more common methods of multiplexing to push the data rates and spectral efficiency up to 2.56 Tbit/s and 95.7 bit/s/Hz [24]. The use of OAM is also not limited to optical frequencies, as Tamburini et al. have demonstrated the use of specifically shaped radio antennas to increase the bandwidth of a radio link, with some promising results [25].

Researchers are not only considering OAM as a method to increase the bandwidth of communications links, it is also being seen as a method to make these links more secure. The larger alphabet which OAM gives has the potential to increase the security of cryptographic keys transmitted with a quantum key distribution (QKD) system [29]. Groblacher et al. demonstrated such a QKD system using three OAM modes, known as qutrits, showing an increased coding density for increased security [30]. Malik et al. recently presented a scheme using 11 OAM modes, increasing the security further and considering the effect of atmospheric turbulence on such a system [31].

12.3 TECHNIQUES FOR OAM GENERATION, MULTIPLEXING/ DEMULTIPLEXING, AND DETECTION

12.3.1 OAM generation

Since Allen et al. from Leiden University recognized that light beam with azimuthal phase dependence carries orbital angular momentum in 1992 [2], different methods have been proposed and developed to generate OAM-carrying light beams.

12.3.1.1 Cylindrical lens mode converter

In the early 1990s, researchers successfully generated OAM-carrying LG modes from Hermite-Gaussian (HG) modes by using a pair of cylindrical lenses [13]. As shown in Figure 12.5, b properly aligning the spatial intensity profile of the HG modes to the cylindrical lens pair, the specific portions of the decomposed modes convert to LG modes after their combination. In principle, this method is quite similar to converting

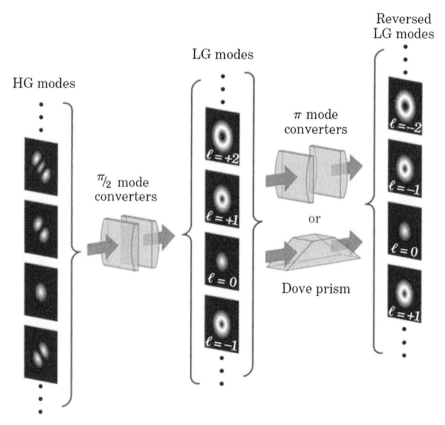

FIGURE 12.5 Hermite-Gaussian (HG) to Laguerre-Gaussian (LG) mode converter using a pair of cylindrical lenses [13].

linearly polarized light to circular polarization using a quarter-wave plate. Moreover, reversed order OAM modes can be obtained by increasing the separation between the two cylindrical lenses or simply a Dove prism [32]. This method has very high efficiency. However, it requires HG mode as the input source. Moreover, giving a HG mode, it can only generate limited LG modes. Therefore, it potentially limits the reconfigurability of the system.

12.3.1.2 Spiral phase plate (SPP)

Gaussian beam is quite common and widely available, thus a laudable goal would be to generate OAM beams from an input beam with Gaussian profile. One straightforward way is to realize such a function using a spiral phase plate [4,33]. From Figure 12.6 we can see the thickness of the spiral phase plate changes according to its azimuthal position. This results in an azimuthally changing optical path length, and gives the input light beam an angle dependent phase-delay. After passing through the spiral phase plate, the Gaussian beam has an azimuthally dependent phase and forms an OAM beam. However, as the spiral phase plate is fixed, it can only perform specific mode transform. It would be highly desirable to have a method that can reconfigurably generate different OAM beams from an input Gaussian beam.

12.3.1.3 Spatial light modulator (SLM)

Nowadays, a commonly used method for generating OAM beams is to use numerically computed holograms. By adjusting the holograms, one can generate any desired OAM beams from the same initial input beam. This approach has largely been facilitated by commercially available spatial light modulators (SLMs). The SLM is a pixellated liquid crystal device that has hundreds by hundreds of pixel arrays. The phase value of each pixel can be individually controlled, and thus imposes spatially varying modulation onto the incident beam. The phase pattern on the SLM can be programmed through the video interface of a computer. By implementing these computer-controlled holograms onto SLMs, the input optical beam can be converted into

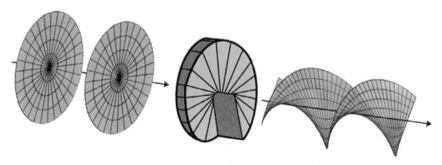

FIGURE 12.6 Gaussian to OAM mode conversion using a spiral phase plate [4].

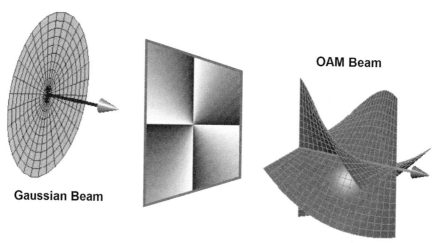

OAM Beam

Gaussian Beam

FIGURE 12.7 Gaussian to OAM mode conversion using a spatial light modulator [24].

a beam with almost any desired phase and amplitude distribution [4,34]. As shown in Figure 12.7, a hologram with four periods of 2π azimuthal phase change can convert an input Gaussian beam to a beam with OAM $l=4$. Furthermore, the hologram on the SLM can be changed hundreds of times per second, and thus provide reconfigurability to the optical system.

12.3.1.4 Fiber mode converter

For OAM-based fiber optical communications system, it is highly desired to have a compact and fiber-based OAM mode converter. Several fiber-based approaches have been proposed and demonstrated. Similar to the cylindrical lens mode converter, stressed fiber has been used to impart OAM to a HG laser mode. The transfer of OAM to the light occurs because of a difference in phase velocity within the fiber for two orthogonal modes that comprise the input beam [35]. Moreover, generating an OAM mode in an optical fiber has been demonstrated by using acoustic-optic interaction [36]. Another promising approach to generate OAM mode for fiber transmission is to use microbend grating [37]. The microbend grating can convert the fundamental mode to OAM mode in the vortex fiber with >99% efficiency. Even after $\sim$1 km fiber transmission, the OAM mode is well maintained.

Recently, mode coupling-enabled OAM mode generation in the fiber was proposed [38]. As shown in Figure 12.8a, a <2 mm fiber coupler consisting of a central ring and four external cores was designed to generate up to 10 OAM modes with >99% mode purity. Four coherent input lights were launched into the external cores and then coupled into the central ring waveguide for the generation of OAM modes. The generated OAM mode can be controlled by choosing the proper size of external cores. This fiber coupler design can also be extended to enable all-fiber spatial-mode multiplexing and demultiplexing.

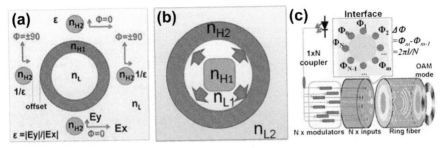

FIGURE 12.8 Fiber cross sections of the mode coupling-enabled OAM mode generation [38–41].

Another fiber structure to perform OAM mode generation was proposed as shown in Figure 12.8b [39]. By breaking the circular symmetry of the waveguide, the input circularly polarized fundamental mode in the square core can be coupled into the ring region to generate higher-order OAM modes. Simulations showed that the generated OAM modes with a topological charge l up to 9 have >96.4% mode purity and 30-dB extinction ratio.

Moreover, an approach was proposed recently to efficiently generate and multiplex OAM modes in a fiber with a ring high-refractive index profile by using multiple coherent inputs from a Gaussian mode [40,41], as shown in Figure 12.8c. By controlling the phase relationship of the multiple inputs, one can selectively generate OAM modes of different states l or generate multiple OAM modes simultaneously without additional loss coming from multiplexing.

12.3.2 OAM multiplexing/demultiplexing

As OAM can provide an additional orthogonal degree of freedom for increasing the capacity and spectrum efficiency of an optical communications link, efficient multiplexing and demultiplexing of OAM modes are thus of great importance. Besides the above-mentioned fiber-based multiplexing/demultiplexing schemes, there are a few approaches developed by using free-space optics and integrated optics.

12.3.2.1 Free-space multiplexing/demultiplexing

Figure 12.9a and b show the generation and back-conversion of multiple data-carrying (e.g. 16-QAM) OAM beams using spatial light modulators loaded with different spiral phase masks. Figure 12.9c illustrates the multiplexing/demultiplexing of OAM beams (PDM is also shown in combination with OAM multiplexing/demultiplexing). The multiplexing of multiple OAM beams with doughnut-like intensity profiles (third column, left panel of Figure 12.9c) can be simply achieved in free space using non-polarizing beam splitters. To demultiplex an OAM beam (ℓ) of interest, a spatial light modulator loaded with an inverse spiral phase mask ($-\ell$) is used to remove

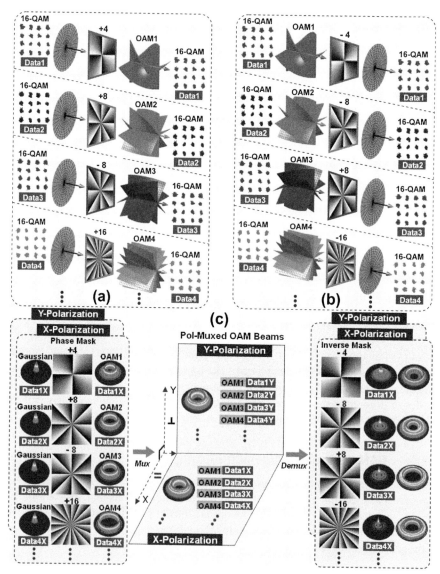

FIGURE 12.9 Data-carrying OAM multiplexing/demultiplexing [24].

the azimuthal phase term $\exp(i\ell\theta)$ of (θ: azimutal angle) of the OAM beam, which is therefore converted back to a beam with a planar phase front. This beam has a high-intensity bright spot at the center (second column, right panel of Figure 12.9c), which is separable from other OAM beams with updated charges and doughnut-like intensity profiles (third column, right panel of Figure 12.9c).

12.3.2.2 Free-space mode sorter

A free-space optical system that is able to sort even and odd OAM modes was demonstrated using an interferometer structure [42]. In principle, this can be further extended to sort an arbitrarily large number of OAM states. This is achieved by cascading additional interferometers with different rotation angles. However, it will require $N-1$ interferometers to sort N OAM states. In such case, the system tends to be quite bulky in terms of scalability. A system that can only sort four OAM states was demonstrated in experiment.

Recently, another compact solution has been proposed by Gregorius C.G. Berkhout from Leiden University to sort OAM states using two static optical elements [43]. The scheme of the optical system is shown in Figure 12.10a. SLM1 is programmed with both phase and intensity information to generate desired LG modes. SLM2 performs a Cartesian to log-polar coordinate transformation, converting the helically phased light beam corresponding to OAM states into a beam with a transverse phase gradient. SLM3 is used to project the phase-correcting element. A subsequent lens then focuses each input OAM state to a different lateral position. Figure 12.10b shows modeled and observed phase and intensity profiles at various places in the optical system for a range of OAM states. In the second row, one can see that an input beam with circular intensity profile is unfolded to a rectangular intensity profile with a 2π phase gradient. The position of the elongated spot changes with the OAM input state. The observed spots are slightly broader than the modeled ones due to aberrations in the optical system. The researchers successfully applied it to separate 11 OAM states. However, there is a 70% light loss associated with the two SLMs that comprise the mode sorter. The transmission efficiency has recently been further improved to 85% by replacing the SLMs with the custom-made refractive optical elements, and 50 OAM modes were efficiently sorted [44].

12.3.2.3 Integrated mode (de)multiplexer

Silicon photonics has attracted a lot of attention due to its great potential for Complementary Metal Oxide Semiconductor (CMOS) compatibility. Other benefits from integrated devices include compact size, low-cost, and low-power consumption.

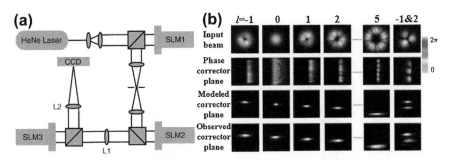

FIGURE 12.10 (a) Schematic overview and (b) phase and intensity profiles at various planes of OAM mode sorter [43].

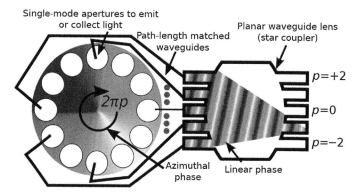

FIGURE 12.11 Concept for multiplexing and demultiplexing OAM modes using integrated mode (de)multiplexer [45].

By taking the above advantages, multiplexing and demultiplexing 9 OAM modes have been realized on integrated silicon chips by researchers from Bell Laboratories [45].

Figure 12.11 depicts the concept for multiplexing and demultiplexing OAM modes using an integrated mode (de)multiplexer. From a demultiplexer point of view, the OAM beam launches into the integrated silicon chips vertically through its surface and is coupled horizontally into an array of M waveguides. The waveguides capture any azimuthal variations of phase and amplitude and direct these variations through path-length matched waveguides toward a $M \times M$ star coupler. The azimuthal phase variations are now linear phase variations enabling a star coupler to steer the beam to a different single-mode output depending on the input OAM. Due to the reciprocity, this device can work as both multiplexer and demultiplexer. By simply increasing the number of waveguide arms, the operational mode-number can be scalable. Figure 12.12 illustrates the scheme used to transmit and receive the multiplexed OAM modes using two integrated silicon chips. Recently, 1-b/s/Hz, 10-Gb/s binary phase-shift keying (BPSK), and 2-b/s/Hz, 20-Gb/s quadrature phase-shift keying (QPSK) data for individual and two simultaneous OAM states have been successfully transmitted [46].

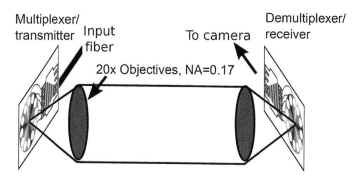

FIGURE 12.12 Scheme for transmitting and receiving OAM beams [45].

12.3.3 **OAM detection**

As discussed in OAM multiplexing/demultiplexing (Figure 12.9), the OAM beam can be converted back to a beam with a planar phase front. The back-converted beam has a bright high-intensity spot at the center which can be separated from other doughnut-like OAM beams via spatial filtering and received by fiber for detection.

During the past years, huge progress has been made on the methods of generation, multiplexing/demultiplexing, and detection for OAM modes since the recognition of OAM-carrying light beams in 1992. We envision that more efficient, compact, cost-effective methods, especially for multiplexing/demultiplexing OAM modes, will be developed in the near future, and will facilitate implementing OAM modes in optical communications systems to boost capacity and spectral efficiency.

12.4 **FREE-SPACE COMMUNICATION LINKS USING OAM MULTIPLEXING**

OAM has recently seen applications in free-space information transfer and communication [23]. Given their inherent orthogonality, OAM-carrying beams with different topological charges (ℓ) are readily distinguishable from one another. As opposed to SAM, which has only two possible values of $\pm\hbar$, the unlimited values of ℓ give, in principle, infinite possibilities of achievable OAMs. Thus, it has the potential to tremendously increase the capacity and spectral efficiency of optical communication systems using two approaches: (i) encode data as OAM states; (ii) multiplex data-carrying OAMs. In the former case, data is encoded as one of multiple OAM states, and the unbounded state-space provides high capacity. In the latter case, light beams with different OAMs serve as carriers of different data streams.

This latter multiplexing approach has similarities to various other multiplexing technologies in optical fiber communications, such as WDM [47–49], OTDM [50], PDM [47–51], and SDM [51,52]. Note that recent optical-communication-systems advances in multi-level amplitude/phase modulation formats, coherent detection, and electronic digital signal processing have enabled tremendous increases in total capacity and overall spectral efficiency [47–53]. Hence, when considering OAM as a new degree of freedom, it is expected to achieve yet another significant increase of capacity and spectral efficiency gained by the combined contributions from multi-level modulation formats, traditional multiplexing technologies (WDM, PDM, SDM, etc.), and OAM multiplexing. Several experiments were carried out in free-space communication links using data-carrying OAM multiplexing.

12.4.1 **OAM+WDM link**

Figure 12.13 illustrates the concept and principle of a typical OAM+WDM link. Two SLMs loaded with different spiral phase masks were employed to convert two data-carrying (WDM signals) Gaussian beams into two OAM beams. Two OAM beams were spatially multiplexed using a non-polarizing beam splitter which

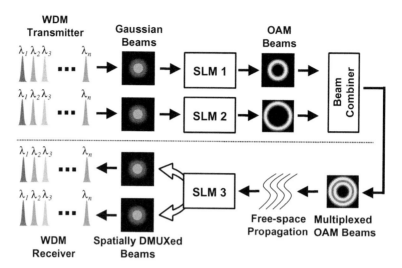

FIGURE 12.13 Concept and principle of an OAM+WDM communication link.

increased twice the capacity. After free-space propagation, the multiplexed OAM beams were demultiplexed using a third SLM and coupled into fiber for detection. The data carried by OAM beams can take different modulation formats. 40-Gbit/s non-return-to-zero (NRZ), 40-Gbit/s non-return-to-zero differential phase-shift keying (NRZ-DPSK), and 10-Gbaud/s QPSK signals were adopted in the experiment. 100-GHz ITU-grid-compatible 25 wavelength channels (from Ch.1: 1537.40 nm to Ch.25: 1556.55 nm) and two OAM beams (OAM_{-8} and OAM_{-16}) were employed. Figure 12.14 plotted measured bit-error rate (BER) performance for OAM+WDM link with 25-channel 40-Gbit/s NRZ-DPSK signals over OAM_{-8} and OAM_{-16}

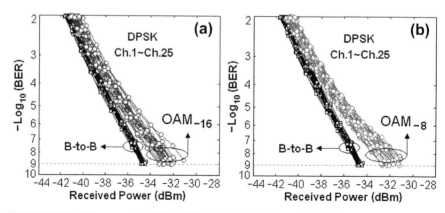

FIGURE 12.14 BER curves for OAM+WDM link with 25-channel 40-Gbit/s NRZ-DPSK signals over OAM_{-8} and OAM_{-16} beams. B-to-B: back-to-back.

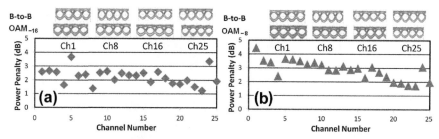

FIGURE 12.15 Power penalties for OAM+WDM link with 25-channel 40-Gbit/s NRZ-DPSK signals over OAM$_{-8}$ and OAM$_{-16}$ beams.

beams (total capacity: 2 Tbit/s). As shown in Figure 12.15, the maximum power penalty of 25 channels was ~3.7 dB for OAM$_{-16}$ beam and ~4.5 dB for OAM$_{-8}$ beam at a BER of 10^{-9}, while the average power penalty of 25 channels was ~2.2 dB for OAM$_{-16}$ beam and ~2.9 dB for OAM$_{-8}$ beam. Shown in insets of Figure 12.15 are measured balanced eye diagrams of demodulated DPSK signals. For 40-Gbit/s NRZ, the observed maximum power penalty was ~2.7 dB for OAM$_{-16}$ beam and ~2.1 dB for OAM$_{-8}$ beam at a BER of 10^{-9}. For 10-Gbaud/s QPSK, the observed maximum power penalty was ~0.5 dB for OAM$_{-16}$ and OAM$_{-8}$ beams at a BER of 2×10^{-3} (enhanced forward error correction (EFEC) threshold).

12.4.2 OAM+PDM link

Figure 12.16 illustrates the concept and principle of a typical OAM+PDM link. Four Gaussian beams (wavelength: 1550.12 nm) with planar phase fronts, each carrying a 16-QAM signal, are converted into four OAM beams (e.g. OAM$_{+4}$, OAM$_{+8}$, OAM$_{-8}$, OAM$_{+16}$) with helical phase fronts by four SLMs loaded with four different spiral phase masks. After the multiplexing of OAM beams via non-polarizing beam splitters and the polarization multiplexing through polarizing beam splitters, a significant increase of capacity and spectral efficiency can be gained owing to the combined contributions from OAM multiplexing, PDM and multi-level modulation format (i.e. 16-QAM). The OAM beams propagate in free-space over meter-length scale. For the demultiplexing, the pol-muxed OAM beams are polarization demultiplexed using a polarizer, and another SLM, loaded with a specified spiral phase mask, is used to demultiplex one of the OAM beams back to a beam with a planar phase front for coherent detection.

We first demonstrated the multiplexing/demultiplexing of four OAM beams (OAM$_{-8}$, OAM$_{+10}$, OAM$_{+12}$, OAM$_{-14}$). The polarization multiplexing stage in Figure 12.16 was not used. Figure 12.17 shows the obtained experimental and theoretical results. The doughnut-like intensity profiles of four OAM beams and their superposition are observed as shown in Figure 12.17a1–a5. The interferograms, i.e. interference between the OAM beam and a Gaussian beam reference, were

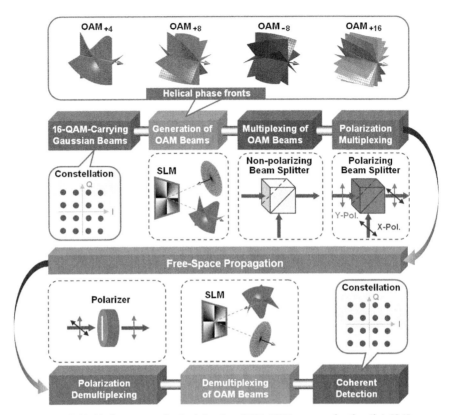

FIGURE 12.16 Concept and principle of an OAM+PDM communication link [24].

measured to verify the charge of the generated OAM beams, as shown in Figure 12.17b1–b4. The number and direction of twists in the interferograms indicated the magnitude and sign of ℓ, showing that four OAM beams for multiplexing were OAM_{-8}, OAM_{+10}, OAM_{+12}, and OAM_{-14}. Figure 12.17c1–c3 shows an example of the demultiplexing of an OAM beam (e.g. OAM_{-8}). Shown in Figure 12.17c1 is the observed intensity profile of the back-converted beam from the OAM_{-8} beam as other OAM beams were blocked (w/o crosstalk). The back-converted beam had a high intensity bright spot at the center. Shown in Figure 12.17c2 is the crosstalk of the demultiplexing as the OAM_{-8} beam was blocked while other OAM beams are on. Negligible intensity was observed near the center. Shown in Figure 12.17c3 is the observed intensity profile as all OAM beams were on (w/crosstalk). The crosstalk surrounding the center of the back-converted beam could be removed by spatial filtering. The multiplexing/demultiplexing of OAM beams was also theoretically analyzed using the angular-spectrum propagation method. The obtained simulation results shown in Figure 12.17d1–d5, e1–e4, and f1–f3 were in good agreement with the corresponding experimental results shown in Figure 12.17a1–a5,

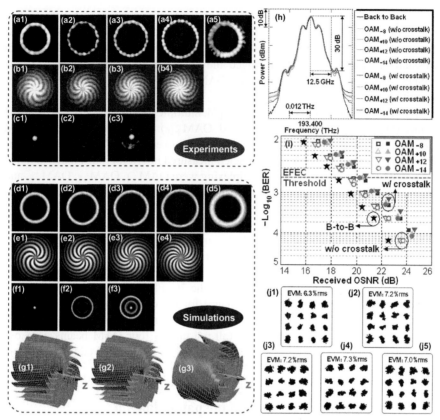

FIGURE 12.17 Experimental and theoretical results of multiplexing/demultiplexing of four OAM beams (OAM$_{-8}$, OAM$_{+10}$, OAM$_{+12}$, OAM$_{-14}$) [24].

b1–b4, and c1–c3. Remarkably, for the demultiplexing of OAM$_{-8}$ beam, only the OAM$_{-8}$ beam was back-converted to a beam with a high intensity bright spot at the center (Figure 12.17c1 and f1), while OAM$_{+10}$, OAM$_{+12}$, and OAM$_{-14}$ beams were transformed into OAM$_{-18}$, OAM$_{-20}$, and OAM$_{+6}$ beams considering the "mirror image" effect of reflective-type SLM with doughnut-like intensity profiles (Figure 12.17c2 and f2), which were verified by the simulated helical phase fronts as shown in Figure 12.17g1–g3. The measured spectra of back-converted beams from different 10.7-Gbaud/s 16-QAM-carrying OAM beams were shown in Figure 12.17h. These spectra overlapped with each other, occupying the same bandwidth. The power suppression of ~30 dB was achieved at an offset frequency of 12.5 GHz from the center. Considering 10.7-Gbaud/s 16-QAM signals over four OAM beams (12.5-GHz grid [48]), a capacity of 171.2 (10.7 × 4 × 4) bit/s together with a spectral efficiency of 12.8 bit/s/Hz was obtained, including the 7% forward error correction (FEC) overhead. The measured BER performance for the demultiplexing of OAM beams (OAM_{-8} and OAM_{-14} as examples) was shown in Figure 12.17i. The optical signal-to-noise ratio (OSNR) penalty at a BER of 2×10^{-3} (EFEC threshold) was measured to be less than 1.2 dB without crosstalk and 2.2 dB with crosstalk. Figure 12.17j1–j5 displayed the observed constellations of 16-QAM (OSNR > 30 dB) for the demultiplexed OAM beams with measured error vector magnitude (EVM).

We then demonstrated the multiplexing/demultiplexing of pol-muxed four OAM beams (OAM_{+4}, OAM_{+8}, OAM_{-8}, and OAM_{+16}). Figure 12.18a1–a5 showed observed doughnut-like intensity profiles of pol-muxed four OAM beams and their superposition. Figure 12.18b1–b4 depicted measured interferograms, indicating that pol-muxed four OAM beams for multiplexing were OAM_{+4}, OAM_{+8}, OAM_{-8}, and OAM_{+16}, respectively. Figure 12.18c showed measured spectra for the demultiplexing (OAM_{+4}, OAM_{+16}) of pol-muxed four OAM beams (8 channels in total), with each channel carrying a 42.8-Gbaud/s 16-QAM signal. A power suppression of $\sim$30 dB was obtained at a 50-GHz frequency offset from the center. Hence, a spectral efficiency of 25.6 bit/s/Hz was achieved for a total capacity of 1369.6 ($42.8 \times 4 \times 4 \times 2$) Gbit/s, i.e. 42.8×4-Gbit/s 16-QAM signals over pol-muxed four OAM beams (50-GHz grid [49]), including the 7% FEC overhead. Figure 12.18d1 and d2 showed measured BER performance for the demultiplexing of pol-muxed OAM beams along X- and Y-polarizations without (only X-/Y-polarization of a pol-muxed OAM beam was on) and with (all pol-muxed four OAM beams were on) crosstalk. The OSNR penalties at a BER of 2×10^{-3} were measured to be less than 1.5 dB without crosstalk and 3.0 dB with crosstalk.

12.4.3 Scalability of OAM+PDM in spatial domain

We finally demonstrated the scalability of the multiplexing/demultiplexing of OAM beams in the spatial domain. As illustrated in Figure 12.19a, two groups of OAM

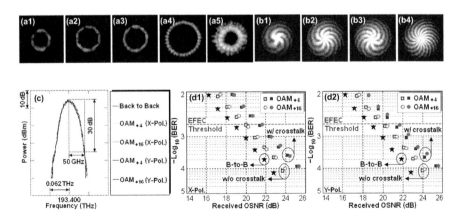

FIGURE 12.18 Experimental results of multiplexing/demultiplexing of pol-muxed four OAM beams (OAM_{+4}, OAM_{+8}, OAM_{-8}, OAM_{+16}) [24].

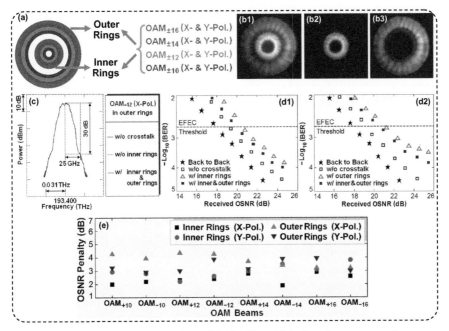

FIGURE 12.19 Experimental results of multiplexing/demultiplexing of two groups of concentric rings of pol-muxed eight OAM beams (OAM$_{\pm 10}$, OAM$_{\pm 12}$, OAM$_{\pm 14}$, OAM$_{\pm 16}$) [24].

beams with the same charges but different beam sizes, i.e. one (outer rings) was expanded compared to the other (inner rings), were spatially multiplexed together as concentric rings. Figure 12.19b1 showed observed intensity profile of the multiplexed two groups of concentric rings of pol-muxed eight OAM beams (OAM$_{\pm 10}$, OAM$_{\pm 12}$, OAM$_{\pm 14}$, OAM$_{\pm 16}$). Spatial filtering was used to enable the demultiplexing of two groups of concentric rings. Figure 12.19b2 and b3 showed observed intensity profiles of demultiplexed inner and outer rings, respectively. Figure 12.19c depicted measured typical spectra for the demultiplexing (OAM$_{-12}$ beam, X-polarization, outer rings) of two groups of concentric rings of pol-muxed eight OAM beams (i.e. 32 channels in total), with each channel carrying a 20-Gbaud/s 16-QAM signal. A power suppression of $\sim$30 dB was achieved at a 25-GHz frequency offset from the center. Thus, a spectral efficiency of 95.7 bit/s/Hz was gained for a capacity of 2560 ($20 \times 4 \times 8 \times 2 \times 2$) Gbit/s, i.e. 20-Gbaud/s 16-QAM signals over pol-muxed eight OAM beams in two groups of concentric rings (25-GHz grid [47]), considering the 7% FEC overhead. Figure 12.19d1, d2, and e plotted measured BER performance for the demultiplexing of two groups of concentric rings of pol-muxed eight OAM beams. All the 32 channels achieved a BER of $<2 \times 10^{-3}$. The average OSNR penalties at a BER of 2×10^{-3} for inner and outer rings were measured to be 2.7 and 3.6 dB, respectively.

From the scalability of OAM multiplexing in spatial domain, one would expect further improvement of capacity and spectral efficiency by increasing the number of OAM beams and groups of concentric rings.

12.5 FIBER-BASED TRANSMISSION LINKS

In the last decade, perhaps the most extensively studied complex beam-shape of light is the class of vortex beams, characterized by a dark hollow center, which possess phase or polarization singularities [54,55]. These beams have several potential scientific and technological applications, such as laser-based electron and particle acceleration [56], single-molecule spectroscopy [57], optical tweezers that can apply torques [58], metal machining [59], higher-dimensional quantum encryption [60], and lastly, but perhaps most importantly, mode-division multiplexing in classical communications systems [24]. Given the diversity of interest in producing these beams, fiber means of generating and propagating vortices would be highly beneficial and, in the context of communications links utilizing them, it would be critical to develop fibers that support these states.

Fiber means of generating both polarization and phase vortex states appear feasible because they are true eigenmodes of cylindrically symmetric optical waveguides [61]. For instance, the radially and azimuthally polarized modes of a fiber are simply the TM_{01} and TE_{01} vector solutions of a cylindrical step-index fiber, while OAM states with $\ell = \pm 1$ can be represented as $\pi/2$-phase-shifted linear combinations of the vector modes HE_{21}^{even} and HE_{21}^{odd} of a fiber. By extension, higher order ℓ states are simply the $\pi/2$-phase-shifted linear combinations of the vector modes $HE_{\ell+1}$ states, yielding an OAM of $\pm\ell$, with their spin (orientation of right- or left-circular polarization) aligned with the sign of ℓ, or the vector combinations of $EH_{\ell-1}$ states with spin anti-aligned with the sign of ℓ.

A laser cavity, defined by a 1.6-m long fiber held straight and rigid, free-space external mirrors, and a conical prism as mode selective element, has yielded lasing in a radially polarized mode [62]. Alternatively, a free-space LG mode has been used to excite a mixture of vortex states in a 35-cm fiber held rigid and straight, following which free-space polarization transformations with a combination of waveplates yield a pure vortex [63]. The lowest loss and compact means to obtain vortices with high modal purity directly from fibers involve inducing the required mode transformations in the fiber itself. This has been demonstrated with fiber gratings [36] or adiabatic mode couplers [64]. In both cases, immediately after the mode-converter (i.e. within less than 3 cm of the rigidly held fiber), a radially or azimuthally polarized beam is obtained. In the context of phase vortices, helecoidal gratings in photonic crystal fibers [65], or phased inputs into multi-core fibers [66] have had limited success in producing states that resemble the first nonzero OAM state in a short (<1 m) fiber. To the best of our knowledge, the purity of excitation of OAM states has not been addressed in any of these approaches, and propagation lengths have been limited to less than a meter, which suggests that fibers supporting OAM states may have fundamental mode instability problems.

12.5.1 Fiber design

The perceived fundamental limitation of vortices in optical fibers has been that they always exist in sets of four (or more) almost degenerate sets. The source of this near degeneracy can be seen by considering the equation [61] satisfied by the time- and longitudinal coordinate-harmonic transverse electric field $\mathbf{e}_t$

$$\left(-\nabla_t^2 - k^2 n^2\right) \mathbf{e}_t - \nabla_t \left(\mathbf{e}_t \cdot \nabla_t \ln n^2\right) + \beta^2 \mathbf{e}_t = 0, \tag{12.4}$$

where $n(r)$ is the refractive index as a function of radius, $k = 2\pi/\lambda$ is the free space wavenumber, β is the propagation constant, and ∇_t is the transverse gradient. If we neglect the index gradient (second) term, then Eq. (12.4) becomes the scalar wave equation and the solutions break up into degenerate subspaces, of dimension two or four, that are spanned by the so-called linearly polarized (LP) mode sets $\mathrm{LP}_{\ell m}$. All modes in one of these subspaces share a common propagation constant $\beta_{\ell m}$. Each such subspace, indexed by the azimuthal and radial numbers ℓ and m, is spanned by a linearly polarized basis set (here chosen to be real):

$$-F_{\ell m}'' - \frac{1}{r} F_{\ell m}' + \left(\frac{\ell^2}{r^2} - k^2 n^2\right) F_{\ell m} = -\beta^2 F_{\ell m}, \tag{12.5}$$

where the radial wavefunction $F_{lm}(r)$ satisfies the radial wave equation

$$\mathbf{e}_{\ell m}^{(\mathrm{LP})}(r, \phi) = F_{\ell m}(r) \begin{Bmatrix} \cos \ell\phi \\ \sin \ell\phi \end{Bmatrix} \begin{Bmatrix} \hat{x} \\ \hat{y} \end{Bmatrix}, \tag{12.6}$$

When $\ell = 0$, the dimension of the degenerate subspace is two, and the corresponding modes cannot form vortices: the intensity does not vanish on the fiber axis. For each value of $\ell > 0$, however, the four LP basis elements may be combined into four orthogonal vortex states. When $\ell = 1$, these states are designated as $\mathrm{TE}_{\ell-1,m}$, $\mathrm{TM}_{\ell-1,m}$, and a pair of even and odd $\mathrm{HE}_{\ell+1,m}$ modes; when $\ell > 1$ the vortex states are a pair of even and odd $\mathrm{EH}_{\ell-1,m}$ modes along with the $\mathrm{HE}_{\ell+1,m}$ pair. The spatial polarization dependence of the $\ell = 1$ case is illustrated in the top row of Figure 12.20.

Inclusion of the second term in Eq. (12.4) partially breaks the degeneracy of the LP mode sets. This term, in contrast to the first term, contains a gradient of the index profile. Fundamentally, it is this index gradient that is responsible for polarization (vector) splitting of the effective indexes. The result is that, for the $\ell = 1$ set, the effective indexes of the TE and TM modes are split from each other, and from the (still) degenerate pair of HE modes, while for $\ell > 1$ those of the HE and EH pairs each remain degenerate, but are split from each other.

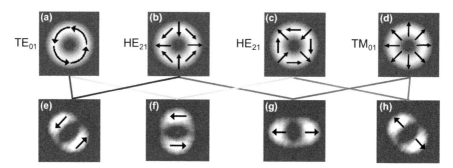

FIGURE 12.20 Modal intensity patterns in optical fibers; (a–d) vortex modes that are true eigenmodes of an optical fiber; $\pi/2$ phase-shifted linear combinations of the HE_{21} modes (b, c) leads to OAM states; (e–d) the resultant, unstable intensity patterns due to mode mixing between the top row of eigenmodes.

After the vector term in Eq. (12.4) is taken into account, the remaining $\ell > 0$ degenerate subspaces all correspond to true vortex states in the fiber. However, in typical fibers the resulting splitting is much too small to isolate the modes from one another, and the elements of each LP set randomly share power with one another due to longitudinal inhomogeneities caused by fiber bends, geometry imperfections during manufacture, and stress-induced index perturbations. Examples of the resulting unstable combinations are illustrated in the bottom row of Figure 12.20. This near-degeneracy, and the associated mode instability problem, has been a ubiquitous feature of fibers. Although the mathematical description above is only for the circularly symmetric fiber case, experimental observations of the vortex states in the case of non-circular symmetric fibers (such are photonic crystal and multi-core fibers) [65,66] confirm the same intermodal coupling problem.

The mode instability problem is addressed by realizing fiber designs that lift the near-degeneracy of the TM_{01} (radially polarized), TE_{01} (azimuthally polarized), and HE_{21} modes. The large splitting in effective indexes minimizes coupling between the states, since mode mixing efficiency decreases exponentially with the square of the difference in their propagation constants [67]. Therefore, such splitting maintains the purity of the vortex modes and avoids the tendency of optical fibers to yield the undesired LP_{11} mode as an output.

The fiber design problem, therefore, is how to enhance the vector splitting by choice of index profile $n(r)$. The key physical intuition behind the solution is the elementary observation that, during total internal reflection, phase shifts at index discontinuities (index steps) critically depend on an incident wave's polarization orientation, and the propagation constant of a mode represents phase accumulation. The translation of this physical reasoning into mathematical expressions amenable to optimization can be achieved with a full vectorial solution of the Maxwell equations. However, more analytical insight is obtained by means of a first-order perturbative analysis [68]. First, we calculate the scalar propagation constants of the LP_{11} mode group (identical in the scalar approximation) following which, the real propagation constants are obtained through a vector correction, given by:

$$\delta\beta_{TE_{01}} = 0\delta\beta_{TM_{01}} = -2\left(\delta_1 + \delta_2\right); \delta\beta_{HE_{21}} = \left(\delta_2 - \delta_1\right), \qquad (12.7)$$

$$\delta_1 = \frac{1}{2a^2 n_{co}\beta_{11}}\int_0^\infty F_{11}\left(r\right)\frac{\partial F_{11}\left(r\right)}{\partial r}\frac{\partial\Delta n\left(r\right)}{\partial r}r\,dr,$$

$$\delta_2 = \frac{1}{2a^2 n_{co}\beta_{11}}\int_0^\infty F_{11}^2\left(r\right)\frac{\partial\Delta n\left(r\right)}{\partial r}dr$$

where r is the radial coordinate, $F_{11}(r)$ is the radial wavefunction for the scalar mode, β_{11} is its unperturbed propagation constant, a is the size of the waveguiding core, n_{co} is a core refractive index, and $\Delta n(r)$ is its refractive index profile relative to the index of the infinite cladding. Equation (12.7) embodies the physical intuition of the relationship between the field profile of a mode and its propagation constant. The search for a solution that substantially separates the propagation constants of the TE_{01}, TM_{01}, and HE_{21} modes boils down to a search for a waveguide that yields high fields ($F_{11}(r)$) and field gradients ($\partial F_{11}(r)/\partial r$) at index steps. Furthermore, this separation, and hence mode stability, grows with the magnitude of the index step Δn.

We now describe a fiber that we designed and fabricated that achieves a large vector separation, and hence stable propagation of vortex modes [69]. We start by first considering a conventional fiber (index profile shown in Figure 12.21a). In light of Eq. (12.7), the problem with obtaining stable vortex modes from such fibers becomes immediately apparent—increasing Δn to increase mode separation does not help because the mode becomes increasingly confined and the field amplitudes dramatically decrease at waveguide boundaries. Using the intuition gathered from Eq. (12.7), we conclude that a waveguide whose profile mirrors that of the mode itself—i.e. an annular waveguide resembling an anti-guide—would be more suitable for maximizing Δn while also maximizing field-gradients at index steps. This is schematically illustrated in Figure 12.21b.

For this class of designs, increasing Δn_{max} does not automatically lead to reductions of $F(r)$ and $\partial F(r)/\partial r$ at index steps. The refractive index profiles shown in Figure 12.21a and b are plotted in normalized units to elucidate the fundamental differences between the two profiles. The simulations assumed an identical Δn_{max} of 0.025 for both cases, and their lateral dimensions were adjusted to obtain similar cutoff wavelengths ($\sim$2600 nm) for the first higher-order antisymmetric modes. This ensures that at the simulated wavelength (1550 nm) the modes (intensity profiles shown as red curves) in both waveguides are similarly well confined or stable, and hence propagate with roughly similar losses. The calculated effective index $n_{eff} = \beta\lambda/2\pi$ of the radially polarized (TM_{01}) mode is separated from its nearest neighbor (the mixed-polarization HE_{21} mode) by $\sim$10^{-5} for the step-index fiber and by $\sim$1.6 $\times$ 10^{-4} for our new design. This is an improvement of 16$\times$, which is significant on two counts: (a) n_{eff}-separation by more than order-of-magnitude is dramatic because mode-mixing in fibers scales *exponentially* as the *square* of this n_{eff} difference; and (b) we

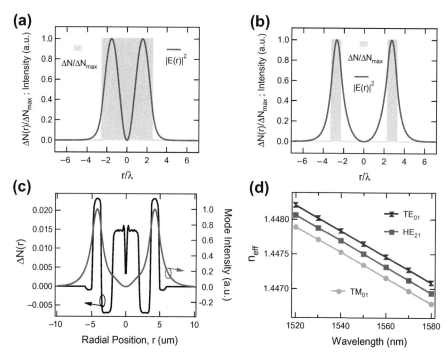

FIGURE 12.21 Normalized refractive index profile (Grey background), and corresponding mode intensity for the scalar LP_{11} mode (red) for (a) a conventional step-index fiber, and (b) the novel ring design. Intensity $|E(r)|^2$, rather than electric field $E(r)$, plotted for visual clarity—field reverses sign at $r/\lambda = 0$; (c) Measured refractive index profile (relative to silica index) for fabricated fiber, and corresponding LP_{11} mode intensity profile; (d) Effective index for the three vector components of the scalar LP_{11} mode for fiber shown in (c). n_{eff} of radially polarized (TM_{01}) mode separated by 1.8×10^{-4} from other modes. (For interpretation of the references to color in this figure legend, the reader is referred to the web version of this book.)

know from the development of conventional polarization-maintaining (PM) fibers that mode separation should ideally be $>10^{-4}$ for mode coupling free long-range signal transmission.

Figure 12.21c shows the measured refractive index profile of the fiber preform we fabricated to test this concept. Note that this fiber possesses the high-index ring as demanded by our design depicted in Figure 12.21b. However, it also has a step-index central core, as do conventional fibers. This core does not sufficiently perturb the spatial profile of the LP_{11} mode of interest, and hence does not detract from the design philosophy illustrated in Figure 12.21b. Instead, it allows for the fundamental mode to be Gaussian-shaped, enabling low-loss coupling.

Figure 12.21d shows the n_{eff} for the TM_{01}, HE_{21}, and TE_{01} modes in this fiber, measured by recording grating resonance wavelengths for a variety of grating periods.

The n_{eff} for the desired TM_{01} mode (radially polarized mode) is separated by at least 1.8×10^{-4} from any other guided mode of this fiber. Note that this value is actually larger than that of the theoretical schematic, primarily because the fabricated fiber effectively had a higher Δn_{max} owing to the down-doped region between the core and the ring. For conventional fibers, the three curves would be indistinguishable in the scale of this plot.

12.5.2 Coupling and controlling OAM in fibers

Practical use of OAM states requires means of multiplexing and demultiplexing these states into and out of the fibers. Methods for estimating the purity of these states in a fiber are necessary for characterizing both the (de)multiplexing devices and the fibers themselves. We describe the pure OAM states, formed via linear combinations of the conventional vector eigenmodes of a fiber, by [70]:

$$
\begin{aligned}
\mathbf{v}_{\ell+1,1}^{(HE+)}(r,\theta) &= \left(\mathbf{HE}_{\ell+1,1}^{(e)} + i\,\mathbf{HE}_{\ell+1,1}^{(o)}\right)/\sqrt{2} = e^{i\ell\theta}\left(\hat{\mathbf{x}} + i\hat{\mathbf{y}}\right) F_{\ell,1}/\sqrt{2}, \\[4pt]
\mathbf{v}_{\ell+1,1}^{(HE-)}(r,\theta) &= \left(\mathbf{HE}_{\ell+1,1}^{(e)} - i\,\mathbf{HE}_{\ell+1,1}^{(o)}\right)/\sqrt{2} = e^{-i\ell\theta}\left(\hat{\mathbf{x}} - i\hat{\mathbf{y}}\right) F_{\ell,1}/\sqrt{2}, \\[4pt]
\mathbf{v}_{\ell-1,1}^{(EH+)}(r,\theta) &= \left(\mathbf{EH}_{\ell-1,1}^{(e)} - i\,\mathbf{EH}_{\ell-1,1}^{(o)}\right)/\sqrt{2} = e^{-i\ell\theta}\left(\hat{\mathbf{x}} + i\hat{\mathbf{y}}\right) F_{\ell,1}/\sqrt{2}, \\[4pt]
\mathbf{v}_{\ell-1,1}^{(EH-)}(r,\theta) &= \left(\mathbf{EH}_{\ell-1,1}^{(e)} + i\,\mathbf{EH}_{\ell-1,1}^{(o)}\right)/\sqrt{2} = e^{i\ell\theta}\left(\hat{\mathbf{x}} - i\hat{\mathbf{y}}\right) F_{\ell,1}/\sqrt{2}.
\end{aligned}
$$

$$(12.8)$$

These vortex states are linear combinations of degenerate pairs of true guided modes, and therefore are themselves true guided modes, or eigenstates of the fiber. Their instability in practice, in conventional fibers, is due to their near-degeneracy with other vortex states that, in combination with small geometric perturbations, causes debilitating crosstalk. An important feature of the solutions in Eq. (12.8) is that the orbital and spin degrees of freedom are decoupled: the phase terms $\exp(\pm i\ell\theta)$ and polarization terms $\mathbf{x} \pm i\mathbf{y}$ always appear in product form. Physically, this means that the phase fronts have the spiral angular dependence of Laguerre-Gaussian modes, which give rise to values of OAM of $\pm\ell\hbar$, and the fields have spatially uniform circular polarization, which give rise to SAM of $\pm\hbar$. Practically, this decoupling of spin and orbital degrees of freedom means that existing mode-sorter technology, for example based on spatial light modulators or specially designed phase masks, that was developed for free-space applications may also be applied to (de)multiplexing with optical fibers.

An alternate means for generating OAM states in fiber involves in-fiber gratings. Since, as described above, OAM states are true eigenmodes of a fiber, we can use a fiber grating to resonantly couple from an incoming fundamental (LP_{01}) mode, excited in these fibers via simply splicing them to single-mode fiber (SMF), to the desired OAM state.

In order to analyze the state present at the fiber output, we may project it onto various polarization states, as well as interfere it with a reference beam. The goal is to use the recorded output spatial intensity patterns to uniquely determine the mode content, as given by the amplitudes $|\gamma_i|^2$, in the expansion of the transverse electric field at the fiber output

$$\mathbf{E}_t(r,\theta,t) = e^{i\omega t}\sum_{i=1}^{6}\gamma_i e_i(r,\theta) \tag{12.9}$$

where the sum is over the mode index i. The six fields e_i represent the vector modes either in the standard basis $HE_{1,1}{}^{(e)}$, $HE_{1,1}{}^{(o)}$, $HE_{2,1}{}^{(e)}$, $HE_{2,1}{}^{(o)}$, $TE_{0,1}$, and $TM_{0,1}$, or alternatively, the vortex basis given in Eq. (12.8). From this representation, it is evident that a spatially uniform circular polarizer of positive (resp. negative) helicity will project onto the three "+" (resp. "−") vortex states.

Figure 12.22 shows the experimental setup that we used to control and analyze the OAM states of a vortex fiber. Using standard SMF, a 50-nm-wide 1550-nm light emitting diode (LED), and a narrowband continuous-wave (CW) tunable laser (Agilent 8168F) were multiplexed into a 20-m-long vortex fiber. Thereafter, using a microbend grating (40-mm length, 475-μm period) [68], with only the LED source turned on, we obtained 18-dB of mode conversion from the input LP_{01} mode to the desired HE_{21}-odd mode. Next, we switched the source to the laser, tuned to the resonant mode-conversion wavelength (1527 nm). The vortex fiber output was imaged onto a camera (VDS, NIR-300, InGaAs).

In order to observe the state present at the fiber output, and to analyze its purity, we projected the fiber output onto the left circular (LC) and right circular (RC) polarization states (Figure 12.22c). In addition, to observe the phase of the beam, the vertical (V) polarization projection with a reference beam. The "V+ref" image in Figure 12.22c shows a clear spiral interference pattern indicative of the output mode

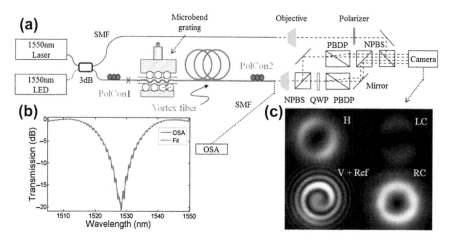

FIGURE 12.22 (a) Experimental setup. (b) Grating resonance spectrum used to deduce HE_{21}^{odd} mode conversion level. (c) Camera image showing $l = 1$ OAM, $s = 1$ SAM state.

carrying OAM. Using a combination of non-polarizing beam splitters (NPBS), quarter wave plates (QWP), and polarizing beam displacing prisms (PBDP), we devised a setup capable of recording these projections in one camera shot.

When most of the light is confined to one of the modes (as is the case in this experiment, where a single OAM state was excited by means of a fiber grating), evaluating the positive and negative SAM states' intensity projections on the ring of peak radial intensity r_0 reveals valuable information about the modal content. The result of the projection onto the, say, positive SAM state, can be written as

$$|P_+\mathbf{E}(r_0,\theta)|^2 \approx DC + \Delta_1 \cos(\theta+\phi_{21,11}) + \Delta_2 \cos(2\theta+\phi_{21,T}), \quad (12.10)$$

where knowledge of the coefficients DC and $\Delta_{1,2}$ are sufficient to recover the mode powers $|\gamma_i^s|^2$. Figure 12.23a shows an example of a measurement of $|P_+\mathbf{E}(r_0,\theta)|^2$; the azimuthal intensity variation of the image that occurs due to interference between the vortex states. Figure 12.23b shows an example of the Fourier series analysis used to recover the DC and $\Delta_{1,2}$ parameters, and Figure 12.23c illustrates the powers of the extracted modes. Note that mode purities exceeding 20 dB may be obtained, and measured, by our coupling and measurement technique.

Next, we show that, once an OAM state is generated in this fiber, we can control its topological charge (ℓ), by simple, off-the-shelf, commercial fiber components. It can be shown that a linear combination of two $\ell=\pm1$ OAM modes will have a total OAM of topological charge that lies between $-1\le\ell\le1$ [71]. By adjusting paddles on a commercial polarization controller mounted on our vortex fiber (PolCon2 of Figure 12.22a), we were able to tune the output OAM state from $\ell=-1$ to $\ell=1$ (Figure 12.22c shows the output of the $\ell=1$ OAM and $s=1$ SAM state). In order to show the ability of the system to control OAM, camera images were observed and analyzed (based on the aforementioned technique illustrated in Figure 12.23) while the polarization controller (PolCon2 on Figure 12.22a) was manually tuned using adjustable paddles. Figure 12.24a shows the modal power distribution as the system was tuned through the different linear combinations of the vortex states. In particular, Figure 12.24b shows observed camera images for the pure positive helicity vortex

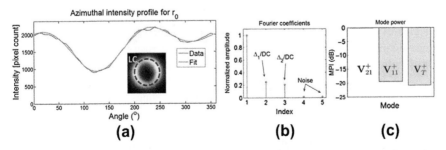

(a)   **(b)**   **(c)**

FIGURE 12.23 (a) Azimuthal intensity profile of LC projection for radius r_0. (b) Fourier series coefficients for profile in (a). (c) Extracted modal power contribution.

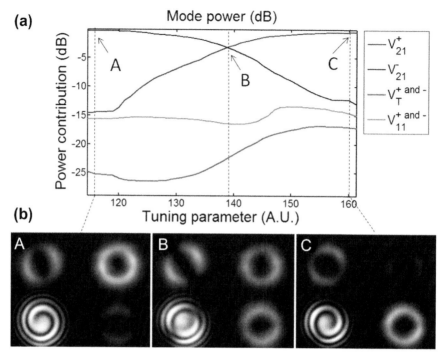

FIGURE 12.24 (a) Mode powers as PolCon2 (see Figure 12.22) is adjusted to obtain the desired superposition of the OAM states. (b) Observed camera images at points (A–C).

state (A), linear combination of the vortex states (B), and the pure negative helicity vortex state (C). Our calculations showed that the combined power of the undesired modes mostly stayed below a level of −14 dB (3%).

12.5.3 Long-length propagation of OAM in fiber

Having demonstrated the ability to excite, with a high level of mode purity, OAM states in our fiber, as well as control the OAM state at the output, we now turn our attention to the primary motivation for developing these fibers—namely, the ability to encode information in OAM states and transmit data over reasonable communications lengths. Here, we describe tests on the ability to maintain the OAM state over km-length fibers. The key question is, while the mode separation designed into vortex fibers enabled pure excitation and control of these states, would inadvertent random perturbation over km lengths serve to couple them to all other existing states in a fiber, or would they maintain their purity? Again, the means of studying this would be to perform mode purity analysis as described in the previous section.

The setup devised to study lengthwise mode purity? is the same as in Figure 12.22. A tunable narrowband laser is spliced to the input of the vortex fiber and conversion

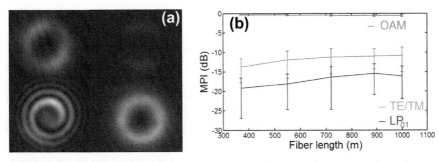

FIGURE 12.25 (a) Mode images at the output of a 1.1-km fiber; (a) cutback mode purity study using images in (a).

from the fundamental mode into the OAM mode is achieved using a microbend fiber grating. Using polarization controllers before the microbend grating, we ensured high mode conversion efficiency (>99%) to the $OAM^{\pm}$ modes. At the other fiber end, after 1.1 km propagation, the mode was imaged or interfered with a reference beam, in order to obtain information about mode's intensity or phase (Figure 12.25a).

When directly imaged, the mode intensity profile had a familiar "doughnut" shape (Figure 12.25a—bottom right) expected of OAM states. Moreover, note that there is a distinct spiral interference pattern when this mode is interfered with an expanded Gaussian beam reference (Figure 12.25a—bottom left), indicating that the output state is clearly in one of the two allowed OAM states. Detailed spatial Fourier analysis, of the kind described in the previous section, enables accurately measuring the relative modal content at the output of our fiber. We may thus define a figure of merit for this fiber in terms of its modal purity, defined as multipath interference level (MPI). MPI is defined, in our case, as the fraction of energy in unwanted modes in comparison to the OAM states. Figure 12.25b shows a cutback study of the mode purity. Our experiments reveal that, even after starting with imperfect input coupling (with mode purities of ~97%), OAM purity decreases only by ~10% over a km of propagation. We showed that OAM states do indeed degrade in purity as they propagate through fibers, but this degradation appears to asymptote at longer lengths, implying that even longer length transmission may be feasible [72].

12.5.4 Fiber-based data transmission using OAM

The previous sections described a novel optical fiber design that addresses a long-standing perceived limitation of optical fibers' ability to support OAM states. We showed that when a fiber is designed such that the scalar fields of OAM carrying ($|\ell|>0$) light encounter large index gradient steps in regions where their field amplitude is high, the almost degenerate subspace of OAM supporting states breaks, thereby making each of the individual OAM states stable, in nature. Armed with this powerful intuition, several fibers were developed, which showed the ability to generate OAM in fibers, as well as the ability to control the output to yield the desired OAM state. Most importantly, we see this fiber design enables stable propagation of these OAM states over length exceeding a km, wherein they become interesting for exploring their potential as additional optical channels to scale the capacity of a single fiber.

Preliminary work toward demonstrating the ability of these fibers to carry data over km lengths has been recently reported [73]—100 Gbit/s QPSK encoded data has been multiplexed on to four distinct modes of the vortex fiber, and successfully transmitted over a km length to yield a data rate of 400 Gbit/s. This represents the first demonstration of data transmission over lengths exceeding a km, without the use of multiple-input multiple-output (MIMO)-based signal processing—indeed, this was critically enabled by a fiber in which the modes experience low levels of cross-talk and MPI.

12.6 OPTICAL SIGNAL PROCESSING USING OAM

As an additional degree of freedom, OAM beam multiplexing has been used to advance optical communications in terms of both capacity and spectral efficiency [24]. To date, communications using multiple OAM beams have been relatively static point-to-point links, such that data on all the modes is transmitted as a unit from transmitter to receiver. Similar to time- and wavelength-based networks, different users in an OAM-based communication system occupy different orthogonal spatial channels, i.e. laser beams with different OAM values. Therefore, it might be interesting to see the capability of processing the data streams carried on randomly selected OAM beams without affecting all other user channels. Actually, some networking functions, such as data exchange, add/drop multiplexing, and multicasting are common in wavelength- and time-multiplexed networks in which different orthogonal data channels sharing the same transmission medium are addressed by wavelengths and time slots, respectively. The presence of these functions could be also helpful to advance the usefulness of OAM for multi-user communication applications. We will discuss the following generally useful networking functions in an OAM-based transmission system: (a) dynamic data exchange among OAM channels; (b) reconfigurable channel add/drop multiplexing of individual OAM channels; (c) all-optical multicasting of data onto multiple orthogonal OAM channels; (d) OAM beam distortion monitoring and compensation.

12.6.1 Data exchange

Data exchange could be a useful function in a multiple-channel communication network. Generally, different users occupy different orthogonal channels in the degree of freedom such as wavelength, time slot, polarization state, or spatial mode. At the network nodes, it is possible that two different channels need to swap their data information. One option would be to detect these two channels first and then re-modulate them with the swapped information. Obviously, in that case optical to electrical (*O/E*) and electrical to optical (*E/O*) conversions are required and the speed is limited. It would be more efficient if we can exchange the data directly in the optical domain. The exchange of data between different wavelengths and time slots has been reported [74–77]. Specifically, in an OAM-based optical communication network, different users may be addressed by different OAM charges. Hence, a laudable goal would be to achieve data exchange between different OAM channels.

The concept and principle of data exchange between OAM channels are shown in Figure 12.26 [24]. Superposed two data-carrying (Signal A, Signal B) OAM channels ($OAM_{\ell 1}$, $OAM_{\ell 2}$) shine at a reflective-type SLM loaded with a spiral phase mask with a charge of $\ell_R = -(\ell_1 + \ell_2)$. After reflecting off the SLM, the phase mask adds an azimuthal phase term $\exp(i\ell_R\theta)$ to the two OAM beams and converts them into $OAM_{-\ell 2}$ and $OAM_{-\ell 1}$, which are further transformed into $OAM\ \ell_2$ and $OAM\ \ell_1$ due to reflection of the SLM which flips the charge sign. As a result, data exchange between two OAM beams is implemented. It is noted that such data exchange scheme using linear optics is simple and low power consumption. For the input of two OAM channels with varied charges, reconfigurable data exchange is available by updating the phase mask loaded into the reflective-type SLM. Shown in Figure 12.26a is an example of data exchange between differential quadrature phase-shift keying (DQPSK)-carrying OAM channels (OAM_{+8}, OAM_{+6}).

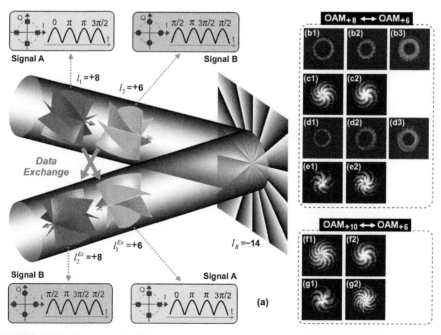

FIGURE 12.26 Concept, principle, and observed intensity profiles of data exchange between OAM channels [24].

The observed intensity profiles of two OAM beams before exchange and their superposition are depicted in Figure 12.26b1–b3. The measured interferograms in Figure 12.26c1 and c2 indicate that the charges of two OAM beams before exchange are +8 and +6. After exchange, the intensity profiles of exchanged OAM beams and their superposition are shown in Figure 12.26d1–d3. The measured interferograms in Figures 12.26e1 and e2 confirm that the charges of two OAM beams after exchange are +6 and +8. Figure 12.26f1, f2 and g1, g2 presents measured results of reconfigurable data exchange between another two OAM channels (OAM_{+10} and OAM_{+6}) by updating the spiral phase mask loaded into the SLM. Figure 12.27a1 and a2 shows measured BER performance for data exchange between two 100-Gbit/s return-to-zero DQPSK (RZ-DQPSK) carrying OAM channels (OAM_{+10} and OAM_{+6}) with power penalty <1.9 dB at a BER of 10^{-9}. The observed temporal waveforms and balanced eyes of demodulated in-phase (Ch. I) and quadrature (Ch. Q) components of 100-Gbit/s RZ-DQPSK signals verified the successful implementation of data exchange between two OAM channels (OAM_{+10} and OAM_{+6}) (Figure 12.27b).

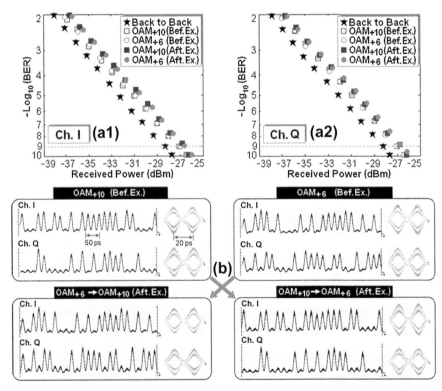

FIGURE 12.27 Measured BER performance, temporal waveforms and balanced eyes for data exchange between two 100-Gbit/s RZ-DQPSK-carrying OAM channels (OAM_{+10} and OAM_{+6}).

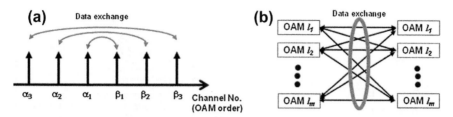

FIGURE 12.28 (a) Concept of data exchange between multiple OAM channels. (b) Envision of a full cross-connection of OAM network by cascading multiple OAM data exchange stages.

From the example provided, a more general rule would be expected: any two channels (with OAM orders of α_i and β_i, respectively) that satisfy $\alpha_1 + \beta_1 = \alpha_2 + \beta_2 = \ldots = m$ (m is a constant integer) can switch their data using OAM mode order conversion, i.e. by reflecting with a single SLM with a phase pattern of $-m$, as depicted in Figure 12.28a. Consequently, simultaneous data exchange between multiple OAM channels can be achieved. If multiple stages can be cascaded, we randomly swap information among arbitrary OAM channels, and a full cross-connect switch can be envisioned, as illustrated in Figure 12.28b.

12.6.2 Add/drop

An add/drop multiplexer is rather useful in a multi-channel, high-performance communication network. For example, a user at an intermediate point may want to access an individual channel among many channels in the link and upload a new channel instead. This requires a function block that can drop any randomly selected channel and add a new channel, without disturbing/detecting the non-selected channels. In wavelength-multiplexed networks, add/drop functionality has been demonstrated to great advantage by selectively dropping a given wavelength channel and adding a new channel at an intermediate point along the link without interrupting pass-through channels [78]. A similar concept can be employed in an OAM-based system where independent channels are addressed by different topological charges of OAM beams.

The concept of OAM-based add/drop function is shown in Figure 12.29. If a spatial channel K (OAM_K) needs to be dropped from OAM channels $1 \sim N$, SLM1 can be programmed with a phase pattern of $\exp(-iK\phi)$. As a result, each channel will be converted to a new OAM mode with azimuthal order $1-K \sim N-K$. Specifically, the channel K (the dropped channel) will transform to the center-focused Gaussian-like beam after passing the SLM1. In principle, the other OAM channels experience a $-K$ charge but are still OAM modes that have no power distributed at the center of the beam's field. By using a double-sided small mirror at the center of the light beam, we can efficiently drop channel K while the other channels bypass the mirror. Again, using the double-sided mirror, a different new added channel K' with a Gaussian field distribution can be directed to SLM2 with the inverse charge of order K. We can also use a specially designed grating pattern to extract the Gaussian beam

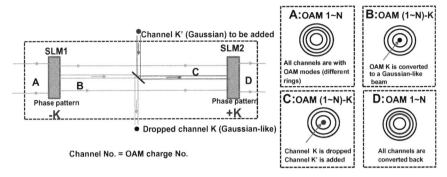

FIGURE 12.29 Concept of OAM channel add/drop.

located at the center from the rest of the ring-shaped beams (i.e. drop), and add a new Gaussian beam (i.e. add). Since SLM1 and SLM2 have conjugate phase masks, the added channel K' will have the same OAM mode order as the original dropped channel K. Additionally, the orders of the other OAM channels are not affected because the phase patterns from SLM1 and SLM2 cancel out each other. Since the SLMs can be arbitrarily programmed, a reconfigurable add/drop multiplexer can be achieved.

Figure 12.30 shows an example of add/drop function in a two-channel OAM link. These two channels (OAM_{+6} and OAM_{-6}) are multiplexed together and then launched onto the first SLM loaded with a phase mask with a charge of $+6$. Because of the OAM mode order conversion by the SLM, OAM_{+6} and OAM_{-6} will then be converted to OAM_{+12} and a Gaussian-like beam, respectively, as shown in Figure 12.30. It can be seen that these two channels are efficiently separated, with only the channel to be dropped at the center. The Gaussian-like beam is then separated out and a new

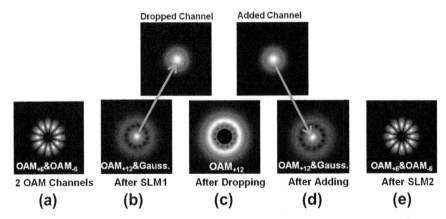

FIGURE 12.30 Simulation results of OAM channel add/drop from a two-channel OAM link (OAM_{+6} and OAM_{-6}).

Gaussian beam is added by using a double-sided mirror or another SLM with a fork-shaped grating pattern. The added beam together with OAM_{+12} is launched onto the second SLM loaded with a phase mask charge of -6. The second SLM will convert the added Gaussian beam and OAM_{+12} into OAM_{-6} and OAM_{+6}.

12.6.3 Multicasting

A multicasting function (i.e. fanout) indicates that data on a single channel can be duplicated onto multiple channels, without the need of detection and remodulation processes. Previously, wavelength and time slots channels have been multicasted in order to replicate the data such that they co-propagate and can be split subsequently for different potential user destinations [79,80]. Following the same concept, it might be desirable to multicast data information from one OAM channel to multiple different OAM channels [81].

The multicasting function can be achieved through a programmable SLM, as shown in Figure 12.31a. The multiple OAM beams after multicasting can be either divergent or collimated, determined by the phase pattern used for the SLM.

To achieve a divergent multicasting, a fork pattern (superposition of a spiral phase profile and a blazed grating) can be loaded to the SLM. Due to the diffraction of the grating, the beam can be split into multiple diffraction orders, each with a different reflecting angle in the incident plane, and therefore with different OAM orders. For example, if the incoming OAM beam has an order of ℓ, and the fork pattern includes a spiral phase pattern in the order of k, then the OAM orders after multicasting become $\ell+nk$ ($n=0, \pm1, \pm2, \pm3, \ldots$ is the diffraction order). The divergent angle of each beam is determined by the blazed grating period. Moreover, by combing two "fork" phase patterns with one perpendicular to the other, we can achieve multicasting at both x-axis and y-axis directions, as shown in Figure 12.31b.

In some cases, the multicasted copies of a data channel need to be multiplexed (i.e. overlapped in the spatial domain). The approach for the "collimated" multicasting is shown in Figure 12.31c. A data stream carried by one OAM beam can be easily duplicated and loaded onto different OAM charges/beams by using a phase-only SLM with a specially designed phase mask. As an example, Figure 12.31c shows the phase pattern to multicast the input OAM beam (OAM_{+15}) to five channels (OAM_{+6}, OAM_{+9}, OAM_{+12}, OAM_{+15}, and OAM_{+18}). After multicasting, the input OAM mode becomes the superposition of multiple OAM modes having a triangle intensity distribution. Figure 12.31d shows the power distribution before and after multicasting. The crosstalk between the three equalized spatial channels of $\ell=9, 12$, and 15 and undesired spatial channels is below -20 dB.

12.6.4 Monitoring and compensation

High mode purity and a low phase distortion are required in many applications that utilize OAM beams. For example, in near-field optical communications, a high capacity and spectral efficiency can be achieved by multiplexing/demultiplexing

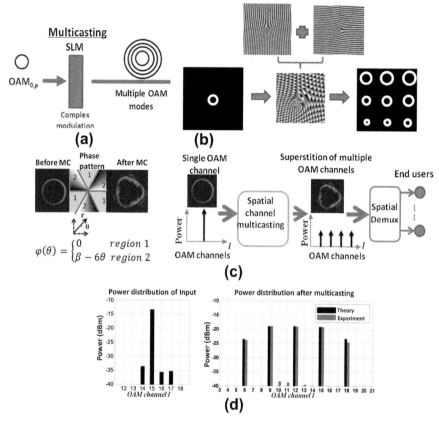

FIGURE 12.31 (a) Concept of data channel multicasting in an OAM-based mode-division multiplexing system. A spatial light modulator is used to distribute energy from a single input OAM channel to multiple OAM channels. (b) Principle of "divergent" OAM channel multicasting. (c) Concept and principle of "collimated" OAM channel multicasting (MC). (left: the designed phase pattern for multicasting, right: concept diagram). (d) Simulated and experimental results: power distribution on each OAM channel before and after multicasting.

many different OAM beams [24]. However, unwanted phase distortions, e.g. due to mild turbulence along the path or imperfect optics, degrade the multiplexing and demultiplexing efficiency [82]. Therefore, it is desirable to monitor the distortions of OAM beams, in order to use a feedback loop to compensate for those distortions.

Commonly used approaches of phase extraction include phase-stepping interferometry [83], in which moving optical components provide a variable phase reference, and the Fourier-transform method [84], in which the fringe pattern from off-axis interference is utilized to estimate the phase. Phase measurements of singularities in a vortex beam have been reported with the latter technique [85,86]. Another method measures the positions of the focal spots behind an array of lenslets, resulting in a piecewise-linear approximation to the phase profile of the incoming beam with spatial resolution determined by the lenslet size [87,88]. In this section, wavefront reconstruction for OAM beams using on-axis interference is discussed [89].

The principle of phase reconstruction is depicted in Figure 12.32a, in which A is an OAM beam to be measured, and B is a reference plane-wave beam. Both A and B are launched into a free space hybrid, in which the OAM beam interferes with the reference beam on one path and with $\pi/2$ phase-shifted reference beam on the other path. The fringe patterns at C and D can be expressed as [84]:

$$|A + B|^2 = |A|^2 + |B|^2 + 2\,|A|\,|B|\cos(\Phi_A - \Phi_B), \qquad (12.11)$$

$$|A + B\exp(j\pi/2)|^2 = |A|^2 + |B|^2 + 2\,|A|\,|B|\sin(\Phi_A - \Phi_B), \qquad (12.12)$$

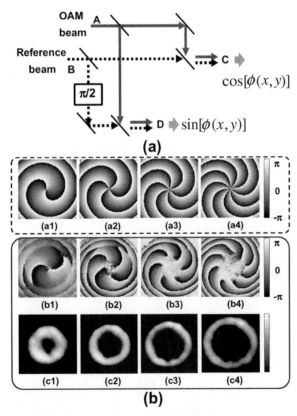

(a)

(b)

FIGURE 12.32 (a) Principle of OAM beam phase reconstruction based on a free space $\pi/2$ hybrid. One copy of OAM beam and the reference beam interfere at C, while the other copy of two beams with a $\pi/2$ phase shift for the reference beam interfere at D. (b) Simulated and measured wavefronts of generated OAM beams. (a1)~(a4): Simulated phase profiles of OAM$_{-2}$, OAM$_{-4}$, OAM$_{-6}$, and OAM$_{-8}$, respectively. (b1)~(b4): Measured phase profiles of OAM$_{-2}$, OAM$_{-4}$, OAM$_{-6}$, and OAM$_{-8}$, respectively. (c1)~(c4): Measured intensity profiles.

where $|A|$ and Φ_A are the amplitude and phase profiles of the OAM beam respectively, and $|B|$ and Φ_B are those of the reference beam. Here all variables are functions of the transverse-plane coordinate with respect to the propagation direction of the beams. All the amplitudes, including $|A|$, $|B|$, $|A+B|$, and $|A+B\exp(j\pi/2)|$, can be measured by a camera, and Φ_B is a constant since the reference beam is a plane wave. Consequently, $\cos(\Phi_A)$ and $\sin(\Phi_A)$ can be determined, and the phase profile of the OAM beam also can be obtained thereafter. Figure 12.32b shows the simulated and experimentally measured wavefronts of generated OAM beams, including OAM$_{-2}$, OAM$_{-4}$, OAM$_{-6}$, and OAM$_{-8}$.

Since OAM beam distortion exists in many cases, it is natural to think about the possibility of compensation. Although there have been many experiments on the compensation of distortions caused by the atmospheric turbulence [90], the demonstrations using OAM beams are rarely reported. One promising method is using the closed-loop adaptive optics, including a wavefront sensor and a bimorph piezoceramic mirror [88]. Another recently reported work employs Gerchberg-Saxton (GS) algorithm for compensation [91]. The concept is shown in Figure 12.33a. A camera captures the images of distorted OAM mode as well as the Gaussian beam incident on SLM1, and then these two images are taken as inputs of the GS algorithm. The GS algorithm can determine the corresponding phase hologram that would generate the observed intensity pattern. After iterating, a hologram $\Phi(x,y)$ which consists of an ideal spiral phase $H(x,y)$, overlaid by the distortion $D(x,y)$, can be produced. Then the distortion $D(x,y)$ can be acquired by subtracting $H(x,y)$ from $\Phi(x,y)$. The phase distortion of OAM mode can then be compensated for with another SLM by loading the conjugation of the distortion calculated from above. The phase pattern and intensity profile at each point of the simulation process can be seen in the inset of Figure 12.33a. The mode purity of corrected OAM modes together with crosstalk between adjacent modes is then analyzed.

Figure 12.33b shows the OAM mode purity as a function of strength of turbulence distortion Cn^2 before and after wavefront correction for different modes [92]. It is observed that the mode purity decreases dramatically with phase distortion

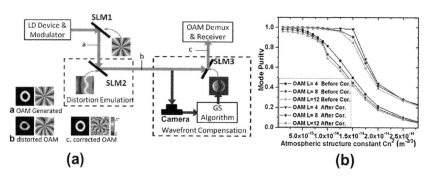

FIGURE 12.33 (a) Principle and simulation setup. (b) OAM mode purity as a function of strength of turbulence distortion before and after compensation for different modes [92].

getting stronger, especially when $Cn^2 > 1.25 \times 10^{-14}$. However, after phase retrieval and compensation using GS algorithm, the mode purity of OAM can be corrected to approach unit (nearly perfect) purity when $Cn^2 < 1.25 \times 10^{-14}$. The simulation results also indicate that the efficiency of the compensation decreases under a stronger turbulence, but still has an improvement of >20% on the OAM mode purity.

12.7 FUTURE CHALLENGES OF OAM COMMUNICATIONS

Moving from traditional applications in physics (e.g. optical manipulation, tweezers, imaging, astronomy, etc.) to optical communications, OAM offers an additional degree of freedom to further dramatically increase the capacity and spectral efficiency of communication systems. The aforementioned works, including free-space communication links, fiber-based transmission links, and OAM-based optical signal processing, have shown preliminary progress to achieve multimode communications using OAM multiplexing. However, there are several key challenges and crucial issues regarding the potential for using OAM in future communication systems.

For free-space communication links, it is of great importance to maintain the spatial phase structure of OAM beams which is sensitive to turbulence in the atmosphere [26,93]. There are possible solutions to overcome these problems using adaptive optics (e.g. phase correction method) [94]. Some alternative schemes, using low-density parity-check (LDPC) codes, together with MIMO and equalization techniques, might be used to deal with the atmospheric turbulence [95]. Consequently, data-carrying OAM multiplexing could potentially be exploited in future deep-space and near-Earth optical communications. Additionally, for situations that require high capacity or spectral efficiency over relatively short distances, OAM multiplexing could be appealing as the harmful effects of atmospheric turbulence can be either tolerated or compensated. Hence, data-carrying OAM multiplexing might be used in practical short-distance free-space propagation systems for high transmission capacity and spectral efficiency combined with other conventional multiplexing techniques such as WDM, PDM, and SDM [24]. Also, there are some other exciting opportunities for long-distance satellite-to-satellite communications in space, where turbulence is not an issue.

For fiber-based transmission links, one laudable goal would be to stably transmit multiple data-carrying OAM beams over a long-distance fiber with negligible mode-coupling crosstalk. One step was achieved in view of recently reported 400-Gbit/s data transmission over four distinct modes along a fiber exceeding a km [73]. Future work in this space would need to address the ability to obtain similar performance from longer-length (>10 km) fibers, and the ability to support and propagate many more OAM modes, so one can truly exploit the theoretically vast number of OAM states available from a cylindrically symmetric optical fiber.

For optical signal processing using OAM, miscellaneous grooming functions are expected to be exploited beyond data exchange, add/drop, multicasting, monitoring and compensation. Simple, scalable, reconfigurable, and multi-functional

OAM-based optical signal processing applications would be desired to increase the flexibility and efficiency of data management and enable robust networking.

Additionally, future green communication systems would benefit greatly by having compact, efficient, and cost-effective methods of generating, multiplexing/demultiplexing, and detecting beams with different OAM values, and the work is progressing on integrated devices and techniques.

The developments in OAM communications are sure to continue, and will potentially lead to OAM becoming an important alternative degree of freedom to achieve high bandwidth multiplexing in future. Although multimode communications using OAM are young together with lots of challenges and issues, the prospects are expected to be exciting [96].

References

[1] P.J. Winzer, G.J. Foschini, MIMO capacities and outage probabilities in spatially multiplexed optical transport systems, Opt. Express 19 (2011) 16680–16696.

[2] L. Allen, M.W. Beijersbergen, R.J.C. Spreeuw, J.P. Woerdman, Orbital angular momentum of light and the transformation of Laguerre-Gaussian laser modes, Phys. Rev. A 45 (1992) 8185–8189.

[3] S. Franke-Arnold, L. Allen, M.J. Padgett, Advances in optical angular momentum, Laser Photon. Rev. 2 (2008) 299–313.

[4] A.M. Yao, M.J. Padgett, Orbital angular momentum: origins, behavior and applications, Adv. Opt. Photon. 3 (2011) 161–204.

[5] J.H. Poynting, On the transfer of energy in the electromagnetic field, Phil. Trans. R. Soc. Lond. 175 (1884) 343–361.

[6] J.H. Poynting, The wave motion of a revolving shaft, and a suggestion as to the angular momentum in a beam of circularly polarised light, Proc. R. Soc. Lond. Ser. A 82 (1909) 560–567.

[7] R. Beth, Mechanical detection and measurement of the angular momentum of light, Phys. Rev. 50 (1936) 115–125.

[8] C.G. Darwin, Notes on the theory of radiation, Proc. R. Soc. Lond. Ser. A 136 (1932) 36–52.

[9] J.F. Nye, M.V. Berry, Dislocations in wave trains, Proc. R. Soc. Lond. Ser. A 336 (1974) 165–190.

[10] M.V. Berry, J.F. Nye, F. Wright, The elliptic umbilic diffraction catastrophe, Phil. Trans. R. Soc. Lond. 291 (1979) 453–484.

[11] P. Coullett, G. Gil, F. Rocca, Optical vortices, Opt. Commun. 73 (1989) 403–408.

[12] V. Bazhenov, M. Vasnetsov, M. Soskin, Laser beams with screw dislocations in their wavefronts, JETP Lett. 52 (1990) 429–431.

[13] M.W. Beijersbergen, L. Allen, H. Vanderveen, J.P. Woerdman, Astigmatic laser mode converters and transfer of orbital angular momentum, Opt. Commun. 96 (1993) 123–132.

[14] L. Allen, M.J. Padgett, M. Babiker, The orbital angular momentum of light, Prog. Opt. 39 (1999) 291–372.

[15] H. He, M. Friese, N. Heckenberg, Direct observation of transfer of angular momentum to absorptive particles from a laser beam with a phase singularity, Phys. Rev. Lett. 75 (1995) 826–829.

[16] A. Ashkin, J. Dziedzic, J. Bjorkholm, S. Chu, Observation of a single-beam gradient force optical trap for dielectric particles, Opt. Lett. 11 (1986) 288–290.

[17] M. Friese, J. Enger, H. Rubinsztein-Dunlop, Optical angular-momentum transfer to trapped absorbing particles, Phys. Rev. A 54 (1996) 1593–1596.

[18] N. Simpson, L. Allen, M.J. Padgett, Optical tweezers and optical spanners with Laguerre-Gaussian modes, J. Mod. Opt. 43 (1996) 2485–2491.

[19] K. Dholakia, N.B. Simpson, M.J. Padgett, L. Allen, Second-harmonic generation and the orbital angular momentum of light, Phys. Rev. A 54 (1996) R3742–R3745.

[20] R.W. Boyd, Nonlinear Optics, Academic Press, 2003.

[21] A. Mair, A. Vaziri, G. Weihs, A. Zeilinger, Entanglement of the orbital angular momentum states of photons, Nature 412 (2001) 313–316.

[22] M.J. Padgett, J. Courtial, Poincare-sphere equivalent for light beams containing orbital angular momentum, Opt. Lett. 24 (1999) 430–432.

[23] G. Gibson, J. Courtial, M.J. Padgett, M. Vasnetsov, V. Pas'ko, S.M. Barnett, S. Franke-Arnold, Free-space information transfer using light beams carrying orbital angular momentum, Opt. Express 12 (2004) 5448–5456.

[24] J. Wang, J.-Y. Yang, I.M. Fazal, N. Ahmed, Y. Yan, H. Huang, Y. Ren, Y. Yue, S. Dolinar, M. Tur, A.E. Willner, Terabit freespace data transmission employing orbital angular momentum multiplexing, Nature Photon. 6 (2012) 488–496.

[25] F. Tamburini, E. Mari, A. Sponselli, B. Thidé, A. Bianchini, F. Romanato, Encoding many channels on the same frequency through radio vorticity: first experimental test, New J. Phys. 14 (2012) 033001.

[26] C. Paterson, Atmospheric turbulence and orbital angular momentum of single photons for optical communication, Phys. Rev. Lett. 94 (2005) 153901.

[27] G.A. Tyler, R.W. Boyd, Influence of atmospheric turbulence on the propagation of quantum states of light carrying orbital angular momentum, Opt. Lett. 34 (2009) 142–144.

[28] B. Rodenburg, M.P.J. Lavery, M. Malik, M.N. O'Sullivan, M. Mirhosseini, D.J. Robertson, M.J. Padgett, R.W. Boyd, Influence of atmospheric turbulence on states of light carrying orbital angular momentum, Opt. Lett. 37 (2012) 3735–3737.

[29] M. Bourennane, A. Karlsson, G. Bjork, Quantum key distribution using multilevel encoding, Phys. Rev. A 64 (2001) 012306.

[30] S. Groblacher, T. Jennewein, A. Vaziri, G. Weihs, A. Zeilinger, Experimental quantum cryptography with qutrits, New J. Phys. 8 (2006) 75.

[31] M. Malik, M. O'Sullivan, B. Rodenburg, M. Mirhosseini, J. Leach, M.P.J. Lavery, M.J. Padgett, R.W. Boyd, Influence of atmospheric turbulence on optical communications using orbital angular momentum for encoding, Opt. Express 20 (2012) 13195–13200.

[32] M. Padgett, J. Courtial, L. Allen, Light's orbital angular momentum, Phys. Today 57 (2004) 35–40.

[33] M.W. Beijersbergen, R.P.C. Coerwinkel, M. Kristensen, J.P. Woerdman, Helical-wavefront laser beams produced with a spiral phaseplate, Opt. Commun. 112 (1994) 321–327.

[34] J.E. Curtis, B.A. Koss, D.G. Grier, Dynamic holographic optical tweezers, Opt. Commun. 207 (2002) 169–175.

[35] D. McGloin, N.B. Simpson, M.J. Padgett, Transfer of orbital angular momentum from a stressed fiber-optic waveguide to a light beam, Appl. Opt. 37 (1998) 469–472.

[36] P.Z. Dashti, F. Alhassen, H.P. Lee, Observation of orbital angular momentum transfer between acoustic and optical vortices in optical fiber, Phys. Rev. Lett. 96 (2006) 043604.

[37] N. Bozinovic, P. Kristensen, S. Ramachandran, Long-range fiber-transmission of photons with orbital angular momentum, in CLEO:2011 - Laser Applications to Photonic Applications, OSA Technical Digest (CD) (Optical Society of America, 2011), paper CTuB1.

[38] Y. Yan, J. Wang, L. Zhang, J.Y. Yang, I.M. Fazal, N. Ahmed, B. Shamee, A.E. Willner, K. Birnbaum, S. Dolinar, Fiber coupler for generating orbital angular momentum modes, Opt. Lett. 36 (2011) 4269–4271.

[39] Y. Yan, L. Zhang, J. Wang, J.-Y. Yang, I.M. Fazal, N. Ahmed, A.E. Willner, S.J. Dolinar, Fiber structure to convert a Gaussian beam to higher-order optical orbital angular momentum modes, Opt. Lett. 37 (2012) 3294–3296.

[40] Y. Yan, J. Yang, Y. Yue, M.R. Chitgarha, H. Huang, N. Ahmed, J. Wang, M. Tur, S. Dolinar, A.E. Willner, High-purity generation and power-efficient multiplexing of optical orbital angular momentum (OAM) modes in a ring fiber for spatial-division multiplexing systems, in CLEO: Science and Innovations, OSA Technical Digest (online) (Optical Society of America, 2012), paper JTh2A.58.

[41] Y. Yan, Y. Yue, H. Huang, J.Y. Yang, M.R. Chitgarha, N. Ahmed, M. Tur, S.J. Dolinar, A.E. Willner, Efficient generation and multiplexing of optical orbital angular momentum modes in a ring fiber by using multiple coherent inputs, Opt. Lett. 37 (2012) 3645–3647.

[42] J. Leach, M.J. Padgett, S.M. Barnett, S. Franke-Arnold, J. Courtial, Measuring the orbital angular momentum of a single photon, Phys. Rev. Lett. 88 (2002) 257901.

[43] G.C.G. Berkhout, M.P.J. Lavery, J. Courtial, M.W. Beijersbergen, M.J. Padgett, Efficient sorting of orbital angular momentum states of light, Phys. Rev. Lett. 105 (2010) 153601.

[44] M.P.J. Lavery, D.J. Robertson, G.C.G. Berkhout, G.D. Love, M.J. Padgett, J. Courtial, Refractive elements for the measurement of the orbital angular momentum of a single photon, Opt. Express 20 (2012) 2110–2115.

[45] N.K. Fontaine, C.R. Doerr, L. Buhl, Efficient multiplexing and demultiplexing of free-space orbital angular momentum using photonic integrated circuits, in Optical Fiber Communication Conference, OSA Technical Digest (Optical Society of America, 2012), paper OTu1I.2.

[46] T. Su, R.P. Scott, S.S. Djordjevic, N.K. Fontaine, D.J. Geisler, X. Cai, S.J.B. Yoo, Demonstration of free space coherent optical communication using integrated silicon photonic orbital angular momentum devices, Opt. Express 20 (2012) 9396–9402.

[47] A.H. Gnauck, P.J. Winzer, S. Chandrasekhar, X. Liu, B. Zhu, D.W. Peckham, Spectrally efficient long-haul WDM transmission using 224-Gb/s polarization-multiplexed 16-QAM, J. Lightwave Technol. 29 (2011) 373–377.

[48] X. Zhou, J. Yu, M.-F. Huang, Y. Shao, T. Wang, L. Nelson, P. Magill, M. Birk, P.I. Borel, D.W. Peckham, R. Lingle, B. Zhu, 64-Tb/s, 8 b/s/Hz, PDM-36QAM transmission over 320 km using both pre- and post-transmission digital signal processing, J. Lightwave Technol. 29 (2011) 571–577.

[49] A. Sano, H. Masuda, T. Kobayashi, M. Fujiwara, K. Horikoshi, E. Yoshida, Y. Miyamoto, M. Matsui, M. Mizoguchi, H. Yamazaki, Y. Sakamaki, H. Ishii, Ultra-high capacity WDM transmission using spectrally-efficient PDM 16-QAM modulation and C- and extended L-band wideband optical amplification, J. Lightwave Technol. 29 (2011) 578–586.

[50] T. Richter, E. Palushani, C. Schmidt-Langhorst, R. Ludwig, L. Molle, M. Nölle, C. Schubert, Transmission of single-channel 16-QAM data signals at terabaud symbol rates, J. Lightwave Technol. 30 (2012) 504–511.

[51] X. Liu, S. Chandrasekhar, X. Chen, P.J. Winzer, Y. Pan, T.F. Taunay, B. Zhu, M. Fishteyn, M.F. Yan, J.M. Fini, E.M. Monberg, F.V. Dimarcello, 1.12-Tb/s 32-QAM-

OFDM superchannel with 8.6-b/s/Hz intrachannel spectral efficiency and space-division multiplexed transmission with 60-b/s/Hz aggregate spectral efficiency, Opt. Express 19 (2011) B958–B964.

[52] R. Ryf, S. Randel, A.H. Gnauck, C. Bolle, A. Sierra, S. Mumtaz, M. Esmaeelpour, E.C. Burrows, R.-J. Essiambre, P.J. Winzer, D.W. Peckham, A.H. McCurdy, R. Lingle, Mode-division multiplexing over 96 km of few-mode fiber using coherent 6×6 MIMO processing, J. Lightwave Technol. 30 (2012) 521–531.

[53] D. Hillerkuss, R. Schmogrow, T. Schellinger, M. Jordan, M. Winter, G. Huber, T. Vallaitis, R. Bonk, P. Kleinow, F. Frey, M. Roeger, S. Koenig, A. Ludwig, A. Marculescu, J. Li, M. Hoh, M. Dreschmann, J. Meyer, S. Ben Ezra, N. Narkiss, B. Nebendahl, F. Parmigiani, P. Petropoulos, B. Resan, A. Oehler, K. Weingarten, T. Ellermeyer, J. Lutz, M. Moeller, M. Huebner, J. Becker, C. Koos, W. Freude, J. Leuthold, 26 Tbit s^{-1} line-rate super-channel transmission utilizing all-optical fast Fourier transform processing, Nature Photon. 5 (2011) 364–371.

[54] K.S. Youngworth, T.G. Brown, Focusing of high numerical aperture cylindrical-vector beams, Opt. Express 7 (2000) 77–87.

[55] Q. Zhan, Cylindrical vector beams: from mathematical concepts to applications, Adv. Opt. Photon. 1 (2009) 1–57.

[56] Y.I. Salamin, Electron acceleration from rest in vacuum by an axicon Gaussian laser beam, Phys. Rev. A 73 (2006) 043402.

[57] L. Novotny, M.R. Beversluis, K.S. Youngworth, T.G. Brown, Longitudinal field modes probed by single molecules, Phys. Rev. Lett. 86 (2001) 5251–5254.

[58] A.T. O'Neil, I. MacVicar, L. Allen, M.J. Padgett, Intrinsic and extrinsic nature of the orbital angular momentum of a light beam, Phys. Rev. Lett. 88 (2002) 053601.

[59] A.V. Nesterov, V.G. Niziev, Laser beams with axially symmetric polarization, J. Phys. D 33 (2000) 1817–1822.

[60] A. Vaziri, J.-W. Pan, T. Jennewein, G. Weihs, A. Zeilinger, Concentration of higher dimensional entanglement: qutrits of photon orbital angular momentum, Phys. Rev. Lett. 91 (2003) 227902.

[61] A.W. Snyder, J.D. Love, Optical Waveguide Theory, Chapman and Hall, London, 1983.

[62] J.-L. Li, K.-I. Ueda, M. Musha, A. Shirakawa, L.-X. Zhong, Generation of radially polarized mode in Yb fiber laser by using dual conical prism, Opt. Lett. 31 (2006) 2969–2971.

[63] G. Volpe, D. Petrov, Generation of cylindrical vector beams with few-mode fibers excited by Laguerre-Gaussian beams, Opt. Commun. 237 (2004) 89–95.

[64] A. Witkowska, S.G. L-Saval, A. Pham, T.A. Birks, All-fiber LP$_{11}$ mode convertors, Opt. Lett. 33 (2008) 306–308.

[65] G.K.L. Wong, M.S. Kang, H.W. Lee, F. Biancalana, C. Conti, T. Weiss, P.St.J. Russell, Excitation of orbital angular momentum resonances in helically twisted photonic crystal fiber, Science 337 (2012) 446–449.

[66] Y. Awaji, N. Wada, Y. Toda, T. Hayashi, World first mode/spatial division multiplexing in multi-core fiber using Laguerre-Gaussian mode, in 37th European Conference and Exposition on Optical Communications, OSA Technical Digest (CD) (Optical Society of America, 2011), paper We.10.P1.55.

[67] A. Bjarklev, Microdeformation losses in SMFs with step-index profiles, J. Lightwave Technol. 4 (1986) 341–346.

[68] S. Golowich, S. Ramachandran, Impact of fiber design on polarization dependence in microbend gratings, Opt. Express 13 (2005) 6870–6877.

[69] S. Ramachandran, P. Kristensen, M.F. Yan, Generation and propagation of radially polarized beams in optical fibers, Opt. Lett. 34 (2009) 2525–2527.

[70] N. Bozinovic, S. Golowich, P. Kristensen, S. Ramachandran, Control of orbital angular momentum of light with optical fibers, Opt. Lett. 37 (2012) 2451–2453.

[71] C.H.J. Schmitz, K. Uhrig, J.P. Spatz, J.E. Curtis, Tuning the orbital angular momentum in optical vortex beams, Opt. Express 14 (2006) 6604–6612.

[72] N. Bozinovic, P. Kristensen, S. Ramachandran, Are orbital angular momentum (OAM/vortex) states of light long-lived in fibers?, in Laser Science, OSA Technical Digest (Optical Society of America, 2011), paper LWL3.

[73] N. Bozinovic, Y. Yue, Y. Ren, M. Tur, P. Kristensen, A.E. Willner, S. Ramachandran, Orbital angular momentum (OAM) based mode division multiplexing (MDM) over a Km-length fiber, in European Conference and Exhibition on Optical Communication, OSA Technical Digest (online) (Optical Society of America, 2012), paper Th.3.C.6.

[74] K. Uesaka, K.K.-Y. Wong, M.E. Marhic, L.G. Kazovsky, Wavelength exchange in a highly nonlinear dispersion-shifted fiber: Theory and experiments, IEEE J. Sel. Top. Quantum Electron. 8 (2002) 560–568.

[75] J. Wang, H. Huang, X. Wang, J.-Y. Yang, A.E. Willner, Multi-channel 100-Gbit/s DQPSK data exchange using bidirectional degenerate four-wave mixing, Opt. Express 19 (2011) 3332–3338.

[76] J. Wang, Z. Bakhtiari, O.F. Yilmaz, S.R. Nuccio, X. Wu, A.E. Willner, 10 Gbit/s tributary channel exchange of 160 Gbit/s signals using periodically poled lithium niobate, Opt. Lett. 36 (2011) 630–632.

[77] J. Wang, O.F. Yilmaz, S.R. Nuccio, X.X. Wu, A.E. Willner, Orthogonal tributary channel exchange of 160-Gbit/s pol-muxed DPSK signal, Opt. Express 18 (2010) 16995–17008.

[78] J. Berthold, A.A.M. Saleh, L. Blair, J.M. Simmons, Optical networking: past, present, and future, J. Lightwave Technol. 26 (2008) 1104–1118.

[79] A. Biberman, B.G. Lee, A.C. Turner-Foster, M.A. Foster, M. Lipson, A.L. Gaeta, K. Bergman, Wavelength multicasting in silicon photonic nanowires, Opt. Express 18 (2010) 18047–18055.

[80] C.-S. Brès, A.O.J. Wiberg, J. Coles, S. Radic, 160-Gb/s optical time division multiplexing and multicasting in parametric amplifiers, Opt. Express 16 (2008) 16609–16615.

[81] Y. Yan, Y. Yue, H. Huang, Y. Ren, N. Ahmed, A.E. Willner, S. Dolinar, Spatial-mode multicasting of a single 100-Gbit/s orbital angular momentum (OAM) mode onto multiple OAM modes, in European Conference and Exhibition on Optical Communication, OSA Technical Digest (online) (Optical Society of America, 2012), paper Th.2.D.1.

[82] J.H. Shapiro, S. Guha, B.I. Erkmen, Ultimate capacity of free-space optical communications, J. Opt. Netw. 4 (2005) 501–516.

[83] C. Joenathan, Phase-measuring interferometry: new methods and error analysis, Appl. Opt. 33 (1994) 4147–4155.

[84] M. Takeda, H. Ina, S. Kobayashi, Fourier-transform method of fringe-pattern analysis for computer-based topography and interferometry, J. Opt. Soc. Am. 72 (1982) 156–160.

[85] T. Ando, N. Matsumoto, Y. Ohtake, Y. Takiguchi, T. Inoue, Structure of optical singularities in coaxial superpositions of Laguerre–Gaussian modes, J. Opt. Soc. Am. A 27 (2010) 2602–2612.

[86] C. Rockstuhl, A.A. Ivanovskyy, M.S. Soskin, M.G. Salt, H.P. Herzig, R. Dändliker, High-resolution measurement of phase singularities produced by computer-generated holograms, Opt. Commun. 242 (2004) 163–169.

[87] F.A. Starikov, G.G. Kochemasov, S.M. Kulikov, A.N. Manachinsky, N.V. Maslov, A.V. Ogorodnikov, S.A. Sukharev, V.P. Aksenov, I.V. Izmailov, F.Y. Kanev, V.V. Atuchin, I.S. Soldatenkov, Wavefront reconstruction of an optical vortex by a Hartmann-Shack sensor, Opt. Lett. 32 (2007) 2291–2293.

[88] F.A. Starikov, G.G. Kochemasov, M.O. Koltygin, S.M. Kulikov, A.N. Manachinsky, N.V. Maslov, S.A. Sukharev, V.P. Aksenov, I.V. Izmailov, F.Y. Kanev, V.V. Atuchin, I.S. Soldatenkov, Correction of vortex laser beam in a closed-loop adaptive system with bimorph mirror, Opt. Lett. 34 (2009) 2264–2266.

[89] H. Huang, Y. Ren, N. Ahmed, Y. Yan, Y. Yue, A. Bozovich, J. Yang, A.E. Willner, K. Birnbaum, B. Erkmen, J. Choi, S. Dolinar, M. Tur, Demonstration of OAM mode distortions monitoring using interference-based phase reconstruction, in CLEO: Science and Innovations, OSA Technical Digest (online) (Optical Society of America, 2012), paper CF3C.4.

[90] A.W.M. van Eekeren, K. Schutte, J. Dijk, P.B.W. Schwering, M. van Iersel, N.J. Doelman, Turbulence compensation: an overview, in: Proc. SPIE 8355, 83550Q, 2012.

[91] R.W. Gerchberg, W.O. Saxton, A practical algorithm for the determination of phase from image and diffraction plane pictures, Optik 35 (1972) 237–246.

[92] Y. Ren, H. Huang, J. Yang, Y. Yan, N. Ahmed, Y. Yue, A.E. Willner, K. Birnbaum, J. Choi, B. Erkmen, S. Dolinar, Correction of phase distortion of an OAM mode using GS algorithm based phase retrieval, in CLEO: Science and Innovations, OSA Technical Digest (online) (Optical Society of America, 2012), paper CF3I.4.

[93] J.A. Anguita, M.A. Neifeld, B.V. Vasic, Turbulence-induced channel crosstalk in an orbital angular momentum-multiplexed free-space optical link, Appl. Opt. 47 (2008) 2414–2429.

[94] S.M. Zhao, J. Leach, L.Y. Gong, J. Ding, B.Y. Zheng, Aberration corrections for free-space optical communications in atmosphere turbulence using orbital angular momentum states, Opt. Express 20 (2012) 452–461.

[95] I.B. Djordjevic, Deep-space and near-Earth optical communications by coded orbital angular momentum (OAM) modulation, Opt. Express 19 (2011) 14277–14289.

[96] A.E. Willner, J. Wang, H. Huang, A different angle on light communications, Science 337 (2012) 655–656.

Transmission Systems Using Multicore Fibers

13

Yoshinari Awaji[a], Kunimasa Saitoh[b], and Shoichiro Matsuo[c]

[a]National institute of information and communications technology (NICT), Japan,
[b]Hokkaido University, Japan,
[c]Fujikura Ltd, Japan

13.1 EXPECTATIONS OF MULTICORE FIBERS

To overcome the issue of capacity limits in the existing optical fiber communications infrastructure, increasing the spatial efficiency within the available fiber cross-section is the most effective solution. Multicore fibers (MCFs), in conjunction with space-division multiplexing (SDM), are one of the most promising technologies for realizing such an increase in spatial efficiency. In practice, the available number of spatial channels is also important.

The parallelism of MCFs makes them a strong candidate for realizing LAN-PHY networks to take Ethernet beyond 100 GE. Most MCFs can provide reliable independent transmission channels for short-haul applications. In the same manner that 100 GE LANPHY was realized by using four lanes of 25-Gbps WDM channels, LANPHY terabit Ethernet networks can be realized by using seven cores accommodating six lanes of 25-Gbps WDM channels, for example. A dramatic reduction in the footprint of network equipment can be expected by using MCF technology.

On the other hand, in addition to common short-haul optical links, MCFs are also useful for long-haul trunk lines to avoid nonlinear signal impairment that originates from excessive power concentration inside the core. In this case, the important features of MCFs are the aggregated core cross-section and the extremely high transmission capacity for a given cladding diameter.

This chapter mainly considers uncoupled MCFs and gives details of transmission systems that have been demonstrated in recent years. In such systems, the choice of MCF is the most important issue with regard to inter-core crosstalk. Hence, different types of MCFs and theoretical designs will be reviewed first. Second, MCF coupling methods will be introduced, including fan-in/fan-out couplers for connecting with standard single-mode fiber (SSMF) and splicing techniques for connecting with other MCFs.

13.2 MCF DESIGN

13.2.1 Types of MCFs

A multicore structure has been considered as a candidate for improving the efficiency of space-division multiplexing (SDM) [1,2]. In particular, after the successful fabrication of an MCF with a holey structure in 2008 [3], research on MCFs accelerated, with the aim of increasing the transmission capacity per fiber [4].

Figure 13.1 shows the types of MCFs reported so far. MCFs can be classified into uncoupled-type and coupled-type fibers. In uncoupled MCFs, each core has to be suitably arranged to keep the inter-core crosstalk sufficiently small for long-distance transmission applications. A number of core arrangements have been reported for uncoupled MCFs, including homogeneous MCFs with multiple identical cores [5–7], quasi-homogeneous MCFs with several kinds of slightly different cores [8], and heterogeneous MCFs with several kinds of different cores [9,10]. In addition, each core can be designed to support not only a single mode but also a few modes or even multiple modes. These are called multi-core single-mode fiber (MC-SMF) [11–18], multi-core few-mode fiber (MC-FMF) [19], and multi-core multi-mode fiber (MC-MMF) [20], respectively.

In coupled MCFs on the other hand, several cores are placed so that they strongly and/or weakly couple with each other. Coupled MCFs supporting a single transverse mode and multiple transverse modes have been investigated for high-power fiber laser applications [21,22], and coupled MCFs supporting a few super-modes can be used as few-mode fibers for large-capacity transmission experiments with mode-division multiplexing (MDM) [23–26].

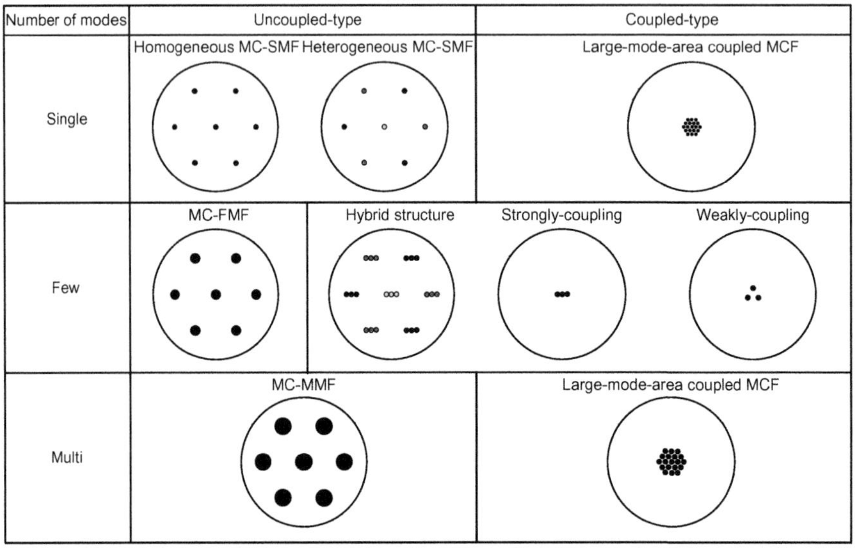

FIGURE 13.1 Types of MCFs [5–26].

Table 13.1 summarizes recently fabricated uncoupled MC-SMFs [5,6,8,11–18]. In the uncoupled MCFs used as transmission fibers, a high-core density, low attenuation, and large effective area are the important characteristics for improving the optical signal-to-noise ratio and SDM efficiency [27,28]. In order to compare the core density of MCFs, a core multiplicity factor (CMF) was proposed [29]:

$$\text{CMF} = \frac{n A_{\text{eff}}}{\pi (CD/2)^2},\qquad(13.1)$$

where n is the number of cores each having effective area A_{eff} in a cladding and CD is the cladding diameter. The CMF indicates the ratio occupied by the core area in the cladding. In Table 13.1, the relative CMF (RCMF) of MCFs with various A_{eff} and CD values is shown. The RCMF is the ratio of the CMF of an MCF and that of a standard single-core single-mode fiber with A_{eff} of $80\,\mu\text{m}^2$ at $1.55\,\mu\text{m}$ and a cladding diameter of $125\,\mu\text{m}$. The maximum reported RCMF is 6.66 [13], and the maximum reported number of cores is 19 [18].

13.2.2 Inter-core crosstalk in homogeneous uncoupled MCFs

The number of cores that can be multiplexed in a fiber is determined by the core-to-core distance for a fixed outer cladding diameter. However, a small core-to-core distance results in a large crosstalk between neighboring cores in uncoupled MCFs. The suppression of crosstalk between cores is one of the critical issues for practical use of uncoupled MCFs. Figure 13.2a shows a schematic cross-section of an MCF with a step-index homogeneous core arrangement (SI-MCF), where seven identical cores having a step-index profile are arranged hexagonally. The relative refractive index difference between the core and cladding is Δ, and the core-to-core distance is D.

Table 13.1 Characteristics of reported uncoupled MC-SMFs.

Reference	Number of Cores	CD [μm]	Attenuation [dB/km]	A_{eff} [μm^2]	Crosstalk/ Length [dB/km]	RCMF
OFC2009 [5]	7	~180	2.38	41.8	−60	1.76
OFC2010 [6]	7	215	0.205	101	−23	2.99
OFC2010 [8]	7	125.9	0.43	40.1	−39	3.46
OFC2011 [11]	7	217	0.213	110	−43	3.19
OFC2011 [12]	7	150	0.18	80	−92	4.86
OFC2011 [13]	7	125.4	0.21	76.6	−45	6.66
Opt. Expr. [14]	7	186.5	0.23	~75	−60	2.95
ECOC2011 [15]	7	240	0.2	100	−50	2.37
ECOC2011 [16]	7	181.3	0.198	112.4	−59	4.67
Opt. Lett. [17]	10	204.4	0.242	116	−46	5.42
OFC2012 [18]	19	200	0.227	71.5	−42	6.63

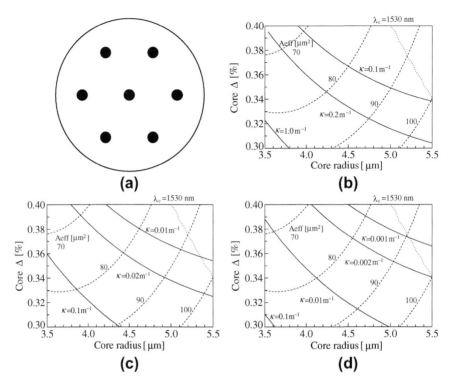

FIGURE 13.2 (a) Cross-section of an SI-MCF with homogeneous core arrangement, where the relative refractive index difference between the core and cladding is Δ, and the core-to-core distance is D. The coupling coefficient, κ, between neighboring cores as a function of core radius and Δ at a wavelength of 1550 nm for (b) $D = 35\,\mu m$, (c) $D = 40\,\mu m$, and (d) $D = 45\,\mu m$, where the dashed lines represent the effective area values, and the dotted lines correspond to the cutoff wavelength of 1530 nm.

If the core-to-core distance D and the cutoff wavelength λ_c of each core are fixed, the effective area has to be decreased to decrease the coupling coefficient between neighboring cores, resulting in higher fiber nonlinearity. Therefore, there is a trade-off relationship between inter-core crosstalk, fiber nonlinearity, and core density while maintaining the single-mode condition.

In order to estimate the inter-core crosstalk in MCFs numerically, a coupled-mode theory (CMT) [10,30] and a coupled-power theory (CPT) [31] have been introduced. In CMT, in order to obtain sufficiently accurate average values of crosstalk, a large number of samples must be simulated. In CPT, on the other hand, average crosstalk values can be obtained by only one simulation. Figure 13.3 shows the first demonstrated length dependence of measured crosstalk in two fabricated seven-core MCFs [8], where the core-cladding relative refractive index difference (Δ), the core diameter, and the core-to-core distance are 0.7%, 6.1 μm, and 40.4 μm, respectively, in Fiber A,

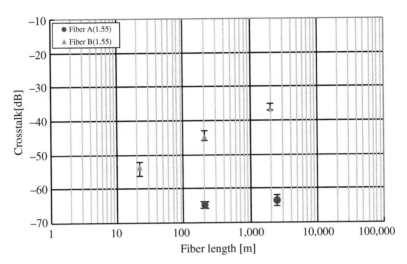

FIGURE 13.3 Length dependence of measured crosstalk in fabricated MCFs [8].

and 0.7%, 5.3 μm, and 35.4 μm, respectively, in Fiber B. The measured results show that the crosstalk degrades monotonically with increasing fiber length. This length dependence of crosstalk in the seven-core fiber was successfully simulated using the CPT with a conventional power-coupling coefficient (PCC), defined as the ratio of maximum power conversion efficiency to coupling length derived from CMT [31].

More recently, a novel PCC, $h = \kappa^2 S(\Delta\beta)$, has been derived based on exponential, Gaussian, and triangular autocorrelation functions [32], where κ is the mode-coupling coefficient between neighboring cores, $\Delta\beta$ is the local propagation-constant difference between adjacent cores, and $S(\Delta\beta)$ is the power spectrum density, which is the Fourier transform of the autocorrelation function. For homogeneous MCFs with small bending radius R, the average value of the PCC based on an exponential autocorrelation function is given by $h_{avg} = 2\kappa^2 R/(\beta D)$, where β and D are the propagation constant and the core-to-core distance, respectively. Therefore, the average value of the statistical distribution of the crosstalk XT_{ave} in homogeneous MCFs is given by [28,32]

$$XT_{ave} = 2\frac{\kappa^2}{\beta}\frac{R}{D}L, \tag{13.2}$$

where L is the fiber length. For example, if we target a crosstalk level of $-30\,\mathrm{dB}$ after 100 km propagation with $D < 45\,\mu\mathrm{m}$ and $R < 300\,\mathrm{mm}$, the coupling coefficient κ has to be lowered to around $2 \times 10^{-3}\,\mathrm{m}^{-1}$ or less. Figure 13.2b, c, and d shows the coupling coefficient κ between neighboring cores of SI-MCFs as a function of core radius and core Δ at a wavelength of 1550 nm for $D = 35\,\mu\mathrm{m}$, $D = 40\,\mu\mathrm{m}$, and $D = 45\,\mu\mathrm{m}$, respectively, where the dashed lines represent the effective area values and the dotted lines correspond to the cutoff wavelength of 1530 nm. When the core-to-core distance is 35 μm, the coupling coefficient is on the order of $0.1\,\mathrm{m}^{-1}$ for a reasonable A_{eff} size with single-mode operation. Even if the core-to-core distance becomes 40 μm, the coupling coefficient is on the order of $0.01\,\mathrm{m}^{-1}$, and a coupling coefficient on the order of $0.001\,\mathrm{m}^{-1}$ is obtained by increasing the core-to-core distance to 45 μm. These results indicate that a core-to-core distance of least 45 μm is required for low crosstalk transmission with large A_{eff} and single-mode operation in SI-MCFs with a homogeneous core arrangement. This limitation on the core-to-core distance is a high barrier to high-core-density SDM transmission using MCFs with practical cladding diameters. A small cladding diameter is preferable not only for achieving a dense core arrangement but also for maintaining high mechanical reliability for bending.

In order to realize low crosstalk and a dense core arrangement simultaneously in MCFs, trench-assisted MCFs (TA-MCFs) [12,13] and hole-assisted MCFs (HA-MCFs) [33] have been proposed. Figure 13.4a shows a schematic cross-section of a TA-MCF with a homogeneous core arrangement, where seven identical cores with a low-index trench profile are arranged hexagonally. A schematic diagram of an index profile with a trench-assisted structure is also presented in Figure 13.4a. The core radius is r_1, the distance from the core center to the trench position is r_2, and the trench width is W. The relative refractive index difference between the core and the cladding is Δ_1, the relative refractive index difference between the trench region and the cladding is Δ_2, and the core-to-core distance is D. The trench-assisted structure can reduce not only macro-bending loss [34] but also micro-bending loss [35]; therefore, it is effective for reducing the outer cladding thickness [16], resulting in a higher core density with a limited cladding size.

Figure 13.4b, c, and d shows the coupling coefficient κ between neighboring cores of a TA-MCF as a function of the core radius and the core Δ_1 at a wavelength of 1550 nm for $D = 35\,\mu\mathrm{m}$, $D = 40\,\mu\mathrm{m}$, and $D = 45\,\mu\mathrm{m}$, respectively, where the relative trench position r_2/r_1 and the relative trench width W/r_1 are assumed to be, as an example, 2.0 and 1.0, respectively, and Δ_2 is fixed at –0.70%. The dashed lines represent the effective area values, and the dotted lines correspond to the cutoff wavelength of 1530 nm in an isolated trench-assisted core. When the core-to-core distance is 45 μm, the coupling coefficient κ is on the order of $0.0001\,\mathrm{m}^{-1}$ with single-mode operation. Even if the core-to-core distance is reduced to 35 μm, the coupling coefficient κ can be on the order of $0.002\,\mathrm{m}^{-1}$ while maintaining a large A_{eff} of $>100\,\mu\mathrm{m}^2$. These results indicate that we can reduce the core-to-core distance in TA-MCFs to around 35 μm by appropriately setting the fiber parameters and the core arrangement.

Figure 13.5a shows a cross-sectional view of a recently proposed trench-assisted 10-core fiber with a large effective area of about $120\,\mu\mathrm{m}^2$ [17]. It has a two-pitch layout, which was designed to increase the number of cores with a limited cladding diameter in order to maintain mechanical reliability. The core-to-core distance between the outer cores is 40.5 μm, and the core-to-core distance between the center core and the outer cores is 59.2 μm. Figure 13.5b shows the measured crosstalk at 1550 nm for a fiber length of 3962 m, where "2–3" denotes the crosstalk between core 2 and core 3, and so forth. The upper graph in Figure 13.5b represents the crosstalk of adjacent

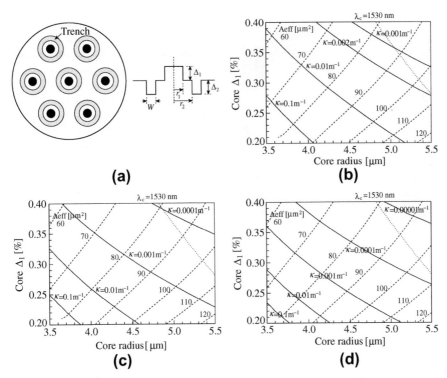

FIGURE 13.4 (a) Cross-section of a TA-MCF with a homogeneous core arrangement, where the relative refractive index difference between the core and the cladding is Δ_1, the relative refractive index difference between the trench region and the cladding is Δ_2, and the core-to-core distance is D. The coupling coefficient κ between neighboring cores as function of the core radius and core Δ_1 at a wavelength of 1550 nm for (b) $D = 35\,\mu m$, (c) $D = 40\,\mu m$, and (d) $D = 45\,\mu m$, respectively, where r_2/r_1 and W/r_1 are assumed to be 2.0 and 1.0, respectively. The dashed lines represent the effective area values, and the dotted lines correspond to the cutoff wavelength of 1530 nm.

outer cores, whereas the lower graph shows the crosstalk between the center core and the outer cores. The crosstalk between outer cores is about −40 dB. On the other hand, the crosstalk between the center and outer cores is less than −70 dB. The very small center-outer core crosstalk will be effective for avoiding crosstalk degradation of the center core in the worst case where all of the outer cores carry equal power.

13.2.3 Inter-core crosstalk in heterogeneous uncoupled MCFs

Another approach for realizing a low-crosstalk uncoupled MCF is to introduce different kinds of cores in the fiber [9], which is called a heterogeneous MCF. The heterogeneous MCF consists of several kinds of cores whose propagation constants are

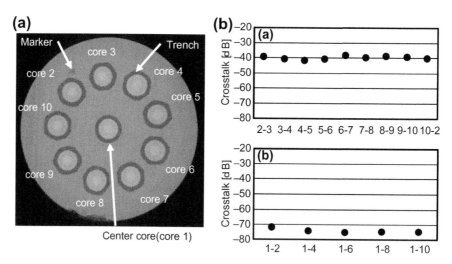

FIGURE 13.5 (a) Cross-sectional view of a fabricated trench-assisted ten-core fiber and (b) its measured crosstalk for a fiber length of 3962 m [17].

different from each other. Figure 13.6a shows a schematic cross-section of a heterogeneous MCF, where three kinds of cores are arranged hexagonally. In heterogeneous MCFs, the maximum normalized power transferred between the non-identical cores (that is, the power-conversion efficiency) is given by $F = 1/[1+\Delta\beta^2/(2\kappa)^2]$, where $\Delta\beta$ is the propagation constant difference of the fundamental modes between cores and κ is the coupling coefficient. It is known that the power-conversion efficiency

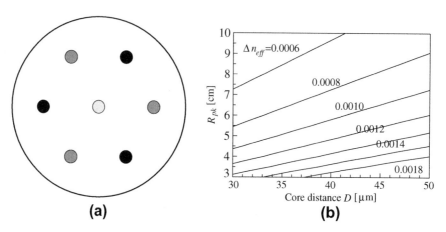

FIGURE 13.6 (a) Cross-section of a heterogeneous MCF with three kinds of cores arranged hexagonally. (b) Required effective index difference Δn_{eff} as a function of the core-to-core distance D and R_{pk}.

can be effectively suppressed by introducing slightly different cores, if there are no perturbations, such as bending, twisting, and so on, along the fiber length. However, in a practical situation, this is not the case, and bending strongly affects the crosstalk characteristics in heterogeneous MCFs [10].

Figure 13.7a shows a cross-sectional view of a fabricated heterogeneous MCF [36]. The outer cores, whose diameters were categorized into two groups, were alternately arranged in the circumferential direction. Figure 13.7b shows the measured core diameter of the fabricated MCF. The core-to-core distances were about 39.2 μm for all cores. The relative refractive index deference of the cores was about 0.4%. Mode field diameters at 1550 nm ranged from 9.52 μm to 9.77 μm for all cores. Figure 13.7c shows the bending diameter dependence of the measured crosstalk of the outer cores. Similar tendencies were observed at smaller bending diameters of less than 500 mm, namely, a linear relation between the bending diameter plotted on a logarithmic scale and the crosstalk. On the other hand, at bending diameters larger than 500 mm, the crosstalk decreased rapidly at a particular bending diameter, especially in cores with even core IDs. This phenomenon originates from the variation of the equivalent propagation constant in the outer cores due to bending and twisting effects [10]. Considering the fiber bending and twisting effects, the propagation constant difference between neighboring cores, $\Delta\beta$, is a function of propagation distance and becomes zero at many positions if the bending radius is smaller than a specific threshold value of R_{pk}, which is given by

$$R_{pk} = n_{\text{eff}}D/\Delta n_{\text{eff}}, \tag{13.3}$$

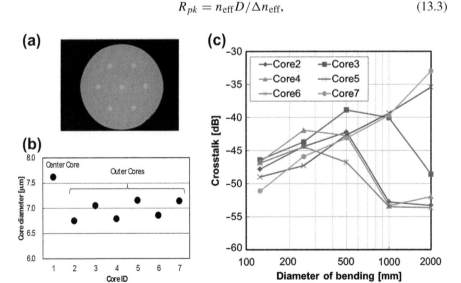

FIGURE 13.7 (a) A cross-sectional view and (b) the measured core diameters of a fabricated heterogeneous MCF [36]. (c) Measured bending diameter dependence of crosstalk of a 100-m MCF. Crosstalk between outer cores (core IDs = 2–6) and a center core (core ID = 1) was measured at a wavelength of 1550 nm.

where n_{eff} and Δn_{eff} are the effective index of a core and the effective index difference between non-identical cores, respectively, and D is the core-to-core distance. The large crosstalk occurs at bending radii less than R_{pk} due to index matching between non-identical cores [10,36]. In this phase-matching region, the bend perturbations are crucial. In the non-phase-matching region at bending radii greater than R_{pk}, on the other hand, the crosstalk is dominated by the statistical properties [32]. Therefore, a large effective index difference between cores will be required for pushing the value of R_{pk} toward a sufficiently small range. Figure 13.6b shows the required effective index difference Δn_{eff} as a function of the core-to-core distance D and R_{pk}. It is found that, if we try to shift R_{pk} to a bending radius smaller than 5 cm with a core-to-core distance of less than 40 μm, a Δn_{eff} value larger than 0.001 would be needed between non-identical cores in a heterogeneous MCF. Figure 13.8 shows the effective index value of the fundamental mode at a wavelength of 1550 nm as a function of the core radius and core Δ, where the dashed lines represent the effective area values, and the dotted lines correspond to the cutoff wavelength of 1530 nm. The shaded region in Figure 13.8 represents a high-bending-loss structure, where the bending loss of the fundamental mode is larger than 0.5 dB/100 turn at a wavelength of 1625 nm with a bending radius of 30 mm. This result indicates that it is difficult to select three kinds of cores while keeping Δn_{eff} between non-identical cores larger

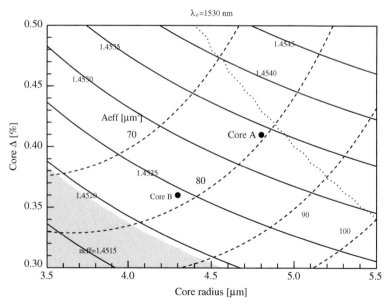

FIGURE 13.8 Effective index value of the fundamental mode at a wavelength of 1550 nm as a function of the core radius and core Δ, where the dashed lines represent the effective area values, the dotted line corresponds to the cutoff wavelength of 1530 nm, and the shaded region represents a high-bending-loss structure.

than 0.001 under single-mode, large A_{eff} ($>80\,\mu\text{m}^2$), and low-bending-loss conditions. For example, if we select a core with a radius of $4.8\,\mu\text{m}$ and $\Delta = 0.41\%$ (Core A in Figure 13.8), another non-identical core with similar A_{eff}, a core radius of $4.3\,\mu\text{m}$ and $\Delta = 0.36\%$ (Core B in Figure 13.8) could be selected to realize $R_{pk} = 5$–$5.5\,\text{cm}$ with $D = 35$–$40\,\mu\text{m}$.

Based on the relationship between R_{pk}, Δn_{eff}, and D, we evaluated the crosstalk characteristics of a heterogeneous MCF with sufficiently small R_{pk} under bending conditions. Figure 13.9a shows a schematic cross-section of a heterogeneous MCF with two kinds of cores considered here (MCF-1), where one of the cores has a radius of $4.8\,\mu\text{m}$ and $\Delta = 0.41\%$ (Core A in Figure 13.8), and the other one has a radius of $4.3\,\mu\text{m}$ and $\Delta = 0.36\%$ (Core B in Figure 13.8). Core A and Core B are arranged alternately with a core-to-core distance of $35\,\mu\text{m}$, and a total of six cores are provided. For comparison, a homogeneous MCF with six identical cores shown in Figure 13.9b (MCF-2) is also considered, where the core-to-core distance is $35\,\mu\text{m}$, the core radius is $4.8\,\mu\text{m}$, and the core Δ is 0.41%, which are same as those of Core A in MCF-1.

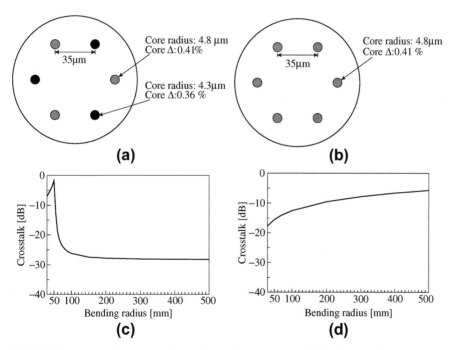

FIGURE 13.9 Schematic cross-sections of (a) a heterogeneous MCF with two kinds of cores (MCF-1) and (b) a homogeneous MCF with six identical cores (MCF-2). Numerically calculated average crosstalk between neighboring cores after 100-km propagation as a function of bending radius for (c) MCF-1 and (d) MCF-2, where the operating wavelength is 1550 nm, and the fiber twisting rate is assumed to be 5 turns per 100 m.

Figure 13.9c and d shows the numerically calculated mean value of the inter-core crosstalk between neighboring cores after 100-km propagation as a function of the bending radius for MCF-1 and MCF-2, respectively, where the operating wavelength is 1550 nm, and the fiber twisting rate is assumed to be 5 turns per 100 m [32,36]. For evaluating the crosstalk characteristics in MCF-1 and MCF-2, the CPT [36] was used, where the PCC based on an exponential autocorrelation function with a correlation length of 0.05 m was adopted [36]. We can see that, in the heterogeneous MCF-1, the inter-core crosstalk after 100-km propagation is quite large for bending radii R_{pk} of less than 5 cm, since the bend perturbations are crucial in this range, whereas it is greatly suppressed for bending radii of larger than 6 cm, even if the core-to-core distance is 35 μm. In the homogeneous MCF-2, on the other hand, the inter-core crosstalk is lower than −18 dB for bending radii of less than 3 cm due to bending and twisting effects; however, it increases with increasing bending radius, as shown in Figure 13.9d. These results indicate that the heterogeneous MCFs with sufficiently large effective index difference Δn_{eff} between the non-identical cores can be used as bending-radius-insensitive, low-crosstalk MCFs in practical situations.

13.3 METHODS OF COUPLING TO MCFs

13.3.1 Lens coupling systems

Currently, almost all transmitters and receivers were connected to SMF, hence the MCF transmission system has to have a fan-in/fan-out (FI/FO) device to utilize all of the core in the MCF. The function of the FI/FO device is to efficiently couple light from each core of the MCF into individual SMFs and vice versa. Figure 13.10 depicts the coupling principle for 19 cores of MCF allocated hexagonally as one center core, six inner cores, and 12 outer cores. A single lens placed in front of the MCF translates the laterally displaced diverging beams emerging from the MCF facet into a bundle of collimated beams each propagating in a slightly different direction. Assuming a core-to-core distance of 35 μm and a focal length, f_{MCF}, of 2 mm, the inner and outer core beams will propagate along two cones with angles of 1° and 2°, respectively. Propagation over a few centimeters is thus sufficient to spatially separate and focus the beams onto the facets of individual SMFs.

The preceding prototype of the coupling devices was a seven-core type. After the demonstration in principle of seven-core, a 19-core type was designed and fabricated. Generally, the seven-channel coupling device can be considered a simplified version of the 19-channel device. To facilitate coupling on the SMF side, the beams carrying the light from the inner and outer cores are deflected by two layers of circularly arranged prisms, as illustrated in Figure 13.11. In this way, the SMF fiber collimators need not be placed in a conical arrangement but can be simply placed in two separate planes, denoted as the top and middle layers in Figure 13.11. A photograph of the six-prism array in the top layer is shown in Figure 13.12a. The window diameter at the center is about 2 mm, leaving enough room for the 0.5 mm-wide collimated center

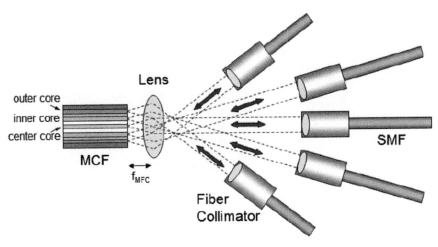

FIGURE 13.10 Coupling principle of lens coupling system [37].

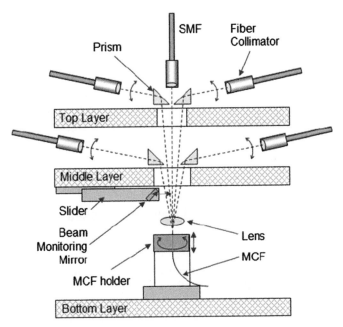

FIGURE 13.11 Layout of 19-core MFC coupling system [37].

core beam to pass through unimpeded. An image of all 19 collimated beams before hitting the prism array is shown in Figure 13.12b. The 7-channel device occupies a cylindrical volume of 20 cm in diameter and 35 cm in height, whereas the 19-channel device is a little larger, at 30 cm in diameter and 43 cm in height. Compared with our

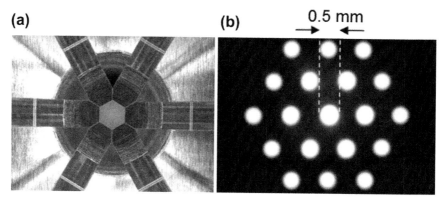

FIGURE 13.12 (a) Top view of the 6-prism array, and (b) collimated beams [37].

7-channel prototype used previously [37], this is an improvement in compactness by a volume factor of 11 for the 7-channel device and a volume factor of 4 for the 19-channel device.

The evaluation index of the coupling device is mainly characterized by two parameters: insertion loss and crosstalk. We evaluated a pair of coupling devices connected by a short piece (approx. 2 m) of 19-core MCF. The total insertion loss of the 19-channel MUX/DEMUX system was measured to be 1.3 dB on average with a variation of ±0.2 dB across all 19 channels. This value also included losses due to one connector (less than 0.3 dB) and the two uncoated MCF facets (0.3 dB). Other contributions stemmed from the combined loss of the AR-coated optical surfaces inside both coupling devices, which was estimated to be in the range of 0.2–0.3 dB, and the loss due to mode field mismatch at both fiber facets, which was estimated by numerical calculations to be on the order of 0.5 dB in total.

13.3.2 Fiber-based systems and waveguide-based systems

Figure 13.13 shows examples of a fiber-based FI/FO device called a Tapered Multicore Coupler (TMC), which has been used in many transmission experiments [14]. One end of the TMC consists of single-core fibers. The other end of the TMC has the same structure as the MCF to be spliced with it. A TMC that is designed to be connected with an MCF whose mode field diameters (MFDs) are about 9.6 μm at 1550 nm showed a crosstalk of less than −45 dB at 1550 nm and an insertion loss that ranged from 0.45 dB to 2.77 dB at 1550 nm [14]. The high insertion loss was considered to originate from slight core misalignment, mode field mismatch, and slight core asymmetry. The characteristics of the TMC, such as the effective core area, were not reported. Another embodiment of a TMC is a FI/FO device for a multicore erbium-doped fiber (EDF) [38]. The TMC was designed to realize adiabatic MFD conversion from 9.04 μm to 6.1 μm to minimize the insertion loss and the splice loss between EDFs with a small MFD and SMFs.

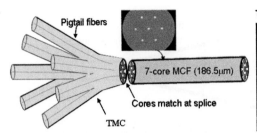

Table 1 insertion loss and crosstalk of TMCs

Core #	TMC#1		TMC#2	
	Loss (dB)	Crosstalk (dB)	Loss (dB)	Crosstalk (dB)
Center core	1.11		0.55	
Outer core1	0.75	−48.0	2.10	−49.0
Outer core2	2.77	−47.5	0.90	−46.5
Outer core3	1.95	−45.0	0.45	−46.0
Outer core4	0.98	−48.0	1.13	−47.0
Outer core5	1.42	−48.0	2.05	−48.5
Outer core6	1.37	−47.8	1.61	−45.5
average	1.48	−47.4	1.26	−47.1

FIGURE 13.13 Schematic diagram of tapered multicore fiber connector (TMC) and examples of insertion loss and crosstalk at 1550 nm of two fabricated TMCs [14].

The TMC has a similar structure to a tapered fiber bundle (TFB) and a photonic lantern. The TFB is used as a cladding pump device of EDFs and ytterbium-doped fibers with a double-clad structure. The combined pump power through the TBF is introduced to an inner cladding of the double-clad fiber. A photonic lantern is designed to be used in astronomy to transport light from a telescope to a distant device. Figure 13.14 shows the structure of a photonic lantern [39]. Seven single-core single-mode fibers are bundled into a capillary tube. The capillary tube is fused and tapered down to a certain diameter. The collapsed region after the tapered section functions as a large-core multi-mode fiber. In the case of TMCs, the individual core structure should be preserved at both ends, and the crosstalk between cores should be suppressed. Satisfying these requirements requires careful design of the relative refractive index of the cores. When the single-core fibers are collapsed into a multi-core fiber, the radius of the individual cores decreases. If they use a conventional step-like core structure, the MFD of the fiber excessively expands as the core radius decreases, which results in excessive loss in the device, large crosstalk between cores, and large splice loss with an MCF. To overcome this problem, a two-step core structure with a trench region has been proposed in order to maintain the MFD at both ends of a FI/FO device and in the tapered region, as well as to reduce the insertion loss of the TMC [40]. Suppression of excessive loss due to the change of the core radius is an important issue with TMCs.

Another example of a fiber-based FI/FO is a connector-based device. Figure 13.15 illustrates an example of a connecter-based FI/FO device that was used in a seven-core fiber transmission experiment using Raman amplification [41]. Small-diameter fibers, whose diameter was adjusted to match the core-to-core distance of an MCF, were assembled into a ferrule. The A_{eff} values at 1550 nm of the MCF and the small-diameter fibers were $110\,\mu m^2$ and $88\,\mu m^2$, respectively. A very low crosstalk of −60 dB at 1550 nm and an insertion loss from 0.5 dB to 1.9 dB were realized. About 0.5 dB of the insertion loss originated from the field mismatch between the MCF and the small-diameter fibers. This contribution to the insertion loss will be reduced by optimizing the MFD of the fibers. The rest of the insertion loss was due to misalignment.

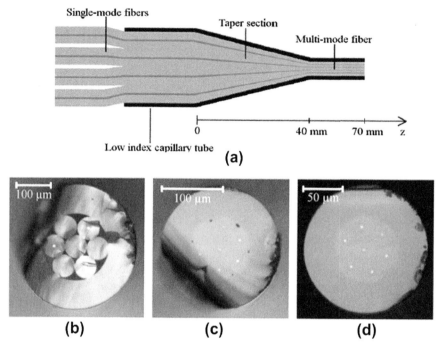

FIGURE 13.14 An example of a photonic lantern [40]. (a) Schematic illustration of a photonic lantern. (b)–(d) Microscope images of the cross-section of the taper section of the lantern at different postions.

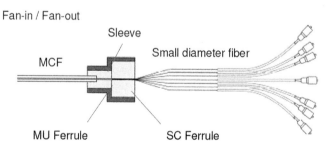

FIGURE 13.15 Schematic diagram of a connector-based FI/FO device [6].

A common issue with fiber-based FI/FO devices is excessive loss in the outer cores due to misalignment of the cores. Improvement of the fabrication process in order to achieve precise core alignment will be indispensable for reducing the insertion loss of fiber-based FI/FO devices.

Another embodiment of a FI/FO device is a waveguide-based device. Figure 13.16 schematically shows two waveguide-based FI/FO devices. Figure 13.16a is

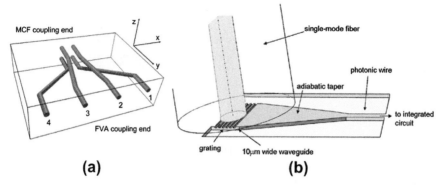

FIGURE 13.16 Examples of waveguide-based FI/FO devices: (a) A horizontal coupling device [42]. (b) A vertical coupling device with a grating coupler [43].

an example of a horizontal-coupling device [42]. A femtosecond pulsed laser was used to generate a waveguide structure in a silica glass substrate. The average insertion loss was 5 dB at 1550 nm. One reason for the high insertion loss was the field mismatch between the waveguide and the MCF. Figure 13.16b illustrates a vertical-coupling device using a grating coupler [43]. The MFD of the waveguide is enlarged to the MFD of a SMF in the adiabatic taper section. Higher power is coupled into the SMF through a grating section. The grating-based device exhibits bandwidth limitation due to the nature of grating. However, the device has other benefits compared with the horizontal-coupling device: flexibility in placing the device on an integrated tip, wafer-scale testing, and elimination of polishing and other surface preparation steps for connection. Figure 13.17 shows a grating-coupler-based FI/FO device fabricated for a seven-core MCF [44]. The seven-point triangular lattice was slightly rotated with respect to a hexagonal end face, as shown in Figure 13.17a and b to avoid interference between adjacent cores. Figure 13.17c shows the waveguide layout of the MCF coupler. All seven-core centers are spread out and are connected to an array of waveguides arranged at equal intervals. The measured insertion loss was less than 9.8 dB. This relatively high insertion loss was dominated by loss in the grating coupler, whose structure was not optimized. A device with a grating coupler is also useful for mode mux/demux devices for mode division multiplexing [44,45].

Another interesting approach is direct coupling between an MCF and devices such as vertical-cavity surface-emitting lasers (VCSELs) and photodetectors (PDs). Figure 13.18 shows an example of a direct coupling device [46]. VCSELs and PDs were arrayed. The device was designed to couple with a multimode MCF having a hexagonal array. The MCF had seven cores: a center core and six outer cores. However, the device had only six VCSELs and six PDs due to the limitation of the wiring density.

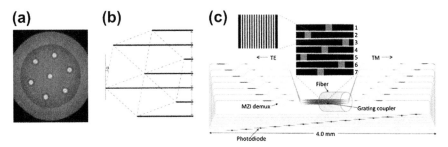

FIGURE 13.17 An example of a grating-coupler based FI/FO device [44]: (a) Photograph of a seven-core MCF. (b) Rotation of grating couplers that allows coupling to all cores. (c) An example of the waveguide layout.

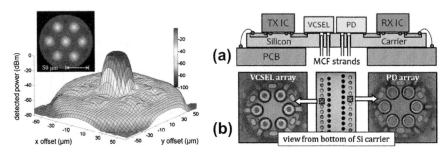

FIGURE 13.18 An example of a VCSEL array and a photodiode array that couple to a multi-core fiber without other coupling devices [46]. (Left figure) A cross-sectional image of an MCF and core-to-core crosstalk for a 100-m MCF. (Right figure) (a) A schematic of a transceiver with a VCSEL array and a photodetector array. (b) Selected image of a VCSEL array and a photodetector array from the underside of the package.

13.3.3 Splicing techniques

Splicing is an important technique for constructing optical transmission systems with MCFs. Two splicing techniques are generally required: detachable splicing using an optical connector and permanent splicing with a fusion splice.

Optical connectors are formed of precisely fabricated and aligned components. In the case of MCFs, rotational alignment is additionally required to align the outer cores, and therefore the optical connectors should also have rotation degree-of-freedom. On the other hand, the freedom should be limited for fixing the fibers after alignment. An optical connector using an Oldham's shaft coupling has been proposed for polarization-maintaining fibers, which require connectors that reduce the rotation degree-of-freedom while ensuring enough space for floating [47]. Further development will be required to realize both flexibility for alignment and reliability for maintaining the fiber positions.

Fusion splicing is also an important method for connecting fibers. Alignment is an issue for fusion splicing of MCFs because MCFs have not only a center core but also outer cores. One of the essential functions of a fusion splicer for MCFs is rotational alignment for the outer cores. A conventional fusion splicer aligns cores by moving the fibers in the lateral directions only. A rotational alignment system is a popular function of fusion splicers designed for polarization-maintaining fibers. Fusion splicing of an MCF by using a fusion splicer designed for polarization-maintaining fibers has been demonstrated [48]. After optimizing the splicing parameters through careful calibration using the optical power, automatic alignment was achieved. The average splice loss in a seven-core MCF with a cladding diameter of $130\,\mu m$ was about $0.1\,dB$. However, the individual splice loss varied widely. The maximum splice loss exceeded $0.3\,dB$.

Another essential function is detection of markers. Recently reported MCFs have markers for identifying the core number [6,11–13,15–18]. Two types of markers have been proposed: solid markers that have a slightly different refractive index from the cladding, and hollow makers. A manual core alignment technique for an MCF with hollow makers has been proposed [49]. Clear contrast of the hollow markers in a side-view image enables manual alignment by using the images. The average splice loss in a seven-core MCF with a cladding diameter of $208\,\mu m$ was $0.119\,dB$ at $1550\,nm$. The maximum splice loss was $0.198\,dB$. This alignment method is only applicable to MCFs having the same end-face structure with hollow markers.

A precise alignment system and a marker detection system are required for fusion splicing of MCFs. Recently, a fusion splicer with an end-view function has been proposed [50]. The end-view function was originally developed for aligning specialty fibers such as a polarization-maintaining fibers and noncircular-cladding fibers. The direct detection of cores and markers by using an end-view image will be a powerful approach to automatically align MCFs including marked cores. Figure 13.19 shows a schematic diagram of the mechanical structure used to realize the end-view function. A mirror is located between the fibers to be spliced. The end of the fiber is illuminated by laterally injected light. The illuminated end-view image is reflected by the mirror and is monitored with an imaging device through a lens. After completing alignment using the end-view image, the mirror is moved down, and the fibers are spliced. The cores of the MCFs are aligned by processing the end-view images.

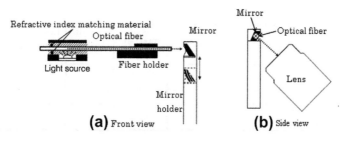

FIGURE 13.19 Schematics of an end-view system [50].

The end-view method was employed as an early alignment system for polarization-maintaining fibers. Recent fusion splicers align polarization-maintaining fibers with side-view images. The development of more innovative and simpler core-alignment methods is expected in future.

13.4 TRANSMISSION EXPERIMENTS WITH UNCOUPLED CORES

13.4.1 Early demonstrations

The first practical trial of MCF transmission was demonstrated in May 2010 as shown in Figure 13.20 [48]. The seven-core MCF designed and fabricated for that demonstration had a length of 11.3 km, a core diameter of 8 μm, a cladding diameter of 130 μm, a core-to-core distance of 38 μm, and mean crosstalks of −39.0 dB at 1310 nm and −24.8 dB at 1490 nm. Connection to the MCF was realized with help of a tapered MCF connector (TMC) made from seven special single-mode fibers by using a taper process. The TMC had an insertion loss of 0.38–1.6 dB and crosstalk of −39.32 to −43.8 dB at the best. At a central office (CO), seven optical line terminals (OLTs), each comprising PON transceivers (1490 nm downstream (DS) signal transmitter and 1310 nm upstream (US) receiver), were connected to the TMC via seven single-core fibers, and the TMC was connected via a fusion splice to the 11.3 km seven-core MCF. At the remote node (RN), a second TMC was connected via a fusion splice to the MCF, and seven pigtail fibers at the other end of the TMC

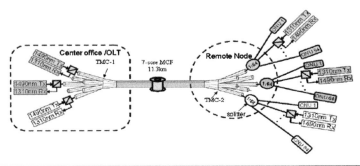

	Fiber loss (dB/km)	Loss (dB) 11.3 km MCF	Loss (dB) TMC #1	Loss (dB) TMC#2	Loss (dB) 1:64 splitter	Total link loss (dB)
Center core (1310 nm)	0.39	4.41	1.64	0.38	21.0	27.4
Outer cores-averaged (1310 nm)	0.41	4.63	1.48	1.16	21.0	28.3
Center core (1490 nm)	0.30	3.39	1.64	0.38	21.0	26.4
Outer cores-averaged (1490 nm)	0.53	5.99	1.48	1.16	21.0	29.6

FIGURE 13.20 Seven-core MCF transmission of PON signals [48].

were connected to 1:64 optical fiber splitters. Each splitter was then connected to optical network units (ONU) at the subscriber premises. The line rate was 2.5 Gb/s, achieving seven-parallel bi-directional transmission of 35 Gb/s aggregated capacity.

Because inter-core crosstalk is the most serious issue in MCFs, short-reach applications such as data communications are suitable for immediate commercialization. Multi-mode seven-core fiber transmission has been demonstrated as shown in Figure 13.21 [51,52]. This MCF had a core diameter of 26 μm, a cladding diameter of 125 μm, and a core-to-core distance of 39 μm. The average crosstalk was −42.8 dB over 550 m transmission.

In that demonstration, two types of MCF coupling techniques, using a TMC and a VCSEL array, were attempted in the transmission experiment. In the TMC case, 70 Gb/s aggregated capacity transmission through 100 m and 550 m of MCF was demonstrated. In the VCSEL array case on the other hand, 120 Gb/s aggregated capacity transmission through 100 m of MCF was demonstrated. In the latter case, only six cores were utilized in the seven-core fiber because of layout limitations of the IC. Such a transmission scheme can be separated from long-haul transmission technology because of the ease of coupling, lower cost, and high demand.

A trial for huge capacity transmission through an MCF started in 2011 [53]. In order to accommodate advanced modulation formats, the crosstalk requirement becomes much more severe. A record capacity of over 100 Tb/s was achieved for the first time (Figure 13.22) [53] by using a trench-assisted seven-core fiber whose maximum crosstalks were −77.6 dB at 1550 nm and −67.7 dB at 1625 nm for a 17.1 km length [12]. A seven-core lens coupling system was adopted for SDM-MUX/DEMUX. The maximum crosstalk was less than −53 dB, and losses were 5.1–5.4 dB, including

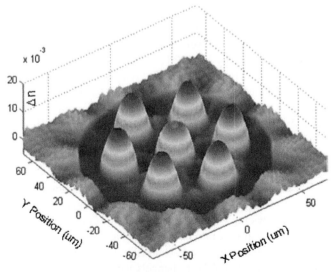

FIGURE 13.21 Refractive-index profile of multimode seven-core fiber [52].

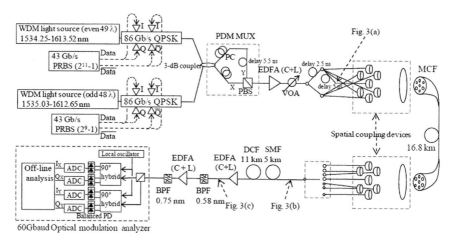

FIGURE 13.22 Transmission setup with 109 Tb/s capacity [53].

MCF and MUX/DEMUX losses. Ninety-seven wavelength channels with 100 GHz spacing were modulated by an 86 Gb/s PDM-QPSK signal and were launched into 16.8 km seven-core MCF after a tellurite-based C+L EDFA and SDM-MUX. The total signal power measured at each input port of the SDM-MUX was adjusted to approximately +3.5 dBm/core by using a variable optical attenuator (VOA). After transmission and SDM-DEMUX, the signals went through a dispersion compensation fiber (DCF) for partial compensation of the accumulated chromatic dispersion of the MCF. Residual chromatic dispersion of the MCF (−20–−60 ps/nm, depending on the WDM channel) was digitally compensated for on a 60-Gbaud optical modulation analyzer (Agilent Technologies, N4391A). The aggregated capacity was 109 Tb/s after subtracting the FEC overhead.

Huge capacity transmission over a longer distance has also been demonstrated. An aggregated capacity of 112 Tb/s through a 76.8 km seven-core MCF was achieved as shown in Figure 13.23 [14], giving a capacity-distance product as high as 8.6 Pb/s·km. The MCF had a core diameter of 9 μm, a core-to-core distance of 46.8 μm, and a cladding diameter of 186.5 μm. The losses in the center core were 0.23 dB/km at 1550 nm and 0.37 dB/km at 1300 nm. The average losses for the six outer cores were 0.26 dB/km at 1550 nm and 0.40 dB/km at 1300 nm. The 76.8 km seven-core MCF span consisted of two spools (23.5 km and 53.3 km) of MCF, which were spliced together using a commercially available polarization-maintaining fiber splicer. A TMC was used for SDM-MUX/DEMUX. The maximum total link crosstalk was less than −36.6 dB across the C- and L-bands. One hundred and sixty DWDM channels with 50 GHz spacing were modulated by a 107 Gb/s PDM-QPSK signal and separated by a C/L band splitter into the C-band channels and the L-band channels, which were separately amplified by a C-band EDFA and an L-band EDFA, respectively. For each band, the 80 DWDM channels were split by a 1:8 power splitter, and seven of the outputs were amplified by seven EDFAs. After that, the amplified

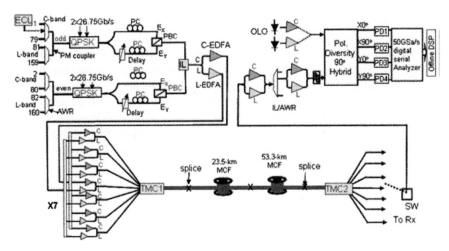

FIGURE 13.23 Setup for 112 Tb/s transmission through 76.8 km seven-core MCF [14].

C- and L-band channels were combined by using C-/L-band couplers and launched into each core of the MCF.

In order to explore the potential of MCF-based SDM for long-haul transmission, net aggregation of the spectral efficiency of seven-core fiber transmission was demonstrated as shown in Figure 13.24 [54]. The seven cores were utilized as seven fold links for one spatial channel. The total length of the seven fold links was 76.8 km × 7 = 537.6 km, and the distance was extended to 1075.2 km, 1612.8 km, and so on by using seven separate re-circulating loops running synchronously. These loops shared a common load switch that launched identical copies of the DWDM

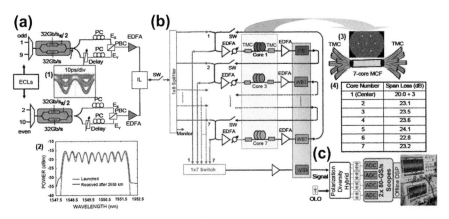

FIGURE 13.24 Transmission setup for total length of 2688 km, re-circulated with seven fold links [54].

channels into each of the loops via a 1×8 power splitter. The signals at each of the seven loop inputs were amplified by an EDFA before being launched into the 76.8 km MCF through a TMC. A second TMC was used to couple out the signals after transmission. The signals were amplified to compensate for the fiber loss, and the channel powers were equalized using a wavelength blocker (WB) array module that incorporated eight independent 96-channel 50 GHz WBs. One WB was used per core of the MCF, and the eighth WB was used at the receiver. After each WB, the DWDM signals from one core were sent to the re-circulating loop input of the next core, in a cyclic fashion. Finally, a transmission distance of 2688 km was achieved below the FEC limit for 10×128 Gb/s PDM-QPSK signals.

It is important to strike a balance between high spectral efficiency and high spatial efficiency. A 1.12 Tb/s 32-QAM-OFDM superchannel was successfully transmitted through a seven-core fiber without significant impairment due to inter-core crosstalk as shown in Figure 13.25 [55]. A 5-comb generator driven by a 25.94 GHz sinusoidal wave generated five carriers with a spacing of 25.94 GHz. Additionally, a 4-comb generator, whose two nested branches were respectively driven by 3.24 GHz and 9.73 GHz sinusoidal waves, quadrupled the number of frequency-locked carriers

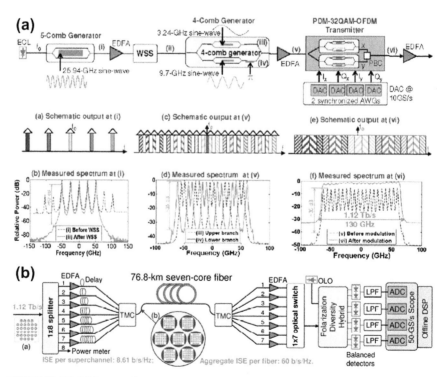

FIGURE 13.25 (a) Schematic of the experimental setup of (a) 1.12 Tb/s superchannel transmitter, (b) superchannel transmission and detection [55].

to 20 with a carrier spacing of 6.48 GHz. The x- and y-polarization components of the PDM signal were independently modulated to better emulate a real transmitter. Four independent drive patterns were stored in two synchronized arbitrary waveform generators (AWGs), each having two 10 GS/s DACs. The IFFT size used for OFDM was 128, and the guard interval overhead was only 1.56%. Each polarization component of an OFDM symbol contained seventy-eight 32-QAM data subcarriers (SCs), four pilot SCs, one unfilled DC SC, and forty-five unfilled edge SCs. The spectral bandwidth of each modulated subchannel was 6.48 GHz, and the total bandwidth of 20 frequency-locked carriers was 130 GHz. Three correlated dual-polarization training symbols were used for every 697 payload OFDM symbol. The net payload data rate of the superchannel was 1.12 Tb/s ($=10$ GHz $\times$ 10 b/s/Hz, 78/130 $\times$ 697/700 $\times$ 20/1.07) without 7% FEC overhead. This corresponds to a net intrachannel spectral efficiency (ISE) of 8.61 b/s/Hz (1.12 Tb/s/129.7 GHz). By using a seven-core fiber, the aggregated per-fiber ISE became 60 b/s/Hz. These signals were injected into 76.8 km MCF via TMF and transmitted by using seven pairs of in-line single-core EDFAs.

13.4.2 Scalability of core number

A common question is how many cores can be fabricated in an MCF? Seven-core MCFs are the most popular, and many of them have been fabricated. Ten-core MCFs have been proposed and fabricated with different core-to-core distances which could suppress crosstalk effectively [17]. The development of 19-core MCFs was predicted by extending seven-core MCFs; however, mechanical strength will be an issue when the cladding diameter increases. The first demonstration of 19-core MCF fabrication (Figure 13.26 [56]) and transmission (Figure 13.27 [57]) was presented just 1 year after the demonstration of 109 Tb/s transmission using a seven-core MCF as shown in Figure 13.27.

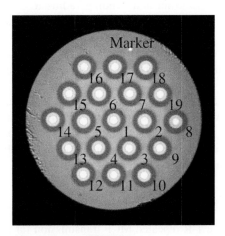

FIGURE 13.26 Cross section of 19-core MCF [56].

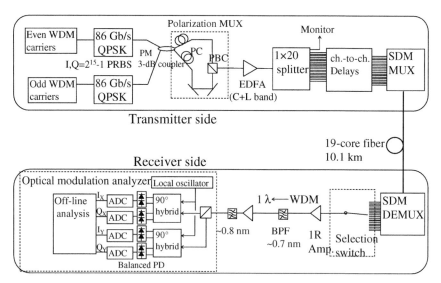

FIGURE 13.27 Experimental setup for 305 Tb/s, 19-core MCF transmission [57].

In that demonstration, the core-to-core distance, cladding diameter, and coating diameter were 35 μm, 200 μm, and 345 μm, respectively. The attenuation was less than 0.26 dB/km (average 0.23 dB/km) at 1550 nm, and A_{eff} was about 72 μm^2 for all cores. The cable cutoff wavelengths (λ_{cc}) of the center core and inner cores were longer than those of the outer cores, which was considered to be due to the influence of the trench layer in neighboring cores. The maximum value of λ_{cc} was 1528 nm, and so single-mode transmission through the C-/L-band can be expected. The crosstalk properties of the fabricated 19-core MCF after 10.4 km transmission were measured at $\lambda = 1550$ nm and bending radius $R = 90$ mm. For example, at the center core, the mean crosstalk from all other cores was measured. It was confirmed that the crosstalk from the inner cores was about –31 dB, whereas that from the outer cores was much smaller, at less than –50 dB. This means that the crosstalk from next-nearest neighboring cores hardly affects the total crosstalk from all cores.

The lens-coupled SDM-MUX/DEMUX described in Section 13.4.1 was used for the transmission experiment. One hundred wavelength channels were modulated by an 86 Gb/S PDM-QPSK signal and were divided into 20 branches. Of these, 19 branches were launched into the corresponding cores of the MCF after a tellurite-based C+L EDFA. The aggregated capacity reached 305 Tb/s.

13.4.3 1-R repeated demonstrations

It is essential to achieve 1-R repeated MCF transmission for practical applications. A seven-core distributed Raman amplifier (DRA) has been developed, and 1-R

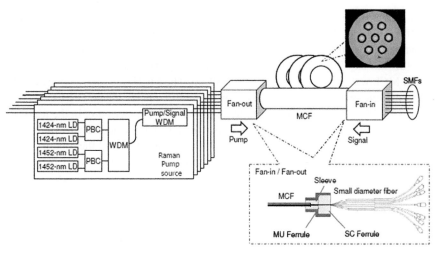

FIGURE 13.28 Configuration of DRA on MCF [41].

repeated MCF transmission was demonstrated for the first time as shown in Figures 13.28 and 13.29 [41].

A 75-km seven-core MCF transmission line with a DRA including FI/FO and a Raman pump source (RPS) was fabricated. The MCF had a trench-assisted structure. The core-to-core distance was 49 μm, and the cladding diameter was 195 μm. In

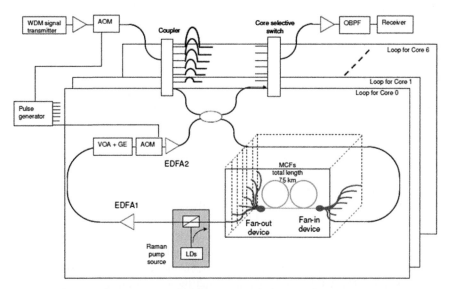

FIGURE 13.29 Experimental setup of 1000 km transmission by using DRA on MCF [41].

order to achieve high Raman gain and suppress other optical nonlinearities, the cable cutoff wavelength, λ_{cc}, was set to be less than 1400 nm, and A_{eff} was about 110 μm^2 on average. The attenuation and dispersion at a wavelength of 1550 nm were 0.190–0.199 dB/km and 20.5–20.8 ps/nm/km, respectively. The measured crosstalk of the 75-km MCF was –65 dB on average at 1550 nm. The MCF was fusion spliced with a FI/FO module, which was a bundle of seven small-diameter fibers. DRA gains from 9.2 to 12.9 dB were obtained at a pumping power of 1.1 W per core. In the transmission experiment, a total power of 6.5 W per fiber was launched. There were seven re-circulating loops corresponding to each core of the MCF, and the total transmission distance was 1050 km. Ten wavelength channels were modulated by 96 Gb/s PDM-16-QAM signals. Q-factors of all 10 channels of the seven cores exceeded the Q-limit of 8.3 dB.

13.5 LAGUERRE-GAUSSIAN MODE DIVISION MULTIPLEXING TRANSMISSION IN MCFs

Orbital angular momentum (OAM) modes have attracted much attention as one candidate for a new class of space-division multiplexing for overcoming the capacity crunch which has been becoming a serious issue in fiber telecommunications. Until recently, free-space transfer of information encoded with multiple Laguerre-Gaussian (LG) modes was demonstrated using a phase mask [58]. Ultrafast switching of OAM conversion was also realized by photoinduced OAM of semiconductor electrons [59]. Therefore, the stable transfer of OAM via optical fiber is the only remaining hurdle for telecommunication applications. So far, few-mode fibers have usually been employed for the propagation of multiple LG modes [60]. In this case, however, the guided modes are actually LP modes; thus, additional conversion processes are necessary for extracting the LG modes. MCFs are promising for the direct propagation of multiple LG modes, where the relative phase among the cores can preserve the OAM [61].

In Figure 13.30, l LG modes are spatially overlapped (multiplexed) and coupled with several cores in the MCF. The output from the MCF is an aggregation of beams corresponding to each core. If OAM is preserved after propagation through the MCF, a combination of appropriate beams reconstitutes the LG modes. Thus, the proper LG mode can be extracted (demultiplexed) by discriminating the OAM [62]. In this sense, the role of an MCF is similar to an image fiber; however, each core carries a single mode and can deliver a high-speed signal.

Figure 13.31 shows the setup used for MCF transmission [62]. Two tunable semiconductor lasers (Santec, TSL510) were used as light sources, and both center wavelengths were 1550 nm. These beams (CH1 and CH2) were independently modulated by PRBS:2^{23}-1 at 10 Gbit/s. Both CH1 and CH2 were launched as Gaussian-mode (TEM$_{00}$) beams from fiber collimators, but the beam diameters were adjusted to be different. The CH1 beam was diffracted by a two-dimensional spatial light modulator (PAL-SLM1; Hamamatsu Photonics, PPM X8267, 1024 × 768 pixels).

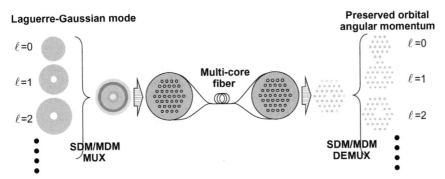

FIGURE 13.30 Conceptual concept of MDM/SDM of Laguerre-Gaussian mode in MCF [62].

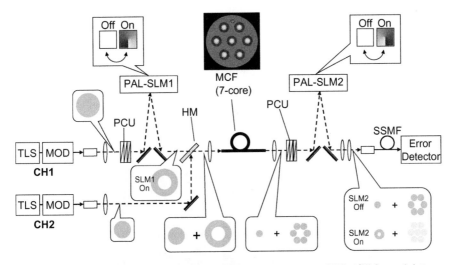

FIGURE 13.31 Experimental setup [62] TLS: Tunable laser source, MOD: LiNbO₃ modulator, PCU: Polarization control unit composed by λ/2, λ/4, and polarizer, HM: Half mirror.

The mode conversion from the Gaussian ($\ell = 0$) to LG modes was achieved by using PAL-SLM1, and the mode conversion from the LG to Gaussian modes was achieved using a second SLM (PAL-SLM2; Hamamatsu Photonics, PPM X7550, 640×480 pixels). The PAL-SLMs were used not only for generating the LG mode but also for mode switching. When the PAL-SLM applies no modulation, the injected mode will be maintained. When a phase pattern corresponding to $\ell = +1$ is applied, the beam is converted to a first-order LG-mode beam. After that, CH1 in the LG mode and CH2 in the Gaussian mode were mode/space division multiplexed by a half mirror (HM) when PAL-SLM1 was on. The multiplexed beam was focused at the end of the MCF. Unlike conventional focusing conditions into standard single-mode fiber (SSMF), the beam waist was adjusted so as to overlap all seven cores. This beam waist was estimated to be ∼110 μm. The length of the MCF was 80 cm. When the CH1 LG_{+1} ($\ell = +1$) beam was introduced onto all the six outer cores, the output showed circular located six Gaussian beams, which could be resolved into $\ell = +1$ and $\ell = +1 \pm 6$ LG modes. The CH2 Gaussian ($\ell = 0$) beam was injected into the center core. Both modes maintained the OAM even after propagation through the MCF.

The output from the MCF was collimated by an aspheric lens and introduced to PAL-SLM2. A phase pattern corresponding to $\ell = -1$ was prepared and applied to PAL-SLM2. The diffracted beam from PAL-SLM2 was coupled into an SSMF. Basically, only the fundamental Gaussian mode was coupled into the SSMF, and the LG mode was blocked by the aperture of the SSMF. Therefore, the multiplexed CH1 ($\ell = +1$ and $+1+6$) and CH2 ($\ell = 0$) from the MCF could be discriminated by using the SSMF as a spatial filter. Also, it was possible to change the coupling conditions of CH1 and CH2 into the SSMF by altering the modulation pattern on PAL-SLM2, which worked as a mode switch, because $\ell = 0$ was generated from the $\ell = +1$ component of CH1, whereas CH2 was converted from $\ell = 0$ to $\ell = -1$ when PAL-SLM2 was turned on.

Figure 13.32 shows beam profiles before and after the MCF. It seems that CH1 and CH2 have different beam sizes after the MCF, which may be considered a simple form of SDM. However, the most notable feature of the MDM based on the LG mode is preservation of OAM, and this was confirmed by switching the OAM by using

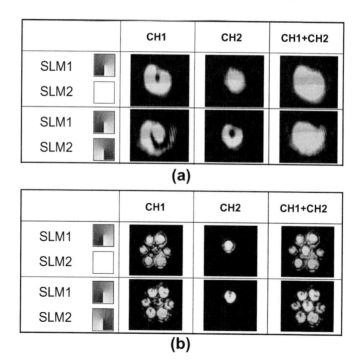

FIGURE 13.32 Beam profiles [62] (a) Before MCF, (b) After MCF.

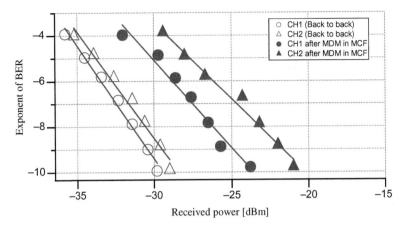

FIGURE 13.33 Measured bit error rate.

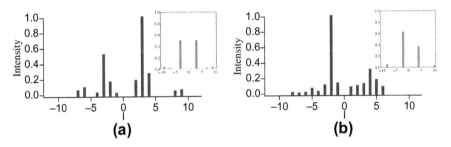

FIGURE 13.34 OAM spectra (a) $\ell=+3$, (b) $\ell=+4$.

PAL-SLM2 for demultiplexing. Thanks to the low inter-core crosstalk of the MCF, the OAM preservation was fair, and the mode extinction ratio between CH1 and CH2 was less than 15 dB.

Figure 13.33 shows the measured bit error rate. The typical power penalty of MDM/SDM was about 6–9 dB, which mainly came from power loss at the surface of the MCF because of rejected energy at the cladding. Therefore, it should be possible to improve the power loss of MCF injection by increasing the core density, which is also beneficial for OAM preservation of higher-order LG modes. In other words, the proposed MDM/SDM method is more suitable for MCFs with denser cores.

Also, OAM spectra after propagation through an MCF have been successfully observed [63]. Figure 13.34a shows the case of Tx: $\ell=+3$. We can see the $\ell=+3$ component and the $\ell=-3$ component with significant side-mode suppression ratios. The spectrum shown in the inset is a numerical calculation. Since the output beam group from the MCF has six separated beams, the intensity oscillation in the azimuthal direction results in satellite modes with shifts of ± 6 from the main component. Therefore, it is predicted that the $\ell = +3$ component will be accompanied by the $\ell = -3$ and $\ell = +9$ components. The reason why the $\ell = +9$ component is small may be due to its higher index, which can be confirmed in the case of Tx: $\ell = +4$ shown in Figure 13.34b, where the $\ell = +10$ and $\ell = -2$ components are small. It was confirmed that the intensity oscillation in the azimuthal direction results in satellite modes with shifts of ± 6 from the main component on OAM spectra. It was also found that independent rotation of the polarization in each core can cause fluctuations in the propagation conditions. Nevertheless, the OAM spectrum was successfully observed by eliminating background noise from the diffracted beams.

References

[1] S. Inao, T. Sato, H. Hondo, M. Ogai, S. Sentsui, A. Otake, K. Yoshizaki, K. Ishihara, N. Uchida, High density multicore-fiber cable, in: Proceedings of the International Wire & Cable Symposium IWCS, 1979, pp. 370–384.

[2] G. Le Noane, D. Boscher, P. Grosso, J.C. Bizeul, C. Botton, Ultra high density cables using a new concept of bunched multicore monomode fibers: a key for the future FTTH networks, in: Proceedings of the International Wire & Cable Symposium IWCS, 1994, pp. 203–210.

[3] K. Imamura, K. Mukasa, R. Sugisaki, Y. Mimura, T. Yagi, Multi-core holey fibers for ultra large capacity wide-band transmission, in: ECOC 2008, Sept. 2008, P.1.17.

[4] T. Morioka, New generation optical infrastructure technologies: "EXAT Initiative" towards 2020 and beyond, in: Proceedings of the OECC 2009, Paper FT4.

[5] K. Imamura, K. Mukasa, Y. Miura, T. Yagi, Multi-core holey fibers for the long-distance (>100 km) ultra large capacity transmission, in: OFC/NFOEC 2009, Mar. 2009, OTuC3.

[6] K. Imamura, K. Mukasa, T. Yagi, Investigation on multi-core fibers with large A_{eff} and low micro bending loss, in: OFC/NFOEC 2010, Mar. 2010, OWK6.

[7] B. Zhu, T.F. Taunay, M.F. Yan, J.M. Fini, M. Fishteyn, E.M. Monberg, F.V. Dimarcello, Seven-core multicore fiber transmission for passive optical network, Opt. Express 18 (2010) 11117–11122.

[8] K. Takenaga, S. Tanigawa, N. Guan, S. Matsuo, K. Saitoh, M. Koshiba, Reduction of crosstalk by quasi-homogeneous solid multi-core fiber, in: OFC/NFOEC 2010, Mar. 2010, OWK7.

[9] M. Koshiba, K. Saitoh, Y. Kokubun, Heterogeneous multi-core fibers: proposal and design principle, IEICE Electron. Express 6 (2009) 98–103.

[10] T. Hayashi, T. Nagashima, O. Shimakawa, T. Sasaki, E. Sasaoka, Crosstalk variation of multi-core fiber due to fiber bend, in: ECOC 2010, Sept. 2010, We.8.F.6.

[11] K. Mukasa, K. Imamura, Y. Tsuchida, R. Sugizaki, Multi-core fibers for large capacity SDM, in: OFC/NFOEC 2011, Mar. 2011, OWJ1.

[12] T. Hayashi, T. Taru, O. Shimakawa, T. Sasaki, E. Sasaoka, Ultra-low-crosstalk multi-core fiber feasible to ultra-long-haul transmission, OFC/NFOEC 2011, Mar. 2011, PDPC2.

[13] K. Takenaga, Y. Arakawa, S. Tanigawa, N. Guan, S. Matsuo, K. Saitoh, M. Koshiba, Reduction of crosstalk by trench-assisted multi-core fiber, in: OFC/NFOEC 2011, Mar. 2011, OWJ4.

[14] B. Zhu, T.F. Taunay, M. Fishteny, X. Liu, S. Chandrasekhar, M.F. Yan, J.M. Fini, E.M. Monberg, F.V. Dimarcello, 112-Tb/s space-division multiplexed DWDM

transmission with 14-b/s/Hz aggregate spectral efficiency over a 76.8-km seven-core fiber, Opt. Express 19 (2011) 16665–16671.

[15] K. Imamura, K. Mukasa, R. Sugizaki, Trench assisted multi-core fiber with large A_{eff} over 100 μm^2 and low attenuation loss, in: ECOC 2011, Sept. 2011, Mo.1.LeCervin.1.

[16] K. Takenaga, Y. Arakawa, Y. Sasaki, S. Tanigawa, S. Matsuo, K. Saitoh, M. Koshiba, A large effective area multi-core fiber with an optimized cladding thickness, in: ECOC 2011, Sept. 2011, Mo.1.LeCervin.2.

[17] S. Matsuo, K. Takenaga, Y. Arakawa, Y. Sasaki, S. Tanigawa, K. Saitoh, M. Koshiba, Large-effective-area ten-core fiber with cladding diameter of about 200 μm, Opt. Lett. 36 (2011) 4626–4628.

[18] J. Sakaguchi, B.J. Puttnam, W. Klaus, Y. Awaji, N. Wada, A. Kanno, T. Kawanishi, K. Imamura, H. Unaba, K. Mukasa, R. Sugizaki, T. Kobayashi, M. Watanabe, 19-core fiber transmission of 19x100x172-Gb/s SDM-WDM-PDM-QPSK signal at 305 Tb/s, in: OFC/NFOEC 2012, Mar. 2012, PDP5C.1.

[19] K. Takenaga, Y. Sasaki, N. Guan, S. Matsuo, M. Kasahara, K. Saitoh, M. Koshiba, A large effective area few-mode multi-core fiber, in: IEEE Photonics Society Topical Meeting on Space Division Multiplexing for Optical Systems and Networks, July 2012, TuC1.2.

[20] B. Zhu, T.F. Taunay, M.F. Yan, M. Fishteyn, G. Oulundsen, D. Vaidya, 70-Gb/s multicore multimode fiber transmission for optical data links, IEEE Photon. Technol. Lett. 22 (2010) 1647–1649.

[21] M.M. Vogel, M. Abdou-Ahmed, A. Voss, T. Graf, Very-large-mode-area multicore fiber, Opt. Lett. 34 (2009) 2876–2878.

[22] Y. Huo, P.K. Cheo, G.G. King, Fundamental mode operation of a 19-core phase-locked Yb-doped fiber amplifier, Opt. Express 12 (2004) 6230–6239.

[23] Y. Kokubun, M. Koshiba, Novel multi-core fibers for mode division multiplexing: proposal and design principle, IEICE Electron. Express 6 (2009) 522–528.

[24] C. Xia, N. Bai, I. Ozdur, X. Zhou, G. Li, Supermodes for optical transmission, Opt. Express 19 (2011) 16653–16664.

[25] R. Ryf, A. Sierra, R.-J. Essiambre, A.H. Gnauck, S. Randel, M. Esmaeelpour, S. Mumtaz, P.J. Winzer, R. Delbue, P. Pupalaikis, A. Sureka, T. Hayashi, T. Taru, T. Sasaki, Coherent 1200-km 6×6 MIMO mode-multiplexed transmission over 3-core microstructured fiber, in: ECOC 2011, September 2011, Th.13.C.1.

[26] Y. Kokubun, T. Komo, K. Takenaga, S. Tanigawa, S. Matsuo, Mode discrimination and bending properties of fore-core homogeneous coupled multi-core fiber, in: ECOC 2011, Sept. 2001, We10.P1.08.

[27] K. Imamura, Y. Tsuchida, K. Mukasa, R. Sugizaki, K. Saitoh, M. Koshiba, Investigation on multi-core fibers with large A_{eff} and low micro bending loss, Opt. Express 19 (2011) 10595–10603.

[28] T. Hayashi, T. Taru, O. Shimakawa, T. Sasaki, E. Sasaoka, Design and fabrication of ultra-low crosstalk and low-loss multi-core fiber, Opt. Express 19 (2011) 16576–16592.

[29] K. Takenaga, Y. Arakawa, Y. Sasaki, S. Tanigawa, S. Matsuo, K. Saitoh, M. Koshiba, A large effective area multi-core fiber with an optimized cladding thickness, Opt. Express 19 (2011) B542–B550.

[30] J.M. Fini, B. Zhu, T.F. Taunay, M.F. Yan, Statistics of crosstalk in bent multicore fibers, Opt. Express 18 (2010) 15122–15129.

[31] K. Takenaga, Y. Arakawa, S. Tanigawa, N. Guan, S. Matsuo, K. Saitoh, M. Koshiba, An investigation on crosstalk in multi-core fibers by introducing random fluctuation along longitudinal direction, IEICE Trans. Commun. E94-B (2011) 409–416.

[32] M. Koshiba, K. Saitoh, K. Takenaga, S. Matsuo, Multi-core fiber design and analysis: coupled-mode theory and coupled-power theory, Opt. Express 19 (2011) B102–B111.

[33] K. Saitoh, T. Matsui, T. Sakamoto, M. Koshiba, S. Tomita, Multi-core hole-assisted fibers for high core density space division multiplexing, in: OECC 2010, July 2010, 7C2-1.

[34] S. Matsuo, M. Ikeda, K. Himeno, Low-bending-loss and suppressed-splice-loss optical fibers for FTTH indoor wiring, in: OFC 2004, Mar. 2004, ThI3.

[35] P. Sillard, S. Richard, L.-A. de Montmorillon, M. Bigot-Astruc, Micro-bend losses of trench-assisted single-mode fibers, in: ECOC 2010, Sept. 2010, We.8.F.3.

[36] S. Matsuo, K. Takenaga, Y. Arakawa, Y. Sasaki, S. Tanigawa, K. Saitoh, M. Koshiba, Crosstalk behavior of cores in multi-core fiber under bent condition, IEICE Electron. Express 8 (2011) 385–390.

[37] W. Klaus, J. Sakaguchi, B.J. Puttnam, Y. Awaji, N. Wada, Free-space coupling optics for multi-core fibers, WC3.3, in: IEEE Photonics Society Summer Topicals 2012.

[38] K.S. Abedin, T.F. Taunay, M. Fishteyn, M.F. Yan, B. Zhu, J.M. Fini, E.M. Monberg, F.V. Dimarcello, P.W. Wisk, Amplification and noise properties of an erbium-doped multicore fiber amplifier, Opt. Express 19 (17) (2011) 16715–16721.

[39] D. Noordegraaf, P.M.W. Skovgaard, M.D. Nielsen, J. Bland-Hawthorn, Efficient multi-mode to single-mode coupling in a photonic lantern, Opt. Express 17 (3) (2009) 1988–1994.

[40] J.M. Fini, T.F. Taunay, M.F. Yan, B. Zhu, Techique and devices for low-loss, modefield matched coupling to a multicore fiber, US Patent Application No. 2011/0280517, 2011.

[41] H. Takara, H. Ono, Y. Abe, H. Masuda, K. Takenaga, S. Matsuo, H. Kubota, K. Shibahara, T. Kobayashi, Y. Miyamoto, 1000-km 7-core fiber transmission of 10×96-Gb/s PDM-16QAM using Raman amplification with 6.5 W per fiber, Opt. Express 20 (9) (2012) 10100–10105.

[42] R.R. Thomson, H.T. Bookey, N.D. Psaila, A. Fender, S. Campbell, W.N. MacPherson, J.S. Barton, D.T. Reid, A.K. Kar, Ultrafast-laser inscription of a three dimensional fan-out device for multicore fiber coupling applications, Opt. Express 15 (18) (2007) 11691–11697.

[43] F.V. Laere, G. Roelkens, M. Ayre, D. Taillert, D.V. Thourhout, T.F. Krauss, R. Baets, Compact and Highly Efficient Grating Couplers Between Optical Fiber and Nanophotonics Waveguides, J. Lightwave Technol. 25 (1) (2007) 151–156.

[44] C.R. Doerr, T.F. Taunay, Silicon photonics core-, wavelength-, and polarization-diversity receiver, IEEE Photon. Technol. Lett. 23 (9) (2001) 597–599.

[45] N.K. Fontaine, C.R. Doerr, M.A. Mestre, R.R. Ryf, P.J. Winzer, L.L. Buhl, Y. Sun, X. Jiang, R. Lingle Jr., Space-division multiplexing and all-optical MIMO demultiplexing using a photonics integrated circuit, in: OFC/NFOEC 2012, 2012, PDPB5B1.

[46] B.G. Lee, D.M. Kuchta, F.E. Doany, C.L. Schow, P. Pepeljugoski, C. Baks, T.F. Taunay, B. Zhu, M.F. Yan, G. Oulundsen, D.S. Vaidya, W. Luo, N. Li, End-to-End Multicore Multimode Fiber Optic Link Operating up to 120 Gb/s, J. Lightwave Technol. 30 (6) (2012) 886–892.

[47] R. Nagase and S. Mitachi, MU-type PANDA fiber connector, in: Proc. SOFM '96, 1996, pp. 53–56.

[48] B. Zhu, T.F. Taunay, M.F. Yan, J.M. Fini, M. Fishteyn, E.M. Monberg, F.V. Dimarcello, Seven-core multicore fiber transmissions for passive optical network, Opt. Express 18 (11) (2010) 11117–11122.

[49] K. Watanabe, T. Saito, K. Imamura, Y. Nakayama, M. Shiino, Study of Fusion Splice for Single-mode Multicore Fiber, in: 17th Microoptics Conference (MOC'11), 2011, H-8.

[50] K. Yoshida, A. Takahashi, T. Konuma, K. Yoshida, K. Sasaki, Fusion Splicer for Speciality Optical Fiber with Advanced Functions, Fujikura Tech. Rev. 41 (2012) 10–13.

[51] B.G. Lee, D.M. Kuchta, F.E. Doany, C.L. Schow, C. Baks, R. John, P. Pepeljugoski, T.F. Taunay, B. Zhu, M.F. Yan, G.E. Oulundsen, D.S. Vaidya, W. Luo, N. Li, 120-Gb/s 100-m transmission in a single multicore multimode fiber containing six cores interfaced with a matching VCSEL array, IEEE Summer Topical Meeting 2010, 2010, TuD4.4, 223–224.

[52] B. Zhu, T.F. Taunay, M.F. Yan, M. Fishteyn, G. Oulundsen, D. Vaidya, 70-Gb/s Multicore Multimode Fiber Transmissions for Optical Data Links, IEEE Photon. Technol. Lett. 22 (22) (2010) 1647–1649.

[53] J. Sakaguchi, Y. Awaji, N. Wada, A. Kanno, T. Kawanishi, T. Hayashi, T. Taru, T. Kobayashi, M. Watanabe, 109-Tb/s ($7 \times 97 \times 172$-Gb/s SDM/WDM/PDM) QPSK transmission through 16.8-km homogeneous multi-core fiber, in: Optical Fiber Communication Conference, OSA Technical Digest (CD), Optical Society of America, 2011, Paper PDPB6.

[54] S. Chandrasekhar, A. Gnauck, X. Liu, P. Winzer, Y. Pan, E.C. Burrows, B. Zhu, T. Taunay, M. Fishteyn, M. Yan, J.M. Fini, E. Monberg, F. Dimarcello, WDM/SDM Transmission of 10×128-Gb/s PDM-QPSK over 2688-km 7-Core Fiber with a per-Fiber Net Aggregate Spectral-Efficiency Distance Product of 40,320 km.b/s/Hz, in: 37th European Conference and Exposition on Optical Communications, OSA Technical Digest (CD) Optical Society of America, 2011, Paper Th.13.C.4.

[55] X. Liu, S. Chandrasekhar, X. Chen, P.J. Winzer, Y. Pan, T.F. Taunay, B. Zhu, M. Fishteyn, M.F. Yan, J.M. Fini, E.M. Monberg, F.V. Dimarcello, 1.12-Tb/s 32-QAM-OFDM superchannel with 8.6-b/s/Hz intrachannel spectral efficiency and space-division multiplexed transmission with 60-b/s/Hz aggregate spectral efficiency, Opt. Express 19 (2011) B958–B964.

[56] K. Imamura, H. Inaba, K. Mukasa, R. Sugizaki, 19-core multi core fiber to realize high density space division multiplexing transmission, in: IEEE Photonics Society Summer Topicals 2012, TuC4.3.

[57] J. Sakaguchi, B.J. Puttnam, W. Klaus, Y. Awaji, N. Wada, A. Kanno, T. Kawanishi, K. Imamura, H. Inaba, K. Mukasa, R. Sugizaki, T. Kobayashi, M. Watanabe, 19-core fiber transmission of $19 \times 100 \times 172$-Gb/s SDM-WDM-PDM-QPSK signals at 305Tb/s, in: Optical Fiber Communication Conference, OSA Technical Digest, Optical Society of America, 2012, Paper PDP5C.1.

[58] R. Celechovsky, Z. Bouchal, Optical implementation of the vortex information channel, New. J. Phys. 9 (2007) 328.

[59] Y. Ueno, Y. Toda, S. Adachi, R. Morita, T. Tawara, Coherent transfer of orbital angular momentum to excitons by optical four-wave mixing, Opt. Express 17 (2009) 20567–20574.

[60] G. Volpe, D. Petrov, Generation of cylindrical vector beams with few-mode fibers excited by Laguerre-Gaussian beams, Opt. Commun. 237 (2004) 89.

[61] Y. Awaji, N. Wada, Y. Toda, T. Hayashi, Propagation of Laguerre-Gaussian mode light through multi-core fiber at telecom wavelength, in: CLEO2011, 2011, CThGG2.

[62] Y. Awaji, N. Wada, Y. Toda, T. Hayashi, World first mode/spatial division multiplexing in multi-core fiber using Laguerre-Gaussian mode, ECOC2011, 2011, We.10.P1.55.

[63] Y. Awaji, N. Wada, Y. Toda, Observation of orbital angular momentum spectrum in propagating mode through seven-core fibers, in: CLEO2012, 2012, JTu2K.3.

Elastic Optical Networking

14

Ori Gerstel[a] and Masahiko Jinno[b]

[a]*Principal Engineer, Cisco, Israel*

[b]*Senior Research Engineer, Supervisor, NTT Network Innovation Laboratories, Japan*

14.1 INTRODUCTION

14.1.1 The only constant in the future network is change

Service provider (SP) networks are undergoing major changes. Traffic continues to grow at an exponential rate—around 40% per year globally [1] and much faster in some cases—such as cellular networks. At the same time, a growing percent of the direct and indirect revenues from the services are going to "over the top" (OTT) service providers, such as Google and Netflix, leaving SPs with almost flat revenues. This in itself strains the business model of SPs, as the gap between the cost of the network and the revenues from it shrinks. But more fundamental changes are becoming apparent:

- Consumer traffic—mostly video—is now much larger than business traffic—skewing the required technologies toward more dynamic IP-based technologies.
- The number of main bandwidth sources of this traffic is becoming much smaller. For example, in the US, Netflix traffic represents 32.7% of the downstream traffic and YouTube represents another 11.3% [2].
- This implies that a change in peering arrangement for one of the major sources of content, or an addition of a new data center, will dramatically affect the demand pattern—all the way to the optical layer.
- Another implication is that the disparity between small demands and large ones in the optical layer will grow: some data centers will require "elephant" demands between them, while demands on links closer to the consumer will be much smaller.
- The emerging cloud computing paradigm will make it easy to mobilize an application from one server to another based on power savings considerations, proximity to the users of the application and commercial considerations, further increasing traffic dynamism.
- It will thus be increasingly hard to predict what the type of traffic will be and its behavior: 3 years ago Netflix did not generate any significant traffic, today it is 1/3 of the traffic in the US, and there are already questions whether the peak is not already behind us [2].

Optical Fiber Telecommunications VIB. http://dx.doi.org/10.1016/B978-0-12-396960-6.00014-6

These changes imply that the optical layer will have to be low cost, flexible, configurable, and reconfigurable:

Low cost: the increased pressure on SP margins implies that the future network must be as spectrally efficient as possible, and as streamlined as possible—one cannot afford many layers as in today's network, in fact most networks may only have two layers: a transport layer and a service layer.

Flexible: the lack of ability to forecast how traffic will evolve implies that the network will have to be as flexible as possible—the "one size fits all" approach of previous generation DWDM systems will no longer work as some connections will require much larger bandwidth than other connections, and the same gear will have to accommodate these varying demands via software configurability.

Configurable: unpredictable traffic patterns imply that today's static DWDM layer will not do. If connections cannot be set up between any source/destination, making use of available resources, then the network will not react quickly enough to unpredictable growth.

Reconfigurable: since traffic patterns will change more frequently, the network will have to support graceful release, redeployment, and reoptimization of resources. Without these capabilities, resource will sit idle and the cost of the network will grow well beyond the required cost. For example, fragmentation of wavelength resources implies that the effective utilization of DWDM link will deteriorate over time.

In the rest of this chapter, we will see how elastic optical networking can optimally address these requirements.

14.1.2 Why "business as usual" is not an option for DWDM

Let's start by reviewing the architecture of DWDM networks today. These networks are based on the International Telecommunication Union (ITU) wavelength grid, which splits the useful spectrum in a fiber into fixed spectrum slots—typically 50 GHz wide for long-haul networks. The use of the grid is quite rigid today: reconfigurable add/drop multiplexers (ROADMs) are becoming more ubiquitous, however, a surprising number of large networks still don't use them, or use very limited versions of ROADMs that do not allow for fully automated setup of lightpaths. Even networks that do use fully flexible ROADMs (which allow for directionless and colorless switching) do not have a control plane that understands the availability of resources in the network and its constraints, requiring offline planning tools to determine which connections can be set up. However, these tools base their decisions on network information that is often dated and inaccurate. This means that the network will not have the flexibility to adjust itself in case of significant demand changes. While the above shortcoming can be fixed even with the current fixed grid, a key issue that is inherent to the 50 GHz grid is that in many cases it is unlikely that it will support bit rates beyond 100–200 Gb/s. This does not automatically point to a flexible grid, but given that a new grid is needed, the question arises: why not make it flexible?

Even if sufficiently broad spectrum is available per channel, high data-rate signals become increasingly difficult to transmit over long distances at high spectral efficiency (i.e. how many bits per second fit into a GHz worth of spectrum). Therefore, it becomes beneficial for the network to measure the actual conditions along the link and adapt to these conditions, giving rise to software programmable (or adaptive) transceivers. Moreover, if resources are constrained, it may not make sense to limit the network to transporting fixed amounts of bandwidth per connection: transporting 100 Gb/s for a client demand that only requires 60 Gb/s may prove overly expensive, especially if 100 Gb/s requires regeneration, while 60 Gb/s could get by without it. Therefore it may be important to adapt the transported data rate to the client layer demands.

To properly address this challenge, one needs flexible and adaptive networks equipped with flexible transceivers and network elements that can adapt to the actual traffic needs. Fortunately, the same technologies that are being considered for achieving very high bit rates, of 100 Gb/s and beyond, can also provide this added flexibility (for example, coherent detection and increased reliance on digital signal processing). The combination of adaptive transceivers, a flexible grid, and intelligent client nodes enables a new "elastic" networking paradigm [3], allowing SPs to address the increasing needs of the network without frequently overhauling it.

100 Gb/s-based transmission systems have been commercialized in the recent two years. Since they are compatible with the 50 GHz ITU grid already deployed, the need for replacing the grid did not arise. Both the telecom and datacom industries are now considering data rates beyond 100 Gb/s, and 400 Gb/s is receiving a lot of attention as a possible next step. Unfortunately, the spectral width occupied by 400 Gb/s using reasonable modulation formats is too broad to fit in the 50 GHz ITU grid, and forcing it to fit by adopting a higher spectral efficiency modulation format would only allow short transmission distances. Figure 14.1 shows an existing ITU

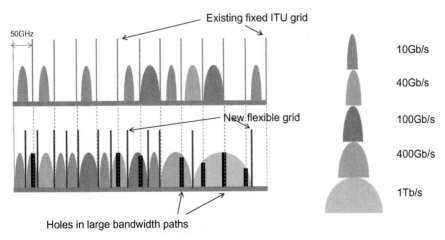

FIGURE 14.1 Use of spectrum for a link with different bit rates [4].

grid (top) vs. a flexible grid (bottom). The fixed grid does not support bit rates of 400 Gb/s and 1 Tb/s at standard modulation formats, as they overlap with at least one 50 GHz grid boundary.

Figure 14.2 shows different bit rate demands interconnecting node A with B, C, and D. The component that switches the channels arriving at B toward C or D is called a ROADM. If this device conformed to the ITU grid, then it would not be able to switch the broader spectrum channels; as Figure 14.1 shows, the optical spectrum coinciding with an ITU grid boundary (marked in black) will not be transmitted through the ROADM. Therefore, in order to build a flexible network, a new kind of ROADM is required that allows flexible spectrum to be switched from the input to the output ports.

Figure 14.2 shows several features that will help us define some important terms. This new approach is called Elastic Optical Networking (EON). The term "elastic" refers to two key properties: (i) the optical spectrum can be divided up flexibly and ROADMs can aggregate and switch this flexible spectrum and (ii) the transceivers can generate Elastic Optical Paths (EOP)—i.e. paths with variable bit rates and spectrum needs. These new ROADMs and transceivers are called Flexible Spectrum ROADM (FS-ROADM) and Bandwidth Variable Transceivers (BVT), respectively.

The drivers for developing the EON paradigm are listed below.

Clearly, the main driver is the need to support high bit rate demands, beyond 100 Gb/s. Indeed, one does not need a flexible grid to carry such demands: they can be implemented on a fixed grid network by demultiplexing the demand to smaller 100 Gb/s channels, using a technique called "inverse multiplexing" in the context of TDM networks, or a related technique for packet networks called "link bundling," but such an approach will not yield additional spectrum efficiencies. Further improvements require closer spacing of channels, giving rise to "superchannels" that use an amount of spectrum which is proportional to their bandwidth. Consider Figure 14.3, which provides a historic perspective of spectral efficiency in

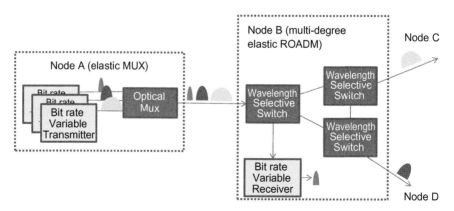

FIGURE 14.2 Example for 3 demands propagating through an EON [4].

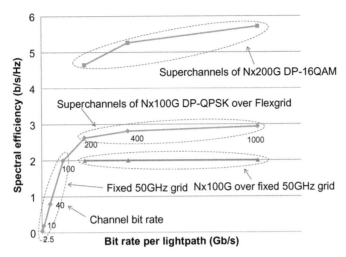

FIGURE 14.3 Spectral efficiency improvements for existing systems and for future EON.

commercial fixed systems, as well as the achievable spectral efficiency for EON (assuming 33 GHz per each channel in a superchannel and an additional 10 GHz of guard-band between superchannels). If we keep the 50 GHz grid and do not assume superchannels, the graph would have flattened at a spectral efficiency of 2 b/s/Hz, as shown in the figure. Note that if the demand spans more than approximately 500 km, then even if we use superchannels, it will follow the DP-QPSK graph and will not experience significant spectral efficiency improvement, however the overall savings will be more significant if one takes a network level perspective as discussed later in this chapter.

Another key driver for EON is the disparate bandwidth needs that demands in a typical network exhibit. In the past, the disparity was lower and it was dealt with by either allocating excessive bandwidth to small demands, or by using an extra layer of TDM cross-connects to allow multiplexing of small demands onto a higher bandwidth wavelength—thereby decoupling the actual demands from the wavelength bandwidth, but the cost of both approaches seems less acceptable for higher bit rates. This point is discussed in more detail in the comparison section below. Since EOPs can be right-sized to match the client layer links, EON will enable a much closer mapping between the two layers. This is especially true for a client layer that requires significant bandwidth over the transport layer, like an IP core network.

A related driver is the geographic disparity of demands. In the future, many large demands may span short distances, as a result of cloud computing and video caching technologies. If an EOP is short in distance, the BVT can adjust to a more efficient modulation format—say DP-16QAM, and the connection will still perform error-free even if this format provides more limited reach. This will improve the spectral efficiency considerably as shown in the upper graph in Figure 14.3. It is becoming cost prohibitive to ignore such savings. At the same time, some demands will require longer reach, and regenerating them frequently will be less cost effective and less conducive to a dynamic behavior, than using a less efficient yet better performing modulation format such as DP-QPSK or even DP-BPSK.

Finally, the need for more dynamic response to traffic changes will also drive to an EON. If demands change dramatically, then the network should allow for changes in connections capacities, which is more difficult with fixed DWDM systems.

The development of EON will require innovations in both hardware and software. New components will need to be developed, and will often be more complex than their fixed grid counterparts. Also challenging will be the control and management of the network, including setting up EOPs and changing their properties over time. To make this development worthwhile, it is important to understand whether the benefits of EON outweigh its disadvantages, but before we do so, we'll survey the technologies that enable EON, and outline the vision that could be enabled, should the technology be embraced to its full extent.

14.2 ENABLING TECHNOLOGIES

The main technologies that enable EON are flexible bandwidth transmitters and receivers that can adapt to the needs of the network and create superchannels, and ROADMs, typically based on wavelength selective switches, that can manipulate arbitrary spectrum slices. No less important is the control of these flexible resources via network management systems and control planes. This is the focus of this section.

14.2.1 Flexible Spectrum ROADM

A ROADM typically comprises several interconnected wavelength selective switches (WSS), which take multiple wavelengths on an input port, and can select which of these wavelengths will be routed to an output port—see Figure 14.2 for a simple three-way ROADM that requires 1×3 WSS devices. ROADMs also include amplifiers and other components which are less sensitive to the existence of the ITU grid and are not discussed herein. The capabilities of the WSS device determine to a large extent the capabilities of the ROADM insofar as the manipulation of the spectrum is concerned. While mature WSS technologies are specific to the ITU grid, newer WSS technologies that allow switching almost arbitrary spectrum slices (in 3.125–6.250 GHz steps) have been productized recently—enabling FS-ROADMs. These devices are based on one of several technologies: optical MEMS, Liquid Crystals on Silicon (LCOS), or silica Planar Lightwave Circuits (PLCs)—see [5] for a good survey.

We describe below the optical MEMS approach, but the other technologies are conceptually similar. The device comprises the following elements, schematically depicted in Figure 14.4:

- The input fiber launches the light into free space (inside the device).
- A diffractive element separates the incoming spectrum over space.

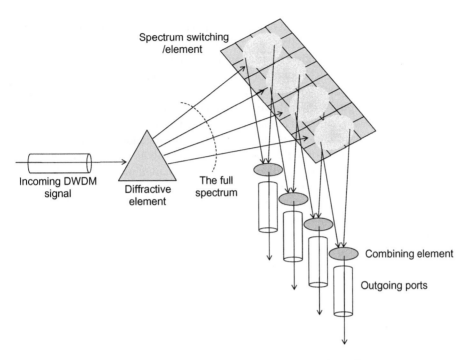

FIGURE 14.4 Conceptual depiction of a flexible wavelength selective switch.

- An optical switching array, capable of redirecting the various chunks of the spectrum toward different output ports.
- A combining element (e.g. lenses) recombines the light to the different output ports.

The unique feature of a flexible spectrum WSS, compared to the conventional fixed bandwidth WSS, is that the switching array consists of a large number of two-dimensionally arranged pixels, which are much smaller than the light beam. Therefore, the light spectrum can be contiguously changed by adjusting the number of pixels.

Due to the low cost overhead for these devices, we expect them to become common in next-generation ROADMs, and to be deployed irrespective of whether the larger EON vision will materialize, simply because they provide assurance that the network will be robust against future uncertainty, without paying a significant premium.

14.2.2 Bitrate Variable Transceiver

High-speed transmission is based on four techniques, all of which aim at reducing the bit rate for the electro-optical components and analog-to-digital converters, as well as baud rate sent over the network [6,7]: first, higher order modulation formats help transmit more bits per symbol and thus increase the bit rate while keeping the baud rate low: from BPSK, QPSK, and 16-QAM to perhaps 64-QAM in the future. While higher order modulation formats allow transmission of a high bit rate over a single carrier, they severely limit the reach due to the increased sensitivity to OSNR and transmission impairments, for example, a 16-QAM signal is expected to enable transmission of a 100 Gb/s signal over a few hundred km at best.

A second technique, which has seen deployment even for 40 Gb/s and is widely used for 100 Gb/s transmission, is to exploit the two polarizations of light to transmit two independent signals over the same wavelength. This is denoted by the prefix "DP" or "PM" (for "dual polarization" or "polarization multiplexed"). Third, coherent receivers allow for electronic processing of the signal and greatly improve the resiliency of the system to impairments such as chromatic dispersion and polarization mode dispersion. Coherent receivers also allow for easier processing of a polarization multiplexed signal in the electronic domain.

However, the above three techniques together are not likely to enable transmission of 400 Gb/s or higher bit rates over long-haul distances in real-world networks. The only solution to this problem seems to be to use multiple channels in parallel, using the aforementioned superchannel approach [7]. Superchannels are different from a group of individual channels in that they are very closely spaced without guard-bands between them, traverse the network as a single entity, and share the same endpoints. Several superchannel technologies have been discussed in the literature: coherent wavelength division multiplexing (CoWDM) [8], coherent optical orthogonal frequency division multiplexing (CO-OFDM) [9], Nyquist-WDM [10], as well as dynamic optical arbitrary waveform generation (OAWG) [11].

We will not attempt to review the differences between these approaches here. The reader is referred to [12] as well as to other chapters in this book for an in-depth description of these approaches. We will devote a bit more attention to Nyquist-WDM, as this seems the leading approach for the implementation of superchannels in commercial systems at this point in time. Nyquist-WDM attempts to minimize the spectral utilization of each channel and to reduce the guard-bands required between WDM channels generated from independent lasers. By filtering the channel spectrum in a square shape that approaches a Nyquist filter, the channel bandwidth is minimized, to a value that approaches the channel baud rate. The channels are then packed closely together such that the subcarrier spacing is equal to or slightly larger than the baud rate. This poses a challenge of how to separate the channels at the receiver. Luckily, with coherent detection, each receiver can select a separate channel from the aggregate without explicit demultiplexing.

Figure 14.5 demonstrates how a DWDM system is conceptually put together, from the incoming bits on the left, which are mapped on to symbols (16-QAM in the example) creating a modulated signal, which is combined with another modulated signal via polarization multiplexing; this forms a channel, which is combined with other channels to form a superchannel; finally, multiple superchannels are combined to form the aggregate DWDM signal.

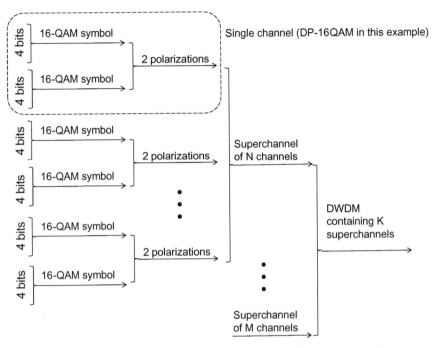

FIGURE 14.5 The parallel nature of high-speed transmission: different stages in the construction of a DWDM signal.

14.2.3 The extended role of network control systems

If EON is to be widely deployed, it must be reasonably easy to operate. This is not a simple feat, given the extra degrees of flexibility that EON introduces. Some of this complexity can be automated (for example, automatic decision on the optimal modulation format for a connection), but some of it will have to be visible to the operator, if only for trouble-shooting purposes (for example, understanding why the system has picked a specific modulation format). These functions are part of the networks control system, which comprises network management functions (centrally operated, GUI-based systems under operator control) and control plane (typically distributed and automated). We briefly cover these functions below.

Figure 14.6 shows a connection-management model of an EON, which applies to both functions. It includes the client layer demands as well as the optical layer details. Beyond the usual information needed in a DWDM network, this model must be aware of the channel bit rate, modulation format, and achievable optical reach. The figure also shows the concept of a "frequency slot," which is defined as a frequency range where an optical channel is allowed to occupy. The frequency

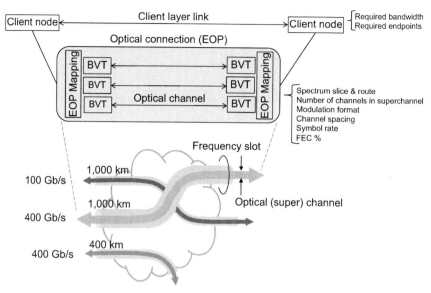

FIGURE 14.6 Simple connection model for EON.

slot is characterized by the filtering and switching window of the FS-ROADMs on the route.

14.2.3.1 Management plane aspects

For wide-scale deployment, it must be possible to initially operate the EON in a centralized and manual fashion, without relying on a control plane. This is because the operations of large transport networks are not adapted to using control planes. While control planes are becoming more acceptable to such operators, and will greatly enhance the value of EON, one should first examine how the technology can be operated without such automation.

In its most basic form, the network could be operated in similar fashion to today's DWDM networks: the decision on connection parameters will be based on rigid rules, such as: "QPSK for all long haul links, 16QAM for metro links," and could even be taken in offline planning tools rather than based on actual network parameters. In such a system, the use of superchannels may not be allowed, or they may be restricted to, say, 400 Gb/s demands implemented as 4×100 Gb/s superchannels. Spectrum allocation could be based on fixed spectrum slots, say 50 GHz for 100 Gb/s and 100 GHz for 400 Gb/s, where each type is allocated in first fit fashion from opposite ends of the spectrum to avoid fragmentation. Clearly, the value of such a system to the operator will be lower than the full potential that EON holds, but it will still be more flexible than today's DWDM systems.

More advanced management systems will understand how to best map a given demand to an optical connection. To this end they will understand which of the available options for implementing the connection is most suitable given the required reach and bandwidth for the demand. This requires understanding of transmission considerations, a capability that typical management systems do not have today. In addition, they will track the use of spectrum and find a suitable frequency slot for a new connection. We later expand on how this should be done.

Another important new functionality is to allow the operator to understand the current usage of spectrum in the network and its availability for future EOPs. The system could even enable a reoptimization of the use of the spectrum, a concept similar to defragmentation of hard-disk resources.

14.2.3.2 Control plane aspects

The control plane is most suitable to fully unlock the value of EON, because a truly adaptive network must base its decisions on actual measured data from the network and not on offline information in planning tools, or stale information in management systems (both of which are common practice today). The control plane could be fully distributed, or rely on a centralized path computation engine (PCE). Either way, it must be aware of the network topology, the different fiber characteristics, the spectrum utilization, and the design rules that dictate how to select connection parameters. Conceivably, this decision can be refined via trial and error: the control plane decides on some rough connection parameters and establishes the connection; the receivers at both endpoints measure connection performance (BER) and other metrics (CD, PMD, SNR, etc.) and decide how to tweak the connection parameters: if the BER is too high to guarantee the quality of the connection, it is improved, say, by changing the modulation format to a more robust one; on the other hand, if the BER is very low, then it may be possible to improve the use of spectrum by moving to a more efficient modulation format, while still keeping the BER below the required maximum. Once the two endpoints decide on first-order parameters like modulation format, they can refine second-order parameters, like the spacing between channels inside a superchannel. This fine-tuning process is likely to occur using control plane signaling.

The main control plane protocols that are likely to play a role for EON are:

- Distributed topology dissemination protocols such as OSPF or ISIS will be used to disseminate impairment information and spectrum availability and to allow the first node on the path to decide on the route, frequency slot, and connection parameters.
- Connection setup signaling protocols such as RSVP will be used to set up the EOP, reserve the resources, and possibly to fine-tune the connection parameters after it has been established. This class of protocols will also play a role during network optimization processes, such as defragmentation.
- Protocols for communicating with PCE, such as PCEP, will be used to convey topology and connection information to PCE.
- Link and fault management protocols, such as LMP, will be used for autodiscovery of BVTs and for fault isolation.

More details on the information carried in control plane signaling messages and on how spectrum should be allocated in EON will be given in subsequent sections.

14.2.4 **EON trials and other proof points**

In the past two years, as the enabling technologies for EON have started to mature, EON research has transitioned from theory to experimentation. Rate-adaptive spectral allocation has been experimentally shown to allow scaling of the optical path capacity from 40 to 440 Gb/s [3] by using no-guard-interval optical OFDM signals [14] and bandwidth-variable wavelength cross-connects based on flexible wavelength selective switches as shown in Figure 14.7 from [13]. The same architecture was also used to demonstrate optical aggregation of multiple hundreds of Gb/s optical channels into a single continuous superchannel in an EON [15]. Distance adaptive spectrum allocation, where the minimum necessary spectrum resources is adaptively allocated to an optical path according to end-to-end physical network condition, was demonstrated by using 16-APSK and QPSK modulation formats [16]. A recent field trial of EON-based OFDM transmission has demonstrated over 620 km distance with 10 G/40 G/100 G/555 G with defragmentation [17], and an EON network testbed with real-time automated adaptive control plane and sliceable transceivers has been demonstrated in [18]. We expect more substantial testbeds to be built in the near term.

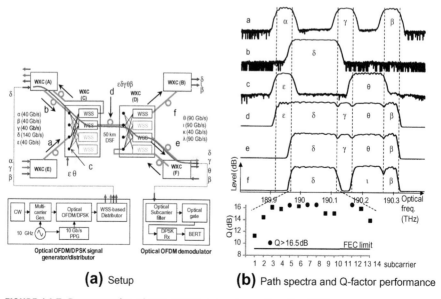

(a) Setup

(b) Path spectra and Q-factor performance

FIGURE 14.7 Demonstration of rate-adaptive spectrum allocation [13].

14.3 THE EON VISION AND SOME NEW CONCEPTS

We now provide a vision for an ideal, fully automated EON, based on a control plane that is aware of transmission impairments. We also assume that the clients are smart enough to exploit the flexibilities that EON provides:

- In such a network, each client link will typically be mapped to a single EOP which is right-sized for the link, eliminating the need for an extra layer of multiplexing. This mapping is cost effective, assuming that the transceivers are sliceable and can serve multiple client links, when each link does not consume the full transceivers resources.

- When the client link requires more bandwidth, it signals the EON via the control plane, and EON extends the EOP hitlessly in the most efficient way to accommodate the demand. The client can now increase the usage on the link.

- When the bandwidth decreases on a client link, the client reduces the capacity of its flexible interface to EON at both ends of the link, and then signals to EON to reduce the bandwidth of an EOP. This can be done, for example, by removing a channel from a superchannel.

- When a new link is needed, EON picks the most optimal solution for the EOP, based on knowledge of the client layer needs, the actual transmission conditions along the path, eliminating the inaccurate and labor intensive collection process of this data as is done today, and reducing the cost due to the reduced need for margins to account for these inaccuracies.

- Upon failure, the network load shifts from the failed links to the surviving ones, necessitating the augmentation of their capacity. Today, this is handled by over-provisioning links for failure scenarios, but in the future, if links can increase capacity dynamically, such over-provisioning will not be needed. Another course of action for the network is to heal failed links automatically, by rerouting them over a different optical path. Such rerouting will be more flexible than in today's networks, since different connection parameters could be used to overcome the longer reach of a protection path (for example, moving from QPSK to BPSK, and ensuring the client layer reduces the use of the link accordingly).

- Periodically, the network will consider reoptimizing the use of resources. This may be needed due to the aging of the fiber plant—rendering certain paths less attractive than in the past, or due to the desire to improve the allocation of spectrum, or due to the addition of new DWDM layer links that allow some EOPs to be routed more optimally. This will be done in a hitless manner via coordination with the client layer.

- The network will be flexible enough to make use of general purpose resources that are not tied to a particular direction or connection. For example, if BVTs are sliceable as explained below, they can be exploited to connect to multiple remote peers as needed. All the operator needs to do is provide sufficient BVTs per site to ensure traffic can be terminated there. The need for accurate forecast will be significantly reduced.[1]

In such a network, the planning process is significantly simplified: no longer does the process rely on manual entry of often inaccurate measurements of the fiber plant, or on inaccurate traffic forecast information. Instead, the network will adjust itself to optimally meet the demands with the current transmission conditions (with appropriate margins for aging, etc.), and readjust as needed if demands or conditions change. The role of planning reduced to ensuring sufficient BVTs are allocated per site, based on aggregate (and hence less error-prone) forecast information.

Below we describe several concepts that are unique to EON and provide more detail on the above vision.

14.3.1 Flexible choice of EOP parameters

In fixed DWDM networks, there is typically one way to implement a given lightpath: the wavelength bit rate is fixed, the optical reach is fixed, and the spectrum is fixed. Depending on the bandwidth needs, the demand will require one or more lightpaths, and depending on the required reach it may require one or more regenerators. In EON, multiple choices may be feasible when implementing a connection:

1. A straightforward approach to adapting an EOP to the bit rate of a demand is to change the transmitted symbol rate. However, a higher symbol rate implies lower resiliency to most impairments, so this approach will mostly apply for shorter reaches. Also note that from a practical standpoint it is hard to dramatically change the symbol rate for a given transceiver; however, this is certainly possible within 10–20%.
2. A given demand can be assigned a modulation format which gives sufficient performance to reach the required distance, while minimizing the spectral bandwidth occupied by the optical path [19].
3. Today the ratio between the amount of Forward Error Correction (FEC) and payload is fixed, but it could be made adaptive in EON to enable greater distances to be reached when the required bandwidth is lower—see [20] for details.
4. Whenever a connection passes through a ROADM, the ROADM reduces the optical bandwidth for the (super) channel. When this happens over and over, the resulting bandwidth may be too narrow, affecting the quality of the signal and limiting the reach. With EON, the spectrum allocated for longer EOPs can be adjusted to account for such bandwidth narrowing, increasing the number of ROADMs the connection can go through [19].

---

[1]Note that this capability is not specifically tied to EON and useful (but lacking) in today's networks as well.

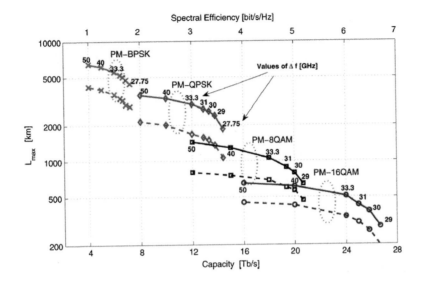

FIGURE 14.8 The performance of a superchannel with different channel spacing (Δf) at BER $\leq 4 \times 10^{-3}$, for SSMF fiber (solid line) and NZDSF (dashed line) [21].

5. Figure 14.8 demonstrates another level of flexibility: the spacing between channels in a superchannel can be modified to achieve either better performance or better spectral efficiency.

   Finally, if the required distance is too long and the required bandwidth is too high for an unregenerated EOP, one could always resort to regeneration of the EOP, but this is just one of several alternatives that can be considered, unlike the fixed DWDM case, where regeneration was the only option if the reach was too long.
   A summary of the parameters that characterize the EOP is shown in Figure 14.9.

14.3.2 Sliceable transceiver

In some transceiver designs, a single BVT could be "sliced" into several "virtual transceivers" that serve separate EOPs as in Figure 14.10b, where a sliceable 400 Gb/s transceiver is sliced into three EOPs: 100 Gb/s, 100 Gb/s, and 200 Gb/s. This flexibility is key to the economic justification of EON since it is hard to justify "wasting" a 400 Gb/s BVT on, say, a 100 Gb/s EOP alone (as in Figure 14.10a). If the transceiver is not sliceable, it may make more sense to use a standard rigid 400 Gb/s transceiver and electrical sub-wavelength grooming to fill up the remaining 300 Gb/s—but this introduces another layer and eliminates some of the cost gains of EON [22]. This flexibility seems feasible in some next-generation BVT designs,

Flexibilities

(a) Symbol rate

(b) Modulation type

(c) Inter-channel spacing

(d) Inter-superchannel spacing

(e) FEC vs payload

FIGURE 14.9 Different flexibilities for an EOP.

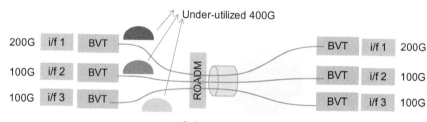

(a) Fixed transceivers

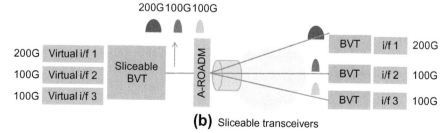

(b) Sliceable transceivers

FIGURE 14.10 Support for 3 sub-wavelength demands using fixed or sliceable transceivers (the ROADMs are assumed to support EON) [1,4].

and has been demonstrated in [18]; however, challenges still exist. For example, on the receiver side, one might need N independently tunable lasers as local oscillators to enable coherent reception of N EOPs that come from difference sources and use different parts of the spectrum.

14.3.3 Flexible client interconnect

Since different EOPs may have different bandwidth, one key question is how to connect the client (such as an IP router) to a BVT. From a protocol perspective, extensions of the Optical Transport Network (OTN) and Ethernet standards seem appropriate, similar to the concept of ODUflex in OTN. This client interface is further challenged if the BVT is sliceable and if the bandwidth for each virtual interface must expand/contract over time. In this case, the interface between the client and the BVT becomes a flexible pool of channels that can be grouped in different ways. This also affects the client line card architecture. A simpler option may be to eliminate the client-EON interface altogether by integrating the BVT into the client box, but this creates a separate set of challenges which are outside the scope of this chapter.

14.3.4 Spectrum allocation and reallocation

Let us assume that the system has figured out how to best implement a given demand and what the required spectrum width for it will be. Now it must find an available contiguous frequency slot along the path. This problem is exacerbated in a dynamic environment, where EOPs must have room to expand without affecting other EOPs.

In conventional DWDM systems, a key problem is calculating a route while maintaining the same wavelength end to end for a lightpath. This problem is called the routing and wavelength-assignment (RWA) problem. Spectrum allocation in EONs introduces another constraint to wavelength-continuity, which is finding a contiguous piece of the spectrum along the way. This is a more general routing and *spectrum*-assignment problem (RSA).

The RSA problem can be divided into two stages for a simpler (suboptimal) solution. The first stage is to identify candidate routes for the EOP. These routes satisfy various constraints for the path but do not consider spectrum usage—see Figure 14.11b. The second stage allocates contiguous spectrum to the route, by sliding a window from lowest frequency to highest and checking whether the spectrum is available on all the links along the path, as shown in Figure 14.11c. Finally (outside the RSA problem) the resources are locked as in Figure 14.11d. Some more details on this scheme can be found in [19].

When considering a dynamic network, in which EOPs can be added or removed, as well as contract or expand, it is important to leave sufficient available spectrum on both sides of the frequency slot for an EOP, to allow it to grow either up or down in spectrum. However, leaving such room for growth impacts the promised savings of EON. While some solutions have been suggested (e.g. [23]), they only serve to delay the need to reallocate the spectrum at a future point in time.

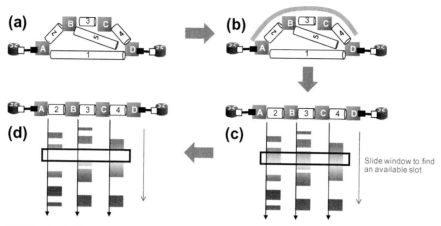

FIGURE 14.11 The spectrum assignment process.

Thus there will be a need to reallocate the spectrum over time. The required mechanism will remove stranded fragments of spectrum between EOPs and redistribute them to allow for further growth. In the context of a superchannel comprising different channels, this hitless spectrum shift can be done by adding an adjacent channel to the EOP in the desired direction of the shift and re-mapping the data sent over the channel that will be released to the new channel using either a transport mechanism called Link Capacity Adjustment Scheme (LCAS), or an IP layer mechanism called "link bundling." Both mechanisms allow for such changes to occur in a hitless manner. Note that this mechanism requires coordinating the changes in spectrum at the client device, the transmitter, and the optical switching device, but this can be done slowly and hence seems achievable.

A related problem is spectrum fragmentation: since the spectrum per EOP varies, as EOPs come and go, over time there will be small parts of the available spectrum that are unusable. This is called "fragmentation," and is analogous to how a computer hard disk becomes fragmented. The same spectrum reallocation process can be used for this case as well.

14.3.5 Managing a connection per demand instead of managing wavelength

In traditional DWDM networks, a connection corresponds to a single wavelength that is managed individually: it appears as a separate entity to management systems, has its separate alarms under failure conditions, and is switched individually through the network. If multiple wavelengths are needed to realize a larger demand, then their bundling occurs at the client layer (via OTN, Ethernet, or IP layer bundling) without reducing the complexity of the network. This is not the case with EON, where an optical layer demand will be mapped directly to a single EOP. If the demand is smaller than the capacity of a BVT, then part of the BVT will be (ideally) carved out to carry it independently, while the rest of the BVT resources can be used for other purposes. If the demand is larger than the capacity of a BVT, then channels from multiple BVTs can be optically bundled into a single superchannel that implements the demand. Either way, the demand will be accommodated in a single frequency slot, switched as one entity through ROADMs, and managed as a single entity in management systems. A side effect of this is that the number of managed entities in the optical layer will be dramatically reduced—at least for large IP networks, in which a large number of channels are typically bundled. This has a positive impact on the scale of ROADMs— no longer does a ROADM have to add and drop many dozens of separate connections; instead it will only need to add/drop a handful of EOPs. Since the nodal degree of a large IP core router is merely 2–8, the number of EOPs for a typical core site that contains two core routers will be typically less than 16, which is much lower than the number of wavelengths added and dropped by a typical ROADM today.

14.3.6 Adaptive restoration

The ability of EON to adapt the capacity and reach per connection enhances survivability of optical paths in a case of widespread serious disasters. This will be demonstrated on the network example in Figure 14.12. Under normal circumstances, the two routers in the figure connect via the primary path (#1). If this path fails, a protection scheme moves the traffic quickly to a secondary path (#2). This can be done at the client layer (e.g. via MPLS fast reroute) or in the optical layer, if optical

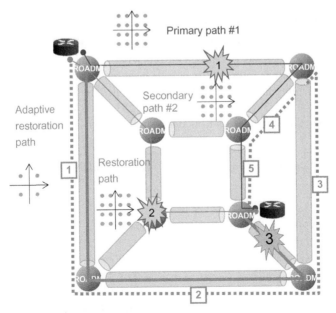

FIGURE 14.12 Optical layer adaptive restoration.

protection is used. If the secondary path fails, then a conventional DWDM system may find a restoration path, and switch traffic to it by reconfiguring the ROADMs along the path. However, this third path may not be feasible due to the required extended reach, or it could fail as well, due to its proximity to the disaster area. Here is where EON provides a significant advantage as it can find a long path that "weaves its way" between failed resources. However, such surviving detour routes may not have sufficient spectral resources to transport the original data rate, and/or the length of the detour route could exceed the optical reach of the original optical signal. The ability of the elastic optical path to adapt its reach and offer a lower bandwidth path guarantees that at least minimum connectivity is achieved for high-priority traffic. One should note that this assumes that the client layer is intelligent enough, and could choose to drop low priority traffic, or reroute some of the traffic over other paths in the client topology (via traffic engineering).

The decision to trigger such a restoration process could be taken by the control plane or by management systems as explained above.

14.4 A COMPARISON OF EON AND FIXED DWDM

Let's start with a basic comparison of EOP to fixed DWDM networks—just looking at the spectrum efficiency of a point to point link. Later in the section, we will look at the impact of EON in the context of an entire optical network, followed by considerations that include the client layer as well.

14.4.1 A point-to-point comparison

It is tricky to compare EON to fixed DWDM even for the simplest network of a single link. The comparison depends on the link length and transponders used for both cases. Questions like the following ones must be answered to provide a fair comparison: are we assuming a fixed DWDM system is limited to using a single modulation format? Is a fixed DWDM system limited to a 50 GHz grid? What is the extra cost for a BVT compared to a regular transponder? Are there any limitations to how a BVT

Table 14.1 Comparison between fixed DWDM and EON for a 400 Gb/s demand.

Reach (km)	Fixed network			EON		
	Solution	Regen	Spectrum (GHz)	Solution	Regen	Spectrum (GHz)
300	4 × DP-QPSK	0	200	2 × 200G DP-16QAM	0	85
2000	4 × DP-QPSK	0	200	4 × 100G DP-QPSK	0	142
5000	4 × DP-QPSK	4	200	8 × 50G DP-BPSK	0	274
				4 × 100G DP-QPSK	4	142

can be sliced? What FS-ROADM granularity is assumed? etc. One attempt to compare these options was provided in [4]. A different perspective is provided in Table 14.1 below. The table assumes superchannels comprise 33 GHz channels and that these are 10 GHz guard-bands between superchannels. The fixed DWDM network is assumed to only use DP-QPSK transponders (reflecting today's reality). As can be seen, EON saves significant spectral resources for short links (57%), and medium links (~30%), while for long links several options exist. One option is to use 37% more spectrum but eliminate regenerators, while another option is to use the same approach as in the fixed DWDM case, which will require regenerators, but still save 30% of spectrum.

Note that the cost savings depend on whether BVTs are sliceable and to what extent. In the ideal case, a BVT can handle any channel combination as long as the total bandwidth it handles is constant, say 400 G. In this case a single BVT can support all the required combinations (2×200 G, 4×100 G, and 8×50 G) at no cost premium for EON; while for long links, there are savings due to the elimination of regenerators. On the other hand, in a non-ideal case, there may be need for more BVTs for, say, the 8×50 G case, in which case the regenerator savings are offset by the extra cost at the endpoints.

14.4.2 A network level comparison

We now turn our attention to a network level perspective. Such a viewpoint differs from the point-to-point perspective provided above, because inefficient use of spectrum may increase the blocking probability of demands, thus exacerbating the impact of such inefficiencies and increasing the relative value of EON.

Figure 14.13 shows a comparison of the required total spectrum at the worst case link in a 12 node ring network for fixed grid and flexible grid with distance-adaptive spectrum allocation [19]. This assumes some client-layer 100 G demands are protected using a pair of diverse optical paths (clockwise and counterclockwise), while other 100 G demands are unprotected and transported using an optical path between source-destination node pair (see Figure 14.13b). Required spectrum resources are obtained from a spectrum allocation map as a function of number of node-hops (see Figure 14.13a). The map is created by numerical transmission simulation of 112 Gb/s DP-QPSK and DP-16QAM signals considering OSNR degradation and frequency clipping due to the filtering effect of cascaded ROADMs. The required total spectrum was evaluated by using a heuristic RSA algorithm with the spectrum-continuity constraint. The longest and shortest paths, which are paired route-diverse paths between adjacent nodes, have 11 and 1 hop, respectively. As can be seen from Figure 14.13c, unlike the fixed grid, which requires 100 GHz spectrum for every optical path, the flexible grid requires 37.5 GHz for optical paths shorter than five hops and 62.5 GHz for optical paths longer than 10 hops. This results in 45% spectrum saving in flexible grid for the paired route-diverse optical paths and even higher spectrum saving of 56% for the unprotected optical paths where most optical paths use spectrally efficient 16QAM modulation format.

We now turn to a different network comparison in [23]. Figure 14.14 compares the required spectrum for a mesh network for fixed grid and flexible grid with flexible

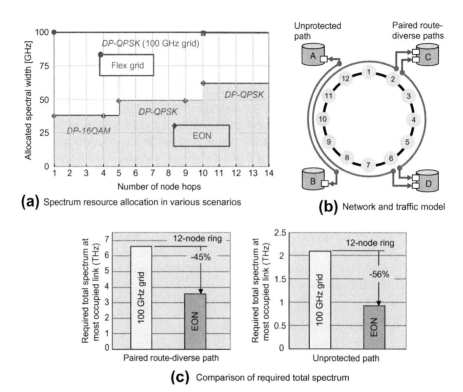

(a) Spectrum resource allocation in various scenarios

(b) Network and traffic model

(c) Comparison of required total spectrum

FIGURE 14.13 Comparison of required total spectrum for fixed grid and flexible grid with distance-adaptive spectrum allocation [19].

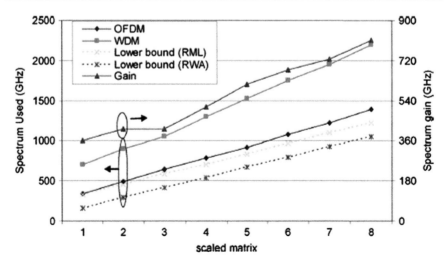

FIGURE 14.14 Spectrum utilization of a fixed 50 GHz network with spectral efficiency of 0.8 bits/s/Hz and an OFDM-based elastic network [23].

line-rate and distance-adaptive spectrum allocation. The fixed grid WDM network uses 40 Gb/s wavelengths with QPSK in a 50 GHz grid. The flexible grid network employs multiple subcarriers each modulated with BPSK, QPSK, 8-QAM, and 16-QAM to provide flexible line rate and adaptive optical reach. Each subcarrier is assumed to be aligned with 5 GHz subcarrier spacing and modulated at a symbol rate of 2.5 Gbaud so that 8 QPSK subcarriers (40 Gb/s in total) with 10 GHz guard-band fill a 50 GHz slot. These values were chosen so as to have the same spectrum efficiency for fairly evaluating the gain of flexible line-rate and distance-adaptive spectrum allocation. The network used in the evaluation is a generic Deutsche Telecom-like network topology with the reference traffic matrix with the 15 Gb/s average demand between nodes. As can be seen from the figure, the flexible grid network has better spectrum utilization due to the rate-adaptive spectrum allocation at light load and the distance-adaptive higher level modulation at heavy load. It should be noted that this result is applicable to any multi-carrier based systems including spectrum-efficient OFDM and Nyquist WDM.

Comparisons were also performed for dynamic networking environment, where optical paths are set up and torn-down according to the client connection requests [24,25]. The flexible grid network exhibited better blocking probability and more than twofold average spectrum efficiency enhancement with respect to the fixed grid, fixed modulation format network [25].

14.4.3 A comparison that includes the client network

Let's now take an even wider viewpoint, and consider how the network will be architected from a client layer perspective. Specifically we look at the most common case for such a client layer, namely an IP core network [26].

The most natural way to grow a network is to keep adding more wavelengths in parallel to existing ones and achieve high capacity. If taken to the extreme, this requires a large number of wavelengths in the network—say, 128×10 Gb/s wavelengths on a 25 GHz grid. Such a High Density DWDM (HD-DWDM) system is schematically shown in Figure 14.15a. The bandwidth granularity benefits of EON also exist with HD-DWDM—in fact, the grid-based approach may provide an even denser topology since there is no added penalty for fine-grain connections, while in EON, fine-grain EOP connections may not be supported via sliceable transceivers. However, the spectral efficiency of HD-DWDM is low if most connections require more than 10 Gb/s capacity: namely, a total of 1.28 Tb/s per fiber vs. 8 Tb/s if 80 channels of 100G each are used. In addition, HD-DWDM requires a very large number of add/drop ports in ROADMs (W*D ports for 100% add/drop of all W wavelengths in a node with D outgoing directions), as well as very large cross-connects. The number of managed entities to keep track of is also proportionally high. Another disadvantage of HD-DWDM is that the narrow passband per connection on a 25 GHz grid implies it will suffer from filter cascading effect when going through many hops, while this is not the case for EON. Finally, from a router perspective, this approach may have additional disadvantages as the density of low-speed interfaces on high capacity routers may be more limited, and the efficiency of large link bundles may also be lower.

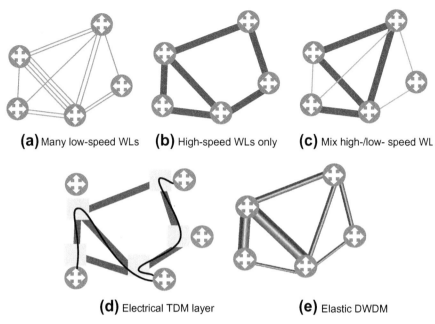

(a) Many low-speed WLs   **(b)** High-speed WLs only   **(c)** Mix high-/low- speed WL

(d) Electrical TDM layer   **(e)** Elastic DWDM

FIGURE 14.15 Network architecture comparison.

A more common way to address high bandwidth needs is to settle on a single high-speed wavelength—say, 100 Gb/s today or 400 Gb/s in the future—and build out the network with such wavelengths. The problem with this approach is that typically many links will not have sufficient capacity to require high-speed wavelengths, resulting in low link utilization and resource waste. As a result, the client network design will be typically changed to a sparser topology, which has fewer links that are more utilized as shown in Figure 14.15b. However, the network cannot be allowed to be too sparse, since a certain level of connectivity is required to enable sufficient alternate paths for use under failures. Note that this approach will increase the average number of IP hops from source to the destination, which increases the transit traffic in routers and drives to excessive use the router resources and transponders.

A third approach, depicted in Figure 14.15c, is a hybrid of the above two approaches. It uses a mix of high-speed links and low-speed links to address the different demands in the network. In practice a network may evolve to a hybrid model since they start out with lower-speed wavelengths and add higher-speed wavelengths when they become commercially viable; however, large carriers rarely continue to add the low-speed wavelengths once they have switched to the higher-speed ones, as this increases operational overheads (e.g. increasing the chance of human error, increasing inventories, etc.). This approach also requires accurate traffic forecasts, since getting the forecast wrong implies that the wrong hardware is installed in the network.

Table 14.2 Comparison of EON and fixed grid architectures.		
Architecture	**Pros**	**Cons**
(a) Many low-speed wavelengths	Low cost per transceiver, pay as you grow	Low spectral efficiency. Inefficient link bundling
(b) High-speed wavelengths only	Scalable	Low utilization → very high capex. Straining router scale
(c) Mix of high- and low-speed wavelenghts	Optimal cost and spectral efficiency	Complex design, inflexible to traffic changes, higher opex
(d) Electrical TDM layer (OTN)	Fine grain bandwidth allocation	Still high capex Very high opex
(e) Elastic DWDM	Good cost, spectral efficiency, and flexibility	Immature, non-standard Dynamic network challenges

Another common solution to this problem is to rely on high-speed wavelengths only—as in the second solution, but add a layer of time-division-multiplexed (TDM) cross-connects (most likely based on OTN standards) to provide sub-wavelength grooming as shown in Figure 14.15d. This architecture will tend to a similar number of high-speed wavelengths as in option b (for the same reasons), and will rely on OTN connections between client nodes instead of full wavelengths, as shown by the black links in Figure 14.15d. The optical layer cost of this solution will be higher than that of option b (but not significantly, since most of the cost is in transponders—not the OTN layer), but the cost of the client layer will be reduced compared to option b, due to the reduced amount of transit traffic at that layer. Traffic will still have to transit cross-connects, but they are assumed to be lower cost than the client gear. However, since the cost of the future network will be heavily skewed toward the transponders— even as compared to router ports—this solution will still have the same fundamental cost structure as option b. Beyond equipment cost considerations, this solution will be burdened by the added complexity of managing an additional TDM layer.

Compare these solutions to EON—depicted in Figure 14.15e, in which a single EOP is needed between the adjacent client nodes, right-sized based on the direct traffic between them. In this case, the average number of hops in the IP layer is reduced (ideally to a single hop), and transit traffic at the client layer is also reduced, driving down the required client resources compared to option b, as well as the required transponders compared to options b and d.

The comparison between these options is summarized in Table 14.2.

14.5 STANDARDS PROGRESS

Since the transition from current rigid networks to EON will be a significant leap forward, early initiatives by the standardization bodies will be indispensable. Clarifying what should be inherited from today's technology, what should be extended, and what should be created is imperative as the starting point for studying the possible extension of Ethernet, OTN, and ASON/WSON/GMPLS standards to support EON. In the following, potential study items and some candidate proposals from a standards viewpoint are presented.

14.5.1 DWDM network architecture

ITU-T Recommendation G.872 "Architecture of optical transport networks" specifies the functional architecture of DWDM from a network level viewpoint. G.872 defines an optical network layered structure that comprises an Optical Channel (OCh), Optical Multiplex Section (OMS), and Optical Transmission Section (OTS). Although the data rate, modulation format, and spectral width of an optical path in an EON may change according to the user demand and network conditions, an elastic optical path is naturally mapped into the OCh of the current OTN layered structure. Therefore, there will be no significant impact on the current G.872 when introducing the EON concept.

14.5.2 OTN mapping and multiplexing

The interfaces and mappings of OTN are specified in ITU-T Recommendation G.709 "Interfaces for the optical transport network (OTN)." The standard can accommodate various client signals and transport them over long distances. Originally, the OTN specified client signal mapping into ODUk ($k = 1, 2, 3$), which have bit rates of approximately 2.5 Gb/s, 10 Gb/s, and 40 Gb/s, and their multiplexing to ODUk with a higher bit rate if necessary. The multiplexed ODUk signal is then transported as an OTUk signal with a forward error correction (FEC) code. A new ODU was recently specified in G.709 called ODUflex, which can approximate any bit rate (down to 1.25 G granularity), to accommodate any client signal efficiently.

One possibility toward a rate-flexible OCh is to use the current standard of virtual concatenation (VCAT), where multiple ODUk(s) are concatenated to provide higher capacity and each ODUk is transported independently. The other possibility is to introduce rate-flexible OTUs (OTUflex) as well as rate-flexible HO ODUs (HO ODUflex).

14.5.3 Control plane: ASON, WSON, and GMPLS

The ITU-T Recommendations on automatically switched optical network (ASON) provide requirements, architecture, and protocol-neutral specifications for automatically switched optical networks with a distributed control plane. The goal is not to define new protocols but to provide mappings between abstract protocol specifications and the existing candidate protocols, and development of new protocols such as GMPLS has been tasked by the Internet Engineering Task Force (IETF) and Optical Internetworking Forum (OIF). More recently, extensions of the control plane that are specific to DWDM networks have been defined under the collective name of WSON.

The ASON network resource model is based on a generic functional model for transport networks defined in G.805, functional models for Synchronous Digital

Hierarchy (SDH) defined in G.803, and OTN defined in G.872. As examined in the previous subsection, an elastic optical path is naturally mapped into the OCh layer of the current OTN layered structure defined in G.872, which implies that there will be no significant impact on the current ASON standards when introducing a distributed control plane into EONs, although further studies are still necessary.

As for the technology-specific aspects of routing and signaling in elastic and adaptive optical networks, discussions on possible extension of GMPLS protocols in the IETF, the OIF, and the ITU-T are needed. A new switching type in GMPLS architecture may be introduced, representing "spectrum switching" capabilities. In order to establish a necessary frequency slot, a new label must be defined to represent a variable amount of spectrum. This can be based on center frequency and frequency slot width, or on lowest and highest frequency of frequency slot. The label will be used in various signaling constructs, such as the Upstream Label, Explicit Route Label, and Record Route Objects in the resource reservation protocol—traffic engineering (RSVP-TE) protocol. In addition, new parameters are needed in order to configure the symbol rate, number of sub-carriers, and modulation level in transponders at the endpoints of an optical path according to the required data rate and optical reach. This will likely be defined in RSVP-TE objects called "Sender TSpec" and "FlowSpec."

However, the standardization effort will have to extend to tighter coordination with the client layer, to enable the specification of client layer bandwidth needs and coordination between the layers during processes such as restoration and spectrum defragmentation.

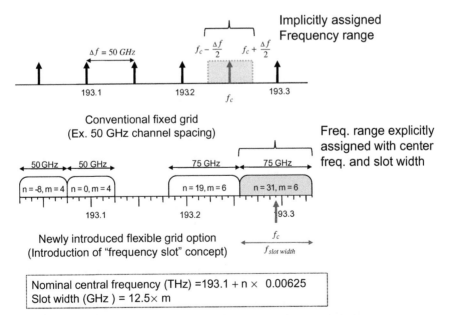

FIGURE 14.16 Introduction of frequency slot into ITU-T G.694.1 recommendation.

14.5.4 **Standardizing on Flexible Spectrum**

The current ITU-T frequency grid specified in G.694.1 "Spectral grids for WDM applications: DWDM frequency grid" is anchored to 193.1 THz, and supports various channel spacings of 12.5 GHz, 25 GHz, 50 GHz, and 100 GHz. In conventional optical networks, when a nominal central frequency is assigned to an optical channel, a fixed frequency range between plus and minus half the channel spacing from the central frequency is implicitly assigned to the channel. In order to utilize fully the spectrally efficient and scalable nature of EON, we need to consider some extension of G.694.1. Early standardization initiatives have also started, such as revisiting the notation of frequency resources assigned to an optical channel. Recently, ITU Study Group 15 introduced a flexible DWDM grid into its G.694.1 recommendation, where the allowed frequency slots have a nominal central frequency (THz) defined by $193.1 + n \times 0.00625$ (n is a positive or negative integer including 0), and a slot width defined by $12.5\,\text{GHz} \times m$ (m is a positive integer). Depending on the application, a subset of the possible slot widths and central frequency defined in the flexible DWDM grid can be selected (see Figure 14.16).

14.6 **SUMMARY**

We started this chapter by looking at the challenges the optical layer will face in the future, fueled by the insatiable appetite for more bandwidth, coupled with reduced ability to forecast and plan for such growth. A conclusion of this is that it will become increasingly expensive to continue to develop the optical network architecture along its current trajectory.

One of the first principles of traditional DWDM systems that must break is the fixed 50 GHz grid. Whether this will mean adoption of a fully flexible grid, or just a slightly more flexible grid, which supports a few fixed slot options, still remains to be seen, but the required FS-ROADM technologies for both options will enable significant flexibility.

Another principle that is breaking with next-generation transceivers is the fixed modulation format and bit rate of the transceiver. On one hand, it seems fairly easy to support multiple formats such as BPSK, QPSK, and 16QAM using common hardware and optics, while on the other hand, it seems necessary to provide flexible support for higher bandwidth at reduced reach and vice versa. Since such transceivers mostly use coherent technology, the use of superchannels seems natural and will help improve spectral efficiency—even if not at the same rate as we've experienced in the past.

A third principle that is no longer a given is the lack of control planes in the DWDM layer and more generally the lack of automation in this layer. The optical layer is changing from a fixed collection of point to point "pipes" with very limited ability to change (even when ROADMs enable such changes in theory), to a true network layer. Impairment-aware control plane technology is being productized even for non-elastic DWDM networks, but it will become almost essential for realizing the potential of the elastic network, as it will allow picking from the rich set of configuration knobs based on actual network conditions.

Elastic optical networking is not ready for prime time yet. There are still a lot of open questions to be figured out for EON to become a widely adopted technology. For example, the architecture of fully sliceable BVTs is still unclear. Control plane extensions for EON must be standardized and commercialized. One needs to better understand the dynamic behavior of the network, such as the allocation of spectrum under changing transmission conditions and defragmentation. EON will also greatly benefit from research and development of the relationship between the client layer and the optical layer. However, it seems quite clear that many of the pieces of the elastic network puzzle are becoming a reality.

Acknowledgments

This chapter is based in part on our paper [4]. We would like to acknowledge the other authors of this paper, Andrew Lord and Ben Yoo.

References

[1] Cisco, Visual network index (VNI), http://www.cisco.com/en/US/netsol/ns827/networking_solutions_sub_solution.html.

[2] Sandvine, Global Internet Phenomena Report, Fall 2011, http://www.sandvine.com/downloads/documents/10-26-2011_phenomena/Sandvine%20Global%20Internet%20Phenomena%20Report%20-%20Fall%202011.PDF.

[3] M. Jinno et al., Spectrum-efficient and scalable elastic optical path network: architecture, benefits, and enabling technologies, IEEE Commun. Mag. 47 (2009) 66–73.

[4] O. Gerstel et al., Elastic optical networking: a new dawn for the optical layer? IEEE Commun. Mag. 50 (2) (2012).

[5] T.A. Strasser, J.L. Wagener, Wavelength-selective switches for ROADM applications, IEEE J. Sel. Top. Quant. Electron. 16 (2010) 1150–1157.

[6] K. Roberts et al., 100G and beyond with digital coherent signal processing, IEEE Commun. Mag. 48 (July) (2010) 62–69.

[7] X. Liu, S. Chandrasekhar, High spectral-efficiency transmission techniques for systems beyond 100 Gb/s, Signal Processing in Photonic Communications, 2011, Paper # SPMA1.

[8] P. Frascella et al., Unrepeatered field transmission of 2 Tbit/s multi-banded coherent WDM over 124 km of installed SMF, Opt. Express 18 (2010) 24745–24752.

[9] W. Shieh et al., Coherent optical OFDM: theory and design, Opt. Express 16 (2008) 841–859.

[10] G. Gavioli et al., Investigation of the impact of ultra-narrow carrier spacing on the transmission of a 10-carrier 1 Tb/s superchannel, in: Proc. OFC 2010, March 21–25, 2010, Paper OThD3.

[11] D.J. Geisler et al., Bandwidth scalable, coherent transmitter based on the parallel synthesis of multiple spectral slices using optical arbitrary waveform generation, Opt. Express 19 (2011) 8242–8253.

[12] S. Gringeri, E.B. Basch, T.J. Xia, Technical considerations for supporting data rates beyond 100 Gb/s, IEEE Commun. Mag. 50 (2) (2012).

[13] M. Jinno, H. Takara, B. Kozicki, Y. Tsukishima, T. Yoshimatsu, T. Kobayashi, Y. Miyamoto, K. Yonenaga, A. Takada, O. Ishida, S. Matsuoka, Demonstration of novel spectrum-efficient elastic optical path network with per-channel variable capacity of 40 Gb/s to over 400 Gb/s, in: Proc. ECOC, 2008, Th3F6.

[14] T. Kobayashi, A. Sano, E. Yamada, E. Yoshida, Y. Miyamoto, Over 100 Gb/s electro-optically multiplexed OFDM for high-capacity optical transport network, IEEE J. Lightwave Technol. 27 (16) (2009) 3714–3720.

[15] B. Kozicki, H. Takara, Y. Tsukishima, T. Yoshimatsu, T. Kobayashi, K. Yonenaga, M. Jinno, Optical path aggregation for 1 Tb/s system in spectrum-sliced elastic optical path network, Photon. Tecnol. Lett. 22 (17) (2010) 1315–1317.

[16] B. Kozicki, H. Takara, T. Tanaka, Y. Sone, A. Hirano, K. Yonenaga, M. Jinno, Distance-adaptive path allocation in elastic optical path networks, IEICE Trans. Commun. E94-B (7) (2011) 1823–1830.

[17] N. Amaya, et al., Gridless optical networking field trial: flexible spectrum switching, defragmentation and transport of 10G/40G/100G/555G over 620-km field fiber, in: Proc. ECOC, Geneva, Switzerland, 2011.

[18] D.J. Geisler, et al., The first testbed demonstration of a flexible bandwidth network with a real-time adaptive control plane, in: ECOC 2011, Geneva, Switzerland, 2011.

[19] M. Jinno et al., Distance-adaptive spectrum resource allocation in spectrum-sliced elastic optical path network, IEEE Commun. Mag. (2010) 138–145.

[20] G.-H. Gho et al., Rate-adaptive coding for optical fiber transmission systems, J. Lightwave Technol. 29 (2011) 222–233.

[21] G. Bosco et al., On the performance of Nyquist-WDM terabit superchannels based on PM-BPSK, PM-QPSK, PM-8QAM or PM-16QAM subcarriers, J. Lightwave Technol. 29 (1) (2011) 53–61.

[22] O. Gerstel, Flexible use of spectrum and photonic grooming, in: Photonics in Switching 2010, PMD3.

[23] K. Christodoulopoulos, I. Tomkos, E.A. Varvarigos, Elastic bandwidth allocation in flexible OFDM-based optical networks, J. Lightwave technol. 29 (9) (2011) 1354–1366.

[24] T. Takagi, et al., Dynamic routing and frequency slot assignment for elastic optical path networks that adopt distance adaptive modulation, in: Proc. OFC/NFOEC, March 2011, OTuI7.

[25] C.T. Politi, V. Anagnostopoulos, C. Matrakidis, A. Stavdas, A. Lord, V. López, J.P. Fernández-Palacios, Dynamic operation of flexi-grid OFDM-based networks, in: Proc. OFC/NFOEC, March 2012.

[26] O. Gerstel, Realistic approaches to scaling the IP network using optics, in: Proc. OFC/NFOEC, March 2011.

ROADM-Node Architectures for Reconfigurable Photonic Networks

15

Sheryl L. Woodward[a], Mark D. Feuer[a], and Paparao Palacharla[b]

[a]*AT&T Labs-Research, 200 Laurel Ave, South Middletown, NJ, USA,*
[b]*Fujitsu Laboratories of America, Inc. 2801 Telecom Parkway, Richardson, TX, USA*

SUMMARY

The history of optical communications is often organized into generations, defined by technology innovations such as Wavelength-Division Multiplexing (WDM) and the Erbium-Doped-Fiber Amplifier (EDFA), which together enabled a rapid reduction in cost per bit and stimulated decades of explosive demand growth. Arguably the most epochal change was the commercial deployment of Reconfigurable Optical Add/Drop Multiplexers (ROADMs), which is gradually transforming an electronic network of optical "wires" into a highly interconnected, reconfigurable photonic mesh. Reconfigurable photonics have already contributed significantly to improved network efficiency, by enabling traffic to grow gracefully without a priori knowledge of future traffic demands, while minimizing the use of expensive optoelectronic regenerators.

To date, the widespread use of ROADMs has been driven by the cost savings and operational simplicity they provide to *quasi-static* networks (i.e. networks in which new connections are frequently set up, but rarely taken down). However, new applications exploiting the ROADMs' ability to *dynamically* reconfigure a photonic mesh network are now being investigated. In this chapter we review the attributes and limitations of today's ROADMs and other node hardware, and survey proposals for future improvements. We also discuss new features and applications being developed for reconfigurable networks, with emphasis on the needs of the backbone network of a major communications service provider (carrier). Finally, we assess which of these new developments are most likely to bring added value in the near-term and long-term future.

To place this work in context, the chapter begins with a brief description of how optical networks have progressed since their first deployments, and possible directions of the on-going evolution. In Section 15.2 we discuss various ROADM architectures and the elements that make up the ROADM node. After reviewing ROADM attributes widely accepted as necessities, we discuss newer features that may prove valuable in the future. Section 15.3 focuses on applications enabled by ROADMs with these enhanced features. We speculate on the future of ROADM-enabled networks in Section 15.4, and conclude in Section 15.5.

Optical Fiber Telecommunications VIB. http://dx.doi.org/10.1016/B978-0-12-396960-6.00015-8

15.1 **INTRODUCTION**

Optical communication systems have evolved dramatically, from simple point-to-point links to mesh networks. This evolution has been driven by the need to improve the efficiency (as measured in cost per bit) of the network. Photonic networks rely on ROADMs to provide mesh interconnection of nodes on a wavelength-by-wavelength basis. By enabling wavelengths to optically bypass nodes, ROADMs minimize the number of optoelectronic regenerators required in a network, without requiring dedicated fiber pairs between each terminal pair. They also reduce the operational expense and failure risk associated with manual setup of intermediate nodes. Because ROADMs are (by definition) reconfigurable, each wavelength can follow an individualized path, enabling networks to grow efficiently without perfect knowledge of future traffic demands. This improvement in efficiency has led to large-scale deployment of ROADMs in metro, regional, and long-haul optical networks [1].

However, the agility of today's networks is still limited—new circuits are added over time, but once provisioned, circuits are virtually permanent. Resources such as transponders and regenerators that have been deployed are dedicated to a particular circuit, and cannot be applied to new traffic demands without manual replacement of fiber jumpers. One great advantage of a truly dynamic photonic network is the efficient sharing of these scarce resources. This requires a node design that is reconfigurable from the DWDM transport, through the transponders, all the way to the client connections—an infrastructure that extends well beyond the ROADM itself [2]. Not only will the physical hardware need to include additional capabilities, but new management systems, modes of operation, and applications will all need to be developed. This will not happen overnight, but will require a gradual evolution during which networks become increasingly dynamic.

Although this chapter focuses on the physical layer, we note that the control plane must also be significantly enhanced to support dynamic services. References [3–7] provide overviews of multiple efforts in this area. While these all explore the challenges and benefits of creating a dynamic network, they have largely focused on control plane issues, and most utilize electronic switching, rather than switching at the ROADM layer ([5] and [6] are exceptions, addressing optical layer limitations in the algorithms used by the control plane). Arnaud et al. [7] address the benefits of developing a control plane that enables a private network (with either a single or small number of customers) to be reconfigurable. This effort has largely focused on the challenges of managing such a network, and does not address issues of scale which arise in a major carrier's backbone network. In [5] Doverspike and Yates focus on optical network management and control challenges faced by a major carrier in order to provide bandwidth on demand.

One ambitious vision of the future of dynamic photonic networks is DARPA's CORONET program, in which near-instantaneous bandwidth-on-demand services are provided at the wavelength layer. Although the CORONET program focused primarily on network design, operations, and management issues, including a three-way handshake protocol for fast path setup [8], major hardware developments would also be required to implement the vision. Some of these include millisecond-scale optical switching in the ROADM node, fast-tuning optical transponders (including fast frequency matching for coherent transponders), and optical amplifier chains that are tolerant to rapidly changing channel counts.

Other applications are closer to realization. For example, it has long been known that wavelength grooming in ROADM-enabled networks can maximize the throughput of the network while minimizing the amount of wavelength conversion required (thus minimizing the number of optoelectronic regenerators) [9]. However, today's manual wavelength grooming interrupts customer traffic, making it difficult to implement. It has recently been demonstrated in a testbed built with commercial equipment that by implementing "bridge and roll" at the photonic layer, wavelength circuits can be moved without disrupting higher layers [10]. This not only enables traffic to be groomed for improved network efficiency, it also allows circuits to be moved for scheduled maintenance, thus improving network availability. If networks with mixed channel spacing (flex-grid networks, see below) are deployed in the future, dynamic wavelength grooming will become essential, to avoid stranded bandwidth caused by the ebb and flow of unequal-width channels.

Rapidly reconfigurable photonic hardware could also have a major impact on network resilience. If the ROADM provides colorless, non-directional add/drop ports (cf. Section 15.2.3), wavelengths can be re-routed after a fiber cut, using the original optical transponder pair. Reconfigurability in the client-to-transponder connections (i.e. client-side crossconnect or C-XC) could enable 1:N protection against failures of transponders or client interface cards, bringing substantial cost savings compared to a network with 1:1 protection. Such a client-side crossconnect might be all-optical, all-electrical, or hybrid optical-electrical, but whatever option is chosen, it must scale gracefully from small to large port counts. Also, the C-XC implementation must include enough redundancy to assure that it does not detract from overall network availability.

Another dimension to network reconfiguration is flexible spectrum allocation (flex-grid), which would allow optical channels of unequal spectral width to coexist within a fiber [13–16]. Flex-grid has hardware implications for the ROADMs and the transponders. It also presents challenges/opportunities for the network as a whole, including network management, stranded bandwidth, and comprehensive system testing/verification across a vast number of possible configurations. Research in the coming years will clarify the conditions under which flex-grid's benefits justify the substantial cost of overcoming these challenges.

The future of photonic networks will be driven by the need to constantly improve the efficiency of our communication infrastructure. This will of course require that photonic and electronic resources be used as efficiently as possible, and may also push additional functionality into the photonic layer, to improve overall network efficiency. The latter will be driven by new applications which bring new requirements to the ROADM-node architecture. In Section 15.2 we discuss various features that nodes must, should, or may possess, and how these can be realized in hardware.

15.2 THE ROADM NODE

Whereas the term "ROADM" is commonly defined to include the components and subsystems needed to interconnect two or more WDM transmission fibers and a number of fiber add/drop ports [1], we will also discuss the broader system that includes transponders, regenerators, and the C-XC, an assemblage which we denote as a "ROADM node." The ROADM node connects to client-side interfaces on one side, and to inter-node fiber transmission lines on the other. A high-level view of a ROADM node is shown in Figure 15.1. In addition to the elements mentioned above, ROADM nodes are likely to have ancillary equipment such as node controllers, optical amplifiers, performance monitors, transceivers for an optical supervisory channel, etc.

15.2.1 Features—from necessities to luxuries

Decisions on ROADM node requirements will be driven by the network operator's needs and budget. For example, a major communications service provider (carrier) must aggregate connections to millions of customers at thousands of points of presence, while an enterprise network might comprise only a handful of locations. The former's network is likely to be hierarchical, heterogeneous, and large, emphasizing the importance of operations and management, while the latter may be more concerned with minimizing the number of fibers that must be leased. A large carrier will also have different requirements for different parts of the network—metro systems are more cost-sensitive, while ultra-long-haul systems place a higher premium on

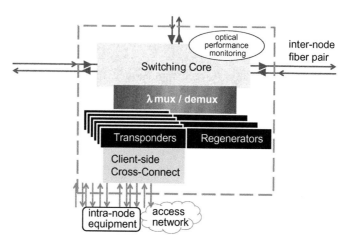

FIGURE 15.1 A block diagram of a 3-degree ROADM node. Note that it includes equipment beyond the ROADM, as all the equipment between the three inter-node fiber pairs and the ports for the client signals is included.

remote reconfigurability. In this section we enumerate many of the features which describe a ROADM node, and indicate when they are most needed.

- *Connectivity:* The basic function of a ROADM is to provide connectivity between all the inter-node fiber pairs, in addition to the node's transponders. A *full* ROADM is one that provides add/drop (de)multiplexing of any arbitrary combination of wavelengths supported by the system with no maximum, minimum, or grouping constraints. A *partial* ROADM only has access to a subset of the wavelengths, or the choice of the first wavelength introduces constraints on other wavelengths to be dropped. The "routing power" is a measure of the connectivity of a ROADM [1,11]. Full connectivity is preferred, though it is sometimes compromised in very cost-sensitive applications.
- *Hitless operation:* There is consensus within the industry that changing a ROADM's configuration (e.g. adding a new wavelength to the network) should not introduce errors on the wavelengths that are already in operation.
- *Directional separability:* A network element is said to be *directionally separable* if there is no single failure that will cause loss of add/drop service to any two of its line ports. This insures that a circuit's working and protection paths will not fail simultaneously. Separability implies not only that the initial failure affects only a limited fraction of the traffic, but also that it must be possible to repair the failed element without disturbing the protection paths.
- *High degree for span relief:* The fiber degree of a ROADM node is the number of inter-node fiber pairs served by the node. For high-capacity networks it is desirable to be able to "over-build" congested routes, to use multiple fiber pairs between nodes along those routes (this is sometimes called span relief). Therefore, even though most maps of nodes' physical locations indicate they are directly connected to no more than $\sim$4 other nodes, it is desirable for the ROADM node to be able to grow to a larger fiber degree so that the capacity of some of those routes can be upgraded.
- *Cascadeability:* Because ROADMs are designed to be able to separate wavelength channels from one another, they must contain wavelength-selective elements. When an optical signal passes through a cascade of ROADMs, the clear passband available to it decreases, inducing inter-symbol interference (ISI) that limits the number of ROADMs that a signal can pass through. This is one reason ultra-long-haul networks have ROADMs placed at large intervals, and regional or metro-networks are used to serve smaller nodes. The squareness of a ROADM's passband and the accuracy of its center wavelength both contribute to its maximum cascade limit.
- *Channel conditioning:* The wavelength-selective element within a ROADM can be used to perform channel equalization, thus improving the reach of long-haul transmission systems. Performance monitoring equipment within the ROADM node can be used to generate the feedback signal. Some

ROADM types may also be able to shape each channel's amplitude and phase transfer characteristic, though such capability is not yet used in deployed systems.

- *Scaleability:* A scaleable ROADM node architecture allows a node to grow, so that the cost of a large ROADM node can be spread over time, and its ultimate size need not be known at the beginning of life. A scaleable architecture also allows the same basic architecture to be used at both large and small ROADM locations, minimizing development and testing costs. Finally, a ROADM that can gracefully grow over time enables congested routes to be overbuilt for span relief, as discussed above. This effectively removes bottlenecks within the network.

- *100% add/drop capability:* A ROADM node is 100% add/drop capable if all wavelengths from all fiber degrees are able to be added/dropped at the node. For example, a ROADM node with D fiber degrees and W wavelengths per fiber is 100% add/drop capable if it can support $D \times W$ add/drop ports to access all the traffic at the node. Just as economic arguments make it desirable for a node to be scaleable, it is best if the node's add/drop capability can grow seamlessly (without any hits to existing traffic) from a small drop fraction to 100% as needed. Although some studies have estimated the average add/drop fraction to be 25%, actual networks show a wide variation in the add/drop fraction at different locations, so the ROADM node must be capable of supporting these divergent requirements.

All of the features listed above are readily available, and widely deployed in today's networks. The following features are either more recent developments, or still on the "wish-list" of features. These features are associated with more rapidly reconfigurable networks, either because they make such networks economically feasible, or because they become more practical once rapid reconfiguration is an accepted practice.

- *Colorless and non-directional[1] add/drop ports:* One key requirement of a dynamic network is that the transponders must be shared; otherwise the cost of underutilized transponders will make the network too expensive. This translates into the requirement that any unused transponder in a ROADM node must be able to serve almost any traffic demand. Therefore, it is desirable for the ROADM to be *colorless* (any given add/drop port is capable of handling any wavelength) and *non-directional* (any given add/drop port can connect to any inter-node fiber link via the switching core). Colorless and non-directional designs are discussed in more detail in Section 15.2.3. The desire to enable transponder sharing also motivates using a client-side cross-connect, which is discussed at more length in Section 15.2.4.

---

[1]"Non-directional" is also referred to as "steerable" [12] or "directionless" [14].

- *Spectral flexibility:* Today's wavelength-division multiplexed networks utilize a standardized grid, with equally spaced wavelength channels (50 GHz channel spacing is typical). Going to a network that utilizes a more flexible grid has been proposed [13]. Such a major change would have implications for the switching core, the add/drop section of the node, and the transponders, as well as network management software, routing protocols, etc. A flex-grid network not only places requirements on the optical components (the hardware implications are further discussed in Sections 15.2.2, 15.2.3, and 15.2.5), it also needs more sophisticated planning tools. Some researchers advocate flex-grid ROADMs as a means to ensure that today's systems can support higher rate channels (e.g. 400 Gbps or 1 Tbps channels) and thus be "future-proof" [14]. Some are motivated by the promise of improved spectral efficiency achieved in mixed-rate networks when channel spacing is optimized on a per-channel basis [13–16]. A blocking analysis of a dynamic network has shown tangible improvements, but demonstrated that unless additional features (such as inverse multiplexing) are incorporated, the fairness of the network is compromised [15]. Other simulations show roughly a 30% improvement in capacity; however, these simulations focused on the static network design problem [16]. This implicitly assumes that all the traffic forecasts are perfect, or that grooming is possible. The ability to groom wavelengths is even more crucial in a flex-grid network than in a fixed-grid network, since the evolution of the traffic demand pattern can create stranded bandwidth slices that are too narrow to support later-arriving demands. Note that if grooming is performed, the 30% improvement in network capacity is still far less than the growth enabled by span relief using a scaleable ROADM (of course, a flex-grid ROADM design should also be fully scaleable).
- *Flexible transponders:* In recent years software-defined transponders have been developed. Some have software-defined client protocols [17] so that the same transponder can support FibreChannel, SONET, or GbE customer interfaces, while others can adjust the modulation format and/or symbol rate employed on the line side so as to optimize reach or spectral efficiency [18,19]. This is discussed further in Section 15.2.5.
- *Switching speed:* To date, no network operator has placed a premium on switching speed, and so little effort has been made by system vendors to provide fast switching. While tens of seconds may seem fast today, many see advantages in reducing this to tens of milliseconds, as this would enable optical layer protection [20] and virtually instantaneous bandwidth-on-demand services [8]. An improvement of more than three orders of magnitude will not be done in a single step—system developers will address delays in the slowest facet of system design until they are no longer the limiting factor, and then attention will turn to other components and sub-systems. This is discussed further in Section 15.4.

15.2.2 Evolution of the switching core

At the heart of the ROADM is the switching core, where the devices that perform the wavelength switching reside. Figure 15.2 shows two examples of early ROADM cores—both are designed for linear transmission systems where the nodes have a fiber degree of two (i.e. they serve two inter-node fiber pairs, though only one direction for each fiber pair is shown in the figure). In the first example, all wavelengths are demultiplexed, then each 2×2 optical switch is set so that its associated wavelength channel either terminates at the node, or optically bypasses the node. Wavelength signals are added using the same 2×2 switches, and a multiplexer combines the signal for transmission to the next node. These components can be integrated in a planar lightwave circuit (PLC). Cascadeability of these nodes is limited by crosstalk associated with finite extinction ratio of the switches, multiplexers, and demultiplexers, as well as by signal distortion (ISI) due to passband narrowing. Nonetheless, this design was widely deployed in early ring-based ROADM systems. For national-scale, mesh-based backbone networks, the design pictured in Figure 15.2b was the early favorite.

The second ROADM core (Figure 15.2b) employs a wavelength blocker. Wavelength blockers efficiently handle 80–100 wavelength channels, and have a passband that is more square than that found in most planar waveguide demultiplexers. In this core the wavelength channels are sent (via an optical power splitter) to both a demultiplexer and the wavelength blocker. This blocker either allows wavelengths to express through the node, or blocks wavelengths that terminate at the node. Wavelength channels can be added using a power combiner located after the wavelength blocker. Wavelength blockers based on liquid crystal technology have been widely deployed. They not only have improved cascadeability due to their broad passband,

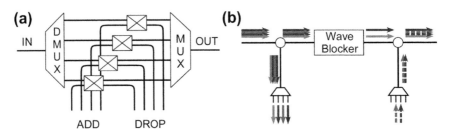

FIGURE 15.2 Two early ROADM core designs. (a) An array of 2×2 switches is nested between arrayed waveguide gratings serving as demultiplexer and multiplexer. In this design, the switching core and the mux/demux section are integrated. This ROADM type has been used for optical bypass in linear and ring systems. (b) A power-splitter transmits the incoming signals to both the demultiplexer and to a wave blocker; wavelengths to be terminated at the node are blocked, so that an added signal can reuse that wavelength channel without interference. The wavelength blocker also provides variable attenuation, to equalize the power in each wavelength channel. The channel mux and demux are shown here for clarity; they are not integral to the switching core.

they also serve as channel power equalizers, as they can attenuate each channel independently.

Neither of these designs scales well to multi-degree ROADMs (nodes with fiber degree >2), which are central to photonic mesh networks.

Figure 15.3 depicts the core of a multi-degree ROADM based on wavelength-selective switches (WSS). Unlike wavelength blockers, WSS enable the ROADM to smoothly scale to high degree. This ROADM utilizes a broadcast and select architecture—incoming signals are split and transmitted to each WSS serving an out-bound inter-node fiber. The $N \times 1$ WSS then selects the appropriate wavelength signals to be transmitted to the next node. In large degree nodes the optical power splitters may be replaced with $1 \times N$ WSS in a route and select topology—the $1 \times N$ WSS has lower optical loss than a large optical power splitter, and also improves the optical isolation. (Note that the route and combine architecture, with WSS at the input and power combiners at the output, is not directionally separable under certain WSS failure modes.)

In a flex-grid ROADM each wavelength-selective element must meet additional specifications. Such a ROADM does not use a fixed wavelength grid of equally spaced wavelength channels, but can adapt a wavelength channel's allocated bandwidth to meet demand. This can be achieved by having wavelength-selective elements that can slice spectrum on a finer grid (e.g. 12.5 GHz) and which do not have a lossy notch between slices. For the 12.5 GHz example, adjacent bandwidth slices can be merged to form a 25 GHz, 37.5 GHz, or wider channel. However, the 37.5 GHz channel will not have the same center frequency as the 25 GHz channel, so the system's transmit

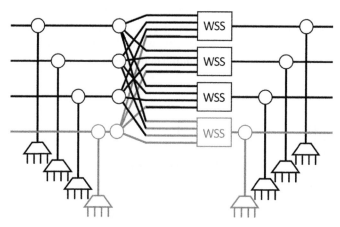

FIGURE 15.3 $N \times 1$ wavelength-selective switches (WSS) are the foundation of today's multi-degree ROADMs (M-ROADMs). A typical ROADM has full interconnection of up to $N + 1$ inter-node fiber pairs, and is in-service upgradeable. This ROADM utilizes power splitters and WSS in a broadcast and select topology. With fixed demultiplexers and multiplexers deployed before and after each WSS, each transponder is locked to a specific wavelength channel and a specific direction; the node is neither colorless nor non-directional.

lasers would need to be settable at 6.25 GHz resolution. Although some WSS already support flex-grid, the feature is not actually used in today's networks.

15.2.3 The mux/demux section of the ROADM node

The mux/demux section of the ROADM node is used to combine/separate individual wavelength channels transmitted from/to the transponders. Most ROADM deployments to date have used arrayed waveguide gratings (AWGs) as the mux/demux connected to each fiber degree (see Figure 15.3) to provide fixed wavelength access at each add/drop port. Therefore, these ROADM nodes are colored and directional.

A colorless ROADM has add/drop ports capable of handling any wavelength, and three design options for the demultiplexer section of such a colorless ROADM node are shown in Figure 15.4. These design options are also suitable for the multiplexer section of the ROADM nodes (e.g. reverse the arrows in Figure 15.4). The first option is to use an optical power splitter (PS) followed by large port count WSSs (e.g. $1 \times 20/1 \times 23$) to provide adjustable wavelength filtering. The second option is to use an AWG with an optical fiber cross-connect (e.g. for an 80 wavelength system, an 80×80 optical fiber cross-connect would be needed) to provide colorless capability. The third option is to use a power splitter with optical tunable filters or coherent receivers to select wavelengths. The coherent receivers allow electronic filtering instead of optical filtering by tuning the local oscillator to the required wavelength [21]. Depending on the coherent receiver implementation, the impact of the power levels and number of channels on filtering performance needs to be carefully considered; similar factors must also be taken into account on the transmit side to avoid optical crosstalk. Optical tunable filters are needed for use with direct-detection receivers (e.g. 10 Gbps NRZ, 40 Gbps DPSK). The first and third design options allow modular growth of add/drop capacity, while the second option needs the full multiplexer and fiber cross-connect at initial deployment. In addition, the first design

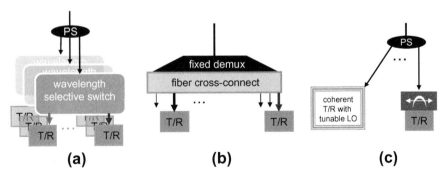

FIGURE 15.4 Three configurations for the demultiplexer section of the ROADM node—
(a) demux using WSS, (b) fixed AWG and fiber cross-connect, (c) power splitter with coherent receiver or tunable filter.

option with flex-grid WSS and the third design option are applicable for flex-grid colorless ROADM nodes.

An add or drop port is said to be non-directional if it can be set to address any of the line ports (e.g. East or West for a 2-degree ROADM). The ROADM node shown in Figure 15.5 is non-directional, since any transponder can serve any direction (inter-node fiber pair). This non-directional capability is provided by the PSs and WSS connected to the colorless mux/demux blocks. Each add/drop WSS selects the desired wavelength signals from among the incoming wavelengths on each inter-node fiber and passes the selected signals to the colorless demux. Meanwhile, the PS receives the wavelengths from each colorless mux and distributes them to all of the transit WSSs for switching to an out-bound inter-node fiber. The colorless mux/demux blocks, based on one or a mix of the design options in Figure 15.4, organize the transponders into *banks*, each of which is connected to the core by a single fiber pair. This ROADM node architecture offers smooth in-service growth of the number of fiber degrees as well as the add/drop capacity. Its primary limitation is intra-node wavelength contention: it is possible to add/drop at most a single instance of any given wavelength channel to transponders in a given mux/demux bank.

A contentionless ROADM is a *full* ROADM that never blocks traffic demands within its add/drop structure (an arbitrary add/drop port can serve a valid demand for any degree at any wavelength regardless of what other demands are already present).

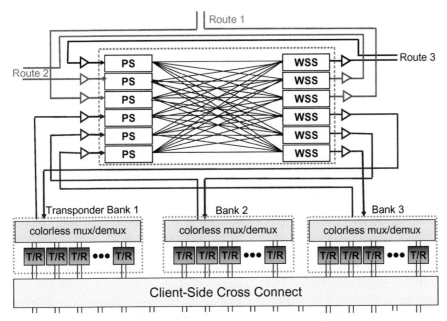

FIGURE 15.5 Architecture of a colorless, non-directional ROADM node with a client-side cross-connect [2].

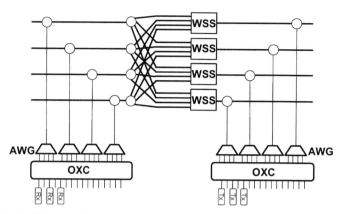

FIGURE 15.6 Architecture of a colorless, non-directional, contentionless ROADM node using large fiber cross-connects.

Two types of colorless, non-directional, contentionless (CNC) ROADM architectures have been proposed: (i) the conventional ROADM architecture of Figure 15.3 with the addition of a large fiber cross-connect that interconnects all of the transmitters/receivers and all of the multiplexer/demultiplexer ports, and (ii) a WSS-based ROADM core with a mux/demux section using $M \times N$ multicast switches or $M \times N$ WSSs. The first architecture, shown in Figure 15.6, uses the colored, directional ROADM architecture of Figure 15.3 with a large fiber cross-connect (e.g. 704×704 ports for 8-degree with 88 wavelengths) to provide non-blocking connectivity between transponders and inter-node fiber pairs. This architecture is not modular since the large cross-connect is required from day one. In addition, this cross-connect represents a single point of failure, thus requiring fully redundant hardware, which is problematic for both cost and optical loss. Recently, variations of this architecture to provide modular growth were proposed [22].

The second architecture is shown in Figure 15.7 using $M \times N$ multicast switches (MCS). The express path of the ROADM is based on a route and select architecture with large port count (1×24) WSSs on the ingress and egress ports. Each $M \times N$ MCS in the mux/demux section of the ROADM is a (non-wavelength-selective) broadcast and select switch with M $1 \times N$ splitters mesh connected to N $M \times 1$ switches [23]. Following the path through the demultiplexer, each inter-node fiber pair (up to M fiber degrees) is connected to one of the M input ports. A MCS is non-directional because each of the N output ports can select the signals from any direction. Each of the N output ports can be connected to a coherent receiver or tunable filter (e.g. third design option in Figure 15.4) to create a colorless drop port. A MCS is also contentionless since it is capable of dropping the same wavelength from different directions to different ports. A typical MCS of 8×16 using planar lightwave circuit (PLC) or MEMS technology is currently available. Scaling the MCS architecture to obtain 100% add/drop for an 8-degree node is difficult. For example, the design

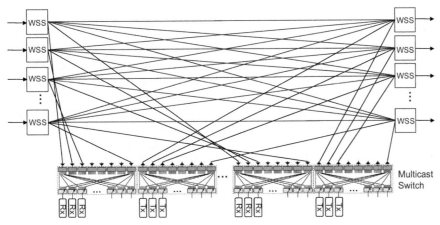

FIGURE 15.7 Architecture of a colorless, non-directional, contentionless ROADM node using M × N multicast switches.

of Figure 15.7 would require 88 (=2 × 704/16) modules, as well as 1 × 52 (52 × 1) WSS at the ingress (egress). The WSS degree could be reduced if additional power splitters were introduced between the WSS and MCS ports, but that would add to the already high loss of the add/drop paths. In general, many amplifiers (such as EDFA arrays) are needed to accommodate the high loss of the MCS-based mux/demux, contributing to the cost, power dissipation, and failure rate of the ROADM. Overall, this contentionless ROADM architecture employs a larger number of EDFAs and WSS ports than the architecture shown in Figure 15.5, resulting in higher cost as well as space and power issues. It may be most suitable for specialized networks in which a low add/drop fraction can be guaranteed. An alternative CNC ROADM architecture using M × N WSS instead of M × N MCS has been proposed [24]. The optical loss of M × N WSS could be lower than M × N MCS, but the M × N WSS is not yet commercially available. Simulations exploring the value of deploying contentionless ROADMs are covered in Section 15.3.

15.2.4 Client-side switching

Transponders represent a significant cost in long-haul optical systems, so it is important that they be efficiently utilized. In a dynamic network this drives a requirement that they be "shareable." In order for transponders to be shared among multiple clients, there must be some switching on the client-side of the transponder. In addition to enabling the statistical multiplexing of transponders, having a C-XC also enables 1:N protection of optical transponder units and client interface cards. This can improve the reliability of the network at a cost much lower than for 1:1 protection, provided the C-XC is sufficiently reliable.

Figure 15.5 depicts the C-XC as a single large device, and no design details are provided. Just as the ROADM needs to be reliable, hitless, and scaleable, so does the C-XC. A fully connected, scaleable cross-connect can be built using a Clos architecture, but that approach requires a significant multiplication of the total number of switch ports consumed. Alternatively, a modular design utilizing smaller C-XC arranged in parallel has been studied. Simulations on the blocking performance demonstrate that even with partitioning limiting the connectivity of the C-XC, the impact on the blocking performance in a dynamic optical network is small [25,26]. These studies assumed that the C-XC modules were fiber-cross-connects, so that the granularity of the switch equaled the full payload of a transponder.

Note that the C-XC can be either optical or electronic; while fiber-cross-connects have lower port costs, electronic switching (e.g. an OTN switch) can provide sub-wavelength grooming and aggregation. Therefore the most favorable solution may prove to be a balance of the two, a fiber-cross-connect for moving full wavelength channels, and an electronic switch used to handle sub-wavelength traffic demands [27,28]. Optimizing the architecture of the C-XC will be challenging as it will need to provide a balance of cost, reliability, and functionality.

15.2.5 Flexible transponders

In most optical transponders both the client and line-side interfaces are hardware defined, and cannot be changed. The client-side interface will be a well-established standard interface (e.g. 10 GbE) while the line-side interface may employ proprietary forward-error correction, and customized modulation formats. In the past decade more flexible transponders have become available. Some have software-defined client ports [17] so that the same transponder can support FibreChannel, SONET, or GbE customer interfaces. In a static network such flexibility greatly simplifies ordering and sparing, while in a dynamic network it improves the efficiency of sharing transponders, as the transponders are not partitioned into multiple pools.

Transponders which adjust the modulation format employed on the line side so as to optimize reach or spectral efficiency have been demonstrated [18,19]. Such transponders utilize more spectrally efficient modulation formats when long reach is not needed, but employ more robust formats when the circuit must traverse a long distance, and spectrum is available. To do this, not only must the ASICs within the transponder support both formats, but the optical components must meet the specifications of both. This implies that the lasers must have the narrow linewidths required of high-order modulation formats, while the modulators must have sufficient bandwidth to support the less spectrally efficient formats. For lower-rate transponders this might not be challenging, but for transponders operating at state-of-the-art rates this may have significant implications. The added cost of employing better components may be offset by savings due to larger production volumes.

15.3 NETWORK APPLICATIONS: STUDIES AND DEMONSTRATIONS

The widespread deployment of colorless, non-directional ROADMs will be a significant step toward enhancing our ability to exploit reconfigurable photonic networks. Such ROADMs immediately enable carriers to improve operational efficiency, and the addition of a client-side cross-connect will further improve the flexibility of the photonic layer. Eventually, the improved flexibility provided by an advanced ROADM layer will enable new services, such as bandwidth on demand. Some features, such as colorless, non-directional ROADMs, offer significant operational advantages and their adoption is imminent, while others, such as contentionless ROADMs, may not offer enough value to justify their cost.

In this section we provide some concrete examples of how the capabilities of advanced ROADMs can be utilized. We also point out some of the hurdles that must be overcome before these applications can be widely implemented. Expected performance is examined either through simulations [29–35] or testbed demonstrations [10,28,36]. The majority of the simulation results presented below are based on the CORONET CONUS network topology, shown in Figure 15.8 [8,37], and the node architecture shown in Figure 15.5 [2].

Because this node architecture is not contentionless, the first simulations we present compare the performance of this ROADM with an ideal colorless, non-directional, contentionless (CNC) ROADM. We follow this with results on the performance of various applications. We chose these applications because each helps build the case for investment in an increasingly dynamic photonic layer.

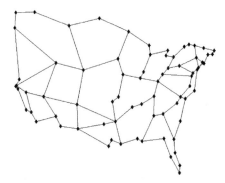

FIGURE 15.8 The continental United States (CONUS) portion of the CORONET topology, with 75 nodes and 99 links. Although this topology is loosely based on existing commercial networks, it does not reproduce any actual network in use today.

15.3.1 **CN-ROADMs and CNC-ROADMs in dynamic optical networks**

CN-ROADM nodes provide full interconnection for transit traffic among all fiber degrees and full interconnection between fiber degrees and add/drop structures grouped in transponder banks (Figure 15.5). Each transponder bank connects to the ROADM core through a single fiber pair, so a bank can only support a single instance of a particular wavelength. This restriction can cause intra-node contention when two lightpaths request the same wavelength from transponders located in the same transponder bank. To quantify the impact of such intra-node contention, the blocking rate was estimated using Monte Carlo simulations of dynamic traffic in a continental-scale network (CORONET CONUS topology shown in Figure 15.8) [29]. In these simulations, the lightpath requests for each source-destination pair were generated using a Poisson arrival distribution, with hold times according to a Gaussian distribution. Three types of ROADM nodes were considered: (i) CN-ROADM node without C-XC, (ii) CN-ROADM node with C-XC, and (iii) CNC-ROADM node. For each CN-ROADM node, the number of transponder banks was equal to the number of inter-node fiber pairs. Two routing and wavelength assignment (RWA) schemes were simulated: simple RWA (S-RWA) and Enhanced RWA (E-RWA). S-RWA assigns the first-fit wavelength for a route without any consideration of intra-node contention, while the E-RWA scheme selects a route and assigns first-fit wavelength with awareness of the intra-node contention at source and destination nodes [29]. Figure 15.9 shows the blocking rate versus offered load. The blocking rate for CN-ROADM nodes using S-RWA is high, due to the intra-node contention, but by using E-RWA the blocking rate approaches the performance of CNC-ROADM nodes. Similar improvement in blocking rate was shown for CN-ROADM nodes with C-XCs using S-RWA because the C-XC allows flexible assignment of transponders. These results demonstrated that the impact of intra-node contention in CN-ROADMs on blocking performance in dynamic networks is negligible with enhanced hardware (C-XC) or enhanced software (E-RWA). Thus, this study demonstrated that CN-ROADMs can be efficiently used while CNC-ROADMs may not add sufficient value in dynamic optical networks to justify their higher complexity

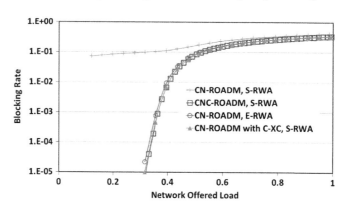

FIGURE 15.9 Simulation results comparing the blocking performance of CN-ROADMs and CNC-ROADMs in dynamic optical networks.

and cost. Other recent studies of contention-aware RWA schemes confirm that blocking rates and port utilizations of CN-ROADM networks can be very close to those of networks with contentionless nodes [30–32]. For some topologies and traffic patterns, excellent network performance can be achieved with fewer transponder banks than inter-node fiber pairs [31,32]. In [31], an E-RWA method has been implemented with only minor changes to existing distributed GMPLS/RSVP control plane protocols.

15.3.2 Predeployment of regenerators for faster provisioning and lower MTTR

One of the most attractive applications enabled by deploying advanced ROADMs is to reduce the time it takes to provision new circuits in the network by pre-deploying regenerators. Such pre-deployment of optoelectronic equipment has never been practical before, because CN-ROADMs are strictly necessary to achieve reasonable utilization of the pre-deployed resources. This rapid service provisioning mode is referred to as service velocity, where new connections are remotely provisioned without requiring visits (e.g. truck rolls) and manual operations (e.g. changing fiber jumpers) at any intermediate nodes [33]. Using Monte Carlo simulations, we have studied the performance of CN-ROADM networks in terms of number of active and idle regenerators deployed for traffic growth. Each regenerator connects to two distinct transponder banks in CN-ROADM nodes of Figure 15.5 to allow same-wavelength regeneration. We assumed traffic is quasi-static, in which demands arrive randomly but persist indefinitely. Two network scenarios were studied—(i) all nodes are regenerator sites, and (ii) a limited subset of nodes are regenerator sites (see [38] for details on regenerator site selection). Figure 15.10 shows

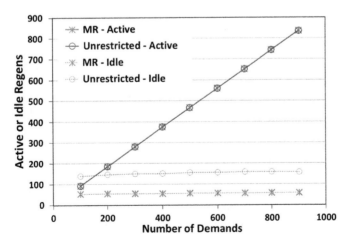

FIGURE 15.10 Simulation results of active and idle regenerators in the **CORONET CONUS** network. Two scenarios were studied. The first allowed regenerators to be placed at any node ("unrestricted"). In the second scenario, regenerator placement was restricted to 22 nodes, carefully selected to enable minimum-regenerator routing for each source-destination node pair ("MR").

the simulation results for the number of active regenerators and the number of idle regenerators deployed in the CORONET CONUS network (Figure 15.8), for systems with 2000 km optical reach [34]. The first network scenario is labeled as unrestricted in Figure 15.10 since all 75 nodes are used as regenerator sites. For the second scenario, 22 regenerator sites were selected to achieve minimum-regenerator (MR) routing for each source-destination node pair. Although the count of active regenerators grows linearly with the traffic, the count of idle regenerators remains nearly constant for both network scenarios. The second scenario (labeled as MR) has the same number of active regenerators as the first scenario, but a much lower number of idle regenerators, offering substantial cost savings. These results show that pre-deployment of regenerators can be a cost-effective way to provide high-velocity wavelength services in CN-ROADM networks [33,34].

Another benefit of service velocity is that it enables 1:N sparing of regenerators, thereby reducing the mean-time-to-repair (MTTR). When a regenerator fails, the service connection can be automatically restored by using a pre-deployed regenerator at the same location, or by establishing a new path through an unused regenerator at a different location.

15.3.3 Wavelength grooming and traffic re-routing

Wavelength grooming (also referred to as network defragmentation) is another application that we believe will be rapidly adopted. With colorless, non-directional ROADMs, circuits can be re-routed, and/or tuned to another wavelength channel, without needing to move fiber jumpers. On protected circuits, this can be done by switching to the protection circuit, reconfiguring the primary circuit, and then switching back to the primary circuit.

Grooming unprotected wavelengths is more challenging, as service interruption must be short enough that it does not disrupt higher network layers (such as the IP-Layer). SONET has a 50-ms restoration requirement, so higher layer networks typically will not respond to an outage with a duration below that. With today's ROADMs and transponders, such short outage times are best accomplished by a bridge and roll operation, wherein a second wavelength circuit is established (the bridge) before the traffic is rolled to that circuit. A client-side cross-connect may be used to implement such a bridge-and-roll operation, enabling the customer's received traffic to quickly switch from the original wavelength channel to the new one. We have assembled a GRIPhoN (Globally Reconfigurable Intelligent Photonic Network) testbed and demonstrated this wavelength bridge and roll application [10]. The outage due to the final switch from the original circuit to the bridge took under 50 ms—fast enough to prevent higher layers from reacting to it. This bridge and roll capability would also be useful in rerouting traffic for planned route maintenance.

15.3.4 Automated wavelength restoration

The GRIPhoN testbed has also been used to demonstrate automatic wavelength restoration on a time scale far faster than manual repair [36]. In this demonstration a centralized GRIPhoN controller used standard alarms to detect a cable cut, correlated this information with all impacted traffic, and dynamically re-provisioned each wavelength connection using the existing optical transponders. By reusing the same transponders, this promises to be much more cost-effective than $1+1$ optical protection. The average outage time was under 3 min—orders of magnitude faster than cable repair. This time could be further reduced by streamlining the element management software (the software used was not designed for this application). Once the original cable has been repaired, a spare transponder can be used to perform "bridge and roll" operations to move the traffic back to its original route. Because bridge and roll operations can be performed sequentially, this traffic normalization requires only a single spare transponder at each terminal, rather than a dedicated spare for each wavelength connection. While dynamic wavelength restoration is slower than $1+1$ protection, it should also be far less expensive, and therefore be an attractive alternative for many customers.

15.3.5 Bandwidth on demand

The previously mentioned applications are all focused on improving the efficiency with which the network can deliver today's services. However, advanced ROADMs offer the promise of enabling new services, such as wavelength-on-demand. Bandwidth-on-demand services are already offered to large customers, but they are limited to lower rates, and provided via electronic switching [39]. With an advanced ROADM, such services can be offered at higher rates, as network capacity can be dynamically reconfigured [28].

15.4 TWO COMPATIBLE VISIONS OF THE FUTURE

15.4.1 Vision 1: highly dynamic network

This first vision foresees a future in which optical networks have gradually evolved to become highly dynamic. Since this evolution is unlikely to reduce the cost of building and operating the optical layer, it must be justified by enhancing the optical layer's capabilities in a way that provides cost savings in other layers or enables new services that increase revenue.

For example, with a highly dynamic optical layer it is possible to integrate restoration across the IP and optical layers. This has been explored as part of DARPA's CORONET program, which investigated the development of protocols and network management to enable optical connections to be established in roughly 50 ms. Such fast connection times enable the integration of IP and optical restoration. An economic study revealed that although the integration results in an increase in the required wavelength-miles in the network, it can reduce the overall cost by ~20% as compared with a restoration strategy that performs all IP restoration within the IP layer [35]. A significant fraction of the savings come from the efficient reuse of transponders; the transponders used to support a service connection can be reused for restoration of that connection. This is only possible because the ROADM-node architecture is both colorless and non-directional, so that transponders can serve an alternate path. It also relies on the presence of a client-side cross-connect between the routers and the transponders, so that transponders can provide alternate connections in the event of a router failure.

Another capability provided by a dynamic optical layer is the ability to provide bandwidth on demand at the optical layer. Research consortia clearly desire the ability to share large datasets [7,40] while a commercial enterprise might incorporate such a service into its disaster recovery strategy. To provide bandwidth on demand at a cost lower than that of dedicated circuits, the network must be able to share resources among multiple clients. This service will become more cost-effective as the number of clients increases, which should improve the utilization of the shared resources. Efficient sharing is only possible with a fully reconfigurable network, which includes a client-side cross-connect. Utilization will also improve as the idle time of resources is minimized—an argument in favor of rapid connection set-up times.

Realization of this futuristic dynamic optical layer depends on overcoming a number of technical hurdles. It will be important that each step is economically justified, and that the solutions employed do not introduce future roadblocks.

The first step will be the development of streamlined management systems, so that they do not limit the speed with which new circuits can be provisioned. This is not a fundamental limitation, but will require that resources be devoted to minimizing unnecessary delays. Once the management software is no longer the limiting factor, then the delays in other components and subsystems will become apparent.

One area that is of significant concern is the response time of transmission lines' control loops (used to control the gain of the optical amplifiers and the power in each wavelength channel). The sudden addition or deletion of multiple channels is known to cause power transients that can build up along a transmission path. Since the sudden deletion of multiple wavelength channels (e.g. due to a cable cut) is already a recognized concern for network operators, builders of optical transmission systems are already tackling the problem of fast transient control to handle the simultaneous loss of multiple optical signals. Various solutions have been developed, and some are publicly disclosed [41], but whether the current solutions will also work when many channels are suddenly *added* has yet to be studied.

The tuning speed of the transmitting laser (and the local oscillator, when coherent detection is employed) is another significant source of delay. While fast tunable lasers have been demonstrated, combining fast tuning with narrow linewidth requirements remains challenging.

Although not currently a limiting factor, the speed of the wavelength selective components within the mux/demux and ROADM core also needs to be considered. While the individual switches within a MEMS or liquid crystal based WSS are adequately fast, the ROADM must be designed so that wavelengths can be switched in parallel, rather than having to sequentially address wavelengths (if ~100 wavelengths need to be moved, then sequentially addressing wavelengths can adversely affect switching times).

15.4.2 Vision 2: space-division multiplexed systems

As researchers ponder ways to extend cost-per-bit reduction into the Terabit era, space-division multiplexing (SDM) has emerged as a leading candidate for the next technology revolution [42–45]. SDM boosts the maximum capacity of a single fiber by using multiple modes in multimode fiber (MMF) [46–49] or multiple cores in multicore fiber (MCF) [45,50–52] to carry independent data streams or subchannels on a single wavelength. To sustain the advantages of a reconfigurable network, such SDM systems will need a new generation of ROADMs. Potentially, SDM ROADMs could independently route each wavelength and each spatial mode, improving the granularity of each fiber multiplex section. However, in practice, fiber transport is likely to induce some mode mixing, especially for MMF options. Such mode mixing can be unraveled by MIMO techniques in the receiver, but only if all of the modes reach the receiver. Although one can imagine placing mode detanglers at each ROADM node, in analogy to dispersion-compensation in non-coherent long-haul systems, no such mode detangler has yet been demonstrated.

As a more practical alternative, the use of spatial superchannels, in which all of the modes at a given wavelength make up a single data channel that is routed together at each ROADM node, has been proposed. Besides avoiding the need for mode detanglers, spatial superchannels could facilitate cost-reduced integrated transmitters with shared lasers and cost-reduced receivers with lower DSP complexity [52]. ROADMs suitable for spatial superchannels could be implemented with WSSs that have the same number of addressable elements as current WSSs, though a redesign of the imaging optics would be required. Considering the importance of photonic-layer reconfiguration in modern networks, we expect advanced ROADMs to become an important part of SDM research.

15.5 CONCLUSIONS

In this chapter we have outlined how today's ROADM-enabled backbone networks can evolve into rapidly reconfigurable networks capable of providing highly dynamic services. ROADMs in future networks will need to maintain the characteristics that we have come to depend on: they must scale gracefully to large or small nodes, sustain signal quality through many cascaded nodes, enhance network availability, and support high fiber degree for span relief as traffic grows. The first, critical step will be the deployment of colorless, non-directional ROADMs that optimize utilization of both electronic and optical resources. Additional advanced features, such as contentionless architectures and flex-grid designs, will follow when (and if) they can justify the costs associated with their introduction. Flex-grid, in particular, goes far beyond the ROADM hardware, implying a complete ecosystem of flexible transponders and regenerators, automated wavelength and bandwidth grooming, and new algorithms for network routing and operations. To complete the dynamic photonic layer, a client-side cross-connect (C-XC) will extend reconfigurability down to the subtending clients, enabling efficient utilization and protection of transponders and client interfaces.

We have also considered some proposed applications of rapidly reconfigurable networks. Rapid provisioning schemes, such as service velocity, and automated wavelength restoration both depend on pre-deployment of optoelectronic equipment that is not dedicated to a single customer; as such, they represent a major departure from prior practice. We have reviewed simulations and lab demonstrations that point to the most promising directions. Finally, although the focus of this chapter has been on the hardware, development of the network control plane must happen at the same time. Driven by the need for networks that are bigger, faster, smarter, and more efficient, we expect to see continued advancement in reconfigurable photonic networks in the years to come.

Acknowledgments

We are indebted to our colleagues at AT&T and Fujitsu for years of fruitful collaboration.

References

[1] M.D. Feuer, D.C. Kilper, S.L. Woodward, ROADMs and their system applications, in: I.P. Kaminow, T. Li, A. Willner (Eds.), Optical Fiber Telecommunications VB, Academic Press, 2008 (Chapter 8).

[2] S.L. Woodward, M.D. Feuer, J.L. Jackel, A. Agarwal, Massively-scaleable highly-dynamic optical node design, OFC/NFOEC 2010, Paper JThA18.

[3] A. Jajszczyk, Automatically switched optical networks: benefits and requirements, IEEE Commun. Mag. 43 (2) (2005) S10–S15.

[4] X. Zheng, M. Veeraraghavan, N.S.V. Rao, Q. Wu, M. Zhu, CHEETAH: circuit-switched high-speed end-to-end transport architecture testbed, IEEE Commun. Mag. 43 (8) (2005) S11–S17.

[5] R.D. Doverspike, J. Yates, Optical network management and control, Proc. IEEE 100 (5) (2012) 1092–1104.

[6] T. Lehman, J. Sobieski, B. Jabbari, DRAGON: a framework for service provisioning in heterogeneous grid networks, IEEE Commun. Mag. 44 (3) (2006) 84–90.

[7] B.S. Arnaud, J. Wu, B. Kalali, Customer-controlled and -managed optical networks, J. Lightwave Technol. 21 (11) (2003) 2804–2810.

[8] A.L. Chiu et al., Architectures and protocols for capacity-efficient, highly-dynamic and highly-resilient core networks, IEEE J. Opt. Commun. Netw. 4 (1) (2012) 1–14.

[9] G.R. Hill, I. Hawker, P.J. Chidgey, Applications of wavelength routing in a core telecommunication network, in: International Conference on Integrated Broadband Services and Network 1990, 15–18 October, 1990, pp. 63–67, URL: <http://ieeexplore.ieee.org/stamp/stamp.jsp?tp=&arnumber=114155&isnumber=3371>.

[10] X.J. Zhang, M. Birk, A. Chiu, R. Doverspike, M.D. Feuer, P. Magill, E. Mavrogiorgis, J. Pastor, S.L. Woodward, J. Yates, Bridge-and-roll demonstration in GRIPhoN, in: OFC/NFOEC 2010, March 22–25, 2010, Paper NThA1, San Diego, CA.

[11] M.D. Feuer, D. Al-Salameh, Routing power: a metric for reconfigurable wavelength add/drops, in: Proceedings of the Optical Fiber Communication Conference, 2002, pp. 156–158.

[12] S.L. Woodward, M.D. Feuer, J. Calvitti, K. Falta, J.M. Verdiell, A high-degree photonic cross-connect for transparent networking, flexible provisioning & capacity growth, in: Proceedings of the European Conference on Optical Communications, Paper Th1.2.2, 2006.

[13] Masahiko Jinno, Takuya Ohara, Yoshiaki Sone, Akira Hirano, Osamu Ishida, Masahito Tomizawa, Elastic and adaptive optical networks: possible adoption scenarios and future standardization aspects, IEEE Commun. Mag. 49 (10) (2011) 164–172.

[14] S. Gringeri, B. Basch, V. Shukla, R. Egorov, T.J. Xia, Flexible architectures for optical transport nodes and networks, IEEE Commun. Mag. (2010) 40–50.

[15] S. Thiagarajan, M. Frankel, D. Boertjes, Spectrum efficient super-channels in dynamic flexible grid networks—a blocking analysis, in: OFC/NFOEC 2011, Paper OTuI6.

[16] T. Takagi, H. Hasegawa, K. Sato, T. Tanaka, B. Kozicki, Y. Sone, M. Jinno, Algorithms for maximizing spectrum efficiency in elastic optical path networks that adopt distance adaptive modulation, in: ECOC 2010, 19–23 September, Torino, Italy, Paper We.8.D.5.

[17] E. Fung, L.J. Nociolo, M. Nuss, S.A. Surek, T.K. Woodward, Apparatus for implementing transparent subwavelength networks, US Patent No. 7,570,660 B1, August 4, 2009.

[18] B. Kozicki, H. Takara, Y. Sone, A. Watanabe, M. Jinno, Distance-adaptive spectrum allocation in elastic optical path network (SLICE) with bit per symbol adjustment, in: OFC/NFOEC 2010, paper OMU3.

[19] D. McGhan, W. Leckie, C. Chen, Reconfigurable coherent transceivers for optical transmission capacity and reach optimization, in: OFC/NFOEC 2012, Paper OW4C.7.

[20] J. Strand, A. Chiu, Realizing the advantages of optical reconfigurability and restoration with integrated optical cross-connects, J. Lightw. Technol. 21 (2003) 2871–2882.

[21] L. Nelson, S. Woodward, P. Magill, S. Foo, M. Moyer, M. O'Sullivan, Real-time detection of a 40 Gbps intradyne channel in the presence of multiple received WDM channels, in: OFC/NFOEC 2010, Paper OMJ1.

[22] Rich Jensen, Andrew Lord, Nick Parsons, Highly scalable OXC-based contentionless ROADM architecture with reduced network implementation costs, in: OFC/NFOEC 2012, Paper NW3F.7.

[23] Winston I. Way, Optimum architecture for MxN multicast switch-based colorless, directionless, contentionless, and flexible-grid ROADM, in: OFC/NFOEC 2012, Paper NW3F.5.

[24] B.C. Collings, Wavelength selectable switches and future photonic network applications, in: Photonics in Switching Conference, 2009.

[25] T. Zami, Contention simulation within dynamic, colorless and unidirectional/ multidirectional optical cross-connects, We.8.K.4, ECOC 2011.

[26] I. Kim, P. Palacharla, X. Wang, D. Bihon, M.D. Feuer, S.L. Woodward, Performance of colorless, non-directional ROADMs with modular client-side fiber cross-connects, in: OFC 2012, Paper NM3F.7.

[27] S.L. Woodward, M.D. Feuer, Toward more dynamic optical networking, in: OptoElectronics and Communications Conference (OECC), 2010, 5–9 July 2010, pp. 114–115.

[28] A. Mahimkar et al., Bandwidth on demand for inter-data center communication, in: Proceedings of the 10th ACM Workshop on Hot Topics in Networks, Article No. 24, November 14–15, 2011.

[29] Paparao Palacharla, Xi Wang, Inwoong Kim, Daniel Bihon, Mark D. Feuer, Sheryl L. Woodward, Blocking performance in dynamic optical networks based on colorless, non-directional ROADMs, in: OFC/NFOEC2011, Paper JWA8, 2011.

[30] G. Shen, Y. Li, L. Peng, How much colorless, directionless and contentionless (CDC) ROADM help dynamic lightpath provisioning? in: OFC/NFOEC 2012, Paper NW3F.1.

[31] P. Pavon-Marino, M. Bueno-Delgado, Distributed online RWA considering add/drop contention in the nodes for directionless and colorless ROADMs, in: OFC/NFOEC 2012, Paper NW3F.4.

[32] S. Thiagarajan, S. Asselin, Nodal contention in colorless, directionless ROADMs using traffic growth models, in: OFC/NFOEC 2012, Paper NW3F.2.

[33] S.L. Woodward, M.D. Feuer, I. Kim, P. Palacharla, X. Wang, D. Bihon, Service velocity strategies in optical ROADM networks, J. Opt. Commun. Netw. IEEE/OSA 4 (2) (2012) 92–98.

[34] M.D. Feuer et al., Simulations of a service velocity network employing regenerator site concentration, in: OFC/NFOEC 2012, Paper NTu2J.5.

[35] A.L. Chiu, G. Choudhury, M.D. Feuer, J.L. Strand, S.L. Woodward, Integrated restoration for next-generation IP-over-optical networks, J. Lightw. Technol. 29 (6) (2011) 916–924.

[36] A. Mahimkar et al., Outage detection and dynamic re-provisioning in GRIPhoN—a Globally Reconfigurable Intelligent Photonic Network, in: OFC/NFOEC 2012, Paper NM2F.5.

[37] Available online at: <http://monarchna.com/topology.html>.

[38] B.G. Bathula et al., On concentrating regenerator sites in ROADM networks, in: OFC/NFOEC 2012, Paper NW3F.6.

[39] See, for example, AT&T's Optical Mesh Service (OMS). <http://www.business.att.com/wholesale/Service/data-networking-wholesale/long-haul-access-wholesale/optical-mesh-service-wholesale/>.

[40] O. Yu, A. Li, Y. Cao, L. Yin, M. Liao, H. Xu, Multi-domain lambda grid data portal for collaborative grid applications, Future Gener. Comput. Syst. 22 (8) (2006) 993–1003.

[41] X. Zhou, M. Feuer, M. Birk, Fast control of inter-channel SRS and residual EDFA transients using a multiple-wavelength forward-pumped discrete Raman amplifier, in: OFC/NFOEC 2007, March 25–29, 2007, OMN4.

[42] A.R. Chraplyvy, The coming capacity crunch, in: ECOC Plenary Talk, Vienna, Austria, September 20–24, 2009, Paper 1.0.2.

[43] T. Morioka, New generation optical infrastructure technologies: "EXAT initiative" towards 2020 and beyond, in: OECC2009, Paper FT4.

[44] P.J. Winzer, Energy-efficient optical transport capacity scaling through spatial multiplexing, Phot. Technol. Lett. 23 (2011) 851–853.

[45] Transmission Systems using Multicore Fibers, chapter 36 in OFT VI.

[46] An Li, Abdullah Al Amin, Xi Chen, William Shieh, Reception of mode and polarization multiplexed 107-Gb/s CO-OFDM signal over a two-mode fiber, in: OFC/NFOEC2011, Paper PDPB8.

[47] M. Salsi et al., Transmission at 2×100Gb/s, over two modes of 40km-long prototype few-mode fiber, using LCOS-based mode multiplexer and demultiplexer, in: OFC/NFOEC2011, Paper PDPB9.

[48] R. Ryf et al., MIMO-based crosstalk suppression in spatially multiplexed 3 56-Gb/s PDM-QPSK signals for strongly coupled three-core fiber, Phot. Technol. Lett. 23 (2011) 1469–1471.

[49] R. Lingle, Few-Mode Fiber Technology for Spatial Multiplexing, Chapter 11 in this book; J. Kahn, R.K.-P. Ho, Fundamentals of Propagation in Multimode Fibers, Chapter 40 in this book.

[50] B. Zhu et al., Seven-core multicore fiber transmissions for passive optical network, Opt. Express 18 (11) (2010) 11117–11122.

[51] Refer to Chapter 9 in Optical Fiber Telecommunications VI-A (on MCF).

[52] M.D. Feuer, L.E. Nelson, X. Zhou, S.L. Woodward, R. Isaac, B. Zhu, T.F. Taunay, M. Fishteyn, J.F. Fini, M.F. Yan, Demonstration of joint DSP receivers for spatial superchannels, in: IEEE Photonics Society Summer Topical Meeting on Space-Division Multiplexing, Seattle, 2012.

Convergence of IP and Optical Networking

16

Kristin Rauschenbach and Cesar Santivanez

Raytheon BBN Technologies, Cambridge, MA, USA

16.1 INTRODUCTION

Increasing demands on capacity, service delivery speed, and variable quality of service are stressing both IP (Internet Protocol) and optical networks. Methods are being sought to provide intelligent provisioning, efficient restoration and recovery from failures, and effective management schemes that reduce the amount of "hands-on" activity to plan and run the network. Over the past 10 years, great strides have been made in integrating the service-friendly IP layer together with the efficient transport capabilities of the optical layer. Converged IP-optical networks are being demonstrated in large multi-carrier and multi-vendor venues. Research is continuing on making this convergence more efficient, flexible, and scalable.

New challenges are emerging in the face of an increasingly heterogeneous future of networking. Certainly any convergence of the IP and optical layers must be done within a context that recognizes the complexity and heterogeneity of all the network services and resources that use and support these layers. In this chapter, we review the current key technologies that contribute to the convergence of IP and optical networks, describing control and management plane technologies, techniques, and standards in some detail. We also illustrate current research challenges and discuss future directions for research. While we focus primarily on the IP and optical layer, we also provide perspectives on research toward an even more heterogeneous future.

Much of the technology and methods described in this chapter apply to networks of both small and large geographic scale, however most activity on converged networks is aimed at large-scale core networks with global reach.

This chapter is organized as follows. First, we begin with a description of the services, architectures, and technologies that are driving toward converged networks. We then provide the background on the existing and emerging technologies that will contribute to the establishment of widespread converged network implementation and deployment. We discuss relevant standards and industry activities that are facilitating this deployment. We then describe a novel integrated control framework that enables both network convergence and scalable dynamic network control. Finally, we discuss some research that will both drive and enable future highly flexible and heterogeneous networks.

Optical Fiber Telecommunications VIB. http://dx.doi.org/10.1016/B978-0-12-396960-6.00016-X

16.2 MOTIVATION

16.2.1 Network services

Cloud-based services, mobility, and video-based content delivery are driving increasingly unpredictable network traffic that is stressing network management and control. Application innovations are driving changes in the make-up of network services. For example, the commercial Amazon™ Elastic Compute Cloud service provides resizable compute capacity to general consumer market users via a simple, quick-response interface. Combining compute and connection resources would appear a simple extension of this service. Likewise, the rapid proliferation of handheld technologies and performance improvement of wired devices such as high definition television and high-density storage systems have caused dramatic changes in how consumers and businesses receive and deliver content. This drives not only tremendous bandwidth growth, but also presents challenges of supporting mobility and quality on-demand video delivery.

Along with these exciting new services, the traditional challenges of growing network capacity remain. In the past, reducing capital cost of network equipment has been the major solution to address capacity growth. However, today, network scale is demanding that operational expense reduction play a larger role in technology decisions than in the past. This directs more attention to automation and interoperability of network capabilities.

16.2.2 Network architectures

Today's deployed optical core network architectures rely exclusively on static point-to-point transport infrastructure. Higher-layer services operate according to their place in the traditional OSI (Open Systems Interconnection) network stack (see Section 3.1). The stack helps to confine conceptually similar functions into layers, invoking a service model between them to implement the services using multiple layers. However, this practice has led to stovepiped management, creating multiple parallel networks within a single network operator's infrastructure.

This type of rigidly layered architecture is expensive to build and operate, and will not react quickly to variable traffic and service types. As such, the industry has been calling for "network convergence" to simplify network management and provisioning and ultimately save operational and capital costs.

IP services now dominate network traffic. However, IP, or layer 3, networks utilize stateless per-node forwarding that is costly at high data rates, prone to jitter and packet loss, and ill suited to global optimization. Layer 2 switching mechanisms are more deterministic, but they lack fast signaling, which hinders service setup time. GMPLS (generalized multiprotocol label switching) attempts layer 2 and 3 coordination but is not yet mature enough for optical layer 1 and layer 2 shared protection over wide areas. Today's Synchronized-Optical-Network-based (SONET) methods of provisioning protected routes for critical services consume excessive resources, which drive down utilization, increase cost, and limit the use of the more efficient route protection schemes.

There is a move to integrate multiple layer 1 and layer 2 functions to reduce cost and minimize space and power requirements. These efforts also aim to minimize the costly equipment (router ports, transponders, etc.) in the network by maximizing bypass at the lowest layer possible. These coordination efforts require a control plane that supports dynamic resource provisioning across the layers to support scalable service rates (from 100 Mb/s to 100 Gb/s) and multiple services, e.g. Time Division Multiplexing (TDM), Storage Area Networking (SAN), and IP services. Such a control plane also enables automated service activation and dynamic bandwidth adjustments, reducing both operational and capital costs. Surmounting these challenges requires a re-think of core network architectures to overcome the limitations of existing approaches and better leverage emerging technologies.

16.2.3 Network technologies

Increased capabilities of the switching and transmission systems, underlying component commoditization, improved and open-sourced software are providing substantial improvements in accessibility, agility, and extensibility of the network elements, such as switches, routers, and transport equipment. The latest software and control protocols are represented in reference [1].

Advances in optical technology now allow practical reconfigurable wavelength networks to be constructed. These networks use wavelength-switching components to dynamically route wavelengths across mesh topologies, and provide a level of flexibility and scalability previously not possible [2–4].

Optical switch architectures that combine Wavelength Selective Switching (WSS) and optical cross-connect switching (OXC) technologies like micro-electro-mechanical switches (MEMs) are emerging to support dynamic routing at the optical layer [5,6]. Currently in deployed networks, WSSs are used to remotely configure optical wavelength bypass in an optical transport node. WSSs can be reconfigured to automatically route wavelengths through mesh networks while minimizing the amount of optical-to-electrical conversion because of this bypass function.

One challenge of the WSS technology is that it is not colorless, because wavelength sensitive elements are used to provide the spatial separation in the switch that routes colors to different output ports. That is, with WSS only certain wavelengths are available at a particular port. Colorless switching is enabled through MEMs devices and other technologies that spatially redirect optical signals independently of the signal wavelength. While colored switching was not an issue with fixed-wavelength transceivers, the advent of tunable transponders has in turn demanded colorless architectures. Cost is an important factor, and recently technology advances have led to the creation of cost-effective colorless switch architectures, just emerging commercially [7].

In addition to the optical switches, other components are needed to enable more agility in optical networks. These include fast-tunable transmitters and receivers, low-noise optical amplifiers, electronic dispersion compensators, and advanced, programmable modulation techniques. These technologies have existed for several years in research and prototype form, and most of them are also now beginning to appear in commercial subsystems [8].

16.3 BACKGROUND

Converged networking builds off a long history of network technology development. To understand the challenges and opportunities of network convergence, some background in the methods and architectures that provide efficient and reliable network provisioning is useful. In this section, we start with a description of the data and transport network stacks. We then describe the functions and architecture of the management, control, and data planes. Traffic management is the key to efficient network operation, and methods for routing, wavelength assignment, and grooming are presented in this context. We also discuss restoration- and protection-based recovery methods to achieve high network availability. Finally, we discuss multi-domain network architecture and optimization approaches. In all cases, the focus is on IP, optical, and the emerging converged IP/Optical networks.

16.3.1 Network stack

In a multi-layer network, layered models are useful to track how traffic is generated and carried across a heterogeneous infrastructure. Both the data community and the transport community have similar, but not identical, stacks that represent the layered model. Figure 16.1 shows both the telecommunications standard OSI (Open Systems

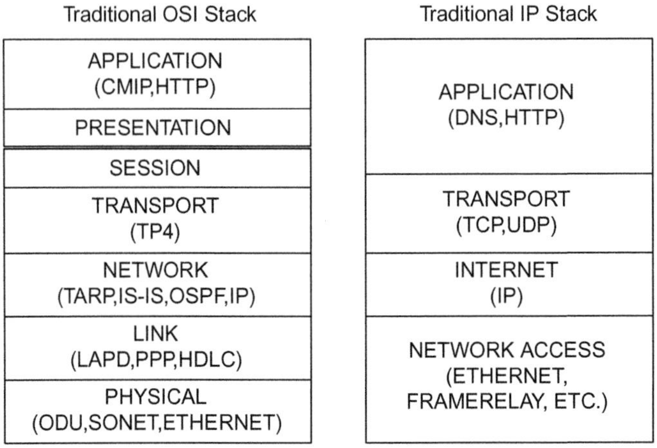

Traditional OSI Stack	Traditional IP Stack
APPLICATION (CMIP,HTTP)	APPLICATION (DNS,HTTP)
PRESENTATION	
SESSION	
TRANSPORT (TP4)	TRANSPORT (TCP,UDP)
NETWORK (TARP,IS-IS,OSPF,IP)	INTERNET (IP)
LINK (LAPD,PPP,HDLC)	NETWORK ACCESS (ETHERNET, FRAMERELAY, ETC.)
PHYSICAL (ODU,SONET,ETHERNET)	

FIGURE 16.1 Side-by-side comparison of the traditional OSI and IP stacks with some common protocols in parentheses.

Interconnection) stack and the data communications TCP/IP (Transport Control Protocol/Internet Protocol) stack. In general, each layer of the stack interacts with the adjacent layers to create an end-to-end connection for the application. While many variations are possible, the boxes that implement the technology in a layer above select, or are assigned, paths from the box technology that implements the lower layer.

As an information packet or circuit passes through a network, it traverses several network nodes, and may at each node pass through interfaces at different layers of the network as shown in Figure 16.2. The specific equipment (or layer) traces the packet or circuit travels is determined by the routing requirements for that particular information flow.

Various protocols are used to manage the interaction that directs a flow or traffic demand on its path through the network. As an example, an email from the application layer server that distributes email would become encapsulated in a TCP/IP protocol within the server. The TCP/IP packet would then be encapsulated in an Ethernet frame in the local area network to which the server is attached, that Ethernet frame is then encapsulated into a SONET-formatted stream from the enterprise network that connects to the next hop in the IP route as established by the TCP/IP protocol. At this hop, the stream is demultiplexed, and the IP packet is extracted and sent to the local router for processing and retransmission to the next IP node. This process of encapsulation, multiplexing, and de-encapsulation and demultiplexing continues through multiple hops through the IP network, which is typically connected via SONET or OTN (Optical Transport Network) circuits in the network core. Ultimately, the IP

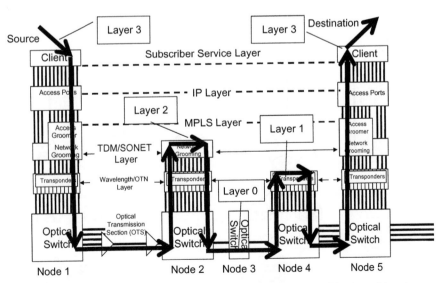

FIGURE 16.2 Trace of an information packet or circuit through a multi-layer, multi-node network. The path crosses interfaces at different layers in different nodes based on routing requirements.

packet that contains the mail message is received by the recipient's mail server, and the message can be unpacked at the application layer.

Each layer in the network performs various management functions, and attempts to operate the functions of the layer efficiently and quickly. However, these layer functions are often redundant, and the packing and unpacking process results in unnecessary processing, latency, and loss of efficiency. This has prompted investigation into multi-layer network management and control, which is currently an active area of research [9–13].

16.3.2 Management, control, and data planes

As shown in Figure 16.3a, originally networks had data planes that carried the traffic, and management planes to conduct the operations of the network. This approach leads to a proliferation of management systems as the network scale and scope increase. Also, because management plane response times are typically slow, especially given the large amount of information they manage, service changes in these architectures are also slow. Increasingly today, there are three distinct planes in a network architecture: the management plane, the control plane, and the data plane as shown in Figure 16.3b. This architecture yields faster and more efficient management and service instantiation.

The management plane typically handles high-level network management and operations including network monitoring and customer billing. The telecommunication

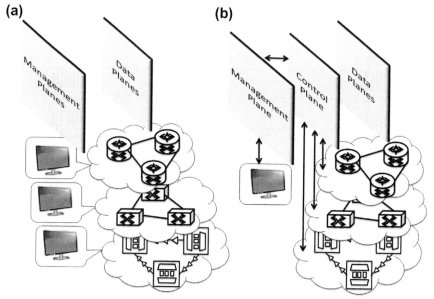

FIGURE 16.3 (a) Traditional management and data plane architecture and **(b)** emerging management, control, and data plane architecture.

FCAPS (Fault, Configuration, Accounting, Performance, Security) model describes the functions of network management. The data plane carries the user data, and employs interfaces to network transport equipment such as transponders, optical amplifiers, and optical and electronic switches and switch ports. The control plane operates between these two layers. The role of the control plane is evolving, but generally, the control plane handles automated network provisioning, some fault recovery, and other administrative control functions like topology management and liveness verification. As shown in Figure 16.4 illustrations of example control plane architectures, the control

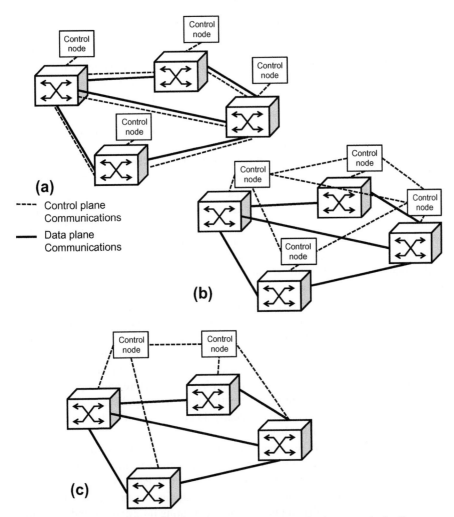

FIGURE 16.4 Control plane architectures. (a) In-band communications, topologically isomorphic, (b) out-of-band communications, topologically isomorphic, and (c) out-of-band non-isomorphic topology.

plane data communications can be provided in-band or out-of-band. The control plane data communications topology can also be isomorphic to the communications data plane, or not, as the use case and technology demand.

The advantage of a control plane is that it can operate much more quickly than a typical management plane, enabling bandwidth to be provisioned on-demand (carrier or customer initiated), novel scheduled services, and more efficient automated restoration and recovery. There is a major push by standards organizations to ensure that the optical control plane is multi-vendor and multi-carrier capable. This will ensure interoperability that will hasten its introduction and use.

16.3.3 Control plane functions

While the definition of a control plane continues to evolve, in general, the control plane serves to decouple services from service delivery mechanisms and to decouple quality-of-service from its realization mechanisms. It also provides boundaries for policy and information sharing based on trust, business models, and need for scale. It provides end-to-end signaling path and maintains topology and path information across heterogeneous platforms.

A typical control plane handles functions of network element discovery, routing, path computation, and signaling and connection setup. Discovery is the process of learning the links and nodes that are adjacent to a given node. Discovery protocols manage the finding and verification of nodes and links, including real-time updates for liveliness. Discovery is enabled by exchanging address or naming information over either the in-band or out-of-band control channel.

Routing is the process of establishing all link and node connectivity in the network topology, the reachability of a node to other nodes in the network, and resource information such as link bandwidth and node switching capacity and connectivity. The routing information is used to construct a full network topology of the available links and nodes in the network, and is either stored locally in the network elements, maintained in management database, or in some combination of these. The management of the topology information database, including where the information is stored, how it is updated, and how it is used for network optimization and service creation, drives the control and management plane architecture and implementation. In today's optical networks routing information is typically gathered infrequently, as network topology does not change frequently. In contrast, in IP networks, routing is handled on a packet-by-packet basis.

To achieve scale, not all resource information is included in the routing protocol. Abstraction, or minimal resource representation, is used to filter information and help scale, albeit at the sacrifice of optimization potential. Path computation provides available source-destination paths to either the endpoints or the traffic-engineering engine in the network. Path computation takes in routing information, and adds other optimization features, such as cost functions, desired protection schemes, or diverse routing requirements to determine viable network paths.

Signaling serves to implement a desired path setup or tear-down, as well as maintaining liveness information of an active path.

16.3.4 Traffic management

The objective of traffic management is to efficiently assign service demands to available network resources. Grooming, routing, and wavelength assignment are methods for assigning traffic to network resources in a multi-layered, multi-node network. Regardless of the underlying transmission method, traffic demands are said to flow from a "source" to a "destination" over a "path" (e.g. see Figure 16.5). There are different "paths" available in a typical network, including fiber paths (the specific fiber optic strand a signal transits), wavelength paths (the specific wavelength channel a demand is assigned to), SONET paths, etc.

In the recent past, a significant amount of human intervention was involved in the planning of the network capacity and the allocation of wavelengths between network nodes. As the network migrated from simple SONET-based rings, to singly- and dually-honed multiple-rings, to today's more common mesh architectures, more automated methods have been introduced, because the complexity associated with path selection is significantly higher.

While the terms are often used interchangeably, in general grooming describes the multiplexing of lower rate signals into higher rate signals for subsequent routing and path assignment. The wavelength assignment process determines how to assign desired service paths to specific wavelengths. This is usually constrained to ensuring wavelength continuity (e.g. wavelength remains the same) along the path.

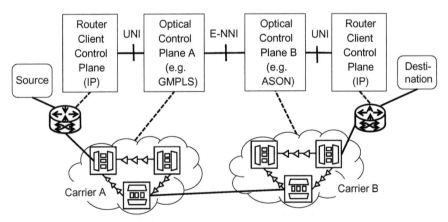

FIGURE 16.5 Multi-domain network and control plane. Source and destination traffic traverses two routed domains and two optical transport domains in carrier A and carrier B. The use of UNI and E-NNI interfaces allows interoperability of different control planes unique to each domain (e.g. GMPLS or ASON).

In many studies, routing and wavelength assignment is considered separately from grooming, because wavelengths are considered to be fully packed from source to destination. However, to achieve the best efficiency, the problems should be considered together. Overall, the objective is to either maximize throughput, minimize blocking, or minimize network costs (equipage) for a specific throughput of blocking requirement. Routing and wavelength assignment and grooming are each described in more detail in the following sections.

16.3.4.1 Routing and wavelength assignment

Wavelength assignment involves determining a path in the network between the two nodes and allocating a free wavelength on all of the links on the path. An all-optical path is commonly referred to as a lightpath and may span multiple fiber links without any intermediate electronic processing, while using one wavelength channel per link. The entire bandwidth on the lightpath is reserved for this connection until the call is ended and service terminated, at which time the associated wavelengths become available on all the links along the route.

In a typical wavelength-division multiplexed network, a fixed number of fixed color wavelengths are available on an optical fiber. Today, wavelength conversion occurs only at transponders. Therefore, routing and wavelength assignment of a demand consists of finding a path with available capacity, and then selecting a single wavelength used along the entire path. This is known as the wavelength continuity constraint. If wavelength conversion is allowed (either via back-to-back optical/electrical conversions or with an all-optical wavelength converter) then different colors can be used on different segments along the path. While wavelength conversion would appear to be a major advantage for the efficiency of traffic assignment, especially since a limited number (<100) of wavelengths are available on a given fiber, in general the gains to be realized by using wavelength conversion are relatively small [14,15].

Routing and wavelength assignment can be divided into pre-planned and on-demand. In pre-planned assignment, the traffic is known, and wavelengths are assigned to minimize blocking of the known demand set. This is analogous to the generic bin-packing problem. The solution is conducive to an integer linear program which is NP-complete (NP stands for Non-deterministic Polynomial time), and heuristic approaches are utilized to minimize blocking [16–18]. In on-demand wavelength assignment, paths are selected as the demand appears in the network. Routing paths are typically selected based on shortest path, or least-loaded path criteria. Methods to approach this problem include layered graphs and logical-link representation [19,20].

Routing and wavelength assignment becomes even more complex in an optical mesh network where optical cross-connects are used to reroute wavelengths dynamically onto different fibers. In these cases, the use of a centralized path computation element (e.g. the PCE Section 3.3) allows for a near-optimal, deterministic, and stable solution [21,22].

In ASONs (Automatically Switched Optical Networks), GMPLS-based (Generalized Multi-Protocol Label-Switched) resource reservation schemes are used for wavelength assignment and resource reservation based on either forward or backward reservation protocols. In both cases, signaling establishes available wavelengths on the path from source to destination. At the destination, a single continuous wavelength is selected. In forward reservation protocol, all available wavelengths are reserved on the forward path, and those not used are released. In backward reservation protocol, the backward signal reserves the selected path on the way back to the destination. These distributed algorithms do not guarantee a path will be available until two signaling passes on the path, and instability can arise if multiple reservations compete for a resource during setup.

16.3.4.2 Grooming

As compared to wavelength routing and assignment, the grooming problem is typically focused on multiple layers in the network. This is because grooming involves the aggregation of multiple lower rate streams onto a single higher rate stream. Often, the highest layers in the network, like layer 3, consist of highly granular flows with small data rates that are groomed together either at layer 3 or at lower layers to create higher rate streams. Grooming is the optimization of network transmissions that span multiple distinct transmission channels or methods. Grooming can occur within multiple layers of the same technology or between technologies. Grooming is performed at the edge of the network, where tributaries are merged into long-haul connections, and also at intermediate nodes where the electronic transport, switching, and routing equipment is capable of converting signals between different wavelengths, channels, or time slots. This intermediate grooming is only advantageous when multiple input ports at the node combine toward common output ports.

Many demands do not fill a full wavelength. If one such demand is uniquely assigned to a full wavelength, without sharing it with other demands, it will result in wasting bandwidth and long-reach transponders. To alleviate this problem, demands can be aggregated into larger flows at the source node. They can also be combined with other nodes' demands at intermediate nodes so that wavelength utilization at the core is close to 100%. However, not all nodes are capable of the required aggregation or disaggregation. Once demands are fully groomed onto a wavelength, the resulting channel can take advantage of optical bypass at intermediate nodes. This can reduce network capital cost, because optically bypassed signals do not require an optical-to-electrical-to-optical conversion (OEO), saving the cost of the electronics required for this function in the node.

Deciding where and when to groom demands is a difficult optimization problem. It must take into account different tradeoffs among capacity available, the cost (both capital and operational) of the grooming ports and transponders, and the fact that constantly adding or removing demands will unavoidably result in fragmentation inside a wavelength. What may appear to be a good grooming decision in the short term may hurt performance in the future. Grooming decisions, then, must balance medium- to long-term resources tradeoffs and be based on medium-term traffic patterns. Like wavelength assignment, grooming is typically represented as a mixed-integer linear-programming problem, and heuristics and partitioning are used to achieve near-optimal results. Several good papers have been published regarding grooming solutions for multi-layer networks, including real-time grooming via online algorithms, and static integer linear programming approaches [23–25].

16.3.5 Recovery

Another major function of the control and management planes in distributed systems is recovery. Network recovery deals with the ability of a network to recover from the failure of a technology within the network. This includes link failures, transponder failures, node failures, and others. The control and management planes play a role in recovery by identifying faults, notifying appropriate resources about that fault, and establishing the eligible pool of resources used to recover from the fault, either pre-planned or on-demand. Particular implementation methods vary, depending on the speed and determinism demanded of the recovery process. Usually, the bottlenecks for recovery latency are round-trip delays for the signaling messages and the queuing delays for requests at the failover switching node. There are schemes that address these issues, which are particularly acute in mesh architectures [26].

Fault recovery can be broadly classified into protection and restoration. Protection relies on dedicated resources that are reserved in advance during connection setup. Restoration takes place immediately after failure detection to discover and reserve capacity and reroute the signal using this capacity. Compared with protection switching, restoration is more efficient in terms of resource utilization, but usually requires longer restoration time.

Typically, the lowest layers of the network (fibers, wavelengths, SONET, OTN) use protection-based schemes to recover from failures, while higher layers in the network (IP, TCP, Application) use restoration-based schemes. Also, because protection schemes are simpler to implement on a link basis, or using ring-based topologies typical of SONET infrastructure, restoration was also implemented in telecommunications digital cross-connect networks.

16.3.5.1 Recovery approaches

There are three basic approaches to recover an established connection in the face of network node and link failures: link-based, segment-based, and path-based. For purposes of this discussion the *interior* of a path consists of all links on the path and all nodes except the path's endpoints; two paths are said to be *interior-disjoint* when there is no node or link that is in the interior of both paths.

In link-based recovery, for each interior element along the primary path, a backup route is found by omitting the failed element from the network topology and recalculating the end-to-end path. Thus for each working path there is a set of n backup paths where n is the number of interior elements on the primary path. These paths need not be (and usually are not) interior-disjoint from the primary path or from one another. For a single failure, link-based recovery may give an efficient alternate route; however, the approach faces combinatorial explosion when protecting against multiple simultaneous failures.

In segment-based recovery, a working path is associated with a set of n backup paths, one for each interior link or node. A given backup path is associated with one of these interior elements; it is not based on the end-to-end service requested but simply defines a route around that element. A classic example of segment-based recovery can be found in SONET rings, where any one element can fail and the path is rerouted the other way round the ring. Because segment-based backup paths are independent of any particular working path, they may be defined per failed element instead of per path. However, they can also be highly non-optimal from the perspective of a specific service request, and are ill suited to protect against multiple simultaneous failures.

Path-based recovery defines one or more backup paths for each working path. A working path with one backup path is said to be *singly protected*; a working path with two backup paths is *doubly protected*; and similarly for higher numbers. Each backup path for a primary is interior-disjoint with the primary path and interior-disjoint with each of the working path's other backup paths (if any). Practical algorithms exist for jointly optimizing a working path and its backup path(s). Relative to link- and segment-based methods, path-based recovery maximizes bandwidth efficiency, provides fast reaction to partial failures, and is readily extended to recover from multiple simultaneous failures. Its main drawbacks are a high signaling load if the path contains services with many different endpoints. MPLS and dynamic optical mesh networks both tend to favor the use of path-based recovery for these reasons.

16.3.5.2 Restoration

There are many restoration schemes used today. In general, these schemes use some form of signaling after a failure is detected to determine an alternative path. The different schemes use different mechanisms to flood failure state information, to calculate alternative paths, and to switch the data to the new path [27–30]. In general, it takes on the order of seconds for a restoration event to stabilize, and the network to recover. In large IP/MPLS networks, where IP forwarding tables must be updated, this can take 10s of seconds. In cases where network instability occurs, and/or a large amount of fragmentation of the topology results, the network may not recover without human intervention.

IP networks typically rely on restoration for failure recovery. IP networks are stateless, so in IP restoration, control plane and forwarding plane convergence time contributes to overall traffic restoration times. Control plane protocol updates (e.g. IGP, EGP, PIM) are required to exchange information after a topology change. Control plane convergence is completed when all network elements reflect the updated topology. The forwarding plane allows routers and switches to forward traffic from ingress to egress. The forwarding plane convergence is completed when affected traffic flows are restored.

To help speed recovery in IP-based networks, MPLS (Multi-Protocol Label Switching) is useful. MPLS was designed to provide a connection-oriented framework for the connectionless IP networks running at the layer above. MPLS utilizes labels, attached to flows of multiple IP packets headed to the same destination, to establish virtual paths (called label-switched paths or LSPs) between the MPLS-enabled router nodes.

MPLS Fast Reroute (also called MPLS local restoration or MPLS local protection) is a local restoration network resiliency mechanism. It is a capability represented in the RSVP Traffic Engineering (RSVP-TE) standard. In this case, each LSP passing through a facility is protected by a backup path that originates at the node immediately upstream to that facility.

In theory, labels can be assigned to any of a variety of underlying layer 2 technologies, such as ATM (Asynchronous Transfer Mode) virtual circuits, SONET circuits, or even wavelengths. These paths can be managed like the any other circuit, and so the MPLS layer can be used for traffic engineering, and protection and restoration. Since the MPLS paths are often running over underlying SONET infrastructure, which is already protected, hold-off mechanisms are used to prevent the MPLS layer from reacting to a failure that the lower layer protocol will address.

16.3.5.3 Protection

In protection schemes, a pre-planned alternative, or backup, path is either pre-provisioned and standing by idle, or reserved, and data is automatically routed to the backup path in the event a failure is detected. The SONET standard for back-up path switching time is 50 ms. Protection can be implemented so as to be "hitless," if the back-up path is up and providing redundant data to the receiver (so-called $1 + 1$ protection). Slightly more efficient protection is afforded by schemes where a backup link is used to protect more than one path, as in $1:N$ SONET protection, or shared mesh protection. As long as only one failure occurs at a time, these shared schemes provide the same level of determinism as the dedicated protection.

Automatic protection switching is the capability of a transmission system to detect a failure on a working link or lightpath and to switch to a standby to recover the traffic. There are two commonly used types of protection for the links in SONET and OTN transport networks. They are differentiated by how they reserve backup resources. One-plus-one protection provides a continuously active backup path for each working path. At the source, the optical signal is split into two signals and sent over both the working and the protection facilities simultaneously, producing a working signal and a protection signal that are identical and always on. At the destination, both signals are monitored independently for failures. The receiving equipment selects either the working or the protection signal. Extending SONET-like protection schemes to WDM mesh architectures involves different algorithms to select both working and disjoint protection paths, but otherwise the switching mechanism is similar [31–34].

In one-for-N protection ($1:N$), there is one backup path for several working paths (the range is from 1 to 14). In the $1:N$ protection architecture, all communication from the source to destination is carried out over the signaling channel. Because this represents a shared protection scheme, all traffic reverts to the working facility as soon as the failure has been corrected.

This approach to sharing protection paths can be extended to meshes. In these shared protection approaches, for a given failure or set of failures, only some primary paths are affected, and only some of their protection paths (in the case of multiple failures) are affected. A protection resource can be reserved for use by an entire set of protection paths if none of the failures under consideration can simultaneously require use of that resource by more than one path in the set. Several methods to determine shared protection paths have been reported [35–37].

In general, path-based shared protection requires significantly less reservation of network resources than does dedicated protection. However, in real-world optical networks, a fundamental tradeoff arises between protection sharing and electrical port count: more sharing (shorter protection links) requires more electrical ports (regenerators, OEO converters, electrical switching). These practical considerations imply that not all nodes or regions of the network need the same strategy. Regions of the network that are bandwidth-rich or port-scarce will likely benefit from dedicated protection, while regions that are bandwidth-poor or port-rich will likely benefit from shared protection. Optimization schemes with a broad purview can consider a combination of shared and dedicated protection resource assignment strategies for those areas of the network that benefit from one or the other.

The control plane must be appropriately engineered to implement shared protection in a mesh, especially when fast (sub-100-ms) recovery is needed, because careful synchronization between the network resources and the control system is required to trigger a protection switch, and signaling to the distributed resources involved in the failover is required [38].

16.3.6 Multi-domain

Because of many technological and business reasons, today's global core networks are heterogeneous. Multiple global carriers operate using different network control and management practices. Within a typical carrier, there are again multiple networks operating that serve various purposes including local, metropolitan, and wide area or backbone connections, as well as different services such as public IP, private IP, and circuit networks. These networks are outfitted with equipment from different box vendors that each utilize proprietary box control (so-called element management), single-box-vendor network control (so-called network management), and interfaces to carrier management systems (so-called north-bound management interfaces). It is useful to delineate separate control regions into "domains," and control and management across different domains is then called multi-domain management. Typically within a domain, there is a common method for resource representation and the control and management functions that enable service creation on the equipment within that domain.

Multi-domain network management and control strives to achieve management and administration to enable functions such as service setup and restoration end-to-end between a source and a destination that may lie in separate domains. These functions are achieved at domain boundaries through interfaces that contain typically limited information about the internal networks, a so-called abstract representation.

The challenge for multi-domain management is that the abstraction must contain enough information for rapid, efficient, and complete satisfaction of the service request, but not so much information to limit scale or sacrifice the privacy concerns of the information within the domain.

The control plane for multi-domain networks connecting a user source and destination is described by a user network interface (UNI) and an external network-to-network interface or E-NNI (External-Network-to-Network Interface), as shown in Figure 16.5. It is important to note that the control plane connections are logical diagrams. The control planes may communicate via the underlying transport networks they control or by a completely separate communication channel, like the control planes for single-domain networks described in Section 3.2. Thus, we refer to intra-network control plane "reachability" as the ability of one control plane to communicate with another over an available data communication network. This data communications network used for control signaling may be separate for reasons of practicability, security, survivability, availability, and other quality of service considerations or scale.

Most of today's applications require end-to-end delivery of information, and so provisioning in optical networks must address the multi-domain issues. Recent research activities have started to consider multiple domain scenarios with a focus on improving routing and signaling protocol to increase network utilization [39,40]. Most multi-domain routing protocols use abstract representations of the local resource representations to advertise domain information to other domains in the network. Since each domain receives only abstract information about the other domains, calculating an optimal end-to-end path is challenging task.

Multi-domain network optimization proceeds much like single-domain optimization, and aims to optimize capacity utilization, reduce blocking, provide low, or guaranteed latency for an end-to-end path that crosses multiple domains. The challenge is the lack of complete information across all domains. There are two extreme approaches to resolving resource conflicts, a distributed method that probes the domains on a demand-by-demand basis to establish resource availability, and a more centralized approach that uses control elements in each domain, with full intra-domain knowledge. These control elements then manage resource allocation across domains.

To illustrate a distributed approach, recently, a dynamic optimal end-to-end path computation algorithm for multi-domain optical transport networks was reported [41]. This algorithm determines an *a priori* optimal path using domain abstraction information. Then, the end-to-end path is re-optimized by dynamically re-assessing the inter-domain paths and domain's egress node after visiting intermediate domains in the path. The algorithm updates the inter-domain paths and domain's egress node after visiting all intermediate domains in the path, so egress can be changed to avoid congestion in a mid-path domain. The update decision considers several parameters such as available bandwidth, shortest hop count, and link failures. The update process occurs as the circuit is set up from source to destination. In simulation of a 100+ node, 4-domain United States-based network topology, this "two-pass" approach showed improvement in blocking at high loads and no increase in blocking at light loads owing to path computation overhead.

The more centralized approach is covered in the next section.

16.3.6.1 Path computation element

In a multi-domain scenario, visibility among different domains is usually limited. So, to improve accuracy, determinism, efficiency, and scale, an architectural element that provides purview within and across different domains is needed. This is the function of the Path Computation Element (PCE) introduced by the IETF (Internet Engineering Task Force) in 2006 and described in RFC 4655 [42]. The PCE incorporates the special computation control elements needed to coordinate path selection in multi-domain networks. The IETF PCE model defines elements for computation (centralized or distributed), synchronization, discovery and load balancing, and liveness. It also describes control communications among these elements. It also supports coordination among multiple, distributed PCEs (both stateful and stateless) with functions including synchronization and monitoring. Finally, it establishes need for policy, confidentiality, and evaluation metrics.

Centralized path computation allows CPU-intensive (control processor unit) calculations to be dedicated to a high-end processor as opposed to the limited computation available in network equipment. It also improves determinism and optimality of resource assignment.

For multi-domain networks with limited visibility, a hybrid of centralized and distributed architecture is envisioned. A collection of PCEs, each with full information of a particular topology, coordinate at domain boundaries to establish end-to-end paths. This semi-distributed approach helps the system scale.

A PCE-based architecture also helps with extensibility, because it eases inclusion of a network node, which may or may not have its own control plane, which lies outside the original domain. The PCE also enables multi-domain protection and restoration schemes, as well as providing an insertion point for cross-domain policy implementation.

The IETF and ITU-T, which addresses PCE functionality in G.7715.2 (ASON routing architecture and requirements for remote route query), have an ongoing collaboration to align routing requirements for large multi-domain networks. The OIF is already working toward implementation and testing in carrier networks.

Work continues on standardization of the following key elements of the PCE architecture:

- Methods for communication between PCEs for policy updates, and between resources and PCE.
- Protocols on support of PCE discovery and signaling of inter-domain paths.
- Metrics to evaluate path quality, scalability, responsiveness, robustness, and policy to support path computation algorithms.
- Management modules related to communication protocols, routing and signaling extensions, metrics, and PCE monitoring information.

16.4 STANDARDS

Networks have traditionally been built by two distinct communities, the data community and the transport community, each with their own respective standards organizations: the IETF for the data community and the ITU-T for the transport community. One place these communities converge is in the area of multi-layer network control. There is currently much activity focused on converging the IETF control plane model, so-called GMPLS, or generalized multi-protocol label switching, and the ITU control plane model, so-called ASON for automatically switched optical network. These models address large-scaled switched optical networks that would typically contain both data-centric network equipment (IP routers) and transport-centric equipment (OTN-based WDM transport and optical switching).

Merging of the control schemes for the data world and the transport world is a major challenge, because the starting points for the two network approaches are quite different. In traditional data, or IP networking, router nodes function as stateless per-node forwarding engines. There is no separate control plane, all control information is contained within the packet header, and packet forwarding governed by a forwarding table residing in each router. The forwarding table is periodically updated when the underlying topology changes. Generally, there is no management function, and traffic is routed on a "best effort" basis. This kind of approach supports low-cost, on-demand (though not guaranteed) service over a heterogeneous transport infrastructure, and works particularly well when the underlying router connections are low loss and statically connected (so that the router tables do not have to be frequently updated). These attributes have served to make IP the "service layer" of choice for most of today's applications, including computer interconnection, but also voice and video.

In contrast, transport networks were built to support very high-efficiency circuit-oriented connections between telecommunications switching and aggregation points. There is a nominally centralized, hierarchical management plane that handles call setup and connection control, which is the term given to the setup, tear down, and management of connections through the transport network. The data transport plane of these networks operates deterministically, using well-defined and well-timed frames to handle multiplexing and demultiplexing functions cost effectively. These connection-oriented networks operated with a very high level of fidelity (the 5–9s or 99.999% availability standard), and utilized pre-planned protection schemes to achieve this availability in the face of typical failure modes (back-hoe link outages and equipment failures). The traditional transport network is deterministic and highly efficient. The equipment is able to cost effectively pack data into transport channels (more than a factor of five times lower cost than the equivalent speed router interface) because the interface cards do not have to examine every packet for the control and routing information. However, these networks are not as agile or flexible to accommodate unplanned growth and new service creation as the IP networks.

Building off the traditional SONET circuit-based standards, the ITU has introduced the OTN standard, which addresses multiple wavelength and multi-service features as compared to the SONET standard. OTN introduces containers, or optical data units (ODU) with different rates (2.5–100 Gb/s) into which not only traditional SONET-framed data, but others such as Gigabit Ethernet, Fibre Channel, FICON, and ESCON can also be conveniently packed. These containers can be configured in a multiplexing hierarchy for grooming and aggregation, and the OTN control plane (G.709) supports discovery, signaling, and routing of the connections established for containers in the multiplexing hierarchy. There is also recently added an ODU-flex channel that allows for client-specified data rates. The ODUflex-type containers necessitate the need for more signal information (data rate and frame size), and the ability to adjust the frame size, which adds significant flexibility to the otherwise conventional transport method.

Packet-based network control is based on multi-protocol label switching protocol (MPLS). MPLS-TP (MPLS-Transport Profile), introduced in 2008, is emerging as a method to converge IP and optical TDM transport control. This standard adds traditional operations and management functions common in ITU-based SONET and OTN standards to the MPLS protocol. The standard is evolving with both IETF and ITU activity.

For IP networks, the IETF has specific RFCs (request for comments) that govern the control plane for heterogeneous, e.g. IP, MPLS, TDM, and WDM optical networks. RFC 3471 describes extensions to Multi-Protocol Label Switching (MPLS) signaling required to support Generalized MPLS [43]. Generalized MPLS extends the MPLS control plane to encompass time-division (e.g. Synchronous Optical Network and Synchronous Digital Hierarchy, SONET/SDH), wavelength (optical lambdas), and spatial switching (e.g. incoming port or fiber to outgoing port or fiber). The GMPLS control plane ensures traffic-grooming capability on edge nodes by operating on a two-layer model; that is, an underlying pure optical wavelength routed network and an electronic grooming layer built over it (MPLS or TDM). In the wavelength routed layer, operating exclusively at lambda granularity, when a transparent lightpath connects two physically adjacent or distant nodes, these nodes will seem adjacent for the upper layer. The upper layer can perform multiplexing of different traffic streams into these wavelengths. The GMPLS control plane essentially facilitates routing, resource discovery, and connection management and recovery.

In GMPLS, lightpaths are established by exchanging control information among nodes, distributing labels, and reserving resources along the path to route appropriately labeled flows. In practice, the signaling protocol is closely integrated with the routing and wavelength assignment protocols. Typical GMPLS signaling protocols include Resource Reservation Protocol (RSVP) and Constraint-Based Label Distribution Protocol (CR-LDP). GMPLS also uses the Link Management Protocol (LMP) to communicate proper cross-connect information between the network elements. LMP runs between adjacent systems for link provisioning and fault isolation. It can be used for any type of network element, particularly in natively photonic switches.

An emerging area of control plane standards is that for managing purely optical layer capability, including optical impairments and, potentially, dynamic optical layer topologies realized through optical switching and reconfigurable add/drop function.

Activities for this standard include routing and wavelength assignment methods, methods to include impairments, as well as the required signaling extensions to the emerging optical components and their performance monitoring test points. The Wavelength Switched Optical Network (WSON) standard emerging from the IETF provides a framework for applying Generalized Multi-Protocol Label Switching (GMPLS) and the Path Computation Element (PCE) architecture to the control of wavelength switched optical networks [44].

The relevant ITU (International Telecommunication Union) standards for optical control plane are part of both the architecture for optical transport networks (G.805), the ASON (Automatically Switched Optical Network) standards to govern the architecture for switched optical networks (G.8080), transport network functions (G.807), and various call setup and connection management (G.7713) and discovery (G.771) [45].

The layered model of the ITU-T standards includes a client-server model, and is recursive such that any particular layer is a server to the layer above, and a client to the layer below. Links consist of a set of ports that connect the edge of a subnetwork to another. Link connections are static, but subnetwork connections are flexible and managed by the management plane. Links and subnetwork connections are delimited by connection points (CPs) in the client layer. The network connection in client layer is delimited by a terminal connection point (TCP). A link connection is represented in the server layer by a pair of adaptation functions and a trail. In the management plane, these reference points are represented by objects called connection termination points and trail termination points (CTP and TTP) for connection points and trails [46]. In G.8080, these concepts are extended to switched network topologies. A subnetwork point (SNP) is an abstraction that represents a connection point or a terminal connection point, and a set of SNPs that are grouped together for routing purposes is called a subnetwork point pool (SNPP). The SNPs may be static, an SNP link connection, or dynamic, and SNP subnetwork connection.

ASON maintains separation of the control plane from the transport plane. The control plane can be assigned link connections without the link being physically connected. Thus, there are two steps to network discovery. In the first, transport plane discovery, a discovery agent maintains the transport connections for later binding to the associated control plane connections. The second step, control plane discovery, is handled by a link resource manager (LRM) that holds the SNP-SNP link connection information. A termination adapter performer (TAP) maintains the relationship of the control plane and transport plane resource names, which is necessary with the separate control planes.

The OIF (Optical Internetworking Forum), launched in 1998, is an industry group that sponsors internetworking activities and demonstrations, and has forged development of user network interface implementations (UNI), and networking interfaces (E-NNI for intra-carrier and I-NNI and N-NNI for internal networks). The work of the OIF has allowed multi-carrier, multi-domain demonstrations of control plane interoperability and early implementations of end-to-end circuit setup and restoration functions in an automated fashion via an optical control plane. This kind of

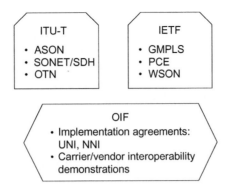

FIGURE 16.6 Network communication standards bodies (ITU-T, IETF) and industrial forum (OIF) and their associated purview for optical control plane evolution.

interoperability demonstration has been ongoing since 2004. A diagram that illustrates the purview of the IETF, ITU-T, and OIF is shown in Figure 16.6.

16.5 NEXT-GENERATION CONTROL AND MANAGEMENT

16.5.1 Drivers

Much of the industry work to date has focused on optical control plane for integration of the dynamic, flexible IP layer over a static, circuit-oriented wavelength-division-multiplexed optical transport layer. Research, however, has addressed such challenges as making the optical layer dynamic and responsive to traffic changes [47–51], and also including non-traditional resources into the purview of the network control plane such as compute, storage, and sensor resources [52]. This increase in scope puts additional stress on the control plane to handle more and more dynamism and heterogeneity going forward. Fortunately, there is a sound framework, building off the current optical control plane research and development, for managing and implementing services across these large, complex systems. Below we describe one of these frameworks, and then we describe work toward future control and management framework that supports an even greater degree of heterogeneity.

16.5.2 Novel framework

A functional architecture that can manage and control the instantiation of services across a large, heterogeneous infrastructure must address several key requirements. It must be technology agnostic, allowing not only for introduction of new generations of traditional optical and routing equipment, but also the ability to add higher layer capabilities from processing and storage and lower layer functions, such as energy and other supporting infrastructure, into the model. The architecture must be able to optimize and manage flexibly across a heterogeneous subset of resources and constraints. Finally, it must be scalable and deterministic or stable.

A functional architecture that meets these objectives was developed in 2008 called PHAROS (Petabits Highly Agile Robust Optical System) [38]. While the PHAROS functional architecture is general, and applies to any large-scale dynamic system, certain specific design choices were made based on the desire to apply the architecture to control of a global-scale (~100-node) dynamic optically switched wavelength-division multiplexed system as part of DARPA's CORONET project. Here we describe the general framework, and provide some of the specific implementation decisions that apply to a global-scale wavelength-division multiplexed optical network.

16.5.2.1 Governance, decision, action

The PHAROS functional architecture explicitly separates *governance*, *decision making*, and *action* as three key roles in control and management of multi-layer, multi-domain networks. Their functional relationship is illustrated in Figure 16.7. They are analogous to the traditional management plane, emerging control plane and existing data plane functions.

The governance function controls the behavior of the full system, establishing which actions and parameters will be performed automatically and which require human intervention. Governance establishes policy and reaction on a human scale. It is not on the critical path for service instantiations. This function contains the

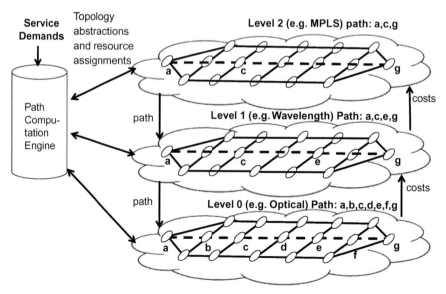

FIGURE 16.7 A functional breakdown of control and management for a multi-layered distributed system.

primary repository of nonvolatile governance information and is the primary interface between human operators and the network.

The decision function applies the policies established by the governance function to effectively allocate resources to meet service demands. It is highly time-critical. The decision process is on the critical path for realizing each service request on-demand: the decision process is applied to each service request and creates directives for control of network resources. The PCE in the IETF architecture addresses the mechanisms required to carry out the decision role. In PHAROS, the decision process is unitary. That is, one and only one decision maker is assigned to a given resource. Minimizing the negotiations required to make a decision improves both the optimality of the decision and the consistency of the state it was made from, ensures deterministic decision times (without backtracking or thrashing), and enhances speed and resilience by reducing the number of entities on the critical path that need to reach a consensus. This results in globally consistent resource allocation, with consistently fast service setup. However, as is the case with the PCE, the decision function may also be implemented in a distributed fashion.

The challenge of allocating and assigning communications resources across multiple technological layers, rapidly and efficiently, requires careful attention to the functionality of the decision role. The key characteristics of the role are to minimize negotiations while maximizing the horizon of resource allocation decisions: that is, making each decision with the widest feasible awareness of the total resources in the network and the total demands upon it. Maximizing the horizon of a resource allocation decision allows consideration of the potential uses of a resource for local as well as for transit and protection functions.

The action function implements decisions made by the decision function quickly and reports any changes in the state of the system. The action function is time-critical. The responsibility of the action role is limited to implementing directives. The network element controllers in a typical router or switch device would be the primary implementation components responsible for carrying out the action function.

16.5.2.2 Signaling network

The PHAROS functional architecture includes a signaling network, a closed system connecting all the components required for connecting the elements that perform governance, decision, and action. The signaling communication network can be implemented in-band with the transmission network, either with a separate signaling channel like OTN, or within a packet header as in IP, or out-of-band or even on a separate network. Cost, congestion, and delay are all factors affecting this design decision. There is also the option to implement signaling on a topology that is isomorphic to the data plane topology. Isomorphism has the advantage that it simplifies routing and speeds up signaling. However, this choice may result in additional delay because it dictates the link distances between control nodes. In the PHAROS functional architecture, the desire to manage resources dynamically, with sub-100-ms-class response times, drove a design choice of a data-plane-isomorphic signaling network with dedicated in-fiber, separate wavelength channel, bandwidth.

16.5.2.3 Resource representation

Current systems employ some degree of abstraction in managing network resources, using interface adapters that expose a suite of high-level parameters describing the functionality of a node. Such adapters, however, run the risks of obscuring key blocking and contention constraints for a specific node implementation, and/or tying their interfaces and the system's resource management algorithms too tightly to a given technology.

A more generalized and extensible framework for resource representation is based on topology abstraction. Topology abstraction is used to track network resources and route demands. For a multi-layer system, multiple topology abstractions are used, each representing a set of like resources or addressing a set of like services. Topology abstractions represent these collections of resources by graphs with edges based on their connectivity. Each abstraction then presents a view of resources at different "levels" of the network (these levels may be the traditional network optical, MPLS layers, as shown in Figure 16.8, but may also be a more complex arrangement of resources) and are tuned for the optimization calculations required for specific tasks.

For example, within an optical wavelength-division multiplexed network, there may be a topology abstraction that represents the transparent, non-electrically regenerated, network connections within a particular network configuration. As another example, protection resources may be represented in a "shared protection" topology abstraction that describes the resources available for protection, and may, for example, put lower costs for shorter routes, as opposed to other topology abstractions that may favor lightly loaded links. As a final example, an optical cross-connect within a network node may be represented by a topology abstraction that provides all the

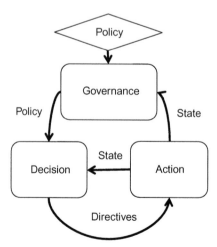

FIGURE 16.8 Illustration of topology abstraction for MPLS, wavelengths, and fiber transmission along with a path computation element that can optimize resource utilization for a set of service demands via appropriate path assignment algorithms. Each demand is routed at its corresponding topology abstraction. A higher level edge contains all relevant information from lower layers.

input output port connections available per wavelength. The topology representation can communicate constraints such as wavebanding through its topology. By using abstract topological representations for all levels of the network, representations extend down to an abstract network model of the essential contention structure of a node, and extend upward to address successive (virtual) levels of functionality across the entire network, as shown in Figure 16.9. This method is highly extensible because it uses one approach common to all levels of resource representation and allocation.

Constraints are incorporated by considering the edges available in a particular topology abstraction. Levels are layered as appropriate given the network configuration, and the standard "client-server" model is used for higher layer topology abstraction nodes to select particular "paths" from the layer below. Cost information is passed from lower layers to higher layers, and routing information is passed from higher to lower, as shown earlier in Figure 16.8.

The critical advantage of using topology abstractions is efficiency and agility. For efficiency, the optimization of resource allocation becomes truly global, with cross-layer and cross-network properties evaluated jointly. For agility, the technology agnosticism within the abstractions ensures that legacy and new technologies are readily incorporated and the requirements of emerging service classes addressed.

Note that topology abstractions keep track of resource usage, and adapt to resource utilization, and each topology abstraction keeps track of constraints at its "level." For example, optical reach and wavelength continuity would be captured in

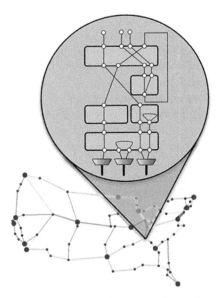

FIGURE 16.9 Illustration of recursive nature of topology abstraction representation. Topologies may represent links and nodes in a wide area optical mesh network, and extend down to the connection of technologies within a specific node. Copyright Cambridge Press.

the "available resources" attribute of the edges in the optical transparent topology abstraction described above. Policy and "learning" about network behavior can also be inserted into the topology abstractions to affect routing decision. For example, as traffic patterns are discovered, expertise can be captured by adding/removing edges from the respective topology abstraction to influence routing decisions from the layer/level above.

16.5.2.4 *Optimization strategies*

Whether the objective is efficient grooming or efficient routing and wavelength assignment on static link resources, efficient assignment of dynamic traffic to a dynamic circuit layer, or agile bandwidth assignment to a best-effort routing, layer optimization is achieved via a combination of a resource assignment strategy (centralized, distributed, or some hybrid) and underlying optimization algorithms that drive that strategy.

Thus, a key element of a multi-layer resource management system is its choice of setup strategy for allocating resources to a new service request. There are three broad classes of strategy for doing resource allocation when setting up a service: pure centralized (the single master), path threading, and predistributed resources. In addition, hybrid strategies are available, including the one selected for the PHAROS project: *unitary resource allocation* strategy. The engineering tradeoffs of the various strategies are summarized in Table 16.1, and described further below.

16.5.2.4.1 Single-master setup

The *single-master strategy* entails a single control node that receives all setup requests and makes all resource allocation decisions for the network. This approach allows, in principle, global optimization of resource allocation across all network resources. It has the further virtue of allowing highly deterministic setup times: it performs its resource calculation with full knowledge of current assignments and service demands, and has untrammeled authority to directly configure all network resources as it decides. The challenge for this strategy is that a single processing node with sufficient capacity for communications, processing, and memory has to encompass the entire network's resources and demands. This node becomes a single point of failure, a risk typically ameliorated by having one or more additional, equally capable standby nodes. Moreover, each service request must interact directly with the master allocator, which not only adds transit time to service requests (which may need to traverse the entire global network) but also can create traffic congestion on the signaling channel, potentially introducing unpredictable delays and so undercutting the consistency of its response time.

16.5.2.4.2 Path-threading setup

The *path-threading strategy* goes to the other extreme: each node controls and allocates its local resources, and a setup request traces a route between source and destination(s). When a request reaches a given node, it reserves resources to meet the request, based on its local knowledge, and determines the next node on the request's

Table 16.1 Comparison of the setup strategies for resource allocation in a distributed network.

Strategy	Advantages	Disadvantages
Single master	• Optimal allocation • Deterministic latency • Fast decision algorithm	• Worst case: adds round trip to master to setup latency • Can cause focused traffic loads • Potentially limited scalability • Vulnerable to single node failure • Vulnerable to network partition
Path threading	• Fastest latency (most of the time)	• If high call-setup rates, high chance of long setup • Potential thrashing behavior at high setup rates • Additional state distribution (flooding) • Did not work well in practice • No global optimality
Predistributed	• Fastest latency (most of the time)	• Less optimal • If high setup rates and utilization, maybe long setup • Potential thrashing at high setup rates and utilization • Additional state distribution (flooding)
Unitary	• Optimal allocation • Fast and deterministic latency • Robust and scalable	• Additional state distribution (flooding) • Requires governance (to set parameters controlling assignment of scopes and resources) • More complex implementation than single master

path. If a node has insufficient resources to satisfy a request, the request backtracks, undoing the resource reservations, until it fails or reaches a node willing to try sending it along a new candidate path. This strategy can yield very fast service setup, provided enough resources are available and adequately distributed in the network. There is no single point of failure; indeed, any node failure will at most render its local resource unavailable. Similarly, there is no single focus to the control traffic, reducing the potential for congestion in the signaling network.

However, the strategy has significant disadvantages. Setup times can be highly variable and difficult to predict; during times of high request rates, there is an exceptionally high risk of long setup times and potential thrashing, as requests independently reserve, compete for, and release partially completed paths. Because backtracking is more likely precisely during times when there are already many requests being set up, the signaling network is at increased risk of congestive overload due to the nonlinear increase in signaling traffic with increasing request rate. The path-threading strategy is ill suited to global optimization, as each node makes its resource allocations and next-hop decisions in isolation. This drawback may be ameliorated by global state flooding, though this adds to the risk of congestive overload during times of many service requests. At all times optimization decisions may suffer as a given node's model of the network is neither internally consistent nor consistent with that of other nodes along a request's path. In practice, the path-threading strategy has not worked well owing to these limitations.

16.5.2.4.3 Predistributed-resources setup

The *predistributed-resources strategy* is an alternative approach to distributed resource allocation. In this strategy, each node "owns" some resources throughout the network. When a node receives a setup request, it allocates resources that it controls and, if they are insufficient, requests other nodes for the resources they own. This strategy has many of the strengths and weaknesses of path-threading. Setup times can be very quick, if sufficient appropriate resources are available, and there is no single point of failure or a focus for signaling traffic. Under high network utilization or high rates of service requests, setup times are long and highly unpredictable; thrashing is also a risk. Most critically, resource use can be quite suboptimal. Not only is there the issue of local knowledge limiting global optimization, there is also an inherent inefficiency in that a node will pick routes that use resources it owns rather than ones best suited to global efficiency. In effect, every node is reserving resources for its own use that might be better employed by other nodes setting up other requests.

16.5.2.4.4 Unitary resource management

To resolve the tradeoffs among these strategies, the PHAROS functional architecture relies on a resource allocation strategy called *unitary resource management*. This approach involves running the optimization algorithm and resource management from a resilient hierarchy of control nodes. For each service request there is exactly one control node responsible at any time; and for each network resource there is exactly one control node controlling its allocation at any time. Each control node has an assigned scope that does not overlap with that of any other control node. Scope consists of a service context and a suite of assigned resources. The service context defines the service requests for which the control node will perform setup. For example, a service context may be a set of tuples, each consisting of a service class, a source node, and one or more destination nodes. A scope would typically be based on a meaningful network service region, for example, all service requests whose endpoints fall in the continental US. The unitary strategy allows a high degree of optimization and highly consistent setup times, as a control node can execute a global optimization algorithm that takes into account all resources and all service demands within its scope. There is no backtracking or thrashing, and no risk of nonlinear increases in signaling traffic in times of high utilization or of high rates of setup requests. There is some risk of suboptimal resource decisions.

The unitary strategy uses multiple control nodes to avoid many of the problems of the single master strategy. Thus, a standard distributed system failover mechanism is used to manage information updates, and handoff in the event of failure of a control node.

16.5.2.5 Shared protection

Responding as quickly to a detected fault is an important capability of the network. At the same time, it is important to do so efficiently, using the current network state where feasible. With PHAROS, speed and efficiency are achieved by using a fault notification scheme that relies on flooding, ensuring that every node is aware of the event as quickly as possible and taking action in parallel across the network to implement service recovery. The data plane nodes responsible for the failed element send a status change to all the adjacent control nodes and this is forwarded along all interfaces that have not acknowledged failover. The switching actions required to respond to a given failure are pre-determined at service setup, and loaded into the switching nodes in the form of playbooks.

Upon a service request, the control node calculates both the primary route (the path taken when there are no failures affecting the service) and the required disjoint protection route(s). The network resources required to instantiate each protection route are identified in terms of pools of reserved resources.

Disjoint routes for shared protection are established by an algorithm running in the control node. If two desired paths have no links or nodes in common, then no single failure can break both of them, and therefore they can be protected using some of the same backup nodes and links. The requirement is that no single failure will cause them to contend for these resources. Resources for the joint protection can be calculated straightforwardly by determining a set of all paths that simultaneously fail when at least one link or node along those paths fails. The capacity of services on this "jointly protected set" then defines the capacity required to protect the set. The protection actions for each failure, along with the required reserve capacity pools, are then sent to the local nodes in playbooks that define the actions to take upon failure notification.

Upon receipt of a failure message, each local node will check its playbook to determine if it has protection actions to take due to the status change. If not, then it is done. If a protection action is required, the local node selects from the pool, implements the switch, and signals to its adjacent nodes that the resource is now assigned to the specific service request. Once all connections have been made, subscriber traffic may flow on the connection. The benefit of this solution is that the establishment of the resource pools was made by the element that implements the decision function (the "decision node"), but no decision node needs to be consulted to take action. Each local node will send a message to the decision node and to its neighbors acknowledging the connection; the decision node informs all local nodes of the specific resources to monitor for the flow and the abstracted "failure handler" to report if any failure occurs. Playbook semantics are based on these "failure handlers," which reside in the local nodes. Use of shared protection substantially reduces the amount of spare capacity required in a large core network as compared to dedicated protection, and is a major driver for the economic advantage of mesh-based architectures.

16.5.2.6 Optimal resource assignment

There are many published approaches to optimal wavelength assignment and grooming, as described earlier in this chapter. To illustrate how a subset of these approaches might come together in a consistent framework, the optimal service assignment approach for the PHAROS project is described further here.

PHAROS integrated routing, wavelength assignment, grooming, and protection strategy is guided by the topology abstractions described above. Each topology abstraction keeps track of resource availability and constraints (e.g. optical reach and wavelength continuity for level 1 (L1) and wavelength topology abstraction edges) at its level.

Each demand is routed at its associated level, as shown in Figure 16.8. Note that the corresponding topology abstraction contains all required information for its level as well as all levels below. This information is abstracted into edge attributes (e.g. availability, cost, latency, intersection) that are used for routing and resource allocation decisions. Conversely, once a route is chosen, this route can be mapped into actual resources and cross-connects decisions. For example, a level 2 (L2) edge in a path implies that grooming is performed at the edge's endpoints only. Furthermore, the L2 edge lower level path (e.g. L1 path) carries information of exactly where regeneration and wavelength conversion (at the full wavelength level) is taking place. Similarly, a L1 edge in a path implies OEO at the edge's endpoints and optical bypass along the edge's physical (level 0) path.

For example, Figure 16.8 shows an IP demand is routed at level 2 and the lowest-cost path obtained is the (level 2) path a–c–g. Thus, grooming is performed at the source and destination nodes (nodes a and g) as well as in node c. Furthermore, expanding level 2 edge c–g reveals its associated level 1 path to be c–e–g. Thus, OEO conversion must also be done at node e (along with nodes a, c, and g). Finally, nodes b, d, and f are optically bypassed.

PHAROS routing algorithm is based on successive Dijkstra computation [53]: it first computes the working path, then removes those links and computes the first protection path and so forth. Successive Dijkstra was chosen to provide very fast (sub-100-ms-class) setup times. It should be noted that standard algorithms such as Bhandari are not applicable in our cross-layer setting since they require non-overlapping edges. To minimize the impact of "trap links" and very long protection paths, we used the concept of "forbidden node/forbidden links." Basically, an edge will not be used for routing if its corresponding level 0 path contains a forbidden element. Standard algorithms such as Bhandari can be used to detect trap links and add them to the forbidden set without compromising the cross-layer optimization [54].

The above formulation allows for trading off different costs at different layers (from bandwidth to transponders to MPLS ports). Also, the costs are not fixed but they can be expressed as a function of resource availability. This allows for not only load balancing at a single layer (e.g. bandwidth congestion) but to change the relative cost of different resources as the network operating point changes (say, from bandwidth rich to bandwidth poor). It also allows for an optimization based on equipment or operational (e.g. energy) costs or total-cost-of-ownership. For example, Figure 16.10 shows an example of cross-layer decision making. The best path (either the one with 2 OEOs or the one with 1 MPLS port) will depend on the selected cost function. This cost function can be set to optimize cost and/or to minimize the likelihood of future blocking. In any case, the level 2 topology abstraction contains all the required information.

Grooming, i.e. the action of joining several sub-lambda flows into a single wavelength with the purpose of maximizing wavelength utilization, is a relatively expensive operation since it operates at the electrical (and sometimes packet) level. To be worthwhile, grooming must save more resources (e.g. bandwidth or ports) than it costs, and so are made based on the aggregate traffic traversing a link.

Determination of the cost/benefit of grooming is relatively simple when the traffic is static, but when the traffic is dynamic, as is the PHAROS case, then the decision is more complicated. Grooming determinations need to be made on-the-fly at the moment a demand arrives (and for both working and protection paths) *before* the system knows the total traffic that will flow through a link in the future.

PHAROS addresses this issue by decoupling the timescales. Level 2 edges are created *a priori* based on expect traffic patterns (e.g. using traffic matrices derived from past history or current expectations). The effect of the existence of a long level 2 edge in the algorithm is that it makes available an "express link" that the algorithm can use to pack small flows into. When a demand arrives, the control node uses the L2 TA to decide the path to use. This decision is made based on network-wide, cross-layer criteria (such as resource availability, congestion, etc., as described above). If the path includes a long L2 edge, the flow will be packed into it. Thus, the TA uses medium-term to long-term information to decide whether to add an edge and the decision to actually use the edge and groom traffic is decided by the routing algorithm on-the-fly.

During path setup, protection resources are pre-staged across the network and assigned to protection pools. The information about when and how to use a protection pool's resourcs is maintained by local nodes in playbooks provided and updated by the control node. Both playbooks and protection pools are expressed in terms

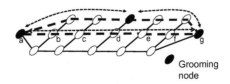

Grooming
node

FIGURE 16.10 Example of cross-layer decision making. Path A includes two all-optical paths and one grooming node, and path B includes two OEOs, but no intervening grooming. Routing decisions at the next higher layer are based on the cost comparison of these two paths' costs which are shared with the layer above.

of edges in the TA representation. The playbooks map failure identification(s) and affected services onto the protection pools to be used to protect those services (i.e. protection pools associated with the edges in the protection path). The playbook does not specify the resource within a protection pool to use. Instead, the assignment is efficiently made only after a failure occurs. At recovery time, each node on the protection path selects the resources for protecting the "outgoing" direction of a service independently and informs its neighbor nodes. Based on this distributed selection, each intermediate node can complete protection actions by interconnecting the outgoing resources it selected for protection with those incoming resources it was informed about by the neighbor nodes. Note that since each node only chooses resources in one direction ("outgoing"), and since the resources have been preassigned and sized for the worst failure event, no conflict occurs. This process is successively repeated across layers. That is, when a L2 resource is needed, first the underlying L1 path is established, and then the L2 resource (e.g. timeslot, label, etc.) is grabbed. Once again, since there is no need to coordinate between endpoints, the outgoing timeslot/label is selected shortly after the outgoing L1 resource is selected (no need to wait for a path-length round-trip period).

One unique feature of the PHAROS system is that it allows the sharing of protection resources across layers. E.g. the same wavelength can provide protection to a wavelength service and to a set of L2 demands, as shown in Figure 16.11. Each demand, d, whose protection path includes a protection pool, PP e, has an associated "Failure resource allocation matrix" (FRAM_e^d). Each entry (i, j, k, ...) in FRAM_e^d represents the number resources from PP e that are required if the failure sequence

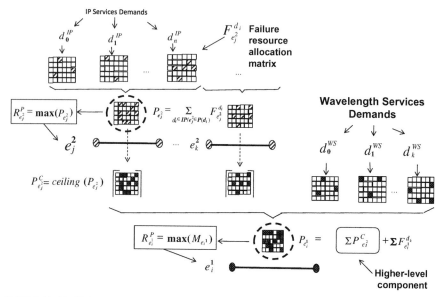

FIGURE 16.11 Sizing protection pools in multilayer shared protection.

$f_i f_j f_k, \cdots$ occurs. A failure f_i may represent a node failure, a span cut, etc. The size of the FRAM matrices is $|F|^K$, where $|F|$ is the number of failure events and K is the maximum level of protection LoP of any demand. All the FRAMs associated with a PP are combined (simple matrix addition) into the pool's Protection Matrix (M_e). The maximum entry (worst failure sequence) in M_e determines the number of resources needed in the protection pool (i.e. max sum instead of sum max). Note that there is no need to individually track each demand's ($FRAM_e^d$), instead their sum (M_e) can be updated each time a demand arrives/departs. The rules determining which entries of M_e to update for a given demand are dependent on the protection policy (i.e. reversion mandatory or not) and the control plane capabilities (i.e. can it distinguish between span and node failures?). Therefore, PHAROS is flexible and can support different update rules. For example, when the maximum LoP is 2, and under certain conditions, the update rule for a PP e in the protection path of a singly protected demand d of bit rate r is equal to $M_e = M_e + FRAM_e^d$, where $FRAM_e^d = r \cdot \delta_{I_e^d}, \delta_s(x) = 1$ iff $x \in S$ (zero otherwise), and $I_e^d = P_1 \times P_2^C$, referred to as the "protected index set," is a set containing all failure events (ordered failure sequences) for which the demand d will need protection resources from the PP e. P_1 and P_2 represent the working and protection paths, respectively, P_2^C represents P_2s complement, and x represents the Cartesian product.

Figure 16.11 provides an illustration of sizing L1 protection resources due to both wavelength services and "L2 protection edges" which serve IP services. For the L1 edge under consideration, each L2 edge traversing it is a "client" (i.e. a demand) the same as a wavelength service demand. The FRAMs associated with each of these L2 edges are derived directly from the L2 edge Protection Matrix M by applying the generalized "ceiling" function, which round up each entry in P into the units of the lower level. Then, the L1 edge's Protection Matrix M is the sum of all the FRAMs of the WS as well as the rounded-up Protection Matrices of all L2 edges including the L1 edge in its lower level path. This relatively simple methodology allows for sharing of protection resources between different services classes at different layers (i.e. IP and WS).

16.5.3 Research extensions: highly heterogeneous networks

Going forward, the trend toward converging "layers" to improve efficiency and improve service delivery times while improving the richness of the service offering is extending beyond the IP and optical layers, as shown in Figure 16.12. Recently, research has been focusing on dynamic resource management "higher up the stack" as well. This research represents an important step toward large-scale distributed system management that will reduce the amount of human intervention required to establish and use a complex array of resources. The ability to realize complex yet agile "systems-of-systems" autonomously will improve efficiency, and ultimately lead to simpler methods to solve larger and more distributed problems. To illustrate the extensions to the multi-layer and multi-domain network control and management strategies described earlier in this chapter, we describe an emerging approach to heterogeneous resource management being forged in the National Science Foundation's GENI project, the Global

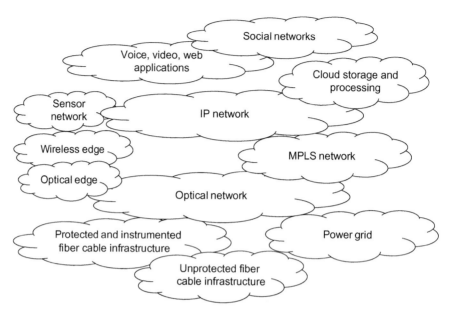

FIGURE 16.12 Pieces of a large and heterogeneous networked system-of-systems. Part or all of these individual network "clouds" may be jointly optimized through the use of a generic control plane and optimization strategy.

Environment for Network Innovation. GENI is a deeply programmable infrastructure suite for performing computer science and networking experimentation at scale [55].

GENI allows programmable configuration and control of all aspects of a computation network (computation, storage, and communication), and provides transparent access to a federation of shareable and sliceable resources in programmable topologies. A federated architecture such as GENI provides a set of mechanisms to allow organizations and users to share and collaborate across a set of separately owned and operated resources. Details of the GENI architecture can be found at http://groups. geni.net/geni/wiki/GeniArchitectTeam.

The GENI functional architecture, illustrated in Figure 16.13, brokers the capabilities of resource owners to the needs of experimenters (users) who access slices of resources that may span several different resource owners. Single-owner resources are managed as aggregates via a GENI aggregate manager. Examples of aggregates are regional network, backbone network, campus networks, or computer clusters. As the resource broker, the GENI functional architecture is designed to ensure accountability of the actions taken by experimenters on an aggregate's resources, and manage authentication and authorization services in the GENI clearinghouse. The GENI meta-operations center (GMOC) oversees the health and operations of the aggregates, though not as a replacement for individual aggregated management and operations functions. Experimenters access and create slices via experiment tools. They obtain an identity via a third-party trusted authority that authenticates them

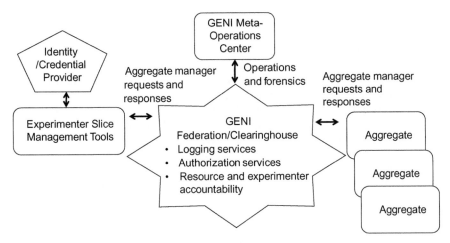

FIGURE 16.13 A functional architecture for the Global Environment for Network Innovation, a system which provisions services across a highly heterogeneous array of computer and communication network resources (aggregates).

to the GENI federation, and credentials that authorize particular actions needed to create slices on GENI aggregates. A picture that illustrates the data flow among the principal actors in the GENI functional architecture is shown below.

At the network layer, GENI relies on an emerging technology called OpenFlow (http://www.openflow.org/). OpenFlow provides an emerging standard to open the internal flow tables in an Ethernet switch together with an interface to add and remove entries in the table. More generally, OpenFlow is an attempt to unify control across both packets and circuits, which can both be considered "flows." OpenFlow provides a unified method of assigning flows to an underlying switch and transport infrastructure through the OpenFlow interface. To date, the interface is well defined for data center switching, and multiple Ethernet switch vendors are supporting the OpenFlow specification. Research and development are continuing on extending OpenFlow concepts and standards to optical layer. OpenFlow per se does not address the optimization and management function for these open networks; however, the OpenFlow architecture stresses the need for a separate control plane, and enables a nominally centralized resource management strategy, which manifests a trend of the past decade of network research.

Acknowledgments

The authors acknowledge the important contributions of the PHAROS architecture team for valuable input to this chapter. In particular, Ilia Baldine, Alden Jackson, Will Leland, Walter Milliken, Ram Ramanathan, and Dan Wood contributed to the PHAROS framework and functional architecture described herein.

References

[1] B. Ramamurthy, G.N. Rouskas, K.M. Sivalingam, Next Generation Internet: Architectures and Protocols, Cambridge University Press, 2011.

[2] A. Saleh, J. Simmons, Evolution toward the next-generation core optical network, J. Lightwave Technol. 24 (9) (2006) 3303.

[3] A. Gladisch, R.-P. Braun, D. Breuer, A. Ehrhardt, H.-M. Foisel, M. Jaeger, R. Leppla, M. Schneiders, S. Vorbeck, W. Weiershausen, F.-J. Westphal, Evolution of terrestrial optical system and core network architecture, Proc. IEEE 94 (5) (2006) 869–891.

[4] J. Strand, A. Chiu, Realizing the advantages of optical reconfigurability and restoration with integrated optical cross-connects, J. Lightwave Technol. 21 (11) (2003) 2871–2882.

[5] W.J. Tomlinson, Requirements, architectures, and technologies for optical cross-connects, 13th Annual Meeting on Lasers and Electro-Optics Society, LEOS 2000, vol. 1, IEEE, 2000., pp. 163–164.

[6] R. Shankar, M. Florjanczyk, T.J. Hall, A. Vukovic, H. Hua, Multidegree ROADM based on wavelength selective switches: architectures and scalability, Opt. Commun. 279 (1) (2007) 94–100.

[7] R.A. Jensen, Optical switch architectures for emerging colorless/directionless/contentionless ROADM networks, in: Optical Fiber Communication Conference and Exposition (OFC/NFOEC), 2011 and the National Fiber Optic Engineers Conference, March 2011, 1–3, pp. 6–10.

[8] E.B. Basch, R. Egorov, S. Gringeri, S. Elby, Architectural tradeoffs for reconfigurable dense wavelength-division multiplexing systems, IEEE J. Sel. Top. Quant. Electron. 12 (4) (2006) 615–626.

[9] B.J. Wilson, N.G. Stoffel, J.L. Pastor, M.L. Post, K.H. Liu, L. Tsanchi, K.A. Walsh, J.Y. Wei, Y. Tsai, Multiwavelength optical networking management and control, J. Lightwave Technol. 18 (2000) 2038–2057.

[10] D.C. Blight, P.J. Czezowski, Management issues for IP over DWDM networks, Opt. Netw. Mag. 2 (1) (2001) 81–91.

[11] L. Raptis, G. Hatzilias, An integrated network management approach for managing hybrid IP and WDM networks, IEEE Netw. 17 (3) (2003) 37–43.

[12] J. Schonwaelder, A. Pras, J. Martin-Flatin, On the future of internet management technologies, IEEE Commun. Mag. 41 (10) (2003) 90–97.

[13] C. Pinart, G.J. Giralt, On managing optical services in future control-plane-enabled IP/WDM networks, J. Lightwave Technol. 23 (10) (2005) 2868–2876.

[14] R.A. Barry, P.A. Humblet, Models of blocking probability in all-optical networks with and without wavelength changers, IEEE J. Sel. Area. Commun. 14 (5) (1996) 858–867.

[15] E. Karasan, E. Ayanoglu, Effects of wavelength routing and selection algorithms on wavelength conversion gain in WDM optical networks, IEEE/ACM Trans. Netw. 6 (2) (1998) 186–196.

[16] R. Ramaswamy, K.N. Sivarajan, Routing and wavelength assignment in all-optical networks, IEEE/ACM Trans. Netw. 3 (5) (1995) 489–500.

[17] A. Brzezinski, E. Modiano, Dynamic reconfiguration and routing algorithms for IP-over-WDM networks with stochastic traffic, J. Lightwave Technol. 23 (10) (2005) 3188–3205.

[18] A.E. Ozdaglar, D.P. Bertsekas, Routing and wavelength assignment in optical networks, IEEE Trans. Netw. 2 (2003) 259–272.

[19] S. Xu, L. Li, S. Wang, Dynamic routing and assignment of wavelength algorithms for multi-fiberwavelength division multiplexing networks, IEEE J. Sel. Areas Commun. 18 (10) (2000) 2130–2137.

[20] B. Zhou, M. Bassiouni, G. Li, Routing and wavelength assignment in optical networks using logical link representation and efficient bitwise computation, Photonic Netw. Commun. 10 (3) (2005) 333–346.

[21] J. Hu, B. Lieda, Traffic grooming, routing, and wavelength assignment in optical WDM mesh networks, in: IEEE INFOCOM, 2004.

[22] Y. Zhau, J. Zhang, Y. Ji, W. Gu, Routing and wavelength assignment problem in PCE-based wavelength-switched optical networks, J. Opt. Commun. Netw. 2 (4) (2010) 196–205.

[23] K. Zhu, B. Mukherjee, Traffic grooming in an optical WDM mesh network, IEEE J. Sel. Areas Commun. 20 (1) (2002) 122–133.

[24] R. Dutta, G.N. Rouskas, Traffic grooming in WDM networks: past and future, IEEE Netw. Mag. 16 (6) (2002) 46–56.

[25] P. Iyer, R. Dutta, C.D. Savage, On the complexity of path traffic grooming, in: Second International Conference on Broadband Networks, vol. 2, 3–7 October 2005, pp. 1231–1237.

[26] C. Assi, W. Hou, A. Shami, N. Ghani, Improving signaling recovery in shared mesh optical networks, J. Comput. Commun. 29 (2005) 59–68.

[27] L. Zhou, P. Agrawal, C. Vijaya Saradhi, V.F.S. Fook, Effect of routing convergence time on lightpath establishment in GMPLS-controlled WDM optical networks, in: 2005 IEEE International Conference on Communications, vol. 3, 16–20 May 2005, pp. 1692–1696.

[28] E. Bouillet, J.-F. Labourdette, R. Ramamurthy, S. Chaudhuri, Enhanced algorithm cost model to control tradeoffs in provisioning shared mesh restored lightpaths, in: Optical Fiber Communication Conference and Exhibit, 2002, OFC 2002, 17–22 March, 2002, pp. 544–546.

[29] S. Yuan, B. Wang, Highly available path routing in mesh networks under multiple link failures, IEEE Trans. Reliab. 60 (4) (2011) 823–832.

[30] M. Kodialam, T.V. Lakshman, Dynamic routing of bandwidth guaranteed tunnels with restoration, Proceedings of the Nineteenth Annual Joint Conference of the IEEE Computer and Communications Societies, INFOCOM 2000, vol. 2, IEEE, 2000, pp. 902–911.

[31] C. Xin, Y. Ye, S. Dixit, C. Qiao, A joint working and protection path selection approach in WDM optical networks, Global Telecommunications Conference, 2001, GLOBECOM '01, vol. 4, IEEE, 2001, pp. 2165–2168.

[32] C. Ou, B. Mukherjee, H. Zang, Sub-path protection for scalability and fast recovery in WDM mesh networks, in: Optical Fiber Communication Conference, OFC 2002, 17–22 March 2002, pp. 495–496.

[33] L. Guo, H. Yu, L. Li, Path protection algorithm with trade-off ability for survivable wavelength-division-multiplexing mesh networks, Opt. Express 12 (24) (2004) 5834.

[34] P.-H. Ho, H.T. Mouftah, A framework for service-guaranteed shared protection in WDM mesh networks, IEEE Commun. Mag. 40 (2) (2002) 97–103.

[35] Ou. Canhui, J. Zhang, Z. Hui, L.H. Sahasrabuddhe, B. Mukherjee, New and improved approaches for shared-path protection in WDM mesh networks, J. Lightwave Technol. 22 (5) (2004) 1223–1232.

[36] P.-H. Ho, H.T. Mouftah, Shared protection in mesh WDM networks, IEEE Commun. Mag. 42 (1) (2004) 70–76.

[37] F. Dikbiyik, L. Sahasrabuddhe, M. Tornatore, B. Mukherjee, Exploiting excess capacity to improve robustness of WDM mesh networks, IEEE/ACM Trans. Netw. 20 (1) (2012) 114–124.

[38] I. Baldine, A. Jackson, J. Jacob, W. Leland, J. Lowry, W. Miliken, P. Pal, R. Ramanathan, K. Rauschenbach, C. Santivanez, D. Wood, PHAROS: an architecture for next-generation core optical networks, in: Byrav Ramamurthy, G.N. Rouskas, K.M. Sivalingam (Eds.), Next-Generation Internet Architectures and Protocols, Cambridge University Press, 2011, pp. 54–179.

[39] L. Guanglei, C. Ji, V.W.S. Chan, On the scalability of network management information for inter-domain light-path assessment, IEEE/ACM Trans. Netw. 13 (1) (2005) 160–172.

[40] J. Berthold, L. Ong, Next-generation optical network architecture and multidomain issues, Proc. IEEE 100 (5) (2012) 1130–1139.

[41] D. Benhaddou, S. Dandu, N. Ghani, J. Subhlok, A new dynamic path computation algorithm for multi-domain optical networks, in: Mediterranean Winter, 2008, ICTON-MW 2008, 2nd ICTON, 11–13 December 2008, pp. 1–6.

[42] A. Farrel, J.P. Vasseur, J. Ash (Eds.), A Path Computation Element (PCE)-based Architecture, RFC 4655, August 2006.

[43] L. Berger (Ed.), Generalize Multi-Protocol Label Switching (GMPLS) signaling functional description, RFC 3471, January 2003.

[44] Y. Lee, G. Bernstein, W. Imajuku (Eds.), Framework for GMPLS and Path Computation Element (PCE) Control of Wavelength Switched Optical Networks (WSONs), RFC6163, April 2011.

[45] ITU-T, Architecture for the Automatically Switched Optical Network (ASON), Recommendation G.8080/Y.1304, November 2001 (Revision, January 2003).

[46] P. Tomsu, C. Schmutzer, Next Generation Optical Networks, Prentice Hall PTR, 2002.

[47] R. Mahalati, R. Dutta, Reconfiguration of traffic grooming optical networks, in: Proceedings of the Broadband Networks, 2004, BroadNets 2004, First International Conference, 25–29 October 2004, pp. 170–179.

[48] D. Simeonidou, R. Nejabati, G. Zervas, D. Klonidis, A. Tzanakaki, M.J. O'Mahony, Dynamic optical-network architectures and technologies for existing and emerging grid services, J. Lightwave Technol. 23 (10) (2005) 3347–3357.

[49] Z. Pandi, M. Tacca, A. Fumagalli, L. Wosinska, Dynamic provisioning of availability—constrained optical circuits in the presence of optical node failures, J. Lightwave Technol. 24 (9) (2006) 3268.

[50] J. Berthold, A. Saleh, L. Blair, J. Simmons, Optical networking: Past, present, and future, J. Lightwave Technol. 26 (9) (2008) 1104–1118.

[51] N. Sambo, N. Andriolli, A. Giorgetti, L. Valcarenghi, I. Cerutti, P. Castoldi, F. Cugini, GMPLS-controlled dynamic translucent optical networks Network, IEEE Netw. 23 (3) (2009) 34–40.

[52] R. Nejabati, E. Escalona, P. Shuping, D. Simeonidou, Optical network virtualization, in: 15th International Conference on Optical Network Design and Modeling (ONDM), 8–10 February 2011, pp. 1–5.

[53] E.W. Dijkstra, A note on two problems in connexion with graphs, Numer. Math. 1 (1956) 269–271.

[54] R. Bhandari, Survivable Networks: Algorithms for Diverse Routing, Kluwer Academic Publishers, 1999.

[55] <www.geni.net>.

Energy-Efficient Telecommunications

17

Daniel C. Kilper[a] and Rodney S. Tucker[b]

[a]Bell Labs, Alcatel-Lucent, USA

[b]Center for Energy-Efficient Telecommunications, University of Melbourne, Australia

17.1 INTRODUCTION

Energy has long held a central role in the physics of optical communications and more generally in communication theory. We communicate through symbols that are generated and manipulated using energy to transition physical systems through different energetic states at a cost that is determined by losses to the environment. Information bits in fiber optic systems are conveyed using photons, each carrying an energy equal to the optical frequency times Planck's constant, and when converted to electronic signals, these bits are typically represented in the form of energy stored in electrons on capacitances.

The Shannon channel capacity is precisely the minimum entropy required to receive a bit of information error free within a communication channel and prescribes the minimum energy per bit at a given temperature. In practice, however, commercial fiber optic telecommunication systems typically operate far from the Shannon limit. Furthermore, the electrical power drawn by a telecommunications equipment rack is many orders of magnitude above the power set by these fundamental energy limits [1] and indeed also far above the power in the optical signals that are launched into the fiber. An optical line card might output an optical signal at a power level of 10 mW, while drawing electrical power typically above 10 W. Historically, the energy use in such systems has been dominated by the control and management electronics, which are largely independent of the bit rate [2].

This points to the historical scalability of optical systems: as the bit rate has increased, the energy and consequently the footprint of the equipment has exhibited only modest increases. Efficient system design was good practice and per-bit energy efficiency was a consequence of capacity expansion. Longer transmission reach and increasing optical spectrum likewise improved the overall system energy per bit, contributing further to the scaling [2]. Figure 17.1 shows the historical energy per bit of 1000 km transmission for Trans-Atlantic communications comparing wireless, telegraphy, coax, and optical systems [3]. These systems have enjoyed a 20% per annum efficiency improvement since the early radio frequency and telegraphy systems. Optical systems are at the leading end of this trend and have experienced

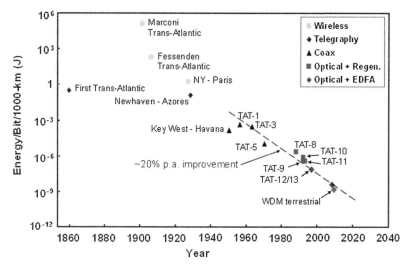

FIGURE 17.1 Energy per bit per 1000 km reach for Trans-Atlantic communication systems [3].

roughly three orders of magnitude improvement in the last two decades. Sharing a similar technology base, we expect that terrestrial systems have seen similar efficiency improvements.

Energy per bit is a widely used metric for energy efficiency in optical communication systems. As seen in Figure 17.1, however, in some cases one must account for the scale of the system or network and thus metrics such as energy/bit/distance may be relevant. The energy/bit metric often requires a more exact definition. For an optical device such as a modulator, one can simply calculate the energy used for each bit of information generated in the modulator. Transceivers or line cards under development today might make use of sleep modes or rate adaptation. Therefore, either multiple energy/bit values need to be reported depending on the mode of operation or an average is taken over a period of time and for a given traffic pattern. In networks, there is not only application data that is ultimately delivered to and used by the end user (often referred to as goodput data) but also other overhead data associated with the internal operation of the network.

For good network energy performance the quantity of interest to minimize is typically the mean energy used to support an average goodput traffic level characteristic of services in the network. The energy/bit can also be specified using the traffic capacity of the equipment rather than the mean amount of traffic carried by the equipment when in service. This metric is useful for comparing different individual elements, but can be misleading since systems are often underutilized for much of their life. An element that is highly efficient at full capacity may have very poor efficiency at low utilization due to a large overhead or base unit (traffic independent) energy requirement. The energy per bit is sometimes specified as power per data rate because the power and data rate are quantities that are normally measured in practice.

Other important energy-related metrics include the carbon footprint or greenhouse gas (GHG) impact, total energy or power, and the energy or power per user. Methods for determining the carbon footprint of network equipment are described in the GHG protocol guidelines [4]. This can be calculated on a per-user or per-service basis to determine the GHG emissions associated with using telecommunications. It is important to understand that the carbon footprint per user varies widely depending on the region of the world due to (a) the extent to which the electric grid utilizes renewable energy, (b) the extent to which telecom equipment makes use of local renewable energy, and (c) the state of technological development and practice of the telecom equipment in the region. For example, data-based telephony is far more efficient than the old POTS systems, so the extent to which the POTS network has been replaced by data networks will dictate the efficiency and carbon impact of telephony. In some parts of the world such as British Columbia, Canada, more than 90% of the electrical grid operates off renewable energy [5] and therefore the carbon impact of telecom equipment connected to that grid is 10 times lower than when used in other regions.

Due to the massive aggregation of signals in commercial data networks, the power per user has been small in comparison with other information technologies. A single 10 Gb/s wavelength might carry the data from 10,000 individual computers. The average power used by the optical equipment to serve a user browsing the Internet has been estimated at below 1 W [6]. Therefore from an overall carbon footprint perspective, optical systems have not been a great concern.

List I. Recent trends bring focus to energy in fiber optic telecommunication systems:

1. Increasing use of optical systems in carbon-sensitive applications and access networks; ubiquitous use of optical systems to enable smart or green technologies.
2. Rapidly increasing power needed to support high-capacity optical systems resulting in an end to traditional optical system scaling.
3. Thermal hot spots in central offices and data centers.
4. Slowing of power scaling in electronics as a result of challenges related to continued scaling according to Moore's law.

A number of technological and application issues have emerged in recent years that are resulting in a changing perspective on energy use in optical communication systems (see List I). Currently data traffic is increasing at approximately 40–50% per annum in mature markets [7,6]. This rate of change is much larger than the historical system efficiency improvement rate of 20% shown in Figure 17.1. In a scalable system, the capacity can be increased over time with only a modest increase in power, cost, or footprint. Thus, the rate of efficiency increase in the equipment must be similar to the traffic increase. Recent estimates suggest that the telecommunications network consumes approximately 1% of the world's electricity supply [8]. Extrapolations of the energy consumption of the network, taking into account projections for increases in traffic growth and user numbers [9], suggest that this could rise to 10%

or more of the world's electricity supply by 2020 if there are not significant improvements in the energy efficiency of the network.

If we write the total mean network power as $P = T/\eta$, where η is the equipment or network efficiency (with units of mean traffic load per mean operating power—written as the inverse of the energy per bit) and T is the mean traffic volume, then a higher rate of growth in T compared to η will result in higher power. In Figure 17.2, the traffic volumes for the Northern America region (US and Canada) and equipment efficiencies projected to 2020 are normalized to their respective 2007 magnitudes. As these quantities diverge, the required power for communication increases.

In addition to the challenge of keeping up with the rapid growth of traffic, telecom equipment is also contending with technology power scaling issues that are expected to slow the rate of efficiency improvement [10]. Most of the power in telecom equipment is used by the electronic processing elements, which are tied to Moore's law. Moore's law is expected to continue for 2–3 more decades through a variety of innovations including new materials and 3D structures. However, the efficiencies have not kept pace with increases in both speed and density. Consequently processors have moved to parallel processing in multiple cores rather than increasing the clock frequency. The associated slowing in energy per bit for processing in telecom gear has resulted in higher thermal densities, larger multi-rack systems, and higher total power.

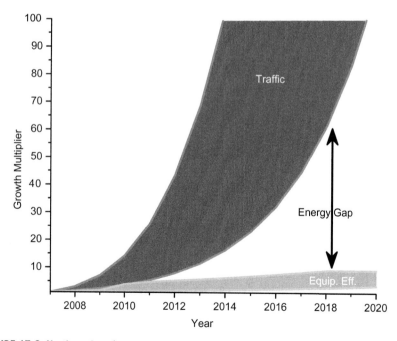

FIGURE 17.2 Northern America core network traffic growth, mobile and wireline, and telecom equipment efficiencies normalized to a start in 2007 and projected to 2020 [6].

The costs associated with rising energy use in the equipment are also a concern. The additional power required due to the gap between the growth of traffic and equipment efficiency impacts the network operating budget. With rising energy costs and increased regulations on GHG emissions, there is more pressure on controlling energy use in optical systems. As mentioned above, the total energy use associated with core networks is relatively small. However, it is increasing rapidly and network operators may not be prepared to see their energy bills double in a period of a few years. Including both wireline and wireless networks, telecommunications equipment amounts to an estimated 0.25% of the worldwide carbon footprint [8]. BT and Telecom Italia are reported to be among the top two energy consumers for their respective countries [11]. In other regions such as North America, the energy costs account for only a few percent of the total network operating costs, so the importance of energy costs as seen by network operators varies regionally.

At the same time that network energy use is gaining attention, communication systems in general are receiving interest as a potential platform to enable carbon offsetting strategies for other sectors of the economy [8,12]. Information and communication technologies (ICT) can be used to build so-called smart systems that use automation to control energy use. Such systems might regulate heating and cooling systems in buildings or control automobile traffic signals to reduce idling and transit times. The goal of such measures is to realize reductions in greenhouse gas (GHG) emissions and therefore one seeks low or zero carbon solutions as the communication systems are built out to provide this ubiquitous control capability. Fiber optic systems are an attractive solution in this environment due to their good energy efficiency. However, this efficiency is realized through high capacity and many applications do not have high throughput. Fiber to the home has only recently become economical as home access rates start to approach optical signal speeds, although significant aggregation is still used in the passive optical network (PON) systems widely used today. As optics moves into the home and new application areas, greater focus is placed on the power and energy efficiency of these systems.

A number of studies have looked in detail at these telecom energy trends to try to understand the long-term implications [13,6,14]. In Figure 17.3 we show one such analysis that considered various state-of-the-art commercial technology models to determine business-as-usual energy trends to 2020. Figure 17.3a shows the estimated power per user for a Northern America model network. Since these are state-of-the-art models, the power associated with deployed networks will be higher than these curves by a factor estimated to be between 2 and 5 times. This is due to the use of less-efficient legacy equipment and non-state-of-the-art network practices or constraints. The analysis includes service-dependent traffic models with a range of traffic growth rates shown by the increasing spread in the curves. Figure 17.3b shows the same trends broken out by equipment type. Although the WDM/optical core network equipment consumes by far the lowest power, it increases rapidly over the decade. Such increases would be difficult to sustain from a business perspective.

In Figure 17.3a we also show a scenario of "optimistic improvements," in which various efficiency measures were taken to the ideal limits. For example, excess

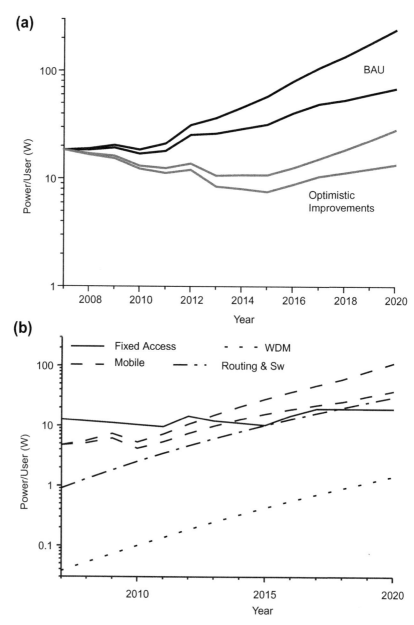

FIGURE 17.3 Energy projections for a Northern America model network assuming state-of-the-art equipment and practices. (a) Total business-as-usual trend for a range of traffic growth models and a best-case "optimistic improvements" estimate; (b) the corresponding BAU trends broken down by equipment type [6].

capacity provisioning is taken to zero for a perfect load proportional network. Such measures cannot be achieved in practice, so these curves should be taken as bounds on the potential efficiency gains with current technology, not predictions.

The main conclusion from this analysis is that even in this ideal situation, the network power can at best be kept roughly flat over the next decade. Note also that efficiency measures, such as sleep mode operation, provide a fixed "one-time" improvement that does not continue into the future, unlike the benefits of electronic scaling. Therefore, new network architectures or practices, or transformations in the underlying technology base are needed to address the long-term energy trends. An example at the component level of such change that is being investigated today would be to bring the optical signals all the way to the chip and possibly even on-chip in order to reduce the interconnect losses in the electronics (which are starting to dominate the energy use in high speed and high density electronics) and eliminate redundant components in the transmission hardware [15]. At the architectural level, research is under way to break through the current layered network structure so that services and applications can use awareness of the physical layer technologies to streamline connections and avoid inefficient use of the underlying equipment [16]. Such transformative measures are bounded not by the curves in Figure 17.3, but instead by the practical or physical limits of the technology. In Sections 17.3 and 17.4 we will investigate these more fundamental limits.

17.2 ENERGY USE IN COMMERCIAL OPTICAL COMMUNICATION SYSTEMS

17.2.1 Long reach and core transmission systems

State-of-the-art long-reach and core transmission systems use coherent techniques with advanced modulation formats to achieve high spectral density and high capacity in a single fiber. Much of the focus in these systems has been around addressing the so-called (single-) fiber capacity crunch. The latest commercial systems use 100-Gb/s interfaces on a 50-GHz grid in the C-band. Recent field trials and lab experiments have shown greater than 1 Tb/s in a single interface using multi-carrier techniques [17]. With the use of coherent receivers and very high spectral density, optical orthogonal frequency division multiplexing has become a candidate for future optical systems [18,19]. Recent studies have examined the various contributions to energy use in these high capacity and high spectral density transceivers. Figure 17.4 shows such an analysis for a 100 Gb/s quadrature phase-shift keyed transceiver that is characteristic of the state of the art in commercial systems [20].

Roughly 25% of the transceiver line card power is dissipated in the digital signal processing electronics used in the coherent reception. The management electronics including thermal management (mostly fans) accounts for 20% of the total. The 10-Gb/s short-reach client interfaces and the framing electronics account for about another 25% of the total. Forward error correction consumes less than 5% of the total

	Component	Units	Unit Power (W)	Power (W)
Client side	Client-card	10	3.5	35
	Framer	1	25	25
FEC	FEC	1	7	7
E/O Modulation	Drivers	4	9	36
	Laser	1	6.6	6.6
O/E Receiver	Local oscillator	1	6.6	6.6
	Photodiode + TIA	4	0.4	1.6
	ADC	4	2	8
	DSP	1	100	100
FEC	FEC	1	7	7
Client Side	Deframer	1	25	25
	Client-card	10	3.5	35
	Management power	–	20% of above	58
	Total Power			351

FIGURE 17.4 Breakdown of energy use in a 100 Gb/s long haul line card [20], courtesy of Annalisa Morea.

and the remainder (60 W) is consumed by the optical transceiver itself: laser, modulators, and receivers.

The power profile in Figure 17.4 is a departure from earlier generations of line cards. For earlier line cards operating at 10 Gb/s and lower rates, the line card power was dominated by the management electronics, which is largely independent of the bit rate [2]. Within the current trend, line card power is strongly dependent on the bit rate and thus the power will increase with increasing capacity. Figure 17.5 shows the recent line card power trends and extrapolates this forward. Ten gigabytes per second is seen as a crossing point at which the bit rate dependent "line hardware" power is roughly equal to the (bit rate independent) management or common equipment power. If this trend continues, 1-Tb/s line cards will require more than 1 kW, which is roughly the power used by an entire shelf today. With greater reliance on electronic processing, this trend will benefit from efficiency improvements in the electronics over time. However, the move to higher capacity interfaces typically requires additional components. For 40-Gb/s, tunable dispersion compensators and optical pre-amplifiers were added. At 100 Gb/s, coherent receivers and digital signal processing were added. These additional components have fueled the trend in Figure 17.5.

At high bit rates, the transceiver line cards typically consume the largest fraction of total power in an optical transmission system. However, other line elements are important when considering the overall picture. Single-band Erbium-doped fiber amplifiers (EDFA) are the most commonly used line amplifiers and consume roughly 50 W. Of this, 10–20 W is the amplifier module itself including the high optical power pump lasers and their thermoelectric coolers. The energy per bit of the amplifier depends on the number of channels it can support. A typical C-band amplifier supports approximately 90 channels at 50 GHz spacing. Fully loaded with 100-Gb/s

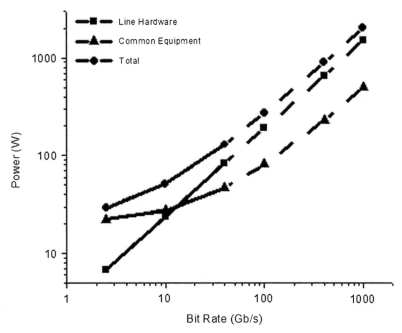

FIGURE 17.5 Transceiver line card power trends showing bit rate dependent line hardware and bit rate independent common equipment separately. Values for 100 Gb/s and above are extrapolated from commercial values for other interface rates [2].

channels, an individual C-band EDFA can achieve 5 pJ/b. A long haul system will utilize two amplifiers every 50–100 km, plus roughly two amplifiers per degree at each add-drop node. Thus a long haul system might include 10–100 or more amplifier line cards. Expressing the efficiency in terms of the distance, the state-of-the-art energy efficiency in commercial systems is around 100 fJ/b/km.

Chromatic dispersion can be compensated in-line passively using dispersion compensating fibers (DCF) or using electronic techniques at the receiver. A recent study has compared the energy use for these two approaches [21]. If only a few channels are present, then the multichannel benefit of the DCF is small as well as the per-channel electronic dispersion compensation power and therefore electronic techniques may be advantageous. As the system is more heavily loaded, then DCF becomes more attractive. A key issue is to what extent the DCF requires additional amplification in the line. Since the amplifier power is usually dictated by the output channel power, additional mid-stage loss for DCF may not incur significant impact on the power. On the other hand, if one can move to a single stage amplifier, then the savings can be significant.

Today transmission line cards do not support power adaptation modes [22]. Recent studies have examined the potential for transitioning line cards to use sleep modes.

Powering transceivers down to an idle mode that leaves the lasers on was estimated to enable 67% energy savings for 100-Gb/s transmission [20]. In another study, rate adaptation combined with power adaptation was shown to enable as high as 90% energy savings [23].

In transmission system design there are a number of interesting trade-offs to consider [24,14]. A transmission system employs engineering rules that include performance margins to account for component variations, varying operating conditions, and uncertainties in the network design. Large margins ensure that new connections will meet the bit error rate requirements and retain reliable communication over the life of the system [25]. A consequence of large margins is that more transceivers are needed in order to regenerate signals that cannot fulfill a connection. These regenerators increase the energy used by the system. Smaller margins will therefore enable lower energy use, but with a greater likelihood of a failure or other transmission problems. Thus one can trade transmission performance for energy by adjusting the margins. Different services may have very different performance requirements and thus may provide opportunities for optimization in this regard. Furthermore, the wavelength blocking probability may be affected by the choice of routes based upon energy optimization, which in turn decreases the efficiency of the network [26].

Another transmission trade-off that has received much interest in research is bit rate and modulation format adaptation balanced with transmission reach or impairment tolerance [27–30]. Different combinations of bit rate and modulation format will have different reach and impairment tolerance performance. Higher bit rates and advanced modulation formats can achieve higher spectral efficiencies and potentially save energy, but they will usually have a much shorter transmission reach. If a system only requires short reach, then this may be an efficient solution. However, if long distances are required, then the frequent regeneration will drive up the energy use. Such network design considerations from an energy perspective are quite complicated and will require extensive study. Until recently such design aspects have only been considered from a cost or performance perspective, and not energy.

17.2.2 Access networks

The public access network is the least energy efficient of the networks that comprise the Internet. Figure 17.6 shows the power consumption per user against the access bit rate for a range of access technologies, including wireless (LTE), fiber to the node (FTTN), hybrid fiber coax (HFC), digital subscriber line (DSL), passive optical network (PON), and point-to-point optical (PtP) [31]. As can be seen in Figure 17.6, the power per user of wireless access networks rises rapidly as the access rate increases. This is primarily because as the data rate increases, fewer users can share the resources of a wireless cell. There is also an energy cost associated with mobility and ubiquitous connectivity afforded by cellular systems.

Unlike wireless access, the power consumption per user in wired access technologies is generally independent of the access rate, as indicated by the lower band

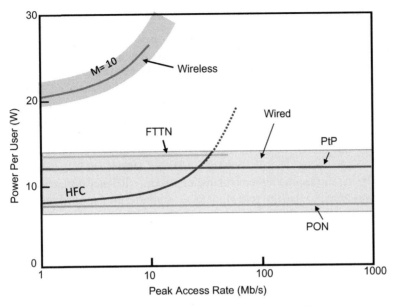

FIGURE 17.6 Comparison of the power per user for wireless and wireline access technologies [31].

in Figure 17.6. PON is the most energy-efficient access technology over a range of access rates. The inefficiency of PON relative to other transmission technologies (as shown in Figure 17.3b) is partly due to the fact that PON equipment is provisioned for peak traffic and transmission rates, which at the edge of the network can be several orders of magnitude above the mean traffic. Provisioning well above the mean is important in order to ensure a high quality of service for the end user experience. Sleep mode operation can be especially effective in this environment, provided that the transitions are fast enough so that they do not impact the user experience. Fiber-to-the-premises (FTTP) access networks using PON technologies such as EPON and GPON are being rolled out in many countries, and a number of research groups and consortia have established research programs aimed at further improving the energy efficiency of PONs, including today's technologies such as GPON, next-generation XGPON technologies, and future yet-to-be-standardized technologies.

Most of the energy consumed in PON networks is attributed to electronic circuitry associated with processing of the data and managing the various signaling protocols. Approximately 20% of the energy is consumed in the optical transmitters and receivers [32]. Sleep modes are being aggressively developed to reduce the energy use in the electronics, and advanced switching techniques such as bit-interleaved de-multiplexing [33] can further improve energy efficiency.

17.2.3 Switching and routing equipments

Energy use in switching and routing equipment used in optical networks is largely dictated by the efficiency of the electronic processing elements. These in turn are tied to Moore's law-related semiconductor trends. In a recent study [10], the energy efficiency of commercial optical networking equipment was analyzed starting from data in 2005 and then projected forward to 2020. One complication with comparing such equipment from different vendors is that the efficiency varies widely depending on the available semiconductor process technology at the time of product introduction. In this study, an average efficiency was taken for equipment across the industry in each year; however, the efficiency of equipment introduced in earlier years was adjusted by the corresponding semiconductor process efficiency difference—as if the product was introduced in the year of interest. Similarly, the expected semiconductor efficiency trends out to 2020 were used to obtain extrapolated values. The efficiency of commercial equipment ranges from 0.5 nJ/b to 8 nJ/b for 2012. Note that this is for a bi-directional system. The rate of efficiency improvement is slowing to approximately 10% per annum as is evident from the curves in Figure 17.7 and this is mostly due to the slowing of the semiconductor energy efficiencies. Since core routers are already heavily constrained due to thermal density limits in central offices, this trend will bring even greater pressure on energy efficiency measures.

Recent studies have considered the potential of optical circuit, flow, or packet switching as alternatives to electronic techniques. These optical techniques suffer from two main limitations: (1) inefficiency of optical storage and processing and (2)

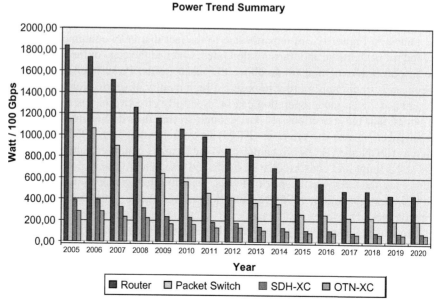

FIGURE 17.7 Power trend for optical networking commercial equipment [10].

complications associated with wavelength switching and transmission. Both of these limitations are closely tied to energy use. If more energy-efficient methods of optical storage and processing can be identified or methods of rapid and efficient wavelength switching introduced, then these solutions will likely find use in the marketplace. Several recent approaches to optical switching have received commercial interest [34]. We address these methods below in more detail from a fundamental energy consumption perspective.

Optical switching can be a very efficient means of moving large volumes of traffic within a network. However, there is also energy used in configuring and controlling such systems. In order to switch a wavelength over long distances the network must first determine the path, accounting for transmission performance, provide signaling to set the elements along the path, control the different elements for stable provisioning without interfering with other signals, and finally communicate with the higher layer elements that will use the new connection [25]. The energy associated with this overhead is difficult to estimate, but must be traded off against any benefit from the optical switching itself.

With optical circuit or flow switching, one also has the issue that latency may be impacted due to either the switching event or due to the wavelength occupancy. As mentioned above, there are many operations that need to be performed to switch a wavelength and these may introduce considerable latency. Once a connection is established and a large flow or transaction is introduced, then the channel is occupied until this flow or transaction finishes. Therefore small flows or transactions may incur large delays—a log-jam effect [35]. In Figure 17.8, this effect is studied considering different interface rates. The delay decreases due to such "log-jam" effects with increasing bit rate. However, the energy per bit shows an optimum due to setup and teardown delays that have a larger impact at higher rates. Breaking up the signals into smaller either statistical or time multiplexed pieces can alleviate this problem,

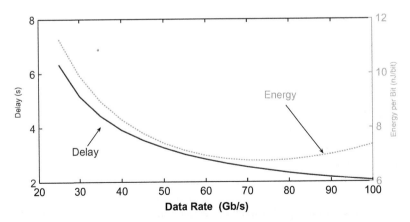

FIGURE 17.8 Energy and delay in an optically switched system considering different interface rates and including 1 s setup and teardown delays [23].

but then more processing is required, which may remove any benefit from the optical switching.

One important use of wavelength switching is to by-pass routers and other higher layer and processing intensive hardware. Today this is done at a node to provide patch panel capability—"glassing through" connections that don't need to be processed [25]. When switching is used, the efficiency is dependent on the persistence of the connection or the transaction size relative to the time required to set up the dynamic connection. The minimum transaction size can be written as: $B > CT^0 p_{wdm}/p_r$, where C is the wavelength capacity (e.g. 10 Gb/s), T^0 is the setup (and teardown) time, p_{wdm} is the energy per bit of the transmission equipment, and p_r is the energy per bit of the router or other processing hardware [36]. Assuming the transmission equipment is 10 times more efficient than the routing equipment, for a 1 s setup time, switching of a 10-Gb/s wavelength would only be efficient for transactions of 1 Gb or larger. In systems today, wavelength provisioning over long haul distances might require a setup time of 1000 s or longer.

From an architectural perspective one may trade off transport energy versus processing or storage energy. For the case of content delivery, numerous caches of content close to the end users will reduce the transmission distance to a given copy, but will increase the energy due to storage [37]. Some types of data processing, such as compression coding or video processing, can reduce the number of bits that need to be transported for a given content element, but will use energy to do this. A variant of the transport and processing or storage energy trade-off is to consider the use of renewable energy. Processing and storage locations, particularly in terms of data center facilities, can be migrated (and turned on and off) around a network depending on the local availability of renewable energy [38,39]. In this way, data and telecom centers can participate in the energy marketplace. The main cost associated with this approach is the adaptation of the transport system to the changing connections and in some cases moving necessary data between centers—resulting in potential large volumes of data being transported over large distances.

17.2.4 Overhead energy and common equipment constraints

Commercial communication gear includes common equipment elements associated with operation, maintenance, and regulatory requirements. These include electrical power regulation, thermal management, safety systems, operator communication interfaces, logging and debugging capabilities, and alarm management systems, among others. Furthermore, a line card will include elements that directly support the communication functions, but do not use energy in proportion to the energy of the communication signal itself. Figure 17.9 illustrates the flow of energy in a practical subsystem (or system), and illustrates how the total input energy per bit period E_{Total} into the subsystem is the sum of two components: E_f and E_{ovhd} [3]. E_f is the energy per bit period required to perform the key active function of the subsystem and E_{ovhd} is the energy per bit period consumed by ancillary functions that are not essential to the key function but are required in order for the subsystem to operate. The power

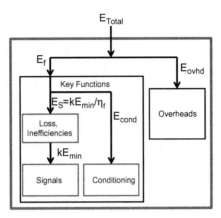

FIGURE 17.9 Model of energy overheads in commercial subsystems.

P_{ovhd} consumed by the overhead ancillary functions is independent of the number of wavelengths k and the bit rate B_r on each wavelength channel. The overhead energy per bit is related to the bit period B_r by: $E_{ovhd} = P_{ovhd}/kB_r$.

The key function energy includes contributions from all elements that are bit rate or signal energy dependent and is further divided into the energy associated with the conditioning elements E_{cond} and the signal elements E_S. The minimum energy used in the optical signals is given by kE_{min} (just E_{min} when considering a single wavelength sub-system such as an optical transceiver), which is related to the signal energy Es by the system efficiency η_f. The conditioning elements include digital signal processing electronics in coherent receivers, signal framing and multiplexing electronics, and modulator drivers. In commercial optical transceivers, the overall subsystem efficiency is $\eta_{subsystem} = E_{min}/E_{Total}$ is approximately 0.001 or less and in commercial optical line amplifiers, $\eta_{subsystem}$ is around 0.01. As described above, historically, optical transceivers have been characterized by $E_{ovhd} \gg E_f$. From Figures 17.4 and 17.5, we see that E_{ovhd} for current 100G transceiver line cards is now roughly 20% of the total. However, signal efficiency remains small due to the large conditioning component energy. For line amplifiers, the key functions make up roughly 10% of the total. Thus for transceivers, the focus should be on reducing the energy associated with the signal conditioning elements, whereas for optical amplifiers the emphasis should be placed on reducing the overhead power.

17.3 ENERGY IN OPTICAL COMMUNICATION SYSTEMS

Energy is a fundamental cost that applies to all telecommunications. Generally, this cost increases with performance and features. A major effort in optical fiber telecommunications research and development today is to understand the details of the energy cost functions and to use this information in optimizing system efficiency and performance. These cost relationships are often cast in the form of a trade-off between energy or power and the quantity of interest. In many cases these relationships are straightforward. Generally, a higher bit rate modulator will require more power. Quite frequently, however, an accurate accounting of these costs is highly complex and either requires a full end-to-end network picture or involves architectural and functional changes that make a comparison difficult. As an example, consider a higher bit rate modulator that uses more power than devices designed for lower bit rates. Depending on the cost function, it might be more efficient to use two lower bit rate modulators. However, the modulators are driven by amplifiers and rely on other electronics that might use more energy than the modulators themselves and duplicate sets of these electronics, even at lower rates, might in fact result in a much higher power for the two-modulator case than the single-modulator case. Thus, the single higher rate modulator, though using more power, results in an overall higher efficiency.

This picture may change again if we consider the possibility of turning off modulators and their associated electronics. If we can put the components to sleep, then the two-modulator configuration might allow for significant energy savings during periods of low traffic since half the electronics are turned off and the system is operating at a lower rate and as a result lower power. This benefit, however, depends on the traffic statistics, and sleep-mode operation might incur wake-up delays or other performance degradations that also need to be taken into account. Although these performance issues might make sleep-mode operation impractical today, with some technology development it might quickly become viable. This chain of dependencies makes energy-efficient optical system design a difficult practice with no clear optimal solution. It also opens the potential for multiple competing architectures, each achieving advantage based upon technology development. An end-to-end perspective, although difficult to come by, is essential for making real progress.

For a given network with a range of specified functional criteria, in principle one can map out the space of energy-efficient technology solutions. An example of this is illustrated in Figure 17.10, which shows a trade-off between functionality (vertical axis) and energy per bit E_b (horizontal axis). Here we only consider energy associated with the key functions as shown in Figure 17.9. The network is assumed to include transmission distance and capacity requirements that result in optical transmission being the more efficient transmission platform, such as in a long haul network. Increasing functionality, expressed in terms of support of heterogeneous services, dynamic capacity/traffic engineering (e.g. packet switching), flexible service provisioning, advanced monitoring and control capabilities, and other processing rich features, requires more energy per bit.

There is a fundamental lower-bound physical limit on the energy cost that defines the minimum energy required for a given functionality. There will be similar cost functions for the state-of-the-art optical and electronic technologies, where the commercial system state of the art is given by the minimum energy for either class of technologies. Indeed, communication systems today are a hybrid of optical and electronic devices, making best use of each depending on the performance and cost requirements. Opaque or electronic technologies are more energy efficient for systems that

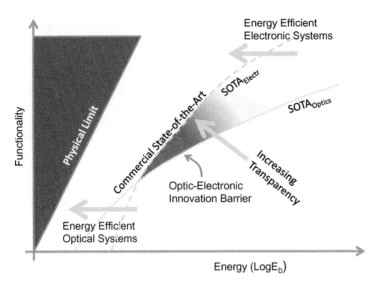

FIGURE 17.10 Energy cost space and opportunities for innovation in optical communication networks. System functionality is expressed in terms of the energy, given as the logarithm of the energy per bit.

require higher functionality [40]. Transparent systems are more energy efficient for systems needing less functionality. Different networks with different size, capacity requirements, topology, etc. will yield a different set of curves, but the same qualitative features usually apply. Note that a related set of curves can often also be defined from a pure cost perspective, in part because cost is related to efficiency.

The multi-layered networks in use today take advantage of these trends (primarily for cost reasons, not energy) by using opaque systems in the higher layers that need advanced functionality and using transparent systems in the lower layers with less functionality. Research to achieve greater energy efficiency in commercial systems can focus on efficiency improvements in either the optical systems or the electronic systems as shown in Figure 17.10 by the horizontal arrows. Since the electronic systems, due to higher functionality, typically use more energy in the end-to-end network transport process than optical systems, they are where the most focus is placed in terms of efficiency improvements (compare the switching and routing trend with the WDM trend in Figure 17.3) [2].

Much research today is also focused on increasing transparency, which effectively moves the state of the art for optics either vertically to achieve more "intelligence" in the optical layer or diagonally to achieve advanced functionality at lower energy (diagonal arrow in Figure 17.10). Note, however, that such efforts to achieve energy savings through transparency must first demonstrate efficiency improvements relative to electronic systems before they will impact the commercial state of the art. Thus there is a barrier, which we refer to as the optic-electronic innovation barrier.

This barrier must be overcome before commercial systems will benefit. Also note that it is possible to move the SOTA$_{\text{Optical}}$ curve without providing any useful benefit to commercial systems, since more efficient electronic solutions already may exist for a particular level of functionality.

The optic-electronic innovation barrier is an important consideration for research on optical systems, particularly those targeting advanced functionality or transparency. Any such solutions will not impact commercial systems until they first demonstrate capabilities that exceed that of electronic systems. An example might be header extraction and processing to determine the output port for a flow or packet of information. This process is normally done using electronics, but optical methods have been demonstrated for simple headers and small numbers of ports. For sufficiently long-lived flows or low operations per bit, some advanced switching or processing functions may be more efficiently implemented using optical techniques. Figure 17.11 shows the results of a recent analysis in which the energy per bit for simple processing operations is compared for optical and electronic methods [40]. As seen in the figure, signal processing functions that involve fewer than 20 operations per bit have the potential to be more energy efficient using optical processing elements compared to electronic processing, at a bit rate of 100 Gb/s and using 2010 technologies. At more than a few 100 operations, optical processing is less efficient even at 1 Tb/s.

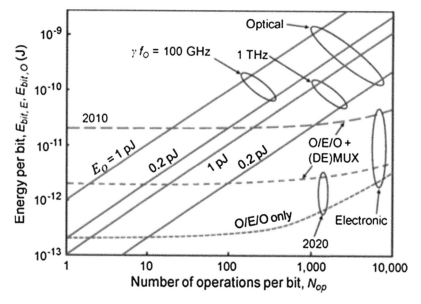

FIGURE 17.11 Energy efficiency of optical and electronic processing functions for different optical device efficiencies (E_0) and generations of electronic devices (2010 and 2020), including energy of optical to electronic to optical (O/E/O) conversion with and without electrical multiplexing. Native optical processing rates (γf_0) of 100 GHz and 1 THz are considered [40].

If opportunities exist to provide functionality with only a few operations per bit, then optically based processing techniques have the potential to provide efficiencies that are 1–2 orders of magnitude better than that obtained through electronic techniques. Advanced packet processing, typical of any modern electronic switch or router, performs functions that far exceed the number of operations per bit at which optical methods become feasible.

So far, we have focused on trends in state-of-the-art telecommunications equipment. In the following sections we examine the basic energy relations that are important for optical telecommunications and further use these to estimate the lower-bound physical limits referred to in Figures 17.10 and 17.11. These sections pertain to the key function elements in Figure 17.9.

17.4 TRANSMISSION AND SWITCHING ENERGY MODELS

17.4.1 Transmission system energy model

Our analysis considers a WDM transmission system, comprising m identical optically amplified stages, as shown in Figure 17.12 [3]. The system is terminated with a WDM optical transmitter at the input and a WDM optical receiver at the output. The transmitter, receiver, and amplifiers support k wavelengths. Each stage has a length (i.e. amplifier spacing) L_{stage}. The total length of each link is $L = mL_{stage}$.

Each stage in Figure 17.12 incorporates a fiber with a fractional power loss of $D = e^{\alpha L_{stage}}$, where α is the power attenuation per unit length of the fiber, and an amplifier gain block with power gain G which is equal to the loss per stage (i.e. $G = D$). There are $(m-1)$ line amplifiers and one amplifier preceding the receiver, giving a total of m amplifiers. The system has an optical bandwidth B_0. Each amplifier contains optical filters (not shown in Figure 17.12) that match the optical bandwidth of the link to the optical spectrum of the optical transmitter. We assume that the optical gain G is the same at all signal wavelengths and that the optical signal power is the same at all wavelengths. Note that the analysis here focuses on basic energy relationships. Additional complexity may be introduced in considering system impairments due to dispersion and other non-ideal behavior of the transport medium

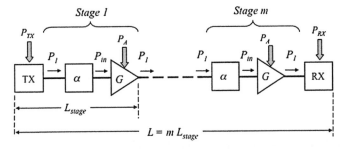

FIGURE 17.12 WDM transmission system of length L, comprising an optical transmitter, m identical stages of optical gain, and an optical receiver. Each stage has length L_{stage} [3].

as appropriate for the system under consideration. Different fiber types introduce different levels of dispersion. This dispersion requires compensation and possibly introduces a need for additional amplification or signal processing.

The total power P_{out} consumed by the transmission system in Figure 17.12 is given by

$$P_{tot} = m P_A + P_{TX/RX}, \tag{17.1}$$

where P_A is the supply power to each amplifier and $P_{TX/RX} = P_{TX} + P_{RX}$ is the supply power to each WDM transmitter/receiver pair and P_{TX} and P_{RX} are the transmitter and receiver supply powers. At each wavelength, the signal input power and output power at the input and output of each amplifier are P_{in} and P_1, respectively, as shown in Figure 17.12. The signal input power at each wavelength can be written as $P_{in} = P_1 e^{-\alpha L_{stage}}$ and $k(P_1 - P_{in}) = \eta_{PCE} P_p$, where η_{PCE} is the amplifier power conversion efficiency and P_p is the amplifier pump power [41].

The pump power P_p is related to the total amplifier supply power P_A via the expression, $P_p = \eta_E(P_A - P_{ovhd})$, where η_E is the power conversion efficiency of the amplifier control and management circuitry, including the temperature controllers and the pump laser. Neglecting the overheads and combining these equations, we obtain $P_1 = P_{in} + \eta_{EPCE} P_A / k$, where $\eta_{EPCE} = \eta_E \eta_{PCE}$ is the overall electrical power conversion efficiency of the amplifier [41].

17.4.2 Lower bound on energy consumption of optically amplified transport

We now focus on determining the lower limit or lower bound on transport energy using the power model of the optically amplified transmission link. Here we use a semiclassical model that is appropriate for fiber optic systems with high losses and phase-insensitive amplification [42]. Fully quantum treatments are covered elsewhere [43]. In this model, the minimum signal power is limited by the magnitude of the noise in the optical amplifiers. At each wavelength, the optical signal-to-noise ratio (OSNR) of the output of stage m in each optically amplified link is given by [3,41]:

$$OSNR = \frac{P_1}{2n_{sp}m(e^{\alpha L_{stage}} - 1)h\nu B_0}, \tag{17.2}$$

where P_1 is the average signal power at each wavelength at the output of the transmitter and each of the amplifiers, n_{sp} is the spontaneous emission factor of each of the amplifiers, h is Planck's constant, and ν is the optical frequency.

The signal-to-noise ratio per bit SNR_{bit} is given by $SNR_{bit} = 2\tau_{bit} B_0 OSNR$, where τ_{bit} is the bit period [44]. Thus the energy per bit of the optical signal at the output of the mth amplifier is given by

$$E_1 = P_1 \tau_{bit} = SNR_{bit} n_{sp} m (e^{\alpha L_{stage}} - 1)h\nu. \tag{17.3}$$

Combining (17.1) and (17.3), the total energy per bit per wavelength consumed by all active devices in the optically amplified link is $E_{bit} = P_{tot} \tau_{bit}/k$:

$$E_{bit} = E_{AMP} + E_{TX/RX}, \tag{17.4}$$

where

$$E_{\text{AMP}} = m P_A \tau_{\text{bit}}/k \cong SNR_{\text{bit}} n_{\text{sp}} m^2 (e^{\alpha L_{\text{stage}}} - 1) h\nu / \eta_{\text{EPCE}} \qquad (17.5)$$

is the total energy per bit per wavelength in the amplifiers and $E_{\text{TX/RX}} = P_{\text{TX/RX}} \tau_{\text{bit}}/k$ is the energy per bit per wavelength in the transmitter/receiver pair.

The theoretical lower limit $E_{\text{AMP–min}}$ on the amplifier energy per bit E_{AMP} is obtained by setting the power conversion efficiency η_{PCE} to its theoretical maximum value of ~1.0 for an EDFA with a pump wavelength near 1480 nm, and signal wavelengths around 1550 nm. In addition, we set $\eta_E = 1.0$, the spontaneous emission factors n_{sp} of the amplifiers to unity, and assume zero coupling loss at the input and output of the amplifier.

Figure 17.13 is a plot of $E_{\text{AMP–min}}$ as a function of the amplifier spacing L_{stage} for a total transmission distance $L = 2000$ km and with a fiber loss of 0.2 dB/km, a wavelength of 1.55 μm and for on-off keying (OOK) and differential binary phase-shift keying (DBPSK) with a bit error rate (BER) of 10^{-9} [3]. For this error rate, the required signal-to-noise ratio per bit is 16.1 dB for OOK and 13.4 dB for DBPSK [44]. Not surprisingly, the well-known ~3-dB system sensitivity advantage of DBPSK over OOK [44], as indicated by these SNR_{bit} values translates into DBPSK achieving a lower bound on total system energy per bit. An even more energy-efficient modulation format is 8-level dual-polarization quaternary phase-shift keying

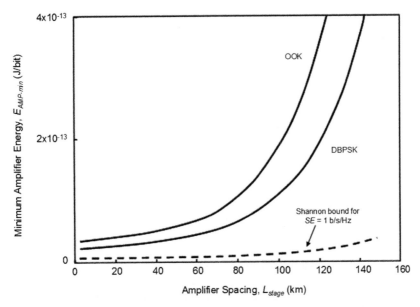

FIGURE 17.13 Minimum total amplifier energy per bit for 2000-km optically amplified system as a function of amplifier spacing. The fiber loss is 0.2 dB/km and the SNR per bit is 16.1 dB for OOK and 13.4 dB for DBPSK [3].

DP-QPSK [45]. This modulation format gives a 1-dB improvement over DBPSK. Also shown is the ultimate limit placed by the Shannon information capacity. The Shannon limit sets the minimum energy per bit required for error-free transmission for an arbitrary modulation format in a dissipative system. Non-dissipative systems [1] have the potential for lower energy transmission, but are not feasible in transmission systems today.

Figure 17.14 shows $E_{AMP-min}$ as a function of the total transmission distance, for OOK systems with amplifier spacing of 50 and 100 km. As before, the SNR per bit at the receiver is 16.1 dB. For a 1000-km link, the transport energy $E_{AMP-min}$ with a 100-km amplifier spacing is about 0.2 pJ/b and for a 10,000-km system it is 5 pJ/b.

Figure 17.14 also shows the minimum amplifier energy per bit $E_{AMP-dist}$ for a system using distributed amplification and the minimum amplifier energy per bit per km of transmission distance for distributed amplification. Note that the amplifier energy per bit decreases as the amplifier spacing is decreased and reaches a minimum for distributed amplification.

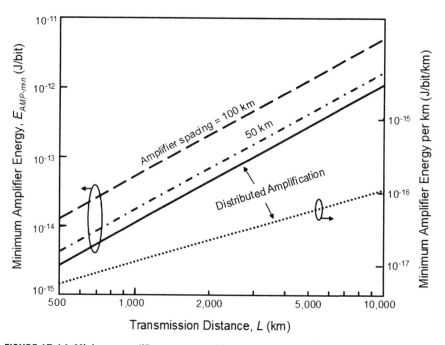

FIGURE 17.14 Minimum amplifier energy per bit for optically amplified system with OOK and amplifier spacing of 50 km, 100 km, and with distributed amplification. Also shown is the minimum amplifier energy per bit per km of transmission distance for distributed amplification. The energy per bit per km for lumped amplifiers has the same slope as this curve, but the absolute values increase with amplifier spacing. The fiber attenuation is 0.2 dB/km [3].

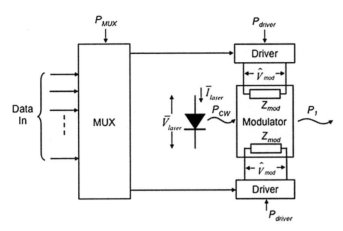

FIGURE 17.15 Optical transmitters using a dual-drive electro-optic modulator [3].

Commercial amplifiers consume much more energy than the lower bound. The intrinsic power conversion efficiency η_{PCE} for 980-nm pumps is $\sim$63%, and for 1480-nm pumps it is $\sim$95%. Including coupling and other losses, the actual η_{PCE} of commercial devices is typically in the range of 40% and the electrical power conversion efficiency η_{PCE} of commercial amplifiers is less than 1% [2]. Therefore, practical systems using today's technology typically consume at least two orders of magnitude more power than the lower bound.

As pointed out earlier, the data in Figures 17.13 and 17.14 are based on an ideal linear WDM system, in which there are assumed to be no nonlinear effects in the fiber or other impairments that require additional transmission margins and therefore higher transmission power. In practice, additional limitations caused by fiber non-linearities and other practical constraints may reduce the actual reach in commercial systems to distances below those indicated in Figure 17.14 [46].

17.4.3 Energy consumption in optical transmitters and receivers

17.4.3.1 Optical transmitters

Figure 17.15 shows a laser transmitter incorporating a multiplexer (MUX) used to combine a number of incoming data streams into data streams that are applied to an electro-optic modulator. The modulator in Figure 17.15 has two electrical ports, but in general more ports may be used, depending on the modulation format [3]. A driver at each port of the modulator supplies an instantaneous voltage of V_{mod} at each port. The input impedance of the modulator, as seen by the driver, is Z_{mod}. The details of this impedance depend on the type of modulator and affect the design of the driver and the interface between the driver and the modulator. These impedance matching considerations influence the energy consumption of the modulator and the modulation speed [3]. The energy consumption of the transmitter in Figure 17.15 depends on the energy consumption of the laser, the drivers, and the MUX.

Table 17.1 provides some examples of energy consumption per bit in optical transmitters. Data are presented for representative examples of 2010-era technology and "target" future device performance that may be achievable through aggressive improvements in technology over the next 10–15 years. These target figures are speculative, but help to identify the kinds of improvements in energy consumption that may be achievable in the 2020–2025 timeframe. The bit rate is taken to be 40 Gb/s for 2010-era technology and 100 Gb/s for the "Target" column. It is possible that some devices will operate at bit rates above 100 Gb/s in the 2020–2025 timeframe. At higher bit rates, the energy per bit figures may be similar to or somewhat higher than at 100 Gb/s.

The data in Table 17.1 includes estimates of energy use for key electronic circuits in transmitters: MUXs and drivers. In making these estimates it is important to recognize the differences between state-of-the-art laboratory results and commercial product designs. State-of-the-art laboratory results often push devices to the limit of voltage and temperature. On the other hand, commercial products need to incorporate large safety margins on device operating conditions.

While speculative, the 2020 figures reflect the performance of laboratory devices today. For example, an 8:1 multiplexer using 130-nm SiGe bipolar-CMOS (BiCMOS) technology has an energy consumption of 10 pJ/b at a bit rate of 87 Gb/s [47], and an integrated 100-Gb/s transmitter module using InP double-heterojunction bipolar transistors (DHBT) consumes 15 pJ/b [48]. In Table 17.1, we have selected an intermediate figure of 10 pJ/b for 2010-era technology at 40 Gb/s. In this context, we note that it is expected that the energy consumption for next-generation 100-Gb/s BiCMOS [48] MUX/DEMUX chipsets could be around 20 pJ/b [49]. High-speed electronic device technologies are improving rapidly and it is not unreasonable to anticipate that with future generations of Silicon-Germanium (SiGe) technology the energy consumption of high-speed multiplexers could be reduced to around 2–5 pJ/b. We use a figure of 2-pJ/b fj in Table 17.1 for the "target" energy consumption for future MUX devices operating at 100 Gb/s.

Table 17.1 Energy consumption in optical transmitters.

Component	Energy per Bit	
	2010-era Technology (40 Gb/s)	**2020-era Target (100 Gb/s)**
MUX	10 pJ	2 pJ
Laser ($\overline{V}_{laser}\overline{I}_{laser}/B_r$)	1.5 pJ	500 fJ
$2\hat{V}^2_{mod}/Z_0 B_r$	4 pJ	1.6 pJ
Driver (P_{driver}/B_r)	25 pJ	3 pJ
Laser + 4 Drivers	~100 pJ	12 pJ

In many current-day commercial optical transmitters and in state-of-the-art laboratory demonstrations of advanced transmitter circuits, the electronic drivers are often conventional commercial RF amplifiers with 50-Ω coaxial inputs and outputs terminations. This kind of amplifier is convenient for laboratory demonstrations, but is not necessarily the most energy-efficient solution. For example, the most energy-efficient circuit for an externally modulated transmitter with a resistively-terminated lumped modulator would be a low-impedance driver (ideally $<<50\,\Omega$) located in very close proximity to the modulator. Alternatively, low-capacitance modulators integrated with high-impedance drivers can achieve high efficiency [3].

A typical state-of-the-art wideband driver for 40 Gb/s systems, capable of delivering voltage swings of up to around 6 V into a 50-Ω load [50], consumes about 2 W of supply power, or about 50 pJ/b, which is much larger than either the laser or the modulator. There is scope for considerable reduction in this energy using advanced circuit designs such as distributed amplifiers. For example, a 180-nm CMOS distributed amplifier with an output voltage swing of 1.6 V into 50 Ω has been demonstrated at 40 Gb/s, with an energy consumption of 6.5 pJ/b [51] and a 50-Ω driver amplifier energy consumption of 5 pJ/b is reported in [48] for a 100-Gb/s driver using DHBT technology. In Table 17.1, we have chosen 25 pJ/b as being representative of the energy consumption of today's drivers.

17.4.3.2 Optical receivers

Figure 17.16 shows a circuit diagram of a balanced optical receiver. The two photo diodes are connected to a transmission line with characteristic impedance Z_0 (typically 50 Ω) and are shunted with an impedance-matching resistor R_S. The transmission line is terminated at the input of an electronic clock and data recovery (CDR) and demultiplexer (DEMUX) circuit with a matched input resistance $R_{in} = Z_0$. The CDR/DEMUX circuit may also include analog-to-digital converters (ADCs). The configuration in Figure 17.16 enables the photodiodes to be located some distance from the electronic components and to use a low-loss transmission line to convey the detected signal to the electronic circuitry. The signal input voltage to the electronic demultiplexers is V_{in}.

The photodiodes in Figure 17.16 are reverse-biased with a voltage V_{Bias}. If an EDFA preamplifier (not shown in Figure 17.16) is employed in front of the receiver, it is possible to achieve an input voltage swing V_{in}^{p-p} on the order of 0.5–1 V. This signal level is generally adequate to directly drive the CDR/DEMUX, thereby eliminating the need for a preamplifier. The power consumption of the clock recovery and

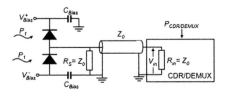

FIGURE 17.16 Photodiode and bias circuitry of balanced receiver front end, with transmission line connecting photodiodes to CDR/DEMUX [3].

Table 17.2 Energy consumption in optical receivers.

Component	Energy per Bit	
	2010-era Technology (40 Gb/s)	**2020-era Target (100 Gb/s)**
CDR/DEMUX	20 pJ	2 pJ
Optical preamplifier power and bias power	35 pJ	2 pJ

de-multiplexer circuits (CR/DEMUX) in Figure 17.16 is $P_{CDR/DEMUX}$. Table 17.2 shows the dominant contributions to power consumption in the optical receiver. In calculating the figures in Table 17.2 we have made the fairly optimistic assumption that the power conversion efficiency of the optical preamplifier can be improved to around 25% by 2020.

17.4.4 Transmission system lower bounds

We have shown earlier in this chapter that the lower bound on amplifier energy consumption is approximately 50 fJ/b in a 1000-km amplified system with a 100-km repeater spacing. Table 17.2 shows that the lower bound on energy in a 2010-era externally modulated laser transmitter with four drivers is about 100 pJ/b (dominated by the drivers). Table 17.2 shows that the lower bound on energy consumption in a 2010-era high-speed optical receiver is about 55 pJ/b. Therefore, the lower bound of the total energy consumption for a 2010-era 1000-km system is around 165 pJ/b with roughly equal contributions from each of the main elements described here.

It is instructive to compare the lower-bound figures in the previous paragraph with the energy consumption data for a typical commercial system as presented in Section 17.2. It turns out that the 110-pJ/b lower-bound figure in the previous paragraph for a 1000-km system is about 2% of the ~5-nJ/b energy consumption in a typical commercial systems. However, the commercial systems are typically only 10–50% utilized due to practical constraints such as survivability, allowances for capacity expansion, and wavelength blocking. Note that the lower bound on total energy consumption in the optical amplifiers is about 0.01% of the ~100-pJ/b in a typical commercial system.

Much research and development today is focused on optical interconnect technologies including intra-core/core to core (chip level), core to memory (board level), and rack to track (inside the data center). These applications require very lower power optical solutions for integration with the electronics and to be competitive with other electronic interconnect solutions. However, due to the short distances the signal conditioning and overhead elements of Figure 17.9 can often be minimized and one can approach the lower bounds of the key function signal components. Figure 17.17 shows current generation electrical and optical interconnect link efficiencies, not including the multiplexing/serialization (serdes) [52]. Directly modulated vertical

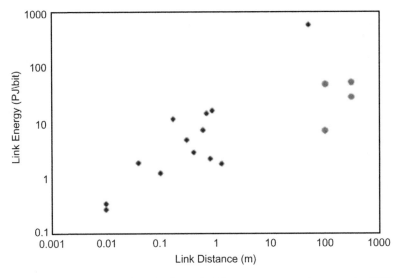

FIGURE 17.17 Transceiver efficiencies for optical interconnects, serdes to serdes [52].

cavity surface emitting lasers (VCSEL) can be used for links of a few 100 m and achieve transceiver efficiencies of tens of pJ/bit (circles in figure). These reflect laser, modulator, and receiver efficiencies achievable today. Future on-chip optical interconnect technologies are targeting a few hundred fJ per bit. These targets are below those given for 2020 in Table 17.1, consistent with the lower optical power requirements for short distances and the reduced conditioning elements/requirements.

17.5 NETWORK ENERGY MODELS

In this section, we consider the energy use in optical networks—combining switching and transmission. We analyze the energy performance of a range of switching devices and technologies and develop quantitative models of the lower bounds on energy consumption in these devices. These models are then incorporated into a simple model of a global switched network and the lower bound on total network energy use is estimated.

17.5.1 Network energy model

The mathematical modeling of modern telecommunications is well advanced, but there is a lack of a comprehensive theory of energy use in interconnected communication networks. Therefore, to obtain an estimate of a lower bound on the energy consumption of networks, we focus on a simple analysis that provides a broad guide to energy use and the corresponding lower bounds in a network of many interconnected users and devices [53].

The telecommunications network provides an end-to-end connection mechanism enabling data from any device connected via an access network to communicate with any other device connected to a network. The lower bound on energy consumption of the network is the minimum energy required for data to traverse the network between the two connected devices. This minimum energy is the energy consumed as each bit of data passes each switching device and the energy used in the interconnecting transport systems between the switching devices. This minimum depends on many implementation or boundary conditions such as the number of end nodes, distances between nodes, amount of traffic, and both time and space traffic statistics, among others. In this section, we estimate the energy consumption for two illustrative examples: (a) a global network with 10 billion attached devices, with the end-user devices operating at bit rates of 10 Mb/s and (b) a future global network with 100 billion attached devices with end-user bit rates of 1 Gb/s [53].

To keep our analysis simple, we model the network as a "minimal" array of switching devices connected using a configurable non-blocking multi-stage Clos architecture [54]. A schematic of the network model is shown in Figure 17.18. In Figure 17. 18, the users are connected through the access network to a stage of multi-port switches that is connected by optical transport systems to further stages of multi-port switches. Any bit of data in an end-to-end connection traverses s stages of switches, $s-1$ stages of core optical transport, and two stages of access network transport.

In general, the switches in Figure 17.18 can use either optical or electronic technologies. To determine the total energy consumption, it is necessary not only to determine the energy consumption of the key functional blocks in Figure 17.18, but also to consider the energy consumption in the interfaces between each port of the switches and the transport medium. The total (end-to-end) energy per bit E_{NET} in the

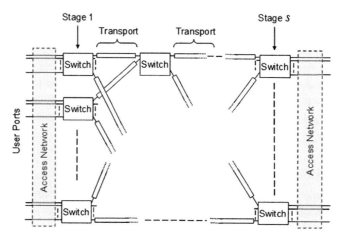

FIGURE 17.18 Simple model of global network, comprising an array of multi-port switches interconnected by optical transport systems and connected to user devices via an access network [53].

network in Figure 17.18 is the energy consumed by each bit in each stage of switching and in each transport system. Thus, E_{NET} is given by:

$$E_{NET} = \sum_{i=1}^{s} E_{switch,i} + \sum_{i=1}^{s-1} E_{core\text{-}transport,i} + E_{access\text{-}transport}, \qquad (17.6)$$

where $E_{switch,i}$ is the energy per bit in stage i of switching in Figure 17.18, $E_{core\text{-}transport,i}$ is the energy per bit in a transport system at the output of a switch in stage i of the network in Figure 17.18, $E_{access\text{-}transport}$ is the two-way transport energy per bit in the access network (i.e. two one-way access transport paths), and s is the number of stages of switching. Note that in (17.6) the energies $E_{switch,i}$ and $E_{transport,i}$ include terms that account for the particular configuration of the interface between the switches and the transport systems. For example, if the switch fabric is all-optical and there are no O/E or E/O converters at the switch ports, the $E_{core\text{-}transport,i}$ term for the transport systems that interface to the switch should reflect this [53].

17.5.2 Switching devices and fabrics

In this section we develop estimates of the energy $E_{switch,i}$ in (17.6) for a variety of different high-speed device technologies. We will combine this with estimates of $E_{transport,i}$ from Section 17.3 and estimate the total network energy per bit E_{NET} using (17.6). This will enable us to determine how device technologies influence total energy consumption.

Figure 17.19 shows a switching device or circuit with j signal input ports and k signal output ports. The switch is powered from a supply (either optical or electrical) and the operation of the switch is via a control input (either optical or electrical). In general, the input ports, the supply and the control ports receive energy over some period (typically a bit period or a switching period) and the output ports deliver energy over the same period. The individual energies are defined in Figure 17.19.

The total energy E_{Total} used by the switch in Figure 17.19 is given by [55]:

$$E_{Total} = \sum_{i=1}^{j} E_{Ii} + E_{Supply} + E_{Control} - \sum_{j=1}^{k} E_{Oi}. \qquad (17.7)$$

Note that *all* input energies contribute to the total energy use of the switch. This point might seem obvious and not worth stating, but the importance of including all contributions to energy consumption is not always acknowledged in the literature relating to optical switching devices. Furthermore, Eq. (17.7) can be expressed as the energy per bit or the energy per switching cycle for cases in which the switch is operated with a cycle time of multiple bit periods. This lower rate switching operation is often represented by a multiplicative activity factor, which proportionally reduces the energy per bit.

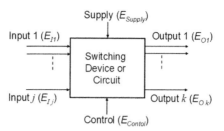

FIGURE 17.19 Energy in switching devices and circuits.

We now consider the relative importance of the various terms in (17.7) for a number of optical and electronic switching devices. We consider two distinct classes of switches: (a) linear analog switches and (b) digital switches. Linear analog switches pass the input signal to the appropriate output, without altering the waveform of the signal (except perhaps for the addition of some noise and/or crosstalk). These switches are linear in the sense that the properties of the device are independent of the input optical signal level. Digital switches, on the other hand, operate at the bit level and generally incorporate highly nonlinear devices such as logic devices.

17.5.2.1 Analog switches: electro-optic devices

A basic building block for linear analog switches is the traveling-wave electro-optic (E-O) phase modulator. Phase modulators form the basis of a number of optical components such as Mach-Zehnder interferometer switches [56] and other more complex switching devices [57]. Electro-optic phase modulators typically incorporate a resistively terminated matched transmission line and present a resistive load at their control terminals.

For an optical switch with a resistively terminated control port, the control energy $E_{\text{Control-}R}$ associated with the switch control function is given by

$$E_{\text{Control}-R} = \tau_s V_\pi^2 / Z_0, \tag{17.8}$$

where τ_s is the period over which the energy is measured, V_π is the switching voltage, and Z_0 is the matched load impedance. For a device with a switching voltage of 3 V and a switching rate of 100 GHz, $E_{\text{Control-}R}$ is 1.8 pJ. At a switching rate of 1 GHz, $E_{\text{Control-}R}$ is 180 pJ. If the drive amplifier is matched to the termination impedance, an additional 180 pJ is dissipated in the driver.

Attenuation is an often-ignored contribution to energy use in analog optical switching. Consider an analog optical switch with an input-to-output loss of 3 dB. If the input optical signal level is 2 mW, the optical power lost in the switch is 1 mW. If $\tau_s = 100$ ns, the energy per switch cycle would be 100 pJ, which is similar in magnitude to $E_{\text{Control-}C}$ (180 pJ).

17.5.2.2 Analog switches: semiconductor optical amplifier gate arrays

Semiconductor Optical Amplifiers (SOAs) are useful optical switching devices because of their fast switching speed (~1 ns) and the ability to achieve a high extinction (on/off) ratio [58,59]. Arrays of integrated high-extinction-ratio SOA gates potentially provide a platform for large cross-connect arrays with sufficient switch-ing speed for potential use in optical packet switching and burst switching [60,61].

A basic 2×2 SOA gate array will employ four SOAs, each of length L_A. This structure can form the basis of a variety of larger gate arrays [58,62]. When an SOA is biased into its on state, the device gain overcomes the losses in the splitters and combiners as well as other waveguide losses in the system. When an SOA is turned off (i.e. at zero bias), the signal is blocked through that path. A critically important parameter affecting crosstalk in large switch arrays is the extinction ratio of the gates between the on and off states [58]. Typically, the required extinction ratio is at least 50 dB.

To achieve high extinction ratio, the intrinsic attenuation of the amplifier needs to be large when the device is unbiased [59]. This large intrinsic loss is relatively easy to achieve in practice. But it means that large on-level bias current is needed to overcome this intrinsic loss when the device is biased to its on level [58]. This extinction ratio constraint on the minimum power consumption of SOA gate arrays is easily ignored in small switch arrays, but it is likely to have significant impact on the scalability of SOA gate arrays. To achieve a reasonable on-level gain in practical SOAs with length of around 1 mm, it is necessary to pump the device with a current of around 200 mA [59]. Therefore the on-level power consumption P_{SOA} is around 400 mW for a device with a forward-bias voltage of 2 V. At a bit rate of 100 Gb/s, this translates to an energy consumption of 4 pJ/b per SOA.

17.5.2.3 Digital switches: CMOS

An example of an electronic digital switch is the simple single-input single-output CMOS inverter shown in Figure 17.20. This inverter has a supply voltage V_{DD}, a leakage current $I_{leakage}$, a combined gate capacitance of C_{gate}, and an output wire of length L_W and capacitance $C_W L_W$. Neglecting the leakage current, the total energy per bit consumed by the inverter and the output wire is:

$$E_{bit} = \frac{1}{2}(C_{gate} + C_W L_W)V_{DD}^2. \tag{17.9}$$

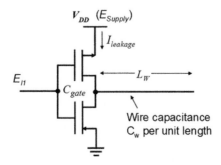

FIGURE 17.20 CMOS inverter.

The energy $1/2C_{\text{gate}}V_{\text{DD}}^2$ term in (17.9) is sometimes referred to as the *switching energy* of the device. For the CMOS inverter in Figure 17.20 the switching energy is simply the input energy required to charge the two gate capacitances. Figure 17.21 shows the switching energy of a single CMOS transistor through different process technologies as a function of time. Figure 17.21 uses historical data and future predictions based on the ITRS roadmap [63]. The feature size of the devices has been shrinking over time, leading to a decrease in CMOS gate capacitance and thus from (17.9) a decrease in switching energy. The smaller feature size, together with better manufacturing process control, has allowed a reduction in operating voltage, leading to further switching energy reductions. It is worth pointing out here that Figure 17.21 assumes continued reductions in CMOS operating voltage beyond today's technology [53]. In recent years, reductions in operating voltage have slowed and if this continues then the rate of reduction of device energy will be less than that shown in Figure 17.21. Note also that Eq. (17.9) and Figure 17.21 do not include leakage currents, which are becoming increasingly significant as CMOS device sizes shrink.

The upper curve in Figure 17.21 represents the total energy per transition, as given by (17.9). This data has been estimated for a 2010-era circuit based on 32-nm

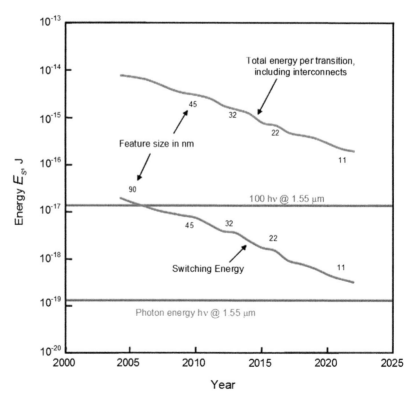

FIGURE 17.21 Switching energy in CMOS gates [53].

devices and projected to 2022 based on the ITRS roadmap [64]. The energies in the upper curve in Figure 17.21 are dominated by device-to-device interconnects [53,65].

For reference, Figure 17.21 also shows the energy of a single photon at a wavelength of $1.55\,\mu m$ and the energy of 100 photons. Note that the switching energy of a single 45-nm CMOS transistor is less than the energy of 100 photons.

17.5.2.4 Switch fabrics: 2×2 crosspoints

We now consider the energy consumption of switch fabrics constructed using an array of 2×2 crosspoints [53]. Figure 17.22a shows a schematic of a $p \times p$ Benes switch [66] based on 2×2 crosspoints. The number of stages r in the Benes structure in Figure 17.22a is

$$r = \lceil 2 \log_2 p - 1 \rceil. \tag{17.10}$$

Crosstalk is a potential barrier to scaling arrays of 2×2 crosspoints, but this crosstalk issue is sometimes overlooked in the literature on the scaling properties of optical switches. One approach to the mitigation of crosstalk is to use the dilated crosspoint structure illustrated in Figure 17.22b [67]. This dilated crosspoint uses four elemental 2×2 crosspoints. It improves the crosstalk significantly, but doubles the optical losses and increases the control energy by a factor of 4. Dilated versions of larger switch architectures are also feasible, but they all come at the cost of significant increase in energy consumption and optical loss.

If the switches in Figure 17.22 are constructed using 2×2 crosspoints in the Benes architecture in Figure 17.22a, the energy $E_{switch,i}$ in (17.3) is given by

$$E_{switch,i} = \sum_{i=1}^{r} E_{crosspoint,i}, \tag{17.11}$$

where $E_{crosspoint,i}$ is the energy per bit in a crosspoint in stage i of the Benes switch in Figure 17.22. If all crosspoints consume about the same energy, (17.11) can be simplified to $E_{switch,I} = r E_{crosspoint}$.

We consider below a network based on core switches with 1024 input ports and 1024 output ports. For $p = 1024$, the number of stages is $r = 19$. Thus, the energy per bit in a 1024×1024 switch is approximately 19 times the energy per bit in a single crosspoint. This logarithmic scaling is an important feature of such systems that contributes to the scalability of communication networks. A wavelength-routed switch fabric can be implemented using optical wavelength converters combined with arrayed-waveguide gratings (AWGs) [68]. An energy analysis of a three-stage switch fabric shows the energy consumption of a switch with 2500 input and output ports to be around $1\,pJ/b$, or about $300\,fJ/b$ per stage.

17.5.2.5 Switching packets

Analog optical switches potentially offer energy-saving advantages when used for switching packets [69]. When switching on a packet-by-packet basis rather than a bit-by-bit basis, it is not necessary to separately operate the switches on each individual bit [70]. This is illustrated in Figure 17.23, which shows a linear analog switch

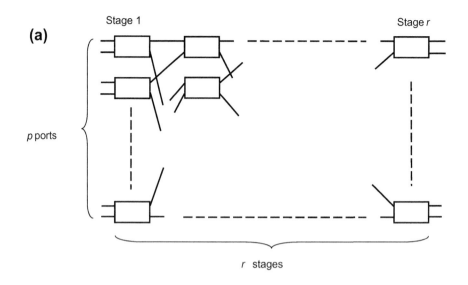

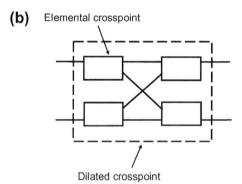

FIGURE 17.22 (a) $p \times p$ switch using 2×2 crosspoint building blocks. (b) Low-crosstalk dilated 2×2 crosspoint comprising four elemental crosspoints [53].

array operating as a packet switch. An incoming packet is shown in Figure 17.23 with a bit period of τ_b and a packet length of τ_p. The switch includes a switch array and an amplifier to overcome the losses in the switch array (not necessary in SOA gate arrays). The amplifier energy per bit is E_{AMP} and the control energy per packet is E_{Control}. The total energy per bit in the optical switch in Figure 17.23 is:

$$E_{\text{bit}} = E_{\text{AMP}} + \frac{E_{\text{Control}}}{N_b}, \qquad (17.12)$$

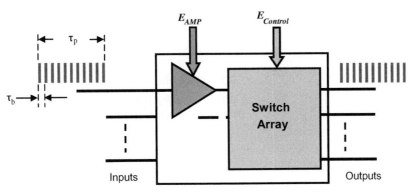

FIGURE 17.23 Packet switching using a linear analog switch array [70].

where $N_b = \tau_p/\tau_b$ is the number of bits in a packet [70]. The second term in (17.12) represents the control energy per bit. In some switch implementations the control energy per bit can be much smaller than the control energy per switching operation. This neglects any energy required to maintain the switching state, such as thermal management or switch holding voltages. In such cases, as the number of bits in the packet is made larger, the control energy per bit becomes smaller.

This reduction in the control energy per bit is a potential advantage of optical packet switches using linear analog optical switch devices. However, this theoretical reduction in control energy per bit cannot always be realized in practice. For example, in higher-speed switches such as SOA gate arrays and electro-optic switch arrays using traveling-wave switches with resistive terminations, the array consumes power continuously. In such cases, the control energy per packet $E_{control}$ is proportional to N_b. In comparison, electronic packet switches incur a fixed energy cost associated with the O/E and E/O conversions in addition to the switching energy [71,73].

17.5.3 Switching sub-system energy

Table 17.3 provides some indicative figures for device and switch energies [55]. The E-O and SOA-based switches in Table 17.3 have optical inputs, optical outputs, and the signal remains in optical form as it passes through the switch. Therefore, we refer to these switches as O/O/O devices. The switching rate is one operation per packet, which is equivalent to an average switching rate of 10 MHz. The E-O switches are classified as lumped (i.e. low switching speed and capacitance limited) and TW (i.e. resistance limited).

The O/E/O switches in Table 17.3 are devices in which the signal is converted to electronic form at some point internal to the switch. The wavelength-routed switch is based on an AWG switch fabric, with optoelectronic (O/E/O) wavelength converters. The CMOS switch incorporates O/E converters at the inputs and E/O converters at the outputs.

Table 17.3 Energy consumption in switch devices and a 1024×1024 switch fabric.

Technology			2 × 2 Switch	10³ × 10³ Switch	2 × 2 Switch	10³ × 10³ Switch
			2010-era Technology		**Target**	
O/O/O	E-O	Lumped	20 fJ	–	20 fJ	–
		TW	1.8 pJ	–	1 pJ	20 pJ
	SOA Gate Array		8 pJ	–	4 pJ	75 pJ
O/E/O	Wavelength-routed		–	–	–	10 pJ
	CMOS		200 fJ (+61 pJ)	4 pJ (+61 pJ)	20 fJ (+5.1 pJ)	400 fJ (+5.1 pJ)

The numbers included in Table 17.3 under "2×2 Switch" refer to a single crosspoint switch and are based on data at a bit rate of 100 Gb/s and an average packet length of 10^4 bits, which is typical of an Internet Protocol (IP) packet. The data under "$10^3 \times 10^3$ Switch" refer to a $10^3 \times 10^3$ Clos architecture switch fabric at the same bit rate and average packet length. We have not included an estimate for optical $10^3 \times 10^3$ switches using 2010-era technology because practical high-speed optical switches suitable for packet switching are yet to be demonstrated.

The energies for the E-O switches in Table 17.3 are based on a switching voltage of 3 V and a resistive termination for TW devices of 50Ω [53]. For lumped electro-optic switches, we assume a device propagation time τ_{prop} of 1 ns. Lumped switches with capacitive inputs do not scale easily to large size because of the need for long electrical lines, which ultimately need to be terminated. Therefore, we have not included data for a $10^3 \times 10^3$ switch with lumped E-O devices. The insertion loss of the electro-optic switches is taken to be 3 dB per crosspoint. At this attenuation level, the E_{AMP} term in (17.12) is much smaller than the E_{Control} term.

The CMOS switch circuits in the bottom row of Table 17.3 incorporate electronic multiplexers and de-multiplexers. This requires multiple parallel switch fabrics to be used, each operating below the line rate. Note that while multiple switch fabrics are used in these CMOS examples, the energy per bit in the switch fabrics is independent of the number of parallel switch fabrics, highlighting the fact that electronic switching is readily adaptable to parallel processing of data. We have assumed that each 2×2 crosspoint in the switch fabrics uses 16 transistors (see [53]).

The "target" energy figures in Table 17.3 are estimates based on expectations of technology improvements. Improvements in electro-optic devices, SOA gate arrays, and wavelength-routed switches are expected to be limited, given the mature status of these technologies. Table 17.3 confirms earlier findings [68] that suggest optical switch fabrics are generally uncompetitive with CMOS from an energy efficiency point of view. Wavelength-routed switches may eventually become competitive [13]. However, the energy advantages or wavelength-routed optical switches over CMOS are, at best, marginal and there are many practical problems to be solved, including the much larger footprint of optical switches [68].

It is instructive to compare the estimated switching energies in the $10^3 \times 10^3$ switch fabric in Table 17.3 with the data in Figure 17.7. As shown in Figure 17.7, today's large routers and Ethernet switches consume around 5–10 nJ/b. This is between two and three orders of magnitude larger than the energy consumption of the 2010-era CMOS $10^3 \times 10^3$ switch in Table 17.3 (70 pJ/b) and between three and almost four orders of magnitude larger than the target energies per bit in the right-hand column of Table 17.3.

There are two key reasons for these differences: (a) In a modern router the switch fabric accounts for only about 10% of the total power consumption [71], and as mentioned already, other ancillary components also consume significant power, and (b) today's network devices typically use older CMOS technologies (e.g. 130-nm or 90-nm devices rather than the 45-nm or 33-nm devices in state-of-the-art technology and the 11-nm devices that have been used in estimating the CMOS target in Table 17.3). The difference in the energy consumption between the 90-nm and 11-nm technology generations is around two orders of magnitude (see Figure 17.21).

Table 17.3 highlights the fact that CMOS switching is expected to improve significantly. This suggests that CMOS and other electronic technologies may offer greater potential for low-energy packet switching than optical switching technologies [70].

17.5.4 End-to-end network energy models

We now estimate the energy consumption in an end-to-end system, using a network model based on a core network comprising an array of $10^3 \times 10^3$ switches as building blocks. This size of switch is representative of the scale of practical network switches. The arrangement of the $10^3 \times 10^3$ switches in the end-to-end network model is shown in Figure 17.24 [53].

The network is connected to N_{user} user devices, each operating at a bit rate of B_{user}. The core network operates at a bit rate of B_{core}. User data is aggregated in an array of aggregation switches that multiplex the user data at bit rate B_{user} to the core bit rate of B_{core}. The number of input ports on each aggregation switch is B_{core}/B_{user} and there is one output port on each aggregation switch. As shown in Figure 17.22, the total number of aggregation switches is $N_{user}B_{user}/B_{core}$. The core of the network comprises a $N_{user}B_{user}/10^3B_{core} \times N_{user}B_{user}/10^3B_{core}$ array of $10^3 \times 10^3$ switches, connected in a Clos architecture. The core network feed an array of disaggregation switches that demultiplex data back to the user device bit rate B_{user}. The number of stages in the $N_{user}B_{user}/10^3B_{core} \times N_{user}B_{user}/10^3B_{core}$ core array of $10^3 \times 10^3$ switches is q. The number q of core stages in Figure 17.24 is given by

$$q = \lceil 2\log_{1024} t - 1 \rceil, \tag{17.13}$$

where $t = N_{user}B_{user}/10^3B_{core}$ is the number of $10^3 \times 10^3$ switches in each stage of the core switch network.

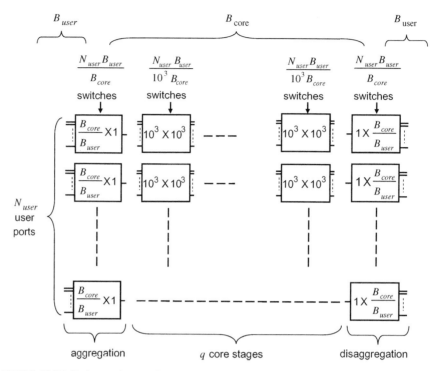

FIGURE 17.24 End-to-end network model, showing the number and port counts on the aggregation, core, and disaggregation switches. The access network and the optical transport systems are omitted, for simplicity [53].

To estimate the lower bound on energy consumption trends in the network, we now consider two examples. Scenario 1 is a global network with 10 billion attached devices operating at bit rates of 10 Mb/s, and Scenario 2 is a global network with 100 billion attached devices with end-user bit rates of 1 Gb/s. The first of these two networks has an aggregate data rate of 10^{17} b/s (100 Pb/s) and represents a scenario that might become reality in the 2020–2030 timeframe. (For comparison, today's Internet operates at an average aggregate bit rate of about 100 Tb/s [72].) The second scenario has an aggregate data rate of 10^{20} b/s (100 Exabits per second) and represents a much more futuristic scenario that may perhaps become a reality in the 2050 timeframe or beyond.

The energy per bit for both network scenarios is given in Table 17.4. To calculate the energies in Table 17.4, we have used the energy per bit for the $10^3 \times 10^3$ switch fabrics in Table 17.3 for each of the switch technologies considered in Table 17.3. In all entries in Table 17.4 except the bottom row, the data for switch energies is taken from the "Target" column in Table 17.3. The bottom row of Table 17.4 includes, for comparison, data based on 2010-era technologies. The energy figures in the "all switch fabrics" column in Table 17.4 represent the $\sum_{i=1}^{s} E_{switch,i}$ term in (17.6) and

Table 17.4 Minimum energy per bit E_{net} in two network scenarios. All entries in this table except the bottom row refer to "target" energies in Table 17.3.

Switch Technology	Scenario 1			Scenario 2		
	N_{user}	B_{user} (Mb/s)		N_{user}	B_{user} (Mb/s)	
	10^{10}	10		10^{11}	10^3	
	Energy per Bit, E_{NET} (pJ)			Energy per Bit, E_{NET} (pJ)		
	All Switch Fabrics	All Transport	Total	All Switch Fabrics	All Transport	Total
E-O TW (Target)	72	2	74	83	2.5	86
SOA Array (Target)	290	2	292	340	2.5	343
AWG/TWC (Target)	36	2	36	42	2.5	45
CMOS (Target)	20	10	30	31	24	55
CMOS (2010-era)	225	160	385	372	210	582

include the energy per bit in the core stages (see Table 17.3) and the energies per bit in the aggregation and disaggregation switches, which have been estimated using the data in Section 17.3. For Scenario 1, $q=1$ and $s=3$, and for Scenario 2, $q=3$ and $s=5$.

The energy figures in the "all transport" columns of Table 17.4 represent the $\sum_{i=1}^{s-1} E_{\text{core-transport},i} + E_{\text{access-transport}}$ terms in (17.6). The figures for transport energy in the core come from Tables 17.1 and 17.2. We assume an average transport system span (i.e. between switches) of 3000 km for Scenario 1 and 2000 km for Scenario 2. In both scenarios, the amplifier spacing is 50 km and the total end-to-end transmission distance is approximately 9000 km. We have included a figure of 3.6-pJ for the energy of each TX/RX pair (including MUX/DEMUX) in the transport systems for the CMOS switch example (bottom row of Table 17.3) and we have assumed that no transmitters or receivers are used at the input and outputs of the transport systems for the optical switch examples. This is a fairly generous assumption for optical switching because, in practice, packet switches require optical receivers and transmitters, even if the switch fabric technology is optical [70,71]. However, as can be seen from Table 17.4, this assumption does not have a significant impact on the energy because the network energy consumption is dominated by switching energy rather than by the energy associated with O/E/O conversion. We assume that the access network uses optical transmitters and receivers at the user interface. Because the access network operates a lower bit rate than the core network, there is no need for MUX/DEMUX circuitry in the access network. Therefore, we have used a figure of 5 pJ for the $E_{\text{access-transport}}$ term.

Table 17.4 shows that for both scenarios, the energy consumed in the switch fabrics dominates over the energy consumption in optical transport. The table also illustrates how well, in principle, the network can scale to larger size and capacity. There are 10 times as many users in the network in Scenario 2 as in Scenario 1, and the user bit rate is larger. But the energy per bit in Scenario 2 is less than twice the energy per bit in Scenario 1. Note that in this ideal scenario, we presume an effective overall management of the configuration of the international switched network. Note also that we have assumed an ideal Clos network architecture. Ad hoc switch and router deployment could dramatically increase the numbers in Table 17.4.

17.5.5 Comparison of energy projections with network-based data

It is instructive to compare the data in Table 17.4 from the analysis in Figure 17.3. The key difference between the data in Table 17.4 and Figure 17.3 is that Table 17.4 provides projections of the ideal practical lower bound on network energy. On the other hand, Figure 17.3 is based on a network model using commercial equipment data and extrapolated to future equipment performance based on a projection of current trends in improvements in technology over time [6,13]. The optimistic improvements trend in Figure 17.3 is a best estimate on such equipment within the current architectural constraints. Table 17.4 represents bottom-up extrapolations based on device-level estimates of practical lower-bounds on energy use. The difference between the current trends analysis and the practical lower bound analysis represents a measure of the gap between trends in current practice and the ideal performance that could be achieved if all inefficiencies and all energy consumption associated with ancillary functions could be eliminated. The lower-bound analysis also assumes that semiconductor trends will continue at near historical rates—based on the expectation that new technology innovations will overcome the current challenges. This is clearly an optimistic scenario because while significant slowing is already observed in the rate of improvement of CMOS technology and no clear solutions to these scaling challenges exist [15].

Figure 17.25 shows the end-to-end router/switch and transport energy per bit from Figure 17.3 and from Table 17.4, plotted against time. The data in Figure 17.25 are based on CMOS switch technologies in Table 17.3 rather than optical switch technologies. However, it is important to note that the "Target" data from Table 17.3 plotted at the 2020 time point would be almost identical to the CMOS data if AWG-based or E-O-based switch technologies are used in place of CMOS. For simplicity, Figure 17.25 does not include energy consumption in the access network.

The energies in Figure 17.25 based on the lower bounds on energy (broken lines) are approximately four orders of magnitude smaller than the estimates of energy consumption based on Figure 17.3 (solid lines). There are five main reasons for this:

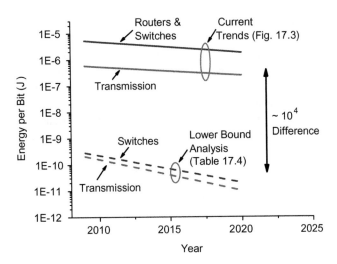

FIGURE 17.25 End-to-end switch and transport network energy consumption based on business-as-usual analysis using current trends as developed in [6] and minimum end-to-end network switch and transport energy calculated in this paper using an analysis based on lower bounds of energy consumption of CMOS switches and optically-amplified transport systems. Access network energy consumption is omitted, for simplicity.

1. Today's network is much more complicated than the "minimal" network model in Figure 17.24, which forms the basis for Table 17.4. In today's network, an end-to-end connection typically traverses up to 20 switches and routers [38]. This is almost an order of magnitude larger than the number of switches used in the network model in Figure 17.24.

2. The lower-bound data (broken lines) are for ideal subsystems with no inefficiencies, overheads, and conditioning (see Figure 17.9).

3. As pointed out, routers typically consume about an order of magnitude more energy than the internal cross-connect. Other functions in routers that consume considerable energy are the pattern matching for QoS and security, forwarding engine, and buffers. Similarly, Ethernet switches and other network devices have substantial energy overheads and advanced functionality. Overhead functions in routers, switches, and other devices including SDH cross-connects are discussed in detail in [10].

4. In today's network, many optical links and many routers operate well below maximum load. When operating below maximum load, the energy per bit increases.

5. The components in a production router are generally one or two generations behind the leading-edge technology of the day and well behind the "target" technology states used in Table 17.4.

17.6 CONCLUSION

The six installments of this series of books have highlighted the importance of optical fiber technologies in advanced telecommunications systems. Combined with advances in electronic switching and signal processing, and with advanced protocols and network management systems, today's telecommunications network has reached high levels of capacity, reach, reliability, flexibility, and affordability to users. For many years, the prime drivers behind advances in telecommunications have been considerations of capacity and cost. But recently, concerns about the rising energy use of telecommunications networks have brought the issue of energy efficiency into the mix, both for equipment vendors and for network operators.

We have identified several reasons for this recent increase in interest in energy efficiency. First, as networking equipment such as optical transceivers and network switches and routers grow in capacity, there is a need to increase the density of active devices in order to maintain an acceptably small footprint for the equipment. This expanding density of devices has resulted in challenges associated with heat dissipation from equipment racks. Improved thermal engineering can help to alleviate some of these problems, but ultimately there is a need to improve the energy efficiency of the active devices. Secondly, operational expenses (OpEx) associated with energy consumption of equipment are becoming an increasingly important part of the total OpEx in network operators. In the past, energy costs were such a small portion of an operator's total OpEx that many operators paid little or no attention to energy. But this is now rapidly changing. Third, as the telecommunications network continues to expand to satisfy the ever-increasing demand for new services, new applications, and to accommodate an increasing user base, the energy consumption of the network has a small but growing impact on global GHG emissions.

We further provided a detailed analysis of the energy use of the different core elements of a telecommunication system, including switching and transport. The basic energy relationships were used to describe a lower bound on the energy use of a minimal network based upon practical technologies today and anticipated in the next decade. This result was discussed relative to estimates for energy use in networks based on technology projections for commercial systems. The four orders of magnitude separating these trends depend not only on the technology efficiencies, but also on the many functional and performance requirements on commercial systems today. Thus, progress on energy-efficient telecommunications will require a combination of technology improvements together with new, intelligent, service aware capability that can realize essential performance or functionality at lower energy.

References

[1] S. Lloyd, Ultimate physical limits to computation, Nature 406 (2000) 1047.
[2] D. Kilper, K. Guan, K. Hinton, R. Ayre, Energy challenges in current and future optical transmission networks, Proc. IEEE 100 (5) (2012) 1168–1187.
[3] R. Tucker, Green optical communications—part I: energy limitations in transport, IEEE J. Sel. Top. Quant. Electron. 17 (2) (2011) 261–274.

[4] Green House Gas Protocol (online). <www.ghgprotocol.org>.

[5] Ministry of Energy, Mines and Petroleum Resources (2007) The BC Energy Plan (online). <http://energyplan.gov.nc.ca>.

[6] D. Kilper et al., Power trends in communication networks, IEEE J. Sel. Top. Quant. Electron. 17 (2) (2011) 275.

[7] Cisco, Hyperconnectivity and the Approaching Zettabyte Era, 2009.

[8] The Climate Group, SMART 2020: enabling the low carbon economy in the information age, Global e-Sustainability Initiative (GeSI), 2008.

[9] T. Asami, S. Namiki, Energy consumption targets for network systems, in: ECOC, 2008, p. Tu4A3.

[10] O. Tamm, C. Hermsmeyer, A.M. Rush, Eco-sustainable system and network architectures for future transport networks, Bell Labs Tech. J. 14 (2010) 311–328.

[11] S. Roy, Energy logic: a road map to reducing energy consumption in telecommunications networks, in: IEEE 978–1-4244-2056-8/08, 2008.

[12] World Wildlife Federation, The Potential Global CO_2 Reductions from ICT Use, Identifying and Assessing the Opportunities to Reduce the First Billion Tonnes of CO_2, 2008.

[13] J. Baliga, R. Ayre, K. Hinton, W. Sorin, R. Tucker, Energy consumption in optical iP networks, J. Lightwave Technol. 27 (13) (2009) 2391–2403.

[14] Y. Zhang, P. Chowdhury, M. Tornatore, B. Mukherjee, Energy efficiency in telecom optical networks, in: IEEE Communication Surveys & Tutorials, 2010.

[15] R.G. Beausoleil, P.J. Kuekes, G.S. Snider, S.-Y. Wang, R.S. Williams, Nanoelectronic and nanophotonic interconnect, Proc. IEEE 96 (2) (2008) 230–247.

[16] A. Ahmad et al., Power-aware logical topology design heuristics in wavelength routing networks, in: Proceedings of ONDM, 2011.

[17] S. Chandrasekhar, X. Liu, Terabit superchannels for high spectral efficiency transmission, in: Proceedings of European Conference on Optical Communications, 2010, p. Tu3C5.

[18] K. Christodoulopoulos, I. Tomkos, E. Varvarigos, Elastic bandwidth allocation in flexible OFDM-based optical networks, J. Lightwave Technol. 29 (9) (2011) 1354–1366.

[19] W. Shieh, OFDM for flexible high-speed optical networks, J. Lightwave Technol. 29 (10) (2011) 1560–1577.

[20] A. Morea et al., Power management of optoelectronic interfaces for dynamic optical networks, in: Proceedings of European Conference on Optical Communications, Geneva, 2011.

[21] B.S.G. Pillai, B. Sedighi, W. Shieh, R.S. Tucker, Chromatic dispersion compensation—an energy consumption perspective, in: Proceedings of OFC/NFOEC, 2012, p. OM3A8.

[22] L. Chiaraviglio, D. Ciullo, M. Mellia, M. Meo, Modeling sleep modes gains with random graphs, in: INFOCOM Workshop on Green Communications and Networking, 2011.

[23] K.C. Guan, D. Kilper, Y. Pan, O. Rival, A. Morea, Energy efficient file transfer over rate adaptive optical network, in: IEEE Green Communications Conference, 2012.

[24] W. Vereecken et al., Optical networks: how much power do they consume and how can we optimise this? in: European Conference on Optical Communications, 2010, p. Mo1D1.

[25] M.D. Feuer, D.C. Kilper, S. Woodward, ROADMs and their system applications, in: I.P. Kaminow, T. Li, A.E. Willner (Eds.), Optical Fiber Communications V B, San Diego, Academic Press, 2008.

[26] P. Wiatr, P. Monti, L. Wosinska, Power savings versus network performance in dynamically provisioned WDM networks, in: IEEE Communication Magazine, 2012, pp. 48–55.

[27] A. Nag, M. Tornatore, B. Mukherjee, Optical network design with mixed line rates and multiple modulation formats, J. Lightwave Technol. 28 (4) (2010) 466–475.

[28] O. Rival, A. Morea, J.-C. Antona, Optical network planning with rate-tunable NRZ transponders, in: European Conference on Optical Communications, 2009.

[29] P. Chowdhury, M. Tornatore, B. Mukherjee, On the energy efficiency of mixed-line-rate networks, in: Optical Fiber Communication Conference, 2010.

[30] K. Christopdoulopoulos, K. Manousakis, E. Varvarigos, Adapting the transmission reach in mixed line rate WDM Transport Networks, in: International Conference on Optical Network Design and Modeling, 2011.

[31] J. Baliga, R. Ayre, W. Sorin, K. Hinton, R.S. Tucker, Energy consumption in access networks, in: Proceedings of OFC/NFOEC, 2008.

[32] K.-L. Lee, B. Sedighi, R.S. Tucker, H. Chow, P. Vetter, Energy efficiency of optical transceivers in fiber access networks, J. Opt. Commun. Netw. 4 (9) (2012) A58–A68.

[33] D. Suvakovic et al., Low energy bit-interleaving downstream protocol for passive optical networks, in: IEEE Green Communications Conference, 2012.

[34] D. Chiaroni et al., Packet OADM for the next generation of ring networks, Bell Labs Tech. J. 14 (4) (2010) 243–264.

[35] S. Namiki et al., Ultrahigh-definition video transmission and extremely green optical networks for future, IEEE J. Sel. Top. Quant. Electron. 17 (2) (2011) 446–457.

[36] K. Guan, D. Kilper, G. Atkinson, Evaluating the energy benefit of dynamic optical bypass for content delivery, in: INFOCOM Workshop on Green Communication and Networking, 2011.

[37] J. Baliga, R. Ayre, K. Hinton, R. Tucker, Architectures for energy-efficient IPTV networks, in: OFC/NFOEC, 2009, p. OTHQ5.

[38] S. Figuerola, M. Lemay, V. Reijs, M. Savoie, B.S. Arnaud, Converged optical network infrastructures in support of future internet and grid services using IaaS to reduce GHG emissions, J. Lightwave Technol. 27 (12) (2009) 1941–1947.

[39] X. Dong, T. El-Gorashi, J. Elmirghani, IP over WDM networks employing renewable energy sources, J. Lightwave Technol. 29 (1) (2011) 3–14.

[40] R.S. Tucker, K. Hinton, Energy consumption and energy density in optical and electronic signal processing, IEEE Photon. J. 5 (3) (2011) 821–833.

[41] D.E.D. Bayart, B. Desthieux, S. Bigo, Erbium-Doped Fiber Amplifiers: Device and System Developments, Wiley, New York, 2002.

[42] Y. Yamamoto, H. Haus, Preparation, measurement, and information capacity of optical quantum states, Rev. Mod. Phys. 58 (1986) 1001–1020.

[43] T. Li, M.C. Teich, Bit error rate for a lightwave communication system incorporating an erbium doped fibre amplifier, Electron. Lett. 27 (1991) 598–600.

[44] X. Liu, S. Chandrasekhar, A. Leven, Self-coherent optical transport systems, Optical Fiber Telecommunications V B: Systems and Networks, Elsevier, 2008.

[45] M. Karlsson, E. Agrell, Which is the most power-efficient modulation formats in optical links? Opt. Express 17 (2009) 10814–10819.

[46] R.-J. Essiambre, G. Kramer, P. Winzer, G.J. Foschini, B. Goebel, Capacity limits of optical fiber networks, J. Lightwave Technol. 28 (4) (2010) 662–701.

[47] S.P. Voinigescu et al., Towards a sub-2.5V, 100-Gb/s serial transceiver, in: IEE CICC, 2007, pp. 471–478.

[48] M. Chacinski et al., 100 Gb/s ETDM transmitter module, IEEE J. Sel. Top. Quantum Electron. 165 (5) (2010) 1321–1327.

[49] M. Moeller, High-speed electronic circuits for 100 Gb/s transport networks, in: Proceedings of OFC/NFOEC, 2010.

[50] SHF. SHF communication technologies (online). <www.shf.de>.

[51] J.-C. Chien, L.-H. Lu, 40-Gb/s high-gain distributed amplifiers with cascaded gain stages in 0.18-micrometer CMOS, IEEE J. Solid-State. Circ. 42 (2007) 2715–2725.

[52] A.V. Krishnamoorty et al., Progress in low-power switched optical interconnects, IEEE J. Spec. Top. Quant. Electron. 17 (2) 357–376.

[53] R. Tucker, Green optical communications—part 2: energy limitations on networks, J. Sel. Top. Quant. Electron. 17 (2) (2011) 261.

[54] C. Clos, A study of non-blocking switching networks, Bell Sys. Tech. J. 32 (1953) 406–424.

[55] K. Hinton et al., Switching energy and device size limits on digital photonic signal processing technologies, IEEE J. Sel. Top. Quant. Electron. 14 (3) (2008) 938.

[56] E.J. Murphy et al., 16×16 strictly nonblocking guided-wave optical switching system, J. Lightwave Technol. 14 (1996) 352–358.

[57] I.M. Soganci et al., High-speed 1×16 optical switch monolithically integrated on InP, in: European Conference on Optical Communication, 2009.

[58] R.F. Kalman, L.G. Kazovsky, J.W. Goodman, Space division switches based on semiconductor optical amplifiers, IEEE Photon. Technol. Lett. 4 (1992) 1048.

[59] S. Tanaka et al., Monolithically integrated 8:1 SOA gate switch with large extinction ratio and wide input power dynamic range, IEEE J. Quant. Electron. 45 (2009) 1155–1162.

[60] Y. Kai, et al., A compact lossless 8×8 SOA gate switch subsystem for WDM optical packet interconnections, in: European Conference on Optical, Communications, 2008.

[61] M. Renaud, M.E. Bachmann, Semiconductor optical space switches, IEEE J. Sel. Top. Quant. Electron. 2 (1996) 277–288.

[62] N. Sahri et al., A highly integrated 32-SOA gates optoelectronic module suitable for IP multi-terabit optical packet routers, in: Proceedings of OFC/NFOEC, 2001, pp. PD32-1–PD32-3.

[63] International Technology Roadmap for Semiconductors, 2010, 2009 Edition (online). <http://public.itrs.net>.

[64] International Technology Roadmap for Semiconductors, 2005 Edition (online). <http://public.itrs.net>.

[65] D.A.B. Miller, Device requirements for optical interconnects to silicon chips, in: Proceedings of IEEE, 2009, p. 1166.

[66] V.E. Benes, Mathematical Theory of Connecting Networks and Telephone Traffic, Academic Press, 1965.

[67] G.H. Song, M. Goodman, Asymmetrically-dilated cross-connect switches for low-crosstalk WDM optical networks, in: IEEE LEOS Annual Meeting, 1995, pp. 212–213.

[68] R.S. Tucker, The role of optics and electronics in high-capacity routers, J. Lightwave Technol. 24 (2006) 4655–4673.

[69] D.K. Hunter, I. Andonovic, Approaches to optical Internet packet switching, IEEE Commun. Mag. 28 (2000) 116–122.

[70] R. Tucker, Optical packet switching: a reality check, Opt. Switch. Netw. 5 (2008) 2.

[71] R.S. Tucker, Scalability and energy consumption of optical and electronic packet switching, J. Lightwave Technol. 29 (2011) 2410–2420.

[72] R.S. Tucker et al., Evolution of WDM optical IP networks: a cost and energy perspective, J. Lightwave Technol. 27 (2009) 243–252.

[73] Minnesota Internet Traffic Studies, 2010 (Online). <www.dtc.umn.edu/mints/home.php>.

Advancements in Metro Regional and Core Transport Network Architectures for the Next-Generation Internet

18

Loukas Paraschis

Cisco, 170 W. Tasman Dr., San Jose, CA 95134, USA

18.1 INTRODUCTION

The expanding availability of fast and reliable network connectivity has been enabling applications to transition increasingly to an Internet-based service delivery model. At the same time, the underlying infrastructure consists of data-centers (DC), often of massive, pooled, and "virtualized" compute and storage resources, commonly referred to as "cloud." Networking is crucial in the delivery of services to the end-users (consumers or enterprises), as well as the DC interconnection for cost-performance optimization. Moreover, this "cloud" traffic has become the fastest growing part of the Internet, and of the capacity demand in its transport networks. As a result, the structure of the Internet has started to change toward a flatter hierarchy with more densely interconnecting networks (Figure 18.1) [1]. In this chapter, we explore the implications of this change in the metro regional and core transport network architectures, which are expected to cost-effectively scale globally to more than a Zettabyte in 2015. We particularly focus on the important advancements in the optical, routing, and traffic engineering technologies enabling this evolution.

Network architecture evolutions are typically motivated by a combination of significant new services, or evolving business needs, and important technology advancements that serve better the new requirements. An excellent example of such an evolution has been the worldwide introduction over the last 10 years of reconfigurable DWDM as the converged multi-service transport architecture of choice for the evolution of metro regional networks, which we discussed in the previous OFT edition 5 years ago [2]. In the current OFT edition, we review the regional and core transport network evolution currently under way, motivated by the changes in the structure of the Internet outlined in the initial paragraph. Section 18.2 starts by reviewing today's network architecture. We summarize the main functional characteristics, and application requirements that motivated these networks to scale leveraging IP/MPLS, and DWDM transport. We then discuss the evolving requirements, and business needs arising from the proliferation of DC-based service delivery

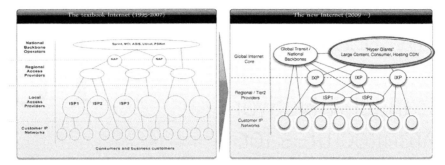

FIGURE 18.1 The evolution of the Internet to a flatter hierarchy with more densely interconnecting networks (right), motivated by the increasingly important role of the large data-center-based, "cloud," service providers [1].

models. In Section 18.3, we shift our attention to the technology innovations that best serve the required transport network evolution. We specifically review the advancements in 100 Gb/s coherent transmission, reconfigurable DWDM systems, optical transport, routing, and traffic engineering technology. Other chapters in this book provide a much more detailed analysis of some of the related photonics innovations. The goal of this chapter is to highlight the increasing role photonics have been playing in the evolution of transport architectures. We argue that photonics technology innovation is today even more important than the initial glory days of DWDM transmission, or the subsequent adoption of optical-add-drop (OADM) networking. As a new generation of DWDM system offers flexible optical switching and spectral efficiencies aiming to exceed 2 b/s/Hz, the embedded fiber infrastructure can scale to multi-Tb/s transport, albeit at a significantly higher proportion of the overall network cost. Section 18.4 specifically explores the related network value and remaining challenges of these photonics technologies. Section 18.5 then reviews the emerging convergence of packet (IP/MPLS) and optical transport in network architectures that combine photonics, system advancements, multi-layer network control-plane, and interoperability to improve overall network operational performance, and capital cost. We discuss the benefits in capacity provisioning optimization, and optical restoration, resulting from these more dynamically configurable transport networks. Two other chapters of this book discuss in much further detail rapidly reconfigurable networking (by AT&T, Sheri Woodward et al.), and convergence of IP and optical networking including recovery (by MIT, Kristin Rauschenbach et al.). As such, both chapters provide important, complementary material related to this chapter. Finally, in Section 18.6 of this chapter, we provide an outlook of important research and development efforts, most notably including adaptive rate transport networks, and future novel architectures toward a fully automated transport infrastructure optimized based on software-defined network operation.

18.2 NETWORK ARCHITECTURE EVOLUTION

Internet-protocol (IP) traffic has been dominating the worldwide capacity and growth demands in telecommunication transport networks for more than 10 years. IP global traffic is expected in 2012 to exceed an Exabyte per day (an average of 100 Tb/s), and continues to grow at around 30% compounded annual rate (CAGR) [3,37]. Over the past 20 years, network architectures have evolved to accommodate the IP services, and cost-effectively scale to a global network with an annual capacity forecasted to exceed a Zettabyte by 2015. Moreover, the increasingly important high-bandwidth, predominantly consumer video-related, applications, which often have diverse quality-of-service requirements, have been skewing the required technologies towards more dynamic IP based technologies, and have thus motivated a fundamental shift from circuits to packets. This has been the most significant evolution of transport networks in recent history, and is commonly also referred to as the IP Next-Generation Network (IPNGN) architecture. IPNGN has been extensively analyzed, e.g. [2,4]. Figure 18.2, taken from [2], provides a good summary of the IPNGN architecture functional characteristics. By offering an access agnostic aggregation, leveraging IP/MPLS, and Ethernet packet transport, IPNGN allowed a converged and scalable infrastructure for all Internet services. More specifically, for metro regional transport networks, the IPNGN architecture offers a single converge infrastructure for both wireline and wireless services. (Note that wireless, particularly WiFi, traffic is growing currently at around 100% CAGR, and is expected to surpass wireline traffic in 5 years.)

Today, IPNGN is adopted worldwide by operators as the most flexible and cost-effective architecture for scaling their networks (e.g. [2,4]). Packet transport enables significant efficiencies due to its crucial ability for statistical multiplex of bursty traffic flows. At the same time, the advancements in optical transport have been crucial in enabling IPNGN to meet the Internet growth needs. More specifically, the fiber capacity of a DWDM system has been doubling every couple of years. Moreover, new photonics technologies like ROADM, tunable transmitters, pluggable optical modules, combined with systems advancements like self-calibrating transmission systems with automatic power control, and alarm correlation management,

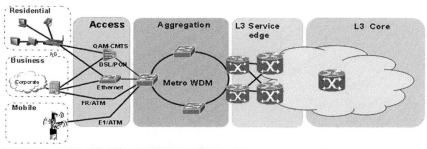

L0-L1	SONET, CWDM	CWDM, DWDM, OTN	N/A	DWDM, OTN
L2-L3	Ethernet	Ethernet, IP, MPLS	IP, MPLS	IP, MPLS

FIGURE 18.2 The IPNGN architecture typical functional characteristics [2].

have been improving DWDM capacity planning, service flexibility, and operational robustness [5]. Note that operational cost (OpEx) usually dominates the total network cost (TCO), and can therefore be more critical than capital cost (CapEx). These TCO benefits led to the wide deployment of metro regional networks with converged WDM transport for both the IP/Ethernet services and traditional TDM traffic. The analysis in [2] provided a detailed account of the DWDM adoption in metro regional networks, and the significantly reduced operational cost it has enabled. Core transport has also been leveraging the DWDM with particular emphasis in scaling from 10 Gb/s per channel (which is still the channel capacity in many DWDM networks around the world), to 40 Gb/s, or recently 100 Gb/s at the largest operators.

Figure 18.3 [6,7] summarizes the transport building blocks of the typical IPNGN architecture deployed currently by network operators worldwide. Generalized routers with increasingly sophisticated, robust IP/MPLS control-plane functionality have been the basic building block. Most Internet traffic would flow from the centralized peering nodes in the IP core through metro regional transport networks to access the different end-users, as described in [2]. The role of the core is to connect the metro networks together through the regional (edge) nodes. This architecture allows the Internet to serve millions of endpoints, in a highly dynamic and scalable way, employing several hierarchically organized autonomous networks (or autonomous systems) internetworked through pre-computed IP routes which statistically multiplex traffic flows. The hierarchical nature of service delivery offers multiple levels of statistical multiplexing. Generally, more aggregation and statistical multiplexing leads to more predictable and efficiently utilized routes. However, a hierarchical architecture may inevitably terminate at each core node a certain amount of traffic that is intended as pass-through traffic. To resolve this issue, an intermediate circuit switching

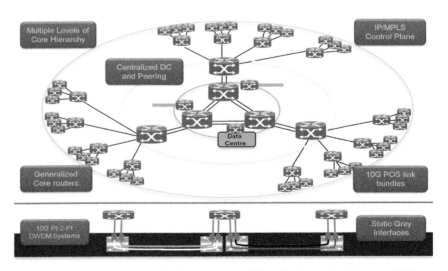

FIGURE 18.3 The typical current IPNGN transport building blocks and architecture [6,7].

(TDM-based) transport layer has been proposed to interconnect directly the outer metro/regional routers in a full mesh, and replace the core routers with lower cost systems, thus saving CapEx. However, this alternative "hollow-core" architecture was not widely adopted due to its operational limitations. More specifically, any full mesh of N endpoints requires that operators accurately provision and operate $N(N-1)/2$ separate connections. In most cases, such a mesh would imply that network operators would need to accurately forecast and maintain an order of magnitude more core connections, a task considered prohibitively expensive operationally. Instead, core routers have been directly interconnected through point-to-point links in DWDM systems, to which routers connect through a short-reach (SR) interconnection.

Recently, network operators have started considering how to re-optimize the IPNGN transport architecture based on the changing dynamics of the Internet [4,7]. New services and business models have been the main drivers, but new technologies with important implications in the network architecture have also provided strong motivation for an evolution. We review the main business aspects here, before we focus on the role of related technology innovations in the next sections. The changing dynamics of the Internet traffic flows has been the primary reason for the IPNGN evolution. Increasingly, the majority of the content being delivered to the Internet end-users (consumers and enterprises alike) has been originating from a handful of large content providers. The combined proportion of the 10 largest such content providers has grown to more than 40% of the global Internet traffic [1]. At 40 Exabytes per month, 40% corresponds to more than 40 Tb/s of transport capacity. The majority of this capacity originates, or terminates, within the DCs that these large content providers operate, usually in multiple locations around the world.

Note that in this "cloud" infrastructure, the DC interconnection would generate additional, usually large (also referred to as "elephant") traffic flows, which are generally treated separately from the Internet (end-user) traffic because they have different requirements [39]; for example, they can be particularly sensitive to latency or other connection quality metrics. As we mentioned in the introduction, the inter-DC network connectivity is crucial for the cost-performance optimization of the participating "virtualized" computing and storage infrastructure pool. Hence, the inter-DC traffic could be often served through a dedicated transport network, especially when the related capacity needs are significant [8,39].[1] In many cases, however, inter-DC traffic with moderate capacity needs would be part of the same transport network infrastructure that serves the Internet traffic. As such, the inter-DC capacity requirements have become increasingly important for the future metro regional and core transport networks [25]. We discuss the implications of this use-case further in Section 18.5.

Here, it is important to also highlight that most of the Internet content today is employing an "over-the-top" (OTT) delivery model,[2] which means that it uses the

---

[1]For example, Google has reported building such a dedicated inter-DC transport network, which in 2011 was actually employing the largest reported single deployment of 100 Gb/s DWDM channels.

[2]A good OTT example is Facebook which has enjoyed the fastest growth of end-users these last few years.

transport networks of other service providers (e.g. the Comcast or ATT broadband network) to reach end-users. From a pure technical perspective, the transport of an OTT bit-stream may not differ from other bit-streams carried over a transport network. From a business perspective, however, the difference can be big, because network operators monetize OTT bit-streams at much lower rates than their own traffic. Consequently, the network TCO has been growing faster than revenues, and new investment in transport networks has faced with increased scrutiny to justify its Return-on-Investment (RoI). This business challenge has been a significant motivation of the evolution to transport network architectures that scale with proportionally lower TCO, leveraging technology innovations.

These new "cloud" service delivery and operational models, and the emergence of the large OTT content providers, have been motivating an evolution to the flatter Internet architecture (Figure 18.1). Consequently, network operators have evolved to bring services closer to end-users, in order to improve delivery, or lower transport cost. This evolution is mainly affecting the metro regional and core transport networks, which is the focus of this chapter. At the same time, given that practically all new services, and traffic growth is packet based, and statistical multiplexing efficiencies remain very important, the existing IPNGN transport architecture remains the foundation for this evolution. An increasing amount of recent analysis has discussed the important operational characteristics and new technologies that would best enable this IPNGN evolution [4–9]. In addition to the need for scale to meet the Internet traffic growth, the value of a flatter hierarchy with denser connectivity has been emphasized. The combination calls essentially for greater operational flexibility and ability to respond to forecast uncertainty. Statistical multiplexing of packet transport offers the most efficient way to deal with bursty traffic as long as the traffic variations remain relatively small, typically an order of magnitude less than the capacity interconnecting the two routers involved. The IPNGN control plane offers sufficient robustness and flexibility also with large, but slow (time-of-day type of) traffic variations, as long these variations remain within the available overall network capacity. However, the introduction of a new DC, or a change in a OTT content provider, can disrupt traffic patterns significantly. Essentially, the advent of "cloud" has introduced significant variance also at the large flows. A low traffic node can fast become a large traffic node, demanding additional capacity that exceeds the available router interconnection bandwidth. Figure 18.4 provides evidence of this behavior; by documenting traffic growth by much more than 1000% in just days, for an ISP who adds a new consumer. Thus, more flexible capacity provisioning also becomes an important motivation for the evolution of the transport architecture.

Provisioning $10\times$ the capacity in few hours, or even days, is far from what a typical network can do today. There are many operational limitations that prohibit fast capacity provisioning and optical transport networks are among the most important limiting factors [23]. Strong consensus is emerging that an expanded role for optical transport, and a closer integration with IP/MPLS transport, would best enable this network evolution. At the same time, the connection granularity does not usually match the client links that use these connections. The capacity of a wavelength is a

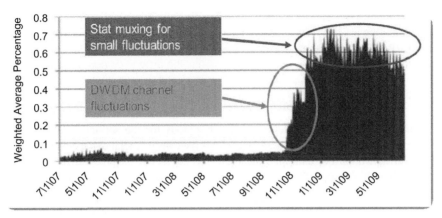

FIGURE 18.4 The traffic dynamics of an ISP (Carpathian Hosting) when adding a new consumer [6,7].

function of the transport technology at hand, and can be too small or too large for the bandwidth needs of the application layer. This creates a mismatch between the topology of the client layer and the transport layer, and motivate the addition of a sub-wavelength grooming layer to provide a better match. The optimal combination between IP and DWDM, the role of OTN, and the best use of the new photonic technologies has been extensively debated. A good synopsis of this debate occurred during the related 2011 and 2012 IEEE OFC workshops, and more interesting discussions are expected also in upcoming 2013 OFC [8]. The 2011 workshop [4] (organized by Deutsche Telecom) focused on the NGI transport optimization, the role of optical transport in this flatter hierarchy, the value of distributed peering and optimal content placement, and the role of control-plane innovations. The 2012 workshop [7] (organized by AT&T) focused more on the convergence of routing with optical transport, and particularly the optimal amount of traffic bypass for core nodes. The rest of this chapter summarizes the increasing value of photonics and optical transport in this network evolution. We start by reviewing the main transport technology innovations in the next section.

18.3 TRANSPORT TECHNOLOGY INNOVATIONS

Transport has been experiencing significant technological innovation in practically all[3] network layers. Photonics innovations in coherent transmission and non-blocking switching have enhanced significantly the DWDM layer performance. New development, and some standardization, of WSON promises better DWDM

---

[3]SONET/SDH and ATM are the only exceptions, as neither has advanced for many years.

system operations, and eventually improved network provisioning. The optical transport network (OTN) standard has introduced much needed OAMP. Meanwhile, IP/MPLS transport has continued to advance operational flexibility, and system scale.

18.3.1 **IP/MPLS transport**

Because IP/MPLS transport has become ubiquitous for the services that drive the traffic growth, its layer-3 (or L3) advancements have been crucial to the core and metro transport networks. For example, the router evolution has become the main driving force behind higher optical channel capacity, and optical transport economics. More specifically, core router ports were the first to use 40 Gb/s, and then 100 Gb/s as soon as the IEEE 802.3ba standard became available. Consequently, the analysis of transport network evolution has to account for the interplay between routing and optical transport [5].

IP/MPLS transport has been leveraging the significant advancements in ASIC of network processors (NPU) and electronic switching fabrics, and highly available operating systems, to achieve impressive system scale. Routers are already approaching Tb/s capacity at single-chassis systems, with densities of Gb/s/cm^2 and less than 2 W/Gb/s. (Note that in addition to lowering operational cost, lower power consumption is crucial because power density is usually the least flexible system design constraint.) Moreover, currently deployed core routing systems can scale to tens of Tb/s of non-blocking capacity in multi-chassis configurations.[4] Such high capacity transport systems allow the design of large hierarchical packet networks that would maximize the statistical multiplexing benefit, and could meet the forecasted capacity needs of even the most demanding inner core multi-degree nodes for many years. At the same time, robust implementations of IP/MPLS control plane with advancements in MPLS Traffic Engineering (TE), and protection mechanisms like Fast-ReRoute (FRR), IPv6, and proximity routing, improve the packet transport flexibility, scalability, and availability. The combined data plane and control-plane IP/MPLS transport innovations (some in hardware and many in software) have enabled the adoption of the IPNGN transport architecture (Figure 18.2). Moreover, leveraging at its best the electronics technology innovations which collectively we refer commonly to as "Moore's law," each new generation of core IP/MPLS routers, since their introduction in the mid-1990s, has been able to improve the routing system cost-performance by an order of magnitude every 4–5 years. Continuing this progress, the next-generation systems currently under development leverage NPU ASIC designs with hundreds of multi-threaded processors to enable systems that can scale beyond 100 Tb/s in multi-chassis configurations. These new systems offer additional CapEx optimization by transitioning from general to more specialized NPU ASICs that

---

[4]Multi-chassis systems use also optical intra-system interconnects, which is another very interesting photonics innovation, albeit beyond the scope of this chapter.

match the distinct functional requirement of inner core,[5] peering, or regional (edge) nodes, all in single flexible system architectures. Also, further innovations in operating system and control plane have advanced operational robustness and flexibility (OpEx). We account for the network value of these router innovations, and especially the significant implications in the network economics in the next sections. Meanwhile, in the rest of this section, we review the optical transport innovations that have accommodated the massive scale achieved by IP/MPLS transport.

18.3.2 **100 Gb/s interconnections and coherent DWDM transmission**

High capacity optical channels have been very useful for efficiently interconnecting high capacity routers, especially in the core. The channel bandwidth is particularly important for the operational efficiency of a router interconnection because the gains from statistically multiplexing bursty traffic flows improve when fewer, higher capacity links are employed. For example, it has been reported that the same IP/ MPLS "logical" link can be 10–30% more efficient when it is implemented with a single 40 Gb/s physical link instead of link-bundling four separate 10 Gb/s physical channels. (The 30% gain would occur for the least uniform, bustier traffic flow statistics.) These operational efficiencies have led core routers to offer ports operating at 40 Gb/s and 100 Gb/s as soon as it was technologically possible, even though the cost per bit for such higher rate links has been initially higher. Nevertheless, early adoption has been crucial in motivating the transition of the enabling photonics technology to volume manufacturing which in turn enabled cost reduction, and worldwide adoption of the higher rate links. Currently, most new deployments would use 100 Gb/s core router links as soon as the transport fiber infrastructure can support it.

Core routers have also driven higher rate DWDM transmission. Initial deployments leveraged the improved spectral efficiency to scale the most capacity constrained core links. Even though 40 or 100 G/s DWDM channels have generally been more expensive ($ per b/s), their deployments have led to significantly lower network cost by prolonging the life of the existing DWDM system. For operators without their own fiber plant additional savings also arise from postponing the leasing of an additional fiber pair. However, to achieve these savings required that the higher rate DWDM channels shall operate over the existing systems and fiber infrastructure. For such interoperability, the higher rate DWDM channels need to meet the transmission system design constraints of: (1) the OSNR for each path given the existing optical amplification, (2) the predeployed optical filter pass-bands and channel spacing, and (3) the higher impairments due to fiber (chromatic and polarization) dispersion. Advancements in DWDM channel modulation successfully addressed these requirements [10]. Optical duo-binary (ODB) modulation was the first departure from the traditional NRZ modulation, and

---

[5]Routers with lower buffer needs optimized for inner core nodes have been a very active area of innovation; e.g. G. Appenzeller, I. Keslassy, and N. McKeown. Sizing router buffers. In Proceedings of the ACM SIGCOMM, pp. 281–292, 2004.

enabled the initial deployment of 40 Gb/s DWDM channels [11]. ODB was helpful in doubling the channel spectral efficiency, meeting the required optical filter pass-bands and improving chromatic dispersion tolerance. However, ODB did not offer any significant OSNR benefits, thus limiting the reach of such 40 Gb/s channels. As a result, phase-shift-key (PSK) formats soon dominated, allowing also for lower OSNR than an amplitude modulation. Eventually, photonics innovations in new transmitter and receiver designs with QPSK with coherent detection, combined with advanced electronic signal processing, and new stronger FEC algorithms have allowed the most recent generation of 100 Gb/s DWDM systems to meet (or exceed in the case of dispersion tolerance) the performance of previously deployed 10 Gb/s DWDM systems [12].

18.3.3 Optical transport networking (ITU G.709 standard)

In addition to the important advancements in 100 Gb/s photonics technology, the development of the IEEE 802.3ba and ITU G.709 optical transport network (OTN) standards [13] has also helped the successful adoption of 40 Gb/s and 100 Gb/s DWDM systems. More specifically, OTN has defined the encapsulation, and multiplexing hierarchy, to allow for two important, and much needed, networking functions:

First, OTN has introduced standards-based Operations, Administration, Management and Provisioning (OAM&P) functionality, through an encapsulation format that wraps the data payload with well-defined OAM&P overhead bytes that provide information about the section, link, and path of the optical connection. The encapsulation also defines a separate overhead section for use by FEC bytes. (The G.709 has defined a specific FEC, but in practice alternative, higher gain FEC algorithms can also be employed.) The OTN OAM&P can support all services offering a much needed replacement of the traditional SONET/SDH OAM&P.[6]

Second, OTN has also defined a TDM-based mechanism for aggregating and switching lower rate payloads within an higher rate optical channel. More specifically, G.709 has defined payloads (ODU-n) at 1.25 Gb/s, 2.5 Gb/s, 10 Gb/s, 40 Gb/s, and 100 Gb/s (and also an ODU-flex function similar to VCAT in SONET/SDH). So, OTN allows for the efficient aggregation of multiple lower rate flows into a higher rate optical channel. In the most complete implementation of this functionality, combined with the ITU G.782 framework, OTN could become a network wide TDM circuit switching layer. We review the cost benefit analysis, potential value, and limitation of an OTN-based switching layer as part of the transport architecture evolution in Section 18.5.

In general, the OTN digital-wrapper encapsulation has become widely accepted as the OAM&P layer for practically all new DWDM transmission deployments,

---

[6]For completeness, it may be useful to refer here also to the MPLS-TP IETF standard that offers OAM&P functionality for MPLS paths. MPLS-TP operates essentially at the IP/MPLS transport layer we discussed in Section 18.3.1, and does not provide for any DWDM link level and FEC functionality. Hence, MPLS-TP is practically complementary to OTN digital wrapper, and typically applicable to metro links as they migrate from SONET/SDH to MPLS transport, and may not employ DWDM transmission.

at 10, 40 and 100 Gb/s. On the other hand, the use of OTN switching has been mostly considered as part of future 100 Gb/s DWDM deployments.

18.3.4 Fully flexible DWDM add-drop multiplexing and switching

At the same time advancements in 100 Gb/s coherent DWDM transmission, FEC, and OTN encapsulation, have been scaling the fiber capacity, significant photonics innovations in wavelength add, drop, and switching technologies, leveraging MEMS and liquid crystals, have been enabling a generation of system designs with fully flexible DWDM node switching [38]. The new systems allow for:

* add-drop flexibility of any channel at any wavelength in any port or direction, also referred to as "colorless" and "directionless" port, and
* different channels using the same wavelength to be dropped in the same node (which is often referred to in the marketing literature as "contention-less").

This new functionality is removing the two remaining capacity planning constraints of the previous generation of wavelength-selective ROADM systems that enabled the initial multi-degree switching [5]. This progress is even more impressive when compared to the first generation DWDM systems that offered only point-to-point transmission (essentially no networking), or the second generation DWDM systems with fixed OADM that did not allow for nontraffic-affecting change in the network provisioning after deployment [5].

This is a particularly important advancement in the sense that the DWDM layer can now offer fully non-blocking transport in terms of wavelength provisioning and subsequent redeployment—provided of course that channel OSNR meets the optical path requirements (or sufficient signal regeneration is employed along the path). This new generation of DWDM systems also uses wavelength tunable transmitters (and receivers in the case of coherent detection), making feasible the vision of an agile DWMD transport layer, which we discuss in detail in the next section.

18.3.5 WSON and GMPLS control-plane advancements

The advent of an agile DWDM transport layer has motivated also a renewed focus and significant development of control-plane implementations that would advance the automation of capacity provisioning. We can distinguish these efforts between the Wavelength Switched Optical Network (WSON), and Generalized MPLS (GMPLS).

WSON refers to a collection of new control-plane development efforts focused on the DWDM layer. They generally aim to automate wavelength provisioning. In addition to the DWDM node switching technology, the WSON implementations may leverage optical system innovations like wavelength performance monitoring and automatic power controls to better control the DWDM channel transmission performance. A particularly noteworthy component of WSON has been the path routing optimization. Employing the significant advancements in DWDM transmission simulation tools and network planning algorithms, WSON aims to optimize, in sufficiently short time, the routing of each WDM channel given the traffic matrix, channel capacity, and OSNR of each path, or even the available regenerators [14]. Although most components of a WSON, like the path feasibility calculations, are proprietary, with each implementation serving a specific DWDM system, WSON standards are currently under development in IEFT to define the user-network-interfaces (UNIs) with higher layers, most notably IP/MPLS transport, that would enable multi-layer network provisioning. We generally refer to these multi-layer control-plane (MLCP) efforts as GMPLS.

Most GMPLS proposals have been around for many years, with frameworks generally distinguished into two broad categories: the peer and the overlay models [15]. Nevertheless, the recent optical transport innovations have provided a renewed interest for optimized multi-layer control-plane implementations [16]. We discuss the network value of WSON and GMPLS in Section 18.5, after accounting for the main network values of the photonics technology innovations.

18.4 THE NETWORK VALUE OF PHOTONICS TECHNOLOGY INNOVATION

The *100 Gb/s optical transmission* is arguably the most valuable recent photonics innovation from a network perspective. As Internet traffic growth leads network operators to employ multiple DWDM channels to connect the same two routers, higher capacity optical channels become desirable. Higher capacity optical channels can simplify the network operations, enable higher statistical multiplexing gains of bursty traffic, and improve the spectral efficiency of DWDM transmission prolonging the "life" of the fiber infrastructure. Even at high initial cost per b/s, 100Gb/s channels have enabled significant, typically more than 20%, savings in the transport TCO. Figure 18.5 summarizes the

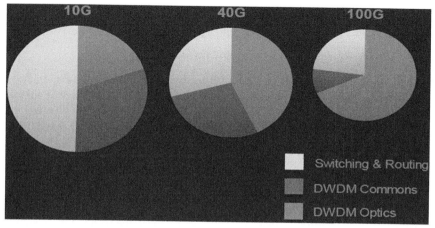

FIGURE 18.5 CapEx comparison for different generations of transport based on modeling of the network of a large operator [17].

results of such a cost study modeling the network and capacity forecast of a large (tier-1) American operator [17]. More specifically, the CapEx of routers and DWDM was compared when building the same network using the previously deployed 10 Gb/s technology, or the available (at the time of the study) 40 Gb/s, or the emerging at the time 100 Gb/s technology (which is currently available). There are two important conclusions from this study which are both widely applicable, beyond the details of the specific network:

- The network cost gets lower for higher rate channels. The savings are typically substantial enough to justify the fast adoption of 100 Gb/s.
- For networks operating at 100 Gb/s, the DWDM transponders[7] dominate the overall network cost. This is radically different from the relative costs at 10 Gb/s where routing, and to a lesser extent DWDM common equipment (like optical amplifiers and ROADMs), are proportionally more expensive.
- Note that other studies have reported very similar results, validating the consistency of this conclusion [7,9].

We have already discussed the main technology innovations that explain both conclusions. The most important savings result from the "Moore's law" scaling of the IP/MPLS transport (Section 18.3.1). More specifically, the latest generation of IP/MPLS routers (which correspond to 100 Gb/s operation) can typically achieve more than 40% lower cost per b/s than the previous generation routers deployed at 10 Gb/s operation. This router cost reduction explains the significant shift in the relative network costs because during the same time the cost per b/s of 100 Gb/s DWDM has not reduced much relatively to the previous generation DWDM.[8] As a result, transponders have become an increasingly higher proportion of the overall CapEx (Figure 18.5). Although the majority of this cost arises from DWDM, the high-speed short-reach (SR) interconnections (between routers and DWDM systems) also contribute. Particularly at 100 Gb/s, the first generation of SR modules (CFP) that implemented the IEEE802.3ba standard was substantially more $ per b/s than a same reach 10 GE solution. A new generation of SR optics is leveraging parallelism, and new technologies like silicon photonics, to improve the SR cost-effectiveness. On the other hand, the CapEx of the DWDM common equipment is proportionally much lower at 100 Gb/s because, in any multi-fiber network deployments, we would need much less EDFAs or ROADMs to build the same capacity. In the case of fiber leasing, the increase of spectral efficiency of 100 Gb/s DWDM would offer additional OpEx savings. OpEx is not included in Figure 18.5 because it is harder to generalize being usually too sensitive to operational details (e.g. leasing fiber cost, or man-hour

---

[7]Transponder is the typical industry term to refer to the combined transmitter and receiver DWDM subsystem, which includes all the photonics and analog-to-digital electronics technologies that generate and terminate the optical digital signals.

[8]As we discussed in Section 18.3, the 40 and 100 Gb/s optical modules were more expensive in $ per b/s until volume manufacturing. Even then, because much of the photonics cost scales proportionally to volume, the cost curves of the highest rate optics have not improved faster than the previous generation optics.

cost). Generally, however, OpEx studies have also showed significant savings for higher rate channel operation due to lower footprint, power, and less frequent provisioning or re-provisioning events. Particularly at 100 Gb/s DWDM, the increased performance robustness achieved by coherent detection, PSK modulation, and DSP has simplified also the optical design.[9]

The adoption of **IP/MPLS transport directly over DWDM** (Figure 18.3) has motivated also the convergence of these two network layers in order to simplify operations and reduce cost [18]. Most notably, the OTN encapsulation has made it easy to incorporate an elaborate OAM into routers, thus eliminating the need in the IPNGN architecture for an intermediate dedicated SONET/SDH layer for just the OAMP functionality [18]. Convergence of L3 with L1 can occur both at network level and through system level integration. There are a few advantages and certain limitations in a system level integration. The most important benefit of system integration where the IP/MPLS router incorporates DWDM transponders, has been the elimination of the SR optical interconnections [18]. Additional benefits arise from the corresponding power and footprint savings. For these reasons, the current generation of core and edge (regional) IP/MPLS routers has typically integrated 10 Gb/s, 40 Gb/s, and 100 Gb/s DWDM when possible. This system level integration was made possible to a great extent because current DWDM transponders have been compatible in power and footprint with the current generation of IP/MPLS routers. However, as the next generation of IP/MPLS routers scale to NPUs that can serve multiple 100 Gb/s channels per line-card (Section 18.3.1), coherent DWDM transponders cannot scale to the same density levels with today's technology (Section 18.3.2). Research and development efforts are under way to improve the DWDM density; for example, photonic integrated circuits and silicon photonics are two important complementary related innovation areas. Meanwhile, the next-generation multi-Tb/s transport could be better served by the network L3 and L1 convergence based on optimizing the cross-system architecture and improving network provisioning, scale, and operational flexibility.

The convergence of IP/MPLS with DWDM has already enabled some important network level benefits. Most notably, the integration of OTN encapsulation as part of the router has led to the widely accepted innovation of proactively protecting IP/MPLS traffic, using L3 protection mechanisms like FRR, triggered by the pre-FEC bit-error rate of each DWDM channel [19]. Figure 18.6 illustrates the basic idea, and compares it to the previous operational paradigm. The initial evaluations of proactive protection schemes show substantial network performance improvements, particularly for slower developing impairments. For example, in the case of gradual optical noise injection from a degrading EDFA, proactive IP-based protection reduced the average packet loss from 2.7 s to 0.076 ms, a performance improvement of five orders of magnitude [19]. The same basic principle, namely the better exchange of information between the IP/MPLS and DWDM transport, enables other important network

---

[9]The design rules can be more complicated in the case of DWDM systems with channels employing both phase and amplitude modulation.

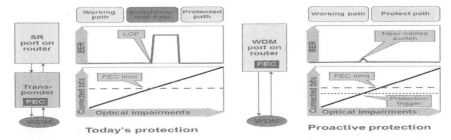

FIGURE 18.6 The operational steps of proactive IP/MPLS protection (right) based on pre-FEC bit-error rate monitoring of a DWDM channel are compared with the traditional protection scheme (left) [19].

benefits, such as better coordination in network provisioning or restoration. This could take also into account information about the shared-risk link-groups (SRLG) between L1 and L3 [18]. Additional value proposition could arise from the elimination of duplicate management inefficiencies, e.g. alarm correlation using LMP [18]. However, many such benefits from the L3 and L1 convergence depend on, and have so far been often constrained by limitations from deployed legacy systems and uncoordinated L1 and L3 organizations and operational practices.

The initial deployments of DWDM interfaces in routers have been motivated by the drive to operate higher rate physical links over the predeployed 10 Gb/s DWDM systems [20]. The increased operational robustness from innovations in modulation formats and dispersion compensation has enabled this interoperability, typically referred to as "alien-wavelength" transmission [21]. This application has also reinforced the "dark-link" standard [22] efforts toward an open DWDM transport layer that allows for fully interoperable deployment of "alien-wavelengths."

At the same time, *the new flexible DWDM transport layer* with fully reconfigurable OADM and multi-directional switching (Section 18.3.4) is also contributing to this vision, by allowing carriers to optimize a "wrongly" forecasted wavelength provisioning, and adapt to the evolving network needs. More specifically, the DWDM flexibility has enabled the introduction of wavelength level bypass as a valid optimization to the IPNGN transport hierarchy. Figure 18.7 [6] illustrates this IPNGN evolution. Note that wavelength level bypass today remains a carefully planned provisioning step to optimize the underlying hierarchical IPNGN topology, based on high capacity, stable in time, traffic needs. (So, the process does not create any network stability concerns by dynamically changing routing adjacencies.) Basically, if the traffic between two previously nonadjacent nodes, say two peering locations, approaches the capacity of the DWDM channel, a wavelength can now be provisioned directly between these two nodes, utilizing the non-blocking ROADM layer flexibility. In this sense, the new ROADM layer enables cost savings from minimizing the unnecessary intermediate routing. This is particularly useful for the IPNGN core nodes where transient traffic can grow beyond 50% [6]. It could also be appropriate for

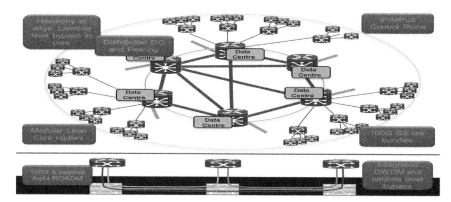

FIGURE 18.7 The new IPNGN transport architecture, including wavelength level bypass and distributed peering [6,7].

the increasingly important DC-to-DC interconnections of large (elephant) demands discussed in Section 18.2.

Wavelength level bypass has also been considered as the first step in a much more radical evolution of the transport network architecture with fully automated DWDM capacity provisioning and restoration, utilizing a dynamic multi-layer (GMPLS) control plane. However, this vision of dynamic wavelength provisioning is not supported by today's DWDM layer which remains limited by the legacy systems still in operation [23]. We review the current challenges and opportunities of the optical transport innovation in the next section.

18.5 THE NETWORK VALUE OF OPTICAL TRANSPORT INNOVATION

The introduction of DWDM nodes with flexible ROADM and multi-directional switching (Section 18.3.4), combined with the 100Gb/s DWDM robust transmission (Section 18.3.2), has enabled wavelength level bypass to successfully optimize the IPNGN transport architecture, as we summarized in the previous paragraph. From this starting point, a more dynamic optical transport provisioning, protection, and restoration has been often advocated to provide additional benefits. Although more feasible technologically, this vision has also to address carefully the questions of what are the key services, the main savings, and the optimal technology adoption during each implementation step of the IPNGN evolution.

From a services perspective, a more dynamic transport network would be justified by requirements for faster provisioning. More specifically, a dynamic DWDM layer would imply the requirement for on-demand wavelength services [23]. With the exception of special purpose, mostly research, networks, there has been very little need for dynamic wavelength services [23]. On the contrary, service providers are increasingly relying on IP/MPLS transport for dynamic provisioning and protection of packet-based services. IP/MPLS (L3) transport (Section 18.3.1) enables network operators to achieve fast, cost-effective, network utilization leveraging stat-multiplexing with the ability for elaborate QoS prioritization and TE. At the same time, OTN has provided a transport layer for non-IP/MPLS traffic, mainly including the migration of traditional SONET/SDH private-line services. In this sense, OTN has increasingly served the provisioning of such TDM-based, connection oriented services, and their related shared-meshed protection [24]. Figure 18.8a summarizes the current consensus around the transport layer for each network service. Specifically for Ethernet private line (EPL), an interesting debate has emerged as either IP/MPLS or OTN could be valid, and have been adopted, depending on the operational (and organizational) details of each specific network operator and their preferred method for offering the bandwidth reservation required for the specific application carried over the EPL. The bandwidth reservation mechanism has also been increasingly important for the inter-DC interconnections, and their sub-wavelength and wavelength level provisioning [25].

From a network cost savings perspective, the extensive bypass of core routers has been the most commonly proposed, and debated, value for dynamic optical transport [7]. In this proposal, capacity is provisioned by an intermediate sub-wavelength switching layer based on OTN (or alternatively MPLS-TP or Ethernet) and a dynamic DWDM layer, which interconnect directly the service (edge) routers. However, such a mesh router architecture presents a few challenges, related to efficiently forecasting, provisioning, and maintaining many more (typically an order of magnitude more—see Section 18.2) point-to-point links, and layer-3 adjacencies. Recent, detailed CapEx studies [26] have established that an OTN-based router bypass would usually increase more than 30% the overall transport cost for typical IP networks. The exact amount of extra cost depends on the proportion of transit traffic, and the relative cost of OTN vs. the cost of the bypassed routers.

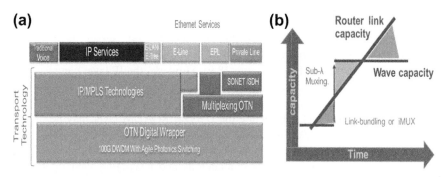

FIGURE 18.8 (a) The transport layer for the different network service (left) [6,7] and (b) the value of sub-wavelength multiplexing (right) [7].

Network cost could increase further when OTN protection of IP traffic is also taken into consideration [27], and when OpEx inefficiencies due to the additional layer are included. The most important limitation of any future architecture based on router bypass, however, may be that such an IPNGN transport evolution is focused on minimizing the IP/MPLS router cost which is a decreasingly important proportion (Figure 18.5) of the future networks [7,9].

As the cost of 100 Gb/s DWDM transponders starts dominating CapEx, efficient wavelength utilization may be the most important network optimization target. For efficient wavelength utilization, statistical multiplexing makes hierarchical packet transport very useful and extensive router bypass potentially less efficient. Wavelength bypass remains valid for stable traffic in fully utilized wavelengths. Sub-wavelength multiplexing (Figure 18.8b) also becomes useful for aggregating many small stable flows into a single wavelength, especially when flows have slow growth thus minimizing the need for re-provisioning. The OTN digital hierarchy allows efficient sub-wavelength multiplexing when each flow is not much less than 1 Gb/s (the lowest OTN granularity). OTN multiplexing could also combine IP/MPLS and non-IP traffic into the same wavelength, making it able to better utilize the deployed capacity. In this sense, OTN sub-wavelength multiplexing would be beneficial in the parts of the network with traffic flows that are small proportionally to the wavelength capacity, have slow growth, and may include a significant proportion of non-IP services [37].

On the other hand, for high capacity, fast-growing IP/MPLS traffic, the convergence of DWDM and IP/MPLS transport layers is considered the most cost-effective IPNGN evolution [4–7]. We have already discussed (Section 18.4) the benefits in OAM&P based on the OTN encapsulation, including proactive protection from pre-FEC monitoring (Figure 18.6). Improvements in network efficiency could also be achieved from coordinated L3 and L1 provisioning or restoration based on multi-layer control-plane (Section 18.3.5) and wavelength bypass. Recent advancements in multi-layer control-plane implementations have been a promising transport innovation. An innovative such framework was reported in [16]. A new set of functions that improve network operations has been built around extensions to GMPLS standards [28]. The aim is to create a new standard that balances between the two previous, diametrical GMPLS models: the "peer model" which required too much information exchange leading to complexity, and the "overlay model" which enabled too little exchange resulting in inefficiency. To this end, the DWDM "server" layer is sharing, through a UNI, the required information about the optical paths that constitute the links of the IP "client" layer (Figure 18.9). Similarly, the IP layer can also share requirements with the optical layer. At the same time, the participating layers remain independent and reasonably decoupled; as such, for example the IP/MPLS layer may continue to run multi-level ISIS, while the DWDM ROADM layer may be running OSPF. This decoupling makes the definition of multi-layer control-plane particularly powerful for allowing the two layers to more easily scale independently while maintaining the chosen organizational segmentations and leveraging their operational expertise. It also fosters multi-vendor solutions. In the most general case, there could be more than two layers in the network, with more than one client-server (UNI)

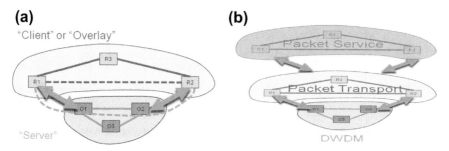

FIGURE 18.9 The multi-layer control-plane UNI between (a) IP/MPLS and DWDM (left), and (b) in the case of more than two layers, with more than one tier of client-server (UNI) interactions (right) [16].

interaction, as in the example depicted in Figure 18.9b, however generally the functionality becomes more complex with more layers.

This information awareness could eliminate multi-layer inefficiencies, and support improved network planning and operation. The goal is to maintain the SLA of the IP/MPLS services at reduced overall network cost. Conversely, other GMPLS features could allow improving the SLA at equal total network cost. For example, improved network availability can be achieved by leveraging DWDM SRLG information to guide IP/MPLS routing decisions. Operational benefits could stem from optimization based on cost, or latency, or improve handling of catastrophic failures. More specifically, a successful GMPLS implementation can enable network optimization during:

- Normal operation, by sharing optical layer information (e.g. SRLGs or latency) with the client layer.
- New connection setup, by improving the QoS, e.g. selecting the lowest latency path.
- Traffic or network changes, through path re-optimization, or by rerouting traffic appropriately (e.g. during a maintenance window).
- Restoration, by requesting the optical layer to restore a connection upon physical layer failure, and subsequent network re-optimization when the network recovers from this failure.

From an operational perspective, adoption in current networks may initially be easier during normal operation, since this does not require operational adjustments to connection setup or rerouting. This also requires smaller extensions to the existing GMPLS UNI [28]. The adoption of DWDM routing based on IP requirements will be more challenging, as this requires less mature automation in ROADM and impairment-aware WSON [29]. Eventually, however, MLCP network optimization during traffic and network changes could offer significant TCO savings, particularly since currently most IP links operate normally at less than 50% utilization due to the requirement for sufficient protection capacity against failures at the IP layer.

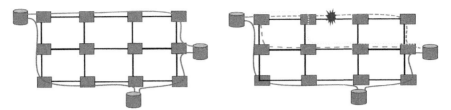

FIGURE 18.10 Normal operation (left), and DWDM restoration of a failed link over alternative paths (right) [28].

More specifically, in the event of a link failure, an advanced MLCP implementation combined with a flexible DWDM transport layer can offer over 20% CapEx savings [28], when (A) the IP/MPLS layer is designed assuming node failures is a low risk event, and (B) the SLAs of some best-effort traffic allows for a few seconds of recovery time under failure. In such a case, best-effort traffic without strict failure-related SLA guarantees remains only partially protected in the IP layer until the DWDM layer restores the failed links over alternative DWDM paths, as depicted in Figure 18.10. At this point, peak traffic will be supported again for all traffic classes. The network savings would increase proportionally to the percentage of best-effort traffic, because minimal additional protection bandwidth must be pre-provisioned for such traffic in the IP/MPLS layer.

Link failure restoration is indeed becoming increasingly important for future network operations, because next-generation routers (with multi-chassis configurations and distributed operating systems) would achieve significantly higher system availability, thus making node failures less common. However, the practical limitations of the deployed, mostly legacy, DWDM systems prevent currently the wide adoption of optical restoration [23]. The new flexible DWDM transport layer, once widely deployed, and combined with advanced WSON and GMPLS implementations could enable optical restoration within the required time-scales, typically no more than several 10s of seconds [14,28,29]. Note that, unlike optical protection, the GMPLS restoration scheme kicks in after the IP layer has quickly protected (e.g. FRR) some (or all) of the traffic, relying on QoS to ensure best use of the potentially lower L3 capacity until L1 fully recovers the links. Multiple restoration path computation options could be explored, including trying a pre-planned path first, followed by calculating a dynamic path.

More generally, network optimization based on a centralized path computation element (PCE) [30], and implemented in combination with a distributed, open, multivendor, MLCP, has been proposed as an additional advancement of the converged (L1 and L3) transport. In this vision, PCE network optimization continuously maximizes the IP/MPLS and optical transport, preventing the network from getting gradually less optimal (as is common in today's networks). In today's networks, optimization is typically performed using offline modeling tools, because network wide traffic matrix cannot be collected real time, and manual input of future traffic forecast is typically required. The benefits from continuously optimized network operation would generally increase as the traffic becomes more dynamic and the network more adaptable. However, the current limitations, particularly in the DWDM layer automation, and from legacy network management systems (NMS, OSS), prevent a practical implementation, or a realistic account of the network value of this vision.

Given the remaining limitations from legacy deployments, and DWDM automation, a radically alternative approach to L3 and L1 convergence is also being actively explored based on a very simple transport architecture with next-generation multi-Tb/s routers directly interconnected using the highest possible capacity DWDM point-to-point links, e.g. [31]. In this approach, network automation occurs only at the IP/MPLS layer, thus simplifying operations and minimizing cross-layer coordination. At the same time, many of the synergies from IP and DWDM convergence, like proactive protection, still benefit network operations. Also, the simplification of the DWDM layer could allow for lower cost DWDM systems by reducing OADM and optical amplifier complexity. The inherit simplicity of this architecture, at least from an optical transport perspective, makes it more readily available for scaling IPNGN to hundreds of Tb/s. The main challenges here arise from cost-effectively scaling the IP/MPLS routers and high-speed DWDM transmission. We explore some of the more promising future technologies that could enable this evolution in the next section.

18.6 OUTLOOK

The recent innovations in optical and IP/MPLS transport have improved the network cost-effectiveness, scale, and operational efficiency. Emerging technology and system advancements are promising to improve further these metrics. Before elaborating on the benefits from future innovations, however, it is important to highlight the significant, often-untold, value of the evolution to a transport architecture that is increasingly "open." "Openness" has been a key enabler to network scale, service flexibility, operational efficiency; including interoperability and early adoption of important new technologies. The advancements in DWDM coherent transmission and reconfigurability, optical transport, and routing have certainly contributed to this "openness," and this is the expectation also from future innovation.

The Internet needs to scale to Zettabytes of capacity, while addressing the traffic unpredictability in terms of service type, origin, destination, and duration. As the majority of traffic growth is increasingly coming from IP/MPLS services, IP/MPLS routers are expected to continue to play an important role in the future transport network, based on systems that cost-effectively scale to hundreds of Tb/s of capacity in multi-chassis configurations. Another particularly interesting development for the future routing systems is the emergence of service-agnostic electronic switching that would enable the convergence of IP, MPLS, and OTN (or any other future transport standard) into a single system. This convergence at the system level would improve significantly the cost and operational efficiency of future transport networks, and accommodate future unanticipated services and protocols without affecting the installed hardware infrastructure.

Photonics innovation in short-reach optical interconnections will also be increasingly important for the next-generation (multi-chassis) transport systems. Silicon photonics has become a very promising technology platform for this application, for its ability to improve density and power (both much needed attributes in the future multi-Tb/s chassis).

Future DWDM innovations are also promising significant network benefits. For example, Tb/s super-channels (see related chapter) aim to increase 2−4× the DWDM transmission spectral efficiency, scaling fiber capacity beyond 30 Tb/s. At the same time, advanced DSP (see related chapter) aims to increase the channel bandwidth and its robustness to physical layer impairments. From the network architecture perspective, the most intriguing innovation may be the proposed "flexible-spectrum" [32] DWDM systems that depart from the traditional fixed ITU wavelength grid convention. These systems aim to enable a future "Elastic Optical Networking" (EON) architecture that would offer more efficient and flexible utilization of the fiber capacity, when compared to today's DWDM. Another chapter in this book discusses in detail the EON architecture, its value, potential implementations, and enabling technologies. Although EON "is not ready for prime time," some related technologies are getting closer to play an important role also in the evolution of the current transport networks. Most notably, "variable bit-rate" transmission could soon allow DWDM channels to adapt their data rate, trading off channel bandwidth (and spectral width) for OSNR and hence unregenerate reach. One practical implementation is leveraging different phase modulations over the coherent DWDM channel to adapt transmission from 50 Gb/s with BPSK for the longest reach, to current 100 Gb/s DQPSK, and eventually 400 Gb/s with multi-QAM that requires though a much higher OSNR path [32]. Various other designs have also been proposed, e.g. leveraging different FEC [33], or new photonics super-channel technology [34] or even arbitrary optical wave-forms [35]. Assuming a cost-effective implementation becomes possible, this new generation of "adaptive transmission" DWDM systems would enable a useful cost-performance transport optimization. In the simplest network implementation of a static point-to-point optical transport, like in [31], adaptive DWDM would be optimized for the required transmission distances, which could differ by more than an order of magnitude in different carrier networks. In the most elaborate network implementation of a fully reconfigurable, automated, converged (L1 and L3) transport architecture with dynamic MLCP, adaptive DWDM could participate in the overall network PCE-based optimization, essentially becoming an enabling technology in a broader vision of software-defined network operation [36,39].

It is hard to predict the exact adoption of future innovations, as this will be determined by many important implementation details which remain yet undefined. For example, the exact amount of reconfigurability in future optical transport will depend on the extra cost of making the optical layer flexible in comparison to the cost savings enabled by such flexibility. Nevertheless, drawing analogies from the recent optical transport evolution, and more specifically the successful adoption of metro DWDM multi-service transport [2], suggests that DWDM will continue to be the main technology for scaling the capacity of future networks, and photonic switching and wavelength bypass will continue to help scale the electrical, dominantly IP/MPLS, transport layer.

18.7 SUMMARY

The expanding role of Internet-based service delivery, and its underlying "cloud"-based infrastructure of Internet worked data-centers, is motivating the evolution to an IPNGN architecture with a flatter hierarchy of more densely interconnecting networks (Figure 18.7). This next-generation Internet is required to cost-effectively scale to Zettabytes of bandwidth with improved operational efficiency, in an environment of increasing traffic variability, dynamisms, forecast unpredictability. In this chapter, we have explored the implications of this change in the metro regional and core transport network architectures, and the important advancements in optical, routing, and traffic engineering technologies that are enabling this evolution. We have particularly accounted for the increasingly important role of optical transport and photonics technology innovations. More specifically, a new generation of coherent and flexible DWDM systems with more than 2 b/s/Hz spectral efficiency is enabling the existing fiber infrastructure to scale to multi-Tb/s capacity, albeit at a significantly higher proportion, typically more than 50%, of the total transport network cost. The convergence of IP/MPLS and optical transport has been suggested as the most promising next step in cost-efficient transport evolution, in open architectures that combine advancements in photonics, routing, multi-layer control-plane and management coordination, with interoperability, to improve operation, automate provisioning and restoration, and optimize network utilization.

Acknowledgments

I would like to acknowledge numerous insightful discussions on this topic with many colleagues, and especially O. Gerstel, S. Spraggs, A. Clauberg, A. Gladish, B. Koley, R. Doverspike, and P. Magill.

References

[1] C. Labovitz et al., ATLAS Internet Observatory Annual Report. <http://www.nanog.org/meetings/nanog47/presentations/Monday/Labovitz_ObserveReport_N47_Mon.pdf>.

[2] L. Paraschis, O. Gerstel, M. Frankel, Metro networks: services and technologies, in: I.P. Kaminow, T. Li, A. Willner (Eds.), Optical Fiber Telecommunications V B, Systems and Impairments, Academic Press, 2008, pp. 477–509 ISBN: 978-0-12-374172-1 (Chapter 12).

[3] Cisco, Visual Network Index (VNI), 2011. <http://www.cisco.com/en/US/netsol/ns827/networking_solutions_sub_solution.html>.

[4] F.J. Westphal, A. Gladisch, Next-generation network convergence: How will the architectures of mega-data-centers and traditional telecommunication networks evolve towards a future ICT infrastructure, in: Workshop in the OSA/IEEE Optical Fiber Communication Conference, Los Angeles, California, March 7, 2011.

[5] L. Paraschis, Photonics enabling the Zettabyte network evolution, in: IEEE LEOS 2009 Topical Meeting, Newport Beach, California, USA, July 21, 2009.

[6] S. Spraggs, Combining a universal service-rich edge with cost-optimized core transport to create the optimal service provider infrastructure, in: Ethernet World Congress 2011, October 12, 2011.

[7] P. Magill, Core router bypass: via ROADM, OTN, Ethernet, MPLS or not at all? in: Workshop in the OSA/IEEE Optical Fiber Communication Conference, Los Angeles, California, March 12, 2012.

[8] B. Koley, V. Vusirikala, 100 GbE and beyond for warehouse scale computing interconnects, Opt. Fiber Technol. 17 (2011) 363–367.

[9] R. Doverspike, Can transport equipment cost reductions keep pace with carrier internet growth? in: Workshop in the OSA/IEEE Optical Fiber Communication Conference, Anaheim, California, March 17, 2013.

[10] K. Roberts et al., 100G and beyond with digital coherent signal processing, IEEE Commun. Mag. 48 (2010) 62–69.

[11] M. Birk et al., Field trial of 40 Gb/s PSBT channel upgrade to an installed 1700 km 10 Gb/s system, in: OSA/IEEE 2005 Optical Fiber Communication Conference Technical Digest Series, Paper OtuH3 Anaheim, California, March 2005.

[12] C. Fludger et al., Coherent equalization and POLMUX-RZ-DQPSK for robust 100-GE transmission, IEEE J. Lightwave Technol. 26 (1) (2008) 64–72.

[13] S. Gorshe, A Tutorial on ITU-T G.709 Optical Transport Networks (OTN) Technology White Paper, 2010. <http://www.elettronicanews.it/01NET/Photo_Library/775/PMC_OTN_pdf.pdf>.

[14] C. Saradhi et al., Traffic independent heuristics for regenerator sites election for providing any-to-any optical connectivity, in: Proceedings of the Optical Fiber Communication Conference, Los Angeles, CA, March 2010.

[15] P. Ashwood-Smith, Y. Fan, A. Banerjee, J. Drake, J. Lang, L. Berger, G. Bernstein, K. Kompella, E. Mannie, B. Rajagopalan, D. Saha, Z. Tang, Y. Rekhter, V. Sharma, Generalized MPLS Signaling Functional Description, IETF Internet Draft, 2000.

[16] C. Filsfils, iOverlay Framework, Keynote 2 MPLS & Ethernet World Congress 2012, Paris, France, February 8, 2012.

[17] J. Maddox, Market watch, in: the OSA/IEEE Optical Fiber Communication Conference, Los Angeles, California, March 7, 2011.

[18] R. Batchelor, O. Gerstel, et al., IP over DWDM architecture value proposition, in: OSA/IEEE 2006 Optical Fiber Communication Conference Technical Digest Series, Paper PD42, Anaheim, California, March 2006.

[19] O. Gerstel et al., Proactive protection of IP/MPLS traffic based on pre-FEC bit-error rate monitoring, in: IEEE/OSA 2008 Conference on Optical Fiber Communications (OFC) '08, Paper NWD4, San Diego, CA, February 2008.

[20] D. Ventori et al., Demonstration and evaluation of IP-over-DWDM networking as "alien-wavelength" over existing carrier DWDM infrastructure, in: IEEE/OSA Conference on Optical Fiber Communications (OFC) '08, Paper NME3, San Diego, CA, February 2008.

[21] I. Leung et al., 40 Gb/s Transmission over a reconfigurable 10 Gb/s WDM system scaling beyond 1000 km of G.652 fiber, in: IEEE Lasers and Electro-Optics Society Annual Meeting 2006, Paper WP2, Montreal, November 2006.

[22] Dark-link IETF Draft. <http://tools.ietf.org/html/draft-kunze-black-link-management-framework-00>.

[23] R. Doverspike, J. Yates, Optical network management and control, Proc. IEEE 100 (5) (2012).

[24] Wayne D. Grover, Mesh-based Survivable Networks: Options and Strategies for Optical, MPLS, SONET and ATM Networking, Prentice Hall PTR, Upper Saddle River, New Jersey, 2003.

[25] A. Mahimkar, R. Doverspike, J. Yates, M. Feuer, S. Woodward, P. Magill, A. Chiu, Bandwidth on demand for inter-data center communication, in: ACM SIGCOMM Workshop on Hot Topics in Networks, 2011.

[26] P. Belotti et al., Multi-layer MPLS network design: the impact of statistical multiplexing, Comput. Networks 52 (6) (2008).

[27] O. Gerstel, R. Batchellor, Increased IP layer protection bandwidth due to router bypass, in: ECOC 2010.

[28] O. Gerstel et al., GMPLS-based IP/optical integration—benefits and case studies, MPLS & Ethernet World Congress 2012, Paris, France, February 10, 2012.

[29] draft-ietf-ccamp-wson-impairments-10.txt. <http://www.rfc-editor.org/rfc/rfc6566.txt>.

[30] IETF, Path Computation Element (PCE) Subgroup. <http://datatracker.ietf.org/wg/pce/charter/>.

[31] A. Clauberg, Revolutionizing carrier service delivery using a software defined native IP network, in: Open Networking Summit 2012 Program, Santa Clara, CA, April 18, 2012. <http://opennetsummit.org/>.

[32] O. Gerstel et al., Elastic optical networking: a new dawn for the optical layer? IEEE Commun. Mag. (2012).

[33] G. Gho, L. Klak, J. Kahn, Rate-adaptive coding for optical fiber transmission systems, J. Lightwave Technol. 29 (2) (2011) 222.

[34] M. Chitgarha et al., Flexible, reconfigurable capacity output of a high-performance 64-QAM optical transmitter, in: ECOC 2012.

[35] D.J. Geisler, R. Proietti, Y. Yin, R.P. Scott, X. Cai, N.K. Fontaine, L. Paraschis, O. Gerstel, S.J.B. Yoo, Experimental demonstration of flexible bandwidth networking with real-time impairment awareness, Opt. Express 19 (2011) B736–B745.

[36] D. Ward, Service provider approach to SDN, in: Open Networking Summit 2012 Program, Santa Clara, CA, April 18, 2012. <http://opennetsummit.org/>.

[37] A. Gerber, R. Doverspike, Traffic types and growth in backbone networks, in: IEEE/OSA 2011 Conference on Optical Fiber Communications (OFC), Los Angeles, CA, March 2011.

[38] T.A. Strasser, J.L. Wagener, Wavelength-selective switches for ROADM applications, IEEE J. Sel. Top. Quant. Electron. 16 (September–October) (2010) 1150–1157.

[39] U. Hoelzle, in: Open Networking Summit 2012 Program, Santa Clara, CA, Tuesday, April 17, 2012. <http://opennetsummit.org/>.

Novel Architectures for Streaming/Routing in Optical Networks

19

Vincent W.S. Chan

EECS, Joan & Irwin Jacobs Professor, MIT, US

19.1 INTRODUCTION AND HISTORICAL PERSPECTIVES ON CONNECTION AND CONNECTIONLESS ORIENTED OPTICAL TRANSPORTS

Most future growth of Internet data volume will not merely result from more users with the same application rate demands, but also from the dramatic increase in data rate demand of emerging applications, such as a new computing and storage paradigm using the network as a service bus as depicted in Figure 19.1. A significant portion of these new applications will be comprised of bursty and unscheduled large transactions that can make up most of the traffic volume of the Internet with its traffic exhibiting heavy tail behavior. This growth has set the stage for optical networking technology to make significant contributions in next-generation data network architectures [24].

In the early days of the Internet, the most precious resource was long haul transmission capacity. The electronic packet switching (EPS) architecture was designed to use this resource as efficiently as possible. The Achilles' heel of the EPS architecture is its scalability: the difficulty, due to the complexity of electronic packet processing at routers, to keep pace with the unfolding exponential growth in network data rate demand. Indeed, given the trajectory of present-day demands, even electronic processing advancing with the pace of Moore's law—arguably optimistic owing to the super-linear complexity of switching and routing computation at the network processing units of routers—will not be able to avert a bottleneck in electronic-based switching. Figure 19.2 illustrates the cost trend of the different generations of data networks. Present-day network cost is dominated by the processing cost at network nodes and not long-haul transport cost. Even with passive optical access networks, the processing of which are electronically based, the cost and thus user data rate evolution will be limited by Moore's law. Optical networking technology is capable of dramatic increases in data rates ($\sim$3 orders of magnitude) in the next decade and only a new network architecture can unleash its potential [19,21,24].

There is no doubt that computing power will increase in the near future with the development of advanced multi-core processors and cloud computing and storage.

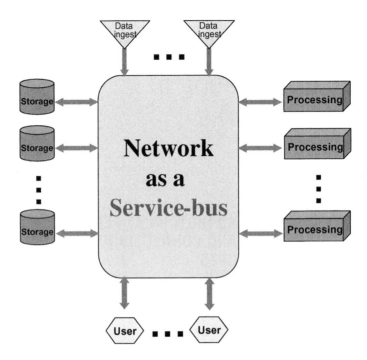

FIGURE 19.1 Network as a service bus, supporting both small and large transactions particularly of bursty unscheduled nature.

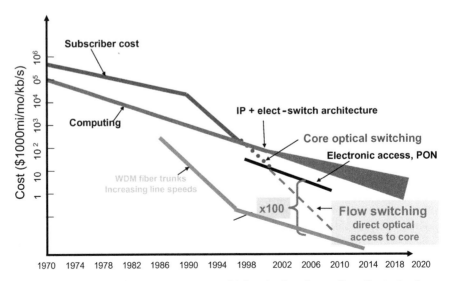

FIGURE 19.2 Cost evolution of data networks with breakpoints due to disruptive technology insertions and possible effect of introduction of optical flow switching.

Coupled with processing power increase, the I/O speeds of the new generations of processors will also increase substantially.[1] The limiting factor on how fast new applications will develop is the availability of high network speeds and much better quality of service at reasonable costs. To continue to reap the full benefits of optical networking technology, there has to be some significant network architectural changes—not mere substitution of optical components for electronic ones. The consequence of such a shift toward optical networking technology is that most architectural elements of networks—from the Physical Layer to the higher network layers, as well as network management and control—must be re-examined at the most fundamental level. All-optical switching in various forms such as packet switching, burst switching, and flow switching are new architecture constructs that show significant promise.

19.2 ESSENCE OF THE MAJOR TYPES OF OPTICAL TRANSPORTS: OPTICAL PACKET SWITCHING (OPS), OPTICAL BURST SWITCHING (OBS), AND OPTICAL FLOW SWITCHING (OFS)

Since the characteristics of optical devices are very different from their electronic counterparts, and in some cases may not even have an electronic analog, it is safe to assume that the optimum optical network architecture will not be the same as the current electronic network architecture of the Internet. To highlight this thesis, the following is a list of the major differences between optical and electronic technologies relevant to networks:

1. *Abundance of bandwidth in the fiber*—especially in the access network (important not to be confused with the data rate supported by the transmission system, which is typically much less and more expensive).
2. *Lack of viable and cost-effective optical random access memory*—an important building block of routers.
3. *Cost of SiCMOS logic gates is about 10^{-8}/gate compared to the current cost of optical logic gates at 10^4–10^5/gate*—a difference of 12 orders of magnitude.
4. *The fundamental limit of minimum switching energy of an optical logic gate is $\sim h\nu$ which is significantly larger than the corresponding limit for electronic, $\sim kT$.* Thus, optics will generate more heat than electronics for the same number of logical operations. Given the huge number of logical operations required in a packet switch, heat management will be a big problem.

---

[1]The new generation of 128 core Intel processors will have 2×10 Gbps I/O speeds using "Lightpeak," an optical USB size I/O interface.

5. *Electronic time slot interchangers are a lot cheaper (~6–7 orders of magnitude) than their dual in the optical WDM domain (wavelength converters).*
6. *Optical switches are more efficient when switching in bulk,* such as entire wavelengths, for longer configuration durations (>10 ms); whereas electronic switches can switch small segments of data efficiently at much higher reconfiguration rates.
7. *Optical switches are inherently a better broadcast medium than electronics* and may lend to simpler architectures for multicasting, narrow-casting, multiple access as a shared medium with contention resolution, and conferencing.
8. *Laser sources are significantly more expensive than the corresponding electronic signal sources.*

To keep reaping the cost benefits of optical technologies, long duration (quasi-static) optical circuit switching will be used in the next cycle of buildup in the core (known as GMPLS, Generalized Multi-Protocol Label Switching). At the access network, electronic routers and switches are currently used for aggregation to deal with bursty computer data communications. This will eventually slow down the trend of cost reductions at which point optical access and some form of optical switching and routing for large transactions must be used to maintain the same cost reduction slope. Thus it is important to explore *architectures that will exploit optical switching and routing technology as well as all-optical transport, in order to continue lowering cost faster than Moore's law, and to make high-rate services accessible.*

The optical transport mechanisms used will have first-order effects—effects on the orders of magnitude—on network cost. Thus, it is important to understand the throughput-cost tradeoffs of several leading optical network switching architectures. The various categories of optical transport mechanisms that have been considered are listed below[2]:

1. *Optical packet switching (OPS)* [1–7,30]: OPS is an analog of electronic packet switching (EPS) using optical buffering in place of electronic buffers, or variable fiber delay loops to order optical packets in the right timing order for switching.
2. *Tell-and-go (TaG)/optical burst switching (OBS)* [27–29]: IP packets from IP routers are aggregated for a few microseconds (or longer) at the edge of the network and bursts of common-destination packets, with appropriate "just-in-time-switching" or "deflection-routing" headers or control packets, are sent through the network without waiting for end-to-end acknowledgments. Burst loss due to contention and collisions is allowed to occur (unless the optical switches employ "optical buffers" of microseconds or longer durations).

---

[2]*Generalized multi-protocol label switching (GMPLS)* not addressed here is a signaling-based optical circuit provisioning technique for the core network and is already being deployed in the current generation of networks. GMPLS provisions full line-rate circuits (e.g. SONET or Ethernet) in response to GMPLS-based automatic signaling by client equipment with service holding times of minutes to days or longer.

3. *Optical flow switching (OFS)* [1–7]: Establishing connections in response to flow-based requests by client-layer IP routers for direct access by individual users, with service holding times of a several milliseconds or longer.

Figure 19.3 organizes the different forms of all-optical data networking architectures into a logical framework [1–7]. OPS is a connectionless transport as in traditional EPS except the switching functions are done via optical means. OBS and OFS are connection-oriented services. What sets them apart from traditional circuit switching as in GMPLS and MPLS are the dynamic nature of the circuit assignments and the per transaction resource allocation.

Figure 19.4 shows a generic router architecture. Packets are stored temporarily in a buffer and their headers are processed and scheduled by the network processing unit (NPU) for switching across the router switch fabric to the designated output ports. Packets may also be temporarily stored in output buffers. Headers are typically processed at

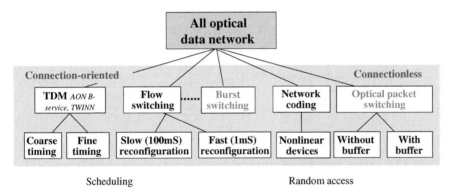

FIGURE 19.3 Candidate electronic and optical transport mechanisms.

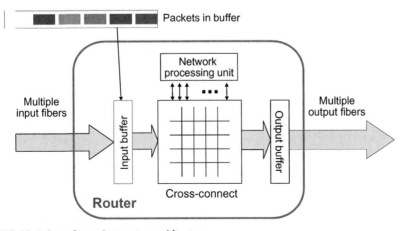

FIGURE 19.4 Generic packet router architecture.

line rates. The ultimate all-optical version of OPS has all the functions of a traditional packet router implemented in optical forms [30]. In its most modest form all computations are done electronically and only the switching and buffering are done optically, see Figure 19.5. The technology of optical buffering, especially random access memory, and computation is yet to mature and the cost and power consumption of optical digital processing as this juncture is still orders of magnitude from realizing a cost-effective network architecture, what with the current router NPUs having upwards of 100s of million of gates. While many promising new technologies involving silicon and nanophotonics, especially in the power consumption and foot print areas, may completely change the realizable architectures, time will tell when a low cost and low power network can be built using OPS. Nonetheless, today this is vibrant area of continued interesting research.

Both optical burst switching (OBS) and optical flow switching (OFS) use an electronic control plane which is separated from the optical data plane for management and control of the transport mechanisms [16]. While OBS emphasizes fast, almost instantaneous transmissions suitable for lightly loaded networks, OFS uses scheduling to improvement network utilization and lower costs. Except for the scheduling difference, both use the essentially the same optical hardware as shown in Figure 19.6.

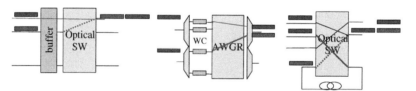

FIGURE 19.5 Several common realizations of optical packet switching with: (a) optical buffers, (b) wavelength converters (WC), and (c) fiber delay lines.

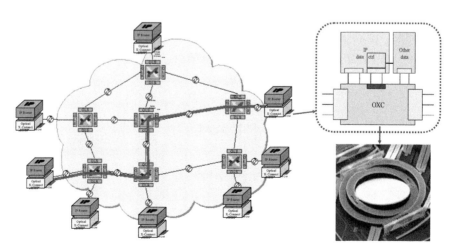

FIGURE 19.6 Optical connection oriented network based on switching of all-optical lightpaths and off-band electronic control plane.

For OBS [27–29], the most common and purest form of implementation aggregates packets at ingress nodes/routers according to destination to form bursts of collections of packets for end-to-end switching to the egress nodes/routers. Prior to transmission, control packets are end to intermediate nodes to set up lightpaths for the burst transfers. After a prescribed time delay a burst will be sent into the network without acknowledgment. There is a chance that the burst can be discarded due to contention of resources inside the network. The probability of collision occurrence is typically kept loaded by only lightly loading the network. While this may seem inefficient compared to OFS with scheduling, OBS can be very effective for networks with plenty of bandwidths such as those in a data center and inter-data-centers. OBS has the advantage that it has almost no delay due to access protocol. Currently under research are mutations of OBS and OFS with features that are in between the two types of closely related transports.

The OFS concept was conceived in 1989 at the inception of the "All-Optical-Network" (AON) Consortium [2–4] and was called the D-service at the time. For OFS, Figure 19.7, the scheduling is per session and the network control must be dynamic and responsive. When a transaction finishes, the network resources are immediately relinquished to other users. There will consequently be a tradeoff among three network performance parameters: delay, blocking probability, and wavelength utilization. This tradeoff allows the network to accommodate multiple levels of service quality to co-exist in the same network. The key to high utilization of backbone wavelength channels—a precious network resource owing to the necessary use of optical amplifiers and dispersion management—is statistical multiplexing of large flows from many users in a *scheduled* fashion. Dynamically assigned multi-access all-optical broadcast groups are arranged for multiple transaction durations using the LAN and MAN node architecture. In a conservative cost comparison [18–21], it has been suggested that OFS provides significant cost savings over other transport mechanism for large transactions.

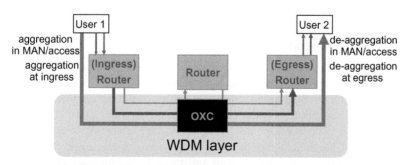

FIGURE 19.7 OFS (an end-to-end optical service) and conventional IP and IP over GMPLS (Generalized Multi-Protocol Label Switching) packet switching services [23].

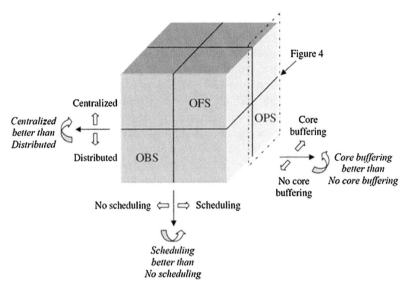

FIGURE 19.8 Taxonomy of optical network transport architectures [15].

Figure 19.8 depicts the architectural differences among the three connection and connnectionless oriented transport mechanisms based on their unique attributes on several key architecture features:

1. With or without buffering.
2. With or without scheduling.
3. Using centralized or distributed control.

These architecture features directly affect the capacity, response time, delay and blocking performance, cost and power consumption of the network. While the values of these metrics are very much dependent on the state-of-the-art of the underlying technologies, OPS is less mature at this juncture and thus will have less coverage in this chapter except for capacity considerations. OFS and OBS are primed for usage and will be treated in more detail architecturally.

19.2.1 A brief history of OFS

In 1988, MIT initiated a research project to explore all new optical network architectures that can make full use of the properties of optical devices and not be tied to the IP packet switching architecture that evolved from accommodating expensive long-haul transports with the EPS architecture that first used copper cables and then optical fibers used merely as high-speed replacements of copper cables. It was apparent immediately at the time that bypassing electronic processing at intermediate nodes has the potential of substantially saving node cost, provided

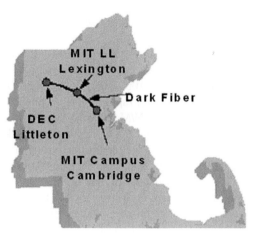

FIGURE 19.9 All-Optical-Network (AON) testbed [1–3].

the transaction size is large (flows that occupy an entire wavelength for a fraction of a second or longer). The companion benefit of saving node power consumption was not noted until much later when energy consumption becomes a great concern [25,26]. The above study led to the AON Consortium (1990–1997) formed among MIT, AT&T, and the Digital Equipment Corporation funded by DARPA. There were four different types of services built in the AON testbed that was a LAN/MAN combination of 100 km extent in Massachusetts, Figure 19.9 [1–5]. Except for the electronic packet service (C-service) used for network management and control all the other services were all-optical end-to-end. The B-service was a time slotted wavelength service to provide users periodic subwavelength capacity on demand. The A service was usually known by its slow provisioning form and it was similar to subsequent SONET and G709 services providing long duration WDM circuit switching services. In its most aggressive and dynamic form, which was called the D-service internally (not used externally since the difference between A and D services is just the flow duration and the fast setup time), the testbed could set up WDM circuits at 10 ms time scales using the C service, for short duration flows (>100 ms). The A and D services used wavelength tuned optical switching with an array grating router and the D service was the first realization of OFS in the LAN/MAN environment.

In 1997–1998 an architecture study ORAN, Optical Regional Network Study performed by MIT and AT&T, led to two coupled experimental programs ONRAMP, Optical Network for Regional Access with Multi-Protocols and BossNet, Boston-Southern Network. One of the major findings of that study is the realization that *to achieve the ultimate benefits of optical technologies, all-optical-network architectures need to consider both access and long-haul networks together as one network in their architecture construct and especially for the very dynamic OFS service.*

The ONRAMP Program (1999–2004), with MIT, AT&T, JDS, HP/Compac, and Nortel as partners, deployed OFS as its center-piece. Using BossNet (an MIT only program), as the long-haul network (~1000 km), Figure 19.10, OFS technology was demonstrated in 2001 as an on-demand rapid optical wavelength service from end user to end user over an all-optical LAN/MAN/WAN with setup times <100 ms and holding times of 1 s or longer, Figure 19.6. Comparison to packet switching using TCP/IP confirmed the benefit of OFS for large transactions in terms of much better delay performance [6,7]. While these testbed demonstrations were valuable in showing the potential of OFS and that optical and electronic technologies were fast enough to perform at the time scale of ms to enable OFS, there were only a few users and nodes in the network and a small number of alternative paths. These demonstrations in no way answered the important questions of scalability to large numbers of users and a large complex WAN/MAN/LAN network. To that extent these were technology "stunts" similar to many hero experiments. A multi-faceted architecture study was done in the last few years to create a sensible scalable and cost-effective optical network at all layers of the network from the Physical Layer to the Application Layer [14,15,17].

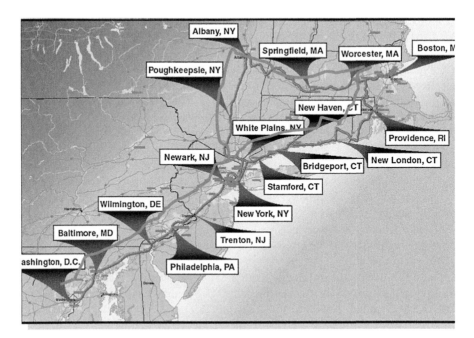

FIGURE 19.10 ONRAMP and BossNet with two alternate all-optical paths between Boston and Washington DC.

19.3 NETWORK ARCHITECTURE DESCRIPTION AND LAYERING

19.3.1 The need for new architecture constructs for optical networks

Left to incremental developments, the current Internet architecture will not be able to meet heterogeneous network applications over heterogeneous networks at affordable costs. The principal reason is that the current partitioning of networks into layers for research has run its course and approaches the saturation point for further major improvements. However, breaking down the layer structure entirely for optimization is not the correct path for future network research. As an amorphous system with no boundaries, the network is too complex for global optimization as a single large problem. Some form of partitioning is important for modeling and thinking through the right architecture for each network function. The current partitioning may or may not be the best one for the research of radically different and better network architectures. Indeed, recent developments of physical layer communication systems substantially transcend the original assumptions of the properties of the physical layer. There are now many more physical layer techniques that can be put to bear on network architectures. A good example is agile optical circuits for large elephant transactions in fiber networks. Few if any of the upper layer network architectures in recent years have taken advantages of this new space of possibilities of the physical layer. Substantial network performance can be realized if one exploits the interplay of the physical layer and the higher network layers including applications. Due to the heterogeneity of the subnets and applications, a fundamental question to ask is: whether a single suite of protocols can perform in these networks at high efficiencies or an adaptive hybrid architecture is needed. With current tradition network research running against performance/cost barriers, to realize better network performance, network architectures should be re-examined from a fresh set of viewpoints. In particular, the network control plane needs much more attention with better modeling and quantitative analysis and the current highly suboptimum architecture designed by black art needs to be turned into a science. Hitherto network management and control is an underinvestigated area because sessions are small and statistical multiplexing smoothes over large short duration fluctuations, slowing down the control plane substantially. In the future this paradigm will not hold with elephant sessions taking over the granularity of a single wavelength in optical networks at multi-Gbps. The dynamic nature of these applications will take the speed of the control plane from minutes between adaptations to seconds and subseconds. With such magnitude of changes, scalability and network state sensing and propagation are big concerns. How the network should deal with such dynamics and still contain the cost of the network is a wide open problem. This is more than an isolated problem in the control plane, for the physical/routing/transport layers architectures can be designed and tuned to relieve the pressures on the control plane. Thus, a key facet of network research is a scientific treatment of network management and control in conjunction with how it interacts with network layer architectures. Only with a new architecture can the network finally have the response and cost structure to match and utilize the capability of the already existing device technologies.

19.3.2 OFS architectural principles

The dynamic OFS service is shown in an explicit LAN/MAN/WAN context in Figure 19.11. There are three important architecture attributes that are necessary for any viable new optical networks:

1. Scalability in number of users, data rates, and geographic extent.
2. Good quality of service, QoS, in terms of easy access, low blocking probability, and low delay.
3. Low cost and low power consumption.

The basic construct of the network architecture uses the following features to solve the scalability, cost/power, and QoS requirements:

1. End-to-end OFS lightpaths, without intervening electronic regeneration and processing, assigned on demand for large user transactions.
2. A slowly changing quasi-static WAN that only responds to traffic trends and not fast per session scheduling which are decoupled from long-haul resource allocation.
3. An optimum reconfigurable MAN physical topology minimizing hop counts with medium switching speeds responding to short-term traffic demands.

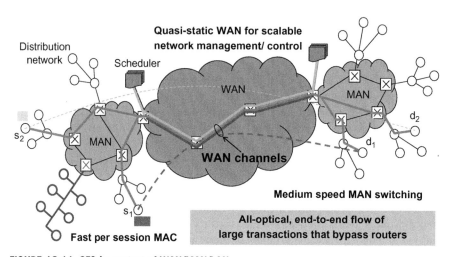

FIGURE 19.11 OFS in context of WAN/MAN/LAN.

4. A remotely pumped (at head-end) passive optical LAN to allow more users to statistically share efficiently expensive MAN switching resources and WAN lightpaths.
5. Fast per flow OFS media access control, used only for wavelength assignments of end user transceivers and time intervals scheduled for transmissions.
6. A slow centralized control plane for the WAN and fast but distributed control plane for the MAN/LAN combination.
7. For applications with extreme time deadlines, an ultra-fast OFS service that allows flow setups essentially in one roundtrip propagation time plus switch settling times between users.
8. A new network architecture that spans the Physical Layer to the Application Layer and at the same time accommodating traditional IP service as a hybrid coexistence, Figure 19.12.

The major driver that led to the above network architecture feature is cost, as will be more fully discussed later. Note in this architecture the network management and control system is still in full control of the network resources. Upon user requests, wavelength and time of transmission assignments are issued. Users cannot change scheduler policies, algorithms, and directly usurp resources. This is an important network management and security feature that ensures contending and, at moderate and high loads, conflicting user requests are arbitrated fairly and efficiently. As such, OFS is different from OBS and other network architectures in which users have access to and can change the control plane routing algorithms that can result in unresolved contentions and unfair resource allocations and open up security vulnerabilities.

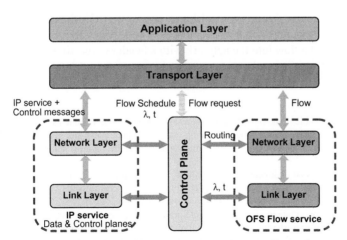

FIGURE 19.12 Multi-layer network architecture that supports both OFS and IP services [23].

19.4 DEFINITION OF NETWORK "CAPACITY" AND EVALUATION OF ACHIEVABLE NETWORK CAPACITY REGIONS OF DIFFERENT TYPES OF OPTICAL TRANSPORTS

For the considerations of capacity of a network which is significantly more complex than the capacity of a communication channel, this section will make some idealized assumptions of neglecting propagation delays and consider only unicast transactions. *The capacity region of a network is defined as the set of exogenous traffic rates for which the system of queues in the network is rate-stable for some routing under its operational constraints.*[3]

Not surprisingly, due to the ability to buffer data and wait for open time slots, OPS achieves the highest capacity region of all three transport mechanisms: *The capacity region of an OPS network is the convex hull of the union (over all possible routings) of the admissible set of traffic rates. That is, OPS networks achieve the maximum possible capacity region.*

OBS and OFS both do not have core network buffering and thus their capacities are less than that of OPS. In [16] using conflict graph[4] techniques, the capacities of both transport mechanisms were determined: *The capacity region P of a network without core buffering is the convex hull of the union (over all possible routings) of the stable set polytopes[5] of the conflict graphs.*

The reader is referred to [16] for a detailed exposition of the determinations of the capacity regions. Here the relative merits of the different transport architectures, from the perspective of *capacity alone*, are summarized and notionally depicted in Figure 19.13. Electronic/Optical Packet Switching, EPS/OPS, achieves the largest capacity region followed by Optical Flow Switching, OFS; then IP packet switching at the edge using a form of optical circuit in the core network such as Generalized Multi-Protocol Label Switching (GMPLS), and finally Tell-and-Go/Optical-Burst-Switching, TaG/OBS. The random access TaG/OBS allows the user to send a flow into the network with a header requesting resources without any prior notification and agreement by the network and has the least capacity

---

[3]A queue is rate-stable if and only if the time-averaged queue length is finite as time evolves. A set of exogenous traffic rates is admissible if a routing exists for which every link in the network is offered a rate of traffic which is strictly less than its link capacity. A more rigorous definition can be found in [16].

[4]A conflict graph is an undirected graph in which vertices represent the set of flows to be served in the network. An edge exists between vertices i and j if the flows corresponding to nodes i and j cannot simultaneously exist in the network (i.e. the flows share at least one link). Note that, for a fixed routing, there is a one-to-one mapping between feasible network states and stable sets of the conflict graph. Thus, by means of time-sharing, the stable set polytope of the conflict graph is achievable. This fully characterizes the capacity region of a network without core buffering [16].

[5]A stable set is a set of vertices in which no two vertices have an edge connecting them, and the convex hull of the incidence vectors of stable sets is the stable set polytope [16].

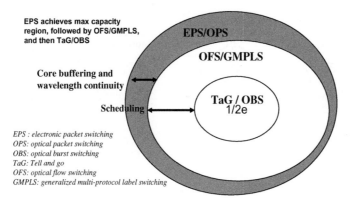

FIGURE 19.13 Notional capacity regions of different optical transport architectures [15].

performance because of its unscheduled nature with collisions/conflicts resulting in loss of rates especially at high utilizations. In its purest form the efficiency is upper-bounded by 1/2e, the throughput of unslotted ALOHA multiple access. Nonetheless, capacity utilization is not the only important differentiator of optical network transport mechanisms. Ultimately cost is the important differentiator, though cost is intimately related to throughput efficiency. The OFS architecture construct presented in this chapter uses cost as well as delay, blocking probability, and network throughput as metrics for architecture optimization. IP packet switching is processing intensive and is shown to be not cost effective for very large transactions (>Gbyte) compared to OFS. The throughput efficiency of the pure form of OBS is too low for long-haul transmission because of the high cost of amplifiers, dispersion, and polarization compensation though OBS may have a distinct advantage in local area networks (e.g. data centers, where bandwidth is plentiful) due to its distinctly low media access delay.

19.5 PHYSICAL TOPOLOGY OF FIBER PLANT AND OPTICAL SWITCHING FUNCTIONS AT NODES AND THE EFFECTS OF TRANSMISSION IMPAIRMENTS AND SESSION DYNAMICS ON NETWORK ARCHITECTURE

A sensible network architecture must exhibit a non-increasing—and preferably decreasing—cost/user/data rate as the number of users and user data rates increase. In all-optical networks, the manner in which fiber is connected to network nodes is a key design element that has significant leverage. For low access costs, the following are important key design features for the access network (LAN and MAN), Figure 19.14 [9–11,18,20], and the core network:

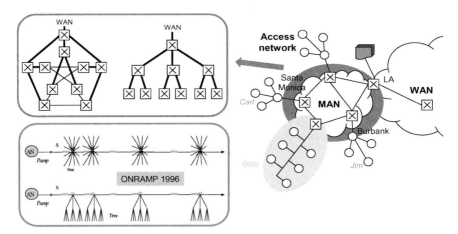

FIGURE 19.14 Access network physical architecture with remotely-pumped broadcast PON [17].

1. Minimize average lightpath lengths, in terms of the number of optical nodes and thus optical cross-connects, OXCs,[6] traversed. In the MAN, any connected fiber plant topology can be configured to any lightpath topology using OXCs and bypass patch-panels (at much lower cost than OXCs) if path agility is not required. To accommodate dynamic traffic demands the physical topology must be capable of realizing a range of possible lightpath topologies via routing and wavelength assignment (RWA) algorithms by using OXCs and transceiver tuning.
2. Use of head-end pumped erbium amplifiers in passive optical access networks (PONs) to increase the number of users in a broadcast group, for efficient statistical multiplexing without per flow switching of the node equipment [9–11].
3. Quasi-static MAN broadcast groups that adjust based on mid-term time scale traffic trends to minimize fast, <100 ms, per flow network reconfigurations.

Owing to the absence of buffering and optical-electronic-optical (OEO) conversions in the interior of OFS networks, the economic viability of OFS hinges largely on cost-effective deployment of all-optical components in the metro and access networks to carry out optical aggregation of data—while respecting the stringent physical layer constraints imposed by the architecture. In the MANs, reconfiguration via (expensive) optical cross-connects (OXCs) is economically justifiable, owing to the large number of end-users supported; whereas (less expensive) broadcast architectures, coupled with wavelength and time reservation/scheduling, are

---

[6]In this section OXCs are assumed to use all-optical switching without any OEO conversion. It includes switches that are wavelength sensitive and capable of switching wavelength by wavelength.

appropriate for access networks, where the number of supported end-users is significantly smaller. For the MAN, considerations on arbitrary mesh networks led to the conclusion that networks that are based upon Generalized Moore Graphs with optical amplifiers compensating for OXC losses are the most efficient [20,21].

Figure 19.15 summarizes the average minimum hop distance between all node pairs in the network and the network diameter D, defined as the longest shortest path over all node pairs. It is clearly a strong function of the degree Δ of the network. The larger the degree, the shorter will be the average min-hop distance and diameter, and the fewer number of optical switching ports needed. It is evident that some familiar topologies, such as rings and Hamilton graphs, scale poorly with the number of nodes, N, in the network. More sophisticated topologies, such as the Shuffle-Net and deBruijn graphs, that come close to the theoretical limit known as the Moore bound, scale favorably as log N, keeping the lightpaths short and hence using fewer optical

Graph type	Ring	P nearest neighbor	Symmetric Hamilton Graph	Shuffle-Net	deBrujin Graph	Moore Graph (Bound)
Average min hop H_{min}	N/4	N/2p	N/4p	$N = kp^k$ $H = \dfrac{3k}{2}$	$N=p^D$ ~D	$\sim D - \dfrac{1}{p}$
Diameter D	N/2	N/p	N/2p	$2k \sim 2\log_p N$	$D \sim \log_p N$	$\sim \log_p N$

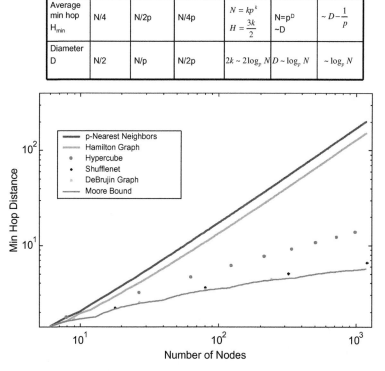

FIGURE 19.15 Average min-hop distances between nodes as a function of number of nodes *N* and degree *p* for different classes of symmetric regular graph physical lightpath topologies [17].

amplifiers and optical switching ports [17]. There are dramatic differences between good and pedestrian network topologies as shown in Figure 19.15. The differences in min-hop distances can be over an order of magnitude. Even if the reach of the lightpath is not exceeded, the order of magnitude difference in min-hop distance will result in a similar difference in cost of the corresponding networks, owing to the usage of optical switching ports at the nodes. Another way to appreciate this fact is to examine the amount of pass-through traffic at each node. The good topologies have a lower fraction of pass-through traffic, and hence only require small OXCs. Thus, from the point of view of switching cost at nodes, a good topology is crucial. These results are for regular topologies with uniform traffic (deterministic and random).

Figure 19.16 shows that the normalized MAN node cost can be brought down substantially with the right connection topology and all-optical switching at the nodes. While this has a first-order affect on the cost of the MAN, the same connection architecture is important for high performance computing where high-speed switching cost can be the important cost driver. In reality, traffic is seldom symmetric, nor is it regular or regularizable. In [17], generalization to irregular and non-uniform traffic shows the notion of minimizing lightpaths hops and using network graphs with almost full spanning trees, such as the generalized Moore graphs, are still good reference physical architectures and can be used to guide operating network topologies.

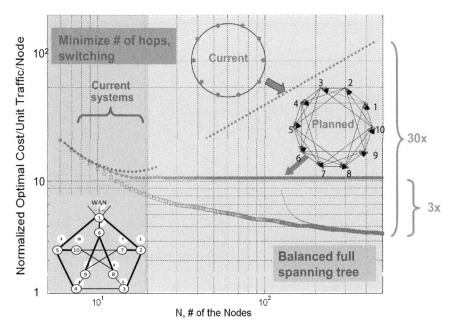

FIGURE 19.16 Scalability of optical switch and connection graph architectures as a function of the number of nodes in a MAN [18].

In the access environment, conventional passive optical network, PON designs are inadequate for distribution networks, owing to the stringent constraints of a maximum power (~3 mW/wavelength) that can be injected into a single mode fiber at the head-end yielding an upper limit of ~100 users/PON. A family of passive all-optical distribution networks that employ multiple Erbium-doped fiber (EDF) segments (without pumps) *within* the network, Figure 19.12 [9–11,20], can be used to increase the number of users attached to a single passive optical network head-end node. In contrast to conventional PON schemes which employ a single, lumped amplifier at the head-end, this design employs a remotely pumped configuration in which a single pump laser at the head-end supplies power to all of the EDFA fiber segments keeping the LAN totally passive without the need of electrical power and active optical components. The net effect is a substantial increase of the number of users in a single broadcast group to $10^{3\text{-}4}$. This allows thousands of bursty users to efficiently share expensive MAN/WAN resources via time multiple access without per flow active optically switching, Figures 19.17 and 19.18.

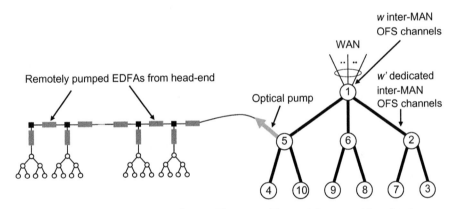

FIGURE 19.17 Access network physical architecture with remotely pumped broadcast network to increase efficient time statistical sharing of MAN/WAN resources [9–11,20].

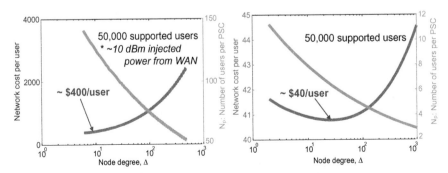

FIGURE 19.18 Cost of flow-based optical services with optical switching in the MAN and without/with remotely pumped amplified PONs [20].

19.6 NETWORK MANAGEMENT AND CONTROL FUNCTIONS AND SCALABLE ARCHITECTURES

An inherent tradeoff exists between centralized and distributed scheduling schemes. Centralized schemes provide solutions which can be globally optimum and no worse than those of distributed schemes. However, centralized schemes require the network to collect global state information and run algorithms for resource allocations that are typically NP complete on a large data set rendering such algorithms impractical albeit optimum. Thus, any implementable centralized algorithms will be suboptimal. Distributed schemes are more scalable, faster acting, and resilient. In the OFS architecture described here is a hybrid framework comprising both centralized and distributed components, whose complementary advantages allow us to support fast flow setups with manageable network state sensing and scalable scheduling algorithms.

There are two major extreme classes of flow services based on their QoS properties [20,22,23][7] (and many other gradations in between):

1. *Basic per flow services requiring admission control and queuing at the access network.* On a per flow basis, the ingress scheduler at the gateway between the ingress MAN and WAN communicates over the control plane with the egress scheduler at the gateway between the WAN and the egress MAN to determine a time schedule and open wavelength for the transmitter and the receiver. Blocked requests are scheduled and wait in a queue until resources open up. The number of sessions scheduled to wait may be finite for ease of algorithm implementation. Overflowed sessions are considered blocked. Considering transaction sizes that are beneficial by using OFS being approximately 1 s or longer, the waiting time in the queue is at most a few seconds [20,24]. This service will be sufficient for applications that are not interactive and can tolerate delays of the order of a few transaction times, such as big file transfers and backups.

2. *Ultra-fast flow service with one round-trip time scheduling and low blocking probability.* For sessions requiring subsecond setup times, routing and wavelength assignment must be completed immediately. Here the architecture uses a hybrid centralized (slow processes) and distributed (fast processes) scheme relying on up-to-date local and slightly stale global information to set up these sessions. A source node desiring ultra-fast service sends out multiple pre-computed (centralized) path[8] requests over the control plane (electronic probes over the control plane without interfering with the on-going optical flows in the data plane) to the nodes residing on these multiple paths. If the network states are updated and broadcast on time scales less than the shortest

---

[7]These two classes were picked based on the big differences in QoS requirements. There can be many more service variants that can be constructed.

[8]These paths are precomputed in an offline, centralized fashion and only have to change slowly based upon long-term traffic changes.

session durations, the state information is likely to be very accurate. If the resources requested at a node are available, then the node forwards an ACK to the next node along the path en route to the destination. If the resources at a node are not available, a NACK is sent upstream to release all booked resources in that path. Available paths are temporarily booked until use or release. In the event that two or more simultaneous requests for the same resources arrive at a node, the node adjudicates the dedication of the resources according to the priorities of the requests, how many nodes each lightpath request has successfully passed through up to that point (longest survivor will have priority everything else being equal), as well as its own version of slightly stale global network state information. Immediately after forwarding an ACK, a path node temporarily reserves the relevant resources in case the source/ destination nodes choose to use these resources for the session. In the event that multiple ACKs have arrived at the destination node at the egress network, each corresponding to different lightpaths, the destination node decides which path to use. Upon making this decision, the destination node notifies the source node of the chosen path, and the source node begins data transmission immediately thereafter. Simultaneously, the destination node sends release messages along all lightpaths that have ACKed but that will not be used for data transmission, so that temporarily booked network resources may be released for other purposes. Also, the centralized management system is notified so that it will refresh its lists of available resources in the next update. In this fashion, lightpaths can be set up in as little time as one round-trip time plus hardware reset times (<5 ms) [22–24]. Since OFS is for sessions that use the entire wavelength for at least a second and typically more, the time the ultra-fast reservation protocol tying up candidate links during probing is a small fraction of the transaction times. Moreover, it is assumed the special ultra-fast service is a small fraction of all OFS sessions; the basic services that rely on booking ahead of time do not suffer any interruptions. The performance of this ultra-fast scheme is further discussed in Section C.

Global WAN wavelength assignment on a session-by-session basis—due to cross coupling globally of all links—are deemed too complex, not scalable, and hence not cost effective. Instead, the architecture slows down the control plane to adaptive provisioning among ingress and egress MANs wavelengths over the WAN, Figure 19.19. The wavelengths between MANs act as high data rate limited access highways or shunts with no incoming and outgoing traffic at intermediate nodes in the WAN. They are allocated in response to changes in time-averaged traffic load over many sessions that can be as fast as minutes but not faster. This approach will allow the network to always meet a target queueing and overall delay performance. In this architecture, MAN and LAN physical connection topology using the programmability of optical switches and optical amplifiers also only react to medium-term traffic patterns. The only fast per session process in the control plane is the MAC protocol which does not involve any network hardware setup and switching and only IP packet

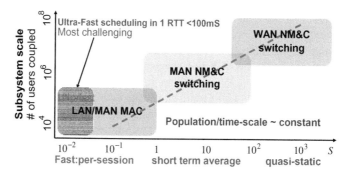

FIGURE 19.19 The time scale of action and number of users within each control region of the network management and control architecture for LAN, MAN, and WAN [24].

communication, algorithm processing, and transceiver tuning. Any MAN pairs that do not have enough traffic for at least one long-haul wavelength shunt are not a candidate for OFS.[9] In that event, multiple adjacent MANs can be merged as far as management and control to make up enough traffic for the provisioning of at least one wavelength in the core.

The ultra-fast service is a specialty service that costs more for instantaneous response and occupies only a small fraction of all flows. Factors that contribute to costs include the probing effort which is dependent on the average load of the network and the state reporting and dissemination efforts to provide node pairs current information on candidate lightpaths for probing.

The OFS network needs an integrated control plane that cuts across the entire LAN/MAN/WAN for end-to-end scheduling. The next section describes the access control and routing algorithm for the basic OFS service and the ultra-fast OFS service.

19.7 MEDIA ACCESS CONTROL (MAC) PROTOCOL AND IMPLICATIONS ON ROUTING PROTOCOL EFFICIENCY AND SCALABILITY

Basic OFS service [20,24]. Since the architecture uses quasi-static WAN highways between ingress and egress MANs, the access control problem is substantially reduced to scheduling within the pair of access networks in which the source and destination nodes reside.[10] Also the user nodes in the ingress access networks have access to wavelengths that are routed to the egress network via a

---

[9]This is highly unlikely when data rates are increased by 2–3 orders of magnitudes in future networks.
[10]Access network pairs that cannot fill at least one wavelength of traffic have too little traffic to use OFS and should use conventional IP service. Higher up in the aggregation chain in the MAN, the routers themselves may become an OFS user. In that case, the use of OFS is not per session but similar to GMPLS.

broadcast structure configured by the MAN network management systems in response to mid-term traffic demands between pairwise MANs. When a session allowing a setup time on the order of several seconds and beyond arrives at a source node, it is unwise to assign a route and wavelength immediately. This is because the network state may evolve considerably between this time and when the session actually begins, thus rendering the initial RWA suboptimal. Moreover, these pre-assigned resources will yield unnecessarily higher blocking probabilities for the Ultra-Fast Services and will result in low utilization. Hence, while it is necessary to acknowledge a session's request for resources immediately, it is prudent to defer the final RWA to as late a time as possible. Indeed, between the time of the arrival of the session request and its actual commencement, global network state information will be refreshed at every node many times via network-wide broadcasts. Thus, the scheduling system only has to ensure that there exists at least one surviving lightpath for each of the scheduled sessions, and to assign these lightpaths only at the last possible instance before use (<50 ms). To guarantee the low blocking probability of Ultra-Fast Services, the network will only admit or book longer-horizon sessions up to a certain level of occupancy (can be as high as 80%), as dictated by an admission control policy. Furthermore, booked basic services before initiation can be pre-empted by the Ultra-Fast Services, provided that the negotiated blocking probability of the booked services can be satisfied. The combination of this basic service with the ultra-fast service results in excellent utilization of network resources.

With the special structure of WAN highways and broadcast tree in the access networks, the access control and routing problem on a per flow basis are reduced to a manageable scheduling problem (without any network hardware switching) of the three networks: ingress, long-haul and egress networks, and the two user terminals. An optimum algorithm would search for a jointly optimized schedule, though further reduction in scheduling complexity can be implemented by finding an open schedule for the most precious resource, the WAN, first and then schedule the ingress and egress networks afterwards sequentially. This would sacrifice some performance as shown in Figure 19.20 but the less than 10% of throughput loss is well worth the simplification of the sequentially scheduling algorithm over the much more complicated jointly optimized algorithm.

Ultra-fast service [22–24]. For special sessions requiring fast service, routing and wavelength assignment must be done within ~50 ms. The Ultra-Fast Service, a hybrid scheme relying on up-to-date local and slightly stale global[11] information, will set up these sessions without asking the network management system to update network state information at that rate! Here the architecture applies the design precept that only entities at the network edge—MAC, transceiver wavelength tuning and transmission times—are fast controlled processes, while maintaining internal core network node switching quasi-static (changing only with traffic pattern shifts and not per flow).

---

[11]It has been estimated that even for update intervals of one second, the network control plane needs a management network overlay capable of 10 Gbps between nodes!

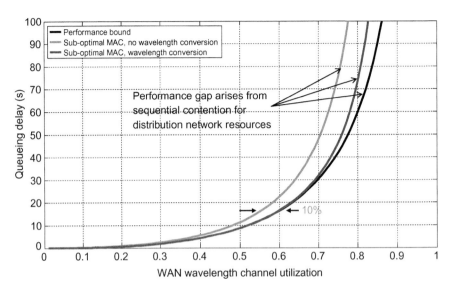

FIGURE 19.20 Queueing delay performance of access protocol of the basic OFS service [20].

Figure 19.21 depicts the concept of designing into the network physical connection topology multiple independent paths, $\{p\}$, between source and destination. The (slow) centralized network management system periodically broadcasts the path states at regular intervals. The user probes $K_p(t) = \min\{K_K, K_o(t)\}$ paths to achieve a desired blocking probability β. The number of paths probed will be an increasing function of t, the time since the last update. For manageability the number of paths to be probed should be ≤ 5. The information at the beginning of the update interval is accurate and yields the lowest blocking probabilities. Whereas, toward the end

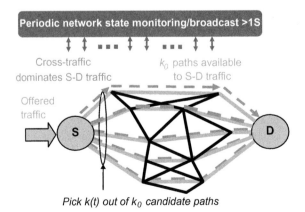

FIGURE 19.21 Multi-path probing for open path [21].

of the interval between updates, new traffic may have joined the network and result in higher blocking. Some simplifying assumptions are made to get a sense on the requirement on how frequently the network states have to be updated and broadcast. Figure 19.22 shows that the update interval should be ~0.5 of the expected transaction service time. Note that this algorithm can also be used in networks with merging and disappearing traffic in the core WAN.

The complexity of the Ultra-Fast scheduling [11] can still be daunting but can be substantially improved using the evolution of the state of each network domain grouped into one scalar parameter: the average entropy. The entropy of the network states in a network region or domain is defined as $-\sum p_i \log p_i$, where $p_i = 1/0$ if the wavelength link is used/free. In the control plane, the entropy evolution and the set of available paths are measured using life network traffic and broadcast periodically in two different time scales: at a period of its coherence time (that can range from several minutes to hours, depending on the actual traffic statistics), and the set of available paths at a period of ~half of the average transaction time (~1 s). With the updated information of entropy evolution, it is possible to get a close approximation of the average entropies at any time. In [23], it was shown that the number of paths to probe is monotonically increasing with increasing entropy. The algorithm chooses paths from different network domains by selecting the ones with the least total average entropy or uncertainty, Figure 19.23.

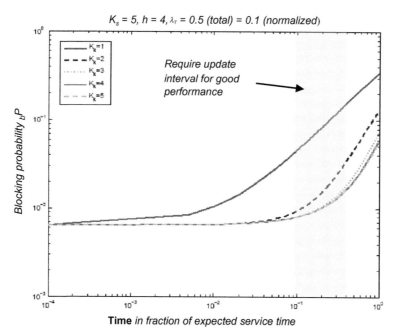

FIGURE 19.22 Blocking probability versus update period [22].

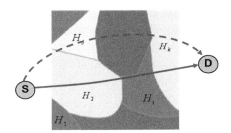

FIGURE 19.23 Entropy assisted routing across multiple domains.

Within one network domain, as the information about its internal states (e.g. the availabilities of each individual links) are aggregated into one single parameter, the entropy, the network control plane will lose some of the detailed statistics of the arrival and departure processes, resulting in probing more paths than is necessary. However, this algorithm is not dependent on detailed assumptions of the statistical model of the network and thus is much more robust. Moreover, since in the coarse updating time scale, the network control system broadcasts the entropy evolution of the group of paths in one network domain instead of the detailed statistics of each individual path, the management system has reduced the amount of control traffic by orders of magnitude. When the network state has just been updated, there is no uncertainty about the path states and thus the entropy is zero. However, as time evolves between updates, the information will become somewhat stale and the uncertainty and thus the entropy increases, Figure 19.24. The algorithm will probe a few more (2–3) paths in exchange for management and control plane simplicity.

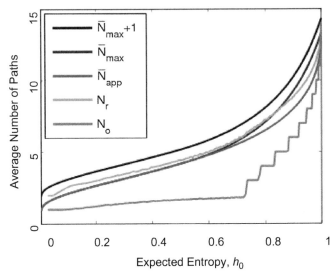

FIGURE 19.24 Number of paths to probe for entropy-assisted routing [23].

19.8 TRANSPORT LAYER PROTOCOL FOR NEW OPTICAL TRANSPORTS

The Transport Layer in the current Internet performs several important functions in conjunction with the Network Layer:

1. End-to-end data delivery reliability.
2. Flow control to prevent congestion in the network.
3. "Fair" resource allocation.

TCP carries out these functions for the current IP-based Internet rather successfully using the following major design features:

1. Segmentation of files into packets and reassembly.
2. Error recovery via sliding window ARQ.
3. Window flow control.
4. Use of time-outs to terminate dead sessions by estimating round-trip time, RTT and delay standard deviation, σ.
5. Work together with Layer 3 to perform congestion control in a distributed fashion with no direct knowledge of network congestion.

While this works fine in most cases for packet switching with routers, this protocol is not best suited for flows. If the TCP feature of window flow control is used, its slow start mechanism (even with acceleration) will prevent the full capacity of the optical path being used until many round trips (can be in the 100s) between source and destination have elapsed. Often by the time the big transaction is finished TCP has yet to allow the source node to use the maximum rate resulting in extremely inefficient use of network resources. In this OFS architecture flows are scheduled or reserved, Figure 19.25, the function of congestion control is performed by the MAC and the scheduling algorithm, and thus the function of fair resource allocation via windowing is not necessary and will be eliminated. What is left is the end-to-end reliability requirement which in the case of TCP is provided by ARQ. The OFS architecture

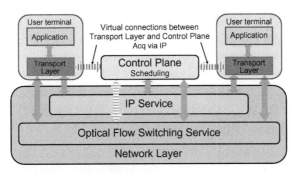

FIGURE 19.25 Transport layer protocol for OFS.

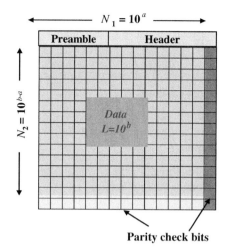

FIGURE 19.26 Low density parity check code for large OFS files.

needs to have a new reliable transport protocol, Figure 19.26. The idea is to arrange the file to be transmitted in a multiple N-dimensional, "N-cube." At each surface of the cube the protocol uses a parity check plane to check for errors. Each binary digit in the parity check plane is the parity of the corresponding row/column of the cube. The number of parity check planes must be large enough to determine if there are any errors incurred during transmission of the file. The dimension of the cube N is chosen based on the error performance of the underlying optical transport. For example, if the link is very error-prone, N will be large. If the network is error-free, N will be set to 1. The code can also be made to be robust against burst errors which in dynamic optical networks will be the toughest problem to tackle. This detection mechanism is of very low rate and the computation effort is low. Instead of the complication of requesting specific packets to be transmitted upon error in transmission, retransmission of the whole file or large segments is done if there are any errors. The size of the segmentation (found to be >1 Mbytes) is optimized based on delay and resources utilization. With the reliability of optical networks today, retransmissions will not occur very often and the slight sacrifice in theoretically maximum throughput is more than compensated by the simplicity of not having to segment a large file into many small IP packets and use per erroneous packet ARQ retransmission.

19.9 COST, POWER CONSUMPTION THROUGHPUT, AND DELAY PERFORMANCE

Throughput-cost comparison of the architectures was considered in [19,21,31]. Four cases were treated—EPS, EPS with GMPLS, OFS, and OBS/TaG—in the context of the simple hierarchical network: two large groups of users, located in different MANs, communicating over a quasi-static WAN. Common practice architectures are

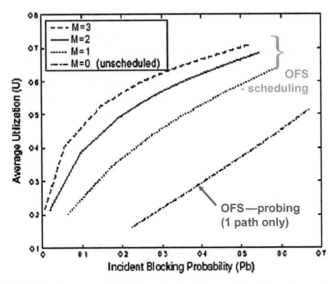

FIGURE 19.27 Utilization versus blocking probability for number of sessions scheduled into the horizon, showing good performance for only several scheduled transactions in OFS [24].

chosen for EPS and GMPLS, and sensible architectures that seem to perform well are chosen for TaG/OBS and OFS. The network cost model employed focused on initial capital expenditure: transceiver, switching, routing, and amplification costs. The results of the throughput-cost comparison are given in Figures 19.27 and 19.28.

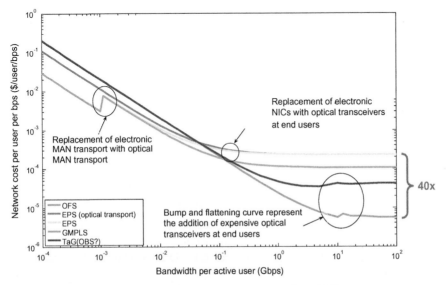

FIGURE 19.28 Cost comparison of different transport mechanisms [21].

Figure 19.28 shows the cost-optimal optical network architecture as a function of sustained data rate per user in the network. OFS is the most scalable architecture of all when the average user data rate is high; EPS is most sensible at low to moderate data rates; the GMPLS architecture, which is conceptually intermediate to EPS and OFS, is optimal at moderate user data rates with a moderate to large number of users; and, finally, there does not exist a regime of optimality for TaG/OBS since the low cost scheduling in OFS yields greater throughput performance benefit relative to the otherwise identical TaG architecture.

The above results suggest that a hybrid switching architecture may be optimal, where high-rate users would employ OFS for their large transactions, and low-rate users would employ the IP/EPS architecture for their small transactions. Based on the above analysis, using cost-optimal hybrid architectures [14] is expected to result in more than an order of magnitude cost reduction over present-day architectures. Moreover, only a small fraction ($\sim$1%) of "elephant" users need to be equipped with OFS hardware and a substantial fraction ($\sim$95%) of data traffic volume would be drained from the network by OFS [20].

Figure 19.29 shows the results of an analysis [25,26] of the US IP backbone to provide power minimizing design heuristics for WAN architecture. Optical networks using lightpaths for end-to-end transports bypassing electronic switches and routers are the most scalable architecture design with respect to power consumption, especially when quality of service, network flexibility, reliability, and protection are considered. The power consumption of the standard optical bypass design can be

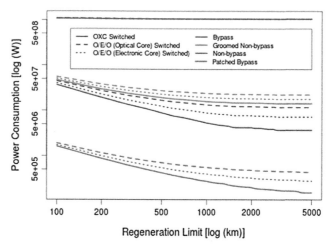

Baseline Power Consumption (Log-Log)

FIGURE 29 Comparison of the network power consumption for network designs implemented with available switches. The group of curves in red use all-optical by pass and WDM circuit switching, [24,25]. (For interpretation of the references to color in this figure legend, the reader is referred to the web version of this book.)

further improved through a hybrid patched bypass/bypass network. Under the hybrid scheme, whole wavelengths of core, stable traffic between the node pairs are routed via direct, fixed lightpaths using patch paneling to avoid lightpath switching at intermediate nodes and all other unexpected, fluctuating, and/or bursty traffic is switched on a standard optical bypass network. The hybrid network can provide services with low power, exceptional latency, transmission, and reliability performance. Just as in cost optimization, shortest path and minimum hop routing is power optimal and traffic balanced routing should be avoided when a majority of the traffic demand between any pair of nodes is sufficiently large to warrant whole wavelengths, as is the case when network capacity grows from the current 30-Terabit/sec to a projected 300-Terabit/sec capacity.

19.10 SUMMARY

OFS exploits the strengths of optics to serve large (sometimes called "elephant") transactions that can be a substantial fraction of the total traffic by providing end-to-end user all-optical access and will enable orders of magnitude in cost reductions and at the same time is energy efficient. This feature separates OFS from quasi-static circuit switching such as GMPLS where optical circuits are not scheduled per user sessions. The per-session scheduling substantially changes the dynamics of the control plane by speeding it up by orders of magnitude. The major hurdle of the architecture is the scalability in network management and control and session scheduling complexity. Thus, the shift toward OFS requires some architectural elements of the network, from the physical to the higher network layers, as well as network management and control, to be redesigned at the fundamental level. The architecture construct in this chapter shows how the physical architecture coupled with a matched media access control protocol can help slow down the control plane and still can operate the network at high efficiency which is critical for low cost and low power operations. The proposed architecture addresses this very important issue by using quasi-static WAN/MAN provisioning to slow down the wide/metro area network control planes by decoupling session-by-session scheduling of cross-flow traffic interior to the core network. The only fast per session process is the scheduling of transmission wavelength and time of transmission of the users, without necessitating any fast reconfigurations of optical switches. Thus, OFS only uses transmitter/receiver wavelength tuning and time of transmission for statistical multiplexing to achieve efficient use of resources, instead of using packet switching for multiplexing in the current IP architecture. With much higher data rates and more agile optical networks on the horizon, it is inevitable that the current network architecture including the control plane will have to be updated. Some simple changes to the network architecture can be implemented then to support OFS with significant cost benefits, especially to large transaction users.

While OFS is more efficient in the WAN, Figure 19.30, and in general for the future Internet architecture, for intra-data center and cloud interconnection, OBS

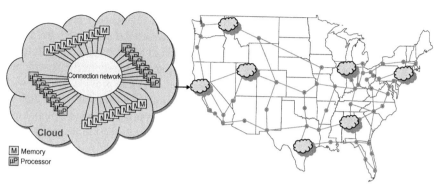

FIGURE 19.30 Cloud and data center applications of all-optical networks.

may be more favorable in terms of delays if the fiber plant has enough bandwidths and the transceivers have broad enough tuning range to exploit them. In those cases the utilization of the wavelengths must be kept way below 1/2e. Finally, the future Internet will be very heterogeneous in transaction sizes. While OFS and OBS are suitable for large elephant files, electronic packet switching will, for the foreseeable future, be the transport of choice for small messages. Figure 19.31 attempts to roughly scope the ranges of the preferred transport mechanism for different transaction sizes. The boundaries are sensitive to technologies and architectures they allow.

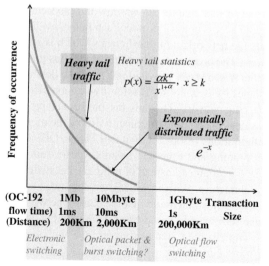

FIGURE 19.31 Ranges of the preferred transport mechanism for different transaction sizes.

References

[1] V.W.S. Chan, All-Optical Networks, LEOS Summer Topical Meeting, Santa Barbara, California, 3–5 August 1992, IEEE 1993 Optical Fiber Communications Conference, San Jose, California, 21–26 February 1993.

[2] S.B. Alexander, R.S. Bondurant, D. Byrne, V.W.S. Chan, S. Finn, R. Gallager, B. Glance, H.A. Haus, P. Humblet, R. Jain, I.P. Kaminow, M. Karol, R.S. Kennedy, A. Kirby, H.Q. Le, A.A.M. Saleh, B. Schofield, J. Shapiro, N.K. Shankaranarayanan, R.W. Thomas, R. Williamson, R.W. Wilson, A precompetitive consortium on wideband all-optical networks, IEEE J. Lightwave Technol. 11 (5) (1993) 714–735.

[3] V.W.S. Chan, A.J. Kirby, A. Saleh, AT&T/DEC/MIT Consortium on Wideband All-Optical Networks, San Jose, CA, November 1993 LEOS 1993, 15–18.

[4] V.W.S. Chan, A.J. Kirby, A. Saleh, AT&T/DEC/MIT Precompetitive Consortium on Wideband All-Optical Networks, New Orleans, LA, November 1993 GOMAC 1993 Conference, 1–4.

[5] I.P. Kaminow, C.R. Doerr, C. Dragone, T. Koch, U. Koren, A.A.M. Saleh, A.J. Kirby, C.M. Ozveren, B. Schofield, R.E. Thomas, R.A. Barry, D.M. Castagnozzi, V.W.S. Chan, B.R. Hemenway, Jr., D. Marquis, S.A. Parikh, M.L. Stevens, E.A. Swanson, S.G. Finn, R.G. Gallager, A wideband all-optical WDM network, IEEE J. Selected Areas Commun. 14 (5) (1996) 780–799.

[6] V.W.S. Chan, K.L. Hall, E. Modiano, K.A. Rauschenbach, Architectures and technologies for high-speed optical data networks, IEEE J. Lightwave Technol. 16 (12) (1998) 2146–2168.

[7] V.W.S. Chan, All-Optical Networks, Sci. Am. (JA-7272) (1995) 72–75.

[8] M. Kuznetzov, N.M. Froberg, S.R. Henion, H.G. Rao, J. Korn, K.A. Rauschenbach, E.H. Modiano, V.W.S. Chan, A Next-Generation Optical Regional Access Network, IEEE Commun. Mag. 38 (1) (2000) 66–72.

[9] S. Mookherjea, V.W.S. Chan, Remotely-pumped optical distribution networks, Conference on Lasers and Electro-Optics (CLEO'01), Technical Digest, May 2001, pp. 497–498.

[10] V.W.S. Chan, Optical access networks, Thirty-Ninth Annual Allerton Conference on Communication, Control, and Computing, 2001.

[11] V.W.S. Chan, S. Chan, S. Mookherjea, Optical distribution networks, Opt. Netw. Mag. 3 (1) (2002) 25–33.

[12] G. Liu, C. Ji, V. Chan, On the scalability of network management information for inter-domain light path assessment, IEEE/ACM Trans. Netw. 13 (1) (2005) 160–172.

[13] C. Guan, V.W.S. Chan, Topology design of OXC-switched WDM mesh networks, IEEE J. Selected Areas Commun. Opt. Commun. Netw. 23 (8) (2005) 1670–1686.

[14] V.W.S. Chan, Editorial: Optical network architecture from the point of view of the end user and cost, IEEE J. Selected Areas Commun., Opt. Commun. Netw. 24 (12) (December 2006) 1–2.

[15] V.W.S. Chan, Editorial: hybrids, IEEE J. Selected Areas Commun., Opt. Commun. Netw. 25 (4) (2007) 1.

[16] G. Weichenberg, V.W.S. Chan, M. Medard, On the capacity of optical networks: a framework for comparing different transport architectures, Opt. Commun. Netw. Suppl. IEEE J. Selected Areas Commun. 25 (6) (2007) 84–101.

[17] V.W.S. Chan, Editorial: Near-term future of the optical network in question? IEEE J. Selected Areas Commun. Opt. Commun. Netw. 25 (9) (2007) 1.

[18] G. Weichenberg, V.W.S. Chan, Access network design for optical flow switching, in: Proceedings of IEEE Global Telecommunications Conference (Globecom 2007), Washington, D.C., November 2007.

[19] Guy Weichenberg, Vincent W.S. Chan, Muriel Medard, Performance analysis of optical flow switching, IEEE/ICC Dresden Germany, June 2009.

[20] Kyle C. Guan, Vincent W.S. Chan, Cost-efficient fiber connection topology design for metropolitan area WDM networks, IEEE/OSA J. Opt. Commun. Netw. (June 2009).

[21] Guy Weichenberg, Vincent W.S. Chan, Design and analysis of optically flow switched networks, IEEE/OSA J. Opt. Commun. Netw. (2009)

[22] Anurupa R. Ganguly, Vincent W.S. Chan, Guy Weichenberg, Optical flow switching with time deadlines for high-performance applications, IEEE Globecom December, 2009 Honolulu.

[23] Zhang Lei, Vincent Chan, Fast scheduling for optical flow switching, IEEE Globecom December 2010, Miami.

[24] Vincent W.S. Chan, Optical flow switching, Invited Plenary Talk, IEEE Globecom December 2010, Miami.

[25] Katherine Lin, Vincent Chan, Power optimization of wide area and metropolitan area networks, ICTON July 1, 2012, Coventry, England.

[26] Katherine Lin, Green optical network design: power optimization of wide area and metropolitan area networks, MIT MEng Thesis EECS, June 2011.

[27] C. Qiao, M. Yoo, Optical burst switching—a new paradigm for an optical Internet, J. High Speed Netw. 8 (1) (1999) 69–84 (special issue on Optical Networks)

[28] J.S. Turner, Terabit burst switching, J. High Speed Netw. 8 (1) (January 1999) 3–16.

[29] J.P. Juc, V.M. Vokkarane, Optical Burst Switched Networks, Optical Networks Series, Springer, 2005 ISBN: 0-387-23756-9

[30] R.A. Barry, V.W.S. Chan, K.L. Hall, E.S. Kintzer, J.D. Moores, K.A. Rauschenbach, E.A. Swanson, L.E. Adams, C.R. Doerr, S.G. Finn, H.A. Haus, E.P. Ippen, W.S. Wong, M. Haner, All-optical network consortium—ultrafast TDM networks, IEEE J. Sel Areas Commun. 14 (5) (June 1996) 999–1013.

[31] Vincent W.S. Chan, Optical Flow Switching Networks, Invited Paper, Special Issue, in: Proceeding of IEEE, March 2012.

Recent Advances in High-Frequency (>10 GHz) Microwave Photonic Links

20

Charles H. Cox, III and Edward I. Ackerman

Photonic Systems, Inc., MA, USA

20.1 INTRODUCTION

The development of microwave photonic links, like their more common relative—digital photonic links—has been an active area for more than three decades. Research and development on this type of link has been driven largely by the dual objectives of meeting application needs and developing a deeper understanding of the limits of their performance.

The primary commercial application for early microwave photonic links was the distribution of analog TV signals for CATV systems. Meeting this need focused development on bandwidths up to about 1 GHz. Also, at least up until recently, CATV distribution involved only one-way signal flow—from a central location to remote users. Consequently microwave links were almost all (exclusively?) unidirectional links.

There was some early work on higher frequency links, driven mainly by the application of microwave links to remoting the signals from antennas. For these applications the microwave links needed to convey signals at the common radar bands, such as S-, C-, X-, Ka, Ku, etc. Again the focus was on links for receive-only, since remoting the transmit signal from the single, high-power amplifier—such as a traveling wave tube (TWT)—to an antenna required conveying high power, which photonic links were incapable of doing.

For microwave photonic links to be used in a particular application, they had to offer advantages over competing methods of conveying microwave signals at a competitive price. In practice this meant that photonic links had to offer a performance advantage over coaxial cable at a price that was commensurate with their performance advantage.

Microwave photonic links still need to earn their place vs. an alternative. But recently the technical requirements for such links have changed radically. Driven largely by the explosion in the demand for wireless networks, microwave links have needed to keep pace with conveying signals at ever higher frequencies. Of particular interest are frequencies above 10 GHz, such as, for example, 60 GHz [1], where the attenuation of coax would limit the distance to a few meters and where low-attenuation waveguide is too bulky and cumbersome. Further, networks routinely involve bi-directional signal flow. The need for bi-directional links is also being driven by the development of solid-state power amplifiers, which in turn permit the power amplifier to be moved deeper into the antenna array, with the ultimate goal of locating a power amplifier at each antenna element of the array.

Section 20.2 of this chapter describes several recent developments that have greatly improved the high-frequency (up to and beyond 10 GHz) microwave performance of *two-port* photonic links for *receive-only* antenna remoting applications. These evolutionary advances paved the way for the revolutionary new development of *three-port* photonic links, described in Section 20.3, enabling signal remoting from antennas that simultaneously *transmit and receive* over multiple decades of bandwidth.

Whereas this chapter reports microwave photonic link performance using conventional microwave figures of merit like gain and noise figure, the next chapter (21), which focuses on their application to remoting signals for networks (commonly referred to as "radio over fiber") concentrates on correspondingly appropriate performance parameters such as bit error rate (BER).

20.2 PHOTONIC LINKS FOR RECEIVE-ONLY APPLICATIONS

In a receive-only application, possibly the most important figure of merit for any two-port component in the signal chain is its noise figure (NF). This has been a particularly significant parameter in the historical development of microwave photonic links, because early link demonstrations earned them a reputation for having unattractively high noise figure (e.g. 30 dB or greater, even at frequencies below 1 GHz)—a reputation that they have not fully shaken off despite much more impressive results reported in recent years (e.g. 3.5 dB at 2 GHz [2] and <6 dB at frequencies up to 12 GHz [3]). These recent results are described in more detail further on in this section of the chapter.

In addition to demonstrations of links with improved noise figure, recent years have also seen improvements to how the noise figure of an analog link is modeled. These recent modeling improvements are reviewed first, because they help to explain what made the recently improved results possible.

For any analog component with two or more RF ports, noise figure is defined as the extent to which the signal-to-noise ratio degrades between the input and output ports—i.e.

$$\text{NF (in dB)} = 10 \log \left[\frac{(s_{\text{in}}/n_{\text{in,th}})}{(s_{\text{out}}/n_{\text{out}})} \right]_{n_{\text{in,th}}=kT_0 B} = 10 \log \left[\frac{n_{\text{out}}}{kT_0 B g_i} \right], \qquad (20.1)$$

where g_i is the component's small-signal gain, and where the input noise, $n_{\text{in,th}}$, is specified as only the noise arising from thermal sources. As shown in (20.1), this input noise is the product of k (Boltzmann's constant), T_0 (room temperature, standardized by the IEEE as 290 K [4]), and B (the instantaneous bandwidth of the electronic receiver, sometimes called the "resolution bandwidth" or "noise bandwidth"). Because the output noise term n_{out} is also proportional to B, a component's noise figure is independent of B.

For an external modulation link, the g_i term in (20.1) is easily calculated from the square of the product of small-signal slope efficiencies of the external modulator and the detector. The modulator's slope efficiency, s_m, is the change in its optical output power for a given change in input voltage. Specifically,

$$s_m \equiv \left| \frac{dp_{M,O}(v_M)}{dv_M} \right|_{v_m=0},$$

(20.2)

where $p_{M,O}$ is the modulator's output optical power, which is a function of v_M, the total voltage on the modulator, which is in turn the sum of a DC bias voltage V_M and the modulation signal voltage v_m. If the modulator's transfer function (i.e. the mathematical relationship between its output optical power and its input voltage) is known, (20.2) can be used to derive an explicit expression for s_m. A Mach-Zehnder interferometric (MZI) modulator, for example, has the transfer function

$$p_{M,O} = \frac{T_{FF}\, P_I}{2} \left[1 + \cos\left(\frac{\pi v_M}{V_\pi(f)} \right) \right],$$

(20.3)

where T_{FF} is the modulator's fiber-to-fiber optical insertion loss, P_I is the CW optical power supplied to its input fiber, and

$$V_\pi(f) = \frac{\alpha L}{1 - e^{-\alpha L}} V_\pi(DC),$$

(20.4)

is the voltage change necessary to shift the optical phase in one of the arms of the MZI by π radians relative to the other. In (20.4), L is the length of the modulator electrodes and α is their frequency-dependent loss coefficient [5]. Given that $v_M = V_M + v_m$, substituting (20.3) into (20.2) yields

$$s_m = \frac{\pi\, T_{FF}\, P_I\, R_M}{2 V_\pi(f)} \sin\left(\frac{\pi V_M}{V_\pi(f)} \right),$$

(20.5)

where R_M is the modulator's resistance. The detector's slope efficiency is simply

$$s_d = r_d\, |H_d(f)|,$$

(20.6)

where r_d is its responsivity and $H_d(f)$ represents the frequency response of the detector circuit (including any resistive or reactive matching circuitry). Therefore,

$$g_i = \left[\frac{\pi\, r_d\, T_{ff}\, P_I\, R_S}{2 V_\pi(f)} |H_d(f)| \sin\phi \right]^2,$$

(20.7)

where we have defined the ratio $\pi\, V_M/V\pi$ as the MZI modulator's "bias angle" ϕ. Equation (20.7) shows that an MZI modulator-based external modulation link gain is maximum at the so-called "quadrature" bias angle (90°), and that one can increase slope efficiency by minimizing $V\pi$ and increasing P_I, such that it is possible to achieve gi > 1 [6]. Figure 20.1 (after [7]) shows the measured noise figures for nine different links without RF amplifiers, as reported in eight different publications [2,3,5,8–12].

Figure 20.2 (after [6]) shows an equivalent circuit for an external modulation link, including noise sources representing the photodetected laser relative intensity noise

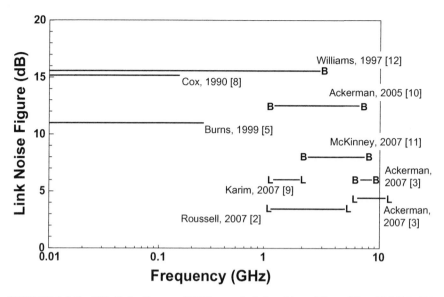

FIGURE 20.1 (after [7]) Noise figures of RF fiber optic links without RF amplifiers [2,3,5,8-12]. Characters at line endpoints indicate links using the Low-biasing and Balanced differential modulation and detection techniques, both of which are discussed further on in this chapter.

(RIN), the shot noise, and several sources of thermal noise that contribute to the total output thermal noise $\overline{n}_{out,th}$. The noise figure of this link is

$$\mathrm{NF} = \frac{\overline{n}_{out,th} + \overline{n}_{out,rin} + \overline{n}_{out,shot}}{kT_0\,g_i}\bigg|_{\overline{n}_{in}=kT_0}, \tag{20.8}$$

where $\overline{n}_{out,RIN}$ and $\overline{n}_{out,shot}$, the output noise spectral densities due to the laser's relative intensity noise (RIN) and the shot noise of the detection process, respectively, are defined in [13] and elsewhere. Recall from (20.1) that noise figure is defined for input noise equal to the thermal noise generated at $T_0 = 290\,\mathrm{K}$. This input thermal noise is amplified (or attenuated) by the link's gain (or loss) g_i, and if g_i is sufficiently large that the amplified input thermal noise term makes the dominant contribution to the total noise at the link output, then there is virtually no degradation in signal-to-noise ratio from the link's input to its output, and NF will be minimized. This thermal-noise-limited situation yields the lower limit to NF, which we can express as NF_{min}:

$$\mathrm{NF} = \frac{\overline{n}_{out,th}}{kT_0 g_i}\bigg|_{\overline{n}_{in}=kT_0} = 10\log\left[1 + x + \frac{1}{g_i}\right], \tag{20.9}$$

where the three terms in the latter expression are the individual contributions of the input thermal noise source, of the modulation device and its optional matching circuit, and of the detector and its optional matching circuit. If we assume that the

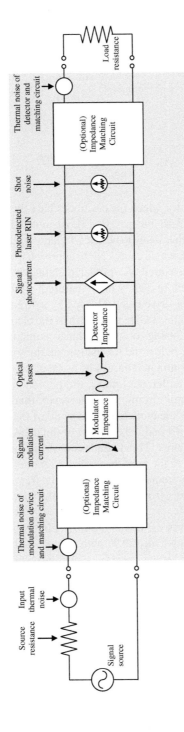

FIGURE 20.2 Equivalent circuit of an amplifier-less RF fiber optic link using external intensity modulation and direct detection. (*Analog Optical Links: Theory and Practice*, by Charles H. Cox, III, Copyright © 2004 Cambridge University Press. Reprinted with permission.)

modulator electrodes are *lossless* and that the RF and optical velocities have been successfully matched to one another in the modulator, then

$$x = x_{\text{lossless}} = \frac{\sin^2(\beta_{\text{RF}}L)}{(\beta_{\text{RF}}L)^2}, \tag{20.10}$$

where

$$\beta_{\text{RF}} = \frac{2\pi}{\lambda_{\text{RF}}} = \frac{2\pi f_{\text{RF}} n_{\text{opt}}}{c}. \tag{20.11}$$

At low RF frequencies, for which the electrodes have an electrical length $<<360°$, $x_{\text{lossless}} \to 1$, so that $NF \to 3\,\text{dB}$ for $g_i >> 0$ [13]. At most RF frequencies of interest, however, the electrode length is likely to be at least a large fraction of an RF wavelength, and in this case $x_{\text{lossless}} \to 0$, which causes NF to approach a value less than 3 dB.

Figure 20.3 shows measured vs. modeled data for the 1–12 GHz external modulation link described in [2], which featured an extremely low-V_π MZI modulator with 14-cm dual-drive electrodes fed by a directional coupler. Even for RF frequencies as low as $f_{\text{RF}} = 1\,\text{GHz}$ the length of the electrodes in this modulator was greater than 360°, causing x_{lossless} to have a magnitude so small that its contribution would appear below the lowest line on the y-axis of Figure 20.3a. As shown in the figure, the dominant sources of noise at the link output were due to loss in the input hybrid coupler, shot noise, the photodetected laser *RIN*, and the thermal noise of the detector circuit. Note, however, that the uppermost curve in Figure 20.3a, which shows the modeled effect of all of the above-mentioned noise sources, falls between 1.1 dB and 2.3 dB short of the measured noise figure across the 1–12 GHz frequency range. The degradation of the lossless model's accuracy as frequency increases led to the hypothesis that it ignores a source of noise whose magnitude increases with frequency, and that this frequency-dependent source of noise is that which arises from ohmic loss in the modulator's traveling-wave electrodes [14].

Including electrode loss in the model has *three* effects on the link's noise figure, so that rather than replacing x in (20.9) with x_{lossless} as in (20.10), we replace it instead with the sum $x_{1,\text{lossy}} + x_{2,\text{lossy}} + x_{3,\text{lossy}}$. The first term, $x_{1,\text{lossy}}$, expresses the effect of the thermal noise generated by the electrode termination resistance, which we derive from [15] for lossy electrodes:

$$x_{1,\text{lossy}} = \frac{\left(1 - e^{-\alpha L}\right)^2 + 4e^{-\alpha L}\sin^2(\beta_{\text{RF}}L)}{(\alpha L)^2 + 4(\beta_{\text{RF}}L)^2} \frac{(\alpha L)^2}{\left(1 - e^{-\alpha L}\right)^2}, \tag{20.12}$$

where α is the electrodes' RF frequency-dependent electrode loss coefficient. [Note that, for $\alpha = 0$, (20.12) reduces to (20.10), as it should.] The second and third terms, $x_{2,\text{lossy}}$ and $x_{3,\text{lossy}}$, express the effect of co- and counter-propagating thermal noise generated by the ohmic loss in the traveling-wave electrodes themselves, respectively.

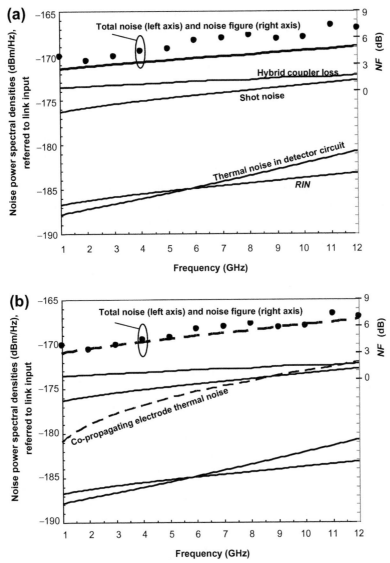

FIGURE 20.3 (after [14]) (a) Modeled sources of noise in the link of Figure 20.2, referred back to its input (left axis) assuming *lossless* modulator electrodes, and the measured link noise figure, NF (right axis). Noise due to the electrode termination resistance, giving rise to the $x_{lossless}$ term, falls below −190 dBm/Hz. (b) Same model, assuming lossy modulator electrodes. The modified effect of termination resistance (accounted for in the $x_{1,lossy}$ term) is again negligible, as is counter-propagating thermal noise arising from the electrode loss ($x_{3,lossy}$). The effect of co-propagating thermal noise from the electrode loss ($x_{2,lossy}$) is shown by the additional dashed line. With electrode loss accounted for, the modeled link NF more nearly matches the measured data.

For electrodes designed for minimum RF loss, it has been shown that the extent of this effect can be approximated as follows [14]:

$$x_{2,\text{lossy}} \approx \frac{2}{3}\alpha L \tag{20.13}$$

and

$$x_{3,\text{lossy}} \approx 0. \tag{20.14}$$

The effect of modifying the noise figure model to account for loss in the modulator electrodes is shown in Figure 20.3b. The unlabeled solid curves in this plot are unchanged relative to Figure 20.3a, but the three additional effects of electrode loss factor into the total shown by the bold dashed curve. Two of the three effects are so small as to never reach the lowest y-axis value of $-190\,\text{dBm/Hz}$, but the third effect—co-propagating electrode thermal noise—does increase the expected noise figure to the point where the modeled data of Figure 20.3b much more nearly match the measured data.

20.2.1 Effect of modulator bias point

One effective way to improve a receive link's performance is to reduce its noise figure using a technique that was discovered independently by three groups in 1993 [16-18] and that has generally become known as "low-biasing" the external modulator. The benefits of this technique are easiest to quantify in the case of an MZI modulator in a linear electro-optic material like lithium niobate, because its transfer function and slope efficiency can be expressed as simple functions of its dc bias voltage V_M using (20.3) and (20.5), respectively. These simple expressions yield straightforward dependence of the external modulation link gain, g_i, on the bias point ϕ as given in (20.7), which shows, for example, that gain is maximum at the "quadrature" bias point $\phi = 90°$. Equation (20.8) would, at first glance, seem to imply that noise figure is always minimum at this same bias point. In actuality, because the three terms in the numerator of (20.8) each depend on ϕ to different extents, external modulation link noise figure is a relatively complicated function of modulator bias. Specifically,

$$\overline{n_{\text{out,th}}} = kT_0 g_i (1 + x) + kT_0, \tag{20.15}$$

$$\overline{n_{\text{out,rin}}} = \langle I_D \rangle^2 \text{RIN} |H_d (f)|^2 R_S, \text{ and} \tag{20.16}$$

$$\overline{n_{\text{out,shot}}} = 2q \langle I_D \rangle |H_d (f)|^2 R_S, \tag{20.17}$$

where

$$\langle I_D \rangle = \frac{r_d T_{\text{FF}} P_I}{2} (1 + \cos \phi). \tag{20.18}$$

Figure 20.4 (after [13]) shows, for a specific set of assumed values for the components in Eqs. (20.4)–(20.18), the effect of the low-biasing technique on the NF of an MZI modulator-based external modulation link. From the curves showing the intrinsic link gain (g_i) and the total output noise power density $\overline{n}_{out,total} = \overline{n}_{out,thermal} + \overline{n}_{out,RIN} + \overline{n}_{out,shot}$, it is evident that, over a large range of modulator bias points ϕ, the noise decreases more quickly than the signal as the modulator bias is increased from 90° toward the light-extinguishing bias of 180°. At some optimum low-biasing point between 90° and 180° that depends on component parameters such as the laser RIN, NF is minimized. If the bias point is moved further toward 180°, the signal gain begins to decrease more quickly than the noise, and therefore NF begins to increase relative to its value at the optimum low-biasing point.

The lowest noise figure ever demonstrated for an RF fiber optic link at X-band frequencies (6–12 GHz) was described in [3] and is repeated in Figure 20.5. The link consisted of a high-power (2.5 W), low-RIN (~–175 dB/Hz) master oscillator power amplifier (MOPA), a dual-drive MZI modulator with the lowest RF V_π available—less than 1.4 V at 12 GHz when measured from the delta port of a broadband directional coupler connected to its two RF inputs, and a p-i-n photodetector with r_d of nearly 1.0 A/W at $\lambda = 1.55\,\mu m$. The modulator was biased at about 165°, resulting in an average photocurrent $<I_D>$ of only 6.2 mA. For the 6–12 GHz band, the square data points in Figure 20.5 show the measured link gain of >12.7 dB and

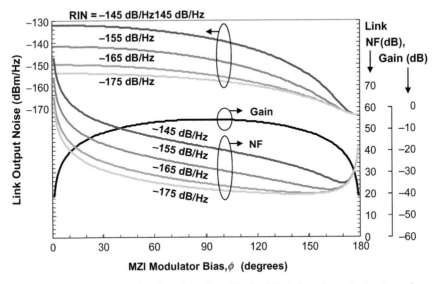

FIGURE 20.4 (after [13]) Illustration of the "low-biasing" technique for reducing the noise figure of an external modulation link that uses a MZI modulator (assumptions: $V_\pi = 3$ V, $<I_D> = 10$ mA at $\phi = 90°$). The link's output noise (left axis) decreases more quickly than its gain (right axis), causing the optimum NF (right axis) to occur at a bias point between 90° and 180°.

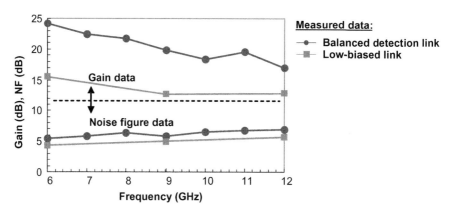

FIGURE 20.5 (after [3]) Measured gain and noise figure of links using a single photodetector to detect one output of a low-biased MZI modulator (curves with square data points), and balanced photodetectors to differentially detect two anti-phase outputs of a quadrature-biased MZI modulator (curves with circular data points).

record low NF—only 4.4 dB at 6 GHz and <5.7 dB at 12 GHz—of this "low-biased" link.

20.2.2 Effect of balanced photodetection

Another way to reduce a link's NF is to use an external modulation link architecture that employs a balanced differential photodetector configuration to cancel the CW laser's RIN. Figure 20.6 shows such a link configuration and why it has improved NF. This link uses a quadrature-biased MZI modulator in which the y-branch at the output end of the modulator is replaced by an optical directional coupler that produces the necessary interferometry and balanced outputs. The two optical outputs from this coupler are conveyed via equal-length optical fibers to two detectors whose photocurrents are made to subtract from one another at the link output port—e.g. by connecting the output port to the cathode of one photodiode and the anode of the other. Because the two modulated outputs from the MZI modulator are complementary to one another—i.e. 180° out of phase with one another—subtracting them in a differential detector configuration doubles the output signal (thereby increasing g_i) while cancelling the common-mode component, which is the laser's RIN [19].

As was reported in [19], cancellation of the laser RIN using the balanced photodetection link architecture shown in Figure 20.6 also yielded impressively low NF. The circular data points in Figure 20.5 show the measured gain and NF for a link using a balanced photodetector to demodulate both anti-phase outputs of the same MZI modulator used in the "low-biased" link that had the measured performance shown by the red data points.

Note from Figure 20.5 that the balanced photodetection link exhibited much higher gain than the low-biased link because its modulator must be biased at quadrature

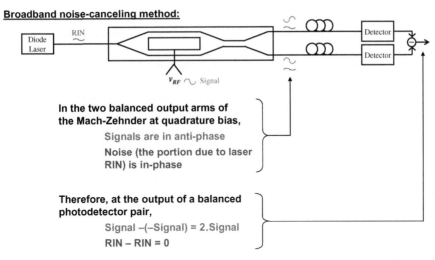

Broadband noise-canceling method:

In the two balanced output arms of the Mach-Zehnder at quadrature bias,

Signals are in anti-phase

Noise (the portion due to laser RIN) is in-phase

Therefore, at the output of a balanced photodetector pair,

Signal −(−Signal) = 2.Signal

RIN − RIN = 0

FIGURE 20.6 External modulation link configuration using a dual-output MZI modulator and balanced photodetector to cancel the laser's RIN.

($\phi = 90°$, where g_i is maximum) for the RIN cancellation to succeed. Its measured NF, however, ranging from 5.5 dB at 6 GHz to 6.9 dB at 12 GHz, is not quite as low as that enabled by the low-biased configuration because low-biasing reduces both the RIN and shot noise components of the output noise, whereas balanced detection suppresses only the RIN component, as shown in (20.16) and (20.17).

20.3 PHOTONIC LINKS FOR TRANSMIT AND RECEIVE APPLICATIONS

With the evolution of solid-state power amplifiers, it is now possible to place a power amplifier at the element level in an array. More recently array designers have begun to consider the advantages of using separate links for conveying receive signals from—and transmit signals to—the T/R module that is located with each antenna element. The conventional way to do this involves using two completely separate links, one for the receive signal and one for the transmit signal.

Suppose instead we combine these two links such that the resulting new link has three ports vs. the two ports of a conventional link. This new link would have one port where the transmit signal goes in, one port where the receive signal comes out and one port that connects to the antenna. The antenna port[1] would need to be capable of

---

[1]It is of course possible to use separate antennas for transmit and receive. While this does simplify some aspects of the design, it can prove difficult at best—and impossible at worst—to achieve the required level of isolation between the transmit and receive signals. Hence for the purposes of this discussion we will assume the goal is to interface both transmit and receive signals with a single antenna.

bi-directional operation: conveying the receive signal detected by the antenna to the receive port while at the same time conveying the transmit signal to the antenna for transmission.

There are two key figures of merit for this new type of link. One is that it has sufficient sensitivity to permit achieving the desired noise figure on receive without a conventional low noise amplifier (LNA), because with the bi-directional signal flow it is not possible to place an LNA at the antenna as would conventionally be done. In the previous section we showed that photonic link technology has evolved to the point that it is indeed possible to achieve the required sensitivity.

The other key figure of merit is the isolation of the transmit signal from the receive path, and to achieve high isolation over a wide bandwidth. We will discuss two techniques for achieving high isolation over a broad bandwidth in this section.

A circulator is a non-reciprocal RF component that can be used to interface separate transmit and receive paths to a common antenna. A circulator achieves its operation by launching two, counter-propagating waves from one port around the circumference of a circle and arranging these two waves to add constructively at a second port and destructively at a third port. The different phase shifts that are applied to the two waves occur due to the non-reciprocal propagation in a ferrite material that is subjected to a dc magnetic bias field.

Any device that relies on the phase of a signal to achieve its performance is an inherently narrow bandwidth device. Consequently the typical bandwidth of ferrite-based circulators is at most an octave, i.e. 2:1 in bandwidth. However, for some applications such as the 60 GHz wireless band, a ferrite circulator has sufficient bandwidth.

A critical parameter when trying to detect a weak receive signal—simultaneously—in the presence of a much stronger transmit signal is the transmit-to-receive (TR) isolation. The required TR isolation, which depends on a number of system parameters, can range from as low as 40 dB to as high as 120 dB. Compare this range of required TR isolation with the TR isolation of a ferrite circulator, which is typically 15–20 dB. It is possible to improve a ferrite circulator's isolation by adding an auxiliary circuit to intentionally leak a small portion of the transmit signal, with the proper phase and magnitude, into the receive path so as to cancel the signal leaking through circulator. TR isolations as high as 60–75 dB have been achieved via this approach, but the improved isolation has only been achieved over an extremely narrow bandwidth measured in MHz.

Photonics, which is an intrinsically broad bandwidth technology, offers the ability to realize the circulator function over an extremely broad bandwidth—four decades has been demonstrated to date—*and* high isolation—greater than 40 dB has been demonstrated to date. Below we will present two versions of a photonics-based link that accomplishes the circulator function. However, we choose not to call this a photonic circulator. The reasons are twofold. Perhaps the key reason is that the photonics approach can have RF power gain—for reasons that were discussed in Section 20.2—between the antenna and receiver ports, which means that it can replace the low noise amplifier (LNA) as well as the circulator. A second reason is that the ports of the photonics approach cannot be permuted as the ports of a ferrite circulator can be. Consequently what we really have is a new component that combines the functions of both an LNA and a circulator. Therefore we have given this configuration a new name: TIPRx, which is short for Transmit-Isolating Photonic Receiver.

Since both versions of the TIPRx are built on the operation of a Mach-Zehnder modulator, which is a common type of electro-optic modulator, we briefly describe its operation; for a more detailed description, see [6] for example.

Figure 20.7(a) is a sketch of the layout of a Mach-Zehnder electro-optic modulator. Single mode optical waveguides, shown in gray, are fabricated in an electro-optic material, most commonly lithium niobate. An electro-optic material has the property that an applied electric field, E, causes a change in the refractive index of the material, Δn. For a modulator, the electric field is created across the optical waveguides by metal electrodes that are deposited on the surface of the lithium niobate; the electrodes are shown by the textured rectangles in the figure. The electric field is proportional to the voltage, V, applied to the electrodes and the separation distance between the electrodes, d; i.e. $E \propto \frac{V}{d}$.

Figure 20.7(b) is a plot of the response of the modulator as a function of the voltage that is applied to its electrodes. Assume for the moment that the optical waveguides are of equal length. Then with zero voltage applied to the electrodes the light will split at the input end of the modulator, travel along the equal-length optical waveguides, and recombine in phase at the output end of the modulator. The result will be the maximum output from the modulator. When a voltage is applied, the resulting electric field will induce changes in the refractive index in the optical waveguides. For the electrode configuration shown there will be opposite changes in each waveguide: $+\Delta n$ in one guide and $-\Delta n$ in the other. Hence when the light from the two waveguides recombine, they will no longer be in phase. The result will be at least a partial cancellation. If the applied voltage is sufficient to induce a $180°$—or π radians—phase shift between the two optical paths, then there will be, at least ideally, complete cancellation at the modulator output. The voltage at which this occurs is called V_π. For digital modulation, a voltage of V_π is applied resulting in 100%—i.e. full on/off—modulation of the optical carrier. As was discussed in conjunction with equation (20.7) for analog modulation, the modulator is typically biased at $V_\pi/2$ and the modulation applied about this bias point; the optical modulation depth is typically only a few %, which is considerably less than is common with digital modulation.

20.3.1 Broad bandwidth TIPRx

Figure 20.8 shows the basic configuration of the broad bandwidth TIPRx [20,21], which consists of the same basic components as a conventional link: laser, modulator, and photodiode. The increased functionality of aTIPRx, as compared to a conventional receive-only link, comes from the connections that are made to the modulator.

To make the necessary connections, the modulator electrodes are arranged in a balanced or differential drive configuration: the center electrode is the common,

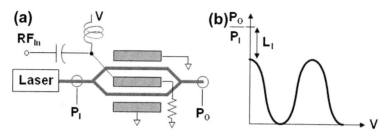

FIGURE 20.7 (a) Sketch of Mach-Zehnder modulator (MZM) showing single-mode optical waveguides (gray) and RF electrodes (textured rectangles); (b) plot of MZM optical transfer function vs. electrode voltage.

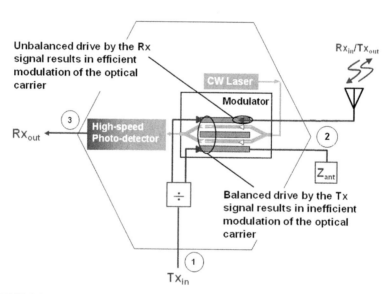

FIGURE 20.8 (after [20]) Block diagram of broad bandwidth TIPRx showing its key components.

or ground electrode, and each of the two outside electrodes can be driven independently. In an ideal modulator, if the same voltage is applied to both outside electrodes, then the same refractive index change is imposed in each optical waveguide. Consequently when the light in these two waveguides recombine, the modulation on each optical carrier cancels out resulting in no net intensity modulation of the optical wave. Hence there is no intensity modulation for the photodiode to detect. If a voltage is applied to only one of the electrodes—and not to the other—then there is a net intensity modulation of the light out of the modulator, which the photodiode can detect.

First consider the TIPRx receive function.[2] The receive signal is connected to one of the modulator's outside electrodes. Since the receive signal modulates the light in only one arm of the MZ modulator, the receive signal produces a corresponding intensity modulation at the modulator output, which the photodiode detects to reproduce the receive signal. Thus the TIPRx receive function is basically that of a conventional receive-only link.

Next consider the TIPRx transmit function. As shown in Figure 20.8 the transmit signal is split and applied equally to the modulator's two outside electrodes. The Tx power on the electrode that is connected to the antenna is radiated; the Tx power on the other electrode is dissipated by a load whose impedance vs. frequency, Z_{ant}, replicates the impedance of the antenna. As was discussed above, applying the same signal to both outside electrodes results in no net intensity modulation of the optical carrier. Consequently none of the transmit signal is conveyed to the photodiode, which is the desired result.

One of the key figures of merit for a TIPRx is the amount of isolation that can be achieved between the transmit and receive (T and R) signals. Figure 20.9 is a plot of TR isolation vs. frequency with the antenna and balance ports connected to a 50 Ω load. The plot shows more than 40 dB of isolation over nearly four decades of bandwidth—which is limited by the measuring equipment (Agilent PNA-L, N5230A 4-port network analyzer), not the TIPRx.

20.3.2 High frequency TIPRx

Figure 20.10 shows the basic configuration of the high frequency TIPRx [20,22]. Note that for this version of the TIPRx the antenna and transmit amplifier connections are to the center electrode.

Key to understanding the high frequency TIPRx is the following: for a modulator to achieve efficient modulation at high frequency (e.g. $> \sim 2\,\text{GHz}$) the velocity with which the RF signal travels along the electrode should be matched to the velocity with which the light travels along the optical waveguide. This is the so-called "velocity match condition," which maximizes the time that the RF signal has to interact with the light, thereby achieving the most efficient modulation possible for a given modulator design.

With this as background, first consider the receive signal operation of the high frequency TIPRx. The receive signal from the antenna is connected to the modulator,

---

[2]Since the TIPRx is a network with two independent sources—i.e. the transmit and receive sources—the formal way to present this discussion, and the following discussion on the high frequency TIPRx, would be to invoke the principle of superposition, which facilitates the analysis of networks with more than one independent source. We have found that the details of the formal approach can obscure the intuitive understanding of the TIPRx we are trying to convey. We invite readers familiar with superposition to confirm, as we have, that the TIPRx operation is indeed on a firm theoretical foundation. For the rest of the readers, rest assured that the intuitive descriptions of the present discussions are rigorously correct.

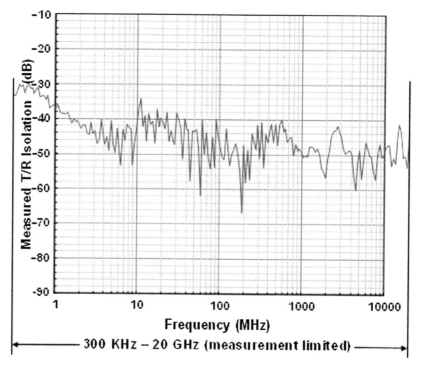

FIGURE 20.9 Plot of T/R isolation vs. frequency for the broad bandwidth TIPRx.

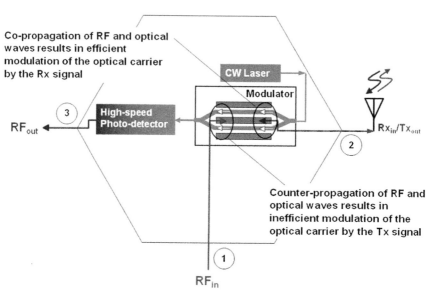

FIGURE 20.10 (after [20]) **Block diagram of high frequency TIPRx showing its key components.**

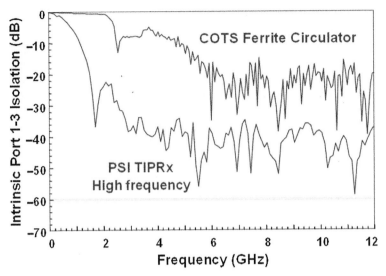

FIGURE 20.11 (after [20] Plot of T/R isolation vs. frequency for the high frequency TIPRx (black curve) and COTS ferrite circualtor (gray curve).

exactly as it would be in a receive-only link. The modulated light is conveyed to the photodetector, which recovers the receive signal from the optical modulation. Since the receive signal co-propagates with—and is maximally velocity matched to—the light, the modulation efficiency of the receive signal is high, which is as desired.

In a receive-only application, the other end of the modulator electrode would be terminated in a 50 Ω load. In the case of the TIPRx, the termination of the electrode is via the output impedance of the transmit power amplifier.

This configuration provides a point to connect the transmit signal to the modulator electrode and hence on to the antenna from which it is radiated. However, note that by injecting the transmit signal from what would normally be considered the output end of the modulator, the transmit signal is traveling along the modulator electrode in the opposite direction with respect to the light. Hence the transmit signal counter-propagates with respect to—and is minimally velocity matched to—the light. This means that the modulation efficiency of the transmit signal is low, which is also as desired.

Figure 20.11 is a plot of TR isolation vs. frequency with a 50 Ω load at the antenna port. At high frequencies the isolation the TIPRx, as shown by the black curve, is better than ~40 dB over almost a decade, which is similar to the isolation of the broad bandwith TIPRx. As the frequency decreases, the distinction between co- and counter-propagating waves diminishes, which results in a corresponding degradation of the TR isolation. For comparison, the TR isolation of a commercial ferrite circulator is also plotted as the grey curve in the figure. The ferrite circulator isolation is ~20 dB over one octave.

20.4 SUMMARY

Section 20.2 of this chapter described recent developments that have enabled amplifierless microwave photonic receive-only links having gain >20 dB at 12 GHz and low noise figure, <6 dB at frequencies up to 12 GHz, much like commercially available microwave LNAs. The record-setting performance—especially the noise figure—required advances in the ability to model the signal-to-noise performance mathematically, which were also discussed in Section 20.2.

The improvements to microwave photonic link performance—especially the newly achieved low noise figures—have opened the door to a completely new method of designing a photonically-remoted front-end for an antenna that simultaneously transmits and receives. As Section 20.3 described, the enabling component for this new front-end design is called a Transmit-Isolating Photonic Receiver (TIPRx), which substitutes as the functional equivalent of a receive-channel LNA, a photonic link for microwave signal remoting, and an RF ferrite circulator. Measured TIPRx performance data provided in Section 20.3 showed that the inherently broad bandwidth of its microwave photonic components enable both its TR isolation (>40 dB) and its bandwidth (nearly four decades) to greatly exceed that of a ferrite circulator. The ability to simultaneously transmit and receive over very broad bandwidths is an entirely new capability enabled by the high TR isolation that the TIPRx imparts to an antenna front-end.

References

[1] Y.-T. Hsueh, Z. Jia, H.-C. Chien, A. Chowdhury, J. Yu, G.-K. Chang, Generation and transport of independent 2.4 GHz (Wi-Fi), 5.8 GHz (WiMAX), and 60-GHz optical millimeter-wave signals on a single wavelength for converged wireless over fiber access networks, in:OFC/NFOEC, OTuJ1, 2009.

[2] H. Roussell, M. Regan, J. Prince, C. Cox, J. Chen, W. Burns, E. Ackerman, J. Campbell, Gain, noise figure, and bandwidth-limited dynamic range of a low-biased external modulation link, in: Proceedings of the IEEE International Topical Meeting on Microwave Photonics, October 2007, pp. 84–87.

[3] E. Ackerman, G. Betts, W. Burns, J. Campbell, C. Cox, N. Duan, J. Prince, M. Regan, H. Roussell, Signal-to-noise performance of two analog photonic links using different noise reduction techniques, in: IEEE MTT-S Int. Microwave Symp. Dig., June 2007, pp. 51–54.

[4] H. Haus et al., IRE standards on methods of measuring noise in linear twoports, Proc. IRE 48 (1959) 60–68.

[5] W. Burns, M. Howerton, R. Moeller, Broad-band unamplified optical link with RF gain using a LiNbO$_3$ modulator, IEEE Photon. Technol. Lett. 11 (1999) 1656–1658.

[6] C. Cox, Analog Optical Links: Theory and Practice, Cambridge University Press, 2004.

[7] E. Ackerman, C. Cox, Microwave photonic links with gain and low noise figure, in: Proc. IEEE Lasers Electro-Opt. Soc. Annu. Meet. (Orlando, Florida), October 2007, pp. 38–39.

[8] C. Cox, D. Tsang, L. Johnson, G. Betts, Low-loss analog fiber-optic links, in: IEEE MTT-S Int. Microwave Symp. Dig., pp. 157–160, May 1990.

[9] A. Karim J. Devenport, Low noise figure microwave photonic link, in: IEEE MTT-S Int. Microwave Symp. Dig., June 2007, pp. 1519–1522.

[10] E. Ackerman, G. Betts, W. Burns, J. Prince, M. Regan, H. Roussell, C. Cox, Low noise figure, wide bandwidth analog optical link, in: Proc. of the IEEE International Topical Meeting on Microwave Photonics, October 2005 pp. 325–328.

[11] J. McKinney, M. Godinez, V. Urick, S. Thaniyavarn, W. Charczenko, K. Williams, Sub-10-dB noise figure in a multiple-GHz analog optical link, IEEE Photon. Technol. Lett. 19 (2007) 465–467.

[12] K. Williams, L. Nichols, R. Esman, Externally-modulated 3 GHz fibre-optic link utilizing high current and balanced detection, Electron. Lett. 33 (1997) 1327–1328.

[13] C. Cox, E. Ackerman, G. Betts, J. Prince, Limits on the performance of RF-over-fiber links and their impact on device design, IEEE Trans. Microwave Theory Tech. 54 (2006) 906–920.

[14] E. Ackerman, W. Burns, G. Betts, J. Chen, J. Prince, M. Regan, H. Roussell, C. Cox, RF-over-fiber links with very low noise figure, J. Lightwave Technol. 26 (2008) 2441–2448.

[15] G. Gopalakrishnan, W. Burns, R. McElhanon, C. Bulmer, A. Greenblatt, Performance and modeling of broadband LiNbO3 traveling wave optical intensity modulators, J. Lightwave Technol. 12 (1994) 1807–1819.

[16] G. Betts F. O'Donnell, Improvements in passive, low-noise-figure optical links, in: Proc. of the Photonic Systems for Antenna Applications Conf., Monterey, California, USA, 1993.

[17] M. Farwell, W. Chang, D. Huber, Increased linear dynamic range by low biasing the Mach-Zehnder modulator, IEEE Photon. Technol. Lett. 5 (1993) 779–782.

[18] E. Ackerman, S. Wanuga, D. Kasemset, A. Daryoush, N. Samant, Maximum dynamic range operation of a microwave external modulation fiber-optic link, IEEE Trans. Microwave Theory Tech. 41 (1993) 1299–1306.

[19] E. Ackerman, S. Wanuga, J. MacDonald, J. Prince, Balanced receiver external modulation fiber-optic link architecture with reduced noise figure, in: IEEE MTT-S Int. Microwave Symp. Dig., Atlanta, Georgia, USA, June 1993, pp. 723–726.

[20] C.H. Cox, E.I. Ackerman, Transmit isolating photonic receive links : a new capability for antenna remoting, in: IEEE/OSA Optical Fiber Communications Exhibition and National Fiber Optic Engineers Cong., Los Angeles, CA, USA, March, 2011.

[21] C.H. Cox, III, Bi-directional Signal Interface and Apparatus Using Same, US Patent 7,809,216, October 5, 2010.

[22] C. Cox, E. Ackerman, Bi-directional Signal Interface, US Patent 7,555,219, June 30, 2009.

Advances in 1-100 GHz Microwave Photonics: All-Band Optical Wireless Access Networks Using Radio Over Fiber Technologies

Gee-Kung Chang, Yu-Ting Hsueh, and Shu-Hao Fan

Georgia Institute of Technology, Atlanta, GA 30332-0250, USA

21.1 INTRODUCTION

The evolution of fiber-optics communications has been thriving for the last several decades. All Internet service providers are eager to provide broadband data and video services through optical nodes as close to the users as possible to maximize the communication bandwidth. For access networks, which are deployed within the last miles or the last meters near consumers, it is a popular trend to develop high-speed access networks integrated with wireless services. From old-fashioned 2.5G/3G "Global System for Mobile" (GSM) to modern technologies, such as Long-Term Evolution (LTE), WiMAX, or Wireless Fidelity (Wi-Fi), the wireless transmission rate has been growing from a few kilobits per second (Kb/s) to one gigabit per second (Gb/s), but the highest data rate of these microwave bands is restricted by their available bandwidth of a few tens MHz. However, there is a multitude of multimedia applications calling for wireless transmission, and their required data rates are larger than hundreds of megabits per second. For instance, the uncompressed HDTV 1080i requires a transfer rate of 1.485 Gb/s while the HDTV 1080p requires 2.97 Gb/s. Obviously, the existing wireless services cannot accommodate these very-high-throughput (VHT) applications, and a new wireless technology should be developed to support future demands.

Extremely high frequency band (EHF), defined as a range of electromagnetic waves with frequencies between 30 GHz and 300 GHz, has a wavelength of 1 to 10 mm; therefore, it is also called millimeter wave (mmW; mmWave). Compared to lower bands with spectral congestion, the mmW can offer a massive amount of spectral space to support high-speed applications such as wireless local area networks (WLAN) and HD video distribution [1]. In particular, the 60-GHz band has gained much attention and has been viewed as the most promising carrier in next-generation wireless networks because of its widest unlicensed bandwidth among all the wireless channels. The large atmosphere absorption restricts the applications of

Optical Fiber Telecommunications VIB. http://dx.doi.org/10.1016/B978-0-12-396960-6.00021-3

60-GHz mmW for long-distance transmission. Nevertheless, 60-GHz bands still hold great potential for indoor and short-range transmission.

In 2009, multi-gigabit 60-GHz wireless transceivers were already off-the-shelf products. The first 60-GHz standard is available through the WirelessHD Consortium[2], which is promoted by many major electronics companies, such as Broadcom Corporation, Intel Corporation, LG Electronics Inc., NEC Corporation, Samsung Electronics, Co. Ltd., Silicon Image who acquired SiBEAM in 2011, Hittite Microwave, Sony Corporation, and Toshiba Corporation. The 60-GHz transceivers enable the point-to-point wireless gigabit transmission over tens of meters, intending to replace conventional wired links (such as USB and HDMI cables). Beyond point-to-point wireless links, the WirelessHD Consortium has announced its specification 1.1, providing up to 28-Gb/s transmission speed over its wireless video area network (WVAN) [2]. Not limited by only HD video transmission, the IEEE 802.11ad task group [3] and the WiGig Alliance [4] are searching for a solution to the convergence of Wi-Fi and 60-GHz for WLAN, while the IEEE 802.15.3c group specializes in development of 60-GHz wireless personal area networks (WPAN) [5].

In order to accomplish mmW wireless communications, a technology called radio-over-fiber (RoF) has recently been proposed [6–9]. The key idea of this technology is the integration of optical transmission systems and wireless access networks. For a mmW RoF system, the mmW signals are carried by optical carriers and then distributed to user terminals through fibers and antennas. Compared with the electrical mmW generation method, optical up-conversion methods offer definitive advantages. First of all, optical fiber provides immunity to electromagnetic interference. Secondly, it can be seamlessly applied to the deployed optical transport networks. Thirdly, some optical mmW generation methods are capable of generating multiples of a reference local oscillator (LO), while relaxing the bandwidth requirement of electronic devices and saving costs. Therefore, the converged RoF technology with high transmission capacity, large coverage, and good cost efficiency has been considered as the most promising solution for future 60-GHz wireless access networks.

The outline of second half of this chapter is as follows. In Section 21.2, the overview of common methods for optical RF generation is presented. The types of optical RF waves are also shown in this section. Section 21.3 focuses on the multi-band generation over one optical platform based on RoF technology. Various schemes are presented for simultaneously delivering mmW, microwave, baseband, and commercial low-RF wireless services. Finally, Section 21.4 provides a summary and presents the potential research topics in the future.

21.2 OPTICAL RF WAVE GENERATION
21.2.1 Overview of optical RF signal generation

Most technologies of optical RF signal generation can be classified into two categories: one laser or two lasers needed in a system. The generation schemes with one laser are based on direction modulation, external modulation, and dual-mode DFB [10,11], while the schemes with two lasers are based on nonlinear effects in the nonlinear media [9,12], heterodyne mixing of independent lasers [13], and injection locking [14,15].

The direct modulation by laser is a simple way to generate optical RF signals, but it is limited by the modulation bandwidth of laser source, which is around 30 GHz. The realization of mmW generation by nonlinear effects such as four-wave mixing (FWM) and cross-phase modulation (XPM) [9] in the nonlinear media has the advantages of high conversion efficiency and large conversion bandwidth. However, it requires high input power and has uneven response for the large bandwidth. Furthermore, the heterodyne mixing of two independent lasers has been viewed as the most simple method to optically generate mmW signal. Unfortunately, the large jitter and phase noise induced by the wide signal linewidth and the need of accurate polarization control are the disadvantages for the heterodyne mixing scheme. Although the dual-mode DFB laser can generate two phase-locked modes with specific frequency difference and is without any requirement for electronic feedback systems [11], this kind of laser is not a commercial product and is absent of system flexibility. For the injection locking method, the large injection signal power and only few GHz locking range make it impractical for system implementation.

For the optical RF signal generation using external optical modulation, it takes advantage of the intrinsic nonlinear response of modulators to generate high-order sidebands and then obtain the corresponding RF carrier through the optical heterodyning. The external modulation can be intensity or phase modulation. Among all-optical RF signal generation methods, this one is most flexible and easy to integrate with WDM system. The main drawback of this method is the insertion loss of external modulators, so it usually requires additional optical amplifiers to compensate the power loss.

21.2.2 Types of optical RF waves

There are four types of optical RF waves introduced, including optical double-sideband (ODSB), optical single-sideband-plus-carrier (OSSB+C), homodyne optical-carrier-suppression (homoOCS), and heterodyne optical-carrier-suppression (heteroOCS). The RF carrier of interest is focused on 60-GHz mmW because of its high potential of multi-gigabit wireless transmission and high-frequency related constraints.

21.2.3 ODSB millimeter wave

An ODSB mmW is generated in the most straightforward way. Any RF carrier linearly modulated to the intensity of a coherent lightwave can be represented by an ODSB signal. As shown in Figure 21.1, the optical spectrum of an ODSB mmW signal is represented by an optical central carrier and two complex-conjugated data sidebands. ODSB signals can be generated straight from an analog directly modulated laser (DML) or an optical intensity modulator (MZM: Mach-Zehnder modulator) biased at the linear point with an RF-signal input. The structure and principle of operation of single-arm and dual-arm MZM are described in Chapter 20 of this book and in Reference [28]. Being detected by a photodiode (PD), the beating between the optical

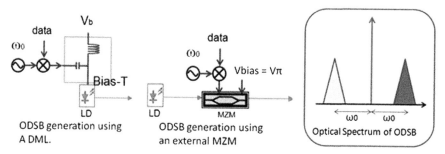

FIGURE 21.1 Setup of ODSB generation and optical spectrum.

central carrier and the data sideband generates the electrical signal at the RF band. Assuming the angular frequency of the RF is ω_0 and the complex data symbol is $x(t)$, the generated electrical signal is expressed by

$$S(t) \propto |1 + \alpha \mathrm{Re}\{x(t)\exp(j\omega_0 t)\}|^2 \sim 1 + 2\alpha \mathrm{Re}\{x(t)\exp(j\omega_0 t)\}, \quad (21.1)$$

where α represents the ratio of the sideband to the optical central carrier. And the beating frequency at $2\omega_0$ is neglected here. Therefore, the amplitude and phase information is preserved by the beating of the PD.

21.2.4 OSSB+C millimeter wave

The optical spectrum of OSSB+C mmWs is shown in Figure 21.2. It is similar to the ODSB optical spectrum except that one of the sideband is suppressed. OSSB+C mmWs can be generated by a dual-arm MZM, whose two RF inputs have a 90-degree phase shift and two bias inputs have $V_\pi/2$ difference. Otherwise, the sideband can also be suppressed by an optical bandpass filter (OBPF). Neglecting out-of-band signals, the electrical signal detected by a PD can be approximated by

$$S(t) \propto |1 + \alpha x(t)\exp(j\omega_0 t)\}|^2 \sim 2\alpha \mathrm{Re}\{x(t)\exp(j\omega_0 t)\}, \quad (21.2)$$

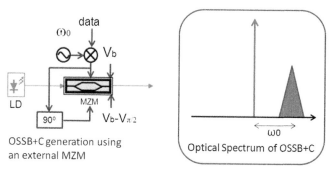

FIGURE 21.2 Setup of OSSB+C generation and optical spectrum.

Therefore, the resulting electrical signal is proportional to the original RF electrical signal.

Both ODSB and OSSB+C mmW signals are readily generated by electrical up-conversion, which assures the sensible convenience of downlinks and uplinks. However, the bandwidth of the required o/e conversion interface has to manage mmW frequency. 60-GHz mmWs, for example, cannot be modulated directly into 60-GHz ODSB or OSSB+C mmW signals. Using indirect optical up-conversion to double or triple the carrier can be achieved [16]. However, it complicates the optical mmW generation even more, and the problem of uplinks has still not been resolved.

21.2.5 OCS millimeter wave

Instead of suppressing one of the sideband, OCS generates optical mmWs by suppressing the optical central carrier of an ODSB mmW. OCS can be realized by using a MZM biased at the null point or by using an optical stop-band filter to suppress the optical central carrier. When using an optical filter to suppress the optical central carrier, the two sidebands of OCS mmW can be generated either by a MZM or by a phase modulator (PM). The OCS optical spectrum is shown in Figure 21.3. Using electrical up-conversion (Figure 21.3a), the RF signal after a PD is given by

$$S(t) \propto |\text{Re}\{x(t)\exp(j\omega_0 t)\}|^2 \sim \text{Re}\{x^2(t)\exp(j\omega_0 t)\}, \qquad (21.3)$$

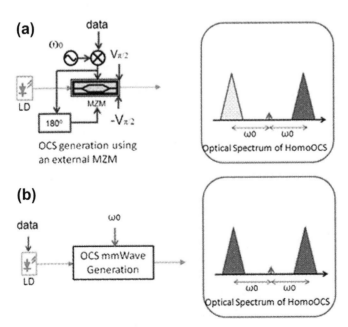

FIGURE 21.3 Setup of HomoOCS generation with the electrical (a) and optical (b) up-conversion and optical spectrum.

where the low-frequency term is neglected. Notice that the RF frequency is doubled because of the OCS technique. Even if the bandwidth of the E/O interface in the transmitter is not enough to generate $2\omega_0$, the $2\omega_0$ mmW can be still generated by the square law of a PD. Also, notice that the complex data symbol $x(t)$ is squared. This indicates that any modulation format except OOK will be distorted by the OCS mmW generation. It also means that the RF signals are generated indirectly. Therefore, the problem of uplinks is unsolved.

When using the optical up-conversion (Figure 21.3b), the baseband signal $x(t)$ is modulated separately from the OCS mmW generation. The RF signal generated by the PD is

$$S(t) \propto |x(t)\text{Re}\{\exp(j\omega_0 t)\}|^2 \sim \frac{1}{2}|x(t)|^2 \cos(2\omega_0 t). \tag{21.4}$$

Therefore, the main difference between the electrical and optical up-conversion is that the phase information of the complex data symbol $x(t)$ is completely lost in the optical up-conversion. Because the data symbol is directly modulated onto the RF carrier ($2\omega_0$), this generation method is also known as homodyne OCS (homoOCS).

To preserve the phase information in the optical up-conversion, we can use an intermediate frequency (IF) to carry the baseband data symbol by $\text{Re}\{x(t)\exp(j\omega_{\text{IF}}t)\}$ in electrical circuits, where the IF angular frequency ω_{IF} is usually much lower than RF angular frequency $2\omega_0$. Therefore, the IF signals can be directly modulated onto laser lightwaves (either ODSB or OSSB+C). Then OCS mmW generation can be used as the optical up-conversion method to generate the optical mmWs. The RF signal after a PD becomes

$$S(t) \propto |E_{\text{ocs}'}(t)|^2 \sim \frac{1}{2}\cos(2\omega_0 t) + \frac{\alpha}{2}\text{Re}\{x(t)\exp(j(2\omega_0 + \omega_{\text{IF}})t)\}, \tag{21.5}$$

Here we neglect all the terms that are lower than $2\omega_0$ since low-frequency components are unable to pass through RF amplifiers. The RF carrier carries the complex data symbol at ($2\omega_0 + \omega_{\text{IF}}$) without any distortion. Another RF carrier at $2\omega_0$ is also generated without carrying any data, which should not cause any interference in the wireless channel. As a result of different IF generation methods, there are variants for the IF-carried OCS mmW generation methods, as shown in Figure 21.4.

21.2.6 Conversion efficiency

To compare the optical mmW generation, it would be interesting to know the optical-to-RF conversion efficiency of the various kinds of optical mmWs. That is, when the optical power detected by a PD is the same, what is the RF power that the optical mmW can produce? The optical power, in terms of Poynting vectors, varies with time for RoF signals. However, we can calculate the average optical power by

$$P_{\text{optical}} = \frac{1}{T}\int_T \frac{|E(t)|^2}{2\eta}dt \times \text{Area}. \tag{21.6}$$

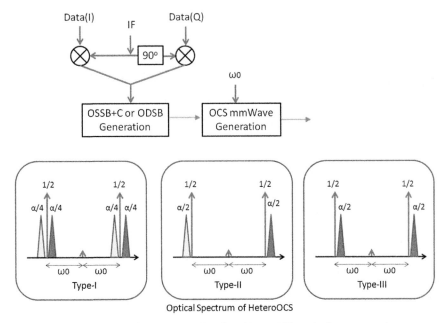

FIGURE 21.4 Setup and optical spectra of various HeteroOCS optical up-conversion schemes.

Here η represents the characteristic impedance of optical fibers and T is the period of mmWs. Area is the effective area of the fiber. Assuming that the PD responsivity (ampere/watt) is one, the average RF power generated by a PD at the specific mmW frequency is

$$P_{RF} = \frac{1}{T} \int_T \left| \frac{s(t)^2}{2\eta} \right| Z \, dt, \qquad (21.7)$$

where Z is the transimpedance of the PD electronics. Without considering data symbols, the optical-to-RF conversion efficiency can be approximated by $P_{RF}/P_{optical}$ as

(a) ODSB

$$\mu_{ODSB} = C_0 \frac{4\alpha^2}{1 + \alpha^2/2}. \qquad (21.8a)$$

(b) OSSB+C

$$\mu_{OSSB+C} = C_0 \frac{4\alpha^2}{1 + \alpha^2}. \qquad (21.8b)$$

(c) HomoOCS

$$\mu_{OCS} = C_0 \frac{1}{2}. \qquad (21.8c)$$

(d) HeteroOCS Type-I

$$\mu_{OCS'}^{I} = C_0 \frac{\alpha^2}{2 + \alpha^2}.$$
(21.8d)

(e) HeteroOCS Type-II

$$\mu_{OCS'}^{II} = C_0 \frac{\alpha^2}{2 + 2\alpha^2}.$$
(21.8e)

(f) HeteroOCS Type-III

$$\mu_{OCS'}^{III} = C_0 \frac{\alpha^2}{1 + \alpha^2}.$$
(21.8f)

The parameter C_0 is represented by

$$C_0 = \frac{Z E_0^2}{2\eta \text{Area}}.$$
(21.9)

Notice that the parameter α can be individually adjusted for each kind of optical mmW by the modulation indices. To ensure the modulation linearity, α is usually much smaller than 0.5. Under this circumstance, HomoOCS is the most efficient optical-mmW generation scheme. However, it is also the only optical mmW generation method that is unable to carry phase information (or vector signals) without distortions by a PD. In addition, part of the HeteroOCS power is dedicated to the RF carrier, so the RF conversion efficiency of HeteroOCS mmWs is low.

21.3 CONVERGED ROF TRANSMISSION SYSTEM
21.3.1 Generation and transmission of multiple RF bands

With the rapid growth of bandwidth demand for all kinds of emerging data and video-intensive wired and wireless services, converging traditional baseband systems (EPON and GPON) and wireless systems on the same optical platform based on RoF techniques is foreseen as a promising way to provide multi-band services. This is because the optical transport of radio signals based on RoF technology that can offer wireless services with an extended coverage that is beyond the purely RF range, and have format transparency to integrate multiple wireless services on an optical platform. Figure 21.5 shows the optical-access network architecture, which can simultaneously provide wired and wireless services. The service providers send the data to the central office (CO), an RF/optical interface. Based on the RoF technology, the data are optically up-converted to different RF bands, including baseband, Wi-Fi, WiMAX, MW, mmW, and so on. In the remote node, the up-converted multi-band signals with different wavelengths are multiplexed and demultiplexed by AWGs before they are routed to access points (APs) over fiber. The APs then locally distribute wireless services to antennas for broadcasting, and wired services (baseband) to local

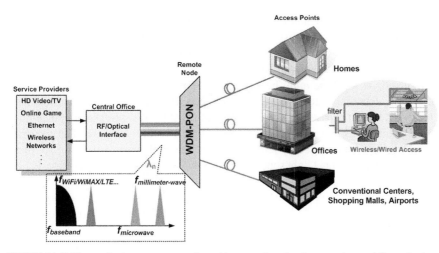

FIGURE 21.5 The optical access network architecture for simultaneously providing wired and wireless services.

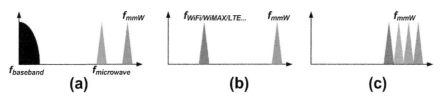

FIGURE 21.6 Three multi-band types: (a) baseband, MW, and mmW, (b) mmW and wireless services at lower RF regions, and (c) mmW (60 GHz) sub-bands.

area networks such as fiber-to-the-home (FTTH). The APs can be conference centers, airports, hotels, shopping malls, and ultimately homes and small offices.

In this section, some system experiments involving the multi-band generations and transmissions will be discussed. These architectures can be divided into three types according to their generated bands as Figure 21.6 shows. The first type can generate and transmit mmW, MW, and baseband signals at the same time (Figure 21.6a), and the second one can simultaneously offer mmW and commercial wireless services at lower RF regions such as Wi-Fi and WiMAX (Figure 21.6b). Not like the previous two ones to view the whole 60-GHz mmW as one band, the third type is capable of transmitting multiple services at 60-GHz sub-bands (Figure 21.6c).

21.3.2 Baseband, microwave, and millimeter wave

Recently, the RoF systems with simultaneous multi-band modulation and transmission, including baseband, MW, and mmW, have been demonstrated [17–19]. However, the multi-band signal generation in [17] is implemented with an expensive and complicated modulator, and the EAM nonlinearity, residual chirp and the crosstalk among three signals would limit the transmission performance in [18] and [19]. Additionally, these methods require high-frequency clock source to generate mmW signal. More importantly, the transmission limitation arising from optical fiber chromatic dispersion was not resolved yet for these multi-band systems. Here a novel full-duplex RoF system transmitting downlink wireless 60-GHz mmW, 20-GHz MW, and wired baseband data is presented [20]. It is based on a single dual-arm MZM and a subsequent optical filter (OF). The high dispersion tolerance is achieved by the OSSB+C format. The upstream data transmission is obtained by reusing the wavelength downstream, therefore, there is no need of an additional light source and wavelength management at the base station (BS), which significantly reduces the cost and increases the system stability.

Figure 21.7 shows the schematic diagram of this multi-band RoF system. It is achieved by only one MZM. The continuous wave (CW), which is represented by $E_{in} = E_0 \exp(j\omega_c t)$, is modulated by two driving signals, $V_1(t)$ and $V_2(t)$, via a dual-arm MZM. $V_1(t)$, consisting of a baseband data mixed with a 40-GHz sinusoidal clock, can be written as $V_1(t) = A(t) \cos(\omega_1 t)$, where $A(t)$ represents the 2.5-Gb/s baseband data and ω_1 is the angular frequency of the sinusoidal signal. $V_2(t)$, consisting of a 20-GHz RF sinusoidal clock and a DC bias voltage, is written as $V_2(t) = V_b + B \cos(\omega_2 t)$. Here V_b is the DC bias voltage, and B and ω_2 are the amplitude and the angular frequency of the 20-GHz sinusoidal signal. We assume the power splitting ratio of two arms of the MZM is 1/2. The electrical field at the output of the MZM is given by

$$E_{out} \approx \frac{\sqrt{\alpha} \cdot E_0}{2}[(J_0(m_A) + J_0(m_B)e^{j\pi V_b/V_\pi}) \cdot e^{j\omega_0 t} + J_1(m_A) \cdot e^{j[(\omega_0+\omega_1)t+\pi/2]}$$
$$+ J_1(m_A) \cdot e^{j[(\omega_0-\omega_1)t+\pi/2]} + J_1(m_B) \cdot e^{j\pi V_b/V_\pi} \cdot e^{j[(\omega_0+\omega_2)t+\pi/2]}$$
$$+ J_1(m_B) \cdot e^{j\pi V_b/V_\pi} \cdot e^{j[(\omega_0-\omega_2)t+\pi/2]}. \tag{21.10}$$

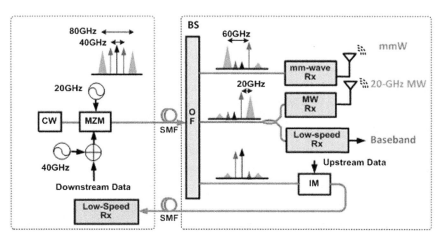

FIGURE 21.7 Schematic diagram of the multi-band generation (baseband, 20 GHz, and 60 GHz) based on a dual-arm MZM.

Here, α is the insertion loss of the MZM. Equation (21.10) shows that carrier and four subcarriers are equally spaced by 20 GHz, and two inner subcarriers are without any data information. After a fiber of length l, an OF is used to separate these subcarriers to generate 60-GHz and 20-GHz signals. As shown in Figure 21.7, the 60-GHz mmW and 20-GHz MW both belong to the optical OSSB+C scheme, which has high tolerance to CD. Moreover, the carrier and the first-order subcarrier with lower frequency are remodulated by the upstream data, and then sent back to the CO where a low-speed receiver is utilized for the O/E conversion. For more details of experimental setup, please refer to [20].

The BER measurements for the DL wired signal and the DL 60-GHz signal with a 4-m wireless transmission are shown in Figure 21.8. The receiver sensitivity is relatively high because an optical pre-amplifier is utilized. At the given BER of 10^{-9}, the power penalties after a 50-km SSMF transmission are about 0.3 dB and 1.4 dB for the wired and wireless services, respectively. The penalty for the wireless transmission is mainly due to the amplified spontaneous emission (ASE) from EDFA and the noise from EAs at the BS. The BERs for the 2.5-Gb/s remodulated UL are also shown in Figure 21.8, and its power penalty is less than 0.5 dB over the same transmission distance. The electrical eye diagrams after the 50-km SSMF transmission for the DL and UL signals are also provided in the insets of Figure 21.8.

The unique advantages of this multi-band RoF system are to utilize simple components to realize dispersion-tolerant transmission and without the lasers in the BS for the uplink transmission. The experimental results prove that this scheme is suitable to simultaneously offer high bandwidth, multi-services for the future optical-access networks.

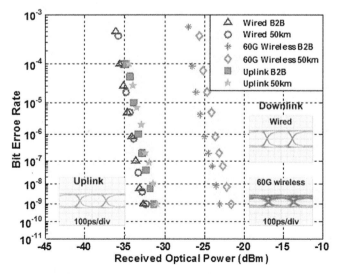

FIGURE 21.8 BER measurements and electrical eye diagrams for the UL and DLs of the wired and 60-GHz signal with a 4-m wireless transmission.

21.3.3 Millimeter wave with wireless services in low RF regions

As discussed in Section 21.1, optical-wireless access technologies have been considered the most practical and efficient solution to increase the capacity, coverage, bandwidth, and mobility for the last mile and last meter. Various RoF systems with commercial wireless services at lower RF regions (ex. Wi-Fi at 2.4 GHz and WiMAX at 5.8 GHz) that deliver multi-megabit signal over in-building fiber network have been proposed and demonstrated [21–23]. On the other hand, the RoF system architecture operating at 60-GHz has recently gained much attention for its availability to achieve multi-gigabit data rate. In order to exploit such bandwidth advantage of 60-GHz mmW and have the backward compatibility of legacy wireless services, a RoF architecture design that accommodates advanced mmW and existing wireless services simultaneously for in-building broadband access systems is discussed here [24].

In Figure 21.9, the scheme based on a PM and subsequent filtering can simultaneously provide Wi-Fi, WiMAX, and 60-GHz mmW services. The multiple sidebands are generated by a PM, which is driven by a 30-GHz clock source. The different subcarriers, which are separated by an OF, are modulated by different baseband data from Wi-Fi, WiMAX, and mmW services, respectively. After transmitting over a length of fiber, the combined signal is separated by the other OF. The baseband data for Wi-Fi and WiMAX services are up-converted to their belonging frequency bands by using the commercial transceivers, and then broadcast to users. The 60-GHz optical signal are converted to the electrical domain and broadcast by an antenna. Therefore, all-optical up-conversion in the CO is for the high-data-rate 60-GHz signal while the remote electrical up-conversion in the BS for Wi-Fi and WiMAX services. The wired service can also be achieved by using a low-speed receiver to accomplish direct detection as Figure 21.9 shows. Of course, if the input signals for Wi-Fi and WiMAX services from service provider are already at their corresponding RF bands, the IMs can be directly modulated by these RF signals. At this situation, the Wi-Fi and WiMAX transceivers are unnecessary. For more details of experimental setup, please refer to [24].

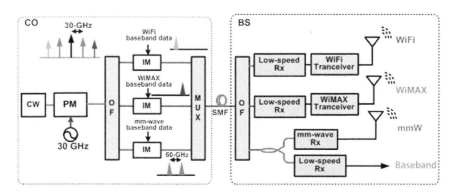

FIGURE 21.9 Schematic diagram of a multi-band RoF system with mmW, Wi-Fi, and WiMAX.

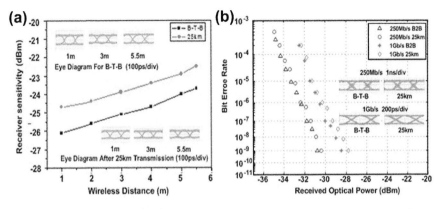

FIGURE 21.10 (a) Receiver sensitivities and electrical eye diagrams for the 2.5-Gb/s signal with different wireless distances. (b) BER measurements and electrical eye diagrams for the 250-Mb/s and 1-Gb/s signal after transmitted over a 25-km SSMF and 4-m wireless transmission.

The receiver sensitivities and electrical eye diagrams for the 2.5-Gb/s signal at 60 GHz with different wireless distances are shown in Figure 21.10a. After transmitting over a 25-km SSMF, the power penalties at the given BER of 10^{-9} for different air distances are less than 1.5 dB. The penalties result from the CD for the two subcarriers with 60-GHz spacing and the nonlinear modulation in the modulator. Moreover, Figure 21.10a also shows that the receiver sensitivities degrade with the increase of the air distance because the power loss is proportional to transmission distance. The BERs and corresponding electrical eye diagrams for the 250-Mb/s and 1-Gb/s signal at the air distance of 4-m are shown in Figure 21.10b. After a 25-km SSMF, for these two signals, there are almost no power penalties as the CD effects are negligible at these rates. The distance of wireless transmission in the experiment is limited by our lab's space, and our scheme mainly focuses on the functional verification of the multi-services generation and transmission with only one light source at the CO. To our knowledge, this is the first experimental demonstration for simultaneously feeding three independent wireless signals on a single wavelength to provide last mile and last meter wireless access services.

21.3.4 60-GHz sub-bands generation

The wide 60-GHz unlicensed band can be divided into several sub-bands to increase system flexibility. As mentioned in ECMA-387 [25], the unlicensed 8-GHz band is actually divided into four sub-bands centered at 58.32 GHz, 60.48 GHz, 62.64 GHz, and 64.80 GHz with a frequency separation of 2.160 GHz and the symbol rate of 1.728 GS/s on each channel to lower the bandwidth requirement of mmW circuits and devices. To address the generation and transmission issues of a converged optical-wireless network with multiple 60-GHz sub-bands, [26] based on the ODSB technique generated 60-GHz and 64-GHz mmW signals simultaneously but with an unwanted signal at 62 GHz produced and short distance of fiber transmission. Thus, a better RoF system based on the OSSB+C format and cascaded modulation was proposed, which is with high tolerance to CD and without unwanted RF signals [27].

In Figure 21.11, the optical carrier with angular frequency ω_c is first modulated by two independent data at different RF signals ($\omega_{A1} = 36\pi$ GHz and $\omega_{A2} = 40.32\pi$ GHz) via a MZM. An OF_1 with a sharp passband window is used for vestigial sideband (VSB) filtering. The frequency of an RF clock, ω_B, is set at 40.32π GHz to further generate higher and lower subcarriers. Another optical filter (OF_2) is used to eliminate unwanted peaks. The remaining three subcarriers of interest, the left one without data and the right twos with different data information, are employed for the system demonstration. The coherent beating between the resultant optical subcarriers ω_1 ($=\omega_c - \omega_B$) and ω_2 ($=\omega_c + \omega_{A1} + \omega_B$) and the beating between ω_1 and ω_3 ($=\omega_c + \omega_{A2} + \omega_B$) would produce two mmW signals: one is at 50.32 GHz, and the other one is at 60.48 GHz. It can be seen if ω_A is equal to ω_B, one generated signal at frequency $\omega_A + 2\omega_B$ ($=3\omega_B$) to form exact frequency tripling. However, if ω_B is larger than ω_A by few GHz, it is approximate frequency tripling. The frequency tripling technique for the mmW generation makes these systems cost-efficient, which is with low bandwidth requirement for electronic components compared with the traditional mmW RoF systems using frequency doubling technique with the OCS format.

At the BS, the optical signal is detected by a high-speed receiver and broadcast by a 60-GHz mmW broadband antenna, and finally, the end users can select the appropriate band based on their need from the broadcast multi-band signals. Based on this proposed scheme, [27] demonstrated the generation and transmission of two independent 1-Gb/s data on 55-GHz and 60-GHz carriers over a 50-km SSMF without any dispersion compensation. The reason why it chose two bands over the four ones in ECMA-387 is the limitation of available components and equipments.

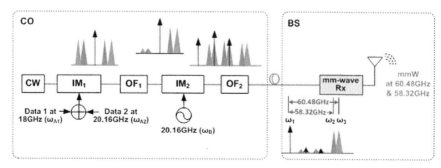

FIGURE 21.11 Schematic diagram of the RoF scheme based on cascaded modulation for the generation of multiple 60-GHz sub-bands.

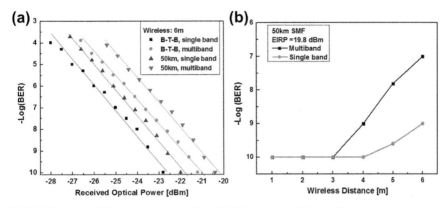

FIGURE 21.12 (a) BER measurements of the 60-GHz service with or without the presence of the 55-GHz service after a 6-m wireless transmission. (b) BER versus wireless distance for the 60-GHz service with or without the presence of the 55-GHz service with EIRP $= 19.8\,dBm$ and after a 50-km SSMF transmission.

Figure 21.12a shows the BER of the 60-GHz band with or without the presence of 55-GHz band after a 6-m wireless transmission. The 60-GHz signal in the multi-band case needs 2 dB more power to achieve BER $= 10^{-9}$ compared with the single-band case. Moreover, the penalty from the 50-km SMF transmission is 0.8 dB at the BER of 10^{-9} for both single-band and multi-band cases. Figure 21.12b shows the BER against the wireless distance of the multi-band transmission is 4 m compared to 6 m in the single-band case. The penalty mainly arises from the lower EIRP for each band signal in the distance of free-space propagation with EIRP $= 19.8\,dBm$, under the FCC regulations for in-building 60-GHz radio frequency applications. At the BER of 10^{-9}, the achieved multi-band case compared with the one in the single-band case, and some interference from the 55-GHz signal. Therefore, the performance of the multi-band case can be improved by adding a band pass filter to separate the two bands signals before the down-conversion.

21.4 CONCLUSIONS

With the growing bandwidth demand for providing broadband access services in the last mile and last meter, RoF technology at mmW band has been viewed as the most promising solution to provide ubiquitous multi-gigabit wireless services with simplified and cost-effective BSs and low-loss, bandwidth-abundant fiber optic networks. In such a system, the technical challenges on optical mmW generation, transmission impairments compensation, and converged multi-band services are vital for the successful deployment in the future. In this chapter we first concludes the general methods and types of optical mmW generation, and summarizes their advantages and disadvantages.

Then, the numerical analysis suggests that HomoOCS is the most efficient optical-mmW generation scheme but it is unable to carry vector signals without phase distortion.

Owing to ultra-wide bandwidth and protocol transparent characteristics, advanced RoF system can be utilized to simultaneously deliver wired and multi-band wireless services for both fixed and mobile users. In Section 21.3, several multi-band 60-GHz RoF systems are categorized in three types: mmW with baseband and MW, mmW with commercial wireless services in low RF regions, and 60-GHz sub-bands; they are designed to realize lightwave centralization, frequency doubling/tripling, and long transmission distance.

For the RoF systems used for the wireless-coverage extension, the point-to-point fiber-wireless transmission has been thoroughly studied and experimentally demonstrated before. However, RoF systems are designed for multi-user applications, such as pico-cells or femto-cells. Multiple-access controlling schemes are rarely designed and implemented for RoF systems, and multi-user scenarios are rarely studied and experimentally tested for mmWave end-to-end transmission systems. To promote the RoF wireless-coverage extension systems to more practical systems, link-layer issues, such as multi-services and multi-users, together with higher layer protocols require further studies and experimental tests in the future.

References

[1] P. Smulders, Exploiting the 60 GHz band for local wireless multimedia access: prospects and future directions, IEEE Commun. Mag. 40 (1) (2002) 140–147.

[2] WirelessHD Specification ver. 1.1, May 2010. Available: <http://www.wirelesshd.org>.

[3] IEEE P802.11ad Task Group. Available: <http://www.ieee802.org/11/>.

[4] WiGig Alliance. Available: <http://wirelessgigabitalliance.org>.

[5] IEEE 802.15 WPAN Millimeter Wave Alternative PHY Task Group 3c (TG3c). Available: <http://www.ieee802.org/15/pub/TG3c.html>.

[6] G.-K. Chang, A. Chowdhury, Z. Jia, H.-C. Chien, M.-F. Huang, J. Yu, G. Ellinas, Key technologies of WDM-PON for future converged optical broadband access networks, J. Opt. Commun. Network 1 (4) (2009) C35–C50.

[7] A. Koonen, L. Garcia, Radio-over-MMF techniques—part II: Microwave to millimeter-wave systems, J. Lightwave Technol. 26 (15) (2008) 2396–2408.

[8] A. Chowdhury, H.-C. Chien, Y.-T. Hsueh, G.-K. Chang, Advanced system technologies and field demonstration for in-building optical-wireless network with integrated broadband services, J. Lightwave Technol. 27 (22) (2009) 1920–1927.

[9] Z. Jia, J. Yu, G. Ellinas, G.-K. Chang, Key enabling technologies for optical-wireless networks: optical millimeter-wave generation, wavelength reuse, and architecture, J. Lightwave Technol. 25 (2007) 3452–3471.

[10] U. Gliese, S. Nørskov, T.N. Nielsen, Chromatic dispersion in Fiber-Optic Microwave and millimeter-wave links, IEEE Trans. Microwave Theory Tech. 44 (10) (1996) 1716–1724.

[11] D. Wake, C.R. Lima, P.A. Davies, Transmission of 60-GHz signals over 100 km of optical fiber using a dual-mode semiconductor laser source, IEEE Photon. Technol. Lett. 8 (4) (1996) 578–580.

[12] J. Yu, Z. Dong, W. Jian, M.-F. Huang, Z. Jia, X. Xin, G.-K. Chang T. Wang, All-optical up-conversion 10-Gb/s signal in 60-GHz ROF system using 2-m Bismuth Oxide-based fiber, OFC/NFOEC 2010, OTu06.

[13] I. González Insua, D. Plettemeier, C.G. Schäffer, Simple remote heterodyne radio-over-fiber system for gigabit per second wireless access, J. Lightwave Technol. 28 (16) (2010) 2289–2295.

[14] J. Liu et al., Efficient optical millimeter-wave generation using a frequency-tripling Fabry-Pérot laser with sideband injection and synchronization, IEEE Photon. Technol. Lett. 23 (18) (2011) 1325–1328.

[15] L.A. Johansson, A.J. Seeds, Millimeter-wave modulated optical signal generation with high spectral purity and wide-locking using a fiber-integrated optical injection phase-lock loop, IEEE Photon. Technol. Lett. 12 (6) (2000) 690–692.

[16] Z. Jia, J. Yu, Y.-T. Hsueh, A. Chowdhury, H.-C. Chien, J.A. Buck, G.-K. Chang, Multiband signal generation and dispersion-tolerant transmission based on photonic frequency tripling technology for 60-GHz radio-over fiber systems, IEEE Photon. Technol. Lett. 20 (17) (2008) 1470–1472.

[17] C. Qingjiang, F. Hongyan, S. Yikai, Simultaneous generation and transmission of downstream multiband signals and upstream data in a bidirectional radio-over-fiber system, IEEE Photon. Technol. Lett. 20 (3) (2008) 181–183.

[18] K. Ikeda, T. Kuri, K. Kitayama, Simultaneous three-band modulation and fiber-optic transmission of 2.5-Gb/s baseband, microwave-, and 60-GHz-band signals on a single wavelength, J. Lightwave Technol. 21 (12) (2003) 3194–3202.

[19] J.J.V. Olmos, T. Kuri, K. Kitayama, Reconfigurable radio-over-fiber networks multiple-access functionality directly over the optical layer, J. Lightwave Technol. 58 (11) (2010) 3001–3010.

[20] Y.-T. Hsueh, Z. Jia, H.-C. Chien, A. Chowdhury, J. Yu, G.-K. Chang, A novel bidirectional 60-GHz radio-over-fiber scheme with multiband wignal generation using a single intensity modulator, IEEE Photon. Technol. Lett. 21 (18) (2009) 1338–1340.

[21] R. Llorente, S. Walker, I.T. Monroy, Triple-play and 60-GHz radio-over-fiber techniques for next-generation optical access networks, NOC 2011, IT-5.

[22] M. Morant, T. Quinlan, S. Walker, R. Llorente, Real world FTTH optical-to-radio interface performance for bi-directional multi-format OFDM wireless signal transmission, OFC/NFOEC 2011, NTuB6.

[23] M. Sauer, A. Kobyakov, J. George, Radio over fiber for picocellular network architectures, J. Lightwave Technol. 25 (11) (2007) 3301–3320.

[24] Y.-T. Hsueh, Z. Jia, H.-C. Chien, A. Chowdhury, J. Yu, G.-K. Chang, Generation and transport of independent 2.4 GHz (Wi-Fi), 5.8 GHz (WiMAX), and 60-GHz optical millimeter-wave signals on a single wavelength for converged wireless over fiber access networks, OFC/NFOEC 2009, OTuJ1.

[25] Standard ECMA-387, High rate 60 GHz PHY, MAC and HDMI PAL, December 2008—under ballot in JTC 1 as ISO/IEC DIS 13156.

[26] A. Chowdhury, H.-C. Chien, S.-H. Fan, J. Yu, G.-K. Chang, Multi-band transport technologies for in-building host-neutral wireless over fiber access systems, J. Lightwave Technol. 28 (16) (2010) 2406–2415.

[27] Y.-T. Hsueh, Z. Jia, H.-C. Chien, A. Chowdhury, J. Yu, G.-K. Chang, Multiband 60-GHz wireless over fiber access system with high dispersion tolerance using frequency tripling technique, J. Lightwave Technol. 29 (8) (2011) 1105–1111.

[28] C. Cox, Analog Optical Links: Theory and Practice, Cambridge University Press, 2004.

PONs: State of the Art and Standardized

22

Frank Effenberger

VP of Access R&D Futurewei Technologies, Inc. 400 Crossing Blvd, Bridgewater, NJ, USA

22.1 INTRODUCTION TO PON

Ever since low-loss telecom grade fibers were available in the mid-1970s, researchers began to look for ways to apply them to the access network, and particularly fiber to the home (FTTH). At the beginning, the techno-economic situation was very challenging, as the fiber was much more expensive than copper, the optoelectronics that would light up the fiber were expensive, and the labor to install the cables (and the associated civil works) was costly. This situation played a big part in guiding the mainstream of research in fiber access.

The simplest way to apply fiber to the access is to simply replace the copper wires with fiber strands, resulting in a point-to-point fiber network (Figure 22.1a). This has the advantages of simplicity, very high ultimate capacity, and service and provider flexibility. Since this network provides a raw fiber path to each home, the capacity is limited only by the capability of the optical equipment at the endpoints. Also, since each home subscriber has his own link, this could be used for any of a range of communications protocols, and it could be routed to any of a set of service providers. However, this network tends to be the maximum of cost: it uses the most fiber, and each home will require two transceivers (one at the home and one at the local exchange). Also, the traffic capacity of such a network (dedicated bandwidth per endpoint) does not fit well with the traffic profile of typical access networks (which is bursty and requires statistical gain). For these reasons, a better architecture was desired.

The next type considered was to implement electronic concentration at a remote terminal near the subscribers (Figure 22.1b). This scheme solves the large fiber count problem in the longer feeder portions of the network. For example, a typical distribution area serves 300–500 homes and could be served with only a handful of fibers. However, it does nothing to reduce the number of optoelectronics (still 2 per home), and what is worse, it creates a requirement for electrical power at the remote node. Local powering at these sites is costly to install, and presents operational issues to maintain back-up batteries, and to provide power in the case of sustained electrical outage. From a strategic view, most network operators would like to reduce the number of active sites in their network, not increase them. So, with a few exceptions, this architecture is not widely used.

Optical Fiber Telecommunications VIB. http://dx.doi.org/10.1016/B978-0-12-396960-6.00022-5

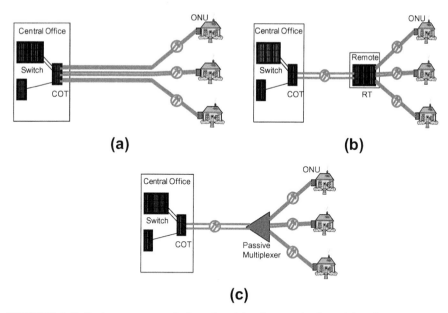

FIGURE 22.1 Optical access networks based on (a) point-to-point fiber, (b) active remote terminals, and (c) passive optical networks.

In the 1980s, a new form of optical access topology was invented, based on passive optical networking (PON) [1]. The key idea was that the shared feeder fiber would be connected to the distribution fiber using a purely optical component (e.g. a splitter or wavelength multiplexer). This achieves the fiber savings without requiring electrical power at the remote terminal. What's better, the optical components tend to be compact, rugged, and long lived. In many cases, it also reduces the number of optical transceivers by 50% (roughly 1 per home, rather than 2). This network also pushes the traffic multiplexing close to the users, so it minimizes stranded capacity. Thus, for these very important reasons, the PON architecture has remained the preeminent optical access design ever since the 1980s.

There is some basic terminology shown in Figure 22.1c. The equipment on the network end of the network is the optical line terminal (OLT). The equipment near the user is the optical network unit (ONU). Note, some prefer to describe an ONU that serves only one user as an optical network termination (ONT); however, from a technical view, ONT and ONU are equivalent. From the network to the user is termed the "downstream direction," while the reverse is the "upstream direction." The optical network that connects the OLT and the ONU is the optical distribution network (ODN).

Another important fact to consider is that once a topology is implemented, it tends to stay in the network. Since about 2000, PONs have begun mass deployment, with nearly 100 million lines being deployed by 2012. Such an infrastructure represents a huge investment, and one that will not be soon replaced. Therefore, even if the initial motivations to concentrate on PONs subside, the establishment of the network that supports PON will retain its relevance indefinitely. PON is here to stay.

22.2 TDM PONs: BASIC DESIGN AND ISSUES

The simplest form of PON is that based on optical splitters. At an optical level, the splitter provides directional optical coupling between one or more common fibers and multiple distribution fibers. Signals traveling downstream are split to all the users. Signals traveling upstream are combined. Due to the conservation of power and of luminance, the splitter exhibits a loss proportional to the log of the split ratio in both directions. This loss typically becomes the major component of the loss budget of the PON. While all the descriptions here assume a single stage of splitting, in fact this is equivalent to any multi-stage splitting arrangement.

Ideally, the downstream and upstream channels should be independent; however, optical reflections will cause some cross coupling of the two directions. These reflections can jam the transmission, and therefore require a means of duplexing the two directions on a single fiber. There are two methods used to eliminate this problem. The first is using time division duplexing (TDD), wherein first the OLT transmits, then the channel is allowed to clear (due to the round-trip time of flight of the signals), and then the ONUs are allowed to transmit (Figure 22.2a). The second method is wavelength division duplexing (WDD) that uses a different color of light for each direction. Any reflections are optically rejected by the duplexing filter, and thus the protocol need not be concerned with them (Figure 22.2b). This method is far simpler than TDD, and it avoids the channel clearing guard time. Except for one very early PON system, all PONs use the WDD method.

The simplest media access control (MAC) protocol is time division multiplex. In the downstream, each ONU receives a copy of the OLT's output. The transmission protocol on the PON provides some form of tag on each data frame to enable the

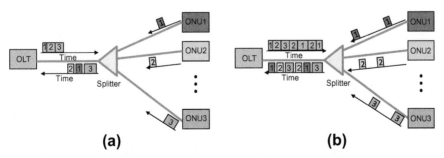

(a)   **(b)**

FIGURE 22.2 TDM-PON duplex/MAC methods: (a) TDD+TDMA, and (b) WDM+TDMA.

ONU to select its own data and discard the rest. However, in the upstream, all the ONU's transmissions are combined passively, so if two ONUs transmit at the same time, their transmissions will be jammed. The OLT must coordinate the transmissions of the ONUs so that they arrive at the splitter at different times. Because the ONUs are at different distances from the splitter, this requires some means to equalize all the ONUs to a common time frame.

The PON upstream is rather unique from previous optical transmission systems in that there are multiple transmitters facing a single receiver. Even when the MAC protocol is working well to avoid collisions, each of the transmissions will have different power levels, and have a different clock phase. This is known as the "burst mode problem" [2–4]. The OLT's receiver must be able to cope with this. There are different strategies to deal with this. The first conceived was the so-called "memory-based receiver," which built up a profile on each ONU's transmission based upon previous observations. This could be faster on each reception, because it could be prepared for each incoming burst; but it is complicated to implement. The second is the so-called "instant receiver," which deals with each burst as it arrives. This is a more challenging problem, as there is no aide of the past memory; but it is simple. In practice, all modern OLTs use the latter approach.

The burst mode receiver must first identify the decision threshold (optimally located halfway between the zero level and the one level). In the normal continuous mode receiver, this is usually accomplished using AC-coupling between the preamplifier and the limiting (thresholding) amplifier (Figure 22.3a) [5]. If this is used for burst mode, then the AC-coupling circuit requires a relatively long time to recover from one burst to the next, particularly when a quiet burst follows a loud

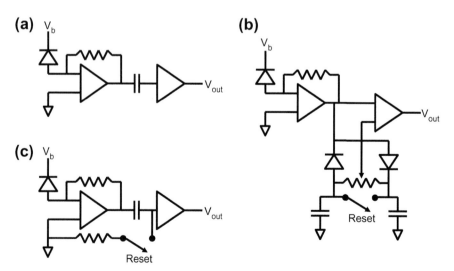

FIGURE 22.3 Burst mode level recovery, via (a) AC-coupling, (b) DC-coupling, and (c) dynamic AC-coupling.

one. The alternative to this is to maintain DC-coupling throughout the preamplifier stages, and then implement peak detectors to determine the maximum and minimum levels of the signal [6]. The decision threshold is then placed at their average value. This is a very fast technique, but is difficult to implement. An intermediate method is to use AC-coupling, but to create a variable time constant for that coupling (Figure 22.3c). In this way, the AC-coupling capacitor time constant can be made fast at the beginning of the burst, but slower during the data portion of the burst.

The second task of the burst mode receiver is the identification of the signal's bitwise clock phase. In a typical continuous mode receiver, this is accomplished via a phase-locked loop (PLL), where the difference between the local clock and the incoming signal is computed, and that is used as a control signal to adjust the local clock (Figure 22.4a). Once again, this is usually too slow to handle the burst mode problem. The brute force method to attack this problem is to oversample the incoming signal at 8–16 times the bit rate. Logic circuits can then select the best phase from that lot (Figure 22.4b). This is very fast, but difficult to implement without doing specialized logic device development. An intermediate method is to use a PLL, but to provide it with "dummy data" in between the bursts, so that the control loop does not drift away from the desired frequency range (Figure 22.4c). This allows the reuse of continuous mode integrated circuits with minimal change.

The bottom line of burst mode recovery is illustrated in Figure 22.5. On the left is shown the incoming optical signal, with a very strong burst followed by a guard time, and then a weak burst that is 20 dB smaller. On the right-hand side, the burst mode recovery circuit sets the bursts to a common amplitude and a common clock.

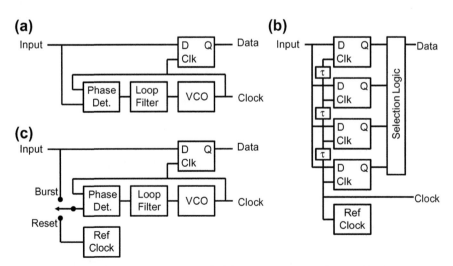

FIGURE 22.4 Burst mode clock recovery, via (a) PLL, (b) oversampling, and (c) dummy data + PPL.

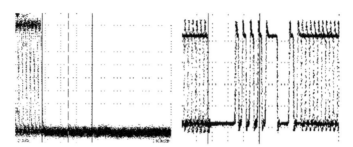

FIGURE 22.5 Burst mode data and clock recovery.

Any TDM-PON system has the issue of bandwidth allocation. Static allocation of bandwidth to each ONU is simple, but it does not allow the system to respond to the users' varying demand levels. It is far more efficient to give each ONU a bandwidth that is somewhat proportional to its traffic demand. The process of doing this is termed dynamic bandwidth allocation (DBA) and is a feature of all modern PONs. The procedure must begin with the OLT finding out what each ONU's demand is. This can be done implicitly, by observing the fraction of idle responses from each ONU. If the ONU reacts to a certain bandwidth allocation with transmissions that are half idle, then that ONU's demand can be estimated to be half of his allocation. Iterating over time, the OLT can settle on the appropriate allocation.

Alternatively, the OLT can determine the traffic demands by having the ONUs send traffic reports to it [7]. These reports state the amount of traffic waiting in the ONUs traffic queues, and coupled with the allocations sent previously, the OLT can use them to determine the rate at which traffic is arriving at the ONUs. The DBA procedure must then calculate the correct allocation of bandwidth to each service on each ONU. This is a function of the offered load and the contracted service level agreements. The service agreements basically amount to contracting a certain amount of traffic at one or more priority levels. The OLT then must allocate bandwidth fairly to each class according to a queue service discipline. Lastly, the OLT sends the allocations to the ONUs via the line protocol. All of this DBA procedure must occur in about 1 ms cycles. If done properly, DBA can result in greatly improved system performance, as shown in Figure 22.6 [8]. This shows the packet loss as a function of utilization load factor with and without DBA. One can see that without DBA, packet loss sets in at 40% utilization, while with DBA it can be raised to about double that.

The last common aspect of all TDM-PONs is protection. In fact, most access systems are not protected, and PON is not an exception, as nearly all PONs deployed today are unprotected simplex systems. But, as the PON carries more and more traffic, it is likely that some applications will require some protection. There are two basic topologies of PON protection, termed type-B and type-C, as shown in Figure 22.7a and b. Type-B protection duplicates the OLT and feeder fiber portions of the network, while leaving the splitter, distribution, and ONU portions alone. This is

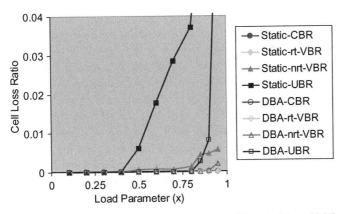

FIGURE 22.6 Packet loss as a function of load factor both with and without DBA for a variety of traffic types constant, real-time and non-real-time variable, and unspecified bit rates (CBR, rt-VBR, nrt-VBR, and UBR, respectively).

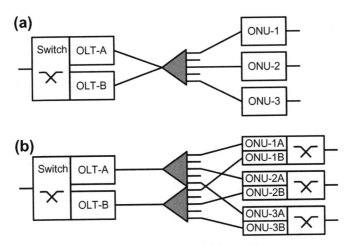

FIGURE 22.7 PON protection schemes (a) type-B and (b) type-C.

cost effective, but provides only partial protection against failures. Type-C protection duplicates the entire PON section, providing full protection, but at double the cost of the unprotected system.

22.2.1 Brief review of TDM PON standards

Given the basic technology and approach of TDM PON, there have been several practical implementations standardized. These have included PONs based on synchronous transfer mode (STM), Asynchronous transfer mode (ATM), and Frame oriented (G-PON and EPON). For the most part they fulfill the same requirements and do so in a similar way. The major difference is the networking protocol family upon which they are based.

The first PON system was described in G.982 [9], also known as pi-PON. This was developed in the early 1990s, and the optical component industry was still immature. At that time, 1550 nm laser sources were still a challenge to make, and 1310 nm lasers were considerably cheaper. This was the major factor leading to the selection of TDD, as both the OLT and ONU could use 1310 transmitters. Also, the Internet did not exist at this time as a major consumer application, and so the primary service was seen to be ISDN (a 64 kb/s channelized voice and data service). Pi-PON implemented a static TDM multiplexing scheme, where each ONU could be given a number of 64 kb/s timeslots in a fixed length frame.

The bandwidth of this system was relatively modest; the line rate was about 50 Mb/s, and the usable symmetric bandwidth was around 25 Mb/s. While this was seen as adequate for interactive services, it was obvious that video was unsupportable with such low bandwidth. Thus, a video overlay wavelength was added to this system, operating at 1550 nm. As mentioned, this was expensive to transmit, but the video transmitter would be highly shared, and so the overall system cost could be controlled. There will be more discussion about video overlays in Section 22.3.

The G.982 system was implemented by a couple of system vendors and deployed in Japan on a trial basis. However, it never reached full scale, because it was overtaken by the rapid evolution of both the optical technology and the service offerings in the market. The technical choices that were made in its design were based on assumptions that rapidly changed, and the system became obsolete. While the industry moved on to the next system, there are still some remnants of the G.982 system that live with us today. The ODN loss budget classes (A: 5–20 dB, B: 10–25 dB, C: 15–30 dB) were first laid out here.

The next system to be developed was based on ATM and described in the G.983 series of standards. This was originally referred to as ATM-PON (A-PON), but over time more and more system features were added, and the term "ATM" took on a negative marketing connotation. Hence, a new name was adopted, broadband-PON (B-PON). This system was based on wavelength duplexing, using the 1480–1580 nm window in the downstream, and 1260–1360 nm in the upstream. The range of loss budget classes was defined, as were several data rate options (155 Mb/s symmetric, 622/155 Mb/s asymmetric, and later, a 622 Mb/s symmetric was added).

The line protocol was based very heavily on formats and concepts borrowed from ATM. The protocol is illustrated in Figure 22.8. Note that in the downstream, the signal is composed entirely of ATM cells. The downstream cell delineation scheme could therefore be reused from the I.432 series of recommendations. The PON-specific parts of the downstream are located in the physical layer operation administration and maintenance (PLOAM) cells. These cells appear every 28th cell, and they have a special cell header address; it is these factors that enable the ONUs to identify them. An important feature of this scheme is that the remaining payload

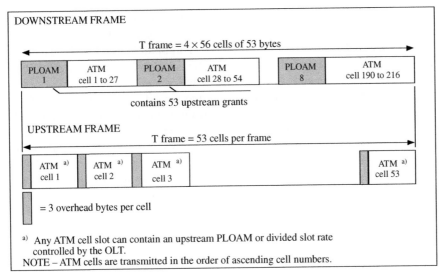

FIGURE 22.8 B-PON line protocol.

(27/28 of the line rate) is sufficient to carry a full synchronous digital hierarchy (SDH) ATM payload.

Inside the PLOAM cells are several important PON management functions. Some framing and frequency alignment features are present. A messaging channel is included, to pass asynchronous control messages between the OLT and ONU. But most important, the PLOAM cells contain the bandwidth grants that give each ONU the permission to transmit in the correct upstream timeslots. Because ATM is cell based, each upstream timeslot is a fixed size, and so each grant is for a fixed unit of time. In the case of B-PON, this was set to be 56 bytes of time. Each 53-byte upstream cell is transmitted on its own, with 3 bytes of physical layer overhead time. Even if the same ONU transmits consecutively, the 3-byte overhead is transmitted before each cell. While not perfectly efficient, it is the simplest scheme.

The important process of range equalization is done during activation. New ONUs are allowed to transmit during an unallocated ranging time. The OLT measures the response time of the ranging transmission. The OLT then calculates how much extra delay the ONU should use to equalize its response time to all the other ONUs on the PON. The OLT then sends this equalizing delay back to the ONU. The ONU then delays its future responses by the equalizing delay. This scheme results in all the ONUs having a shared timing reference frame, which then allows the bandwidth grants to be simply defined.

Over its life, the G.983 series of standards added several important features that have become common specifications for nearly all PON systems. An integrated ONU management and control interface (OMCI) was defined in G.983.2, and this has been reused in every ITU PON since. The video overlay was covered in G.983.3

(including the narrowing of the downstream wavelength to 1480–1500 nm), and this is widely used in many PONs today. The detailed mechanics of DBA were laid out in G.983.4, and as discussed above DBA is a key part of all PONs. Finally, the details of PON protection were described in G.983.5. So, while B-PON is no longer in active deployment, its technical influence can still be felt.

The next PONs to be standardized were the gigabit rate systems, G-PON and E-PON. These were developed contemporaneously in the early 2000s [10]. The gigabit-capable PON (G-PON) is specified by the ITU-T G.984 series. Since then, a few amendments were consented by the ITU-T on most of the documents in the series. The Ethernet PON is specified by the IEEE 802.3ah standard, approved in 2004. As is the style in the 802.3 working group, no significant modifications to this work have been made, but other groups have made their own extensions of the EPON technology to suit their needs. It is interesting to note that both of these systems began from a common point and had many of the same requirements. For this reason, the two systems are quite similar from the physical layer point of view. They diverge more and more the higher one goes in the stack, as those layers are more determined by the standardization group involved than by the fundamental technology requirements.

The physical network architecture for both G-PON and E-PON is shown in Figure 22.9, and supports a two-wavelength WDM scheme for bidirectional digital services. The systems use 1490 nm for downstream, and 1310 nm for upstream. While not explicitly defined, spectrum near 1555 nm is set aside for analog video service. The network supports up to 20 km reach, and the theoretical split ratio supported is up to 128. Practical deployments typically would have lower reach and split ratios, limited by the optical budget

On the issue of bit rates, the G-PON system initially specified a wide range of rates (seven combinations in all!). As network operators' requirements evolved, the preferred G-PON bit rate was selected to be 2.488 Gbps downstream, 1.244 Gbps upstream. This focus has then allowed the definition of best-practice optical parameters for G-PON, which was documented as an amendment to G.984.2. The parameters, known as Class B+, support a loss budget of 28 dB.

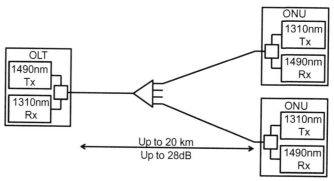

FIGURE 22.9 G-PON and EPON physical network architecture.

In contrast, the EPON system had immediately locked into the choice of 1 Gb/s symmetric, as this was the only Ethernet rate that was practical and was previously standardized for other physical layers. Two loss budgets (PX10 and PX20) were specified. The details of these budgets do not allow them to be expressed as a single loss number, but they are approximately equivalent to 20 dB and 24 dB of loss, respectively. In practice, these loss budgets have not been used, and the deployed systems implement budgets closer to the B+ levels.

Each of these PON systems defines a transmission protocol. These two systems accomplish the same functions, but they do so in decidedly different ways. The G-PON Transmission-Convergence (GTC) layer specified by G.984.3 performs the adaptation of user data onto the PMD layer, and the control of the TDMA transmissions in the upstream. Additionally, the GTC layer provides basic physical layer management. This is done by defining a light-weight framing-based protocol that provides a precise timing reference, and that carries three control elements. The first is field-based signaling that allows the very lowest layer functions of the PON to be realized in a compact form. This includes items like the bandwidth map and the DBA reports. The second is a message-based signaling to handle more advanced and less frequent functions to be done. This borrowed heavily from the B-PON legacy of the PLOAM channel. GPON also reused the activation and range equalization system from B-PON. The third is the user data transport, which uses G-PON encapsulation method (GEM). GEM was a replacement for ATM, and has as its major advantage the capacity to carry arbitrary size datagrams. GEM also provides for user data logical multiplexing (port identification), as well as frame fragmentation.

The EPON protocol was designed in a different way. The objective in this design was to reuse the 1 Gb/s optical Ethernet protocols (8b10b coding) with a minimum of modifications. Of course, all the basic functions mentioned for G-PON are still needed, so the art of the EPON standard is finding ways to support these things in an existing protocol. The essential signaling of the EPON (bandwidth allocation, reporting, and activation) were all handled using the MAC control channel, which carries multi-point control protocol (MPCP) packets.

The exact timing reference needed by the PON was implemented by inserting time stamps on each MPCP packet. The OLT stamps the time of transmission of every packet, and when the ONU receives these, it sets its local clock to match these stamps. The ONU then stamps the time of transmission of each of its MPCP packets. When these arrive back at the OLT, they will appear to be delayed by just the round-trip time of the PON. The OLT can use this information to build an efficient upstream schedule. This is done implicitly by the OLT advancing each of the bandwidth allocations by the round-trip delay of the ONU in question.

The logical multiplexing in EPON is done by adding a special logical link identifier (LLID) into the preamble of each Ethernet frame. The preamble is an 8-byte pattern that was originally designed to accommodate multi-point transmission on coaxial media. Such networks are not commonly used now, and so the preamble is substantially unused in most Ethernet systems. For EPON, those bytes were re-purposed, and a 16-bit identifier is added before each frame. The LLID is used by the

ONU to select which downstream frames belong to it, and also to mark the upstream frames for easy handling at the OLT.

Above the transmission convergence layer, these PON systems require a full set of management controls, to provide configuration, fault detection, performance monitoring, accounting, and security functions. In the case of G-PON, this is provided by the OMCI standard (described in G.984.4). In the case of EPON, this is provided via a combination of enhanced OAM and proprietary standards developed by the major users of EPON.

22.2.2 Generation 4: 10 Gbit/s PONs

The IEEE 802.3 working group took up the study of a next generation of EPON in 2006 [11]. The main objective of the new system was to address what was seen as a coming shortage of bandwidth in the current 1 Gb/s GE-PON systems. The natural solution for this requirement in the 802.3 family of standards was then 10Gb/s PON systems. Very early in the process, it was seen that 10G symmetric systems would be quite difficult, so two options were allowed: an asymmetric 10G down/1G up system (eventually labeled 10/1G Base-PRX) and a symmetric 10G system (labeled 10G Base-PR). It is interesting to note that the choice of these bandwidth numbers was not driven by requirements or technical considerations; rather, they were dictated by tradition within the 802.3 group. This has left the 10GE-PON system with a sort of large bandwidth gap between the 10/1 system and the 10/10 system.

The second major requirement that was agreed was one of backward compatibility. Since the GE-PON system has scaled to millions of deployed endpoints, any new PON system should be capable of coexisting with this large base. In addition to the basic GE-PON system, many deployments also have a video overlay, and this system was included in the coexistence framework. Coexistence in the 10GE-PON context was achieved by using a combination of WDM techniques for the downstream wavelengths, and multi-rate TDM for the upstream. The basic system is shown in Figure 22.10. Note that due to the TDMA upstream, the existing GE-PON OLTs will need to be replaced with a new hybrid OLT. While this is a disruption, it is manageably small compared to any ONU or ODN replacement.

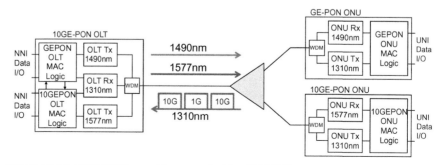

FIGURE 22.10 WDM/TDMA coexistence method for 10GE-PON.

With this architecture settled, the details of the system were specified. The most important aspect of the PMD is the loss budget, and this is largely dictated by the coexistence with the existing GE-PON systems. The existing 802.3 standards defined two loss budgets: PX10 (5–20 dB) and PX20 (10–24 dB). On top of this, the largest deployment of GE-PON in the world (at that time, at least) used a budget of 15–29 dB. To properly coexist with these systems, the new PRX and PR systems were given matching loss budgets. The loss budgets specified in the 802.3 PON standard are given in Figure 22.11. Note that for the high budgets, it is assumed that an avalanche photodiode (APD) is used at the ONU, while for the lower budgets a less costly positive-intrinsic-negative (PIN) detector can be used.

The rest of the 10G EPON system operates exactly as the 1GE-PON system does. This is to say that the media control still uses the MPCP protocol, and the rest of the system architecture is not specified in 802.3 documents. How this has manifested itself in the real world is that each major deployer of EPON technology has undertaken to define their own system architecture specification. There is an NTT specification, a KT specification, a CCSA specification, and a Cablelabs specification.

Starting in 2010, a new IEEE standards project P1904.1 embarked to address this situation. This project's goal is to document all the things that 802.3 left undone, and to describe all the system integration parameters for EPON systems. One of the big difficulties of this project is that it is starting 6 years too late, and there are already multiple large deployments of EPON with fundamentally incompatible specifications.

The way this was handled was to define a set of major technical features. For each of these major features, one or more profiles are defined that describe all the technical information needed to implement it in a certain style. Some features have only one profile, but most have more than one. Then, packages of profiles were defined, where each package selects which profile is to be used for that system. The three packages (A, B, and C) roughly correspond to America, Japan, and China, respectively. While this is not the ideal situation for a standard, it is the best that can be done given the situation on the ground.

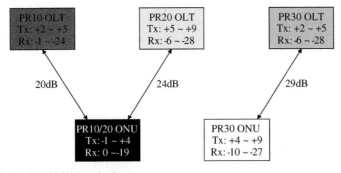

FIGURE 22.11 10G-EPON loss budgets.

Looking forward, additional projects are being founded to develop conformance test plans and other aspects to verify actual equipment against the newly written specifications. Since the specifications were actually written around as-built equipment, the results would appear to be self-fulfilling.

The full services access network (FSAN) and ITU-Q2/15 communities began the consideration of next generation PON (NG-PON) in 2007 [12–14]. Rather than approach the problem with a preconceived notion of the solution, the group took a complete stock of all the possibilities, and worked step-by-step to find the best solution. The first thing that became obvious was that there were two great classes of new systems: those that required the ODN to be changed, and those that did not. These two classes were of such a difference that the NG-PON project was split into two phases: NG-PON1 would consider the ODN-compatible systems, and NG-PON2 would consider systems that were either non-ODN-compatible or required technologies that were not mature. While not explicitly restricted, the general feeling was that the NG-PON1 system would see the field sooner than the NG-PON2 systems.

It was determined that 10 Gb/s would be the next reasonable step in bandwidth from the current 2.5 Gb/s of G-PON, given the very large commercial development of 10G optics, and due to the relative cost insensitivity of the downstream link. The upstream is far more cost sensitive, however, and the requirement for bandwidth could not be set so easily. The accumulated experience of the operators told them that the upstream was in fact not as heavily used as the downstream in residential applications, by a significant margin. For these reasons, the upstream was set to be 2.5 Gb/s, to leverage the significant base of optics at that well-known SDH rate. This 10G down, 2.5G up system was termed XG-PON, and is described in the ITU G.987 series.

Coexistence was again a major driving force for the architecture of XG-PON. In this respect G-PON had an advantage over GE-PON in that a recent addition to the G.984 series defined three new features that made G-PON equipment forward compatible. The first was the definition of a blocking filter that would protect the G-PON ONU's receiver from the new XG-PON downstream wavelengths. The second was the definition of a WDM1 filter that combines/separates the G-PON signals from the XG-PON signals at the CO end. The third was the narrowing of the G-PON upstream wavelength band. With these in place, the way was clear to implement a full WDM/WDMA coexistence method, as shown in Figure 22.12. This allows an operator to leave both the G-PON ONUs and also the OLT's in place when upgrading to XG-PON.

A wide range of loss budgets were defined for XG-PON, ranging from 29 to 35 dB, as shown in Figure 22.13. All of these budgets can be achieved with a single ONU type, because all of the budget effecting changes are made at the OLT side. Since it is far easier to stock a range of OLT circuit packs rather than a range of ONUs, this is an optimal scheme for budget flexibility. The lower capability OLTs assume relatively conventional and conservative optics are used. The mid-range of capability could be achieved by doing component selection. The highest capability OLTs would

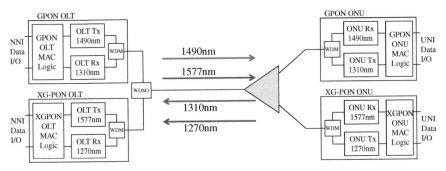

FIGURE 22.12 WDM/WDMA coexistence method for XG-PON.

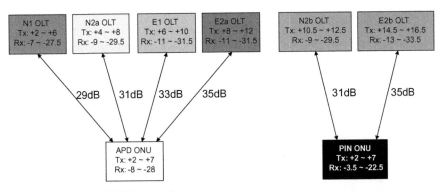

FIGURE 22.13 The XG-PON loss budgets.

require more advanced (and expensive) technology such as optical preamplifiers and boosters. In addition to the multiple budgets, there are also two options for the ONU receiver design to achieve the 31 and 35 dB links. One is based on APD components, and the other uses PIN components. This provides the market with the flexibility to fully explore the cost optimization problem, and arrive at the best answer. When the XG-PON system begins to achieve significant volumes, it is expected to revisit these budgets and down-select to those that have proven their economic viability.

Just as with 10GE-PON, XG-PON finds the solutions to all its other design issues by reusing those from G-PON. The transmission convergence (TC) layer is highly similar to G-PON, with some streamlining of the protocols, and expansion of some of the fields to expand the address range (maximum split ratio) among others. Unlike 10GE-PON, XG-PON could draw heavily on the accumulated system integration progress of G-PON, and fully inherited the OMCI, the BBF-156/167 specification, and the G-PON interoperability and conformance work that is ongoing. The OMCI for both G-PON and XG-PON are now defined by the same single document (G.988), which virtually assures that the same software implementation that over 20 companies have implemented for G-PON will work perfectly for XG-PON. The Broadband Forum (BBF) documents provide the guidance on how to apply the TR-101 model of data internetworking to (X) G-PONs. This gives designers the concrete instructions they need to build OLT and ONU equipment that will connect seamlessly with backbone networks and CPE, respectively.

The interoperability and conformance work is what puts all of the standards and specifications into action, because it involves testing real implementations. Interoperability involves testing OLTs and ONUs together, in a double-blind fashion, where the ONU and OLT do not know which vendor is on the other side. This mimics the real-world case of full plug-and-play interoperability. Conformance involves testing and ONU against a test equipment that issues a known good test protocol sequence. Proper response and behavior to this sequence confirm that the ONU has conformed to the standards. These efforts are already complete for G-PON, and as mentioned above, they port directly to XG-PON. In that sense, XG-PON is fully interoperable on day 1.

22.2.3 **40G serial**

The next step beyond 10G would seem to be 40 Gb/s transmission, presuming that the PON systems would follow in the footsteps of the optical transmission systems. Figure 22.14 shows one proposed system that explores this possibility [15]. The basic factor is that 40 Gb/s serial NRZ transmission is quite difficult to manage. The first problem is that of dispersion. Assuming ideal Fourier-limited transmission, the chromatic dispersion impact grows worse as the link speed squared. 10 Gb/s is already on the boundary of usability, and so 40 Gb/s requires special measures. Using the traditional C- or L-band wavelengths for 40G would result in very short reach. More reasonable reach can be achieved by operating in the O-band.

In addition, 40 Gb/s optoelectronics are very expensive and difficult to manufacture. The bandwidth demands can be reduced by using a bandwidth limited line code, such as duobinary. This permits the use of optics that have a 20 GHz bandwidth, rather than the 30 Gb/s bandwidth required by 40 Gb/s NRZ signals. Taken

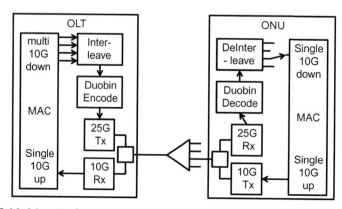

FIGURE 22.14 A hypothetical 40 Gb/s TDM PON.

together, the use of duobinary code at 1350 nm should produce workable penalties at 40 km reach.

Another concern is the high processing power required at the ONU. In a conventional TDM PON system, the ONU receives and processes every downstream packet, and then determines based on its Port-ID or LLID whether it should be forwarded to the downstream customers. This becomes burdensome at 40 Gb/s, and even worse, it is unlikely that 40 Gb/s would ever be destined for any single ONU. The highest speed user interface that is likely to be seen is 10 Gb/s. Thus, requiring the ONU data processing to be rated for 40 Gb/s is a waste of resource. One approach to avoid this is to implement an interleaving approach, wherein the downstream is composed of several bitstreams. Each bitstream would be a self-contained downstream protocol, and any particular ONU would only need to process one such bitstream to obtain its data. The number, size, and membership of each bitstream could be static or dynamic, depending on the design objective. This kind of design begins the departure from pure packet-based TDM systems, and toward more channelized approaches.

The commercial suitability of 40 Gb/s transmission remains in doubt. At present, the component technology seems too costly, and the design difficulties require too many trade-offs against other requirements. Furthermore, there seem to be other architectures (WDM-PON, FDM-PON, or hybrids) that appear easier to implement. So, 40 Gb/s PON systems may remain a theoretical curiosity, at least for now.

22.3 VIDEO OVERLAY

From the first TDM-PON system, there was a recognition that broadcast video was an important service that was missing from the offering. At that time, video compression was not very effective, and the data rates on the digital PON were relatively modest. Also, the entire video service infrastructure was oriented toward simple broadcast delivery, and so any efforts to provide video over the digital path were difficult. To meet this need, a video overlay scheme was devised that used a third wavelength channel at around 1550 nm, as shown in Figure 22.15 [16]. The system consists of a signal head end and transmitter, some number of stages of optical amplification, and then insertion onto the PON. At the ONU, an optical filter isolates the 1550 nm wavelength, and feeds it into a linear receiver and amplifier for distribution inside the user's residence. These signals can then be used by any of the millions of existing televisions or set-top boxes.

This overlay channel uses analog transmission of subcarrier multiplexed video. There are a few modulation systems that have been carried in this way, including terrestrial analog (AM-VSB) video, satellite FM format, and digital cable QAM channels. All of these signal formats are carried over the overlay channel unchanged (except for up-/downconversion to suit the spectrum plan of the operator). Such systems are specified in ITU-T J.186.

One of the issues with analog transmission is that it is relatively sensitive to noise and other impairments. The primary issue is that of noise, which arises in the

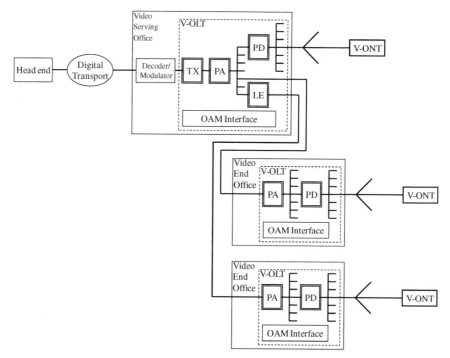

FIGURE 22.15 A video overlay system, showing the typical arrangement of network elements.

transmitter, receiver, and the intrinsic shot noise of the carrier light. This noise sets the required signal level to achieve acceptable signal-to-noise ratio, and in common practice this tends to be rather high. One method that was explored to reduce this was the use of FM modulation of the entire video spectrum. This results in a signal that has perhaps five times bandwidth enhancement, which in turn can improve sensitivity by a factor of 10 or more. However, it does require very large (5 GHz) bandwidth receivers and FM detector circuits. Such systems are described in J.185.

Another important impairment is the Raman crosstalk between the digital channels and the analog overlay [17]. There are two aspects to this interaction. The first is the continuous wave transfer of power from the shorter wavelength to the longer wavelength. For example, The G-PON 1490 nm downstream loses power to the 1550 nm video signal, and the 1550 nm video loses power to the XG-PON 1577 nm downstream. The total level of this CW effect is relatively small (no more than 1 dB effective loss for G-PON), and it has little practical effect. The second aspect is the transfer of the modulation from one wave to another. This has a more pronounced effect on the video signals, as they have higher signal-to-noise ratio requirements. These effects have been well analyzed, and while they are definite and observable, they are at a level that is manageable using various remediation techniques.

22.4 WDM PONs: COMMON ELEMENTS

Following TDM-PONs, the next alternative that was investigated was wavelength division multiplexing [18,19]. There are many PON architectures that utilize multiple wavelengths, but here we define WDM-PON as that which uses WDM as the primary method of multiplexing multiple ONUs onto a shared ODN. This implies either one or two wavelengths per endpoint, and this is the key differentiator from other PONs that use WDM as a subfeature (such as TDM-PON, which uses WDM for duplexing, or hybrid PONs, which has a pair of wavelengths per sub-PON).

With that distinction in mind, there are two major types of WDM-PONs: splitter-based and grating-based (Figure 22.16a and b). The splitter-based WDM-PON uses a wavelength blind ODN, and a copy of every wavelength is routed to every ONU. The system relies on the ONUs to tune to the correct downstream and upstream wavelengths using some kind of electronic control. The grating-based WDM-PON uses a wavelength selective passive component to route one or more wavelength paths to each ONU [20]. In this case, the ONU can blindly receive the downstream channel; however, it must still have the ability to transmit at the correct upstream wavelength to gain connection to the OLT. The grating-based system has many advantages, in that the ODN has a lower intrinsic loss, and the signals are intrinsically protected from eavesdropping. However, it is less flexible, as the wavelength assignments are essentially hard wired.

In many WDM-PONs, each ONU uses two wavelengths for diplexing purposes, to avoid reflections and other impairments. Since the ODN is wavelength selective, it must route not just one, but two wavelengths to each ONU. Fortunately, most grating devices can be arranged to support more than one order of diffraction. In this way, one order can be used to route the downstream wavelengths, while the next order can be used to route the upstream wavelengths out of the same set of ports.

The most critical problem of any WDM-PON is to produce the so-called "colorless ONU." Since a WDM-PON has many colors (typically 32–64), it is impractical to produce so many different ONUs. A practical system would have only a single

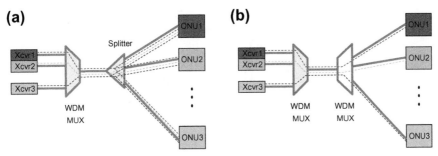

FIGURE 22.16 WDM-PONs based on (a) splitters or (b) gratings. Note the difference of connectivity.

ONU type that would automatically adapt to whatever color is needed where it is installed. There are a variety of methods of implementing this, which drive the five basic types of design below. These descriptions are kept at a very high level, mainly to motivate the standardization efforts that we see for each of these. However, it is impossible to review the 30 years of research that has gone into the WDM-PON field. For more technical detail, please consult the companion chapter of this book.

22.4.1 Injection locked

In the recent phase of research, the first popular method of colorless ONU technology was that of injection locking [21,22]. This system requires the use of a grating-based ODN (Figure 22.17). For the downstream, the OLT transmits an array of modulated wavelengths, which are routed to the appropriate ONUs. For the upstream, first the OLT transmits a strong broadband unpolarized optical seed signal downstream, and the ODN slices this spectrum into many colors—one per drop fiber. This seed light enters the ONU and is conducted to the injected semiconductor light source. This source is induced to operate at a color substantially similar to that of the injection. By modulating the electrical drive of the semiconductor, it can be modulated, and thereby used to transmit the upstream signal.

The injected laser source at the ONU can have a range of properties. The simplest is a reflective semiconductor optical amplifier (RSOA). This operates as a simple modulated amplifier, and returns the same spectrum that is delivered to it. There are two major issues with this style of device. First, the RSOA center wavelength depends on temperature (roughly 0.5 nm per degree variation). With a temperature range of 120°C, the gain shifts by 60 nm. If the system uses 30 nm of spectrum for its upstream transmission, then the RSOA needs 90 nm of gain bandwidth, which is a challenge. Second, the spectrum sliced by the AWG is quite broad (a fair fraction of the channel spacing). This line width results in a very large dispersion penalty for the distances and rates that are interesting for future access systems.

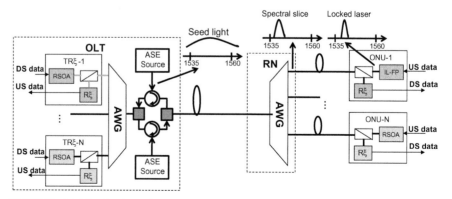

FIGURE 22.17 Injection locked WDM-PON.

An alternative to the RSOA is an injection locked Fabry-Perot (F-P) laser. This is effectively an RSOA with a non-zero front-facet reflectivity. The advantages of this device are that it is easier to fabricate than an RSOA, and it operates on a single cavity mode, so its line width is far smaller than the injected spectrum. This can reduce the dispersion difficulties to a manageable level. That said, the dynamics of injection locking are not so well controlled. If the injection light is insufficiently strong, or at a wavelength far from the gain peak of the laser, then the injection will fail, and the laser will be essentially free-running.

Taken together, all of these optical effects tend to limit the loss budget and distance that the system can achieve. The current state-of-the-art systems operate at 1.25 Gb/s over distances of 20 km of G.652 fiber. Some newer systems operate at 2.5 Gb/s, but at a significant penalty of link margin. Recently, such systems have been standardized in metro transport applications as ITU recommendation G.698.3.

22.4.2 Wavelength reuse

One of the drawbacks of the injection system is that the OLT must provide the broadband seed signal in addition to the usual downstream data signals. The seed source is a significant cost and power dissipation for the OLT. One possible scheme to reduce it is to reuse the downstream transmission channels as the injection source [23,24]. Such a system is shown in Figure 22.18, and by comparing with the previous system, one can see the simplification. Essentially, one gets two functions for the price of one. What's even better, the injection is now a laser-like narrow band optical signal, which is more effective at locking an F-P laser, and which has far better dispersion characteristics.

However, this system has its own issues. First, most conventional transmitters are polarized, and so the injection signal will have a high degree of polarization. The ODN is not polarization-maintaining, so the state of polarization at the ONU

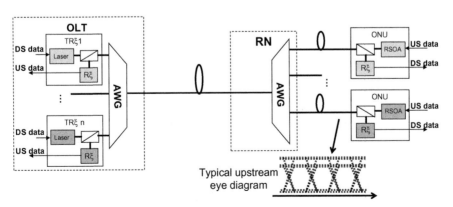

FIGURE 22.18 A wavelength reuse WDM-PON.

will be a randomly fluctuating one. At some times, the polarization may not be at the desirable direction, and it will be ineffective in locking. The alternative is to use a polarization-insensitive component (such as an RSOA); however, making such a device that operates over a wide temperature range is very challenging.

Second, the downstream wavelength is modulated with the user's data. In the simplest case where the ONU's modulator is linear, then this pattern will leak into the upstream remodulated signal. There are a variety of solutions that can attempt to reduce this. One is to use an SOA modulator that is operated in saturation. This device will compress the difference between the downstream zero and one levels. Coupled with using a downstream signal that does not have a very high extinction ratio, this can produce usable signals. Another method is to use the detected downstream signal to feed forward to the modulator driver. This gives the zero level inputs more gain, equalizing them. A final method would be to use a different modulation scheme in the downstream than the upstream, whereby the two methods have substantially different frequency or amplitude characteristics. In this method, the signal leakage has less of an impact on the signal reception.

22.4.3 Self-seeded

The ultimate solution to eliminating the seed source in such systems is to use the ONU transmitter as its own seed source [25,26]. This type of system is sketched in Figure 22.19. The wavelength plan and data transmission are similar to the externally seeded system. The key addition to this scheme is an intentional reflector in the PON, just upstream of the WDM multiplexer. The reflector most popular to use is a Faraday rotating mirror, for reasons that will be explained below. The concept of operation of this system is that the ONU's RSOA device will emit amplified spontaneous emission (ASE) into the PON. The AWG will spectrally slice this emission, and only a single channel's color will be returned to each ONU. Within a few round-trip times, the cavity that is formed by the ONU gain element, drop fiber, and reflective AWG will reach steady state, and sustained transmissions at the appropriate color are produced.

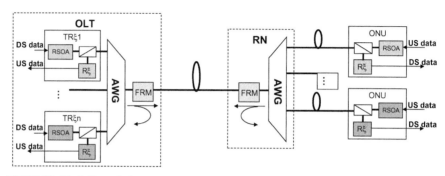

FIGURE 22.19 Self-seeded WDM-PON.

The self-injection system's biggest benefit is that the seeding power only travels from the ONU to the AWG and back again. Since the drop is much shorter than the feeder fiber, the total loss that the seed power must withstand is lower than the previous schemes. Also, the seed source is distributed to the ONUs, which makes each ONU self-sufficient on an active component basis.

An interesting aspect has to do with the laser action of the very long, very low resonance cavity in the system. In typical lasers, the cavity round-trip time is very short compared to the attempted modulation frequency, and the mode photon population is effectively following the modulation for the most part. This is not the case in the long-cavity case. In the self-seeded laser, the cavity is being operated in a quite different regime, where the modulation is much faster than the cavity round-trip time. Also, the round-trip gain of the laser is quite high (approximately 15 dB). In this case, the laser is not operating in a steady state, but rather in a mode dynamic situation. This is what enables the system to be modulated at rates much higher than the natural cavity relaxation time.

Self-injection could have polarization issues, as the ONU device will have residual polarization dependence. As mentioned for the wavelength reuse system, the arbitrary birefringence of the optical fiber could result in undesirable feedback states. A particularly useful way to overcome this is to use a Faraday rotator placed before the reflective element. This element rotates the linear component of the light 90°. This has the effect that any polarization dependence is averaged out, and the resulting laser emission is largely unpolarized and it is relatively stable.

Of course, the feedback returning from the Faraday rotator mirror is modulated with the upstream signals that were sent a short time previously (delayed by the time of flight in the fiber). This is a similar situation as with the wavelength reuse system. Most of the same schemes of pattern erasure apply to this system as well, with the exception of using different modulation schemes, of course. Lastly, as will all injected schemes, this system has the problem with temperature dependence of the ONU gain device.

22.4.4 Tunable

The previous three systems were based on physical-layer methods of controlling the ONUs wavelength. Each of those had advantages and disadvantages. The tunable transmitter ONU WDM-PON aims to avoid all of these by using a more straightforward active tuning approach [27]. A typical wavelength routed tunable WDM-PON is shown in Figure 22.20. Each ONU is equipped with a tunable transmitter, and this device is under electronic control to operate at the correct wavelength. The tunable approach avoids all the optical issues, but it brings its own set of challenges. The first and foremost is the cost and availability of tunable transmitters. Even in the optical transport field, tunable transmitters are not used so widely due to their increased cost with respect to fixed transmitters. Considering that the access field is much more cost sensitive, this is particularly worrisome. Thus, most systems feature a new design of a tunable transmitter device that can be manufactured at lower cost.

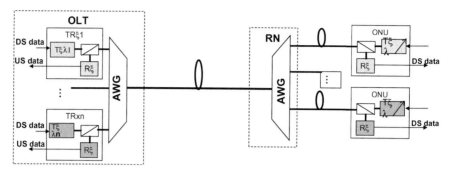

FIGURE 22.20 Tunable transmitter ONU WDM-PON.

One of the common themes of lower cost tunables is the elimination of tempera-ture control. As mentioned previously, all semiconductor devices have a temperature dependence of their wavelength, and this must be compensated. When cost is no object, the simplest method is to thermostat the laser; however, the thermo-electric cooler (TEC) typically used to do this control is undesirable. It has direct cost, and it also requires the use of types of packages that are more expensive. TECs also require a fair amount of power, usually drawing about three times as much power as the dis-sipation of the device being controlled. If the tunable device can be made to tolerate the natural temperature swings on its own, it would be better. The challenge is that the tuning control (whatever that is) must be made to cover a much wider effective range of wavelengths. As analyzed in Section 22.4.1, the gain spectrum must be a very broad 90 nm. On top of that, most grating stabilized devices drift at 0.1 nm/C. With a full temperature range of 120°C, this is 12 nm of wavelength shift on top of the 30 nm or so of spectrum coverage, requiring a tunability of about 45 nm. This becomes more and more difficult.

As to the actual mechanism of tunability, there are several concepts being explored currently. On one end of the spectrum are the "electronic" methods, where a special semiconductor laser is controlled via injection of currents into its multiple sections. These currents can set the device's gain, Bragg grating constant, and cav-ity phase. Typical devices also contain a separate output amplifier for power boost, and a separate modulator device for high-quality chirp-free modulation. In principle, such a device should not cost much more to fabricate than a conventional DFB laser. But, device calibration requires a lot of time, since there is a wide operating space of injection current combinations to test.

At the other end of the design spectrum are the "mechanical" methods, where an external cavity laser is used. The external cavity laser can be constructed of athermal materials and designed such that the wavelength forcing is very strong. Thus, these transmitters can be relatively robust against optical disturbance. How-ever, the opto-mechanical nature of these lasers can be a reliability issue, and the long-term cost of manufacture of such lasers may not be as low as the electronic methods.

22.4.5 **Coherent**

Continuing in the path of more advanced optics, if the ONU can have a tunable laser, then the next step would be to use this tunable laser as a local oscillator for a coherent receiver [28]. A characteristic system is shown in Figure 22.21. The downstream data is modulated using QPSK or similar code. The most common channel rate imagined is 1 Gb/s, and so the actual width of each channel is around 1 GHz. About 10 such carriers can be modulated onto a single optical carrier at the OLT, using a channel spacing of perhaps 3.125 GHz. This requires a Mach-Zehnder modulator bandwidth of about 25 GHz, and a sampling rate of 50 Gsamp/s on the OLT side. The OLT receiver requires similar capabilities, and requires polarization diversity since the fiber channel is not polarization maintaining.

On the ONU side, a tunable local oscillator is positioned adjacent to one of the downstream transmitted channels. This is mixed with the incoming signal, and routed to a polarization diverse coherent receiver. The bandwidth of this signal chain can be much less than that of the OLT, as it is only handing a single ~1 GHz carrier. The sampling rate of the ONU is about 5 Gsamp/s, and the raw samples are fed into a DSP for post-processing. As the local oscillator may not have a stable frequency control, the DSP will need to track the frequency mismatch, and process the received signals as needed. In the upstream the DSP will need to similarly calculate the appropriately offset modulation. A portion of the local oscillator is used to drive a Mach-Zehnder modulator, and this carries the upstream data to the OLT.

The big advantages of this approach are the very high optical capabilities: loss budgets of about 45 dB are possible, and the spectrum efficiency is quite good (bidirectional 1.25 Gb/s in 3.125 GHz of spectrum). Due to the large loss budget and intrinsic tunable receiver, the coherent WDM-PON can be operated over a splitter-based ODN. This makes the system backward compatible with previous generations of TDM-PONs. Taken together, these characteristics make the coherent WDM-PON very attractive, and indeed it is a leading contender for future standardization.

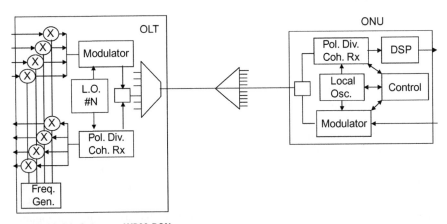

FIGURE 22.21 Coherent WDM-PON.

The big challenge for this sort of system is its complexity. One look at the diagram in Figure 22.21 reveals that there are a lot of components and subsystems. Over time, more and more of these may be integrated into optical circuits. Indeed, there is a rising tide of silicon photonics interest that aims to turn all the optical components shown above into a single small chip. However, such devices are still not available at the quality and robustness level required for commercial application. Moreover, silicon still faces the fundamental issue of a lack of a native optical emitter. Various solutions to this problem are being researched (and have been for many years). When all of this basic device research will finally bear fruit is unknown. Until these technologies have been proven, it is likely that the coherent PON will remain a research system and not move toward formal standardization.

22.5 FDM-PONs: MOTIVATION

For most of the preceding systems, the optical transmission of data has used simple NRZ coding. In fact, it can be shown that NRZ is the most power efficient code possible if the channel bandwidth is "free." In the early stages of optical transmission (until quite recently!), this has been the case. But now, the transmission speeds are getting to the point where the channel does have significant bandwidth limitations. Opto-electronic devices become increasingly difficult to fabricate beyond 10 GHz bandwidth, and 40 GHz is the current practical limit. More fundamentally, the fiber channel's dispersion acts to limit the usable bandwidth for each channel. For these reasons, the optical transmission field has begun to explore the use of modulation formats beyond NRZ. In particular, frequency division multiplexing with multi-level modulation is an attractive way to increase channel capacity [29].

22.5.1 Pure FDM

The first and most fundamental use of FDM is to use it as the multiplexing method of the PON, that is to say, different frequency signals are used to support different service bundles or different ONUs [30]. Such a system is illustrated in Figure 22.22. In the downstream, the OLT transmits several frequency bands over a linear transmission signal chain. In the modern situation, the synthesis of these multiple frequency carriers would be done in a digital signal processor, and transformed to analog using a DAC. The signals pass over the power splitting PON, and are received by every ONU. The ONUs contain a linear optical receiver, which is fed into an ADC and DSP. The processing inside of the DSP is equivalent to an RF tuner and demodulator. In this way, the ONU electronically isolates the one carrier that it is assigned. The extreme case would be that each ONU would get its own carrier, but this would only be suitable if that user was receiving a constant bit rate service. What is more likely is that a carrier would be assigned to a group of ONUs that are sharing the common pool of bandwidth via packet-based TDM.

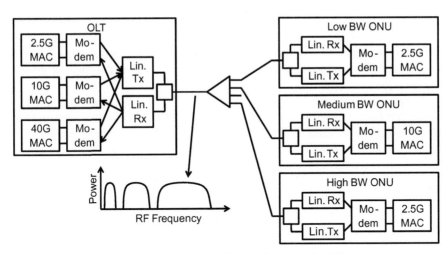

FIGURE 22.22 A (non-orthogonal) FDM-PON, with multiple capacity ONUs.

In the upstream, the ONUs have a similar arrangement, with a DSP, DAC, and linear transmitter. The OLT has a linear receiver, ADC, and DSP demodulator. However, in this case, the important issue is that of optical beat noise. Since FDM is being used for multiplexing, this implies that more than one ONU is transmitting at the same time (unlike TDMA-PONs). This will have multiple optical carriers falling on the single OLT receiver simultaneously. If the wavelengths of the ONU transmitters are far enough apart, there is no issue. However, if the wavelengths are very close together, then the difference in frequency signal will be generated, and this can effectively jam the transmissions of the ONUs that are using that frequency. So, either the wavelengths must be engineered to be different enough, or to be essentially equal, wherein the beat note will be at zero frequency which is typically not a usable frequency. Various schemes have been devised to avoid this issue, such as controlling the wavelength, or remodulating a carrier provided from the OLT. In that sense, many of the FDM schemes begin to resemble some of the WDM-PON wavelength reuse or seeding concepts. However, in contrast to WDM-PON, the goal here is to stabilize the wavelengths of the ONUs to be exactly the same.

22.5.2 Incoherent OFDM

In systems that use classical frequency division, there are several limitations and impairments that arise. The first is that any channel has a pass band with appreciable energy, and a guard band to isolate it from the adjacent channels. The ratio of pass or guard band impacts the spectral efficiency of the system. Filters of reasonable complexity can only go so far in delivering good isolation. But even more fundamentally, the pass band must also have a flat response or the modulated signal will be distorted. It is possible to equalize a non-ideal channel, but that itself has limits. The orthogonal FDM method works to address these problems by modulating all the carriers in an organized way such that guard bands are minimized, and the effective pass band of each carrier can be made quite narrow, minimizing the effect of channel distortion [31].

A typical arrangement is shown in Figure 22.23. The signal chain in both directions is substantially the same as in the simple FDM case (and the traditional problem of optical beat interference is still there). However, in this case, the signal format is OFDM, so the pass band is divided into a large number of (e.g. 1024) subcarriers. These carriers are all harmonically related and modulated in a coordinated fashion. This has two big advantages. The first is that the interference between channels is largely eliminated (note the nulls in the spectra). The second is that the entire modulated multi-carrier can be generated in a single step using the fast Fourier transform algorithm. This is what makes it possible to go to such large subcarrier count with a reasonable processing capability.

The OFDM system in the downstream is relatively straightforward, as all the ONUs receive the composite signal and demodulate it in bulk. Then, the actual use of carriers can be dynamically allocated, both in frequency and time. On aspect of the demodulation is that the receiver must align the FFT frame to that of the incoming signal. This is typically done using a cyclic prefix code. In the upstream, the subcarriers can be shared in a similar time and frequency way. However, there is an extra aspect, in that the ONUs are at different distances from the OLT, so the propagation times are different. There has to be a protocol that works to equalize the ONUs, just as in a TDMA-PON, so that their symbols arrive at the OLT location at the same time.

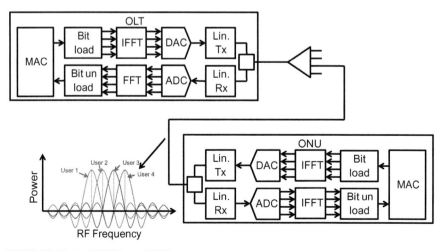

FIGURE 22.23 An OFDM-based PON.

22.5.3 Optical OFDM

In the previous two schemes, the optical carrier was being used on an intensity basis, that is, the power of the light is the signal. Better schemes become possible when the optical carrier's electric field is used as the signal [32]. In this case, the system can use both the amplitude and phase of the light to carry information, with obvious benefits of spectral efficiency. Such a system is illustrated in Figure 22.24. In the downstream, the OFDM signal processing is as before, but now those signals drive a modulator that is capable of controlling the phase and amplitude of the light. At the ONU, a coherent receiver is used to detect the signals. This gives the advantage that if a banded spectrum plan is used, then the receiver only needs to have the bandwidth of the band of interest, rather than the whole spectrum. Add to that the fact that the coherent detector will have very good sensitivity, and the system becomes very capable, albeit very complicated.

In the upstream, coherent transmission provides similar benefits of spectral efficiency and sensitivity. Also, since the local oscillator is locked from the central source, all the LOs can be arranged to have a known frequency relationship. This can effectively control the optical beat interference problem.

An additional enhancement in such systems is to use polarization multiplexing to double the capacity of the system [33]. When coherent reception is used, the random

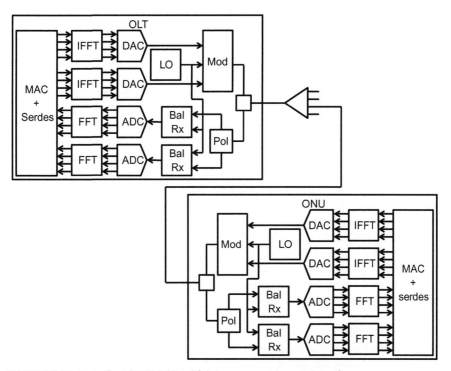

FIGURE 22.24 An optical OFDM-PON, which leverages coherent detection.

polarization received due to the optical channel can be analyzed and digitally rotated back to the originally transmitted states. Of course, this requires both the transmitter's and the receiver's signal chains to be doubled. There has also been work to explore polarization multiplexing even with direct detection, using multiple-input-multiple-output techniques. However, such schemes would face the issue of channel fading, when the two polarization lose their diversity.

It should be noted that there are also differing combinations of all the FDM methods. It is possible to use OFDM in the downstream, but just simple TDMA in the upstream. Or, intensity modulated OFDM is used in the downstream, while coherent OFDM is used in the upstream. The number of possibilities is quite large.

22.6 HYBRID TWDM-PON

Now that the preceding sections have gone though the entire gamut of possible PON systems, we consider the hybrid solution [34,35]. In fact, some of the systems mentioned above could already be considered hybrids, in that the use of TDM/TDMA is usually an option. But those systems do not feature their hybrid nature as their key element. In the scope of standardization, at least, the hybrid of TDMA and WDM seems to gather the most interest. The basic motivation for this system is to find the most cost-effective way to move to higher transmission speeds. As mentioned above, 10 Gb/s is the most cost-effective speed, with the next option (40 Gb/s) being costly. As in the long-distance transmission market, using multiple wavelengths at the best speed available becomes the cheaper option. This system has been given the name time and WDM multiplexed PON (TWDM-PON).

Figure 22.25 shows the architecture of TWDM-PON, in this case with four pairs of wavelengths [36]. Note that the system could have more wavelengths than that. In the simplest concept, each pair of wavelengths would carry a TDM-PON protocol, such as XG-PON. For simple network deployment and inventory management purposes, the ONUs use colorless tunable transmitters and receivers. The transmitter is tunable to any of the upstream wavelengths. The receiver is tunable to any of the downstream wavelengths. In order to achieve power budget higher than that of XG-PON1, optical amplifiers are employed at the OLT side to boost the downstream signals as well as to preamplify the upstream signals. ODN remains passive since both the optical amplifier and WDM Mux/DeMux are placed at the OLT side.

Coexistence with previous generations of PONs in the legacy ODN relies upon the TWDM-PON wavelength plan. Many possibilities exist. One option is to reuse the XG-PON1 wavelength bands. It defines a finer grid inside of the previously defined bands. There happens to be just about enough bandwidth to support 8 wavelengths at 100 GHz spacing. This wavelength plan leverages the development work that has gone into XG-PON1 optics and coexistence filters. It is compatible with G-PON and the 1555 nm RF video overlay channel, but blocks standardized XG-PON. However, it is difficult to meet the 40 km passive reach objectives with such a system, as it uses the higher loss 1270 nm band. Other options are being considered such as the

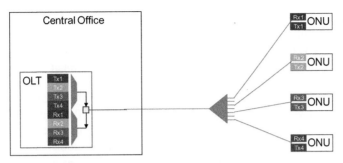

FIGURE 22.25 The TWDM-PON in the pure splitter-based ODN.

O-plus band at 1350 nm, or the guard band at 1535 nm. These would possibly allow coexistence with XG-PON1 or 10GE-PON. If the TWDM system could use only the longer wavelength bands (1480 nm and above), its long reach capability would be far easier to achieve.

Beyond the physical layer changes, TWDM could completely reuse without changing all the material that was developed for XG-PON1. The TC layer, OMCI, and interoperability are all the same. All that is needed are some small additions to specify the methods to manage and control the multiple wavelengths in the system. This should make the standardization of this system rather quick.

In addition to the basic system that delivers extra bandwidth, there are other applications for the multiple wavelengths. The first one to consider is for local loop unbundling. This scheme is shown in Figure 22.26. Each operator would have their own OLT, each of which would contain some set of wavelength channels. A wavelength selective device would be used to multiplex the OLT ports onto a single PON. The wavelength selective device could be as simple as a filter-based demultiplexer, or as shown in the figure, it could be an arrayed waveguide router kind of device. This scheme offers the possibility of every operator's OLT being the same (containing all the wavelengths), and allowing pay-as-you-grow features, in that even a single operator could add OLT resources as they want.

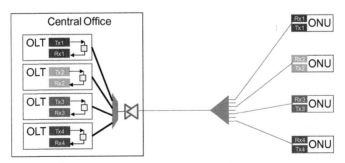

FIGURE 22.26 A TWDM-PON arrangement with multiple OLTs suitable for unbundling.

Another application is for PON reach extension [37,38]. The typical case is that the OLT is located in a main central office that is a significant distance from the satellite offices. The more classical PONs emanate from the satellite offices, and serve the users. By placing a WDM at the satellite office, the fiber count going between the offices can be reduced by a factor of 8.

The fact that the TWDM-PON has a pool of dynamically allocated wavelengths can be used for yet other purposes. Protection against OLT optical failures is possible. If one of the OLT optics stops working, the ONUs that were using that channel can be re-tuned to use other wavelengths, and thereby come back into service. Power saving at the OLT becomes possible, as underutilized wavelengths can be deactivated. The ONUs would then be concentrated onto the remaining channels. Lastly, rogue ONU behavior can be better controlled, as ONUs can be moved to other wavelengths to avoid the uncontrolled ONU.

22.7 SUMMARY AND OUTLOOK

This chapter has given a review of all the technologies that have been suggested for use in the PON field, beginning with the TDM family of PON (most of which have been standardized). The remaining systems (from the 40 Gb/s serial up to TWDM-PON) have been considered for the next step of standardization, NG-PON2. The story of this recent study is outlined here.

Immediately after the XG-PON1 project was spun out of FSAN and into the ITU, work began on NG-PON2. At the beginning of this work project, the basic requirements were for a system with at least 40 Gb/s of capacity and 40 km of reach at 64-way split, but not necessarily backward compatibility with existing ODN or even previous PON systems like the video overlay. Based on that very loose scope of requirements, many different systems were proposed, including 40 Gb/s TDM-PON, TWDM-PON, five flavors of WDM-PON, and three types of OFDM-PON. Importantly, some of the WDM-PONs required wavelength selective ODNs, while others did not.

After some initial study, it was obvious that the broad-brush approach had gathered together systems that were in several ways incomparable. Some systems could maintain backward compatibility, while others couldn't. Some systems were quite close to practical implementation, while others were further away from deployment. Some systems had moderate performance, while others were superlative.

Given this solution space, the network operator members of FSAN had to reconsider what were their true objectives for the NG-PON2 project. The ability to operate on the existing fiber ODN was the first firm requirement to crystallize, followed by the support of up to eight independent operators, and the compatibility with the video overlay. The compatibility with XG-PON was also required. Also, the time frame for practical availability was set to be 2015.

These new or shifted requirements had a huge effect. Some systems were practically eliminated due to the ODN compatibility requirement. Some could be eliminated due to the time-frame requirement. In a short time, all eyes turned to the TWDM-PON system, and indeed this was selected in April 2012 as the architecture for NG-PON2. This architecture will be standardized over the next few years, but in somewhat non-trival document structure. The requirements, physical layer, and TC layer transmission speed enhancements of the NG-PON2 system will be contained in a new series of recommendations, G.ngpon2. The wavelength parts of the TC layer will be contained in G.multi—a recommendation that aims to cover all multi-wavelength PON systems. The OMCI layer aspects will be contained in the G.988 document.

This structure has a couple of advantages. First, the dedicated NG-PON2 material is minimized, which certainly reduces the scale of those documents. Generally speaking, smaller standards documents are better, because there is less chance of making errors. Second, the multi-wavelength materials are generalized, which works to broaden the coverage of several PON systems. For example, the TWDM architecture could just as easily be based on 10GE-PON as XG-PON. The G.multi recommendation can (and likely will) contain a part that covers this possible implementation.

In closing, the progress of PON technology and standards has a long and varied history. Each generation of PON has had an effect on subsequent generations, as the transmission bandwidth was increased time over time. NG-PON2 will place the capability at 40–80 Gb/s aggregate bandwidth standardized at the year 2014. One can hypothesize that there could be an NG-PON3 perhaps at the end of the decade. This may be true. However, the ultimate goal of optical access should be considered, and the trends in the network as a whole. For example, the current trend of cloud computing puts all the processing and storage in the network, and the user would have only a user interface device. The usable and effective bandwidth of the human user is certainly limited, perhaps at around 1 Gb/s as an upper bound. Thus, once access systems get close to delivering this service level, the demand will likely saturate. In fact, NG-PON2 substantially reaches this point for branching levels that are practical with its optical technology. So, an NG-PON3 will probably not increase the per-user bandwidth. However, there are still possibilities to increase the branching ratio (optically) and this will then require the aggregate PON bandwidth to be increased. That is probably the last frontier of PON.

References

[1] J.R. Stern, J.W. Balance, D.W. Faulkner, S. Hornung, D.B. Payne, K. Oakely, Passive optical local networks for telephony applications and beyond, Electron. Lett. 23 (24) (1987) 1255–1256.

[2] Y. Ota, R.G. Swartz, V.D. Archer, S.K. Korotky, M. Banu, A.E. Dunlop, High-speed, burst-mode, packet-capable optical receiver and instantaneous clock recovery for optical bus operation, J. Lightwave Technol. 12 (2) (1994) 325–331.

[3] R.G. Swartz, Y. Ota, M.J. Tarsia, V.D. Archer, A burst mode, packet receiver with precision reset and automatic dark level compensation for optical bus communications, In: Digest of Technical Papers Symposium on VLSI Circuits, 1993, pp. 67–68

[4] Chao Su, Lian-Kuan Chen, Kwok Wai Cheung, Theory of burst-mode receiver and its applications in optical multiaccess networks, J. Lightwave Technol. 15 (4) (1997) 590–606.

[5] Quan Le, Sang-Gug Lee, Yong-Hun Oh, Ho-Yong Kang, Tae-Hwan Yoo, Burst-mode receiver for 1.25 Gb/s Ethernet PON with AGC and internally created reset signal, in: IEEE International Solid-State Circuits Conference, Paper 26.3, 2004.

[6] Cedric Mélange, Xin Yin, Bart Baekelandt, Tine De Ridder, Xing-Zhi Qiu, Johan Bauwelinck, Jan Gillis, Pieter Demuytere, Jan Vandewege, Fully DC-Coupled 10 Gb/s Burst-Mode PON Prototypes and Upstream Experiments with 58 ns Overhead, OFC/NFOEC, Paper OWX2, 2010.

[7] G. Kramer, B. Mukherjee, G. Pesavento, IPACT: a dynamic protocol for an Ethernet PON (EPON), IEEE Commun. Mag. 40 (2) (2002) 74–80.

[8] F. Effenberger, H. Ichibangase, H. Yamashita, Advances in broadband passive optical networking technologies, IEEE Commun. Mag. 39 (12) (2001) 118–124.

[9] All ITU-T recommendations can be downloaded at <www.itu.int>.

[10] F. Effenberger, D. Cleary, O. Haran, G. Kramer, R.D. Li, M. Oron, T. Pfeiffer, An introduction to PON technologies, IEEE Commun. Mag. 45 (3) (2007) 517–525.

[11] K. Tanaka, A. Agata, Y. Horiuchi, IEEE 802.3av 10G-EPON standardization and its research and development status, J. Lightwave Technol. 28 (4) (2010) 651–661.

[12] J. Kani, F. Bourgart, A. Cui, A. Rafel, M. Campbell, R. Davey, S. Rodrigues, Next-generation PON—Part I: Technology roadmap and general requirements, IEEE Commun. Mag. 47 (11) (2009) 43–49.

[13] F. Effenberger, H. Mukai, S. Park, T. Pfeiffer, Next-generation PON—Part II: Candidate systems for next-generation PON, IEEE Commun. Mag. 47 (11) (2009) 50–57.

[14] F. Effenberger, H. Mukai, J. Kani, M. Rasztovits-Wiech, Next-generation PON—Part III: System specifications for XP-PON, IEEE Commun. Mag. 47 (11) (2009) 58–64.

[15] Ed Harstead, Technologies for NGPON2: why I think 40G TDM PON (XLG-PON) is the clear winner, in: OFC/NFOEC, March 2012.

[16] R. Olshansky, V.A. Lanzisera, P.M. Hill, Subcarrier multiplexed lightwave systems for broad-band distribution, J. Lightwave Technol. 7 (9) (1989) 1329–1342.

[17] Michael Aviles, Kerry Litvin, Jun Wang, Barry Colella, Frank J. Effenberger, Feng Tian, Raman crosstalk in video overlay passive optical networks, in: OFC'04, February 2004.

[18] N.J. Frigo, A survey of fiber optics in local access architectures, in: I.P. Kaminow, T.L. Koch (Eds.), Optical Fiber Telecommunications IIIA, Academic Press, 1997 (Chapter 13).

[19] E. Harstead, P.H. van Heyningen, Optical access networks, in: I.P. Kaminow, T. Li (Eds.), Optical Fiber Telecommunications IVB, Academic Press, 2002 (Chapter 10).

[20] S.S. Wagner, H.L. Lemberg, Technology and system issues for a WDM-based fiber loop architecture, J. Lightwave Technol. 7 (1989) 1759–1768.

[21] M.D. Feuer, J.M. Wiesenfeld, J.S. Perino, C.A. Burrus, G. Raybon, S.C. Shunk, N.K. Dutta, Single-port laser-amplifier modulators for local access, Photon. Technol. Lett. 8 (1996) 1175–1177.

[22] H. D Kim, S.-G. Kang, C.H. Lee, A low-cost WDM source with an ASE injected Fabry-Perot semiconductor laser, Photon. Technol. Lett. 12 (2000) 1067–1069.

[23] N. Calabretta, M. Presi, R. Proietti, G. Contestabile, E. Ciaramella, A bidirectional WDM/TDM-PON using DPSK downstream signals and a narrowband AWG, IEEE Photon. Technol. Lett. 19 (16) (2007) 1227–1229.

[24] Wai Hung et al., An optical network unit for WDM access networks with downstream DPSK and upstream remodulated OOK data using Injection-locked FP Laser, IEEE Photon. Technol. Lett. 15 (10) (2003).

[25] E. Wong, K. Lee, T. Anderson, Directly modulated self-seeding reflective semiconductor optical amplifiers as colorless transmitters in wavelength division multiplexed passive optical networks, J. Lightwave Technol. 25 (2007) 67–74.

[26] N. Cheng, Z. Xu, H. Lin, D. Liu, 20 Gb/s Hybrid TDM/WDM PONs with 512-split using self-seeded reflective semiconductor optical amplifiers, in: National Fiber Optic Engineers Conference, OSA Technical Digest, Optical Society of America, 2012, Paper NTu2F.5.

[27] Markus Roppelt, Felix Pohl, Klaus Grobe, Michael Eiselt, Jörg-Peter Elbers, Tuning methods for uncooled low-cost tunable lasers in WDM-PON OFC/NFOEC, Paper NTuB1, 2011.

[28] S. Smolorz, E. Gottwald, H. Rohde, D. Smith, A. Poustie, Demonstration of a Coherent UDWDM-PON with Real-Time Processing, Post Deadline PDPD4, OFC, 2011.

[29] N. Cvijetic, OFDM for next-generation optical access networks, IEEE/OSA Journal of Light wave Technology 30 (4) (2012) (Invited Tutorial).

[30] T. Duong, N. Genay, P. Chanclou, B. Charbonnier, A. Pizzinat, R. Brenot, Experimental demonstration of 10 Gbit/s upstream transmission by remote modulation of 1 GHz RSOA using adaptively modulated optical OFDM for WDM-PON single fiber architecture, in: Proc. Eur. Conf. Opt. Communications. Conf. ECOC, September 2008, Brussels, Belgium.

[31] D. Qian, J. Hu, J. Yu, P.N. Ji, L. Xu, T. Wang, M. Cvijetic, T. Kusano, Experimental demonstration of a novel OFDM-A based 10 Gb/s PON architecture, in: Proc. Eur. Conf. Opt. Communication. Conf. ECOC, September 2007, Berlin, Germany.

[32] N. Cvijetic, M. Cvijetic, M.F. Huang, E. Ip, Y.K. Huang, T. Wang, Terabit optical access networks based on WDM-OFDMA-PON, J. Lightwave Technol. 30 (4) (2012).

[33] D. Qian, N. Cvijetic, J. Hu, T. Wang, 108 Gb/s OFDMA-PON with polarization multiplexing and direct detection, J. Lightwave Technol. 28 (4) (2010) 484–493.

[34] F.T. An, K.S. Kim, D. Gutierrez, S. Yam, E. Hu, K. Shrikhande, L.G. Kazovsky, SUCCESS: a next-generation hybrid WDM/TDM optical access network architecture, J. Lightwave Technol 22 (11) (2004).

[35] G. Talli, P.D. Townsend, Hybrid DWDM-TDM long reach PON for next generation optical access, J. Lightwave Technol. 24 (2006) 2827–2834.

[36] Yuanqiu Luo, Xiaoping Zhou, Frank Effenberger, Xuejin Yan, Guikai Peng, Yinbo Qian, Yiran Ma, Time and wavelength division multiplexed passive optical network (TWDMPON) for Next Generation PON Stage 2 (NG-PON2), J. Lightwave Technol. 31 (4) (2012) 587–593.

[37] R.P. Davey, D.B. Grossman, M. Rasztovits-Wiech, D.B. Payne, D. Nesset, A.E. Kelly, A. Rafel, Sh. Appathurai, Sheng-Hui Yang, Long-reach passive optical networks, J. Lightwave Technol. 27 (2009) 273–291.

[38] P. Ossieur, C. Antony, A.M. Clarke, A. Naughton, H.-G. Krimmel, Y. Chang, C. Ford, A. Borghesani, D. Moodie, A. Poustie, R. Wyatt, R. Harmon, I. Lealman, G. Maxwell, D. Rogers, D.W. Smith, D. Nesset, R.P. Davey, P.D. Townsend, A 135 km, 8192-split, carrier distributed DWDM-TDMA PON with $2 \times 32 \times 10$ Gb/s capacity, J. Lightwave Technol. 29 (4) (2011) 463–474.

Wavelength-Division-Multiplexed Passive Optical Networks (WDM PONs)

23

Y.C. Chung and Y. Takushima[1]

Korea Advanced Institute of Science and Technology (KAIST),
Department of Electrical Engineering, Daejeon, South Korea

23.1 INTRODUCTION

Since the advent of optical fiber, there have been many attempts to utilize it for the broadband access services [1–7]. It is interesting to note that the first fiber-to-the-home (FTTH) trial was planned as early as in 1976, almost at the same time when the low-loss (<1 dB/km) optical fiber was realized for the first time [1,2,8]. However, it remained as a research curiosity in laboratories for a long time due to the high cost and uncertain service demands at that time and became a commercial reality only about 15 years ago. FTTH is now experiencing fast growth and the global FTTH subscribers have already exceeded 67 million [9].

The most prevalent optical access solution deployed around the world at present is the time-division-multiplexed passive optical network (TDM PON) capable of providing up to ∼100 Mb/s service to each subscriber. However, due to the continual growth of various multimedia services, it is evident that we will use more bandwidth in the future. In fact, many service providers are already offering 1-Gb/s access service to their residential subscribers [10]. In addition, it has been forecast that 10-Gb/s access service will be introduced by ∼2020 [11,12]. Also, considering the historical trends of the optical access network development, it has even been speculated that we may need to introduce 100-Gb/s service by ∼2030 [13]. However, it will be unrealistic to provide such broadband access services by using the conventional TDM PON.

The idea of using wavelength-division-multiplexing (WDM) technology in the optical access network had also started in the late 1970s [14–16]. At this early stage, its primary goal was to develop an integrated broadband network that could provide various services by using several wavelengths [16]. In the late 1980s, some pioneering attempts were made to demonstrate the WDM PON in double-star architecture

---

[1]Presently with Spectronix Corp., Japan.

Optical Fiber Telecommunications VIB. http://dx.doi.org/10.1016/B978-0-12-396960-6.00023-7

for the distribution of broadband services [17–20]. However, even at that time, it was well recognized that WDM PON could be very expensive due to the use of wavelength-specific light sources. There had been some efforts to overcome this problem by spectrum-slicing the output of the light-emitting diode (LED) and utilizing it as a "colorless" light source [19–21]. However, its performance was limited due to the small output power of LED and large spectrum-slicing loss. In the mid-1990s, several innovative results have been reported on the development of colorless light sources and revitalized the research interests on WDM PON [22–26]. Since then, there have been numerous efforts to develop practical WDM PONs and improve their performances including the large-scale field trials held in Korea [27–29].

The advantages of WDM PON can be summarized as follows.

1. *High capacity:* In WDM PON, each subscriber is assigned an individual wavelength and it is not shared with other subscribers. As a result, each subscriber can in principle utilize the full capacity achievable by using a single wavelength (which is larger than 100 Gb/s at present). Also, unlike in TDM PON, there is no need to utilize the transceivers operating much faster than the speed each subscriber is serviced at.

2. *Security:* WDM PON provides excellent network security as each subscriber operates on its own wavelength and there is no need to share it with other subscribers.

3. *Enhanced reach:* Unlike TDM PON, WDM PON does not suffer from large power-splitting loss. As a result, the maximum reach can be extended without using remote optical amplifiers.

4. *Simplified operation:* Each subscriber in WDM PON has a virtual point-to-point connection to the central office (CO). Thus, unlike in TDM PON, it does not require use of the ranging and dynamic bandwidth allocation (DBA) algorithms. Also, the medium access control (MAC) layer is simplified.

5. *Upgradability:* To increase the operating speed of a particular subscriber in TDM PON, it is necessary to upgrade all the equipment attached to the network. However, in WDM PON, this speed can be increased on a needs basis by replacing only the equipment relevant to that particular subscriber since each wavelength can operate at a different speed and with a different protocol.

6. *Service transparency:* WDM PON can be used as a platform for various services due to its bit rate and protocol transparency.

Because of these advantages, WDM PON has long been considered as an ultimate solution for the future optical access network capable of providing practically unlimited bandwidth to each subscriber. However, it is still considered to be too expensive for the massive deployment. On the other hand, WDM PON becomes increasingly important with the ever-increasing demand for bandwidth. Thus, it is necessary to improve the competitiveness of WDM PON for its commercial success in the coming years. For this purpose, it is critical to further increase the operating speed and maximum reach of WDM PON. However, the increased speed and reach may not be helpful unless these objectives are achieved cost-effectively. Thus, for the development of the next-generation WDM PON, it is vital to use colorless optical network units (ONUs) and bidirectional single-fiber systems, while restraining the use of expensive external modulators and optical amplifiers [13].

This chapter reviews the current status and future directions of the WDM PON technologies with a special emphasis on the techniques required for the realization of the next-generation high-speed WDM PON. In Section 23.2, various light sources developed for use in WDM PON are reviewed. Section 23.3 describes several exemplary network architectures proposed for WDM PON. The progresses in the long-reach WDM PON are reviewed in Section 23.4. Section 23.5 discusses the recent efforts to realize the high-speed WDM PON operating at the per-wavelength speed faster than 10 Gb/s cost-effectively. The fault monitoring and protection techniques developed for use in WDM PON are reviewed in Section 23.6. Finally, this chapter is summarized in Section 23.7.

23.2 LIGHT SOURCES FOR WDM PON

Although WDM PON has many advantages, it has been considered to be prohibitively expensive for the practical deployment. This is mainly because of the extra costs involved in the installation and maintenance of the wavelength-specific light sources required in WDM PON. In particular, it is not a simple task to install all different transmitters (operating at different wavelengths) at the subscriber's premises located everywhere. In addition, if one of these transmitters deviates from the assigned wavelength, it can deteriorate the performances of its neighboring channels as well as its own. Thus, the wavelengths of these transmitters should be tightly controlled. To overcome these problems and implement WDM PON by using identical light sources (especially for the ONUs), there have been numerous efforts to develop the so-called "colorless" light sources. In particular, the spectrum-sliced broadband light sources and reflective semiconductor optical amplifiers (RSOAs) have attracted significant attention. These light sources can simplify the ONU installations at the subscriber's premises and mitigate the inventory problem of stocking many wavelength-specific light sources. In this section, we review the characteristics of various light sources proposed for use in WDM PON and discuss their potential problems and possible solutions.

23.2.1 Distributed feedback (DFB) laser

At the early stage of the WDM PON development in the late 1980s, it had been mostly demonstrated by utilizing the distributed feedback (DFB) lasers [19,30]. Thus, it was necessary to pick and choose specific lasers operating at the required wavelengths. Figure 23.1 shows a typical configuration of the WDM PON implemented by using such wavelength-selected DFB lasers. This configuration is conceptually simple and there are many advantages of using DFB lasers. For example, DFB lasers have a field-proven record of reliability and are operable at the high speed faster than 10 Gb/s,

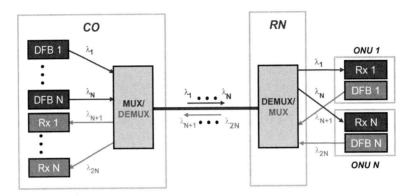

FIGURE 23.1 Typical configuration of WDM PON implemented by using wavelength-selected DFB lasers (DFB: Transmitter based on DFB laser, Rx: Receiver).

which is critical for future upgradability. In addition, by using DFB lasers, we can in principle realize the long-reach WDM PON without the technical problems caused by using the colorless light sources in loopback configuration such as the limited power budget and reflection-induced power penalties. However, it is considered to be too expensive for use in access networks due to not only the laser cost itself (i.e. DFB lasers are usually much more expensive than Fabry-Perot lasers) but also the difficulties in managing a large number of different wavelength lasers. In fact, it is impractical if a different laser should be installed at every subscriber's site as it increases the installation complexity and requires a large inventory of many different wavelength lasers. Also, it has been reported that the wavelength of the DFB laser can be shifted in long term by aging [31]. Thus, to ensure the operation of the DFB laser at a specific wavelength during the system's lifetime, it is indispensable to use a temperature controller (which usually consumes a large amount of electric power) as well as a wavelength-locker. Nevertheless, at least at the optical line terminal (OLT), it may still be necessary to consider the use of DFB lasers (or other type of single-frequency lasers) to avoid the performance limitations caused by the colorless light sources. Recently, an integrated-optic WDM transmitter based on the DFB laser array has been reported [32–34]. This type of the integrated device is well suited for the downstream transmission as it can significantly reduce the footprint and cost of the OLT equipment placed at the CO and negates the inventory issue associated with using many different wavelength lasers. To avoid the use of many lasers operating at all different wavelengths, a multi-frequency laser has also been developed for use in WDM PON by using an arrayed waveguide grating (AWG) and optical amplifier arrays [35,36]. In these lasers, it is possible to modulate each WDM channel by directly modulating its corresponding amplifier. This solution also has a potential for use as downstream WDM transmitters and reduce the size and cost of the OLT equipment.

23.2.2 **Tunable laser**

There are many potential advantages of using tunable lasers for the realization of colorless ONUs. For example, by using tunable lasers in WDM PON, we can avoid the difficulty of installing a different laser at every subscriber's premise and the necessity of stocking many lasers operating at different wavelengths. Also, in comparison with the WDM PONs implemented by using the reflective type of colorless light sources in loopback configuration (described later in this section), the operating speed and maximum reach can be increased easily due to the large modulation bandwidth and power budget achievable by using tunable lasers, respectively. The reflection-induced power penalties, which are often problematic in the WDM PON implemented in loopback configuration, can also be mitigated. In addition, extra functionalities such as network protection can be easily achieved by using tunable lasers.

Various types of tunable lasers have been developed for telecom applications including the external cavity lasers, distributed Bragg reflector (DBR) lasers, and micro-electromechanical vertical cavity surface emitting lasers (MEM-VCSELs), etc. [37–40]. It is interesting to note that, even in the 1980s, there have been attempts to realize a dense WDM PON by using external cavity tunable lasers [17,18,41]. However, such a conventional external cavity laser based on the diffraction grating is bulky and expensive, lacks long-term reliability, and requires an external modulator for high-speed (>1 Gb/s) operation. There have been some efforts to overcome these problems by fabricating an integrated external cavity laser on a polymer-waveguide platform [42–45]. This external cavity laser is small (i.e. it can be mounted in a standard 14-pin butterfly package), can be tuned over 26 nm by separately heating the Bragg grating and phase control section mounted on the polymer-waveguide platform, and directly modulated at 2.5 Gb/s. Recently, it has also been demonstrated that the modulation speed of these lasers can be increased to 10 Gb/s by shortening the cavity length and optimizing the polymer-based grating design [46]. It has been claimed that these lasers can be fabricated cost-effectively enough for use in WDM PONs [42–45]. On the other hand, there have been significant progresses in the development of monolithically integrated tunable lasers during the last decade [47–49]. As a result, a large variety of tunable transponders based on these integrated devices are now available in the market for metro and core applications. However, it has been considered that these lasers are still too expensive for use in WDM PONs [45,50]. In the long run, however, it is evident that the monolithically integrated tunable lasers will be more cost-effective than the external cavity lasers. In fact, it has been reported that DBR lasers can now be manufactured with high yield (>99%) despite their complicated structure [51]. Thus, with the future improvement in low-cost packaging technology, these lasers may eventually satisfy the stringent cost requirement needed for use in WDM PONs [51–53]. However, the use of tunable lasers in WDM PON can incur additional costs beside the laser cost itself. For example, it is not simple to operate the tunable laser at a specific wavelength since it usually has more than three tuning sections. Thus, a microprocessor is often used to control the injection currents to these tuning sections. In addition, to ensure that these lasers operate at their assigned wavelengths during the system's lifetime without significant drifts or mode hopping, it is necessary to control their operating temperature and utilize wavelength-lockers, which can also add extra cost to the WDM PONs implemented by using tunable lasers. To solve this problem, there have been attempts to utilize the tunable laser without the thermoelectric cooler and wavelength locker, and stabilize its lasing wavelength simply by monitoring the temperature and adjusting the tuning currents in accordance with the lookup table [54,55]. However, it is not clear if this method can indeed guarantee the operation of the tunable laser at a specific wavelength during the system's lifetime. There have also been several attempts to minimize these extra costs by using only one wavelength locker in the OLT and sharing it with all the tunable lasers used in WDM PON instead of installing it in every tunable laser module [50,56,57].

23.2.3 Spectrum-sliced incoherent light source

From the very beginning of WDM PON development, it is well recognized that the wavelength-selected lasers can be too costly for use in optical access networks. To solve this problem, it has been proposed to slice the output spectrum of a broadband light source such as the light-emitting diode (LED), superluminescent diode (SLD), and amplified spontaneous emission (ASE) source (i.e. based on the Erbium-doped fiber amplifier (EDFA)) by using a narrow optical filter, and utilize it as a wavelength-specific light source for each WDM channel [19,20,22,26,58–63]. Thus, this technique can be used to realize the colorless light sources and, consequently, to avoid the difficulties of managing many different-wavelength lasers used in WDM PON. Figure 23.2 shows the operating principle of the spectrum-slicing technique [22]. In WDM PON, spectrum-slicing can be achieved automatically by the AWGs installed at the remote node (RN) and CO [26,62]. Thus, there is no need to utilize additional optical filters for spectrum-slicing. It has been demonstrated that this technique can be used for the WDM system operating at the per-wavelength speed faster than 1 Gb/s [22,59,60]. However, the performance of this technique is limited

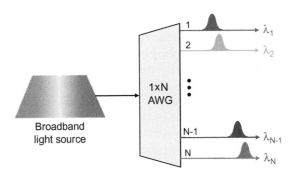

FIGURE 23.2 Operating principle of spectrum-slicing technique [22].

by the excess intensity noise arising from the beat noises between the ASE components within the spectrum-sliced signal and the limited power budget caused by the inherently large splicing loss [22]. For example, due to the excess intensity noise, the signal-to-noise ratio (SNR) of the received spectrum-sliced signal is determined approximately by the ratio between the optical bandwidth of the spectrum-sliced signal and the electrical bandwidth of the receiver (which depends on the signal's operating speed) [22]. As a result, to increase the operating speed, it is necessary to increase the optical bandwidth of the spectrum-sliced signal. However, the use of the large optical bandwidth can inevitably increase the dispersion penalty and decrease the maximum number of channels achievable by using the spectrum-slicing technique in WDM PONs. There have been many efforts to overcome this problem by using the optoelectronic feed-forward technique, gain-saturated semiconductor optical amplifiers, and optical preamplifier receivers [64–67]. It has also been reported that this problem can be mitigated by spectrum-slicing the coherent light having a broad optical spectrum such as the femtosecond laser and supercontinuum generator (instead of the broadband incoherent light) [68–70]. However, this light source has been used mostly as a WDM transmitter in the high-speed optical time-division-multiplexed (OTDM) transmission systems and is not suitable for use in ONUs. On the other hand, to overcome the large slicing losses, it has been attempted to increase the output power of the broadband light sources [71–73] and utilize the forward error-correction (FEC) codes [62,74,75]. Figure 23.3 shows the upstream link in an exemplary WDM PON demonstrated in an effort to mitigate these limitations [62]. In this network, the large splicing loss was reduced by using the multiple peaks of the spectrum-sliced light (generated by the cyclic property of AWG) and the receiver sensitivity was improved by use of the FEC technique. The large dispersion penalty (caused by using multiple peaks of the spectrum-sliced light) was reduced by using 1.3-μm LEDs and simple dispersion pre-compensation circuits. However, the

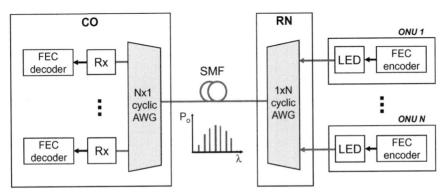

FIGURE 23.3 Upstream link in WDM PON based on the spectrum-slicing technique [62]. In this network, the cyclic property of AWG is used together with the FEC technique to mitigate the large slicing loss.

operating speed of the upstream signal in this network was limited to be 622 Mb/s per channel due to the relatively slow modulation speed of the LED. Recently, it has been reported that it is possible to develop an efficient LED having 10-GHz modulation bandwidth [76].

23.2.4 Reflective light sources

The colorless ONU can be realized by utilizing a reflective device such as the RSOA [25,77,78]. In fact, the WDM PON realized by using such ONUs has already been commercially deployed [79]. Figure 23.4 shows an exemplary configuration of the WDM PON implemented by using RSOAs. In this network, it is necessary to provide the seed light from the CO to each ONU. At the ONU, the cw seed light is amplified and modulated by the RSOA and sent back to the CO. Thus, there is no need to control the wavelength of the upstream transmitter at each ONU and the complexity of stabilizing the signal's wavelengths is handled completely at the CO. However, due to the loopback configuration used for the upstream transmission in this network, the upstream signals experience the link loss twice. As a result, the performance of this network is usually limited by the upstream signals. This type of reflective light source can also be used for implementation of the colorless OLT equipment, if necessary.

The seed light can be obtained by using either a coherent light source such as the DFB laser array and multi-frequency laser (MFL) or an incoherent light source such as the spectrum-sliced ASE light. However, it is advantageous to utilize the coherent seed source for increasing the operating speed and maximum reach of WDM PON due to its higher SNR and optical power than the incoherent source, respectively. On the other hand, use of the incoherent seed source can be advantageous for mitigating

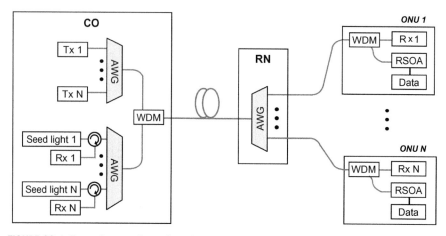

FIGURE 23.4 Exemplary configuration of WDM PON implemented by using RSOAs.

the reflection-induced noises, which is often problematic in this type of network implemented in loopback configuration [80,81].

RSOA is a potentially low-cost device due to its simple structure and is capable of uncooled operation over a wide range of temperature [82,83]. However, its modulation bandwidth is typically limited to be 1–3 GHz. As a result, most previous demonstrations based on RSOAs have been achieved at the moderate speed in the range of 155 Mb/s–5 Gb/s [82,84]. Thus, for the implementation of the next-generation WDM PON operating at the per-wavelength speed of >10 Gb/s by using RSOAs, it is essential to utilize the electronic equalization techniques and/or multi-level modulation formats [85–91]. This complexity can be significantly reduced if the modulation bandwidth of the RSOA can be increased [87]. Recently, it has been reported that the RSOA's bandwidth can be substantially increased by increasing its device length and, as a result, reducing the carrier's lifetime [92]. On the other hand, the 10-Gb/s colorless WDM PON can also be implemented without this complexity by using the photonic integrated circuit (PIC), which consists of a reflective electro-absorption modulator (REAM) and a semiconductor optical amplifier (SOA) [93,94]. It is not clear yet which technique will be most cost-effective for the realization of the WDM PON operating at 10 Gb/s and beyond.

Another type of the reflective light source proposed for the implementation of the colorless ONU is the ASE-injected Fabry-Perot (FP) laser [95]. In this scheme, the spectrum-sliced ASE light is injected into the FP laser and forces it to operate at the specific wavelength determined by the incident ASE light. However, although the ASE-injected FP laser has been widely considered as a different colorless light source than RSOA, its fundamental operating principles are the same [96]. When the spectrum-sliced ASE light is injected into the FP laser, it no longer lases due to the increased threshold and merely amplifies the injected ASE light (with possible noise suppression due to the saturated operation, as in the case of using RSOA). However, since the FP laser is not an ideal optical amplifier, a relatively large incident optical power is needed to suppress its side modes sufficiently. Also, unlike in the case of RSOA, the coherent light (obtained from a single-frequency laser) cannot be used as the seed light for the FP laser since it can cause unstable behaviors [97,98]. Thus, the maximum operating speed of the ASE-injected FP laser is limited not only by the modulation bandwidth of the FP laser, but also the intensity noises arising from the injected ASE light. To overcome this problem and secure the sufficient SNR required for high-speed operation, it is essential to utilize the ASE light having a broad optical bandwidth [22]. However, this will increase the channel spacing and, consequently, decrease the number of WDM channels (i.e. subscribers). Due to these effects, the highest operating speed of the ASE-injected FP laser reported to date has been limited to 1.25 Gb/s in the 100-GHz spaced WDM PON [99,100]. Recently, it has been reported that this speed can be increased to 2.5 Gb/s by using the polarization-insensitive FP laser having an improved front-facet reflectivity of 0.1% [101]. It is interesting to note that, for high-speed operation, the specifications of the FP laser become similar to those of RSOAs. In any case, due to the large intensity noises inherent in the spectrum-sliced ASE light, it is difficult to further increase the operating speed of the ASE-injected FP laser to >10 Gb/s. To solve this problem, there have been some efforts to utilize the FP laser injection-locked to the coherent seed light. Although such an injection-locked FP laser can certainly operate at a speed faster than 10 Gb/s, it cannot be considered as a colorless light source since its locking range is extremely narrow (~0.03 nm) [102].

23.2.5 Re-modulation scheme for upstream transmission

The colorless ONU can also be realized by re-modulating the downstream signal (instead of the cw seed light) by using an external modulator or an RSOA for the upstream transmission [103–119]. Figure 23.5 shows an exemplary WDM PON architecture implemented by using this re-modulation technique in single-fiber loop-back configuration and RSOAs. This technique is considered to be simple and cost-effective since there is no need to use additional light sources for the seed light. In addition, the system's capacity (i.e. the number of WDM channels) can be doubled as the upstream and downstream signals operate at the same wavelength. However, since the downstream wavelengths are reused for the upstream transmission, the performance of the upstream signal can be seriously deteriorated by the residual downstream signal, as shown in Figure 23.6. Various techniques have been proposed to mitigate this problem caused by the residual downstream signal (which is often referred as the re-modulation noise). For example, to mitigate this problem, it has been proposed to utilize a small extinction ratio for the downstream signal and re modulate it at a substantially slower rate with high extinction ratio for the upstream transmission [105,106]. In this case, the extinction ratio of the downstream signal should be optimized to balance the performances of the upstream and downstream signals [107]. This re-modulation noise problem can also be mitigated by erasing

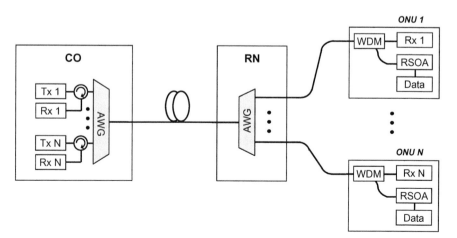

FIGURE 23.5 An exemplary WDM PON architecture implemented by using the re-modulation technique in single-fiber loopback configuration and RSOAs.

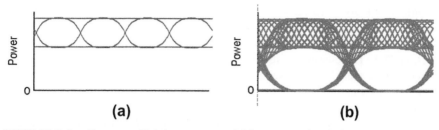

FIGURE 23.6 Eye diagrams of (a) downstream and (b) upstream signals in the WDM PON implemented by using the re-modulation technique. The "mark" level of the upstream signal is seriously contaminated by the residual downstream signal.

the downstream signal by using a gain-saturated RSOA [108,109]. However, when this technique is used, the maximum reach of the WDM PON can be limited by the incident optical power required for the gain saturation. Other techniques proposed to solve this problem include the use of subcarrier-multiplexing (SCM), time-division-multiplexing (TDM), and different modulation formats and coding for the downstream signals such as the inverse return-to-zero (IRZ), frequency-shift-keying (FSK), differential phase-shift-keying (DPSK) formats, and Manchester coding, etc. [107,110–115]. The performances of these techniques have been reviewed in several literatures [116–118].

When the re-modulation technique is utilized, especially in the WDM PON implemented in single-fiber loopback configuration, the system's performance is bound to be sensitive to the unwanted external reflections (caused by the fiber connectors, splices, and Rayleigh backscattering, etc.) since both the upstream and downstream signals operate at the same wavelength and propagate along the same fiber [119–121]. This problem can be somewhat mitigated by broadening the signal's spectrum or separating the upstream and downstream signals in the frequency domain using the SCM and Manchester coding [111,122,123]. However, it still remains the most serious limiting factor for the re-modulation technique.

23.3 WDM PON ARCHITECTURES

WDM PON is typically implemented in double-star architecture consisting of a shared feeder fiber and dedicated drop fibers connected to RN, although its details can be different from each other depending on the type of light sources, network size, required services, and necessity of protection, etc. For example, the WDM PON architecture can be significantly affected by the type of upstream light sources used in the network (i.e. some colorless light sources require use of the seed light). To increase the number of subscribers supported by a feeder fiber in a WDM PON, various network architectures have been proposed including the ultra-dense WDM PONs and multi-stage WDM PONs [124–127] as well as hybrid WDM/TDM, WDM/SCM, and

WDM/CDM PONs [110,128–132]. The maximum reach of WDM PON can be substantially increased by installing various types of optical amplifiers (such as EDFAs, remotely pumped EDFAs, and Raman amplifiers) in the network, as described in Section 23.4. The commonly used WDM PON architecture (i.e. implemented by using AWG at the RN) is not suitable for delivering the broadcast signals due to its virtual point-to-point connectivity. To solve this problem, it is necessary to modify the network architecture by using an additional passive splitter or an additional light source having a broad optical spectrum [133–135]. The WDM PON architecture can also be modified in various ways to have the protection capability against fiber failures, as described in Section 23.6. Among these various options for the WDM PON architecture, the optimal option should be selected considering the cost and required performances (such as the bandwidth, splitting ratio, and maximum reach). In particular, the cost is a critical issue for the realization of the practical WDM PON. In this respect, it is highly desirable to implement the WDM PON by using the colorless light sources (especially for the upstream signals) and the single-fiber configuration (i.e. using the bidirectional transmission technique).

This section briefly reviews several representative architectures used for the implementation of WDM PON including the wavelength-routing, broadcast-and-select, and ring/bus architectures.

23.3.1 WDM PON in wavelength-routing architecture

Figure 23.7 shows an exemplary WDM PON implemented in wavelength-routing architecture. This is the most common architecture used for the realization of WDM PON at present, although there are many variants depending on the type of light sources, network size, required services, and necessity of protection, etc. This network utilizes AWG at the RN instead of the power-splitting coupler. As a result, the power budget of this network does not suffer from the large splitting loss (i.e. the insertion loss of AWG is in the range of only 3–6 dB regardless of the number of channels). In addition, unlike the WDM PON implemented in broadcast-and-select architecture, there is no need to utilize the wavelength-specific receivers at the ONUs since their corresponding signals are automatically selected by the AWG placed in the RN.

Several WDM PONs implemented in the wavelength-routing architecture have already been described in Section 23.2. In this section, the major issues in this WDM PON architecture are discussed by using the exemplary network shown in Figure 23.7. In this network, the downstream signals are combined by an AWG at the CO and then coupled into the feeder fiber, while the upstream signals are separated by an AWG and sent to the receivers. The downstream signals can be generated by using various types of light sources at the CO including the wavelength-selected DFB lasers, MFLs, tunable lasers, and colorless light sources. However, the use of expensive external modulators should be avoided for cost-effectiveness. On the other hand, it will be desirable if an integrated-optic WDM transmitter array can be used to reduce the footprint of the OLT equipment, assuming that such a device can eventually be

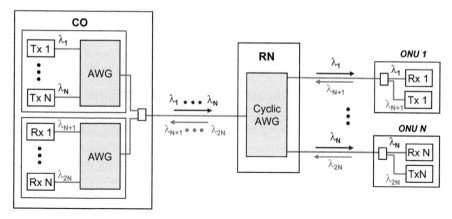

FIGURE 23.7 An exemplary WDM PON implemented in wavelength-routing architecture. In this network, a cyclic AWG is used at the RN and the wavelengths of the upstream and downstream signals are separated by the free-spectral range of the AWG used at the RN.

manufactured cost-effectively [32–34]. An integrated-optic WDM receiver array can also be used at the OLT for the same purpose [136,137]. It should also be noted that, recently, there have been many efforts to develop the coherent detection techniques suitable for use in WDM PONs, especially for the long-reach applications [126,138–141]. At the CO, the downstream and upstream signals can be combined and separated by using a 3-dB coupler, a coarse WDM filter, or an optical circulator for the bidirectional transmission along the feeder fiber. At the RN, a cyclic AWG is commonly used to minimize the AWG's port count by utilizing the same AWG port for both upstream and downstream signals [142,143]. For this purpose, the upstream and downstream signals should operate at the different wavelength bands separated by the free-spectral range (FSR) of this AWG. It has been reported that the temperature at the outside plant of a passive optical network can be changed by as much as 125 °C [144]. Such a large temperature variation can cause the spectral misalignment between the WDM sources and the AWG at the RN (or between the AWGs at the CO and RN) and result in the significant power loss and crosstalk. Thus, it is crucial for the AWG (usually placed at the unpowered RN) to have athermal characteristics [145]. Also, if necessary, wavelength-tracking techniques should be used to avoid this problem [144,146,147]. A remote optical amplifier can be added at the RN to extend the maximum reach of WDM PON, as described in Section 23.4. For the practical WDM PON, it is indispensable to utilize the colorless ONUs. It can be realized by various types of colorless light sources described in Section 23.2. However, to utilize some of these light sources, it is required to modify the network architecture significantly to provide the seed light from the CO.

It is sometimes necessary to deliver the broadcast video signals over the PON. However, the WDM PON implemented in the wavelength-routing architecture is not suitable for this purpose due to the wavelength-selective AWG used at the RN.

Various techniques have been developed to solve this problem by installing a power splitter near the AWG at the RN, modifying the AWG configuration, and using a broadband light source such as LED, etc. [133,135,148–151].

23.3.2 WDM PON in broadcast-and-select architecture

The broadcast-and-select architecture has been widely used for various applications especially at the early stage of the WDM network development in the late 1980s including access networks, computer networks, distribution of video signals, and optical cross-connects, etc. [16,18,152–155]. Figure 23.8 shows a representative WDM PON implemented in this architecture. The most distinctive feature of this architecture is use of the $1 \times N$ fiber coupler (instead of AWG) in the RN as in TDM PON. In this network, the downstream signals are combined by the $N \times 1$ coupler at the CO and then broadcast to every ONU via the $1 \times N$ coupler placed at the RN. Thus, at each ONU, it is necessary to select its corresponding downstream signal by using an optical bandpass filter (or a coherent receiver). The upstream signals are combined by the $1 \times N$ coupler at the RN and sent to the CO. At the ONUs and CO, the upstream and downstream signals are combined and separated by using 3-dB couplers or coarse WDM filters. By using this architecture, the existing TDM PON can be upgraded to WDM PON without changing the outside plant. In addition, this architecture is well suited for providing the broadcast and multicast services since it can efficiently distribute the downstream signals to every subscriber (although it may require the use of a tunable receiver at the ONU). However, it is problematic to realize the colorless ONU as it requires not only the wavelength-specific transmitter but also the wavelength-specific receiver. Also, this architecture suffers from the limited power budget as in TDM PON due to the use of the power-splitting coupler at the RN. As a result, this architecture is not extensively exploited at present for the realization of WDM PONs.

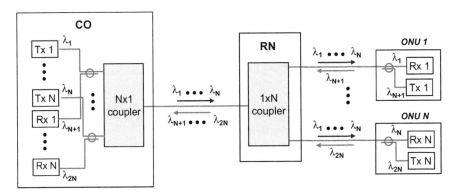

FIGURE 23.8 An exemplary WDM PON implemented in broadcast-and-select architecture.

23.3.3 **WDM PON in ring/bus architectures**

The ring architecture is widely used in the metro area network due to its superb restoration capability. By using this architecture, the total fiber length required for interconnecting each node in the network can be minimized. There have been some efforts to exploit these advantages, along with other objectives achievable by using the ring architecture, in WDM PON [156–168]. Figure 23.9 shows a conceptual schematic diagram of the WDM PON implemented in single-fiber ring architecture. In this network, every ONU is connected to the CO through the feeder fiber and a portion of the ring. At the ONU, its corresponding upstream and downstream signals are added to and dropped from the ring, respectively, by using an optical add/drop multiplexer (OADM). In this network, when a fiber failure is occurred in the ring, the signals are rerouted to the opposite directions by using the optical switches installed in the ONUs. Thus, this network is robust against the fiber failures. However, the maximum number of ONUs that can be attached to this ring is usually very limited due to the large insertion loss of the ONU (mostly caused by the OADM). It is certainly possible to overcome this limitation by installing optical amplifiers along the ring (or in the ONUs), but it can seriously deteriorate the cost-effectiveness of the network. Thus, to enhance the practicality of this network, it is crucial to develop an inexpensive OADM integrated with optical amplifiers.

When it is necessary to deploy WDM PON in a sparsely populated rural area, the linear bus topology can be useful to save the fiber cost. However, this network can accommodate only a limited number of ONUs due to large insertion losses of

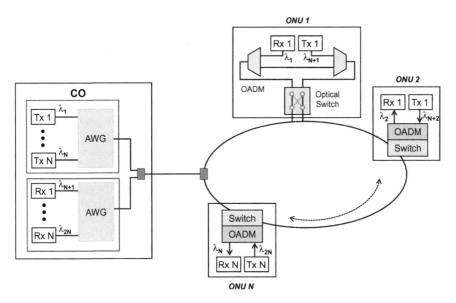

FIGURE 23.9 A conceptual schematic diagram of WDM PON implemented in ring architecture.

OADMs, as the WDM PON implemented in ring architecture. Nevertheless, it has been demonstrated that the WDM PON in this architecture can be implemented to support up to 16 ONUs by using a simple WDM filter as an OADM and RSOAs at the ONUs [169,170]. To further increase the number of ONUs supported in this architecture, it is necessary to reduce the insertion loss of OADM and/or utilize optical amplifiers in the link.

23.4 LONG-REACH WDM PONs

There have been many efforts to develop a long-reach access network for the purpose of extending the coverage of the CO. This is because of the fact that the network's cost can be reduced significantly through the simplification of the network [171]. Figure 23.10 illustrates this network simplification by using a typical hierarchical structure of the current optical networks consisted of the backbone, metro, and access networks. The backbone network transports huge traffic between different cities and provides connectivity among the metro networks. Its size can be national or international scale. The metro networks consist of core and edge rings, and deliver traffic to service points of presence. Also, these networks serve as the gateways for both the backbone and access networks. The access networks connect the "last-mile" between the subscribers and the CO (which plays a role of the gateway between access and metro networks). In the access network, each CO covers a small area within a radius of $\sim 20\,\mathrm{km}$ (which is limited by the maximum reach of the standard PON). As a result, a large number of COs are required to cover the entire service region, as shown in the left side of Figure 23.10b. However, if the maximum reach of PON is increased, the coverage of CO is expanded (hence the number of the required COs is reduced), as shown in the center of Figure 23.10b. In fact, if the maximum reach is increased to $>100\,\mathrm{km}$, a single CO can cover the entire region served by a metro network. In other words, the access and metro networks can be combined into a single system, as shown in the right side of Figure 23.10a and b. As a result, the number of equipment interfaces, network elements, and network nodes can be reduced significantly. In fact, it has been reported that, by using long-reach access networks, the number of exchange buildings in the United Kingdom's network can be reduced from ~ 5700 to a few hundred [171]. Because of these advantages, there have been numerous efforts to increase the maximum reach of the access network, not only for WDM PONs, but also for TDM PONs [171] and hybrid WDM/TDM PONs [158,172]. This section reviews the techniques proposed to extend the maximum reach of WDM PON and hybrid WDM/TDM PON, and discuss the issues related to the use of colorless light sources.

23.4.1 Fundamental limitations on the reach of WDM PON

WDM PON has much larger power margin than TDM PON due to its use of AWG at the RN instead of the power-splitting coupler. For example, the class B+ G-PON

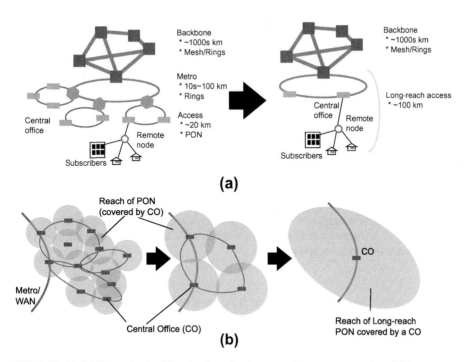

FIGURE 23.10 (a) Network simplification by using long-reach access network and (b) illustration of the reduction in the required number of COs achieved by using long-reach access network. (For interpretation of the references to color in this figure legend, the reader is referred to the web version of this book.)

specifications, which are conformed in most commercial systems, dictate the system to support 32 subscribers and physical reach up to 20 km. In its maximum optical loss budget (specified to be >28 dB), the splitter's loss takes up the largest portion of >16 dB. On the other hand, the insertion loss of the AWG used in the WDM PON is typically 3–6 dB only, and is almost independent of the number of ports (i.e. wavelength channels). Thus, the maximum reach of WDM PON can be easily increased to >60 km without using any optical amplifiers or reducing the split ratio [107,173].

In WDM PON, the maximum reach is determined by the maximum transmission distances of both the upstream and downstream signals. However, in most cases, it is settled by the upstream signal since the output power of the colorless light source used in the ONU is usually much lower than that used in the OLT. Thus, we focus our discussions on the fundamental limitation of the upstream transmission, while assuming that the signal distortion caused by the chromatic dispersion and fiber non-linearities can be ignored (which is valid in most WDM PONs operating at the speed of <2.5 Gb/s). Figure 23.11 shows the received powers of the upstream signals in the WDM PONs implemented in two different architectures: (a) one implemented without using loopback configuration (by using non-reflective type colorless light

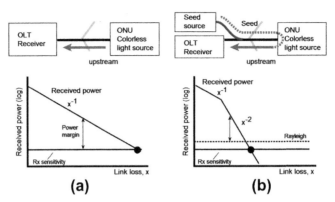

FIGURE 23.11 Power margin vs. link loss in the long-reach WDM PON (a) implemented in non-loopback configuration (e.g. by using tunable lasers), (b) implemented in loopback configuration (e.g. by using RSOAs).

sources such as tunable lasers and spectrum-sliced ASE sources at the ONU) and (b) the other implemented in loopback configuration (using reflective type colorless light sources such as RSOAs, REAMs, and ASE-injected FP lasers). To simplify the notation, the reach is described by the link loss between the ONU transmitter and OLT receiver. In case of (a), the output power of the ONU transmitter is constant regardless of the link loss. Thus, the maximum reach is determined simply by the output power of the ONU transmitter and the sensitivity of the OLT receiver. On the other hand, in the case of (b), the seed light is provided from the CO to the ONU, and then sent back to the CO after being modulated with the upstream signal by the colorless light source. The colorless light source usually has an optical gain. For example, Figure 23.12 shows the gain characteristics of an RSOA used in WDM PON [121]. This RSOA has the constant small-signal gain of 17 dB, when the input power is lower than −20 dBm. Also, it has the saturation output power of ∼5 dBm (for the input power: >−2 dBm). Thus, when the link loss is small and the input power of the seed light to the colorless light source is high enough to saturate the gain, the output power of the ONU remains almost constant. In this region, the received power decreases in proportional to the link loss. This situation occurs in the short-reach WDM PONs. On the contrary, as the reach increases, the link loss is also increased and the input power of the seed light incident on the colorless light source becomes reduced. As a result, the optical gain of the colorless light source becomes almost constant (i.e. it becomes almost same as the small-signal gain). In this case, the received power is decreased in proportion to the link loss squared (because the seed light experiences the link loss twice). Thus, the impact of the link loss is more critical in the loopback system, as shown in Figure 23.11b.

The maximum reach of WDM PON can also be limited by the Rayleigh backscattering of the fiber as well as the discrete reflections caused by fiber connectors and splices, etc. [107,119–121,174,175]. As shown in Figure 23.13a, the seed light

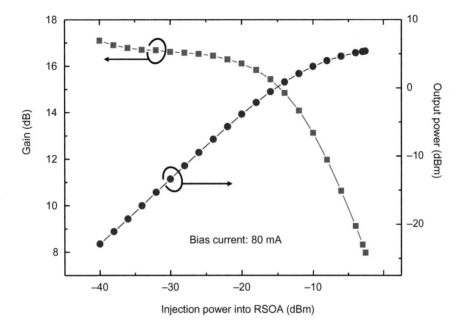

FIGURE 23.12 Measured gain characteristics of RSOA [121]. In this figure, the vertical axes in the left- and right-hand sides represent the fiber-to-fiber gain and the fiber output power, respectively. (For interpretation of the references to color in this figure legend, the reader is referred to the web version of this book.)

can be reflected back to the CO by the Rayleigh backscattering occurring in the fiber. The backscattered light co-propagates with the upstream signal, and causes the in-band crosstalk (since the seed light and upstream signal operate at the identical wavelength). In particular, when a coherent single-frequency laser is used to generate the seed light, it can result in the severe interferometric noise (i.e. optical beat interference (OBI) noise) [174]. It should be noted that the crosstalk level caused by the Rayleigh backscattering (i.e. the ratio of the Rayleigh backscattered power to the received power of the upstream signal) increased as the link loss is increased. This is because of the fact that the Rayleigh backscattered power remains almost constant when the fiber length is longer than $1/\alpha$ (where α is the fiber loss per unit length), while the received signal is in proportion to the link loss squared, as shown in Figure 23.11b. The crosstalk level cannot be improved by increasing the power of the seed light since it also increases the backscattered power proportionally. The discrete reflections caused by the fiber connectors and splices can also induce similar impairments (i.e. OBI noises). The impacts of the discrete reflections as well as the Rayleigh backscattering on the performances of the upstream signals have been systematically investigated by classifying the effects of the reflection into two categories as shown in Figure 23.13 [119,121]:

- Reflection-I: the reflection of the seed light (which interferes with the upstream signal).
- Reflection-II: the reflection of the upstream signal (which returns to the ONU is amplified and modulated by the colorless light source again at the ONU, and then is sent back to the CO and interferes with the upstream signal).

To ensure the error-free operation, it is important to maintain the crosstalk level to be sufficiently small. If only Reflection-I is taken into account, it is preferable to increase the upstream power by increasing the ONU gain (provided by the colorless light source). However, the crosstalk caused by Reflection-II can also be amplified by the ONU gain. Thus, the ONU gain should be optimized to minimize the effects of the Rayleigh backscattering and discrete reflections [119]. The impacts of the Rayleigh backscattering occurred in the feeder fiber can also be mitigated by placing the seed light source in the RN [172]. This technique not only negates the Rayleigh backscattering of the seed light in the feeder fiber but also reduces the required optical power of the seed light, although it may complicate the configuration of RN. In the case of the WDM PON utilizing the re-modulation scheme, it becomes more complicated to analyze the effects of the reflection since Reflection-II can deteriorate the performances of both the upstream and downstream signals [121].

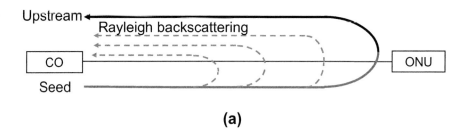

(a)

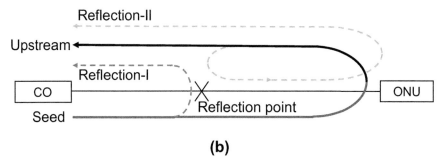

(b)

FIGURE 23.13 Effects of reflections in the WDM PON implemented in loopback configuration. (a) Effects of Rayleigh backscattering, (b) classifications of the effects of back-reflection and Rayleigh backscattering.

Various techniques have been proposed to improve the reflection tolerance of WDM PON. Many of these techniques utilize the fact that the reflection-induced OBI noise can be suppressed effectively by broadening the optical spectrum of the signal (by dithering the bias current of its transmitter [176] or using a phase modulator [177]). When the WDM PON implemented by using the re-modulation technique, the reflection tolerance can be improved by using the Manchester coding or SCM for the downstream signals [111,178]. This is mainly because the upstream and downstream signals can be separated in frequency by using these modulation formats (i.e. the upstream signals operate in the baseband, while the downstream signals operate at the high-frequency region). Thus, the effects of reflection can be easily filtered out at the receivers. On the other hand, when the WDM PON is implemented in single-fiber loopback configuration without using the re-modulation technique (i.e. external seed light is used), the reflection tolerance can be improved drastically by using the incoherent light (such as ASE light) as the seed light. In fact, it has been reported that, when this technique is used, the RSOA-based WDM PON can even tolerate the large reflection caused by an open connector (~16 dB) [80].

23.4.2 Long-reach WDM PON using remote optical amplifiers

To extend the maximum reach of WDM PON, a remote amplifier is often used to compensate for the losses of the transmission fiber and RN. Figure 23.14 shows an exemplary long-reach WDM PON implemented by installing a remote amplifier at the RN. For this purpose, EDFAs have been most widely used in the previous experiments. Figure 23.14a shows the typical configuration of the bidirectional amplifier implemented by using two unidirectional EDFAs. One of the difficulties in using EDFAs in WDM PON arises from the fact that the upstream and downstream signals are usually transmitted in packets. If such a burst signal is amplified by the EDFA,

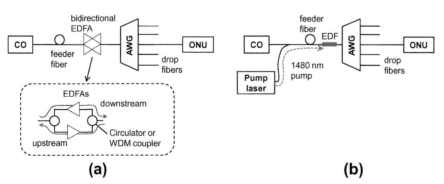

(a)   **(b)**

FIGURE 23.14 Long-reach WDM PONs implemented by using remote optical amplifiers. (a) WDM PON using bi-directional remote amplifier, (b) WDM PON using remotely pumped amplifier.

the waveform is severely distorted due to its transient since the leading edge of the burst signal can be amplified excessively [179]. Also, the EDFA's gain can be varied when the number of wavelength channels (subscribers) is changed. Thus, it may be necessary to utilize the dynamic gain control technique for the EDFA to achieve the stable amplification [180,181]. The SOA has also been widely used as a remote optical amplifier for the demonstration of the long-reach WDM PON [171]. The SOA is compact and has a small gain transient due to its short carrier lifetime. Thus, it is suitable for the amplification of the burst-mode signals. However, the SOA can suffer from the gain variation due to the intermodulation among WDM channels. To overcome this problem, it has been proposed to use the dynamic gain control technique [182] or utilize an SOA per each WDM channel (i.e. use SOA as a single-channel amplifier after demultiplexing WDM channels) [172].

For the use in the hybrid WDM/TDM PON capable of supporting a large number of subscribers (>1000), the configuration of the remote amplifier can be complicated. For example, Figure 23.15 shows the network configuration of the 10-Gb/s hybrid WDM/TDM PON capable of supporting 8291 subscribers with the reach of 124 km [183,184]. In order to avoid the crosstalk between WDM channels and provide the high gain needed to compensate for the large splitting loss of the passive couplers used in the TDM part as well as the loss of the feeder fiber, automatic gain-controlled (AGC) EDFAs are used for all 32 WDM channels before and after the (de)multiplexing by the AWGs. Thus, the configuration of the RN becomes extremely complex, although the cost of these EDFAs can be shared by a large number of users supported in the network.

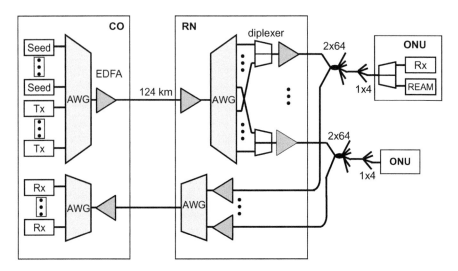

FIGURE 23.15 Example of the remote amplifications used in large-scale hybrid WDM/TDM PON [183].

To utilize a remote optical amplifier, it is requisite to install a power supply in the RN since the optical amplifier is not a passive device. In fact, strictly speaking, it may not be appropriate to use the term "passive optical network" when the remote amplifier is used in the RN, although it has been widely used in the previous literatures despite this contradiction. In any case, the use of a power supply in the RN could increase the installation and operation costs of WDM PON. To solve this problem, it has been proposed to utilize the remote pumping technique shown in Figure 23.13b. In this technique, a short section of Erbium-doped fiber (EDF) is attached to the feeder fiber at the RN, and the intense pump light operating at $1.48\,\mu$m is launched from the CO to remotely pump the EDF [185–187]. Since the fiber loss at $1.48\,\mu$m is relatively small, the sufficient pump power can be delivered to the remotely located EDF from the CO. When the feeder fiber is very long, a distributed Raman amplifier can also be used to compensate for the link loss without the necessity of installing any active fibers in the outside plant [188].

Although these remote amplification techniques can compensate for the transmission losses of the feeder and drop fibers as well as the splitting loss at the RN in the long-reach and high-split WDM PON, the effect of the back-reflections becomes serious since the back-reflected light can also be amplified by the remote amplifier. Thus, to mitigate the impact of back-reflection in the remotely amplified WDM PON, the location and gain of the remote amplifier should be properly optimized according to the distribution of losses in the system [189–192].

23.4.3 Long-reach WDM PON using coherent detection technique

The coherent detection technique provides the excellent receiver sensitivity as well as the frequency selectivity. Also, due to recent advances of digital signal processing technique, linear and nonlinear impairments can be compensated without the knowledge of the transmission line [192]. Because of these features, the coherent detection technique together with advanced modulation formats and digital signal processing are the key enablers for the optical communication systems operating at 100 Gb/s and beyond. These advantages are also very attractive for the access networks. For example, the excellent receiver sensitivity can enhance the power budget and enables to extend the maximum reach of WDM PON [138,140,193]. Also, by exploiting the excellent frequency selectivity of the coherent receiver as well as the high spectral efficiency of the advanced modulation format, it is possible to accommodate a large number of WDM channels within the limited bandwidth and increase the number of subscribers supported by a WDM PON [194]. However, the coherent detection technique has been considered to be prohibitively expensive for the use in the access network. For example, the coherent receiver needs a local oscillator (LO) and its wavelength control circuitry to track the signal's wavelength precisely. In addition, it requires 90-degree optical hybrids, which are usually made of bulk optics or complex optical waveguides. Furthermore, for polarization-independent operation, it is usually needed to utilize two sets of 90-degree optical hybrids and optical receivers. Thus, for use in access networks, it is essential to develop a cost-effective coherent detection technique. Recently, there have been several attempts to achieve this objective [138,140,192,193]. For example, a cost-effective self-homodyne coherent receiver has been developed for use in the long-reach WDM PON [140]. Figure 23.16 shows the schematic diagram of the upstream link of a long-reach WDM PON implemented by using this coherent receiver. In this coherent receiver, a portion of the seed light (cw light in the figure) is used as a LO for the coherent detection of the upstream signal. Thus, there is no need to utilize an additional laser for the LO as well as the optical frequency-locking loop associated with it. Also, the state-of-polarization (SOP) of the upstream signal is automatically stabilized simply by placing a Faraday rotator (FR) in front of the RSOA [140]. Accordingly, there is also no need to utilize the costly polarization-diversity technique (or the polarization-tracking technique). In addition, to further enhance its cost-effectiveness, the coherent receiver is realized by using an inexpensive 3×3 fiber coupler as a 120-degree optical hybrid instead of using the expensive 90-degree optical hybrid. In the ONU, the upstream data is modulated in the binary phase-shift keying (BPSK) format by utilizing the frequency chirp of the directly modulated RSOA (refer to Section 23.5) [140]. In this way, the expensive optical components used in the conventional coherent receiver are eliminated or replaced with cost-effective passive components. Thus, this coherent receiver has a potential to be cost-effective enough for use in the access network.

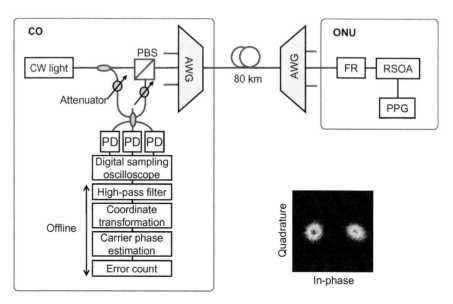

FIGURE 23.16 Configuration of long-reach WDM PON employing RSOAs and self-homodyne coherent receivers [140].

The cost-effective coherent receiver shown in Figure 23.16 can still provide the excellent sensitivity close to the shot-noise limit. For example, the sensitivities of this receiver at 2.5 Gb/s are measured to be −46.4 dBm in back-to-back condition and −45.6 dBm at the bit-error-rate (BER) of 10^{-4} after 80-km-long transmission [140]. Thus, the excess penalties from the shot-noise limit are only 3.4–4.2 dB. This receiver can also mitigate the impacts of Rayleigh backscattering and discrete reflections can also be mitigated by using the coherent detection. For example, Figure 23.17a shows the upstream signal and the in-band crosstalk caused by the backscattered seed light generated in the loopback system implemented by the direct detection technique. In this case, the upstream signal and the in-band crosstalk interfere with each other and induce large beat noise. However, in the case of using the homodyne detection technique shown in Figure 23.17b, this beat noise component becomes much smaller than the detected signal component, since the signal component is the beat between the high-power LO and the received signal. Also, the beat noise caused by the interference between the crosstalk and LO can be effectively eliminated simply by using a high pass filter [138,140]. Thus, the system's performance is almost immune to the back-reflection. The effectiveness of this coherent receiver has been successfully demonstrated by using the installed fibers [140].

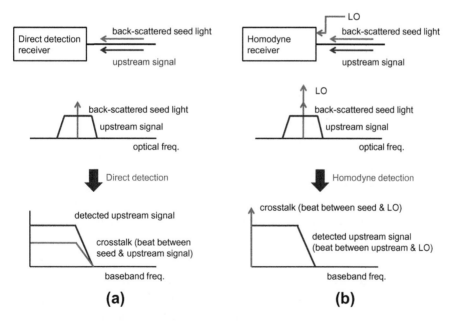

FIGURE 23.17 Improvement of reflection tolerance by using the coherent detection technique. (a) The optical and baseband spectra in the case of using direct detection technique. (b) The optical and baseband spectra in the case of using coherent detection technique. In this case, the crosstalk component falls at around the zero frequency. Thus, it can be easily eliminated by using a high pass filter.

23.5 NEXT-GENERATION HIGH-SPEED WDM PON

It has been forecast that 10-Gb/s access service will be needed by ∼2020 to accommodate the ever-increasing traffic demand [12,13]. Also, if the current development trends of the optical access network persist in the future, it may be necessary to provide even 100-Gb/s service to each subscriber by ∼2030 [14]. The WDM PON has long been considered as an ultimate solution for the next-generation optical access network capable of providing practically unlimited bandwidth to each subscriber, and is well suited for these next-generation high-speed optical access networks. However, until recently, the maximum operating speed of WDM PON has been limited to ∼1 Gb/s in most demonstrations mainly due to the limited modulation bandwidth of the colorless light sources. To overcome this problem and increase the per-wavelength operating speed of WDM PON to >10 Gb/s, it has been proposed to utilize various equalization techniques and/or the advanced modulation formats [85–87,89,195–198]. For example, it has been proposed to increase the operating speed of the colorless light source (such as RSOA) to >10 Gb/s by using the electronic equalizers such as the decision-feedback equalizer (DFE), maximum likelihood sequence estimation (MLSE) equalizer [85–87] and the optical equalizers based on the optical offset filtering and interferometer-based techniques [195,196]. In addition, there have been many attempts to overcome the bandwidth limitation of the colorless light source by using the advanced modulation formats such as the direct duobinary, quadrature phase shift keying (QPSK), and optical orthogonal-frequency-division-multiplexing (OFDM) formats, etc. [89–91,141,197,198]. The 10-Gb/s operation of WDM PON can also be achieved by using high-speed external modulators (such as REAMs and externally modulated tunable lasers) [93,94,199,200]. All these techniques have some advantages and disadvantages. Thus, for the realization of the WDM PON operating at the speed >10 Gb/s, it is necessary to select the proper technical options considering the required costs and performances. In this section, we first discuss the limitations on the high-speed operation of the colorless light sources. We then describe several techniques proposed to increase its operating speed to >10 Gb/s by using the electronic equalization technique and the advanced modulation format together with the coherent detection technique. We also introduce the potential technical solutions proposed to provide an extremely broadband (∼100 Gb/s) service to each subscriber cost-effectively.

23.5.1 Limitation on the operating speed of colorless light sources

As discussed in Section 23.2, various types of light sources can be used for the implementation of the WDM PON. When the WDM PON is implemented by using directly modulated DFB lasers or externally modulated tunable lasers, there is no serious technical difficulty in operating them at >10 Gb/s, although their costs may be a big concern. On the other hand, when the colorless light sources are used for the implementation of the WDM PON, its operating speed can be limited by various causes. For example, in the case of using the spectrum-sliced ASE light described in Section 23.2.3, it is necessary to increase its optical bandwidth significantly to achieve the SNR required for the high-speed operation [22]. In particular, for the 10-Gb/s operation, the optical bandwidth of the spectrum-sliced ASE light should be as wide as 480 GHz (to maintain the power penalty caused by the SNR degradation to be less than 2 dB). However, it can inevitably cause a large dispersion penalty and consequently limit the maximum reach of the WDM PON significantly. This limitation (i.e. imposed by the excess intensity noise of the spectrum-sliced ASE light) is also applicable to the WDM PON implemented by using the ASE-injected FP lasers. On the other hand, this limitation becomes irrelevant when the coherent single-frequency laser is used to provide the seed light instead of the ASE light (i.e. WDM PON is implemented by using the reflective-type colorless light sources based on the coherent seed light). In this case, however, the operating speed can be limited by the relatively slow electro-optic response of the reflective light source (such as RSOA). This problem can of course be resolved simply by utilizing a high-speed external modulator such as the LiNbO$_3$ modulator at the output of the colorless light source [201]. However, it may not be realistic to use such an external modulator in the access networks due to not only its high cost but also the technical difficulties (such as the large insertion loss and polarization dependence). The high-speed WDM PON can also be realized by using REAM as a colorless light source [93,94,202]. Previously, this compact device has been successfully used for the demonstrations of 10-Gb/s WDM PONs [202]. There have also been many efforts to realize the 10-Gb/s WDM PON by using the inexpensive colorless light sources such as RSOA. For this purpose, various techniques have been developed to overcome the limited modulation bandwidth of the RSOA (which is typically 1–3 GHz) by using the electronic equalization, optical equalization, and advanced modulation formats (such as the duobinary, QPSK, and OFDM formats) [85–91,195–198,203]. However, it is not clear yet which technique will be most cost-effective for the realization of the high-speed WDM PON operating at the speed at 10 Gb/s and beyond.

23.5.2 Modulation bandwidth of RSOA and its equalization technique

The modulation bandwidth of the RSOA is typically in the range of 1–3 GHz. This bandwidth is in principle determined by the carrier lifetime [204]. However, it usually becomes smaller than this maximum value due to the electrical parasitics associated with the packaging. For example, Figure 23.18 shows the photographs of the RSOAs in TO-can and butterfly packages and their measured frequency responses. This RSOA is fabricated in a planar buried hetero-structure (PBH) with a bulk InP active layer [87]. The intrinsic modulation bandwidth of this RSOA, limited by the carrier lifetime, is measured to be ~3.2 GHz by optically modulating its carrier density (to avoid the effects of packaging) [205]. However, when this RSOA is packaged in a TO-can, its modulation bandwidth is reduced to ~2.2 GHz due to the electrical parasitics. On the other hand, when this RSOA is mounted in a butterfly package, there is almost no reduction in the modulation bandwidth, as shown in Figure 23.18c.

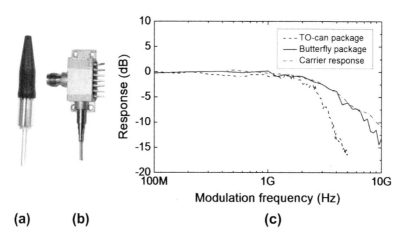

(a) **(b)** **(c)**

FIGURE 23.18 (a) TO-can packaged RSOA, (b) butterfly-packaged RSOA, and (c) measured frequency responses of RSOA (dotted curve: TO-can packaged RSOA, solid curve: butterfly-packaged RSOA, dashed curve: intrinsic carrier response of RSOA) [87].

Figure 23.18c shows that the frequency response of the RSOA has smooth roll-off characteristics with no relaxation oscillation peak or spectral zero, indicating that it is well suited for the electronic equalization technique based on the feed-forward equalizer (FFE) and the decision-feedback equalizer (DFE) [206]. Thus, by using this electronic equalization technique, it is possible to operate the RSOA at >10 Gb/s despite its narrow modulation bandwidth. The performance achievable by using the electronic equalization technique can be estimated by calculating the penalty induced by the equalizer (PIE), which represents the power penalty when the ideal equalizer (with infinite tap length) is applied. Figure 23.19 shows the calculated PIEs as a function of the bit rate for the cases of using ideal FFE and DFE in comparison with the simulated power penalty when no equalizer is used. In this calculation, it is assumed that the butterfly-packaged RSOA shown in Figure 23.18b is modulated in the non-return-to-zero (NRZ) format. The result shows that, when no electronic equalization technique is used, this RSOA can be operated at the speed of up to ~5 Gb/s. However, by using FFE and DFE, this RSOA can be operated at 10 Gb/s with a small power penalty (<1 dB). This result also indicates that it is possible to operate this RSOA even at >20 Gb/s if a large penalty can be tolerated (e.g. by using FEC code).

By using the electronic equalization technique described above, it has been demonstrated that even the TO-can packaged RSOA (having the modulation bandwidth of only 2.2 GHz) can be used for the 10-Gb/s upstream transmission in a WDM PON [85]. Figure 23.20a shows an eye diagram obtained by directly modulating this TO-can packaged RSOA with a 9.95-Gb/s NRZ signal. This figure shows that, due to the limited modulation bandwidth of this RSOA, the eye is completely closed. However, the eye opening can be recovered by applying the electronic equalization technique

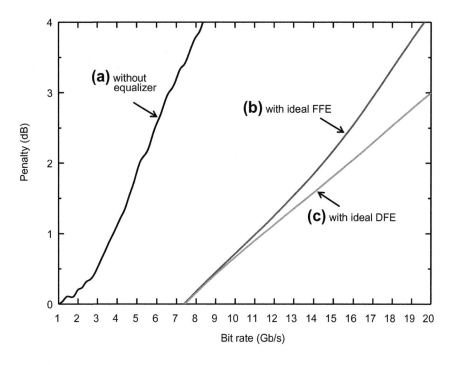

FIGURE 23.19 Calculated power penalties by using the frequency response curve of the butterfly-packaged RSOA shown in Figure 23.18 for the cases (a) without equalizer, (b) with ideal FFE, and (c) with ideal DFE.

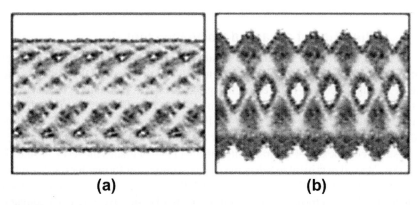

FIGURE 23.20 Eye diagrams of the TO-can packaged RSOA operating at 10 Gb/s: (a) before equalization and (b) after equalization [85].

to the received signal. For example, Figure 23.3b shows the eye diagram obtained by using an electronic equalizer consisting of half-symbol-spaced 17-tap FFE and 3-tap DFE [85]. A clear eye opening is observed, which enables the error-free transmission with the help of a proper FEC code [203]. This result indicates that it is possible to realize the high-speed colorless light source operating at 10 Gb/s cost-effectively by using the RSOA and inexpensive package.

There are electronic equalization techniques more effective than the FFE and DFE in mitigating the effects of the inter-symbol interference (ISI) caused by the insufficient modulation bandwidth such as the MLSE based on the Viterbi algorithm [206]. This performance enhancement is achieved at the cost of the increased complexity. However, due to the recent advancements in the digital signal processing (DSP) technique, it is now possible to realize the 16-state MLSE at 10 Gb/s, capable of compensating for the ISI of four neighboring bits in real time [207]. Accordingly, there have been some efforts to utilize the MLSE equalizer for increasing the operating speed of the bandwidth-limited RSOA [86,208,209]. For example, by using an 8-state MLSE receiver, it has been demonstrated that the RSOA having a modulation bandwidth of only 2.5 GHz can be used to generate the 10-Gb/s upstream signal in a bidirectional WDM PON with the maximum reach of 18 km [209]. The performance of the MLSE receiver can be further improved by using the blind partial-response maximum likelihood (PRML) equalizer, which consists of the partial response (PR) equalizer in addition to the MLSE equalizer [210]. By using this PRML equalizer, 20-Gb/s upstream transmission has been demonstrated in a 20-km reach bidirectional WDM PON using an RSOA having a modulation bandwidth of only 1 GHz. It has been reported that this equalizer is also effective in mitigating the effects of the reflection noise. The electronic equalization technique can also be utilized at the transmitter's side instead of the receiver's side [211]. For example, the electrical signal applied to the RSOA can be pre-distorted in a way that the high-frequency components in the modulation signal are emphasized. It has been reported that, to maximize the system's performance, it is necessary to utilize both the pre-emphasis technique at the transmitter and the electronic equalization technique at the receiver [211].

23.5.3 Utilization of advanced modulation formats

To overcome the limited modulation bandwidth of the inexpensive colorless light sources such as the RSOAs and utilize them in the high-speed WDM PON, the baud rate of the signal can be reduced by using the advanced modulation formats. Various modulation formats have already been examined for this purpose including the 4-ary pulse amplitude modulation (PAM) [90], direct duobinary [89], OFDM [198], and QPSK [141,197]. However, the use of these multi-level modulation formats can increase the complexity of the transmitter/receiver and degrades the receiver sensitivity [206]. Thus, it is critical to select the proper modulation format for the cost-effective implementation of the high-speed WDM PON. In this sense, it is desirable to avoid the modulation formats which require the use of expensive external modulators.

When the 4-ary PAM format is used, the baud rate of the driving signal is reduced to half. Thus, this format can alleviate the impacts of the limited modulation bandwidth of the RSOA as well as the sensitivity to the chromatic dispersion significantly. In addition, it is relatively simple to generate the 4-ary PAM signal (since it only requires modulating the RSOA with four-level electrical signal). It has been demonstrated that this format can be used to generate the 11-Gb/s signal (5.5-Gbaud) by using an RSOA having a bandwidth of merely 2.2 GHz [90]. This signal can be transmitted over >20 km with a relatively small penalty induced by the chromatic dispersion.

The duobinary signal has also been used to generate the 10-Gb/s signal from the bandwidth-limited RSOA [89]. For this purpose, the 10-Gb/s driving signal is filtered out by using a low-pass filter having a cutoff frequency of 2.5 GHz (i.e. 1/4 of the bit rate). As a result, the filtered signal has three levels $(-1, 0, +1)$. When this direct duobinary signal is applied to the RSOA under a certain bias condition, it is transmitted as a three-level PAM signal. At the receiver, the dc component in the three-level signal is eliminated, and then the binary signal $(0, 1)$ is recovered by rectification. By using this format together with the pre-emphasis technique, the 10-Gb/s transmission has been demonstrated by using an RSOA having a bandwidth of only 1.5 GHz.

There have been some attempts to utilize the directly modulated OFDM format in the RSOA-based WDM PON [198]. In this demonstration, RSOA is directly modulated with an OFDM signal at the ONU. Thus, the intensity of the RSOA is modulated with the OFDM signal and there is no need to utilize the expensive external modulator. In this OFDM signal, its subcarriers are modulated in the quadrature amplitude modulation (QAM) format. As a result, the spectrum of the driving electrical signal becomes very compact. It has been demonstrated that, by using this direct OFDM signal, the 10-Gb/s transmission can be achieved by using an RSOA having a modulation bandwidth of only 1 GHz [198].

In the previous demonstrations described above, the RSOAs have been used to generate the intensity-modulated multi-level signals. However, when the RSOA is directly modulated, its refractive index is modulated as well as the gain due to the modulated carrier density. In fact, it is well known that the semiconductor optical amplifier usually has a large chirp (i.e. large linewidth enhancement factor) [204,212]. Thus, the RSOA can also be used to generate the multi-level phase-modulated signal such as the QPSK signal, as shown in Figure 23.21 [141,197]. This is attractive since the phase-modulated signals have much superior sensitivities to the intensity-modulated signals. For example, the receiver sensitivity of the QPSK signal is better than that of the 4-level PAM signal by >8 dB [213]. However, for its detection, it is necessary to utilize the coherent detection or the differential detection technique. It is also interesting to note that the receiver sensitivities of the QPSK and BPSK signals are the same (when the coherent receivers are used). Thus, by modulating the RSOA for the QPSK signal, it is possible to achieve the high-speed operation without sacrificing the receiver sensitivity. Recently, there have been many efforts to develop the coherent detection technique cost-effective enough for use in the WDM PON, as described in Section 23.4.3. By using such a coherent receiver and the QPSK signal generated by directly modulating an RSOA having a bandwidth of 2.2 GHz, it has been demonstrated that 80-km-reach, 10-Gb/s WDM PON can be realized without using any remote optical amplifiers [141].

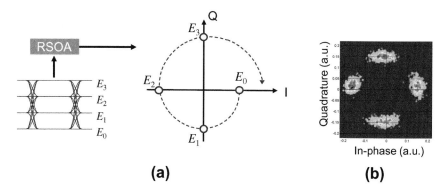

(a)   **(b)**

FIGURE 23.21 (a) Generation of the QPSK signal by directly modulating an RSOA with 4-level PAM signal and (b) the measured constellation diagram of the generated QPSK signal [197]. (For interpretation of the references to color in this figure legend, the reader is referred to the web version of this book.)

23.5.4 Ultrahigh-speed WDM PON

Due to the recent standardization activities of the 100 Gigabit Ethernet (100 GbE), there have been growing interests in the PON capable of providing 100-Gb/s service to each subscriber [214]. In some sense, it is difficult to imagine the benefits of providing such an extremely broadband service in the access network. On the other hand, it has been forecast that the next-generation PON capable of providing 10-Gb/s service to each subscriber can be deployed by the year 2020 [12,13]. Thus, it is expected to take about 10 years from the introduction of the PON capable of providing 1-Gb/s service to a subscriber to that of the PON capable of providing 10-Gb/s service. In fact, it appears that the capacity of PON has been increased by an order of magnitude every ~10 years. Considering this fact, it has been speculated that the future PON capable of providing 100-Gb/s service to each subscriber can be emerged by 2030 [14].

If necessary, the WDM PON capable of providing 100-Gb/s service can be realized, in principle, by using the polarization-multiplexed QPSK signals (e.g. generated by using tunable lasers and external LiNbO$_3$ modulators), and the digital coherent detection technique, as in the current 100-Gb/s long-haul networks. However, this solution will be prohibitively expensive even for use in future access networks. As pointed out in [215], it usually becomes unrealistic when the core transport technology is directly applied to an access system.

Recently, there have been some efforts to increase the per-wavelength operating speed of the RSOA-based WDM PON to be much faster than 10 Gb/s. For example, it has been reported that the RSOA can be directly modulated at 25.78 Gb/s [87]. For this purpose, a butterfly package is used instead of the TO-can package for the RSOA. As a result, its modulation bandwidth is improved by >40% (i.e. from 2.2 GHz to 3.2 GHz). By using this butterfly-packaged RSOA together with the electronic equalization and FEC techniques at the receiver, the error-free transmission

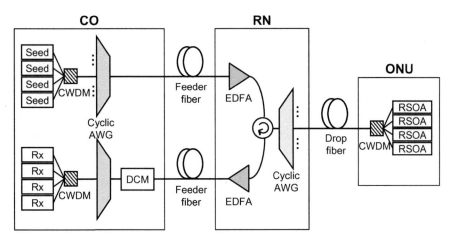

FIGURE 23.22 Upstream link of the 103-Gb/s per subscriber, long-reach WDM PON [216].

of 25.78-Gb/s signal (including the FEC overhead) is achieved. However, for the transmission of such high-speed signals, the influence of the chromatic dispersion cannot be ignored. Thus, for the demonstration of the 20-km long SMF transmission of 25.78-Gb/s signal, a dispersion-compensating module (DCM) is used at the CO to mitigate the distortion caused by the accumulated dispersion [87]. This technique has also been used for the demonstration of the ultrahigh-speed WDM PON capable of providing 100-Gb/s service to each subscriber [216]. Figure 23.22 shows the schematic diagram of the upstream link used in this demonstration. For the generation of the 108-Gb/s upstream signal, four RSOAs operating at 25.78 Gb/s are used at each ONU. The seed light for these RSOAs operates at four different wavelengths separated by the FSR of the cyclic AWG used at the RN. As the result, all the seed light assigned for a specific ONU are aggregated at the same output port of the AWG. At the ONU and the OLT, a CWDM filter is used to separate the wavelength channels. It should be noted that this ONU is still operating colorlessly. Another plausible solution for the implementation of 100-Gb/s WDM PON is the use of REAM, since it has a much wider modulation bandwidth than RSOA. In fact, it has recently been demonstrated that REAM can be used for the implementation of the high-speed WDM PON capable of providing 40-Gb/s service to each subscriber [217]. Thus, it may be possible to realize the 100-Gb/s WDM PON by using two REAMs at each ONU.

23.6 FAULT MONITORING, LOCALIZATION AND PROTECTION TECHNIQUES

Previously, the real-time fault monitoring and the network protection techniques have been considered only as optional mechanisms in the optical access networks due to their stringent cost requirements. However, it can be indispensable to have these capabilities for the future high-capacity WDM PONs, especially for use in business and mobile backhaul applications. This section reviews various techniques proposed for fault localization and network protection in WDM PONs.

23.6.1 Fault localization techniques for WDM PON

For the practical deployments of WDM PON, it is highly desirable to have the capability of detecting and localizing the fiber failures without delay. The optical time-domain reflectometry (OTDR) is the most powerful technique for this purpose. It can detect the distribution of the fiber losses as well as the fiber breaks by launching a short optical pulse into the fiber and analyzing the back-reflected light [218]. However, it is not simple to utilize the OTDR in the passive optical networks (both TDM PON and WDM PON). For use in the TDM PON, the OTDR must have a very high dynamic range to overcome the large loss occurred at the power-splitting RN [219]. In addition, it is difficult to localize the fiber failures occurred in the drop fibers, since all the OTDR traces are overlapped when the OTDR is used at the CO [219–220]. On the other hand, when the conventional OTDR is used in the WDM PON, the OTDR pulse can be blocked by the wavelength-sensitive AWG placed at the RN (Figure 23.23a). Thus, the conventional OTDR cannot localize the failures occurred in the drop fibers. Several techniques have been proposed to solve this problem by implementing an additional $1 \times N$ coupler to bypass the AWG at the RN (Figure 23.23b), using a tunable OTDR (Figure 23.23c), or generating the OTDR pulse for a specific drop fiber by using its corresponding WDM transmitter (Figure 23.23d), etc. The details of these fault localization techniques are described below.

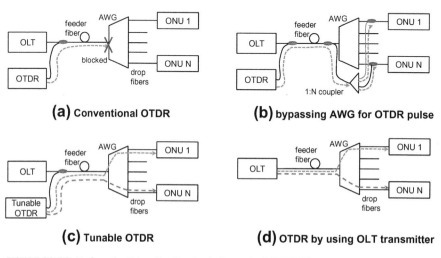

(a) Conventional OTDR

(b) bypassing AWG for OTDR pulse

(c) Tunable OTDR

(d) OTDR by using OLT transmitter

FIGURE 23.23 Various fault localization techniques for WDM PON.

23.6.1.1 Technique based on conventional OTDR and additional couplers at the RN to bypass AWG

Although the conventional OTDR can certainly be used for monitoring the failures occurring in the feeder fiber of WDM PON, it is not suitable for monitoring the failures in the drop fibers due to the wavelength-sensitive AWG placed at the RN. To overcome this problem, it has been proposed to implement additional optical paths at the RN by using a $1 \times N$ coupler and a number of WDM couplers, as shown in Figure 23.23b [221]. Thus, the OTDR pulses, operating at the wavelength different from the signal's wavelengths, can bypass the AWG through the $1 \times N$ coupler. As a result, this technique enables the conventional OTDR to monitor the failures of drop fibers even in WDM PON. This technique may be useful in the WDM PON where the additional paths are already installed at the RN for other purposes (such as the broadcast overlay) [133–135].

23.6.1.2 Technique based on tunable OTDR

Technically, a tunable OTDR is well suited for monitoring the fiber failures in WDM PON [222–226]. Figure 23.23c shows the schematic diagram of the WDM PON implemented with a tunable OTDR. In this network, it is assumed that the AWG installed at the RN has a cyclic property and the tunable OTDR operates in the different wavelength band from the signal (i.e. it is separated from the signal's wavelength by at least one FSR of the AWG used at the RN). Thus, the tunable OTDR can monitor the status of each drop fiber by tuning its wavelength without interrupting the data traffic. However, for this purpose, it is needed to install an optical bandpass filter at each ONU to block the OTDR pulse. There have also been some efforts to develop the tunable OTDR by using a broadband light source and a tunable filter instead of using a tunable laser [224–226].

23.6.1.3 Fault localization by utilizing the downstream signals

In WDM PON, instead of using the tunable OTDR, the downstream transmitters can be reused as the OTDR pulse sources whenever the fiber failure occurs in the network [227]. Figure 23.24 shows the experimental setup used to evaluate the performance of this technique. For the detection of fiber failures, the statuses of the upstream signals are monitored at the CO by using the upstream receivers. When a failure is detected, the downstream transmitter of the failed channel is automatically switched to transmit the OTDR pulses instead of data. Thus, this technique can detect and localize the failures in the drop fibers as well as in the feeder fiber without disturbing the operation of other WDM channels. Also, this technique can be cost-effective since it does not require any additional light sources or modifications in the RN and ONUs.

Another technique proposed to localize the fiber failures by using the downstream signals utilizes the correlation between the downstream signal and its reflected signal [228,229]. Figure 23.25 shows the schematic diagram of the OLT equipment implemented by using this technique. In this technique, the amplitude of the downstream signal is slightly modulated by using a low-speed pattern and the OTDR trace

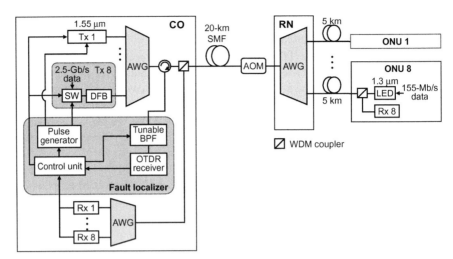

FIGURE 23.24 Fault localization technique by reusing the downstream transmitters as OTDR pulse sources [227].

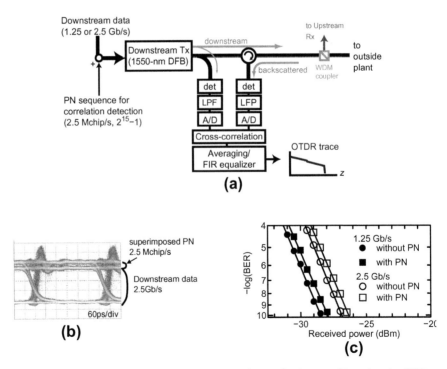

FIGURE 23.25 (a) Schematic diagram of the OLT equipment implemented by using the OTDR based on the correlation-detection technique, (b) eye diagram of the downstream signal with a superimposed PRBS pattern, and (c) measured receiver sensitivities of the downstream signals (operating at 1.25 and 2.5 Gb/s) with and without the superimposed PRBS pattern [228].

is obtained by calculating the cross-correlation between this PRBS pattern and the received backscattered light. The amplitude of the PRBS pattern is set to be small enough to minimize the degradation in the sensitivity of the downstream receiver, as shown in Figure 23.25b and c. A sufficient dynamic range is obtained by averaging out the noise components for a long period of time during the cross-correlation operation.

23.6.2 Survivable WDM PONs

Various protection and restoration techniques are currently being used to enhance the survivability of the long-haul and metro optical networks [230]. On the other hand, it has been considered only as an optional mechanism for the optical access networks due to the cost problem (as it can drastically increase the capital expenditure) [231]. However, it is bound to be more important in the future optical access networks as their capacity and split ratio are continued to be increased. In particular, it can be a critical issue for the future high-capacity WDM PON capable of supporting a large number of subscribers. For example, for the hybrid WDM/TDM PON (e.g. capable of supporting ~1000 subscribers), it will be highly desirable to have the protection capability at least for the feeder-fiber failures. The protection capability can also be a critical issue if the high-capacity WDM PON is used for the business applications or mobile backhaul applications. Thus, it is needed to develop the protection techniques cost-effective enough for the use in WDM PON. For this purpose, several techniques have been proposed to provide a protection mechanism against the fiber and/or equipment failures in WDM PON [167,232–246]. This section briefly reviews some of these techniques.

23.6.2.1 Network protection by duplicating the fiber links

One of the simplest protection techniques against the fiber failure is duplicating the relevant fiber link. This technique has been commonly used in various optical networks, and is certainly applicable to the PON. Figure 23.26a shows an exemplary protection architecture for TDM PON described in an ITU-T recommendation [231]. In this configuration, both the feeder and drop fibers are duplicated as well as the OLT and ONU equipment. Thus, this network can restore the data traffic seamlessly even in the event of a fiber or equipment failure. This technique can also be utilized in WDM PON [233–235]. For example, Figure 23.26b shows a survivable WDM PON implemented by using this technique [233]. The topology of this network is practically identical to the TDM PON shown in Figure 23.26a. However, it is interesting to note that a 3-dB coupler is used at the RN to feed the signals into the working and protection fibers in both networks shown in Figure 23.26. Thus, the power margin is inevitably reduced by 3 dB in these networks. However, this reduction in the power margin can limit the maximum reach, especially in the WDM PON implemented by using the reflective type colorless light sources in loopback configuration (since the impact of the insertion loss is doubled, as described in Section 23.4.1). To avoid this problem, an optical switch can be used instead of the 3-dB coupler, but then it is necessary to utilize a power supply at the RN and the network becomes no

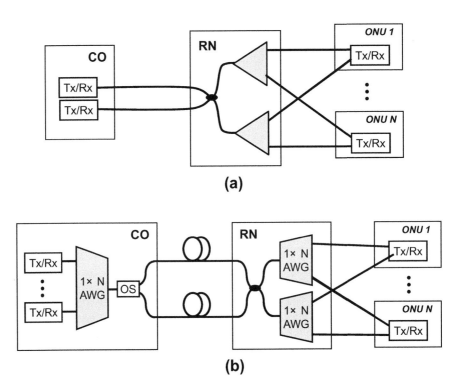

FIGURE 23.26 Network protection by duplicating fiber links in access network. (a) TDM PON [231], (b) WDM PON [233]. (For interpretation of the references to color in this figure legend, the reader is referred to the web version of this book.)

longer passive. This problem can also be avoided by using a $(2N+1)\times(2N+1)$ AWG instead of two $1\times N$ AWGs at the RN, where N is the number of ONUs [235].

23.6.2.2 Network protection by rerouting the disrupted traffic via adjacent ONU

Any fiber failures in WDM PON can be protected by duplicating every fiber connection, as described above. However, it can be too costly for the practical deployment in the access networks. To mitigate this problem, there have been some attempts to reduce the number of fiber connections required for the protection by rerouting the disrupted traffic via an adjacent ONU [236–238]. Figure 23.27 shows the schematic diagram of the proposed network architecture. In this network, the feeder fiber is protected simply by duplicating the fiber links. However, when the failure occurs in the drop fiber, the disrupted traffic is rerouted via the adjacent ONU within the same group. For this purpose, additional fiber links are installed between the neighboring

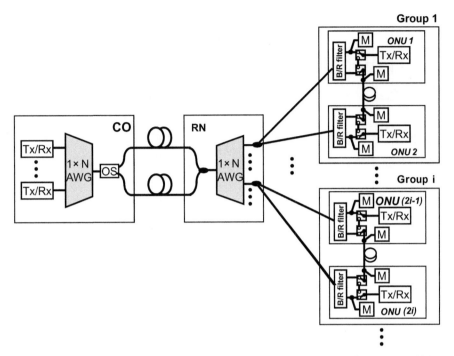

FIGURE 23.27 Network protection by rerouting the disrupted traffic via adjacent ONU [236].

ONUs (instead of duplicating the entire drop fiber links). Also, each output port of the AWG placed at the RN is split and sent to two neighboring ONUs within the same group, which operate at the different wavelength bands (i.e. "blue" and "red" bands) separated by the FSR of the AWG. Thus, each ONU always receives two signals (one for itself and the other for the neighboring ONU, which is separated in wavelength by the FSR of the AWG). An optical bandpass (blue/red) filter is used at the ONU to select only the intended signal. During the normal operation, the signal for the neighboring ONU is not used. However, when a fiber failure occurs between the RN and ONU, this unused signal is transmitted to the neighboring ONU through the optical switches and the extra fiber connection installed between the neighboring ONUs. Thus, this technique can protect any failures occurred in the drop fibers by utilizing the extra fiber links installed between the neighboring ONUs.

23.6.2.3 Network protection by rerouting the disrupted traffic via adjacent RN

In a survivable WDM PON, the most important feature is the capability of protecting the network from the feeder fiber failure. It has been proposed that this objective can be achieved without duplicating the feeder fiber installations by rerouting the disrupted traffic via the RN of the adjacent WDM PON [239]. Figure 23.28 shows an exemplary architecture of the WDM PON implemented by using this technique.

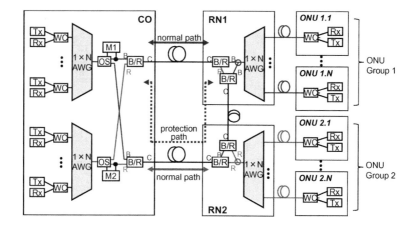

FIGURE 23.28 WDM PON architecture capable of protecting the feeder fiber failure [239].

In this network, the neighboring WDM PONs are designed to operate in the different wavelength bands (i.e. "blue" or "red" band) separated by the FSR of the AWG used at the RN. Also, instead of duplicating the feeder fiber of every WDM PON, the RNs of the neighboring WDM PONs are connected with each other by using an additional strand of fiber for the protection path. Thus, in normal operation, the downstream and upstream signals of each ONU group are transmitted through their corresponding feeder fiber. However, in the case of a fiber cut in the feeder cable, the power monitor (M1 or M2) at the CO detects a transmission failure by monitoring the optical powers of the upstream signals, and then triggers the optical switches to transfer the disrupted signals to the other port. For example, when a fiber failure occurs between the CO and RN1, the downstream signals of the ONU group 1 are transmitted through the feeder fiber between the CO and RN2 (which is already in use for the transmission of red-band signals). The blue-band signals are then separated from the red-band signals at RN2 by the B/R filters, sent to RN1 via the protection fiber installed between these two RNs, and delivered to their corresponding ONUs through the B/R filter, 3-dB coupler, AWG, and drop fibers. The upstream signals of ONU group 1 can also be transferred to CO by the opposite procedure. Thus, in this network, the protection against the feeder fiber failure can be achieved with the minimum amount of protection fibers.

This protection technique can be extended to recover the failures in drop fibers as well as feeder fibers. Figure 23.29 shows an example. At the CO, the optical path to each RN is divided by using a 3-dB coupler, and then connected to two feeder fibers of the neighboring WDM PONs through the B/R filters. The RN consists of a $1 \times N$ cyclic AWG and N B/R filters. The blue port of each B/R filter is connected to the corresponding ONU of group 1 and the red port is connected to the corresponding ONU

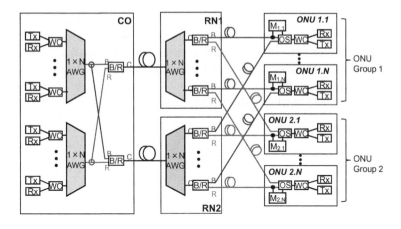

FIGURE 23.29 WDM PON architecture capable of protecting both the feeder and drop fiber failures [239].

of group 2. Thus, every ONU has two optical paths (one connected to the CO via RN1 and the other connected to the CO via RN2). One path is used in normal operation and the other path is used in case of a fiber failure (either feeder or drop fiber). When a fiber cut occurs, the power monitor at each ONU detects the transmission failure by monitoring the optical power of the downstream signal and triggers the optical switch to redirect the disrupted signals to the protection path. Since a cyclic AWG is used at the RN, all the signals (i.e. the upstream and downstream, working and protection) operating in different wavelength bands can be carried by the same AWG. Thus, this architecture can provide the self-protection capability against the failures of drop fibers as well as feeder fibers by duplicating only the drop fibers.

23.6.2.4 Network protection by using ring topology for WDM PON

It is well known that the ring topology provides excellent restoration capability for the optical network. As described in Section 23.3, there have been some efforts to utilize the self-healing ring architecture also for WDM PON [164–168,240–246]. In these networks, the CO is connected with ONUs through single- or dual-fiber rings. At the ONU, an OADM is used to add/drop its corresponding upstream/downstream signals to and from the ring. When the WDM PON is implemented in a dual-fiber ring topology, the disrupted signals are restored through the protection ring in case of a fiber failure [240–242]. However, this technique can be too costly for use in access networks. Thus, there have been several attempts to implement the WDM PON in a single-fiber ring topology [167,243–246]. In these networks, the protection is achieved by rerouting the disrupted signals to the opposite direction by using the optical switches installed at the ONUs. However, when the WDM

PON is implemented in a ring topology, the maximum number of ONUs that can be supported by this network is usually very limited due to the large insertion loss of the ONU. Thus, to overcome this limitation, it is necessary to install an optical amplifier at every ONU and compensate for its large insertion loss. However, to make this approach cost-effective enough for use in the WDM PON, it is essential to develop inexpensive photonic integrated circuits consist of an OADM and optical amplifiers.

23.7 SUMMARY

During the last three decades, the per-subscriber data rate of the access networks (including both copper- and fiber-based networks) has been increased roughly by two orders of magnitude in every 10 years. If this trend continues, it will be necessary to provide 10-Gb/s service to each subscriber by ~2020. However, it is unrealistic to provide such a broadband service by using the conventional TDM PON. Thus, WDM PON will be increasingly important for the future broadband optical access networks. However, WDM PON is still considered to be too expensive for the massive deployment. Thus, the competitiveness of WDM PON is yet to be improved for its commercial success in the coming years. For this purpose, it is essential to develop the cost-effective methods to further increase the operating speed and maximum reach of WDM PON. In this chapter, we have reviewed the fundamental issues critical to achieve these objectives and introduced the recent progresses reported for the realization of the high-speed and long-reach WDM PONs.

Acknowledgments

The authors wish to acknowledge the contributions of their current and former students at KAIST to many materials described in this chapter including K.Y. Cho, U.H. Hong, H.K. Shim, S.P. Jung, E.S. Son, K.H. Han, K.J. Park, and D.K. Jung. The authors also would like to thank Dr. M. Suzuki and Dr. A. Agata of KDDI R&D Laboratories for their helpful discussions on this subject. This work was supported in part by the National Research Foundation of Korea (2012R1A2A1A01010082).

Appendix: Acronyms

AGC	automatic gain-controlled
ASE	amplified spontaneous emission
AWG	arrayed waveguide grating
BPSK	binary phase-shift keying
CDM	code-division-multiplexing

CO	central office
DBA	dynamic bandwidth allocation
DBR	distributed Bragg reflector
DCM	dispersion-compensating module
DFB	distributed feedback
DFE	decision-feedback equalizer
DPSK	differential phase-shift-keying
DSP	digital signal processing
EDF	Erbium-doped fiber
EDFA	Erbium-doped fiber amplifier
FEC	forward-error correction
FFE	feed-forward equalizer
FP	Fabry-Perot
FR	Faraday-rotator mirror
FSK	frequency-shift-keying
FSR	free-spectral range
FTTH	fiber-to-the-home
IRZ	inverse return-to-zero
ISI	inter-symbol interference
LED	light-emitting diode
LO	local oscillator
MAC	medium access control
MEM-VCSEL	micro-electromechanical vertical cavity surface emitting lasers
MFL	multi-frequency laser
MLSE	maximum likelihood sequence estimation
NRZ	non-return-to-zero
OADM	optical add/drop multiplexer
OBI	optical beat interference
OFDM	optical orthogonal-frequency-division-multiplexing
OLT	optical line terminal
ONU	optical network unit
OTDM	optical time-division-multiplexed
OTDR	optical time-domain reflectometry
PAM	pulse amplitude modulation
PBH	planar buried hetero-structure
PIC	photonic integrated circuit
PIE	penalty induced by the equalizer
PR	partial response
PRBS	pseudo-random bit sequence
PRML	partial-response maximum likelihood
QAM	quadrature amplitude modulation
QPSK	quadrature phase shift keying

REAM	reflective electro-absorption modulator
RN	remote node
RSOA	reflective semiconductor optical amplifier
SCM	subcarrier-multiplexing
SLD	superluminescent diode
SNR	signal-to-noise ratio
SOA	semiconductor optical amplifier
SOP	state-of-polarization
TDM PON	time-division-multiplexed passive optical network
WDM	wavelength-division-multiplexing
WDM PON	wavelength-division-multiplexed passive optical network

References

[1] T. Nakahara, H. Kumamaru, S. Takeuchi, An optical fiber video system, IEEE Trans. Commun. COM-26 (7) (1978) 955–961.

[2] M. Kawahata, Optical visual information systems—some first results, in: Proc. OFC, Paper TuB2, Mar. 1979.

[3] K.Y. Chang, Fiberguide systems in the subscriber loop, Proc. IEEE 68 (10) (1980) 1291–1299.

[4] K. Sakurai, K. Asatani, A review of broad-band fiber system activity in Japan, IEEE J. Sel. Areas Commun. SAC-1 (3) (1983) 428–435.

[5] J. Kanzow, BIGFON: preparation for the use of optical fiber technology in the local network of the Deutsche bundespost, IEEE J. Sel. Areas Commun. SAC-1 (3) (1983) 436–439.

[6] K.Y. Chang, E.H. Hara, Fiber-optic broad-band integrated distribution—Elie and beyond, IEEE J. Sel. Areas Commun. SAC-1 (3) (1983) 439–444.

[7] P.W. Shumate, Fiber-to-the-home: 1977–2007, J. Lightwave Technol. 26 (9) (2008) 1093–1103.

[8] M. Horiguchi, H. Osanai, Spectral losses of low-OH-content optical fibres, Electron. Lett. 12 (12) (1976) 310–312.

[9] Global FTTH/B subscribers reach 67 million at mid-2011, Telecompaper, Feb. 13, 2012, Available at <http://www.telecompaper.com/news/global-ftthb-subscribers-reach-67-mln-at-mid-2011>.

[10] J. Savage, M. Render, Residential Gigabit Subscribers Services, Applications and Attitudes—A Joint Report Prepared for the FTTH Council, Telecom TinkTank Whitepaper, North America, Mar. 2012

[11] J.P. Elbers, K. Grobe, Optical access solutions beyond 10G-EPON/XG-PON, in: Proc. OFC, Paper OTuO1, Mar. 2010.

[12] R. Heron, Next-generation optical access networks, in: Proc. ANIC, Paper AMA2, June 2011.

[13] Y.C. Chung, Recent advancement in WDM PON technology, in: Proc. ECOC, Paper Th.11.C.4, Sept. 2011.

[14] T. Miki, H. Ishio, Viabilities of the wavelength-division-multiplexing transmission system over an optical fiber cable, IEEE Trans. Commun. COM-26 (7) (1978) 1082–1087.

[15] K. Asatani, R. Watanabe, K. Nosu, T. Matsumoto, F. Nihei, A field trial of fiber optic subscriber loop systems utilizing wavelength-division multiplexers, IEEE Trans. Commun. 30 (9) (1982) 2172–2184.

[16] E.-J. Bachus, R.-P. Braun, W. Eutin, E. Großmann, H. Foisel, K. Heimes, B. Strebel, Coherent optical-fiber subscriber line, Electron. Lett. 21 (25/26) (1985) 1203–1205.

[17] B. Glance, K. Pollock, C.A. Burrus, B.L. Kasper, G. Eisenstein, L.W. Stulz, Densely spaced WDM coherent star network, Electron. Lett. 23 (17) (1987) 875–877.

[18] Y.C. Chung, K.J. Pollock, P.J. Fitzgerald, B. Glance, R.W. Tkach, A.R. Chraplyvy, WDM coherent star network with absolute frequency reference, Electron. Lett. 24 (21) (1988) 1313–1315.

[19] S.S. Wagner, H. Kobrinski, T.J. Robe, H.L. Lemberg, L.S. Smoot, Experimental demonstration of a passive optical subscriber loop architecture, Electron. Lett. 24 (6) (1988) 344–346.

[20] M.H. Reeve, A.R. Hunwicks, W. Zhao, S.G. Methley, L. Bickers, S. Hornung, LED spectral slicing for single-mode local loop applications, Electron. Lett. 24 (7) (1988) 389–391.

[21] A.R. Hunwicks, L. Bickers, P. Rogerson, A spectrally sliced, single-mode, optical transmission system installed in the UK local loop network, in: Proc. IEEE Globecom, 1989, pp. 1303–1307.

[22] J.S. Lee, Y.C. Chung, D.J. DiGiovanni, Spectrum-sliced fiber amplifier light source for multichannel WDM applications, IEEE Photon. Technol. Lett. 5 (12) (1993) 1458–1461.

[23] N.J. Frigo, P.P. Iannone, P.D. Magill, T.E. Darcie, M.M. Downs, B.N. Desai, U. Koren, T.L. Koch, C. Dragnone, H.M. Presby, G.E. Bodeep, A wavelength-division multiplexed passive optical network with cost-shared components, IEEE Photon. Technol. Lett. 6 (11) (1994) 1365–1367.

[24] A.M. Hill, S. Carter, L. Blair, D.J. Pratt, G. Sherlock, R.A. Harmon, S. Culverhouse, M. Searle, J.J. O'Reilly, C. Appleton, Broadband upgrade of optical access networks by HDWDM, in: Proc. ECOC, 1994, pp. 777–780.

[25] M.D. Feuer, J.M. Wiesenfeld, J.S. Perino, C.A. Burrus, G. Raybon, S.C. Shunk, N.K. Dutta, Single-port laser-amplifier modulators for local access, IEEE Photon. Technol. Lett. 8 (9) (1996) 1175–1177.

[26] D.K. Jung, S.K. Shin, C.H. Lee, Y.C. Chung, Wavelength-division-multiplexed passive optical network based on spectrum-slicing techniques, IEEE Photon. Technol. Lett. 10 (9) (1998) 1334–1336.

[27] A. Banerjee, Y. Park, F. Clarke, H. Song, S. Yang, G. Kramer, K. Kim, B. Mukherjee, Wavelength-division-multiplexed technologies for broadband access: a review, J. Opt. Netw. 4 (11) (2005) 737–758.

[28] C.-H. Lee, S.-M. Lee, K.-M. Choi, J.-H. Moon, S.-G. Mun, K-T. Jeong, J.H. Kim, B. Kim, WDM-PON experiences in Korea, J. Opt. Netw. 6 (5) (2007) 451–464.

[29] K. Grobe, J.-P. Elbers, PON in adolescence: from TDMA to WDM-PON, IEEE Commun. Mag. (2008) 26–34.

[30] H. Kobrinski, R.M. Bulley, M.S. Goodman, M.P. Vechhi, C.A. Brackett, Demonstration of high capacity in the LAMDANET architecture: a multiwavelength optical network, Electron. Lett. 23 (16) (1987) 824–826.

[31] Y.C. Chung, J. Jeong, L.S. Cheng, Aging-induced wavelength shifts in 1.5-mm DFB lasers, IEEE Photon. Technol. Lett. 6 (7) (1994) 792–795.

[32] R. Nagarajan, C.H. Joyner, R.P. Schneider, Jr J.S. Bostak, T. Butrie, A.G. Dentai, V.G. Dominic, P.W. Evans, M. Kato, M. Kauffman, D.J.H. Lambert, S.K. Mathis, A. Mathur, R.H. Miles, M.L. Mitchell, M.J. Missey, S. Murthy, A.C. Nilsson, F.H. Peters, S.C. Pennypacker, J.L. Pleumeekers, R.A. Salvatore, R.K. Schlenker, R.B. Taylor, H.S. Tsai, M.F. Van Leeuwen, J. Webjorn, M. Ziari, D. Perkins, J. Singh, S.G. Grubb, M.S. Reffle, D.G. Mehuys, F.A. Kish, D.F. Welch, Large-scale photonic integrated circuits, IEEE J. Sel. Top. Quantum Electron. 1 (1) (2005) 50–65.

[33] T. Schrans, G. Yoffe, Y. Luo, R. Narayan, S. Rangarajan, D. Hui, F. Kusnadi, A. Hanjani, B. Pezeshki, 100 Gb/s 10 km link performance of 10×10 Gb/s hybrid approach with integrated WDM array of DFB lasers, in: Proc. NFOEC, Paper NThA4, Mar. 2009.

[34] J. Li, X. Chen, N. Zhou, J. Zhang, X. Huang, L. Li, H. Wang, Y. Lu, H. Zhu, Monolithically integrated 30-wavelength DFB laser array, in: Proc. ACP (SPIE-OSA-IEEE Vol. 7631), Paper 763104, Nov. 2009.

[35] M. Zirngibl, C.H. Joyner, C.R. Doerr, L.W. Stulz, H.M. Presby, An 18-channel multifrequency laser, IEEE Photon. Technol. Lett. 8 (7) (1996) 870–872.

[36] K. Lawniczuk, R. Piramidowicz, P. Szczepanski, P.J. Williams, M.J. Wale, M.K. Smit, X.J.M. Leijtens, 8-Channel AWG-based multiwavelength laser fabricated in a multi-project wafer run, in: Proc. IPRM (Inter. Conf. on Indium Phosphide and Related Materials), May 2011, pp. 1–4.

[37] T.L. Koch, U. Koren, Semiconductor lasers for coherent optical fiber communications, J. Lightwave Technol. 8 (3) (1990) 274–292.

[38] L.A. Coldren, Monolithic tunable diode lasers, IEEE J. Sel. Top. Quantum Electron. 6 (6) (2000) 988–999.

[39] L.A. Coldren, G.A. Fish, Y. Akulova, J.S. Barton, L. Johansson, C.W. Coldren, Tunable semiconductor lasers: a tutorial, J. Lightwave Technol. 22 (1) (2004) 193–202.

[40] M. Jiang, C.-C. Lu, P. Chen, J.-H. Zhou, J. Cai, K. McCallion, K.J. Knopp, P.D. Wang, M. Azimi, D. Vakhshoori, Error free 2.5 Gb/s transmission over 125 km conventional fiber of a directly modulated widely tunable vertical cavity surface emitting laser, in: Proc. OFC, Paper TuJ3-1, Mar. 2001.

[41] B. Glance, C.A. Burrus, L.W. Stulz, Fast frequency-tunable external-cavity laser, Electron. Lett. 23 (3) (1987) 98–99.

[42] J.H. Lee, M.Y. Park, C.Y. Kim, S.H. Cho, W. Lee, G. Jeong, B.W. Kim, Tunable external cavity laser based on polymer waveguide platform for WDM access network, IEEE Photon. Technol. Lett. 17 (9) (2005) 1956–1958.

[43] G. Jeong, J.H. Lee, M.Y. Park, C.Y. Kim, S.H. Cho, W. Lee, B.W. Kim, Over 26-nm wavelength tunable external cavity laser based on polymer waveguide platforms for WDM access networks, IEEE Photon. Technol. Lett. 18 (20) (2006) 2102–2104.

[44] Y.O. Noh, H.J. Lee, J.J. Ju, M.S. Kim, S.H. Oh, M.C. Oh, Continuously tunable compact lasers based on thermo-optic polymer waveguides with Bragg gratings, Opt. Express 16 (22) (2008) 18194–18201.

[45] K.H. Yoon, S.H. Oh, K.S. Kim, O.K. Kwon, D.K. Oh, Y.O. Noh, H.J. Lee, 2.5-Gb/s hybridly-integrated tunable external cavity laser using a superluminescent diode and a polymer Bragg reflector, Opt. Express 18 (6) (2010) 5556–5561.

[46] B.-S. Choi, S.H. Oh, K.S. Kim, K.-H. Yoon, H.S. Kim, M.-R. Park, J.S. Jeong, O.-K. Kwon, J.-K. Seo, H.-K. Lee, Y.C. Chung, 10-Gb/s direct modulation of polymer-based tunable external cavity lasers, Opt. Express 20 (18) (2012) 20358–20375.

[47] Y.A. Akulova, G.A. Fish, P.-C. Koh, C.L. Schow, P. Kozodoy, A.P. Dahl, S. Nakagawa, M.C. Larson, M.P. Mack, T.A. Strand, C.W. Coldren, E. Hegblom, S.K. Penniman,

T. Wipiejewski, L.A. Coldren, Widely tunable electroabsorption-modulated sampled-grating (DBR) laser transmitters, IEEE J. Sel. Top. Quantum Electron. 8 (6) (2002) 1349–1357.

[48] A.J. Ward, D.J. Robbins, G. Busico, E. Barton, L. Ponnampalam, J.P. Duck, N.D. Whitbread, P.J. Williams, D.C.J. Reid, A.C. Carter, M.J. Wale, Widely tunable DS-DBR laser with monolithically integrated SOA: design and performance, IEEE J. Sel. Top. Quantum Electron. 11 (1) (2005) 149–156.

[49] J. Buus, E.J. Murphy, Tunable lasers in optical networks, J. Lightwave Technol. 24 (1) (2006) 5–10.

[50] M. Roppelt, M. Eiselt, K. Grobe, J.-P. Elbers, Tuning of an SG-Y branch laser for WDM-PON, in: Proc. OFC, Paper OW1B.4, Mar. 2012.

[51] M.J. Wale, Technology options for future WDM-PON access systems, in: Proc. OECC, Paper TuH4, July 2009.

[52] M.J. Wale, Photonic integration challenges for next-generation networks, in: Proc. ECOC, Paper 1.7.4, Sept. 2009.

[53] M.J. Wale, Options and trends for PON tunable optical transmitters, in: Proc. ECOC, Paper Mo.2.C.1, Sept. 2011.

[54] Y. Liu, J.D. Ingham, R.G.S. Plumb, R.V. Penty, I.H. White, D.J. Robbins, N.D. Whitbread, A.J. Ward, Directly-modulated DS-DBR tunable laser for uncooled C-band WDM system, in: Proc. OFC, Paper OThG8, Mar. 2006.

[55] C.C. Renaud, M.J. Fice, I. Lealman, P. Cannard, L. Rivers, A.J. Seeds, 100 GHz spaced 10 Gbit/s WDM over 10°C to 70°C using an uncooled DBR laser, in: Proc. OFC, Paper OWI76, Mar. 2006.

[56] K.J. Park, S.K. Shin, H.C. Ji, H.G. Woo, Y.C. Chung, A multi-wavelength locker for WDM system, in: Proc. OFC, Paper WE4-1, Mar. 2000.

[57] M. Roppelt, F. Pohl, K. Grobe, M. Eiselt, J.-P. Elbers, Tuning methods for uncooled low-cost tunable lasers in WDM-PON, in: Proc. NEOEC, Paper NTuB1, Mar. 2011.

[58] S.S. Wanger, T.E. Chapuran, Eiselt an high-density WDM transmission using superluminescent diodes, Electron. Lett. 26 (11) (1990) 696–697.

[59] Y.C. Chung, J.S. Lee, R.M. Derosier, D.J. DiGiovanni, 1.7 Gbit/s transmission over 165 km of dispersion-shifted fiber using spectrum-sliced fiber amplifier light source, Electron. Lett. 30 (17) (1994) 1427–1428.

[60] G.J. Pendock, D.D. Sampson, Transmission performance of high bit rate spectrum-sliced WDM systems, J. Lightwave Technol. 14 (10) (1996) 2141–2148.

[61] K.Y. Liou, U. Koren, E.C. Burrows, J.L. Zyskind, K. Dreyer, A WDM access system architecture based on spectral slicing of an amplified LED and delay-line multiplexing and encoding of eight wavelength channels for 64 subscribers, IEEE Photon. Technol. Lett. 9 (4) (1997) 517–519.

[62] K.H. Han, E.S. Son, H.Y. Choi, K.W. Lim, Y.C. Chung, Bidirectional WDM PON using light-emitting diodes spectrum-sliced with cyclic arrayed-waveguide grating, IEEE Photon. Technol. Lett. 16 (10) (2004) 2380–2382.

[63] M. Oksanen, O-P. Hiironen, A. Rervonen, A. Pietlainen, E. Gotsonoga, H. Jarvinen, H. Kaaja, J. Aarnio, A. Grohn, M. Karhiniemi, V. Moltchanov, M. Oikkonen, M. Tahkokorpi, T. Wallenius, Spectral slicing passive optical access network trial, in: Proc. OFC, Paper ThH2, Mar. 2002.

[64] A.J. Keating, W.T. Hollway, D.D. Sampson, Feedforward noise reduction of incoherent light for spectrum-sliced transmission at 2.5 Gb/s, IEEE Photon. Technol. Lett. 7 (12) (1995) 1513–1515.

[65] V. Arya, I. Jacobs, Optical preamplifier receiver for spectrum-sliced WDM, J. Lightwave Technol. 15 (4) (1997) 576–583.

[66] F. Koyama, T. Yamatoya, K. Iga, Highly gain-saturated GaInAsP/InP SOA modulator for incoherent spectrum-sliced light source, in: Proc. International Conf. on Indium Phosphide and Related Materials, Paper WP2.26, May 2000.

[67] A.D. McCoy, B.C. Thomsen, M. Ibsen, D.J. Richardson, Filtering effects in a spectrum-sliced WDM system using SOA-based noise reduction, IEEE Photon. Technol. Lett. 16 (2) (2004) 680–682.

[68] E.A. De Souza, M.C. Nuss, M. Zirngibl, C.H. Joyner, Spectrally sliced WDM using a single femtosecond source, in: Proc. OFC, Paper PD16-1, Feb. 1995.

[69] S. Kawanish, H. Takara, K. Uchiyama, I. Shake, and K. Mori, 3 Tbit/s (160 Gbit/s × 19 ch) OTDM/WDM transmission experiment, in: Proc. OFC/IOOC, Paper PD-1, Feb. 1999.

[70] L. Boivin, S. Taccheo, C.R. Doerr, P. Shiffer, L.W. Stulz, R. Monnard, W. Lin, 400 Gbit/s transmission over 544 km from spectrum-sliced supercontinuum source, Electron. Lett. 36 (4) (2000) 335–336.

[71] D.D. Sampson, W.T. Holloway, 100 mW spectrally-uniform broadband ASE source for spectrum-sliced WDM systems, in: Proc. OAA, Paper PD2-1, Aug. 1994.

[72] K.Y. Liou, B. Glance, U. Koren, E.C. Burrows, G. Raybon, C.A. Burrus, K. Dreyer, Monolithically integrated semiconductor LED-amplifier for applications as transceivers in fiber access systems, IEEE Photon. Technol. Lett. 8 (6) (1996) 800–802.

[73] F. Koyama, High power superluminescent diodes for multi-wavelength light sources, in: Proc. LEOS Annual Meeting, Paper TuY2, Nov. 1997.

[74] S. Kaneko, J. Kani, K. Iwatsuki, A. Ohki, M. Sugo, S. Kamei, Scalability of spectrum-sliced DWDM transmission and its expansion using forward error correction, J. Lightwave Technol. 24 (3) (2006) 1295–1301.

[75] J. Briand, F. Payoux, P. Chanclou, M. Joindot, Forward error correction in WDM PON using spectrum slicing, Opt. Switching Networking 4 (2007) 131–136.

[76] D. Fattal, M. Fiorentino, M. Tan, D. Houng, S.Y. Wang, R.G. Beausoleil, Design of an efficient light-emitting diodes with 10 GHz modulation bandwidth, Appl. Phys. Lett. 93 (2008) 243501-1–243501-3.

[77] A.M. Hill, S. Carter, D.J. Pratt, G. Sherlock, R.A. Harmon, S. Culverhouse, M. Searle, J.J. O'Reilly, C. Appleton, Broad-band upgrade of optical access networks by HDWDM, in: Proc. ECOC, Sept. 1994, pp. 777–780.

[78] P. Healey, P. Townsend, C. Ford, L. Johnston, P. Tonwley, I. Lealman, L. Rivers, S. Perrin, R. Moore, Spectral slicing WDM-PON using wavelength-seeded reflective SOAs, Electron. Lett. 37 (19) (2001) 1181–1182.

[79] J.H. Lee, S.-H. Cho, H.-H. Lee, E.-S. Jung, J.-H. Yu, B.-W. Kim, S.-H. Lee, J.-S. Koh, B.-H. Sung, S.-J. Kang, J.-H. Kim, K.-T. Jeong, S.S. Lee, First commercial deployment of a colorless gigabit WDM/TDM hybrid PON system using remote protocol terminator, J. Lightwave Technol. 28 (4) (2010) 344–351.

[80] H.C. Jeon, K.Y. Cho, Y. Takushima, Y.C. Chung, High reflection tolerance of 1.25-Gb/s RSOA-based WDM PON employing spectrum-sliced ASE source, in: Proc. SPIE, vol. 7136, pp. 71360N-1–71360N-11 (presented at APOC 2008), Oct. 2008.

[81] Y. Takushima, K.Y. Cho, Y.C. Chung, Design issues in RSOA-based WDM PON, in: Proc. IEEE Photonics Global Conference, pp. 1–4, Dec. 2008.

[82] H.S. Shin, D.K. Jung, D.J. Shin, S.B. Park, J.S. Lee, I.K. Yun, S.W. Kim, Y.J. Oh, C.S. Shim, 16 × 1.25 Gbit/s WDM-PON based on ASE-injected R-SOAs in 60 oC temperature range, in: Proc. OFC, Paper OTuC5, Mar. 2006.

[83] K.Y. Cho, Y. Takushima, Y.C. Chung, Enhanced operating range of WDM PON implemented by using uncooled RSOAs, IEEE Photon. Technol. Lett. 20 (18) (2008) 1536–1538.

[84] P. Chanclou, F. Payoux, T. Soret, N. Genay, R. Brenot, F. Blache, M. Goix, J. Landreau, O. Legouezigou, F. Mallecot, Demonstration of RSOA-based remote modulation at 2.5 and 5 Gbit/s for WDM PON, in: Proc. OFC, Paper OWD1, Mar. 2007.

[85] K.Y. Cho, Y. Takushima, Y.C. Chung, 10-Gb/s operation of RSOA for WDM PON, IEEE Photon. Technol. Lett. 20 (18) (2008) 1533–1535.

[86] A. Agata, K.Y. Cho, Y. Takushima, Y.C. Chung, Y. Horiuchi, Study on ISI mitigation capability of MLSE equalizers in RSOA-based 10 Gbit/s WDM PON, in: Proc. ECOC, Paper 9.5.5, Sept. 2009.

[87] K.Y. Cho, B.S. Choi, Y. Takushima, Y.C. Chung, 25.78-Gb/s operation of RSOA for next-generation optical access networks, IEEE Photon. Technol. Lett. 23 (8) (2011) 495–497.

[88] Q. Guo, A.V. Tran, C.-J. Chae, 10-Gb/s WDM-PON based on low-bandwidth RSOA using partial response equalization, IEEE Photon. Technol. Lett. 23 (20) (2011) 1442–1444.

[89] M. Omella, V. Polo, J. Lazaro, B. Schrenk, J. Prat, 10 Gb/s RSOA transmission by direct duobinary modulation, in: Proc. ECOC, Paper Tu.3.E.4, Sept. 2008.

[90] K.Y. Cho, Y. Takushima, Y.C. Chung, Enhanced chromatic dispersion tolerance of 11 Gbit/s RSOA-based WDM PON using 4-ary PAM signal, Electron. Lett. 46 (22) (2010) 1510–1512.

[91] K.Y. Cho, U.H. Hong, A. Agata, T. Sano, Y. Horiuchi, H. Tanaka, M. Suzuki, Y.C. Chung, 10-Gb/s, 80-km reach RSOA-based WDM PON employing QPSk signal and self-homodyne receiver, in: Proc. OFC, Paper OW1B.1, Mar. 2012.

[92] G. de Valicourt, D. Make, C. Fortin, A. Enard, F. Van Dijk, R. Brenot, 10 Gbit/s modulation of reflective SOA without any electronic processing, in: Proc. OFC, Paper OTh2, Mar. 2011.

[93] A. Garreau, J. Decobert, C. Kazamierski, M.-C. Cuisin, J.-G. Provost, H. Sillard, F. Blache, D. Carpentier, J. Landreau, P. Chanclou, 10 Gbit/s amplified reflective electroabsorption modulator for colorless access networks, in: Proc. International Conf. on Indium Phosphide and Related Materials, Paper TuA2.3, May 2006.

[94] C. Kazmierski, Remote amplified modulators: key components for 10 Gb/s WDM PON, in: Proc. OFC, Paper OWN6, Mar. 2010.

[95] H.D. Kim, S.-G. Kang, C.-H. Lee, A low-cost WDM source with an ASE injected Fabry-Perot semiconductor laser, IEEE Photon. Technol. Lett. 12 (8) (2000) 1067–1069.

[96] M. Fujiwara, H. Suzuki, K. Iwatsuki, M. Sugo, Noise characteristics of signal reflected from ASE-injected FP-LD in loopback access network, Electron. Lett. 42 (2) (2006) 111–112.

[97] P.J. De Groot, Range-dependent optical feedback effects on the multimode spectrum of laser diodes, J. Modern Optics 37 (7) (1990) 1199–1214.

[98] G. Yabre, H. de Waardt, H.P.A. van den Boom, G-D. Khoe, Noise characteristics of single-mode semiconductor lasers under external light injection, IEEE J. Quantum Electron. 36 (3) (2000) 385–392.

[99] D.J. Shin, D.K. Jung, H.S. Shin, J.W. Kwon, S. Hwang, Y. Oh, C. Shim, Hybrid WDM/TDM-PON with wavelength-selection-free transmitters, J. Lightwave Technol. 23 (1) (2005) 187–195.

[100] S.-G. Mun, J.-H. Moon, H.-K. Lee, J-Y. Kim, C-H. Lee, A WDM-PON with a 40 Gb/s (32 × 1.25 Gb/s) capacity based on wavelength-locked Fabry-Perot laser diodes, Opt. Express 16 (15) (2008) 11361–11368.

[101] H.-K. Lee, H.-S. Cho, J.-Y. Kim, C.-H. Lee, A WDM-PON with an 80 Gb/s capacity based on wavelength-locked Fabry-Perot laser diode, Opt. Express 18 (17) (2010) 18077–18085.

[102] Z. Xu, Y.J. Wen, W.-D. Zhong, C.-J. Chae, X.-F. Cheng, Y. Wang, C. Lu, J. Shankar, High-speed WDM PON using CW injection-locked Fabry-Perot laser diodes, Opt. Express 15 (6) (2007) 2953–2962.

[103] N.J. Frigo, P.P. Iannone, P.D. Magill, T.E. Darcie, M.M. Downs, B.N. Desai, U. Koren, T.L. Koch, C. Dragone, H.M. Presby, G.E. Bodeep, A wavelength-division-multiplexed passive optical network with cost-shared components, IEEE Photon. Technol. Lett. 6 (11) (1994) 1365–1367.

[104] L.Y. Chan, C.K. Chan, D.T.K. Tong, F. Tong, L.K. Chen, Upstream traffic transmitter using injection-locked Fabry-Perot laser diode as modulator for WDM access networks, Electron. Lett. 38 (1) (2002) 43–45.

[105] J.J. Koponen, M.J. Soderlund, A duplex WDM passive optical network with 1:16 power split using reflective SOA remodulator at ONU, in: Proc. OFC, Paper MF99, Feb. 2004.

[106] F. Payoux, P. Chanclou, T. Soret, N. Genay, R. Brenot, Demonstration of a RSOA-based wavelength remodulation scheme in 1.25 Gbit/s bidirectional hybrid WDM-TDM PON, in: Proc. OFC, Paper OTuC4, Mar. 2006.

[107] S.Y. Kim, S.B. Jun, Y. Takushima, E.S. Son, Y.C. Chung, Enhanced performance of RSOA-based WDM PON by using Manchester coding, J. Opt. Netw. 6 (6) (2007) 624–630.

[108] H. Takesue, T. Sugie, Wavelength channel data rewrite using saturated SOA modulator for WDM networks with centralized light sources, J. Lightwave Technol. 21 (11) (2003) 2546–2556.

[109] W. Lee, M.Y. Park, S.H. Cho, J. Lee, C. Kim, G. Jeong, B.W. Kim, Bidirectional WDM-PON based on gain-saturated reflective semiconductor optical amplifiers, IEEE Photon. Technol. Lett. 17 (11) (2005) 2460–2462.

[110] J.-M. Kang, S.-K. Han, A novel hybrid WDM/SCM-PON sharing wavelength for up- and down-link using reflective semiconductor optical amplifier, IEEE Photon. Technol. Lett. 18 (3) (2006) 502–504.

[111] K.Y. Cho, A. Murakami, Y.J. Lee, A. Agata, Y. Takushima, Y.C. Chung, Demonstration of RSOA-based WDM PON operating at symmetric rate of 1.25 Gb/s with high reflection tolerance, in: Proc. OFC, Paper OTuH4, Feb. 2008.

[112] J. Prat, C. Arellano, V. Polo, C. Bock, Optical network unit based on a bidirectional reflective semiconductor optical amplifier for fiber-to-the-home networks, IEEE Photon. Technol. Lett. 17 (1) (2005) 250–252.

[113] G.-W. Lu, N. Deng, C.-K. Chan, L.-K. Chen, Use of downstream inverse-RZ signal for upstream data re-modulation in a WDM passive optical network, in: Proc. OFC, Paper OFI8, Mar. 2005.

[114] J. Prat, V. Polo, C. Bock, C. Arellano, J.J.V. Olmos, Full-duplex single-fiber transmission using FSK downstream and IM remote upstream modulations for fiber-to-the-home, IEEE Photon. Technol. Lett. 17 (3) (2005) 702–703.

[115] Y. Tian, Y. Su, L. Yi, L. Leng, X. Tian, H. He, X. Xu, Optical VPN in PON based on DPSK erasing/rewriting and DPSK/IM formatting using a single Mach-Zehnder modulator, in: Proc. ECOC, Sept. 2006.

[116] C. Arellano, C. Bock, J. Prat, K.-D. Langer, RSOA-based optical network units for WDM-PON, in: Proc. OFC, Paper OTuC1, Mar. 2006.

[117] S.Y. Kim, E.S. Son, S.B. Jun, Y.C. Chung, Effects of downstream modulation formats on the performance of bidirectional WDM-PON using RSOA, in: Proc. OFC, Paper OWD3, Mar. 2007.

[118] J.-H. Yu, N. Kim, B.W. Kim, Remodulation schemes with reflective SOA for colorless DWDM PON, J. Opt. Netw. 6 (8) (2007) 1041–1054.

[119] M. Fujiwara, J. Kani, H. Suzuki, K. Iwatsuki, Impact of backreflection on upstream transmission in WDM single-fiber loopback access networks, J. Lightwave Technol. 24 (2) (2006) 740–746.

[120] Y.J. Lee, K.Y. Cho, A. Murakami, A. Agata, Y. Takushima, Y.C. Chung, Reflection tolerance of RSOA-based WDM PON, in: Proc. OFC, Paper OTuH5, Feb. 2008.

[121] K.Y. Cho, Y.J. Lee, H.Y. Choi, A. Murakami, A. Agata, Y. Takushima, Y.C. Chung, Effects of reflection in RSOA-based WDM PON utilizing remodulation technique, J. Lightwave Technol. 27 (10) (2009) 1286–1294.

[122] P.J. Urban, A.M.J. Koonen, G.D. Khoe, H. De Waardt, Mitigation of reflection-induced crosstalk in a WDM access network, in: Proc. OFC, Paper OthT3, Feb. 2008.

[123] A. Murakami, H.C. Jeon, K.Y. Cho, A. Agata, Y. Takushima, Y.C. Chung, Y. Horiuchi, Reflection tolerance enhancement of RSOA-based WDM PON by using optical frequency dithering, in: Proc. OFC, Paper ONM4, Mar. 2009.

[124] C. Bock, J.M. Fabrega, J. Prat, Ultra-dense WDM PON based on homodyne detection and local oscillator reuse for upstream transmission, in: Proc. ECOC, Sept. 2006.

[125] H. Rohde, S. Smolorz, E. Gottwald, K. Kloppe, Next generation optical access—1 Gbit/s for everyone, in: Proc. ECOC, Paper 10.5.5, Sept. 2009.

[126] S. Smolorz, E. Gottwald, H. Rohde, D. Smith, A. Poustie, Demonstration of a coherent UDWDM-PON with real-time processing, in: Proc. OFC, Paper PDPD4, Mar. 2011.

[127] G. Mayer, M. Martinelli, A. Pattavina, E. Salvadori, Design and cost performance of the multistage WDM PON access networks, J. Lightwave Technol. 18 (2) (2000) 121–142.

[128] C.R. Giles, M. Zirngibl, C. Joyner, 1152-subscriber WDM access PON architecture using a sequentially pulsed multifrequency laser, IEEE Photon. Technol. Lett. 9 (9) (1997) 1283–1284.

[129] G. Talli, P.D. Townsend, Hybrid DWDM-TDM long-reach PON for next-generation optical access, J. Lightwave Technol. 24 (7) (2006) 2827–2834.

[130] J. Yu, M. Lee, Y. Kim, J. Park, WDM/SCM MAC protocol suitable for passive double star optical network, in: Proc. CLEO/Pacific Rim 2001, Paper ThB4-4, July 2001.

[131] T. Pfeiffer, J. Kissing, J.-P. Elbers, B. Deppisch, M. Witte, H. Schmuck, E. Voges, Coarse WDM/CDM/TDM concept for optical packet transmission in metropolitan and access networks supporting 400 channels at 2.5 Gb/s peak rate, J. Lightwave Technol. 18 (12) (2000) 1928–1938.

[132] G.C. Gupta, M. Kashima, H. Iwamura, H. Tamai, T. Ushikubo, T. Kamijoh, Over 100 km bidirectional, multi-channel COF-PON without optical amplifier, in: Proc. OFC, Paper PDP51, Mar. 2006.

[133] U. Hilbk, Th. Hermes, J. Saniter, F.-J. Westphal, High capacity WDM overlay on a passive optical network, Electron. Lett. 32 (23) (1996) 2162–2163.

[134] P.P. Iannone, K.C. Reichmann, N.J. Frigo, High-speed point-to-point and multiple broadcast services delivered over a WDM passive optical network, IEEE Photon. Technol. Lett. 10 (9) (1998) 1328–1330.

[135] E.S. Son, K.H. Han, J.K. Kim, Y.C. Chung, Bidirectional WDM passive optical network for simultaneous transmission of data and digital broadcast video service, J. Lightwave Technol. 21 (8) (2003) 1723–1727.

[136] M. Zirngibl, C.H. Joyner, L.W. Stulz, WDM receiver by monolithic integration of an optical preamplifier, waveguide grating router and photodiode array, Electron. Lett. 31 (7) (1995) 582–583.

[137] F. Tong, C.-S. Li, G. Berkowitz, A 32-channel tunable receiver module for wavelength-division multiple-access networks, IEEE Photon. Technol. Let. 9 (11) (1997) 1523–1525.

[138] S.P. Jung, Y. Takushima, Y.C. Chung, Transmission of 1.25-Gb/s PSK signal generated by using RSOA in 110-km coherent WDM PON, Opt. Express 18 (14) (2010) 14871–14877.

[139] D. Lavery, M. Ionescu, S. Makovejs, E. Torrengo, S.J. Savory, A long-reach ultra-dense 10Gbit/s WDM PON using a digital coherent receiver, Opt. Express 18 (25) (2010) 25855–25860.

[140] K.Y. Cho, K. Tanaka, T. Sano, S.P. Jung, J.H. Chang, Y. Takushima, A. Agata, Y. Horiuchi, M. Suzuki, Y.C. Chung, Long-reach coherent WDM PON employing self-polarization-stabilization technique, J. Lightwave Technol. 29 (4) (2011) 456–462.

[141] K.Y. Cho, U.H. Hong, S.P. Jung, Y. Takushima, A. Agata, T. Sano, Y. Horiuchi, M. Suzuki, Y.C. Chung, Long-reach 10-Gb/s RSOA-based WDM PON employing QPSK signal and coherent receiver, Opt. Express 20 (14) (2012) 15353–15358.

[142] H. Takahashi, K. Oda, H. Toba, Y. Inoue, Transmission characteristics of arrayed waveguide $N \times N$ wavelength multiplexer, J. Lightwave Technol. 13 (3) (1995) 447–455.

[143] K. Okamoto, T. Hasegawa, O. Ishida, A. Himeno, Y. Ohmori, 32×32 arrayed-waveguide grating multiplexer with uniform loss and cyclic frequency characteristics, Electron. Lett. 33 (22) (1997) 1865–1866.

[144] D. Mayweather, L. Kazovsky, M. Downs, N. Frigo, Wavelength tracking of a remote WDM router in a passive optical network, IEEE Photon. Technol. Lett. 8 (1996) 1238–1240.

[145] S. Kamei, Recent progress on athermal AWG wavelength multiplexer, in: Proc. OFC, Paper OWO1, Mar. 2009.

[146] R. Monnard, M. Zirngible, C.R. Doerr, C.H. Joyner, L.W. Stulz, Demonstration of a 12×155 Mb/s WDM PON under outside plant temperature conditions, IEEE Photon. Technol. Lett. 9 (1998) 1655–1657.

[147] D.K. Jung, S.K. Shin, H.G. Woo, Y.C. Chung, Wavelength-tracking technique for spectrum-sliced WDM passive optical network, IEEE Photon. Technol. Lett. 12 (3) (2000) 338–340.

[148] P.P. Iannone, K.C. Reichman, N.J. Frigo, Broadcast digital video delivered over WDM passive optical network, IEEE Photon. Technol. Lett. 8 (7) (1996) 930–932.

[149] D.K. Jung, H. Kim, K.H. Han, Y.C. Chung, Spectrum-sliced bidirectional passive optical network for simultaneous transmission of WDM and digital broadcast video signals, Electron. Lett. 37 (5) (2001) 308–309.

[150] J.-H. Moon, K.-M. Choi, C.-H. Lee, Overlay of broadcasting signal in a WDM PON, in: Proc. OFC, Paper OThK8, Mar. 2006.

[151] Y. Zhang, N. Dentg, C.-K. Chan, L.-K. Chen, A multicast WDM-PON architecture using DPSK/NRZ orthogonal modulation, IEEE Photon. Technol. Lett. 20 (17) (2008) 1479–1481.

[152] B.S. Glance, K. Pollock, C.A. Burrus, B.L. Kasper, G. Eisenstein, L.W. Stulz, WDM coherent optical star network, J. Lightwave Technol. 6 (1) (1988) 67–72.

[153] E. Aurthors, M.S. Goodman, H. Kobrinski, M.P. Vecchi, HYPASS: An optoelectronic hybrid packet switching system, IEEE J. Sel. Areas Commun. 6 (9) (1988) 1500–1510.

[154] H. Kobrinski, R.M. Bulley, M.S. Goodman, M.P. Vechhi, C.A. Brackett, Demonstration of high capacity in the LAMBDANET architecture: a multiwavelength optical network, Electron. Lett. 23 (16) (1987) 824–826.

[155] E. Authrs, J.M. Cooper, M.S. Goodman, H. Kobrinski, M. Tur, M.P. Vecchi, Multiwavelength optical crossconnect for parallel-processing computers, Electron. Lett. 24 (2) (1988) 119–120.

[156] P.P. Iannone, K.C. Reichmann, N.J. Frigo, A.H. Gnauck, L.H Spiekman, A flexible metro WDM ring using wavelength-independent subscriber equipment to share bandwidth, in: Proc. OFC, Paper PD38, Mar. 2000.

[157] N. Takachio, H. Suzuki, M. Fujiwara, J. Kani, K. Iwatsuki, H. Yamada, T. Shibata, T. Kitoh, Wide area gigabit access network based on 12.5 GHz spaced 256 channel super-dense WDM technologies, Electron. Lett. 37 (5) (2001) 309–311.

[158] F.-T. An, K.S. Kim, D. Gutierrez, S. Yam, E. Hu, K. Shrikhande, L.G. Kazovsky, SUCCESS: a next-generation hybrid WDM/TDM optical access network architecture, J. Lightwave Technol. 22 (11) (2004) 2557–2569.

[159] J.A. Lázaro, R.I. Martínez, V. Polo, C. Arellano, J. Prat, Hybrid dual-fiber-ring with single-fiber-tree dense assess network architecture using RSOA-ONU, in: Proc. OFC, Paper OTuG2, Mar. 2007.

[160] J.A. Lázaro, C. Bock, V. Polo, R.I. Martinez, J. Prat, Remotely amplified combined ring-tree access network architecture using reflective RSOA-based ONU, J. Opt. Netw. 6 (6) (2007) 801–807.

[161] H. Erkan, A.D. Hossain, R. Dorsinville, M.A. Ali, A. Hadjiantonis, G. Ellinas, A. Khalil, A novel ring-based WDM-PON access architecture for the efficient utilization of network resources, in: Proc. ICC, May 2008. pp. 5175–5181.

[162] P.J. Urban, E.G.C. Pluk, M.M. de Laat, F.M. Huijskens, G.D. Khoe, A.M.J. Koonen, H. de Waardt, 1.25-Gb/s transmission over an access network link with tunable OADM and a reflective SOA, IEEE Photon. Technol. Lett. 21 (6) (2009) 380–382.

[163] P.J. Urban, B. Huiszoon, R. Roy, M.M. De Laat, F.M. Huijskens, E.J. Klein, G.D. Khoe, A.M.J. Koonen, H. de Waardt, High-bit-rate dynamically reconfigurable WDM-TDM access network, J. Opt. Commun. Netw. 1 (2) (2009) A143–A158.

[164] X. Sun, Z. Wang, C.-K. Chan, L.-K. Chen, A novel star-ring protection architecture scheme for WDM passive optical access networks, in: Proc. NFOEC, Paper JWA53, Mar. 2005.

[165] P.-C. Peng, K.-M. Feng, H.-Y. Chiou, W.-R. Peng, J. Chen, H.-C. Kuo, S.-C. Wang, S. Chi, Reliable architecture for high-capacity fiber-radio systems, Opt. Fiber Technol. 13 (3) (2007) 236–239.

[166] H. Erkan, G. Ellinas, A. Hadjiantonis, R. Dorsinville, M. Ali, Native Ethernet-based self-healing WDM-PON local access ring architecture—a new direction for supporting simple and efficient resilience capabilities, in: Proc. ICC, May 2010.

[167] Y. Zhou, C. Gan, L. Zhu, Self-healing ring-based WDM PON, Opt. Commun. 283 (9) (2010) 1732–1736.

[168] U.H. Hong, K.Y. Cho, Y.C. Chung, Multi-ring architecture for survivable WDM PON, in: Proc. OECC, Paper 6A2-4, July 2012.

[169] K.T. Kim, WDM PON based on bus architecture, ME thesis, KAIST, 2007.

[170] H.-H. Lee, S.-H. Cho, E.-G. Lee, S.-S. Lee, Demonstration of RSOA-based 20 Gb/s linear bus WDM-PON with simple optical add-drop node structure, ETRI J. 32 (2) (2010) 248–254.

[171] R.P. Davey, D.B. Grossman, M. Rasztovits-Wiech, D.B. Payne, D. Nesset, A.E. Kelly, A. Rafel, S. Appathurai, S. Yang, Long-reach passive optical networks, J. Lightwave Technol. 27 (3) (2009) 273–291.

[172] G. Talli, P.D. Townsend, Hybrid DWDM-TDM long reach PON for next generation optical access, J. Lightwave Technol. 24 (7) (2006) 2827–2834.

[173] S.M. Lee, S.G. Mun, M.H. Kim, C.H. Lee, Demonstration of a long-reach DWDM-PON for consolidation of metro and access networks, J. Lightwave Technol. 25 (1) (2007) 271–276.

[174] M.D. Feuer, M.A. Thomas, L.M. Lunardi, Backreflection and loss in single-fiber loopback networks, IEEE Photon. Technol. Lett. 12 (8) (2000) 1106–1108.

[175] J.H. Moon, K.M. Choi, S.G. Mun, C.H. Lee, Effects of back-reflection in WDM-PONs based on seed light injection, IEEE Photon. Technol. Lett. 19 (24) (2007) 2045–2047.

[176] P.J. Urban, A.M.J. Koonen, G.D. Khoe, H. de Waardt, Mitigation of reflection-induced crosstalk in a WDM access network, in: Proc. OFC, Paper OThT3, Feb. 2008.

[177] A. Murakami, K.Y. Cho, Y.J. Lee, A. Agata, Y. Takushima, Y.C. Chung, Y. Horiuchi, Enhanced reflection tolerance of SCM signal in an RSOA-based WDM-PON system by using phase modulation, in: Proc. IEICE Society Conference, Paper B-10-46, Sept. 2008 (in Japanese).

[178] A. Murakami, Y.J. Lee, K.Y. Cho, Y. Takushima, A. Agata, K. Tanaka, Y. Horiuchi, Y.C. Chung, Enhanced reflection tolerance of upstream signal in a RSOA-based WDM PON by using Manchester coding, in: Proc. SPIE vol. 6783, pp. 67832I-1–67832I-5 (presented at APOC 2007), Nov. 2007.

[179] C.R. Giles, E. Desurvire, J.R. Simpson, Transient gain and crosstalk in erbium-doped fiber amplifiers, Opt. Lett. 14 (1989) 880–882.

[180] T. Shinozaki, M. Fuse, S. Morikura, A study of gain dynamics of Erbium-doped fiber amplifiers for burst optical signals, in: Proc. ECOC, Paper P4.2, Sept. 2002.

[181] H.G. Krimmel, T. Pfeiffer, B. Deppisch, L. Jentsch, Hybrid electro-optical feedback gain-stabilized EDFAs for long-reach wavelength-multiplexed passive optical networks, in: Proc. ECOC, Paper 9.5.3, Sept. 2009.

[182] M.S. Lee, B.T. Lee, B.Y. Yoon, B.T. Kim, Bidirectional amplification with linear optical amplifier in WDM-PON, in: Proc. OFC, Paper JThB73, Mar. 2006.

[183] C. Antony, P. Ossieur, A.M. Clarke, A. Naughton, H.-G. Krimmel, Y.F. Chang, A. Borghesani, D.G. Moodie, A. Poustie, R. Wyatt, W.B. Harmon, I. Lealman, G. Maxwell, D. Rogers, D.W. Smith, D. Nesset, R.P. Davey, P.D. Townsend, Demonstration of a carrier distributed 8192-split hybrid DWDM-TDMA PON over 124 km field installed fibers, in: Proc. OFC, Paper PDPD8, March 2010.

[184] P. Ossieur, C. Antony, A.M. Clarke, A. Naughton, H.-G. Krimmel, Y.F. Chang, C. Ford, A. Borghesani, D.G. Moodie, A. Poustie, W.B. Harmon, I. Lealman, G. Maxwell, D. Rogers, D.W. Smith, D. Nesset, R.P. Davey, P.D. Townsend, A 135-km 8192-split carrier distributed DWDM-TDMA PON with 2×32 10 Gb/s capacity, J. Lightwave Technol. 29 (4) (2011) 463–474.

[185] J.P. Blondel, F. Misk, P.M. Gabla, Theoretical evaluation and record experimental demonstration of budget improvement with remotely pumped erbium-doped fiber amplification, IEEE Photon. Technol. Lett. 5 (12) (1993) 1430–1433.

[186] M.D. Feuer, R.D. Feldman, J.L. Zyskind, S.C. Shunk, J. Sulhoff, T.H. Wood, Remotely-pumped self-amplified star network for local access, in: Proc. OFC, Paper WI5, Feb. 1996.

[187] J.M. Oh, S.G. Koo, D.H. Lee, S.J. Park, Enhancement of the performance of a reflective SOA-based hybrid WDM/TDM PON system with a remotely pumped Erbium-doped fiber amplifier, J. Lightwave Technol. 26 (1) (2008) 144–149.

[188] J.H. Lee, Y.-G. Han, S.B. Lee, C.H. Kim, Raman amplification-based WDM-PON architecture with centralized Raman pump-driven, spectrum-sliced erbium ASE and polarization-insensitive EAMs, Opt. Express 14 (20) (2006) 9036–9041.

[189] M.J.L. Cahill, G.L. Pendock, M.A. Summerfield, A.J. Lowery, D.D. Sampson, Optimum optical amplifier location in spectrum-sliced WDM passive optical networks for customer access, in: Proc. OFC, Paper FD5, Feb. 1998.

[190] U.H. Hong, K.Y. Cho, Y. Takushima, Y.C. Chung, Effects of Rayleigh backscattering in long-reach RSOA-based WDM PON, in: Proc. OFC, Paper OThG1, Mar. 2010.

[191] U.H. Hong, K.Y. Cho, Y. Takushima, Y.C. Chung, Maximum reach of long-reach RSOA-based WDM PON employing remote EDFA, in: Proc. OFC, Paper OMP1, Mar. 2011.

[192] D.-S. Ly-Gagnon, S. Tsukamoto, K. Katoh, K. Kikuchi, Coherent detection of optical quadrature phase-shift keying signals with carrier phase estimation, J. Lightwave Technol. 24 (1) (2006) 12–21.

[193] S.J. Park, Y.B. Choi, S.P. Jung, K.Y. Cho, Y. Takushima, Y.C. Chung, Hybrid WDM/TDMA-PON using self-homodyne and differential coding, IEEE Photon. Technol. Lett. 21 (7) (2009) 465–467.

[194] H. Rohde, S. Smolorz, J.S. Wey, E. Gottwald, Coherent optical access networks, in: Proc. OFC, Paper OTuB1, Mar. 2011.

[195] I. Papagiannakis, M. Omella, D. Klonidis, A.N. Birbas, J. Kikidis, I. Tomkos, J. Prat, Investigation of 10-Gb/s RSOA-based upstream transmission in WDM-PONs utilizing optical filtering and electronic equalization, IEEE Photon. Technol. Lett. 20 (24) (2008) 2168–2170.

[196] H. Kim, 10-Gb/s operation of RSOA using a delay interferometer, IEEE Photon. Technol. Lett. 22 (18) (2010) 1379–1381.

[197] S.P. Jung, Y. Takushima, Y.C. Chung, Generation of 5-Gbps QPSK signal using directly modulated RSOA for 100-km coherent WDM PON, in: Proc. OFC, Paper OTuB3, Mar. 2011.

[198] T. Duong, N. Genay, P. Chanclou, B. Charbonnier, A. Pizzinat, R. Brenot, Experimental demonstration of 10 Gbit/s upstream transmission by remote modulation of 1 GHz RSOA using adaptively modulated optical OFDM for WDM-PON single fiber architecture, in: Proc. ECOC, Paper Th.3.F.1, Sept. 2008.

[199] L. Xu, K. Padmaraju, L. Chen, M. Lipson, K. Bergman, First demonstration of symmetric 10-Gb/s access networks architecture based on silicon microring single sideband modulation for efficient upstream signal re-modulation, in: Proc. OFC, Paper OThK2, Mar. 2011.

[200] L. Xu, H.K. Tsang, WDM-PON using differential-phase-shift-keying remodulation of dark return-to-zero downstream channel for upstream, IEEE Photon. Technol. Lett. 20 (10) (2008) 833–835.

[201] H. Suzuki, H. Nakamura, J. Kani, K. Iwatsuki, A wide-area WDM-based passive optical network (WDM-PON) accommodating 10 gigabit Ethernet-based VPN services, in: Proc. ECOC, paper Tu4.6.3, Sept. 2004.

[202] D. Smith, I. Lealman, X. Chen, D. Moodie, P. Cannard, J. Dosanjh, L. Rivers, C. Ford, R. Cronin, T. Kerr, L. Johnston, R. Waller, R. Firth, A. Borghesani, R. Wyatt, A. Poustie,

Colourless 10 Gb/s Reflective SOA-EAM with Low Polarization Sensitivity for Long-reach DWDM-PON Networks, in: Proc. ECOC, Paper 8.6.3, Sept. 2009.

[203] K.Y. Cho, A. Agata, Y. Takushima, Y.C. Chung, Performance of forward-error correction code in 10-Gb/s RSOA-based WDM PON, IEEE Photon. Technol. Lett. 22 (1) (2010) 57–59.

[204] G.P. Agrawal, N.K. Dutta, Long-Wavelength Semiconductor Lasers, Van Nostrand Reinhold, 1986.

[205] Y.C. Chung, J.M. Wiesenfeld, G. Raybon, U. Koren, Y. Twu, Intermodulation distortion in a multiple-quantum-well semiconductor optical amplifier, IEEE Photon. Technol. Lett. 3 (2) (1991) 130–132.

[206] J.G. Proakis, Digital Communications, fourth ed., McGraw-Hill, New York, 2001.

[207] D. Fritzsche, L. Schürer, A. Ehrhardt, D. Breuer, H. Oeruen, C.G. Schäffer, Field trial investigation of 16-states MLSE equalizer for simultaneous compensation of CD, PMD and SPM, in: Proc. OFC, Paper OWE4, Mar. 2009.

[208] J. Zhao, L.-K. Chen, C.K. Chan, Electronic equalization for unsynchronized modulation and chromatic dispersion compensation in supercontinuum-based WDM-PON, in: Proc. ECOC, Paper We3.P.159, Sept. 2006.

[209] I. Cano, M. Omella, J. Prat, P. Poggiolini, Colorless 10 Gb/s extended reach WDM PON with low BW RSOA using MLSE, in: Proc. OFC, Paper OWG2, Mar. 2010.

[210] Q. Guo, A.V. Tran, 20-Gb/s single-feeder WDM-PON using partial-response maximum likelihood equalizer, IEEE Photon. Technol. Lett. 23 (23) (2011).

[211] W. Rosenkranz, J.V. Hoyningen-Huene, Nonlinear compensation and equalization in access networks, in: Proc. OECC, Paper 5B3-2, July 2012.

[212] A.J. Zilkie, J. Meier, M. Mojahedi, A.S. Helmy, P.J. Poole, P. Barrios, D. Poitras, T.J. Rotter, C. Yang, A. Stintz, K.J. Malloy, P.W.E. Smith, J.S. Aitchison, Time-resolved linewidth enhancement factors in quantum dot and higher-dimensional semiconductor amplifiers operating at 1.55 μm, J. Lightwave Technol. 26 (11) (2008) 1498–1509.

[213] A.D. Ellis, J. Zhao, D. Cotter, Approaching the non-linear Shannon limit, J. Lightwave Technol. 28 (4) (2010) 423–433.

[214] H. Kimura, N. Iiyama, Y. Sakai, K. Kumozaki, A WDM-based future optical access network and support technologies for adapting the user demands' diversity, IEICE Trans. Commun. E93-B (2) (2010) 246–254.

[215] P.P. Iannone, K.C. Reichmann, Optical access beyond 10 Gb/s PON, in: Proc. ECOC, Paper Tu.3.B.1, Sept. 2010.

[216] K.Y. Cho, U.H. Hong, Y. Takushima, A. Agata, T. Sano, M. Suzuki, Y.C. Chung, 103-Gb/s long-reach WDM PON implemented by using directly modulated RSOAs, IEEE Photon. Technol. Lett. 24 (3) (2012) 209–211.

[217] Q. Guo, A.V. Tran, Demonstration of 40-Gb/s WDM-PON system using SOA-REAM and equalization, IEEE Photon. Technol. Lett. 24 (11) (2012) 951–953.

[218] D. Derickson, Fiber Optic Test and Measurement, Prentice Hall, 1998 (Chapter 11).

[219] Y. Enomoto, H. Izumita, M. Nakamura, Over 31.5 dB dynamic range optical fiber line testing system with optical fiber fault localization function for 32-branched PON, in: Proc. OFC, Paper ThAA3, Mar. 2003.

[220] J. Laferriere, M. Sagat, A. Champavere, Original method for analyzing multipath networks by OTDR measurement, in: Proc. OFC, Paper TuT4, Feb. 1997.

[221] U. Hilbk, M. Burmeister, B. Hoen, T. Hermes, J. Saniter, F.J. Westphal, Selective OTDR measurements at the central office of individual fiber link in a PON, in: Proc. OFC, Paper Tuk3, Feb. 1997.

[222] K. Tanaka, H. Izumita, N. Tomita, Y. Inoue, In-service individual line monitoring and a method for compensating for the temperature-dependent channel drift of a WDM-PON containing an AWGR using a 1.6 μm tunable OTDR, in: Proc. ECOC, Paper WE4A4, Sept. 1997.

[223] S. Hann, J.-S. Yoo, C.-S. Park, Monitoring technique for a hybrid PS/WDM-PON by using a tunable OTDR and FBGs, Meas. Sci. Technol. 17 (2005) 1070–1074.

[224] J.H. Lee, J.H. Park, J.G. Shim, H. Yoon, J.H. Kim, K.M. Kim, J.O. Byun, N.K. Park, In-service monitoring of 16 port × 32 wavelength bi-directional WDM-PON systems with a tunable, coded optical time domain reflectometry, Opt. Express 15 (11) (2007) 6874–6882.

[225] J. Park, J. Baik, C.H. Lee, Fault-detection technique in a WDM-PON, Opt. Express 15 (4) (2007) 1461–1466.

[226] M. Thollabandi, T.-Y. Kim, S. Hann, C.-S. Park, Tunable OTDR based on direct modulation of self-injection-locked RSOA for in-service monitoring of WDM-PON, IEEE Photon. Technol. Lett. 20 (15) (2008) 1323–1325.

[227] K.W. Lim, E.S. Son, K.H. Han, Y.C. Chung, Fault localization in WDM passive optical network by reusing downstream light sources, IEEE Photon. Technol. Lett. 17 (12) (2005) 2691–2693.

[228] Y. Takushima, Y.C. Chung, In-service OTDR for passive optical networks, in: Proc. NFOEC, Paper NWC2, Mar. 2010.

[229] H.K. Shim, K.Y. Cho, Y. Takushima, Y.C. Chung, Correlation-based OTDR for in-service monitoring of 64-split TDM PON, Opt. Express 20 (5) (2012) 4921–4926.

[230] D. Zhou, S. Subramaniam, Survivability in optical networks, IEEE Network (2000) 16–23.

[231] ITU-T Recommendation G983.1, Broadband optical access systems based on passive optical networks (PON), Jan. 2005.

[232] C.K. Chan, Protection architectures for passive optical networks, Passive Optical Networks: Principles and Practice, Academic Press, 2007, pp. 243–266 (Chapter 6).

[233] A.H. Gnauck, A.A.M. Saleh, S.L. Woodward, Restorable architectures for fiber-based broadband local access networks, US Patent Number: H2075, Aug. 2003.

[234] H. Nakamura, H. Suzuki, J. Kani, K. Iwatsuki, Reliable wide-area wavelength division multiplexing passive optical network accommodating gigabit Ethernet and 10-Gb Ethernet services, J. Lightwave Technol. 24 (5) (2006) 2045–2051.

[235] S.B. Park, D.K. Jung, D.J. Shin, H.S. Shin, S. Hwang, Y.J. Oh, C.S. Shim, Bidirectional wavelength-division-multiplexing self-healing passive optical network, in: Proc. OFC, Paper JWA57, Mar. 2005.

[236] T.K. Chan, C.K. Chan, L.K. Chen, F. Tong, A self-protected architecture for wavelength-division-multiplexed passive optical networks, IEEE Photon. Technol. Lett. 15 (11) (2003) 1660–1662.

[237] C.M. Lee, T.J. Chan, C.K. Chan, L.K. Chen, C. Lin, A group protection architecture (GPA) for traffic restoration in multi-wavelength passive optical networks, in: Proc. ECOC, Paper Th.2.4.2, Sept. 2003.

[238] X.F. Sun, C.K. Chan, L.K. Chen, A survivable WDM PON architecture with centralized alternate-path protection switching for traffic restoration, IEEE Photon. Technol. Lett. 18 (4) (2006) 631–633.

[239] E.S. Son, K.H. Han, J.H. Lee, Y.C. Chung, Survivable network architectures for WDM PON, in: Proc. OFC, Paper OFI4, Mar. 2005.

[240] B. Glance, C.R. Doerr, I.P. Kaminow, R. Montagne, Optically restorable WDM ring network using simple add/drop circuitry, J. Lightwave Technol. 14 (11) (1996) 2453–2456.

[241] H. Toba, K. Oda, K. Inoue, K. Nosu, T. Kitoh, An optical FDM-based self-healing ring network employing arrayed waveguide grating filters and EDFA's with level equalizers, IEEE J. Sel. Areas Commun. 14 (5) (1996) 800–813.

[242] C.H. Kim, C.H. Lee, Y.C. Chung, Bidirectional WDM self-healing ring network based on simple bidirectional add/drop amplifier modules, IEEE Photon. Technol. Lett., 10 (9) (1998) 1340–1342.

[243] Y. Joo, G. Lee, R. Kim, S. Park, K. Song, J. Koh, S. Hwang, Y. Oh, C. Shim, 1-fiber WDM self-healing ring with bidirectional optical add/drop multiplexers, IEEE Photon. Technol. Lett. 16 (2) (2004) 683–685.

[244] S.B. Park, C.H. Lee, S.G. Kang, S.B. Lee, Bidirectional WDM self-healing ring network for hub/remote nodes, IEEE Photon. Technol. Lett. 15 (11) (2003) 1657–1659.

[245] Z. Wang, C. Lin, C.K. Chan, Demonstration of a single-fiber self-healing CWDM metro access ring network with unidirectional OADM, IEEE Photon. Technol. Lett. 18 (1) (2006) 163–165.

[246] X. Sun, C.K. Chan, Z. Wang, C. Lin, L.K. Chen, A single-fiber bi-directional WDM self-healing ring network with bi-directional OADM for metro-access applications, IEEE J. Sel. Areas Commun. 25 (4) (2007) 18–24.

FTTX Worldwide Deployment

24

Vincent O'Byrne[a], Chang Hee Lee[b], Yoon Kim[b], and Zisen Zhao[c]

[a]Verizon, USA,
[b]KAIST, Korea,
[c]Wuhan Research Institute, Fiberhome, South America

24.1 INTRODUCTION

The copper network has been deployed since the late 1800s across the world in order to meet the needs of the residential and business communities. However, with the increase in subscriber demands for internet speeds and broadband in general over the last 10–15 years the underlying copper technologies, based on DSL (Digital Subscriber Loop) variants such as ADSL (Asymmetric DSL) and VDSL (Very high DSL), have been found to have limiting capacity especially at longer loop lengths and thus operators across the world have been looking to fiber-based technologies to provide the added bandwidth independent of distance. Operators have considered their options in terms of invested capital and overall long-term strategies and determined how much fiber to deploy and how close to the customer to deploy this fiber.

The copper network is illustrated in Figure 24.1. There is typically a cross-connect box, sometimes called a SAI (Serving Area Interface), located typically several thousand feet from the customer's location and which terminates the larger copper cables from the central office. There are then copper distribution cables running from these boxes to a distribution box typically a few hundred feet from the customer and then finally a copper drop from these terminals to the NID (Network Interface Device) at the side of the customer's house.

When considering a fiber overlay there are many different locations where the Optical-to-Electrical (O/E) conversion can occur. Thus Fiber-to-the-X (FTTX) has many meanings within the industry depending on their reference point. It can mean Fiber to the Node (FTTN), Fiber to the Curb (FTTC), Fiber to the Building/Business (FTTB), and Home/Premises (FTTP) as illustrated in Figure 24.1. As we approach the home or business the amount of fiber deployed grows and the bandwidth available to the customer grows due to the rather limited bandwidth of the copper cable in comparison to the fiber optic cable. In the case of multiple dwelling units some operators deploy to the building and then use the inside copper wiring to gain access to the customer. In some cases operators deploy all the way to the living unit.

However, this comes at a cost as it is generally more costly to deploy fiber deeper into the network, but the advantage is the future proofness of that architecture given this

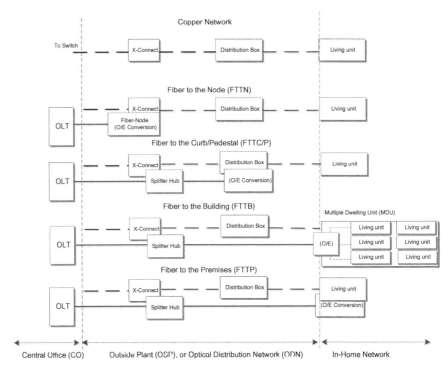

FIGURE 24.1 FTTX, where X denotes the positioning of the optical-electrical conversion.
(Note: Fiber _____, Copper _ _ _ _.)

added bandwidth. This trade-off between how far to deploy the fiber versus bandwidth achieved is one of the most important decisions an operator makes in deploying FTTX.

In the following sections we cover other architectural choices that can be made as well as the standing of the various technologies for the access space. This will then be a lead-in to a summary of the status of FTTX and some of the issues operators are facing around the world.

24.2 BACKGROUND OF FIBER ARCHITECTURES

There are basically two types of architectures deployed around the world. One is a shared network among many users and the other is a point-to-point network. These are outlined in the following sections together with the main benefits and disadvantages as well as their relative acceptance by the operators.

24.2.1 Passive optical networks (PONs)

The PON (Passive Optical Network) is a passive optical network that is typically deployed in a point-to-multipoint fashion similar to a star network. The single fiber leaving the central office is typically split, using a power splitter or many power splitters distributed along the fiber. The power split level ranges typically from 1 by 64 down to 1 by 4, or 1 by 8. These basic architectures are shown in Figure 24.2.

The most appropriate level and placement of the split design used can depend on many operational issues such as:

- Customer density. There is a trade-off in the need for more feeder fiber versus the need for increased distribution fiber. As the density reduces studies [1] have shown that it may become more cost effective to place the split closer to the customer and thus use a distributed, or multiple stage split as outlined in Figure 24.2b.
- Desire for simplicity of design and increased utilization of CO (Central Office) equipment. For example, deploying a single split helps in ensuring that the utilization of CO equipment and equipment upstream is fully utilized and one is less sensitive to localized take rate variations.
- Desire for increased operational support. NTT (see Section 24.4.1.2) in Japan, for example, deploys dual splitting in their architecture and adds an OTDR (Optical Time Domain Reflectometer) after the first stage of splitting in order to be able to monitor their fiber plant [2].

Some of the advantages of the PON architecture can be summarized as:

- The passive nature of the PON where the fiber is relatively impervious to electric interference.
- More reliable than its copper equivalent.
- Increased bandwidth available at the end point ("X").
- Support of a single network without having to supply powering in the field and its associated battery backup in some variants of FTTX.
- Powering at the customer location provided by the customer in the case of FTTH.

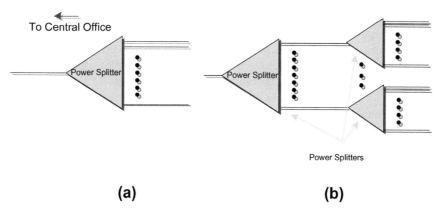

(a)   **(b)**

FIGURE 24.2 Basic PON architecture. (a) Single splitter stage and (b) double stage splitting.

In the following section we will briefly discuss an alternate architecture to those described above in Figure 24.2 called point to point.

24.2.2 Point to point

Another architecture alternative is a point-to-point architecture where there is a dedicated fiber from the CO all the way to the customer's location as illustrated in Figure 24.3. Because of the dedicated fiber this architecture is seen as the most future proof and secure architecture as each customer gets their own fiber and connection to the CO. However, it results in a large deployment of fiber in the ODN (Optical Distribution Network) and an increased need for space and powering in the CO.

Its advantages and disadvantages can be summarized as:

Advantages:

- Future proofness of the network, in that it entails a direct fiber to the customer and they have access to all the bandwidth that is transmitted along their fiber.
- Higher security as it is a dedicated fiber to every user.
- Each user's equipment is only operating at their maximum rate, as opposed to the aggregate rate as in the case of a shared PON. So for example if the subscriber is on a 1 Gbps link the ONU (Optical Network Unit)/ONT (Optical Network Terminal)[1] receiver operates at that rate whereas in a shared PON network the ONT receives at the full downstream rate of the aggregate rated required by the ONTs. So if each ONT required 1 Gbps then on a 32-way split it would require 32 Gbps downstream unless some level of oversubscription is utilized at the CO.
- Increased robustness:
  - The point-to-point network is more tolerant to a single rogue [3] ONT as any ONT that goes bad and starts to cause interference only impacts its own link, rather than the situation in a shared PON, or shared network where it is possible to interfere with other customers sharing the same fiber.
  - Ease of determining the location of a fault on the fiber using an OTDR as there is sufficient link budget to identify clearly the position of the fault.
- Lower cost ONUs: Generally the ONUs can be run at the customer's rate and thus do not need to operate at the aggregate speed of the technology. Thus if the customer only wants a 100 Mbps service the ONU can be customized to only support that rate at a lower cost. In a shared PON network each ONU must demodulate and process the signal at the total downstream rate.

---

[1]Note that the terms ONU and ONT are used interchangeably in the industry and in this chapter. Historically the reference used to be that an ONT was where the ONU and ONT (network termination device, or NID) were combined, which typically occurred in FTTP deployments where the ONT became the effective NID and where the interfaces from the ONT were customer interfaces.

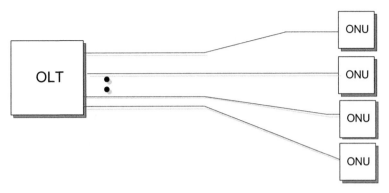

FIGURE 24.3 Point-to-point network.

Disadvantages:
- Increased cost in the ODN and in the CO: The cost of deployment is typically higher largely due to the additional fiber and facilities needed per customer. Though fiber is relatively cheap, in larger COs the cable size becomes too unwieldy and requires additional conduit as well as new deployment techniques [3]. In addition there are added space requirements in the central office as well as increased powering requirements. Some studies have shown that for an average-sized CO the additional requirements can result in an order of magnitude increase in real estate and powering requirements with associated increases in cost [3,4].

WDM-PON (see Section 24.3.4.2) provides virtual point-to-point connectivity by assigning dedicated wavelength and/or wavelength pair to each ONT, while maintaining the point-to-multipoint PON architecture. However, some transparencies (wavelength and bit rate) need to be sacrificed.

Operators around the world have looked at PON networks and also point-to-point networks and the majority have chosen to go forward with a PON deployment [4] as outlined in Figure 24.4.

24.3 TECHNOLOGY VARIANTS

In the following sections we describe the technology options that operators across the world have investigated for possible deployment to meet the customer's demands and their present status.

The Access network in general consists of several components as outlined in Figure 24.5.

- OLT (Optical Line Terminal), which manages the ONUs and interfaces the Access network to the Services networks.
- If chosen a V-OLT (Video Optical Line Terminal), which transports the video broadcast signal between 1550 nm and 1560 nm in the downstream to the ONU.

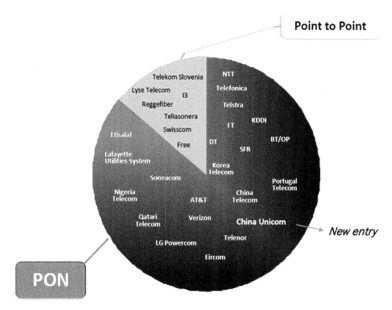

FIGURE 24.4 FTTX deployment around the world [4].

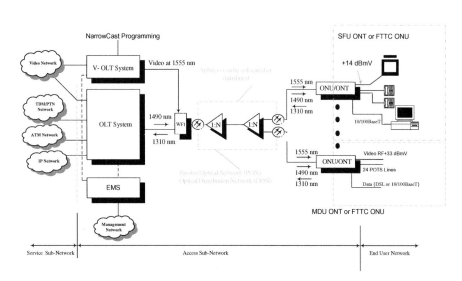

FIGURE 24.5 Illustrative example of the FTTP architecture with sample reference levels.

This is typically made up of several stages of EDFA (Erbium-Doped Fiber Amplifier) optical amplifiers. Some operators have chosen this format for video while others use the 1490 nm downstream signal to offer an IPTV (Internet Protocol TV) video offering.

- ODN (Optical Distribution Network), which connects to the OLT and provides the optical paths over which they communicate. There can be different stages of splitting as mentioned previously, but typically a single stage of splitting is defined in order to make better use of the central office equipment.
- ONU/ONT, which interfaces the Access network to the End User network. Within the house different technologies could be used to distribute the data and video, with some examples being MOCA (Multimedia over Coax Alliance [5]) and HPNA (Home Phoneline Networking Alliance, [6]) using the already embedded coaxial or phone lines in order to reduce the overall installation costs.
- At the ONT Powering can be either Customer or Network provided depending on whether it is deployed at the home or remote from it.
- EMS (Element Management System) provides the PON network system interface to standard industry core operations network.

24.3.1 B-PON

B-PON (Broadband Passive Optical Network) was originally developed by FSAN (Full Service Access Network) working body in 1995 and was standardized by the ITU (International Telecommunications Union) in 2001. Prior to this there were several providers of FTTH equipment though using proprietary protocols. The specific workings of B-PON are defined by the standards ITU G.983.x, where x equates to different standards.

- G.984.1: Broadband optical access systems based on passive optical networks (PON), which was updated in 2005 [7], with amendment 1. Includes the definition of the physical layer optical power classes (see for example Tables 24.1–24.3), and basic operating wavelengths for upstream and downstream. It specifies a maximum range of the technology of 20 km which was judged to be sufficient for most applications initially for deployment.
- G.984.2: ONT management and control interface specification for B-PON [8].
- G.984.3: A broadband optical access system with increased service capability by wavelength allocation, 2001 [9], with amendments 1 and 2.
- G.984.4: A broadband optical access system with increased service capability using DBA (Dynamic Bandwidth Assignment), 2001 [10], with amendment 1 and corrigendum 1.
- G.984.5: A broadband optical access system with enhanced survivability, 2002 [11]. This covers at a high level the requirements for different levels of protection of the ODN all the way to the ONU. Since B-PON is typically used for residential applications, the deployments have largely been deployed without protection.

B-PON is based on a clock rate of 8 kHz and employs Asynchronous Transfer Mode (ATM) protocol at layer 2 and has a typical downstream rate of 622.08 Mbps and an upstream rate of 155.52 Mbps, though other rates are specified in the standards [7]. Transmission is continuous, using TDM (Time Division Multiplexing) in

Table 24.1 Classes of B-PON power budgets [7].

Items	Unit	Specification
Fiber type	–	ITU-T Rec. G.652
Attenuation range (ITU-T Rec. G.982)	dB	Class A: 5–20 Class B: 10–25 Class C: 15–30
Differential optical path loss	dB	15
Maximum optical path penalty	dB	1
Maximum differential logical reach	km	20
Maximum fiber distance between S/R and R/S points	km	20
Minimum supported split ratio	–	Restricted by path loss and ONU addressing limits PON with passive splitters (16- or 32-way split)
Bidirectional transmission	–	1-fiber WDM or 2-fiber

Table 24.2 Optical interface parameters for 622 Mbps downstream [7]. Note that references to figures, appendices, etc. are as outlined in the reference.

Items	Unit	Single Fiber			Dual Fiber		
		OLT Transmitter (optical interface o_{ld})					
Nominal bit rate	Mbit/s	622.08			622.08		
Operating wavelength	nm	1480–1500			1260–1360		
Line code	–	Scrambled NRZ			Scrambled NRZ		
Maximum reflectance of equipment, measured at transmitter wavelength	dB	NA			NA		
Minimum ORL of ODN at O_{lu} and O_{ld} (Notes 1 and 2)	dB	More than 32			More than 32		
ODN class		Class A	Class B	Class C	Class A	Class B	Class C
Mean launched power MIN	dBm	–7	–2	–2	–7	–2	–2
Mean launched power MAX	dBm	–1	+4	+4	–2	+3	+3

Table 24.2 (Continued)

Items	Unit	Single Fiber			Dual Fiber		
Launched optical power without input to the transmitter	dBm	NA			NA		
Extinction ratio	dB	More than 10			More than 10		
Transmitter tolerance to incident light power	dB	More than −15			More than −15		
If MLM Laser— Maximum RMS width	nm	NA			1.4		
If SLM Laser— Maximum −20 dB width (Note 3)	nm	1			1		
If SLM Laser— Minimum side mode suppression ratio	dB	30			30		
ONU receiver (optical interface o$_{rd}$)							
Maximum reflectance of equipment, measured at receiver wavelength	dB	Less than −20			Less than −20		
Bit error ratio	–	Less than 10^{-10}			Less than 10^{-10}		
ODN class		Class A	Class B	Class C	Class A	Class B	Class C
Minimum sensitivity	dBm	−28	−28	−33	−28	−28	−33
Minimum overload	dBm	−6	−6	−11	−7	−7	−12
Consecutive identical digit immunity	bit	More than 72			More than 72		
Tolerance to the reflected optical power	dB	Less than 10			Less than 10		

Note 1—The value of "minimum ORL of ODN at point O$_{ru}$ and O$_{rd}$, and O$_{lu}$ and O$_{ld}$" should be more than 20 dB in optional cases which are described in Appendix I.
Note 2—The values on ONU transmitter reflectance for the case that the value of "minimum ORL of ODN at point O$_{ru}$ and O$_{rd}$, and O$_{lu}$ and O$_{ld}$" is 20 dB are described in Appendix II.
Note 3—Values of maximum −20 dB width, and minimum side mode suppression ratio are referred to in ITU-T Rec. G.957.

Table 24.3 Optical interface parameters for 155 Mbps upstream [7]. Note that references to figures, appendices, etc. are as outlined in the reference.

Items	Unit	Single Fiber		Dual Fiber	
		ONU Transmitter (optical interface O_{ru})			
Nominal bit rate	Mbit/s	155.52		155.52	
Operating wavelength	nm	1260–1360		1260–1360	
Line code	–	Scrambled NRZ		Scrambled NRZ	
Maximum reflectance of equipment, measured at transmitter wavelength	dB	Less than –6		Less than -6	
Minimum ORL of ODN at O_{ru} and O_{rd} (Notes 1 and 2)	dB	More than 32		More than 32	
ODN class		Class B	Class C	Class B	Class C
Mean launched power MIN	dBm	–4	–2	–4	–2
Mean launched power MAX	dBm	+2	+4	+1	+3
Launched optical power without input to the transmitter	dBm	Less than Min sensitivity –10		Less than Min sensitivity –10	
Extinction ratio	dB	More than 10		More than 10	
Transmitter tolerance to incident light power	dB	More than –15		More than –15	
If MLM Laser— Maximum RMS width	nm	5.8		5.8	
If SLM Laser— Max. –20 dB width (Note 3)	nm	1		1	
If SLM Laser— Minimum side mode suppression ratio	dB	30		30	
Jitter generation from 0.5 kHz to 1.3 MHz	UI p-p	0.2		0.2	
		OLT receiver (optical interface O_{lu})			
Maximum reflectance of equipment, measured at receiver wavelength	dB	Less than –20		Less than –20	

Table 24.3 (Continued)

Items	Unit	Single Fiber		Dual Fiber	
Bit error ratio	–	Less than 10^{-10}		Less than 10^{-10}	
ODN class		Class B	Class C	Class B	Class C
Minimum sensitivity	dBm	–30	–33	–30	–33
Minimum overload	dBm	–8	–11	–9	–12
Consecutive identical digit immunity	bit	More than 72		More than 72	
Jitter tolerance	–	NA		NA	
Tolerance to the reflected optical power	dB	Less than 10		Less than 10	

Note 1 — The value of "minimum ORL of ODN at point O_{ru} and O_{rd}, and O_{lu} and O_{ld}" should be more than 20 dB in optional cases which are described in Appendix I.

Note 2 — The values of ONU transmitter reflectance for the case that the value of "minimum ORL of ODN at point O_{ru} and O_{rd}, and O_{lu} and O_{ld}" is 20 dB are described in Appendix II.

Note 3 — Values of maximum -20 dB width, and minimum side mode suppression ratio are referred to in ITU-T Rec. G.957.

the downstream and TDMA (Time Division Multiple Access) in the upstream as shown in Figure 24.6. The ONUs are ranged and synchronized to arrive at the OLT in an orderly fashion and the OLT receiver makes adjustments in its decision levels to handle different received powers between near in ONTs and those further out, in line with the maximum limits defined within the specification.

In June of 2003 in the USA an RBOC Consortium [AT&T (formerly SBC and BS) and Verizon] chose B-PON as the technology of choice for its deployment. As a result Verizon has deployed B-PON in its network. B-PON is typically deployed as shown in Figure 24.5, where a WDM device is employed at the CO to optically

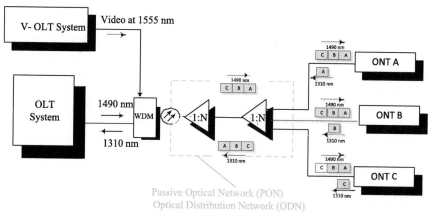

FIGURE 24.6 Illustration of the communication between the OLT and ONTs A, B, and C.

combine the video overlay signal at 1550 nm with the data and voice signals on the 1490 nm wavelength. Verizon used B-PON as part of its initial PON deployment starting in 2004. POTS (Plain Old Telephony Service), data, and IP video were offered on the 1490 nm and 1310 nm was used for the return channel and the video overlay at 1550 nm carried the broadcast, or linear video.

Care has to be taken in overlaying the video signal in order to ensure that the Stimulated Raman Scattering or Raman Crosstalk [12–16] as well as SBS (Stimulated Brillouin Scattering) [17,18] are kept to a minimum. B-PON is still being deployed in the US within the verizon network and has proven efficiency and is robust. However, since the initial deployment of PON technology the demand for more bandwidth from the consumer has resulted in speed offerings beyond the ability of B-PON to efficiently handle it and this resulted in the need for higher bandwidth technologies as described in the following sections.

24.3.2 GE-PON

GE-PON [19] is Gigabit Ethernet passive optical network that is designed to carry Ethernet signal over an optical power splitter based PON. This should be compared with ATM-based fixed cell signal format for B-PON. The transport signal is basically an Ethernet frame although there exist some modifications in the preamble area for clock recovery. To transmit upstream signal, GE-PON uses the clock recovered from the downstream signal. Since GE-PON is a TDM-PON-like B-PON, it still has a time domain multiple access protocol for the upstream, while the downstream is broadcasting to all users. The multiple access protocol for GE-PON defined as MPCP (Multi-Point Control Protocol) is a 64-byte control message. When new ONUs are connected to a PON, auto-discovery and ranging algorithms register the ONU for use. The DBA is one of the most important features for GE-PON as in the generic TDM-PON, since upstream data from each user in a PON is not constant. However, the implementation of DBA is vendor specific.

The physical layer specification of GE-PON was standardized under IEEE 802.3 ah study group in 2004 [20]. To make a low-cost system, the timing control of GE-PON is less strict than that of B-PON. The transmission length target has two versions, 10 km and 20 km. The specifications of the ONUs for the two versions are almost identical in order to lower the cost by increasing the volume. The split ratio starts from 16. The GE-PON emulates point-to-point connection and shared medium by using LLID (Logical Link IDentification) that has the maximum number of 32,768 (15 bits). The LLID also provides a minimum security for downstream data selection. It is also possible to use RF video overlay as in B-PON, although the standard IEEE 802.3 ah does not define it.

The data rate (line rate) of 1.25 Gbps employs 8 B/10 B code resulting in a 1 Gbps usable payload. However, actual payload capacity will depend on many parameters. The physical layer specifications are summarized in Table 24.4. The GE-PON uses transmission wavelength band similar to B-PON and G-PON in order to ensure use of similar lasers and receivers and gain from economies of scale. The upstream transmission can be done with F-P (Fabry-Perot) LD for cost-effective implementation.

Table 24.4 GE-PON and 10 GE-PON physical layer specifications.

Items	Unit	GE-PON				10 GE-PON				
		Downstream		**Upstream**		**Downstream**		**Upstream**		
Class		PX10-D	PX20-D	PX10-U	PX20-U	PR-D1, PR-D3 PRX-D1 PRXD3	PR-D2 PRX-D2	PR-U1	PR-U3	PRX-U3
Nominal bit rate	Gbit/s	1.25±100ppm				10.3125±100ppm				1.25±100ppm
Wavelength range	nm	1480–1500	1260–1360			1575–1580		1260–1280		1260–1360
RMS spectral width	nm	Dep. on wavelength								Dep. on wavelength
Side mode sup. ratio	dB	NA				30		30		NA
Average launched power Max	dBm	+2	+7	+4	+4	5	9	4	9	5.62
Average launched power Min	dBm	−3	+2	−1	−1	2	5	−1	4	0.62
Average launched power off TX	dBm	−39		−45		−39		−45		−45
Extinction ratio	dB	6				6		6		6
RIN Max	dB/Hz	−118	−115	−113	−115	−128		−128		−115
Launch OMA min	mW	0.6	2.8	0.95	0.95	2.46	4.91	0.95	3.01	1.38
Ton Max	Ns	NA		512		NA		512		512
Toff Max	Ns	NA		512		NA		512		512

(Continued)

Table 24.4 (Continued)

Items	Unit	GE-PON		10 GE-PON				
		Downstream	Upstream	Downstream		Upstream		
				PR-D1	PR-D2 PR-D3	PRX-D3	PR-U1 PRX-U1 PRX-U2	PR-U3 PRX-U3
Optical return loss	dB	15		15		15		15
Tx reflectance max	dB			−10		−10		−10
Tx and dispersion penalty	dB	1.3, 2.3	2.8, 1.8	1.5			3	1.4
BER max		10^{-12}			10^{-3}	10^{-12}		10^{-3}
Average received power Max	dBm	−1	−3	−1	−6	−9.38	0	−10
Receiver sensitivity	dBm	−24	−24	−24	−28	−29.78	−20.50	−28.50
Receiver reflectance	dB	−12			−12		−12	−12
Stressed receiver sensitivity	dBm	−22.3	−21.4, −22.1	−21	−25	−28.38	−19	−27
Vertical eye closure penalty	dB	1.2	2.2, 1.5	2.99		1.4	1.5	
Rx setting time max	ns	400	NA	800			NA	

It is also possible to have vendor-specific FEC (Forward Error Correction) in order to increase the link budget.

To secure the transmission data, GE-PON uses vendor-specific encryption especially for the downstream. The encryption is based on AES (Advanced Encryption Standard) with a block cipher mode. The interoperability of GE-PON is currently in discussion and is being standardized by project SIEPON [21].

To provide more bandwidth to users 10 GE-PON was also standardized under IEEE 82.3av in 2009 [22]. The downstream speed is 10 Gbps with two versions of upstream speed, 1 Gpbs and 10 Gbps. For 10 Gbps transmission, FEC (Reed Solomon (255,223)) is mandatory. Table 24.4 represents the specification of 10 GE-PON. The link budget was specified up to 29 dB with a transmission length of 20 km. It may be noted that the timing control of the 10 GE-PON is also less tight than G-PON for low cost implementation.

To support coexistence of GE-PON and 10 GE-PON ONUs with a single ODN, the OLT can be operated in dual rate mode, both at 10 Gbps and 1 Gbps.

22.3.3 G-PON

ITU-T Recommendation G.984 is a family of recommendations that defines G-PON (Gigabit-Passive Optical Networks). It comprises four basic recommendations which are numbered similarly as B-PON and base a lot of their structure on B-PON except that all references to ATM have been removed. It also increased the number of splits supported from 32 in B-PON to 64. G-PON is defined by the specifications as summarized below [23–26]:

- G.984.1: G-PON: General characteristics.
- G.984.2: G-PON: PMD (Physical Media Dependent) layer specification, 2003, with amendments 1 and 2. The various classes are the same as B-PON in Table 24.1, with an additional class of C+ defined in an amendment [24] to the standard. However, the maximum logical reach was extended to 60 km from 20 km of B-PON.
- G.984.3. G-PON: Transmission convergence layer specification, 2008, with amendments 1, 2, and erratum 1.
- G.984.4: G-PON: OMCI (ONT management and control interface) specification, 2008, with amendments 1, 2, 3, corrigendum 1, erratum 1, and an implementer's guide.

The G-PON OMCI recommendation G.984.4 draws on its B-PON counterpart, which defines the B-PON management model. However, G.984.4 removed all references to ATM. G-PON supports the need for greater splitting in the ODN as well as a longer distance out at a distance of 60 km as compared with 20 km for B-PON. In addition, further amendments allowed the deployment out to a differential reach of 40 km [27] as opposed to 20 km as outlined in Table 24.5. This can facilitate the technician following the exact same procedures as they do today when deploying on a PON out to 40 km without the systems, or methodology having to change (see Table 24.6).

Table 24.5 Optical interface parameters of 2488 Mbit/s downstream direction. Note that references to figures and tables within the table refer to references in [24].

Items	Unit	Single Fiber			Dual Fiber		
		OLT Transmitter (Optical Interface O_{ld})					
Nominal bit rate	Mbit/s	2488.32			2488.32		
Operating wavelength	nm	1480–1500			1260–1360		
Line code	–	Scrambled NRZ			Scrambled NRZ		
Maximum reflectance of equipment, measured at transmitter wavelength	dB	NA			NA		
Minimum ORL of ODN at O_{lu} and O_{ld} (Notes 1 and 2)	dB	More than 32			More than 32		
ODN Class		A	B	C	A	B	C
Mean launched power MIN	dBm	0	+5	+3 (Note 4)	0	+5	+3 (Note 4)
Mean launched power MAX	dBm	+4	+9	+7 (Note 4)	+4	+9	+7 (Note 4)
Launched optical power without input to the transmitter	dBm	NA			NA		
Extinction ratio	dB	More than 10			More than 10		
Tolerance to the transmitter incident light power	dB	More than –15			More than –15		
If MLM Laser— Maximum RMS width	nm	NA			NA		
If SLM Laser— Maximum –20 dB width (Note 3)	nm	1			1		
If SLM Laser—- Minimum side mode suppression ratio	dB	30			30		
		ONU receiver (optical interface O_{rd})					
Maximum reflectance of equipment, measured at receiver wavelength	dB	Less than –20			Less than –20		

Table 24.5 (Continued)

Items	Unit	Single Fiber			Dual Fiber		
Bit error ratio	–	Less than 10^{-10}			Less than 10^{-10}		
ODN Class		A	B	C	A	B	C
Minimum sensitivity	dBm	–21	–21	–28 (Note 4)	–21	–21	–28 (Note 4)
Minimum overload	dBm	–1	–1	–8 (Note 4)	–1	–1	–8 (Note 4)
Consecutive identical digit immunity	bit	More than 72			More than 72		
Tolerance to the reflected optical power	dB	Less than 10			Less than 10		

Note 1 — The value of "minimum ORL of ODN at point O_{ru} and O_{rd}, and O_{lu} and O_{ld}" should be more than 20 dB in optional cases which are described in Appendix I/G.983.1.
Note 2 — The values on ONU transmitter reflectance for the case that the value of "minimum ORL of ODN at point O_{ru} and O_{rd}, and O_{lu} and O_{ld}" is 20 dB are described in Appendix II/G.983.1.
Note 3 — Values of maximum –20 dB width, and minimum side mode suppression ratio are referred to in ITU-T Rec. G.957.
Note 4 — These values assume the use of a high-power DFB laser for the OLT Transmitter and of an APD-based receiver for the ONU. Taking future developments of SOA technology into account, a future alternative implementation could use a DFB laser + SOA, or a higher power laser diode, for the OLT Transmitter, allowing a PIN-based receiver for the ONU. The assumed values would then be (conditional to eye-safety regulation and practice):
Mean launched power MAX OLT Transmitter: +12 dBm
Mean launched power MIN OLT Transmitter: +8 dBm
Minimum sensitivity ONU Receiver: –23 dBm
Minimum overload ONU Receiver: –3 dBm

Table 24.6 Optical interface parameters of 1244 Mbit/s upstream direction. Note that references to figures and tables within the table refer to references in [24].

Items	Unit	Single Fiber	Dual Fiber
		ONU Transmitter (Optical Interface O_{ru})	
Nominal bit rate	Mbit/s	1244.16	1244.16
Operating wavelength	nm	1260–1360	1260–1360
Line code	–	Scrambled NRZ	Scrambled NRZ
Maximum reflectance of equipment, measured at transmitter wavelength	dB	Less than –6	Less than –6

(Continued)

Table 24.6 (Continued)

Items	Unit	Single Fiber			Dual Fiber		
Minimum ORL of ODN at O_{ru} and O_{rd} (Notes 1 and 2)	dB	More than 32			More than 32		
ODN Class		A	B	C	A	B	C
Mean launched power MIN	dBm	−3 (Note 5)	−2	+2	−3 (Note 5)	−2	+2
Mean launched power MAX	dBm	+2 (Note 5)	+3	+7	+2 (Note 5)	+3	+7
Launched optical power without input to the transmitter	dBm	Less than Min sensitivity −10			Less than Min sensitivity −10		
Maximum Tx Enable (Note 3)	bits	16			16		
Maximum Tx Disable (Note 3)	bits	16			16		
Extinction ratio	dB	More than 10			More than 10		
Tolerance to transmitter incident light power	dB	More than −15			More than −15		
MLM Laser— Maximum RMS width	nm	(Note 5)			(Note 5)		
SLM Laser— Maximum −20 dB width (Note 4)	nm	1			1		
If SLM Laser— Minimum side mode suppression ratio	dB	30			30		
Jitter generation from 4.0 kHz to 10.0 MHz	UI p-p	0.33			0.33		
		OLT Receiver (optical interface O_{lu})					
Maximum reflectance of equipment, measured at receiver wavelength	dB	Less than −20			Less than −20		
Bit error ratio	−	Less than 10^{-10}			Less than 10^{-10}		
ODN class		A	B	C	A	B	C
Minimum sensitivity	dBm	−4 (Note 6)	−28	−29	?24 (Note 6)	−28	−29

Table 24.6 (Continued)

Items	Unit	Single Fiber			Dual Fiber		
Minimum overload	dBm	−3 (Note 6)	−7	−8	−3 (Note 6)	−7	−8
Consecutive identical digit immunity	Bit	More than 72			More than 72		
Jitter tolerance	−	NA			NA		
Tolerance to the reflected optical power	dB	Less than 10			Less than 10		

Note 1 — The value of "minimum ORL of ODN at point O_{ru} and O_{rd}, and O_{lu} and O_{ld}" should be more than 20 dB in optional cases which are described in Appendix I/G.983.1.
Note 2 — The values of ONU transmitter reflectance for the case that the value of "minimum ORL of ODN at point O_{ru} and O_{rd}, and O_{lu} and O_{ld}" is 20 dB are described in Appendix II/G.983.1.
Note 3 — As defined in 8.2.6.3.1.
Note 4 — Values of maximum -20 dB width, and minimum side mode suppression ratio are referred to in ITU-T Rec. G.957.
Note 5 — While MLM laser types are not applicable to support the full ODN fiber distance of Table 24.2 a, such lasers can be used if the maximum ODN fiber distance between R/S and S/R is restricted to 10 km. The MLM laser types of Table 24.2 e can be employed to support this restricted fiber distance at 1244.16 Mbit/s. These laser types are subject to the same conditions as indicated in Note 5 of Table 24.2 e.
NOTE 6 — These values assume the use of a PIN-based receiver at the OLT for Class A. Depending on the amount of ONUs connected to the OLT, an alternative implementation from a cost point of view could be based on an APD-based receiver at the OLT, allowing it to use more economical lasers with lower fiber-coupled emitted power at the ONUs. In this case, the values for Class A would be:
Mean launched power MIN ONU Transmitter: −7 dBm
Mean launched power MAX ONU Transmitter: −2 dBm
Minimum sensitivity OLT Receiver: −28 dBm
Minimum overload OLT Receiver: −7 dBm

24.3.4 Next-generation PON technologies

Since the definition of G-PON and GE-PON the standards bodies have been working on the evolution of the access technologies given the continued growth in residential and business services which have been nearly doubling in bandwidth [28] every 2–3 years for residential traffic alone. In the following sections we outline some of the continuing work on the latest technologies being defined to handle the growing bandwidth requirements for both residential and business services.

24.3.4.1 XG-PONs

The bandwidth needs for the customer continues to increase. In order to meet the future bandwidth demand, next-generation PONs have been under investigation. Among them, 10 Gbit capable TDM PONs are considered as the first choice for next NG-PON (XG-PON, where the "X" denotes 10 in Roman numerals) and have been standardized by ITU-T [29,30].

The most important requirements for XG-PONs were higher bandwidth and seamless migration from G-PON to protect operator's investment. Thus mainly the technologies that can support existing power splitter based ODNs were considered. Approaches that require a new ODN based on wavelength division multiplexing devices were not considered for XG-PON, since existing G-PON users who do not want to upgrade to XG-PON should be able to remain with G-PON services, while other users can be upgraded to XG-PON services. In other words, minimization of service interruption during the migration was one of the major requirements for XG-PONs. Finally, XG-PONs need to support/emulate G-PON services in case of full migration. With these constraints, two XG-PONs were defined. XG-PON 1 supports 10 Gbps downstream and 2.5 Gbps upstream, while XG-PON 2 supports symmetric 10 Gbps [31]. However, XG PON 2 is not standardized yet due to a lack of a clear need and burst mode receiver that operates at 10 Gbps.

To use the same ODN, the spectrum (or wavelength) allocation is critical. Coexistence of two PON systems with the same ODN requires a WDM filter at the CO to combine the G-PON signal and XG-PON signal. At each ONU, a blocking filter is needed to block signal from the one of the unintended PON technologies [32]. The WDM filter and the blocking filter should be preinstalled to enable migration to XG-PON without service interruption. Details of coexistence with wavelength band will be discussed in Section 24.3.5.

The power budgets of the XG-PON 1 specification have two classes defined. Since the power budgets were designed with the coexistence with G-PON in mind, the nominal budget corresponds to G-PON class B+ and the other class is to account for an extended budget G-PON C+. Table 24.7 summarizes these classes of XG-PON 1 power budgets. The transmission distance was specified as 20 km and 40 km with support of an optical split up to 256 for better cost sharing compared with G-PON. The transmitter eye-masks shown in Figure 24.7 were also defined to allow the testing of the optical transmitter without actual incorporating it into the full end-to-end system. This may help to reduce testing time and cost. As optics alone were not seen as being able to achieve the full link budget economically the XG-PON1 specification incorporates mandatory FEC. It should be noted that the nominal transmission speed for downstream is different from 10 GE-PON (see Table 24.8).

Table 24.7 Class of XG-PON 1 power budgets.

Fiber Type	ITU-T G.652 or Equivalent
Attenuation range	N1 class: 14–29 dB
	N2 class: 16–31 dB
	E1 class: 18–33 dB
	E2 class: 20–35 dB
Maximum fiber distance	DD20: 20 km
	DD40: 40 km
Minimum fiber distance	0 km
Bidirectional transmission	1-fiber WDM

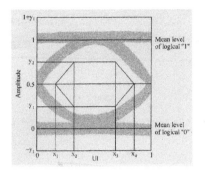

	9.95328 Gbit/s
x3-x2	0.2
y1	0.25
y2	0.75
y3	0.25
y4	0.25

	2.48832 Gbit/s
x3-x2	0.2
y1	0.25
y2	0.75

FIGURE 24.7 Eye-masks for XG-PON 1.

Table 24.8 Optical interface parameters of 9.95328 Gbit/s downstream direction.

Item	Unit	Value					
OLT Transmitter (optical interface O_{ld})							
Nominal line rate	Gbit/s	9.95328					
Operating wavelength (Note 1)	nm	1575–1580					
Line code	-	NRZ					
Mask of the transmitter eye diagram	-	See Figure 24.7					
Maximum reflectance at S/R, measured at transmitter wavelength	dB	NA					
Minimum ORL of ODN at O_{lu} and O_{ld}	dB	More than 32					
ODN Class		N1	N2 N2a	N2b	E1	E2 E2a	E2b
Mean launched power MIN	dBm	+2.0	+4.0	+10.5	+6	+8	+14.5
Mean launched power MAX	dBm	+6.0	+8.0	+12.5	+10	+12	+16.5

(Continued)

Table 24.8 (Continued)

Item	Unit	Value					
OLT Transmitter (optical interface O_{ld})							
Launched optical power without input to the transmitter	dBm	NA					
Minimum extinction ratio	dB	8.2					
Transmitter tolerance to reflected optical power (Note 7)	dB	More than −15					
Dispersion range	ps/nm	0 –400 (DD20) 0 –800 (DD40)					
Minimum side mode suppression ratio	dB	30					
Maximum differential optical path loss	dB	15					
Jitter generation	–						
ONU receiver (optical interface O_{rd})							
Maximum optical path penalty (Note 6)	dB	1.0					
Maximum reflectance at R/S, measured at receiver wavelength	dB	Less than −20					
Bit error ratio reference level	–	10^{-3}					
ODN Class		N1	N2 / N2a	N2b	E1	E2 / E2a	E2b
Minimum sensitivity at BER reference level (Note 5)	dBm	−28.0	−28.0	−21.5	−28.0	−28.0	−21.5
Minimum overload at BER reference level	dBm	−8.0	−8.0	−3.5	−8.0	−8.0	−3.5
Consecutive identical digit immunity	bit	More than 72					
Jitter tolerance	–						
Receiver tolerance to reflected optical power (Note 8)	dB	Less than 10					

The TC (transmission convergence) layer of XG-PON 1 is based on the G-PON TC protocol with some modifications to support low complexity and more users [33] (see Table 24.9).

Table 24.9 Optical interface parameters of 2.48832 Gbit/s upstream direction.

Item	Unit	Value			
ONU transmitter (optical interface O$_{ru}$)					
Nominal line rate	Gbit/s	2.48832			
Operating wavelength	nm	1260–1280			
Line code	–	NRZ			
Mask of the transmitter eye diagram	–	See Figure 24.7			
Maximum reflectance at R/S, measured at transmitter wavelength	dB	Less than −6			
Minimum ORL of ODN at O$_{ru}$ and O$_{rd}$	dB	More than 32			
ODN Class		N1	N2	E1	E2
Mean launched power MIN	dBm	+2.0	+2.0	+2.0	+2.0
Mean launched power MAX	dBm	+7.0	+7.0	+7.0	+7.0
Launched optical power without input to the transmitter (Note 3)	dBm	Less than "Minimum sensitivity at BER reference level" −10			
Maximum Tx enable (Note 3)	bits	32			
Maximum Tx disable (Note 3)	bits	32			
Minimum extinction ratio	dB	8.2			
Transmitter tolerance to reflected optical power (Note 8)	dB	More than −15			
Dispersion range (Notes 4 and 5)	ps/nm	0–140 (DD20) 0–280 (DD40)			
Minimum side mode suppression ratio	dB	30			

(Continued)

Table 24.9 (Continued)

Item	Unit	Value			
ONU transmitter (optical interface O_{ru})					
Jitter transfer	–				
Jitter generation	–				
OLT receiver (optical interface O_{lu})					
Maximum optical path penalty (Note 7)	dB	0.5			
Maximum reflectance at S/R, measured at receiver wavelength	dB	Less than –20			
Bit error ratio reference level	–	10^{-4}			
ODN Class		N1	N2	E1	E2
Minimum sensitivity at BER reference level	dBm	–27.5	–29.5	–31.5	–33.5
Minimum overload at BER reference level	dBm	–7.0	–9.0	–11	–13
Consecutive identical digit immunity	bit	More than 72			
Jitter tolerance	–				
Receiver tolerance to reflected optical power (Note 9)	dB	Less than 10			

The envisioned XG-PON services typically include POTS and VoIP (Voice over Internet Protocol), IPTV and digital TV broadcasting, leased lines, high-speed internet, mobile back haul, VPN, and IP services. See G.987.1 for more details [29].

24.3.4.2 NG-PON 2

Consumer appetite for increased bandwidth is continuing to increase, especially due to the rapid growth of video traffic which is expected to occupy 90% of internet traffic by 2015. As video traffic is becoming more unicast in nature, the need for bandwidth is growing dramatically. In addition, the evolution of large-scale display technology brings about video traffic requiring very high bandwidth [34].

From the operator's point of view, reducing the total cost of ownership is very important while supporting the expected explosive bandwidth growth. It is envisioned that this can be achieved by defining an optical access network that utilizes large transmission capacity shared over many users. With this in mind, the NG-PON 2 technologies are being investigated in terms of increased number of users per PON, CO consolidation with long reach transmission, together with a single platform for residential, business, and mobile backhaul.

To initiate the various NG-PON 2 technologies, operators came to a consensus on the general set of requirements. These requirements are for expected deployment in the 2015 timeframe. Table 24.10 is a good summary based on a recent publication [34] from FSAN.

Based on operator's requirements, many technologies have been proposed as candidate of NG-PON 2. The technologies can be categorized based on multiple access

Table 24.10 Operator's requirements for NG-PON 2.

Feature	Requirement	Remark
Capacity	Per feeder fiber >40 Gb/s for downstream >10 Gb/s for upstream per ONU sustainable 1 Gb/s for down, 0.5 Gb/s for up	Business and mobile back haul applications require sustained symmetric bandwidth
Reach	40 km (passive) 60 km (RE is acceptable)	RE should remotely manageable through an OLT
Maximum differential reach	Up to 40 km (either 20 km or 40 km, configurable)	
End user termination (split ratio)	>64	Higher split ratio is preferable with WDM
Coexistence and migration	Reuse deployed power splitter Support migration from GPON over XGPON to NGPON2	Legacy ONU and OLT must remain unchanged. No additional filter is allowed
Service-specific requirements	Support legacy services Must support G.987.1 Section 7 Meets latency limits for mobile backhaul	Better delay and jitter performance
Resiliency	Support feeder fiber redundancy including hardware redundancy of oLT	Restoration time <50 ms
Operational requirements	ONU management by OMCI (G.988) Early identification of faults in physical layer and service layer	OTDR for ODN monitoring end-to-end performance monitoring up to Ethernet layer
Operating temperature	Outdoor (−45 °C to +45 °C) for ONU Extended outside temperature of OLT (optional)	

methods in time domain, frequency domain (subcarrier domain), and wavelength domain [35]. It is also possible to use the mixture of these technologies. To have narrow channel spacing and high link budget, coherent detection is also investigated [36]. The orthogonal frequency division technique can be considered as a more efficient version of the frequency division method [37]. It is interesting to note that there was no code division multiplexing. The colorless operation (simply a single type of ONU) can be supported regardless of the technologies with the help of a tunable laser and/or tunable filter. However, complexity and cost of the system will be dependent on the technologies. In addition, different technologies have their own preferred ODN architectures. The studied technologies are summarized in Table 24.11. The WDMA

Table 24.11 NG-PON 2 technologies.

Multiple Access Method	# of WDM ch.	Variations	Signal Format	ONU BW (sustained) (Gb/s)	Scalability of ONU BW (Gb/s)	Challenges
TDMA	1		NRZ Duo-binary DQPSK	40/N		Dispersion comp. Link budget 40 Gb/s burst mode receiver
TDMA+ WDMA	N/M		NRZ	10/M	~2.5 G with higher wavelength	10 Gb/s burst mode receiver Tunable LD/ Tunable filter
WDMA	N	Injection seeding	NRZ	1	5–10 (with tunable LD)	Tunable filter for legacy ODN support
		Wave-length reuse	IRZ	1	5	Tunable filter for legacy ODN support
		Tunable LD	NRZ	1	10	Low cost tunable LD
		UDDWM	DQPSK	1	10 with reduced ONU	Photonic integration
FDMA(OFDM)	1	Coherent detection	QAM	1		Low cost ADC Low cost Tx/Rx
		Direct detection	QAM	1		

N: number of ONUs per PON (default is 64)
M: number of TDMA users (8 or 16).

technique based on injection seeding was standardized recently under ITU-T (ITU-T document G.698.3) [38]. It may be noted that WDMA prefers WDM filter at ODN, except for UDWDM with coherent detection [36].

The TDMA PON is an extension of XG-PONs in terms of transmission bit rate. For downstream, 40 Gbps transmission is needed to meet NG-PON 2 requirements. The most important question is achievable link budget at 40 Gbps, since impairment due to fiber dispersion is very serious and an intrinsic degradation of receiver sensitivity is inevitable. For NRZ transmission in low loss regions, the maximum reach was limited to a few km. Thus a potential solution is to use transmission near the dispersion zero wavelength with a semiconductor optical amplifier. In this case, however, optical nonlinearity induced impairment should be investigated. In addition, the bandwidths of the currently available modulators and APDs (Avalanche Photo Diode) are not sufficient for 40 Gbps NRZ signal. A new modulation format (e.g. duobinary [39]) that occupies narrow spectral width can be considered as a possible solution including coherent detection. However, a low cost implementation is still questionable.

There are varieties of WDM-PON technologies that can meet FSAN requirements for capacity. For WDM-PONs, two types of optical branching devices can be used, a wavelength division multiplexer (AWG: arrayed waveguide grating) and an optical power splitter. Since colorless operation of ONU (wavelength independent operation) is essential to reduce inventory and operational costs, an injection seeding with a reflective modulator or tunable laser is desirable for ONUs. The colorless operation with injection seeding can be achieved by using broadband reflective modulators [40] that can be realized by either anti-reflection-coated F-P LD or RSOA (Reflective Semiconductor Optical Amplifier). In this configuration, the AWG in the ODN determines the wavelengths for communication. The seed light located at OLT can be either a broadband light source or a coherent multi-wavelength light source. For a transmission speed up to 2.5 Gbps per channel, incoherent light can be used for seed light to mitigate any coherent interference induced penalty (e.g. optical back reflection). With the help of FEC, the achieved link budget at 1.25 Gbps per channel at 25 GHz channel spacing was 27 dB [41].

The seed source can be eliminated by using the downstream signal as the seed light [42]. However, this approach suffers from an extinction ratio penalty and extra insertion loss. In addition, back reflection induced penalty limits transmission performance. It is possible to reduce the extinction ratio penalty and back reflection induced penalty with a special modulation format, such as IRZ [42]. However, the power penalty (or reduction of link budget) more than 5 dB may not be avoided compared with the case of an independent seeding source.

A WDM-PON with a tunable laser may be simple to realize. In addition, it may provide good performance with a high link budget. However, the cost of the tunable laser at present time may not be affordable for access application.

Since the legacy ODN support (OND with optical power splitter) is one of the mandatory requirements by the operators, the WDM-PON technologies need to provide colorless operation with the legacy ODN. At the OLT, the use of AWG for multiplexing and demultiplexing is desired. At the ONUs, different wavelengths can be transmitted to the OLT with tunable transmitters. However, a tunable filter is needed at each ONU to receive only the desired channel. For injection seeded WDM-PON, a tunable filter is also needed in front of the reflective modulator of each ONU to select a single seed channel for transmission. In these systems, the transmission distance can be limited by the insertion loss of optical components, AWG, optical power splitter (a main loss element), tunable filter, and the fiber.

Longer reach and high split ratio can be achieved by employing coherent detection at both ends. The sensitivity gain with coherent detection can be as high as 17 dB (sensitivity of −46 dBm at 1.25 Gbps) compared with a PIN detector case. However, an optical source with a narrow optical linewidth is needed for transmission of the coherent signal. Based on the fact that the channel spacing or the frequency spacing can be comparable to the transmission bit rate, it may be possible to have a 1000 split ratio (1000 channels) with a limited transmission distance [36]. Although this method has many interesting features, the realization of this system requires extensive photonic integration at the OLT. This is currently the topic of much research and development within the industry.

The multiplexing and/or multiple access can be done in the frequency domain (classically subcarrier domain) instead of the wavelength domain. This method can be distinguished with coherent detection in terms of electrical domain processing of frequency division multiplexed signal. An extensive digital signal processing including ADC (Analog-to-Digital Converter) and DAC (Digital-to-Analog Converter) are essential building blocks. A well-known example is orthogonal frequency division multiplexing [37] that gives the narrowest frequency spacing based on the orthogonality between the modulated subcarriers. A single optical carrier is used to transmit the OFDM signal. The different number of subcarriers (RF carriers) to ONU that can be assigned depends on the bandwidth from/to ONU. This corresponds to the dynamic bandwidth allocation in a TDM-PON system. However, as a direct result of the significant digital signal processing a large amount of electrical power is consumed. In addition, the cost of high-speed DSP is still relatively expensive for access applications and will take several years before it is seen as a viable alternative.

The high capacity access network can be realized by combining the technologies explained above. The most common and straightforward method is time and wavelength division multiplexing PON. In other words, some ONUs with a single wavelength can be shared in the time domain [44]. It is called a TDM WDM hybrid PON. This technology can be used for many different ODN architectures. However, the ODN with only an optical power splitter is preferred to support coexistence with legacy PON deployments. At the OLT, a WDM MUX and DMUX are used for the downstream signal and upstream signal, respectively. Burst mode optical receivers are used for each channel. At the ONU, a tunable transmitter and tunable receiver are required for colorless operation and flexible bandwidth allocation. This type of PON is called particularly TWDM-PON inside of the FSAN. To support coexistence with legacy PONs (G-PON and XG-PON) the wavelength should be carefully allocated. Although it looks like a straightforward extension of XG-PON, the usable channel can be determined by the tunable laser and tunable filter. In addition, the costs of these devices are not economical at present time for access application, though there is active research in this area.

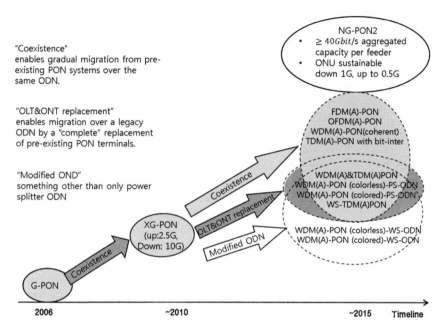

"Coexistence"
enables gradual migration from pre-existing PON systems over the same ODN.

"OLT&ONT replacement"
enables migration over a legacy ODN by a "complete" replacement of pre-existing PON terminals.

"Modified OND"
something other than only power splitter ODN

NG-PON2
• ≥ 40*Gbit*/s aggregated capacity per feeder
• ONU sustainable down 1G, up to 0.5G

FDM(A)-PON
OFDM(A)-PON
WDM(A)-PON(coherent)
TDM(A)-PON with bit-inter

WDM(A)&TDM(A)PON
WDM(A)-PON (colorless)-PS-ODN
WDM(A)-PON (colored)-PS-ODN
WS-TDM(A)PON

WDM(A)-PON (colorless)-WS-ODN
WDM(A)-PON (colored)-WS-ODN

XG-PON
(up:2.5G, Down: 10G)

G-PON

Coexistence
OLT&ONT replacement
Modified ODN

2006   ~2010   ~2015   Timeline

FIGURE 24.8 Migration and coexistence scenario for access network.

As stated NG-PON 2 is targeting high speed at longer distances and higher split ratio with a resulting increase of the total coupled power into the transmission fiber, regardless of technology choice. This brings about signal impairments induced by optical nonlinearities in the transmission fiber. Particularly, when RF video overlay is present, Raman interaction with NG-PON 2 signals should be carefully investigated to mitigate signal quality degradation [14,15].

The operators participated in the FSAN (Full Service Access Network) will select technologies for NG-PON 2. For the first step, the operators published potential solutions and a migration path roadmap for NG-PON 2 as shown in Figure 24.8 [34].

It should be noted that the authors are presently participating in FSAN activities for NG-PON 2. FSAN operators have made consensus for TWDM-PON as primary NG-PON 2 with WDM-PON overlay if needed. Currently, many technical contributions are ongoing including the wavelength plan for coexistence. The target for ratification of the NG-PON 2 standard is in the 2013 timeframe.

24.3.5 Coexistence and wavelength plan

At the beginning of the FTTX development, transmission wavelength was not a major concern. For B-/GE-/G-PON upstream wavelength, a wide spectral range around 1310 nm (1260–1360 nm) was selected to take advantage of a low cost multimode F-P LD at the ONUs and zero dispersion [7]. This was a very reasonable choice to minimize the cost of ONUs. For these PON systems, asymmetric bandwidth between upstream and downstream was generally accepted considering traffic patterns in access network at that time. Then, the downstream transmission wavelength was selected in the low loss wavelength range in a narrow spectral range suitable for uncooled DFB lasers. The selected wavelength band was 1480–1500 nm. This is also one of the CWDM (Coarse WDM) wavelength bands. It should be noted that separation between the wavelength bands for the upstream and downstream is sufficiently wide to use a simple low cost WDM filter for a diplexer.

ITUT-T document G.983.3 also defined other wavelength bands [9]. The enhancement band (1539–1565 nm) was for video services and wavelength division multiplexing services. Two additional wavelength bands were reserved for future uses, i.e. 1380–1460 nm and long wavelength side. (Specific numbers were not defined for further study.) Since the digital video was not popular at that time (2001), most video services were delivered via analog signals that required high-power transmission to achieve high-quality video. Then, it was natural to select an EDFA amplification band of 1550–1560 nm within the enhancement band for video overlay. Figure 24.9 shows wavelength plan in G. G983.3 with a typical loss curve of the transmission fiber.

Coexistence became one of the major concerns for next-generation PONs, since a considerable amount of legacy PONs (B-/GE-/G-PONs) have been and will be deployed before these technologies are readily available. The coexistence requirement makes possible gradual migration from existing PONs with a deployed ODN. The coexistence was actively discussed in the IEEE standard group 802.3av for standardization of 10 GE-PON [22]. At the same time, FSAN discussed the blocking filter for G-PON ONU for coexistence with XG-PON. The 10 GE-PON standard approved in 2009 defined upstream transmission wavelength within the existing upstream band by shrinking it to 1290–1330 nm [22]. Then, upstream wavelength of both XG-PON and 10GE-PON was assigned to 1260–1280 nm. The downstream G-PON wavelength was untouched by assigning the XG-PON downstream wavelength to 1575–1580 nm (outdoor case 1575–1581 nm) which should be outside of the G-PON blocking filter pass band. Figure 24.10 shows wavelength allocation for G-PON and XG-PON. See ITU-T Document G.987.1 for details [29].

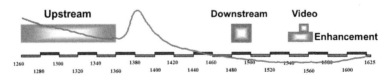

FIGURE 24.9 Old wavelength band allocation.

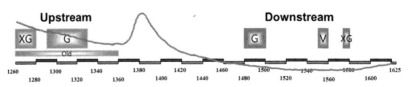

FIGURE 24.10 New wavelength allocation after XG-PONs.

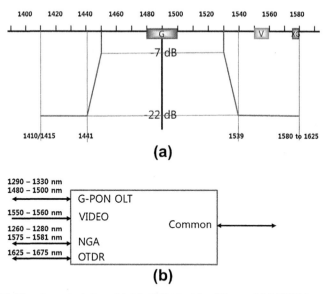

(a)

(b)

FIGURE 24.11 The characteristics of (a) G-PON blocking filter and (b) WDM 1r.

For coexistence with G-PON and XG-PON, the OLT should be equipped with a WDM filter (WDM 1r) to combine and to separate the G-PON and XG-PON signal. The G-PON ONU should block the XG-PON signal by using the blocking filter, and vice versa. These filters should be preinstalled to minimize service interruption during the service migration. Figure 24.11 shows defined characteristics of the G-PON blocking filter [46] and the WDM1r [47]. Then, the XG-PON OLT can be installed at the CO without service interruption. At the ONU side, the G-PON ONU can be upgraded to the XG-PON ONU upon a customer's request.

The coexistence issue can be more complex when we consider NG-PON 2. At the OLT, the modified WDM 1r or CE (Coexistence) filter combines G-PON, XG-PONs, and NG-PON 2 signals, while the blocking filter at the ONU blocks the signals from unwanted PONs. Figure 24.12 shows a configuration of the PONs that provides coexistence with three different PON technologies [34]. The definition of filters or wavelength plans of the NG-PON 2 is a major topic for NG-PON 2 standardization. It is envisioned that in the future when all the users are upgraded to the XG-PON or NG-PON 2, the wavelength band used by G-PON can be reserved for the future PONs.

Considering the isolation characteristics of blocking filter at the G-PON ONU, one possible region of downstream band for NG-PON 2 can be 1410–1441 nm. (A shorter wavelength range may be possible. However, isolation characteristic is not defined in standard.) There are some potentially available spectral ranges in the 1539–1575 nm and 1580–1625 nm range. However, the actual wavelength ranges usable depend on use of the video overlay and the guard bands between signals.

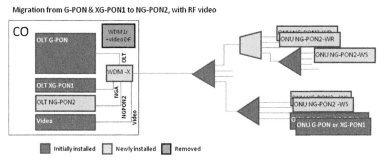

Migration from G-PON & XG-PON1 to NG-PON2, with RF video

FIGURE 24.12 Coexistence of G-PON with XG-PON and NG-PON2.

If we take into account all of these factors, the wavelength band of NG-PON 2 is rather limited.

Coexistence can also be achieved by introducing a WDM filter before the legacy power splitter [48]. The 3-port WDM filter combines/separates legacy PON signals and the NG-PON 2 signals. If existing legacy systems support coexistence between G-PON and XG-PON then the NG-PON 2 can be added by using the WDM filter without disturbing the services of either G-PON or XG-PONs. The advantage of this method is opening of the available spectrum for NG-PON 2 regardless of the current G-PON ONU blocking filter. In addition, the insertion loss of the branching device can be reduced considerably, when an AWG is accepted instead of an optical power splitter for NG-PON 2. It is possible to use 1370 (1340 nm for reduced G-PON band)–1470 nm, 1510–1540 nm, and wavelengths above 1590 nm, assuming a guard band of 10 nm between the bands. If there is no analog video overlay, the available spectrum can be the 1370–1470 nm, 1510–1565 nm, and >1590 nm regions. Unfortunately, a spectrum around 1390 nm is not easy to use due to high attenuation arising from water peak absorption as shown in Figures 24.9 and 24.10. However, recently installed zero water peak fiber with no absorption peak frees use of this band for signal transmission. Again, the new WDM filter should be preinstalled to save on service interruption at a later date.

24.3.6 Extended reach systems

G-PON [23] TC layer was designed to handle a logical limit of 60 km as outlined previously. However, the actual transmission distance can be limited by the fiber loss, split ratio, and transmission impairments. Thus, nominal distance of the G-PON standard was defined as 20 km with split ratios of up to 64. The same transmission length and the split ratio were defined for XG-PONs. It should be noted that the NG-PON 2 requirements specify a nominal reach of 40 km with split ratio of 64 in order to reduce the effective cost of the OLT.

In principle, the transmission length can be increased by reducing the split ratio. However, it is better to have greater transmission length with higher split ratio to reduce cost of the system per subscriber. The addition of link budget will provide great flexibility in reach and split ratio by isolating the trunk loss (loss between the OLT and the ODN) and ODN loss (splitting loss). The RE (Reach Extender) was investigated in FSAN for G-PON and XG-PON. It can be located in between an optical power splitter and the OLT or in between the optical power splitters for some cases.

Several methods can be used for the reach extender [49,50]. A straightforward method is O-E-O (optical-electrical-optical) reach extender. The RE can employ similar optics as used in the OLT and ONU. Another method is to employ optical amplification. A semiconductor optical amplifier that operates at transmission wavelength band can be used. However, a conventional EDFA cannot be used due to wavelength mismatch. A wavelength conversion may be considered before the optical amplifier if the cost is affordable. It is also possible to combine an optical amplifier and O-E-O solution. These RE technologies can increase the maximum transmission length but we are still limited to 60 km due to logical limit of G-PON specifications.

The main drawback of the RE is expected to be the operational cost induced by powering, protection, and management by the OLT, since the RE is an active element. In addition by definition these are orthogonal to the original concept of a PON being passive. To date the RE economics are unproven. Unfortunately, the RE for coexistence requires additional optics for separation/combining of G-PON signal and XG-PON signals [50].

24.3.7 **CO consolidation**

The use of the term CO consolidation has been widely used in the industry to describe the collapsing of the number of COs required and the eventual migration of customers from their present copper services to a fiber-based platform. In general this can be broken down into several stages such as *CO bypass* with the newer fiber-based technology, *CO migration* which covers the migration of customers from copper-based technologies to fiber-based services, and *CO collapse* which represents the closing of the bypassed CO. These different, but inter-related activities are described in more detail below.

24.3.7.1 *CO bypass*

Fiber optic communications have the ability to reach much further than the legacy copper technologies. Thus for companies deploying fiber there is no need to deploy the technology in every central office. Rather it is possible to centralize the aggregation equipment (for example OLTs) and provide service from a centralized location that takes advantage of the fiber's extra reach. This is illustrated in Figure 24.13.

The enablers for this type of deployment have been the extensive reach of PON systems, the availability of higher powered PON systems as described in the standards (C+ Optics [Amendment 2 of [24]), and improved ONT receiver sensitivities and higher powered EDFAs.

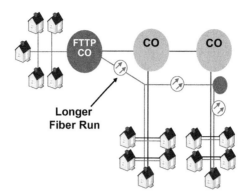

FIGURE 24.13 FTTP serving multiple CO areas while bypassing the incumbent COs.

24.3.7.2 CO migration

The initial deployment of FTTP in many areas of the world was largely driven by the residential market given that that was where the volumes were necessary to justify the costs of deploying fiber. However as more and more geographical areas are covered it makes sense to include the business passed, thus lowering their costs as well.

Copper is aging in the outside plant and it has been found that it can be prone to environment effects so by migrating services over to fiber it is possible to save on operational costs. As a result operators in many parts of the world are looking at the potential of moving services historically carried over copper over to fiber to take advantage of the increased reliability and Opex.

24.3.7.3 CO collapse

Most operators started out just offering telephone services and have an abundance of copper deployed. This copper requires ongoing maintenance and upgrading as its quality properties are very sensitive to rain and other environment effects. As a result maintenance budgets have increased over time and there is a desire to reduce the overall cost while increasing the reliability of the services provided to the customer. Operators worldwide are reviewing the technical as well as regulatory challenges on transitioning over to an all-fiber network and the challenges that this produces.

As a result operators are investigating the possibility of bypassing the COs by the newer technology which reduces the service value of these legacy COs and could reduce the number of COs required to support a given geographical area [51].

24.4 STATUS AND FTTX DEPLOYMENTS AROUND THE WORLD

In the following sections we review the status of FTTX in its many flavors around the world, initially covering Asia where there is the largest market deployment of FTTX technologies in the world as shown in Figure 24.14. Europe as a whole accounts for greater than 15%; the Americas follow with 11%.

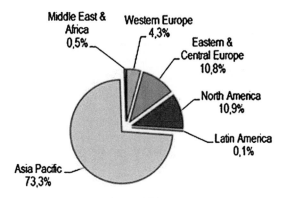

FIGURE 24.14 FTTH around the world. Source [52].

Table 24.12	FTTH/B around world at end of 2011.	
	Subscribers [M]	**Home Passed [M]**
Asia Pacific	58	175.3
North America	8	22.6
South America	0.35	4.2
Europe	10.2	43.7
MENA	0.5	1.6
Total	77.05	247.4

There are various kinds of technologies that use optical fiber to serve subscribers in access network including the fiber-rich solutions such as FTTH/B. The FTTH means connection of houses with optical fiber, while the FTTB means connection of buildings with optical fiber. We focus on FTTH/B deployment status in this section. Around the world FTTH/B subscribers had reached 77 M by 2011, while the homes passed exceeded 247.4 M. Table 24.12 shows the status of FTTH/B and its breakdown by region.

The ranking for FTTH/B penetration is summarized in Figure 24.15 with South Korea leading with a penetration rate of 58%. An interesting example is UAE with above 55% penetration of FTTH to date.

In the following sections we outline the present state of these deployments in various regions, namely Asia, Europe, and the Americas before focusing on some specific deployments by companies which are representative of the issues and choices operators are facing as they choose to deploy FTTX.

24.4.1 FTTX in Asia

Asia recorded 75% (58 M) of the world's total subscribers of FTTH/B at the end of 2011. Within Asia, subscribers in China exceeded 25 M, while in Japan it is 22 M.

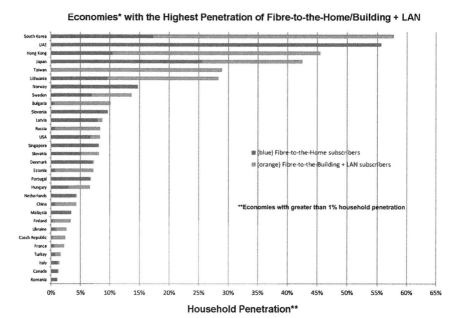

FIGURE 24.15 Rank of FTTH/B penetration at end of 2011. Source [52].

The next highest is Korea at 10 M. China and India show continuous growth, while Japan and Korea show nearly saturated growth rate due to their present very high penetration rate and relatively mature fiber deployments. It may be noted that broadband subscribers in China were 156 M in 2011 and expected to reach 250 M in 2015.

Within the Asia Pacific region, FTTH/B represents greater than 27% of the overall broadband subscribers. In 2011, two important milestones were reached in China. The FTTH subscribers exceeded FTTB subscribers and FTTx subscribers exceeded DSL subscribers [54]. For early adapters like Japan and Korea, FTTH subscribers exceeded DSL subscribers in 2006. In these two countries, DSL subscribers started to decrease in 2006 once the new technology became available. Overall, CAGR of the FTTH/B growth will in Asia Pacific be 21% (FTTH 36%, FTTB 11%), while DSL is expected to decline by 2% [55].

The three countries mentioned above (China, Japan, and Korea) initially deployed GE-PON. However, China slowed down GE-PON deployment in 2010 and has moved to G-PON to provide more bandwidth. As a result, G-PON deployments have increased and by 2011 it occupies about 60% of China's FTTH/B deployments. China will also be stopping deployment of additional GE-PON OLTs in 2012. In contrast GE-PON has been the major selection in Japan and Korea for their FTTH deployment.

Figure 24.16 shows a penetration rate of FTTH/B for Asia Pacific countries in 2011. South Korea, Hong Kong, Japan, and Taiwan had more than 20% penetration, while China had more than about 5%.

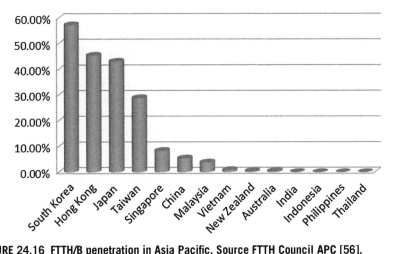

FIGURE 24.16 FTTH/B penetration in Asia Pacific. Source FTTH Council APC [56].

PON is the first choice of FTTH/B in the Asia Pacific area. At the end of 2011, about 92% of all FTTH/B rollouts were PON. Among them, FTTB was 64% according to FTTH Council APAC. The architecture of the ODN depends on the country where specific factors influence the ultimate choice (see Table 24.13).

Table 24.13 Summary of FTTH/B in Asia Pacific.

Country	Architecture	Technologies	Subscribers	Operator	Remarks
China	1 × 16 to 1 × 64	GE-PON → G-PON PtP	25 M	China Telecom China Unicom China Mobile	Target FTTH passed: 100 M at 2015
Japan	Two stages 1 × 4 (CO) and 1 × 8 at field	GE-PON	22 M	NTT	Video overlay service 10 G E-PON is under test
Korea	Two stages 1 × 8 and 1 × 4	GE-PON (major) PtP WDM-PON	10 M	KT SK broadband	XG-PON under test
Taiwan					
Singapore					
Malaysia			~ 0.3 M	Telecom Malaysia	1.2 M passed
Indonesia		G-PON	16 k	Telecom Indonesia	Plan: 8.3 M at 2014
Hong Kong		G-PON PtP	0.6 M	HKBN	

24.4.1.1 KT's (Korea Telecom's) deployment

Korea is leading in the penetration of broadband access and FTTH. In the early 1990s, KT, the leading ISP in Korea, had carried out a small field test for FTTH with TDM-PON (A-PON) with video overlay. However, KT ended up with deploying ADSL due to higher cost of PON at the time. In 2004, KT deployed newly developed WDM-PON [57] in a small scale of a few hundred thousand subscribers for the first time in the world. However, KT selected lower cost GE-PON for mass deployment that began in 2005 and currently has about 3.1 M (2012.3) FTTH subscribers. For video services, KT uses IP-based video including the IPTV. As seen in Figure 24.17, KT employs several different configurations for FTTx with pure FTTH services for residential areas. However, for apartment complexes with installed UTP (Unshielded Twisted Pair) cables, FTTB is the preferred choice. FTTC configurations and direct fiber connections have also been used for some individual homes and big apartment complexes.

It may be noted that Korea is running a Gigabit Internet project which started in 2009. It will serve approximately 6000 households in 19 cities in 2012. For this service, WDM-PON, 10G/1G E-PON, G-PON, RFoG+E-PON will be deployed [58].

24.4.1.2 NTT's deployment (Japan's deployment)

Japan has been very active in deploying FTTH led primarily by NTT. They started the development of FTTH systems as early as 1991 and deployed commercial FTTH in 1997 with POTS/ISDN and RF video overlay services. In 2002, NTT started to deploy B-PON with a downstream bit rate of 622.08 Mbps and an upstream bit rate of 155.52 Mbps. The deployment of B-PON lasted until 2006 with subscribers totaling 3.5 M and the deployment of GE-PON began in 2005. As seen in Figure 24.18,

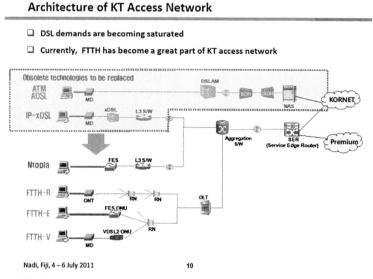

FIGURE 24.17 Korea Telecom's access network architecture. Source [58].

FTTH System Configuration today

Since 2004, GE-PON has deployed for FTTH.
IP-based services: Internet access, PSTN-quality VoIP and IP-TV
RF Digital video broadcasting service over GE-PON

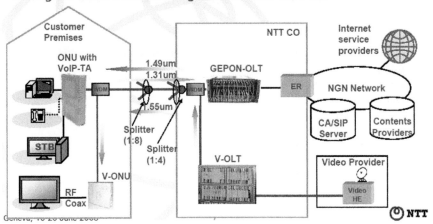

FIGURE 24.18 FTTH deployment in NTT. Source [59].

NTT uses a two-stage power splitter configuration, 1×4 at CO and 1×8 in the field. They also use an OTDR for monitoring the ODN at 1620 nm band. Although there is no video overlay standard in GE-PON, NTT is using video overlay at the 1550 nm band. NTT has about 25 M (2012.03) FTTH subscribers after massive deployment of GE-PON since 2005.

24.4.1.3 China's deployment

Although China is relatively new to FTTX deployment, China has a huge potential market. The recent growth rate in China is remarkable. Three operators, China Telecom, China Unicom, and China Mobile, are the major service providers. Main driving forces for FTTX deployment are national strategy, competition, market demand for bandwidth from video services, and cost reduction of FTTH. China operators deployed GE-PON for FTTH/B at the beginning. However, current deployments are with G-PON for FTTH, while promoting 10 GE-PON for FTTB. For village areas, major architecture is FTTN. An example of deployment architecture for Shanghai Telecom which is one of the subsidiaries of China Telecom is shown in Figure 24.19.

China MIIT (Ministry of Industry and Information Technology) initialized a national action plan for broadband popularization and speed upgrade. The goals for 2012 are (1) 35 M new additions of FTTH households, (2) 4 Mb/s access speed for more than 50% of broadband users, (3) 20 M new additions of broadband subscribers, (4) increase the coverage of WiFi, and (5) popularize BB applications and reduce cost per Mb/s [60].

3. Technology Innovation for the 1st Difficulty: Network Planning

4) Develop unified scalable duplicable ODN construction template.

Shanghai telecom develop ODN construction standards based on the local characters.

■ ODN construction standards include :
- Optical cable access network topology.
- ODN radius of coverage
- Splitters pattern and location
- Two grade splitting
- Selection of fiber cable and connection pattern
- Cross-connect equipments application according to different scenarios
- Selection of optical apparatus

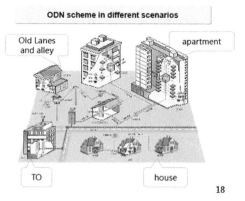

18

FIGURE 24.19 China's deployment architecture. Source [61].

24.4.2 **FTTX in Europe and Africa**

As shown in Figure 24.14, the European portion of FTTH/B is about 15.1% with Russia leading with 4.5 M subscribers (12.3 M homes passed). Sweden (0.64 M) and France (0.63 M) follow as shown in Figure 24.20. Growth in Russia is noticeable [62]. The fiber-based access speed is considerably faster than the copper-based one in Russia and growth rate and take rate are quite high. However, western Europe is relatively slow in deploying FTTH/B even though they have been very active in the standards. Only Sweden and France have subscribers more than 0.6 M with still low increase rate.

For Europe, Ethernet (PtP) is the first choice across EU-39 accounting for 70% of all FTTH/B. The FTTB portion was 59% at the end of 2011.

For the penetration of FTTH/B, Lithuania leads with 28.3%. The second is Norway with 14.7%. Some countries like Norway and Slovenia, Latvia, etc. deployed mainly FTTH as shown in Figure 24.21.

FTTH/B in Africa and the Middle East is at a very early stage. At present they account for less than 1% of global subscribers. However, recent growth in UAE is remarkable. The FTTH penetration is above 55% and this is the second highest penetration after Korea's 58%.

24.4.2.1 *British Telecom's deployment*

British Telecom (BT) is deploying FTTH in green fields with a 100% connection strategy, and the combination of FTTH and FTTC cabinet for brown fields. For BT's

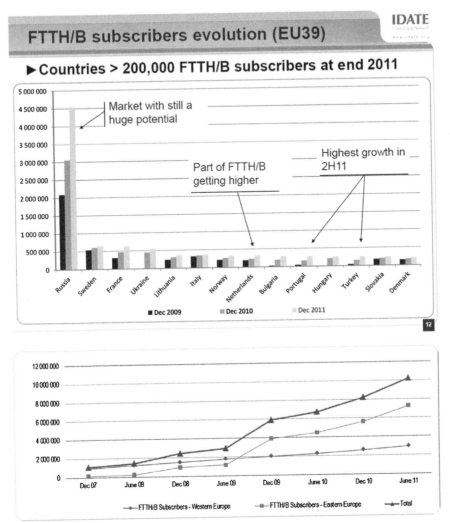

FIGURE 24.20 FTTH/B subscribers in EU39. Source: IDATE for FTTH Council Europe [61].

G-PON, there exist handover points at Layer 2 as shown in Figure 24.22. This can be considered as unbundling services. FTTH in brown field uses a PON for 120 customers with 1×32 split. For the FTTC case, a single gigabit link (or a single OLT port per cabinet) is assigned for NGA services.

24.4.2.2 Deutsche Telecom's deployment

Recently the Federal Government of Germany announced national broadband initiative [65] in 2009. One of the targets is to provide broadband service faster than

European Ranking – December 2011

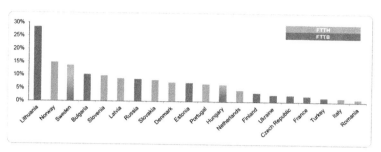

FIGURE 24.21 FTTH/B penetration in Europe. Source: IDATE for FTTH Council Europe [63].

FTTP GEA Network Architecture

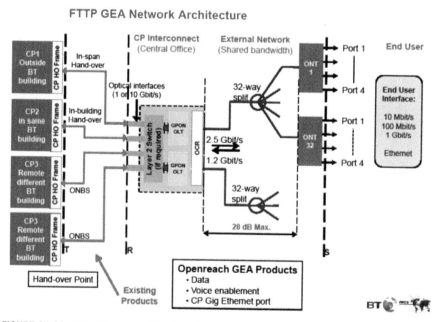

FIGURE 24.22 BT's network architecture. Source [63].

50 Mb/s for 75% of households by 2014. Deutsche Telecom was involved from the beginning and decided to deploy G-PON-based FTTH. Some of the major drivers for G-PON were separation of infrastructure and transport from services, upgradability to XG-PON, and a mature standard.

A FTTB pilot trial started in the city of Dresden from 2008 and covers about 3500 buildings (27,000 households). The pilot area was not served by any broadband available technology before the trial. The VDSL2 ONU was installed in the

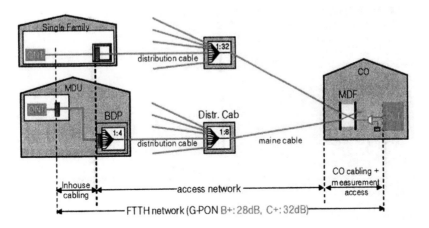

distribution cable

1:32

MDU

BDP

1:4

distribution cable

Distr. Cab

1:8

maine cable

CO

MDF

ONT

CO cabling +
measurement
access

Inhouse
cabling

————access network————

————FTTH network (G-PON B+: 28dB, C+: 32dB)————

FIGURE 24.23 DT's network architecture. Source [65].

basement of the building using existing spare fiber to the building. There also exists a showcase of Gigabit Family that connects with high-speed fiber access up to 1 Gbps download and 500 Mbps upload. The assessment of the trial was that it did not provide sufficient bandwidth to compete with cable network providers due to noise and interference from the in-house network. As a result it was decided to proceed with FTTH-based broadband services instead.

The FTTH pilot trial started in 2010 in the cities of Hennigsdorf and Braunschweig targeting 20,000 households. As shown in Figure 24.23, it has a centralized 1×32 split for a small building and distributed PON splitters (1×8 and 1×4) for multi-dwelling units. There also exists in-service monitoring of the ODN by using 1625 nm or 1650 nm wavelength. The FTTH rollout continues in 2011 to achieve 160,000 homes passed.

24.4.2.3 France Telecom's deployment

France Telecom had a pilot test for FTTX at five cities in the Hauts-de-Seine area and six districts in Paris with 14,000 homes passed. The trial started in 2006 with approximately 2100 customers at the end of 2007. The technologies used for the trial were G-PON and point-to-point. FT identified FTTH services and pricing structure, and selected G-PON from the field trial.

A dedicated fiber plant per operator is not allowed in France. In other words, flexible access must support each living unit for any operator [66]. Then, they introduced a kind of distribution frame (cross-connect point in Figure 24.24) called PM. For dense areas, the PM is located in the building's basement or close to it. By targeting 30% market share, 1×4 final split was usually used as a part of multiple splits. However, for non-dense areas, the size of PM covers at least 300 living units. Thus a single split stage is preferred at the PM.

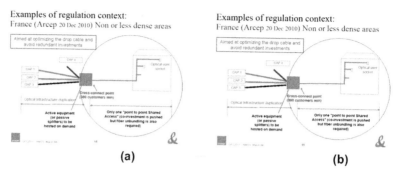

(a)　　　　　　　　　　　**(b)**

FIGURE 24.24 **FT's deployment examples (a) for dense area, (b) for nondense area. Source [66].**

It may be noted that FT does not offer lifeline service with battery back-up. The target of FTTH is 10 M connected homes in France at 2015 and 15 M (60% of all homes) in 2020 [67]. It may be noted that FT has also been rolling out FTTH in Slovakia since 2006.

24.4.2.4 ER Telecom's deployment (Russia's deployments)

ER Telecom is a major broadband service provider who is deploying FTTH/B network. Currently (May 2011) ER telecom has 4.5 M subscribers in more than 31 Russian cities [68]. The network architecture is FTTB for apartment which has 100 households on average. When two thirds of owners voted for services, they installed a network in a block of flats. The ONUs installed in the apartment basement convert digital TV and Internet signals to electrical signals. Then, separate coaxial and FTP (Foiled Twisted-Pair) cables are used to deliver TV and Internet services, respectively.

24.4.3 **FTTX in the Americas**

FTTX took off in America in the early part of this century with several developments by the major operators after much experience looking at proprietary alternatives and several field trials. At present it accounts for nearly 11% of the world's deployments (see Figure 24.14 for example).

Canada has not seen a lot of FTTX deployments to date [52] with the majority of installation occurring in the US. In the following sections we cover the status of the deployments of the main operators as well as some of the nascent deployments to date in South America.

24.4.3.1 AT&T's deployment

In 2002 and later in 2003 AT&T (then SBC and BS) joined with Verizon to try and agree on a base set of requirements for a PON system and architecture. Verizon went forward with B-PON, while AT&T waited for G-PON and a reduction in costs in order for PON to make sense in its network. AT&T utilizes basically two flavors of PON. It utilizes FTTH in greenfield applications and FTTN in brownfield deployments using G-PON with the majority of its deployments using FTTN. Their platform is known as U-verse and uses the telephone lines from a remote node to offer service. Its maximum package offered is 24 Mbps and 3 Mbps on the upstream. Figure 24.25 outlines the basic U-verse architecture for FTTN as deployed by AT&T.

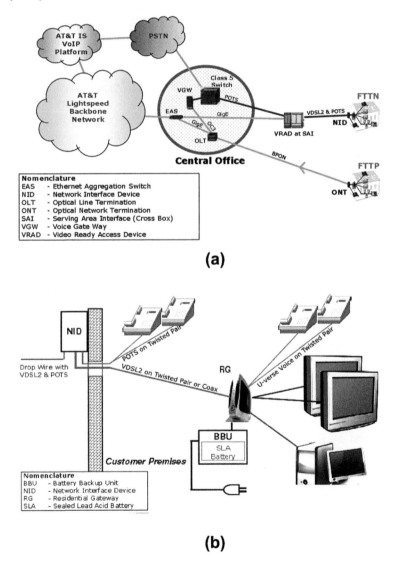

FIGURE 24.25 FTTN as deployed by AT&T [669] showing the high level architecture (a) and the in-home architecture (b).

At the cross-connect AT&T would deploy a Video Ready Access Device (VRAD) that would combine the VDSL and the POTS signals for transmission to the NID (Network Interface Device). From there the POTS would be passed to the inside wiring supporting the legacy PON service (Plain Old Telephone) while the VDSL signal would continue on to the Residential Gateway (RG) over the inside wiring, be it twisted pair or coaxial cable (Figure 24.25b). The RG would then communicate with the STBs and TVs as well as the computer for internet access. The residential gateway also supported U-verse voice service on telephone phones connected to the RG. The RG has a battery backup. In the FTTP case the deployment is similar except that the connection between the ONT and the RG is via an Ethernet cable and the ONT itself requires a battery backup unit. In order to minimize bandwidth needs it uses MPEG-4 for its video delivery.

IPTV was first deployed in 2006 to customers in San Antonio Tx and as of April 2012 U-verse reached a total of 5.9 million Internet users and 4 million subscribers for its video service [70]. AT&T U-verse deployment has resulted in a take rate of just under 17% while reaching over 27% in areas where it has been available for over 42 months [70]. It presently passes around 30 million households.

24.4.3.2 Google's deployment: think big with a gig

One of the more recent entrants to the FTTX story has been Google. Google made an announcement in the open press seeking to get communities that would like to take advantage of the latest technologies and would support the deployment of an all fiber network with the offering of a 1 Gbps speed to its customers. Their initial target was to build a network that would reach between 50,000 and 500 K residential customers and offer 1 Gbps at an affordable rate. Their initial trial area was Stanford University with around 850 customers with two other larger deployments in Kansas City, Kansas and Kansas City, Missouri. The architecture is believed to be a PON system using initially G-PON with an aim to evolving to a WDM-type deployment to take advantage of the point-to-point nature of WDM while minimizing the actual fiber used [71].

The plan is evolving with the expectation that the customer is charged for the installation ($\sim$\$250 [72]) but gets free service for a year, or can opt for a lower fee self-installation kit. The offering is a single play-Ethernet with the technician installing the fiber to an optical NID and having the customer responsible for installing the ONT and the Broadband Home Router (BHR). At present the take rate for this service in Stanford is reported to be very high. The planned deployments in Kansas City, Kansas and Missouri will be in competition with AT&T FTTN deployment and Time Warner's DOCSIS 3.0 which offer lower rates.

24.4.3.3 Verizon's deployment

Verizon's deployment uses both B-PON and GPON together with an RF overlay at 1550 nm. It started deploying PON systems in mid-2004 with POTS and data and then a year later in September 2005 with video overplayed on an additional wavelength at 1550 nm. This entry into the FTTH though had come after many trials of FTTX systems in Cerritos, CA [73], and several other towns like Clearwater,

Somerville, MA, Brooklyn, NY, and Brambleton, VA. The present architecture uses a 1×32 splitter as shown in Figure 24.26 in the ODN through is moving forward with an architecture using also a 2×64 architecture, where the original WDM device is replaced by a 2×2 splitter which combines both the Video Overlay (EDFA output) and the OLT output onto two separate feeder fibers going to two separate 1×32 splitters (see Figure 24.27).

At present Verizon has passed close to 16.5 million homes (4Q/2011) in 12 states, on which around 13.6 million are marketed too. Verizon's FTTX deployment is a fiber all the way to the customer. It provides approximately 8+ hours of battery backup as part of its deployment. In multi-dwelling units Verizon, in the majority of cases,

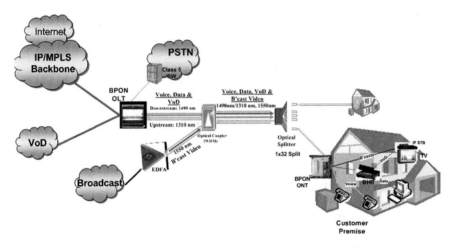

FIGURE 24.26 Baseline architecture as deployed in the Verizon network [74].

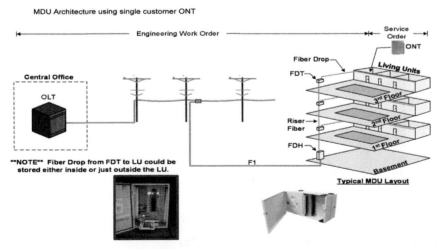

FIGURE 24.27 High level architecture used in MDUs.

deploys splitters within the apartment complex and pulls a fiber passed each living unit in the apartment block. Upon a customer service order a fiber drop is connected to the fiber passing in front of the living units and connected to an inside optical NID which is then connected to the ONT (see Figure 24.26b as an illustrative example).

Verizon is utilizing the advantages of the PON distances to deploy FiOS to remote offices without having to place equipment locally. This can significantly reduce the number of offices that have to be equipped with CO equipment while increasing the utilization of already embedded equipment. Though not shrinking the number of COs in total it is reducing the number of COs that have to be equipped with fiber and the associated electronics.

The present penetration is around 35.5% (31.5%) for FiOS Internet (TV). The maximum speed it offers is at 150 Mbps/35 mbps, but has trialed higher speeds at close to 1 Gbps as well as having done field trials on pre-standardized versions of XG-PON1 [75] and a pre-standardized XG-PON2 at 10 G/2.5 G and 10 G/10 G respectively.

In the case of multiple dwelling units all the engineering work is done in the ODN and within the MDU. Then when an order occurs the final installation of the drop and in-home architecture work goes head.

24.4.3.4 Multiple service operators (MSO) deployments using PON

The cable companies have provided video services in America for many years, with the result that they have many subscribers and pass many homes with their present coaxial network, as illustrated in Figure 24.28.

In the cable TV industry the foundation is a coaxial network, where the main services started out as video and then voice services and data were added later. To meet the customer's need for increased quality, Internet service, and telephony services, these networks have evolved over the last three decades with fiber being deployed deeper and deeper to the customer (e.g. [76,77]) and reducing the node, or serving

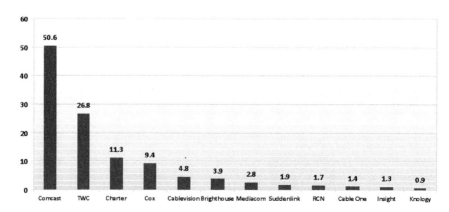

FIGURE 24.28 Homes passed by Cable Companies [78].

area and number of amplifiers between the O/E conversion point and the customers. These networks are known as HFC networks, or Hybrid Fiber Coaxial networks as illustrated in Figure 24.29.

As a result of being able to reduce the number of amplifiers, typically down to around 3, the bandwidth, or spectrum illustrated in Figure 24.30 can be increased steadily allowing the cable companies to meet their customer's requirement for higher throughput/bandwidth as well as the need for increased video channels. The resulting node size typically supports neighborhoods of around 500 homes and work is continuing on reducing these nodes down to around 125 in order to support more concurrent Internet usage while minimizing costs.

PON though has gained traction in the MSO space for many reasons. While DOCSIS 3.0 is seen as an evolution path for its residential customers, more and more cable companies are seeing the advantage of PON, GE-PON specifically, for business applications (for example [79]). In addition many residential estate developers insist on fiber being deployed rather than coaxial cable given the developer's desire for future proofing their investments.

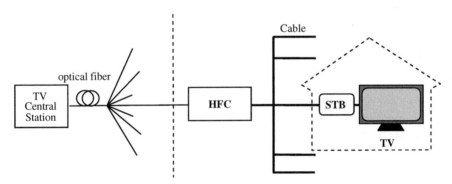

FIGURE 24.29 Illustrative example of the architecture used by cable operators.

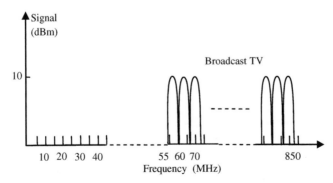

FIGURE 24.30 Typical spectrum used by cable operators.

For residential services the cable operators have been looking at the use of Radio Frequency over Glass (RfoG) [80] which allows the cable industry to support the same service set and protocols and acts as a physical replacement for the coaxial cable. It does, however, allow the cable companies to increase the range and thus facilitates the deployment of fiber deeper into the network. Fiber is less sensitive to environmental issues such as ingress noise and thus the return channel is cleaner allowing increased throughput.

24.4.3.5 Other FTTX deployments in North America

In the rest of the US there has been a lot of interest in FTTX and how this can be used by both municipalities and towns to offer increased bandwidth. However, to date the deployments have been relatively small as illustrated in Figure 24.31.

These deployments are largely made up of Independent Local Exchange Companies (ILEC), municipalities, and towns [78], in order to replace the aging copper networks or the desire to differentiate real estate developments in the case of developers.

24.4.3.6 Deployments in Latin America

Latin America accounts for around 0.1% of the global market share for FTTX [52]. Brazil is the largest Latin America market in the region with approximately 38% [81] of the fixed access subscribers in the region. In Brazil Telefonica is deploying FTTH [82] as a direct result of increased competition in order to increase revenue and decrease customer churn. They plan to ploy to 100K subscribers by the end of 2012 and their deployment is around 90% aerial and they have passed around 1 million homes by 2011.

In Mexico, the second largest market with approximately 17% [81] of the fixed subscribers in the region, Telmex is deploying FTTH in Mexico. Over the coming years this market is expected to increase as a result of the trials being conducted within the region.

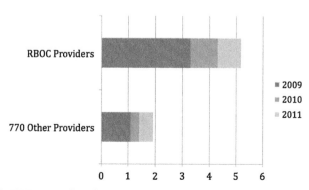

FIGURE 24.31 FTTH connections in millions as of March 30, 2011 [78].

24.5 WHAT'S NEXT?

It is clear that deployment of FTTX, especially FTTH/B, will grow continuously to meet customers' demand for bandwidth and to help reduce total cost of ownership. For the Asia Pacific region, the CAGR of FTTH/B can be as high as 21% according to predictions by Ovum and others and subscribers will reach a total of 140 M by 2015, while DSL will continue to decrease by 2% CAGR [83]. This can be further broken down as a CAGR of 36% for FTTH and 11% for FTTB. Figure 24.32 shows the prediction by Ovum. The prediction of FTTH/B for global customers will be 230 M in 2016 according to Idate, while homes passed will be 520 M [84]. The global growth of FTTH/B will exceed that of VDSL as illustrated in Figure 24.33. In the near future, we anticipate that fiber will be the dominant media for subscribers' networks as we have witnessed to date.

In addition to the growth in the number of subscribers, bandwidth per subscriber is expected to double every 2–3 years [28]. Technologies discussed in NG-PON 2 (see Section 24.3.4.2) are targeting to provide greater than a gigabit per second to subscribers. Though there already exist homes that are connected by gigabit per second bandwidth in some countries, this trend will be more prevalent in the coming years. To support future bandwidth demand, it is clear that we need these new technologies that support greater bandwidth economically in the near future for both residential and business applications.

It is hard to predict future technologies that will be used in access network. As seen in the discussion of NG-PON 2 technology, however, it is clear that we need to

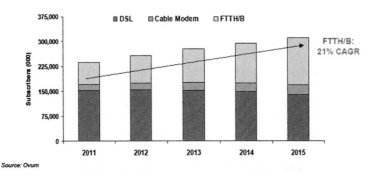

FIGURE 24.32 FTTH/FTTB forecast. Source [81].

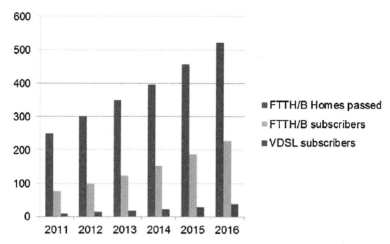

FIGURE 24.33 Forecast of FTTH growth. Source: IDATE, FTTx Watch Service, June 2012 [83].

continue to use the wavelength domain in order to reduce impairments in transmission. For splitter-based ODN, wavelength transparency of the ODN is maintained, which is a key need of the present operators in order to take advantage of legacy deployments. However, the high insertion loss of the splitter inhibits long reach transmission and also tunable transmitters and tunable receivers are required to satisfy the requirements of colorless ONUs. Therefore long reach/high-speed transmission with splitter-based ODN could require the use of optical amplifiers, digital signal processing in the electrical domain to help enhancement of link budget, or better transmitter or receiver technology.

Use of wavelength selective branching devices in the ODN reduces the total losses in the ODN considerably and provides great flexibility for reach and transmission bit rate. It also provides dedicated connectivity (or point-to-point connectivity) by scarifying the wavelength transparency of the ODN. Increased number of wavelength requires greater number of transceivers at OLT, raising issues of power consumption, port density, and cost. The use of PICs (Photonic Integrated Circuits) can solve these issues as ICs have done in electronics.

Coherent detection can also increase link budget by improving both receiver sensitivity and impairment mitigations. The advances of digital signal processing also open the door for higher spectral efficiency transmission for future bandwidth demands.

The next few years are envisioned to produce extensive research within these areas in order to continue to support the residential and business needs as well as to determine how best to synergize the benefits of these technologies with the deployment decisions and technologies already deployed by the operators around the world.

24.6 SUMMARY

This chapter has looked at the evolution of the PON technology and where and how it is being deployed around the world. It is clear that there are many drivers that are continuing to drive the need for higher capacity as well as the more efficient use of the outside plant. These will continue to drive the need for more efficient technologies and continued research in how to address these demands at a reasonable cost. These are exciting times as operators are transforming their networks from their legacy copper-based networks to the more bandwidth and thus future proof fiber-based networks.

References

[1] OFS, FTTP Outside Plant: Considerations and Case Study Analysis. <http://www.gcscte. org/presentations/2008/FTTH%20-%20Swindell%20-%20OFS.pdf>.

[2] Katsuaki Suzuki, Noriyuki Araki, Arata Natsume, Yutaka Kozawa, Development of OTDR that easily locate a fault in an optical Drop Cable below an Optical Splitter, in: Proc. of the 57th IWCS, 2008, pp. 22–26.

[3] Gianfranco Ciccarella, Managing the new generation network transition, in: Telecom Italia, FTTH Council Europe 2011 Milan, 9–10 February 2011.

[4] Paolo Solina, European PON Deployments: a carrier view of the technical, regulatory, and economic challenges, We.8.B.1, in: ECOC 2010, September 19–23, Torino Italy.

[5] <http://www.mocalliance.org/>.

[6] <http://homepna.org/>.

[7] ITU-T Recommendation G.983.1, Broadband optical access systems based on passive optical networks (PON), 2005.

[8] ITU-T G.983.2, ONT management and control interface specification for BPON 2005.

[9] ITU-T G.983.3, A broadband optical access system with increased service capability wavelength allocation, 2001.

[10] ITU-T, G.983.4, A broadband optical access system with increased service capability using dynamic bandwidth assignment, 2001.

[11] ITU-T G.983.5, a broadband optical access system with enhanced survivability, 2002.

[12] David Large, James Farmer, Broadband Cable Access Networks, Morgan Kaufmann, 2009, pp. 139.

[13] C.R.S. Fludger, V. Handerek, R.J. Mears, Pump to signal RIN transfer in Raman fiber amplifiers, J. Lightwave Technol. (19) (2001) 1140–1148.

[14] M.R. Phillips, D.M. Ott, Crosstalk due to optical fiber nonlinearities in WDM CATV lightwave systems, J. Lightwave Technol. 17 (10) (1999) 1782–1792.

[15] M Aviles, K. Litvin, J. Wang, B. Colella, F.J. Effenberger, F. Tia, Raman crosstalk in video overlay passive optical networks, OFC 2004, Paper FE7.

[16] D. Phieler, Appendix I: Nonlinear Optics-Raman Crosstalk and Cross Phase Modulation, 7 August 2009.

[17] R. Smith, Optical power handling capacity of low loss optical fibers as determined by stimulated Raman and Brillouin scattering, Appl. Opt. 11 (1972) 2489–2494.

[18] <http://www.corning.com/docs/opticalfiber/R-00000016.pdf>.

[19] G. Kramer, Ethernet Passive Optical Networks, McGraw-Hill, 2005.

[20] IEEE 802.3 ah 2004, Carrier sense multiple access with collision detection (CSMA/CD) access method and physical layer specification, Amendment: Media access control parameters, physical layer, and management parameters for subscriber access networks.

[21] IEEE P1904.1 Working group, Standard for service interoperability in Ethernet Passive Optical Networks (SIEPON), <http://grouper.ieee.org/groups/1904/1/>.

[22] IEEE 802.3 av 2009, Carrier sense multiple access with collision detection (CSMA/CD) access method and physical layer specifications, Amendment 1: Physical layer specifications and management parameters for 10 Gb/s passive optical networks.

[23] ITU-T G.984.1, Gigabit-capable Passive Optical Network (GPON), General Characteristics, 2003.

[24] ITU-T G.984.2, Gigabit-capable Passive Optical Network (GPON), Physical Media Dependent (PMD) layer specification 2003. Amendment 2 (2008).

[25] ITU-T G.984.3, Gigabit-capable Passive Optical Network (GPON), Transmission convergence layer specification, 2004.

[26] ITU-T G.984.4, Gigabit-capable Passive Optical Network (GPON), ONT management and control interface (OMCI) specification, 2008, with amendments 1, 2, 3, corrigendum 1, erratum 1, and an implementer's guide.

[27] ITU-T G.984.7, Gigabit-capable Passive Optical Network (GPON), Long-Reach, 07/2010.

[28] Cisco VNI forecasts 2011–2016.

[29] G. 987.1 (2010), 10-gigabit-capable passive optical networks (XG-PON): general requirements.

[30] G. 987.2 (2010), 10-gigabit-capable passive optical networks (XG-PON): physical media dependent (PMD) layer specifications.

[31] J.-I Kani, F. Bourgari, A. Cur, A. Rafel, M. Campbell, R. Davey, S. Rodrigues, Next-generation PON, Part 1—Technology roadmap and general requirements, IEEE Commun. Mag. (2009) Nov.

[32] G. 984.5, Gigabit-capable passive optical networks (G-PON): enhancement band for gigabit capable passive optical networks, 2007.

[33] D. Hood, E. Trojer, Gigabit Capable Passive Optical Networks, Wiley, 2012.

[34] P. Chanclou, A. Cui, F. Geilhardt, H. Nakamura, D. Nesset, Network operator requirements for the next generation of optical access networks, IEEE Network (2012) 8–14 March.

[35] Chang-Hee Lee, Wayne V. Sorin, Byoung Yoon Kim, Fiber to the home using a PON infrastructure, IEEE J Lightwave Technol. 24 (12) (2006) 4568–4583.

[36] S. Smolorz, Demonstration of a coherent UDWDM-PON with real time processing, in: OFC 2011, PDPD4, 6–10 March 2011.

[37] N. Cvijetic, OFDM for next-generation optical access networks, IEEE J. Lightwave Technol. (2012).

[38] G. 698.3, Multichannel seeded DWDM applications with single-channel optical interfaces, 2012.

[39] J. Sinsky et al., 39.4-Gb/s Duobinary transmission over 24.4 m of coaxial cable using a custom indium phosphide duobinary-to-binary converter integrated circuit, IEEE Trans. Microw. Theor. 56 (12) (2008) 3162–3169.

[40] Hyun Deok Kim, Seung-Goo Kang, Chang-Hee Lee, A low-cost WDM source with an ASE injected Fabry-Perot semiconductor laser, IEEE Photon. Technol. Lett. 12 (8) (2000) 1067–1069.

[41] Joon-Young Kim, Sang-Rok Moon, Sang-Hwa Yoo, Chang-Hee Lee, DWDM-PON at 25 GHz channel spacing based on ASE injection seeding, European Conference on Optical Communication (ECOC2012), Amsterdam, The Netherlands, September 2012, We.1.B.4.

[42] F. Ponzino, F. Cavaliere, G. Berrettini, M. Presi, E. Ciaramella, N. Calabretta, A. Bogoni, Evolution scenario toward WDM-PON, J. Opt. Commun. Netw. 1 (2009) C25–C34.

[43] I.B. Djordjevic, B. Vasic, Orthogonal frequency division multiplexing for high-speed optical transmission, Opt. Express 14 (2006) 3767–3775.

[44] C. Bock, J. Prat, S. Walker, Hybrid WDM/TDM PON using the AWG FSR and featuring centralized light generation and dynamic bandwidth allocation, IEEE J. Lightwave Technol. 23 (2005) 3981–3988.

[45] G. 983.3, A broadband optical access system with increased service capability by wavelength allocation, 2001.

[46] G. 984.5, Gigabit-capable passive optical networks (G-PON): Enhancement band, 2007.

[47] G. 984.5, Gigabit-capable passive optical networks (G-PON): Enhancement band, Amendment 1, 2009.

[48] Ki-Man Choi, Sang-Mook Lee, Min-Hwan Kim, Chang-Hee Lee, An efficient evolution method from TDM-PON to next-generation PON, IEEE Photon. Technol. Lett. 19 (9) (2007) 647–649.

[49] G. 984.6, Gigabit-capable passive optical networks (G-PON): G-PON reach extension (G.984.re), 2008.

[50] G. 987.4, 10-gigabit-capable passive optical networks (XG-PON): Reach extension, 2012.

[51] D.P. Shea, J.E. Mitchell, Long-reach optical access technologies, IEEE Network (2007) September/October.

[52] Idate, FTTX 2012: Market and Trends, facts and figures, in: Digiworld Summit, Montpellier, November 14, 15 France.

[53] <http://www.ftthcouncil.org/en/knowledge-center/ftth-council-resources/chart-global-ranking-ftth-market-penetration-2012-0>.

[54] L. Wei, View on FTTH deployment: Current status and future trends, in: FTTH Council Asia Pacific Conference, Shanghai, China, May 5–9, 2012.

[55] <http://ovum.com/2012/05/10/ftth-council-asia-pacific-conference-low-pon-prices-inspire-new-applications/>.

[56] <http://www.ftthcouncilap.org/index.php?option=com_content&view=category&layout=blog&id=6&Itemid=36>.

[57] Soo-Jin Park, Chang-Hee Lee, Ki-Tae Jeong, Hyung-Jin Park, Jeong-Gyun Ahn, Kil-Ho Song, Fiber-to-the-home services based on wavelength-division-multiplexing passive optical network, IEEE J. Lightwave Technol. 22 (11) (2004) 2582–2591.

[58] B.-I. Choi, National strategy to drive broadband nationwide Korea, in: FTTH Council Asia Pacific Conference, Shanghai, China, May 5–9, 2012.

[59] Y. Fujimoto, NTT's FTTH deployment status and perspective toward next generation, in: Joint ITU-T IEEE Workshop on Next Generation Access System, Geneva, June 19–20, 2009.

[60] X. Tang, Broadband services and FTTH development of China Unicom, in: FTTH Council Asia Pacific Conference, Shanghai, China, May 5–9, 2012.

[61] Z. Weihua, Practice and innovation: the development of Shanghai MONET, in: FTTH Council Asia Pacific Conference, Shanghai, China, May 5–9, 2012.

[62] <http://www.idate.org/en/News/Twenty-six-percent-increase-in-the-number-of-homes-passed-for-FTTH-B-in-Europe_695.html>.

[63] <http://www.ftthcouncil.eu/documents/Reports/Market_Data_December_2011.pdf>.

[64] P. Barker, FTTH infrastructure: BT pilot deployments, in: Proceedings of the 58th IWCS/ IICIT, 2010, pp. 133–137.

[65] F. Escher, H.-M. Foisel, A. Templin, B. Nagel, M. Adamy, Enabling broadband communication, Deutsche Telecom FTTH deployment architecture, plans, rollout, in: OFC 2012, Los Angeles, March 4–8, 2012.

[66] F. Bourgrat, Regulation environment around the world: impacts on deployments, in: OFC2011, Los Angles, March 6–10, 2011.

[67] <http://www.orange.com/en_EN/networks/network_it/ftth_broadband/>.

[68] <http://www.ftthcouncil.eu/documents/CaseStudies/ER_TELECOM.pdf>.

[69] AT&T, Backup power for voice services in the customer premises, Performance Reliability Standards Workshop, Sponsored by the California Public Utilities Commission (CPUC), San Francisco, CA, February 2 2009.

[70] FierceIPTV Report on AT&T Reports: <http://www.fierceiptv.com/story/att-reports-52-earnings-bump-tops-4m-u-verse-tv-subs/2012-04-24?utm_medium=nl&utm_source=internal>.

[71] Ryohei Urata, Cedric lam, Hong Liu, Chris Johnson, High performance, low cost, colorless ONU for WDM-PON, NTH3E.4, in: OFC/NEOFC Los Angeles, March 4–8, 2012.

[72] <http://www.dslreports.com/shownews/115250>.

[73] R.F. Kearns, V.S. Shukla, P. N. Baum, L.W. Ulbricht, An 80 channel high performance video transport system over fiber using FM-SCM techniques for super trunk applications, NCTA Technical Papers, 1991, pp. 71–76.

[74] F.J. Effenberger, Kent McCammon, Vincent O'Byrne, Passive optical network deployment in North America, J. Opt. Nctw, OSA Web J. (2007) March.

[75] Shweta Jain, Frank Effenberger, Andrea Szabo, Zhishan Feng, Albert Forcucci, Wei Gui, Yuanqiu Luo, Robert Mapes, Yixin Zhang, Vincent O'Byrne, in: Optical Fiber Communication Conference (OFC) 2010 Paper: PDPD6, Los Angeles, 2010.

[76] J.A. Chiddix, D.M. Pangrac, Fiber backbone: a proposal for an evolutionary catv network architecture, NCTA 1988 Technical Papers, 1988, pp. 73–82.

[77] Dean Stoneback, The evolution of hybrid fiber-coaxial cable networks to an all-fiber network, in: OFC, Los Angeles, March 2008.

[78] R.C. Atkinson, I.E. Schultz, T. Korte T. Krompinger, Broadband in America—2nd Edition: where it is and where is it going (according to broadband service providers), an update of the 2009 Report originally prepared for the staff of the FCC Omnibus Broadband Initiative, 2009.

[79] Ed Mallett, Bright House Networks, USA, Mobile Backhaul for PON: a case study, in: OFC 2012, Los Angeles, March 4–8, 2012.

[80] <http://www.cable360.net/technology/strategy/RFoG-for-Business-Services_27351.html>.

[81] Keith Russell, Fiber Deployment in Latin America: An Emerging Market for FTTX, Paper NTH3E3.pdf, OFC, March 4–8, 2012.

[82] Andre Kriger, Telefonical Vivo, FTTH Roll-out Telefonica Brazil, Operational and Commercial Challenges, FTTH LATAM Forum, April 2012.

[83] J. Kunstler, FTTH/FTTB Asia-Pacific: subscriber base and forecast, shipments, worldwide trends, challenges and solutions, in: FTTH Council Asia Pacific Conference, Shanghai, China, May 5–9, 2012.

[84] <http://blog.idate.fr/?p=2489>.

Modern Undersea Transmission Technology

25

Jin-xing Cai, Katya Golovchenko, and Georg Mohs

TE SubCom, 250 Industrial Way West, Eatontown, NJ 07724, USA

This chapter provides an overview over the progress in undersea transmission technology since the last edition of Optical Fiber Telecommunications in 2007. Since then, digital coherent transmission has become available, enabling a ten-fold increase in spectral efficiency and system capacity. Channel data rates have increased from 10 Gb/s to 100 Gb/s and beyond. After a brief general introduction to undersea systems with their unique challenges and design constraints in Section 25.1, Section 25.2 outlines the principles of coherent transmission technology as they apply to undersea systems including polarization multiplexing, linear equalizers and high-order modulation formats. Section 25.3 describes the use of strong optical filtering to help to improve spectral efficiency and reviews techniques to limit the effects of inter-symbol interference. The best trade-off between spectral efficiency and transmission performance is achieved at Nyquist carrier spacing and Section 25.4 discusses different transmission techniques for this condition. Section 25.5 then introduces higher order modulation formats to further increase spectral efficiency by increasing the number of bits per symbol and discusses the implications and mitigation techniques of the receiver sensitivity degradation that goes along with it. Future trends are outlined in Section 25.6 before summarizing the chapter in Section 25.7.

25.1 INTRODUCTION

Undersea communication systems have been providing low latency and high capacity connectivity between the continents of the world for more than 150 years. The first transoceanic telegraph cable became operational in 1858 with a communication speed of about 2 min per character [19], a vast improvement over the 10 days it took to deliver a message by clipper ship. The cable spanned a 4025 km route between Newfoundland and Ireland and lasted only for about a month but was successfully replaced in 1866 with an increased transmission speed of about 8 words per minute [19].

Steady progress has been made ever since to provide faster and faster communication speed or higher and higher bandwidth with submarine cables. Voice traffic on a transatlantic cable became available in 1956 with Transatlantic No. 1 (TAT-1)

Optical Fiber Telecommunications VIB. http://dx.doi.org/10.1016/B978-0-12-396960-6.00030-4

carrying 36 simultaneous telephone channels with 4 kHz bandwidth each replacing radio connections that had been in use between the United States and Europe since 1927 but which were unreliable due to their dependence on atmospheric conditions. In 1988 the first fiber optic transatlantic cable was put into service ultimately carrying 40,000 telephone circuits equivalent to 560 Mb/s of aggregate traffic on two fiber pairs [1]. The latest generation of systems in the Atlantic between the United States and Europe can carry multiple Tb/s of aggregate capacity and the next generation is actively being discussed with multiple tens of Tb/s capacity [26,20]. Figure 25.1 shows the lit undersea communication capacity around the globe and used capacity by country as of December 2011. Undersea communication cables are nearly ubiquitous in all international waters connecting the countries and continents of the world and making the web world wide.

25.1.1 Reach, latency and capacity

From Figure 25.1 the defining challenge of under sea communication systems becomes apparent. The transmission paths can be very long. Typical transatlantic

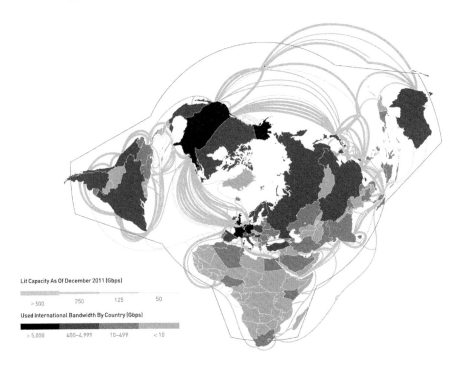

FIGURE 25.1 Lit undersea capacity in 2011. (Courtesy of TeleGeography) [http://www. telegeography.com/].

type systems have a length of about 6,000 km or more. Distances in the Pacific are even longer. A typical connection between the United States and Japan is about 9,000 km and between the United States and South East Asia can be up to 11,000 km or more. This is much longer than any terrestrial system. However, there is also some advantage compared with terrestrial systems [4,5]. A submarine system owner does not have to install a new system onto an existing fiber route as is common for terrestrial systems but can freely design the transmission path and choose fiber types and amplifier spacing for the best solution of a new system. Undersea amplifiers are often called "repeaters" and are based on Erbium doped fiber amplifiers (EDFA) that amplify the optical signals without any retiming or reshaping.

Many people believe that their international internet data and telephone connections are carried by satellites. For a great many connections, however, that is not true. Most intercontinental traffic is carried by undersea cables. Compared with satellite connections undersea cables have a significant advantage. The signals travel along a much more direct and therefore shorter path. A geostationary satellite orbits earth at about 36,000 km above the equator such that a roundtrip signal between two points on the globe would travel at least 4 times that distance or 144,000 km. At the speed of light it takes 0.48 s to travel this distance, a significant delay that is noticeable and irritating in a telephone conversation. Low latency is even more important in modern applications. Round trip latencies as low as sub-60 ms have been announced between New York and London [26], a significant achievement considering that the great circle distance is about 5600 km and therefore the minimum latency between New York and London is 54 ms using a refractive index of 1.45 for optical fiber.

Undersea systems have another advantage: massive capacity. A modern cable can carry about 100 channels at 100 Gb/s each per fiber depending on distance. Since a cable can typically contain up to 8 fiber pairs (one fiber for each direction in a pair) this equates to 80 Tb/s in each direction. Satellite communication links use microwave radio for transmission and are typically limited to several Gb/s capacity.

25.1.2 The capacity challenge

Typically, the higher the capacity per fiber pair of an undersea communication system the lower the cost per transported bit since the total system cost can be amortized over a larger capacity. This simple fact explains the drive towards higher and higher capacity in submarine systems. Figure 25.2 summarizes the experimentally demonstrated transmission capacity over transoceanic distances (>6000 km) since 2000, when the first >1 Tb/s transoceanic transmission was demonstrated [18].

Since fiber pair capacity is simply the product of the repeater bandwidth (BW) and the spectral efficiency of the transmitted signals, fiber pair capacity can be increased in two ways. One, by increasing the repeater bandwidth and two, by increasing the spectral efficiency of the transmitted signals. Much effort has been devoted to maximizing both of these parameters. Repeater bandwidths up to 80 nm have been demonstrated as early as 2002 [21] but so far commercial applications in the undersea space are limited to the full C-band of about 40 nm or 5 THz for various reasons.

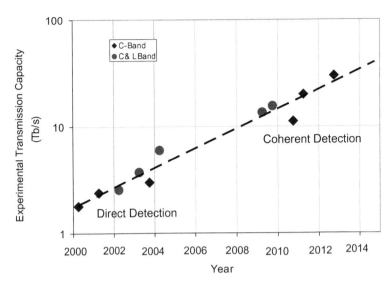

FIGURE 25.2 Experimentally demonstrated single-fiber transoceanic capacity records.

Coherent techniques on the other hand have enabled dramatic progress in spectral efficiency over the last few years.

Progress in ASIC (application specific integrated circuit) technology enables powerful digital signal processing (DSP) which makes coherent transmission technology practical for commercial applications. Coherent transmission and DSP provide much of the advancement in undersea transmission technology over the past few years allowing for higher spectral efficiency and capacity. Spectral efficiency (SE) describes how well bandwidth is utilized in terms of transmitted information. In optical communication systems it can be expressed by the ratio of the channel information rate (R_I) and the channel spacing Δf (see Figure 25.3). For example, in a direct detection system just a few years ago the channel information rate was 10 Gb/s and

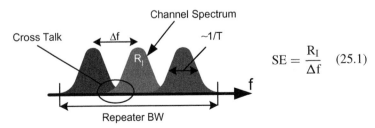

$$SE = \frac{R_I}{\Delta f} \quad (25.1)$$

FIGURE 25.3 Spectral efficiency is the ratio of information rate R_I to channel spacing Δf. Cross talk limits the achievable spectral efficiency.

the channel spacing 25 GHz for a spectral efficiency of 0.4 b/s/Hz. To increase spectral efficiency the information rate per channel must be increased and/or the channel spacing decreased. However, the spectral width of a channel is proportional to the inverse symbol pulse width 1/T which is related to the symbol rate. Once the channel spectra start to overlap, there will be channel crosstalk and transmission performance rapidly decreases which limits the spectral efficiency that can be achieved.

Coherent detection can help us in several ways to further increase spectral efficiency. First, single mode fiber actually supports two degenerate orthogonal polarization modes that can be separated effectively with coherent detection and used nearly independently by polarization multiplexing doubling the information rate of a channel without changing its spectral width. Second, coherent transmission also allows us to depart from binary transmission formats and transmit more than one bit per symbol without incurring fundamental implementation penalties. Since the spectral width of a channel is determined by the symbol rate not the bit rate, its spectral width is unchanged and we can increase the information rate for the same channel spacing. The basics of coherent detection including polarization multiplexing and demultiplexing as well as quadrature phase shift keying (QPSK) with 2 bits per symbol are discussed in the following section (25.2). These two steps alone allow us to increase spectral efficiency fourfold compared with a single polarization binary format.

In the last few years even higher spectral efficiencies have been achieved enabled by coherent techniques. One approach is bandwidth constraint by strong optical filtering. The spectral width of a channel can be reduced significantly by optical filtering which then enables smaller channel spacing values and therefore higher spectral efficiency. However, when reducing the spectral width of a channel by filtering, the signal pulses in the time domain broaden and pulses start to overlap causing intersymbol interference (ISI) rapidly decreasing transmission performance. Coherent detection allows us to process the received optical field and introduce advanced filters and algorithms that can correct ISI and mitigate the penalty associated with strong optical filtering. This will be discussed in detail in Section 25.3. Another approach to increasing spectral efficiency is to go to higher order modulation formats with combinations of amplitude and phase modulation and more bits per symbol increasing the information rate and/or decreasing the spectral width per channel. Coherent detection makes these formats practical. However, receiver sensitivity decreases with higher order modulation formats requiring a higher optical signal-to-noise ratio (OSNR) and limiting the transmission distance and spectral efficiency that can be achieved. The lower receiver sensitivity can be countered by moving to more sophisticated coded modulation techniques that will be discussed in Section 25.5.3.

25.2 COHERENT TRANSMISSION TECHNOLOGY IN UNDERSEA SYSTEMS

This section reviews coherent transmission technology with a focus on undersea applications. Section 25.2.1 introduces the basic concepts and takes us from direct detection to polarization multiplexed QPSK with coherent detection. Linear impairment compensation for chromatic dispersion, polarization evolution and polarization mode dispersion are discussed in Section 25.2.2. At the end, we review nonlinearity accumulation and compensation in dispersion uncompensated transmission in Section 25.2.3

25.2.1 Introduction to coherent detection

Before coherent transmission, optical communication systems were mostly based on binary formats. The simplest of these formats is on-off keying (OOK) where light is transmitted for a mark and no light for a space. Since the light is not turned off between two consecutive marks, this format is often referred to as Non-Return-to-Zero or NRZ. To improve the nonlinear tolerance, in undersea systems this format was often modified by sending a light pulse for a mark and no light pulse for a space such that light intensity would drop significantly between two consecutive marks and hence the name Return-to-Zero or RZ. The top line in Figure 25.4 schematically shows the light field for this case. In the figure the period of the optical carrier has been greatly exaggerated for illustration purposes. For a 10 Gb/s signal there are about 19,000 cycles per symbol.

With Differential Binary Phase Shift Keying (DBPSK) or often simply abbreviated as DPSK enhanced receiver sensitivity can be achieved by using a combination of balanced detection and phase modulation [3,22]. In this case the information is no longer encoded in the intensity of the light field but in its phase as shown on the second line of Figure 25.4 for binary phase shift keying (BPSK). More specifically the information is usually encoded in the difference between the phase of two consecutive symbols. To make the phase information accessible with a direct detection

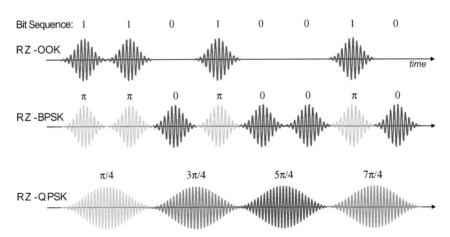

FIGURE 25.4 Schematic representation of the RZ-OOK, RZ-BPSK and RZ-QPSK modulation formats.

receiver an interferometer is used that interferes a received symbol with the previous symbol. The interferometer has the property that if the phase of the light field of the two symbols is the same, light appears at one output port of the interferometer and if it is reversed (180° shift) at the other output port. When the two output ports of the interferometer are connected to a balanced photo detector, a positive photo current can be detected when the phase of the light field between two consecutive symbols is the same and a negative photo current if the phase is reversed. The interferometer and intensity detectors in direct detection remove the optical carrier phase leaving only the encoded information. In this scheme the carrier phases of two subsequent symbols are compared and the phase reference is therefore noisy.

The concept can be extended to even higher level modulation formats with phase changes other than 180°, as shown on the third line of Figure 25.4 for the case of four phase states. This format is known as quadrature phase shift keying (QPSK). Since each symbol now has four possible states QPSK carries two bits per symbol reducing the symbol rate by half compared with BPSK decreasing the spectral width of the channel and enabling higher spectral efficiency. However, with direct detection technology significant implementation penalties are incurred.

Coherent detection allows us to detect arbitrary amplitude and phase information of the received light field by interfering the received signal with a local oscillator. The local oscillator is simply a continuous wave laser tuned roughly to the carrier frequency of the received signal providing a less noisy phase reference compared with direct detection schemes which results in improved receiver sensitivity. The interferometer is replaced by a 90° optical hybrid that interferes the light field of the received signal with the light field of the local oscillator fourfold with phase shifts in 90° increments. The resulting four optical output signals are detected in pairs by two balanced photo detectors. The signals from the photo detectors are then processed in DSP. This description applies to single polarization coherent detection. Polarization multiplex requires a polarization diverse receiver as described in Section 25.2.2.2.

A schematic of the single polarization 90° optical hybrid and coherent receiver is shown in Figure 25.5. The received signal is split into two parts by a 3 dB coupler.

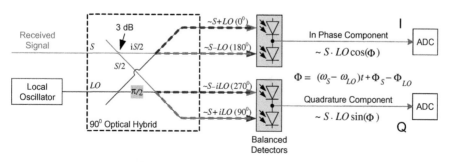

FIGURE 25.5 Single polarization 90° optical hybrid and coherent receiver (ADC: analog to digital converter).

The 3 dB coupler imparts a 90° phase shift between the two output signals. Each of the two parts of the received signal is then mixed with the local oscillator in two other 3 dB couplers where the couplers again impart 90° phase shifts between the outputs. Before this mixing an additional 90° phase shift is applied to one part of the local oscillator such that the resulting four signals interfere the received signal and local oscillator with 90° phase steps. The four signals are then detected in pairs by balanced photo detectors in an arrangement where the phase difference between the received signal and local oscillator changes by 180° for each pair. The resulting signal from the balanced photo detectors is proportional to the product of the signal and local oscillator field amplitudes. In addition the output signal from one balanced detector is proportional to the cosine of the phase difference Φ between received signal and local oscillator. This is the in-phase component. The output of the other balanced detector is proportional to the sine of Φ and is called the quadrature component.

The in-phase and quadrature components for a received symbol can be plotted in a constellation diagram as shown in Figure 25.6. The distance from the origin is proportional to the signal amplitude and often normalized such that average power is 1. Each modulation format has a number of target amplitude and phase combinations for each symbol that constitute the constellation points. The number of constellation points determines the number of bits per symbol as the number of constellation points is 2^m where m is the number of bits per symbol (for each polarization). The constellation diagrams for some example constellations are also shown in Figure 25.6. For an OOK format such as RZ or NRZ the amplitude of all marks is the same but the phase Φ is undetermined such that the constellation diagram is a circle with a dot at the center for the spaces that have no amplitude. For a phase shift keying (PSK) format the amplitude of all symbols is the same, however for BPSK for instance there are two possible phase states 180° apart such that the constellation diagram has two points that are point symmetric to the origin. Since the in-phase and quadrature component are orthogonal we can superimpose two BPSK signals with a 90° phase shift and arrive at four constellation points. This format is called quadrature phase

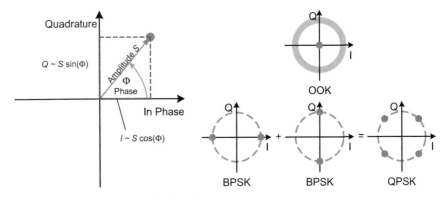

FIGURE 25.6 Schematic constellation diagrams.

shift keying (QPSK) and carries two bits per symbol (and per polarization), i.e., each constellation point represents a combination of two bits.

Coherent detection increases receiver sensitivity. Figure 25.7 shows the theoretical noise loaded back-to-back performance as a function of OSNR for three different coherent detection methods compared with direct detection (DD) DBPSK. For the same OSNR all coherent detection methods achieve a higher performance, i.e., better Q-factor, than DD DBPSK. This is due to the fact that a coherent receiver uses a better phase reference (the local oscillator) than a direct detection receiver where the phase of a symbol is compared to a noisy reference (another symbol).

The outputs of the 90° optical hybrid yield a complex signal with a phase Φ given by:

$$\Phi = (\omega_S - \omega_{LO})t + \Phi_S - \Phi_{LO} \qquad (25.2)$$

where ω_S and ω_{LO} are the angular frequencies of the transmitted signal and local oscillator and Φ_S and Φ_{LO} are the phase of the signal and local oscillator, respectively. At this point the constellation is rotated by Φ_{LO} and spinning with the local oscillator offset frequency $\omega_S-\omega_{LO}$. This has to be taken into account when recovering the transmitted information and results in three different detection methods, each with advantages and drawbacks. The simplest method to recover the transmitted information is to encode it differentially such that the transmitted information is in the difference between two consecutive symbols $\Phi_n-\Phi_{n-1}$, where n is the symbol index in time. In this case the local oscillator offset and phase cancel as long as they vary slowly on the timescale of the symbol rate. This method is called differential demodulation and is the simplest coherent detection method (triangles in Figure 25.7).

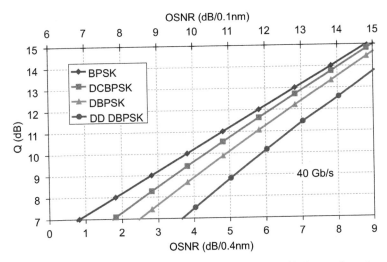

FIGURE 25.7 Noise loaded back-to-back performance for BPSK with three coherent detection methods compared with direct detection (DD) DBPSK.

In order to improve on the differential demodulation method the local oscillator offset and phase must be determined. The local oscillator offset can be established by finding the peak in the Fourier transform of the output signal of the 90° optical hybrid. A popular method to find the local oscillator phase for a QPSK signal uses the Viterbi-Viterbi algorithm [54] taking advantage of the fact that only four phase states separated by 90° are allowed for the signal. Taking the signal to the fourth power multiplies Φ by 4 turning Φ_S into multiples of 2π and removing the signal modulation. The arctangent of 4Φ after removing the local oscillator offset yields Φ_{LO} except for a fourfold ambiguity that has to be resolved by other means, e.g., known framing symbols. Once resolved, Φ_{LO} must be tracked over time until the next time known symbols arrive that can be used for the Φ_{LO} ambiguity resolution. Since both the local oscillator offset and phase are now known, the signal phase can be determined. This method is usually referred to as absolute phase detection and provides the best receiver sensitivity (diamonds in Figure 25.7). However, when an error occurs in the local oscillator phase tracking, all subsequent signal phase values will be incorrect by multiples of 90° leading to symbol errors until the Φ_{LO} ambiguity can be resolved again and the Φ_{LO} tracking error corrected. This is referred to as cycle slip and can present a high risk for errors with absolute phase detection.

The third method is usually referred to as differential coding and has a receiver sensitivity between differential demodulation and absolute phase detection (squares in Figure 25.7). This method avoids the dire consequences of cycle slips by encoding the transmission information differentially in the signal phase but still determines local oscillator offset and phase to decide the signal phase of each received symbol individually. By taking the difference of the decided signal phase of two consecutive symbols a cycle slip is recovered after two symbols. However, each symbol error now leads to two incorrectly decoded differential symbols resulting in a bit error ratio that is twice that obtained with absolute phase detection. Most first commercial implementations of coherent detection use this method.

25.2.2 Linear impairment compensation with coherent detection

Besides increased receiver sensitivity, coherent detection has other advantages compared with direct detection. The fact that both amplitude and phase of the optical field are recovered affords the opportunity to undo linear effects that acted on the optical field during transmission. The following section describes conceptually how linear impairments can be compensated with a special emphasis on undersea applications.

25.2.2.1 Dispersion compensation

Group velocity dispersion due to chromatic dispersion is one of the leading linear effects that limit the bandwidth-distance product of single-mode fibers. The effect is shown schematically in Figure 25.8. The frequency dependence of the effective mode index causes frequency dependent group velocity or group velocity dispersion. Different spectral components of a pulse travel at different velocity which leads to pulse broadening. Without dispersion compensation a 10GBd signal will start to be significantly

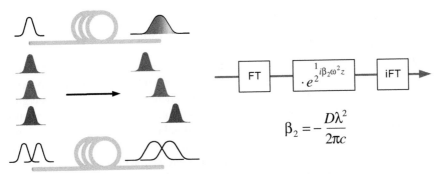

FIGURE 25.8 Pulse broadening due to group velocity dispersion and dispersion compensation in the Fourier domain. β_2 is the group velocity dispersion parameter, ω is the angular frequency of the carrier, z is the fiber length, λ is the wavelength of the carrier, D is the dispersion parameter and c is the speed of light [2].

impaired after only a few hundred ps/nm dispersion. The fiber dispersion is the result of both material dispersion and waveguide dispersion and the latter can be varied widely by altering the waveguide parameters making it possible to design dispersion compensation fibers. Optical dispersion compensation using these fibers has been successfully used commercially for many years to limit the effect of dispersion on transmission performance. However, dispersion compensation fibers have higher loss values than what can be achieved for transmission fibers with relatively large positive dispersion.

With a coherent receiver, dispersion can be compensated in the digital domain using DSP. In the Fourier domain, dispersion is just a phase factor, i.e. an all pass filter. By using a Fourier transform (FT) to convert the received signal into the Fourier domain, applying a phase factor related to the fiber dispersion D (in ps/nm/km) as shown in Figure 25.8 and converting back into the time domain with an inverse Fourier transform (iFT), dispersion can be compensated. The phase factor shown in Figure 25.8 assumes one fiber type with constant dispersion. In general D (and β_2) are location dependent such that the phase factor must be calculated by integrating over the transmission path. Of course there are practical limitations on the size of the Fourier transform that can be accommodated in DSP which determines the range of dispersion compensation that can be achieved. For a 100G channel running at 30 GBd (PDM QPSK including FEC overhead) on a 6,500 km system using a transmission fiber with $D = 21$ ps/nm/km dispersion at $\lambda = 1550$ nm, approximately 1000 symbols overlap that must be covered in the DSP dispersion compensation block.

Having the ability to compensate fiber dispersion digitally with a coherent receiver has a significant advantage. Negative dispersion fibers typically have higher loss than the best positive dispersion transmission fibers. For the same span length a system with inline dispersion compensation will therefore have a lower OSNR than a comparable system with no inline dispersion compensation. For an 80 km span the difference in span loss can be as much as 1.5 dB with a similar impact on receive

OSNR. Negative dispersion fibers also have a smaller effective area such that the optical power per channel must be reduced compared with positive dispersion fibers further lowering the receive OSNR.

While dispersion management using carefully arranged positive and negative dispersion fiber also known as dispersion maps was important for direct detection systems, a new class of dispersion unmanaged systems for coherent detection is now emerging. The contrast between two such maps is shown in Figure 25.9. For the direct detection system map (solid red line in Figure 25.9) the accumulated dispersion is managed such that it is non-zero to avoid phase matching reducing nonlinear impairments but at the same time does not reach high values limiting the interaction between nonlinear phase and dispersion. For the ideal coherent system the undersea transmission path consists of only positive dispersion fiber creating a so-called D+ or dispersion unmanaged map (dotted-dash blue line in Figure 25.9) with very high values of accumulated dispersion of up to more than 200,000 ps/nm depending on system length. As outlined above, the large accumulated dispersion leads to significant pulse broadening and overlap of many signal symbols such that nonlinear interaction between symbols and wavelengths tends to randomize [36] allowing for higher channel power. More details on this can be found in Section 25.2.3. However, such high values of accumulated dispersion are still a challenge for coherent transponders today as they require significant DSP resources. A compromise dispersion map is also shown in Figure 25.9 (dashed green line). Maximum values of accumulated dispersion are reached while keeping the map symmetric and zero accumulated dispersion at both ends [34].

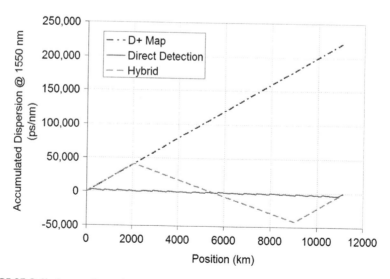

FIGURE 25.9 Undersea dispersion maps for a coherent system without any optical dispersion management (D+), a direct detection system and a hybrid system compatible both with direct detection and coherent transmission. (For interpretation of the references to color in this figure legend, the reader is referred to the web version of this book.)

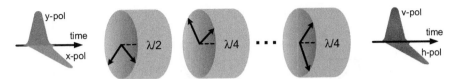

FIGURE 25.10 Evolution of a polarization multiplexed signal. The transmission fiber acts as a collection of wave plates with random strengths and orientations.

25.2.2.2 Polarization multiplexing and demultiplexing

Single mode optical fiber actually supports two orthogonal polarization modes. In typical transmission fiber the two modes are nearly degenerate and couple strongly such that power is easily transferred from one mode to the other. Small random variations of the fiber birefringence make it act like a collection of wave plates with random strengths and orientations (see Figure 25.10). As a result the polarization of a signal launched into the fiber quickly reaches a state of arbitrary polarization that also fluctuates over time. However, the two components of a polarization multiplexed signal remain mostly orthogonal under this transformation. It is therefore possible to send and recover two signals on the same wavelength with orthogonal polarization if the polarization transformation of the fiber can be undone on the receive side. In direct detection systems this can be achieved with a polarization controller before a polarization beam splitter. The polarization controller has to continuously track the temporal evolution of the polarization of the received signal. This turns out to be a significant technical challenge and widespread commercial use was never achieved.

Fortunately, the polarization transformation function of the transmission fiber is mostly linear and can therefore be reversed to a large degree with DSP in a coherent receiver. First, a polarization diversity receiver is required to detect both orthogonal polarization modes. This is simply done by using two 90° optical hybrids, one for each polarization mode, in combination with a polarization beam splitter (see Figure 25.11).

The polarization beam splitter separates the received signal into a horizontal and vertical component and the 90° optical hybrids recover the in-phase and quadrature portion of each component. Due to the polarization evolution in the fiber the signal in the horizontal or vertical channel of the receiver is a mixture of the original two orthogonal polarization states sent at the transmitter which leads to large amplitude variations in each channel. A popular method to recover the original polarization states for a polarization multiplexed QPSK signal is the constant modulus algorithm (CMA) which makes use of the fact that only the phase state is modulated and therefore all symbols have the same average intensity or modulus. By applying a butterfly filter (see Figure 25.12a) where the coefficients are set to achieve constant modulus at the output of the filter, the original polarization can be recovered [45]. The signals detected in the horizontal and vertical polarization channels of the coherent receiver are a mix of the originally transmitted x- and y-polarization states as indicated by the dual colored dots in Figure 25.12b. This mix causes the amplitude of the detected signal to vary considerably. By using the butterfly filter the amplitudes can be restored

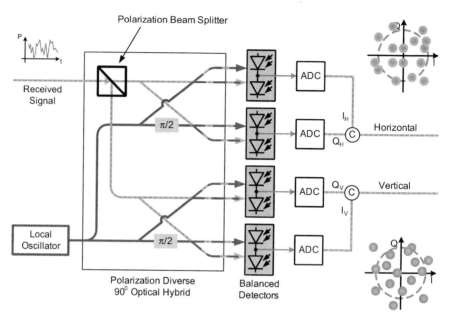

FIGURE 25.11 Polarization diverse 90° optical hybrid and coherent receiver.

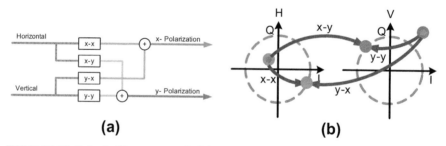

(a) **(b)**

FIGURE 25.12 Butterfly filter structure in (a) and polarization recovery schematic with a constant modulus algorithm in (b). (For interpretation of the references to color in this figure legend, the reader is referred to the web version of this book.)

to constant modulus which maps the horizontal and vertical polarization channels of the receiver to the original x- and y-polarization states of the transmitted signal. The coefficients of the butterfly filter have to be updated fast enough to track any temporal polarization changes caused by the transmission path.

Sending 2 bits per symbol using the QPSK format increases spectral efficiency twofold compared with a binary transmission format. In addition, the two polarization states of the fiber are nearly independent, again doubling spectral efficiency when using polarization multiplexing. A schematic of a typical polarization multiplexed

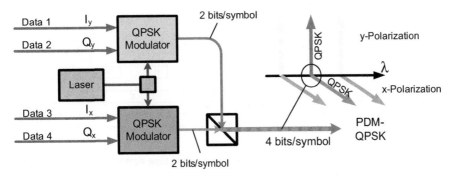

FIGURE 25.13 Schematic of a typical polarization multiplexed QPSK transmitter.

QPSK transmitter is shown in Figure 25.13. The light of a laser source is split into two equal parts and each part is modulated with a QPSK signal created out of two binary inputs (Data1 through Data4). The two parts are then combined to a polarization division multiplexed (PDM) QPSK signal with 4 bits/symbol before wavelength division multiplexing. The spectral efficiency of this transmitter is four times larger than a direct detection based DBPSK transmitter without polarization multiplexing.

25.2.2.3 PMD compensation

An additional advantage of coherent detection is the ability to compensate polarization mode dispersion (PMD) which has been a limiting effect for high bit rate transmission with direct detection. The birefringence of the transmission fiber not only leads to different phase velocities for orthogonal polarization modes (and thus polarization evolution) but together with the frequency dependence of the refractive index also causes a difference in group velocity. This is referred to as PMD and leads to temporal broadening of a signal as schematically depicted in Figure 25.14.

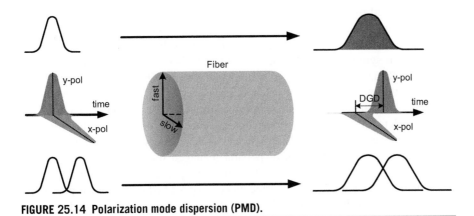

FIGURE 25.14 Polarization mode dispersion (PMD).

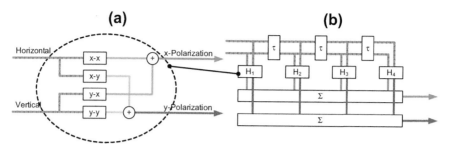

FIGURE 25.15 Generalized butterfly structure with multiple time domain taps for PMD compensation using CMA.

PMD is a statistical process since it depends on the random distribution of birefringence in the transmission fiber and the differential group delay (DGD) between the two orthogonal polarization states follows a Maxwell distribution.

To compensate PMD in a coherent receiver the CMA concept can be expanded. Instead of using a single butterfly structure, a series of multiple butterfly filters can be used where each filter is separated by a certain time delay from the previous. This is schematically shown in Figure 25.15. The range of PMD compensation is determined by the number of taps and the tap delay which is typically chosen to be half the symbol period.

25.2.3 Nonlinearity accumulation in dispersion uncompensated transmission

Coherent detection provides the opportunity to compensate chromatic dispersion in the digital domain using digital signal processing which enables a new class of undersea systems that no longer employ any optical dispersion management (see Section 25.2.2.1). As a result very large values of accumulated dispersion are encountered by the signal during transmission which leads to strong symbol broadening in the time domain and overlap of many symbols. This overlap tends to randomize the time domain field which effectively randomizes the interaction between symbols of the same channel and also neighboring channels with profound effects on the nonlinear system performance [6,14,17,22,36,49,53]. In particular, the randomization of the nonlinear interaction leads to a slower accumulation of nonlinear interference compared to dispersion managed systems used for direct detection. One important consequence is an observable difference in the dependence of the optimum power per channel launched into the transmission fiber on system length. In dispersion managed transmission the nonlinear impairment is related to the value of the nonlinear phase which is proportional to the system length and the path average power. For a constant nonlinear phase, the optimum fiber launch power therefore decreases with system length. In dispersion

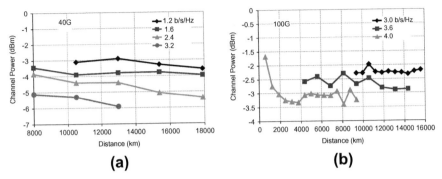

FIGURE 25.16 Optimum channel fiber launch power for 40 G coherent QPSK in (a) and 100 G coherent QPSK in (b) on a D+ transmission path for several spectral efficiencies as indicated in the legend.

uncompensated transmission, however, after a minimum distance at which the randomization takes effect, the optimum fiber launch power becomes independent of system length [48].

Figure 25.16 shows the optimum channel launch power into the transmission fiber for two different symbol rates and at several spectral efficiencies as a function of system length on a D+ transmission path. In particular at 28 GBd (100 G) the optimum channel launch power is independent of system length within the experimental uncertainty after about 2,000 km transmission distance. To achieve a similar level of randomization at 12 GBd (40 G) about 4 times as much dispersion is required explaining the small residual dependence on system length at 40 G.

In essence, the nonlinear interaction on a dispersion uncompensated path with sufficient accumulated dispersion can be treated as a noise term [6,17,36]. The total noise accumulated during propagation is no longer just due to amplified spontaneous emission (ASE) noise but is now the sum of ASE noise and nonlinear interference noise P_{NLI}:

$$P_{Noise} = P_{ASE} + P_{NLI} \tag{25.3}$$

The data of Figure 25.16 can be analyzed further to calculate the optimum power spectral density for each spectral efficiency and symbol rate. The result is shown in Figure 25.17. All calculated values are within 1.5 dB representing a good match with the theory considering the broad range of experimental transmission parameters. In our case the average value of the optimum power spectral density is 0.018 mW/GHz applicable to our transmission path of 52 km spans of D+ fiber with 150 μm^2 effective area, $2.37 \cdot 10^{-20}$ m^2/W nonlinear refractive index, 20.6 ps/nm/km chromatic dispersion and 0.183 dB/km attenuation.

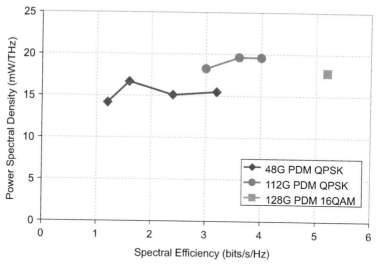

FIGURE 25.17 Optimum power spectral density (PSD) for 40 G and 100 G coherent QPSK on a D+ transmission path.

25.3 INCREASING SPECTRAL EFFICIENCY BY BANDWIDTH CONSTRAINT

To increase spectral efficiency either the information rate of the channel must be increased or the channel spacing decreased. The previous section described how coherent transmission techniques allow us to increase the channel information rate fourfold compared with a direct detection binary phase modulated signal like BPSK while keeping the spectral width constant by using polarization multiplexing and two bits/symbol. In this section we describe techniques to further increase spectral efficiency by bringing the channels closer together. Reducing the channel spacing is limited by crosstalk that occurs when channel spectra start to overlap (Figure 25.18a). Given appropriate optical filters the bandwidth of a channel can be constrained reducing the spectral width of the channel (Figure 25.18c) and enabling smaller channel spacing values and therefore higher capacity (Figure 25.18e). However, when reducing the spectral width of a channel by filtering, the signal pulses in the time domain broaden and pulses start to overlap causing inter-symbol interference (ISI) rapidly decreasing transmission performance (Figure 25.18b and d). Coherent detection allows us to process the received optical field and introduce advanced filters and algorithms that can correct this type of ISI and mitigate the penalty associated with strong optical filtering (Figure 25.18f).

25.3.1 Inter-symbol interference compensation by linear filters

The spectral width of a channel is directly related to its symbol rate, not its bit rate and as we will see in Section 25.4.1, the minimum spectral width for the ISI free case

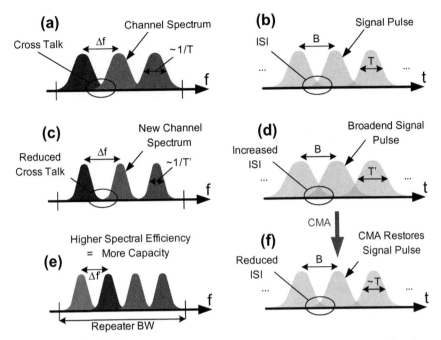

FIGURE 25.18 Increasing spectral efficiency by bandwidth constraint. Frequency domain on the left and corresponding time domain on the right.

is equal to the Baud rate. When comparing different filtering and ISI compensation techniques it makes sense to use Baud rate efficiency rather than spectral efficiency as a metric. Here we define Baud rate efficiency as the ratio of Baud rate over channel spacing such that the maximum Baud rate efficiency that can be achieved without any ISI is unity. Typical commercial 100G systems operate at a Baud rate of 30 GBd and a channel spacing of 50 GHz which yields a Baud rate efficiency of 0.6 Bd/Hz.

When the natural spectrum of a channel is reduced by filtering the signal pulses in the time domain broaden (Figure 25.18a–d), A coherent receiver includes a powerful linear equalizer that performs the polarization demultiplexing and PMD compensation operations. This linear equalizer is based on a butterfly filter structure and often uses a CMA as described in 25.2.2.2 and 25.2.2.3. The resulting filter transfer function is adaptive and can boost the parts of a spectrum that were reduced by optical filtering, thus restoring the time domain signal pulse shape and reducing ISI. Figure 25.19 shows the amplitude transfer function of the bandwidth constraining filter (solid blue) along with the overall filter shape with CMA equalizer (dashed red) [10]. The high frequency components that are attenuated by the bandwidth constraining filter are boosted by the CMA filter response. For Figure 25.19 the CMA response was calculated for a 28 GBd QPSK signal filtered by a 33 GHz optical interleaving filter for the noise free case to enhance the effect. In the presence of noise the CMA filter response will also result in increased high frequency noise components eventually limiting the effectiveness of the technique.

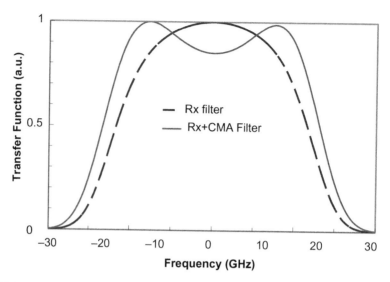

FIGURE 25.19 Bandwidth constraining filter shape (solid blue,) and the overall filter shape with CMA equalizer (dashed red, 15 taps) for a 28 GBd WDM QPSK signal at a Baud rate efficiency of 0.85 Bd/Hz (33 GHz channel spacing). (For interpretation of the references to color in this figure legend, the reader is referred to the web version of this book.)

Figure 25.20 shows the noise loaded back-to-back performance of a 28 GBd PDM QPSK signal with single-tap (red squares) and multi-tap (green triangles) CMA at a constant 18 dB/0.1 nm OSNR. A single tap CMA filter cannot correct ISI. With multi-tap CMA a significant performance boost is observed increasing with Baud rate efficiency (blue diamonds). However, beyond a Baud rate efficiency of unity even with a large number of taps the performance penalty is significant. This region requires multi-symbol detection (10-tap MLSE) that will be discussed on the next section (Section 25.3.2).

A similar linear filtering technique can also be used on the transmit side avoiding much of the noise enhancement effect encountered when implementing the filter on the receive side. This transmitter side filter can either be introduced in the optical domain [43] (however with limitations on implementable filter shapes) or in the electrical domain using digital signal processing and digital to analog converters. The latter is discussed in more detail in Section 25.4.1.

25.3.2 Multi-symbol detection

Linear filters cannot compensate ISI effectively, if the constrained bandwidth is less than the Baud rate (Figure 25.20). In this case the optical filter must constrain the bandwidth to less than the Baud rate to avoid strong linear crosstalk impairment. However, in order for linear filters to be able to restore the ISI free case, frequency content up to at least the Baud rate is required. Since this frequency content

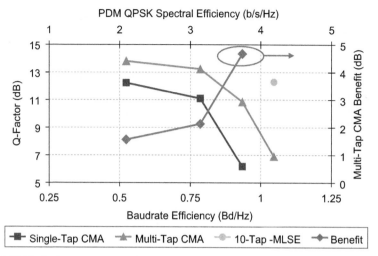

FIGURE 25.20 Performance comparison of 1-tap and multi-tap CMA at different spectral efficiency (received OSNR is fixed at 18 dB/0.1 nm). (For interpretation of the references to color in this figure legend, the reader is referred to the web version of this book.)

is suppressed by the optical filters and no longer present, linear filters are unable to restore the signal pulses and mitigate the ISI impairment.

Figure 25.21 demonstrates the effect of bandwidth constraint at a Baud rate efficiency of 1.12 Bd/Hz on the QPSK constellation [12]. Each one of the four constellation clouds breaks up into 3×3 sub-clouds. Close observation of the bandwidth constrained constellation also indicates that the probability of the sub-clouds is not equal. The center sub-cloud still has the highest probability whereas the corner sub-clouds have the lowest. This can be understood when considering the nearest neighbor ISI. Consider a symbol with $\pi/4$ phase (the (1,1) constellation point). If the previous symbol (N−1) and the next symbol (N+1) also have a $\pi/4$ phase and all three symbols overlap at the decision point of symbol N, then the interference is constructive and the amplitude of symbol N is increased without any phase change. If both

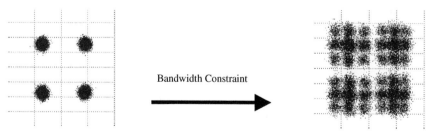

FIGURE 25.21 Effect of bandwidth constraint on the QPSK constellation at a Baud rate efficiency of 1.12 Bd/Hz.

neighboring symbols have a $5\pi/4$ phase (the $(-1,-1)$ constellation point), the neighbors interfere destructively with the symbol reducing its amplitude without changing the symbol phase. If the two neighbors have a phase difference of π between each other, their interference contributions cancel and the symbol constellation point remains unchanged. Similar arguments can be made for all nine sub-clouds with four possible phase combinations of the neighboring symbols to yield the center point, two for the edge points and only one for the corner points.

The effect of bandwidth constraint on a symbol is therefore dependent on its neighboring symbols and introduces correlation among symbols or memory into the channel. This symbol correlation or memory can be used with multi-symbol detection algorithms to mitigate the performance impairment encountered at Baud rate efficiencies larger than one. One such algorithm is Maximum a Posteriori probability (MAP) decoder where the receiver uses a lookup table based on system knowledge or a training sequence and selects the symbol combination with the smallest Euclidean distance. Depending on the signal constellation and the number of symbols considered in the decision process (the number of taps), the size of the lookup table and the complexity of this algorithm can become large quickly.

Figure 25.22 shows the benefit of MAP based detection over CMA only as a function of Baud rate efficiency [12]. Very large performance benefits can be realized for Baud rate efficiencies greater than one and even at a Baud rate efficiency of one and below, residual ISI impairment can be mitigated. Using this technique 10 Tb/s capacity was demonstrated over 10,600 km at a Baud rate efficiency of 0.85 Bd/Hz [8] and 12.5 Tb/s over 9360 km at a Baud rate efficiency of 1.0 Bd/Hz [7].

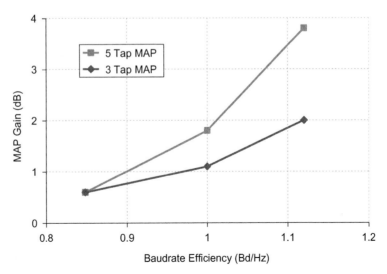

FIGURE 25.22 Benefit of the Maximum a Posteriori probability algorithm over CMA. After [12].

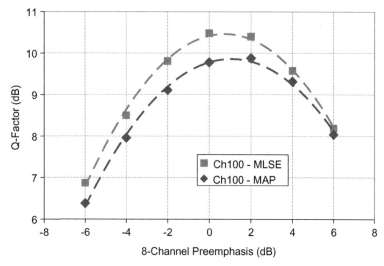

FIGURE 25.23 MLSE benefit over MAP in a 6860 km transmission experiment at a Baud rate efficiency of 1.12 Bd/Hz. After [9]

At a Baud Rate efficiency of 1.12 Bd/Hz the transmission distance was limited to 4370 km [8].

Multi-symbol detection is also possible with Maximum Likelihood Sequence Estimation (MLSE) which allows for efficient utilization of correlation between more distant symbols. Figure 25.23 shows the benefit of MLSE processing with a correlation length of 10 symbols over MAP with 5 taps at a Baud rate efficiency of 1.12 Bd/Hz after 6860 km. Note that the observed benefit is OSNR dependent. Using this technique 20 Tb/s capacity was demonstrated in 40 nm repeater bandwidth over 6860 km at a Baud rate efficiency of 1.12 Bd/Hz or spectral efficiency of 4.2 b/s/Hz [9].

Even though multi-symbol detection is effective in mitigating performance impairment caused by bandwidth constraint, some residual penalty remains. This is shown in Figure 25.24 for MAP detection [7]. Up to a Baud rate efficiency of unity the required OSNR for a constant Q-factor in a noise loaded back-to-back experiment increases only slowly with increasing Baud rate efficiency. If the Baud rate efficiency exceeds unity, the required OSNR increases sharply.

25.4 NYQUIST CARRIER SPACING

The conclusion from Section 25.3 is that the best performance and spectral efficiency is achieved at a Baud rate efficiency of one where the carrier spacing is equal to the

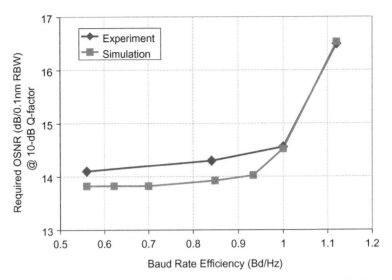

FIGURE 25.24 Noise loaded back-to-back WDM experiment and Monte Carlo simulation with bandwidth constraint and linear filtering.

Baud rate, a condition known as Nyquist carrier spacing. This section reviews techniques to achieve Nyquist carrier spacing with low inter-channel crosstalk and intra-channel inter-symbol interference. The time domain implementation is discussed in 25.4.1 where DSP and digital to analog converters (DACs) are used to implement a linear filter at the transmitter that limits the width of the channel to approximately the Baud rate enabling Nyquist channel spacing. The frequency domain equivalent is orthogonal frequency division multiplexing (OFDM) and discussed in Section 25.4.2 where channel spectra overlap significantly but are orthogonal and can be separated using a Fourier transformation avoiding inter-channel crosstalk. This property of the OFDM signal can be combined with single carrier techniques to create super-channels with very high aggregate channel capacities. Super-channels are discussed in Section 25.4.3.

25.4.1 Spectral shaping for single channel modulation formats

Recent improvements in the bandwidth and sampling rates of DACs have enabled the use of digital signal processing of the transmitted signal. This allows for precise control of the transmitted signal's time domain waveform and spectral characteristics. One well-known class of signal waveforms used to increase spectral efficiency are raised cosine shapes [37]. In the limiting case (beta = 0) the channel's bandwidth can be reduced to equal the symbol rate without introducing ISI. Although this limiting case would theoretically require infinitely long digital filters, the abilities of today's

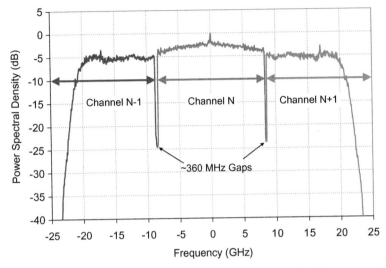

FIGURE 25.25 Sample spectrum of three adjacent channels as acquired by our sampling oscilloscope.

leading edge technology allow for a practical implementation with a minimum of penalty. A demonstration of raised cosine spectral shaping of a 16.64 GBd signal is shown in Figure 25.25. In this case an optical 16QAM, 16.64 GBd, dual polarization signal was produced using a combination of DSP raised cosine filtering and a DAC. The plot shows the power spectral density of channels of a portion of a WDM transmitted spectrum. The data was measured using a laboratory version of a coherent receiver, and the data for both polarizations was used to create the figure. A key feature of the digital filtering is shown by the steep "brickwall" edges of the channels with $\beta=0.001$, which allowed spacing the 16.64 GBd channels at 17 GHz. The "gap" between channels in this case is only 360 MHz. In principal, this concept is similar to the optical filtering discussed in Section 25.3.1 only that the filter here is implemented in DSP which enables much sharper filter shapes. This technique has been used to transmit 30 Tb/s over a distance of 6,630 km at a spectral efficiency of 6.1 bits/s/Hz [33].

25.4.2 Orthogonal frequency division multiplexing (OFDM)

The previous section focused on single carrier modulation formats where a laser source is modulated with a serial information signal in the time domain. This is currently the prevailing method for most commercial fiber optic transmission systems. OFDM is an alternative modulation format which has been used extensively in commercial radio frequency (RF) systems and is now generating significant interest for use in optical communication systems [47]. In this method the data to be transmitted is demultiplexed into a number of parallel streams and the bits of each stream are

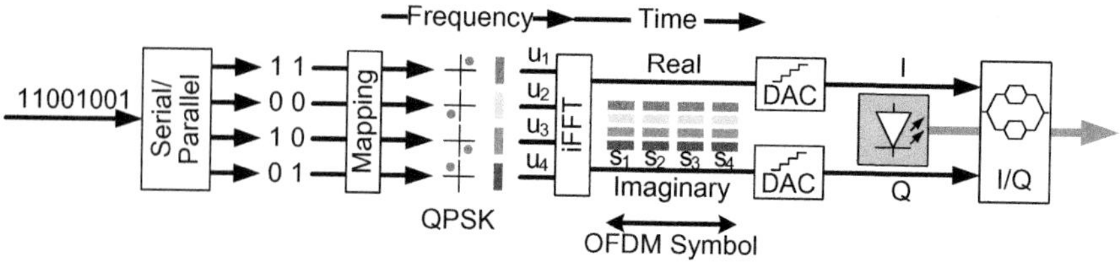

FIGURE 25.26 OFDM transmitter schematic for 4 subcarriers and QPSK mapping. I/Q indicates an I/Q modulator, DAC indicates a digital to analog converter.

then mapped to a symbol of a particular modulation format. The resulting complex values (amplitude and phase) are interpreted as a spectrum in the frequency domain and inverse Fourier transformed. The Fourier transformed complex values are then used as a time domain signal that is imparted onto a laser source with DACs. A schematic representation of this method is shown in Figure 25.26. In this example the chosen degree of parallelization and therefore number of subcarriers is 4 and the mapping is QPSK. The four time domain samples S_1 to S_4 represent one OFDM symbol.

OFDM is very popular in radio transmission since it results in high spectral efficiency and OFDM can be more resilient to fading than a single carrier. The generated subcarriers have a spacing that corresponds to the inverse of the OFDM symbol rate (i.e., Nyquist spacing) and sinc function spectra such that the subcarriers are orthogonal in the frequency domain. At the receiver, Fourier transforming an OFDM symbol demultiplexes the subcarriers and recovers the original mapped symbols for each subcarrier. The challenge is to find and synchronize to the OFDM symbols in the received data stream where the subcarriers are scrambled through transmission effects such as dispersion, polarization evolution and PMD. In contrast to single carrier coherent receivers where these effects can be compensated step by step, in OFDM the transfer function of the channel must be estimated as a whole and a reverse linear filter applied to the incoming data. Pilot tones and a cyclic prefix with a length corresponding to the interaction length encountered during transmission must be applied to the data to enable this channel estimation process. Compared with single carrier transmission, additional overhead may therefore be required slightly reducing the overall spectral efficiency that can be achieved.

Aside from practical aspects such as coherent algorithm implementation, OFDM and the limiting case of a raised cosine spectrum with infinitely steep edges are complementary. One has a rectangular pulse in the time domain and sinc spectra in the frequency domain (OFDM) and the other has sinc pulses in the time domain and rectangular spectra in the frequency domain. This is schematically depicted in Figure 25.27. High spectral efficiency transmission with OFDM has been successfully demonstrated in several experiments [29,38] including spectral efficiency of 4.7 b/s/Hz over 10,181 km [57].

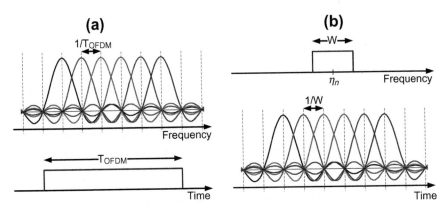

FIGURE 25.27 (a) OFDM signal with rectangular time domain pulses and a sinc spectrum for each subcarrier, (b) Single carrier signal with a rectangular spectrum and sinc pulses in the time domain.

25.4.3 Super-channels

A variant of OFDM, often referred to as super-channel [15], uses multiple frequency-locked continuous wave carriers (either individual lasers or a demultiplexed frequency comb generated from a single laser source) that are modulated individually with the mapped symbols avoiding the inverse Fourier transform operation at the transmitter and not requiring DACs. In this case DAC bandwidth limitations do not apply and each subcarrier can operate at a high Baud rate. The subcarrier spacing is again equal to the Baud rate. Very high aggregate channel capacities have been demonstrated with this method; however, the hardware complexity is similar to multiple WDM channels since each subcarrier requires its own modulation path plus typically its own coherent receiver since current analog to digital converter (ADC) technology cannot support the bandwidth required for these super-channels. Figure 25.28 shows a schematic representation of a super-channel transmitter. Note the subcarrier multiplexing similar to wavelength-division multiplexing (WDM). On the receive side the Fourier operation to de-multiplex the subcarriers is often implemented with a simple tap-delay filter after which single carrier techniques are

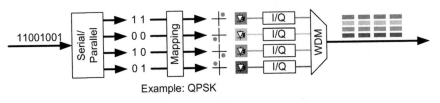

FIGURE 25.28 Schematic representation of a super-channel transmitter.

used for equalization and carrier phase recovery. To preserve the orthogonality of the subcarriers after detection, the bandwidth of the coherent receiver must typically cover more than one subcarrier and has therefore a larger bandwidth requirement than its single carrier counterpart.

Super-channels have been used to demonstrate impressive channel capacities at high spectral efficiency. For example, the first transoceanic Tb/s super-channel transmission over 7,200 km was reported in [15]. Each super-channel consisted of 24 individually modulated subcarriers with PDM QPSK mapping at 12.5 GBd and with 12.5 GHz subcarrier spacing. Assuming 20% FEC overhead this corresponds to an intra-channel spectral efficiency of 3.3 b/s/Hz. While these results are impressive, super-channels do not increase the spectral efficiency compared to OFDM or single carrier channels at the Nyquist spacing with the same mapping. However, photonic integration may eventually provide a cost advantage to super-channels.

25.5 INCREASING SPECTRAL EFFICIENCY BY INCREASING THE CONSTELLATION SIZE

At the Nyquist spacing spectral efficiency can only be increased by increasing the number of bits conveyed by each symbol. This is accomplished by increasing the constellation size or alphabet of the transmitted signals. The spectral efficiency that can be achieved here is fundamentally limited by the overall signal-to-noise ratio including linear and nonlinear noise. This is described in Section 25.5.1. However, increasing the constellation size to create higher order modulation formats results in lower receiver sensitivity such that the achievable distance for the same optical signal-to-noise ratio and forward error correction algorithms rapidly decreases with increasing spectral efficiency. This challenge is outlined in Section 25.5.2. The decreasing receiver sensitivity with higher order modulation formats can be countered by using coded modulation described in Section 25.5.3.

25.5.1 Higher order modulation formats

Optical communication systems have traditionally used binary optical formats such as OOK or BPSK (see Section 25.2.1). These formats have two possible states per symbol and carry one bit of information. Current optical communication systems are being deployed using QPSK which increases the number of states (also known as the constellation size) to four which results in two bits per symbol. The number of bits per symbol can be further increased by increasing the size of the constellation. Common techniques used in the communication industry are phase shift keying (PSK), amplitude shift keying (ASK), and a combination of PSK and ASK which is commonly referred to as quadrature amplitude modulation (QAM). Figure 25.29 shows examples of common constellations or higher modulation formats. The number of bits per symbol is equal to the log base 2 of the number of constellation points. The choice of modulation format is influenced by a number of variables including the

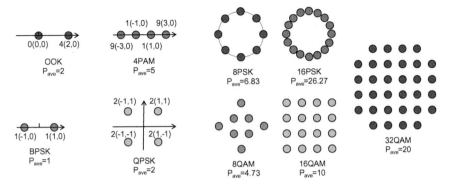

FIGURE 25.29 Constellation diagrams and average symbol energy for some popular higher order modulation formats.

required receiver sensitivity and practical limits for the complexity of the transmit and receive implementations.

Also shown in Figure 25.29 is the average energy per symbol for a minimum Euclidean distance of 2 (i.e., the minimum distance between two constellation points). Although the exact relationship between average energy per symbol and receiver sensitivity depends on the details of the constellation, in general the receiver sensitivity will decrease as the average energy increases.

The lowest energy per symbol is achieved for BPSK where each of the constellation points carries an energy of 1 and therefore the average energy per BPSK symbol is 1. For OOK one constellation point carries an energy of $2^2 = 4$ whereas the other has no energy and therefore the average energy is 2 (assuming equal probability for marks and spaces). For QPSK all constellation points again have the same energy ($\sqrt{2}^2$) and the average energy per symbol is 2. Note that for QPSK each symbol carries two bits such that the average energy per bit is the same as for BPSK, i.e., 1. The average energy values per symbol or per bit for other constellations can be calculated in a similar fashion.

Figure 25.30 shows the maximum spectral efficiency that can be achieved for several QAM constellations assuming polarization multiplex and no channel memory. In the high OSNR case the maximum spectral efficiency asymptotically corresponds to the number of bits per symbol multiplied by two (to account for the two polarizations) since the minimum spectral width is equal to the Baud rate for memory free modulation formats. When the OSNR degrades, the maximum spectral efficiency decreases. In Figure 25.30 the OSNR is normalized to the number of information bits. To get the OSNR in 0.1 nm resolution bandwidth for a 100G channel multiply by 104 Gb/s (the OTN 100G bit rate) and divide by 12.5 GHz (0.1 nm resolution bandwidth in GHz at 1550 nm). Also shown in Figure 25.30 is the upper bound for spectral efficiency in a noise limited channel by modifying Shannon's limit [46] to account for polarization multiplexing. Note that an OSNR per bit of at least −4.6 dB (with polarization multiplexing) is required for any reliable communication to take place.

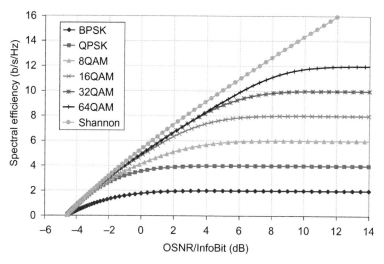

FIGURE 25.30 Maximum spectral efficiency for some popular higher order modulation formats along with the Shannon limit.

25.5.2 Receiver sensitivity

As already indicated in the previous section, when the constellation size increases the average energy per symbol increases if the minimum Euclidean distance is fixed which leads to a reduced receiver sensitivity. The actual OSNR required for a certain bit error ratio (BER) depends on the shape of the constellation with their varying number of inner, edge and corner constellation points. Figure 25.31a shows the maximum Q-factor for several QAM modulation formats in a noisy channel. The corresponding minimum BER calculated from the Q-factor in (a) is shown in Figure 25.31b. The OSNR here is normalized to the line bit rate (i.e., the information

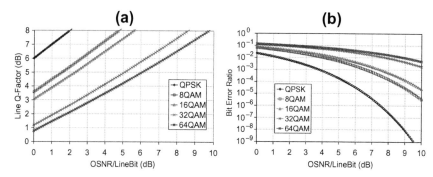

FIGURE 25.31 Receiver sensitivity plots for some popular higher order modulation formats. OSNR/LineBit corresponds to the OSNR normalized to the line bit rate (i.e., the information bit rate plus overhead).

bit rate plus overhead) to make the result independent of bit rate and overhead. Note that the receiver sensitivities for BPSK and QPSK are identical.

It is apparent from Figure 25.31 that increasing the spectral efficiency by increasing the constellation size comes at a cost. For example, when switching from QPSK with 2 bits/symbol to 16QAM with 4 bits/symbol which increases the spectral efficiency twofold, the required OSNR per line bit increases by roughly 4 dB for the same Q-factor or BER. This cuts the achievable transmission distance by more than half and is the reason for the significant challenge facing undersea communication systems to increasing spectral efficiency.

The fundamental loss in receiver sensitivity when moving to higher order modulation formats can be mitigated by advanced forward error correction (FEC) algorithms with increased net coding gain. Since with coherent receivers the signal is already sampled and digitized, integration of a soft decision FEC into the receiver becomes very attractive. Compared to their hard decision counterparts common in direct detection receivers, soft decision can decrease the FEC threshold by more than 1 dB [37,56].

25.5.3 **Coded modulation**

With binary transmission formats error correcting algorithms are limited to codes that are independent of the modulation format. The maximum spectral efficiency is determined by the code rate. Coherent transmission techniques make higher order modulation formats practical and allow us to introduce coded modulation which is the joint design of an error correcting code and a signal constellation. Groups of coded bits are mapped to points of the signal constellation in a way that enhances the distance properties of the code [35]. Typically coded modulation expands the signal constellation size and then applies error correcting coding to increase the minimum Euclidean distance between modulated coded sequences [51,52,28]. A pragmatic approach to coded modulation is bit interleaved coded modulation (BICM) where binary encoding is followed by a pseudorandom bit interleaver [13]. A wide variety of binary codes can be used in this case increasing the flexibility with which this technique can be implemented.

Figure 25.32a shows a schematic for one such implementation as used in [55]. A binary low density parity check (LDPC) code is used for error correction followed by

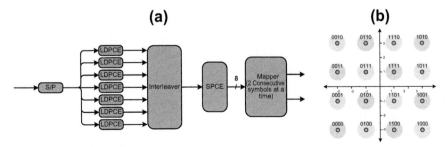

FIGURE 25.32 (a) Implementation schematic for Single Parity Check Bit Interleaved Coded Modulation (SPC BICM). S/P: Serial to parallel converter, LDPCE: Low density parity check encoder, SPCE: Single parity check encoder. (b) 16QAM constellation set partitioning.

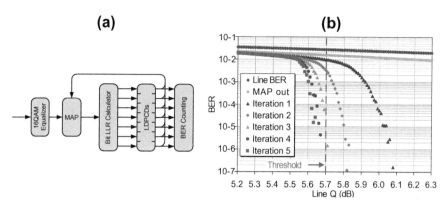

FIGURE 25.33 (a) Receiver schematic and (b) SPC BICM waterfall curve.

the bit interleaver. In addition, the resulting bit stream is further encoded with a 7/8 rate single parity check (SPC) encoder before the bits are mapped in groups of 8 onto two consecutive 16QAM symbols using Gray mapping. The SPC encoding results in set partitioning where the available points of the 16QAM constellation are restricted to 8 possible states in the second symbol depending on the constellation point of the first symbol in the pair. This is schematically shown in Figure 25.32b.

On the receive side a coherent receiver is used to recover the 16QAM constellation. After equalization a soft-input/soft-output (SISO) maximum a posteriori probability (MAP) decoder takes advantage of the SPC bit and calculates the symbol log likelihood ratios (LLRs) of the two consecutive symbols based on the SPC codeword book. The symbol LLRs are then converted to bit LLRs and passed on to the LDPC decoders which calculate the extrinsic information. In the next iteration the MAP decoder uses this extrinsic information as a priori information. A schematic of the receiver is shown in Figure 25.33a. Shown in Figure 25.33b is the waterfall curve of this coded modulation scheme for up to five iterations. The LDPC codes in this particular example use code rate 0.93, codeword length 32,000, girth 8, column weight 4 and 10 inner iterations. The total overhead is 23%.

Other coded modulation formats based on this scheme have also been reported. A particular variant uses SPC encoding for only two out of four 16QAM symbols and stronger LDPC codes with rate 0.833 for a total overhead of 28%. The threshold in this case is 4.9 dB after five iterations [33].

25.6 **FUTURE TRENDS**

At the end, we speculate technologies that are promising to further enhance the capacity of undersea communication systems; however, more progress is needed before commercialization.

25.6.1 **Nonlinearitycompensation**

It is straightforward to achieve higher spectral efficiency and therefore higher capacity if we can increase the OSNR at the receiver. Higher SNR is possible simply by increasing the signal power. However, the total noise power that enters into the SNR calculation is not only given by the linear noise generated by the repeaters (amplified spontaneous emission or ASE) along the transmission path but also the "noise" generated by nonlinear interaction among the signals through the Kerr effect (see Section 25.2.3). Fortunately, the leading contributions to this nonlinear interference noise are generated by signal-signal interaction which is not truly random but deterministic. This provides the opportunity for compensation as long as sufficient knowledge of the signal and physical properties of the transmission system exist. Only the contributions to the nonlinear interference noise that stem from signal-ASE and ASE-ASE interaction cannot be predicted in a deterministic way and will therefore ultimately limit the transmission performance, spectral efficiency and capacity.

A popular method for compensation of nonlinear impairments is digital back propagation (DBP) [30]. The received signal is propagated backwards through the transmission path by solving the inverse nonlinear Schrödinger equation (NLSE) to arrive at an approximation of the transmitted signal before fiber impairment. This technique is computationally intense since the NLSE is solved numerically by the split step Fourier method where dispersion and nonlinear phase shift are treated separately and a small step size is required for accurate results. Much effort has been devoted to find good approximations that allow for large step sizes and therefore fast data processing [32]. However, the complexity for algorithms that can be applied universally is still prohibitively high for commercial applications.

Besides active nonlinearity compensation, passive compensation by decreasing the nonlinear interaction strength on the fiber is also very attractive. The nonlinear interaction strength is proportional to the ratio of nonlinear refractive index and effective area. To decrease the nonlinear interaction strength fiber manufacturers have been able to decrease the nonlinear refractive index slightly by removing dopants from the core glass and moving to pure silica core fiber types. More importantly, advanced fiber designs now enable effective area up to $150 \mu m^2$, a more than twofold improvement compared with legacy dispersion managed undersea transmission fiber. Hollow core fibers can further improve the nonlinear refractive index by about two orders of magnitude; however, at this point in time fiber loss of these photonic crystal fibers is still too high to be attractive.

25.6.2 **Multi-core and multi-mode fiber**

The previous sections of this chapter all focused on single mode and single core transmission. However, instead of increasing capacity by increasing the repeater bandwidth or the spectral efficiency, it is also possible to increase capacity simply by providing a greater number of transmission paths. This can be accomplished by space division multiplexing. Transmission fibers with multiple cores and limited inter-core cross-talk have been demonstrated to be compatible with ultra long haul transmission [25].

However, the number of cores that can fit into a fiber without increasing its cladding diameter is limited to about 7. Beyond 7 cores the outside diameter of the fiber has to be increased such that the core density can only be increased by about one order of magnitude compared with conventional fiber bundles [25]. For applications where space is limited such as in an undersea cable this may be a compelling reason for using multi-core fibers to increase capacity. On the other hand, all the cores in the cable have to be fitted with EDFAs when the cable is first installed. It is not practical to install EDFAs at a later stage in the life of the system to upgrade the cable capacity since involved marine operations would be required. This is in contrast to terrestrial applications where EDFAs are installed along a new path only when it is being lit for the first time. The number of unlit paths in a submarine cable therefore has to balance the initial system cost and final capacity. This scenario assumes that separate amplifiers are required for each core which is the current situation. In the future, multi-core EDFAs may become available such that multi-core fiber can be used in a similar fashion as single mode fiber today.

Another option to increase cable capacity is mode division multiplexing. Multi-mode fiber has a very large number of modes that each can be used for transmission in principle. As pointed out in Section 25.2.2.2 even single mode fiber supports two orthogonal polarization modes that we have learned to use to double the spectral efficiency. The two polarization modes couple strongly in transmission fiber but a coherent receiver is able to unscramble the two modes as long as the differential mode delay from polarization mode dispersion does not exceed the length of the linear equalizer. The linear equalizer in this case is a 2×2 multi-input multiple-output (MIMO) equalizer (two polarization streams in and two demultiplexed polarization streams out). The same principles apply to transmission on multi-mode fiber, only now the number of modes can be very large. This is very attractive to increase the fiber capacity; however, each mode also increases the complexity of the required MIMO equalizer considerably. The modes also experience different group delay that can be very large, even though few mode fibers with limited differential group delay have been demonstrated [24]. Recently, multi-mode amplifiers have been demonstrated in transmission experiments including gain equalization among modes [40]. Another approach uses multi-core fiber with strong coupling among the cores to support a few spatial modes [39].

25.7 SUMMARY

Much progress has been made over the last few years in undersea optical fiber telecommunication systems. Most importantly coherent transmission formats have become practical enabling polarization multiplexing and higher order modulation formats with much increased spectral efficiency. A selection of some high capacity and high spectral efficiency demonstrations over transoceanic distances is provided in Table 25.1 sorted by ascending capacity-distance product. Over the past four years the capacity and capacity-distance product are more than doubled. This progress was enabled by an almost threefold increase in spectral efficiency. The first experimental demonstration of a capacity distance product in excess of 100 Pb/s km relied on C+L band EDFAs and Raman amplifiers to provide a combined repeater bandwidth of approximately 62 nm and transmitted

Table 25.1 Recent demonstrations with large capacity-distance and/or large SE-distance product.

Capacity (Tb/s)	SE (b/s/Hz)	Distance (km)	SE Distance Product (b/s/Hz·km)	Capacity Distance Product (Pb/s·km)	Modulation Format	Year	Reference
1.20	3.60	7200	25,920	8.6	Super-Channel	2009	15
3.00	3.03	8040	24,361	24.1	QPSK	2010	50
4.00	4.70	10,180	47,846	40.7	OFDM 16QAM	2012	57
4.30	3.58	10,180	36,444	43.8	Super-Channel	2011	27
7.54	2.09	7040	14,714	53.1	QPSK	2009	16
8.37	2.09	9000	18,810	75.4	QPSK	2010	42
14.00	2.07	6250	12,938	87.5	OFDM QPSK	2009	31
8.37	2.09	12,000	25,080	100.5	QPSK	2011	44
10.05	3.14	10,610	33,315	106.6	QPSK	2011	8
11.72	3.74	9360	35,006	109.7	QPSK	2010	7
16.65	5.20	6860	35,672	114.3	16QAM	2012	55
9.99	2.78	11,680	32,470	116.7	QPSK	2011	43
16.22	2.09	7200	15,048	116.8	QPSK	2009	41
11.55	4.02	10,180	40,924	117.6	OFDM 8QAM	2011	38
20.73	4.19	6860	28,743	142.3	QPSK	2012	9
26.00	5.20	5530	28,756	143.8	16QAM	2012	11
30.58	6.12	6630	40,576	202.7	16QAM	2012	33

16 Tb/s capacity over 7,200 km [41]. In comparison, the current state-of-the-art uses only the standard C-band with about 40 nm bandwidth and transmits 30 Tb/s capacity over 6,600 km achieving nearly double the capacity and capacity-distance product [33]. The enabling technologies for this result are 16QAM coded modulation and raised cosine spectral shaping using DACs. Note that the highest spectral efficiency-distance product is achieved with OFDM at 4 b/s/Hz over 10,180 km [38]. Super-channels in excess of 1 Tb/s capacity have been transmitted over 7,200 km as early as 2009 [15].

Acknowledgments

The authors wish to thank all of their colleagues in optical fiber telecommunications around the world for their significant achievements in undersea transmission. We also express our heartfelt thanks to the TE SubCom team for their valuable contributions to this work and for kindly providing the material and fruitful discussions that made this chapter possible.

List of Acronyms

ADC	Analog to Digital Converter
ASE	Amplified Spontaneous Emission
ASK	Amplitude Shift Keying
BER	Bit Error Ratio
BICM	Bit Interleaved Coded Modulation
BPSK	Binary Phase Shift Keying
BW	Bandwidth
CMA	Constant Modulus Algorithm
DAC	Digital to Analog Converter
DBPSK	Differential Binary Phase Shift Keying
DD	Direct Detection
DGD	Differential Group Delay
DSP	Digital Signal Processing
EDFA	Erbium Doped Fiber Amplifier
FEC	Forward Error Correction
FT	Fourier Transform
iFT	inverse Fourier Transform
ISI	Inter-Symbol Interference
LDPC	Low Density Parity Check
LLR	Log Likelihood Ratio
MAP	Maximum a posteriori Probability
MIMO	Multi-Input Multiple-Output
MLSE	Maximum Likelihood Sequence Estimation
NLSE	Nonlinear Schrödinger Equation
OFDM	Orthogonal Frequency Division Multiplexing
OOK	On-Off Keying
OSNR	Optical Signal to Boise Ratio
PDM	Polarization Division Multiplexing
PMD	Polarization Mode Dispersion
PSD	Power Spectral Density
PSK	Phase Shift Keying
QAM	Quadrature Amplitude Modulation
QPSK	Quadrature Phase Shift Keying
RF	Radio Frequency
SE	Spectral Efficiency
SISO	Soft-Input/Soft-Output
SPC	Single Parity Check

References

[1] S. Abbott, Review of 20 years of undersea optical fiber transmission system development and deployment since TAT-8, in: Proc. ECOC 2008, Mo.4.E.1.

[2] G.P. Agrawal, Fiber-Optic Communication Systems, John Wiley and Sons, Inc., 2002.

[3] W.A. Atia, R.S. Bondurant, Demonstration of return-to-zero signaling in both OOK and DPSK formats to improve receiver sensitivity in an optically preamplified receiver, in: LEOS'1999, Paper TuM3, San Francisco, California, November 1999, pp. 226–227.

[4] Neal S. Bergano, Undersea communications systems, in: Ivan Kaminow, Tingye Li (Eds.), Optical Fiber Telecommunications IVB, Academic Press, 2002 (Chapter 4)

[5] S. Bigo, Undersea communications systems, in: Ivan Kaminow, Tingye Li (Eds.), Optical Fiber Telecommunications VB, Academic Press, 2007 (Chapter 14)

[6] A. Bononi, E. Grellier, P. Serena, N. Rossi, F. Vacondio, Modeling nonlinearity in coherent transmission with dominant interpulse four wave mixing, in: Proc. ECOC 2011, We.7.B.4.

[7] J.-X. Cai Y. Cai, Y. Sun, C.R. Davidson, D.G. Foursa, A. Lucero, O. Sinkin, W. Patterson, A. Pilipetskii, G. Mohs, N.S. Bergano, 112×112 Gb/s transmission over 9,360 km with channel spacing set to the baud rate (360% spectral efficiency), in: Proc. ECOC 2010, PD2.1.

[8] J.-X. Cai, Y. Cai, C.R. Davidson, D.G. Foursa, A. Lucero, O. Sinkin, W. Patterson, A. Pilipetskii, G. Mohs, N. Bergano, Transmission of 96×100 Gb/s bandwidth-constrained PDM–RZ–QPSK channels with 300% spectral efficiency over 10,610 km and 400% spectral efficiency over 4,370 km, J. Lightwave Technol. 29 (4) (2011) 491.

[9] J.-X. Cai, C.R. Davidson, A. Lucero, H. Zhang, D.G. Foursa, O.V. Sinkin, W.W. Patterson, A.N. Pilipetskii, G. Mohs, N.S. Bergano, 20 Tbit/s transmission over 6,860 km with sub-Nyquist channel spacing, J. Lightwave Technol. 30 (4) (2012) 651–657.

[10] J.-X. Cai, O. Sinkin, H. Zhang, Y. Sun, A. Pilipetskii, G. Mohs, N.S. Bergano, ISI compensation up to Nyquist channel spacing for strongly filtered PDM RZ QPSK using multi-tap CMA, in: Proc. OFC/NFOEC 2012, JW2A.47.

[11] J.-X. Cai, H. Batshon, H. Zhang, C. Davidson, Y. Sun, M. Mazurczyk, D. Foursa, A. Pilipetskii, G. Mohs, N. Bergano, 25 Tb/s transmission over 5,530 km using 16QAM at 5.2 b/s/Hz SE, in: Proc. ECOC 2012, Mo.1.C.1.

[12] Y. Cai, J.-X. Cai, C.R. Davidson, D. Foursa, A. Lucero, O. Sinkin, A. Pilipetskii, G. Mohs, N.S. Bergano, High spectral efficiency Long-Haul transmission with pre filtering and maximum a posteriori probability detection, in: Proc. ECOC 2010, We.7.C.4.

[13] G. Caire, G. Taricco, E. Biglieri, Bit-interleaved coded modulation, IEEE Trans. Inform. Theory 44 (3) (1998) 927–946.

[14] A. Carena, G. Bosco, V. Curri, P. Poggiolini, M. Tapia Taiba, F. Forghieri, Statistical characterization of PM-QPSK signals after propagation in uncompensated fiber links, in: Proc. ECOC 2010, P4.07.

[15] S. Chandrasekhar, X. Liu; B. Zhu, D.W. Peckham, Transmission of a 1.2 Tb/s 24-carrier no-guard-interval coherent OFDM superchannel over 7,200 km of ultra–large–area fiber, in: Proc. ECOC 2009, PD2.6.

[16] G. Charlet, M. Salsi, P. Tran, M. Bertolini, H. Mardoyan, J. Renaudier, O. Bertran-Pardo, S. Bigo, 72×100 Gb/s transmission over transoceanic distance, using large effective area fiber, hybrid Raman-Erbium amplification and coherent detection, in: Proc. OFC/NFOEC 2009, PDPB6.

[17] Xi Chen, W. Shieh, Closed-form expression for nonlinear transmission performance of densely-spaced coherent optical OFDM systems, Opt. Express 18 (18) 19039–19054.

[18] C.R. Davidson, C. Chen, M. Nissov, A. Pilipetskii, N. Ramanujam, H. Kidorf, B. Pedersen, M. Mills, C. Lin, I. Hayee, J.-X. Cai, A. Puc, P. Corbett, R. Menges, H. Li, A. Elyamani, C. Rivers, N. Bergano, 1800 Gb/s transmission of one hundred and eighty 10 Gb/s WDM channels over 7,000 km using the full EDFA C-band, in: Proc. OFC 2000, PD25.

[19] Bern Dibner, The Atlantic Cable, Burndy Library Inc., Norwalk, Conn., 1959.

[20] Emerald Express press release, Houston, TX and Morristown, NJ, July 19, 2011.

[21] D.G. Foursa, C.R. Davidson, M. Nissov, M.A. Mills, L. Xu, J.-X. Cai, A.N. Pilipetskii, Y. Cai, C. Breverman, R.R. Cordell, T.J. Carvelli, P.C. Corbett, H.D. Kidorf, N.S. Bergano, 2.56 Tb/s (256 × 10 Gb/s) transmission over 11,000 km using hybrid Raman/EDFAs with 80 nm of continuous bandwidth, in: OFC'2002, Post-Deadline Paper FC3, Anaheim, California, March 2002.

[22] A.H. Gnauck, G. Raybon, S. Chandrasekhar, J. Leuthold, C. Doerr, L. Stulz, A. Agarwal, S. Banerjee, D. Grosz, S. Hunsche, A. Kung, A. Marhelyuk, D. Maywar, M. Movassaghi, X. Liu, C. Xu, X. Wei, D.M. Gill, 2.5 Tb/s (64 × 42.7 Gb/s) transmission over 40 × 100 km NZDSF using RZ-DPSK format and all-Raman-amplified spans, in: Proc. OFC 2002, FC2.

[23] E. Grellier, Jean-Christophe Antona, S. Bigo, Revisiting the evaluation of non-linear propagation impairments in highly dispersive systems, in: Proc. ECOC 2009, 10.4.2.

[24] L. Grüner-Nielsen, Y. Sun, J. W. Nicholson, D. Jakobsen, R. Lingle Jr, B. Pálsdóttir, Few mode transmission fiber with low DGD, low mode coupling and low loss, in: OFC/NFOEC 2012, PDP5A.1.

[25] T. Hayashi, T. Sasaki, E. Sasaoka, Multi-core fibers for high capacity transmission, in: Proc. OFC/NFOEC 2012, OTu1D.4.

[26] Hibernia Atlantic press release Summit, NJ and Dublin, Ireland, September 30, 2010.

[27] Y.-K. Huang, M.-F. Huang, E. Ip, T. Inoue, Y. Inada, T. Ogata, Y. Aoki, 4 × 1.15 Tb/s DP-QPSK superchannel transmission over 10,181 km of EDFA amplified hybrid large-core/ultra low-loss fiber spans with 2 dB FEC margin, in: Proc. ACP 2011, PD3.

[28] H. Imai, S. Hirakawa, A new multilevel coding method using error-correcting codes, IEEE Trans. Inform. Theory 23 (3) (1977) 371–377.

[29] S.L. Jansen, I. Morita, T.C.W. Schenk, N. Takeda, H. Tanaka, Coherent optical 25.8 Gb/s OFDM transmission over 4,160 km SSMF, J. Lightwave Technol. 26 (1) (2008) 6–15.

[30] X. Li, X. Chen, G. Goldfarb, E. Mateo, I. Kim, F. Yaman, G. Li, Electronic post-compensation of WDM transmission impairments using coherent detection and digital signal processing, Opt. Express 16 (2) (2008) 880.

[31] H. Masuda, E. Yamazaki, A. Sano, T. Yoshimatsu, T. Kobayashi, E. Yoshida, Y. Miyamoto, S. Matsuoka, Y. Takatori, M. Mizoguchi, K. Okada, K. Hagimoto, T. Yamada, S. Kamei, 13.5 Tb/s (135 × 111 Gb/s/ch) no-guard-interval coherent OFDM transmission over 6,248 km using SNR maximized second-order DRA in the extended L-band, in: Proc. OFC/NFOEC 2009, PDPB5.

[32] E. Mateo, M. Huang, F. Yaman, T. Wang, Y. Aono, T. Tajima, Nonlinearity compensation using very-low complexity backward propagation in dispersion managed links, in: Proc. OFC/NFOEC 2012, OTh3C.4.

[33] M. Mazurczyk, D.G. Foursa, H.G. Batshon, H. Zhang, C.R. Davidson, J.-X. Cai, A. Pilipetskii, G. Mohs and Neal S. Bergano, 30 Tb/s transmission over 6,630 km using 16QAM signals at 6.1 bits/s/Hz spectral efficiency, in: Proc. ECOC 2012, Th.3.C.2.

[34] G. Mohs, W.T. Anderson, E.A. Golovchenko, A new dispersion map for undersea optical communication systems, in: Proc. OFC/NFOEC 2007, JThA41.

[35] Robert H. Morelos-Zaragoza, The Art of Error Correcting Coding, second ed., John Wiley & Sons, 2006 ISBN: 0-470-01558-6

[36] P. Poggiolini, A. Carena, V. Curri, G. Bosco, F. Forghieri, Analytical modeling of nonlinear propagation in uncompensated optical transmission links, IEEE Photon. Technol. Lett. 23 (11) (2011) 742–744.

[37] John Proakis, Digital Communications, fourth ed., McGraw Hill, 2001 ISBN 0-07-118183-0

[38] D. Qian, M.F. Huang, S. Zhang, P.N. Ji, Y. Shao, F. Yaman, E. Mateo, T. Wang, Y. Inada, T. Ogata, Y. Aoki, Transmission of $115 \times 100G$ PDM-8QAM-OFDM channels with 4 bits/s/Hz spectral efficiency over 10,181 km, in: Proc. ECOC 2011, Th.13.K.3.

[39] R. Ryf, R. Essiambre, A. Gnauck, S. Randel, Mestre, C. Schmidt, P. Winzer, R. Delbue, P. Pupalaikis, A. Sureka, T. Hayashi, T. Taru, T. Sasaki, Space-division multiplexed transmission over 4200 km 3-core microstructured fiber, in: Proc. OFC/NFOEC 2012, PDP5C.2.

[40] M. Salsi, R. Ryf, G. Le Cocq, L. Bigot, D. Peyrot, G. Charlet, S. Bigo, N.K. Fontaine, M.A. Mestre, S. Randel, X. Palou, C. Bolle, B. Guan, Y. Quiquempois, A six-mode Erbium-doped fiber amplifier, in: Proc. ECOC 2012, Th.3.A.6.

[41] M. Salsi, H. Mardoyan, P. Tran, C. Koebele, E. Dutisseuil, G. Charlet, S. Bigo, 155×100 Gbit/s coherent PDM-QPSK transmission over 7200 km, in: Proc. ECOC 2009, PD2.5.

[42] M. Salsi, C. Koebele, P. Tran, H. Mardoyan, S. Bigo, G. Charlet, 80×100 Gb/s transmission over 9000 km using erbium-doped fibre repeaters only, in: Proc. ECOC 2010, We.7.C.3.

[43] M. Salsi, C. Koebele, P. Tran, H. Mardoyan, E. Dutisseuil, J. Renaudier, M. Bigot-Astruc, L. Provost, S. Richard, P. Sillard, S. Bigo, G. Charlet, Transmission of 96×100 Gb/s with 23% super-FEC overhead over 11,680 km using optical spectral engineering, in: Proc. OFC/NFOEC 2011, OMR2.

[44] M. Salsi, O. Bertran-Pardo, J. Renaudier, W. Idler, H. Mardoyan, P. Tran, G. Charlet, S. Bigo, WDM 200 Gb/s single-carrier PDM-QPSK transmission over 12,000 km, in: Proc. ECOC 2011, Th.13.C.5.

[45] S.J. Savory, Digital filters for coherent optical receivers, Opt. Express 16 (2) (2008) 804–817.

[46] Claude E. Shannon, A mathematical theory of communication, Bell Syst. Tech. J. 27 (3) (1948) 379–423.

[47] W. Shieh, H. Bao, Y. Tang, Coherent optical OFDM: theory and design, Opt. Express 16 (2) (2008) 841.

[48] O.V. Sinkin, J.-X. Cai, D. Foursa, H. Zhang, A. Pilipetskii, G. Mohs, N. Bergano, Scaling of nonlinear impairments in dispersion uncompensated long-haul transmission, in: Proc. OFC 2012, OTu1A.2.

[49] E. Torrengo, R. Cigliutti, G. Bosco, A. Carena, V. Curri, P. Poggiolini, A. Nespola, D. Zeolla, F. Forghieri, Experimental validation of an analytical model for nonlinear propagation in uncompensated optical links, in: Proc. ECOC 2011, We.7.B.2.

[50] E. Torrengo, R. Cigliutti, G. Bosco, G. Gavioli. A. Alaimo, A. Carena, V. Curri, F. Forghieri, S. Piciaccia, M. Belmonte, A. Brinciotti, A. La Porta, S. Abrate, P. Poggiolini, Transoceanic PM-QPSK terabit superchannel transmission experiments at Baud-rate subcarrier spacing, in: Proc. ECOC 2010, We.7.C.2.

[51] G. Ungerboeck, Trellis-coded modulation with redundant signal sets. I: Introduction, IEEE Commun. Mag. 25 (2) (1987) 5–11.

[52] G. Ungerboeck, Trellis-coded modulation with redundant signal sets II: State of the Art, IEEE Commun. Mag. 25 (2) (1987) 12–21.

[53] F. Vacondio, C. Simonneau, L. Lorcy, J.C. Antona, A. Bononi, S. Bigo, Experimental characterization of Gaussian-distributed nonlinear distortions, in: Proc. ECOC 2011, We.7.B.1.

[54] A.J. Viterbi, A.M. Viterbi, Nonlinear estimation of PSK-modulated carrier phase with application to burst digital transmission, IEEE Trans. Inform. Theory IT-29 (4) (1983) 543.

[55] H. Zhang, J.-X. Cai, H.G. Batshon, C.R. Davidson, Y. Sun, M. Mazurczyk, D.G. Foursa, A. Pilipetskii, G. Mohs, N.S. Bergano, 16QAM transmission with 5.2 bits/s/Hz spectral efficiency over transoceanic distance, Opt. Express 20 (11) (2012) 11688–11693.

[56] T. Mizuochi, Recent progress in forward error correction and its interplay with transmission impairments, IEEE J. Sel. Top. Quant. Electron. 12 (4) (2006) 544–554.

[57] S. Zhang, M.F. Huang, F. Yaman, E. Mateo, D. Qian, Y. Zhang, L. Xu, Y. Shao, I.B. Djordjevic, T. Wang, Y. Inada, T. Inoue, T. Ogata, Y. Aoki, 40 × 117.6 Gb/s PDM-16QAM OFDM transmission over 10,181 km with soft-decision LDPC coding and nonlinearity compensation, in: Proc. OFC/NFOEC 2012, PDP5C.4.

Index

Printed and bound by CPI Group (UK) Ltd, Croydon, CR0 4YY

08/05/2025

01864928-0001